CALCULUS

CALCULUS
SECOND EDITION

Lipman Bers
Columbia University

in collaboration with

Frank Karal
New York University

HOLT, RINEHART AND WINSTON
New York Chicago San Francisco Atlanta Dallas Montreal Toronto London Sydney

Library of Congress Cataloging in Publication Data

Bers, Lipman.
   Calculus.

   Includes index.
   1. Calculus. I. Karal, Frank. II. Title.
QA303.B54 1976   515   75-35878
ISBN 0-03-089268-6

Printed in the United States of America
6 7 8 9 032 9 8 7 6 5 4 3 2 1

PREFACE

This is a revised edition of my 1969 *Calculus*. In carrying out the revisions I was fortunate to have had the aid of Professor Frank Karal, who brought to this task his experience as an industrial mathematician, a college teacher, and a researcher in applied mathematics.

Substantial revisions have been made in the presentation of the material, the examples, the problems, and in the design. They all aim at making the book shorter and easier to use. The basic philosophy of the text is, however, unchanged. It is intended for students taking calculus for the first time. The needs of such students, whatever their reason for studying mathematics, are essentially the same. A first course ought to give a grasp of the main ideas of calculus and an ability to use its language and its techniques with ease and understanding. Therefore a first course must be a course in essentials. Calculus is the art of setting up and solving differential equations; this is how it originated, and this is what it is about. Our book is based on this point of view without, of course, attempting to give anything approaching a theory of differential equations.

Applications are emphasized throughout, including applications to mechanics, which played a dominant part in the history of calculus. We have also added some applications to economics and statistics (and eliminated some physical applications which appeared too sophisticated). For the convenience of the user, most applications appear in optional sections and may be omitted without loss of continuity. There are also other optional sections, subsections, and sets of problems which may be omitted at the discretion of the instructor.

V

Intuitive reasoning is used and stressed throughout the book, not as a substitute for but rather as a guide to rigorous thinking. Rigor in mathematics means, first of all, honesty and clarity. The purpose of rigor in a beginning course is to make the concepts easily understood and used. The technique of making definitions and of proving theorems is of little concern to us at this point. All definitions and statements in our book are correct and precise. In most chapters, however, certain key theorems are explained and used without proof. Proofs (some of which appeared in appendixes in the first edition) are collected in a separate Theoretical Supplement at the end of the book.

Like every subject, calculus is learned best with due regard to its history. References to the history of mathematics are therefore made at various places in the book, and historical considerations have influenced the choice of material.

The book presupposes no special preparation beyond the ability to perform ordinary algebraic operations and some rudimentary knowledge of geometry. Though the text centers on calculus of one variable, analytic geometry and calculus of several variables are also developed.

There are over 3500 problems which serve to develop skills and strengthen understanding. Answers are provided for all odd-numbered problems in the main text, and every effort has been made to insure that they are accurate. The number of illustrative examples has been increased, and they have been made more explicit.

The short Chapter 1 on numbers and Chapter 2 on coordinates deal with material familiar to many students; these chapters could be reviewed briefly or assigned for independent reading, except perhaps for some work on inequalities in Chapter 1, §3. The section on conics (§4 in Chapter 2) can be postponed.

Chapter 3 establishes the basic language and main facts about functions, continuity, and limits. The primary aim is to make the student think of continuous functions as those having a graph without breaks and to recognize "at sight" the continuity of simple functions.

The core of one variable calculus is presented in Chapters 4 through 12. The derivative is discussed in Chapters 4 and 6, and its main applications are given in Chapters 5 and 7; some sections in Chapter 7 are optional. A new optional section (§3 in Chapter 6) deals with implicit differentiation. The chapter on integrals (Chapter 8) is followed by a chapter on applications (Chapter 9), some of whose sections are optional. Similarly, the two chapters on transcendental functions (Chapters 10 and 11) are followed by a mostly optional Chapter 12 on applications. This part of the book leads into Chapter 13 on techniques of integration.

The rest of the book consists of three essentially independent units: Taylor's theorem and infinite series (Chapters 14 and 15), curves, vectors, and surfaces (Chapters 16, 17, and 18), and calculus of several variables (Chapters 19, 20, and 21).

We remark that Chapter 16 also contains the continuation of the theory of conics, that in Chapter 18 solid analytic geometry is developed using inner products, and that an optional section on cross products has been added.

November 1975   LIPMAN BERS
New Rochelle, New York

CONTENTS

CALCULUS

NUMBERS

§1 Rational numbers

The invention of calculus in the seventeenth century was a turning point in the history of human thought. This turning point made modern science possible. Two developments prepared the way for the invention of calculus: the extension of the concept of number and the fusion of geometry with algebra. An account of calculus should begin therefore with a discussion of numbers.

1.1 Positive integers and fractions

The simplest numbers are the positive integers used for counting, $1, 2, 3, 4, \ldots$, and the positive fractions used for measuring, such as $\frac{1}{2}$, $\frac{17}{3}$. Positive integers are themselves fractions ($1 = \frac{1}{1}$, $2 = \frac{2}{1}$, and so forth), and every positive fraction is the ratio of two positive integers. The positive integers and fractions are called **positive rational numbers.**

We should, of course, distinguish the numbers themselves from the symbols used to denote them. Thus, instead of the Arabic numerals $1, 2, 3, 4, \ldots$, we could use Roman numerals I, II, III, IV, ... or some self-explanatory symbols like *, **, ***, ****, ...

Positive rational numbers and the arithmetic operations made on them can be represented geometrically.

We choose two distinct points and agree to use the straight segment joining them as a unit of length. Let the unit segment be divided into n congruent segments. Each

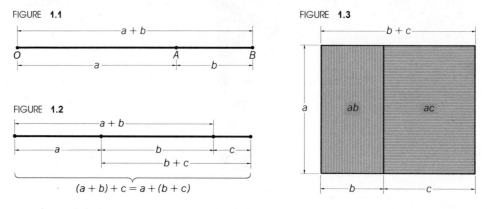

FIGURE **1.1**

FIGURE **1.3**

FIGURE **1.2**

$(a + b) + c = a + (b + c)$

"small" segment is said to have length $\dfrac{1}{n}$. A segment is said to have length $\dfrac{m}{n}$ if it can be divided into m segments, each of which has length $\dfrac{1}{n}$. The length of a segment AB will be denoted by $|AB|$.

Let a and b be positive rational numbers. Let O, A, and B be points on a line such that $|OA| = a$, $|AB| = b$, and A lies between O and B (see Figure 1.1). The sum $a + b$, a positive rational number, is the length $|OB|$. The **commutative law of addition,** $a + b = b + a$, means that the length of the segment OB is the same measured from O to B or from B to O. The geometric meaning of the **associative law of addition** $(a + b) + c = a + (b + c)$ is shown in Figure 1.2. If a is greater than b, the **difference** $x = a - b$ is the positive rational number x such that $b + x = a$.

The **product** ab is again a positive rational number, and so is the **quotient** $x = \dfrac{a}{b}$, that is, the number x such that $bx = a$. $\left(\text{From now on, we often write } a/b \text{ instead of } \dfrac{a}{b}.\right)$ The product ab may be interpreted as the area of a rectangle of length a and width b. The **commutative law of multiplication,** $ab = ba$, means that it does not matter which of the two dimensions of a rectangle we call "length" and which "width." The geometric meaning of the **distributive law,** $a(b + c) = ab + ac$, is shown in Figure 1.3.

The volume of a right parallelepiped ("box") of length a, width b, and height c is $(ab)c$ (= area of base times height). The **associative law of multiplication,** $(ab)c = a(bc)$, follows by noting that a box with base (ab) and height c has the same volume as a box with height a and base (bc).

1.2 Rational numbers and points

Positive fractions were already known in the ancient civilizations of Babylon and Egypt; Hindus are believed to have invented the number **zero. Negative** rational numbers, -1, $-\frac{3}{5}$, and so on, were introduced in Italy during the Renaissance. Positive and negative integers, fractions, and zero form the system of **rational** numbers.

Rational numbers can be represented by points on a straight line. We choose a point on the line (the **origin**) which we label 0, and another point which we label 1. The distance from 0 to 1 is chosen as the unit of length. A positive rational number a is represented by a point on the line whose distance from 0 is a and which lies on the same side of 0 as 1. The point on the other side of 0 and the same distance from it represents the number $(-a)$.

FIGURE **1.4**

It is traditional to draw the line and to choose the points 0 and 1 as in Figure 1.4. We shall often use the expressions "to the right" and "to the left," by which we mean, respectively, "in the direction from 0 to 1" and "in the direction from 0 to -1."

§2 Real numbers

To do calculus we need not only rational numbers but also so-called irrational ones. We explain the need and describe the system of real (rational and irrational) numbers here.

2.1 Incommensurable segments

Rational numbers were invented for measuring, especially for the measurement of length. But there are incommensurable segments, that is, pairs of segments such that if one is chosen as the unit of length, the length of the other cannot be expressed by any rational number. An example is the side of a square and its diagonal. Let the side have length 1. If the length of the diagonal were a rational number x, we should have $x^2 = 2$. This follows from the Pythagorean theorem or by noting that the area of the "big" square in Figure 1.5 is twice that of the "small" square. But *there is no rational number whose square is 2*. We shall recall the proof of this fact in §4 of this chapter.

The existence of incommensurable segments was discovered about 2500 years ago by Greek mathematicians. (There is a legend that the man who announced this discovery was punished by the gods for revealing an imperfection of the universe.) In modern mathematics this difficulty is overcome by introducing a new kind of number. Real numbers, which we discuss in the next section, have the property that (no matter which unit of length we use) every segment has a length expressible as a real number. The Greeks did not have real numbers. They used instead a theory of proportions ▶¹ created by the brilliant mathematician Eudoxos.

The concept of a real number is not the invention of one or several men. It grew, as if by itself, out of a good method for writing down rational numbers.

FIGURE **1.5**

PYTHAGORAS of Samos (6th century B.C.) was the semilegendary founder of a philosophical school. This school was also a religious community which became entangled in politics.

The Pythagoreans are credited with many mathematical discoveries, including the theorem about the square of the hypotenuse (which, however, was known centuries earlier) and the existence of incommensurable segments. They observed the dependence of musical tones on ratios of lengths of strings, which may have led them to the proposition that "all is number." If we interpret this to mean that nature can be described mathematically, this was one of the most successful guesses in the history of thought.

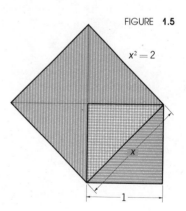

$x^2 = 2$

¹The symbol ▶ signifies that the subject is discussed in the Theoretical Supplement, which begins on page 662. Consult the index, where references to the supplement are given in brackets.

2.2   Positional number systems

There have been three major advances in the art of computing. The third, the development of electronic computers, has occurred in our lifetime. The second, the invention of logarithms, took place during the sixteenth century. The first advance, without which the other two would have been impossible, was the development of the **positional number system.**

The positional number system we use is called **decimal,** or a **base ten system,** since in this system every positive integer is represented as a sum of powers of ten (*decem* in Latin). Thus $203 = 2 \cdot 10^2 + 0 \cdot 10^1 + 3 \cdot 10^0$. This system came to Europe from the Islamic world, after the Crusades; hence the name "Arabic numerals." But the Arabs learned the positional system from the Hindus, who received it from the Hellenistic civilization, and the Greek astronomers borrowed their number system from the Babylonians, who were using it as early as 1500 B.C.

A positional number system has several advantages over the Roman numerals or the Hebrew and Greek method of using the letters of the alphabet as names of numbers. Using only a few digits (in our system 0, 1, . . . , 9) we can name any number, and by remembering the addition table and multiplication table for these digits, we can compute the sum and product of any two integers by a simple procedure.

The choice of a base is arbitrary, and the popularity of ten is an anatomical accident (*digit* means *finger*). The base of the Babylonian system, also used by the Greek astronomer Ptolemy, was 60; this system is preserved in our way of measuring time (1 hour = 60 minutes; 1 minute = 60 seconds). Circuits inside an electronic computer use a base two system, since the two digits 0 and 1 can be represented by the off-on positions of a switch. In principle, however, one base is as good as another, and we restrict ourselves to the familiar decimal notation.

2.3   Decimal fractions

In order to represent positive fractions in the decimal notation, we need an additional symbol, the **decimal point.** For example,

$$1.16 = 1 + \frac{1}{10} + \frac{6}{10^2} = \frac{116}{100} = \frac{29}{25}.$$

The rule is: if a is a nonnegative integer and $\alpha_1, \alpha_2, \ldots, \alpha_m$ are m digits, each of which is one of the numbers 0, 1, 2, . . . , 9, then

$$a.\alpha_1\alpha_2 \ldots \alpha_m = a + \frac{\alpha_1}{10} + \frac{\alpha_2}{10^2} + \cdots + \frac{\alpha_m}{10^m}.$$

(In the example, $a = 1$, $\alpha_1 = 1$, $\alpha_2 = 6$, and $m = 2$.) An expression of this form is called a **terminating decimal.** Its value is not changed if we add one or several zeros on the

EUDOXOS of Cnidus (ca. 403–355 B.C.) lived for a while in Athens where he was connected with Plato's Academy. The theory of proportions is only one of his brilliant achievements. Eudoxos also gave rigorous proofs for various theorems on volumes, and in astronomy he originated the method of describing the visible motion of planets as a superposition of circular motions.

PTOLEMY (ca. 100–187 A.D.) was a mathematician, astronomer, and geographer. His astronomical treatise, known by its Arabic name, *Almagest,* summarized the whole of Greek astronomy including observational data, mathematical methods (in particular spherical trigonometry), and the so-called Ptolemaic world system with the immovable earth at the center of the universe.

right. For instance, $1.16 = 1.160 = 1.1600$ and so forth. By convention, the term a, before the decimal point, may be omitted if $a = 0$. Thus we write .39 instead of 0.39.

We recall some facts about decimals.

If a rational number $x = p/q$, p and q positive integers, can be written as a terminating decimal, the decimal can be obtained by long division; for instance,

$$
\begin{array}{r}
1.16 \\
25{\overline{\smash{\big)}\,29}} \\
25 \\
\hline
40 \\
25 \\
\hline
150 \\
150 \\
\hline
\end{array}
\qquad \text{so that } \frac{29}{25} = 1.16.
$$

If we apply long division to a rational number $x = p/q$ which cannot be represented as a terminating decimal, the process cannot stop, and we obtain the so-called **nonterminating decimal expansion** of x. An example is the number $\frac{1}{3}$; we get

$$\tfrac{1}{3} = .33333 \ldots \qquad \text{(and so on, all 3's)}, \qquad \text{or } \tfrac{1}{3} = .\overline{3}$$

(the bar denoting that 3 is to be repeated indefinitely). This is typical: the *nonterminating decimal expansion of a rational positive number is always repeating*. This means: the same digit, or the same string of digits, is repeated infinitely; thus the repetition may start either immediately after the decimal point or after several digits behind the decimal point. For instance,

$$\tfrac{5}{3} = 1.666 \ldots \text{ (all 6's)} = 1.\overline{6}, \quad \tfrac{11}{6} = 1.8333 \ldots \text{ (all 3's)} = 1.8\overline{3},$$
$$\tfrac{42}{33} = 1.2727 \ldots \text{ (all 27's)} = 1.\overline{27}, \qquad \tfrac{271}{990} = .27373 \ldots \text{ (all 73's)} = .2\overline{73}.$$

Conversely, *every repeating decimal is the decimal expansion of some rational number.*

If a positive fraction *can* be written as a terminating decimal, it can also be written as a nonterminating one. We diminish the last digit by 1 and follow it by a string of 9's. For instance, instead of $\frac{29}{25} = 1.16$, we may write

$$\tfrac{29}{25} = 1.1599 \ldots \text{ (all 9's)} = 1.15\overline{9}.$$

PROBLEMS

Express the following numbers as nonterminating decimal expansions:

1. 3.
2. 15.
3. $\frac{1}{4}$.
4. $\frac{7}{8}$.
5. $\frac{14}{3}$.
6. $\frac{2}{11}$.
7. $\frac{12}{33}$.
8. $\frac{181}{99}$.

2.4 Defining real numbers

There are also nonterminating, nonrepeating decimals (for instance, the decimal $3.12112111211112\ldots$). This suggests one way of defining real numbers—as nonterminating decimals, repeating or nonrepeating. This is how real numbers were first introduced. Mathematicians worked with them long before the concept was formalized and analyzed (by Dedekind, Cantor, and others) in the nineteenth century.

Thus we define: *a positive real number is a nonterminating decimal* $a.\alpha_1\alpha_2\alpha_3 \ldots$.

Here a is a positive integer or zero (and we agree to omit a if $a = 0$), each α is one of the digits $0, 1, \ldots, 9$, and the decimal does not end with a string of zeros.

It may happen that the nonterminating decimal $a.\alpha_1\alpha_2 \ldots$ represents a rational number. (This will be so if the decimal is repeating.) The rational numbers are included among the real numbers. Thus $\frac{1}{3}$ is another name for the real number $.333\ldots = .\bar{3}$, and $\frac{3}{2} = 1.5$ is another name for the real number $1.4999\ldots = 1.4\bar{9}$.

A negative real number is a nonterminating decimal preceded by a minus sign. If the decimal is repeating, this real number is a negative rational number. For instance, $-1.4999\ldots = -1.4\bar{9} = -1.5 = -\frac{3}{2}$.

The positive and negative real numbers *and zero* form the set of **real numbers.** A real number that is not rational is called **irrational.** (For instance $3.12112111211112\ldots$ is irrational since this nonterminating decimal is nonrepeating.)

2.5   The number line

Consider once more (see §1.2) the straight line on which we marked off the points 0 and 1. To *every* point P on this line, which lies to the right of 0, we associate a positive real number which measures the distance from 0 to P; that is, the length $|OP|$ of the segment OP. This is done as follows. We find the largest integer a such that the point a lies to the left of P; this a is positive or zero. Next, let α_1 be the largest of the numbers $0, 1, \ldots, 9$ such that P lies to the right of the point $a.\alpha_1$. (Then P lies either to the left of $a.\alpha_1 + \frac{1}{10}$ or coincides with $a.\alpha_1 + \frac{1}{10}$.) Next we find the largest α_2 among $0, 1, \ldots, 9$ such that P lies to the right of $a.\alpha_1\alpha_2$ and so on. Thus we obtain an infinite sequence of digits $a.\alpha_1\alpha_2\alpha_3 \ldots$; the procedure is shown in Figure 1.6 where $a.\alpha_1\alpha_2\alpha_3 \ldots = 1.345\ldots$. (If the distance $|OP|$ turns out to be a rational number, $a.\alpha_1\alpha_2\alpha_3 \ldots$ will be the nonterminating decimal representation of that number.)

FIGURE **1.6**

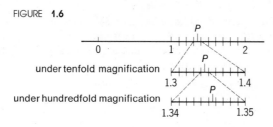

under tenfold magnification

under hundredfold magnification

We call the real number $x = a.\alpha_1\alpha_2\alpha_3 \ldots$ the **coordinate** of P. Each positive real number is the coordinate of some point on the line, to the right of 0, and distinct points have distinct coordinates. We accept this statement as an axiom.

RICHARD DEDEKIND (1831–1916) was primarily an algebraist, one of the originators of so-called "abstract" algebra.

GEORG CANTOR (1845–1918) single-handedly created the theory of infinite sets, perhaps the most revolutionary development in the history of mathematics since calculus. His work provided a common language for most of mathematics. It also led to hitherto unsuspected paradoxes and to logical difficulties, not yet fully resolved. The development of modern mathematical logic is a response to the challenge of Cantor's set theory.

Cantor studied in Berlin and then taught at a mediocre provincial university. His life was troubled by bitter scientific controversies and bouts with insanity.

Let Q be a point to the left of 0. We assign to it the coordinate $-x$, x being the coordinate of a point P, to the right of 0 such that the segment $0Q$ and $0P$ are congruent.

We have associated with each point on the line a unique real number as its coordinate, and each real number is the coordinate of a unique point on the line. We shall often use the expressions real number x, the point with the coordinate x, and the point x interchangeably.

The line on which we marked off the points 0 and 1 is called the **number line.** It is an idealized measuring tape, which can be used to measure the length of *any* segment.

Now we can interpret sums and products or real numbers geometrically, as we did in §1 for rational numbers. This assures us that the real numbers obey the usual rules of algebra ▶.

2.6 Calculating with real numbers

Suppose we want to add two real numbers, say, two positive real numbers, $x = a.\alpha_1\alpha_2\alpha_3 \ldots$ and $y = b.\beta_1\beta_2\beta_3 \ldots$. The natural way to do this is to "round off" both numbers by dropping all digits after, say, the mth and then adding the resulting rational numbers $a.\alpha_1\alpha_2\ldots\alpha_m$ and $b.\beta_1\beta_2\ldots\beta_m$. We get a rational number which, we hope, is close to the true sum of x and y and which will be as close to the true sum as we want, provided we choose m large enough. The same applies to multiplication: we expect that, if m is large enough, the product of $a.\alpha_1 \ldots \alpha_m$ and $b.\beta_1 \ldots \beta_m$ will be as close as we like to the true product of x and y.

But what are the "true" sum and the "true" product of two real numbers? Can they be defined in a precise manner from the properties of rational numbers without recourse to geometry? Can they be computed in the manner indicated? *The answer is yes.*

The reader is advised to accept this on faith for the time being. In doing so, he will be following excellent precedents. The founders of calculus and the great mathematicians who developed and applied calculus considered it self-evident that one could work with real numbers by following the usual rules of algebra. The need for proof, and the proof itself, date from the middle of the nineteenth century ▶.

2.7 Variables. Notations

In a strictly deductive presentation of calculus we would, at this point, list the "usual rules of algebra" which can be applied to real numbers ▶. Instead, we assume that the reader is familiar with these rules, which are the same as the rules for rational numbers.

There are infinitely many numbers, and the only way of making statements about all of them is to use variables; that is, symbols that may be replaced by numbers. We agree that letters stand for (real) numbers. Thus the statement "$a + 0 = a$ for all a" means that $0 + 0 = 0$, $1 + 0 = 1$, $-\frac{5}{7} + 0 = -\frac{5}{7}$, and so forth.

We recall, in particular, that

$$0a = 0 \qquad \text{for all} \quad a.$$

Therefore *division by 0 cannot be defined* without violating the rules of algebra. An important rule is:

$$\text{If } ab = 0, \text{ then either } a = 0 \text{ or } b = 0.$$

Here "or" is the "nonexclusive or" that is always used in mathematics. The statement "$a = 0$ or $b = 0$" means that the following three cases are possible: (1) $a = 0$, $b \neq 0$, (2) $a \neq 0$, $b = 0$, (3) $a = b = 0$, while the case $a \neq 0$, $b \neq 0$ is impossible.

We recall that if n is a positive integer, then a^n means $a \cdot a \ldots a$ (n factors), so that

$$a^1 = a \qquad a^2 = aa, \qquad a^3 = aa^2, \qquad \text{and so forth.}$$

Zero and negative exponents are defined as follows: for $a \neq 0$, we set

$$a^0 = 1, \qquad a^{-1} = \frac{1}{a}, \qquad a^{-2} = \frac{1}{a^2}, \qquad a^{-3} = \frac{1}{a^3}, \qquad \text{and so forth.}$$

No meaning is assigned to expressions such as 0^0, 0^{-1}, 0^{-2}, and so on.

The familiar **laws of exponents** follow:

$$a^1 = a, \qquad a^2 = aa, \qquad a^3 = aa^2, \qquad \text{and so forth.}$$

Here n, m stand for arbitrary integers and a, b are any numbers, except that the values $a = 0$ and $b = 0$ have to be excluded if either of the exponents n or m is negative or zero.

PROBLEMS

Problems 1 to 52 are for review purposes only. (We take the real numbers and the rules for operating with them for granted.)

1. Does $a(b + c + d) = ab + ac + ad$?
2. Does $ab - c = a(b - c)$?
3. Does $(ab + ac)d = a(bd + cd)$?
4. Does $(a + b)(c + d) = (ac + db) + (bc + ad)$?
5. Does $\dfrac{-a}{-b} = \dfrac{a}{b}$ for $b \neq 0$?
6. Does $\dfrac{ab}{ac} = \dfrac{b}{c}$ for $a \neq 0$, $c \neq 0$?
7. Does $\dfrac{a + b}{c} = \dfrac{a}{c} + \dfrac{b}{c}$ for $c \neq 0$?
8. Does $\dfrac{a}{b} + \dfrac{c}{d} = \dfrac{ab + bc}{bd}$ for $b \neq 0$, $d \neq 0$?
9. Does $\dfrac{-a - b - c}{d} = -\dfrac{a}{d} - \dfrac{b}{d} - \dfrac{c}{d}$ for $d \neq 0$?
10. Does $\dfrac{(-a)(-b) - (-b)(-c)}{(-b)} = c - a$ for $b \neq 0$?

Verify the statements in Problems 11 to 22 by showing that one side of the given equation is equal to the other side.

11. $(a + b)^2 = a^2 + 2ab + b^2$.
12. $(a - b)^2 = a^2 - 2ab + b^2$.
13. $(a + b)^3 = a^3 + 3a^2b + 3ab^2 + b^3$.
14. $(a - b)^3 = a^3 - 3a^2b + 3ab^2 - b^3$.
15. $(a - 2b)^2 = a^2 - 4ab + 4b^2$.
16. $a^2 - b^2 = (a + b)(a - b)$.
17. $a^3 + b^3 = (a + b)(a^2 - ab + b^2)$.
18. $a^3 - b^3 = (a - b)(a^2 + ab + b^2)$.

19. $\dfrac{\dfrac{a}{b}}{\dfrac{c}{d}} = \left(\dfrac{a}{b}\right)\left(\dfrac{d}{c}\right) = \dfrac{ad}{bc}.$

21. $\dfrac{\dfrac{1}{a} + \dfrac{1}{b}}{\dfrac{1}{c} + \dfrac{1}{d}} = \dfrac{cd(a+b)}{ab(c+d)}.$

20. $\dfrac{a}{b + \dfrac{c}{d}} = \dfrac{ad}{c + bd}.$

22. $\dfrac{\dfrac{a}{b} + \dfrac{c}{d}}{\dfrac{a}{b} - \dfrac{c}{d}} = \dfrac{ad + bc}{ad - bc}.$

Evaluate the expressions in Problems 23 to 32.

23. $2^2 \cdot 2^4.$

24. $(5)^2(4)^2(3)^2.$

25. $2^0 \cdot 2^1 \cdot 2^2 \cdot 2^3.$

26. $(4^2 \cdot 2^3)^2.$

27. $(2^{-4})^2 \cdot (4^{-2})^{-3}.$

28. $\left(\dfrac{2}{3}\right)^3.$

29. $\dfrac{3^4 \cdot 3^8}{3^5 \cdot 3^7}.$

30. $\dfrac{4^2 \cdot 6^2}{3^3 \cdot 2^3}.$

31. $\dfrac{(10^2)(10^{-4})(10^6)}{(10^0)(10^{-5})(10^8)}.$

32. $3^{2^3}.$

33. Does $2^5 = 5^2$?

34. Does $(2^3)^2 = 2^{3^2}$?

35. Does $(2+3)^2 = 2^2 + 3^2$?

36. Does $2^3 \cdot 3^2 = 6^5$?

37. Does $0^4 \cdot 4^0 = 0$?

38. Does $(0)^{-2}(-2)^0 = 1$?

Simplify the expressions in Problems 39 to 46.

39. $(a^2b^3)(a^3b^2).$

40. $(-2ab^3)(4a^0b^5).$

41. $(a^0b^2c^4)^3.$

42. $(8ab^2)(-2a^2b)^{-2}.$

43. $(-ab^{-2})^3(a^2b^3)^3.$

44. $[(a^{-2})^4]^{-3}.$

45. $[-(-ab^3)^{-3}(a^6b^6)^2]^3.$

46. $\{-(ab^2)^2[-(a^{-2}b^{-1})^{-2}]^3\}^2.$

47. If $ab = 4$, what is $(ab)^3$?

48. If $abc = 2$, what is $a^5b^5c^5$?

49. If $a^{-1}b^{-1}c^{-1} = 2$, what is $(abc)^4$?

50. If $a^2 = 4$, what is $(a^{20})^0$?

51. If $a^{13} = 4$, what is $(a^{-7})^7(a^{-6})^{-6}$?

52. If n is an integer, what is $(-1)^{2n}$?

2.8 Fractional exponents

For every real positive number c and every positive integer $n \neq 1$, there is a unique positive real number x such that $x^n = c$.

We call this x the nth root of c and write $x = \sqrt[n]{c}$. We also set $\sqrt[n]{0} = 0$ and, if n is odd, $\sqrt[n]{-c} = -\sqrt[n]{c}$. Indeed, $0^n = 0$ and, for odd n, $(-\sqrt[n]{c})^n = -c$.

We shall prove the existence of roots later (see Chapter 6, §1.1). If we assume the existence of roots, as we do from now on, we can assign a meaning to the expression c^r, where c is a positive number and r a rational number. The definition reads

$$c^{p/q} = \sqrt[q]{c^p} = (\sqrt[q]{c})^p,$$

where p and q are integers, $q > 0$, and the fraction p/q is reduced to lowest terms. If c is a negative number, c^r is defined as above whenever the rational number r, when reduced to lowest terms $r = p/q$, has an odd denominator. For $c = 0$, we set $c^r = 0$ for all positive rational r.

The definition is chosen so as to make the laws of exponents stated in §2.7 valid for fractional exponents, whenever meaningful.

EXAMPLE   Evaluate $(-1)^{1/3}$. We have $(-1)^{1/3} = \sqrt[3]{-1} = -\sqrt[3]{1} = -1$.

But it would be wrong to write $(-1)^{1/3} = (-1)^{2/6} = \sqrt[6]{(-1)^2} = \sqrt[6]{1} = 1$. We get a wrong result because $\frac{2}{6}$ is not reduced to lowest terms.

PROBLEMS

Evaluate the expressions in Problems 1 to 10.

1. $8^{2/3}$.

2. $(27^{1/3})^2$.

3. $(.49)^{1/2}$.

4. $8^{1/3} \cdot 2^{-2}$.

5. $27^{-4/3}$.

6. $(64^{-1/3})^2$.

7. $(.001)^{-1/3}$.

8. $(\frac{4}{9})^{-1/2}$.

9. $(4^{5/3}) \cdot (2^{2/3})$.

10. $[2^6(4^{-2})^{3/4}]^2$.

Simplify the given expressions in Problems 11 to 20 (a, b, and c are positive).

11. $(a^8 b^{12})^{1/4}$.

12. $(-a^{10} b^{15})^{1/5}$.

13. $(a^{1/2} b^{1/4} c^{1/8})^4$.

14. $[(a^2)^4]^{1/8}$.

15. $[(-a^{-5})^5]^{1/25}$.

16. $(-\frac{9}{64} a^{-1/2} b^2)^{-2}$.

17. $\left(\dfrac{a^{-5/2} b^{-10}}{32} \right)^{-1/5}$.

18. $\left(\dfrac{a^{2n} b^{4n}}{c^{-2n}} \right)^{1/2}$.

19. $\dfrac{a^{(n+1)/2} b^{-n+2}}{a^{-3(n+2)/2} b^{(2n+3)/2}}$.

20. $\dfrac{(a^{-2n/3} - a^{2n/3})}{(a^{-n/3} + a^{n/3})}$.

Simplify the following expressions (x and y are positive).

21. $\sqrt[4]{16x^8}$.

22. $\sqrt[3]{-64x^6}$.

23. $\sqrt[5]{-32x^{10}y^5}$.

24. $\sqrt{\dfrac{81x^8}{25y^4}}$.

25. $\sqrt[4]{\dfrac{4^5 x^{-6} y^2}{2^6 x^2 y^{-6}}}$.

26. $(\sqrt[4]{(x^{1/3})^3})^4$.

27. $(\sqrt{x^{1/3} x^{5/3}})^3$.

28. $\sqrt[3]{(-x^{3/5})^5 y^9}$.

29. $\sqrt[4]{(x^{-8/3})^3 y^2}$.

30. $\sqrt[4]{(x^3)^4 x^2}$.

§3   Inequalities

Since inequalities are indispensable for understanding calculus, we devote a section to this subject.

3.1   Notations

We write "$a < b$" to say that the number a is **less than** the number b. This means the same as "$b > a$" (read: b is **greater than** a). Every **inequality,** that is, a statement that one number is less than another, may be written using either the symbol $<$ or· the symbol $>$. An inequality $a < b$ has a geometric meaning: on the number line, the point with coordinate a (see §2.5) lies to the left of the point with coordinate b.

The statement "$a \le b$" means that either $a < b$ or $a = b$. Thus $-3 \le -2$ and $7 \le 7$, but it is not true that $1 \le \frac{1}{2}$. Similarly, "$a \ge b$" means that either $a > b$ or $a = b$. (The symbols $\le$ and $\ge$ are read "less than or equal to" and "greater than or equal to.") Also, "a is *positive*" and "b is *negative*" is another way of saying that $a > 0$ and $b < 0$, while "*nonnegative*" and "*nonpositive*" may be written as ≥ 0 and ≤ 0, respectively.

If a and b are two distinct numbers, one must be the greater; this is the first basic rule for inequalities (the others will be stated in §3.2). If a and b are both positive and given by nonterminating decimals, we pick the greater number by comparing, successively, the digits. For instance, if $a = 17.3494\ldots\ldots$, $b = 17.3523\ldots\ldots$, then $a < b$. Indeed, the first digit in which a and b differ is the second digit after the decimal point, and that digit is greater for b.

A negative number is less than any positive number. If a and b are both negative, we pick the greater number by remembering that "$a < b$" means the same as "$-b < -a$." For instance, $-3.465\ldots < -3.271\ldots$ since $3.271\ldots < 3.465\ldots\ldots$

REMARK   The nonterminating decimal expansion of a number is an infinite *string of inequalities* that describe the number with greater and greater accuracy; all inequalities taken together determine the number completely. For example, the equation $\frac{1}{3} = .333\ldots = .\overline{3}$ means that

$$0 < \tfrac{1}{3} \le 1, \quad .3 < \tfrac{1}{3} \le .4, \quad .33 < \tfrac{1}{3} \le .34, \quad .333 < \tfrac{1}{3} \le .334, \quad \text{and so forth.}$$

There is no other number, except $\frac{1}{3}$, which satisfies *all* these inequalities ▶. Similarly, the equation $6^{1/3} = 1.81712\ldots$ means that

$$1 < 6^{1/3} \le 2, \quad 1.8 < 6^{1/3} \le 1.9, \quad 1.81 < 6^{1/3} \le 1.82, \quad 1.817 < 6^{1/3} \le 1.818,$$
$$1.8171 < 6^{1/3} \le 1.8172, \quad \text{and so forth.}$$

PROBLEMS

Answer the following questions.

1. Is $7 < 8$?
2. Is $7 \le 8$?
3. Is $5 \ge 5$?
4. Is $-7 > -5$?

5. Is $\frac{1}{2} \ge \frac{1}{3}$?
6. Is $-9 > 1$?
7. Is $-2 \le -\frac{3}{2}$?

8. Is $-5 < 5$?
9. Is $2 - 3 \le -2 + 3$?
10. Is $\frac{1}{2} - \frac{2}{3} < \frac{1}{4} - \frac{1}{3}$?

11. Which number is greater, $x = 2.374512\ldots$ or $y = 2.374506\ldots$?
12. If $x = .334$ and $y = .33\overline{4}$, is $y < x$?
13. Which number is smaller, $x = 1.5$ or $y = 1.4\overline{9}$?
14. Given that $x = .45637070070007\ldots$ and $y = .4563\overline{707}$, which of the two numbers is larger?

3.2   Basic rules

The four basic rules for working with inequalities read:

I. *If a and b are two numbers, then one of the three statements "$a = b$," "$a < b$," "$b < a$" is true and the other two are false.*
II. *If $a < b$ and $b < c$, then $a < c$.*
III. *If $a < b$, then $a + c < b + c$.*
IV. *If $a < b$ and $0 < c$, then $ac < bc$.*
IVa. *If $a < b$ and $c < 0$, then $ac > bc$.*

Rule I was already stated in §3.1. Rule II says that if a is less than b and b less than c, then a is less than c. Rule III asserts that *a true inequality remains true if we add (or subtract) the same number to both sides*. For instance, since $-1 < 3$, we also have that $-1 + \sqrt{2} < 3 + \sqrt{2}$ and $-1 - 6^{1/3} < 3 - 6^{1/3}$. Rule III has an important consequence.

Suppose that

$$x + y < u + v.$$

Adding $(-v)$ to both sides, we obtain $x + y + (-v) < u + v + (-v)$, or

$$x + y - v < u.$$

In other words, *a true inequality remains true if one term is transposed, from one side to the other, with opposite sign.* (Recall that the same rule holds for equations.)

Rule IV says that *a true inequality remains true if both sides are multiplied (or divided) by the same positive number.*

By Rule IVa, *a true inequality remains true if we multiply (or divide) both sides by the same negative number and reverse the direction of the inequality.*

For instance, multiplying the true inequality $-\frac{1}{10} < \frac{1}{2}$ by the positive number 10, we obtain the true inequality $-1 < 5$. Multiplying the true inequality $-\frac{1}{10} < \frac{1}{2}$ by the negative number -100, and reversing the direction of the inequality, we obtain the true inequality $10 > -50$.

For inequalities involving $>$, $\leq$, or $\geq$, similar rules hold.

PROBLEMS

1. If $a \leq b$ and $a \geq b$, what conclusions can you draw? [*Answer*: $a = b$.]
2. If $a \leq b$ and $a \neq b$, what conclusions can you draw?
3. If $a \leq b$, $b \leq c$, and $c \leq a$, what conclusions can you draw?
4. If $a < b$ and $b \leq c$, what conclusions can you draw?
5. If $a < b$ and $b \geq c$, what conclusions can you draw?
6. Is $x + 1 > x$ for all x?
7. Is $x^2 > x$ for all x?
8. Is $x + x > x$ for all x? For all $x > 0$?
9. Is $(x^2 + 1)^{-1} < 1$ for all x?
*10. If $a > 0$ and $b > 0$, show that $a^2 + b^2 \geq 2ab$. [*Hint*: Consider the nonnegative quantity $(a - b)^2$.]
*11. If $x > 0$, show that $x + \dfrac{1}{x} \geq 2$. [*Hint*: set $a = x^{1/2}$, $b = x^{-1/2}$ and use Problem 10.]
*12. Suppose $a_1 > b_1$ and $a_2 > b_2$. Show that $a_1 + a_2 > b_1 + b_2$.

3.3 Solving inequalities

As an illustration of using the rules, we *solve* the inequality

$$3 - 2x < 4x - 5. \tag{*}$$

"To solve" means to find all numbers x for which the statement is true. (The steps are similar to those used in solving an equation of the form $3 - 2x = 4x - 5$.)

Transposing terms we put all terms involving x on the left side, all terms not involving x on the right side:

$$-2x - 4x < -3 - 5 \qquad \text{or} \qquad -6x < -8.$$

To obtain x, multiply both sides by $-\frac{1}{6}$ (that is, divide by -6). Since $-\frac{1}{6}$ is negative, we must reverse the direction of the inequality:

$$(-\tfrac{1}{6})(-6x) > (-\tfrac{1}{6})(-8) \qquad \text{or} \qquad x > \tfrac{4}{3}.$$

---

*Starred problems are nonroutine and occasionally difficult.

Thus (*) is equivalent to $x > \frac{4}{3}$. The solution is a set of numbers (called the *solution set*); it consists of all numbers greater than $\frac{4}{3}$.

EXAMPLES 1. Solve the double inequality $x - 6 < 2x - 5 \leq x - 3$.
ANSWER This means that $x - 6 < 2x - 5$ and $2x - 5 \leq x - 3$. Treating each of these two inequalities separately by the method used above, we obtain

$$
\begin{array}{c|c}
\begin{aligned}
x - 6 &< 2x - 5, \\
x - 2x &< -5 - (-6), \\
-x &< 1, \\
x &> -1.
\end{aligned}
&
\begin{aligned}
2x - 5 &\leq x - 3, \\
2x - x &\leq -3 - (-5), \\
x &\leq 2.
\end{aligned}
\end{array}
$$

Thus our double inequality means that $x > -1$ and $x \leq 2$. The solution set, therefore, consists of all x such that $-1 < x \leq 2$.

2. Solve the double inequality $x + 10 < 2x - 5 \leq x - 3$.
ANSWER The inequality means that $x + 10 < 2x - 5$ and $2x - 5 \leq x - 3$; that is, $x > 15$ and $x \leq 2$. This is not true for any x (so the solution set is empty).

PROBLEMS

Solve the following inequalities.

1. $x - 10 > 2 - 2x$.
2. $6x + 5 \geq x - 5$.
3. $x + 6 \leq 5 - 3x$.
4. $-5 < x - 4 < 2 - x$.
5. $4x + 2 > 5x + 3 > 4x + 4$.
6. $5x - 2 \leq 10x + 8 \leq 2x - 8$.
7. $4 - 3x < 2x + 3 < 3x - 4$.

8. $x^4 < x^2$.
* 9. $2x^2 - 2 \leq x^2 - x$.
*10. $(x^2 - 1)(x + 4) < 0$.
*11. $\dfrac{2x + 1}{x + 1} > 3$.
*12. $1 < \dfrac{3x - 1}{x - 3} < 2$.

3.4 Absolute values

The **absolute value** $|a|$ of a number a is defined by

$$|a| = \begin{cases} a, & \text{if } a \geq 0, \\ -a, & \text{if } a < 0. \end{cases} \tag{1}$$

This $|8| = 8$, $|-8| = -(-8) = 8$. In general, if $|x| = a$, then either $x = a$ or $x = -a$.
Note that

$$|a| > 0, \quad \text{if } a \neq 0, |0| = 0, \tag{2}$$

and

$$|ab| = |a||b|, \quad |a^{-1}| = |a|^{-1}, \quad \left|\frac{a}{b}\right| = \frac{|a|}{|b|}. \tag{3}$$

The absolute value of a number has a geometric interpretation; it is the distance from the corresponding point on the number line to the origin (see Figure 1.7). Therefore

$$|a| < \alpha \quad \text{means that} \quad -\alpha < a < \alpha, \tag{4}$$

as is seen from Figure 1.8. Also, $|a - b|$ **is the distance between points** a **and** b. Two cases illustrating this statement are shown in Figure 1.9. When we say that two

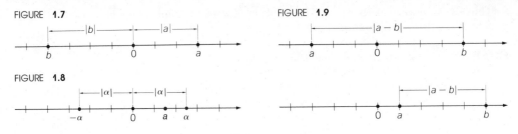

FIGURE **1.7**

FIGURE **1.9**

FIGURE **1.8**

numbers are *close* to each other, we mean that the absolute value of their difference is small. Thus $-\frac{11}{10}$ is closer to -1 than 5 is to 6 for $|-\frac{11}{10} - (-1)| = \frac{1}{10}$, while $|5 - 6| = 1$.

It is geometrically evident that the distance between two points (on a line) is not greater than the sum of their distances from a third point. If we write down this statement for the case in which the two points represent the numbers x and y, and the third point is the origin 0, we obtain $|x - y| \leq |x - 0| + |y - 0|$ or

$$|x - y| \leq |x| + |y|. \tag{5}$$

We often remember this inequality in the form

$$|a + b| \leq |a| + |b|, \tag{6}$$

called the **triangle inequality.** We obtain (6) from (5) by setting $x = a$, $y = -b$.

We shall give an analytic proof of the triangle inequality ("analytic" means "based exclusively on properties of numbers").

If either $a = 0$ or $b = 0$, (6) is obvious. If a and b are positive, so is $a + b$, and we have $a = |a|$, $b = |b|$, and $|a + b| = a + b$. If a and b are negative, so is $a + b$, and we have $a = -|a|$, $b = -|b|$, $a + b = -|a + b|$. In both cases, (6) holds with the equality sign. If $a > 0$ and $b < 0$, then $a = |a|$, $b = -|b|$, and $a + b = |a| - |b|$. Hence either $|a + b| = |a| - |b|$ or $|a + b| = -(|a| - |b|) = |b| - |a|$. Since $|a| - |b| < |a| + |b|$ and $|b| - |a| < |a| + |b|$, we see that (6) holds with the strict inequality sign $<$. The case $a < 0$, $b > 0$ is analogous.

EXAMPLES 1. Solve the equation $|x - 2| = 4$.
ANSWER "To solve" means to find all numbers x for which the statement is true. From the definition of absolute value, $x - 2$ must be 4 or -4. If $x - 2 = 4$, then $x = 6$; if $x - 2 = -4$, then $x = -2$. The solution set consists of two numbers: $x = -2$, $x = 6$.

2. Solve the inequality $|2x - 5| < 9$.
ANSWER In view of (4), the inequality means $-9 < 2x - 5 < 9$. Solving the inequality $2x - 5 < 9$ gives $x < 7$. Solving the inequality $2x - 5 > 9$ gives $x > -2$. Hence the solution set consists of all numbers greater than -2 but less than 7; that is, $-2 < x < 7$.

3. Solve the inequality $\left|\dfrac{2x + 5}{2x - 1}\right| \leq 4$. (We must, of course, exclude the value $x = \frac{1}{2}$, since division by 0 makes no sense.)
ANSWER This means, in view of (4), that

$$-4 \leq \frac{2x + 5}{2x - 1} \leq 4.$$

In order to simplify, we multiply each quantity in the inequality by $(2x - 1)$. However, we must be careful to distinguish between two cases.

Case (i). Assume $2x - 1 > 0$; that is, $x > \frac{1}{2}$. Then the direction of the inequalities is preserved upon multiplication by $(2x - 1)$, and we have

$$-4(2x - 1) \leq 2x + 5 \leq 4(2x - 1).$$

The right inequality gives $x \geq \frac{3}{2}$, and the left inequality gives $x \geq -\frac{1}{10}$. Since $x > \frac{1}{2}$ also, all *three* conditions on x are satisfied if $x \geq \frac{3}{2}$.

Case (ii). Assume $2x - 1 < 0$; that is, $x < \frac{1}{2}$. Then the direction of the inequalities is reversed upon multiplication by $(2x - 1)$, and we have

$$-4(2x - 1) \geq 2x + 5 \geq 4(2x - 1).$$

The right and left inequalities give $x \leq \frac{3}{2}$ and $x \leq -\frac{1}{10}$, respectively. Since $x < \frac{1}{2}$ also, all *three* conditions are satisfied if $x \leq -\frac{1}{10}$. Putting the solutions for Cases (i) and (ii) together, we obtain the following solution set: all numbers x greater than or equal to $\frac{3}{2}$, and all numbers less than or equal to $-\frac{1}{10}$, that is, $x \geq \frac{3}{2}$ or $x \leq -\frac{1}{10}$.

PROBLEMS

In Problems 1 to 8 find the values of the given expressions.

1. $|-3|$,
2. $|(-3)^2|$.
3. $|(-3)^3|$.
4. $|-1 + 2 - 3|$.
5. $|(-4)^2|^2$.
6. $|(-1)(-2)(-3)|$.
7. $\dfrac{|(4)^2(-3)|}{|(6)(-2)^3|}$.
8. $\left|\dfrac{(-2)(-4)(-6)}{(-1)(-3)(-5)}\right|$.

9. Write the "distance from x to 2" in terms of absolute values.
10. Write the "distance from x to -1" in terms of absolute values.
11. Write the "distance from x to a" in terms of absolute values.
12. Write the "distance from x to $\frac{1}{2}(x_1 + x_2)$" in terms of absolute values.

Describe the sets of numbers in Problems 13 to 16 using the absolute value symbol and appropriate inequalities.

13. The set of numbers whose distance from 3 is less than 2.
14. The set of numbers whose distance from the origin is less than 5.
15. The set of numbers whose distance from -1 is greater than 3.
16. The set of numbers whose distance from -2 is less than 4 and greater than 1.

Describe in *words* the inequalities in Problems 17 to 22.

17. $|x - 4| \leq 2$.
18. $|x + 3| < 6$.
19. $|x + 1| \geq 4$.
20. $|x - \frac{1}{2}| < \frac{3}{2}$.
21. $1 < |x + 3| \leq 4$.
22. $|x + 2| < |x - 2|$.

Solve the given equations for x in Problems 23 to 30.

23. $|x + 1| = 2$.
24. $|5 - 2x| = 8$.
25. $|2x - 3| = 4$.
26. $|2x| = |5x + 2|$.
27. $|2x - 1| = |4x + 1|$.
28. $|4x| = |4x + 1|$.
29. $|x^2 - 8| = 1$.
30. $\left|\dfrac{x + 1}{x + 2}\right| = 3$.

In Problems 31 to 38 find the solution set x for the given inequalities.

31. $|x - 1| < 2$.
32. $|2x + 3| < 2$.
33. $|4 - 3x| \leq 3$.
34. $|2x - 1| \geq 4$.

35. $|2 + 3x| > 6.$

36. $\left| \dfrac{2x + 3}{4 - x} \right| \leq 1.$

37. $\left| \dfrac{1 - 2x}{2x - 3} \right| > 2.$

38. $|x^2 + x - 1| > 1.$

3.5 Intervals

This is a good place to introduce some terminology that will be used throughout the book.

Let a and b be numbers such that $a < b$. The set of all numbers x such that $a < x < b$ is called the **open interval** (a,b). "Open" means that the endpoints a and b are not included in the set. The interval (a,b) consists of all points on the segment of the number line with endpoints a and b except for the endpoints themselves. The set of all x such that $a \leq x \leq b$ is called the **closed interval** $[a,b]$; it consists of all points of the above segment, endpoints included. The number $b - a$ is called the **length** of the interval, and the point $(a + b)/2$ is called the **midpoint** of the interval.

The intervals considered above are called **finite.**

The set of all x such that $x > a$ is called the **infinite interval** $(a, +\infty)$. Similarly, $[a, +\infty)$ is the set of all x with $x \geq a$, $(-\infty, b)$ the set of all x with $x < b$, $(-\infty, b]$ the set of all x with $x \leq b$. The whole number line, finally, is denoted by $(-\infty, +\infty)$.

The symbols $+\infty$, $-\infty$ are read: plus infinity and minus infinity. These are *not* names of numbers.

We remark that every interval contains rational and irrational numbers ▶.

PROBLEMS

1. Does 2 lie in the (open) interval $(-3,5)$?
2. Does 5 lie in the (open) interval $(5,6)$?
3. Does 5 lie in the (closed) interval $[5,15]$?
4. Does -1 lie in the (open) interval $(-2,-1)$?
5. Does 0 lie in the (closed) interval $[0,2]$?
6. Does 4 lie in the (open) interval $(-1,5)$?
7. What is the midpoint of the interval $(-3,1)$?
8. What is the closed interval whose midpoint is at $\frac{1}{2}$ and whose length is 2?
9. Find an interval that contains $1 + 10^{-j}$ for $j = 1, 2, 3 \ldots$ but does not contain 1.

§4 The square root of 2 is irrational‡

We recall that every integer is either even, that is, divisible by 2, or odd, that is, not divisible by 2. We remark next that *if m is an integer and m^2 is even, then m is also even.* Indeed, an even integer m can be written as $m = 2n$, and an odd integer m can be written as $m = 2n + 1$ (in both cases, n is an integer). Hence either $m^2 = 4n^2$ (which is even) or $m^2 = (2n + 1)^2 = 4n^2 + 4n + 1$ (which is odd). If m^2 is even, then $m^2 = 4n^2$ and $m = 2n$, so that m is even.

Now let x be a rational number. Then $x = p/q$, p and q integers. We can assume that p and q are not both even, for we can cancel any 2 that appears as a factor

‡Optional section.

in both numerator and denominator. Assume next that $x^2 = 2$. We show that this assumption leads to a contradiction and is therefore untenable.

If $x^2 = 2$, then $p^2/q^2 = 2$ or $p^2 = 2q^2$. Hence p^2 is even. By the remark made above, p is even. Hence $p = 2r$, r a positive integer, and $p^2 = 4r^2$. But $p^2 = 2q^2$. Hence $4r^2 = 2q^2$ or $q^2 = 2r^2$. Therefore q^2 is even, and, by the remark, q is even. Thus both p and q are even, which is contrary to assumption.

This beautiful argument was known to the ancient Greeks.

REMARK   If n and a are positive integers, and if the equation $x^n = a$ has no integral solution, it also has no rational solution. We shall not prove this here.

COORDINATES

§1 Points

While the concept of real numbers developed gradually and was first fully understood only long after calculus had become a flourishing mathematical discipline, the second prerequisite of calculus—analytic geometry—was invented at a definite time independently by two men: Descartes, a philosopher, and Fermat, a judge. Both are among the best mathematicians of all time.

RENÉ DESCARTES (1596–1650) is one of the founders of modern Western thought. He based his philosophy on independent reasoning and research rather than on reliance on authorities. His philosophy also contains an attempt to explain all of natural phenomena, including plant life and animal life, in strictly mechanical terms. Descartes' only mathematical book, *Geometry*, appeared in 1635. This book contains not only the idea of analytical geometry but also important theorems about algebraic equations and some algebraic notations used today.

As a young man, Descartes traveled extensively and was a soldier. Then he settled in Holland, which he left only shortly before his death to go to Stockholm as tutor to Queen Christina.

PIERRE DE FERMAT (1601–1665) was a Counsellor at the Parlement of Toulouse, a judicial position that gave him time for mathematics. He was the co-founder of analytic geometry (with Descartes) and the theory of probability (with Pascal), one of the pioneers of calculus, and a great number theorist. Fermat never published his discoveries.

One of the best-known unsolved mathematical problems concerns the proof, or disproof, of "Fermat's last theorem"; if $n \geq 3$ is an integer and x and y are rational numbers, then $x^n + y^n \neq 1$. "I have found a truly remarkable proof," wrote Fermat in the margin of a book, "but have no room to reproduce it here." (Recent work, utilizing electronic computers, shows that Fermat's theorem is true for all $n < 100,000$.)

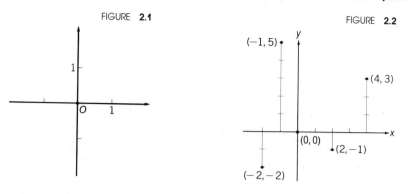

FIGURE **2.1**

FIGURE **2.2**

This chapter contains a brief introduction to analytic geometry; it may be reviewed rapidly by readers familiar with the material.

1.1   Cartesian coordinate system

The basic idea of analytic geometry is as follows: one can specify the position of a point in a plane by two numbers, and one can translate any statement about points into a statement about numbers.

We draw two perpendicular lines in the plane and call one the **horizontal axis,** the other the **vertical axis.** The intersection point O of the axes is called the **origin.** We give each of the axes a direction; the direction of the horizontal axis is called "right" and that of the vertical is called "up"; by "left" we mean the direction opposite to "right," by "down" the direction opposite to "up." We also choose a **unit of length.** (The customary way of drawing the axes is shown in Figure 2.1.) The two axes and the unit of length are called a **Cartesian coordinate system** in the plane. (Cartesius is the Latinized form of the name Descartes.) Instead of "Cartesian coordinate system" we often say "rectangular coordinate system," or, when it is clearly understood from the context, "coordinate system."

Let P be a point in the plane; we associate with P two numbers, called its **coordinates.** The first coordinate is zero if P lies on the vertical axis. If P lies to the right of the vertical axis, the first coordinate is the distance from P to the vertical axis; if P lies to the left of the vertical axis, the first coordinate is the negative of the distance from P to the vertical axis. Similarly, the second coordinate is zero if P lies on the horizontal axis. If P lies above (below) the horizontal axis, the second coordinate is the distance (negative of the distance) from P to the horizontal axis.

We often denote the first coordinate of a point by the letter x, and the second by y. Accordingly, we call the horizontal axis the **x-axis** and the vertical axis the **y-axis.** The first coordinate x is often called the *abscissa;* the second coordinate y is called the *ordinate.*

EXAMPLE   Draw a coordinate system and locate the points with coordinates $(0,0)$, $(4,3)$, $(-1,5)$, $(-2,-2)$, and $(2,-1)$.

ANSWER   The point $(0,0)$ is the origin. The point $(4,3)$ lies four units to the right of the y-axis and three units above the x-axis; the point $(-1,5)$ lies one unit to the left of the y-axis and five units above the x-axis, and so forth. See Figure 2.2.

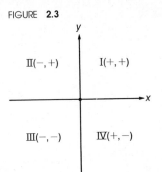

FIGURE   2.3

Here is another way of interpreting the coordinates. The point with coordinates (a,b) is reached from the origin if we first go a units in the x direction and then b units in the y direction. To go a units in the x direction means: not to move at all if $a = 0$, to move the distance $|a|$ in the direction of the x-axis if $a > 0$, to move the distance $|a|$ in the opposite direction if $a < 0$. An analogous explanation holds for the y direction. [For example, the point $(-1,5)$ is reached by going one unit to the left, then five units up; see Figure 2.2.]

A point P is determined by its coordinates, and for any two numbers a and b there is a point P having a as its first coordinate and b as its second. We shall use the ordered pair of coordinates (a,b) as a name for the point P.

When we say that the coordinates of a point form an **ordered pair,** we mean that the order of the two numbers in the pair is important. Thus $(2,3)$ and $(3,2)$ are two distinct ordered pairs. They are names of two distinct points.

The set of points with positive first and positive second coordinates is called the **first quadrant.** The second, third, and fourth quadrants, respectively, are the sets: $x < 0$ and $y > 0$; $x < 0$ and $y < 0$; $x > 0$ and $y < 0$. The four quadrants are shown in Figure 2.3.

PROBLEMS

In Problems 1 to 8 draw a coordinate system, and locate the points with the given coordinates.

1. $(2,3)$.
2. $(-\frac{1}{2},1)$.
3. $(0,4)$.
4. $(-1,-2)$.
5. $(-2,3)$.
6. $(3,-2)$.
7. $(2,-3)$.
8. $(-\frac{9}{2},0)$.

In doing the following problems, make a drawing for each (preferably on graph paper).

9. Find the coordinates of all points in the left half-plane that are at a distance 3 from the x-axis and a distance 2 from the y-axis.
10. Find the coordinates of all points in the upper half-plane that are at a distance 7 from the x-axis and a distance 4 from the y-axis.
11. Find the coordinates of all points with positive abscissas that are at a distance 4 from the x-axis and a distance 6 from the y-axis.
12. Find the coordinates of all points that are at a distance 3 from the x-axis and a distance 4 from the y-axis.

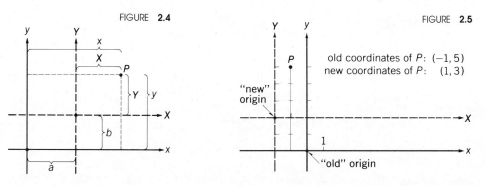

FIGURE **2.4**

FIGURE **2.5**

old coordinates of P: $(-1, 5)$
new coordinates of P: $(1, 3)$

"new" origin

"old" origin

1.2 Translating coordinates

By choosing a Cartesian coordinate system, we establish a one-to-one correspondence between points in the plane and ordered pairs of real numbers. This means: to each point there corresponds a unique pair of numbers, and to every ordered pair of numbers there corresponds a unique point.

But the choice of the coordinate system is arbitrary. The same point will, of course, have different coordinates in different coordinate systems. We consider here the simplest case of this situation.

We choose first one coordinate system and then another "new" system with a different origin but with the axes parallel to the old axes and having the same directions. We say in this case that the new coordinate system is obtained from the old one by a **translation** (see Figure 2.4).

Let a point P have coordinates (x, y) in the old system and coordinates (X, Y) in the new system, and let (a, b) be the coordinates of the new origin in the old system. Then

$$x = a + X \quad \text{and} \quad y = b + Y. \tag{1}$$

This can be read off Figure 2.4 for the case when the new origin lies in the first quadrant and P in the (new) first quadrant. The reader can convince himself that the formulas also hold in all other cases.

EXAMPLE If the coordinates of the new origin in the old system are $(-2, 2)$ and the coordinates of a point P in the old system are $(-1, 5)$, what are the coordinates of P in the new system?
ANSWER Since the coordinates of the new origin in the old system are $(-2, 2)$, we have that $a = -2$, $b = 2$. Using Equation (1), $x = -2 + X$ and $y = 2 + Y$. Substituting for x and y, we find that $-1 = -2 + X$ and $5 = 2 + Y$, that is, $X = 1$, $Y = 3$. See Figure 2.5.

PROBLEMS

In each of the following problems the new coordinate system is obtained from the old one by a translation. Illustrate each problem by a sketch.

1. If the coordinates of the new origin in the old system are $(0, 3)$ and the coordinates of a point P in the old system are $(-1, 4)$, what are the coordinates of P in the new system?
2. If the coordinates of the new origin in the old system are $(-1, 5)$ and the coordinates of a point P in the old system are $(0, 8)$, what are the coordinates of P in the new system?

3. If the coordinates of a point P in the old system are $(1,-2)$ and in the new system are $(3,1)$, what are the coordinates of the new origin in the old system?

4. If the coordinates of a new origin in the old system are $(-1,-4)$ and the coordinates of a point P in the new system are $(-6,2)$, what are the coordinates of P in the old system?

5. If the coordinates of the new origin in the old system are $(3,-2)$ and the coordinates of a point P in the new system are $(\frac{1}{2},\frac{7}{2})$ what are the coordinates of P in the old system?

*6. If the coordinates of the old origin in the new system are $(-3,2)$, what are the coordinates of the new origin in the old system?

1.3 Distance formula

We recall the *Pythagorean theorem*. In a right triangle the area of the square erected on the hypotenuse equals the sum of the areas erected on the legs (see Figure 2.6). This is, by the way, one of the earliest discoveries in mathematics. The discovery was made by the Babylonians; the general statement and proof are due to the Greeks.

FIGURE **2.6**

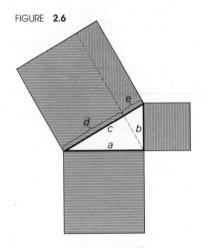

For the Greeks, the Pythagorean theorem was a statement about areas. Having real numbers, we can restate it as a proposition about numbers. If a and b are the lengths of two legs of a right triangle and c is the length of the hypotenuse, then $a^2 + b^2 = c^2$.

(There are many proofs of the Pythagorean theorem. One proof, based on the properties of similar right triangles, follows from Figure 2.6. We have that $d/a = a/c$, $e/b = b/c$, that is, $cd = a^2$, $ce = b^2$. But $c = d + e$. Hence $c^2 = cd + ce = a^2 + b^2$.)

We use the Pythagorean theorem to establish an important formula.

Theorem 1. *The distance d between the points (x_1,y_1) and (x_2,y_2) is*

$$d = \sqrt{(x_1 - x_2)^2 + (y_1 - y_2)^2}.$$ (2)

To prove this distance formula, we consider first a special case: one point, O, is the origin $(0,0)$. The other point, P, may then be denoted by (x,y), and we must show that the distance from $(0,0)$ to (x,y) is $\sqrt{x^2 + y^2}$, that is,

$$|OP|^2 = |x|^2 + |y|^2.$$

This is certainly so if $(x,y) = (0,0)$, or if either $x = 0$ or $y = 0$. In all cases, our assertion is the Pythagorean theorem; see Figure 2.7.

FIGURE **2.7**   FIGURE **2.8**

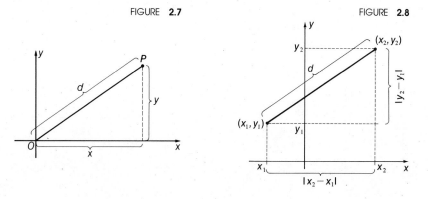

Now we consider the general case. We translate the coordinate system so that the new origin is the point (x_1, y_1). Then the new coordinates of the second point are $(x_2 - x_1, y_2 - y_1)$, as is seen by using (1) with $(a,b) = (x_1, y_1)$ and $(x,y) = (x_2, y_2)$. By the result just proved, the distance between the two points is $\sqrt{(x_2 - x_1)^2 + (y_2 - y_1)^2}$, as asserted. A direct proof of (2) from the Pythagorean theorem is suggested in Figure 2.8.

EXAMPLE   Find the distance between the points whose coordinates are $(-1, 2)$ and $(3, -1)$.
ANSWER   Using Equation (2) with $x_1 = -1$, $y_1 = 2$ and $x_2 = 3$, $y_2 = -1$, we have
$$d = \sqrt{(3 - (-1))^2 + (-1 - 2)^2} = \sqrt{4^2 + 3^2} = \sqrt{25} = 5.$$

PROBLEMS

In Problems 1 to 8 find the distance between the points with the given coordinates.

1. $(0,1)$ and $(1,0)$.
2. $(-1, -2)$ and $(3,1)$.
3. $(-2, 0)$ and $(-1, 1)$.
4. $(2, \frac{1}{2})$ and $(3, -\frac{1}{2})$.

5. $(3, -1)$ and $(-1, 3)$.
6. $(-1, -2)$ and $(-3, -4)$.
7. $(0, -a)$ and $(2a, 0)$, a any number.
8. $(2a, -7a)$ and $(5a, -3a)$.

In doing the following problems make a drawing for each (preferably on graph paper).

9. Find the coordinates of all points in the third quadrant that are at a distance 5 from the origin and a distance 3 from the x-axis. [*Hint:* Use the Pythagorean theorem.]
10. Find the coordinates of all points in the right-half plane that are at a distance 5 from the origin and a distance 5 from the y-axis.
11. Find the coordinates of all points that are at a distance 13 from the point $(1,0)$ and at a distance 5 from the x-axis.
12. Find the coordinates of all points that are at a distance $3\sqrt{5}$ from the point $(3, -6)$ and at a distance $2\sqrt{5}$ from the point $(-2, 4)$.

§2   Lines

In this and the following section we assume that a Cartesian coordinate system in the plane has been selected.

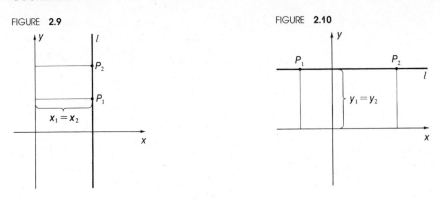

FIGURE **2.9**   FIGURE **2.10**

2.1 Slope

Let l be a straight line in the plane. Let P_1 and P_2 be two distinct points on l, and let them have the coordinates (x_1, y_1) and (x_2, y_2), respectively. If $x_1 = x_2$, then l is parallel to the y-axis, as is seen from Figure 2.9. Such a line is called **vertical**. Conversely, if l is vertical, all points on it have the same x-coordinate.

If l is not vertical, then $x_1 \neq x_2$, and we can compute the number

$$m = \frac{y_2 - y_1}{x_2 - x_1}. \tag{1}$$

The number m will be equal to 0 if $y_1 = y_2$, that is, if l is parallel to the x-axis (such a line is called **horizontal**), and only in this case (see Figure 2.10). In all cases, m is not changed if we interchange the points P_1 and P_2, for then the numerator and denominator in (1) both change sign. Also, m is not changed if we replace the points P_1, P_2 by other points on the line l. To see this, it is enough to verify that if $P_3 = (x_3, y_3)$ is a point on l, distinct from P_1 and P_2, then

$$\frac{y_2 - y_1}{x_2 - x_1} = \frac{y_3 - y_2}{x_3 - x_2}. \tag{2}$$

We assume that $x_1 < x_2 < x_3$; this can be achieved by renaming the points if necessary. From the shaded similar right triangles in Figure 2.11, we see that statement (2) is the same as $|QP_2|/|P_1Q| = |SP_3|/|P_2S|$. This is so since a line crossing parallel lines cuts equal angles, and similar triangles have proportional sides. In our figure the line tilts upward; a similar argument is valid if the line tilts downward.

The number m defined by (1) is called the **slope** of the line l. It is uniquely determined by the line and by the coordinate system used. Furthermore, the slope is not changed if we translate the coordinate system (see §1.2); indeed, if we add one number to both x-coordinates and another to both y-coordinates, the numerator and denominator in (1) remain the same. We do not define the slope of a vertical line.

We multiply both sides of Equation (1) by $x_2 - x_1$ and obtain

$$y_2 - y_1 = m(x_2 - x_1) \tag{3}$$

for any two points (x_1, y_1) and (x_2, y_2) on a line with slope m. Note that Equation (3) holds also if $x_1 = x_2$, $y_1 = y_2$.

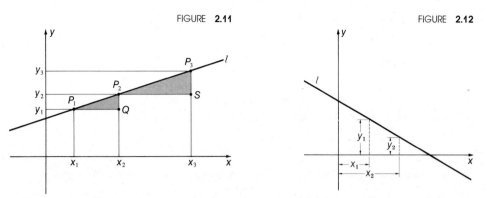

FIGURE 2.11   FIGURE 2.12

This formula shows the geometric meaning of the slope. If $m = 0$, the line is horizontal; any two points on it have the same y-coordinate. If $m > 0$, the line rises as we move from left to right, and the y-coordinate of a point on the line increases m times as fast as the x-coordinate. Thus a small positive slope means a gentle rise, a large positive slope, a steep rise. If $m < 0$, the line falls as we move from left to right; the y-coordinate decreases m times as fast as the x-coordinate (see Figure 2.12).

Let l be a line through the point (x_1, y_1) with slope m. If (x, y) denotes any other point on the same line, it follows from Equation (3) that $y - y_1 = m(x - x_1)$. We reformulate this remark as the following.

Theorem 1. *Let l be a line of slope m through the point (x_1, y_1). A point (x, y) lies on l if and only if*

$$y - y_1 = m(x - x_1). \tag{4}$$

We also say: "Equation (4) is the equation of the line l," or "l is defined by Equation (4)."

PROBLEMS

In each of Problems 1 to 8 find the slope of the line passing through the given pairs of points.

1. $(2, 4)$ and $(5, 7)$.
2. $(-1, 3)$ and $(5, 4)$.
3. $(0, 0)$ and $(-\frac{1}{2}, -7)$.
4. $(-2, -4)$ and $(1, -3)$.

5. $(-2, -4)$ and $(-4, -8)$.
6. $(2, -4)$ and $(2, -6)$.
7. $(a, 3a)$ and $(-4a, 5a)$, $a \neq 0$.
8. $(2a, 2b)$ and $(b, 4a)$, $a \neq 0, b \neq 0$.

9. If the point Q is three units above and two units to the right of the point P, what is the slope of the line running through P and Q?
10. If the point Q is six units above and one unit to the left of the point P, what is the slope of the line running through P and Q?
11. If the point P is two units below and ten units to the right of the point Q, what is the slope of the line running through P and Q?
12. If the point P is four units above and six units to the left of the point Q, what is the slope of the line running through P and Q?
13. A line l has slope -2 and passes through the point $(1, 3)$. Does the point $(4, -3)$ lie on the line?
14. A line l has slope 3 and passes through the point $(1, -2)$. Does the point $(2, 3)$ lie on l?

2.2 Linear equations

The translation of geometric statements into algebraic formulas, that is, into statements about numbers, is based on the distance formula and on another result, which we now state.

Theorem 2. (i) *Let A, B, C be numbers, such that A and B are not both zero. Then the set of all points whose coordinates satisfy the equation*

$$Ax + By + C = 0 \tag{5}$$

is a (straight) line.
  (ii) *Every line is the set of all points satisfying an equation of this form.*

  Proof. (i) Consider a given equation of the form (5), A and B not both zero. Assume first that $A \neq 0$, $B \neq 0$, and rewrite (5) in the form

$$y = -\frac{A}{B}x - \frac{C}{B} \quad \text{or} \quad y - 0 = -\frac{A}{B}\left(x + \frac{C}{A}\right).$$

This is of the form (4) with $m = -(A/B)$, $y_1 = 0$, $x_1 = -(C/A)$, that is, an equation of the line through the point $(-(C/A),0)$, with slope $m = -(A/B)$.
  If $B = 0$, then $A \neq 0$ and Equation (5) says that $x = -(C/A)$; the set of points (x,y) satisfying this equation is a vertical line. If $A = 0$, then $B \neq 0$ and Equation (5) says that $y = -(C/B)$; the set of points (x,y) satisfying this equation is a horizontal line.
  (ii) Now let l be a given line. If l is vertical, it is the set of points (x,y) satisfying the equation $x = a$ for some fixed number a. The equation $x = a$ is of the form (5) with $A = 1$, $B = 0$, $C = -a$.
  If l is not vertical, let m be its slope, and let (x_1,y_1) be some point on l. By Theorem 1, l is the set of points (x,y) satisfying Equation (4); this can be rewritten as

$$y - y_1 = mx - mx_1 \quad \text{or} \quad mx - y + (y_1 - mx_1) = 0.$$

This is an equation of the form (5) with $A = m$, $B = -1$, $C = y_1 - mx_1$.
  Theorem 2 is now completely proved. The argument shows that, given a line l, its equation is determined uniquely, except that it may be multiplied by some number different from zero.
  Equations of the form (5) are called **linear.**
  We list some convenient ways of writing equations of nonvertical lines. If l intersects the x-axis at the point $(a,0)$, a is called the **x-intercept** of l. Similarly, if l intersects the y-axis, the intersection point has coordinates $(0,b)$; b is the **y-intercept** of l.
  Point-slope form. The equation of a line with slope m and containing the point (x_1,y_1) is

$$y - y_1 = m(x - x_1). \tag{4'}$$

This follows from Theorem 1.

  Slope-intercept form. The equation of a line with slope m and y-intercept b is

$$y = mx + b. \tag{6}$$

This follows from Theorem 1 by setting $x_1 = 0$ and $y_1 = b$.

Two-point form. The equation of a line passing through two distinct points (x_1, y_1) and (x_2, y_2) is

$$(y - y_1)(x_2 - x_1) = (y_2 - y_1)(x - x_1). \tag{7}$$

Proof. Equation (7) is a linear equation in the variables x and y, since x_1, y_1, x_2, y_2 are fixed numbers. Hence the solution set of (7) is a line. We can check that (x_1, y_1) and (x_2, y_2) satisfy (7). Since a line is determined by two points, our assertion follows.

Intercept form. The equation of a line with x-intercept a and y-intercept b, $a \neq 0$, $b \neq 0$, is

$$\frac{x}{a} + \frac{y}{b} = 1. \tag{8}$$

Proof. Equation (8) is a linear equation satisfied by $(a, 0)$ and $(0, b)$.

EXAMPLES 1. Draw the line whose equation reads $2x + 3y - 1 = 0$.
SOLUTION In order to draw the line, it suffices to know two distinct points on the line. We first look for a point on the line with $x = 0$. For $x = 0$, the equation gives $3y - 1 = 0$ or $y = \frac{1}{3}$. Thus the point $(0, \frac{1}{3})$ is on our line. Similarly, for $y = 0$, the equation reads $2x - 1 = 0$ or $x = \frac{1}{2}$. Thus the point $(\frac{1}{2}, 0)$ lies on our line. It is now easy to draw it.

2. Does the point $(-3, 2)$ lie on the line $5x + 4y - 7 = 0$?
ANSWER No, since $5(-3) + 4(2) - 7 = -14 \neq 0$,

3. Find the equation of the line through the points $(3,4)$ and $(4,3)$.
FIRST METHOD Using the two-point form (7) with $x_1 = 3$, $x_2 = 4$, $y_1 = 4$, $y_2 = 3$, we obtain $(y - 4)(4 - 3) = (3 - 4)(x - 3)$ or, simplifying, $x + y - 7 = 0$.
SECOND METHOD Using formula (1) we compute the slope $m = (3 - 4)/(4 - 3) = -1$. Now we use the point-slope form (4') and obtain the equation $y - 4 = (-1)(x - 3)$ or $y - 4 = -x + 3$, that is, $x + y - 7 = 0$.
THIRD METHOD The desired equation is of the form $Ax + By + C = 0$ and is to be satisfied by $(3,4)$ and $(4,3)$. Thus we have

$$3A + 4B + C = 0, \qquad 4A + 3B + C = 0.$$

Treat C as given, and solve these two equations simultaneously for A and B. This yields $A = B = -C/7$. The equation of the line reads

$$-\frac{C}{7}x - \frac{C}{7}y + C = 0.$$

The value of C can be chosen by us as long as $C \neq 0$. Choosing $C = -7$, we obtain the same equation as above. Any other choice of C gives the same line. For example, if $C = 1$, we obtain $-\frac{1}{7}x - \frac{1}{7}y + 1 = 0$. Our choice of $C = -7$ avoids fractions.

PROBLEMS

In Problems 1 to 8 find the slope of the given line.

1. $2x + 4y = 1$.
2. $y = 3x + 2$.
3. $2x - 3y = 0$.
4. $4x = 5y + 7$.
5. $8x + 6y + 7 = 0$.
6. $y = -2$.
7. $x = 4$.
8. $\dfrac{x}{a} + \dfrac{y}{b} = 1$, $\quad a \neq 0, b \neq 0$.

9. The equation of line l_1 is $3y + 2x = 6$ and that of line l_2 is $5x - 2 = 2y$. Which line descends as we move from left to right?

10. The equation of line l_1 is $6y - 4x - 2 = 0$, that of line l_2 is $2y - 40x + 7 = 0$, and that of l_3 is $18y - 17x + 51 = 0$. Which line rises most steeply? Which line rises most gently?

In Problems 11 to 20 find the x- and y-intercepts of the given line.

11. $3x + 4y + 1 = 0$.
12. $5x - 2y = 1$.
13. $-2x - 4y = 6$.
14. $4y - 2x = 9$.
15. $-6x + y = 0$.

16. $y = 3$.
17. $x = -5$.
18. $y = mx + b$, $\quad m \neq 0$.
19. $Ax + By + C = 0$, $\quad A \neq 0$, $B \neq 0$.
20. $(y - y_1) = m(x - x_1)$, $\quad m \neq 0$.

21. Draw the line whose equation is $5x - 2y + 10 = 0$. Draw the line whose equation is $5x - 2y + 5 = 0$.
22. Draw the line whose equation is $y - 2 = -3(x - 1)$. Draw the line whose equation is $3x + y = 0$.
23. Draw the line $y = \frac{1}{2}x + 1$. On the same set of axes draw the line $y = -2x + 1$. Where do these lines intersect?
24. Draw the line $y = -\frac{1}{3}x - 2$. On the same set of axes draw the line $y = 3x - 2$. Where do these lines intersect?

In Problems 25 to 41 find the equation of the line satisfying the given requirements.

25. The line passes through the point $(-2, -3)$ and has slope 4.
26. The line passes through the point $(1,7)$ and has slope $-\frac{1}{2}$.
27. The line passes through the point $(100,30)$ and has slope $\frac{3}{4}$.
28. The line has slope 6 and y-intercept 2.
29. The line has slope $-\frac{1}{2}$ and y-intercept 6.
30. The line has slope zero and y-intercept 8.
31. The line passes through the points $(-6,2)$ and $(3,-2)$.
32. The line passes through the points $(-1,-2)$ and $(-4,6)$.
33. The line passes through the points $(0,2)$ and $(-24,0)$.
34. The line has x-intercept 2 and y-intercept 4.
35. The line has x-intercept $-\frac{1}{2}$ and y-intercept 3.
36. The line has x-intercept -8 and y-intercept -8.
37. The line is parallel to the x-axis and passes through the point $(-2,1)$.
38. The line is parallel to the y-axis and passes through the point $(4,6)$.
39. The line has slope 10 and x-intercept 5.
40. The line has x-intercept -3 and passes through the point $(1,4)$.
41. The line has y-intercept 2 and passes through the point $(-2,-4)$.
42. What is the slope of a line which does not pass through the origin and whose x-intercept equals its y-intercept?
43. The line $y = mx + b$ passes through $(1,2)$ and $(3,5)$. Find m and b.
44. Find the equation of the line that cuts off from the third quadrant an isosceles triangle of area 4.
45. Find the point on the line $2x - y - 1 = 0$ that is equidistant from the points $(3,2)$ and $(2,3)$.

2.3   Parallel and perpendicular lines

We consider two distinct lines (in the plane) and ask whether they are parallel. If both lines are vertical, they are. If one line is vertical and the other is not, they intersect and are not parallel. It remains to consider the case of two nonvertical lines.

Theorem 3. *Two distinct nonvertical lines are parallel if and only if they have the same slope.*

FIGURE 2.13

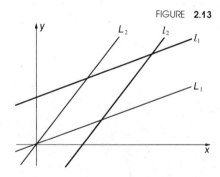

Our statement is rather obvious geometrically. Still we give an analytic proof. Let l_1 and l_2 be two nonvertical lines; let them have slopes m_1 and m_2. The two lines have the equations

$$y = m_1 x + b_1, \qquad y = m_2 x + b_2 \tag{9}$$

(see §2.2). To find an intersection point we must find coordinates (x,y) satisfying both equations, that is, we must solve Equations (9) simultaneously. If (x,y) satisfies both equations, then $m_1 x + b_1 = m_2 x + b_2$ or

$$(m_1 - m_2)x = b_2 - b_1. \tag{10}$$

Assume first that $m_1 = m_2$. Then (10) is possible only if $b_2 = b_1$. In this case l_1 and l_2 coincide. Hence: *distinct* lines with the same slope are parallel.

Assume next that $m_1 \neq m_2$. Then we can solve (10) for x and use this value of x to compute y from either of the Equations (9). We obtain a point with coordinates

$$x = \frac{b_2 - b_1}{m_1 - m_2}, \qquad y = \frac{m_2 b_1 - m_1 b_2}{m_1 - m_2}.$$

It does satisfy both equations and is the intersection point. Hence our two lines are not parallel.

We now ask when two lines (in the plane) are perpendicular. Vertical lines are perpendicular to horizontal lines and no others. For nonvertical lines we have the following.

Theorem 4. *Two nonvertical lines are perpendicular if and only if their slopes, m_1 and m_2, satisfy*

$$m_1 m_2 = -1. \tag{11}$$

Proof. Let l_1, l_2 be two lines. Let L_1 and L_2 be two lines through the origin such that L_1 is parallel to l_1 and L_2 to l_2 (Figure 2.13). Then the lines l_1, l_2 are perpendicular if and only if L_1, L_2 are perpendicular. It is therefore enough to prove the theorem for the two lines passing through the origin.

Consider next two such lines that are neither vertical nor horizontal. Then both have slopes, m_1 and m_2, and $m_1 \neq 0$, $m_2 \neq 0$. Since our lines pass through the origin, their equations are

$$y = m_1 x, \qquad y = m_2 x.$$

FIGURE 2.14   FIGURE 2.15

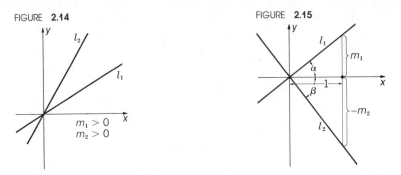

If $m_1 > 0$, $m_2 > 0$, both lines pass through the first quadrant. Then they are not perpendicular (see Figure 2.14), and $m_1 m_2 > 0$ so that (11) does not hold. If $m_1 < 0$, $m_2 < 0$, both lines pass through the second quadrant. Again the lines are not perpendicular, and $m_1 m_2 > 0$.

Assume next that m_1 and m_2 have opposite signs, say $m_1 > 0$ and $m_2 < 0$, as in Figure 2.15. The two lines are perpendicular if and only if the angles α and β are complementary (add up to a right angle). This is so if and only if the two right triangles are similar, and they are similar if and only if $|m_1|/1 = 1/|m_2|$, which is the same as $|m_1 m_2| = 1$. Since m_1 and m_2 have opposite signs, this is equivalent to (11).

EXAMPLE   Find the equation of the line through (1,2) which is perpendicular to the line whose equation is $2x - 4y + 5 = 0$.

SOLUTION   The equation $2x - 4y + 5 = 0$ can be written in the form $y = \frac{1}{2}x + \frac{5}{4}$. Hence the slope of the given line is $\frac{1}{2}$. To find the slope of a line perpendicular to the given line, we use Equation (11) with $m_1 = \frac{1}{2}$, $m_2 = m$. Thus $\frac{1}{2}m = -1$ or $m = -2$. We want a line of slope (-2) passing through (1,2). An equation for this line is obtained from (4) by setting $m = -2$, $x_1 = 1$, $y_1 = 2$. Thus we obtain the equation $y - 2 = -2(x - 1)$ or $y - 2 = -2x + 2$ or $2x + y - 4 = 0$.

PROBLEMS

1. Find the equation of the line through (0,0) which is parallel to the line whose equation is $2x - 3y + 7 = 0$.
2. Find the equation of the line through (0,0) which is perpendicular to the line whose equation is $2x - 3y + 7 = 0$.
3. For which values of α is $\alpha x + 3y - 5 = 0$ the equation of a line perpendicular to the line whose equation is $x - y + 9 = 0$?
4. For which values of u is $2ux - 3y + 6 = 0$ the equation of a line parallel to the line whose equation is $6x + 4y - 3 = 0$?
5. For which values of t are $tx - 2y + 8 = 0$ and $3tx + 6y + 4 = 0$ the equations of perpendicular lines?
6. Find the equation of the line through $(-1,2)$ which is parallel to the line whose equation is $4x + 12y + 3 = 0$.
7. Find the equation of the line through $(6,-3)$ which is perpendicular to the line whose equation is $x - 3y + 12 = 0$.
8. Find the equation of the line through $(\frac{1}{2},-1)$ which is perpendicular to the line whose equation is $18x + 12y - 1 = 0$.

9. Find the equation of the line through the origin that is parallel to the line $3x - 2y - 1 = 0$.
10. Find the equation of the perpendicular bisector of the segment joining $(3, -1)$ and $(-2, -2)$.
11. The points $X = (3, -2)$, $Y = (4,1)$, and $Z = (-3,5)$ are the vertices of a triangle. Find the equation of the line through Y perpendicular to the side XZ.
*12. Show that the altitudes of triangle XYZ in Problem 11 intersect in a point.

§3 Circles

We now give an analytic discussion of circles.

3.1 The equation of a circle

A **circle** is the set of all points (in the plane) that have the same distance, say, r, from a fixed point, say, Q. The point Q is called the **center** of the circle and r the **radius**; the radius is always a positive number. The circle with center at the origin and radius 1 is called the **unit circle.**

Let the center Q have coordinates (a,b). The distance of a point (x,y) from Q is given by $\sqrt{(x - a)^2 + (y - b)^2}$. Hence a circle with center (a,b) and radius $r > 0$ is the solution set of the equation $\sqrt{(x - a)^2 + (y - b)^2} = r$, that is,

$$(x - a)^2 + (y - b)^2 = r^2. \tag{1}$$

If the center of the circle is at the origin, then $a = 0$, $b = 0$, and Equation (1) reduces to

$$x^2 + y^2 = r^2. \tag{1'}$$

The equation of the unit circle is $x^2 + y^2 = 1$.

EXAMPLE 1. Find the equation of the circle with radius 3 and center at $(-1,4)$.

SOLUTION   The coordinates of the center are $a = -1$, $b = 4$, and the radius is $r = 3$. Using Equation (1), we have

$$(x + 1)^2 + (y - 4)^2 = 3^2$$

or

$$x^2 + y^2 + 2x - 8y + 8 = 0.$$

Theorem 1. *Let A, B, and C be numbers. The solution set of the equation*

$$x^2 + y^2 + Ax + By + C = 0 \tag{2}$$

is either a circle, or a point, or the empty set.

The proof will make use of one of the oldest devices in mathematics (known already in ancient Babylon), called **completing the square.** We consider first a special case.

EXAMPLE 2. Describe the solution set of

$$x^2 + y^2 - 2x + 6y - 90 = 0.$$

SOLUTION   First we write our equation in the form

$$(x^2 - 2x) + (y^2 + 6y) - 90 = 0.$$

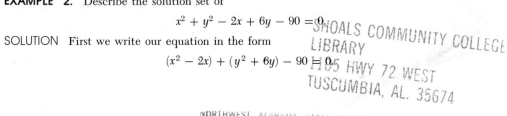

Now we complete the square: we add to and subtract from $x^2 - 2x$ the number 1 (the square of half the coefficient of x), and we add to and subtract from $y^2 + 6y$ the number 9 (the square of half the coefficient of y). The equation becomes

$$(x^2 - 2x + 1) - 1 + (y^2 + 6y + 9) - 9 - 90 = 0$$

or

$$(x - 1)^2 + (y + 3)^2 = 100.$$

The solution set is a circle of radius 10 and center $(1, -3)$.

The proof of Theorem 1 proceeds along the same lines. Equation (1) can be written as

$$(x^2 + Ax) + (y^2 + By) + C = 0$$

or

$$\left(x^2 + Ax + \frac{A^2}{4}\right) - \frac{A^2}{4} + \left(y^2 + By + \frac{B^2}{4}\right) - \frac{B^2}{4} + C = 0;$$

that is,

$$\left(x + \frac{A}{2}\right)^2 - \frac{A^2}{4} + \left(y + \frac{B}{2}\right)^2 - \frac{B^2}{4} + C = 0$$

or

$$\left(x + \frac{A}{2}\right)^2 + \left(y + \frac{B}{2}\right)^2 = D, \tag{3}$$

where $D = \frac{1}{4}(A^2 + B^2) - C$. If $D > 0$, set $r = \sqrt{D}$. Then (3) is the equation of a circle with center $(-A/2, -B/2)$ and radius r. If $D = 0$, Equation (3) says that the sum of two nonnegative numbers, $(x + A/2)^2 + (y + B/2)^2$, is zero. Hence both numbers must be zero. The solution set contains the single point $x = -A/2$, $y = -B/2$. If $D < 0$, finally, (3) cannot hold for any pair of real numbers; the solution set is empty.

PROBLEMS

1. Find the equation of the circle of radius 2 whose center is at $(0,3)$.
2. Find the equation of the circle of radius 3 whose center is at $(-1,4)$.
3. Find the equation of the circle whose center is at $(1,6)$ and which passes through $(-2,2)$.
4. Find the equation of the circle whose center is at $(-3,2)$ and which passes through $(-2,1)$.
5. Find the equation of the circle of radius 4 that passes through $(-3,0)$ and $(5,0)$.
6. Find the equation of the circle of radius 5 that passes through $(0,0)$ and $(-6,8)$.
7. Describe the solution set of $x^2 + y^2 + 2x + 4y + 4 = 0$.
8. Describe the solution set of $x^2 + y^2 + 4x - 4y + 8 = 0$.
9. Describe the solution set of $x^2 + y^2 - x - y + 1 = 0$.
10. Describe the solution set of $x^2 + y^2 + 6x - 2y - 6 = 0$.
11. Describe the solution set of $2x^2 + 2y^2 + 2x - 2y - 1 = 0$.
12. Describe the solution set of $4x^2 + 4y^2 + 8x - 4y + 5 = 0$.

13. Show that for every choice of the number b, the equation $x^2 + y^2 - 2by = 1$ is the equation of a circle passing through the points $(1,0)$ and $(-1,0)$. Where is the center of this circle?

14. Find the equation of the circle that passes through the points $(0,2)$, $(4,0)$, and $(2,-4)$. [*Hint:* Use Theorem 1.]

* 15. Write the equation of all circles passing through the points (a,b) and (c,d). (Assume these points to be distinct.)

* 16. Give an analytic proof of the theorem that through any three points in the plane which do not lie on one line there passes a unique circle.

3.2 Intersection of a line and a circle

We now give two examples of algebraic proofs of geometric theorems. They deal with the relative position of a circle and a line (in the same plane).

Theorem 2. *The intersection of a line l and a circle C is either empty, or consists of one point, or consists of two points.*

Proof. We choose the coordinate system so that the circle C is the unit circle $x^2 + y^2 = 1$, and the line l is horizontal. This is achieved by taking the center of the circle as the origin, the radius of the circle as the unit of length, and the x-axis as parallel to l. Then the equation of the line is $y = b$. A point common to C and l must have coordinates (x,b) with $x^2 + b^2 = 1$ or

$$x^2 = 1 - b^2. \tag{4}$$

If $|b| > 1$, then $1 - b^2 < 0$ so that no x satisfies (4); there are no points common to C and l. [See Figure 2.16(a).] If $|b| = 1$, that is, $b = 1$ or $b = -1$, then (4) means that $x = 0$. There is exactly one intersection point, namely, $(0,b)$. [See Figure 2.16(b).] If $|b| < 1$, then $1 - b^2 > 0$ and (4) means that either $x = \sqrt{1 - b^2}$ or $x = -\sqrt{1 - b^2}$. There are exactly two intersection points, $(\sqrt{1 - b^2}, b)$ and $(-\sqrt{1 - b^2}, b)$. [See Figure 2.16(c).] The theorem is proved.

FIGURE **2.16**

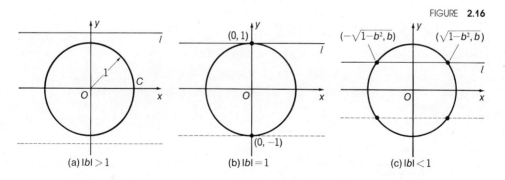

(a) $|b| > 1$ (b) $|b| = 1$ (c) $|b| < 1$

A **diameter** of a circle is that segment of a line through the center of the circle whose endpoints lie on the circle. A **tangent** to a circle is a line that has exactly one point in common with the circle. In Figure 2.17 the segment QP is a diameter, and the line l is a tangent line.

FIGURE 2.17   FIGURE 2.18

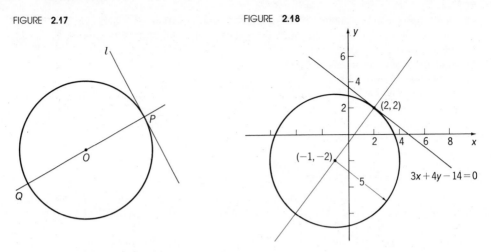

Theorem 3. *Let P be a point on a circle C, and l a line through P. Then l is tangent to C if and only if it is perpendicular to the diameter through P.*

Proof. We choose the coordinate system so that the circle C is the unit circle $x^2 + y^2 = 1$ and the line l is horizontal. Let P have coordinates (a,b). Since l passes through P, the equation of l is $y = b$. Since P is on C, $a^2 + b^2 = 1$. Hence $b^2 \leq 1$ and $-1 \leq b \leq 1$. If l is a tangent line, then, by the argument used in Theorem 2, either $b = 1$ or $b = -1$. This means: either $P = (0,1)$ or $P = (0,-1)$. In both cases OP is vertical, hence perpendicular to l. [See Figure 2.16(b).] If l is not a tangent line, then, by the preceding argument, $|b| < 1$. Hence P is either the point $(\sqrt{1 - b^2},b)$ or the point $(-\sqrt{1 - b^2},b)$. The line OP has either slope $b/\sqrt{1 - b^2}$ or $-b/\sqrt{1 - b^2}$. It is not vertical and thus is not perpendicular to l. [See Figure 2.16(c).]

Note that the proofs of Theorems 2 and 3 were easy because we chose the coordinate system conveniently.

Finding tangent lines to curves other than circles was one of the problems which led to the invention of calculus. (See §1 of Chapter 4.)

EXAMPLE Find the equation of the tangent to the circle of radius 5 with center at $(-1,-2)$, at the point $(2,2)$.

SOLUTION First we check that the distance between $(2,2)$ and $(-1,-2)$ is indeed 5. It is, since

$$\sqrt{(2 + 1)^2 + (2 + 2)^2} = \sqrt{25} = 5.$$

Next, the diameter through $(2,2)$ is the line segment passing through $(2,2)$ and $(-1,-2)$. Its slope is therefore

$$\frac{2 - (-2)}{2 - (-1)} = \frac{4}{3}.$$

The tangent must have slope $-\frac{3}{4}$ and must pass through $(2,2)$. Its equation is therefore $y - 2 = -\frac{3}{4}(x - 2)$ or $4y - 8 = -3x + 6$ or $3x + 4y - 14 = 0$. See Figure 2.18.

PROBLEMS

1. A circle whose center is at $(1,2)$ passes through the point $(-1,1)$. What is the slope of the line tangent to this circle at $(-1,1)$?

2. A line meets a circle at the point $(0,1)$ and no other. If the center of the circle is at $(-1,3)$, what is the slope of the line?

3. The equation of a circle is $x^2 + y^2 - 10x = 28$. Find the equation of the tangent line that meets the circle at the point $(3,7)$.

4. The equation of a circle is $x^2 + y^2 + 4x - 8y - 5 = 0$. What is the equation of the line tangent to the circle at $(1,0)$?

5. The equation of a circle is $x^2 + y^2 - 2x - 24 = 0$. At what point will the line tangent to the circle at $(1,5)$ intersect the line tangent to the circle at $(4,-4)$?

6. If the line $y = x$ is tangent to a circle at $(3,3)$ and the line $y = 2x$ passes through the center of the circle, what is the equation of the circle?

7. Find the equations of all lines that are tangent to the circle $x^2 + y^2 - 2x = 0$ and pass through the point $(4,0)$.

*8. Find the equations of all lines that are tangent to the circle $x^2 + y^2 = 1$ and pass through the point $(3,1)$.

§4 Conics

The power of analytic geometry becomes apparent when we study curves more complicated than lines and circles. In this section we introduce some beautiful curves, discovered by ancient Greek geometers who named them conic sections or **conics.** The Greek theory of conics, summarized in a famous treatise by Apollonius, is one of the most striking achievements of their mathematics. The Greeks studied conics for intellectual and aesthetic reasons. They could not have foreseen the manifold uses of conics in science and technology.

We shall, at this stage, describe only the simplest properties of the conics. In later chapters we shall often return to this subject. The reason for the name "conic sections" will be explained in Chapter 18, §5.1.

4.1 Parabolas

Let l be a line and F a point not on l. The **parabola** with **directrix** l and **focus** F is defined as the set of all points equidistant from l and F (Figure 2.19). This means: a point P is on the parabola whenever

$$\text{distance from } P \text{ to } F = \text{distance from } P \text{ to } l. \tag{1}$$

APOLLONIUS of Perga (ca. 260–170 B.C.) taught in Alexandria and at Pergammon. Of the eight books (that is, parts) of *Conics*, seven survived, some only in an Arabic translation.

Conics was studied for nearly 150 years before Apollonius; his book contains the work of his predecessors as well as his own discoveries. In a sense, the work of Apollonius anticipated analytic geometry. The present names of the conics are also due to him.

The so-called "problem of Apollonius" calls for a compass and ruler construction of a circle tangent to three given circles.

Apollonius also made important contributions to mathematical astronomy.

FIGURE 2.19

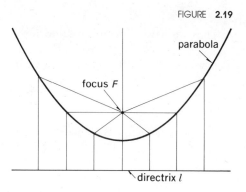

FIGURE 2.20

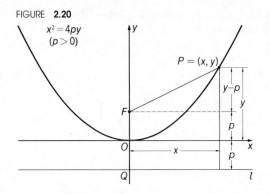

We want to represent the parabola as the solution set of an équation. The equation will depend on the choice of a coordinate system. We try to find a coordinate system that leads to a simple equation.

Drop a perpendicular from F to l; it meets l at a point Q. We choose the midpoint of FQ as the origin, and we choose the horizontal direction parallel to l. Then F is the point $(0,p)$ with $p \neq 0$ and l the line $y = -p$ (see Figure 2.20). Let P be the point (x,y). Then

$$\text{distance from } P \text{ to } F = \sqrt{x^2 + (y - p)^2}$$

by the distance formula in §1.3. Also,

$$\text{distance from } P \text{ to } l = |y + p|$$

as is seen from Figure 2.20. We conclude that Equation (1) says that

$$\sqrt{x^2 + (y - p)^2} = |y + p|,$$

which means (since two nonnegative numbers are equal if and only if their squares are equal)

$$x^2 + (y - p)^2 = (y + p)^2,$$

or

$$x^2 + y^2 - 2py + p^2 = y^2 + 2py + p^2,$$

or

$$x^2 = 4py.$$

Thus we have proved the following.

Theorem 1. *Let $p \neq 0$ be a fixed number. The solution set of*

$$x^2 = 4py \tag{2}$$

is a parabola with focus $(0,p)$ and directrix $y = -p$.

This parabola is shown in Figure 2.20 for a positive value of p. If $p < 0$, the curve "opens downward"; see Figure 2.21. Also, for every fixed $p \neq 0$, the solution set of

$$y^2 = 4px \tag{3}$$

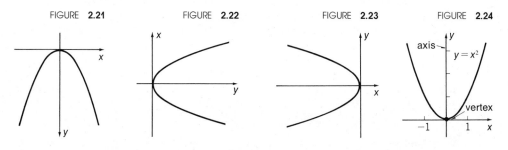

FIGURE 2.21   FIGURE 2.22   FIGURE 2.23   FIGURE 2.24

is a parabola with focus $(p,0)$ and directrix $x = -p$. This is so, since Equation (3) is obtained from (2) by interchanging x and y. The parabola (3) opens to the right or to the left according to whether $p > 0$ or $p < 0$ (see Figures 2.22 and 2.23).

4.2 Properties of the parabola

In deriving Equation (2) of the parabola with focus F and directrix l, we choose the coordinate system in a particular way, namely, we let the focus have coordinates $(0,p)$ and the directrix be the line $y = -p$. Thus the distance between F and l is $2|p|$. We get $p > 0$ if the y direction points from l to F. Assume that we also choose the unit of length so that $p = \frac{1}{4}$. Then Equation (2) becomes

$$y = x^2. \tag{4}$$

This parabola is shown in Figure 2.24. Any other parabola will have the same shape as the parabola (2), although it may have a different location and a different size. In this respect, parabolas are like circles. Given any circle, there is a coordinate system in which the equation of this circle reads $x^2 + y^2 = 1$ (namely, the system for which the center of the circle is the origin and the radius of the circle is the unit of length). Given any parabola, there is a coordinate system in which the equation of this parabola reads $y = x^2$. We express this by saying: all circles are **similar** to each other; all parabolas are similar to each other. Analogously, all squares are similar to each other, whereas two triangles need not be similar. Thus, in studying geometric properties of parabolas, we may restrict ourselves to Equation (4).

The line through the focus perpendicular to the directrix is called the **axis** of the parabola. The intersection point of the parabola with its axis is called the **vertex** of the parabola. In the case of the parabola $y = x^2$, the axis is the y-axis and the vertex is the origin.

If a point (x,y) satisfies $y = x^2$, so does the point $(-x,y)$. But the points (x,y) and $(-x,y)$ are symmetric with respect to the y-axis. This means that the line joining these points is perpendicular to the y-axis and the two points are equidistant from the y-axis. Thus: *a parabola is symmetric with respect to its axis.*

Note that the parabola extends arbitrarily far; every drawing of it shows only a part of the curve.

EXAMPLES   1.   Find the focus and the directrix for the parabola $x^2 = 40y$.
ANSWER   The equation is of the form (2) with $p = 10$. Hence the focus is at $(0,10)$ and the directrix is $y = -10$.

2. Find the equation of the parabola with focus $(0,\frac{1}{3})$ and directrix $y = -\frac{1}{3}$.

FIRST SOLUTION   We can apply Theorem 1 for $p = \frac{1}{3}$. The desired equation is Equation (2) with $p = \frac{1}{3}$; that is, $x^2 = \frac{4}{3}y$.

SECOND SOLUTION   We apply the definition of parabolas. Let (x,y) be a point on the parabola. Its distance from $(0,\frac{1}{3})$ is $\sqrt{x^2 + (y - \frac{1}{3})^2}$; its distance from the line $y = -\frac{1}{3}$ is $|y + \frac{1}{3}|$ (make a drawing to verify this). The point will lie on the parabola if and only if these two distances are equal. This condition reads

$$\sqrt{x^2 + (y - \tfrac{1}{3})^2} = |y + \tfrac{1}{3}|,$$

or (squaring both sides)

$$x^2 + y^2 - \tfrac{2}{3}y + \tfrac{1}{9} = y^2 + \tfrac{2}{3}y + \tfrac{1}{9}.$$

Simplifying we get $x^2 = \frac{4}{3}y$ as before.

PROBLEMS

In Problems 1 to 8 find the coordinates of the focus and the equation of the directrix for each parabola. Sketch each parabola, and show the focus and directrix.

1. $x^2 = 8y$.
2. $x^2 = 2y$.
3. $y^2 = 4x$.

4. $6x + y^2 = 0$.
5. $y = -\frac{1}{4}x^2$.
6. $8y - 3x^2 = 0$.

7. $y + \frac{1}{10}x^2 = 0$.
8. $12x = y^2$.

In Problems 9 to 16 find the equation of the parabola from the given information. The vertex is always at the origin.

9. Focus is at $(2,0)$.
10. Directrix is $x = 4$.
11. Focus is at $(0,-8)$.

12. Directrix is $y = \frac{3}{2}$.
13. Directrix is $x = -\frac{1}{8}$.
14. Focus is at $(-\frac{1}{2},0)$.

15. Directrix is $y + 5 = 0$.
16. Focus is at $(0,\frac{4}{3})$.

17. Find the equation of the parabola whose focus is at the origin and whose directrix is the line $y = -4$. Sketch the curve. [*Hint:* Use the definition of parabola.]

18. Find the equation of the parabola whose directrix is the line $x = 0$ and whose focus is the point $(-4,2)$. What is the vertex and axis of symmetry of this parabola? Sketch the curve.

19. Find the equation of the parabola whose directrix is the line $y = 1$ and whose focus is the point $(-2,-4)$. What is the vertex and axis of symmetry of this parabola? Sketch the curve.

4.3 Ellipses

Let F_- and F_+ be two points, and let a be a positive number such that $2a$ is greater than the distance between the two points. The **ellipse** with **foci** F_-, F_+ and **major semi-axis** a is the set of all points P such that

$$(\text{distance from } F_- \text{ to } P) + (\text{distance from } F_+ \text{ to } P) = 2a \qquad (5)$$

(see Figure 2.25). We want to represent the ellipse as the solution set of an equation; as before, we try to get a simple equation by choosing a convenient coordinate system.

We choose the coordinate system so that the two foci lie on the x-axis and the origin is the midpoint of the foci (Figure 2.26). Then the two foci are $F_+ = (c,0)$ and $F_- = (-c,0)$, where $2c$ is the distance between F_- and F_+. Let $P = (x,y)$ be any point.

FIGURE **2.25**   FIGURE **2.26**

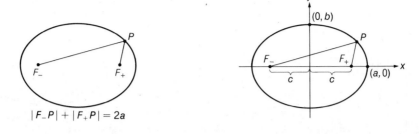

$$|F_-P| + |F_+P| = 2a$$

By the distance formula, $|F_-P| = \sqrt{(x + c)^2 + y^2}, |F_+P| = \sqrt{(x - c)^2 + y^2}.$ Condition (5) reads

$$\sqrt{x^2 + 2cx + c^2 + y^2} + \sqrt{x^2 - 2cx + c^2 + y^2} = 2a,$$

or (squaring both sides)

$$x^2 + 2cx + c^2 + y^2 + x^2 - 2cx + c^2 + y^2 + 2\sqrt{(x^2 + c^2 + y^2)^2 - 4c^2x^2} = 4a^2.$$

Collecting terms, transposing, and dividing by 2, we get

$$\sqrt{(x^2 + y^2 + c^2)^2 - 4c^2x^2} = 2a^2 - (x^2 + y^2 + c^2),$$

or (squaring both sides)

$$(x^2 + y^2 + c^2)^2 - 4c^2x^2 = 4a^4 + (x^2 + y^2 + c^2)^2 - 4a^2 (x^2 + y^2 + c^2).$$

After some manipulations we obtain

$$a^4 - a^2x^2 - a^2y^2 - a^2c^2 + c^2x^2 = 0,$$

which is the same as

$$a^2(a^2 - c^2) = (a^2 - c^2)x^2 + a^2y^2.$$

We assumed that $a > c$; hence $a^2 > c^2$.

Define the **minor semi-axis** b as $b = \sqrt{a^2 - c^2}$. Our equation becomes $b^2x^2 + a^2y^2 = a^2b^2$ or

$$\frac{x^2}{a^2} + \frac{y^2}{b^2} = 1.$$

This equation is equivalent to condition (5); thus we have proved the following.

Theorem 2. *Let $a \geq b > 0$ be fixed numbers, and set*

$$c^2 = a^2 - b^2. \tag{6}$$

The solution set of

$$\frac{x^2}{a^2} + \frac{y^2}{b^2} = 1 \tag{7}$$

is an ellipse with major semi-axis a, minor semi-axis b, and foci $(-c, 0)$, $(c, 0)$.

We did not assume that the foci must be distinct. If they coincide, then $c = 0$, $a = b$, and Equation (7) becomes $x^2 + y^2 = a^2$. The circle is a special case of an ellipse!

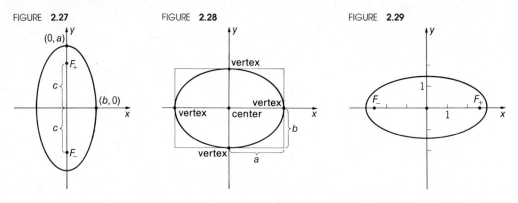

FIGURE 2.27   FIGURE 2.28   FIGURE 2.29

If we interchange x and y in (7), we obtain the equation

$$\frac{x^2}{b^2} + \frac{y^2}{a^2} = 1 \qquad (a \geq b > 0).$$ (8)

It represents an ellipse, but one for which the foci lie on the y-axis (see Figure 2.27).

4.4 Properties of ellipses

We consider first an ellipse which is not a circle. Equation (7) implies that the ellipse is symmetric about the x-axis and about the y-axis [these are called the **axes** of the ellipse (7)]. Indeed, replacing x by $-x$ or y by $-y$ does not affect Equation (7). The intersection point of the axes is called the **center** of the ellipse.

Equation (7) implies that $x^2/a^2 \leq 1$, $y^2/b^2 \leq 1$. Hence $|x|/a \leq 1$, $|y|/b \leq 1$ and $|x| \leq a$, $|y| \leq b$. Thus the ellipse lies completely inside the box formed by the lines $x = \pm a$, $y = \pm b$. The four points at which the ellipse intersects the axes (and touches the box) are called the **vertices** of the ellipse (see Figure 2.28).

A circle (that is, an ellipse with $a = b$) is, of course, symmetric about any line through its center. It has no definite axes and no vertices (or, if we prefer, every point on the circle is a vertex).

EXAMPLES 1. Find the foci and vertices of the ellipse $x^2 + 4y^2 = 9$. Sketch the curve.
SOLUTION   Divide both sides of the equation by 9 to obtain $(x/3)^2 + (y/\frac{3}{2})^2 = 1$. It is of the form (7) with $a = 3$, $b = \frac{3}{2}$. We draw the box formed by the lines $x = \pm 3$, $y = \pm \frac{3}{2}$, and we see that the vertices are $(\pm 3,0)$ and $(0, \pm \frac{3}{2})$; see Figure 2.29. Since $c = \sqrt{a^2 - b^2} = \sqrt{9 - \frac{9}{4}} = \frac{3}{2}\sqrt{3}$, the foci are at $(\pm \frac{3}{2}\sqrt{3}, 0)$.

2. Find the equation of the ellipse with foci at $(\pm 1, 0)$ and major semi-axis 3.
FIRST SOLUTION   We have that $a = 3$, $c = 1$. From (6) we see that the minor semi-axis is $b = \sqrt{a^2 - c^2} = \sqrt{9 - 1} = \sqrt{8}$. By Theorem 2 the desired equation is $x^2/9 + y^2/8 = 1$ or $8x^2 + 9y^2 = 72$.
SECOND SOLUTION   The distances from a point (x,y) to the foci are $\sqrt{(x-1)^2 + y^2}$, $\sqrt{(x+1)^2 + y^2}$. The definition of the ellipse gives the equation

$$\sqrt{(x-1)^2 + y^2} + \sqrt{(x+1)^2 + y^2} = 6,$$

or

$$(x-1)^2 + y^2 + (x+1)^2 + y^2 + 2\sqrt{(x-1)^2 + y^2}\sqrt{(x+1)^2 + y^2} = 36,$$

or

$$\sqrt{(x-1)^2 + y^2}\,\sqrt{(x+1)^2 + y^2} = 17 - x^2 - y^2.$$

Squaring both sides again and simplifying we obtain $8x^2 + 9y^2 = 72$.

PROBLEMS

In Problems 1 to 6 find the coordinates of the vertices and the coordinates of the foci for the given ellipse. Sketch the curve.

1. $\dfrac{x^2}{25} + \dfrac{y^2}{9} = 1.$ 3. $16x^2 + y^2 = 64.$ 5. $x^2 + 4y^2 = 25.$

2. $\dfrac{x^2}{9} + \dfrac{y^2}{4} = 1.$ 4. $\dfrac{x^2}{9} + \dfrac{y^2}{36} = 1.$ 6. $\dfrac{x^2}{49} + y^2 = 1.$

In Problems 7 to 12 find the equation of the ellipse, with center at the origin, from the given information.

7. Vertices at $(5,0)$ and $(0,10)$.
8. Vertices at $(3,0)$ and $(0,-2)$.
9. Focus at $(3,0)$ and vertex at $(5,0)$.
10. Focus at $(0,4)$ and vertex at $(0,-6)$.
11. Focus at $(4,0)$ and vertex at $(0,3)$.
12. Focus at $(0,1)$ and minor semi-axis of length 2.
13. Find the equation of the ellipse with foci at $(0,0)$ and $(4,0)$ and major semi-axis of length $\sqrt{5}$.
14. Find the equation of the ellipse with foci at $(0,0)$ and $(0,8)$ and major semi-axis of length 5.
15. Find the equation of the ellipse with foci at $(-1,0)$ and $(5,0)$ and major semi-axis of length $2\sqrt{3}$.
*16. Find the equation of the ellipse with center at the origin and passing through the points $(-1,4)$ and $(\sqrt{7},-2)$.

4.5 Hyperbolas

The definition of the third kind of conic, the **hyperbola,** is similar to that of the ellipse. We begin with two *distinct* points, F_- and F_+, and a positive number a such that $2a$ is *less* than the distance $2c$ between F_- and F_+. The **hyperbola** with **foci** F_-, F_+ and **major semi-axis** a is the set of all points P with

$$||F_-P| - |F_+P|| = 2a. \tag{9}$$

To translate this into an algebraic equation, we choose the coordinate system as in the case of the ellipse (see Figure 2.30), so that $F_- = (-c,0)$, $F_+ = (c,0)$. If P has coordinates (x,y), then condition (9) reads

$$|\sqrt{(x-c)^2 + y^2} - \sqrt{(x+c)^2 + y^2}| = 2a,$$

or

$$(\sqrt{(x-c)^2 + y^2} - \sqrt{(x+c)^2 + y^2})^2 = 4a^2,$$

or

$$2x^2 + 2c^2 + 2y^2 - 4a^2 = 2\sqrt{(x-c)^2 + y^2}\,\sqrt{(x+c)^2 + y^2}.$$

FIGURE **2.30**

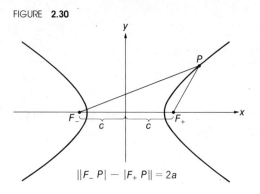

$$\|F_- P\| - \|F_+ P\| = 2a$$

Dividing by 2, squaring both sides, and simplifying, we get

$$(x^2 + y^2 + c^2 - 2a^2)^2 = (x^2 + y^2 + c^2 - 2cx)(x^2 + y^2 + c^2 + 2cx),$$

or

$$(x^2 + y^2 + c^2)^2 - 4a^2(x^2 + y^2 + c^2) + 4a^4 = (x^2 + y^2 + c^2)^2 - 4c^2x^2,$$

or

$$(c^2 - a^2)x^2 - a^2y^2 = a^2(c^2 - a^2).$$

We assumed that $c > a$, so that $c^2 > a^2$.

Define the **minor semi-axis** of the hyperbola by $b = \sqrt{c^2 - a^2}$. Our equation becomes $b^2x^2 - a^2y^2 = a^2b^2$ or

$$\frac{x^2}{a^2} - \frac{y^2}{b^2} = 1.$$

This is equivalent to condition (9); thus we have proved the following.

Theorem 3. *Let $a > 0$, $b > 0$ be given fixed numbers and set*

$$c^2 = a^2 + b^2. \tag{10}$$

The solution set of

$$\frac{x^2}{a^2} - \frac{y^2}{b^2} = 1 \tag{11}$$

is a hyperbola with semi-axes a and b and foci $(-c,0)$, $(c,0)$.

Note that the relations between the semi-axes a, b and the focal distance $2c$ are different for the ellipse and for the hyperbola.

[If we interchange x and y in (11), we obtain the equation

$$\frac{y^2}{a^2} - \frac{x^2}{b^2} = 1 \qquad (a > 0, b > 0). \tag{12}$$

It represents a hyperbola, of course, but one for which the foci lie on the y-axis rather than on the x-axis.]

FIGURE **2.31**   FIGURE **2.32**

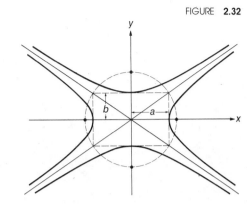

4.6 Properties of hyperbolas

The line through the foci is called the **axis** of the hyperbola. For the hyperbola (11), this is the x-axis. The midpoint between the foci is called the **center** of the hyperbola. For the hyperbola (11), this is the origin. The line through the center, perpendicular to the axis, is called the **conjugate** axis. For the hyperbola (11), this is the y-axis. Like the ellipse, the hyperbola is symmetric about both axes.

Equation (11) shows that for every point (x,y) on the hyperbola, $(x/a)^2 = 1 + (y/b)^2 \geq 1$, so that $|x| \geq a$. Thus the hyperbola consists of two halves, called its **branches**. The hyperbola intersects the axis at two points called **vertices** (see Figure 2.31). Like the parabola, the hyperbola extends arbitrarily far into the plane, and only part of the curve can be shown in a drawing.

The diagonals of the box $|x| \leq a, |y| \leq b$ are the lines $y = (b/a)x$ and $y = -(b/a)x$. These are called **asymptotes** of the hyperbola (11). The asymptotes come arbitrarily close to the hyperbola if we proceed far enough along the curve, but the asymptotes never touch the curve. To verify this, consider a point (x,y) on the upper half of the right branch of the hyperbola (everything we shall say can be repeated, with obvious modifications, for the lower half and for the left branch). For the distance δ shown in Figure 2.31, we have

$$\delta = \frac{b}{a}x - y = \frac{b}{a}x - b\sqrt{\frac{x^2}{a^2} - 1} = \frac{b}{a}(x - \sqrt{x^2 - a^2}) = \frac{ab}{x + \sqrt{x^2 - a^2}}.$$

Hence δ is positive for all x, and, if x becomes large enough, δ becomes as small as we want.

[Equation (12) may be treated similarly. The asymptotes, however, are given by $y = (a/b)x$ and $y = -(a/b)x$, as the reader may verify.]

The equation

$$-\frac{x^2}{a^2} + \frac{y^2}{b^2} = 1, \tag{13}$$

obtained from (11) by interchanging the signs of x^2 and y^2, defines a hyperbola for which the foci lie on the y-axis, and which has the same asymptotes as the hyperbola (11). The two hyperbolas are shown in Figure 2.32. (They are called *conjugate hyperbolas* and have the property that all four foci lie on a circle about the center.)

FIGURE 2.33

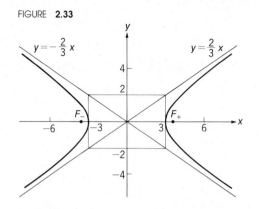

EXAMPLES 1. Find the foci, vertices, and asymptotes of the hyperbola $4x^2 - 9y^2 = 36$.
SOLUTION   Divide both sides of the equation by 36 to obtain $(x/3)^2 - (y/2)^2 = 1$. It is of the
form (11) with $a = 3$, $b = 2$. Hence the vertices are at $(\pm 3,0)$ and the asymptotes are given
by the lines $y = \pm\frac{2}{3}x$. See Figure 2.33. Since $c = \sqrt{a^2 + b^2} = \sqrt{9 + 4} = \sqrt{13}$, the foci are
at $(\pm\sqrt{13},0)$.

2. Find the equation of the hyperbola with vertices at $(\pm 2,0)$ and asymptotes $y = \pm 3x$.
SOLUTION   We have $a = 2$, $b/a = 3$. Hence $b = 6$. By Theorem 3 the desired equation is
$x^2/4 - y^2/36 = 1$ or $36x^2 - 4y^2 = 144$.

PROBLEMS

   In Problems 1 to 8 find the coordinates of the vertices, the coordinates of the foci, and
the asymptotes for the given hyperbola. Sketch the curve.

1. $\dfrac{x^2}{4} - y^2 = 1$.

2. $x^2 - \dfrac{y^2}{4} = 1$.

3. $16x^2 - 9y^2 = 144$.

4. $\dfrac{x^2}{25} - \dfrac{y^2}{4} = 1$.

5. $4y^2 - 9x^2 = 36$.

6. $\dfrac{y^2}{16} - \dfrac{x^2}{4} = 1$.

7. $\dfrac{y^2}{4} - \dfrac{x^2}{16} = 1$.

8. $25y^2 - 9x^2 = 225$.

   In Problems 9 to 14 find the equation of the hyperbola, with center at the origin, from
the given information.

9. Focus at $(5,0)$ and vertex at $(3,0)$.
10. Focus at $(0,-6)$ and vertex at $(0,4)$.
11. Vertex at $(4,0)$ and asymptotes $y = \pm\frac{3}{2}x$.
12. Vertex at $(0,-2)$ and asymptotes $y = \pm 4x$.
13. Focus at $(0,\sqrt{8})$ and asymptotes $y = \pm x$.
14. Focus at $(-\sqrt{5},0)$ and asymptotes $y = \pm\frac{1}{2}x$.
15. Find the equation of the hyperbola with vertices at $(0,0)$ and $(0,4)$ and one focus at $(0,6)$.
16. Find the equation of the hyperbola with foci at $(-2,0)$ and $(8,0)$ and one vertex at $(-1,0)$.
* 17. What is the equation of the hyperbola conjugate to the hyperbola $4x^2 - 16y^2 = 64$? Find
    the equation of the asymptotes. Sketch the conjugate hyperbolas.
* 18. The asymptotes of a set of conjugate hyperbolas are given by $2x \pm y = 0$. If one focus
    is at $(\sqrt{5},0)$, find the equations of the conjugate hyperbolas. Sketch the conjugate hyper-
    bolas.

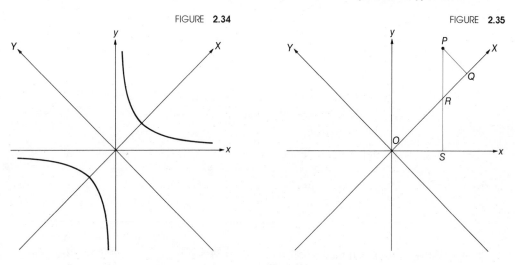

FIGURE 2.34   FIGURE 2.35

4.7   Equilateral hyperbolas

A hyperbola with equal semi-axes is called **equilateral.** Thus such a hyperbola has the equation

$$\frac{x^2}{a^2} - \frac{y^2}{a^2} = 1 \tag{14}$$

or $x^2 - y^2 = a^2$. Choosing the length unit as the semi-axis, we may assume that $a = 1$ and obtain the equation $x^2 - y^2 = 1$. All equilateral hyperbolas are similar!

There is also another convenient way to write the equation of an equilateral hyperbola.

Theorem 4. *Let $p > 0$ be a fixed number. The solution set of the equation*

$$xy = p \tag{15}$$

is an equilateral hyperbola.

Proof.   We introduce a "new" Cartesian coordinate system as shown in Figure 2.34. The old and new system have the same origin and the same unit of length. The new horizontal axis, the X-axis, points into the old first quadrant and bisects the angle between the old axes. The relation between the old coordinates (x,y) of a point P and its new coordinates (X,Y) is obtained from Figure 2.35. We have, by the properties of isosceles triangles, that

$$y = |PS| = |PR| + |RS| = |PQ|\sqrt{2} + |OS| = Y\sqrt{2} + x,$$
$$X = |OQ| = |OR| + |RQ| = |OS|\sqrt{2} + |PQ| = x\sqrt{2} + Y.$$

These may be solved to give

$$x = \frac{X - Y}{\sqrt{2}}, \qquad y = \frac{X + Y}{\sqrt{2}}. \tag{16}$$

Substituting (16) into (15) we obtain $\frac{1}{2}(X - Y)(X + Y) = p$, or

$$\frac{X^2}{a^2} - \frac{Y^2}{a^2} = 1, \qquad a = \sqrt{2p}.$$

This is the equation of an equilateral hyperbola.

We remark that the asymptotes of (15) are the lines $x = 0$ and $y = 0$.

EXAMPLE   Find the foci of the hyperbola $xy = 1$.

SOLUTION   We use the coordinates X, Y defined by (16). In these coordinates the equation $xy = 1$ reads $X^2 - Y^2 = 2$ so that the semi-axes are $a = b = \sqrt{2}$ and the focal distance is $c = \sqrt{a^2 + b^2} = 2$. The foci are $X = \pm 2$, $Y = 0$. Since $x = (X - Y)/\sqrt{2}$ and $y = (X + Y)/\sqrt{2}$, the coordinates of the foci, in the original x,y system, are $(\sqrt{2}, \sqrt{2})$ and $(-\sqrt{2}, -\sqrt{2})$.

PROBLEMS

In Problems 1 to 8 find the coordinates of the vertices, the coordinates of the foci, and the asymptotes for the given equilateral hyperbola. Sketch the curve.

1. $x^2 - y^2 = 9$.   4. $x^2 - y^2 - 25 = 0$.   7. $xy - 2 = 0$.
2. $y^2 - x^2 = 4$.   5. $xy = -4$.   8. $xy + \frac{1}{2} = 0$.
3. $y^2 - x^2 = 8$.   6. $xy = \frac{9}{2}$.

In Problems 9 to 12 find the equation of the equilateral hyperbola, with center at the origin, from the given information.

9. Foci at $(2,2)$ and $(-2,-2)$.
10. Foci at $(2\sqrt{2},0)$ and $(-2\sqrt{2},0)$.
11. One focus at $(-\sqrt{2}, \sqrt{2})$ and one vertex at $(-1,1)$.
12. Vertices at $(0,4)$ and $(0,-4)$.

§5   Geometry and numbers‡

5.1   Axiomatic geometry

The word "geometry" literally means "measuring earth," that is, surveying. Geometry was originally a practical art and, as far as it was a science, experimental. Geometry as a **deductive science** is a creation of the Greeks. Thales, traditionally considered the first Greek philosopher, is also credited with the first proof of a geometric theorem. In deductive geometry certain simple statements about points and lines (and, in the case of solid geometry, also about planes) are assumed as **axioms**; all other statements are **theorems**, that is, logical consequences of the axioms.

Plato, whose philosophy was influenced by Greek mathematics, emphasized that deductive geometry deals with abstract or ideal, rather than physical, figures. The line of the geometer is not the line drawn on paper, a straight rod, or a ray of light. Drawings are only a device to reinforce our memory and stimulate our imagination. Modern mathematicians accept Plato's statement, although perhaps not in his original

---

‡Optional section.

sense. What matters is that no concepts or properties except those explicitly stated in the axioms are to be used in the proofs.

This does not mean, of course, that geometry has nothing to do with the physical world. The axioms are designed so as to give a description of a part of physical reality. The abstract character of the definitions and proofs ensures the validity of the theorems whenever the axioms are assumed. The history of science shows that, by becoming more abstract, mathematics becomes more, rather than less, useful as a tool for understanding nature.

The strictly logical construction of geometry from axioms is the aim of Euclid's *Elements*, a work that summarized the results of a century of Greek mathematics and became the second most influential book of our civilization. It is no disparagement of the stupendous achievement of Euclid and his predecessors to observe that, judged by modern standards of rigor, they did not achieve their aim. A system of axioms really sufficient for Euclidean geometry was created relatively recently. Hilbert's *Foundation of Geometry*, which contains such a system, appeared in 1899.

5.2   Analytic geometry

We are not going to list Hilbert's axioms here. There is no need to do so, because they can be summarized in one sentence: *Euclidean geometry is analytic geometry based on a Cartesian coordinate system.* Let us make this more precise.

We define the **plane** to be the set of all ordered pairs of real numbers (x,y). Every such pair is called a **point.** The **distance** between two points, (x_1,y_1) and (x_2,y_2), is defined to be the number $\sqrt{(x_1 - x_2)^2 + (y_1 - y_2)^2}$. A **straight line** is, by definition, the set of all points (x,y) that satisfy some equation of the form $Ax + By + C = 0$, where A, B, and C are numbers such that $A^2 + B^2 \neq 0$. (We write $A^2 + B^2 \neq 0$ to show that A and B are not both zero.)

Having made these conventions, we can translate every axiom of Hilbert's list, as far as it pertains to the plane, into a statement about numbers, and it turns out that each of these statements can be proved from properties of numbers.

[Here is an example. An axiom of geometry states that through any two points passes a line. Translated, this means that, given two pairs of numbers (x_1,y_1) and (x_2,y_2), there is an equation of the form $Ax + By + C = 0$, with $A^2 + B^2 \neq 0$, which is

---

THALES (6th century B.C.). The reports about Thales' mathematical and astronomical achievements are considered to lack historical credibility.

PLATO (429–348 B.C.), a pupil of Socrates and founder of the idealistic philosophy, was also the inventor of universities. He established the Academy, a school that endured nearly a thousand years.

This great thinker and writer extolled mathematics as a road to true knowledge, and he assigned to it a prominent place in the curriculum of the Academy. Although Plato himself was not a mathematician, and there is no evidence that he was fully abreast of the mathematical achievements of his time, the Platonic tradition played its part in the creation of modern mathematical science.

EUCLID (ca. 365–300 B.C.). Until recently, most high-school texts of geometry were more or less watered-down versions of the *Elements*. But Euclid's book was not intended for elementary instruction. It is a sophisticated treatise dealing with geometry, as well as number theory, and describes contributions by Euclid's predecessors as well as his own work.

DAVID HILBERT (1862–1943), the most influential mathematician of the twentieth century, was a professor at Göttingen, the undisputed world capital of mathematics until its destruction by the Nazis in 1933. Hilbert would devote several years to intensive work in one field of mathematics, and then leave it to go on to a different field. In this way he completely changed the face of many parts of mathematics.

satisfied by $x = x_1$, $y = y_1$ and by $x = x_2$, $y = y_2$. This is indeed so; Equation (7) in §2.2 has the desired property.]

Conversely, starting with Hilbert's axioms, we can construct a Cartesian coordinate system, describe points by coordinates, derive the distance formula, and so forth.

The same procedure applies to solid geometry, as we shall see when we introduce coordinates in space (see Chapter 18).

PROBLEMS

1. Write Equation (7) in §2.2 in the form $Ax + By + C = 0$. Compute $A^2 + B^2$. Show that $A^2 + B^2 = 0$ if and only if $(x_1, y_1) = (x_2, y_2)$.

2. Translate the geometric statement "Two lines (in the plane) either coincide, or are parallel, or have exactly one point in common" into a statement about linear equations.

FUNCTIONS

§1 Functions and graphs

Mathematical analysis, the part of mathematics that includes calculus and all its ramifications, is dominated by the concept of continuity and the related concept of limits. These concepts were cast in their present form in the nineteenth century, primarily by Cauchy and Weierstrass; they were unavailable to the founders of calculus, and the original calculus therefore had a flavor of mystery. To explain continuity and limits, we first recall what is meant by a function.

AUGUSTIN LOUIS CAUCHY (1789–1857). His indefatigable productivity, his inventiveness, and the breadth of his mathematical interests are similar to those of the great mathematicians of the eighteenth century. But Cauchy was a nineteenth-century mathematician and paid great attention to the logical foundation of mathematical analysis.

Cauchy had strong religious and political convictions. After the revolution of 1830, he followed the Bourbons to exile; when he returned to France, he would not accept a university position until the requirement of a loyalty oath to the government was waived.

KARL WEIERSTRASS (1815–1897) began his career as a secondary school teacher. His profound papers, some of which appeared in such unlikely places as a high school graduation program, earned him a professorship in Berlin. From then on, his lectures and seminars had an immense impact on mathematics. The tendency to reduce all mathematics to numbers, and the insistence on complete rigor, are a result, in part, of Weierstrass' influence.

Weierstrass was the first mathematician to sponsor a woman for a Ph.D. (She was Sophie Kowalewski, a distinguished Russian mathematician and a talented writer.)

1.1 Functions

Here are some examples which lead to the concept of function.

I. The volume V of a cube whose sides have length a is $V = a^3$.

II. If a body is released from some height near the surface of the earth, it will fall, and during the first t sec after release, it will traverse a distance s of $s = 16t^2$ ft. (Here we assume that the only force acting on the body is the earth's gravity.)

III. If a wholesaler charges 21 cents for a pound of a product, as well as a handling charge of 2 dollars per order, the cost C of ordering x pounds is $C = (.21)x + 2$ dollars.

What do these examples have in common? In each case one number (side length, time, amount ordered) determines another (volume, distance traveled, cost) by means of a rule (multiply the side length by itself three times, square the time and multiply by 16, multiply the amount by .21 and add 2). In each case we say that the second number is "a function" of the first.

In mathematics a **function** is a rule that assigns to each element of a set an element of the same or of another set. In most of this book we deal with functions that assign to each real number, from some set of numbers, another real number (as in the examples above). The set of numbers, to which other numbers are assigned, is called the **domain of definition** of a function. For the functions we shall consider the domain of definition will be an interval or a collection of intervals. In particular, the domain of definition may be the whole number line. We shall not specify the domain of definition, unless this is necessary. Often it will be obvious from the context.

The rule that assigns one number to another may be of any nature. For example, we may want to assign to every number its square. We write this function in the form

$$x \longmapsto x^2$$

(read: x goes into x^2). Here x is a **variable;** it represents an unspecified real number. The rule assigns a number x^2 to *every* number x; the domain of definition is the whole number line. The function $x \longmapsto x^2$ assigns to 2 the number 4, to -7 the number 49, and so forth. We say the function equals 4 at 2, or the function takes on the value 49 at the point -7.

Often we use a variable, say, y, to denote the number that is assigned to another number by our rule. For the function considered above, we write

$$x \longmapsto y = x^2,$$

or more briefly,

$$y = x^2,$$

and we call x (representing the number to which we assign another) the **independent variable** and y (which represents the number assigned to the first one) the **dependent variable.**

The symbols used for the variables are irrelevant. The formulas

$$t \longmapsto t^2, \qquad \beta = \alpha^2, \qquad w \longmapsto u = w^2, \qquad y \longmapsto x = y^2$$

all represent the same function. In each case the rule is: assign to each number its square.

The function $x \longmapsto 1/x$ assigns to each number, except 0, its reciprocal. (Zero has no reciprocal.) The domain of definition consists of the intervals $(-\infty, 0)$ and $(0, +\infty)$. The same function may be written as $y = 1/x$.

The rule defining the function may be more complicated, for instance,

$$x \mapsto \sqrt{1+x} + \sqrt[4]{2-x} + x^{10}.$$

This function is defined for x such that $1 + x \geq 0$ and $2 - x \geq 0$, that is, for $-1 \leq x \leq 2$.

Here we used a convention: if a function is defined by a formula, then, in the absence of specific instructions to the contrary, the domain of definition is assumed to be the largest set on which the formula makes sense.

The rule defining a function need not be given by a single formula. For instance,

$$x \mapsto \begin{cases} -x, & \text{if } x \leq 0 \\ 1, & \text{if } 0 < x < 3 \\ \sqrt{x}, & \text{if } 3 \leq x \end{cases}$$

is a function, defined for all x.

1.2   Variables denoting functions

Just as we use variables to denote unspecified numbers, we also use variables to denote unspecified functions. When we say "let f be a function," f represents some particular, but not specified, rule assigning numbers to other numbers. If we want f to represent some specified function, say, the function $x \mapsto x^2$, we write $f(x) = x^2$. The symbol $f(x)$ is read "f of x" or "the value of f at x"; it represents the number the rule f assigns to the number x. Thus, if $f(x) = 2x^3$, then $f(0) = 0$, $f(-1) = -2$, $f(t) = 2t^3$, $f(2y) = 2(2y)^3 = 16y^3$, and so on.

If f and g are functions, the statement $f = g$ means that f and g are the same function, which means that f and g are defined on the same set, and $f(x) = g(x)$ for every x in this set.

Strictly speaking, we should distinguish between the function f and the number $f(x)$ which the rule f assigns to a particular number x. In practice, we are not always so careful. We speak, for instance, about the function x^2, that is, the rule $x \mapsto x^2$.

With every number, say, a, there is associated a function $x \mapsto a$ that assigns to every number x the number a. It is called a **constant,** or a constant function.

Let f and g be two functions. The sum $f + g$, the difference $f - g$, the product fg, and the quotient f/g are the functions defined by

$$x \mapsto f(x) + g(x),$$
$$x \mapsto f(x) - g(x),$$
$$x \mapsto f(x) \cdot g(x),$$

and

$$x \mapsto \frac{f(x)}{g(x)},$$

respectively. These function are defined for all those x for which both f and g are defined, except that the quotient f/g is not defined at points x such that $g(x) = 0$.

EXAMPLES   1. What value does the function $u \mapsto \dfrac{2u + 1}{u^2 + 1}$ assign to the number 3?

ANSWER   If $u = 3$, then $(2u + 1)/(u^2 + 1) = (2 \cdot 3 + 1)/(3^2 + 1) = 7/10$. Our function assigns to the number 3 the number $7/10$.

2. If $f(t) = (2t + 1)/(t^2 + 1)$, what is $f(3)$?

ANSWER This is the same function as above, only the names of the variables have been changed. We have $f(3) = 7/10$.

3. For which values of x is the function $x \mapsto \sqrt{x - 1} + 1/\sqrt{2 - x}$ defined?

ANSWER In order to be able to compute $\sqrt{x - 1}$, the number $x - 1$ must be positive or zero. This gives the condition $x \geq 1$. Similarly, $\sqrt{2 - x}$ makes sense if $2 - x \geq 0$, that is, if $x \leq 2$. But in order to compute $1/\sqrt{2 - x}$, we must be sure that the denominator is not zero; thus the value $x = 2$ must be excluded. We collect all conditions obtained: $x \geq 1$, $x \leq 2$, $x \neq 2$. *Conclusion:* our function is defined for $1 \leq x < 2$.

4. Let $f(x) = 2x^2 + 3$, $g(x) = \sqrt{x - 1}$. Then $f + g$, $f - g$, and fg are defined for all x such that $x \geq 1$, and f/g is defined for $x > 1$.

PROBLEMS

1. What value does the function $x \mapsto 2x^2 - 1$ assign to the number 7? To the number $\frac{3}{2}$? To the number -4?

2. What value does the function $\alpha \mapsto (\alpha^2 + 1)(\alpha^3 - 2\alpha)$ assign to the number -1? To the number 1? To the number 2?

3. What value does the function $u \mapsto \left(\dfrac{u + 1}{u^2 + 1}\right)$ take on at the point -1? At the point $\frac{1}{4}$? At the point $\sqrt{2}$?

4. What is the value of the function $f(x) = x^{1/3} + x$ when $x = -8$? When $x = 27$? When $x = 1/125$?

5. Let $f(x) = \sqrt{x + 1} + 2x$. What is $f(0)$? $f(3)$? $f(-1)$?

6. Consider the function $f(x) = 3x^{-2} + x$. What is $f(-3)$? $f(10)$? $f(\frac{1}{3})$?

7. Consider the function $y = 2u^{-2/3} + (u + 1)^2$. What is the value of y when $u = 8$? When $u = -1$? When $u = -8$?

8. Consider the function $f(x) = \dfrac{1}{(x + 1)} - x^3$. What is $f(2)$? $f(\frac{1}{2})$? $f(-\frac{1}{3})$?

9. Let $h(t) = (-t)^{-3/2} + (t - 1)^2$. What is $h(-1)$? $h(-4)$? $h(-2)$?

10. For what values of x is the function $x \mapsto \sqrt{1 - x}$ defined?

11. What is the domain of definition of the function $f(x) = \sqrt{1 - x^2}$?

12. What is the domain of definition of the function $y = (x^2 - 4)^{-1/2}$?

13. For which values of x is the function $x \mapsto (x - 1)^{-1} + (x + 2)^{-1} + (3x - 2)^{1/4}$ defined?

14. For which values of u is the function $f(u) = (4u - 1)^{-1/2} - (1 - u^2)^{1/2}$ defined?

15. Let $f(x) = \sqrt{x + 1}$, $g(x) = x^2 - 9$. For which values of x is $f + g$ defined? f/g?

16. Let $f(x) = \sqrt{x^2 - 4}$, $g(x) = \dfrac{x - 6}{\sqrt{x + 4}}$. For which values of x is $f - g$ defined? f/g?

1.3 Graphs

One way of representing a function f is by a **table,** listing in one row numbers x for which f is defined and in the second row the value $f(x)$ assigned to each x. A partial table for the function $x \mapsto y = x^2$ looks as follows:

x	0	.1	.2	.3	.4	.5
y	0	.01	.04	.09	.16	.25

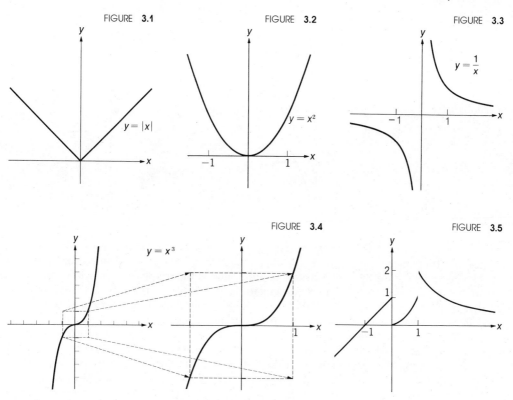

FIGURE **3.1**

$y = |x|$

FIGURE **3.2**

$y = x^2$

FIGURE **3.3**

$y = \dfrac{1}{x}$

FIGURE **3.4**

$y = x^3$

FIGURE **3.5**

We can never write down all possible entries in a table because our functions are defined on intervals, and an interval contains infinitely many numbers.

The **graph** of a function f is the set of all points in the plane with coordinates $(x, f(x))$. The graph is a complete table of the function: all information about the function is contained therein, and, if we draw the graph or a part of it, we can "see" the function.

The simplest functions and some of the most important ones are **linear functions** $f(x) = mx + b$. The graph of a linear function is a line. (See Chapter 2, §2.2.) Conversely, every nonvertical line is the graph of a linear function.

Graphs of nonlinear functions are more complicated and more interesting. Here are a few examples. The graph of $x \mapsto |x|$ is shown in Figure 3.1; it consists of two rays. The graph of $y = x^2$ is shown in Figure 3.2 and is, as we know, a **parabola.** That of $y = 1/x$ is shown in Figure 3.3; this graph is an **equilateral hyperbola.** The graph of $y = x^3$, shown in Figure 3.4, is called a **cubic parabola.** The graph of

$$f(x) = \begin{cases} x + 1, & \text{for } x < 0 \\ x^2, & \text{for } 0 \le x \le 1 \\ 2/x, & \text{for } 1 < x, \end{cases}$$

shown in Figure 3.5, consists of three pieces.

The shape of the graph reflects the properties of the function. We shall encounter many examples of this as we proceed.

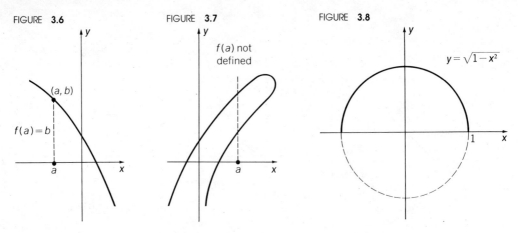

FIGURE **3.6**   FIGURE **3.7**   FIGURE **3.8**

The graph of a function f has the property: a vertical line $x = a$ intersects the graph either once, at the point $(a, f(a))$, or not at all, if f is not defined for $x = a$. The curve in Figure 3.6 satisfies our condition and is the graph of a function. On the other hand, the curve in Figure 3.7 violates our condition; there are vertical lines that meet the curve at two points. The curve is not the graph of a function. Indeed, if it were, what value would we assign to the number a?

The unit circle (center at the origin, radius 1) is not the graph of a function; every line $x = a$, $-1 < a < 1$, intersects the circle at two points. The upper semicircle (see Figure 3.8), the set of points (x,y) such that $x^2 + y^2 = 1$ and $y \geq 0$, is the graph of the function $y = \sqrt{1 - x^2}$. Similarly, the lower semicircle is the graph of $y = -\sqrt{1 - x^2}$. Both functions are defined for $-1 \leq x \leq 1$.

Not every set of points in the plane is the graph of a function. A *necessary and sufficient condition is that every vertical line should meet the set in at most one point.*

PROBLEMS

Sketch the graphs of the following functions over the indicated intervals. Select as *few* points as possible. Only the general shape of the curve is required.

1. $x \mapsto 2x - 4$, $0 \leq x \leq 4$.

2. $x \mapsto \frac{1}{2}(3x + 1)$, $[-\frac{5}{3}, 1]$.

3. $y = \begin{cases} 0, & -1 \leq x < 0, \\ x, & 0 \leq x \leq 2. \end{cases}$

4. $y = \begin{cases} x + 1, & -1 \leq x < 1, \\ 4x - 6, & 1 \leq x \leq 2. \end{cases}$

5. $\phi(x) = \begin{cases} 1 - x, & -2 \leq x \leq 0, \\ 2 + x, & 0 < x \leq 1. \end{cases}$

6. $h(x) = |x| + 1$, $[-4,4]$.

7. $y = 1 - 2|x|$, $[-3,3]$.

8. $f(x) = x + |x|$, $[-\frac{3}{2}, \frac{3}{2}]$.

9. $g(x) = \sqrt{4 - x^2}$, $-2 \leq x \leq 2$.

10. $y = -\sqrt{16 - x^2}$, $|x| \leq 2$.

1.4 Even and odd functions

A function f is called **even** if $f(-x) = f(x)$ and **odd** if $f(-x) = -f(x)$, for all x for which $f(x)$ is defined; in both cases, it is assumed that $f(x)$ is defined whenever $f(-x)$ is. For instance, the function $x \mapsto x^2$ and $x \mapsto |x|$ are even; the functions $x \mapsto x^3$ and $x \mapsto 1/x$ are odd.

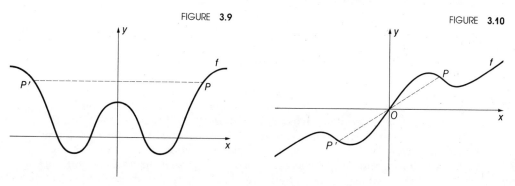

FIGURE 3.9   FIGURE 3.10

The graph of an even function is **symmetric about the y-axis.** [This means that if a point $P = (x,y)$ lies on the graph, so does the point $P' = (-x,y)$; see Figure 3.9.] The graph of an odd function is **symmetric about the origin.** [This means that if a point $P = (x,y)$ lies on the graph, so does the point $P' = (-x,-y)$; see Figure 3.10.]

Note that in general a function will be *neither even nor odd;* the function graphed in Figure 3.5 is such a function. The function $x \mapsto x + x^2$ is also neither even nor odd. Indeed, if $x = 1$, then $x^2 + x = 2$, and if $x = -1$, then $x^2 + x = 0$.

PROBLEMS

In Problems 1 to 16 decide whether the given function is even, odd, or neither.

1. $x \mapsto 1$.
2. $x \mapsto 2x$.
3. $s \mapsto -s$.
4. $y = x + 4$.
5. $t \mapsto t^2 - 2$.
6. $y = 3x^2 - 2x + 1$.

7. $f(x) = x^3 + 2x$.
8. $v \mapsto 4v^4 - 5v^2 + 2$.
9. $\phi(x) = |x|$.
10. $t \mapsto t^{1/2}$.
11. $s \mapsto s^{1/3}$.

12. $g(z) = |z| + z^{10}$.
13. $f(u) = (u^2 + 1)^3 - u^4$.
14. $h(x) = x/(1 + x^2)$.
15. $g(z) = \sqrt{z^3 + 1}$.
16. $v \mapsto \sqrt{v^2 + 1}/|v|$.

17. Prove that if $f(x)$ and $g(x)$ are even, then $f(x) + g(x)$ is even. [*Hint:* Set $\phi(x) = f(x) + g(x)$.]
18. Prove that if $f(x)$ and $g(x)$ are odd, then $f(x) \cdot g(x)$ is even.
*19. Can a function be both even *and* odd? Try to determine all such functions, or at least give an example of one.
*20. Prove that a function $f(x)$ defined for all x can be written in the form $f(x) = \phi(x) + \psi(x)$ with ϕ even and ψ odd. [*Hint:* Set $\phi(x) = \frac{1}{2}f(x) + \frac{1}{2}f(-x)$.]

1.5 Applications of functions

In the application of mathematics, functions are used to represent relationships between measurable, observable quantities as we saw in the examples in §1.1.

Suppose we measure the temperature at some spot during a certain period of time. These measurements can be plotted as a curve; this curve is the graph of a function. We say in this case: the temperature is a function of time. But we should remember that we obtain a mathematical function, that is, a rule assigning one number to another, only after we agree on the **units of measurement** and on **reference points.** If temperature is measured in degrees Fahrenheit and time in minutes, time zero being noon, we obtain one function. If temperature is measured in degrees centigrade and time in seconds, time zero being 2 P.M., the same measurements will be represented by a different function.

Similar remarks apply when the numbers (variables) involved in the definition of a function measure distances, areas, volumes, weights, pressures, intensities, velocities, accelerations, costs, prices, and so forth.

In some cases arising in applications, the functions considered cannot, strictly speaking, assign values to all numbers in some interval. For instance, if $f(x)$ denotes the cost, to some factory, of producing x cans, x must be an integer. Nevertheless it is usually convenient to work with f as if it were a function defined in some interval. A possible cost function is, for instance, a linear function $y = Ax + B$. In working with such a function, say, in graphing it, we pay no attention to the fact that x can take on only integral values.

§2 Polynomial and rational functions

The definition of a function as given in §1.1 is very general. In practice, we use certain special classes of functions. Two such classes (polynomials and rational functions) will be described in the present section. We begin with an important example.

2.1 Quadratic functions

Let a, b, and c be fixed numbers with $a \neq 0$. The function

$$y = f(x) = ax^2 + bx + c \tag{1}$$

is called a **quadratic function.** For any function f, a number α such that $f(\alpha) = 0$ is called a **root,** or a **zero,** of f. A quadratic function has either two distinct roots, or one root, or no roots. (In this book we do not use complex numbers and do not consider complex roots.) If f has one root r, then $f(x) = a(x - r)^2$, and r is called a *double root.* The roots of a quadratic function can be found by "completing the square" (see Chapter 2, §3.1) as we show with some examples.

EXAMPLES   1. Find the roots of the quadratic function $y = 2x^2 - 8x + 4$.
SOLUTION   Completing the square we write

$$y = 2(x^2 - 4x + 2) = 2(x^2 - 4x + 4 - 4 + 2) = 2[(x - 2)^2 - 2] = 2[(x - 2)^2 - (\sqrt{2})^2]$$
$$= 2(x - 2 + \sqrt{2})(x - 2 - \sqrt{2}).$$

Hence $y = 0$ if either $x = 2 - \sqrt{2}$ or $x = 2 + \sqrt{2}$. The roots are $2 - \sqrt{2}$ and $2 + \sqrt{2}$.

2. Find the roots of $f(x) = x^2 - 3x + 3$.
SOLUTION   We complete the square:

$$f(x) = x^2 - 3x + 3 = x^2 - 2\tfrac{3}{2}x + 3 = x^2 - 2\tfrac{3}{2}x + (\tfrac{3}{2})^2 - (\tfrac{3}{2})^2 + 3$$
$$= (x - \tfrac{3}{2})^2 - \tfrac{9}{4} + 3 = (x - \tfrac{3}{2})^2 + \tfrac{3}{4}.$$

This is always positive; there are no roots.

3. For which value of α does the function $x \mapsto x^2 + 2x + \alpha$ have a double root?
ANSWER   We have $x^2 - 2x + \alpha = x^2 - 2x + 4 - 4 + \alpha = (x - 2)^2 - (4 - \alpha)$. This is a square of a linear function of x if $\alpha = 4$.

PROBLEMS

In each of Problems 1 to 6 determine whether the given quadratic function has two distinct roots, one double root, or no roots.

1. $x \mapsto x^2 + 2x - 3$.
2. $x \mapsto x^2 + 4x + 4$.
3. $t \mapsto 2t^2 - t + 8$.
4. $f(x) = 2x^2 + 5x - 3$.
5. $\phi(x) = 5x^2 - 4x + 2$.
6. $g(z) = 3z^2 - 6z + 3$.

7. For which values of α does $x \mapsto x^2 - 2\alpha x + 9$ have two distinct roots?
8. For which values of s does $z \mapsto z^2 + z + 4s$ have no roots?
9. For which values of a does $f(y) = ay^2 + 2y + 1$ have one double root?
10. Write $x \mapsto x^2 + 3x + 1$ as the product of two linear functions.
11. Write $s \mapsto 2s^2 - 4s + 1$ as the product of two linear functions.
12. For which values of z does the graph of $x \mapsto x^2 + 3zx + 5$ intersect the graph of $x \mapsto 1$ in two distinct points?

2.2 Parabolas with vertical axis

We already know from Chapter 2, §4.1, that the graph of the quadratic function $y = ax^2$ is a parabola; indeed, this equation is equivalent to $x^2 = 4py$ with $p = 1/4a$.

It turns out that *the graph of every quadratic function $y = ax^2 + bx + c$ (with $a \neq 0$) is a parabola with vertical axis.*

To verify this we complete the square and write our quadratic function in the form

$$y = ax^2 + bx + c = a\left[x^2 + \frac{b}{a}x + \frac{c}{a}\right] = a\left[\left(x + \frac{b}{2a}\right)^2 + \frac{c}{a} - \frac{b^2}{4a^2}\right]$$

$$= a\left[\left(x + \frac{b}{2a}\right)^2 + \frac{4ac - b^2}{4a^2}\right] = a\left(x + \frac{b}{2a}\right)^2 - \frac{D}{4a},$$

where $D = b^2 - 4ac$. Thus the relation between x and y reads

$$y + \frac{D}{4a} = a\left(x + \frac{b}{2a}\right)^2. \tag{2}$$

Now translate the coordinate system, without changing the unit of length, so as to have as new coordinates

$$X = x + \frac{b}{2a}, \qquad Y = y + \frac{D}{4a}. \tag{3}$$

This means that the new origin $(X = 0, Y = 0)$ is the point with the old coordinates $x = -b/2a$, $y = -D/4a$, and the new Y-axis (the line $X = 0$) is the line $x = -b/2a$. Substituting (3) into (2) we obtain the equation $Y = aX^2$. Note that the parabola lies above the vertex of $a > 0$, below it if $a < 0$.

REMARK The function $y = f(x) = ax^2 + bx + c$ has no roots, one root, or two roots according to whether the number $D = b^2 - 4ac$ is positive, zero, or negative.

We verify this assuming that $a > 0$. (The case $a < 0$ is treated similarly.) If $D < 0$, the y-coordinate of the vertex, $-D/4a$, is positive. The vertex lies above the x-axis, and so does the whole curve. Hence the curve does not intersect the x-axis; the function f has no roots. If $D = 0$, the vertex lies on the x-axis; there is exactly one root of f (called a double root). If $D > 0$, then $-D/4a < 0$; the vertex lies below the x-axis. In this case, the curve intersects the x-axis at two points; f has two distinct roots. The three cases are shown in Figure 3.11.

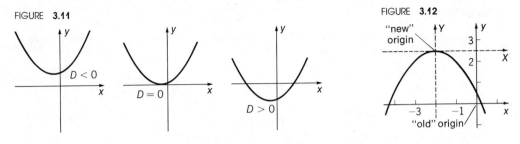

FIGURE **3.11**

FIGURE **3.12**

EXAMPLE   Find the axis and vertex of the parabola $y = -\frac{1}{2}x^2 - 2x + \frac{1}{2}$ and the intersection points of this curve with the x-axis. Sketch the curve.

SOLUTION   Completing the square we have

$$y = -\frac{1}{2}(x^2 + 4x - 1) = -\frac{1}{2}(x^2 + 4x + 4 - 4 - 1)$$
$$= -\frac{1}{2}[(x + 2)^2 - 5] = -\frac{1}{2}(x + 2)^2 + \frac{5}{2}.$$

Introduce new coordinates

$$X = x + 2, \qquad Y = y - \tfrac{5}{2};$$

the equation connecting x and y takes on the form $Y = -\frac{1}{2}X^2$. The axis of the parabola is the line $X = 0$; that is, $x = -2$. The vertex of the parabola is the point $X = 0$, $Y = 0$; that is, the point $x = -2$, $y = \frac{5}{2}$. Finally, $y = 0$ if $-\frac{1}{2}(x + 2)^2 + \frac{5}{2} = 0$. Solving for x, $(x + 2)^2 = 5$; that is, $x = -2 \pm \sqrt{5}$. The curve is sketched in Figure 3.12.

PROBLEMS

In Problems 1 to 6 find the roots, if any, of the given quadratic function. Also find the vertex, axis of the graph, and sketch the graph.

1. $y = x^2 - 4$.
2. $y = x^2 - 2x - 3$.
3. $y = -x^2 + 6x - 9$.

4. $y = \frac{1}{2}x^2 + x - \frac{1}{2}$.
5. $y = \frac{1}{3}x^2 - x + 1$.
6. $y = -\frac{1}{4}x^2 - 2x - 5$.

2.3 Polynomials

Linear and quadratic functions are special cases of **polynomials**. For example, $f(x) = 3x^4 - 2x^2 + 5$ is a polynomial of degree 4, and $f(x) = -7x^6 + x^3 - 2x + 4$ is one of degree 6. In general, a polynomial is a function of the form

$$f(x) = a_0x^n + a_1x^{n-1} + a_2x^{n-2} + \cdots + a_{n-1}x + a_n,$$

where n is a nonnegative integer and $a_0, a_1, \ldots, a_n$ are fixed numbers, with $a_0 \neq 0$. We call n the **degree** of the polynomial, $a_0, a_1, \ldots, a_n$ the **coefficients**, and a_0 the **leading coefficient**. A polynomial of degree 0 is a constant, one of degree 1, a non-constant linear function, and one of degree 2, a quadratic function.

We recall some properties of polynomials.

A sum or a product of two polynomials is again a polynomial.

If r is a root of a polynomial $f(x)$, of degree n, then $f(x)$ is divisible by $x - r$:

$$f(x) = (x - r)P(x),$$

where $P(x)$ is a polynomial of degree $n - 1$. The number r is called a root of multiplicity m if

$$f(x) = (x - r)^m Q(x),$$

where Q is a polynomial (of degree $n - m$) such that $Q(r) \neq 0$. A polynomial of degree n has at most n roots, counting their multiplicities.

EXAMPLE The polynomial

$$f(x) = 2x^3(x + 1)(x^2 + 1) = 2x^6 + 2x^5 + 2x^4 + 2x^3$$

has the roots 0 (with multiplicity 3), -1 (with multiplicity 1), and no others.

PROBLEMS

1. Find a polynomial that has a double root at 1, single roots at -1 and 0, and no others.
   [*Possible answer:* $a(x - 1)^2(x + 1)x$, where a is any number with $a \neq 0$.]
2. Find a polynomial that has a double root at -2, a double root at 3, a triple root at 0, and no others.
3. Find all the real roots (and their multiplicities) of $(x - 2)^{10}(x^4 + 2)$.
4. Find all the real roots (and their multiplicities) of $(x + 3)^3(x - 1)^2(x^4 - 1)^5$.

2.4 Rational functions

A rational function is a function of the form

$$\phi(x) = \frac{f(x)}{g(x)}, \tag{4}$$

where f and g are polynomials, $g \neq 0$.
    For instance,

$$f(x) = \frac{1}{x}, \qquad x \mapsto \frac{x^2 + 1}{x^2 - 1}, \qquad y = \frac{2t}{1 + t^{100}}$$

are rational functions. Every polynomial f is a rational function (since $f = f/1$ and 1 is a polynomial).
    Every rational function can be written as a quotient of two polynomials without common roots.
    Indeed, let f and g be polynomials, $g \neq 0$, and consider the rational function $\phi = f/g$. If f and g have a common root, say, r, then we have (see §2.3) $f(x) = (x - r)f_1(x)$ and $g(x) = (x - r)g_1(x)$. Here f_1 and g_1 are polynomials, their degrees are 1 less than those of f and g, and we have that $\phi = f_1/g_1$. If f_1 and g_1 have a common root, we can repeat the procedure. In this way we finally arrive at a representation of ϕ as a quotient of two polynomials without common roots.

EXAMPLE Consider the function

$$\phi(x) = \frac{x^2 - 2x + 1}{x^2 - 1}. \tag{5}$$

Here the numerator and denominator have a common root, namely, 1. We can write

$$\phi(x) = \frac{(x-1)^2}{(x-1)(x+1)} \tag{5'}$$

or

$$\phi(x) = \frac{x-1}{x+1}. \tag{6}$$

Note that passing from (5) to (6) we have increased the domain of definition of our function. The function given by (5) is not defined at $x = 1$, since $\frac{0}{0}$ is a meaningless symbol. The function given by (6) is defined at $x = 1$ and equals 0 at this point.

PROBLEMS

Write the following rational functions as the quotient of two polynomials without common roots.

1. $x \mapsto \dfrac{x^2 - 5x + 6}{x^2 - 4}.$

2. $x \mapsto \dfrac{x(x-2)}{x^3 - x^2 - 2x}.$

3. $f(x) = \dfrac{x^2 + 1}{x^2 - 2x + 1}.$

4. $f(x) = \dfrac{(x-5)(x^2 + 3x - 10)}{x^2 - 7x + 10}.$

5. $\phi(z) = \dfrac{z^2 - 8z + 16}{z^3 - 12z^2 + 48z - 64}.$

6. $\phi(z) = \dfrac{z^3 - z^2 - 8z + 12}{z^2 + z - 6}.$

§3 Continuous functions

An important property that a function may or may not have is continuity. We shall describe this property, first by examples and then by a definition, and we shall state some basic theorems about continuous functions.

3.1 Examples

Roughly speaking, *a function is* **continuous** *if its graph has no breaks*. Consider the functions whose graphs are shown in Figures 3.13 to 3.20. We note that the graphs in Figure 3.13 and 3.14 each *consist of one piece*. The graphs of the other functions have breaks. For the function in Figure 3.15, the break occurs at $x = 0$, for the one in Figure 3.16, at $x = -1$ and $x = 1$. The functions in Figures 3.17 and 3.18 have breaks at $x = 0$. These functions are not defined at $x = 0$, and for small values of $|x|$ the number $y = f(x)$ becomes very large in absolute value. The first two functions are called continuous. Functions like those in Figures 3.15 to 3.20 are called **discontinuous**. Another example of a discontinuous function is the function whose graph is sketched (or rather, hinted at) in Figure 3.19; its graph has a break at every point.

There is another way of describing the difference between the first two functions and the others. For the functions in Figures 3.13 and 3.14, *if we consider two values of x, say, x_1 and x_2, which are close to each other, the corresponding values of y, $f(x_1)$ and $f(x_2)$, are also close to each other*. This is not always so for the other functions. For the one in Figure 3.15, for instance, if x_1 is negative and x_2 positive, then $|f(x_2) - f(x_1)|$ is close to 1 when x_2 is close to x_1. For the function in Figure 3.17, if x_1 is positive and small, there is always another positive number x_2, such that $0 < x_2 < x_1$ and

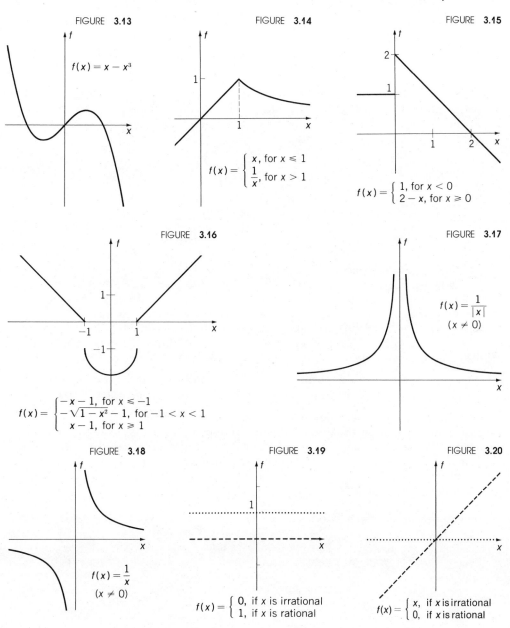

FIGURE **3.13**

$f(x) = x - x^3$

FIGURE **3.14**

$$f(x) = \begin{cases} x, & \text{for } x \leq 1 \\ \dfrac{1}{x}, & \text{for } x > 1 \end{cases}$$

FIGURE **3.15**

$$f(x) = \begin{cases} 1, & \text{for } x < 0 \\ 2 - x, & \text{for } x \geq 0 \end{cases}$$

FIGURE **3.16**

$$f(x) = \begin{cases} -x - 1, & \text{for } x \leq -1 \\ -\sqrt{1 - x^2} - 1, & \text{for } -1 < x < 1 \\ x - 1, & \text{for } x \geq 1 \end{cases}$$

FIGURE **3.17**

$$f(x) = \frac{1}{|x|}$$
$$(x \neq 0)$$

FIGURE **3.18**

$$f(x) = \frac{1}{x}$$
$$(x \neq 0)$$

FIGURE **3.19**

$$f(x) = \begin{cases} 0, & \text{if } x \text{ is irrational} \\ 1, & \text{if } x \text{ is rational} \end{cases}$$

FIGURE **3.20**

$$f(x) = \begin{cases} x, & \text{if } x \text{ is irrational} \\ 0, & \text{if } x \text{ is rational} \end{cases}$$

$|f(x_2) - f(x_1)|$ is as large as we please. The same is true for Figure 3.18. Finally, for Figure 3.19, $|f(x_2) - f(x_1)| = 1$ whenever x_2 is rational and x_1 is irrational, no matter how small the difference $|x_2 - x_1|$ is.

We also note that the functions in Figures 3.15 to 3.18 are "discontinuous" only at certain points ($x = 0$ for Figures 3.15, 3.17, and 3.18, and $x = -1$ and $x = 1$ for Figure 3.16) and continuous elsewhere, while the function in Figure 3.19 is discontinuous everywhere. In Figure 3.20 we show a function which is continuous at one point $(x = 0)$ and discontinuous at all other points.

Continuous functions are of paramount importance in mathematics and its applications. In order to make valid general statements about such functions, we must give a precise definition. In the next subsection we shall state precisely what we mean by saying "the function $f(x)$ is continuous at a point x_0." In doing this, we shall be guided by the examples above. However, for most functions ordinarily encountered in calculus, one needs no formal definition in order to tell where these functions are continuous and where they are discontinuous.

3.2   Definition of continuity

Let x_0 be a point on the number line, that is, a real number. We agree that "**near** x_0" should mean "in some interval with x_0 as midpoint," or, which is the same, "for all x satisfying an inequality $|x - x_0| < \delta$, where δ is some positive number." (Why do we not insist that "near x_0" should mean "in some *small* interval with x_0 as midpoint?" Because what numbers are to be considered "small" depends on the circumstances and cannot be determined once and for all.)

Now let f be a function defined at x_0. We want to give a precise meaning to the statement: the graph of f does not have a break at x_0. Let ϵ be any positive number. If the graph is to have no break at x_0, it cannot immediately jump to a point above the line $y = f(x_0) + \epsilon$ or below the line $y = f(x_0) - \epsilon$. We formulate this property of f as follows: $|f(x) - f(x_0)| < \epsilon$ for all x near x_0, at which f is defined. Thus we are led to the **definition of continuity**:

A function f defined at a point x_0 is called continuous at x_0 *if, for every $\epsilon > 0$, there is a $\delta > 0$ with the property: if $|x - x_0| < \delta$ and $f(x)$ is defined, then $|f(x) - f(x_0)| < \epsilon$.*

The geometric meaning of this definition is shown in Figure 3.21: if we want y to change by less than the distance ϵ, it suffices to restrict x to change by less than δ. The function is continuous at x_0 if we can find an appropriate distance δ for every given distance ϵ. (There is no need to find the largest δ that will do for a given ϵ; the distance δ shown in Figure 3.21, for instance, could be increased.) Briefly speaking: a *sufficiently small change in x produces an arbitrarily small change in y.*

FIGURE **3.21**

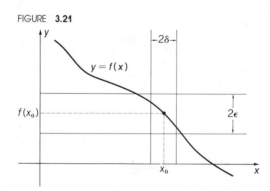

A function is called continuous in an interval *if it is defined and continuous at every point of that interval.*

We give below some examples on using the definition of continuity at a point. (Fortunately, there are more efficient ways of verifying that a function is continuous.)

EXAMPLES 1. The function $f(x) = 3x + 5$ is continuous at $x = 2$.

Proof. We have $f(2) = 11$, and we must show that for a *given* ϵ, we can find *some* δ such that $|(3x + 5) - 11| < \epsilon$ for all x with $|x - 2| < \delta$.

It is convenient to reverse the process. We pick a δ and ask how large $|(3x + 5) - 11|$ can be if $|x - 2| < \delta$. Now if $|x - 2| < \delta$, then

$$-\delta < x - 2 < \delta, \qquad \text{that is,} \qquad 2 - \delta < x < 2 + \delta$$

so that

$$3(2 - \delta) < 3x < 3(2 + \delta), \qquad \text{that is,} \qquad 6 - 3\delta < 3x < 6 + 3\delta,$$

hence

$$6 - 3\delta + 5 < 3x + 5 < 6 - 3\delta + 5,$$

that is,

$$11 - 3\delta < 3x + 5 < 11 + 3\delta \qquad \text{(for } |x - 2| < \delta\text{)}.$$

Thus

$$|(3x + 5) - 11| < 3\delta \qquad \text{if } |x - 2| < \delta.$$

Can we choose δ so that $|(3x - 5) - 11|$ will be less than ϵ? Yes, we may choose, for instance, $\delta = \frac{1}{3}\epsilon$ (or $\delta = \frac{1}{100}\epsilon$ or $\delta = \frac{1}{1000}\epsilon$; there is no need to find the largest δ with the required properties).

2. A constant function is continuous everywhere.

Proof. If $f(x) = c$, a constant, then $|f(x) - f(x_0)| = |c - c| = 0 < \epsilon$ for all x, x_0 and all (positive) ϵ. Hence, for any given ϵ, every δ does the trick.

3. Show that $f(x) = 1/x$ is continuous at $x = 2$.

ANSWER Since $f(2) = \frac{1}{2}$, we are interested in the number

$$\left|\frac{1}{x} - \frac{1}{2}\right| = \left|\frac{2 - x}{2x}\right| = \frac{|2 - x|}{|2x|}. \tag{1}$$

If $|2 - x| < \delta$, then $|x| > 2 - \delta$. [The reader should verify the last statement by a drawing. Here is the analytic proof: $2 = x + (2 - x)$ and by the triangle inequality (Chapter 1, §3.4), $2 \leq |x| + |2 - x| < |x| + \delta$ so that $|x| < 2 - \delta$.] If $\delta < 2$, the inequalities $|2 - x| < \delta$ and $|x| > 2 - \delta$ imply, by (1), that

$$\left|\frac{1}{x} - \frac{1}{2}\right| < \frac{\delta}{4 - 2\delta}.$$

Now let ϵ be given. We need a δ such that $\delta/(4 - 2\delta) < \epsilon$. There is no need to find the largest possible δ. Let us first choose δ so small that $4 - 2\delta > 2$, that is, let us choose $\delta < 1$. Then $\delta/(4 - 2\delta) < \delta/2$. Now it suffices to choose δ so that $\delta/2 < \epsilon$. Then, for a given ϵ, any δ such that $\delta < 1$ and $\delta < 2\epsilon$ has the required property. For this δ, and for any x with $|x - 2| < \delta$, we have $|(1/x) - (1/2)| < \epsilon$.

PROBLEMS‡

1. Let $f(x) = 9x - 5$. Find a $\delta > 0$ such that $|f(x) - 4| < \frac{1}{10}$ for $|x - 1| < \delta$.
2. Let $g(x) = 2x^2 + 3$. Find a $\delta > 0$ such that $|g(x) - g(0)| < \frac{1}{2}$ for $|x| < \delta$.

‡Optional problems.

3. Let $h(x) = x^2 - x$. Find a $\delta > 0$ such that $|h(x) - h(1)| < \frac{1}{5}$ for $|x - 1| < \delta$.
4. Let $f(x) = x + 3$ for $x \leq 2$, $3x - 1$ for $x > 2$. Find a $\delta > 0$ such that $|f(x) - f(2)| < \frac{1}{3}$ for $|x - 2| < \delta$.

Use the definition of continuity to prove that the following functions are continuous at the given points.

5. $f(x) = \frac{1}{5}(2x + 3)$, $\quad x = 1$.
6. $f(x) = 2x - 1$, $\quad x = \frac{1}{2}$.
7. $f(x) = x^2$, $\quad x = 1$.
8. $\phi(x) = x^2 - 2x$, $\quad x = 3$.

9. $\phi(x) = x^2 + 3x + 4$, $\quad x = -4$.
10. $g(x) = 1 + 1/x$, $\quad x = 1$.
11. $g(x) = (x + 2)/(x + 1)$, $\quad x = 1$.
12. $g(x) = \sqrt{x}$, $\quad x = a$, $\quad a > 0$.

3.3 Continuity theorems

We now state a theorem which enables us to recognize that some functions are continuous.

Theorem 1. *Assume that the functions f and g are continuous at x_0. Then so are the functions $f + g$, $f - g$, fg, and, if $g(x_0) \neq 0$, also the function f/g.*

The truth of the theorem is intuitively clear if we think of continuity as meaning that a small change in x results in a small change in the value of the function. The formal proof ▶ of the theorem is somewhat lengthy and will not be given here.

Here are two applications.

Theorem 2. *A polynomial is continuous everywhere.*

Proof. A linear function is continuous since its graph is a line. In particular, constant functions are continuous and so is the function $f(x) = x$. Therefore, by Theorem 1, the functions $x^2 = xx$, $x^3 = x^2x$, and so forth are continuous. By Theorem 1, terms such as cx^n (c a constant, n a nonnegative integer) are continuous. A polynomial is a sum of such terms and hence is continuous, again by Theorem 1.

Theorem 3. *A rational function is continuous at all points at which it is defined.*

Proof. A rational function may be written as f/g where f and g are polynomials without common roots (see §2.3). By Theorems 1 and 2, it is continuous at every point that is not a root of the denominator g.

PROBLEMS

In Problems 1 to 5 assume that $\phi(x)$ and $\psi(x)$ are continuous functions defined for all values of x. Each of the functions listed in 1 to 5 is continuous. Indicate how you know this is so. [*Hint:* Use Theorems 1, 2, and 3.]

1. $f(x) = \phi(x) + \psi(x) + \phi(x)\psi(x)$.
2. $f(x) = x + \phi(x)$.
3. $f(x) = \phi(x)/(1 + x^2)$.

4. $f(x) = \psi(x)/(1 + \phi(x)^4)$.
5. $f(x) = \phi(x)\psi(x)x^{100}/(1 + x^2)$.

6. Is the function $x \mapsto |x|$ continuous?
7. Is the function $x \mapsto -|1 - x|$ continuous?

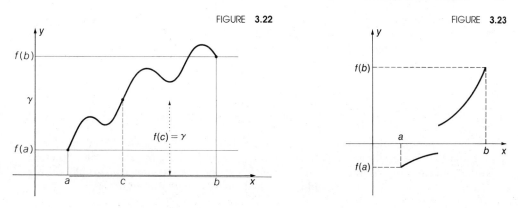

FIGURE **3.22**   FIGURE **3.23**

3.4 Intermediate values

An important property of continuous functions is stated in:

Theorem 4 (the intermediate value theorem). *Let $f(x)$ be continuous for $a \leq x \leq b$ and assume that $f(a) \neq f(b)$. If γ is any number strictly between $f(a)$ and $f(b)$, then there is a point c, $a < c < b$, such that $f(c) = \gamma$.*
     [To say that γ is strictly between $f(a)$ and $f(b)$ means that either $f(a) < \gamma < f(b)$ or $f(a) > \gamma > f(b)$.]

The validity of the theorem is geometrically evident. Since f is continuous, the graph consists of one piece. This curve joins the points $(a,f(a))$ and $(b,f(b))$, one of which lies below the line $y = \gamma$ and one above this line. Hence the curve must cross the line $y = \gamma$ somewhere (Figure 3.22). There is at least one c with $a < c < b$ and $f(c) = \gamma$. An analytic proof ▶ of Theorem 4, however, is not easy and requires a deeper discussion of properties of numbers.
     Theorem 4 is used to locate zeros of continuous functions. If such a function is positive at some point a and negative at a point b, then there is a root of the function between a and b. This need not be so for discontinuous functions; see Figure 3.23.

EXAMPLE Suppose f is continuous in the interval $[-4,4]$ and $f(-4) = -1$, $f(4) = 2$. What is the minimum number of zeros f can have in this interval?
ANSWER Use Theorem 4. Note that $\gamma = 0$ is a point strictly between $f(-4) = -1$ and $f(4) = 2$. Hence there is at least one zero in the interval $[-4,4]$.

PROBLEMS

1. Suppose f is a function continuous in the interval $[0,2]$ and $f(0) = \frac{5}{2}$, $f(2) = -1$. What is the minimum number of zeros f can have in this interval? Make a sketch illustrating the situation.
2. Suppose f is a function continuous in the interval $[-1,2]$ and $f(-1) = 2$, $f(0) = -1$, $f(2) = 3$. What is the minimum number of zeros f can have in this interval? Make a sketch.
3. Suppose g is a function continuous in the interval $[-2,3]$ with $g(-2) = \frac{1}{2}$, $g(-1) = -1$, $g(0) = 2$, $g(1) = 1$, $g(2) = -2$, and $g(3) = \frac{5}{2}$. What is the minimum number of zeros g can have in this interval? Make a sketch.

*4. Define a function on the interval $[-1,1]$, continuous on $[-1,1]$ except at 0, negative at -1, and positive at 1, but which has no zeros. (First try to sketch a graph.) Does this contradict the intermediate value theorem?

*5. Suppose f is continuous in the interval $[0,\frac{5}{2}]$ and $f(0) = 1, f(\frac{5}{2}) = -1$. Can f have an infinite number of zeros in this interval? Make a sketch.

§4   Limits

The concepts of limits and continuity are closely related. Indeed, one can be reduced to the other. We took continuity as the *basic* concept because the notion of a function being continuous in an interval has an intuitive geometric meaning (the graph consists of one piece), and in using calculus one deals mostly with such functions.

4.1   Example

We begin with an example. Consider the function f defined by

$$f(x) = x^2 + x, \qquad \text{for } x \neq 1; \tag{1}$$

the function is not defined at $x = 1$. The graph of (1) is a parabola with one point, the point with $x = 1$, missing (see Figure 3.24). We want to extend the definition of

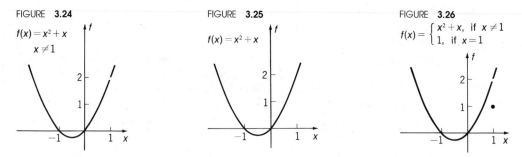

FIGURE **3.24**

$f(x) = x^2 + x$
$x \neq 1$

FIGURE **3.25**

$f(x) = x^2 + x$

FIGURE **3.26**

$f(x) = \begin{cases} x^2 + x, & \text{if } x \neq 1 \\ 1, & \text{if } x = 1 \end{cases}$

f to include $x = 1$. We are, of course, at liberty to define $f(1)$ in any way we want. But there is one "best" way of doing it, namely, setting

$$f(1) = 2. \tag{2}$$

This way of extending the definition of our function is natural, while any other way of defining the value of the function f at 1 would be unnatural. For if we consider a value x that is not 1 but close to 1, then the value $f(x)$ is close to 2. The closer x gets to 1, the closer the value of f gets to 2. For example, $f(.9) = 1.71$, $f(.99) = 1.9701$, $f(.999) = 1.997001$, and so forth; similarly, $f(1.1) = 2.31$, $f(1.01) = 2.0301$, $f(1.001) = 2.003001$, and so forth. Indeed, $f(x)$ will be as close to 2 as we want, provided we take for x numbers that are distinct from, but sufficiently close to, 1.

We see that our way of extending the definition of $f(x)$ to $x = 1$ is natural in the following sense: as the independent variable x "approaches 1," the function value $f(x)$ "approaches 2." We abbreviate this statement by writing

$$\lim_{x \to 1} f(x) = 2. \tag{3}$$

(This is read: the **limit** of $f(x)$, as x approaches 1, is 2.)

Another way of looking at this is as follows. Suppose we graph the function defined by (1) and (2); we obtain the curve shown in Figure 3.25; it has no breaks. Suppose we choose another way of defining f at 1, say, by setting $f(1) = 1$. We obtain a graph (see Figure 3.26) which has a break at $x = 1$. Hence the definition (2) *makes f into a continuous function*, while any other way of assigning a value to $f(1)$ leads to a function which is discontinuous at $x = 1$.

4.2 Definition of limits

The definition is suggested by the example considered above. Let $f(x)$ be a function defined near x_0, except possibly not at x_0 itself. We look for a number α such that, if we define or redefine $f(x_0)$ to be α, then f becomes continuous at x_0. If there is such an α, we call it the **limit** of f at x_0, and we write

$$\lim_{x \to x_0} f(x) = \alpha$$

(read: the limit of f of x as x approaches x_0 is α) or $f(x) \to \alpha$ as $x \to x_0$ [read: $f(x)$ approaches α as x approaches x_0].

If $f(x_0)$ was defined to begin with, and the function was already continuous at x_0, no redefinition is needed; thus

"$\lim_{x \to x_0} f(x) = f(x_0)$" means that "$f$ is defined near x_0 and is continuous at x_0."   (4)

If there is no way to define $f(x_0)$ so as to make the function continuous at x_0, we say that f has **no limit** at x_0 or that the limit of f at x_0 does not exist.

Recalling the definition of continuity (see §3.2), we may restate the **definition of limit** as follows: the statement "$\lim_{x \to x_0} f(x) = \alpha$" means that (1) *the function $f(x)$ is defined for x near x_0, except perhaps at x_0, and (2) for every number $\epsilon > 0$ there is a number $\delta > 0$ such that for all x with $|x - x_0| < \delta$ we have $|f(x) - \alpha| < \epsilon$.*

If we begin with the above ϵ-δ definition of limit, we can define continuity at a point x_0 near which the function $f(x)$ is defined as follows: $f(x)$ *is continuous at x_0 if it has a limit at x_0 and $\lim_{x \to x_0} f(x) = f(x_0)$.*

EXAMPLES   1. If $f(x) = x^3$, what is $\lim_{x \to 3} f(x)$?

ANSWER   The function $f(x)$ is a polynomial, hence continuous everywhere by Theorem 2, §3.3. Thus its limit at $x = 3$ is its value at $x = 3$ [see relation (4) above]. Hence $\lim_{x \to 3} f(x) = f(3) = 3^3 = 27$.

SECOND ANSWER   We show that the desired limit is 27 by using directly the ϵ-δ definition of limits. Set $h = x - 3$ so that $x = 3 + h$. Then

$$|f(x) - 27| = |f(3 + h) - 27 = |(3 + h)^3 - 27|$$
$$= |3^3 + 3 \cdot 3^2 h + 3 \cdot 3h^2 + h^3 - 27|$$
$$= |27h + 9h^2 + h^3| = |h| \, |27 + 9h + h^2|.$$

Let $\delta > 0$ be a given number. We must have $\epsilon > 0$ so that for $|x - 3| = |h| < \epsilon$, we have $|f(x) - 27| < \delta$. Assume that $0 < \epsilon < 1$. Then $|27 + 9h + h^2| < 27 + 9|h| + |h|^2 < 27 + 9 + 1 = 37$ and therefore $|h| \, |27 + 9h + h^2| < 37\epsilon$. If we choose ϵ so that $0 < \epsilon < 1$ and $\epsilon < \delta/37$, then $|f(x) - 27| < \delta$ for $|x - 3| < \epsilon$.

Note that the first answer is shorter and easier!

FIGURE **3.27**

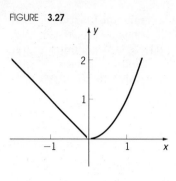

FIGURE **3.28**

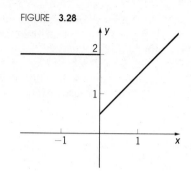

2. If $f(x) = x^3$ for $x \neq 3$, and $f(3) = 26$, what is $\lim_{x \to 3} f(x)$?
ANSWER   The function as defined is *not* continuous at $x = 3$. But if we change the definition of $f(3)$ to 27, the function will be continuous at $x = 3$. Therefore $\lim_{x \to 3} f(x) = 27$.

3. If $f(x) = -x$ for $x < 0$ and $f(x) = x^2$ for $x > 0$, what is $\lim_{x \to 0} f(x)$?
ANSWER   The function is graphed in Figure 3.27. The graph consists of the "left half" of the line $y = -x$ and the "right half" of the parabola $y = x^2$. The value of $f(0)$ which makes the function continuous at $x = 0$ is 0.

4. Assume that $f(x) = 2$ for $x \le 0$ and $f(x) = \frac{1}{2} + x$ for $x > 0$ (see Figure 3.28). Then f has no limit at $x = 0$, since there is no way of redefining $f(0)$ so as to make the function continuous at that point.

PROBLEMS

In Problems 1 to 12 determine if the indicated limits exist, and if so, evaluate them. (Use the definition of limit as given in this subsection, and also use the continuity theorems found in §3.3.)

1. $\lim_{x \to 0} (x^2 + 2x - 1)$.

2. $\lim_{x \to 1} (x^3 - 2x^2 - x + 1)$.

3. $\lim_{x \to 2} f(x)$ if $f(x) = x^2 + x$ for $x \neq 2$ and $f(2) = 10$.

4. $\lim_{x \to -4} f(x)$ if $f(x) = x^2 + 2x - 8$ for $x \neq -4$ and $f(-4) = 2$.

5. $\lim_{x \to 1} f(x)$ if $f(x) = 2x^2 - 4x + 3$ for $x \neq 1$ and $f(1) = 0$.

6. $\lim_{x \to 3} \dfrac{x^2 - 7x}{x + 5}$.

7. $\lim_{x \to -2} \dfrac{x^3 - 2}{x^2 + 1}$.

8. $\lim_{x \to 1} f(x)$ if $f(x) = \dfrac{x^2 + x - 4}{x^2 + 1}$ for $x \neq 1$ and $f(1) = 1$.

9. $\lim_{x \to 2} f(x)$ if $f(x) = x^2 + 4$ for $x < 2$, x^3 for $x > 2$.

10. $\lim_{x \to 1} f(x)$ if $f(x) = |x|$ for $x < 1$, x^2 for $x > 1$.

11. $\lim_{x \to 0} f(x)$ if $f(x) = x - 2$ for $x < 0$, $x^3 - 3$ for $x > 0$.

12. $\lim_{x \to -2} \dfrac{(x + 2)}{|x + 2|}$.

4.3 Computing limits

We now state the basic rules for computing limits.

Theorem 1 (limits of sums, products, and quotients). *Assume that $f(x)$ and $g(x)$ are functions that have limits at a point x_0. Then*

$$\lim_{x \to x_0}(f + g) = \lim_{x \to x_0} f + \lim_{x \to x_0} g, \tag{5}$$

$$\lim_{x \to x_0} (fg) = (\lim_{x \to x_0} f)(\lim_{x \to x_0} g), \tag{6}$$

$$\lim_{x \to x_0} \frac{f}{g} = \frac{\lim_{x \to x_0} f}{\lim_{x \to x_0} g} \qquad \text{if } \lim_{x \to x_0} g \neq 0. \tag{7}$$

In words: *the limit of a sum (product, quotient) is the sum (product, quotient) of limits.*

This statement follows from Theorem 1 in §3.3 and the definition of limits. We prove only the statement about sums, the argument for the other rules being similar. Let $\lim_{x \to x_0} f(x) = \alpha$, $\lim_{x \to x_0} g(x) = \beta$. This means that if we define or redefine $f(x_0) = \alpha$ and $g(x_0) = \beta$, the functions f and g become continuous at x_0. Then the function $f + g$, with the value $\alpha + \beta$ at x_0, is continuous at x_0 (Theorem 1, §3.3) so that $\lim_{x \to x_0} [f(x) + g(x)] = \alpha + \beta$. But this is the statement to be established.

We recall that a constant function is continuous, hence

$$\lim_{x \to x_0} c = c \qquad (c \text{ a constant}) \tag{8}$$

at every point x_0. Combining this with (6) we obtain the rule

$$\lim_{x \to x_0} (cf) = c \lim_{x \to x_0} f \qquad (c \text{ a constant}). \tag{9}$$

Since $f - g = f + (-1)g$, it follows that

$$\lim_{x \to x_0} (f - g) = \lim_{x \to x_0} f - \lim_{x \to x_0} g. \tag{10}$$

REMARK If, at some point x_0, $\lim_{x \to x_0} f = \lim_{x \to x_0} g = 0$, the function f/g may or may not have a limit at x_0, and if it does, Theorem 1 cannot be used to compute it. Nevertheless, such limits will turn out to be particularly important, as we shall see in Chapter 4.

EXAMPLES 1. If $\lim_{x \to 2} f(x) = 3$ and $\lim_{x \to 2} g(x) = 5$, what is $\lim_{x \to 2} \phi(x)$ where $\phi(x) = 3f(x) + x^2 g(x)$?

ANSWER Using Theorem 1 we have $\lim_{x \to 2} \phi(x) = \lim_{x \to 2} [3f(x) + x^2 g(x)] = \lim_{x \to 2} [3f(x)] + \lim_{x \to 2} [x^2 g(x)] = 3 \lim_{x \to 2} f(x) + [\lim_{x \to 2} x^2][\lim_{x \to 2} g(x)] = 3 \cdot 3 + 4 \cdot 5 = 29$.

2. Find $\lim_{x \to 2} \phi(x)$, where $\phi(x) = (x^2 - 4)/(x - 2)$.

ANSWER Note first that we cannot find the limit by substituting $x = 2$ (the ϕ function, as written, is not defined at $x = 2$) or by computing the limits of the numerator and denominator—either procedure leads to the meaningless expression $0/0$. Instead, observe that, for $x \neq 2$, we have

$$\phi(x) = \frac{(x - 2)(x + 2)}{x - 2} = x + 2.$$

The function $x \mapsto x + 2$ is continuous and equal to 4 at $x = 2$. Hence $\lim_{x \to 2} \phi(x) = 4$.

3. Compute $\lim_{x \to 2} (x^2 + 5)/(x - 5)f(x)$, where f is the function in Example 1.

SOLUTION   We use the fact that rational functions are continuous and obtain

$$\lim_{x \to 2} \frac{x^2 + 5}{x - 5} f(x) = \lim_{x \to 2} \frac{x^2 + 5}{x - 5} \lim_{x \to 2} f(x) = \frac{9}{-3} \cdot 2 = -6.$$

PROBLEMS

1. If $\lim_{x \to 0} f(x) = 5$ and $\lim_{x \to 0} g(x) = 7$, what is $\lim_{x \to 0} [f(x) - 2g(x)]$?

2. If $\lim_{x \to 0} f(x) = 3$ and $\lim_{x \to 0} g(x) = -7$, what is the limit, at $x = 0$, of $\phi(x) = [3 + f(x)]g(x)$?

3. If $\lim_{x \to 1} \phi(x) = 6$ and $\lim_{x \to 1} \psi(x) = \frac{1}{2}$, what is $\lim_{x \to 1} \dfrac{[1 + \phi(x)]}{\psi(x)}$?

4. If $\lim_{x \to 0} f(x) = 5$, what is $\lim_{x \to 0} (3x^{10} + f(x))$?

5. If $\lim_{x \to 3} f(x) = 4$ and $\lim_{x \to 3} g(x) = -1$, what is the limit, at $x = 3$, of $\dfrac{[f(x) + x]}{[g(x)^2 - x^2]}$?

6. If $\lim_{t \to 0} \phi(t) = 7$ and $\lim_{t \to 0} \psi(t) = -1$, what is $\lim_{t \to 0} \dfrac{3t + \phi(t)^2}{(2 + t)\psi(t)}$?

7. If $\lim_{x \to 4} f(x) = 2$ and $\lim_{x \to 4} g(x) = 6$, what is $\lim_{x \to 4} \phi(x)$ if $\phi(x) = \dfrac{3f(x)g(x)}{[4 - f(x)^2]}$?

8. Suppose $\lim_{t \to 2} f(t) = -3$ and $\lim_{t \to 2} g(t) = 6$. What is $\lim_{t \to 2} \psi(t)$ if $\psi(t) = \dfrac{2g(t) - 12}{3 + f(t)}$?

In Problems 9 to 14 determine if the indicated limits exist, and if so, evaluate them.

9. $\lim_{h \to 0} \dfrac{h^3 + 2h^2 - h}{h}$.

10. $\lim_{u \to 1} \dfrac{u^2 - 1}{u - 1}$.

11. $\lim_{z \to 2} \dfrac{z^2 + z - 6}{z^2 - 4}$.

12. $\lim_{r \to 3/2} \dfrac{4r^2 - 9}{2r - 3}$.

13. $\lim_{x \to 3} \dfrac{9 + x^2}{3 - x}$.

14. $\lim_{x \to -1} \dfrac{x^2 + 3x + 2}{x^3 + x^2 - 4x - 4}$.

4.4   One-sided limits‡

Consider the function $f(x)$, graphed in Figure 3.29. Note that $|x|/x = 1$ for $x > 0$ and $|x|/x = -1$ for $x < 0$. The function $f(x)$ has no limit at $x = 0$. But for x positive and sufficiently small, $f(x)$ is as close to 1 as we want. For x negative and of sufficiently small absolute value, $f(x)$ is as close to -1 as we want. We say in this case that f has at 0 the **one-sided limits:** 1 from the right and -1 from the left.

The formal **definition of one-sided limits** is very similar to the definition of limit. The statement $\lim_{x \to x_0^+} f(x) = \alpha$ [read: the limit of $f(x)$, as x approaches x_0 from the right, is α] or $f(x) \to \alpha$ as $x \to x_0^+$ [read: $f(x)$ approaches α as x approaches x_0 from the right] means that (1) $f(x)$ *is defined in some interval with x_0 as the left endpoint (that is, for $x_0 < x < b$) and may or may not be defined*

---

‡This subsection may be omitted at first reading.

FIGURE 3.29

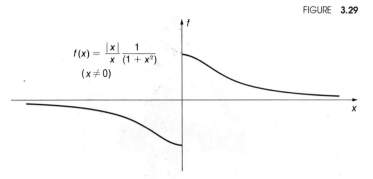

$$f(x) = \frac{|x|}{x} \frac{1}{(1 + x^2)}$$
$$(x \neq 0)$$

at x_0, and (2) if we define or redefine $f(x_0)$ to be α, then f, when considered as defined for $x_0 \leq x < b$, is continuous at x_0.

The equivalent $\epsilon - \delta$ **definition** reads: the statement "$\lim_{x \to x_0^+} f(x) = \alpha$" means that (1) *the function $f(x)$ is defined for $x > x_0$ and near x, and* (2) *for every $\epsilon > 0$, there is a $\delta > 0$ such that if $x_0 < x < \delta$, then $|f(x) - \alpha| < \epsilon$.*

The definition of the limit of f as x approaches x_0 from the left, that is, the meaning of the statement $\lim_{x \to x_0^-} f(x) = \beta$, is completely analogous.

Theorem 1 in §4.3 is valid for one-sided limits.

Suppose that f is defined near x_0, except perhaps at x_0 itself, and that *both* one-sided limits α and β exist at x_0. If $\alpha = \beta$, then $\lim_{x \to x_0} f(x) = \alpha$. If $\alpha \neq \beta$, then there is no way of making f continuous at x_0, and we say that f has a **jump discontinuity** at x_0. A look at Figure 3.29 will explain the name.

EXAMPLES   1.   For the function $f(x) = 1$ for $x < 0$, and $f(x) = 2 - x$ for $x > 0$, $f(0)$ not defined, we have $f(x) \to 2$ as $x \to 0^+$, $f(x) \to 1$ as $x \to 0^-$. Indeed, the function $2 - x$ is continuous at 0 and has the value 2. The constant function 1 is also continuous at 0 and has the value 1. The function f has a jump discontinuity at $x = 0$. See Figure 3.15.

2.   Set $g(t) = 1/t$ for $t < 0$, $g(0) = 4$, $g(t) = 7 + t^2$ for $t > 0$. Then $\lim_{x \to 0^+} g(t) = 7$, and g has no limit as t approaches 0 from the left.

PROBLEMS

In the following problems determine if the one-sided limits exist. If they do, evaluate them.

1. (a) $\lim_{x \to 0^+} x^2$;   (b) $\lim_{x \to 0^-} x^2$.

2. (a) $\lim_{x \to 0^+} \dfrac{|x^3|}{x^3}$;   (b) $\lim_{x \to 0^-} \dfrac{|x^3|}{x^3}$.

3. (a) $\lim_{x \to 2^+} \dfrac{x^3}{1 + x^2}$;   (b) $\lim_{x \to 2^-} \dfrac{x^3}{1 + x^2}$.

4. $\lim_{x \to 2^-} f(x)$, where $f(x) = x^2 - 4$ for $x < 2$, $-x^2 + 4$ for $x > 2$.

5. $\lim_{x \to 1^+} f(x)$, where $f(x) = x^3 - 1$ for $x < 1$, $\frac{1}{2}$ for $x = 1$, and x^2 for $x > 1$.

6. $\lim_{x \to 0^-} h(x)$, where $h(x) = (x^2 + 1)^2$ for $x < 0$, $2x - 8$ for $x > 0$.

7. $\lim_{x \to -1^+} h(x)$, where $h(x) = (1 + x)^2$ for $x < -1$, $(1 + x)^{-2}$ for $x > -1$.

8. $\lim_{t \to 1^-} g(t)$, where $g(t) = (t^2 + 1)(t - 1)^{-1}$ for $t < 1$, $t^2 - 1$ for $t \geq 1$.

DERIVATIVES

§1   The derivative of a function. Tangents

In this chapter we develop the first of the two basic concepts of calculus, the derivative. This concept originated from two seemingly unrelated problems: how to draw a tangent line to a curve and how to compute the velocity of a moving body.

1.1   The derivative, the slope of the tangent line

In the preceding chapter we considered continuous functions, those whose graphs "have no breaks." Now we consider continuous functions whose graphs "have no corners." An example of a graph with corners is shown in Figure 4.1; the graph in Figure 4.2 has no corners. Through every point of the latter graph passes a unique line, the tangent to the curve, which "just touches" the curve. (*Tangere* is a Latin word meaning to touch.)

GOTTFRIED WILHELM LEIBNIZ (1646–1716) spent most of his life in the service of a German ducal family (one of his employers became George I of England). He is equally famous as a philosopher and a mathematician. He was also a lawyer, diplomat, theologian, historian, geologist, economist, librarian, and linguist. Leibniz founded the Berlin Academy of Sciences and one of the first scientific journals. In this journal, he published the first account of calculus. Leibniz developed calculus later than Newton but independently of him.

A central theme in Leibniz' thought was the search for a "universal method" that would reduce all reasoning to calculations. In this sense, Leibniz foresaw mathematical logic. He also invented determinants and designed a computing machine.

FIGURE **4.1**   FIGURE **4.2**

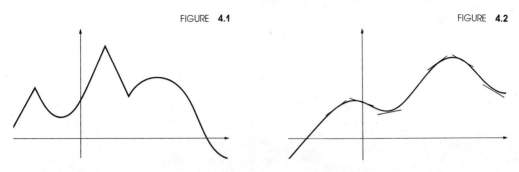

Consider a function $x \mapsto f(x)$. *The derivative of f at x_0 is the slope of the line tangent to the graph of f at the point* $(x_0, f(x_0))$. This, essentially, is the description given by Leibniz in his first published paper on calculus (1684). This paper contains a drawing not unlike our Figure 4.3. The derivative of f at a point x_0 will be denoted by $f'(x_0)$.

In order to use Leibniz's description, we must define what we mean by the tangent. For a circle, a tangent is a line having exactly one point in common with the circle. But such a definition would not work in general. Consider, for example, the graph shown in Figure 4.4. There are several lines (for instance, the lines labeled 1, 2, and 3) passing through the point P on the curve and having only this point in common with the curve. None of these fits the description of the tangent as the line that "just touches" the curve at P. The line labeled 4 does, but this line intersects the curve at other points also.

Our task is therefore to give a precise meaning to the expression: a line just touches the curve at a point. We proceed, as one often does in mathematics, by first assuming that we know what is meant by a tangent to a graph and by asking ourselves how we can compute the slope of this line. Then we shall make the method of computing into a definition.

Suppose we are given a function $x \mapsto f(x)$ and a point $P = (x_0, f(x_0))$; we want to compute the slope m of the tangent l to the graph of f at the point P (Figure 4.5). The difficulty is that we know only one point, namely, P on l, while we need two points on a line in order to compute the slope (see Chapter 2, §2.1). The key observation is that if we pick another point, Q, on the curve, which is close to P, then the slope M of the line L joining P and Q will be close to m. But M can be computed. For, if the coordinates of Q are $(x_1, f(x_1))$, then by the formula for the slope,

FIGURE **4.3**   FIGURE **4.4**

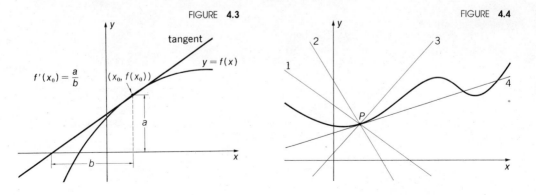

FIGURE **4.5**

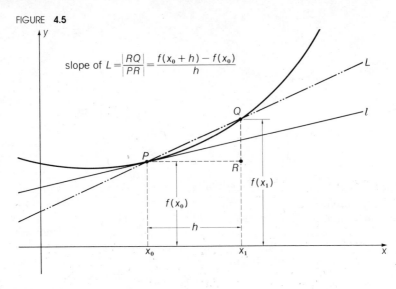

$$\text{slope of } L = \left|\frac{RQ}{PR}\right| = \frac{f(x_0 + h) - f(x_0)}{h}$$

$$M = \frac{f(x_1) - f(x_0)}{x_1 - x_0}.$$

It is convenient to set $h = x_1 - x_0$. Then $x_0 = x_1 + h$ and

$$M = \frac{f(x_0 + h) - f(x_0)}{h}.$$

This number is called the **difference quotient** of f at x_0, for the difference h. For Q close to P, it is an approximation to m.

The point Q will be close to P if $|h|$ is small. We expect that M is close to m if $|h|$ is small, and that the smaller $|h|$ is, the closer M is to m. Let us test this for the function $f(x) = x^2$ (defining a parabola) and the point $P = (1,1)$. First we compute M for various values of h of small absolute value:

h	$f(1 + h) = (1 + h)^2$	$M = \dfrac{(1 + h)^2 - 1}{h}$
.1	1.21	2.1
−.1	.81	1.9
.01	1.0201	2.01
−.01	.9801	1.99
.001	1.002001	2.001
−.001	.998001	1.999

The results show that m should be 2 or close to 2. But we can do better than that. If $f(x) = x^2$, $x_0 = 1$, we have

$$M = \frac{f(1 + h) - f(1)}{h} = \frac{(1 + 2h + h^2) - 1}{h} = 2 + h \qquad (\text{where } h \neq 0).$$

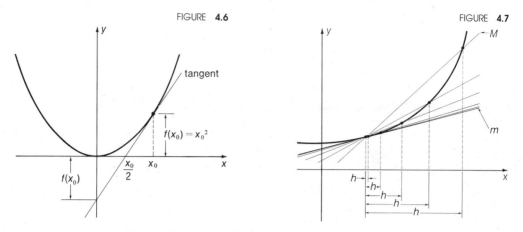

FIGURE 4.6

FIGURE 4.7

We observe that M, as a function of h, has a limit as $h \to 0$ (since $2 + h$ is a continuous function of h). This limit is 2. We conclude that this is the value of the slope m.

More generally, at any point (x_0, x_0^2) on our parabola, we can compute the slope m of the tangent as follows: since $f(x) = x^2$, we have

$$f(x_0 + h) = (x_0 + h)^2 = x_0^2 + 2x_0 h + h^2,$$
$$f(x_0 + h) - f(x_0) = x_0^2 + 2x_0 h + h^2 - x_0^2 = 2x_0 h + h^2,$$

and the difference quotient M is

$$M = \frac{f(x_0 + h) - f(x_0)}{h} = \frac{2x_0 h + h^2}{h} = 2x_0 + h.$$

The limit, as $h \to 0$, is $2x_0$. Thus we conclude that the tangent to the parabola $y = x^2$ passing through (x_0, x_0^2) has slope $2x_0$. The parabola, with a tangent, is shown in Figure 4.6.

We expect that, whenever the graph of a function f has a tangent l at the point $(x_0, f(x_0))$, we can find the slope $f'(x_0)$ of l by first computing the difference quotient M of f at x_0 for a difference h and then finding the limit of M at $h = 0$. The process is illustrated in Figure 4.7.

PROBLEMS

In Problems 1 to 4 tabulate values of the difference quotient M at the given values of x_0 for $h = \pm 1, \pm.1, \pm.01,$ and $\pm.001$. What value does M seem to approach in each case as h gets small?

1. $f(x) = x^2 - 4, \quad x_0 = 2.$
2. $f(x) = 4x^2 + x + 1, \quad x_0 = 1.$
3. $f(x) = 2x^2 - x + 1, \quad x_0 = \frac{1}{2}.$
4. $f(x) = x/(x + 1), \quad x_0 = 0.$

In Problems 5 to 10 compute the difference quotient M at $x = x_0$, and express your answer in terms of h.

5. $f(x) = 5x^2 - 10x + 1.$
6. $f(x) = -6x^2 + 4x + 5.$
7. $f(x) = 2x^3 + 1.$
8. $f(x) = x^3 + 3x.$
9. $f(x) = 1/(4 + x).$
10. $f(x) = 1/(1 + x^2).$

1.2 Analytic definitions of the derivative, the tangent, and the normal

The preceding considerations suggest the following analytic **definition of the derivative.** Let f be a function defined near a point x_0. The derivative of f at x_0 is the number

$$f'(x_0) = \lim_{h \to 0} \frac{f(x_0 + h) - f(x_0)}{h}, \tag{1}$$

provided that this limit exists.

We note that (setting $x_1 = x_0 + h$) we can also write

$$f'(x_0) = \lim_{x_1 \to x_0} \frac{f(x_1) - f(x_0)}{x_1 - x_0}. \tag{1'}$$

The quotient

$$\frac{f(x_1) - f(x_0)}{x_1 - x_0} = \frac{f(x_0 + h) - f(x_0)}{h} \tag{2}$$

is called, as we already noted, the difference quotient (of f at x_0, for the difference h). It is the slope of the line joining the point $(x_0, f(x_0))$ and $(x_1, f(x_1))$ with $x_1 = x_0 + h$, $h \neq 0$. (Such a line is called a chord or a secant of the curve.)

The tangent to the graph of $y = f(x)$, at the point $(x_0, f(x_0))$, is the line through this point with slope $f'(x_0)$. Using the point-slope form of the equation of a line [Chapter 2, §2.2; with $y_0 = f(x_0)$, $m = f'(x_0)$], we obtain the equation of the tangent line in the form

$$y - f(x_0) = f'(x_0)(x - x_0). \tag{3}$$

The line through $(x_0, f(x_0))$ perpendicular to the tangent is called the normal to the graph at this point (see Figure 4.8). Since perpendicular lines have slopes whose product is -1 (see Chapter 2, §2.3), the slope of the normal is $-(1/\text{slope of tangent}) = -1/f'(x_0)$. The equation of the normal is

$$y - f(x_0) = -\frac{1}{f'(x_0)}(x - x_0), \tag{4}$$

provided that $f'(x_0) \neq 0$. If $f'(x_0) = 0$, the normal is vertical. Its equation reads: $x = x_0$.

FIGURE **4.8**

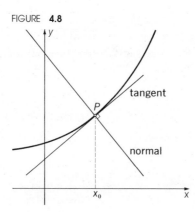

FIGURE **4.9**

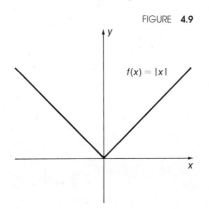

$f(x) = |x|$

It may happen that a function does not have a derivative at a point x_0. This happens if there is no limit (1). In this case, there is no tangent to the graph at the point $(x_0, f(x_0))$.

For instance, the function $f(x) = |x|$ has no derivative at $x = 0$. Indeed, the difference quotient at $x = 0$ is $[f(0 + h) - f(0)]/h = [|0 + h| - |0|]/h = |h|/h$; this is 1 if $h > 0$ and (-1) if $h < 0$. There is no limit as $h \to 0$. The graph of the function $f(x) = |x|$, shown in Figure 4.9, has a corner at the point $(0,0)$.

1.3 Computing the derivative from the definition

In order to compute the derivative of a given function f at a point x_0, we first write down $f(x_0)$ and $f(x_0 + h)$ and form the difference quotient

$$\frac{f(x_0 + h) - f(x_0)}{h}.$$

Here x_0 is a fixed number, òr a symbol representing an unspecified fixed number, while the variable h stands for a number which is close to, but different from, zero. Then we try to transform the difference quotient, by legitimate manipulations, into a form from which the limit (as $h \to 0$) can be read off. (Usually we try to cancel the h in the denominator against an h in the numerator.)

EXAMPLES 1. Compute $f'(2)$ for $f(x) = x^3 - x$.
ANSWER We have

$$f'(2) = \lim_{h \to 0} \frac{f(2 + h) - f(2)}{h} = \lim_{h \to 0} \frac{[(2 + h)^3 - (2 + h)] - (2^3 - 2)}{h}$$

$$= \lim_{h \to 0} \frac{(8 + 12h + 6h^2 + h^3 - 2 - h) - (8 - 2)}{h}$$

$$= \lim_{h \to 0} \frac{11h + 6h^2 + h^3}{h} = \lim_{h \to 0} (11 + 6h + h^2) = 11.$$

2. Find the equations of the tangent and the normal to the curve $y = x^3 - x$ at the point $(2,6)$.
ANSWER First we check that if $x = 2$, $f(x) = x^3 - x$ is 6; this is so. At this point, the slope of the tangent is $f'(2) = 11$, by Example 1. The slope of the normal is $-1/f'(2) = -1/11$.

The equation of the tangent is obtained either by the point-slope formula in Chapter 2, §2.2 (with $x_0 = 2$, $y_0 = 6$, $m = 11$), or directly from (3). It reads

$$y - 6 = 11(x - 2) \quad \text{or} \quad 11x - y - 16 = 0.$$

The equation of the normal is obtained either from the point-slope formula (with $m = -1/11$) or from (4). It reads

$$y - 6 = -\tfrac{1}{11}(x - 2) \quad \text{or} \quad x + 11y - 68 = 0.$$

3. What is the derivative of $g(t) = 1/t$ at $t = 3$?
ANSWER   We have

$$g'(3) = \lim_{h \to 0} \frac{g(3 + h) - g(3)}{h} = \lim_{h \to 0} \frac{1}{h}\left(\frac{1}{3 + h} - \frac{1}{3}\right)$$

$$= \lim_{h \to 0} \frac{1}{h} \cdot \frac{3 - (3 + h)}{(3 + h)3} = \lim_{h \to 0} \frac{-1}{(3 + h)3} = -\frac{1}{9}.$$

4. What is the derivative of $f(x) = 1/x$ at $x = 3$?
ANSWER   This is the same problem as in Example 3; only the names of the variables have been changed.
The answer is, therefore, $f'(3) = -\tfrac{1}{9}$.

5. Find the equation of the normal to the curve $y = f(x) = 3x^3 - 1$ at the point $(0, -1)$.
ANSWER   First we find the slope of the tangent:

$$f'(0) = \lim_{h \to 0} \frac{f(0 + h) - f(0)}{h} = \lim_{h \to 0} \frac{(3h^3 - 1) - (-1)}{h} = \lim_{h \to 0} 3h^2 = 0.$$

Since $f'(0) = 0$, the tangent is horizontal. Its equation reads $x = 0$. The normal is vertical. [Formula (4) is not applicable in this case.]

PROBLEMS

In Problems 1 to 14 the computation of the derivative should be based directly on the definition (as in the foregoing examples).

1. If $f(x) = 3x^2 - 1$, find $f'(-1)$.
2. What is the derivative of $f(x) = -4x^2 + 8x + 3$ at $x = 1$?
3. What is the derivative of $f(x) = \tfrac{1}{2}x^2 + x + 5$ at $x = 4$?
4. If $g(t) = t^3$, what is $g'(0)$?
5. If $f(x) = 1/(x - 1)$, find $f'(2)$.
6. Find the derivative of $g(s) = (s + 1)/(s - 1)$ at $s = 0$. Find also the equation of the tangent to the graph of this function at the point $(0, -1)$.
7. Find the equation of the tangent to the curve $y = 2x^2 - 1$ at the point $(-1, 1)$. [*Hint:* First find the derivative dy/dx at $x = -1$.]
8. If $f(x) = x^2 - x - 1$, find the equation of the normal to the graph at the point $(1, -1)$.
9. Find the equation of the normal to the curve $f(x) = x^3 - 1$ at the point $(2, 7)$.
10. Find the equation of the tangent to the graph of $y = x + (1/x)$ at the point $(1, 2)$.
11. Find the equation of the tangent to the curve $f(x) = x^2/(x + 1)$ at $x = 1$.
12. Find the equation of the tangent to the graph of $z = u^3 + u - 1$ at $u = 2$.
13. Find the equation of the normal to the curve $y = 2x + (3/x^2)$ at $x = 1$.
14. Where does the tangent to the curve $y = x^2 + 2x$ at the point $(2, 8)$ intersect the horizontal and vertical axes?

1.4 The derivative, a new function

Thus far we dealt with the derivative $f'(x_0)$ of a function f at a fixed point x_0. The derivative is a *number*, the slope of the tangent to the graph of $x \mapsto f(x)$ at the point $(x_0, f(x_0))$. We can also define a new *function* which assigns to every relevant x the derivative $f'(x)$ at this point. This function, that is, the rule $x \mapsto f'(x)$, is sometimes called the derived function of f and is denoted by f'.

Strictly speaking, we should distinguish between the derivative $f'(x_0)$ of a function $x \mapsto f(x)$ at some point x_0 and the rule $x \mapsto f'(x)$. It is customary, however, to use the term "derivative" in both cases. Thus, if $f(x) = x^2$, we say "the derivative of this function is the function $x \mapsto 2x$" and "the derivative of $f(x)$ at $x = 2$ is the number 4."

EXAMPLES 1. Set $f(x) = mx + b$, and compute the derivative at some point x.
ANSWER   We have $f(x) = mx + b$, $f(x + h) = m(x + h) + b$ and

$$\frac{f(x + h) - f(x)}{h} = \frac{(mx + mh + b) - (mx + b)}{h} = m.$$

The difference quotient is a constant. Hence

$$f'(x) = \lim_{h \to 0} \frac{f(x + h) - f(x)}{h} = \lim_{h \to 0} m = m.$$

This is not surprising. The graph of $y = mx + b$ is a line of slope m, and a line is to be considered its own tangent.

2. Find the derivative of the function $y = x^2$.
ANSWER   Let $y = f(x) = x^2$. Then $f(x + h) = (x + h)^2 = x^2 + 2xh + h^2$, and the difference quotient becomes

$$\frac{f(x + h) - f(x)}{h} = \frac{x^2 + 2xh + h^2 - x^2}{h} = 2x + h.$$

Hence

$$f'(x) = \lim_{h \to 0} \frac{f(x + h) - f(x)}{h} = \lim_{h \to 0} (2x + h) = 2x.$$

Note that the slope of the tangent to the parabola $f(x) = x^2$ at the point $x = x_0$ is the number $m = f'(x_0) = 2x_0$.

3. Compute $f'(x)$ for $f(x) = 1/x$ (and, of course, $x \neq 0$). In particular, compute $f'(3)$.
ANSWER   We have

$$\frac{f(x + h) - f(x)}{h} = \frac{1}{h}\left(\frac{1}{x + h} - \frac{1}{x}\right) = \frac{1}{h} \cdot \frac{x - (x + h)}{x(x + h)} = -\frac{1}{x(x + h)}.$$

The last expression defines a continuous function of h, at $h = 0$ (since $x \neq 0$). Hence $f'(x) = -1/x^2$. Substituting $x = 3$, we obtain $f'(3) = -\frac{1}{9}$, as in Example 4 in §1.3.

PROBLEMS

In the following problems the computation of the derivative should be based directly on the definition (as in the foregoing examples).

1. If $f(x) = -3x^2 + 1$, find $f'(x)$.
2. If $f(x) = 4x^2 - 3x + 1$, find $f'(x)$.

3. What is the derivative of $f(x) = 2x^2 + 4x + 1$?
4. What is the derivative of $f(x) = -\frac{1}{3}x^2 + 2x - 5$?
5. If $f(s) = s^3$, find $f'(s)$.
6. Find the derivative of $x \mapsto x^3 + 2x$.
7. If $g(x) = 1/(2x - 3)$, find $g'(x)$.
8. If $H(x) = 1/x^2$, find $H'(x)$.
9. Find the derivative of $x \mapsto 1/(1 + x^2)$.
10. If $f(x) = x^2 - (1/x)$, find $f'(x)$.
11. Find the derivative of $G(x) = (2x - 3)/(3x + 2)$.
12. Find the derivative of $t \mapsto (1/t) + (1/t^2)$.

1.5   The Leibniz notation

The notation $f'(x)$ for the derivative of a function $y = f(x)$ was introduced by Lagrange. Leibniz denoted the same derivative by

$$\frac{df(x)}{dx} \quad \text{or} \quad \frac{d}{dx}f(x) \quad \text{or} \quad \frac{df}{dx} \quad \text{or} \quad \frac{dy}{dx}$$

(read "df over dx" or "dy over dx"). Thus we have, from the examples above,

$$\frac{d(mx + b)}{dx} = m, \quad \frac{d(x^2)}{dx} = 2x, \quad \frac{d(1/x)}{dx} = -\frac{1}{x^2}.$$

The Leibniz notation may be explained by a mathematical myth (represented in Figure 4.10). A curve, so the myth goes, consists of infinitely many "infinitesimal" (infinitely small) straight segments. A tangent to the curve is a line containing such an infinitesimal segment. Now let the curve be the graph of a function $y = f(x)$. To compute the slope of the tangent, at the point (x,y), we increase x by an infinitesimal amount dx; this corresponds to a change of $y = f(x)$ by an infinitesimal amount dy such that $f(x + dx) = y + dy$. The piece of the graph between the infinitely close points (x,y) and $(x + dx, y + dy)$ is an infinitesimal segment; its slope is

$$\frac{(y + dy) - y}{(x + dx) - x} = \frac{dy}{dx}.$$

This is how mathematicians thought about calculus before the clarification of the limit concept. They were, of course, aware of the logical deficiencies of this approach, yet they saw in calculus a powerful tool for solving problems. Therefore they followed d'Alembert's advice: "Proceed and faith shall be given to you."

JOSEPH LOUIS LAGRANGE (1736–1813) was a leading mathematician of the eighteenth century. One of his books, the famous *Analytical Mechanics*, is a systematic development of mechanics by the methods of calculus; not a single diagram appears in the book—everything is done by formulas.

Lagrange was under 20 when he became professor of mathematics at the artillery school in his native Turin. Later he was Euler's successor in the Berlin Academy. From 1786 on, Lagrange lived in Paris, and after the French revolution he taught at the newly established École Normale and École Polytechnique.

JEAN LE ROND D'ALEMBERT (1717–1783) was a highly influential scientist and mathematician. He belonged to the group of French Enlightenment philosophers who published the *Encyclopédie*, and he was permanent secretary of the Paris Academy of Sciences. D'Alembert is remembered mostly for his work in mechanics.

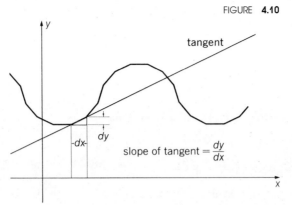

FIGURE **4.10**

Today we do not think of dy and dx as names of infinitely small quantities or of dy/dx as a fraction, but we still use Leibniz's notation. First of all, it is highly suggestive. The symbol dy/dx is a name for the derivative; its form reminds us that the derivative is the limit of the difference quotient (2), which is indeed a fraction, with the increment of x in the denominator and the corresponding increment of $y = f(x)$ in the numerator. We often denote the increment of x by Δx. (It is important to remember that Δx is a single symbol, not the product of Δ and x. This symbol is pronounced "delta x." Previously we denoted this increment by the letter h.) Let us now denote the increment of the function $y = f(x)$ by Δy. Thus $\Delta y = f(x + \Delta x) - f(x)$, and the difference quotient may be written as $\Delta y / \Delta x$. Hence we have

$$\frac{dy}{dx} = \lim_{\Delta x \to 0} \frac{\Delta y}{\Delta x}.$$

This is another "justification" of the Leibniz notation. The principal justification, however, is that the notation is very convenient, as we shall see later on many occasions.

Unfortunately, the Leibniz notation does not indicate the point at which the derivative is computed. To overcome this disadvantage, we indicate this point by a subscript. Thus we write

$$\frac{d(x^2)}{dx} = 2x, \qquad \left(\frac{d(x^2)}{dx}\right)_{x=2} = 4.$$

We shall use this convention whenever convenient.

Another common notation for the derivative of a function $y = f(x)$ is y'.

Other notations for the derivative, which we shall not use, are Df or $(Df)(x)$.

PROBLEMS

In the following problems use the results derived in Examples 1, 2, and 3 in §1.4.

1. If $f(x) = 2x + 3$, what is $\dfrac{d}{dx}(2x + 3)$?   3. If $f(x) = \dfrac{1}{x}$, what is $\dfrac{d}{dx}\left(\dfrac{1}{x}\right)$?

2. If $y = x^2$, what is $\dfrac{dy}{dx}$?   4. If $y = -\tfrac{1}{2}x + 1$, what is $\left(\dfrac{dy}{dx}\right)_{x=2}$?

5. If $f(x) = -3x + 5$, what is $\left(\dfrac{d}{dx}(-3x + 5)\right)_{x=1}$?

6. If $y = x^2$, what is $\left(\dfrac{dy}{dx}\right)_{x=-1}$?

7. If $y = f(x) = 5$ for all x, what is $\left(\dfrac{dy}{dx}\right)_{x=3}$?

8. If $f(x) = \dfrac{1}{x}$, what is $\left(\dfrac{d}{dx}\left(\dfrac{1}{x}\right)\right)_{x=a}$ if a is any number not zero?

1.6 The linear approximation theorem

The mathematical myth told above contains a kernel of truth. To see this, we reformulate the definition of the derivative.

Let the function $f(x)$ have, at x_0, the derivative $f'(x_0)$, and let us denote by $r(x)$ the difference between the difference quotient

$$\frac{f(x) - f(x_0)}{x - x_0}$$

(for $x \neq x_0$) and the derivative $f'(x_0)$. That is, let us set

$$\frac{f(x) - f(x_0)}{x - x_0} - f'(x_0) = r(x). \tag{5}$$

The definition ($1'$) of the derivative (see §1.2) says that

$$\lim_{x \to x_0} r(x) = 0.$$

This means that the function $r(x)$ is continuous at $x = x_0$ if we set $r(x_0) = 0$.

For example, let $f(x) = x^2$, $x_0 = 2$. We know that $f'(2) = 4$. Hence

$$r(x) = \frac{f(x) - f(x_0)}{x - x_0} - f'(x_0) = \frac{x^2 - 4}{x - 2} - 4$$

$$= \frac{(x - 2)(x + 2)}{x - 2} - 4 = x + 2 - 4 = x - 2.$$

This is a continuous function of x with $r(2) = 0$.

For $x \neq x_0$ we may multiply both sides of (5) by $x - x_0$ and obtain the equivalent equation

$$f(x) - f(x_0) - f'(x_0)(x - x_0) = r(x)(x - x_0)$$

or, solving for $f(x)$,

$$f(x) = f(x_0) + f'(x_0)(x - x_0) + r(x)(x - x_0). \tag{6}$$

In the example above [$f(x) = x^2$, $x_0 = 2$, $f'(x_0) = 4$, $r(x) = x - 2$], we have $x^2 = 4 + 4(x - 2) + (x - 2)^2$.

The preceding discussion shows that the following statement is true.

FIGURE 4.11

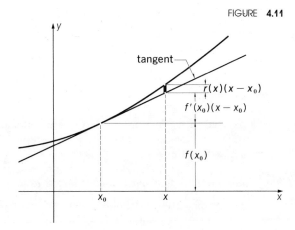

Theorem 1. *Let f be a function defined near x_0. The function has a derivative m at x_0 if and only if*

$$f(x) = f(x_0) + m(x - x_0) + r(x)(x - x_0), \tag{7}$$

where

$$r(x) \text{ is continuous at } x_0 \text{ and } r(x_0) = 0. \tag{8}$$

Indeed, if we set $m = f'(x_0)$, (7) becomes identical with (6). To understand the meaning of the theorem, note that

$$x \mapsto f(x_0) + m(x - x_0), \qquad \text{where } m = f'(x_0),$$

is a linear function whose graph is the tangent line to the graph of f at the point $(x_0, f(x_0))$. The number $r(x)(x - x_0)$ is the error we commit if, in computing the value $f(x)$, for some x close to x_0, we replace the graph of the function f by the tangent at x_0. This is seen from Figure 4.11. The error will be small when $x - x_0$ is small, and it will be *much smaller* than $x - x_0$, as (8) shows. For, if $x - x_0$ is small, the error is $(x - x_0)$ times the small number $r(x)$. *In this sense* we can replace a small arc of a curve by its tangent.

EXAMPLE Let $f(x) = 1/x$, $x_0 = 3$. Write the function $f(x)$, near x_0, as in the linear approximation theorem [Equation (7) above], that is, as a sum of a linear function representing the tangent to the graph of f at x_0 and an error term.

ANSWER   First we note that $f(3) = \frac{1}{3}$. Then we compute $m = f'(3)$ and obtain $m = -\frac{1}{9}$ (see §1.3, Example 4). Substituting into (7), we obtain

$$\frac{1}{x} = \frac{1}{3} - \frac{1}{9}(x - 3) + r(x)(x - 3).$$

To compute $r(x)$, solve the previous equation:

$$r(x)(x - 3) = \frac{1}{x} - \frac{1}{3} + \frac{1}{9}(x - 3)$$

and

$$r(x) = \frac{\dfrac{1}{x} - \dfrac{1}{3} + \dfrac{1}{9}(x - 3)}{x - 3} = \frac{\dfrac{3 - x}{3x} + \dfrac{1}{9}(x - 3)}{x - 3} = -\frac{1}{3x} + \frac{1}{9} = \frac{x - 3}{9x}.$$

Thus

$$\frac{1}{x} = \underbrace{\frac{1}{3} - \frac{1}{9}(x - 3)}_{\text{linear function}} + \underbrace{\frac{(x - 3)^2}{9x}}_{\text{error term}}$$

For $x = 3.1$, the error term is $.01/27.9$, which is less than $.01/20 = .0005$. For $x = 3.01$, the error term is $.0001/27.09 < .000005$.

In discussing the derivative of a function, we have assumed the function to be continuous; we do not expect to have a tangent where the graph has a break. The following result justifies our assumption.

Theorem 2. *If f has a derivative at x_0, then f is continuous at x_0.*

Proof. If $f'(x_0) = m$, then Equation (7) holds. In it, f is represented as a sum of three functions [a constant, a linear function $m(x - x_0)$, and the function $r(x)(x - x_0)$], each of which is continuous at x_0. A sum of continuous functions is continuous (by Theorem 1, in Chapter 3, §3.3).

PROBLEMS

In the following problems write the given function f, near the point x_0, as a sum of a linear function representing the tangent to the graph of f at x_0 plus an error term. State the error term in each case. (The value of m at x_0 is given to facilitate calculation.)

1. $f(x) = x^2 + 1$, $x_0 = 2$, $m = 4$.
2. $f(x) = x^2 - 4x + 5$, $x_0 = -1$, $m = -6$.
3. $f(x) = 3x^2 - x + 6$, $x_0 = 1$, $m = 5$.
4. $f(x) = -4x^2 + 6x + 1$, $x_0 = \frac{1}{2}$, $m = 2$.

5. $f(x) = 2x^2 - 4x + 1$, $x_0 = 0$, $m = -4$.
6. $f(x) = x^3$, $x_0 = 1$, $m = 3$.
7. $f(x) = 1/(x + 1)$, $x_0 = 2$, $m = -\frac{1}{9}$.
8. $f(x) = 1/x^2$, $x_0 = \frac{1}{2}$, $m = -16$.

§2 Differentiation

A function that has a derivative is called **differentiable.** The process of finding the derivative of a function is called **differentiation.** In differentiating functions, we hardly ever go back to the definition of a derivative. Instead, we use a set of rules by which the differentiation of functions commonly encountered in calculus is reduced to a mechanical process.

Rules (1) to (8) derived in this section are among the few rules in calculus that *must be memorized.*

2.1 Constants, sums, and differences

Theorem 1. *Let f and g be functions that have derivatives (at a point) and let c be a constant. Denote derivatives by primes. Then (at the point considered)*

$$c' = 0, \tag{1}$$
$$(cf)' = cf', \tag{2}$$
$$(f + g)' = f' + g', \tag{3}$$
$$(f - g)' = f' - g'. \tag{4}$$

We rewrite the statement in the Leibniz notation:

$$\frac{dc}{dx} = 0 \qquad (c \text{ a constant}), \tag{1'}$$

$$\frac{d(cf)}{dx} = c\,\frac{df}{dx} \qquad (c \text{ a constant}), \tag{2'}$$

$$\frac{d(f + g)}{dx} = \frac{df}{dx} + \frac{dg}{dx}, \tag{3'}$$

$$\frac{d(f - g)}{dx} = \frac{df}{dx} - \frac{dg}{dx}. \tag{4'}$$

One advantage of the Leibniz notation is that Rules (2'), (3'), and (4') look like rules for fractions. These rules read, in words: a constant may be factored out in front of the differentiation sign; the derivative of a sum is the sum of the derivatives; and the derivative of a difference is the difference of the derivatives.

Proof of Theorem 1.   Rule (1) is a special case of Example 1 in §1.4. It is geometrically obvious (a horizontal line has slope 0).

Next we prove Rule (2). The proof uses the definition of derivatives and the rules for computing limits (Chapter 3, Theorem 1 in §4.3). Set $\phi(x) = cf(x)$; we have

$$\phi'(x) = \lim_{h \to 0} \frac{\phi(x + h) - \phi(x)}{h} = \lim_{h \to 0} \frac{cf(x + h) - cf(x)}{h}$$

$$= \lim_{h \to 0} c\,\frac{f(x + h) - f(x)}{h} = c \lim_{h \to 0} \frac{f(x + h) - f(x)}{h} = cf'(x)$$

(since the limit of the product of a constant and a function is the product of the constant and the limit of the function). Thus $\phi' = (cf)' = cf'$, as asserted.

In order to prove the sum rule (3), we set $\phi(x) = f(x) + g(x)$ and obtain

$$\phi'(x) = \lim_{h \to 0} \frac{\phi(x + h) - \phi(x)}{h} = \lim_{h \to 0} \frac{[f(x + h) + g(x + h)] - [f(x) + g(x)]}{h}$$

$$= \lim_{h \to 0} \left(\frac{f(x + h) - f(x)}{h} + \frac{g(x + h) - g(x)}{h} \right)$$

$$= \lim_{h \to 0} \frac{f(x + h) - f(x)}{h} + \lim_{h \to 0} \frac{g(x + h) - g(x)}{h} = f'(x) + g'(x)$$

(since the limit of a sum is the sum of limits). Thus $\phi' = (f + g)' = f' + g'$, as asserted.

Using (2) and (3), we derive Rule (4) by writing

$$(f - g)' = (f + (-1)g)' = f' + ((-1)g)' = f' + (-1)g' = f' - g'.$$

EXAMPLES 1.   Assume that $f(x) = 2x$ and $g(x) = 3$. Verify that Rule (3) holds.
ANSWER   Set $h(x) = f(x) + g(x)$. Then $h(x) = 2x + 3$. Now $h'(x) = 2$, $f'(x) = 2$, and $g'(x) = 0$ by Example 1 in §1.4. Hence $h' = f' + g'$; Rule (3) is verified.

2.   What is the derivative of $100x^2$?
ANSWER   We know by Example 2 in §1.4 that if $f(x) = x^2$, then $f'(x) = 2x$. By Rule (2), the derivative of $100x^2$ is

$$(100x^2)' = 100(x^2)' = 100(2x) = 200x.$$

3. Differentiate the function $f(x) = x^2 - 1/x$, using the results obtained previously.
ANSWER   Since $(x^2)' = 2x$ and $(1/x)' = -1/x^2$ (by Example 3 in §1.4), we have, using Rule (4),

$$\left(x^2 - \frac{1}{x}\right)' = (x^2)' - \left(\frac{1}{x}\right)' = 2x - \left(-\frac{1}{x^2}\right) = 2x + \frac{1}{x^2}.$$

We give next some examples of applying the differentiation rules to more than two functions. We assume that the functions f, g, h have derivatives.

4. What is $(f + g + h)'$?
ANSWER $f' + g' + h'$. Indeed, applying Rule (2) twice, we have that

$$(f + g + h)' = ((f + g) + h)' = (f + g)' + h' = (f' + g') + h'.$$

5. What is $(2f + 3g - 5h)'$?
ANSWER $2f' + 3g' - 5h'$. Indeed, by Rules (1) to (4),

$$(2f + 3g - 5h)' = (2f + 3g)' - (5h)' = (2f)' + (3g)' - (5h)' = 2f' + 3h' - 5h'.$$

PROBLEMS

In Problems 1 to 8 expand the given expressions using differentiation rules (1) to (4), or (1') to (4'). Assume f, g, and h are differentiable.

1. $(f + 8)'$.
2. $(2f + 3g)'$.
3. $(1 + f - 2g)'$.
4. $(-f + 4g - 6h)'$.

5. $\dfrac{d}{dx}(-4 + 3f)$.

6. $\dfrac{d}{dx}(5f - 12g)$.

7. $\dfrac{d}{dx}(-2f + 4g - h)$.

8. $\dfrac{d}{dx}(1 + 2f + 3g + 4h)$.

In Problems 9 to 14 use the following results whenever needed: $(x)' = \dfrac{d}{dx}(x) = 1$, $(x^2)' = \dfrac{d}{dx}(x^2) = 2x$, $\left(\dfrac{1}{x}\right)' = \dfrac{d}{dx}\left(\dfrac{1}{x}\right) = -\dfrac{1}{x^2}$.

9. Suppose $f(x) = 4x$ and $g(x) = \frac{1}{2}x^2$. What is $(f + g)'$?

10. Suppose $f(x) = 1/x$ and $g(x) = 3x^2$. What is $(f - \frac{1}{2}g)'$?

11. Suppose $f(x) = -2x^2$ and $g(x) = 1 + 4x$. What is $\dfrac{d}{dx}(f - g)$?

12. Find $\dfrac{d}{dx}(6 - 2f + g)$ if $f(x) = x$ and $g(x) = x^2$.

13. Find $(2f + g - h)'$ if $f(x) = \frac{1}{2}x$, $g(x) = 1 - 1/x$, and $h(x) = 4x^2$.

14. Find $\dfrac{d}{dx}(3x^2 + f - 4g + 2)$ if $f(x) = 2/x$ and $g(x) = x - 1$.

2.2   The product rule. Derivatives of polynomials

Theorem 2.   *Let f and g be functions which have derivatives (at a point). Then (at the point considered)*

$$(fg)' = f'g + fg'. \tag{5}$$

In the Leibniz notation the rule reads

$$\frac{d(fg)}{dx} = \frac{df}{dx}g + f\frac{dg}{dx}. \tag{5'}$$

In words: the derivative of a product of two functions is the derivative of the first function times the second plus the first function times the derivative of the second. Unfortunately, the Leibniz notation does not help in remembering this rule.

Proof of Theorem 2.   Note first that if $g'(x)$ exists, then $\lim_{h \to 0} g(x + h) = g(x)$. Indeed, by Theorem 2, in §1.6, $g(x)$ is continuous at x so that if $|h|$ is sufficiently small, $g(x + h)$ is as close as we want to $g(x)$. Also $\lim_{h \to 0} f(x) = f(x)$; this is relation (8) in Chapter 3, §4.3. [$f(x)$ is a constant as far as h is concerned.]

Now set $\phi(x) = f(x)g(x)$. Then

$$\frac{\phi(x + h) - \phi(x)}{h} = \frac{f(x + h)g(x + h) - f(x)g(x)}{h}.$$

We use a trick: the numerator to the right is not changed if we subtract and add the number $f(x)g(x + h)$. Hence, using the definition of the derivative and the limit rules from Chapter 3, §4.3, we have

$$\phi'(x) = \lim_{h \to 0} \frac{\phi(x + h) - \phi(x)}{h}$$

$$= \lim_{h \to 0} \frac{f(x + h)g(x + h) - f(x)g(x + h) + f(x)g(x + h) - f(x)g(x)}{h}$$

$$= \lim_{h \to 0} \left(\frac{f(x + h) - f(x)}{h} g(x + h) + f(x) \frac{g(x + h) - g(x)}{h} \right)$$

$$= \lim_{h \to 0} \frac{f(x + h) - f(x)}{h} \lim_{h \to 0} g(x + h) + \lim_{h \to 0} f(x) \lim_{h \to 0} \frac{g(x + h) - g(x)}{h}$$

$$= f'(x)g(x) + f(x)g'(x).$$

This proves (5).

EXAMPLES 1.   Verify that Rule (5) holds if $f(x) = x$ and $g(x) = 1/x$.
ANSWER Set $h(x) = f(x)g(x)$; then $h(x) = 1$. We have that $f'(x) = 1, g'(x) = -1/x^2$, and $h'(x) = 0$, all by Examples 1 and 3 in §1.4. Hence $f'g + fg' = 1(1/x) + x(-1/x^2) = 0 = h' = (fg)'$. Rule (5) holds.

2.   Differentiate the function $x \mapsto x^3$, using the fact that $x^3 = x \cdot x^2$ and Rule (5).
ANSWER Apply Rule (5) with $f(x) = x$, $g(x) = x^2$. We obtain

$$(x^3)' = (x \cdot x^2)' = (x)'(x^2) + (x)(x^2)' = (1)x^2 + (x)(2x) = 3x^2.$$

The last example suggests a general, and very important, differentiation rule.

Theorem 3.   *For every positive integer n, if $f(x) = x^n$, then $f'(x) = nx^{n-1}$; that is,*

$$\frac{dx^n}{dx} = nx^{n-1}.$$

Proof.   For $n = 1$, the theorem says that the derivative of $f(x) = x$ is 1. This is a special case of Example 1 in §1.4, which is geometrically obvious (the graph of the identity function has slope 1). For $n > 1$, we use the product rule (5) and the fact that $x' = 1$. Thus

$$(x^2)' = (xx)' = x'x + x(x)' = x + x = 2x,$$
$$(x^3)' = (xx^2)' = x'x^2 + x(x^2)' = x^2 + x(2x) = x^2 + 2x^2 = 3x^2,$$
$$(x^4)' = (xx^3)' = x'x^3 + x(x^3)' = x^3 + x(3x^2) = x^3 + 3x^3 = 4x^3,$$

and so forth. (To make the argument formally rigorous, we would have to use mathematical induction ▶.)

Consider now a polynomial function, say,

$$f(x) = 3x^5 + 2x^3 - 6x + 1.$$

Using Theorems 1, 2, and 3, we compute its derivative as follows:

$$
\begin{aligned}
f'(x) &= (3x^5 + 2x^3 - 6x + 1)' \\
&= (3x^5)' + (2x^3)' - (6x)' + (1)' \quad \text{by Rules (3) and (4)} \\
&= 3(x^5)' + 2(x^3)' - 6(x)' \quad\quad\;\; \text{by Rules (2) and (1)} \\
&= 3(5x^4) + 2(3x^2) - 6(1) \quad\quad\;\; \text{by Theorem 3} \\
&= 15x^4 + 6x^2 - 6.
\end{aligned}
$$

The same method can be used for any polynomial. In general,

$$(a_0x^n + a_1x^{n-1} + \cdots + a_{n-1}x + a_n)' = na_0x^{n-1} + (n-1)a_1x^{n-2} + \cdots + a_{n-1}.$$

This establishes the following.

Theorem 4. *The derivative of a polynomial of degree $n > 0$ is a polynomial of degree $n - 1$.*

EXAMPLES 3. Find dy/dx if $y = 4x^2 - 2x + 3$.
SOLUTION We have

$$\frac{dy}{dx} = \frac{d}{dx}(4x^2 - 2x + 3) = \frac{d}{dx}(4x^2) + \frac{d}{dx}(-2x) + \frac{d}{dx}(3)$$

$$= 4\frac{d}{dx}(x^2) + (-2)\frac{d}{dx}(x) + 0 = 4(2x) + (-2)(1) = 8x - 2.$$

4. Compute the derivative of $y = (x^3 + 5x^2 + 1)(x^6 + x^3 + 4x)$ at $x = 0$.
SOLUTION Apply Theorem 2 with $f(x) = x^3 + 5x^2 + 1$, $g(x) = x^6 + x^3 + 4x$. We have that

$$\frac{dy}{dx} = (x^3 + 5x^2 + 1)'(x^6 + x^3 + 4x) + (x^3 + 5x^2 + 1)(x^6 + x^3 + 4x)'$$

$$= (3x^2 + 10x)(x^6 + x^3 + 4x) + (x^3 + 5x^2 + 1)(6x^5 + 3x^2 + 4).$$

It is optional whether or not to simplify further. At any rate,

$$\left(\frac{dy}{dx}\right)_{x=0} = (0)(0) + (1)(4) = 4.$$

5. Find the equation of the tangent to the curve $f(x) = x^3 + 3x^2 - 6x$ at the point $(1, -2)$.
SOLUTION We have $f'(x) = 3x^2 + 6x - 6$, so the slope of the tangent at $x = 1$ is $m = f'(1) = 3$. Since the tangent passes through the point $(1, -2)$, the point-slope formula gives us $y - (-2) = 3(x - 1)$; that is, $3x - y - 5 = 0$.

PROBLEMS

In Problems 1 to 12 find the derivatives of the given polynomials using Theorem 1 from §2.1 and Theorem 3 from this subsection.

1. $y = x^5$.
2. $y = 2x^8$.
3. $y = 3x^{10} + 1$.
4. $f(x) = -4x^2 - 3x + 1$.
5. $f(x) = -\frac{1}{2}x^2 + 2x - 1$.
6. $f(x) = \frac{1}{4}(x^2 - 4x + 1)$.

7. $s = -2t^3 + 6t + 1$.
8. $s = t^3 - 3t^2 + 3t - 1$.
9. $s = t^4 + t^3 + t^2 + t + 1$.
10. $g(t) = \frac{1}{2}(t^6 + 3t^4 - 6t^2 + 1)$.
11. $g(t) = t^{12} - 2t^9 + 4t^6 - 6t^3 + 1$.
12. $g(t) = -\frac{1}{3}t^{15} + \frac{1}{5}t^{10} - \frac{1}{5}t^5 + \frac{1}{3}$.

In Problems 13 to 16 compute the derivative of the given function at the indicated point using Theorem 1 from §2.1 and Theorems 2 and 3 from this subsection.

13. $y = (x^3 - 3x^2 + 1)(x^2 + 2x)$, $x = 2$.
14. $y = (x^4 - 2x^2 + 3)(2x^3 + 3x - 1)$, $x = 0$.
15. $f(x) = (x^4 + 4x^2 - 2)(-x^4 - 2x^2 + 1)$, $x = -1$.
16. $f(x) = (x^6 + 2x^3 - 4x)(x^5 + 3x^3 + 2x)$, $x = 1$.

17. Find the equation of the tangent to the curve $y = -3x^2 + 4x + 1$ at the point $(2, -3)$.
18. Find the equation of the normal to the curve $f(x) = \frac{1}{2}(x^2 + 5x - 6)$ at the point $(1,0)$.
19. Find the equations of the tangent and normal to the curve $y = \frac{1}{3}x^3 + 1$ at the point $(3,10)$.
20. Find the point on the curve $f(x) = 2x^2 + 4x - 3$ where the slope of the tangent line is equal to 8.
21. Find the point on the curve $y = x^2 + 8x - 12$ where the tangent line is parallel to the x-axis.
22. Find the points on the curve $f(x) = 2x^3 + 3x^2 - 36x + 12$ where the tangent line is parallel to the x-axis.

2.3   The quotient rule. Derivatives of rational functions

Theorem 5. *Let f and g be functions which have derivatives (at a point), and assume that g $\neq$ 0 at the point considered. Then*

$$\left(\frac{1}{g}\right)' = -\frac{g'}{g^2},\tag{6}$$

$$\left(\frac{f}{g}\right)' = \frac{f'g - fg'}{g^2}.\tag{7}$$

In the Leibniz notation this reads

$$\frac{d}{dx}\left(\frac{1}{g}\right) = -\frac{1}{g^2}\frac{dg}{dx},\tag{6'}$$

$$\frac{d}{dx}\left(\frac{f}{g}\right) = \frac{1}{g^2}\left(\frac{df}{dx}g - f\frac{dg}{dx}\right).\tag{7'}$$

Thus (1) the derivative of a reciprocal is minus the derivative divided by the square of the function, and (2) the derivative of a quotient is the derivative of the numerator times the denominator minus the numerator times the derivative of the denominator, this whole quantity divided by the square of the denominator.

Proof of Theorem 5.   We recall that if $g'(x)$ exists, then $\lim_{h\to 0} g(x + h) = \lim_{h\to 0} g(x) = g(x)$; see the proof of Theorem 2 in §2.2. Next set $\phi(x) = 1/g(x)$, where $g(x) \neq 0$. Then, using the rules for computing limits,

$$\phi'(x) = \lim_{h\to 0}\frac{\phi(x + h) - \phi(x)}{h} = \lim_{h\to 0}\frac{\dfrac{1}{g(x + h)} - \dfrac{1}{g(x)}}{h} = \lim_{h\to 0}\frac{g(x) - g(x + h)}{g(x + h)g(x)h}$$

$$= \lim_{h\to 0}\frac{-\dfrac{g(x + h) - g(x)}{h}}{g(x + h)g(x)} = \frac{-\lim_{h\to 0}\dfrac{g(x + h) - g(x)}{h}}{\lim_{h\to 0} g(x + h)\lim_{h\to 0} g(x)} = -\frac{g'(x)}{g(x)^2}.$$

This proves (6).

Using (5) and (6), we prove the quotient rule (7) by writing

$$\left(\frac{f}{g}\right)' = \left(f\frac{1}{g}\right)' = f'\frac{1}{g} + f\left(\frac{1}{g}\right)' = f'\frac{1}{g} + f\left(-\frac{g'}{g^2}\right) = \frac{f'}{g} - \frac{fg'}{g^2} = \frac{f'g - fg'}{g^2}.$$

Now we can extend the rules for differentiating powers to negative exponents.

Theorem 6. *For every integer n, if $f(x) = x^n$, then $f'(x) = nx^{n-1}$, that is,*

$$\frac{dx^n}{dx} = nx^{n-1}. \tag{8}$$

(If $n < 0$, we assume that $x \neq 0$. For $n = 0$, the right side is interpreted as 0.)

Proof.   For $n > 0$ this is Theorem 3 of §2.2. For $n = 0$, we must show that the derivative of $x^0 = 1$ is 0. This is so, by Rule (1) of Theorem 1. If n is a negative integer, then $m = -n$ is a positive integer. Using Rule (6) of Theorem 5 and Theorem 3 we have

$$\frac{dx^n}{dx} = \frac{dx^{-m}}{dx} = \frac{d(1/x^m)}{dx} = -\frac{1}{x^{2m}}\frac{dx^m}{dx} = \frac{1}{x^{2m}}mx^{m-1} = -mx^{-m-1} = nx^{n-1},$$

as asserted.

EXAMPLE 1.   Differentiate $y = \dfrac{2}{x^3} - \dfrac{3}{x^2}$.

ANSWER   By Rules (4), (2), and by Theorem 6, we have

$$\frac{dy}{dx} = \left(\frac{2}{x^3} - \frac{3}{x^2}\right)' = (2x^{-3} - 3x^{-2})' = (2x^{-3})' - (3x^{-2})' = 2(x^{-3})' - 3(x^{-2})'$$

$$= 2(-3x^{-4}) - 3(-2x^{-3}) = -6x^{-4} + 6x^{-3} = 6\left(-\frac{1}{x^4} + \frac{1}{x^3}\right).$$

The rules derived thus far suffice to differentiate any rational function.

Theorem 7. *A rational function has a derivative wherever the function is defined; the derivative is again a rational function.*

Proof.   Let f be a rational function; then $f = p/q$, where p and q are polynomials without common roots, and f is defined wherever $q \neq 0$. At points where $q \neq 0$ we have, by (7), that $f' = (p'q - pq')/q^2$. By Theorem 4, p' and q' are polynomials. So are $p'q - pq'$ and q^2 (since products and sums and differences of polynomials are polynomials). Hence f' is a quotient of polynomials and thus rational.

EXAMPLES 2.   Find $f'(x)$ if $f(x) = (x^2 + 1)/(x + 1)$.
SOLUTION   We have

$$f'(x) = \frac{(x^2 + 1)'(x + 1) - (x^2 + 1)(x + 1)'}{(x + 1)^2} = \frac{[(x^2)' + 1'](x + 1) - (x^2 + 1)(x' + 1')}{(x + 1)^2}$$

$$= \frac{(2x + 0)(x + 1) - (x^2 + 1)(1 + 0)}{(x + 1)^2} = \frac{2x(x + 1) - (x^2 + 1)}{(x + 1)^2}$$

$$= \frac{2x^2 + 2x - x^2 - 1}{(x + 1)^2} = \frac{x^2 + 2x - 1}{x^2 + 2x + 1}.$$

Note that the derivative is a rational function defined for $x \neq -1$.

3. Compute the derivative of $y = (2x^2 + 5x - 7)/(x^{10} + 1)$ at $x = 1$.

SOLUTION   We have

$$\frac{dy}{dx} = \frac{(3x^2 + 5x - 7)'(x^{10} + 1) - (3x^2 + 5x - 7)(x^{10} + 1)'}{(x^{10} + 1)^2}$$

$$= \frac{(6x + 5)(x^{10} + 1) - (3x^2 + 5x - 7)10x^9}{(x^{10} + 1)^2};$$

it is optional whether or not to simplify. At any rate,

$$\left(\frac{dy}{dx}\right)_{x=1} = \frac{11 \cdot 2 - 1 \cdot 10}{2^2} = 3.$$

PROBLEMS

In Problems 1 to 18 find the derivatives of the given functions.

1. $f(x) = 5x^{-3}$.

2. $f(x) = \dfrac{6}{x^2}$.

3. $f(x) = -x^{-2} - 6x^{-1} + 4$.

4. $y = -6x^{-4} + 2x^{-2} + \frac{1}{2}x^2$.

5. $y = \dfrac{2}{x^4} - \dfrac{4}{x^2} + \dfrac{1}{2}$.

6. $y = \dfrac{1}{x^6} + \dfrac{1}{x^3} + \dfrac{1}{x}$.

7. $g(t) = \frac{1}{2}(t^{-2} + 4t^{-1} + 1 + 4t^2)$.

8. $g(t) = \dfrac{3}{t^4} + \dfrac{2}{t^2} + 1 - 4t^2$.

9. $g(t) = \dfrac{1}{t^8} - \dfrac{2}{t^6} + \dfrac{3}{t^4} - \dfrac{4}{t^2}$.

10. $y = \dfrac{1}{2x + 1}$.

11. $y = \dfrac{x}{x - 1}$.

12. $y = \dfrac{x - 1}{x + 1}$.

13. $f(x) = \dfrac{x + 2}{2x - 1}$.

14. $f(x) = \dfrac{4x}{1 + x^2}$.

15. $f(x) = \dfrac{x^2}{2x + 1}$.

16. $g(t) = \dfrac{t^2 + 1}{t^2 - 1}$.

17. $g(t) = \dfrac{1}{1 + t + 4t^2}$.

18. $g(t) = \dfrac{t + 1}{t^2 + 2t + 3}$.

In Problems 19 to 22 compute the derivatives of the given functions at the indicated points.

19. $f(x) = \dfrac{1}{x^4 + x^2 + 1}$, $x = -1$.

20. $y = \dfrac{x^2 + 4x + 1}{x + 4}$, $x = 0$.

21. $y = \dfrac{x^3 - x + 1}{x^3 + 1}$, $x = 2$.

22. $f(x) = \dfrac{x^3 + 1}{x^6 + 3x^3 + 3}$, $x = 1$.

23. Find the equation of the tangent to the curve $y = x/(1 + x^2)$ at the origin.

24. Find the equation of the normal to the curve $f(x) = (x + 3)/(2x + 1)$ at the point $(2,1)$.

2.4 Higher derivatives

Consider a differentiable function $x \mapsto y = f(x)$ and its derivative $x \mapsto dy/dx = f'(x)$. This rule, which associates with x the slope of the tangent of the curve $y = f(x)$ at the point x, is itself a function. We can compute its derivative if it has one. The derivative of the derivative of f is denoted by $f''(x)$ and is called the **second derivative** of f; f' itself is called the **first derivative**.

This process may be continued. The derivative of $f''(x)$, that is, the derivative of the second derivative or the derivative of the derivative of the derivative, is denoted by $f'''(x)$ and is called the **third derivative** of f. Similarly, the **fourth derivative** of f is defined as the derivative of f''' and could be denoted by f''''. It is customary, however, to denote the fourth derivative by $f^{(4)}$, the fifth by $f^{(5)}$, and so on.

The kth derivative of f, that is, $f^{(k)}(x)$, is also called the **derivative of order** k. The Leibniz notation for the higher derivatives is as follows:

$$f''(x) = \frac{d^2 f}{dx^2} = \frac{d^2 y}{dx^2}$$

(read: "d square of f over dx squared" or "d square of y over dx squared"),

$$f'''(x) = \frac{d^3 f}{dx^3} = \frac{d^3 y}{dx^3}, \qquad f^{(4)}(x) = \frac{d^4 f}{dx^4} = \frac{d^4 y}{dx^4}, \qquad \dots, \qquad f^{(k)}(x) = \frac{d^k f}{dx^k} = \frac{d^k y}{dx^k}.$$

EXAMPLE If $f(x) = \frac{1}{9}x^3 + \frac{1}{4}x^2 + x + 2$, then, by the rules stated in the preceding subsections, $f'(x) = \frac{1}{3}x^2 + \frac{1}{2}x + 1$. Hence, $f''(x) = (\frac{1}{3}x^2 + \frac{1}{2}x + 1)' = \frac{2}{3}x + \frac{1}{2}$, $f'''(x) = (\frac{2}{3}x + \frac{1}{2})' = \frac{2}{3}$, $f^{(4)}(x) = 0$. The fifth, sixth, and all other higher derivatives are also 0.

PROBLEMS

1. If $f(x) = x^7$, find $f''(x)$.
2. If $f(x) = x^4 - 2x^3$, find $f'''(x)$.
3. If $y = x^3 - x^2 + 2x + 1$, find $d^2 y/dx^2$.
4. If $y = x^2 - \dfrac{2}{x} + 3$, find $d^3 y/dx^3$.
5. Find $f'''(1)$ if $f(x) = x^5 - 2x^3 + 4x + 1$.
6. Find $f^{(4)}(-1)$ if $f(t) = t^4 - 2/t^2 + 4$.
7. Compute $d^3 u/ds^3$ for $u = 1 + s + 2s^2 + 3s^3 + 4s^4$.
8. Compute $f^{(5)}(z)$ for $f(z) = z^{10}$.
9. What is $(d^2 y/dx^2)_{x=2}$ if $y = x^4 + 2x^2 - 5$?
10. What is $(d^3 s/dt^3)_{t=-1}$ if $s = 3t^2 - 2t + \frac{1}{2}$?
11. Find the second derivative of $f(x) = 2/(1 + 4x)$.
12. Find the second derivative of the function $u \mapsto (1 + u)/(1 - u)$.

§3 Velocity and acceleration

The problem of tangents was only one of the two questions that led to the concept of the derivative. The other was the description of motion. This problem dominated the work of Newton, who invented calculus as he created the science of dynamics.

3.1 Velocity

We consider a **particle** moving along a straight line (**rectilinear motion**). By a particle we mean a body whose dimensions can be disregarded in the problem considered so that it can be treated as a mathematical point. For instance, in dealing with the motion of the earth around the sun, the earth can be treated as a point.

We choose a point O on the straight line along which our particle moves, and we designate one direction as positive. We also choose a unit of length, a unit of time, and an instant from which we count time. Then the particle moves along a number line and the motion defines a function

$$t \mapsto s(t),$$

which associates to every value of the time t the coordinates of the particle at that time. The graph of this function can be thought of as the picture of the motion. It should be remembered that the graph of the motion depends on the units of length and time. The graph of the same motion will look quite different if we first measure distance in feet and time in hours and then use centimeters and seconds (see Chapter 3, §1.5).

The motion is called **uniform** if the distance traversed by the moving point (particle) is proportional to the time elapsed. Assume that this is so, and let α be a number defined as follows: α is positive or negative according to whether the particle moves in the positive or negative direction; the absolute value $|\alpha|$ is the number of length units traversed by the point during one time unit. We say that our uniformly moving particle has the **velocity**

$$\alpha \, \frac{\text{length units}}{\text{time units}},$$

for instance, α(cm/sec) (read "α centimeters per second"), and it has the **speed** $|\alpha|$ (length units/time units). Assume also that, at time $t = 0$, our point is at the distance β from 0. After t time units, its distance from β will be αt and its distance from 0 will be $\beta + \alpha t$. Thus the motion is represented by the function

$$t \mapsto s(t) = \alpha t + \beta.$$

We see that *the graph of a uniform motion is a straight line and the slope of this straight line is the velocity.*

Consider next the motion of a particle that is not uniform, for instance, the motion of a car accelerating along a straight highway. What is the velocity of the particle at a given time instant? A possible answer is: "the number read on the speedometer." In order to understand motion, however (in particular, in order to build a speedometer), we must reduce the concept of velocity to the more primitive concepts of distance ("the number read on a measuring tape") and time ("the number read on a clock").

Suppose that, at the time t, the particle is located at the point $s = s(t)$, and at the time $t + h$ it is located at $s_1 = s(t + h)$. This means that, in h time units, the particle traversed $s_1 - s$ length units. The "average velocity" of the particle is

ISAAC NEWTON (1642–1727) was educated at Cambridge University and later taught there. During the Great Plague of 1664–65, Newton, who had just received his B.A., withdrew to his native hamlet of Woolsthorpe. In these two years he discovered that white light can be decomposed into rays of different colors, developed calculus, formulated the law of universal gravitation, and then derived from it Kepler's laws of planetary motion. These great discoveries were published much later. For instance, The

Mathematical Principles of Natural Philosophy, containing Newtonian mechanics, appeared in 1687; in this book calculus is not used.

Newton's intellectual interests were not restricted to physics and mathematics, the two fields in which he was never surpassed. He left many manuscripts dealing with theology and alchemy. He was also a successful Master of the Mint (for which he was knighted), and several times represented his university in Parliament.

$$\frac{s_1 - s}{h} = \frac{s(t + h) - s(t)}{h} \tag{1}$$

length units per time units. But we cannot conclude that its velocity at time t was equal to (1), since the particle could have, for instance, moved very fast during the first $h/2$ time units and very slowly during the remaining time. Our experience with motion suggests, however, that, if h is small, the "average velocity" is close to the "true" or "instantaneous" velocity of the particle at the time t. We also feel that the smaller h, the better the agreement between the "average velocity during the time h" and the instantaneous velocity. We therefore *define* the velocity $v(t)$, at the time t, of the moving point as that number (of length units per time unit) to which the average velocity during the time interval from t to $t + h$ is arbitrarily close, provided that h is small enough. In other words,

$$v(t) = \lim_{h \to 0} \frac{s(t + h) - s(t)}{h} = s'(t),$$

or, in Leibniz's notation, $v = ds/dt$.

Thus the *velocity is the derivative of the distance traveled with respect to time, or, which is the same, the slope of the tangent to the graph of the function (time)* $\mapsto$ *(distance traveled)*.

Applying the definition to the uniform motion $s = \alpha t + \beta$, we obtain that $v = ds/dt = \alpha$, as we should.

EXAMPLE   Suppose a particle moves along the x-axis and its coordinate at time t is $s(t) = 2t - t^2$. Then the velocity is $v(t) = s'(t) = 2 - 2t$. For $0 < t < 1$, $v > 0$; the particle moves to the right. For $t > 1$, $v < 0$; the particle moves to the left. At $t = 1$, the particle reverses its direction. At this instant its velocity is 0.

PROBLEMS

In Problems 1 to 10 the given functions describe the motion of a particle moving in a straight line.

1. If $s = t^3 - t$, where the distance s is measured in feet and the time t is measured in seconds, what is the velocity when $t = 10$ sec?
2. If $s = 100 - 16t$, where the distance s is measured in miles and the time t is measured in hours, what is the velocity when $t = 1$ hr?
3. If $s = 20t^2 + 30$, where the distance s is measured in centimeters and the time t is measured in seconds, at what time will the velocity be 80 cm/sec?
4. If $r = t^2 - t$, where the distance r is measured in feet and the time t is measured in seconds, what is the velocity when $t = .25$ sec? When $t = .75$ sec?
5. If $s = t^2 - 3t + 6$, where the distance s is measured in feet and the time t is measured in seconds, what is the velocity when $t = 4$ sec? For what value of t is the velocity zero?
6. If $s = \frac{1}{3}t^3 - \frac{9}{2}t^2 + 18t$, where the distance s is measured in feet and the time t is measured in seconds, what is the velocity when $t = 10$ sec? For what values of t is the velocity zero?
7. If $s = \frac{1}{3}t^3 - 3t^2 + 16t$, where the distance s is measured in centimeters and the time t is measured in seconds, at what time will the velocity be 8 cm/sec?

8. If $s = 10t^2 - 400t + 50$, where the distance s is measured in centimeters and the time t is measured in seconds, what is the velocity when $t = 4$ sec? What is the significance of the sign of the velocity?

9. If $r = 3x^2 - 2x$, where r is distance and x is time, during what time interval is the particle moving in the positive direction?

10. Suppose $s = \frac{2}{3}t^3 - \frac{7}{2}t^2 + 3t + 4$, where s is distance (ft) and t is time (sec). During what time interval is the particle moving in the negative direction?

3.2   Acceleration

We consider next changes in the velocity. Let h be a small, say, positive number; the difference $v(t + h) - v(t)$ is the change in the velocity during the time interval from t to $t + h$; this change is, of course, to be measured in units of velocity, that is, in units of length per unit of time (for instance, in cm/sec). The fraction

$$\frac{v(t + h) - v(t)}{h}$$

is the average change in the velocity during the time interval from t to $t + h$ or the "average acceleration." It is measured in units of length per unit of time per unit of time, for instance, in cm/sec^2 (read "centimeters per second per second"). As we take smaller and smaller values of h, the fraction comes closer and closer to the derivative

$$a(t) = \frac{dv}{dt} = v'(t).$$

We call $a(t)$ the (instantaneous) acceleration and note that

$$a(t) = s''(t) = \frac{d^2s}{dt^2};$$

the acceleration is the second derivative of the distance traveled with respect to the time elapsed.

EXAMPLE   For the particle in the previous example (see §3.1) the acceleration is $a(t) = v'(t) = s''(t) = -2$. When the particle moves to the right ($v > 0$), its velocity decreases and its speed $|v|$ also decreases. When the particle moves to the left ($v < 0$, this happens after $t = 1$), the velocity v decreases and the speed $|v|$ increases.

PROBLEMS

In Problems 1 to 6, s represents the position of a particle at time t moving along a straight line. Assume s is measured in feet, t is measured in seconds, and all motion starts at time $t = 0$ sec. Find the velocity and acceleration for the given time.

1. $s = t^2 - 6t + 4$, $t = 4$ sec.
2. $s = 4t^2 + 3t + 1$, $t = 2$ sec.
3. $s = -\frac{1}{4}t^2 + \frac{1}{2}t + 1$, $t = 1$ sec.

4. $s = t^3 - 3t^2 + 2$, $t = \frac{1}{3}$ sec.
5. $s = 4t^3 - 6t^2 + t - 4$, $t = 4$ sec.
6. $s = -\frac{1}{3}t^3 + \frac{1}{4}t^2 - 2t + 8$, $t = 2$ sec.

7. Suppose a particle moves along a straight line according to the equation $s = t^2/2 - 4t + 1$, where s is the distance measured in centimeters and t is the time measured in seconds. For what interval of time is the particle moving in the positive direction? In the negative direction?

8. Suppose a particle moves along a straight line according to the equation $s = \frac{1}{3}t^3 - 2t^2 + 3t$, where s is distance (ft) and t is time (sec). For what interval of time is the particle moving in the positive direction? In the negative direction? When is the acceleration zero?

9. Suppose a particle moves along a straight line according to the equation $s = \frac{1}{3}t^3 - 4t^2 + 15t + 2$, where s is the distance measured in feet and t is the time measured in seconds. Describe the motion of the particle.

10. Suppose a particle moves along a straight line according to the equation $s = -t^3 + 5t^2 - 3t + 1$, where s is the distance measured in centimeters and t is the time measured in seconds. Describe the motion of the particle.

3.3   Free fall

An example of a nonuniform rectilinear motion is that of a body released at a certain height and permitted to fall to the ground. The discovery of the nonuniformity of this motion was an important event in the history of science.

The ancient Greeks were excellent mathematicians; they also developed a sophisticated statics (that part of mechanics which deals with bodies at rest). But their ideas of dynamics (the mechanics of moving bodies) were naive. Aristotle taught that the motion of falling bodies is uniform, and that the heavier the body, the faster it falls. Since Aristotle's authority acquired an almost religious character, his statement was unchallenged until the sixteenth century. The true law of falling bodies was discovered by Galileo, who thereby founded modern physics. Because his work involved an analysis of the idea of velocity, he became one of the pioneers of calculus.

Galileo's law asserts that *the distance traversed by a freely falling body released from rest is proportional to the square of the time it has been falling, the factor of proportionality being the same for all bodies.* It follows that two bodies, released at the same height, will hit the ground at the same time.

Galileo's law neglects air resistance, so that it cannot be checked on Earth, except in an artificially created vacuum. But the law holds on any planet, and, with the aid of television, millions of people saw it verified on the surface of the airless Moon on August 2, 1971. An astronaut released a hammer and a feather, and both objects reached the ground at the same instant.

ARISTOTLE (384–322 B.C.) was a member of Plato's Academy for 20 years before founding his own school, the Lyceum. For a while, he was tutor to the future Alexander the Great. Aristotle did not contribute to mathematics proper, but he set the pattern for formal logic, as he did for many other subjects (like political science, literary criticism, and descriptive biology). The first significant advance in logic beyond Aristotle occurred in the nineteenth century, with the appearance of symbolic or mathematical logic.

Many of Aristotle's works are preserved. It is believed, however, that these are actually lecture notes written by his students.

A reconciliation between the Platonic-Pythagorean mathematical world view and the Aristotelian emphasis on observing actual phenomena occurred only when calculus made mathematical science possible.

GALILEO GALILEI (1564–1642). The founder of modern science, who said that the book of nature is written in mathematical symbols, was a mathematician, a physicist, an astronomer, a brilliant polemicist, and a talented writer. He formulated the principle of inertia and the laws of motion of the pendulum, of falling bodies, and of projectiles. He built some of the first telescopes and used them to discover the moons of Jupiter and Saturn, the phases of Venus, the rotation of the sun, and the ruggedness of the lunar landscape.

Galileo was professor at Pisa and at Padua, and later "chief philosopher and mathematician" at the court of the Grand Duke of Florence. At the age of 70, Galileo was tried by the Inquisition for his *Dialogue on the Two Great World Systems.* He was forced to renounce the Copernican system and spent the rest of his life under house arrest. Yet he continued to work and to write.

Let us now formulate the law of falling bodies mathematically.

We consider the direction of free fall as negative vertical, choose the point at which the body is released as our reference point O, and count the time from the moment of release. According to Galileo, the motion of the falling body is described by the equation

$$s = -\beta t^2.$$

The graph of this motion is a parabola. We put a minus sign in the formula so that β should be positive. It is customary to set

$$g = 2\beta$$

and to write Galileo's law in the form

$$s = -\tfrac{1}{2}gt^2. \tag{1}$$

The value of g does not depend on the falling body. It does depend on the planet on which the experiment is performed and also on the units used; we note the (approximate) values:

$$g \approx 32 \, \frac{\text{ft}}{\sec^2} \approx 980 \, \frac{\text{cm}}{\sec^2} \text{ on Earth; } g \approx 5.4 \, \frac{\text{ft}}{\sec^2} \approx 160 \, \frac{\text{cm}}{\sec^2} \text{ on the Moon.}$$

Let us compute the velocity at time t. We have

$$v = \frac{ds}{dt} = \frac{d(-\tfrac{1}{2}gt^2)}{dt} = -gt. \tag{2}$$

Thus the velocity and the *speed* (absolute value of the velocity) are proportional to the time elapsed since the release of the body. The acceleration is given by

$$a = \frac{dv}{dt} = \frac{d^2s}{dt^2} = -g. \tag{3}$$

Thus Galileo's law reads: *the acceleration of all freely falling bodies (on the surface of the same planet) has the same constant value.* One calls g the **acceleration of gravity.**

PROBLEMS

1. A ball is dropped from a height of 100 ft. What is its velocity after 2 sec?
2. A stone is dropped from a height of 10 cm. What is its velocity after .1 sec?
3. A book is dropped from a height of 144 ft. In how many seconds will the book strike the ground? What is its velocity at that instant?
4. A brick is dropped from a building. How far has it fallen when its velocity is 128 ft/sec?
5. An astronaut on the Moon drops a hammer and a feather from a height of 2 meters. In how many seconds will they strike the ground?
6. An astronaut on the Moon stands on the edge of a chasm and drops a hammer. If it takes the hammer 1 min to strike the floor of the chasm, how deep is it?

APPLICATIONS
OF DERIVATIVES

§1 Maxima and minima

Having learned how to differentiate, we proceed to apply derivatives to the study of functions. The present chapter contains some basic theorems about derivatives as well as some of the first applications of calculus.

1.1 Largest and smallest values of a function

Finding the largest and the smallest value of a function on an interval is one of the oldest applications of calculus. Indeed, before calculus was cast into anything like its present form, Fermat used a kind of differentiation in order to find the largest value of a polynomial in an interval.

The largest and smallest values of a function f in some interval are called the maximum and minimum of f, respectively. In other words: if $f(x)$ is a function defined in an interval, x_0 a point in this interval such that $f(x) \leq f(x_0)$ for all other points in the interval, we say that f has at x_0 a **maximum** (for the interval), and we call the value $f(x_0)$ **the maximum** of f in the interval. Similarly, the function f has at x_0 a **minimum** if $f(x_0) \leq f(x)$ for all x in the interval; the value of $f(x_0)$ is then called **the minimum** of f.

The maximum of f can occur at an endpoint of the interval considered, or at an inner point, and it can occur at several points. The same is true of the minimum; see Figures 5.1 and 5.2.

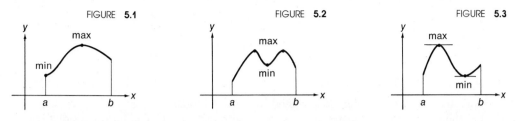

FIGURE **5.1**   FIGURE **5.2**   FIGURE **5.3**

Theorem 1. *A function defined and continuous on a closed (finite) interval has a maximum and minimum.*

This statement is easy to believe, but the proof ▶ is not easy, and it depends on a deeper discussion of numbers.

The following examples show that *all* conditions in the theorem are essential.

EXAMPLES 1. The continuous function $f(x) = x$ has neither a maximum nor a minimum in the open interval $0 < x < 1$. (The function takes on all values y such that $0 < y < 1$, but in the *open* interval $0 < x < 1$ it does not take on the values 0 and 1.) In the closed interval $0 \leq x \leq 1$, the same function has the maximum 1 (at $x = 1$) and the minimum 0 (at $x = 0$).

2. Set $f(x) = x^2$ for $0 < x < 1$, $f(0) = f(1) = \frac{1}{2}$. The function f is defined but not continuous in the closed interval $0 \leq x \leq 1$. In this interval, it has neither a maximum nor a minimum. (The function takes on all values y such that $0 < y < 1$; it does not take on the values 0 and 1.)

3. The continuous function $f(x) = 1/\sqrt{1 - x^2}$ defined in the open interval $-1 < x < 1$ has a minimum 1 at $x = 0$ but no maximum. [For all x considered, $f(x) \geq f(0) = 1$. On the other hand, $f(x)$ becomes as large as we want when x is close enough to the endpoints -1 and 1.]

4. The continuous piecewise monotone function $f(x) = x^3$ has neither a maximum nor a minimum in the infinite interval $-\infty < x < +\infty$. (This function takes on all values y, $-\infty < y < +\infty$.)

Theorem 2. *If the function $f(x)$, defined for a $\leq x \leq b$, has a maximum (or a minimum) at a point x_0, $a < x_0 < b$, and if the derivative $f'(x_0)$ exists, then $f'(x_0) = 0$.*

In geometric language: at a peak or at a trough of a curve the tangent (if there is one) is horizontal, see Figure 5.3.

Proof. Assume that

$$\lim_{h \to 0} \frac{f(x_0 + h) - f(x_0)}{h} = f'(x_0) > 0.$$

If $|h|$ is small enough, the difference quotient $[f(x_0 + h) - f(x_0)]/h$ is as close as we want to the positive number $f'(x_0)$. For such h, the difference quotient is positive. Hence

$$f(x_0 + h) - f(x_0) > 0, \qquad \text{if } h > 0 \text{ and } |h| \text{ is small,}$$
$$f(x_0 + h) - f(x_0) < 0, \qquad \text{if } h < 0 \text{ and } |h| \text{ is small.}$$

But this is absurd! If f has a maximum at x_0, then $f(x_0) \geq f(x_0 + h)$ for all h, and $f(x_0 + h) - f(x_0) \leq 0$. If f has a minimum at x_0, $f(x_0) \leq f(x_0 + h)$ for all h, and $f(x_0 + h) - f(x_0) \geq 0$.

We see similarly that $f'(x_0)$ cannot be negative. Hence $f'(x_0) = 0$.

[The argument works since x_0 is assumed to be an *inner* point of the interval, so that f is defined at $x_0 + h$ for positive h *and* for negative h.]

1.2 Finding maxima and minima

Let f be a continuous function defined in a finite closed interval $[a,b]$, and assume that the derivative $f'(x)$ exists at all points x with $a < x < b$. The maximum of f will be attained at one or several points of the interval, endpoints included. If it is attained at a point x_0, $a < x_0 < b$, then $f'(x_0) = 0$ by Theorem 2. A similar remark applies to the minimum. We conclude that the maximum and minimum of f are found by computing f at the endpoints and at all zeros of f'.

If the function f fails to have a derivative at finitely many points in the interval considered, then these points are also candidates for maxima and minima and must be examined.

EXAMPLES   1.   Find the largest and smallest values of $f(x) = (.1)(x^3 - 12x + 10)$ for $-3 \leq x \leq 3$.

SOLUTION   We have $f'(x) = (.1)(3x^2 - 12) = (.3)(x + 2)(x - 2)$. The zeros of f' are -2 and 2. Since $f(-3) = 1.9$, $f(-2) = 2.6$, $f(2) = -.6$, and $f(3) = .1$, we conclude that the maximum of f is $f(-2) = 2.6$ and the minimum is $f(2) = -.6$.

2.   Find the maximum and minimum of $|x|$ in $-10 \leq x \leq 8$.

ANSWER   There is no derivative of $f(x) = |x|$ at $x = 0$. At all other points $f'(x) = 1$ if $x > 0$ and $f'(x) = -1$ if $x < 0$. We must examine the point 0 and the endpoints -10 and 8. Since $|0| = 0$, $|-10| = 10$, and $|8| = 8$, the minimum is 0 and the maximum is 10.

PROBLEMS

In each of Problems 1 to 7 find the maximum and minimum of the given function over the indicated interval.

1. $f(x) = x^2 - 4x + 12$, $0 \leq x \leq 5$.
2. $f(x) = x^2 - 12x - 64$, $-4 \leq x \leq 4$.
3. $f(x) = x^3 - 12x + 16$, $-3 \leq x \leq 3$.
4. $f(u) = u^3 - 3u^2 - 45u$, $-6 \leq u \leq 6$.
5. $A(x) = x^3 - 9x^2 + 24x + 8$, $0 \leq x \leq 3$.
6. $\phi(t) = (t^2 + 5)/(t + 2)$, $0 \leq t \leq 10$.
7. $\phi(y) = (2 + y - y^2)/(2 - y - y^2)$, $-4 \leq y \leq 4$.

1.3 Applications

Maximum and minimum problems often arise as (or are disguised as) geometric or arithmetic theorems or as practical problems. In such a case, some preliminary work is needed to reformulate the problem into one of finding the largest or smallest value of a function in an interval. We illustrate the procedure by several examples.

EXAMPLES   1.   Find two positive numbers whose sum is 20 and whose product is as large as possible.

SOLUTION   Denote the two numbers by x and y. The product is $P = xy$. At first glance, it does not look as if P is a function of one variable. But it is required that $x + y = 20$. Hence

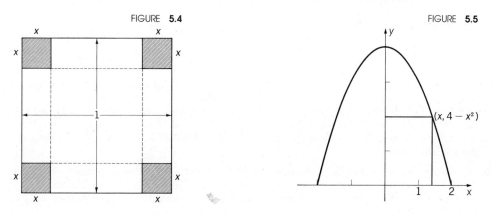

FIGURE 5.4

FIGURE 5.5

$y = 20 - x$ and therefore $P = x(20 - x) = 20x - x^2$. We also need $x > 0$ and $y > 0$, hence $20 - x > 0$ or $x < 20$. Thus we must solve the problem: find the maximum of $f(x) = 20x - x^2$ for $0 \le x \le 20$. (We added the values $x = 0$, $x = 20$ in order to have a closed interval.)

Now $f'(x) = 20 - 2x = 2(10 - x)$. This is 0 for $x = 10$, and for no other x. Since $f(0) = f(20) = 0$ and $f(10) = 100$, the maximum of f is 100. The two required numbers are $x = 10$ and $y = 20 - 10 = 10$.

2. Prove that among all rectangles with given perimeter, the square has the largest area.

SOLUTION   Let x and y be the sides of the rectangle (we mean, of course, the lengths of the sides, but such permissible inaccuracies in language cannot lead to confusion). The area A is given by $A = xy$. The perimeter, call it P, is $2x + 2y$. Since P is given and $2x + 2y = P$, we have $y = \frac{1}{2}P - x$ so that $A = x(\frac{1}{2}P - x) = \frac{1}{2}Px - x^2$. Only values of x between 0 and $\frac{1}{2}P$ are of interest, since for $x > \frac{1}{2}P$, y is negative.

We are looking for the maximum of $A = f(x) = \frac{1}{2}Px - x^2$ in $0 \le x \le \frac{1}{2}P$. Since $f'(x) = \frac{1}{2}P - 2x$, the only solution of $f'(x) = 0$ is $x = \frac{1}{4}P$. Since $f(0) = f(\frac{1}{2}P) = 0$ and $f(\frac{1}{4}P) = \frac{1}{16}P^2$, the maximum of $A = f(x)$ is attained at $x = \frac{1}{4}P$. If $x = \frac{1}{4}P$, then $y = \frac{1}{4}P$ and the rectangle is a square.

(Examples 1 and 2 are not really different!)

3. An open box is made from a square piece of cardboard (of side 1) by cutting out four equal ("small") squares at the corners and then folding (see Figure 5.4). How big should the small squares be, in order that the volume of the box be as large as possible?

SOLUTION   Denote the side length of the squares to be removed by x. The box will have as its base a square of sides $(1 - 2x)$ and will have height x. The volume V will be $V = (1 - 2x)^2 x = (1 - 4x + 4x^2)x = x - 4x^2 + 4x^3$. We see from Figure 5.5 that $2x < 1$, or $x < \frac{1}{2}$. Also $x > 0$. Hence it suffices to consider the closed interval $0 \le x \le \frac{1}{2}$. Now $dV/dx = 1 - 8x + 12x^2$. The zeros of dV/dx are $\frac{1}{2}$ and $\frac{1}{6}$. Now $V = 0$ for $x = 0$ and $x = \frac{1}{2}$, and $V = 2/27$ for $x = \frac{1}{6}$. Thus $x = \frac{1}{6}$ is the required length.

4. A closed container in the form of a right circular cylinder is to be made out of tin. For given surface area, what dimensions should the container have in order to maximize the volume?

SOLUTION   Denote the radius of the cylinder by r, the height by h, the total surface area by A, and the volume by V. Then $A = $ base $+$ lateral surface $+$ top $= \pi r^2 + 2\pi rh + \pi r^2 = 2\pi(r^2 + rh)$, and $V = \pi r^2 h$. Hence

$$h = \frac{A - 2\pi r^2}{2\pi r} \quad \text{and} \quad V = \pi r^2 h = \frac{1}{2}(Ar - 2\pi r^3) = \frac{1}{2}r(A - 2\pi r^2).$$

We are only interested in values of r for which V is not negative, that is, $0 \le r \le \sqrt{A/2\pi}$.

We need the point in this interval at which V has its maximum. At the endpoints we have $V = 0$. Next, $dV/dr = \frac{1}{2}(A - 6\pi r^2)$. The derivative is zero for $r = (A/6\pi)^{1/2}$. For this value of r, $V = \frac{1}{2}r(A - 2\pi r^2) = (A/3)(A/6\pi)^{1/2} > 0$. Hence V has its maximum there. Also, if $r = (A/6\pi)^{1/2}$, then $h = (A - 2\pi r^2)/2\pi r = 2r$. Thus the height must equal the diameter.

5. Find the largest area of a rectangle with vertices at the origin of a Cartesian coordinate system, on the x-axis, on the y-axis, and on the parabola $y = 4 - x^2$ (see Figure 5.5).
SOLUTION   Let the vertex on the parabola be $(x, 4 - x^2)$. The area of the rectangle is $A = (4 - x^2)x = 4x - x^3$. Since $A = 0$ for $x = 0$, $A = 0$ for $x = 2$, the interval in question is $[0,2]$. Since $dA/dx = 4 - 3x^2$, $dA/dx = 0$ for $x = 2/\sqrt{3}$. For this x, $A = (16/3)\sqrt{3}$. Thus the desired largest area is $(16/3)\sqrt{3}$.

6. A rancher has a herd of cows, each weighing 500 lb. It costs 50 cents a day to keep one cow. The cows are gaining weight at the rate of 6 lb a day. The market price for cows is now \$1 per lb and is falling by 1 cent a day. How long should the rancher wait to sell the cows in order to earn the maximum profit? How much will he earn by waiting? [*Note:* Assume cows gain weight uniformly during each day, the cost of keeping them is distributed uniformly throughout the day, and so forth.]
ANSWER   After t days one cow will weigh $500 + 6t$ lb and will sell for $1 - \frac{1}{100}t$ dollars per pound. Thus it will bring in $(500 + 6t)(1 - \frac{1}{100}t)$ dollars. But it will cost the rancher $\frac{1}{2}t$ dollars more to keep the cow for t days. The additional profit per cow is therefore

$$P = (500 + 6t)(1 - \tfrac{1}{100}t) - \tfrac{1}{2}t - 500 = \tfrac{1}{2}t - \tfrac{3}{50}t^2.$$

Now $P = 0$ for $t = 0$, $P < 0$ for $t > \dfrac{25}{3}$. Hence the maximum of P must lie in the interval

$0 \le t \le \dfrac{25}{3}$. The derivative $\dfrac{dP}{dt} = \dfrac{1}{2} - \dfrac{3}{25}t$ is zero when $\dfrac{1}{2} - \dfrac{3}{25}t = 0$, that is, when

$t = \dfrac{25}{6}$ days. For this value of t, $P = \dfrac{1}{2}\left(\dfrac{25}{6}\right) - \dfrac{3}{50}\left(\dfrac{25}{6}\right)^2 = \dfrac{25}{24}$ dollars. Hence by waiting 4 days, approximately, the rancher earns an additional dollar per cow, approximately.

PROBLEMS

1. What positive number plus its reciprocal gives the least sum?
2. What is the minimum value that the square of any number plus the square of its reciprocal can have?
3. Find the number that exceeds its square by the greatest amount.
4. Find two positive numbers whose sum is 100 and whose product is as large as possible.
5. Find two positive numbers whose product is 100 and whose sum is as small as possible.
6. Find two positive numbers whose sum is 1000 and the sum of whose squares is as small as possible.
7. Find two positive numbers x and y such that their sum is 1000 and the product xy^3 is as large as possible.
8. The sum of three positive numbers is 60. The first plus twice the second plus three times the third add up to 120. Find the numbers which maximize (give largest value to) the product of all three numbers.
9. A rectangular cabbage patch 200 sq ft in area is to be laid out so that an adjacent straight wall serves as one of the sides. What are the dimensions of the patch requiring the least amount of fence?
10. A rectangular plot of land is to be surrounded by a fence and then divided into two equal parts by a fence parallel to one side. The area to be enclosed is 1350 sq ft. What are the dimensions of the rectangle that require the least amount of fence?

11. What is the area of the largest rectangle having two vertices on the x-axis and two vertices on the parabola $y = 16 - 3x^2$?

12. A closed box is to be made out of heavy cardboard. If the base of the box is a rectangle twice as long as it is wide, and the box is to hold 9 cu ft, what dimensions require the least amount of cardboard? What dimensions should the box have if there is no top?

13. A sports field consists of a rectangular region with a semicircular region adjoined at each end of the two opposing sides. If the perimeter is to be 1000 ft, find the area of the largest possible field. [*Hint:* Be careful!]

14. A patio is to be built in the shape of a rectangle with a semicircular region adjoined to one side of the rectangle. If the perimeter is to be 100 ft, what is the largest area that can be enclosed?

15. A country is designing a new flag, which is to consist of an orange rectangular region divided by a pink stripe. The perimeter of the entire flag is to be 14 ft, and the orange part is to have an area of 9 sq ft. One faction wants the stripe to be horizontal; another wants the stripe to be vertical. They both agree that the stripe must be as wide as possible. Who wins? (By convention, the standard position for a rectangular flag is with the longest edge horizontal.)

16. An artist decides to paint a picture consisting of a red rectangle surrounded by a white border. If the red rectangle is to have an area of 12 sq in. and the border is to be 1 in. wide along each side and 2 in. wide along the top and bottom, what dimensions should the picture have in order to have the smallest total area?

17. A printer wishes to use 128 sq in. of paper for a handbill. If the top and bottom margins are 2 in. each, and the side margins are 1 in. each, what dimensions should the handbill have if the printed area is to be a maximum?

18. A closed storage tank in the form of a right circular cylinder is to be made out of sheet aluminum. If the tank is to hold V cu ft of liquid, what dimensions should the tank have to require the least amount of material? What dimensions should the tank have if there is no top? [*Hint:* The lateral surface area of a right circular cylinder is $2\pi rh$; the volume is $\pi r^2 h$.]

19. A wire L in. long is cut into two pieces. One piece is bent into a square, the other piece into a circle. How should the wire be cut if the sum of the areas enclosed by the two pieces is to be a minimum?

20. A rectangular storage bin with square base and open top is to have a total surface area of 300 sq ft. Find the dimensions of the bin that will make its volume a maximum.

21. A closed rectangular box is to be constructed from 432 sq in. of wood. If the box is to be twice as long as it is wide, what dimensions will make its volume a maximum?

22. An open box is made from a rectangular piece of tin 5 in. by 8 in. by cutting out equal squares at each corner and then folding up the remaining flaps. What size squares should be cut out so that the box will have maximum volume?

23. Find the volume of the largest right circular cylinder that can be inscribed in a sphere of radius a. [*Hint:* $V_{\text{cylinder}} = \pi r^2 h$.]

24. Find the volume of the largest right circular cone that can be inscribed in a sphere of radius a. [*Hint:* $V_{\text{cone}} = \frac{1}{3}\pi r^2 h$.]

25. A cylindrical drinking cup of circular cross section is to contain 12π sq in. of material. What dimensions should the cup have in order to hold as much liquid as possible?

26. An orange grove now has 40 trees per acre with an average yield of 1000 oranges per tree. It is observed that for each additional tree planted per acre, the yield per tree is reduced by 20 oranges. How many trees should be planted per acre to yield the maximum number of oranges?

27. An owner of a 200-unit apartment complex has full occupancy when the monthly rental per unit is $100. Experience shows that for each $10 increase in rent per unit, 5 units become vacant. How many units should he rent to maximize his total income? What is the monthly rental per unit under these circumstances?

28. (a) A farmer wants to clear a field in the shape of a rectangular plot with a semicircular plot adjoined to one side of the rectangle. The rectangular plot is to be planted in hay, which will yield a profit at 5 cents per sq ft; the semicircular plot is to be planted in rye, which will yield a profit at 6 cents per sq ft. If the perimeter of the field is to be 800 ft, how should the farmer lay out the field to earn the most money?

(b) Suppose the price of rye increases so that a field planted in rye would yield 10 cents per sq ft. How should the farmer lay out the field now?

29. The illumination provided by a point light source is directly proportional to the intensity of the source and inversely proportional to the square of the distance from the source. The illumination provided by several light sources is the sum of the illuminations provided by each one. Suppose there are two light beacons 1 mile apart. If the intensity of the first beacon is eight times the intensity of the second, what point on the straight-line segment joining the beacons receives the least illumination? Generalize this result to the case in which the first beacon is k^3 times stronger than the second.

30. A hay fever sufferer discovers that the amount of pollen that reaches him from a given source is directly proportional to the strength of the source and inversely proportional to the distance from the source. Unfortunately, he is forced to live somewhere on a straight line joining two pollen sources that are 1 mile apart. If one source is four times stronger than the other, where should the man live to suffer the least discomfort? Generalize this result to the case in which one source is k^2 times stronger than the other.

31. The amount of gas produced by a coke oven operating at $1000°F$ is 100 cubic feet per minute (cu ft/min) and will increase by $\frac{1}{5}$ cu ft/min for each degree rise in temperature up to $1500°F$. Above $1500°F$, the amount of gas produced will increase by $\frac{1}{4}$ cu ft/min for each degree rise in temperature. It costs $1000 + \frac{1}{100}(T - 1000)^2$ cents to operate the oven for 1 hr at a temperature of $T°F$ ($T \geq 1000$). If gas can be sold at 1 cent/cu ft, what is the most profitable temperature to run the oven?

32. Previously, before improvements in gas retrieval at high temperatures were introduced, the amount of gas produced by the oven in Problem 31 would only increase by $\frac{1}{10}$ cu ft/min for each degree rise in temperature above $1500°F$. If other factors were the same, what would have been the most profitable temperature to run the oven?

§2   The shape of a graph

In this section we learn how to obtain qualitative information about a function from the knowledge of its derivatives.

2.1   Monotone functions

A function f is called **increasing** if

$$f(x_1) < f(x_2) \qquad \text{whenever } x_1 < x_2.$$

This means that, given any two points on the graph, the one to the right is above the one to the left. The graph rises as one traverses it from left to right. An example is $f(x) = x^3$ graphed in Figure 5.6. A function f is called **nondecreasing** if

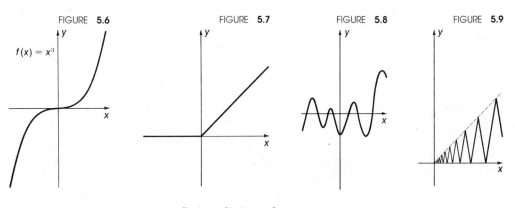

FIGURE 5.6   FIGURE 5.7   FIGURE 5.8   FIGURE 5.9

$f(x) = x^3$

$$f(x_1) \leq f(x_2) \qquad \text{for } x_1 < x_2.$$

This means that, given two points on the graph, the one to the right is not below the one to the left. The graph does not descend. An example is: $f(x) = 0$ for $x \leq 0$, $f(x) = x$ for $x > 0$ (Figure 5.7). An increasing function is also called nondecreasing.

Similarly, f is called **decreasing** if $f(x_1) > f(x_2)$ for $x_1 < x_2$ (the graph descends) or **nonincreasing** if $f(x_1) \geq f(x_2)$ for $x_1 < x_2$ (the graph does not rise).

REMARK If the function f is increasing in an interval, then the function $(-f)$ is decreasing and vice versa. If f is nondecreasing, $(-f)$ is nonincreasing and vice versa.

A function is called **strictly monotone** if it is either increasing or decreasing, **monotone** if it is either nonincreasing or nondecreasing.

The function $f(x) = x^2$ is not monotone. It decreases for $x < 0$ and increases for $x > 0$. This is an example of a **piecewise monotone function**. A function is called piecewise monotone if every finite interval on which it is defined can be divided into *finitely many* intervals, on each of which the function is monotone. A typical case is shown in Figure 5.8. Another example is the function $x \mapsto \sin x$, which we shall discuss later.

Most functions encountered in applying calculus are piecewise monotone. There are, however, continuous functions that do not have this property, for instance, the function graphed in Figure 5.9. Near $x = 0$, the graph of this function changes direction, from "up" to "down," infinitely many times.

PROBLEMS

In Problems 1 to 5 decide whether the given function is increasing, nonincreasing, nondecreasing, decreasing, or none of these. Naturally, an increasing function is automatically nondecreasing; in this case, give the stronger of the two answers.

1. $t \mapsto t^5$.

2. $z = x^3 + x^2$.

3. $g(y) = |y| + y$.

4. $h(u) = |u|/u$.

5. $k(z) = z + 1$ if $z \leq 0$; z^2 if $z > 0$.

In Problems 6 to 9 each of the given functions is a piecewise monotone function. In each case, divide the domain of the function into intervals where it is monotone, and, for each such interval, decide whether the function is increasing, nondecreasing, decreasing, or non-increasing.

6. $A(x) = (x - 1)^4$.

7. $\phi(y) = y^2 + y$.

8. $t \mapsto t^2$ if $t \leq 0$; $1 - t$ if $t > 0$.

9. $g(h) = 1/h$ if $h < 0$; $(h - 1)^2$ if $0 \leq h < 2$; $1/h^3$ if $h \geq 2$.

In Problems 10 to 15 indicate on the horizontal axis the intervals where the function, whose graph is given below, is monotone. For each such interval, indicate whether the function is increasing, nondecreasing, decreasing, or nonincreasing.

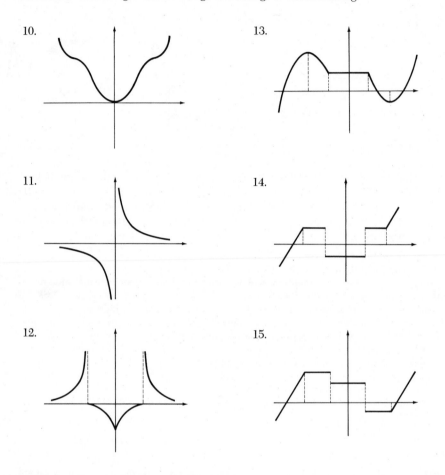

10.

13.

11.

14.

12.

15.

2.2 The sign of the derivative

For a function g defined in an interval, we say that $g > 0$ in this interval if $g(x) > 0$ at *every* point x in the interval. A similar convention holds for inequalities such as $g \geq 0$, $g < 0$, $g \leq 0$.

Let f be a function defined in an interval, and assume that $f' > 0$ in this interval [that is, $f'(x) > 0$ at *every* point of the interval]. Then the tangent to the graph of f points up at every point, and our geometric intuition tells us that the curve rises (see Figure 5.10). We formulate this as the following.

Theorem 1 *(theorem on functions with positive derivatives).* *If $f' > 0$ in an interval, f is increasing in this interval.*

The proof is not immediate, and we postpone it to Chapter 14, §1.4.

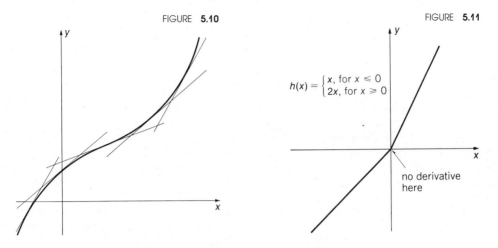

FIGURE 5.10

FIGURE 5.11

$$h(x) = \begin{cases} x, \text{ for } x \leqslant 0 \\ 2x, \text{ for } x \geqslant 0 \end{cases}$$

no derivative
here

Theorem 1 is one of the fundamental statements in calculus.

The converse to Theorem 1 is not valid. A continuous function f may be increasing in an interval and f' may fail to exist at some points (see Figure 5.11), or f' may be zero at some points (the derivative of $x \mapsto x^3$ equals 0 for $x = 0$).

EXAMPLE 1.   Set $f(x) = 3x$. Then $f'(x) = 3 > 0$ for all x. The function is increasing everywhere.

Corollary.   *If $f' < 0$ in an interval, f is decreasing in this interval.*

   Proof.   If $f' < 0$, then $(-f)' > 0$. By Theorem 1, $(-f)$ increases, hence f decreases.

Theorem 2.   *If $f' \geq 0$ in an interval, f is nondecreasing in this interval.*

   This theorem is also geometrically obvious. If the tangent to a graph either points up or is horizontal at every point, the graph cannot descend. The proof is postponed to Chapter 14, §1.4.

EXAMPLE 2.   Let $f(x) = x^3$. Then $f'(x) = 3x^2 \geq 0$ for all x. The function is increasing everywhere.

Corollary.   *If $f' \leq 0$ in an interval, f is nonincreasing in this interval.*

   Proof.   Apply Theorem 2 to the function $(-f)$.

2.3   Changes of sign. Intervals of increase and decrease

The following terminology will be useful in applying Theorem 1. A function g is said to *change sign* at a point x_0 if f is defined near x_0, except perhaps at x_0 itself, and if either (a) $g(x) > 0$ for x near x_0 and $x < x_0$, and $g(x) < 0$ for x near x_0 and $x > x_0$, or (b) $g(x) < 0$ for x near x_0 and $x < x_0$, and $g(x) > 0$ for x near x_0 and $x > x_0$.

   For instance, $g(x) = x^3$ changes sign at $x_0 = 0$ since $x^3 < 0$ for $x < 0$, $x^3 > 0$ for $x > 0$. So does $g(x) = x/|x|$, since $g(x) = -1$ for $x < 0$ and $g(x) = +1$ for $x > 0$.

The intermediate value theorem (Theorem 4 in Chapter 3, §3.4) implies that *a function which is continuous and has no zeros in an open interval cannot change sign at any point in this interval.*

Recall now that the derivative of a polynomial is a polynomial and the derivative of a rational function is a rational function. A polynomial of degree n has at most n distinct zeros or roots (see Theorem 3 in Chapter 3, §2.3); these are the *only* points at which the polynomial can change sign. A rational function p/q, p and q polynomials without common roots, can change sign only at a root of p (where f has a zero) or at a root of q (where f is not defined). Using Theorem 1 and its corollary, we conclude that *polynomials and rational functions are piecewise monotone.*

EXAMPLES   1.   In which intervals is the function $g(x) = x^3 - 2x^2 - 3x$ positive? In which intervals is it negative?

ANSWER   We must first find the zeros of g. This is achieved, in the present case, by factoring: $g(x) = x(x^2 - 2x - 3) = x(x - 3)(x + 1)$. The zeros of f are $-1, 0, 3$; these are the only points at which the function could change sign. The three zeros divide the number line into four intervals: $(-\infty, -1)$, $(-1, 0)$, $(0, 3)$, and $(3, \infty)$, in each of which f has a fixed sign. We check the values of f at convenient points in each interval. We have $g(-2) = (-2)(-5)(-1) = -10 < 0$, $g(-\frac{1}{2}) = (-\frac{1}{2})(-\frac{7}{2})(\frac{1}{2}) = \frac{7}{8} > 0$, $g(1) = (1)(-2)(2) = -4 < 0$, and $g(4) = (4)(1)(5) = 20 > 0$. Thus $g(x) < 0$ for $x < -1$, $g(x) > 0$ for $-1 < x < 0$, $g(x) < 0$ for $0 < x < 3$, and $g(x) > 0$ for $x > 3$.

2.   Find the intervals in which $g(x) = (x - 1)/(x + 1)$ is positive.

ANSWER   The denominator has a zero at $x = -1$, the numerator at $x = 1$. These are the only points where the function could change sign. We must investigate three intervals: $(-\infty, -1)$, $(-1, 1)$, $(1, +\infty)$. We check the value of g at one point in each interval: $g(-2) = 3 > 0$, $g(0) = -1 < 0$, $g(2) = \frac{1}{3} > 0$. Thus $g(x) < 0$ for $-1 < x < 1$ and $g(x) > 0$ for $x < -1$ and for $x > 1$.

3.   Find the intervals of increase or decrease for the function $f(x) = x^3 - 3x^2 + 3x - 1$.

ANSWER   We compute the derivative: $f'(x) = 3x^2 - 6x + 3 = 3(x^2 - 2x + 1) = 3(x - 1)^2$. Thus $f'(x) > 0$ except for $x = 1$. Since f is continuous at $x = 1$, we conclude by Theorem 2 that f is increasing for all x [that is, in $(-\infty, +\infty)$].

4.   Find the intervals of increase and decrease for the function $g(x) = x^3 - 6x^2 + 9x - 5$.

ANSWER   We observe that $g'(x) = 3x^2 - 12x + 9 = 3(x^2 - 4x + 3) = 3(x - 3)(x - 1)$. The zeros of g' are 1 and 3 and $g'(x) > 0$ for $x < 1$ and for $x > 3$, whereas $g'(x) < 0$ for $1 < x < 3$. *Conclusion:* g increases in $(-\infty, 1)$ and in $(3, +\infty)$; g decreases in $(1, 3)$.

5.   Where does the function

$$\phi(s) = \frac{2 + s}{2 - s}$$

increase? Where does it decrease?

ANSWER   We have that $\phi'(s) = 4/(2 - s)^2$. Thus $\phi'(s) > 0$ for $s \neq 2$; for $s = 2$ neither ϕ nor ϕ' are defined.

Conclusion: ϕ increases in $(-\infty, 2)$ and in $(2, +\infty)$. [We cannot conclude that ϕ increases for all x, since ϕ is not continuous for all s. Indeed, $\phi(1) = 3$ and $\phi(3) = -5 < 3$.]

PROBLEMS

In Problems 1 to 10 find the intervals, if any, where the given functions are positive.

1. $f(x) = -x^2 + 4x - 3$.   3. $f(x) = x^2 - 6x + 18$.
2. $f(x) = x^2 - 5x + 4$.   4. $f(x) = x^3 + 4x^2 - 5x$.

5. $f(x) = x^3 - 7x^2 + 10x$.
6. $f(x) = -x^4 + 2x^3 - x^2$.
7. $f(x) = (x - 1)(x - 2)(x - 3)(x - 4)$.
8. $f(x) = \dfrac{x - 5}{x - 7}$.

9. $f(x) = \dfrac{x^2 - 4}{x^2 - 9}$.

10. $f(x) = \dfrac{x^2 + 5x + 4}{x^2 - 1}$.

In Problems 11 to 24 find the intervals of increase and decrease of the given functions. It is understood that these intervals must be contained in the domain of definition of the function and should be maximal (as large as possible).

11. $f(x) = x^2 - 5x + 4$.
12. $f(x) = -x^2 + 4x - 3$.
13. $f(x) = 1 + x^5$.

14. $f(x) = \dfrac{1}{x^2}$.

15. $g(t) = t^{-10}$.
16. $g(t) = t^{-1001}$.
17. $f(x) = 2x^3 - 3x^2 - 36x$.
18. $s \mapsto s^3 + 6s^2 - 15s - 10$.

19. $g(x) = x^3 - 2x^2 + x - 1$.
20. $\phi(s) = s^4 - 2s^2 + 4$.
21. $g(y) = y^4 + 2y^2 - 5$.

22. $h(z) = \dfrac{z}{z - 1}$.

23. $f(t) = \dfrac{t - 2}{t + 1}$.

24. $f(x) = \dfrac{1 + x}{1 + x^2}$.

2.4 The sign of the second derivative

The sign of the second derivative also has a geometric meaning. If $f'' > 0$ in an interval, f' is increasing. Apply Theorem 1 to the function f'. This means that the slope of the tangent to the graph increases as we go along the curve from left to right; the tangent turns counterclockwise (see Figure 5.12). The graph is "bending upward" and "bulging downward." Such a function is called **strictly convex**. It is seen from Figure 5.13 that a strictly convex graph lies "below its chords and above its tangents." An analytic proof ▶ can be given, but it is somewhat complicated. (A function with $f'' \geq 0$ is called **convex** rather than strictly convex.)

Similarly, if $f'' < 0$, then f' decreases, the tangent turns in the clockwise direction, and the graph lies "above its chords and below its tangents" (see Figure 5.14). Such a function is called **strictly concave.**

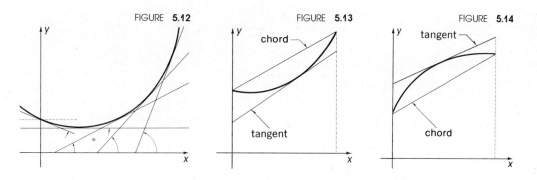

FIGURE **5.12** FIGURE **5.13** FIGURE **5.14**

We note that if a function f is strictly convex, the function $(-f)$ is strictly concave and vice versa.

FIGURE 5.15

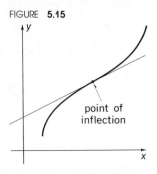

point of
inflection

If $f''(x_0) = 0$ and f'' changes sign at x_0, we say that $(x_0, f(x_0))$ is a **point of infection** of the graph (see Figure 5.15). At such a point the direction of bending changes.

EXAMPLES   1.   Let $f(x) = x^2$. Then $f'(x) = 2x$ and $f''(x) = 2 > 0$ for all x. The function is strictly convex.

2.   Let $f(x) = -x^2$. Then $f'(x) = -2x$ and $f''(x) = -2 < 0$ for all x. The function is strictly concave.

3.   Let $f(x) = x^3$. Then $f'(x) = 3x^2$, $f''(x) = 6x$. For $x > 0$, $f''(x) > 0$ and the function is convex; for $x < 0$, $f''(x) < 0$ and the function is concave. The point $x = 0$ is an inflection point. See Figure 5.6.

PROBLEMS

In the following problems find the intervals where the function is strictly convex or strictly concave. Also, find all inflection points, if any.

1. $f(x) = 2x^2 + 1$.
2. $f(x) = -3x^2 + 4x + 5$.
3. $g(x) = -4x^3$.
4. $f(x) = x^3 - 12x$.
5. $f(x) = -2x^3 + 3x^2 - 12x$.
6. $f(x) = x^3 - 2x^2 + 3x - 4$.
7. $f(x) = \dfrac{1}{x}$.

8. $f(x) = \dfrac{1}{(x-2)^2}$.
9. $g(x) = x^4 - 8x^2 + 16$.
10. $f(x) = -x^6 - x^4 - 10x^2$.
11. $g(s) = s^2 + \dfrac{1}{s}$.
12. $F(t) = \dfrac{1}{1 + t^2}$.

2.5   Local maxima and minima

Consider the function $y = f(x)$ graphed in Figure 5.16. The value of y for $x = 2$ is greater than the values of y for points $x \neq 2$ but close to 2. We say that $f(x)$ has a local maximum at $x = 2$; the value $y = f(2) = 3$ is called a local maximum of $f(x)$. Observe that outside an interval about $x = 2$, our function takes on values greater than $f(2)$; at $x = 4$, for instance, $y = 5$. This possibility is acknowledged by the adjective "local."

Our function also attains a local maximum 6 at $x = 5$. At $x = 3$, the function has a local minimum $f(3) = 2$.

In general, a function f is said to have a **local maximum** at $x = x_0$ if f is defined near x_0 and $f(x_0) \geq f(x)$ for all x near x_0. If $f(x)$ is defined near x_0 and $f(x) \geq f(x_0)$

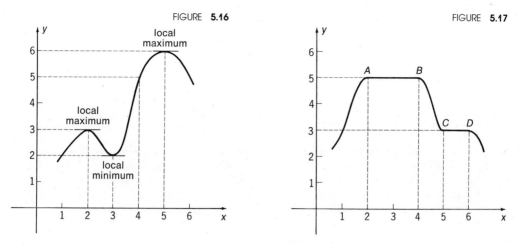

FIGURE **5.16**   FIGURE **5.17**

for all x near x_0, f is said to have a **local minimum** at x_0. (*Remember:* "near x_0" means "in some interval with x_0 as midpoint.")

Consider next the function graphed in Figure 5.17; the graph contains two horizontal segments, AB and CD. The definition above used the symbol $\geq$ (greater than or equal to) rather than $>$ (greater than). Hence we must say that the function graphed in Figure 5.17 has local maxima and local minima at all x such that $2 < x < 4$ and $5 < x < 6$, respectively. The function also has local maxima at $x = 2$, $x = 4$, and $x = 6$, and a local minimum at $x = 5$. In order to distinguish the situation depicted in Figure 5.16 from that of Figure 5.17, we introduce the following terminology.

If f has a local maximum (or minimum) at x_0, and if, near x_0, $f(x) \neq f(x_0)$ for $x \neq x_0$, the local maximum (or minimum) is called **strict.**

In Figure 5.16, the two local maxima (at $x = 2$ and $x = 5$) and the local minimum (at $x = 4$) are strict. In Figure 5.17, there are no strict local maxima or minima.

Theorem 3. *If f has a local maximum or minimum at x_0, and f is differentiable at x_0, then $f'(x_0) = 0$.*

The proof is the same as for Theorem 2 in §1.1.

We should not think, however, that if $f'(x_0) = 0$, then f must have a maximum or minimum at x_0. For the function $f(x) = x^3$, we have $f'(x) = 3x^2$ and $f'(0) = 0$. But there is neither a minimum nor a maximum at $x = 0$; see Figure 5.6. A local maximum or minimum point of a differentiable function is always a zero of the derivative, but a zero of the derivative need not be a local maximum point or minimum point of the function.

On the other hand, a continuous function may have a local maximum or local minimum without having a derivative. For example, the function $f(x) = |x|$ has a strict local minimum at $x = 0$ but no derivative at $x = 0$.

2.6   Tests for local maxima and minima

We now describe two methods for deciding whether a point at which the derivative is zero is a local maximum or a local minimum.

Theorem 4 (first derivative test). *Let $f(x)$ be a continuous function defined near x_0.*

If for x near x_0 and $x < x_0$	and for x near x_0 and $x > x_0$	then f has at x_0 a strict local
$f'(x) > 0$	$f'(x) < 0$	*maximum*
$f'(x) < 0$	$f'(x) > 0$	*minimum*

Proof. We prove the first row only; the second is proved similarly. Let x_1 and x_2 be points near x_0 such that $x_1 < x_0 < x_2$. We must show that $f(x_1) < f(x_0)$ and $f(x_0) > f(x_1)$. Now $f' > 0$ in the interval (x_1,x_0), by hypothesis. Hence f increases in this interval, by Theorem 1, and $f(x_1) < f(x_0)$. Similarly, $f' < 0$ in the interval (x_0,x_1), by hypothesis. Hence f decreases in this interval, by the Corollary to Theorem 1, and $f(x_0) > f(x_1)$.

EXAMPLES 1.   The function $f(x) = |x|$ is continuous everywhere. We have that $f(x) = -x$ for $x < 0$, $f(x) = x$ for $x > 0$. Hence $f'(x) = -1$ for $x < 0$, $f'(x) = 1$ for $x > 0$. Therefore, $f(x)$ has a local minimum at $x = 0$. Note that Theorem 4 does not assume that f is differentiable at $x = x_0$.

2.   Find all local maxima and minima of the function $f(x) = x^4 - 2x^2 + 1$.
ANSWER   The function is a polynomial and hence differentiable everywhere. We compute $f'(x)$ and find that

$$f'(x) = 4x^3 - 4x = 4x(x^2 - 1).$$

The zeros of f' are $x = 0$, $x = +1$, and $x = -1$. By Theorem 3, these are the only candidates for local maxima or minima. For x near 0, x^2 is small and $(x^2 - 1) < 0$. Hence $f'(x)$ is positive for $x < 0$ and negative for $x > 0$. Thus f has a local maximum at $x = 0$. For x near $+1$, $4x$ is positive. Hence $f'(x) < 0$ for $x < 1$, and $f'(x) > 0$ for $x > 1$. Thus f has a local minimum at $x = 1$. For x near -1, $4x$ is negative. Hence $f'(x) < 0$ for $x < -1$, and $f'(x) > 0$ for $x > -1$. Thus f has a local minimum at $x = -1$. Note that all maxima and minima are strict.

Theorem 5 (second derivative test).   *Let f be defined near x_0.*

If	and if	then f has at x_0 a strict local
$f'(x_0) = 0$	$f''(x_0) < 0$	*maximum*
$f'(x_0) = 0$	$f''(x_0) > 0$	*minimum*

Proof.   Assume that $f'(x_0) = 0$ and $f''(x_0) < 0$. Then, noting that $f'' = (f')'$ and

$$f''(x_0) = \lim_{x \to x_0} \frac{f'(x) - f'(x_0)}{x - x_0} = \lim_{x \to x_0} \frac{f'(x)}{x - x_0} < 0,$$

we conclude that $f'(x)(x - x_0) < 0$ for x near x_0, $x \neq x_0$. Therefore $f' > 0$ to the left and near to x_0, $f' < 0$ to the right and near to x_0. By Theorem 4, f has a strict local maximum at x_0.
The second row is proved similarly.

EXAMPLES 3. Find all local maxima and minima of the function $f(x) = 2x^3 - 9x^2 + 12x - 5$.
ANSWER   The function is differentiable everywhere. We compute f' and f''. We have

$$f'(x) = 6x^2 - 18x + 12 = 6(x^2 - 3x + 2) = 6(x - 1)(x - 2),$$
$$f''(x) = 12x - 18.$$

The zeros of f' are $x = 1$ and $x = 2$; by Theorem 3 these are the only candidates for local maxima or minima. We apply the second derivative test (Theorem 5). Since $f''(1) = -6 < 0$, $x = 1$ is a local maximum point; at this point $f(1) = 0$. Since $f''(2) = 6 > 0$, $x = 2$ is a local minimum point. At this point $f(2) = -1$.

4. Find the local maxima and local minima of the function $f(x) = 1 + x^4$.
ANSWER   We have $f'(x) = 4x^3$, $f''(x) = 12x^2$. The only zero of f' is $x = 0$. But $f''(0) = 0$, so the second derivative test tells us nothing. We see directly, however, that $x = 0$ is a strict minimum point since $x^4 > 0$ for $x \neq 0$, so that $f(x) = 1 + x^4 > 1 = f(0)$ for $x \neq 0$.
    We could have also used the first derivative test: $f'(x) = 4x^3$ is negative for $x < 0$ and positive for $x > 0$.

5. Find the local minima and maxima of $f(x) = 1/x$.
ANSWER   The function is differentiable wherever defined. Since $f'(x) = 1/x^2 > 0$, there are no maxima or minima.

PROBLEMS

    In the following problems find all local minima and local maxima of the given function.

1. $f(x) = x^2 + 4x + 4$.
2. $f(x) = 2x^2 - 3x + 5$.
3. $f(x) = -3x^2 + 2x + 1$.
4. $y = x^3 - 12x$.
5. $y = x^3 - 3x^2 + 3x - 1$.
6. $y = x^3 + 3x^2 - 9x + 4$.

7. $g(x) = -2x^3 - 3x^2 + 36x + 36$.
8. $g(x) = 2x^3 - 15x^2 - 36x + 1$.
9. $f(x) = \frac{1}{12}(x - 2)^4$.
10. $f(x) = x^4 - 4x^3 + 6$.
11. $f(x) = x^4 - 18x^2 + 81$.

12. $f(x) = 3x^4 - 16x^3 - 6x^2 + 48x + 2$. [*Hint:* The first derivative contains the factor $(x - 4)$.]
13. $y = 2x^6 - 15x^4 + 24x^2 + 1$.

14. $y = \dfrac{x}{x + 1}$.

15. $f(x) = \dfrac{1}{x^2 + 9}$.

16. $f(x) = \dfrac{x}{x^2 + 1}$.

17. $x \mapsto \dfrac{2 + x^2}{1 + x^2}$.

18. $x \mapsto \dfrac{x^2}{(x^2 + 1)}$.

19. $x \mapsto \dfrac{1}{x^4 + 2x^2 + 1}$.

20. $f(u) = |1 + u|$.
21. $f(u) = |4 - u^2|$.
22. $f(u) = |u^3 - 3u|$.

§3   Graphs of rational functions

    In this section we discuss graphs of rational functions, using derivatives. To do this we need some useful extensions of the limit concept.

3.1   Infinite limits

    Let us consider the function $f(x) = 1/x$, the simplest rational function which is not a polynomial (see Figure 5.18). The function has no limit at $x = 0$, and even no

FIGURE 5.18   FIGURE 5.19

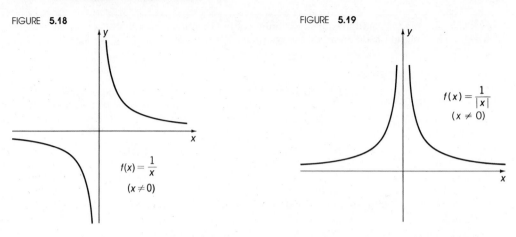

$$f(x) = \frac{1}{x}$$
$$(x \neq 0)$$

$$f(x) = \frac{1}{|x|}$$
$$(x \neq 0)$$

one-sided limits as x approaches 0 from the left or from the right (see Chapter 3, §4.4). For small positive x, $f(x)$ is positive and large; it becomes as large as desired if $|x|$ is small enough. Similarly $f(x)$ is negative for x negative, and $|f(x)|$ can be made as large as desired by choosing x close enough to 0. The statements just made are abbreviated by writing

$$\lim_{x \to 0^+} f(x) = +\infty, \qquad \lim_{x \to 0^-} f(x) = -\infty.$$

[Read: $f(x)$ has the limit plus infinity as x approaches 0 from the right, and the limit minus infinity as x approaches 0 from the left.]

For the function $f(x) = 1/|x|$ we have (see Figure 5.19)

$$\lim_{x \to 0} f(x) = +\infty;$$

we mean by it that $f(x)$ is positive for x close but unequal to 0, and that $f(x)$ is as large as we want for $|x|$ sufficiently small.

These examples suggest the following **definition of infinite limits**: the statement "$\lim_{x \to x_0} f(x) = +\infty$" means that $f(x) > 0$ for x near x_0, $x \neq x_0$, and that $\lim_{x \to x_0} [1/f(x)] = 0$; the statement "$\lim_{x \to x_0^+} f(x) = +\infty$" means that $f(x) > 0$ for x near x_0, $x > x_0$, and that $\lim_{x \to x_0^+} [1/f(x)] = 0$. The meaning of the statements $\lim_{x \to x_0} f(x) = +\infty$, $\lim_{x \to x_0} f(x) = -\infty$, $\lim_{x \to x_0^+} f(x) = -\infty$, and $\lim_{x \to x_0^-} f(x) = -\infty$ are defined similarly.

The symbols $+\infty$ and $-\infty$ are very handy, but the reader should remember that *they are not names of numbers*, or of anything else, but simply a form of shorthand used in certain sentences.

Rational functions have infinite limits at points at which they are not defined, as we show by examples.

EXAMPLES   1.   What limit, or limits, does $f(x) = (1 + x^3)/x^3$ have at $x = 0$?

ANSWER   Since $1/f(x) = x^3/(1 + x^3)$ has limit 0 at $x = 0$, f has at $x = 0$ an infinite limit (or one-sided infinite limits). We must now determine the sign of $f(x)$ for $x \neq 0$ but close to 0. For such x, $1 + x^3$ is close to $1 + 0^3 = 1 > 0$. But $x^3 > 0$ for $x > 0$ and $x^3 < 0$ for $x < 0$. Hence $f(x)$ is positive or negative according to whether x is positive or negative. We conclude: $\lim_{x \to 0^-} f(x) = -\infty$, $\lim_{x \to 0^+} f(x) = +\infty$.

2. What limit, or limits, does $f(x) = (-5 + x^3)/x^3$ have at $x = 0$?

ANSWER   We reason as in Example 1, except that this time the denominator of $1/f(x)$ is, for x close to 0, close to -5, hence negative. Thus $f(x)$ is, for x close to 0, negative if $x > 0$ and positive for $x < 0$. Hence $\lim_{x \to 0^-} f(x) = +\infty$, $\lim_{x \to 0^+} f(x) = -\infty$.

3. What limit, or limits, does $f(x) = (1 + x^3)/|x^3|$ have at $x = 0$?

ANSWER   We note that the denominator $|x^3|$ is positive for $x \neq 0$. Hence $f(x)$ is positive for $x \neq 0$, x close to 0; $\lim_{x \to 0} f(x) = +\infty$.

4. What limit, or limits, does

$$f(x) = \frac{3 - 5x^3}{x^3 - x^2 - x + 1}$$

have at $x = 1$?

ANSWER   First of all, the denominator is 0 at $x = 1$, the numerator is not. Hence $1/f(x)$ has limit 0 at $x = 1$. We must now determine the sign of $f(x)$ for $x \neq 1$, x close to 1. To do this, we first divide the denominator $x^3 - x^2 - x + 1$ by as large a power of $x - 1$ as possible:

$$x^3 - x^2 - x + 1 = (x - 1)(x^2 - 1) = (x - 1)^2(x + 1).$$

Hence

$$f(x) = \frac{3 - 5x^3}{x^3 - x^2 - x + 1} = \frac{1}{(x - 1)^2} \frac{3 - 5x^3}{x + 1}.$$

The factor $1/(x - 1)^2$ is positive for all $x \neq 1$. The factor $(3 - 5x^3)/(x + 1)$ has the limit $-2/2 = -1 < 0$ at $x = 1$; hence this factor is negative for x close to 1. We conclude that $f(x)$ is negative for x close to 1. Therefore, $\lim_{x \to 1} f(x) = -\infty$.

The method of Example 4 is general. Let $f(x)$ be a rational function, so that $f(x) = p(x)/q(x)$, p and q polynomials without common roots. If $q(x_0) = 0$, the infinite limit, or limits, of f at x_0 are found as follows: we write $q(x) = (x - x_0)^m Q(x)$ with Q a polynomial such that $Q(x_0) \neq 0$ (see Chapter 3, §2.3). Thus

$$f(x) = \frac{1}{(x - x_0)^m} \frac{p(x)}{Q(x)}.$$

Now the term $(x - x_0)^{-m}$ has infinite limits at x_0: they are $+\infty$ from the right and $-\infty$ from the left if m is odd, $+\infty$ if m is even. The function f has the same limits as $(x - x_0)^m$ if $p(x_0)/Q(x_0) > 0$, opposite limits if $p(x_0)/Q(x_0) < 0$.

If a function $f(x)$ has an infinite limit or infinite limits at $x = x_0$, then the line $x = x_0$ is called a **vertical asymptote** of its graph. In Figures 5.18 and 5.19, the y-axis is such an asymptote.

REMARK   The statement "f has a limit at x_0" always means that there is a *number* α such that $\lim_{x \to x_0} f(x) = \alpha$. If we want to say that *either* there is a number α that is the limit of f at x_0, or $\lim_{x \to x_0} f(x) = +\infty$, or $\lim_{x \to x_0} f(x) = -\infty$, we say that "$f$ has at x_0 a finite or infinite limit." The same convention applies to one-sided limits.

PROBLEMS

Find the limit, or limits, of the following functions at the indicated points.

1. $f(x) = \dfrac{1}{(x + 2)^2}$, $x = -2$.   2. $f(x) = \dfrac{1 + 2x^2}{x^3}$, $x = 0$.

3. $f(x) = \dfrac{4 + x^2}{|x^3|}$, $x = 0$.

7. $f(x) = \dfrac{x + 1}{|x - 1|}$, $x = 1$.

4. $f(x) = \dfrac{x - 1}{(2x - 1)^2}$, $x = \frac{1}{2}$.

8. $f(x) = \dfrac{x^2}{x^2 - 1}$, $x = -1$.

5. $f(x) = \dfrac{x^2 + 1}{x - 2}$, $x = 2$.

9. $f(x) = \dfrac{2 - x}{x^3 - 8x^2 + 16x}$, $x = 4$.

6. $f(x) = \dfrac{x^3 + 1}{x + 3}$, $x = -3$.

10. $f(x) = \dfrac{x^2 + 4}{x^4 - 6x^3 + 12x^2 - 8x}$, $x = 2$.

3.2   Limits at infinity

The language of limits can be also used to describe the behavior of a function $f(x)$ for large values of $|x|$. For instance, the function $f(x) = 1/x$ has the following properties: $f(x)$ becomes as close as we want to 0 if x is large enough or if x is negative and $|x|$ is large enough. We express this by saying that $f(x)$ has the limit 0 as x approaches $+ \infty$ (read: plus infinity) and also as x approaches $- \infty$ (read: minus infinity). We write

$$\lim_{x \to +\infty} f(x) = 0, \qquad \lim_{x \to -\infty} f(x) = 0.$$

Next, we give an example of a function which has infinite limits at infinity: $f(x) = x^3$. If $x > 0$ and very large, then $f(x) > 0$ and very large, as large as we want provided x is large enough. We express this by writing

$$\lim_{x \to +\infty} f(x) = +\infty.$$

On the other hand, for x negative and $|x|$ sufficiently large, we have that $f(x)$ is negative and $|f(x)|$ is as large as we want; in symbols:

$$\lim_{x \to -\infty} f(x) = -\infty.$$

To give a **definition of limits at infinity** we note that if $x \mapsto f(x)$ is defined for $x > A > 0$, then the function $t \mapsto f(1/t)$, where $t = 1/x$, is defined for $0 < t < 1/A$. As x gets arbitrarily large, $t = 1/x$ gets arbitrarily close to 0 from the right. Therefore, we define

$$\lim_{x \to +\infty} f(x) = \alpha \qquad \text{means that} \qquad \lim_{t \to 0^+} f\left(\frac{1}{t}\right) = \alpha.$$

Similarly,

$$\lim_{x \to -\infty} f(x) = \beta \qquad \text{means that} \qquad \lim_{t \to 0^-} f\left(\frac{1}{t}\right) = \beta.$$

We agree to use the same definitions if α and β are one of the symbols $+\infty$ or $-\infty$.

If $f(x)$ has a finite limit α at $+\infty$ or at $-\infty$, the line $y = \alpha$ is called a **horizontal asymptote** to the graph of f. In Figures 5.18 and 5.19 the x-axis is such an asymptote.

EXAMPLES 1.   Find the limit of $f(x) = \dfrac{3x^3 + 5x - 6}{7x^3 - 4x^2 + 3}$ as $x \to +\infty$.

ANSWER   Replacing x by $1/t$ we have

$$\lim_{x \to +\infty} \frac{3x^3 + 5x - 6}{7x^3 - 4x^2 + 3} = \lim_{t \to 0^+} \frac{\dfrac{3}{t^3} + \dfrac{5}{t} - 6}{\dfrac{7}{t^3} - \dfrac{4}{t^2} + 3} = \lim_{t \to 0^+} \frac{3 + 5t^2 - 6t^3}{7 - 4t + 3t^3} = \frac{3}{7}.$$

2. $\lim_{x \to +\infty} x^n = +\infty$ for every positive integer n.

   Indeed, $\lim_{t \to 0^+} 1/t^n = +\infty$. This is true since $t^n > 0$ for $t > 0$, t^n is continuous at $t = 0$, and $0^n = 0$.

3. $\lim_{x \to -\infty} x^n = +\infty$ for every even positive integer n, and $\lim_{x \to -\infty} x^n = -\infty$ for every odd positive integer n. (Use the same argument as in Example 2.)

The leading term of a polynomial of degree n,

$$p(x) = a_n x^n + a_{n-1} x^{n-1} + a_{n-2} x^{n-2} + \cdots + a_0,$$

is the term $a_n x^n$. For $x \neq 0$ we can write

$$p(x) = a_n x^n \left(1 + \frac{a_{n-1}}{x} + \frac{a_{n-2}}{x^2} + \cdots + \frac{a_0}{x^n} \right).$$

Thus $p(x) = a_n x^n$ times a function which is as close as we want to 1, provided $|x|$ is large enough. This observation leads to the following rule for computing limits at infinity:

   Let f be a rational function, $f = p/q$, where p and q are polynomials. In computing the limits of f at infinity one may replace p and q by their leading terms.

EXAMPLES   4.   Find the limits at infinity of the polynomial $f(x) = -7x^5 + 6x^4 - 3x^2 + 2x - 1$.
ANSWER   The leading term is $(-7x^5)$. Thus

$$\lim_{x \to +\infty} f(x) = \lim_{x \to +\infty} (-7x^5) = -\infty,$$

$$\lim_{x \to -\infty} f(x) = \lim_{x \to -\infty} (-7x^5) = +\infty.$$

5.   Find the limits at infinity of $f(x) = \dfrac{3x^3 + 5x - 6}{7x^3 - 4x^2 + 3}$.

ANSWER   By the rule,

$$\lim_{x \to +\infty} f(x) = \lim_{x \to +\infty} \frac{3x^3}{7x^3} = \lim_{x \to +\infty} \frac{3}{7} = \frac{3}{7}.$$

as in Example 1. As $x \to -\infty$ we obtain the same limit.

PROBLEMS

   In Problems 1 to 20 find the limits of the given functions at plus and minus infinity.

1. $f(x) = x^7 + 25x^2 - 3$.
2. $f(x) = x^{100} + 20x^{10} + 40$.
3. $f(x) = (x^3 - 3)/(x^4 + 25)$.
4. $f(x) = (x^6 - 7x^3 + 2)/(100x^6 - 42x^5)$.
5. $x \mapsto (x^5 - 3)/(x^5 + 3)$.
6. $x \mapsto (x^5 - 3)/(x^6 + 3)$.
7. $x \mapsto (x^6 - 3)/(x^6 + 3)$.
8. $G(s) = (s^5 - 5s^4)/(5s^2 + 1)$.
9. $F(t) = (2t - 1)^2/(3t + 6)^3$.
10. $H(s) = (-s^4 + 2s^2 + 1)^3/(2s^2 + s + 1)^6$.

11. $f(x) = (1 + 6x)/(-2 + x)$.
12. $f(x) = (1 + x + 2x^2)/(10 + 5x - 4x^2)$.
13. $x \mapsto 2x^3/(6 + x + 2x^2 + x^3)$.
14. $x \mapsto 3/(x^3 + x - 1)$.
15. $f(x) = (1 + x - 100x^3)/x^4$.
16. $f(x) = (x^{100} + x^{99})/(x^{101} - x^{100})$.
17. $f(x) = (x^3 + 1)/(x^2 - 1)$.
18. $f(x) = (2x^3 + x^2 - 1)/(x + 5)$.
19. $f(x) = (x^4 - 2x^2 + x)/(1 - x)$.
20. $f(x) = (3 - x^2 + 4x^4)/(3 + x - 2x^2)$.

3.3 Sketching graphs

The information found in the preceding subsections leads to a procedure for sketching quite accurately a graph of a function $f(x)$, while computing only relatively few points. We list the steps below.

 I. Determine whether the function is even, odd, or neither (see Chapter 3, §1.4). If the function is even or odd, only values $x \geq 0$ need to be considered, since the graph is then symmetric.
 II. Find the intervals where the function is defined. Determine the (one-sided) limits of the function at $+\infty$, $-\infty$, and at the endpoints of the intervals of definition. Draw the vertical and horizontal asymptotes, if any.
 III. Locate the zeros of f, if possible. (Mark these points on the graph.)
 IV. Compute f' and locate the zeros of f', if possible. Determine the intervals of decrease and increase of f; for each zero α of f', determine whether it is a local maximum, minimum, or neither, and compute $f(\alpha)$. Draw the points $(\alpha, f(\alpha))$ and the horizontal tangent through them.
 V. Compute f''; locate the zeros of f'', if possible, and determine the intervals of convexity and concavity of f. Determine which zeros β of f'' are points of inflection; for each such β compute $f(\beta)$ and $f'(\beta)$. Draw the points $(\beta, f(\beta))$ and the tangents, with slope $f'(\beta)$, through them.
 VI. Sketch the graph of f, using the results of Steps I to V. Better accuracy can be obtained by computing additional points and slopes.

We note that in any given case not all of Steps I to V are needed or possible.

For polynomials and for rational functions, the limits at $+\infty$ and at $-\infty$ are determined by the rule explained in §3.2. If $f = p/q$ is a rational function, p and q being polynomials without common roots, f becomes infinite (that is, has infinite limit or limits) at every root r of q. Instead of using the procedure explained in §3.1, we may note that at such a point f has the limit $+\infty$ or $-\infty$, from the left, according to whether $f' > 0$ or $f' < 0$ near and to the left of r. The one-sided infinite limit from the right may be determined similarly.

In what follows, we shall write $f(+\infty)$, $f(-\infty)$, $f(x_0^+)$, and $f(x_0^-)$ as abbreviations for the symbols for one-sided limits, $\lim_{x \to +\infty} f(x)$, $\lim_{x \to -\infty} f(x)$, $\lim_{x \to x_0^+} f(x)$, and $\lim_{x \to x_0^-} f(x)$.

EXAMPLES 1. Graph the function $y = 4x^2 + 8x - 5$.
SOLUTION   Let $y = f(x)$. The given function is a polynomial with $f(-\infty) = +\infty$, $f(+\infty) = +\infty$ (by the rule in §3.2). There are no vertical or horizontal asymptotes. Next, we have $f(x) = 4(x^2 + 2x - \frac{5}{4}) = 4(x - \frac{1}{2})(x + \frac{5}{2})$. Hence $x = \frac{1}{2}$ and $x = -\frac{5}{2}$ are zeros of f. Also, $f'(x) = 8x + 8 = 8(x + 1)$ and $f''(x) = 8$. Hence $x = -1$ is a zero of $f'(x)$ and is a candidate for a maximum or minimum. Since $f''(x) = 8 > 0$ for all x, $x = -1$ is a minimum by the second derivative test, and f is convex. At $x = -1$, $f(-1) = -9$. There are no inflection points.
The graph (a parabola) is shown in Figure 5.20.

2. Sketch the graph of the function $f(x) = -\frac{1}{6}x^3 - \frac{1}{6}x + 2$.
SOLUTION   This is a polynomial with $f(-\infty) = +\infty$, $f(+\infty) = -\infty$ (by the rule in §3.2). We have $f'(x) = -\frac{1}{2}x^2 - \frac{1}{6} < 0$ for all x; hence f is decreasing. Also $f''(x) = -x$; hence f is convex for $x < 0$, concave for $x > 0$. There is an inflection point at $x = 0$; at this point $f(0) = 2$, $f'(0) = -\frac{1}{6}$. Also, $f(1) = \frac{5}{3}$, $f(2) = \frac{1}{3}$, $f(3) = -\frac{13}{6}$, so that f has a root between 2 and 3. The function is graphed in Figure 5.21.

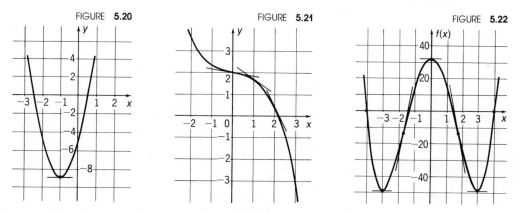

FIGURE **5.20**   FIGURE **5.21**   FIGURE **5.22**

3. Discuss the function $f(x) = x^4 - 18x^2 + 32$ and draw its graph.

SOLUTION   The function is even and therefore symmetric about the y-axis; hence, only values $x \geq 0$ need to be considered. The function is a polynomial with $f(+\infty) = +\infty$. There are no vertical or horizontal asymptotes. We now locate the zeros of f, f', and f'':

$$f(x) = x^4 - 18x^2 + 32 = (x^2 - 2)(x^2 - 16) = (x - \sqrt{2})(x + \sqrt{2})(x - 4)(x + 4),$$
$$f'(x) = 4x^3 - 36x = 4x(x^2 - 9) = 4x(x - 3)(x + 3),$$
$$f''(x) = 12x^2 - 36 = 12(x^2 - 3) = 12(x - \sqrt{3})(x + \sqrt{3}).$$

f has zeros at 0, $\pm\sqrt{2}$, and ±4; f' has zeros at 0 and ±3; and f'' has zeros at $\pm\sqrt{3}$. With this information, we construct the following table for $x \geq 0$:

x	f	f'	f''	Conclusion
0	32	0	$-$	maximum
$\sqrt{2}$	0	$-$	$-$	zero
$\sqrt{3}$	-13	$-24\sqrt{3}$	0	inflection point
3	-49	0	$+$	minimum
4	0	$+$	$+$	zero

We conclude that $x = 0$ is a maximum and $x = 3$ is a minimum, both by the second derivative test. Also, $x = \sqrt{3}$ is an inflection point since $f'' = 0$. At $x = \sqrt{3}$, $f'(\sqrt{3}) = -24\sqrt{3}$. Furthermore, $f(x)$ is concave for $0 < x < \sqrt{3}$ since $f''(x) < 0$, and convex for $x > \sqrt{3}$ since $f''(x) > 0$. [Note that only algebraic signs are needed in the column for f''.] Finally, $f(x) > 0$ for $x > 4$.

The graph of the function is given in Figure 5.22.

4. Discuss the function $f(x) = (x - 1)/(x - 2)$, and draw its graph.

SOLUTION   The function is defined for $x \neq 2$. We note that $f(-\infty) = f(+\infty) = 1$; there is a horizontal asymptote $y = 1$. Since $f'(x) = -1/(x - 2)^2 < 0$ for $x \neq 2$, the function is decreasing for $-\infty < x < 2$ and for $2 < x < +\infty$. The function becomes infinite at $x = 2$, so there is a vertical asymptote. At $x = 2$ we have $\lim_{x \to 2^-} f(x) = -\infty$, $\lim_{x \to 2^+} f(x) = +\infty$. Since $f''(x) = 2/(x - 2)^3$, we have $f''(x) < 0$ (concavity) for $x < 2$ and $f''(x) > 0$ (convexity) for $x > 2$. There are no inflection points.

FIGURE 5.23   FIGURE 5.24

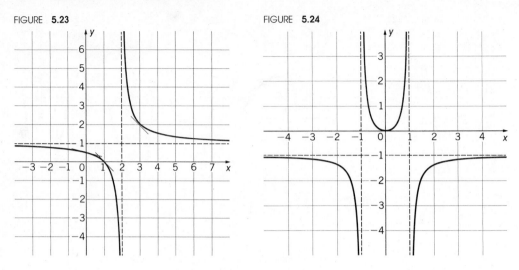

The graph is shown in Figure 5.23 (it is a hyperbola).

5. Graph the function $f(x) = x^2/(1 - x^2)$.

SOLUTION   This is an even function, defined for $x \neq \pm 1$. Since $f(+\infty) = -1$, there is a horizontal asymptote $y = -1$. Since f becomes infinite at $x = 1$, there is a vertical asymptote $x = 1$. We compute the first and second derivatives:

$$f'(x) = \frac{2x}{(1 - x^2)^2}; \qquad f''(x) = \frac{2(1 + 2x^2 - 3x^4)}{(1 - x^2)^4}.$$

We have $f(0) = f'(0) = 0$ and $f''(0) = 2$, hence a local minimum at $x = 0$. Also, $f'(x) > 0$ for $0 < x < 1$ and for $1 < x < +\infty$; the functions are increasing in these intervals. We conclude that $\lim_{x \to 1^-} f(x) = +\infty$, $\lim_{x \to 1^+} f(x) = -\infty$.

In order to determine the sign of f'', we note that

$$2(1 + 2x^2 - 3x^4) = 2(1 - x^2)(1 + 3x^2).$$

Hence we have $f'' > 0$ and a convex graph for $0 < x < 1$, $f'' < 0$ and a concave graph for $1 < x < +\infty$. The graph is shown in Figure 5.24. (It is symmetric about the y-axis, since f is even.)

PROBLEMS

Sketch the graphs of the functions given below. Be sure to find all maxima, minima, inflection points, horizontal asymptotes, and vertical asymptotes. It may occasionally be convenient to use different units of length along the coordinate axes.

1. $y = -x^2 + 6x - 5$.
2. $y = \frac{1}{2}x^2 - x - 4$.
3. $f(x) = x^2 - 4x + \frac{7}{4}$.
4. $f(x) = -4x^2 - 4x + 3$.
5. $y = -x^3 + 3x$.
6. $y = \frac{1}{3}x^3 + x + 2$.
7. $f(x) = x^3 - 3x + 2$.

8. $f(x) = x^3 - 9x^2 + 24x - 18$.
9. $f(x) = x^3 + 3x^2 - 24x + 20$.
10. $F(x) = -x^4 + 6x^2 + 2$.
11. $F(x) = x^4 - 4x^3 + 10$.
12. $g(x) = 3x^5 - 10x^3$.
13. $g(x) = x/(x - 3)$.
14. $y = x + (1/x)$.

15. $y = 8/(x^2 + 4)$.

16. $f(x) = (x^2 - 1)/(x^2 + 1)$.

17. $f(x) = 18(x - 1)/(x^2 + 2x + 1)$.

18. $s(t) = t/(t^2 - 1)$.

19. $s(t) = (t^2 + 2t + 1)/(t^2 + 1)$.

20. $h(u) = (2 - u + u^2)/(2 + u - u^2)$.

21. $h(u) = (u^2 - 7u + 10)/(u - 1)$.

3.4 Remark on odd-degree polynomials

Consider a cubic polynomial

$$f(x) = ax^3 + bx^2 + cx + d$$

with $a \neq 0$. Assume, for the sake of definiteness, that $a > 0$. Then $f(-\infty) = -\infty$, $f(+\infty) = +\infty$. Thus $f(x) > 0$ for x large and positive, and $f(x) < 0$ for $x < 0$ and $|x|$ large. By the intermediate value theorem the graph of f must cross the x-axis. A similar argument shows that *every polynomial of odd degree has at least one (real) root*. But there are polynomials of even degree without (real) roots; for instance, $f(x) = x^2 + 1$.

DERIVATIVES, CONTINUED

§1 Inverse functions. Roots

In the preceding chapter we differentiated only rational functions. We shall now learn techniques for differentiating more complicated functions and also for inverting the differentiation process in some simple cases.

1.1 Existence and continuity of inverse functions

Let $f(x)$ be defined and continuous for $a \leq x \leq b$, and let $\alpha = f(a) < \beta = f(b)$; see Figure 6.1. By the intermediate value theorem (see Theorem 4 in Chapter 3, §3.4), the function f takes on every value γ between α and β somewhere in the interval (a,b). Of course, it may take on the value at several points, even at infinitely many points. Also, f may take on the value α and β not only at the endpoints of the interval but also at inner points (see, for instance, the function graphed in Figure 6.1).

But if $f(x)$ is an *increasing* function, the situation is different. For such a function, if $x_1 < x_2$, then also $f(x_1) < f(x_2)$. Hence f takes on the value α only at a, the value β only at b, and every value γ between α and β at exactly one point between a and b (see Figure 6.2).

Let y be a number such that $\alpha \leq y \leq \beta$. There is a number x with $f(x) = y$ and $a \leq x \leq b$, and only one such number. We write $x = g(y)$ and call g the function **inverse** to f. Note that

$$g(f(x)) = x, \quad \text{for all } x, \qquad a \leq x \leq b.$$

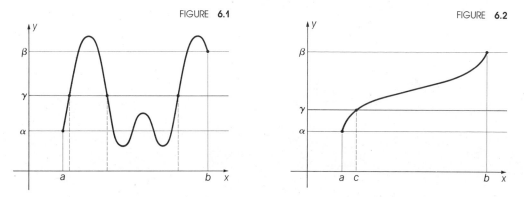

FIGURE 6.1   FIGURE 6.2

Indeed, what is $g(f(x))$? It is the unique number to which f assigns the value $f(x)$; this is x. A similar argument (reminiscent of the question "Who is buried in Grant's Tomb?") shows that

$$f(g(y)) = y, \quad \text{for all } y, \qquad \alpha \le y \le \beta.$$

Indeed, $g(y)$ is that x to which f assigns the value y. Thus f is the function inverse to g. It is evident that g is an increasing function.

A similar argument applies to continuous decreasing functions, except that the inverse of such a function is decreasing.

EXAMPLES   1. The linear function $f(x) = 2x + 3$ is defined and increasing for all x. The inverse function is obtained by "solving the equation $y = 2x + 3$ for x." We get that $x = \frac{1}{2}y - \frac{3}{2}$. Thus $g(y) = \frac{1}{2}y - \frac{3}{2}$ is the function inverse to f. Let us verify that $g(f(x)) = x$. We have

$$g(f(x)) = \tfrac{1}{2}f(x) - \tfrac{3}{2} = \tfrac{1}{2}(2x + 3) - \tfrac{3}{2} = x, \qquad \text{for all } x.$$

2. The function $x \mapsto x^3$ is continuous and increasing for all x. The inverse function is obtained by solving the equation $y = x^3$ for x. Thus $x = \sqrt[3]{y}$. The inverse function is $g(y) = y^{1/3}$ or $y \mapsto y^{1/3}$.

3. The function $f(x) = x^2$, considered for $x \ge 0$ only, is continuous and increasing, and it takes on nonnegative values. The inverse function $g(y)$ is defined as the nonnegative number whose square is equal to y. In other words, $g(y) = \sqrt{y}$.

4. $x \mapsto x^2$ is a decreasing continuous function in the interval $(-\infty, 0]$. The inverse function is now $y \mapsto -\sqrt{y}$.

Theorem 1 (the inverse function theorem). *Let f be a continuous strictly monotone (increasing or decreasing) function defined in the closed interval $[a,b]$. Set $\alpha = f(a)$, $\beta = f(b)$. Then (1) there exists a strictly monotone (increasing or decreasing) function g, called inverse to f, defined in the closed interval with endpoints α and β, such that $g(f(x)) = x$, $f(g(y)) = y$, and (2) this inverse function is continuous.*

Statement (1) follows from what was said at the beginning of the subsection. Statement (2) can be explained geometrically. Note that the graph of g is obtained from the graph of f by interchanging the x- and y-axes, that is, by reflecting ("**flipping**") the graph about the bisector of the first quadrant (see Figure 6.3). The graph of f consists of one piece (has no breaks) since f is continuous. This property of the graph

FIGURE 6.3

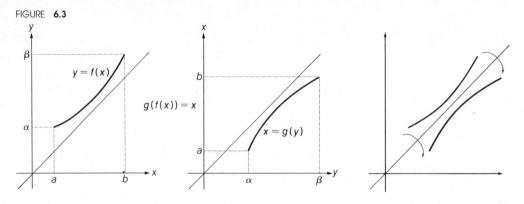

is not destroyed by "flipping." Hence the graph of g consists of one piece, so that g is continuous. [There is also an analytic proof ► of statement (2).]

Let us apply the inverse function theorem to the function $y = x^2$. If c is any positive number, this function is continuous and increasing for $0 \le x \le c$. The inverse function $x = \sqrt{y} = y^{1/2}$ is therefore defined and continuous for $0 \le y \le c^2$. Since we can take c as large as we want, we conclude that $x = \sqrt{y} = y^{1/2}$ is defined and continuous for $0 \le y < +\infty$.

The function $y = x^3$ is increasing and continuous for $-c \le x \le c$, c being any positive number. The inverse function $x = y^{1/3}$ is therefore defined and continuous for $-c^3 \le y \le c^3$. Since c is arbitrary, the function $x = y^{1/3}$ is defined for $-\infty < y < +\infty$. It has limits $-\infty$ and $+\infty$ at $-\infty$ and $+\infty$, respectively.

More generally, if n is any positive integer, the function $y = x^n$ increases for $x \ge 0$, since the derivative $nx^{n-1} > 0$ for $x > 0$. If n is odd, $y = x^n$ increases for all x, since $n - 1$ is even and the derivative $nx^{n-1} > 0$ for $x \ne 0$. The inverse function theorem tells us that the function $x = y^{1/n}$ is defined and continuous for $x \ge 0$ if n is even, for all x if n is odd. (This verifies a statement made in Chapter 1, §2.8.)

For use in later chapters we note the following.

Corollary.　*Let f be a continuous strictly monotone function defined in a (finite or infinite) open interval (a,b), and assume that f has the one-sided (finite or infinite) limits α and β at a and b. Then f has a continuous strictly monotone inverse function g, defined in the open interval with endpoints α and β, and g has one-sided limits a and b at α and β.*

This follows by the reasoning used above for $f(x) = x^2$ (here $a = 0$, $b = +\infty$, $\alpha = 0$, $\beta = +\infty$) and for $f(x) = x^3$ (here $a = -\infty$, $b = +\infty$, $\alpha = -\infty$, $\beta = +\infty$).

PROBLEMS

Each of the functions in Problems 1 to 9 has an inverse on the domain indicated. Find the inverse function.

1. $f(x) = \frac{1}{2}x + 5$,　all x.
   [*Answer:* Let $y = f(x)$.
   Then $x = 2y - 10$.]
2. $f(x) = x^4$,　$x \ge 0$.

3. $f(x) = \sqrt{x}$,　$x \ge 0$.
4. $f(x) = x^{1/5}$,　all x.
5. $y = (x + 1)/(x - 1)$,　$x > 1$.
6. $f(t) = (3t + 1)/t$,　$t > 0$.

7. $g(y) = (1 - y^3)/y^3$, $y > 1$.
8. $h(x) = (1 + \sqrt{x})^3$, $x > 0$.
9. $f(z) = (1 - z^3)^{-1}$, $z > 1$.

In Problems 10 to 15 decide if the given function has an inverse (in the indicated interval). If it does, find it.

10. $x \mapsto x^4$,   all x.
11. $x \mapsto x^4$, $x \geq 0$.
12. $x \mapsto x^4$, $x \leq 0$.

13. $x \mapsto x^2 + x - 1$,   all x.
14. $x \mapsto x^2 + x - 1$, $x \geq -\frac{1}{2}$.
15. $x \mapsto x^2 + x - 1$, $x \leq -\frac{1}{2}$.

In Problems 16 to 21 decide if the function whose graph is given has an inverse, and if so, graph it (make a rough sketch only).

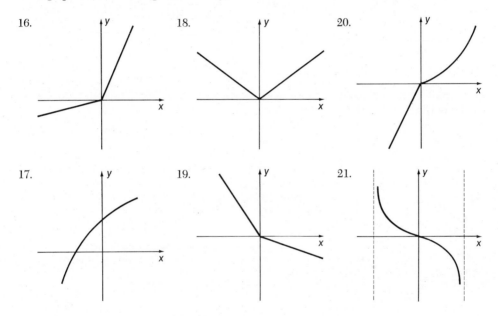

16.

17.

18.

19.

20.

21.

1.2   Derivatives of inverse functions

Theorem 2. *Let $y = f(x)$ and $x = g(y)$ be strictly monotone functions inverse to each other. Assume that f has at x_0 the derivative $f'(x_0) \neq 0$, and set $y_0 = f(x_0)$. Then g has at y_0 the derivative*

$$g'(y_0) = \frac{1}{f'(x_0)}.$$

This rule looks more elegant, and is easier to remember, in the Leibniz notation:

$$\frac{dx}{dy} = \frac{1}{dy/dx}.$$

(If dx and dy were numbers, this would be a true statement about fractions.)

Theorem 2 can be proved analytically ▶ or verified geometrically. The geometric argument follows. Let l be the tangent to the graph of $x \mapsto y = f(x)$ at the point (x_0, y_0).

FIGURE 6.4

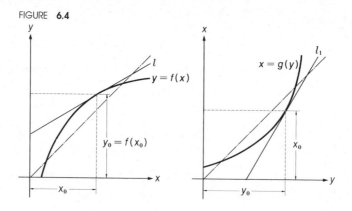

Then l is the graph of a linear function

$$x \mapsto y = mx + b \qquad \text{where} \qquad m = f'(x_0).$$

The graph of the inverse function $y \mapsto x = g(y)$ is obtained by "flipping," that is, by reflecting the graph of f about the line $y = x$ (see Figure 6.4). If we also reflect l, we obtain a line l_1 which is tangent to the graph of g at (y_0, x_0). Now l_1 is the graph of the linear function inverse to $y = mx + b$, that is, of the function

$$x = \frac{1}{m} y - \frac{b}{m}.$$

The slope of l_1 is the derivative $g'(y_0)$. This slope is $1/m$. Therefore $g'(y_0) = 1/m = 1/f'(x_0)$, as asserted.

[If $f'(x_0) = 0$, then l is horizontal and l_1 vertical, and g has no derivatives at y_0.]

EXAMPLE   Find the derivative of the function $x = g(y)$ inverse to the function $y = f(x) = x + x^5$ at $y = 2$, $x = 1$.

ANSWER   Since $f'(x) = 1 + 5x^4 > 0$ for all x, $f(x)$ is increasing. Thus there is an inverse function, $x = g(y)$, although we have no simple formula for it. By Theorem 2,

$$g'(2) = \left(\frac{dx}{dy}\right)_{y=2} = \frac{1}{(dy/dx)_{x=1}} = \frac{1}{(1 + 5x^4)_{x=1}} = \frac{1}{6}.$$

Or, briefer, $g'(2) = 1/f'(1) = \frac{1}{6}$.

PROBLEMS

In Problems 1 to 9 compute the derivatives using the rules that have been proved so far.

1. Let $y = f(x)$ and $x = g(y)$ be strictly monotone functions, inverse to each other. If $f(0) = 3$ and $f'(0) = -\frac{1}{4}$, find $g'(3)$.

2. Let $x \mapsto h(x)$ and $x \mapsto k(x)$ be strictly monotone functions, inverse to each other. If $h(-1) = 2$, $h'(-1) = \frac{1}{3}$, find $k'(2)$. [*Note:* The concept of inverse functions does not depend on what symbols, if any, are used to denote the dependent and independent variables in the inverse function.]

3. Let ϕ and ψ be strictly monotone functions, inverse to each other. If $\phi(0) = 1$ and $\phi'(0) = -\frac{2}{3}$, find $\psi'(1)$.

4. The function $f(x) = x^3 + 2x$ is increasing. If g is the function inverse to f, find $g'(0)$ and $g'(3)$.

5. The function $y = 3 - x - 2x^3$ is decreasing. Find $(dx/dy)_{y=3}$ and $(dx/dy)_{y=0}$.

6. The function $s \mapsto s^5 + 3s^3 - 1$ is increasing. If f is the inverse function, find $f'(-1)$ and $f'(3)$.

7. (a) The function $h(z) = z^2 - z + 1$ is increasing for $z > \frac{1}{2}$. If k is the function inverse to h in this interval, find $k'(3)$ and $k'(7)$.

   (b) The function $h(z) = z^2 - z + 1$ is decreasing for $z < \frac{1}{2}$. If l is the function inverse to h in this interval, find $l'(3)$ and $l'(7)$.

8. (a) The function $y = f(x) = x^4 + 1$ is increasing for $x > 0$. If $x = g(y)$ is the function inverse to f in this interval, find $(dg/dy)_{y=2}$ and $(dg/dy)_{y=82}$.

   (b) The function $y = f(x) = x^4 + 1$ is decreasing for $x < 0$. If $x = h(y)$ is the function inverse to f in this interval, find $(dh/dy)_{y=2}$ and $(dh/dy)_{y=82}$.

   [*Note:* The pure Leibniz notation dx/dy is not really adequate in this situation, because it does not specify whether dg/dy or dh/dy is intended.]

9. (a) The function $f(z) = z^3 - 9z$ is increasing for $z < -\sqrt{3}$. If g is the function inverse to f in this interval, find $g'(0)$.

   (b) The function $f(z) = z^3 - 9z$ is decreasing for $-\sqrt{3} < z < \sqrt{3}$. If h is the function inverse to f in this interval, find $h'(0)$.

   (c) The function $f(z) = z^3 - 9z$ is increasing for $z > \sqrt{3}$. If k is the function inverse to f in this interval, find $k'(0)$.

1.3 Derivatives of roots

Let us apply Theorem 2 to the case $y = f(x) = x^n$, n a positive integer. We have that $f'(x) = nx^{n-1}$, hence $f'(x) \neq 0$ for $x \neq 0$. The inverse function $x = \sqrt[n]{y} = y^{1/n}$ is defined for all y if n is odd, for all $y \geq 0$ if n is even. According to Theorem 2, this inverse function is differentiable for $y \neq 0$, and we have

$$\frac{d\sqrt[n]{y}}{dy} = \frac{dx}{dy} = \frac{1}{dy/dx} = 1 \bigg/ \frac{dx^n}{dx} = \frac{1}{nx^{n-1}}.$$

But $nx^{n-1} = n(\sqrt[n]{y})^{n-1} = n(y^{1/n})^{n-1} = ny^{1-(1/n)} = ny/\sqrt[n]{y}$. Therefore

$$\frac{d\sqrt[n]{y}}{dy} = \frac{\sqrt[n]{y}}{ny} \qquad \text{or} \qquad \frac{dy^{1/n}}{dy} = \frac{1}{n}y^{(1/n)-1}.$$

The name of a variable is of no importance. Writing now x instead of y, we obtain the rule

$$\frac{d\sqrt[n]{x}}{dx} = \frac{\sqrt[n]{x}}{nx} \qquad \text{or} \qquad \frac{dx^{1/n}}{dx} = \frac{1}{n}x^{(1/n)-1}. \tag{1}$$

The case $n = 2$ deserves to be noted:

$$\frac{d\sqrt{x}}{dx} = \frac{1}{2\sqrt{x}} \qquad \text{or} \qquad \frac{dx^{1/2}}{dx} = \frac{1}{2}x^{-1/2}. \tag{2}$$

A consequence of (1) is the rule

$$\frac{d}{dx}\frac{1}{\sqrt[n]{x}} = -\frac{1}{nx\sqrt[n]{x}} \qquad \text{or} \qquad \frac{dx^{-1/n}}{dx} = -\frac{1}{n}x^{(-1/n)-1}. \tag{3}$$

To prove this we combine the reciprocal rule (6') from Chapter 4, §2.3, with rule (1) and obtain

$$\frac{dx^{-1/n}}{dx} = \frac{d}{dx}\frac{1}{x^{1/n}} = -\frac{1}{(x^{1/n})^2}\frac{dx^{1/n}}{dx} = -\frac{1}{x^{2/n}}\frac{1}{n}x^{(1/n)-1} = -\frac{1}{n}x^{(-1/n)-1},$$

as asserted.

The rules (1), (2), (3) are valid for those x for which the formulas make sense. It is easiest to use fractional exponents and to remember only the formula

$$\frac{dx^r}{dx} = rx^{r-1}. \tag{4}$$

For the moment, we proved this when $r = n$ or $1/n$, n an integer. In §2.4 we shall see that the formula is valid for any rational number r.

EXAMPLES 1. Differentiate $f(x) = \sqrt[3]{x} + 2\sqrt[5]{x}$.
ANSWER   First rewrite the given expression with fractional exponents: $f(x) = x^{1/3} + 2x^{1/5}$. Using the rules from Chapter 4, §2.1, and rule (4) above, we have

$$f'(x) = \frac{d}{dx}x^{1/3} + 2\frac{d}{dx}x^{1/5} = \frac{1}{3}x^{-2/3} + \frac{2}{5}x^{-4/5}.$$

Other forms are possible, for instance,

$$f'(x) = \frac{\sqrt[3]{x}}{3x} + \frac{2\sqrt[5]{x}}{5x}.$$

2. Find the slope of the tangent to the curve $y = 3x^{1/3}$ at the point $x = 8$, $y = 6$.
ANSWER   The derivative dy/dx is $x^{(1/3)-1} = x^{-2/3}$. For $x = 8$, this derivative is $8^{-2/3} = (\sqrt[3]{8})^{-2} = 2^{-2} = \frac{1}{4}$. The desired tangent has slope $\frac{1}{4}$ and passes through the point (8,6). By the point-slope formula, the equation of the tangent is $y - 6 = \frac{1}{4}(x - 8)$ or $x - 4y + 16 = 0$.

PROBLEMS

Differentiate the following functions.

1. $y = 3\sqrt{x}$.

2. $y = \dfrac{1}{\sqrt{x}}$.

3. $y = \sqrt[3]{x}$.

4. $y = \frac{1}{3}x^{-1/3}$.

5. $y = 6\sqrt[8]{x}$.

6. $f(x) = 3x^{1/3} + 5x^{1/5}$.

7. $f(x) = \frac{1}{2}\left(\sqrt{x} - \dfrac{1}{\sqrt{x}}\right)$.

8. $f(x) = 4(x^{1/4} + x^{-1/4})$.
9. $f(x) = 3x^{-1/3} + 2x^{-1/6} + 1$.
10. $f(x) = x^{1/2} + 2x^{1/4} + 4x^{1/8}$.

11. $g(t) = \sqrt[3]{t} + 3t + \dfrac{1}{\sqrt[3]{t}}$.

12. $g(t) = 4t^{-1/8} - 2t^{-1/4} + t^{-1/2}$.
13. $g(t) = \sqrt[10]{t} + 5\sqrt[5]{t} + \sqrt{t}$.
14. $g(t) = t^{1/12} + t^{1/6} + t^{-1/6} + t^{-1/12}$.

15. Find the equation of the tangent to the curve $y = \dfrac{2}{\sqrt{x}} + 1$ at the point (4,2).

16. Find the equation of the tangent to the curve $y = \sqrt[3]{x}$ at the point $(-27, -3)$.
17. Find the equation of the tangent to the curve $y = 4\sqrt{x} + x^2$ at the point (4,24).
18. Find the equation of the tangent and normal to the curve $y = x^2 + x + x^{-1/2} + x^{-1/4} - 4$ at the point (1,0).

§2 Composite functions. Radical functions

In this section we consider functions such as $\sqrt{1 + x^2}$ or $\sqrt{x}/(1 - \sqrt{x})$ which are "built up" from other simpler functions.

2.1 Composition of functions

If f and g are two functions, then the composite function, $g \circ f$, is the function

$$x \mapsto g(f(x)).$$

It is defined for numbers x such that $f(x)$ is defined and the number $f(x)$ is in the domain of definition of g. The function $g \circ f$ is the rule: apply first f, then g. It is, in general, distinct from the function $f \circ g$ (apply first g, then f.)

EXAMPLES   1.   Let $f(x) = 1 + x^2$ and $g(y) = 1 - y$. Then $g(f(x)) = 1 - f(x) = 1 - (1 + x^2) = -x^2$. Also, $f(g(y)) = 1 + g(y)^2 = 1 + (1 - y)^2 = 2 - 2y + y^2$. In particular, $g(f(1)) = -1^2 = -1$, $f(g(1)) = 2 - 2 + 1^2 = 1$.

2.   Let $g(x) = x + 1$ and $f(x) = \sqrt{x - 1}$. Then $g \circ f$ is the function $x \mapsto g(f(x)) = f(x) + 1 = \sqrt{x - 1} + 1$. It is defined for $x \geq 1$, not defined for $x < 1$. The function $f \circ g$ is the function $x \mapsto \sqrt{g(x) - 1}$. Since $g(x) = x + 1$, we see that $f \circ g$ is the function $x \mapsto \sqrt{x}$ and is defined for all $x \geq 0$, not defined for $x < 0$. In particular, $g(f(5)) = 3$ and $f(g(5)) = \sqrt{5}$.

Theorem 1.   *Let the function f be continuous at x_0, and set $y_0 = f(x_0)$. Let g be a function continuous at y_0. Then the composite function $g \circ f$ is continuous at x_0.*

Proof.   Let f be a function continuous at x_0, and set $f(x_0) = y_0$. Let g be a function continuous at y_0, and denote the composed function $g \circ f$ by ϕ, so that $\phi(x) = g(f(x))$. We prove that ϕ is continuous at x_0.

Let $\epsilon > 0$ be given. We must exhibit a $\delta > 0$ such that

$$\text{if } |x - x_0| < \delta, \text{ then } |\phi(x) - \phi(x_0)| < \epsilon. \tag{1}$$

Since g is continuous at y_0, there is a $\delta_0 > 0$ such that

$$\text{if } |y - y_0| < \delta_0, \text{ then } |g(y) - g(y_0)| < \epsilon. \tag{2}$$

Since f is continuous at x_0, there is a $\delta > 0$ such that

$$\text{if } |x - x_0| < \delta, \text{ then } |f(x) - f(x_0)| < \delta_0.$$

This δ has the required property (1). Indeed, assume that x is such that $|x - x_0| < \delta$, and set $y = f(x)$. Then $|y - y_0| < \delta_0$, so that $|g(y) - g(y_0)| < \epsilon$, by (2). But $g(y) = \phi(x)$ and $g(y_0) = \phi(x_0)$. Hence $|\phi(x) - \phi(x_0)| < \epsilon$.

Theorem 2   (limits of composite functions).   *If $\lim_{x \to x_0} f(x) = y_0$, $g(y)$ is continuous at y_0, and $g(y_0) = \alpha$, then*

$$\lim_{x \to x_0} g(f(x)) = \alpha,$$

that is,

$$\lim_{x \to x_0} g(f(x)) = g(\lim_{x \to x_0} f(x)).$$

This theorem follows from the definition of limits in terms of continuity and from Theorem 1, which was just proved. One can also prove Theorem 2 using the epsilon-delta definition of limits; that proof would be similar to the proof of Theorem 1. (There is a corresponding theorem about one-sided limits.)

EXAMPLES 3.   What is $\lim_{x \to 3} \sqrt{1 + x^2}$?

ANSWER The function $x \mapsto \sqrt{1 + x^2}$ is continuous since it is the composition of the function $f(x) = 1 + x^2$, which is continuous and positive everywhere, and the function $g(y) = \sqrt{y}$, which is continuous for $y \geq 0$. Therefore the desired limit is $\sqrt{1 + 3^2} = \sqrt{10}$. In other words,

$$\lim_{x \to 3} \sqrt{1 + x^2} = \sqrt{\lim_{x \to 3} (1 + x^2)} = \sqrt{10}.$$

4.   If $\lim_{x \to 2} f(x) = 27$, what is $\lim_{x \to 2} f(x)^{1/3}$?

ANSWER Let $y = f(x)$, $g(y) = y^{1/3}$, and apply Theorem 2. Then

$$\lim_{x \to 2} g(f(x)) = g(\lim_{x \to 2} f(x)) = g(27) = (27)^{1/3} = 3.$$

PROBLEMS

1. Let $f(x) = 3x - 1$, $g(x) = -4x + 3$. What is the function denoted by $f \circ g$? By $g \circ f$?
2. Let $f(x) = x^2 + 3$, $g(x) = 2x + 1$. What is the function denoted by $f \circ g$? By $g \circ f$? By $f \circ f$?
3. Suppose $g(x) = x^2 + 1$, $h(x) = 1/x$. What is the function denoted by $h \circ g$? What value does $h \circ g$ assign to the number -2? What is $h(g(3))$?
4. Suppose $f(x) = x^2 - 5$, $g(x) = x + 5$. What value does $f \circ g$ assign to the number -1? What value does $g \circ f$ assign to the number 10?
5. Let $f(x) = 3x + 1$, $g(x) = x^4 + 2x^2 + 3$. Find the values of $f(g(2))$, $g(f(-1))$, and $g(g(0))$.
6. Let $f(x) = 2x + 2$, $g(x) = x^2 + 2x + 70$. Find those values of x for which $f \circ g = g \circ f$.
7. Let $f(x) = \sqrt{x^2 + 4}$, $g(x) = 2x + 3$. What is the function denoted by $f \circ g$? By $g \circ f$? By $f(f(x))$?
8. Suppose $\alpha(t) = \sqrt{t^2 - 5}$, $\beta(t) = \sqrt{t^2 + 5}$. Find the values of $\alpha \circ \beta$ and $\beta \circ \alpha$ when $t = 4$. Find the values of $\alpha(\alpha(-5))$ and $\beta(\beta(2))$.
9. What is the domain of definition of $f \circ g$ if $f(x) = 1/\sqrt{1 - x^2}$ and $g(x) = 1 - 2x$?
10. What is the domain of definition of $A \circ B$ if $A(s) = \sqrt{s^2 - 4}$ and $B(t) = \sqrt{t - 1}$?

Evaluate the following limits.

11. $\lim_{x \to 1} \sqrt{3x + 1}$.

12. $\lim_{x \to -1} \sqrt{x^2 - 4x + 3}$.

13. $\lim_{x \to 3} \sqrt[3]{x^3 + 3x - 9}$.

14. $\lim_{x \to \frac{1}{2}} \sqrt[3]{x^3 + x^2 + x + \frac{1}{8}}$.

15. $\lim_{x \to 2} \sqrt{\dfrac{x - 1}{2x + 5}}$.

16. $\lim_{x \to 1} \sqrt{\dfrac{2x^2 + 5x + 3}{x^2 + 3x + 1}}$.

17. $\lim_{x \to -2} \sqrt{\dfrac{x^4 - 2x^2 - 1}{x^4 + x^2 + 1}}$.

18. $\lim_{x \to 1} \sqrt{(x^2 + 3x + 2)(x^3 + 4x^2 + x)}$.

19. $\lim_{x \to 4} \sqrt{2 + \sqrt{x}}$. [*Hint:* Use Theorem 2 twice.]

2.2 Radical functions

A rational function may be defined as a function obtained by applying the **rational operations** (addition, multiplication, subtraction, division) to the constants and the identity function $x \mapsto x$. A **radical function** is a function obtained by applying the rational operations *and composition* to the constants and the functions $x \mapsto x^{1/n}$, n a positive integer. In other words, a radical function is one that is defined by a formula involving the symbol (say, x) for the independent variable, numbers, the signs $+$, $-$, $.$, $/$, and rational exponents. The domain of definition of such a function is an interval or several intervals.

EXAMPLE   The function

$$f(x) = \frac{2}{\sqrt{x}} + \sqrt[4]{1 - \sqrt{x}} = 2x^{-1/2} + (1 - x^{1/2})^{1/4}$$

is a radical function. It is defined if $x \geq 0$, and if $x^{1/2} > 0$, and if $1 - x^{1/2} \geq 0$, that is, for $0 < x \leq 1$.

Theorem 3. *A radical function is continuous at all points at which it is defined.*

Proof. As we noted in §1.1, the functions $x^{1/2}$, $x^{1/4}$, $x^{1/6}$, $x^{1/8}$, ..., defined for $x \geq 0$, are inverse to the continuous functions x^2, x^4, x^6, x^8, ... and hence continuous (by the inverse function theorem). The functions $x^{1/3}$, $x^{1/5}$, $x^{1/7}$, ..., defined for all x, are inverse to the continuous functions x^3, x^5, x^7, ... and hence continuous.

Next, $x^{p/q}$, where p and q are integers, $q > 0$, is the composition of two continuous functions, $x^{1/q}$ and x^p [for $x^{p/q} = (x^{1/q})^p$]. Hence this function is continuous, whenever defined (by Theorem 1 above).

Finally, a radical function is continuous, whenever defined, since it is obtained from continuous functions (rational powers) by rational operations and composition, processes that lead from continuous functions to continuous functions.

2.3 The chain rule

We come now to the chain rule for differentiating composite functions. This is one of the most important tools for differentiating.

Theorem 4 (chain rule).   *If the function $u = f(x)$ has at x_0 the derivative $f'(x_0)$, and the function $y = g(u)$ has at $u_0 = f(x_0)$ the derivative $g'(f(x_0))$, then the composed function $\phi = g \circ f$, that is, the function $y = g(f(x))$, has at x_0 the derivative*

$$\phi'(x_0) = g'(f(x_0))f'(x_0). \tag{3}$$

(The derivative of a composition is the product of the derivatives.)

The formulation of this theorem in the Leibniz notation is more elegant:

$$\frac{dy}{dx} = \frac{dy}{du}\frac{du}{dx}. \tag{3'}$$

We are tempted to say "simply cancel du," but this makes no sense, since dy/du and du/dx are not fractions and du is not a number. But formula (3') is easy to remember and convenient to use.

Proof of the chain rule. For the sake of brevity, we assume that $x_0 = 0$, $g(0) = f(0) = 0$, and we set $g'(0) = a$, $f'(0) = b$. We must show that the function $\phi(x) = g[f(x)]$ has, at $x = 0$, the derivative $\phi'(0) = g'[f(0)]f'(0) = g'(0)f'(0) = ab$.

By the linear approximation theorem (Theorem 1 in Chapter 4, §1.6),

$$g(y) = ay + yr(y), \qquad f(x) = bx + xs(x),$$

where the functions r and s are continuous at 0 and $r(0) = s(0) = 0$. In the first equation above, set $y = f(x) = bx + xs(x)$. We obtain

$$\phi(x) = g[f(x)] = a[bx + xs(x)] + [bx + xs(x)]r[f(x)] = abx + xR(x),$$

$$R(x) = as(x) + [b + s(x)]r[f(x)].$$

This R is continuous at 0, being obtained from continuous functions a, b, $s(x)$, $r(x)$, and $f(x)$ by composition, multiplication, and addition. Note that $R(0) = as(0) + [b + s(0)]r(0) = 0$. Hence $\phi'(0) = ab$ by the linear approximation theorem.

The proof for the case when either $x_0 \neq 0$, $f(x_0) \neq 0$, or $g[f(x_0)] \neq 0$ is the same, only the formulas are longer.

Facility in using the chain rule is *very important*.

EXAMPLES 1. Differentiate the function $\phi(x) = (1 + x^3)^2$ using the chain rule.
ANSWER We remark first that here we do not need the chain rule. Since $\phi(x) = (1 + x^3)^2 = 1 + 2x^3 + x^6$, the derivative is $\phi'(x) = 6x^2 + 6x^5$. In order to apply the chain rule, we must write ϕ as $g \circ f$ where f and g are functions which we know how to differentiate. We set $f(x) = 1 + x^3$ and $g(u) = u^2$. Then $\phi(x) = g(f(x)) = (1 + x^3)^2$. Since $f'(x) = 3x^2$ and $g'(u) = 2u$, we have, by (3),

$$\phi'(x) = g'(f(x))f'(x) = (2u)3x^2 = 2(1 + x^3)3x^2 = 6x^2 + 6x^5,$$

as expected.

Now we carry out the same calculation using the Leibniz notation and formula (3′). We must differentiate $y = \phi(x) = (1 + x^3)^2$. We begin by identifying the "parts" from which our function is composed:

$$y = u^2 \qquad \text{where} \qquad u = 1 + x^3.$$

Therefore, by (3′),

$$\phi'(x) = \frac{dy}{dx} = \frac{dy}{du}\frac{du}{dx} = (2u)(3x^2) = 2(1 + x^3)3x^2 = 6x^2 + 6x^5.$$

2. Noting that we know the derivative of $x \mapsto \sqrt{x}$ (see §1.3), differentiate $y = \sqrt{1 - x^2}$.
ANSWER We have

$$y = \sqrt{u} \qquad \text{where} \qquad u = 1 - x^2.$$

Using rule (3′) in §1.3 and the chain rule, we have

$$\frac{dy}{dx} = \frac{dy}{du}\frac{du}{dx} = \frac{1}{2\sqrt{u}}(-2x) = \frac{1}{2\sqrt{1 - x^2}}(-2x) = -\frac{x}{\sqrt{1 - x^2}}.$$

In other words,

$$\frac{d\sqrt{1 - x^2}}{dx} = -\frac{x}{\sqrt{1 - x^2}}.$$

3. Differentiate $f(x) = x^2\sqrt{1 - x^2}$.
ANSWER First we use the product rule (see Chapter 4, §2.2) and write: $f = uv$ where $u = x^2$, $v = \sqrt{1 - x^2}$. Hence

$$f'(x) = \frac{du}{dx}v + u\frac{dv}{dx}. \tag{*}$$

We know that $du/dx = 2x$, and from Example 2 we know that

$$\frac{dv}{dx} = -\frac{x}{\sqrt{1-x^2}}.$$

Substituting the values of du/dx and dv/dx into (*), we obtain

$$f'(x) = 2x\sqrt{1-x^2} + x^2\left(-\frac{x}{\sqrt{1-x^2}}\right) = 2x\sqrt{1-x^2} - \frac{x^3}{\sqrt{1-x^2}}.$$

It is optional whether to simplify. If we do, we get

$$f'(x) = \frac{2x(1-x^2)}{\sqrt{1-x^2}} - \frac{x^3}{\sqrt{1-x^2}} = \frac{2x - 3x^3}{\sqrt{1-x^2}}.$$

4. Differentiate the function

$$y = \sqrt{x + \sqrt{1+x}} = [x + (1+x)^{1/2}]^{1/2}.$$

ANSWER   We try to apply the chain rule and write

$$y = u^{1/2}, \qquad u = x + \sqrt{1+x}$$

so that

$$\frac{dy}{dx} = \frac{dy}{du}\frac{du}{dx} = \frac{1}{2}u^{-1/2}\frac{du}{dx} = \frac{1}{2\sqrt{x + \sqrt{1+x}}}\frac{du}{dx}. \qquad (**)$$

Now we must compute du/dx. This requires that we break up u into its "parts." Thus we write

$$u = x + v, \qquad v = z^{1/2}, \qquad z = 1 + x$$

so that

$$\frac{du}{dx} = 1 + \frac{dv}{dx} = 1 + \frac{dv}{dz}\frac{dz}{dx} = 1 + \frac{1}{2}z^{-1/2}\cdot 1 = 1 + \frac{1}{2\sqrt{1+x}}.$$

Substitution into (**) yields the answer

$$\frac{dy}{dx} = \frac{1}{2\sqrt{x + \sqrt{1+x}}}\left[1 + \frac{1}{2\sqrt{1+x}}\right].$$

5. Compute $f'(1)$ where $f(x) = \sqrt{1 + \sqrt{2 + \sqrt{3 + x^2}}}$.

ANSWER   Here we must use the chain rule repeatedly. We have

$$f = \sqrt{u}, \qquad \text{where } u = 1 + \sqrt{v}, \quad v = 2 + \sqrt{z}, \quad \text{and} \quad z = 3 + x^2$$

so that

$$\frac{df}{dx} = \frac{df}{du}\frac{du}{dv}\frac{dv}{dz}\frac{dz}{dx} = \frac{1}{2\sqrt{u}}\frac{1}{2\sqrt{v}}\frac{1}{2\sqrt{z}}\cdot 2x = \frac{1}{4}\frac{x}{\sqrt{u}\sqrt{v}\sqrt{z}}.$$

Since we need only the value of $f'(x)$ at $x = 1$, there is no need to simplify. For $x = 1$, we have $z = 4$, $\sqrt{z} = 2$, $v = 4$, $\sqrt{v} = 2$, and $u = 3$. Thus $f'(1) = 1/(16\sqrt{3})$.

PROBLEMS

Find the derivatives of the following functions:

1. $y = (1 - 3x^2)^3$.
2. $y = (2x^4 + 1)^6$.
3. $y = (x^2 + 3x - 1)^4$.

4. $f(x) = (x^6 - 3x^4 + 2x^2 - 4)^{12}$.
5. $f(x) = (x^5 - 3x^4 + x^3 - 3x^2 + x)^{100}$.
6. $g(x) = \frac{1}{3}(x^3 - 6x + 2)^{-2}$.

7. $g(x) = (x^4 - 2x^2 + 2x^{-2} - x^{-4})^{-6}$.

8. $\phi(x) = \sqrt{1 - 2x}$.

9. $\phi(x) = \sqrt{1 + 3x^2}$.

10. $\phi(x) = \sqrt{1 + x - 2x^2 + 3x^3 - 4x^4}$.

11. $F(t) = \sqrt{t + t^{-1}}$.

12. $F(t) = \sqrt{2t + t^2 - 2t^{-1} + t^{-2}}$.

13. $F(t) = \dfrac{1}{\sqrt{4 - t^2}}$.

14. $F(t) = (1 - 3t + 5t^3)^{-1/3}$.

15. $y = (1 + 4x)^3(1 - 3x)^4$.

16. $y = (2x^2 + 1)^4(4x + 3)^6$.

17. $y = (3x^2 + 1)\sqrt{x^2 - 4}$.

18. $y = (2x + 1)^3\sqrt{1 - x^2}$.

19. $f(x) = \dfrac{(2x - 1)^2}{(2x + 1)^2}$.

20. $f(x) = \dfrac{(x^2 + 1)^3}{(2x^2 + 3)^4}$.

21. $f(x) = \dfrac{x^2}{\sqrt{1 + 2x}}$.

22. $f(x) = \dfrac{1 - 3x^2}{\sqrt{x^2 - 4x + 1}}$.

23. $\psi(z) = \sqrt{1 + \sqrt{1 - 4z}}$.

24. $\psi(z) = \sqrt{z^2 + \sqrt{1 + 2z^2}}$.

25. $\psi(z) = \dfrac{1 - \sqrt{1 + z}}{1 + \sqrt{1 + z}}$.

26. $\psi(z) = \sqrt{1 + \sqrt{2 + \sqrt{3 + z}}}$.

2.4 Differentiating radical functions

Theorem 5. *Let r be a rational number; then*

$$\frac{dx^r}{dx} = rx^{r-1}$$

(for all x for which x^{r-1} and x^r are defined).

Proof. If r is an integer, this is Theorem 6 of Chapter 4, §2.3. We already proved the result for the case in which $r = 1/n$, n being a positive integer in §1.3.

Assume now that $r = m/n$, where m is an integer and n a positive integer. Set $y = x^r$. Then $y = u^m$ where $u = x^{1/n}$. By rule (1) in §1.3, and Theorem 4, we have

$$\frac{dx^r}{dx} = \frac{dy}{dx} = \frac{dy}{du}\frac{du}{dx} = mu^{m-1}\frac{1}{n}x^{(1/n)-1} = m(x^{1/n})^{m-1}\frac{1}{n}x^{(1/n)-1}$$

$$= \frac{m}{n}x^{(m/n)-(1/n)+(1/n)-1} = rx^{r-1}.$$

Combining Theorem 5 with the chain rule, we obtain the useful *generalized power formula*

$$\frac{d}{dx}[f(x)]^r = rf(x)^{r-1}f'(x)$$

or

$$\frac{dy^r}{dx} = ry^{r-1}\frac{dy}{dx},$$

which holds for all rational numbers r and differentiable functions $y = f(x)$ (for which y^r and y^{r-1} are defined).

The differentiation rules proved thus far suffice to differentiate any radical function and lead to the following statement.

Theorem 6. *A radical function has a derivative wherever it is defined, except perhaps at finitely many points. The derivative is again a radical function.*

EXAMPLES   1.   The function $f(x) = x^{2/3}$ is defined and continuous for all x. By Theorem 6,

$$f'(x) = \tfrac{2}{3}x^{-1/3}.$$

This makes sense only for $x \neq 0$. At $x = 0$, f has no derivatives. (We shall discuss this in more detail in §5.)

2.   Differentiate $x \mapsto (3x^5 - 7x^2 + 4)^{7/9}$.
ANSWER   By the generalized power rule, the derivative is

$$\tfrac{7}{9}(3x^5 - 7x^2 + 4)^{-2/9}(15x^4 - 14x).$$

This is valid for $3x^5 - 7x^2 + 4 \neq 0$.

3.   Find the tangent to the graph of $y = (1 + x^{3/2})^{3/2}$ at $x = 4$.
ANSWER   If $x = 4$, then $y = (1 + 4^{3/2})^{3/2} = (1 + 8)^{3/2} = 27$. Thus the point at which we want to draw the tangent is $(4,27)$. Next, we find the slope of the tangent, that is, the derivative

$$\frac{dy}{dx} = \frac{d(1 + x^{3/2})^{3/2}}{dx} = \tfrac{3}{2}(1 + x^{3/2})^{1/2}\frac{d(1 + x^{3/2})}{dx} = \tfrac{3}{2}(1 + x^{3/2})^{1/2}(\tfrac{3}{2}x^{1/2}).$$

For $x = 4$, $dy/dx = \tfrac{9}{4}(1 + 8)^{1/2}\,2 = 27/2$. Thus the desired tangent line has slope $27/2$ and passes through the point $(4,27)$. By the slope-point formula, the equation of the tangent reads $y - 27 = (27/2)(x - 4)$ or $27x - 2y - 54 = 0$.

4.   What is the shortest distance from the point $(2,\tfrac{1}{2})$ to the parabola $y = x^2$?
ANSWER   The distance between the point $(2,\tfrac{1}{2})$ and a point (x,y) on the parabola is $s = [(x - 2)^2 + (y - \tfrac{1}{2})^2]^{1/2}$. Since $y = x^2$, the distance to be minimized becomes $s = [(x - 2)^2 + (x^2 - \tfrac{1}{2})^2]^{1/2}$. Differentiating with respect to x, we have

$$\frac{ds}{dt} = \tfrac{1}{2}[(x - 2)^2 + (x^2 - \tfrac{1}{2})^2]^{-1/2}[2(x - 2) + 2(x^2 - \tfrac{1}{2})(2x)].$$

The derivative is zero when the numerator is zero. Hence $2(x - 2) + 2(x^2 - \tfrac{1}{2})(2x) = 0$, that is, $2x^3 - 2 = 0$ and $x = 1$. For this x the minimum distance is $s = \sqrt{(1 - 2)^2 + (1 - \tfrac{1}{2})^2} = \sqrt{1 + \tfrac{1}{4}} = \tfrac{1}{2}\sqrt{5}$. (See Figure 6.5.)
   (*Note:* We could have obtained the same result by minimizing the square of the distance.)

FIGURE **6.5**   FIGURE **6.6**

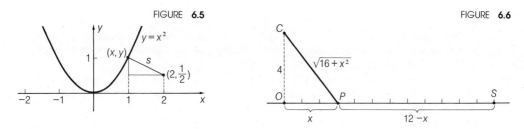

5.   A campsite, located in the woods, is 4 miles from the nearest point on a (straight) road. The distance from this point to a store on the road is 12 miles. If a camper wishes to walk from the campsite to the store in minimum time, what path should he follow if he can walk at the rate of 3 miles per hour in the woods and 5 miles per hour on the road?
SOLUTION   Let C denote the campsite, O the nearest point from the campsite to the road, S the store, and P a point on the road between O and S. (See Figure 6.6.) If x is the distance

from O to P, then $12 - x$ is the distance from P to S, and $\sqrt{16 + x^2}$ is the distance from C to P. We consider only those values for which $0 \le x \le 12$.

The total time to travel from C to P, and from P to S, is given by

$$T = \frac{\sqrt{16 + x^2}}{3} + \frac{12 - x}{5}.$$

To find the minimum time, we differentiate T with respect to x and set the derivative equal to zero. We obtain

$$\frac{dT}{dx} = \frac{x}{3\sqrt{16 + x^2}} - \frac{1}{5} = 0$$

or $\sqrt{16 + x^2} = \frac{5}{3}x$, that is, $16 + x^2 = \frac{25}{9}x^2$. Solving for x, we find that $x = 3$. (The solution $x = -3$ is outside the interval $0 \le x \le 12$.). Hence the camper should walk 5 miles through the woods (from C to P) and 9 miles along the road (from P to S).

PROBLEMS

In Problems 1 to 14 find the derivatives of the given functions.

1. $f(x) = x^{3/4}$.
2. $f(x) = x^{2/3} + 2x^{4/3} + \frac{1}{3}x^2$.
3. $f(x) = x^{2/5} + 2x + x^{-2/5}$.
4. $f(x) = 5x^{-2/7} + 4x^{-3/7} + 3x^{-4/7}$.
5. $g(t) = t^3 \sqrt{t^5}$.
6. $g(t) = \frac{1}{5}t \sqrt[4]{t} - \frac{1}{15}t^3 \sqrt[4]{t^3} + \frac{1}{25}t^5 \sqrt[4]{t^5}$.
7. $g(t) = \dfrac{2\sqrt{t} + 3\sqrt[3]{t} + 4\sqrt[4]{t}}{t^3}$.
8. $y = (2x + 3)^{4/3}$.

9. $y = (x^2 + 2x + 3)^{-3/2}$.
10. $y = \sqrt[3]{(3x^2 + 6x - 1)^2}$.
11. $y = x^3 \sqrt[3]{(8 - x^3)^2}$.
12. $F(x) = \dfrac{3}{2 + \sqrt{x^3}}$.
13. $F(x) = \dfrac{x}{(4 - x^2)^{3/2}}$.
14. $F(x) = \dfrac{1 - x^{2/3}}{1 + x^{2/3}}$.

In Problems 15 to 18 find the equation of the tangent line to the given curve at the indicated point.

15. $y = \sqrt{x^2 + 3x + 5}$, $(1,3)$.
16. $y = \dfrac{x}{\sqrt{x^2 + 1}}$, $(0,0)$.

17. $y = (x^3 + 3x^2 + 4)^{1/3}$, $(1,2)$.
18. $y = (x^2 + 4)^{-2/3}$, $(2,\frac{1}{4})$.

19. If x and y are positive numbers and $x^2 + y^2 = 1$, what is the smallest value that $2x^3 + y^3$ can attain? If k is a positive constant and the same conditions on x and y hold, what is the smallest value that $kx^3 + y^3$ can attain?
20. Find the points on the parabola $y = -\frac{3}{2}x^2 + 11$ that are closest to the origin.
21. What is the shortest distance from the point $(1,0)$ to the curve given by $y = \sqrt{x^2 + 6x + 10}$?
22. Find the shortest and longest distances from the point $(-2,1)$ to the curve given by $y = 1 + \sqrt{18 - 2x^2}$.
23. Find the length of the shortest ladder that will reach from the ground over an 8-ft wall to a building 1 ft beyond the wall.
24. A child 1 yd tall stands 8 yd from the base of a street light. What is the shortest possible distance from the light to the top of the child's shadow?
25. Find the length of the shortest segment intercepted by the coordinate axes and tangent to the circle $x^2 + y^2 = 1$. Assume the segment lies in the first quadrant.

26. A 3-ft-wide corridor intersects a hallway at right angles. A 24-ft-long pole is being pushed along the floor down the corridor. How wide must the hallway be so that the pole will go around?

27. A cowboy riding a thirsty horse is 3 miles from a straight river. Before returning to the ranch, which is on the same side of the river, the cowboy wishes to water his horse. If the ranch is $\sqrt{153}$ miles from the cowboy and 6 miles from the river, what is the shortest distance the cowboy can ride before returning to the ranch?

28. A rectangle is inscribed in the ellipse $x^2/25 + y^2/16 = 1$ with its sides parallel to the axes of the ellipse. What are the dimensions of the rectangle having maximum area?

29. What is the area of the largest rectangle having two vertices on the x-axis and two vertices on the semicircle $y = \sqrt{a^2 - x^2}$? From this result, show that the rectangle of maximum area that can be inscribed in a circle is a square.

30. A rancher wishes to fence off a triangular piece of land adjacent to a straight river. If he uses the river as one side, and the other two sides are to have equal length, what is the maximum area he can enclose if he has 200 ft of fence?

31. A drinking cup in the form of a right circular cone is to be made out of paper. If the cup is to hold 36π cu cm of liquid, what dimensions should the cup have if it is to contain the least amount of paper? [*Hint:* The lateral surface area of a right circular cone is $\pi r \sqrt{h^2 + r^2}$; the volume is $\frac{1}{3}\pi r^2 h$.]

32. A conical tent is to be constructed from $50\pi\sqrt{3}$ sq ft of canvas. What dimensions should the tent have in order to enclose maximum space? Do not count the bottom. [*Hint:* $S_{\text{cone}} = \pi r \sqrt{h^2 + r^2}$, $V_{\text{cone}} = \frac{1}{3}\pi r^2 h$.]

33. A power station is located on one bank of a river, and a factory is located on the opposite bank of the river 1000 ft downstream. The river is 1000 ft wide, and the river banks are parallel. It costs \$12 per foot to lay power lines over land and \$20 per foot under water. What route should be followed from the power station to the factory in order to minimize the cost? Suppose the cost is \$16 per ft over land and \$20 per ft under water. What route should be followed then?

§3   Implicit differentiation. Algebraic functions‡

Implicit differentiation is a technique which, in some circumstances, facilitates the calculation of derivatives. We explain it first in a simple example. Then we apply it to a class of functions—algebraic functions—which contains all functions considered up to now.

3.1   Tangents to circles

The equation of the unit circle (circle of radius 1 with center at the origin) reads

$$x^2 + y^2 - 1 = 0. \tag{1}$$

Solving for y we obtain *two* functions,

$$y = \sqrt{1 - x^2}, \qquad y = -\sqrt{1 - x^2}, \tag{2}$$

whose graphs are the upper and lower semicircles, respectively (see Chapter 3, §1.3). There are several methods for finding the derivatives of these functions.

Appeal to geometry. We know (see Chapter 2, §3.2) that the tangent to the circle (1) at a point (x,y) is perpendicular to the diameter through this point, that is, to the

---

‡This section may be omitted at first reading.

line through $(0,0)$ and (x,y). The slope of this line is y/x (we assume that $x \neq 0$). The slope of the tangent therefore (see Chapter 2, §2.3) is $-x/y$. (This is true also for $x = 0$; then the tangent is horizontal.) But the slope of the tangent is the desired derivative. Thus

$$\frac{dy}{dx} = -\frac{x}{y}.$$ (3)

This holds for both functions (2); substituting we obtain

$$\frac{d\sqrt{1-x^2}}{dx} = -\frac{x}{\sqrt{1-y^2}} \quad \text{and} \quad \frac{d(-\sqrt{1-x^2})}{dy} = \frac{x}{\sqrt{1-y^2}}.$$ (4)

(We assume that $y \neq 0$; at $y = 1$, $x = \pm 1$, the tangent is vertical and has no slope.)

Direct differentiation. We compute the derivative of the functions (2) using the chain rule and the rule for differentiating powers. In fact, we already did this in Example 2, §2.3, and obtained the first Equation (4). The second follows from the first.

Implicit differentiation. We assume that y is a differentiable function of x; then the left side of (1) is a function of x. We differentiate both sides of Equation (1) with respect to x. Keeping in mind that $\dfrac{d}{dx}(y^2) = 2y\left(\dfrac{dy}{dx}\right)$ by the generalized power rule, and that 1 is a constant, we obtain

$$\frac{d(x^2)}{dx} + \frac{d(y^2)}{dx} - \frac{d}{dx}(1) = 0$$

or

$$2x + 2y\frac{dy}{dx} - 0 = 0.$$

Solving this last equation for dy/dx, we obtain (3).

The method can be used also for other circles and, as we shall explain below, for other curves.

EXAMPLE   Find the slope of the tangent to the circle $(x - 3)^2 + (y + 1)^2 = 37$ at the point $x = 2$, $y = 5$, by implicit differentiation.

SOLUTION   We differentiate the given equation (that is, both sides of the given equation) treating y as a function of x and obtain

$$2(x - 3) + 2(y + 1)\frac{dy}{dx} = 0$$

or

$$\frac{dy}{dx} = \frac{3 - x}{y + 1}.$$

For $x = 2$, $y = 5$ the desired slope is $\dfrac{3-2}{5+1} = \dfrac{1}{6}$.

PROBLEMS

In Problems 1 to 6 find the slope of the tangent line to the given circle at the indicated point using implicit differentiation.

1. $(x - 1)^2 + y^2 = 4$, $(1,2)$.
2. $x^2 + (y - 6)^2 = 25$, $(-4,9)$.
3. $(x + 1)^2 + (y - 4)^2 = 8$, $(1,6)$.

4. $(x - 1)^2 + (y + 3)^2 = 34$, $(-2,2)$.
5. $(x + 2)^2 + (y + 1)^2 = 25$, $(1,3)$.
6. $(x - 1)^2 + (y + 5)^2 = 53$, $(-1,-2)$.

3.2   Functions defined implicitly by polynomial equations

A monomial in two variables x and y is an expression of the form $\alpha x^p y^q$ where α is a number and p and q are nonnegative integers. A polynomial $P(x,y)$ in two variables is a sum of monomials. [Examples: $P(x,y) = x^2 + y^2 - 1$, $P(x,y) = x^2 - 3xy^2 + x^2 y^{10}$.] The set of points (x,y) satisfying a polynomial equation

$$P(x,y) = 0 \tag{5}$$

is called a (plane) **algebraic curve**. (Example: a circle.) A point (x_0, y_0) on the curve (5) is called a **simple point with nonvertical tangent** if near (x_0, y_0) the curve coincides with a graph of a differentiable function $y = f(x)$, with $y_0 = f(x_0)$. (Example: every point on $x^2 + y^2 - 1 = 0$ with $y \neq 0$.) If so, the function f is said to be defined implicitly by Equation (5). Functions obtained this way are called **algebraic.**

It can be shown that *all radical functions are algebraic.* However, not all algebraic functions are radical functions. Similarly, all rational functions are radical, but not all radical functions are rational; all polynomials are rational functions, but not all rational functions are polynomials.

Let us *assume* that $y = f(x)$ is a differentiable function defined implicitly by Equation (5) near (x_0, y_0). We want to find the derivative $f'(x_0)$, that is, the slope of the tangent to the curve (5) at (x_0, y_0). To do this we first differentiate both sides of Equation (5), treating y as a function of x and writing y' or dy/dx for the (as yet unknown) derivative $f'(x)$. Then we replace x and y by x_0 and y_0; *if* we obtain an equation for y', its solution is the desired slope. [It can be shown that in this case there *is* a function $y(x)$ satisfying (5), and there was no need to assume its existence. We shall not prove this result (called the implicit function theorem) in this book.]

EXAMPLES   **1.**   Find the slope of the tangent to the curve

$$y^3 + 2yx^2 - 3x - 3 = 0 \tag{6}$$

at the point $(2,1)$.

SOLUTION   First we check that $(2,1)$ lies on the curve: $1^3 + 2 \cdot 1 \cdot 2^2 - 3 \cdot 2 - 3 = 0$. Next we differentiate both sides of (6), using the product rule and the chain rule, and obtain

$$3y^2 y' + 2y'x^2 + 4yx - 3 = 0. \tag{7}$$

For $x = 2$, $y = 1$, this becomes $3y' + 8y' + 8 - 3 = 0$ or $y' = -5/11$. This is the desired slope.

2.   Find dy/dx for the function (or functions) defined implicitly by Equation (6).

ANSWER   Differentiating (6), we obtain (7); solving for y' yields

$$y' = \frac{dy}{dx} = \frac{3 - 4xy}{3y^2 + 2x^2}.$$

FIGURE **6.7**   FIGURE **6.8**   FIGURE **6.9**

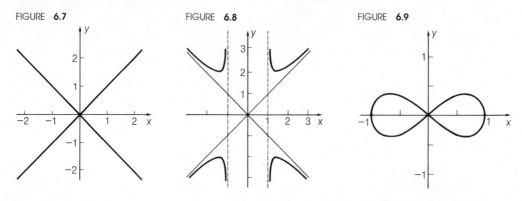

3. Find the derivative, at $x = 0$, $y = 0$, for the function defined near this point by the equation

$$x^2 - y^2 = 0. \tag{8}$$

ANSWER   Differentiation of (8) yields

$$2x - 2yy' = 0. \tag{9}$$

For $x = y = 0$, we obtain $0 = 0$; this yields *no* information on y'. Note that Equation (8) reads $(x - y)(x + y) = 0$; that is, either $y = x$ or $y = -x$. The curve (8) consists of the *two* lines through the origin, of slopes 1 and -1. The origin is *not* a simple point of the curve. (It is a so-called *double point*. For $y \neq 0$, we obtain $y' = x/y$. Since $y = x$ or $y = -x$, we find that $y' = 1$ or $y' = -1$, as we should.) The curve (8) is shown in Figure 6.7.

4. Find the slope of the tangent to the curve

$$y^2(x^2 - 1) - x^4 = 0 \tag{10}$$

at $x = y = 0$.
ANSWER   Differentiating (10), we obtain

$$2yy'(x^2 - 1) + 2y^2x - 4x^3 = 0. \tag{11}$$

For $x = y = 0$, we again obtain the useless equation $0 = 0$. In this case the origin is not a simple point of the curve; it is a so-called *isolated point*. Indeed, near the origin there are no other points of the curve. [For if $x = 0$, we must have $y = 0$; and if $x \neq 0$ and $|x| < 1$, then $x^2 - 1 < 0$ and (10) requires that y^2 be negative, which is not true for any real y.] There is, of course, no tangent through the origin. The curve (10) is shown in Figure 6.8.

5. Find the slope of the tangent to the curve

$$(x^2 + y^2)^2 - x^2 + y^2 = 0 \tag{12}$$

at $x = 0$, $y = 0$, and at $x = 1$, $y = 0$.
ANSWER   Differentiation of (12) yields

$$2(x^2 + y^2)(2x + 2yy') - 2x + 2yy' = 0. \tag{13}$$

Substituting into (13) the values $x = y = 0$, we obtain the useless equation $0 = 0$. Substituting into (13) the values $x = 1$, $y = 0$, we obtain the absurd equation $2 = 0$. The reason why implicit differentiation fails is seen in Figure 6.9 where we graph the curve. The point $(0,0)$ is a *double point*. The point $(1,0)$ is a so-called *simple point with vertical tangent*.

REMARK   Implicit differentiation can also be applied to some nonalgebraic functions, as we shall see later.

PROBLEMS

Find dy/dx for the function (or functions) defined implicitly by the following equations.

1. $x^2 - 4y^2 = 16$.
2. $y^3 - 2x^3 + y = 1$.
3. $y^2 + 2x^2y + x = 0$.
4. $x^2y^4 + 2xy - 4x + 8 = 0$.

5. $y^2(x^2 - 1) = x^2y + 1$.
6. $y^6 + 3x^2y^4 + 3x^4y^2 + x^6 - 8 = 0$.
7. $x^2(y^2 + 3xy) = y^4x + 2$.
8. $(x^2 + y^2)^6 - 2y^3 = 1$.

Find the slopes of the tangent lines to the following curves at the points specified.

9. $x^2 + y^3 + 2x - 5y - 19 = 0$, $(3, -1)$.
10. $x^2 - y^2 + 3xy + 12 = 0$, $(-4, 2)$.
11. $2x^2 - y^3 + 4xy - 2x = 0$, $(1, -2)$.
12. $x^3y^3 - 2x^2y^2 - 4y + x^2 - 2 = 0$, $(1, -1)$.

§4 Primitive functions or antiderivatives

In this section we begin discussing a process inverse to differentiation—a subject which will occupy us on many later occasions.

4.1 Definition of primitive functions

Let f and F be two functions. If $F'(x) = f(x)$ in some interval, we call F a **primitive function** or **antiderivative** of f. For instance, the function $F(x) = 1/x$ is a primitive function of $f(x) = -(1/x^2)$, since we know that

$$\frac{d}{dx}\left(\frac{1}{x}\right) = -\frac{1}{x^2}.$$

We say "a primitive function" and not "the primitive function" because, for example, $F_1(x) = (1/x) + 73$ is also a primitive function of $-(1/x^2)$.

In general, if $F(x)$ is a primitive function of $f(x)$ and C is any constant, then $F(x) + C$ is also a primitive function of $f(x)$ since $(F + C)' = F' + C' = F'$. This is also obvious geometrically. The graph of the function $y = F(x) + C$ is obtained by translating the graph of $F(x)$ in the y direction by the amount C (up, if $C > 0$; down if $C < 0$). This does not change the slope of the tangent (see Figure 6.10).

FIGURE **6.10**

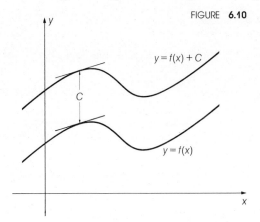

Furthermore, if F is a primitive function of f, then any other primitive function is of the form $F + C$, C a constant. To prove this statement, we make use of the following theorem.

Theorem 1. *If $g'(x) = 0$ for all x in an interval, then $g(x)$ is constant in this interval.*

The theorem asserts that a curve with an everywhere horizontal tangent is a horizontal straight line. Our geometric intuition certainly tells us that this is so. A proof follows from Theorem 2 and its corollary in Chapter 5, §2.2. Let $x_2 > x_1$ be two points in our interval. Since $g' \geq 0$, g is not decreasing; we have $g(x_1) \leq g(x_2)$. Since $g' \leq 0$, g is not increasing; we have $g(x_1) \geq g(x_2)$. Therefore $g(x_1) = g(x_2)$. The function takes on the same value at any two points.

Theorem 2. *If F and F_1 are primitives of the same function in some interval, then $F_1 - F$ is a constant function; that is, there is a number C such that $F_1(x) = F(x) + C$.*

Proof.   Let F and F_1 be primitive functions of f, and set $g(x) = F_1(x) - F(x)$. Then $g'(x) = F_1'(x) - F'(x) = f(x) - f(x) = 0$ for all x in the interval considered. By Theorem 1, g is a constant.

4.2   Finding primitives

It is natural to ask whether every continuous function has a primitive function, that is, whether every continuous function is a derivative of another function. The answer is in the affirmative, as we will see in Chapter 8.

At the moment we can only *guess* primitive functions. Every differentiation rule is at the same time a rule for finding primitive functions. Some results that follow from the differentiation rules established thus far are stated below.

Theorem 3.   (a) *If f has a primitive function F and c is a constant, then cF is a primitive function of cf.* (b) *If, in addition, g has a primitive function G, then $F + G$ is a primitive function of $f + g$, and $F - G$ is a primitive function of $f - g$.*

Indeed, if $F' = f$ and $G' = g$, then $(cF)' = cf$, $(F + G)' = f' + g'$, and $(F - G)' = f' - g'$, by Theorem 1 in Chapter 4, §2.1.

Theorem 4. *If c is a number, r a rational number, $r \neq -1$, then $F(x) = cx^{r+1}/(r + 1)$ is a primitive function for $f(x) = cx^r$.*

This is so, since, by Theorem 5 in §2.4, we have

$$\frac{d}{dx} \frac{cx^{r+1}}{r+1} = \frac{c}{r+1} \frac{dx^{r+1}}{dx} = \frac{c}{r+1}(r+1)x^r = cx^r.$$

These two theorems enable us to find a primitive function for any polynomial and also for any sum of terms of the form cx^r, where r may be any rational number provided $r \neq -1$.

EXAMPLES   1.   Find all primitive functions of $3x^6$.
ANSWER   By Theorem 4, $\frac{3}{7}x^7$ is a primitive function. By Theorem 2, every primitive function is of the form $\frac{3}{7}x^7 + C$, C a constant.

2. Find all primitive functions for $f(x) = x^3 - 2x + 1$.

ANSWER By Theorem 4, the functions x^3, $-2x$, and 1 have primitive functions $\frac{1}{4}x^4$, $-x^2$, and x, respectively. Hence, applying repeatedly Theorem 3(b), we have that $\frac{1}{4}x^4 - x^2 + x$ is a primitive function for f. It is advisable to check every statement about primitive functions by differentiation; we do so: $(\frac{1}{4}x^4 - x^2 + x)' = x^3 - 2x + 1$, as desired. Every primitive function of f is of the form $\frac{1}{4}x^4 - x^2 + x + C$.

3. Find all functions F such that $F'(x) = \sqrt{x}$.

ANSWER We require that F be a primitive function of $\sqrt{x} = x^{1/2}$. By Theorem 4, one primitive function is $\dfrac{1}{\frac{1}{2}+1} x^{(1/2)+1} = \frac{2}{3}x^{3/2}$. Thus all desired F are of the form $F(x) = \frac{2}{3}x^{3/2} + C$.

PROBLEMS

1. $f(x) = 1 - x$.

2. $f(x) = x^3 - 2x$.

3. $f(x) = x^4 + x^2 + 1$.

4. $f(x) = 3x^4 - 7x^3 + 2$.

5. $f(x) = 12x^5 - 8x^3 + 2x - 3$.

6. $f(x) = x^{10} + x^{100}$.

7. $g(t) = \dfrac{1}{t^3}$.

8. $g(t) = -\dfrac{1}{t^3} + \dfrac{4}{t^2} + 2$.

9. $g(t) = \dfrac{2}{t^4} - \dfrac{1}{t^2} + t$.

10. $g(t) = t^{-4} - 2t^{-7}$.

11. $g(t) = 1 + \left(\dfrac{2}{t}\right)^3 - \left(\dfrac{2}{t}\right)^5$.

12. $g(t) = t^{1/3} + 3t^{1/2}$.

13. $g(t) = 2t^{3/2} + t^{5/2}$.

14. $g(t) = -t^{-1/2} + 3t^{-3/2} + 4$.

15. $h(u) = (u + 1)^9$.

16. $h(u) = (3u + 1)^5$.

17. $h(u) = (2u + 3)(u^2 + 3u + 4)^{10}$.

18. $h(u) = \dfrac{u}{\sqrt{u^2 + 1}}$.

4.3   The simplest differential equations

We begin by considering an example. Let us find all functions $\phi(x)$ whose second derivative vanishes identically; that is, for all values of x,

$$\phi''(x) = 0. \tag{1}$$

This means that

$$\frac{d\phi'(x)}{dx} = 0,$$

so that, by Theorem 1, there exists a number α such that $\phi'(x) = \alpha$. Hence ϕ is a primitive function of the constant α. One primitive function is αx; every other is of the form $\alpha x + \beta$, β being some other number. Thus ϕ is a linear function. Conversely, every linear function has an identical vanishing second derivative.

Equation (1) is an example of a differential equation; the function $\phi(x)$ is to be found from a condition imposed on its derivatives. In the case considered, the condition involves a second derivative; the differential equation is said to be of *second order*. Note that the solution is not uniquely determined; condition (1) describes *all* linear functions. To obtain a unique solution we must impose additional conditions. This is characteristic of all differential equations. (The statement "F is a primitive of f" means that "$F' = f$"; this is a first-order differential equation.)

EXAMPLES 1. Find a function $\phi(x)$ such that $\phi''(x) = 0$ for all x, $\phi(1) = 2$, and $\phi'(1) = 3$.

ANSWER We already know that ϕ must be a linear function, say, $\phi(x) = \alpha x + \beta$. Then $\phi'(x) = \alpha$. Our two additional conditions become $\alpha + \beta = 2$ and $\alpha = 3$. Thus $\beta = -1$ and $\phi(x) = 3x - 1$.

2. Find a function g such that $g'(t) = 1/t^2 + 2/t^3$ for all t, and $g(1) = 3$. (This is an example of a differential equation of first order.)

ANSWER   Since $1/t^2 = t^{-2}$, the function $1/t^2$ has a primitive function $t^{-2+1}/(-2 + 1) = -t^{-1} = -1/t$, by Theorem 4. Also, $2/t^3 = 2t^{-3}$ has a primitive function $2t^{-3+1}/(-3 + 1) = -1/t^2$. The most general function g with the required derivative is therefore

$$g(t) = -\frac{1}{t} - \frac{1}{t^2} + C,$$

where C is a constant. The condition $g(1) = 3$ yields

$$3 = -\frac{1}{1} - \frac{1}{1^2} + C,$$

or $C = 5$. Hence $g(t) = -(1/t) - (1/t^2) + 5$.

3. Find all solutions of the differential equation $f'''(x) = 0$.

SOLUTION   Since $f''' = (f'')' = 0$ everywhere, $f''(x)$ is a constant; call it α. Hence $(f')' = \alpha$; $f'(x)$ is a primitive function of α. We conclude that $f'(x) = \alpha x + \beta$, where β is another constant. Now f is a primitive function of $\alpha x + \beta$. One such primitive function is $\frac{1}{2}\alpha x^2 + \beta x$. Hence $f(x) = \frac{1}{2}\alpha x^2 + \beta x + \gamma$, with α, β, γ constants. In other words, f is a quadratic polynomial.

PROBLEMS

1. Find a function f such that $f'(x) = \frac{1}{2}$ for all x.
2. Find a function f such that $f'(x) = x^2$ for all x.
3. Find a function g such that $g''(t) = -1$ for all t.
4. Find a function g such that $g'''(t) = 12t^2 + t - 1$ for all t.
5. Find a solution of the differential equation $f'(x) = x^2 + 1/x^2$.
6. Find a solution of the differential equation $f''(x) = \sqrt{x}$.
7. Find a function g such that $g'(t) = 2t^2 - 3t + 1$ and $g(1) = 0$.
8. Find a function g such that $g'(t) = t(t^2 + 1)^9$ and $g(0) = \frac{1}{10}$.
9. Find a function h such that $h''(u) = 6u + 1$, $h(0) = 2$, and $h(1) = 0$.
10. Find a function h such that $h''(u) = 3u^2$, $h'(0) = 1$, and $h(0) = -1$.
11. What is the equation of the curve whose slope is always equal to half the abscissa and which passes through the point $(4,2)$?
12. What is the equation of the curve whose slope is always equal to $3x^2 - 4x + 1$ and which passes through the point $(2,10)$?

§5   Nondifferentiable functions‡

The derivative of a function f at a point x_0 is the limit of the difference quotient

$$\frac{f(x_0 + h) - f(x_0)}{h}, \tag{1}$$

as $h \to 0$, *provided this limit exists*. Of course, the limit may fail to exist, so the graph need not have a tangent. We know (Theorem 2 in Chapter 4, §1.6) that every differentiable function is continuous, but not every continuous function is differentiable.

---

‡May be omitted at first reading.

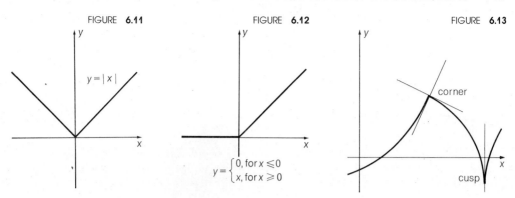

FIGURE 6.11   FIGURE 6.12   FIGURE 6.13

$y = |x|$

$y = \begin{cases} 0, \text{for } x \leq 0 \\ x, \text{for } x \geq 0 \end{cases}$

corner

cusp

5.1 One-sided derivatives. Infinite derivatives

A simple example of a continuous nondifferentiable function is the function $f(x) = |x|$; its graph, shown in Figure 6.11, has a **corner** at $x = 0$. This function is said to have, at $x = 0$, a derivative from the right (equal to 1) and a derivative from the left (equal to -1). We write $f'(0^+) = 1$, $f'(0^{-1}) = -1$. Similarly, the function graphed in Figure 6.12 has at $x = 0$ the one-sided derivatives 0 (from the left) and 1 (from the right). The general definition reads: the **right derivative** of a function f at x_0 is the limit

$$f'(x_0^+) = \lim_{h \to 0^+} \frac{f(x_0 + h) - f(x_0)}{h},$$

whenever it exists. This derivative may exist for a function which is defined only at x_0 and at points near, and to the right of, x_0. The **left derivative** at x_0 is defined similarly:

$$f'(x_0^-) = \lim_{h \to 0^-} \frac{f(x_0 + h) - f(x_0)}{h},$$

if the limit exists.

Let f be a continuous function such that $f'(x_0^+) = \alpha$, $f'(x_0^-) = \beta$. Then if $\alpha = \beta$, f is differentiable at x_0 and $f'(x_0) = \alpha$ as we easily see. If $\alpha \neq \beta$, then f is not differentiable at x_0. We say, in this case, that the graph of f has a **corner** at the point $(x_0, f(x_0))$. A typical case is shown in Figure 6.13.

A continuous function f may fail to have a derivative at a point x_0, because the difference quotient at x_0 can be made larger than any given number provided we make h small enough. This means

$$\lim_{h \to 0} \frac{f(x_0 + h) - f(x_0)}{h} = +\infty.$$

In this case, we write $f'(x_0) = +\infty$. If

$$\lim_{h \to 0^+} \frac{f(x_0 + h) - f(x_0)}{h} = +\infty,$$

we say that $f'(x_0^+) = +\infty$. It is now clear what is meant by statements such as $f'(x_0) = -\infty$, $f'(x_0^+) = -\infty$, $f'(x_0^-) = +\infty$, and $f'(x_0^-) = -\infty$. In all of these cases, f is not differentiable at x_0. If, at x_0, $f' = +\infty$ or $f' = -\infty$, we say that the graph

FIGURE **6.14**

FIGURE **6.15**

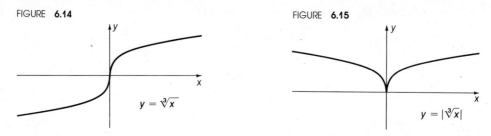

$y = \sqrt[3]{x}$

$y = |\sqrt[3]{x}|$

of f has a *vertical tangent*. If, at x_0, one of two one-sided derivatives is $+\infty$ and the other $-\infty$, the graph of f is said to have a *cusp*, with a vertical tangent (see Figure 6.13).

EXAMPLES **1.** Let $f(x) = x^{1/3}$ (Figure 6.14). Then

$$\frac{f(h) - f(0)}{h} = \frac{h^{1/3}}{h} = h^{-2/3} = \frac{1}{\sqrt[3]{h^2}}.$$

This is nonnegative for $h \neq 0$, and $\lim_{h \to 0} \sqrt[3]{h^2} = 0$. Hence $\lim_{h \to 0} (1/\sqrt[3]{h^2}) = +\infty$, so that $f'(0) = +\infty$. The graph has a vertical tangent at $x = 0$.

2. Set $f(x) = |x^{1/3}|$. Then by a similar argument, $f'(0^+) = +\infty$, $f'(0^-) = -\infty$ (cusp). See Figure 6.15.

PROBLEMS

1. If $g(s) = s^3$ for $s \leq 1$ and s^2 for $s > 1$, find $g'(1^-)$ and $g'(1^+)$.
2. If $f(x) = (1 - x)^{3/2}$, find $f'(1^-)$.
3. Let $h(t) = (t^2 - 4)^{5/2}$. Determine which of the one-sided derivatives $h'(-2^-)$, $h'(-2^+)$, $h'(2^-)$, and $h'(2^+)$ exist; evaluate any that do.
4. If $f(y) = y + 1$ for $y \leq 3$ and $y^2 - 2y$ for $y > 3$, find $f'(3^-)$ and $f'(3^+)$.
5. If $f(x) = x$ for $x \leq 0$ and $x^{5/2}$ for $x > 0$, find $f'(0^-)$ and $f'(0^+)$.
6. If $\phi(u) = (u - 1)^{2/3}$ for $u \leq 0$ and $(u + 1)^{2/3}$ for $u > 0$, find $\phi'(0^-)$ and $\phi'(0^+)$.
7. If $h(s) = s^2 - 1$ for $s \leq 1$ and s for $s > 1$, find $h'(1^-)$ and $h'(1^+)$.

5.2 Limits of derivatives

In computing derivatives (including one-sided derivatives and infinite derivatives), the following rule may be useful.

Theorem 1. *Assume that the function $f(x)$ has a derivative $f'(x)$ for all x near a point x_0, except perhaps at x_0 itself. Assume that f is continuous at x_0, and that the finite or infinite limit $\alpha = \lim_{x \to x_0} f'(x)$ exists. Then $f'(x_0) = \alpha$.*

The corresponding statement involving one-sided limits and one-sided derivatives is also true.

The proof of this result will be given in Chapter 14, §1.4.

EXAMPLE Consider again the function $f(x) = x^{1/3}$. For $x \neq 0$, it has the (finite) derivative $\frac{1}{3}x^{-2/3}$. Therefore $\lim_{x \to 0} f'(x) = +\infty$. Also, f is continuous at $x = 0$. Therefore $f'(0) = +\infty$.

PROBLEMS

In Problems 1 to 7 describe the behavior of the derivative of each function as completely as possible; that is, first determine where the derivative exists. Next, determine where the derivative is infinite, and evaluate it as $+\infty$ or $-\infty$ at these points. Finally, determine any additional points where only a one-sided derivative exists or is infinite, and evaluate the appropriate one-sided derivative(s) at these points.

1. $f(x) = x^{2/3}$.
2. $f(t) = \sqrt{t^2 - 4}$.
3. $f(s) = (s - 1)^{3/7}$.
4. $k(x) = x^{1/3} + x^{2/3}$.
5. $g(u) = (8u - 1)^{2/3}$ for $u \le \frac{1}{8}$; $(8u - 1)^2$ for $u > \frac{1}{8}$.
6. $\phi(y) = (y + 1)^{3/5}$ for $y \le 0$, $\frac{3}{2}(y - 1)^{2/5}$ for $y > 0$.
7. $A(t) = \sqrt{t^{1/3} + 1}$.

5.3   Other nondifferentiable functions

A continuous function may fail to be differentiable in a complicated way. An example is the function graphed in Figure 6.16, which equals 0 at 0. The graph of this function oscillates infinitely often between the lines $y = x$ and $y = -x$. Consider the difference quotient at $x = 0$, that is, $f(h)/h$. There are arbitrarily small values of h for which the difference quotient is 1, as well as arbitrarily small values of h for which the difference quotient is -1. Therefore the difference quotient has no limit as $h \to 0$. There are even no one-sided or infinite limits.

In all examples described thus far, the functions considered failed to have derivatives only at certain points. It would be natural to guess that a continuous function must be differentiable at most points. In this case, however, intuition leads us astray. There are continuous functions that are *nowhere* differentiable, as has been first recognized by Bolzano. To construct such functions is not easy, and it is almost impossible to visualize their graphs. We shall not discuss such functions, since calculus deals primarily with functions that have derivatives.

FIGURE **6.16**

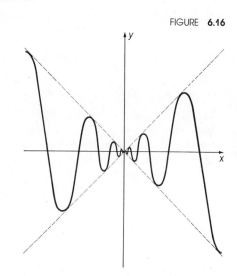

BERNARD BOLZANO (1781–1848), a philosopher, logician and mathematician, anticipated Weierstrass' rigorous approach to calculus and Cantor's set theory. His main mathematical work, *Paradoxes of Infinity*, appeared posthumously.

Bolzano was a priest and a professor of religious philosophy at the University of Prague. In 1820 the Austrian government dismissed him because his sermons to the students were considered subversive (they dealt with war and the conflict between individual conscience and obedience to the government).

The graph of a continuous, nowhere differentiable function oscillates wildly near every point, just as the graph in Figure 6.16 oscillates wildly near $x = 0$. Bolzano's proof of the existence of such functions was not completely convincing. A rigorous proof was published by Weierstrass (and, it seems, was given by Riemann in his lectures).

APPLICATIONS OF DERIVATIVES, CONTINUED

§1 Laws of motion‡

In this chapter we discuss more applications of derivatives. Some of the applications are as old as calculus itself, others are fairly recent. Every section in this chapter may be omitted, with little loss of continuity. But one cannot fully understand calculus without knowing some of its uses.

1.1 Newton's law of motion

Having defined acceleration (in Chapter 4, §3.2), we are able to state one of the most important laws of physics: Newton's law of motion. This law involves two more basic concepts besides acceleration: **force** and **mass.**

The idea of force originates in the subjective feeling of effort that we experience when we change the velocity of a body, for instance, when we throw a stone. In physics, force is defined as that which causes acceleration. In the case of rectilinear motion, we assume that force can be expressed by a number.

It is a common experience that the same effort produces different accelerations in different bodies. We therefore assign to each particle or body a positive number,

‡Optional section.

its mass, which measures its resistance to force. Experiments show that this resistance to acceleration is a property of the body and does not depend on its velocity or on the force applied (this statement must be revised for very fast moving bodies; see §1.4), and that the mass of a body consisting of two parts is equal to the sum of the masses of the parts. We therefore think of the mass of a body as measuring the *quantity of matter* contained in it. In order to measure mass we must, of course, choose a unit; in scientific computations we commonly use grams. We shall mostly use the CGS system of units (centimeters, grams, seconds) in our examples.

Newton's law of motion for a particle of mass m moving along a straight line says that, at every time instant, the acceleration a and the force F acting on the particle are connected by the relation

$$ma = F. \tag{1}$$

This shows that force is measured in units of "mass times acceleration," for instance, in gram $\cdot$ cm/sec^2 (gram centimeter per second per second). The force is positive if $a > 0$, that is, if it pushes the body in the positive direction.

The absolute value of the force that makes a body fall to the ground is called its **weight**. According to Galileo, a freely falling body always has the same acceleration g (see Chapter 4, §3.3). Denoting the weight of our body by W, we have therefore

$$-mg = -W. \tag{2}$$

Thus the weight of the body (on some planet) depends only on its mass. The masses of two bodies can be compared by comparing their weights, that is, by putting them on a scale. The weight of a body of mass 1 gram is a unit of force, called 1 gram weight. A body of a fixed mass will have different weights on Earth and, say, on the Moon.

PROBLEMS

1. A moving particle whose mass is .3 gram has an acceleration of 10 cm/sec^2 at a certain instant. What is the force acting on the particle at this instant?

2. A body whose mass is 500 kilograms (kg) is put on a hydraulic lift and pushed straight up. What force must the lift exert on the body in order to give it a constant acceleration of $\frac{1}{10}$ meter/sec^2?

3. A body whose mass is 10 grams is pushed across a level frictionless table in a straight line. If the velocity of the body is given by $v = (t^2/10) + t$ for $0 \le t \le 10$, where v is measured in centimeters per second and t in seconds, what is the force acting on the body when $t = 1$ sec? When $t = 2$ sec? When $t = 10$ sec?

4. A body whose mass is 100 grams moves in a horizontal plane along a straight line. If the velocity of the body is given by $v = 2t/(1 + t^2)$ for $t \ge 0$, where v is measured in centimeters per second and t in seconds, what is the force acting on the body when $t = 2$ sec? When $t = 3$ sec? What is the force when $t \to +\infty$ sec?

5. A projectile whose mass is 30 grams is shot horizontally into a viscous fluid. Assume that the path is a straight line. If the distance traveled by the projectile is given by $s = 100 - (100 - t)^4$ for $0 \le t \le 100$, where the distance s is measured in centimeters and the time t is measured in seconds, with what force is the fluid resisting the projectile when $t = 10$ sec? When $t = 50$ sec?

1.2   Inertia

We now apply the concept of primitive functions to Newton's law of motion. Since the acceleration a is the first derivative of the velocity $v = v(t)$ and the second derivative of the distance $s = s(t)$, the law of motion may also be written either as

$$m \frac{dv}{dt} = F \qquad (1')$$

or as

$$m \frac{d^2s}{dt^2} = F. \qquad (1'')$$

In order to determine the motion from this **differential equation** (see Chapter 6, §4.3), we must know the force. We shall consider some simple cases.

Assume first that there is *no* force acting on the body. Then $F = 0$ and (1') becomes $m(dv/dt) = 0$ or $dv/dt = 0$. By Theorem 1, in Chapter 6, §4.1, we have that v is constant. This is a special case of the **law of inertia** stated by Galileo. A *body acted on by no force remains at rest or in a state of uniform motion along a straight line.*

1.3   Vertical motion under the influence of gravity

Assume next that the body moves vertically under the influence of its weight. Then $F = -mg$, as we saw in §1.1. The equation of motion can be written as

$$m \frac{d^2s}{dt^2} = -mg \qquad \text{or} \qquad m \frac{dv}{dt} = -mg$$

or, upon canceling m,

$$\frac{d^2s}{dt^2} = -g \qquad \text{or} \qquad \frac{dv}{dt} = -g.$$

This is Galileo's law (see Chapter 4, §3.3). But this time we apply it not only to a body freely released from 0 but to a body thrown upward or downward with velocity v_0 from the height s_0 at the time $t = 0$.

Let $v(t)$ be the velocity at the time t. Then $v(t)$ is a primitive function of the constant function $-g$. Hence $v(t) = -gt + C$, where C is some constant. To find this constant, we note that $v(0) = C$. But we denoted the velocity imparted to our body at $t = 0$ by v_0. Hence $C = v_0$ and

$$v(t) = -gt + v_0. \qquad (3)$$

The velocity is seen to be composed of two parts: the initial velocity v_0, which would be the steady velocity of the body according to the law of inertia if there were no force acting, and the velocity $-gt$ resulting from the acceleration caused by gravity. (We can say that the body simultaneously executes two motions: a uniform motion with velocity v_0 prescribed by the law of inertia and the falling motion with velocity $-gt$ prescribed by Galileo's law.)

What is the height $s(t)$ of the body at the time t? Since $s'(t) = v(t)$, the function $s(t)$ is a primitive function of $-gt + v_0$. It is therefore of the form $s(t) = -\frac{1}{2}gt^2 + v_0t + C$, where C is a constant. To find C, note that $s(0) = C$. Thus $C = s_0$, the height (above a reference point) of the body at the time 0. Therefore

$$s(t) = -\frac{1}{2}gt^2 + v_0t + s_0. \qquad (4)$$

Let us assume, for the sake of definiteness, that $v_0 > 0$ (the body is thrown upward at $t = 0$) and that $s_0 = 0$ (at $t = 0$ the height of the body above the reference point is zero). We also assume that the reference point is at ground level. The motion is now described by

$$s(t) = -\tfrac{1}{2}gt^2 + v_0 t. \tag{4'}$$

At which time will the body hit the ground? We must solve the equation $s(t) = 0$ which gives $t = 0$ (the initial time) and

$$t = \frac{2v_0}{g}. \tag{5}$$

What is the largest height reached by our body? We must find the absolute maximum of $s(t)$ in the interval $[0, 2v_0/g]$. According to the rule for finding maxima, we need the point at which $s'(t) = 0$. Since $s'(t) = -gt + v_0$, this derivative is zero for $t = v_0/g$. For this t, $s''(t) = -g < 0$, so that this is indeed a maximum point (see Theorem 5 in Chapter 5, §2.6), and s takes on the value

$$\frac{v_0^2}{2g}.$$

This is therefore the maximum height achieved. At the instant it is attained, the velocity is zero, since the body reverses the direction of motion. This observation explains again why the derivative ($=$ velocity) is zero at a maximum point.

It is interesting to compute the velocity with which the body hits the ground, that is, the value of $v(t) = s'(t) = -gt + v_0$ at $t = 2v_0/g$. Substitution yields the value $(-v_0)$. The final speed is therefore equal to the initial speed. Later (in Chapter 9, §5.5) we shall recognize in this statement a special case of the law of conservation of energy.

(The description of the motion just given is based on an idealization. We neglect, for instance, the influence of air resistance, which slows down the motion. Also, Galileo's law of instant acceleration applies only to bodies close to the surface of the earth; we discuss this in Chapter 9, §5.6. Such idealizations are necessary whenever one describes an aspect of reality in mathematical terms.)

PROBLEMS

1. A ball is thrown straight up from ground level with an initial velocity of 64 ft/sec. What is the maximum height reached by the ball? How long after release will the ball hit the ground?

2. A stone is thrown vertically upward from a bridge that is 48 ft above the water level below. If the initial velocity of the stone is 32 ft/sec, find the maximum height of the stone above the bridge. What is the time when the stone strikes the water?

3. A stone is thrown straight down from a height of 160 ft with an initial velocity of 8 ft/sec. How long does it take the stone to reach the ground, and what is its velocity then?

4. A stone is dropped from a height of 144 ft. If a second stone is to be thrown down 1 sec later, what velocity must be imparted to it so that the two stones hit the ground at the same time?

5. A man on the ground throws a ball straight up. If he wants the ball to reach a height of at least 50 ft, what is the minimal velocity he must impart to the ball?

1.4 Einstein's law of motion

In the beginning of this century, Einstein discovered that Newtonian physics was not applicable to very high speeds. In the *theory of relativity*, the differential equation of rectilinear motion of a particle, under the influence of a force F, is not $(1')$ but reads

$$m_0 \frac{d}{dt} \frac{v}{\sqrt{1 - (v^2/c^2)}} = F, \tag{6}$$

where m_0 is the mass of the particle measured at rest and c is the speed of light in vacuum $(c = 3 \cdot 10^{10}$ cm/sec$)$. When Einstein proposed this law, its significance was only theoretical, since no one observed bodies moving with velocities v comparable to the velocity of light. And if v/c is very small, the number $\sqrt{1 - (v^2/c^2)}$ is so close to 1 that (6) may be replaced by Newton's law $(1')$. In modern accelerators, however, some subatomic particles reach speeds up to $(.99)c$.

Let us carry out the differentiation in (6) to obtain the acceleration $a = dv/dt$. We have (using the rules in Chapter 6, including the chain rule)

$$\frac{d}{dt} \frac{v}{\sqrt{1 - (v^2/c^2)}} = \left\{ \frac{d}{dv} \frac{v}{\sqrt{1 - (v^2/c^2)}} \right\} \frac{dv}{dt}.$$

This equals

$$a \frac{d}{dv} v \left(1 - \frac{v^2}{c^2}\right)^{-1/2} = a \left\{ \left(1 - \frac{v^2}{c^2}\right)^{-1/2} + v \frac{d}{dv} \left(1 - \frac{v^2}{c^2}\right)^{-1/2} \right\}$$

$$= a \left\{ \left(1 - \frac{v^2}{c^2}\right)^{-1/2} + v \left(-\tfrac{1}{2}\right) \left(1 - \frac{v^2}{c^2}\right)^{-3/2} \left(-\frac{2v}{c^2}\right) \right\}$$

$$= a \left\{ \left(1 - \frac{v^2}{c^2}\right)^{-1/2} + \frac{v^2}{c^2} \left(1 - \frac{v^2}{c^2}\right)^{-3/2} \right\} = \frac{a}{\left(1 - \dfrac{v^2}{c^2}\right)^{3/2}}.$$

Thus we may write (6) in the form

$$\frac{m_0 a}{\left(1 - \dfrac{v^2}{c^2}\right)^{3/2}} = F. \tag{6'}$$

This shows that the ratio F/a depends on the velocity of the body. The greater the velocity, the greater the force needed to produce a given acceleration.

ALBERT EINSTEIN (1879–1955) was working for the Swiss Patent Office in 1905 when he published three papers, each of which initiated a new physical theory. One dealt with Brownian motion; another contained the first application of Planck's quantum hypothesis to a subatomic process. Through papers such as these, Einstein became one of the fathers of quantum theory—yet he never fully accepted this theory. The third paper contained the Special Relativity Theory. This was followed in 1916 by the General Relativity Theory and then by a stubborn search, extending over decades, for a unified theory describing both gravitation and electromagnetism. (There is as yet no such theory.)

Einstein came to the United States as a refugee from Hitler's Germany in 1933 and settled in Princeton, New Jersey. In 1939 he signed a letter alerting President Roosevelt to the possibility of an atomic bomb. Throughout his life, Einstein maintained a strong interest in social causes, particularly world peace, the establishment of a Jewish homeland in Palestine, and civil liberties.

Consider now a particle of rest mass m_0 that moves according to Einstein's equation (6) under the influence of a constant force $F > 0$ and beginning at rest ($v = 0$ at $t = 0$). Then

$$\frac{v}{\sqrt{1 - (v^2/c^2)}}$$

is the primitive function of the constant function F/m_0. Also, this primitive function has the value 0 at $t = 0$. Therefore

$$\frac{v}{\sqrt{1 - (v^2/c^2)}} = \frac{Ft}{m_0}.$$

We square both sides and obtain

$$\frac{v^2}{1 - (v^2/c^2)} = \frac{F^2 t^2}{m_0^2}.$$

Solving for v^2, we get

$$v^2 = c^2 \frac{F^2 t^2}{m_0^2 c^2 + F^2 t^2}$$

or, since $v = v(t)$ is positive,

$$v(t) = c \frac{Ft}{\sqrt{m_0^2 c^2 + F^2 t^2}}.$$

The denominator in the fraction is always larger than the numerator. This shows that, for all t, $v(t) < c$. *The body can never reach the speed of light.* (But the speed of the body will, if the motion continues indefinitely, come as close to the speed of light as we want. In other words, $\lim_{t \to +\infty} v(t) = c$.)

PROBLEMS

In Problems 1 to 4 use the relativistic equations of motion unless otherwise specified. All units are in the CGS system and c denotes the speed of light.

1. A particle whose rest mass is .001 gram has at a certain instant a velocity of $.9c$ and an acceleration of 1 cm/sec^2. What is the force acting on the particle at this instant? Compare with the force that would be predicted, under the same circumstances, by the Newtonian law of motion.

2. A particle whose rest mass is 1 gram has at a certain instant a velocity of $.1c$ and an acceleration of 10 cm/sec^2. What is the force acting on the particle at this instant? Compare with the force that would be predicted by the Newtonian law of motion.

3. A particle whose rest mass is .003 gram is excited, so that its velocity is given by $v = tc/(t + 100)$, $t \geq 0$. When is the velocity equal to $.5c$? What is the force acting on the particle at this instant? When is the velocity equal to $.9c$? What is the force acting on the particle at this instant? Compare with the forces that would be predicted by the Newtonian law of motion.

4. To produce an acceleration of 5 cm/sec^2 in the motion of a certain particle, a force of 10 gm cm/sec^2 is required. If the velocity of the particle is $.8c$, what is its rest mass? Compare with the mass that would be predicted by the Newtonian law of motion.

1.5 Differentiability assumptions in physics

A tacit assumption underlies Newton's (and Einstein's) law of motion: the position of a moving particle is described by a twice differentiable function of time $s(t)$ so that we can talk about the velocity $s'(t)$ and acceleration $s''(t)$. This assumption is amply justified in many cases, but there are some situations in which it is untenable.

A microscopic particle floating in a liquid collides with the much smaller molecules of the liquid, and these innumerable collisions produce an erratic motion of the particle, a phenomenon first noticed by the botanist Brown. In the mathematical theory of *Brownian motion*, the position of the particle is described by a continuous but nondifferentiable function. The particle has therefore no definite instantaneous velocity.

About 50 years ago, physicists realized that subatomic particles, like electrons, protons, and neutrons, do not obey the laws of classical physics of Galileo and Newton. In *quantum mechanics* (the theory of motion of such particles) the velocity is not the derivative of the position, at least not in the same sense as in classical physics. For when we think of velocity as a derivative, we assume that we can measure it by measuring the position $s(t)$ at different times, plotting the curve, drawing the tangent, and measuring its slope. The more accurately we know the position, therefore, the more accurately we know the velocity. But a basic principle of quantum theory asserts that, in any experiment, a greater precision in measuring one of the two quantities, position and velocity, results in a smaller precision in measuring the other. If we measure the position of a particle with the accuracy α and its velocity with the accuracy β, then the Heisenberg *uncertainty principle* states that $\alpha\beta \geq h/m$, where m is the mass of the particle and h is a universal constant (Planck's constant).

This does not mean that calculus is not applicable in quantum theory. On the contrary, the mathematics of quantum mechanics is rooted in calculus.

§2 Rate of change‡

In this section we describe two applications of the derivative which go back to its very definition: the derivative measures how fast a function is changing.

2.1 Derivatives as rates of change

A function $x \mapsto f(x) = y$ describes the relation between two quantities, x and y; the derivative $f'(x) = dy/dx$ is called the **rate of change** of y with respect to x. Thus the

---

MAX PLANCK (1858–1947) proposed in 1900 the "quantum hypothesis," according to which energy can be emitted by electromagnetic radiation only in quantities $Nh\nu$, where N is an integer, ν the frequency of radiation, and h a universal constant.

Planck, a conservative German nationalist, tried to defend Jewish scientists during the Nazi period. His only son was hanged for plotting against Hitler.

ROBERT BROWN (1773–1858).

WERNER HEISENBERG (b. 1906) is one of the inventors of quantum mechanics. He published his uncertainty principle in 1926; it reflects the fact that every measurement affects the body whose position, or velocity, is being measured.

---

‡Optional section.

velocity is the rate of change of distance with respect to time, and the acceleration is the rate of change of the velocity with respect to time. Here is another example.

Let $V = f(x)$ denote the volume of a cube of side length x, so that $f(x) = x^3$. The rate of change of V with respect to x is $dV/dx = 3x^2$. This can be interpreted as follows. Given a cube of side length x_0 (and volume x_0^3), if we increase the sides by a *small* amount h, the volume increases *approximately* by $3x_0^2$ times the amount h, that is, by $3x_0^2 h$.

Indeed, by the linear approximation theorem (Theorem 1 in Chapter 4, §1.6) the volume V for $x = x_0 + h$ equals

$$f(x_0 + h) = f(x_0) + f'(x_0)h + r(h)h, \tag{1}$$

where $r(h)$ is continuous at $h = 0$, and $r(0) = 0$. [This is Equation (7) in Chapter 4, §1.6, where we replaced m by $f'(x_0)$ and $x - x_0$ by h.] For small values of $|h|$, the value of $|r(h)h|$ is *very* small, and writing $\approx$ for "approximately equal," we obtain

$$f(x_0 + h) \approx f(x_0) + f'(x_0)h. \tag{2}$$

In our case this becomes

$$(x_0 + h)^3 \approx x_0^3 + 3x_0^2 h, \tag{3}$$

as asserted.

[Actually, for the simple function x^3, we need no calculus to see that (3) holds. We have $(x_0 + h)^3 = x_0^3 + 3x_0^2 h + 3x_0 h^2 + h^3$. Thus the error committed in replacing $(x_0 + h)^3$ by $x_0^3 + 3x_0^2 h$ is $3x_0 h^2 + h^3$; the absolute value of this error is very small if $|h|$ is small.]

For instance, if $x_0 = 2$ and $h = .1$, we have $(2.1)^3 = 9.261$, whereas the approximation (3) yields $(2.1)^3 \approx 9.2$. For $x_0 = 2$ and $h = .01$, we obtain the exact value $(2.01)^3 = 8.120601$ and the approximation $(2.01)^3 \approx 8.12$. In all of the preceding formulas, h could be negative (a "negative increase" is, of course, a decrease). For instance, using (3) with $x_0 = 2$ and h either $(-.1)$ or $(-.01)$, we obtain $(1.9)^3 \approx 6.8$, $(1.99)^3 \approx 7.88$, whereas the precise values are $(1.9)^3 = 6.859$, $(1.99)^3 = 7.880599$.

PROBLEMS

1. Find the rate of change of the area of a disk (interior of a circle) with respect to its radius.
2. Find the rate of change of the surface area of a cube with respect to side length.
3. Find the rate of change of the volume of a sphere with respect to its radius.
4. Let h denote the height and r the radius of a cylindrical can. If the can is 1 in. high, what is the rate of change of volume with respect to the radius when the radius is $\frac{1}{2}$ in.?
5. Let h denote the height and r the radius of a cylindrical can. If the radius of the can is $\frac{1}{2}$ in., what is the rate of change of volume with respect to height when the height is 1 in.?

2.2   Using derivatives to find function values

The approximate formula (2) can be used, as in the example just discussed, to calculate the value of $f(x_0 + h)$, with a small error and with relatively little calculations, provided $f(x_0)$ and $f'(x_0)$ are easily computable. We shall discuss the accuracy of this formula in a later chapter (see Chapter 14, §1.3).

EXAMPLES 1. Compute, approximately, $\sqrt{1.02}$.
ANSWER We use (2) with $f(x) = x^{1/2}$, $x_0 = 1$, $h = .02$. Since $f'(x) = \frac{1}{2}x^{-1/2}$, we have $f(1) = 1$,
  $f'(1) = \frac{1}{2}$, and $\sqrt{1.02} = (1.02)^{1/2} \approx 1 + \frac{1}{2}(.02) = 1.01$.

2. Compute, approximately, $\sqrt[3]{63}$.
ANSWER We note that 63 is close to $64 = 4^3$. Hence we use (2) with $f(x) = x^{1/3}$, $x_0 = 64$,
  $h = -1$. We obtain $f(64) = 4$, $f'(64) = \frac{1}{3}(64)^{-2/3} = \frac{1}{3}4^{-2} = 1/48 = .0208\ldots$, and $\sqrt[3]{63} = (4^3 - 1)^{1/3} \approx 4 - 1/48 \approx 3.9792$.

PROBLEMS

   In Problems 1 to 10 use the approximation (2).

   1. Compute $(3.05)^2$, approximately.
   2. Compute $(1.08)^2$, approximately.
   3. Compute $(1.988)^2$, approximately.
   4. Compute $(2.95)^4$, approximately.
   5. Compute $\sqrt{49.1}$, approximately.
   6. Compute $\sqrt{35.99}$, approximately.
   7. Compute $1/\sqrt{25.025}$, approximately.
   8. If $g(s) = (s + 1)/(s - 1)$, compute $g(.025)$, approximately.
   9. Find an approximation to $\sqrt{1 + h}$ for $|h|$ small.
   10. Find an approximation to $1/\sqrt{1 + h}$ for $|h|$ small.

2.3   Related rates

The so-called related rate problems refer to two (or sometimes more) quantities which change in time and are also connected by some equation. The rate of change (with respect of time) of one quantity is to be found from that of the other (or others). What kind of problems arise and how they can be solved will be seen from the examples below. Time will always be denoted by the letter t.

EXAMPLES 1. Water is poured into a cylindrical container of radius $\frac{1}{2}$ ft at the rate of 36 cu in./sec. How fast is the level of the water rising?
SOLUTION The two time-dependent quantities involved in this problem are: the volume of the water in the container at time t and the level of the water at time t. Call them V and h, respectively (see Figure 7.1). We are told that $dV/dt = 36$ (in.³/sec), and we are asked to find dh/dt (measured in in./sec). First we need an equation connecting V and h. We assume that we know the formula for the volume V of a right circular cylinder of radius r and height h: $V = \pi r^2 h$. In our case we have $r = \frac{1}{2}$ ft $= 6$ in. (it is necessary to use the same units). Thus

$$V = 36\pi h,$$

so that (remembering that V and h are functions of time t)

$$\frac{dV}{dt} = 36\pi \frac{dh}{dt}.$$

Since $dV/dt = 36$, we have that $dh/dt = 1/\pi = .318\ldots$ Thus the level is rising at the rate of .32 in./sec, approximately.

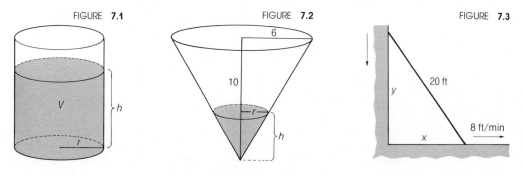

FIGURE 7.1   FIGURE 7.2   FIGURE 7.3

2.   Water is running out of a conical funnel at the rate of 2 in.³/sec. The radius at the top of the funnel is 6 in., and the height of the funnel is 10 in. How fast is the water level dropping when the water is 4 in. deep?

SOLUTION   Let V denote the volume of the water in the funnel, h the water level, and r the radius of the funnel h in. above the vertex. We are told that $dV/dt = -2$ in.³/sec (minus, since the volume is decreasing). We must find dh/dt. The volume of the water in the funnel equals the volume of a right circular cylinder of height h and radius r, so that, by a known geometric formula,

$$V = \tfrac{1}{3}\pi r^2 h. \tag{*}$$

This is not yet the desired relation between V and h, since we do not know r. But from Figure 7.2, and by similar triangles, we see that $r/h = 6/10$ so that $r = \tfrac{3}{5}h$ and (*) becomes

$$V = \frac{3\pi}{25}\, h^3. \tag{**}$$

Differentiating both sides of (**) with respect to t yields

$$\frac{dV}{dt} = \frac{9\pi}{25}\, h^2 \frac{dh}{dt}.$$

For $h = 4$ and $dV/dt = -2$, we obtain $dh/dt = -(25/72\pi) \approx .11$. Thus the water level is dropping at the rate of .11 in./sec, approximately.

3.   A ladder 20 ft long leans against a vertical wall. Suppose the bottom of the ladder is pulled away from the wall at the rate of 8 ft/min. How fast is the top of the ladder moving down the wall when the bottom of the ladder is 12 ft from the wall?

SOLUTION   Denote the height (in feet) of the point at which the ladder touches the wall by y, the distance from the wall to the bottom of the ladder by x (see Figure 7.3). Then $x^2 + y^2 = 400$, and therefore, differentiating both sides with respect to t,

$$2x \frac{dx}{dt} + 2y \frac{dy}{dt} = 0. \tag{*}$$

If $x = 12$, $y = \sqrt{400 - x^2} = \sqrt{256} = 16$. Substituting into (*) these values of x and y and the given value $dx/dt = 8$, we obtain $dy/dt = -6$. The top moves down at the rate of 6 ft/min.

PROBLEMS

1. A swimming pool in the form of a right circular cylinder is 20 ft deep and 100 ft in diameter. If water flows in at the rate of 100 cu ft/min, how fast is the water level rising?

2. A spherical snowball is melting at the rate of $\frac{1}{2}$ cu ft/hr. Assuming the snowball maintains a spherical shape, how fast is the diameter decreasing when the snowball is 4 ft in diameter?

3. A spherical balloon is filled with air in such a way that its volume increases at the rate of 4π cu in./sec. How fast is the surface area increasing when the radius is 8 in.?

4. A stone is cast into a still pond. Ripples, in the form of concentric circles, propagate outward into the undisturbed region. If the radius of the disturbed region increases at the rate of 10 ft/sec, what is the rate of increase of the area of the disturbed region when its radius is 20 ft?

5. A ladder 20 ft long leans against a wall. Suppose the bottom of the ladder slides away from the wall at the rate of 4 ft/min. How fast does the top of the ladder move down the wall when the bottom of the ladder is 12 ft from the wall?

6. A kite at an altitude of 300 ft is moving horizontally at the rate of 20 ft/sec away from the person flying the kite. How fast must the line be let out in order to maintain the same altitude when the kite is 500 ft away?

7. A perfect gas confined within a container and held at constant temperature obeys the law $pv = c$, where p is the pressure, v is the volume, and c is a constant. At a particular instant the pressure is 400 lb/sq in., and the volume is 1000 cu in. If the volume is decreasing at the rate of 5 cu in./min, what is the rate of change of pressure?

8. A sandpile is in the form of a right circular cone with the height equal to the radius of the base. If sand is poured on the top of the sandpile at the rate of $\frac{3}{2}$ cu ft/min, how fast is the height increasing when the height is 5 ft? (Assume the sandpile maintains its original shape.)

9. A barge is pulled toward a dock by means of a taut cable. If the barge is 20 ft below the level of the dock, and the cable is pulled in at the rate of 36 ft/min, how fast is the barge moving when the cable is 52 ft long?

10. A particle moves along the curve $y = \sqrt{x^2 + 16}$ in such a way that its distance from the origin increases at the rate of 6 cm/sec. What is the rate of change of x when the particle is at the point (3,5)?

11. A ship, steaming due north at the rate of 16 knots, observes a fixed buoy 3 nautical miles due west of its course. How fast is the ship receding from the buoy 15 min later? [*Note:* 1 knot = 1 nautical mile per hour; 1 nautical mile = 2000 yd.]

12. A train leaves a station and travels north at the rate of 60 mph. One hour later a second train leaves the same station and travels east at the rate of 45 mph. At what rate are the trains separating from each other 2 hr after the second train leaves the station?

13. A street light hangs 24 ft above the sidewalk. If a man 6 ft tall walks away from the light at the rate of 3 ft/sec, how fast is the tip of his shadow moving?

14. A light is located on the floor of a stage 45 ft in front of a vertical screen. An actress makes an entrance from behind the screen at a point nearest the light and walks toward the light at the rate of 3 ft/sec. If the actress is 5 ft tall, how fast will her shadow be growing on the screen when she is 25 ft from the light?

15. A weather balloon rises vertically at the rate of 30 ft/sec from the top of an 80-ft tower. At the same time a truck leaves the base of the tower at the rate of 37.5 ft/sec. How fast is the distance between the balloon and truck increasing 4 sec later?

16. Oil is leaking from a hemispherical tank at a constant rate of $\pi/8$ cu ft/min. If the diameter of the tank is 16 ft, how fast is the oil level dropping when the oil is 6 ft deep? [*Hint:* The volume of a spherical cap is $V = \pi h^2(a - h/3)$, where a is the radius of the sphere and h is the height of the cap.]

17. A ball is dropped from the roof of a building which is 120 ft high. If a light is located 16 ft away on the edge of the roof, how fast will the ball's shadow be moving on the sidewalk below 2 sec later? Assume that the ball falls according to Galileo's law (see Chapter 4, §3.3).

18. Water is flowing into a leaky cistern at the rate of 31.4 cu ft/min. The cistern is a right circular cone 20 ft deep and 20 ft in diameter at the top. If the water level is rising at the rate of $\frac{1}{4}$ ft/min, how fast will the water be leaking out when the water level is 12 ft deep?

§3 Marginal quantities in economics[‡]

Since its inception, calculus served as a tool of natural sciences. The penetration of mathematics into social sciences is a more recent phenomenon. In this section we give some samples of the uses of calculus in microeconomics, the branch of economics which studies the economic decisions of individual economic units. In particular, we focus on the production and distribution of a single commodity by a single firm.

3.1 Cost, marginal cost, average cost

Consider an enterprise that produces some product. The cost of producing x units of the product will be denoted by $C(x)$. Here x may denote the number of pieces produced, or the number of thousands of pieces, or the number of pounds, or the number of yards, and so forth. The cost $C(x)$ can be measured in dollars, in thousands of dollars, or in other monetary units.

To determine the **cost function** $C(x)$ is a difficult accounting task. Here, however, we take this function as given. We assume that $C(x)$ is a continuously differentiable function. The derivative $C'(x)$ is called the **marginal cost.** In general, economists say "marginal f" where a mathematician would say "derivative of f."

The assumption that the cost function $C(x)$ is continuously differentiable is an idealization of reality. (Such idealizations are needed wherever one uses mathematics. The particle of mass m which we dealt with in §1 is also an idealization.) Let us assume, for the sake of definiteness, that x is the number of pieces produced, hence an integer, and the cost C is measured in dollars. The cost of producing our product can be described by a table listing x in one column and the cost of producing x pieces in another. This cost table can be represented by drawing, for each integer x, the point with first coordinate equal to x and second coordinate equal to the cost of producing x units C (as in Figure 7.4). Now we draw a smooth curve passing through, or nearly through, these points, and we think of this curve as the graph of a continuously differentiable cost function $C(x)$.

The marginal cost $C'(x)$ is the slope of the tangent to the curve $y = C'(x)$. If x is the number of pieces, $C'(x)$ is measured in dollars. If x were measured in, say, tons, $C'(x)$ would be measured in dollars per ton and so forth.

Economists describe marginal cost as "the cost of producing *one more* piece or *one more* unit." This is a good description if the units are small enough. Then the

[‡]Optional section.

FIGURE 7.4   FIGURE 7.5

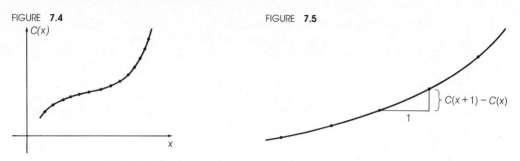

slope of the tangent to $C(x)$ will change very little if x is increased by 1, and the slope $C'(x)$ will be nearly the difference quotient

$$\frac{C(x+1) - C(x)}{(x+1) - x} = C(x+1) - C(x).$$

(See Figure 7.5.) But if we use for x large units, we cannot describe marginal cost as the cost of producing one more unit.

The same applies to concepts such as marginal revenue, marginal profit, and so forth.

A typical shape of the graph of $C'(x)$ is shown in Figure 7.6. The cost function $C(x)$ is increasing (it costs more to produce more). The marginal cost, however, is first decreasing then increasing. (It costs more to produce the first piece than to produce one more piece when many are being produced. But when we produce nearly all we can, it becomes more expensive to increase production by even a small amount.) A typical cost curve $C(x)$ therefore is first concave, then convex. It has an inflection point at the value of x corresponding to the minimum marginal cost. Such a cost curve is shown in Figure 7.7.

The quantity

$$\frac{C(x)}{x} \tag{1}$$

is called the **average cost** of producing x units. A typical average cost curve is shown in Figure 7.8. Since some cost is unavoidable even before a single unit is produced (for instance, the maintenance of the plant), $C(0) > 0$. Therefore $(1/x)C(x)$ has the limit

FIGURE 7.6   FIGURE 7.7   FIGURE 7.8

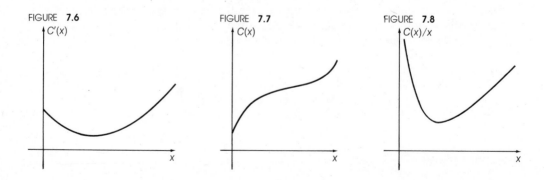

$+\infty$ as $x \to 0^+$. The marginal average cost, that is, the derivative of $(1/x)C(x)$, is

$$\left(\frac{C(x)}{x}\right)' = \frac{xC'(x) - C(x)}{x^2} = \frac{1}{x}\left(C'(x) - \frac{C(x)}{x}\right). \tag{2}$$

This derivative must be zero at the value of x corresponding to the minimum average cost. But when $(C(x)/x)'$ is 0,

$$C'(x) = \frac{C(x)}{x}. \tag{3}$$

Thus: *when average cost is smallest, it equals marginal cost.*

Like most other microeconomical statements, this one could be verified by a verbal argument and bolstered by graphs. But calculus makes the argument brief and lucid.

EXAMPLE   The cost in dollars of producing x units of a product is

$$C(x) = 9000 + 1000x - 10x^2 + \tfrac{1}{3}x^3.$$

What is the marginal average cost? Verify that the smallest value of the average cost is equal to the corresponding value of the marginal cost.

ANSWER   The average cost is

$$\frac{C(x)}{x} = \frac{9000}{x} + 1000 - 10x + \tfrac{1}{3}x^2.$$

The marginal average cost is

$$\left(\frac{C(x)}{x}\right)' = -\frac{9000}{x^2} - 10 + \tfrac{2}{3}x = \frac{1}{x^2}(\tfrac{2}{3}x^3 - 10x^2 - 9000).$$

A zero of this derivative (found by factoring) is $x = 30$. By long division,

$$\tfrac{2}{3}x^3 - 10x^2 - 9000 = (x - 30)(\tfrac{2}{3}x^2 + 10x + 300).$$

The quadratic polynomial $\tfrac{2}{3}x^2 + 10x + 300$ has no (real) zeros. Hence $x = 30$ is the only zero of $(C(x)/x)'$. The second derivative of the average cost is

$$\frac{18{,}000}{x^3} + \frac{2}{3};$$

this is positive for $x = 30$. Thus the average cost attains its minimum at $x = 30$. For this value of x the average cost is

$$\frac{C(30)}{30} = \frac{9000}{30} + 1000 - 10(30) + \tfrac{1}{3}900 = 1300 \text{ dollars.}$$

The marginal cost is

$$C'(x) = 1000 - 20x + x^2;$$

for $x = 30$, $C'(30) = 1000 - 600 + 900 = 1300$ dollars as it should be.

PROBLEMS

1. The cost in dollars of producing x units of a product is $C(x) = 10{,}000 + 10x + x^2$. Find the marginal cost. Compute the value of the marginal cost when $x = 50$. Then compare this value with the cost of producing one more unit "at the level $x = 50$," that is, compute $C(51) - C(50)$. Is the marginal cost a reasonable approximation at this level?

2. The cost in dollars of producing x units of a product is $C(x) = 400 + 20x + x^2$. What is the marginal cost? The average cost? The marginal average cost? For what values of x is the average cost a minimum?

3. The cost function in cents for a commodity is given by $C(x) = 4000 + 10x - 30x^2 + x^3$, where x is the number of pieces produced. Find the marginal cost and the average cost. For what value of x is the average cost a minimum?

4. If the cost in rubles of producing x units of a product is $C(x) = 720{,}000 + 10{,}000x - 100x^2 + \frac{1}{3}x^4$, how many units should be produced so that the average cost is a minimum? What is the average cost of producing this many units?

3.2   Maximizing profit under perfect competition

Let us assume that our enterprise is a so-called perfect competitor. It produces such a small share of the product that it cannot influence the price, and it can sell as much as it produces at a *fixed* market price p. How much should it produce to maximize its profit?

If the output of our firm is x pieces, our total **revenue** $R(x)$ is (price of one piece) times (number of pieces). Thus $R(x) = px$. Our **profit** $\pi(x)$ is (revenue) minus (cost). Therefore

$$\pi(x) = R(x) - C(x) = px - C(x). \tag{4}$$

[We include all marketing costs in the production cost $C(x)$ for the sake of simplicity.] The marginal profit is the derivative

$$\pi'(x) = p - C'(x).$$

This derivative must be 0 at the value of x for which the profit is maximal. And if $\pi'(x) = 0$, then $p = C'(x)$. For a typical case (see Figure 7.9), there will be at most two such values of x. If, at the value of x for which $p - C'(x) = 0$, $C'(x)$ increases, $\pi'(x)$ will change sign from plus to minus, and therefore the profit $\pi(x)$ will indeed have a maximum.

FIGURE **7.9**

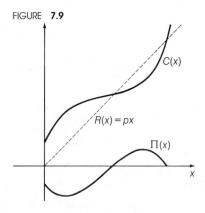

We conclude that *under conditions of perfect competition, a manufacturer maximizes his profits by producing that quantity of the product for which the marginal cost equals the price.*

EXAMPLE   The daily cost to a small firm of producing x units of a product is

$$C(x) = 2002 + 120x - 5x^2 + \tfrac{1}{3}x^3$$

dollars. The market price of 1 unit is 264 dollars. What should be the daily output x in order to maximize the daily profit? What is the daily profit for this x?

ANSWER   The marginal cost is

$$C'(x) = 120 - 10x + x^2.$$

Setting this equal to the price of 264, we obtain the equation $x^2 - 10x - 144 = 0$, which has the roots -8 and 18. The first is, in this context, meaningless. At $x = 18$ the marginal cost increases since $C''(x) = -10 + 2x$ and $C''(18) = 26 > 0$. Hence the profit maximizing daily output is 18 units. For $x = 18$, the daily revenue is $R(18) = 18 \cdot 264 = \$4752$. The daily cost is $C(18) = \$4486$, and the daily profit is $R(18) - C(18) = \$266$.

PROBLEMS

1. The daily cost in dollars of producing x units of a commodity is $C(x) = 1200 + 20x + x^2$. If the market price of 1 unit is 100 dollars, how many units of the commodity should be produced to maximize the profit? What is the daily profit for this x?
2. A product is marketed by a business firm at a price of 50 dollars per unit. The total cost in dollars of marketing x units of the product is $C(x) = 5000 + 650x - 45x^2 + x^3$. How many units should be produced to maximize the profit? What is the daily profit for this x?

3.3   Maximizing the profit of a monopolist

A small producer must sell all he produces at the prevailing market price over which he has no control. But the total sale of a commodity depends on the price at which it is offered. (This is so with "all other things being equal." The term in quotation marks is another idealization made by economists.) The higher the price p, the smaller the amount x which will be sold. Thus x, the amount "demanded," is a decreasing function of the price p of a unit. We assume this function to be continuous. Hence the inverse function is also decreasing and continuous. The function f defined by

$$p = f(x)$$

is called the **demand function.** A typical demand function is graphed in Figure 7.10.

Consider next a firm which has a monopoly and can decide on the total output of its product. (It does not matter here whether or not the firm is state-owned.) The quantity x of the product which will be produced determines the unit price $p = f(x)$. Thus the revenue for the output x is

$$R(x) = xp = xf(x). \tag{5}$$

The profit for this output is

$$\pi(x) = C(x) - R(x),$$

where C is, as before, the cost function. In order that the profit be maximal, we must have $\pi'(x) = 0$, that is,

$$C'(x) = R'(x). \tag{6}$$

FIGURE 7.10

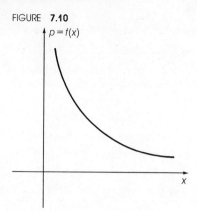

When the profit of a monopolist is maximal, the marginal cost equals the marginal revenue.

This is a necessary condition for maximizing the profit. A sufficient condition, for the same output at which (6) holds, is $\pi''(x) < 0$, that is,

$$C''(x) < R''(x). \tag{7}$$

EXAMPLE The daily demand and cost functions for a commodity are given by

$$p = f(x) = 620 - 8x, \qquad C(x) = 100 + 20x + 2x^2.$$

If the commodity is produced by a monopolist, what is the daily maximal profit? Assume x is the number of units and p and C are in dollars.

ANSWER   The revenue, at output x, is $R(x) = px = 620x - 8x^2$. The daily profit is therefore

$$\pi(x) = R(x) - C(x) = -10x^2 + 600x - 100.$$

The marginal profit, $\pi'(x) = -20x + 600$, is 0 for $x = 30$. Since $\pi''(x) = -20$ is negative for all x, $\pi(x)$ has a maximum at $x = 30$. This maximum is $\pi(30) = -10(30)^2 + 600(30) - 100 = \$8900$.

The reader may check that $R'(x) = C'(x)$ at $x = 30$.

PROBLEMS

1. The demand function for a commodity is given by $p = f(x) = 4000 - 20x$, where x is the number of units produced. If the cost function in dollars is $C(x) = 10x^2 + 1000x + 63,000$, how many units of the commodity should be produced to maximize the profit?
2. Suppose the cost in dollars of making x items of a product is given by $C(x) = 10x^2 + 200x + 5000$. If the demand function is given by $p = f(x) = 2000 - 5x$, how many items should be produced to maximize the profit? What is the maximum profit?
3. The daily demand and cost functions for a commodity are given by $p = 5000 + 25x - x^2$ and $C(x) = 4000 + 1400x + 10x^2$ (dollars), where x is the number of units produced. How many items should be produced to maximize the profit, and what is the maximum profit?
4. Suppose the demand and cost functions for a commodity are given by $p = 1000 - 10x$ and $C(x) = 10x^2 + 200x + 6000$ (dollars). For what values of x will the profit be positive? What value of x gives the maximum profit?

INTEGRALS

§1   The integral of a function. The area under a curve

The "problem of tangents" was one of the stimuli to the development of calculus; it led to the concept of derivatives. Another stimulus, which led to the concept of integral, was the "problem of areas": how to calculate the area of a region with a curved boundary. The fundamental theorem of calculus (see §2) shows that these two, seemingly very different, problems are intimately connected.

1.1   The integral of a positive function

Consider a continuous function $y = f(x)$ defined for $a \leq x \leq b$, which takes on no negative values, so that its graph never lies below the x-axis (Figure 8.1). The curve $y = f(x)$ and the lines $x = a$, $x = b$, and $y = 0$ bound a certain region of the plane, the *region under the curve* $y = f(x)$ *from* a *to* b. This region has a definite area A.

Is this last statement true? Our intuition and our experience tell us that it is. If asked to compute the area, we could, for instance, cover the plane with a mesh of small squares and count the number of squares within our region (Figure 8.2). This would not give us the area precisely, since some squares would lie partly inside and partly outside our region. But we expect that, by making the mesh sufficiently fine, we could compute A with any desired degree of accuracy. That there is indeed a

FIGURE 8.1   FIGURE 8.2   FIGURE 8.3

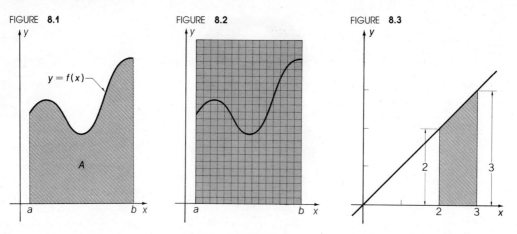

number A that can be so computed is a mathematical statement requiring proof, and we shall discuss this matter later. For the moment, we take the concept of area for granted.

The number A depends on the function f and on the numbers a and b. Following Leibniz, we write

$$A = \int_a^b f(x)\, dx.$$

The right-hand side is read "the **integral** from a to b of the function $f(x)$ with respect to x" or "the integral from a to b of $f(x)\, dx$."

Here is an example. The symbol

$$\int_2^3 x\, dx$$

represents the area indicated in Figure 8.3. The region in question is a trapezoid, so that we know how to compute its area. Thus

$$\int_2^3 x\, dx = \tfrac{5}{2}.$$

The variable x occurring in this formula (the **variable of integration**) is a so-called **dummy variable.** It serves only to identify the form of the function which we "integrate," that is, for which we compute the area. The dummy variable may be replaced by any other symbol. For instance,

$$\int_2^3 y\, dy = \tfrac{5}{2}, \qquad \int_2^3 a\, da = \tfrac{5}{2}, \qquad \int_2^3 \text{$\flat$}\, d\text{$\flat$} = \tfrac{5}{2}.$$

The symbol d serves to pinpoint the variable of integration. For example, for any fixed (positive) number α,

$$\int_1^3 \alpha\, dx = 2\alpha,$$

since the region under the graph of the constant function $y = \alpha$, from $x = 1$ to $x = 3$, is a rectangle of width 2 and height α (see Figure 8.4).

PROBLEMS

In Problems 1 to 5 evaluate the given integrals using knowledge of areas of simple geometric figures. Sketch the graph of the integrand and shade the relevant area.

1. $\int_{1}^{4} 3z\, dz.$ 2. $\int_{-1}^{1/2} (1-s)\, ds.$ 3. $\int_{-1}^{1} |u|\, du.$

4. $\int_{-2}^{1} h(t)\, dt$ if $h(t) = 1$ for $t \leq 0$; $t+1$ for $t > 0$.

5. $\int_{1}^{4} \phi(r)\, dr$ if $\phi(r) = 2 - r$ for $r \leq 2$; $2r - 4$ for $2 < r \leq 3$; 2 for $r > 3$.

1.2   The integral of a function that takes on negative values

We now extend the definition of the integral to functions that take on negative values by agreeing to count areas below the x-axis as negative. Thus, *if* $a < b$, *the number*

$$\int_{a}^{b} f(x)\, dx$$

equals the area below the curve $y = f(x)$ *and above the x-axis, between the lines* $x = a$ *and* $x = b$, *minus the area below the x-axis and above* $y = f(x)$, *between the lines* $x = a$ *and* $x = b$.

Thus, if f is given by the graph in Figure 8.5, the number $\int_{a}^{b} f(x)\, dx$ equals the sum of areas I and III minus the sum of areas II and IV.

As an example, we compute $\int_{-2}^{1} 2x\, dx$ (see Figure 8.6). The triangle under the x-axis has area 4; the one above the axis has area 1. Hence

$$\int_{-2}^{1} 2x\, dx = 1 - 4 = -3.$$

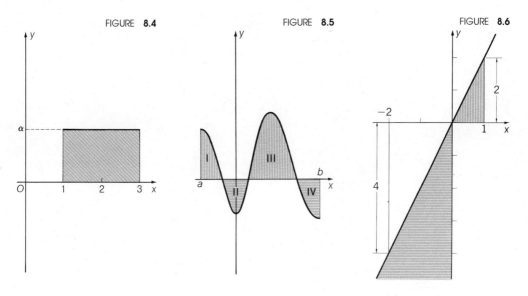

FIGURE **8.4**   FIGURE **8.5**   FIGURE **8.6**

PROBLEMS

In Problems 1 to 5 evaluate the given integrals. Sketch the graph of the integrand and shade the relevant area.

1. $\displaystyle\int_{-2}^{6} (-3)\,dx.$

2. $\displaystyle\int_{0}^{2} (x-2)\,dx.$

3. $\displaystyle\int_{-1}^{4} (1-2x)\,dx.$

4. $\displaystyle\int_{-10}^{10} \phi(z)\,dz$ if $\phi(z) = 2z + 10$ for $-10 \le z \le 0$; $-2z + 10$ for $0 < z \le 10$.

5. $\displaystyle\int_{-1}^{3} f(v)\,dv$ if $f(v) = -2$ for $v \le 0$; $v - 2$ for $0 < v \le 2$; $2v - 4$ for $v > 2$.

1.3 Three basic properties

A special but important case is that of a *constant* function, $f(x) = \alpha$, for all x in the interval considered. We have

$$\int_{a}^{b} \alpha\,dx = (b-a)\alpha, \tag{1}$$

since the rectangle of width $b - a$ and height $|\alpha|$ has area $|\alpha|\,(b-a)$. The formula is illustrated in Figure 8.4 for an $a = 1$, $b = 3$, and $\alpha > 0$. If $\alpha = 0$, then the graph of the function is a segment of the x-axis; there is no actual rectangle, and zero is the only value that makes sense for the area.

A basic and intuitively obvious property of area is: if $a < b < c$, then

$$\int_{a}^{b} f(x)\,dx + \int_{b}^{c} f(x)\,dx = \int_{a}^{c} f(x)\,dx. \tag{2}$$

This formula is illustrated in Figure 8.7: the total shaded area is the sum of the two differently shaded areas. We call relation (2) the *additivity* of the integral with respect to the interval of integration.

Figure 8.7 shows a positive function, but by considering the function shown in Figure 8.8, the reader might be convinced that the argument is general. In this figure, $\int_{a}^{b} f(x)\,dx = \text{I} - \text{II}$, $\int_{b}^{c} f(x)\,dx = -\text{III} + \text{IV}$, and $\int_{a}^{c} f(x)\,dx = \text{I} + \text{IV} - (\text{II} + \text{III})$, so that (2) holds.

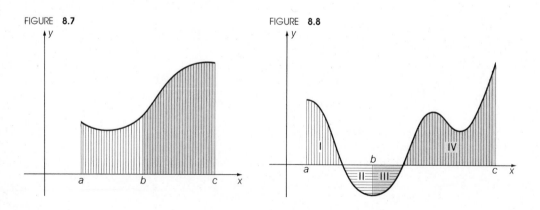

FIGURE **8.7**

FIGURE **8.8**

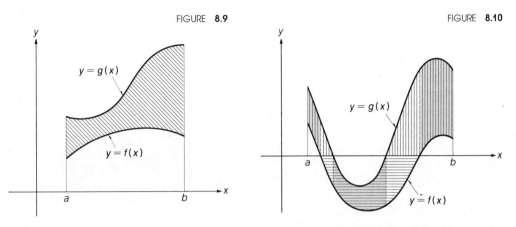

FIGURE 8.9

FIGURE 8.10

Another intuitively obvious property of the area (called the *monotonicity* of the integral) is

$$\text{if } a < b \text{ and } f(x) \leq g(x) \text{ for } a < x < b, \text{ then } \int_a^b f(x)\, dx \leq \int_a^b g(x)\, dx. \qquad (3)$$

In words: "the larger function has the larger integral (over the same interval)." This statement is illustrated in Figure 8.9. The area under f is contained in that under g. A glance at Figure 8.10 might convince the reader that our statement also holds for functions that take on negative values.

1.4 Piecewise continuous, bounded functions

The additivity property (2) suggests a way of defining integrals of certain functions with discontinuities.

Consider, for instance, the function

$$f(x) = \begin{cases} 2, & \text{for } x < 1 \\ x - 1, & \text{for } x > 1 \end{cases}$$

graphed in Figure 8.11; this function is continuous for all $x \neq 1$ and is discontinuous at $x = 1$. What meaning shall we assign to the symbol $\int_0^3 f(x)\, dx$? It is natural to interpret this to mean the sum of the two areas shown in Figure 8.11: $\int_0^1 f(x)\, dx = \int_0^1 2\, dx = $ area of rectangle $= 2$ and $\int_1^3 f(x)\, dx = \int_1^3 (x - 1)\, dx = $ area of triangle $= \frac{1}{2} 2 \cdot 2 = 2$. Thus $\int_0^3 f(x)\, dx = 2 + 2 = 4$. Similarly,

$$\int_{-1}^4 f(x)\, dx = \int_{-1}^1 f(x)\, dx + \int_1^4 f(x)\, dx = \int_{-1}^1 2\, dx + \int_1^4 (x - 1)\, dx$$
$$= 2 \cdot 2 + \tfrac{1}{2} 3 \cdot 3 = \tfrac{17}{2}.$$

The same method works if the interval of integration can be divided, not into two, but into several intervals in each of which the function $f(x)$ is continuous.

A function f, defined in a finite interval $a < x < b$, will be called **piecewise continuous** if f is continuous either at all points of the interval or at all but finitely many points. The function f will be called **bounded** if there is a number M such that

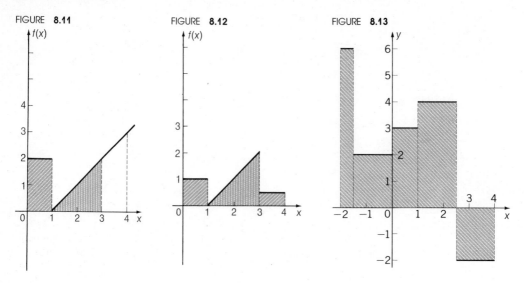

FIGURE 8.11   FIGURE 8.12   FIGURE 8.13

$|f(x)| \le M$ for all x, $a < x < b$. We expect that for such f, the number $\int_a^b f(x)\, dx$ can be defined and that properties (1), (2), (3) remain valid.

EXAMPLE   Let

$$f(x) = \begin{cases} 1, & \text{for } 0 < x < 1 \\ x - 1, & \text{for } 1 < x < 3 \\ \frac{1}{2}, & \text{for } 3 < x < 4. \end{cases}$$

Compute $\int_0^4 f(x)\, dx$. [The values of $f(x)$ at $x = 0, 1, 3$, and 4 have not been defined, they are irrelevant for calculating the integral.]

ANSWER   The piecewise continuous function f is graphed in Figure 8.12. The integral in question is the sum of all shaded areas, hence equals $1 + 2 + \frac{1}{2} = 3.5$; more formally:

$$\int_0^4 f(x)\, dx = \int_0^1 f(x)\, dx + \int_1^3 f(x)\, dx + \int_3^4 f(x)\, dx \qquad \text{[by the additivity property (2)]}$$

$$= \int_0^1 1\, dx + \int_1^3 (x - 1)\, dx + \int_3^4 \tfrac{1}{2}\, dx$$

$$= 1 \cdot 1 + \tfrac{1}{2} \cdot 2 \cdot 2 + 1 \cdot \tfrac{1}{2} \qquad \text{[by the geometric meaning of integrals]}$$

$$= \tfrac{7}{2}.$$

PROBLEMS

Evaluate $\int_a^b f(x)\, dx$ for the following functions $f(x)$ and indicated intervals $[a,b]$.

1. $f(x) = \begin{cases} -2, & -2 \le x < 1, \\ 2, & 1 \le x \le 4. \end{cases}$ 2. $f(x) = \begin{cases} 2x - 1, & -2 \le x < 0, \\ 2, & 0 \le x \le 6. \end{cases}$

3. $f(x) = -1$ for $-2 \le x < 0$; 0 for $0 \le x < 1$; $-\frac{1}{2}x + \frac{5}{2}$ for $1 \le x \le 5$.

4. $f(x) = 1$ for $-2 \le x < 0$; $-x + 2$ for $0 \le x < 2$; -1 for $2 \le x \le 10$.

5. $f(x) = -x$ for $-1 \le x < 0$; -1 for $0 \le x < 2$; $\frac{1}{2}x - 1$ for $2 \le x \le 4$.

1.5   Computing the integral using step functions

We show now how properties (1), (2), and (3) of integrals permit us to compute the integral of a given function.

A function $g(x)$ defined for $a < x < b$ is called a **step function** if the interval (a,b) can be divided into finitely many subintervals in each of which $g(x)$ is constant. An example is the function

$$g(x) = \begin{cases} 6, & \text{for } -2 < x \le -1.5 \\ 2, & \text{for } -1.5 < x \le 0 \\ 3, & \text{for } 0 < x \le 1 \\ 4, & \text{for } 1 < x \le 2.5 \\ -2, & \text{for } 2.5 < x \le 4. \end{cases}$$

This is a step function defined in the interval $(-2,4)$. Its subintervals of constancy are $(-2,-1.5), (-1.5,0), (0,1), (1,2.5)$, and $(2.5,4)$; the values g takes on at the endpoints of the subintervals are of no interest. The graph of the function g is shown in Figure 8.13; its shape explains the name "step function."

The integral of a step function can be computed at once, either using additivity and the formula for the integral of a constant or, even simpler, using the definition of the integral as an area. For the function g defined above, for instance,

$$\int_{-2}^{4} g(x)\,dx = \int_{-2}^{-1.5} 6\,dx + \int_{-1.5}^{0} 2\,dx + \int_{0}^{1} 3\,dx + \int_{1}^{2.5} 4\,dx + \int_{2.5}^{4} (-2)\,dx$$

$$= 6 \cdot \tfrac{1}{2} + 2 \cdot \tfrac{3}{2} + 3 \cdot 1 + 4 \cdot \tfrac{3}{2} + (-2) \cdot \tfrac{3}{2} = 12,$$

as can also be seen in Figure 8.13.

Now let f be a bounded piecewise continuous function defined in a finite interval (a,b), and let ϕ and ψ be two step functions such that $\phi \le f \le \psi$, that is, $\phi(x) \le f(x) \le \psi(x)$ for $a < x < b$. Then, by the monotonicity property (3),

$$\int_{a}^{b} \phi(x)\,dx \le \int_{a}^{b} f(x)\,dx \le \int_{a}^{b} \psi(x)\,dx. \tag{4}$$

The left and right integrals are computable, and knowing them we also know $\int_{a}^{b} f(x)\,dx$, with an error not exceeding $\int_{a}^{b} \psi(x)\,dx - \int_{a}^{b} \phi(x)\,dx$. The geometric meaning of the statement just made is shown in Figure 8.14: the area under the curve $y = f(x)$ is greater than the sum of the areas of the "short" rectangles and less than that of the "tall" ones. Figure 8.15 illustrates the same situation for a function that takes on both positive and negative values.

When f is given, we can find step functions ϕ and ψ such that $\phi \le f \le \psi$: divide the interval (a,b) into subintervals, and in each of the subintervals choose values for ϕ and ψ satisfying our condition. It turns out that in this way *the value of $\int_{a}^{b} f(x)\,dx$ can be computed with any desired degree of accuracy.*

Let us verify this statement for an *increasing* function $f(x)$. We divide the interval (a,b) into N equal subintervals, each of length $(b - a)/N$. In each of these subintervals, we choose the largest possible value for the step function ϕ, namely, the value of f at the left endpoint of the subinterval. Similarly, we choose for ψ the smallest possible value of each subinterval, the values of f at the right endpoint (see Figure 8.16). Now, what is the difference $\int_{a}^{b} \psi(x)\,dx - \int_{a}^{b} \phi(x)\,dx$? It is the sum of the N shaded areas in Figure 8.17. The sum is the tall shaded area in Figure 8.17. It equals $(b - a)/N$

FIGURE 8.14

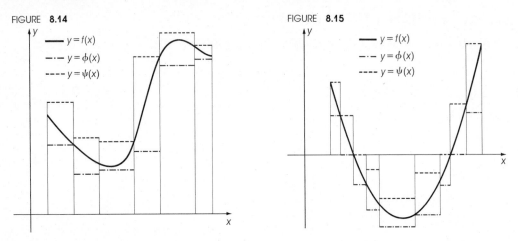

FIGURE 8.15

times the sum of the heights, which is $f(b) - f(a)$. Thus we know $\int_a^b f(x)\, dx$ with an error not exceeding

$$\int_a^b \psi(x)\, dx - \int_a^b \phi(x)\, dx = \frac{1}{N}(b - a)[f(b) - f(a)]. \qquad (5)$$

This number can be made arbitrarily small by choosing N large enough!

The same proof works for nondecreasing functions, and a similar argument applies to nonincreasing functions. [For nonincreasing functions, (5) will be negative; the absolute value of (5) is a bound for the maximum error.] Also, it was only for the sake of convenience that we divided the interval (a,b) into N *equal* subintervals; what matters is only that the subintervals be small enough. In view of the additivity property (2), we conclude that the integral of a *piecewise monotone* function is computable, using step functions, to any prescribed degree of accuracy. This takes care of all functions commonly encountered in applications of calculus.

The method of computing integrals by "trapping" a function between two step functions works also for functions that are not piecewise monotone, but the proof is harder ▶ .

FIGURE 8.16

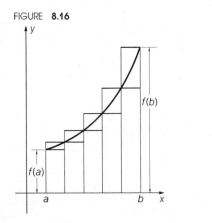

FIGURE 8.17

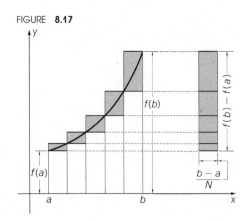

EXAMPLE Compute

$$\int_0^{.5} (1 + x^2)\, dx$$

with an error not exceeding (.03).

ANSWER The function $f(x) = 1 + x^2$ is increasing in the interval considered [since $f'(x) = 2x > 0$ for $0 < x < .5$]. We can use the method just outlined.

First we ask how large an N (number of subdivisions) we need to achieve the desired accuracy. The maximum error is given by (5). In our case $a = 0$, $b = .5$, $b - a = .5$ and $f(b) - f(a) = 1.25 - 1 = .25$. Thus the maximum error (5) equals $(1/N)(.5)(.25) = .125/N$, so that we need an N with $.125/N < .03$ or $N > (.125)/(.03) \approx 4.17$. We choose $N = 5$, so that $.125/N = .025 < .03$. From the table we see that the values of the step function ϕ in the

x	0	.1	.2	.3	.4	.5
$f(x)$	1	1.01	1.04	1.09	1.16	1.25

five subintervals are 1, 1.01, 1.04, 1.09, and 1.16, while the values of the step function ψ are 1.01, 1.04, 1.09, 1.16, and 1.25. The length of each subinterval is $.5/5 = .1$. Therefore

$$\int_0^{.5} \phi\, dx = (.1)(1 + 1.01 + 1.04 + 1.09 + 1.16) = .530,$$

$$\int_0^{.5} \psi\, dx = (.1)(1.01 + 1.04 + 1.09 + 1.16 + 1.25) = .555.$$

Thus

$$.530 \le \int_0^{.5} (1 + x^2)\, dx \le .555.$$

In the following section, we shall learn how to compute this particular integral precisely; we shall find that

$$\int_0^{.5} (1 + x^2)\, dx = .541\overline{6}.$$

PROBLEMS

In Problems 1 to 8 compute the given integrals to within the desired degree of accuracy by using step functions. Whenever possible, compute also the exact value.

1. $\int_0^1 x\, dx$, with an error not exceeding .2.

2. $\int_0^2 (2x + 1)\, dx$, with an error not exceeding .8.

3. $\int_{-1}^2 (2x + 3)\, dx$, with an error not exceeding 2.

4. $\int_0^1 (x^2 + 1)\, dx$, with an error not exceeding .1.

5. $\int_1^3 (x^2 + x)\, dx$, with an error not exceeding 2.

6. $\int_0^9 \sqrt{x}\, dx$, with an error not exceeding 10.

7. $\int_{1/2}^{3/2} \dfrac{2x}{(2x+1)}\, dx$, with an error not exceeding .1.

8. $\int_1^2 \dfrac{dx}{x}$, with an error not exceeding $\dfrac{1}{4}$.

1.6   Analytic definition of integrals

Thus far, we have proceeded as if we knew all about areas, and we have used this knowledge to define integrals. Now we reverse our point of view. We shall give an analytic definition of the integral—a definition based only on properties of numbers.

First we give an analytic definition for step functions: if a step function $\phi(x)$ defined in a finite interval (a,b) has k intervals of constancy, of length $l_1, l_2, \ldots, l_k$, and if ϕ takes on in these intervals the values $\alpha_1, \alpha_2, \ldots, \alpha_k$, then

$$\int_a^b \phi(x)\, dx = \alpha_1 l_1 + \alpha_2 l_2 + \cdots + \alpha_k l_k.$$

We use step functions for defining the integrals of other functions. Our previous discussion suggests that the following statement is true.

Theorem 1 (existence of integrals). *Let f be a bounded, piecewise continuous function defined in the finite interval (a,b). Then there is a number A, and only one such number, with the property: for all step functions ϕ and ψ such that $\phi(x) \le f(x) \le \psi(x)$ for $a < x < b$, we have*

$$\int_a^b \phi(x)\, dx \le A \le \int_a^b \psi(x)\, dx.$$

Definition.   *This number A is denoted by the symbol*

$$\int_a^b f(x)\, dx$$

and is called the integral of f over (a,b).

The reader is invited to accept this theorem without proof ► for the time being. The same applies to the next statement.

Theorem 2 (basic properties of the integral).   *Let (a,b) be a finite interval. Then*

$$\int_a^b \alpha\, dx = \alpha(b-a), \qquad \textit{for every constant } \alpha, \tag{1'}$$

$$\int_a^c f(x)\, dx + \int_c^b f(x)\, dx = \int_a^b f(x)\, dx, \qquad \textit{for any } c \textit{ with } a < c < b, \tag{2'}$$

$$\int_a^b f(x)\, dx \le \int_a^b g(x)\, dx, \qquad \textit{if } f(x) \le g(x) \quad \textit{for } a < x < b, \tag{3'}$$

where f and g are any two bounded, piecewise continuous functions.

This theorem ▶ summarizes the geometrically obvious properties of areas that we already noted in §1.3.

1.7   Riemann sums and the Leibniz notation

The above definition of the integral is equivalent to the one given by Riemann in 1854. We now state Riemann's definition.

Let f be a function defined in the (finite) interval (a,b). Choose an integer $N > 0$, and divide the interval (a,b) into N subintervals with endpoints:

$$a = x_0 < x_1 < x_2 < \cdots < x_N = b. \tag{6}$$

Pick in each subinterval a point ξ_i, so that

$$x_0 \leq \xi_1 \leq x_1 \leq \xi_2 \leq x_2 \leq \cdots \leq x_{N-1} \leq \xi_N \leq x_N, \tag{7}$$

and form the sum

$$S = f(\xi_1)(x_1 - x_0) + f(\xi_2)(x_2 - x_1) + f(\xi_3)(x_3 - x_2) + \cdots + f(\xi_N)(x_N - x_{N-1}). \tag{8}$$

The number S is called the approximating sum, or the **Riemann sum**, for the function f, the interval (a,b), the subdivision (6), and the choice of points (7). The geometric

FIGURE **8.18**

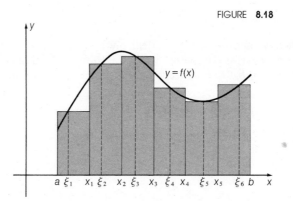

meaning of S is seen in Figure 8.18; the number S is the sum of areas of rectangles of widths

$$(x_1 - x_0), (x_2 - x_1), (x_3 - x_2), \ldots, (x_N - x_{N-1}) \tag{9}$$

and of heights

$$f(\xi_1), f(\xi_2), \ldots, f(\xi_n). \tag{10}$$

(If some of the rectangles would lie below the x-axis, their areas would enter in S with a minus sign.) Riemann defined $\int_a^b f(x)\,dx$ as *the number to which the sums* S *can be made as close as desired by choosing subdivisions with sufficiently small subintervals*, provided such a number exists. It does, whenever f is bounded and piecewise continuous; this statement is equivalent ▶ to Theorem 1.

EXAMPLE   Compute a Riemann sum for the integral $\int_0^{.5} (1 + x^2)\,dx$ and for the subdivision $0 \leq .1 \leq .2 \leq .3 \leq .4 \leq .5$.

ANSWER   We must pick a point in each of the five intervals; we choose .05, .15, .25, .35, and .45. Then the desired sum is

$$(1 + .05^2)(.1 - 0) + (1 + .15^2)(.2 - .1) + (1 + .25^2)(.3 - .2)$$
$$+ (1 + .35^2)(.4 - .3) + (1 + .45^2)(.5 - .4) = (1.0025)(.1)$$
$$+ (1.0225)(.1) + (1.0625)(.1) + (1.1225)(.1) + (1.2025)(.1) = .54125.$$

We expect .54125 to be an approximation to the true value of $\int_0^{.5} (1 + x^2)\,dx$. It is; the true value is .54166 (see the example in §1.5).

Let us consider once more the Leibniz notation for the integral

$$\int_a^b f(x)\,dx.$$

It can be justified by another mathematical myth (see Chapter 4, §1.5). The region under the curve $y = f(x)$ is composed, so the myth says, of infinitely many, infinitely thin rectangles (Figure 8.19). Each rectangle is erected over a point x in the interval

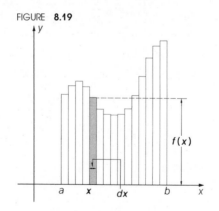

FIGURE **8.19**

$[a,b]$; it has height $f(x)$ and an infinitely small width dx; its area is therefore $f(x)\,dx$. The total area under the curve is the sum of all these areas. The symbol $\int$ means "sum"; it is, in fact, an elongated letter S.

This myth is not literally true, but it contains a kernel of truth. We have just seen what it is. The integral is not equal to a sum of infinitely many areas of infinitely thin rectangles—it is approximately equal to a sum of many thin rectangles!

PROBLEMS

1. Compute two Riemann sums for $\int_0^1 (x + 1)\, dx$. Use five equal subintervals, and use first the left endpoints and then the right endpoints as ξ_i.
2. Repeat Problem 1 using 10 equal subdivisions.
3. Compute a Riemann sum for $\int_0^1 x^2\, dx$ using five equal subintervals and the midpoints of these subintervals as ξ_i.
4. Repeat Problem 3 using 10 equal subdivisions.

1.8   Extending the notations

The number $\int_a^b f(x)\, dx$ has been defined, thus far, only for $a < b$. We want to define it also for $a = b$, so that the additivity rule (2) can remain valid. Therefore we want this to be true:

$$\int_a^a f(x)\, dx + \int_a^c f(x)\, dx = \int_a^c f(x)\, dx.$$

This can be achieved by defining

$$\int_a^a f(x)\, dx = 0. \tag{11}$$

We also want to define $\int_a^b f(x)\, dx$ for $b < a$. Again, we want (2) to remain valid. Hence we must have

$$\int_a^b f(x)\, dx + \int_b^a f(x)\, dx = \int_a^a f(x)\, dx = 0,$$

so that we must define

$$\int_a^b f(x)\, dx = -\int_b^a f(x)\, dx. \tag{12}$$

EXAMPLE Compute $\int_4^0 f(x)\, dx$ when f is the function defined in §1.4, Example 1.
ANSWER By (12) and the answer in the example quoted,

$$\int_4^0 f(x)\, dx = -\int_0^4 f(x)\, dx = -\tfrac{7}{2}.$$

---

§2   The fundamental theorem of calculus

The discovery of calculus is usually ascribed to Newton and Leibniz (there was an ugly and senseless priority dispute among their followers). But it is wrong to assign the sole credit for such a momentous achievement to two men, even to such giants as Leibniz and Newton. The essential content of the crucial fundamental theorem of calculus may have been known to Newton's teacher, Barrow.

Before stating this theorem, we consider an example.

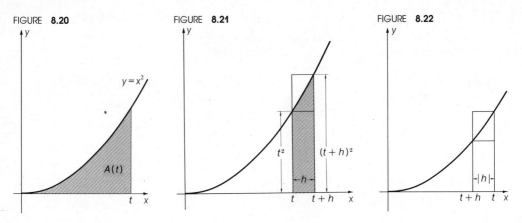

FIGURE **8.20**

$y = x^2$

$A(t)$

t x

FIGURE **8.21**

t^2 $(t+h)^2$

h

t $t+h$ x

FIGURE **8.22**

$|h|$

$t+h$ t x

2.1   Squaring the parabola

The ancient Greeks tried to solve a very difficult problem involving areas. They wanted to "square a circle," that is, construct by ruler and compass a square whose area was precisely equal to that of a given circle. This cannot be done; the impossibility proof was not given until the nineteenth century. But Archimedes, the greatest mathematician of antiquity, succeeded in squaring a region bounded by an arc of a parabola. In modern language, his theorem can be stated as follows. The area $A(t)$ of the region (shown in Figure 8.20) bounded by the parabola $y = x^2$, the x-axis, and the line $x = t$ is $\frac{1}{3}t^3$. In other words,

$$A(t) = \int_0^t x^2 \, dx = \tfrac{1}{3}t^3.$$

We shall prove this using not Archimedes' original method (which was very ingenious but specially adapted for the parabola) but derivatives. (At this point the reader may want to review §4 of Chapter 6.)

It is geometrically obvious that $A(t)$ is a continuous function of t, with $A(0) = 0$. For a fixed t, we try to compute the derivative $A'(t)$. Let $h > 0$ be a small number. Then $A(t + h) - A(t)$ is the area shown in Figure 8.21. It is less than the area

ARCHIMEDES of Syracuse (287–212 B.C.) spent most of his life in his native city, except for a stay in Alexandria, where he studied in the famed Museum, the center of Hellenistic civilization. Many of his works are preserved. Archimedes was a profound mathematician and the founder of statics; he discovered the law of floating bodies and the principle of the lever.

When the Romans besieged Syracuse, Archimedes' inventions were used in defending the city; yet reports by ancient writers about his war machines are probably exaggerated. Archimedes was killed by a Roman soldier during the sack of Syracuse.

ISAAC BARROW (1630–1677) was a theologian and a mathematician. He edited the work of Euclid, Apollonius, and Archimedes. In 1669 Barrow resigned his mathematics chair at Cambridge, and, at his advice, his one-time pupil Newton was appointed his successor.

BONAVENTURA CAVALIERI (1598–1647) was a pupil of Galileo. He belongs to the group of brilliant mathematicians who anticipated many ideas and results of calculus before its formulation by Newton and Leibniz.

$h \cdot (t + h)^2$ of the "taller" rectangle and greater than the area $h \cdot t^2$ of the "shorter" rectangle. Thus

$$ht^2 < A(t + h) - A(t) < h(t + h)^2.$$

Dividing these inequalities by h, we obtain

$$t^2 < \frac{A(t + h) - A(t)}{h} < t^2 + 2th + h^2.$$

For a negative h, of small absolute value, we obtain, from Figure 8.22, the inequalities

$$t^2 + 2th + h^2 < \frac{A(t + h) - A(t)}{h} < t^2.$$

Now the number $t^2 + 2th + h^2$ is as close to t^2 as we want if $|h|$ is small enough. Therefore the last two inequalities show that

$$A'(t) = \lim_{h \to 0} \frac{A(t + h) - A(t)}{h} = t^2.$$

Hence $A'(t) = t^2$. Therefore $A(t)$ is a primitive function of t^2. This means (see Chapter 6, §4, especially Theorem 4) that $A(t) = \frac{1}{3}t^3 + \alpha$, where α is some constant. For $t = 0$, we get $A(0) = \alpha$. But $A(0) = 0$, so that $\alpha = 0$, and we obtain $A(t) = \frac{1}{3}t^3$.

Archimedes' discovery was one of the high points of Greek mathematics. It remained an isolated achievement until nineteen centuries later when Cavalieri showed, by a method similar to that of Archimedes, that the area bounded by the curve $y = x^n$, the x-axis, and the line $x = t$ is equal to

$$\frac{t^{n+1}}{n + 1}.$$

In fact, Cavalieri could prove it only for $n = 3, 4, \ldots, 9$. Today such problems, and even much more complicated problems, are easily solved by anyone who masters the rudiments of calculus. The key to the matter is the connection between two seemingly unrelated problems, measuring areas and drawing tangents. The above example illustrates this connection, which we proceed to develop systematically.

PROBLEMS

1. Use the method of this section to compute the area of the region bounded by the line $y = x$, $x \geq 0$, the x-axis, and the line $x = t > 0$.
2. Use the method of this section to compute the area of the region bounded by the curve $y = x^3$, $x \geq 0$, the x-axis, and the line $x = t > 0$.
3. Use the method of this section to compute the area of the region bounded by the curve $y = \sqrt{x}$, $x \geq 0$, the x-axis, and the line $x = t > 0$.

2.2   First part of the fundamental theorem

Let $t \mapsto f(t)$ be a bounded and piecewise continuous function defined for $a < t < b$. Then, for every number x, $a \leq x \leq b$, we can compute the number $G(x) = \int_a^x f(t)\, dt$. Thus we obtain a new *function*, $x \mapsto G(x)$, defined for $a \leq x \leq b$; note that $G(a) = 0$.

We ask: is this new function continuous, does it have a derivative $G'(x)$, and if so, what is this derivative? Here are the answers.

Theorem 1 (fundamental theorem of calculus, first part). *Let* $t \mapsto f(t)$, $a < t < b$, *·be a bounded, piecewise continuous function, and set*

$$G(x) = \int_a^x f(t)\, dt.$$

Then $x \mapsto G(x)$ *is a continuous function for* $a \leq x \leq b$, *and at all points* x, $a < x < b$, *at which* f *is continuous, we have*

$$G'(x) = f(x).$$

This last equality is the essential point. It can be written as

$$\frac{d}{dx}\int_a^x f(t)\, dt = f(x), \qquad \text{if } f \text{ is continuous at } x. \tag{1}$$

The reasoning behind the fundamental theorem is the one used above in computing the area under the parabola.

We assume that $f > 0$ for the sake of simplicity. Let x be a point in the interval considered and h a small positive number. Then $G(x + h)$ is the area under the curve from a to $(x + h)$, $G(x)$ is the area from a to x, and the difference $G(x + h) - G(x)$ is the whole shaded area in Figure 8.23. It is not greater than Mh if $|f(t)| < M$ everywhere. Hence $|G(x + h) - G(x)|$ is small when h is small. [More precisely, for any small $\epsilon > 0$ we have $|G(x + h) - G(x)| < \epsilon$ if $|h| < \delta = \epsilon/M$.] A similar argument holds for negative h. Thus G is continuous at x. This holds even if f is not continuous at x or if x is an endpoint of our interval.

Next, if f is continuous at x, then $G(x + h) - G(x)$ is approximately the area of

FIGURE **8.23**

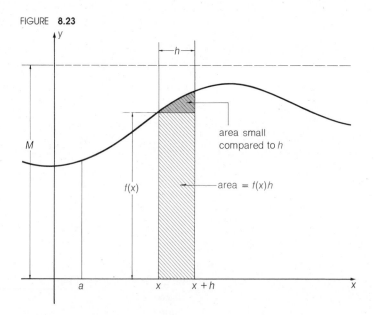

the lightly shaded rectangle, that is, $hf(x)$. Thus

$$\frac{G(x + h) - G(x)}{h} = f(x)$$

plus something as small as we like, if h is small enough. A similar argument holds for negative h. Hence $G'(x) = f(x)$.

The formal proof given in §5 is an elaboration of this intuitive argument.

2.3 Second part of the fundamental theorem

There is another part of the fundamental theorem, and it is this second part that we shall constantly use.

Theorem 2 (*fundamental theorem of calculus, second part*). *Let $f(x)$ be a bounded piecewise continuous function defined for $a < x < b$. Assume that $F(x)$ is a continuous function defined for $a \leq x \leq b$, with*

$$F'(x) = f(x),$$

except perhaps at finitely many points. Then

$$\int_a^b f(t)\, dt = F(b) - F(a).$$

Proof.   We assume first that f is continuous and $F'(x) = f(x)$ for all x. Set

$$G(x) = \int_a^x f(t)\, dt.$$

We must show that $G(b) = F(b) - F(a)$. Set $H(x) = G(x) - F(x) + F(a)$. If we show that $H(b) = 0$, our theorem is proved. Actually, we shall show that $H(x) = 0$ for all x. Indeed, $G'(x) = f(x)$ by Theorem 1, so that $H'(x) = G'(x) - F'(x) + 0 = f(x) - f(x) = 0$ at all points x. By the basic Theorem 1 of Chapter 6, §4.1, we conclude that $H(x)$ is a constant. But $H(a) = G(a) - F(a) + F(a) = G(a) = \int_a^a f(t)\, dt = 0$. Hence $H(x) = 0$ for all x.

The same argument holds if f is only piecewise continuous and $F' = f$ fails to hold at finitely many points. Define H as before. The interval $[a,b]$ can be divided into finitely many intervals in each of which $H'(x) = 0$, and hence H is constant. But since H is known to be continuous over $[a,b]$, it is constant over the whole interval. This completes the proof.

For any function $F(x)$, it is customary to write

$$F(x)\Big|_a^b = F(b) - F(a).$$

With this notation, the second part of the fundamental theorem may be written thus:

$$\int_a^b F'(x)\, dx = F(x)\,\Big|_a^b, \tag{2}$$

provided $F'(x)$ exists, is bounded, and is piecewise continuous. This formula permits us to compute the area under a curve $y = f(x)$ if we know a primitive function for f.

EXAMPLES 1. Find $\int_0^{1/2} (1 + x^2)\, dx$.

ANSWER Since $f(x) = 1 + x^2$ is a polynomial, we know (see §4.2 in Chapter 6) how to find a primitive function. Indeed, $F(x) = x + \tfrac{1}{3}x^3$ is a primitive function for f (as may be checked by differentiation). Hence, by the fundamental theorem, we have

$$\int_0^{1/2} (1 + x^2)\, dx = (x + \tfrac{1}{3}x^3)\,\Big|_0^{1/2} = [\tfrac{1}{2} + \tfrac{1}{3}(\tfrac{1}{2})^3] - 0 = \tfrac{1}{2} + \tfrac{1}{24} = \tfrac{13}{24} = .541\overline{6}.$$

Compare this with the examples in §1.5 and §1.7.

Observe that in applying the fundamental theorem it does not matter which primitive function of f we use. In the example just considered, the result would have been the same if we used instead of $F(x) = x + \tfrac{1}{3}x^3$ the function $x + \tfrac{1}{3}x^3 + 123$.

2. Find the area below the curve $y = x^3$ from $x = 0$ to $x = 2$.
ANSWER We must compute $\int_0^2 x^3\, dx$. A primitive function of x^3 is $\tfrac{1}{4}x^4$. Therefore

$$\int_0^2 x^3\, dx = \tfrac{1}{4}x^4\,\Big|_0^2 = \tfrac{1}{4}2^4 - 0 = 4.$$

3. Let n be a positive integer and t a positive number. Find the area under the curve $y = x^n$ from $x = 0$ to $x = t$. (This is Cavalieri's problem, see §2.1.)
ANSWER We must find $\int_0^t x^n\, dx$. A primitive function of x^n is $x^{n+1}/(n + 1)$, provided $n \neq -1$. Hence

$$\int_0^t x^n\, dx = \frac{x^{n+1}}{n + 1}\,\Big|_0^t = \frac{t^{n+1}}{n + 1} - 0 = \frac{t^{n+1}}{n + 1}, \qquad n \neq -1.$$

Note how easily we obtained the answer. The exceptional case $n = -1$ will be treated when we study logarithms (see Chapter 11).

PROBLEMS

Compute the following integrals using the fundamental theorem.

1. $\int_{-1}^4 3\, dx$.

2. $\int_3^9 \tfrac{1}{2}\, dx$.

3. $\int_4^5 x\, dx$.

4. $\int_1^2 3x^2\, dx$.

5. $\int_{-1}^2 (3x^2 + 1)\, dx$.

6. $\int_{-1}^6 t^2\, dt$.

7. $\int_0^1 (x^2 - x^3)\, dx$.

8. $\int_1^2 (4x^3 - 3x^2 + 1)\, dx$.

9. $\int_{-1}^1 (\tfrac{1}{2}x^3 + x^2 - 7)\, dx$.

10. $\int_{-2}^2 (\tfrac{1}{3}t^5 - t)\, dt$.

11. $\int_1^2 \dfrac{dx}{x^2}$.

12. $\int_1^4 \dfrac{ds}{s^3}$.

13. $\int_{1/2}^1 \left(\dfrac{1}{x^2} + \dfrac{1}{x^3}\right) dx$.

14. $\int_{1/2}^1 \left(\dfrac{1}{s^2} + \dfrac{1}{s^3}\right) ds$.

15. $\int_1^2 nx^n\, dx$, n an integer, $n \neq -1$.

16. $\int_a^{2a} x^n\, dx$, n a positive integer, a some number.

17. Find the area under the curve $y = 1 + x + x^2$ between $x = 0$ and $x = \tfrac{3}{2}$.
18. Find the area under the curve $y = 2 + x^3$ between $x = 0$ and $x = .5$.
19. Find the area under the curve $y = x - 2/x^2$ between $x = \tfrac{1}{2}$ and $x = 1$.
20. Find the area under the curve $y = 2 - 3/x^3$ between $x = 1$ and $x = \tfrac{3}{2}$.

2.4 Indefinite and definite integrals

In Chapter 6, §4, we asked whether every continuous function is a derivative. The fundamental theorem supplies an affirmative answer. *Every continuous function is a derivative of another function, that is, it has a primitive function.* More precisely, the continuous function $f(x)$ is the derivative of the function

$$F(x) = \int_a^x f(t)\, dt.$$

If $f > 0$, $F(x)$ is the area under the graph of f from some fixed point to x. If f takes on both positive and negative values, $F(x)$ is the area above the x-axis minus the area below. If f takes on only negative values, F is minus the area below the x-axis.

A primitive function of f is called an **indefinite integral** of f. Instead of the relation $F'(x) = f(x)$, we often write

$$\int f(x)\, dx = F(x) + C$$

without limits of integration but with an undetermined constant C (the **constant of integration**). The constant reminds us that the primitive function F is determined by f only up to a constant.

The integral

$$\int_a^b f(x)\, dx$$

is called a **definite integral** of f. The numbers a and b are called the lower and upper **limits of integration.** (It is unfortunate that the word "limit" is used in this connection. The limits of integration have nothing to do with the limits considered in Chapter 3, §4.)

If we know the indefinite integral of a function f, that is, a primitive function F of f, we can compute definite integrals, since $\int_a^b f(x)\, dx = F(b) - F(a)$. On the other hand, a definite integral of f with a variable upper limit, $\int_a^x f(t)\, dt = F(x)$, is a primitive function F for f.

EXAMPLES 1. Find $\int x^2\, dx$.
ANSWER We need a primitive function of x^2. One such function is $\frac{1}{3}x^3$. Hence

$$\int x^2\, dx = \frac{1}{3}x^3 + C.$$

2. Find $\int 78x\, dx$.
ANSWER Since $78 \cdot \frac{1}{2}x^2 = 39x^2$ is a primitive function of $78x$,

$$\int 78x\, dx = 39x^2 + C.$$

3. What is $\int 78t\, dt$?
ANSWER This is the same question as above, only the name of the variable has been changed. Thus $\int 78t\, dt = 39t^2 + C$.

4. If $r \neq -1$ is a rational number, then

$$\int x^r\, dx = \frac{x^{r+1}}{r+1} + C.$$

PROBLEMS

Compute the following indefinite integrals.

1. $\int x\,dx$.

2. $\int x^2\,dx$.

3. $\int (1 + x)\,dx$.

4. $\int (1 + x^2)\,dx$.

5. $\int (z + z^5)\,dz$.

6. $\int \dfrac{dz}{z^3}$.

7. $\int (t + t^2 + t^3)\,dt$.

8. $\int \left(\dfrac{2}{x^4} - \dfrac{3}{x^5}\right)dx$.

9. $\int \dfrac{d}{dx}(x^2 + 4x)\,dx$.

10. $\int \dfrac{d}{dx}\sqrt{x^2 + \dfrac{1}{x^2}}\,dx$.

11. If $\int f(x)\,dx = x^7 + c$, what is $f(x)$?

12. If $\int f(x)\,dx = x^4 + 4x^3 - 2x^2 + 8x + c$, what is $f(x)$?

13. If $\int g(t)\,dt = (t^2 + 4t + 1)^{3/2} + c$, what is $g(t)$?

14. If $\int g(t)\,dt = \dfrac{t^2}{(1 + t^2)} + c$, what is $g(t)$?

15. If $F'(t) = t - 1$ and $F(0) = 4$, what is $F(6)$?

16. If $G''(u) = u^2 - 3u$ and $G(0) = 1$, $G(1) = \frac{7}{12}$, what is $G(-1)$?

17. Find a primitive that vanishes at 1 for the function $u \mapsto 2u^{2/3} - u^5$.

18. Find a primitive that attains the value 1 at 2 for the function $z \mapsto z(z^2 - 5)^6$.

19. Find a primitive that takes on the value 0 at 1 for the function $t \mapsto \sqrt{t}\,(t^{3/2} - 1)^4$.

20. If $F(x)$ is a primitive that attains the value -1 at 0 for the function $x \mapsto \int_0^x (s^3 - 1)\,ds$, what is $F(-1)$?

2.5   Inertial navigation‡

As an application, we discuss inertial navigation. Imagine a closed, windowless box moving along a straight line. An observer located in the box has to determine the distance traveled by means of measurements and computation performed inside the box. (Thus a speedometer which must be connected to wheels that touch the ground is not permitted.)

If the box moves with constant speed, the observer inside will be unaware of the motion and will be unable to detect it by any mechanical experiment. This is the *classical principle of relativity* which was known to Galileo. Uniform motion cannot be detected even by experiments inside the box involving electromagnetic phenomena, such as the propagation of light. This is *Einstein's principle of relativity*, the basis of relativity theory.

But anyone who has ridden in a moving vehicle knows that changes in speed, that is, acceleration, can be detected without looking out of a window. This is not merely a subjective feeling. A spring placed in the direction of motion, with one end attached to a wall (Figure 8.24), will contract during (positive) acceleration and expand

‡This subsection may be omitted without loss of continuity.

FIGURE 8.24

position during uniform motion

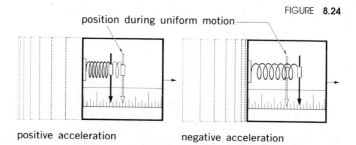

positive acceleration   negative acceleration

during deceleration (negative acceleration). By measuring the length of the spring, our observer will be able to determine the acceleration $a(t)$ of the box at time t.

We recall that the acceleration is the derivative of the velocity $v(t)$, that is, $a(t) = v'(t)$. Hence $v(t)$ is a primitive function of $a(t)$. Assume that at the initial instant $(t = 0)$ the box is at rest. Then the value of $v(t)$ at $t = 0$ is zero. Hence, by the fundamental theorem,

$$v(t) = \int_0^t a(\tau)\, d\tau.$$

(Here τ is, as we know, a dummy variable; it could be replaced by any other symbol or letter, except for v, a, and t, which are used to denote other quantities.) Our observer can therefore compute, for any instant t, the velocity of the box, say, by drawing the curve $a(t)$ and measuring the area under the curve.

Next, the velocity $v(t)$ is the derivative of the distance $s(t)$ traveled from the time $t = 0$. Thus $s(0) = 0$ and $s'(t) = v(t)$. The fundamental theorem implies that

$$s(t) = \int_0^t v(\tau)\, d\tau.$$

In this way, the observer can determine his position at any time t.

The situation just described is far from fanciful. It occurs in nuclear powered submarines, which remain submerged for months at a time, and also in rockets. The principle of inertial navigation is always the same: the accelerations can be measured, and then the velocities and position are computed by integration. Of course, rockets and submarines do not move along straight lines only, and accelerations in various directions must be measured by complicated and highly sensitive instruments. The integrations, that is, the calculations of areas, are in practice performed by automatic computers.

PROBLEMS

1. A rocket is at rest at time $t = 0$. Measurements made inside the rocket show that it experiences an acceleration $a(t) = \frac{1}{64}(t^2 + 10)$ for $t \geq 0$, where time t is measured in seconds and acceleration a is measured in ft/sec^2. What is the velocity at time $t = 64$? How far is the rocket from the starting point at time $t = 64$?

2. Suppose that the rocket in Problem 1 experiences an acceleration $a(t) = 2t + 5$. What is the velocity at time $t = 64$? How far is the rocket from the starting point at time $t = 64$?

3. Suppose that the rocket in Problem 1 experiences an acceleration given graphically by the following drawing. Estimate the velocity at time $t = 1, t = 2, t = 3, \ldots, t = 10$, and

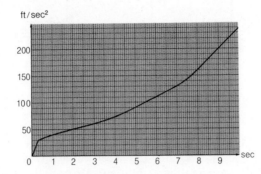

use this information to sketch the graph of the velocity for $0 \le t \le 10$. From your sketch, estimate how far the rocket is from the starting point at time $t = 10$.

§3   Integration

To integrate a function f means to compute its integral, either some **definite integral,** that is, $\int_a^b f(x)\,dx$, or an **indefinite integral,** that is, a primitive function F such that $F'(x) = f(x)$ (see §2.4). There are two ways of doing it: by numerical integration and by formal integration.

3.1   Numerical versus formal integration

Numerical integration is the computing of the number $\int_a^b f(x)\,dx$ using the properties of the integral stated in §1. For instance, we may use two step functions ϕ and ψ such that $\phi \le f \le \psi$ and compute the two integrals $\int_a^b \phi(x)\,dx$ and $\int_a^b \psi(x)\,dx$. Then we know that the desired value $\int_a^b f(x)\,dx$ lies between $\int_a^b \phi(x)\,dx$ and $\int_a^b \psi(x)\,dx$. Later (see Chapter 13, §1) we shall discuss more efficient numerical procedures. Numerical integration involves extensive calculations: today these are usually carried out on electronic computers.

Given a function $f(x)$, one can use numerical integration in order to compute a table of values of a primitive function $F(x) = \int_a^x f(t)\,dt$.

Numerical integration always works. The other method, **formal integration,** works only in some cases, but when it does, it is fast and elegant. The method consists of finding among "known" functions a primitive function F for f. A "known" function is simply a function that has been defined and given a name on some previous occasion, and for which we have either a good table or a method of computing a table. If we know a primitive function F, we can compute $\int_a^b f(x)\,dx$, since this definite integral equals $F(b) - F(a)$.

In finding primitive functions we use certain rules which will be discussed in the present section. We assume that all functions considered are bounded and continuous in the intervals over which we integrate. For a more extensive discussion of formal integration, see Chapter 13, §2 through §5.

3.2   Basic integration rules

In view of the fundamental theorem of calculus, every rule for differentiation can be rewritten as a rule for integration. Of particular importance is the rule

$$\int_a^b [f(x) + g(x)]\, dx = \int_a^b f(x)\, dx + \int_a^b g(x)\, dx. \tag{1}$$

In words: *the integral of a sum is the sum of the integrals.* To prove this, let F and G be primitive functions of f and g, respectively $(F' = f;\ G' = g)$. Then $F + G$ is a primitive function of $f + g$, by the rule for differentiating a sum. The left side of (1) is therefore equal to $[F(b) + G(b)] - [F(a) + G(a)]$, while the right side equals $[F(b) - F(a)] + [G(b) - G(a)]$, which is the same.

By the same token, the rule for differentiating differences implies that

$$\int_a^b [f(x) - g(x)]\, dx = \int_a^b f(x)\, dx - \int_a^b g(x)\, dx; \tag{2}$$

the integral of the difference is the difference of the integrals.

Rules (1) and (2) can be applied to more than two functions. For instance,

$$\int_a^b [f(x) + g(x) - h(x)]\, dx = \int_a^b f(x)\, dx + \int_a^b g(x)\, dx - \int_a^b h(x)\, dx.$$

The rule for differentiating a constant times a function implies that

$$\int_a^b cf(x)\, dx = c \int_a^b f(x)\, dx \qquad (c \text{ a constant}) \tag{3}$$

or, in words: *a constant in front of an integrand may be taken out in front of the integral sign.*

To see this, let F be a primitive function of f. Then $F' = f$, $cF' = cf$ or $(cF)' = cf$. The left side of (3) is therefore equal to $[cF(b) - cF(a)]$, while the right side equals $c[F(b) - F(a)]$, which is the same.

We sometimes write the same results for indefinite integrals, that is, for primitive functions,

$$\int [f(x) + g(x)]\, dx = \int f(x)\, dx + \int g(x)\, dx,$$

$$\int [f(x) - g(x)]\, dx = \int f(x)\, dx - \int g(x)\, dx,$$

$$\int cf(x)\, dx = c \int f(x)\, dx.$$

Since

$$\int_a^b x^r\, dx = \frac{x^{r+1}}{r+1} \Bigg|_a^b \tag{4}$$

for any rational number $r \neq -1$ (in view of Theorem 5 in Chapter 6, §2.4), we can integrate every function that is a sum of terms of the form cx^r ($r \neq -1$ and rational, c arbitrary).

EXAMPLES 1. Compute $\int_1^3 (4x - \frac{10}{9}x^9)\, dx$.

ANSWER

$$\int_1^3 (4x - \tfrac{10}{9}x^9)\, dx = \int_1^3 4x\, dx - \int_1^3 \tfrac{10}{9}x^9\, dx \qquad [\text{by (2)}]$$

$$= 4\int_1^3 x\, dx - \tfrac{10}{9}\int_1^3 x^9\, dx \qquad [\text{by (3)}]$$

$$= 4\frac{x^2}{2}\Big|_1^3 - \frac{10}{9}\frac{x^{10}}{10}\Big|_1^3 \qquad [\text{by (4)}]$$

$$= 2(3^2 - 1^2) - \tfrac{1}{9}(3^{10} - 1^{10}) = -6544.\overline{8}.$$

2. Compute $\int_1^{16} \left(3\sqrt{x} + \dfrac{1}{\sqrt{x}}\right) dx$.

ANSWER It pays to use fractional exponents:

$$\int_1^{16}\left(3\sqrt{x} + \frac{1}{\sqrt{x}}\right) dx = \int_1^{16}(3x^{1/2} + x^{-1/2})\, dx$$

$$= 3\int_1^{16} x^{1/2}\, dx + \int_1^{16} x^{-1/2}\, dx \qquad [\text{by (1) and (3)}]$$

$$= 3\frac{x^{3/2}}{3/2}\Big|_1^{16} + \frac{x^{1/2}}{1/2}\Big|_1^{16} \qquad [\text{by (4)}]$$

$$= 2(16^{3/2} - 1^{3/2}) + 2(16^{1/2} - 1^{1/2})$$

$$= 2(64 - 1) + 2(4 - 1) = 132.$$

PROBLEMS

In Problems 1 to 10 compute the given definite integrals.

1. $\int_1^2 4\, dx$.

5. $\int_0^1 (2x^2 - 4x + 1)\, dx$.

8. $\int_0^4 (7x^{5/2} - 5x^{3/2} + 3x^{1/2})\, dx$.

2. $\int_0^2 x^2\, dx$.

6. $\int_{-1}^1 (x^4 - 4x^2 + 2)\, dx$.

9. $\int_0^1 (z^4 - z^3 + z^2 - z + 1)\, dz$.

3. $\int_4^9 \dfrac{dt}{t^{1/2}}$.

7. $\int_{-2}^{-1}\left(\dfrac{4}{x^2} + 2x + 3\right) dx$.

10. $\int_1^4 (z^2 + \sqrt{z} - z^{-2})\, dz$.

4. $\int_{-2}^0 (2t + 5)\, dt$.

11. If $\int_0^1 f(x)\, dx = 2$ and $\int_0^1 g(x)\, dx = -4$, what is the value of $\int_0^1 [f(x) - g(x)]\, dx$?

12. If $\int_2^3 f(x)\, dx = 3$ and $\int_2^3 g(x)\, dx = -1$, what is the value of $\int_2^3 [3f(x) - 2g(x)]\, dx$?

13. If $\int_a^b \phi(x)\, dx = -4$ and $\int_a^b \psi(x)\, dx = -8$, what is the value of $\int_a^b [6\phi(x) + 4\psi(x)]\, dx$?

14. If $\int_a^b f(x)\, dx = 2$, $\int_a^b g(x)\, dx = -3$, and $\int_a^b h(x)\, dx = 4$, what is the value of $\int_a^b [4f(x) - 3g(x) - h(x)]\, dx$?

15. Find the value of $\int_{-1}^1 [5\phi(x) - 3\psi(x) + x^2]\, dx$ if $\int_{-1}^1 \phi(x)\, dx = 4$ and $\int_{-1}^1 \psi(x)\, dx = -3$.

16. Find the value of $\int_{-2}^{2} [4\phi(x) - 3x^2]\, dx$ if $\int_{-2}^{0} \phi(x)\, dx = 4$ and $\int_{0}^{2} \phi(x)\, dx = -8$.

17. Find the value of $\int_{0}^{1} [f(x) - 2g(x)]\, dx$ if $f(x) = \int_{-2}^{2} 3xt^2\, dt$ and $g(x) = \int_{-2}^{2} (x^2 + 4t)\, dt$.

18. $\int_{0}^{1} \left(\dfrac{1}{\sqrt{x}} - \dfrac{1}{\sqrt[3]{x}} + \dfrac{1}{\sqrt[4]{x}} \right) dx.$ 20. $\int_{0}^{1} \left(\sqrt[3]{x^5} + \dfrac{1}{\sqrt[5]{x^4}} + \sqrt[4]{x^3} \right) dx.$

19. $\int_{0}^{1} \left(\sqrt[3]{x^2} + \sqrt[4]{x^3} + 9\sqrt[5]{x^4} \right) dx.$

3.3   The language of differentials

Some integration rules can be remembered and applied by using an extension of the Leibniz notation, which is described in this subsection.

The Leibniz notation for the derivative of a function $y = f(x)$ is

$$f'(x) = \frac{df}{dx} \quad \text{or} \quad f'(x) = \frac{dy}{dx}. \tag{5}$$

The *symbols df, dy,* and *dx* are called differentials; we need not consider them variables which represent numbers. But sometimes it is convenient to operate with them *as if* they were variables and as if (5) were not a notational convention but a relation between variables. (This leads to no trouble. If we want, we may agree that dx is a variable which may represent any number, except 0, and $dy = df = f'(x)\, dx$. Then the value of dy depends on both the value of x and the value of dx.)

For instance, we may rewrite the first Equation (5) in the form

$$df = f'(x)\, dx, \tag{6}$$

obtained from (5) by multiplying both sides by dx. In the same way, the rules for differentiating sums [see Chapter 4, §2, Equation (3′)] yields the relation

$$d(f + g) = df + dg, \tag{7}$$

while the rule for differentiating the product of a function f by a constant c reads

$$d(cf) = cdf, \qquad c = \text{const.} \tag{8}$$

With this notation, the fundamental theorem can be written as

$$dF(x) = d\int_{a}^{x} f(t)\, dt = f(x)\, dx \tag{9}$$

and

$$\int_{a}^{b} dF(x) = F(x) \Big|_{a}^{b}. \tag{10}$$

The chain rule is built into the language of differentials. Suppose that we are given two functions $y = \phi(u)$ and $u = f(x)$, and set $y = \phi(f(x))$. By (6), $dy = \phi'(u)\, du$ and $du = f'(x)\, dx$. Hence $dy = \phi'(u)f'(x)\, dx$, and therefore $dy/dx = \phi'(u)f'(x)$, as it must be by the chain rule.

For instance, the rule for differentiating a power (Theorem 5 in Chapter 6, §2.4) may be written as

$$du^r = ru^{r-1}\, du. \tag{11}$$

If we consider u as a function of another variable, say, $u = f(x)$, and note that $du = f'(x)\, dx$, (11) may be rewritten as

$$df(x)^r = rf(x)^{r-1}f'(x)\, dx$$

or

$$\frac{df(x)^r}{dx} = rf(x)^{r-1}f'(x),$$

as it must be by the chain rule.

EXAMPLES 1. What is dx^3 (that is, $d(x^3)$)?
ANSWER Since the derivative of x^3 is $3x^2$, we have by (6) $dx^3 = 3x^2\, dx$.

2. If $y = 2x$, what is dy?
ANSWER Since $dy/dx = 2$, we have that $dy = 2\, dx$.

3. If $df = \frac{1}{2}x\, dx$ and $f(0) = 1$, what is $f(x)$?
ANSWER We are told that $f'(x) = \frac{1}{2}x$ (this is the meaning of the equation $df = \frac{1}{2}x\, dx$). Thus $f(x)$ is a primitive function of $\frac{1}{2}x$, that is, $f(x) = \frac{1}{4}x^2 + C$, C a constant. But $f(0) = C$ must be 1. Hence $C = 1$ and $f(x) = \frac{1}{4}x^2 + 1$.

4. If $u = (1 + x^2)^{1/3}$, what is du/dx?
ANSWER By (11),

$$du = d(1 + x^2)^{1/3} = \tfrac{1}{3}(1 + x^2)^{-2/3}d(1 + x^2) = \tfrac{1}{3}(1 + x^2)^{-2/3}2x\, dx,$$

so that $du/dx = \tfrac{2}{3}(1 + x^2)^{-2/3}x$.

5. Suppose $y = w^2 + w^{-2}$, $w = u^4 + 2u^2 - 1$, and $u = x^{3/2} + 4$. Find dy/dx using differentials.
SOLUTION We have

$$dy = 2w\, dw - 2w^{-3}\, dw = 2w(1 - w^{-4})\, dw,$$

$$dw = 4u^3\, du + 4u\, du = 4u(u^2 + 1)\, du,$$

$$du = \tfrac{3}{2}x^{1/2}\, dx.$$

Substitute dw, and then du, into the expression for dy. This gives

$$dy = 2w(1 - w^{-4})4u(u^2 + 1)\, du, \qquad dy = 2w(1 - w^{-4})4u(u^2 + 1)\tfrac{3}{2}x^{1/2}\, dx.$$

Hence

$$\frac{dy}{dx} = 12w(1 - w^{-4})u(u^2 + 1)x^{1/2}.$$

We may, if we wish, express the answer entirely in terms of x by substituting for w in terms of u and then u in terms of x.
The above calculation is equivalent to using the chain rule in the form

$$\frac{dy}{dx} = \frac{dy}{dw}\frac{dw}{du}\frac{du}{dx}.$$

PROBLEMS

1. What is $d(2 - x)$?
2. What is $d(5x^3)$?
3. What is $d\sqrt{x}$?
4. If $y = 3x - 5$, what is dy?
5. If $t = 1 - x^2$, what is dt?
6. If $t = x + 2x^3$, what is dt?

7. If $y = \sqrt{x^3 - 3x^2 + 1}$, what is dy?
8. If $df = x + 1$ and $f(2) = 3$, what is $f(x)$?
9. If $dy = x^{1/2} \, dx$ and $y = 0$ for $x = 9$, write y as a function of x.
10. If $d(x^3 - 2x) = g(x) \, dx$, what is $g(x)$? What is $g(1)$?
11. If $d(5x^2 + x^3) = f(x) \, dx$, what is $f(2)$?
12. If $d(2\sqrt{x} + x^4 - 1) = f(x) \, dx$, what is $f(1)$?
13. If $d[x(x + 1)^5] = g(x) \, dx$, what is $g(-1)$?
14. If $d(t^2/(t - 1)) = h(t) \, dt$, what is $h(2)$?
15. If $d(u/\sqrt{u^2 + 1}) = f(u) \, du$, what is $f(u)$?

16. If $d \int_1^s \sqrt{x^2 + 1} \, dx = h(s) \, ds$, what is $h(s)$? What is $h(1)$?

17. Find a function f such that $df = (x^{3/2} - 2(x + 1)^6) \, dx$.
18. If $f(x) = x^3 - 1$ and $g(x) = 3x - 1$, for what values of x does $df = dg$?
19. If $x^2 \, df = (x^3 - 1) \, dx$ and $f(1) = 0$, what is $f(2)$?
20. If $d(xf) = (2 + 3\sqrt{x}) \, dx$ and f' is defined at 0, what is $f(0)$?
21. If $d(f + g) = x \, dx$, $d(f - g) = (1 - x) \, dx$, and $f(0) = g(0) = 1$, what are $f(x)$ and $g(x)$?

In Problems 22 to 24 assume f and g are differentiable functions of x.

22. Show that $d(fg) = f \, dg + g \, df$.

23. Show that $d\left(\dfrac{1}{f}\right) = -\dfrac{1}{f^2} \, df$.

24. Show that $d\left(\dfrac{f}{g}\right) = \dfrac{g \, df - f \, dg}{g^2}$, $\quad g \neq 0$.

25. Find dy/dx if $y = u^3 + 3u - 1$ and $u = \sqrt{x + 1} + \frac{1}{2}x$.

26. Find dy/dx if $y = \dfrac{u^2 - 1}{u^2 + 1}$ and $u = (x^3 + 6x + 1)^{1/3}$.

27. Find dy/dx if $y = 2\sqrt{1 + u^2} + \sqrt{1 + u^4}$ and $u = \sqrt{x} + 1/\sqrt{x}$.

28. Find dy/dx if $y = w^5 + 10w^3 + 15w$, $w = \sqrt{u^2 + 4}$, and $u = x^2 + 4x + 1$.

29. Find dy/dx if $y = \dfrac{w - 1}{\sqrt{w + 1}}$, $w = u^{3/2} - 3u^{1/2} + 2$, and $u = x^4 + 2x^2 + 3$.

30. Find dy/dx if $y = \sqrt{w + 1} + \dfrac{1}{\sqrt{w + 1}}$, $w = \dfrac{\sqrt{u} - 1}{\sqrt{u} + 1}$, and $u = (x^2 + 1)^3$.

3.4 The substitution rule

If we rewrite the chain rule (Chapter 6, §2.3) as an integration rule, we obtain the basic substitution rule for changing the variable of integration. The rule reads: *if f is a continuous function and ϕ a function with a continuous derivative, then*

$$\int_a^b f(\phi(x))\phi'(x) \, dx = \int_{\phi(a)}^{\phi(b)} f(t) \, dt. \tag{12}$$

In order to prove (12), note first that, by the fundamental theorem, f has a primitive function. Denote a primitive function by F, and form the composed function $G(x) = F(\phi(x))$. Then

$$F'(x) = f(x) \qquad \text{(by the definition of primitives),} \tag{*}$$
$$G'(x) = F'(\phi(x))\phi'(x) \qquad \text{(by the chain rule),}$$

so that

$$G'(x) = f(\phi(x))\phi'(x). \tag{**}$$

The left side of (12) is, by (**) and the fundamental theorem,

$$\int_a^b f(\phi(x))\phi'(x)\, dx = \int_a^b G'(x)\, dx = G(b) - G(a).$$

The right side of (12) is, by (*) and the fundamental theorem,

$$\int_{\phi(a)}^{\phi(b)} f(t)\, dt = \int_{\phi(a)}^{\phi(b)} F'(t)\, dt = F(\phi(b)) - F(\phi(a)) = G(b) - G(a),$$

which is the same as the left side. (We remark explicitly that the function ϕ need not be monotone.)

The *substitution rule for indefinite integrals* reads

$$\int f(\phi(x))\phi'(x)\, dx = \int f(t)\, dt \qquad \text{where} \qquad t = \phi(x). \tag{12'}$$

In other words: if $F(t)$ is a primitive function of $f(t)$, then $F(\phi(x))$ is a primitive function of $f(\phi(x))\phi'(x)$. This follows from (12) or, directly, from the chain rule.

The substitution rule can be used to evaluate an integral $\int g(x)\, dx$ *if* we succeed in writing $g(x)$ in the form $f(\phi(x))\phi'(x)$ where f is a function which we know how to integrate. It is advisable to use the language of differentials, as we do in the examples below.

Indeed, the substitution rule, like the chain rule of which it is the reflection, is built into the language of differentials. For instance, let us rewrite relation (4) in §3.2 for an indefinite integral and replace the letter x by u:

$$\int u^r\, du = \frac{u^{r+1}}{r+1} + C. \tag{13}$$

We may, if we need to, consider u as a function of another variable, say, $u = \phi(x)$. Since $du = \phi'(x)\, dx$, (13) becomes

$$\int \phi(x)^r \phi'(x)\, dx = \frac{\phi(x)^{r+1}}{r+1} + C. \tag{13'}$$

This is a special case of (12'), for $f(t) = t^r$.

EXAMPLES 1. Evaluate the indefinite integral

$$\int (1 - 2x)^3\, dx,$$

that is, find a primitive function of $(1 - 2x)^3$.

ANSWER   It is not immediately clear how to do it, but we see "something to a third power," and we know how to find a primitive for u^3. So let us denote the "something," namely, $1 - 2x$, by a new symbol, say, u, and let us proceed, trusting the Leibniz notation.

We set
$$u = 1 - 2x.$$
Then
$$du = -2\,dx$$
and (solving for dx)
$$dx = -\tfrac{1}{2}\,du.$$
We substitute into the integral and obtain
$$\int (1 - 2x)^3\,dx = \int u^3 \left(-\tfrac{1}{2}\,du\right) = \int \left(-\tfrac{1}{2}\right) u^3\,du$$
$$= -\frac{1}{2} \int u^3\,du = -\frac{1}{2}\frac{u^4}{4} + C = -\frac{1}{8}(1 - 2x)^4 + C.$$

We could have also used (13), with $u = 1 - 2x$, $r = 3$. Since $du = -2\,dx$, we have that $dx = -\tfrac{1}{2}\,du$ and, by (13),
$$\int (1 - 2x)^3\,dx = -\frac{1}{2}\int u^3\,du = -\frac{1}{2}\frac{u^4}{4} + C = -\frac{1}{8}(1 - 2x)^4 + C.$$

The reader should check by differentiation that $x \mapsto -\tfrac{1}{8}(1 - 2x)^4$ is indeed a primitive function for $x \mapsto (1 - 2x)^3$.

2. Compute
$$\int_0^2 (1 - 2x)^3\,dx.$$

ANSWER We use the same substitution as before but pay attention to the limits of integration. Thus
$$u = 1 - 2x, \qquad du = -2\,dx, \qquad dx = -\tfrac{1}{2}\,du.$$
Note that if $x = 0$, then $u = 1$; if $x = 2$, then $u = -3$. Hence
$$\int_0^2 (1 - 2x)^3\,dx = \int_1^{-3} u^3\left(-\tfrac{1}{2}\,du\right) = -\tfrac{1}{2}\int_1^{-3} u^3\,du$$
$$= -\tfrac{1}{2}\tfrac{1}{4}u^4 \Big|_1^{-3} = -\tfrac{1}{8}[(-3)^4 - 1^4] = -10.$$

3. Evaluate
$$\int (3 + 2x)^{1/2}\,dx.$$

ANSWER Set
$$u = 3 + 2x, \qquad du = 2\,dx \qquad \text{so that} \qquad dx = \tfrac{1}{2}\,du.$$
Then
$$\int (3 + 2x)^{1/2}\,dx = \int u^{1/2}\tfrac{1}{2}\,du = \tfrac{1}{2}\int u^{1/2}\,du = \tfrac{1}{2}\tfrac{2}{3}u^{3/2} + C = \tfrac{1}{3}(3 + 2x)^{3/2} + C.$$

4. Evaluate
$$\int_0^{\sqrt{3}} \sqrt{1 + x^2}\, x\,dx.$$

ANSWER Set
$$u = 1 + x^2, \qquad du = 2x\,dx \qquad \text{so that} \qquad x\,dx = \tfrac{1}{2}\,du.$$
If $x = 0$, then $u = 1$; if $x = \sqrt{3}$, then $u = 4$. Thus
$$\int_0^{\sqrt{3}} \sqrt{1 + x^2}\, x\,dx = \int_1^4 u^{1/2}\tfrac{1}{2}\,du = \tfrac{1}{2}\int_1^4 u^{1/2}\,du = \tfrac{1}{2}\tfrac{2}{3}u^{3/2}\Big|_1^4 = \tfrac{1}{3}(8 - 1) = \tfrac{7}{3}.$$

One should remember that the method of substitution works only occasionally. For example, while it enables us to compute a primitive function for $x\sqrt{1 + x^2}$ [namely, $\frac{1}{3}(1 + x^2)^{3/2}$, see Example 4 above], it does not help in finding a primitive function for $\sqrt{1 + x^2}$.

PROBLEMS

Evaluate the following indefinite and definite integrals. When an indefinite integral is required, check your answer by differentiation.

1. $\int (1 - 2x)^4 \, dx$.

2. $\int_{-1}^{1} (1 + x)^4 \, dx$.

3. $\int \sqrt{2 - 3x} \, dx$.

4. $\int_{-1}^{1} \sqrt{10 - 6x} \, dx$.

5. $\int_{-13}^{0} (1 - 2x)^{1/3} \, dx$.

6. $\int (3 + 4x)^{-1/3} \, dx$.

7. $\int \dfrac{dx}{(2 + x)^2}$.

8. $\int_{0}^{1/2} \dfrac{dx}{(1 + 2x)^3}$.

9. $\int \sqrt{2 - x^2} \, x \, dx$.

10. $\int (1 - 4s^3)^{10} s^2 \, ds$.

11. $\int_{1}^{2} \dfrac{x \, dx}{(1 + x^2)^2}$.

12. $\int_{0}^{1} \dfrac{x \, dx}{(3 - 2x^2)^3}$.

13. $\int_{0}^{\sqrt{7}} (1 + x^2)^{1/3} x \, dx$.

14. $\int \sqrt[3]{(3x - 4)^4} \, dx$.

15. $\int \dfrac{x \, dx}{\sqrt{a^2 + x^2}}$, $a \neq 0$.

16. $\int_{2a}^{3a} \dfrac{x \, dx}{(x^2 - a^2)^2}$, $a \neq 0$.

17. $\int \dfrac{x \, dx}{(1 + ax^2)^n}$;

$a \neq 0, n \neq 1$.

18. $\int \sqrt{1 + 3x^3} \, x^2 \, dx$.

19. $\int (2 - x^4)^{1/3} x^3 \, dx$.

20. $\int_{0}^{2} \dfrac{x^2 \, dx}{\sqrt{1 + x^3}}$.

21. $\int \sqrt{1 + 2x + 3x^2} \, (2 + 6x) \, dx$.

22. $\int_{1/2}^{1} \dfrac{(1 + 3x) \, dx}{(1 + 2x + 3x^2)^2}$.

23. $\int_{0}^{1} \dfrac{x + 1}{(x^2 + 2x + 3)^3} \, dx$.

24. $\int \dfrac{(4x + 10) \, dx}{\sqrt[3]{x^2 + 5x + 2}}$.

25. $\int \sqrt{x^2 - \dfrac{1}{x}} \left(2x + \dfrac{1}{x^2}\right) dx$.

26. $\int_{1}^{4} \dfrac{1 + \sqrt{x}}{\sqrt{x}} \, dx$.

27. $\int (1 + \sqrt{x})^r \dfrac{dx}{\sqrt{x}}$,

$r \neq -1$.

28. $\int \dfrac{g'(x) \, dx}{g(x)^2}$.

3.5 The number π

We give two geometric applications of the substitution rule. First, let us compute the area of the region bounded by the circle $x^2 + y^2 = r^2$. The area, call it A, is four times the shaded area in Figure 8.25. The upper semicircle is the graph of $y = \sqrt{r^2 - x^2}$. Therefore

$$A = 4 \int_{0}^{r} \sqrt{r^2 - x^2} \, dx.$$

FIGURE **8.25**

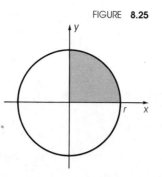

Now make the substitution $x = rt$. Then $dx = r\,dt$. Also, $t = 0$ for $x = 0$; $t = 1$ for $x = r$. Hence

$$A = 4 \int_0^1 \sqrt{r^2 - r^2 t^2}\; r\,dt = 4r^2 \int_0^1 \sqrt{1 - t^2}\; dt. \qquad (14)$$

We *define* the number π to be the area of a circular disk of radius 1. Thus we have

$$\pi = 4 \int_0^1 \sqrt{1 - t^2}\; dt. \qquad (15)$$

Note that, since the integral can be defined analytically, the number π is defined by the above formula, without reference to geometry. That π is also the ratio of the length of the circumference to that of the diameter will be proved later (see Chapter 9, §3.2).

We can now interpret the result (14) as follows: *the area of a disk of radius r is $A = \pi r^2$.*

The number π is irrational; its decimal expansion

$$\pi = 3.1415926536\ldots$$

never terminates and never becomes periodic. (We shall not prove this.) Actually, more is true. In 1882 Lindemann (1852–1939) proved that π is a transcendental number; it is not the root of a polynomial with integral coefficients. This showed that the Greek problem of squaring the circle by compass and straight edge (see §2.1) is unsolvable. Indeed, any geometric construction with compass and straight edge can be imitated algebraically, using coordinates, and this involves only solving linear and quadratic equations.

With the aid of electronic computers it is quite easy to calculate π to thousands of digits. It was not so before the advent of automatic computers and before the discovery of logarithms. Archimedes established that

$$3\frac{10}{17} < \pi < 3\frac{10}{70},$$

and when Ludolph (about 1600) computed 35 digits for π, this achievement so impressed mathematicians that π was named "Ludolph's number."

Next we show that *the area bounded by the ellipse* (see Chapter 2, §4.3)

$$\frac{x^2}{a^2} + \frac{y^2}{b^2} = 1 \tag{16}$$

is πab.

Indeed, the upper half of the curve (16) is the graph of the function

$$y = b\sqrt{1 - \frac{x^2}{a^2}}; \qquad \bullet \tag{17}$$

this relation is obtained by solving (16) for y. The desired area A is four times the area under (17) from $x = 0$ to $x = a$, or

$$A = 4 \int_0^a b\sqrt{1 - \frac{x^2}{a^2}} \, dx.$$

Introduce a new variable of integration t by $t = x/a$. Then

$$x = at; \qquad dx = a\,dt; \qquad t = 0 \quad \text{for} \quad x = 0; \qquad t = 1 \quad \text{for} \quad x = a.$$

Hence

$$A = 4 \int_0^1 b\sqrt{1 - t^2}\, a\, dt = ab\, 4 \int_0^1 \sqrt{1 - t^2}\, dt = \pi ab,$$

where we use (15).

3.6   Integrals of even and odd functions

We give one more application of the substitution rule. Recall that a function $f(x)$ is called *even* if $f(-x) = f(x)$, *odd* if $f(-x) = -f(x)$. A function is even if and only if its graph is symmetric with respect to the x-axis, odd if and only if the graph is symmetric with respect to the origin (see Chapter 3, §1.4). We claim that

$$\int_{-a}^{a} f(x)\, dx = 2 \int_0^a f(x)\, dx, \qquad \text{if } f \text{ is even} \tag{18}$$

$$\int_{-a}^{a} f(x)\, dx = 0, \qquad \text{if } f \text{ is odd.} \tag{19}$$

This is obvious geometrically (see Figure 8.26). If f is even, the areas under the curve from $-a$ to 0 and from 0 to a are equal. If f is odd, the area above the axis and the area below cancel each other.

FIGURE **8.26**

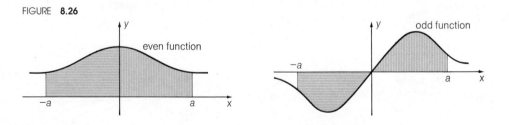

even function

odd function

To prove (19) analytically we note that the integral on the left equals $\int_{-a}^{0} f \, dx + \int_{0}^{a} f \, dx$, so that it suffices to show that $\int_{0}^{a} f \, dx = -\int_{-a}^{0} f \, dx$. We use the substitution $x = -t$. Then $dx = -dt$, $t = 0$ for $x = 0$, $t = -a$ for $x = a$, and, for f odd, we obtain

$$\int_{0}^{a} f(x) \, dx = \int_{0}^{-a} f(-t)(-dt) = -\int_{0}^{-a} f(-t) \, dt = \int_{0}^{-a} f(t) \, dt = -\int_{-a}^{0} f(t) \, dt,$$

as asserted. A similar argument proves (18), as the reader should verify.

EXAMPLE What is $\displaystyle\int_{-5}^{5} (x^{17} - 32x^7 + x^3) \, dx$?

ANSWER Since $f(x) = x^{17} - 32x^7 + x^3$ is odd, we conclude, without any calculations, that the integral equals zero.

PROBLEMS

Evaluate the following definite integrals making use of the properties of even and odd functions.

1. $\displaystyle\int_{-2}^{2} (x^3 - 2x) \, dx.$

2. $\displaystyle\int_{-2}^{2} (x^2 + 1) \, dx.$

3. $\displaystyle\int_{-5}^{5} (\tfrac{1}{5}x^4 - 3x^2) \, dx.$

4. $\displaystyle\int_{-4}^{4} (3x^2 + x) \, dx.$

5. $\displaystyle\int_{-1}^{1} (4x^5 - 6x^3 + 3x + 1) \, dx.$

6. $\displaystyle\int_{-8}^{8} (7x^{4/3} + 4x^{1/3}) \, dx.$

7. $\displaystyle\int_{-a}^{a} (x^4 + 3x^3 - 6x) \, dx, \quad a \neq 0.$

8. $\displaystyle\int_{-a}^{a} \sqrt{4 + x^2} \; x \, dx, \quad a \neq 0.$

§4 Improper integrals‡

The number $\int_{a}^{b} f(x) \, dx$ has been defined thus far only for bounded (and piecewise continuous) functions f and for finite intervals (a,b). In some cases, however, there is also a natural way of assigning a meaning to the integral when the function f is unbounded and for infinite intervals (a,b) as well. We shall use such **improper integrals** only in the case of nonnegative functions. Instead of giving a general definition, we consider some typical examples.

4.1 Unbounded integrands

The graph of the function $y = 1/\sqrt{x}$ defined for $x > 0$ is shown in Figure 8.27. If ϵ is a small positive number, we have

$$\int_{\epsilon}^{1} \frac{dx}{\sqrt{x}} = 2\sqrt{x} \,\Big|_{\epsilon}^{1} = 2 - 2\sqrt{\epsilon}.$$

---

‡This section may be omitted at first reading.

FIGURE 8.27

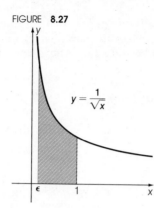

Since $\sqrt{\epsilon}$ is a continuous function of ϵ for $0 \le \epsilon$, and $\sqrt{0} = 0$, we have

$$\lim_{\epsilon \to 0^+} \int_\epsilon^1 \frac{dx}{\sqrt{x}} = 2. \tag{1}$$

(Recall that "$\epsilon \to 0^+$" means "as ϵ approaches zero through positive values.") Therefore we write

$$\int_0^1 \frac{dx}{\sqrt{x}} = 2.$$

This is to be understood as an abbreviated form of Equation (1). We also say that the improper integral $\int_0^1 x^{-1/2}\, dx$ **converges**, and that it converges to the value 2. It is natural to say that the area enclosed between the y-axis, the x-axis, the line $x = 1$, and our curve has the value 2, even though the region considered extends arbitrarily far upward.

Consider next the function $y = 1/x^2$ (defined for $x \ne 0$). We know that $(-1/x)$ is a primitive function. If ϵ is a small positive number,

$$\int_\epsilon^1 \frac{dx}{x^2} = -\frac{1}{x}\bigg|_\epsilon^1 = \frac{1}{\epsilon} - 1.$$

This quantity will be arbitrarily large positive if ϵ is sufficiently small; the statement just made is abbreviated by writing

$$\lim_{\epsilon \to 0^+} \int_\epsilon^1 \frac{dx}{x^2} = +\infty.$$

(Recall that "∞" is read "infinity"; it is *not* the name of a number.) We also say that the improper integral $\int_0^1 x^{-2}\, dx$ **diverges**. We will not ascribe any numerical value to the symbol

$$\int_0^1 \frac{dx}{x^2},$$

and we say that the area enclosed by our curve, the coordinate axes, and the line $x = 1$ is infinite.

PROBLEMS

In Problems 1 to 12 determine if the given improper integrals exist (that is, can be assigned a finite value), and evaluate them if they do.

1. $\int_0^1 \dfrac{dx}{x^{1/3}}$.

2. $\int_0^8 x^{-2/3}\, dx$.

3. $\int_{-1}^0 \dfrac{dt}{t^{4/3}}$.

4. $\int_0^4 \dfrac{dt}{t^3}$.

5. $\int_0^2 \dfrac{dx}{\sqrt{2-x}}$.

6. $\int_{-1}^0 (s+1)^{-5/4}\, ds$.

7. $\int_0^2 \dfrac{s\, ds}{\sqrt{4-s^2}}$.

8. $\int_3^5 \dfrac{s\, ds}{(s^2-25)^3}$.

9. $\int_0^{\sqrt{8}} \dfrac{x\, dx}{(x^2-8)^{1/3}}$.

10. $\int_{-1/2}^0 \dfrac{x^2\, dx}{(1+8x^3)^{1/4}}$.

11. $\int_0^a \dfrac{1}{x^r}\, dx, \quad r < 1,\ a \neq 0$.

12. $\int_0^a \dfrac{1}{x^r}\, dx, \quad r > 1,\ a \neq 0$.

4.2 Infinite intervals of integration

We compute next the area under the curve $y = 1/x^2$ from 1 to a large positive number a. We have that

$$\int_1^a \frac{dx}{x^2} = -\frac{1}{x}\Big|_1^a = 1 - \frac{1}{a}.$$

Therefore

$$\lim_{a \to +\infty} \int_1^a \frac{dx}{x^2} = 1$$

or, in an abbreviated notation,

$$\int_1^{+\infty} \frac{dx}{x^2} = 1. \tag{2}$$

We also say that the improper integral $\int_1^{+\infty} x^{-2}\, dx$ converges (to 1). Similarly,

$$\int_{-\infty}^{-1} \frac{dx}{x^2} = 1, \tag{3}$$

which means that

$$\lim_{a \to -\infty} \int_a^{-1} \frac{dx}{x^2} = 1.$$

On the other hand, the improper integral

$$\int_1^{+\infty} \frac{dx}{\sqrt{x}}$$

diverges, so that no numerical value is attached to the symbol above. To verify this, we note that, for $a > 0$, we have

$$\int_1^a \frac{dx}{\sqrt{x}} = 2\sqrt{x}\Big|_1^a = 2\sqrt{a} - 2,$$

and the limit of this expression for $a \to +\infty$ is $+\infty$.

PROBLEMS

In Problems 1 to 10 determine if the given improper integrals exist (that is, can be assigned a finite value), and evaluate them if they do.

1. $\displaystyle\int_1^\infty \frac{dx}{x^3}$.

2. $\displaystyle\int_1^\infty x^{-1/3}\,dx$.

3. $\displaystyle\int_{-\infty}^{-1} \frac{dt}{t^{4/3}}$.

4. $\displaystyle\int_0^\infty \frac{dt}{\sqrt{t+1}}$.

5. $\displaystyle\int_0^\infty \frac{dx}{(2x+1)^2}$.

6. $\displaystyle\int_{-\infty}^0 \frac{x\,dx}{(x^2+4)^3}$.

7. $\displaystyle\int_2^\infty \frac{(2u+1)\,du}{(u^2+u+2)^{4/3}}$.

8. $\displaystyle\int_{-\infty}^0 u^2(u^3+1)^{-2/3}\,du$.

9. $\displaystyle\int_a^\infty \frac{1}{x^r}\,dx,\quad r<1,\ a>0$.

10. $\displaystyle\int_a^\infty \frac{1}{x^r}\,dx,\quad r>1,\ a>0$.

4.3 Other examples

In the example above, the "impropriety" always occurred at one endpoint of the interval of integration. This need not be so in all cases. Here is an example. The equation

$$\int_{-1}^1 \frac{dx}{\sqrt{|x|}} = 4$$

is an abbreviation of the statement: if ϵ_1 and ϵ_2 are small positive numbers, then

$$\int_{-1}^{-\epsilon_2} \frac{dx}{\sqrt{|x|}} + \int_{\epsilon_1}^1 \frac{dx}{\sqrt{|x|}}$$

is close to 4, as close as we like, that is, provided ϵ_1 and ϵ_2 are small enough. Another way of writing this is as follows: the improper integrals

$$\int_{-1}^0 \frac{dx}{\sqrt{|x|}}, \qquad \int_0^1 \frac{dx}{\sqrt{|x|}}$$

converge, and

$$\int_{-1}^1 \frac{dx}{\sqrt{|x|}} = \int_{-1}^0 \frac{dx}{\sqrt{|x|}} + \int_0^1 \frac{dx}{\sqrt{|x|}} = 2 + 2 = 4.$$

REMARK   The rules for integration derived in §3 remain valid for integrals of piecewise continuous functions and also for improper integrals. The proof is almost obvious and need not be spelled out.

PROBLEMS

In Problems 1 to 10 determine if the given improper integrals exist (that is, can be assigned a finite value), and evaluate them if they do.

1. $\displaystyle\int_{-1}^1 \frac{dx}{x^2}$.

2. $\displaystyle\int_{-1}^1 \frac{dx}{x^{2/3}}$.

3. $\displaystyle\int_{-1}^1 \frac{dy}{(2y+1)^2}$.

4. $\displaystyle\int_0^2 \frac{dy}{(1-y)^{2/3}}$.

5. $\displaystyle\int_0^4 \frac{x\,dx}{(x^2-4)^3}$.

6. $\displaystyle\int_{-\sqrt{10}}^{-1} x(x^2-9)^{-1/3}\,dx$.

7. $\displaystyle\int_1^\infty \frac{v\,dv}{(v^2-1)^{1/3}}.$ 9. $\displaystyle\int_0^\infty \frac{1}{x^r}\,dx,\quad r<1.$

8. $\displaystyle\int_{-\infty}^\infty \frac{v\,dv}{(v^2+4)^2}.$ 10. $\displaystyle\int_0^\infty \frac{1}{x^r}\,dx,\quad r>1.$

§5 Proof of the fundamental theorem[‡]

In §2.2 we gave only a plausability argument for the first part of the fundamental theorem of calculus. Now we give a formal proof.

Let $f(x)$ be a bounded, piecewise continuous function defined for $a < x < b$. We set $G(x) = \int_a^x f(t)\,dt$ and must show that $G(x)$ is continuous for $a \le x \le b$, and that $G'(x_0) = f(x_0)$ if f is continuous at x_0, $a < x_0 < b$.

Let x_0 and x be two distinct points in the interval $[a,b]$. By the additivity of the integral [Theorem 2, part (2) in §1.6]

$$G(x) - G(x_0) = \int_a^x f(t)\,dt - \int_a^{x_0} f(t)\,dt = \int_{x_0}^x f(t)\,dt. \tag{1}$$

Since f is bounded, there is a number M such that $|f(t)| \le M$ for all t considered, that is, $-M \le f(t) \le M$. Assume that $x > x_0$. By the monotonicity of the integral [Theorem 2, part (3), in §1.6], we have

$$\int_{x_0}^x (-M)\,dt \le \int_{x_0}^x f(t)\,dt \le \int_{x_0}^x M\,dt$$

or $-M(x - x_0) \le G(x) - G(x_0) \le M(x - x_0)$ which is the same as

$$|G(x) - G(x_0)| \le M|x - x_0|. \tag{2}$$

We prove in the same way that (2) holds also if $x < x_0$. This inequality shows that G is continuous at x_0: if x lies in an interval of length 2δ with x_0 as midpoint, $G(x)$ differs from $G(x_0)$ by at most $\epsilon = M\delta$. (We note explicitly that x_0 could have been one of the endpoints a or b.)

Next, let x_0 be a point such that $a < x_0 < b$, and let f be continuous at x_0. Let ϵ be a positive number. Then, by the definition of continuity, there is a $\delta > 0$ such that

$$f(x_0) - \epsilon \le f(t) \le f(x_0) + \epsilon, \tag{3}$$

for $x_0 - \delta < t < x_0 + \delta$. Let x be so close to x_0 that $|x - x_0| < \delta$. Then (3) holds for t between x_0 and x. By the monotonicity property of the integral [Theorem 2, part (3) in §1.6], we conclude that

$$\int_{x_0}^x [f(x_0) - \epsilon]\,dt \le \int_{x_0}^x f(t)\,dt \le \int_{x_0}^x [f(x_0) + \epsilon]\,dt, \qquad \text{if } x > x_0$$

and

$$\int_x^{x_0} [f(x_0) - \epsilon]\,dt \le \int_x^{x_0} f(t)\,dt \le \int_x^{x_0} [f(x_0) + \epsilon]\,dt, \qquad \text{if } x < x_0.$$

[‡]Optional section.

Evaluating the integrals of constants [Theorem 2, part (1), in §1.6], we see that

$$[f(x_0) - \epsilon](x - x_0) \leq \int_{x_0}^{x} f(t)\, dt \leq [f(x_0) + \epsilon](x - x_0), \qquad \text{if } x > x_0$$

and

$$[f(x_0) - \epsilon](x_0 - x) \leq \int_{x}^{x_0} f(t)\, dt \leq [f(x_0) + \epsilon](x_0 - x), \qquad \text{if } x < x_0.$$

Dividing the first inequality by the positive number $x - x_0$, we obtain

$$f(x_0) - \epsilon \leq \frac{1}{x - x_0} \int_{x_0}^{x} f(t)\, dt \leq f(x_0) + \epsilon. \tag{4}$$

The same result is obtained by dividing the second inequality by $x_0 - x$, since

$$\frac{1}{x_0 - x} \int_{x}^{x_0} f(t)\, dt = \frac{1}{x - x_0} \int_{x_0}^{x} f(t)\, dt.$$

By (1) and (4),

$$f(x_0) - \epsilon \leq \frac{G(x) - G(x_0)}{x - x_0} \leq f(x_0) + \epsilon. \tag{5}$$

We have shown that, given any $\epsilon > 0$, there is a $\delta > 0$ such that (5) holds for $|x - x_0| < \delta$. This means that

$$\lim_{x \to x_0} \frac{G(x) - G(x_0)}{x - x_0} = f(x_0)$$

or $G'(x_0) = f(x_0)$, as asserted.

REMARK   A similar argument applies to the case in which x_0 is one of the endpoints of $[a,b]$. We conclude that $G'(a^+) = f(a)$ if f is continuous at $x = a$, and $G'(b^-) = f(b)$ if f is continuous at $x = b$. Here $G'(a^+)$ and $G'(b^-)$ are one-sided derivatives; see Chapter 6, §5.1.

APPLICATIONS OF INTEGRALS

§1   Area

In this chapter we describe some applications of integrals to geometry, mechanics, and statistics.

1.1   Computing areas

The area of a region bounded by the x-axis, two lines $x = a$ and $x = b$, and the graph of a nonnegative function $y = f(x)$ is

$$\int_a^b f(x)\, dx.$$

There are two ways of interpreting the statement just made. We may feel that we know what area is and regard the statement above as a definition of the integral, or we may consider the integral to have been defined analytically, as in Chapter 8, §1.5, and take the statement above as the *definition of area* for a region below the graph of a function.

It turns out that areas of more general regions can also be expressed as integrals.

Let R be a "reasonable" region in the plane, that is, a set of points lying inside a smooth closed curve (Figure 9.1) or between several such curves (Figure 9.2); a precise definition need not concern us at this point. We want to compute A, the area of R. (We assume that we know what area is and are concerned only with computing it.)

FIGURE 9.1

FIGURE 9.2

FIGURE 9.3

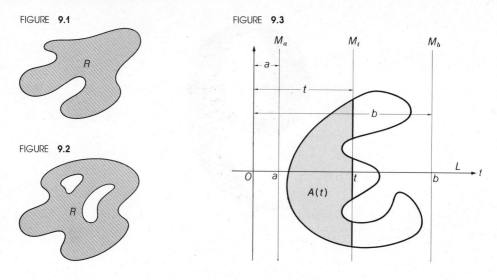

Choose a straight line L and on it a reference point O and a positive direction. We call this direction "to the right." For every number t, let M_t denote the straight line perpendicular to L which intersects L at the (positive or negative) distance t from O. We assume that our region R lies entirely between the lines M_a and M_b with $a < b$ (see Figure 9.3). Let us denote by $A(t)$ the area of that part of R which lies to the left of M_t. Then $A(a) = 0$, since all of R lies to the right of M_a, and $A(b) = A$, since all of R lies to the left of M_b. Therefore $A = A(b) - A(a)$, and, by the fundamental theorem,

$$A = \int_a^b A'(t)\, dt.$$

[Here we assume that $A(t)$ is a continuous function and has a bounded piecewise continuous derivative.] But what is $A'(t)$?

The **intersection** of R and the line M_t is the set of points lying in R and on M_t. (In Figure 9.3, it consists of three disjoint segments.) The **length** $l(t)$ of the intersection is defined whenever the intersection consists of one segment or of several disjoint segments and points: $l(t)$ is the sum of the lengths of the segments. The length $l(t)$ is 0 if $l(t)$ is empty or consists of finitely many points.

Let h be a small positive number. The quantity $A(t + h) - A(t)$ is the area of the part of R between M_t and M_{t+h}. If the intersection of R and M_t consists of one or several segments, then the area $A(t + h) - A(t)$ differs little from the area of one or several rectangles of total height $l(t)$ and width h, as is seen from Figures 9.4 and 9.5. Hence $A(t + h) - A(t)$ is approximately $l(t)h$, so that

$$\frac{A(t + h) - A(t)}{h} \approx l(t).$$

(Recall that "$\approx$" means "approximately equal.") This suggests that $A'(t) = l(t)$, and therefore

$$A = \int_a^b l(t)\, dt. \tag{1}$$

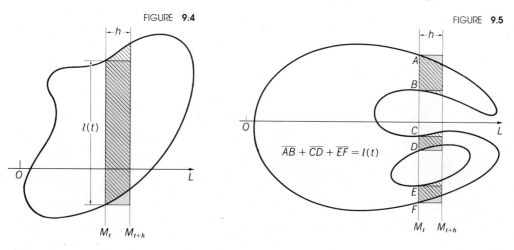

FIGURE 9:4

FIGURE 9.5

$\overline{AB} + \overline{CD} + \overline{EF} = l(t)$

We call (1) the **area formula**.

Let us show that the area formula gives the expected result in simple cases. Suppose R is the region bounded by the x-axis, the lines $x = a$ and $x = b > a$, and the graph of a continuous positive function $y = f(x)$; see Figure 9.6. For L we choose the x-axis. Then $l(t) = f(t)$, and the area formula yields $A = \int_a^b f(t)\, dt$, as expected.

Another example is the area between two graphs. Let the functions $f(x)$ and $g(x)$ be bounded and piecewise continuous for $a \leq x \leq b$, and assume that $g(x) \leq f(x)$. Let R denote the region between the graphs of these functions and the lines $x = a$, $x = b$, and let A denote the area of R. Then

$$A = \int_a^b f(x)\, dx - \int_a^b g(x)\, dx,$$

as we see from Figure 9.7 (where f and g are both positive) or from Figure 9.8 (where f is positive and g negative) by using our intuitive ideas about area. This can be written as

$$A = \int_a^b [f(x) - g(x)]\, dx.$$

If we choose for L the x-axis, we have $l(t) = f(t) - g(t)$. Hence the relation above is equivalent to the area formula.

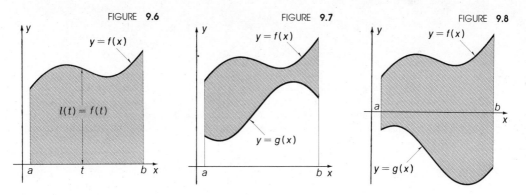

FIGURE 9.6

FIGURE 9.7

FIGURE 9.8

FIGURE 9.9   FIGURE 9.10   FIGURE 9.11

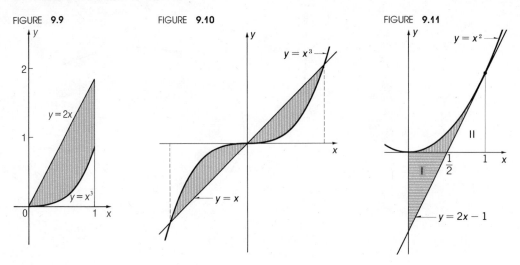

EXAMPLES   1.   Find the area enclosed between the graph of $y = x^3$ and $y = 2x$, from $x = 0$ to $x = 1$ (see Figure 9.9).

FIRST SOLUTION   In the interval considered, the second graph lies above the first. Hence the desired area is

$$A = \int_0^1 2x\,dx - \int_0^1 x^3\,dx = x^2 \Big|_0^1 - \frac{x^4}{4}\Big|_0^1 = 1 - \frac{1}{4} = \frac{3}{4}.$$

SECOND SOLUTION   We use the area formula, with L the x-axis, and with $a = 0$, $b = 1$, $l(x) = 2x - x^3$. This yields

$$A = \int_0^1 (2x - x^3)\,dx = \left(x^2 - \frac{x^4}{4}\right)\Big|_0^1 = \frac{3}{4}.$$

2.   Find the area A of the region enclosed between the curve $y = x^3$ and the line $y = x$ (see Figure 9.10).

SOLUTION   The curve and the line intersect when $x^3 = x$, that is, for $x = -1, 0, 1$. The region consists of two pieces. We choose for L the x-axis. Then our region lies between $x = -1$ and $x = 1$, and

$$l(x) = \begin{cases} x^3 - x, & \text{for } -1 \le x \le 0, \\ x - x^3, & \text{for } 0 < x \le 1, \end{cases}$$

as is seen from the figure. Hence, by the area formula,

$$A = \int_{-1}^1 l(x)\,dx = \int_{-1}^0 (x^3 - x)\,dx + \int_0^1 (x - x^3)\,dx$$
$$= \left(\frac{x^4}{4} - \frac{x^2}{2}\right)\Big|_{-1}^0 + \left(\frac{x^2}{2} - \frac{x^4}{4}\right)\Big|_0^1 = \frac{1}{2}.$$

3.   Find the area enclosed between the parabola $y = x^2$, the y-axis, and the tangent to this parabola at the point $(1,1)$; see Figure 9.11.

FIRST SOLUTION   The equation of the tangent is $y = 2x - 1$. The region is contained between $x = 0$ and $x = 1$. Hence, if L is the x-axis, we have $l(x) = x^2 - 2x + 1$. By the area formula,

$$A = \int_0^1 (x^2 - 2x + 1)\,dx = \left(\frac{x^3}{3} - x^2 + x\right)\Big|_0^1 = \frac{1}{3}.$$

SECOND SOLUTION   The area under the parabola from 0 to 1 is $\frac{1}{3}$. The area of triangle I is $\frac{1}{4}$, and so is the area of triangle II. Therefore $A = (\frac{1}{3} - \frac{1}{4}) + \frac{1}{4} = \frac{1}{3}$.

PROBLEMS

In Problems 1 to 15 find the area of the region enclosed by the given set of curves. Make a sketch of the region in question before applying the area formula. In sketching these graphs, it may occasionally be convenient to use different units of length along the two coordinate axes.

1. $y = x^2$ and $y = 4$.
2. $y = x^2$ and $y = x + 2$.
3. $y = \frac{1}{2}x^2$ and $y = -x^2 + 6$.
4. $y = x^2$ and $y^2 = x$.
5. $y = x^3$ and $y = x^2$.

6. $y = |x|$ and $y = x^2 - 2$.
7. $y = 5x$ and $y = x^3 - 4x$.
8. $y = \sqrt{x}$, $y = x - 2$, and $x = 0$.
9. $y^2 = x$ and $y = 2 - x$.
10. $y = x$, $y = x^2 + 1$, $x = 0$, and $y = 2$.

11. $y = \sqrt{x}$, $y = 2\sqrt{x}$, and the line $y = x$.
12. $y = x^3$, $y = 0$, and the tangent to $y = x^3$ at the point $(1,1)$.
13. $y = x^3 - x$ and the tangent to $y = x^3 - x$ at $x = -1$.
14. $y = x^3 - 12x$ and $y = x^2$.
15. $y = x^2$, $y = (x - 1)^2$, and $y = \frac{1}{9}$. [*Hint:* There are three separate regions.]

1.2 Another derivation of the area formula

We shall show now how the area formula could have been obtained from the definition of the integral without recourse to the fundamental theorem of calculus. We consider again a plane region R and an axis L, with an origin O and a positive direction. Denote by M_t the line perpendicular to L and crossing L at the distance t from O. We assume that R is contained between the lines M_a and M_b. Now we divide the interval $a < t < b$ into many, say, n small subintervals

$$a = t_0 < t_1 < t_2 < \cdots < t_n = b \tag{2}$$

and pick a value, call it τ_j, in each interval $[t_{j-1}, t_j]$. We recall that $l(\tau_j)$ is the length of the intersection of R with M_{τ_j}. The area of the part of R contained between $M_{t_{j-1}}$ and M_{t_j} is approximately $l(\tau_j)(t_j - t_{j-1})$, since it differs little from the area of one or several rectangles of width $t_j - t_{j-1}$ and total height $l(\tau_j)$; see Figure 9.12 where $n = 6$. The total area therefore differs little from the sum

$$l(\tau_1)(t_1 - t_0) + l(\tau_2)(t_2 - t_1) + l(\tau_3)(t_3 - t_2) + \cdots + l(\tau_n)(t_n - t). \tag{3}$$

This sum, however, is a Riemann sum for the integral

$$\int_a^b l(t)\, dt. \tag{4}$$

Thus the area A of R is nearly (3), and our geometric intuition tells us that the difference will be arbitrarily small if we choose a subdivision (2) with sufficiently small subintervals. On the other hand, for such a subdivision the Riemann sum (3) is as close as we like to the integral (4). Therefore A equals the integral, and the area formula (1) holds!

FIGURE 9.12

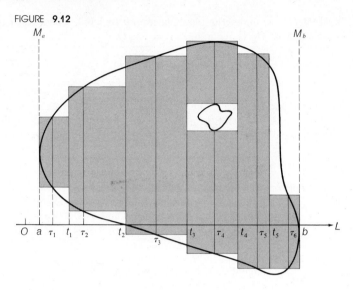

The two ways in which we reached the area formula, via the fundamental theorem and via Riemann sums, are typical for expressing geometric and other quantities by integrals. Which method is simpler depends on the problem considered.

1.3 Defining areas

Now we change our point of view and give an **analytic definition of area.** We simply make (1) into a definition. (Recall that in Chapter 8, §1.6, we defined the integral using only properties of numbers.) A set R of points in the plane will be called a region with area A if the area formula (1) is applicable. This means that there is a line L such that the integral (1) is defined.

In order to justify the definition of area given by (1), we must show that if we compute the area of the same region R using two different lines L, we get the same result. This can be proved, using properties of numbers; it is easier to do this in the context of calculus of several variables ▶.

As an illustration of the fact that the result does not depend on the choice of L, we compute in two different ways the area bounded by the parabola $y = x^2$ and

FIGURE 9.13

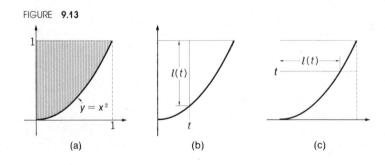

(a)   (b)   (c)

the lines $x = 0$ and $y = 1$ [Figure 9.13(a)]. If we choose for L the x-axis [Figure 9.13(b)], then $l(t) = 1 - t^2$ and the area equals

$$A = \int_0^1 (1 - t^2)\, dt = \left(t - \frac{t^3}{3} \right) \Big|_0^1 = 1 - \frac{1}{3} = \frac{2}{3}.$$

If we take for L the y-axis [Figure 9.13(c)], then $l(t) = \sqrt{t}$ and

$$A = \int_0^1 \sqrt{t}\, dt = \frac{2}{3} \sqrt{t^3} \, \Big|_0^1 = \frac{2}{3},$$

as before. This is not surprising, but nevertheless reassuring.

There are very complicated regions to which the area formula is not applicable, but they do not occur in the usual applications of calculus.

§2 Volume

We now apply integrals to the computation of volumes of solids.

2.1 Volumes of cylinders

Let R be a plane region with area A. At each point of R, erect a segment of length H, perpendicular to the plane of R; let all segments lie on the same side of R. The solid body formed by all these segments is called the **right cylinder** with base R and height H (see Figure 9.14). In elementary geometry, we are taught that the volume of such a cylinder is

$$V = AH, \tag{1}$$

at least in the case in which R is a polygon or a circular disk. We shall show how we can arrive at this formula, using the formula for the volume of a right parallelepiped (box), our intuitive concept of volume, and the basic ideas of calculus.

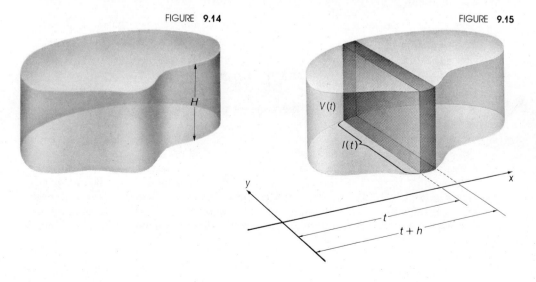

FIGURE **9.14**

FIGURE **9.15**

Let C be a right cylinder, of height H, whose base R lies in the x, y plane, between $x = a$ and $x = b$, and let $V(t)$ denote the volume of that part of C that lies to the left of $x = t$ (see Figure 9.15). Then $V(a) = 0$ and $V(b) = V$ is the total volume of C. If the function $V(t)$ has a bounded piecewise continuous derivative $V'(t)$, we have $V = \int_a^b V'(t)\, dt$.

Let h be a small positive number. Then $V(t + h) - V(t)$ is the volume of the slice of C contained between $x = t$ and $x = t + h$. In Figure 9.15 we drew the case in which the line $x = t$ intersects R in one segment of length $l(t)$. The slice of C between $x = t$ and $x = t + h$ is nearly a parallelepiped of base area $h \cdot l(t)$ and height H. The volume of this slice is approximately $hl(t)H$. Hence $\dfrac{V(t + h) - V(t)}{h} \approx Hl(t)$.

This suggests that $V'(t) = Hl(t)$, and therefore $V = \int_a^b Hl(t)\, dt = H\int_a^b l(t)\, dt$. Since $\int_a^b l(t)\, dt = A$ by the area formula, we have that $V = AH$. In particular, if the base R is a circular disk of radius r, we obtain the familiar formula $V = \pi r^2 H$.

2.2   Volume formula

We now derive a formula for the volume of general solids, proceeding as with areas. First we use freely our geometric intuition: we assume we know what volume means and try to compute it. Then we transform the method of computing into a definition.

Consider a "reasonable" solid body B in space; let V denote its volume. In order to compute V, we choose a straight line L, a reference point O on the line, and a positive direction. For the sake of simplicity, we assume that the intersection of any plane P perpendicular to L with B consists of a single piece. Let P_t be such a plane, t denoting the (positive or negative) distance from O to the intersection point of the plane and the line L. We denote the area of the intersection of P_t and B by $A(t)$ (see Figure 9.16).

Assume that B lies between the planes P_a and P_b, $a < b$. Let $V(t)$ denote the volume of the part of B contained between P_a and P_t. Then $V(a) = 0$ and $V(b) = V$.

The difference $V(t + h) - V(t)$ is the volume of a thin slice that is nearly a right cylinder of height h and base area $A(t)$ (see Figure 9.16). Thus $V(t + h) - V(t)$ is close to $A(t)h$, the fraction $\dfrac{V(t + h) - V(t)}{h} \approx A(t)$, and we may expect that $V'(t) = A(t)$. If so, the fundamental theorem of calculus gives the **volume formula**

$$V = \int_a^b A(t)\, dt. \tag{2}$$

Up to now, we have regarded the concept of volume as known, and we have arrived at Equation (2) by informed guessing. Now we change the point of view and *define the volume by the volume formula*. The formula is applicable to sets B in space such that, if the line L is chosen properly, the intersection of P_t with B is a region with area $A(t)$ and the integral (2) is defined.

The volume formula (2) shows that to compute the volume of a solid, we need to know the area $A(t)$ of a cross section perpendicular to a line L, at the distance t from some reference point. Therefore, if two different solids have cross sections of equal areas (such as the two solids in Figure 9.17), they also have the same volume. This statement is known as **Cavalieri's principle;** it was asserted by Cavalieri before calculus was discovered.

FIGURE **9.16**

FIGURE **9.17**

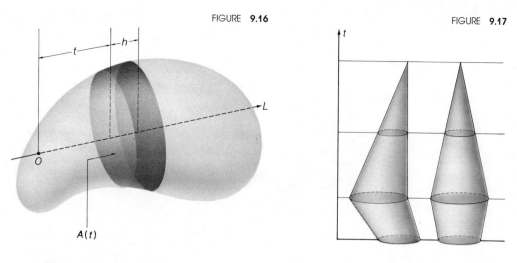

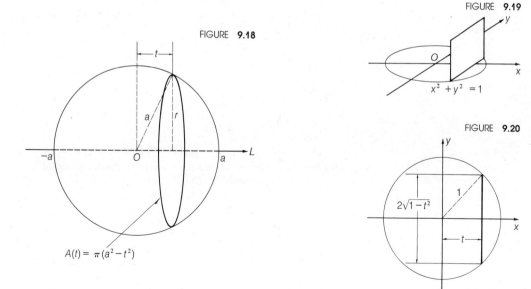

$A(t)$

EXAMPLES 1. Find the volume of a sphere of radius a.

ANSWER Let the line L be any diameter, and take the reference point O at the center of the sphere. The intersection of the plane P_t and the sphere is a circle of area πr^2, or $A(t) = \pi(a^2 - t^2)$, where we use the Pythagorean theorem. (See Figure 9.18.) Since t varies from $-a$ to a, the desired volume is given by

$$V = \int_{-a}^{a} \pi(a^2 - t^2)\, dt = \pi \left(a^2 t - \frac{t^3}{3} \right) \Big|_{-a}^{a} = \frac{4}{3}\pi a^3.$$

2. The base of a solid is the circle $x^2 + y^2 < 1$. Each cross section perpendicular to the x-axis is a square. Find the volume of the solid (see Figure 9.19).

FIGURE **9.19**

$x^2 + y^2 = 1$

FIGURE **9.18**

$A(t) = \pi(a^2 - t^2)$

FIGURE **9.20**

$2\sqrt{1 - t^2}$

FIGURE 9.21

FIGURE 9.22

FIGURE 9.23

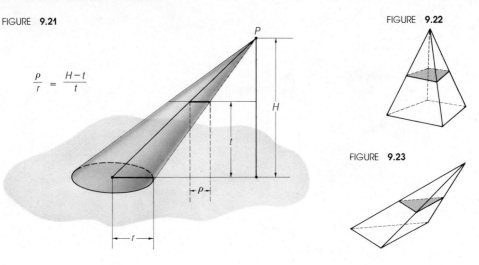

$$\frac{\rho}{r} = \frac{H-t}{t}$$

ANSWER   Let us use the x-axis as the line L. We have $a = -1$, $b = 1$, and $A(t)$ is the area of a square whose side is (see Figure 9.20) $2\sqrt{1 - t^2}$. Thus $A(t) = 4(1 - t^2)$, and, by the volume formula, the desired volume is

$$\int_{-1}^{1} 4(1 - t^2)\, dt = \left(4t - \frac{4}{3}t^3\right)\Big|_{-1}^{1} = \frac{16}{3}.$$

3.   Let P be a point H units above a plane, and let there be given a circle of radius r in the plane. We draw all segments joining P to points inside or on the circle (see Figure 9.21); these segments form a solid, called a **circular cone** of radius r and height H. Find the volume of this cone.

ANSWER   We use the line through P, perpendicular to the plane of the circle as the line L. The cross section of the cone at the height t is a circle of radius $\rho = r(1 - t/H)$, as is seen from Figure 9.21 (by noting similar triangles). Hence $A(t) = \pi\rho^2 = \pi r^2 (1 - t/H)^2$, and the volume of the cone is, by (2),

$$V = \int_{0}^{H} \pi r^2 \left(1 - \frac{t}{H}\right)^2 dt,$$

or, setting $1 - t/H = s$, $dt = -H\,ds$,

$$V = \pi r^2 H \int_{0}^{1} s^2\, ds = \tfrac{1}{3}\pi r^2 H. \tag{3}$$

4.   If, in the construction in Example 3, we replace the circular disk by an arbitrary plane region R of area A, we obtain a **cone with base** R. A cone with a square base is called a **pyramid** (see Figure 9.22); one with a triangular base, a **tetrahedron** (see Figure 9.23).

The volume of a cone is

$$V = \tfrac{1}{3}AH, \tag{4}$$

where A is the area of the base and H the height.

The proof is analogous to the one given above for circular cones, but we shall not give it here.

PROBLEMS

In Problems 1 to 5 use the volume formula.

1. Compute the volume of a right circular cylinder of radius R and height H, taking for L a line perpendicular to the axis of the cylinder.
2. Find the volume of a right circular cone whose height is H and which has a base of radius R.
3. Compute the volume of a spherical cap of thickness h. Assume the radius of the sphere is a. (See figure.)
4. Find the volume of the frustum of a square pyramid if the base and top have sides b and a, respectively, and the altitude is H. (See figure.)
5. Compute the volume of a frustum of a cone of revolution if the base and top have radii R and r, respectively, and the altitude is H. (See figure.)

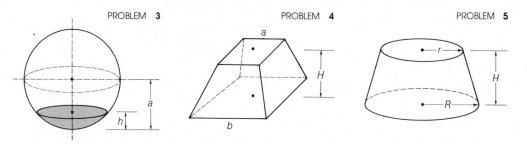

PROBLEM 3   PROBLEM 4   PROBLEM 5

In Problems 6 to 12 we use the following notation: The symbol $\{(x,y): \ldots\}$ denotes the set of points (x,y) in the x, y plane satisfying the condition(s). $\ldots$ For instance, $\{(x,y): y = 0\}$ is the x-axis.

6. The base of a solid is the circle $\{(x,y): x^2 + y^2 \leq 4\}$. Each cross section perpendicular to the x-axis is an isosceles triangle with its altitude equal to the base. Compute the volume of the solid.
7. The base of a solid is the parabolic region $\{(x,y): y^2 \leq x, 0 \leq x \leq 1\}$. Each cross section perpendicular to the x-axis is a square. Find the volume of the solid.
8. The base of a solid is the parabolic region $\{(x,y): y^2 \leq 4x, 0 \leq x \leq 1\}$. Each cross section perpendicular to the x-axis is an equilateral triangle. Compute the volume of the solid.
9. The base of a solid is the parabolic region $\{(x,y): y^2 \leq x, 0 \leq x \leq 2\}$. Each cross section perpendicular to the x-axis is a semicircle. Compute the volume of the solid.
10. The base of a solid is the circle $\{(x,y): x^2 + y^2 \leq 9\}$. Each cross section perpendicular to the x-axis is an isosceles right triangle with one of the legs a chord of the circle. Find the volume of the solid.
11. The base of a solid is the triangle $\{(x,y): 0 \leq x \leq 1, 0 \leq y \leq x\}$. Each cross section perpendicular to the x-axis is a square. Compute the volume of the solid.
12. The base of a solid is the triangle $\{(x,y): 0 \leq x \leq 1, 0 \leq y \leq x\}$. Each cross section perpendicular to the y-axis is a semicircle. Find the volume of the solid.

2.3   Solids of revolution. Disk method

Consider a plane region R between the graph of a function $y = f(x)$ and the x-axis from a to b (Figure 9.24); we assume f to be nonnegative. When R is rotated about

FIGURE **9.24**

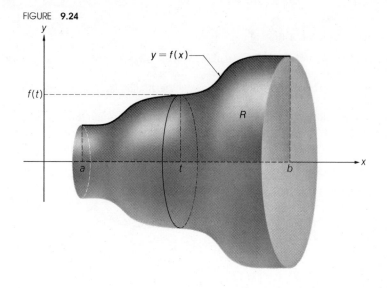

the x-axis by 360° (one full revolution), it sweeps out a solid B called a **solid of revolution.**

In order to compute the volume V of B, we choose for L the x-axis. For $a < t < b$, the intersection of the plane P_t with B is a circular disk of radius $f(t)$. Hence $A(t) = \pi f(t)^2$, and the general volume formula yields the result: *the volume of the solid swept out by rotating the region under $y = f(x)$, $a \le x \le b$, about the x-axis is*

$$V = \pi \int_a^b f(t)^2 \, dt. \tag{5}$$

The formula may be applied also to bodies of revolution for which the axis of rotation is not the x-axis. Assume, for instance, that R is the region bounded by a curve $x = f(y)$, $a \le y \le b$, the y-axis, and the lines $y = a$, $y = b$, and that B is the body swept out by R when it is rotated about the y-axis by 360° (such a body is considered in Example 2 below). The volume of B is again given by (5). Indeed, the volume does not depend on the names of the axes.

EXAMPLES 1. Find the volume obtained by rotating the area under the parabola $y = x^2$ from 0 to 2 about the x-axis (see Figure 9.25).
ANSWER   We have $a = 0$, $b = 2$, and $f(t) = t^2$, so

$$V = \pi \int_0^2 t^4 \, dt = \pi \cdot \frac{t^5}{5} \Big|_0^2 = \frac{32}{5} \pi.$$

2. Find the volume obtained by rotating the region bounded by the arc of the parabola $y = x^2$ from $x = 0$ to $x = 2$, the y-axis, and the line $y = 4$ about the y-axis. (See Figure 9.26.)
ANSWER   We have $a = 0$, $b = 4$, $f(t) = \sqrt{t}$. Therefore, by (5),

$$V = \pi \int_0^4 t \, dt = \pi \frac{t^2}{2} \Big|_0^4 = 8\pi.$$

3. Find the volume obtained by rotating the area under the parabola $y = x^2$ from 0 to 2 about the y-axis. (See Figure 9.27.)

FIGURE 9.25

FIGURE 9.26

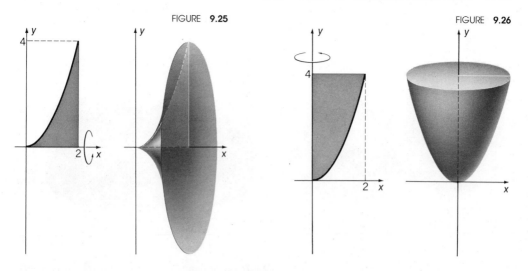

FIRST SOLUTION   We use Equation (5) with the y-axis as the reference line L. For $0 < t < 4$, the intersection P_t with our volume is a circular annulus with radii 2 and $\sqrt{t}$. (See Figure 9.28.) Hence $A(t) = \pi 2^2 - \pi t = \pi(4 - t)$, and

$$V = \int_0^4 A(t)\, dt = \int_0^4 \pi(4 - t)\, dt = \pi \left(4t - \frac{t^2}{2}\right)\bigg|_0^4 = 8\pi.$$

SECOND SOLUTION   The volume considered, together with the volume computed in Example 2, form a circular cylinder of base radius 2 and height 4. The volume of this cylinder is 16π. Hence the desired volume is $16\pi - 8\pi = 8\pi$.

4.   The **sphere** of radius r is obtained by rotating the region under $y = \sqrt{r^2 - x^2}$ from $-r$ to r about the x-axis. The **volume of the sphere** can be computed from Equation (5), setting $a = -r$, $b = r$, $f(t) = \sqrt{r^2 - t^2}$. We obtain

$$V = \pi \int_{-r}^r (r^2 - t^2)\, dt = \pi \left(r^2 t - \frac{t^3}{3}\right)\bigg|_{-r}^r$$

or

$$V = \tfrac{4}{3}\pi r^3, \tag{6}$$

in conformity with Example 1 in §2.2.

FIGURE 9.27

FIGURE 9.28

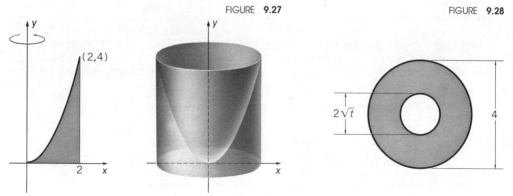

FIGURE **9.29**   FIGURE **9.30**

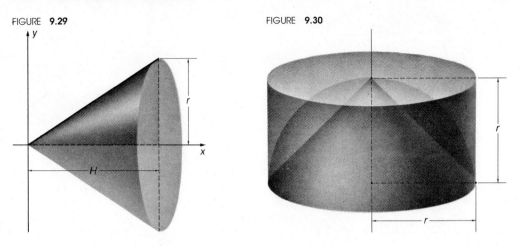

5. The **right circular cone** of radius r and height H is obtained by rotating the region under the straight line $y = (r/H)x$ from $x = 0$ to $x = H$ about the x-axis (see Figure 9.29). From (5) we obtain, for the **volume of the cone**,

$$V = \pi \int_0^H \frac{r^2}{H^2} t^2 \, dt = \frac{\pi r^2}{H^2} \left(\frac{t^3}{3}\right) \bigg|_0^H$$

or

$$V = \frac{\pi}{3} r^2 H. \tag{7}$$

This is a special case of Example 3 in §2.2.

Consider now a circular cylinder whose radius is equal to its height, a hemisphere inscribed in the cylinder, and a cone inscribed in the hemisphere (Figure 9.30). It follows from the formulas just derived that the ratios of the three volumes are like $3:2:1$. This beautiful relationship occurs in a work of Archimedes.

[The ancient Greeks always expressed formulas for lengths, areas, and volumes as statements about ratios of "like" quantities. They rightly counted the formulas for the volumes of spheres and circular cones among the glories of their mathematics. These formulas were first discovered by the "materialist" philosopher Democritus, but a rigorous proof was given only by the mathematician Eudoxus, a friend of the "idealist" Plato. (The words "idealist" and "materialist" are used in their philosophical sense; they do not imply a moral superiority of Plato over Democritus.) The ideas of Plato and his onetime pupil Aristotle triumphed in Greek thought and dominated Western culture for centuries, but no works by Democritus have been preserved. Modern mathematicians learned of Democritus' discovery only when a manuscript of a long-lost book by Archimedes was accidentally found in 1906. From this book we also know that Democritus' method was based on a mathematical myth not unlike the one we retold in Chapter 8, §1.7, and his method was somewhat akin to calculus.]

6. Suppose we are given two functions, $g(x)$ and $f(x)$, such that $0 < g(x) < f(x)$, $a < x < b$, and we want to compute the volume V of the solid of revolution obtained by rotating about the

DEMOCRITUS of Abdera (ca. 400–370 B.C.) founded Greek "atomism," probably together with the philosopher Leucippus. He contributed to most fields of human knowledge, but of his many writings, only a few fragments and several titles of treatises have reached us (among them *On Numbers, On Geometry, On Tangencies, On Irrationals*).

The atomist had the bold idea of explaining the world by the motion of unchangeable, indivisible atoms through empty space.

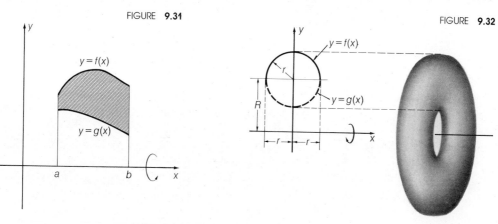

FIGURE 9.31

FIGURE 9.32

x-axis the region bounded by the lines $x = a$, $x = b$, and the graphs of $y = f(x)$, $y = g(x)$; see Figure 9.31. The desired volume is the *difference* of the volumes obtained by rotating the regions under $y = f(x)$ and under $y = g(x)$. Using (5), we obtain

$$V = \pi \int_a^b f(t)^2 \, dt - \pi \int_a^b g(t)^2 \, dt = \pi \int_a^b [f(t)^2 - g(t)^2] \, dt. \tag{8}$$

7. Let r and R be numbers such that $0 < r < R$. If we rotate the circular disk of radius r, with center at $(0,R)$, about the y-axis, we obtain a doughnutlike body called the **solid torus** (see Figure 9.32).

In order to compute the volume V of the solid torus, we may interchange x and y and use formula (8) with $a = -r$, $b = r$, $g(t) = R - \sqrt{r^2 - t^2}$, and $f(t) = R + \sqrt{r^2 - t^2}$. We obtain

$$V = \pi \int_{-r}^r [f(t)^2 - g(t)^2] \, dt$$

$$= \pi \int_{-r}^r \{[R + \sqrt{r^2 - t^2}\,]^2 - [R - \sqrt{r^2 - t^2}\,]^2\} \, dt$$

$$= 4\pi R \int_{-r}^r \sqrt{r^2 - t^2} \, dt = 4\pi R r^2 \int_{-1}^1 \sqrt{1 - s^2} \, ds$$

$$= 2\pi^2 R r^2,$$

where we used the substitution $t = rs$ and the fact that $\int_{-1}^1 \sqrt{1 - s^2} \, ds = \pi/2$ (see Chapter 8, §3.5).

The formula (5) for the volume of a solid of revolution is said to be obtained by the **disk method.** We give another derivation of this formula, which also explains the name.

We want to compute the volume V of the solid obtained by rotating the region under $y = f(x)$, $a \le x \le b$, about the x-axis. We divide the interval $[a,b]$ into many small subintervals, $a = x_0 < x_1 < x_2 < \cdots < x_n = b$ (see Figure 9.33 where $n = 5$), and choose in each subinterval $[x_{j-1}, x_j]$ a value ξ_j. Now, V is the sum of the n (small) volumes obtained by rotating the regions under $y = f(x)$, $x_{j-1} \le x \le x_j$, about the x-axis. Each of these small volumes is nearly the volume of the solid obtained by rotating the rectangle $0 \le y \le f(\xi_j)$, $x_{j-1} \le x \le x_j$, about the x-axis. The latter solid

FIGURE 9.33

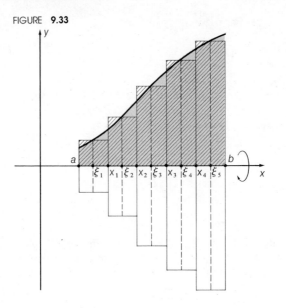

is a right circular cylinder of radius $f(\xi_j)$ and (small) height $x_j - x_{j-1}$. The volume of this "disk" is $\pi f(\xi_j)^2(x_j - x_{j-1})$. Thus V is nearly

$$\pi f(\xi_1)^2(x_1 - x_0) + \pi f(\xi_2)^2(x_2 - x_1) + \cdots + \pi f(\xi_n)^2(x_n - x_{n-1}).$$

But this is the Riemann sum for the integral $\int_a^b \pi f(t)^2\, dt$, and we conclude that

$$V = \pi \int_a^b f(t)^2\, dt.$$

The mathematical myth behind this formula, which goes back to Democritus, says that the solid of revolution consists of infinitely many infinitely thin disks.

PROBLEMS

1. Find the volume of the solid formed by rotating the region under the graph of $y = 1 - x^2$ from $x = 0$ to $x = 1$ around the x-axis.
2. Find the volume of the solid formed by rotating the region under the graph of $y = 1 - x^2$ from $x = 0$ to $x = 1$ around the y-axis.
3. Find the volume of the solid formed by rotating the region under the graph of $y = x^3$ from $x = 1$ to $x = 2$ around the x-axis.
4. Find the volume of the solid formed by rotating the region under the graph of $y = 1/x$ from $x = 1$ to $x = a > 1$ around the x-axis. (What happens to this volume as $a \to \infty$?)
5. Find the volume of the solid formed by rotating the region bounded by the curves $y = x^2 + 2$ and $y = -x^2 + 10$ around the x-axis.
6. Find the volume of the solid formed by rotating the region under the graph of $y = 1 + \frac{1}{2}\sqrt{x}$ from $x = 0$ to $x = 4$ around the y-axis.
7. Find the volume of the solid formed by rotating the region bounded by the curves $y = \sqrt{x}$, $x = 0$, and $y = 1$ around the x-axis.
8. Find the volume of the solid formed by rotating the region bounded by the curves $y = \sqrt{x}$, $x = 0$, and $y = 1$ around the y-axis.

9. Find the volume of the solid formed by rotating the region bounded by the curves $y = \sqrt{1 - x^2}$, $y = x$, and $y = 0$ around the x-axis. (Note that the curve $y = \sqrt{1 - x^2}$ is just the upper half of the unit circle $x^2 + y^2 = 1$.)

10. Find the volume of the solid formed by rotating the region bounded by the curves $y = \sqrt{1 - x^2}$, $y = x$, and $y = 0$ around the y-axis.

11. Find the volume of the solid formed by rotating the region bounded by the curves $y = x^2 + 1$, $x = 0$, and the tangent to $y = x^2 + 1$ at $x = 1$ around the x-axis.

12. Find the volume of the solid formed by rotating the region bounded by the curves $y = x^{2/3}$, $x = 0$, and the tangent to $y = x^{2/3}$ at $x = 1$ around the y-axis.

13. Find the volume of the solid formed by rotating the region bounded by the curves $y = 4 - x^2$ and $y = 3$ around the line $y = -1$. [*Hint*: Translate coordinates.]

14. Find the volume of the solid formed by rotating the region bounded by the curves $y = 1/(x + 1)$, $x = 0$, $x = 1$, and $y = 0$ around the line $x = -1$.

15. Find the volume of the solid formed by rotating the region bounded by the curves $y = x^3$ and $y = x$ around the line $x = -2$.

2.4 Solids of revolution. Shell method‡

There is also another way of computing volumes of solids of revolution called the **shell method.**

The volume obtained by rotating, about the y-axis, the region $0 < a \le x \le b$, $g(x) \le y \le f(x)$ (see Figure 9.34), is

$$V = 2\pi \int_a^b t[f(t) - g(t)]\, dt. \tag{9}$$

To obtain the formula, we denote the volume obtained by rotating the region between $x = a$ and $x = t$, and between the graphs of $g(x)$ and $f(x)$, by $V(t)$; see Figure 9.35. For $h > 0$, the difference $V(t + h) - V(t)$ is the volume swept out by the region

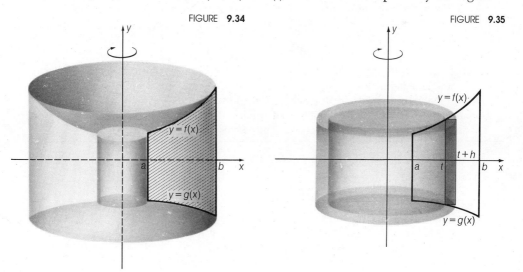

FIGURE **9.34**   FIGURE **9.35**

‡Optional subsection.

$t \leq x \leq t + h$, $g(x) \leq y \leq f(x)$. For small h, this is nearly the volume of the "shell" swept out by the rectangle $t \leq x \leq t + h$, $g(t) \leq y \leq f(t)$. The volume of the shell is the difference: [volume of cylinder of radius $t + h$, height $f(t) - g(t)$] minus [volume of cylinder of radius t, same height]. Thus

$$V(t + h) - V(t) \approx \pi(t + h)^2[f(t) - g(t)] - \pi t^2[f(t) - g(t)],$$

$$\frac{V(t + h) - V(t)}{h} \approx (2\pi t + \pi h)[f(t) - g(t)].$$

This suggests that $V'(t) = \lim_{h \to 0} \{[V(t + h) - V(t)]/h\} = 2\pi t[f(t) - g(t)]$. Since $V(a) = 0$ and the desired volume V is $V(b)$, we have

$$V = \int_a^b V'(t)\, dt = 2\pi \int_a^b t[f(t) - g(t)]\, dt.$$

[The reader may want to derive (9) using Riemann sums.]

EXAMPLE Compute the volume of the solid torus in §2.3, Example 7, by the shell method.
ANSWER The equation of the circle with center $(0,R)$ and radius r is $(x - R)^2 + y^2 = r^2$. Hence we use (9) with $a = R - r$, $b = R + r$, and $f(t) = \sqrt{r^2 - (x - R)^2}$, $g(t) = -\sqrt{r^2 - (x - R)^2}$, and obtain

$$V = 2\pi \int_{R-r}^{R+r} t[\sqrt{r^2 - (t - R)^2} + \sqrt{r^2 - (t - R)^2}\,]\, dt$$

$$= 4\pi \int_{R-r}^{R+r} t\sqrt{r^2 - (t - R)^2}\, dt = 4\pi \int_{-r}^{r} (R + u)\sqrt{r^2 - u^2}\, du$$

$$= 4\pi r \int_{-1}^{1} (R + rv)\sqrt{r^2 - r^2 v^2}\, dv = 4\pi r^2 \int_{-1}^{1} (R + rv)\sqrt{1 - v^2}\, dv.$$

Here we used first the substitution $t - R = u$, then the substitution $u = rv$. Continuing, we see that

$$V = 4\pi R r^2 \int_{-1}^{1} \sqrt{1 - v^2}\, dv + 4\pi r^3 \int_{-1}^{1} v\sqrt{1 - v^2}\, dv.$$

The second integral is 0, since $v\sqrt{1 - v^2}$ is an odd function of v; see Chapter 8, §3.6. [We can also compute the second integral by the substitution $1 - v^2 = w$, $-2v\, dv = dw$; it becomes $-\frac{1}{2}\int_0^0 w^{1/2}\, dw = 0$. Or note that $v\sqrt{1 - v^2}$ is the derivative of $-\frac{1}{3}(1 - v^2)^{3/2}$.] The integral $\int_{-1}^{1} \sqrt{1 - v^2}\, dv$ equals $\pi/2$. Thus we conclude that $v = 2\pi^2 R r^2$.

PROBLEMS

In Problems 1 to 12 use the shell method.

1. Find the volume of the solid formed by rotating the region under the graph of $y = x^2$ from $x = 0$ to $x = 2$ around the y-axis. Repeat, but rotate around the x-axis.
2. Find the volume of the solid formed by rotating the region under the graph of $y = \frac{1}{4}x^2$ from $x = 2$ to $x = 4$ around the y-axis. Repeat, but rotate around the x-axis.
3. Find the volume of the solid formed by rotating the region bounded by the curves $y = 4x - x^2$ and $y = 0$ around the y-axis.

4. Find the volume of the solid formed by rotating the region under the graph of $y = \sqrt{x}$ from $x = 0$ to $x = 4$ around the x-axis.

5. Find the volume of the sphere of radius a. [*Hint:* Rotate the semicircle $y = \sqrt{a^2 - x^2}$ about the x-axis to obtain the sphere of radius a.]

6. Find the volume of the right circular cone with altitude H and base of radius R.

7. Find the volume of the solid formed by rotating the region bounded by the parabolas $y = x^2$ and $y^2 = x$ about the y-axis.

8. Find the volume of the solid formed by rotating the region bounded by the curves $xy = 5$ and $y = -x + 6$ around the y-axis.

9. Find the volume of the solid formed by rotating the region bounded by the curves $xy = 2$, $xy = 4$, $x = 1$, and $x = 2$ about the y-axis.

10. Find the volume of the solid formed by rotating the region bounded by the parabola $y^2 = 8x$ and the line $x = 2$ around the line $x = 2$. [*Hint:* Draw a figure and construct a typical cylindrical shell.]

11. Find the volume of the solid formed by rotating the region in the first quadrant bounded by the curves $y = 1 + x^2$, $y = 2$, and $x = 0$ around the line $x = 1$.

12. A circular hole of diameter a is bored through the center of a sphere of radius a. What is the volume of the remaining solid?

§3 Length

The calculating of lengths of curves also leads to integrals, as we shall see presently. We assume first that we know what is meant by the length of a curve and use this knowledge to make an educated guess about a method for computing it. Then we define length by the formula obtained. This is the procedure we also followed for defining areas and volumes.

3.1 The length formula

Consider the graph of a smooth function $y = f(x)$, $a \le x \le b$, and denote its length by l. Let $l(t)$ denote the length of that part of the curve which corresponds to the values of x between a and t. Thus $l(a) = 0$, $l(b) = l$, and if $l(t)$ is a "nice" function, then by the fundamental theorem of calculus,

$$l = \int_a^b l'(t)\, dt.$$

To find $l'(t)$, let h be a small positive number, and let us consider the difference quotient $[l(t + h) - l(t)]/h$. The numerator is the length of the small arc of the curve lying above the interval $[t, t + h]$. We now make an assumption: if we replace the curve, for values of x between t and $t + h$, by its tangent at the point $(t, f(t))$, the resulting error in computing the difference quotient will be very small (see Figure 9.36). By replacing the curve by its tangent we mean that we replace the length of the arc PQ in the figure by the length of the segment PQ', which equals $\sqrt{h^2 + (mh)^2} = h\sqrt{1 + m^2}$, where m is the slope of the tangent. But $m = f'(t)$. Thus the ratio $[l(t + h) - l(t)]/h$ is approximately

$$\frac{h\sqrt{1 + m^2}}{h} = \sqrt{1 + f'(t)^2}.$$

FIGURE 9.36   FIGURE 9.37

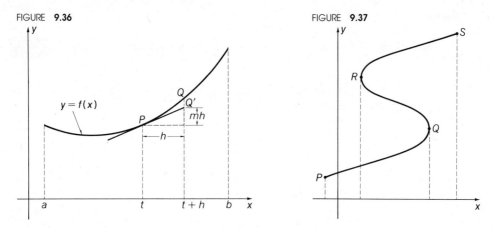

We surmise that $l'(t) = \sqrt{1 + f'(t)^2}$ and obtain the **length formula**

$$l = \int_a^b \sqrt{1 + f'(t)^2} \, dt. \tag{1}$$

Now we change our point of view and *define the length* of the graph of the function $y = f(x)$, $a < x < b$, by the length formula. The formula is applicable whenever the function considered has a piecewise continuous derivative.

If a curve is not the graph of a function (such as the curve shown in Figure 9.37) but can be decomposed into several arcs that are graphs of functions (such as the arcs PQ, QR, and RS in Figure 9.37), we define the length of the curve as the sum of the lengths of these arcs. (Our definition of length does not depend on the position of the coordinate system. This will be shown in Chapter 17, §2.4.)

EXAMPLES 1. Compute the length of the straight segment joining the points $(-2,4)$ and $(3,5)$.
FIRST SOLUTION   By the distance formula, $l = \sqrt{(-2 - 3)^2 + (4 - 5)^2} = \sqrt{26}$.
SECOND SOLUTION   The equation of the line joining the two points (see Chapter 2, §2.1) is

$$(y - 5)(-2 - 3) = (4 - 5)(x - 3) \qquad \text{or} \qquad y = \frac{1}{5}x + \frac{22}{5}.$$

By the length formula,

$$l = \int_{-2}^3 \sqrt{1 + \left(\frac{dy}{dx}\right)^2}\, dx = \int_{-2}^3 \sqrt{1 + \frac{1}{25}}\, dx = \frac{\sqrt{26}}{5}\int_{-2}^3 dx = \sqrt{26},$$

as it should.

2. Find the length of the graph $x \mapsto y = x^{3/2}$ between $x = 1$ and $x = 2$.
SOLUTION   By the length formula,

$$l = \int_1^2 \sqrt{1 + \left(\frac{dy}{dx}\right)^2}\, dx = \int_1^2 \sqrt{1 + \frac{9}{4}x}\, dx.$$

Setting $t = 1 + \frac{9}{4}x$, we get

$$\int_{13/4}^{22/4} t^{1/2} \frac{4}{9}\, dt = \frac{4}{9}\left(\frac{2}{3} t^{3/2}\right)\Big|_{13/4}^{22/4} = \frac{(22^{3/2} - 13^{3/2})}{27} \approx 2.1.$$

PROBLEMS

In Problems 1 to 10 use the arc-length formula to find the length of the given graphs between the given points.

1. $f(x) = x^{3/2}$ between $x = 0$ and $x = 5$.
2. $f(x) = 6x^{3/2}$ between $x = \frac{1}{3}$ and $x = \frac{5}{3}$.
3. $f(x) = \frac{1}{3}(x^2 + 2)^{3/2}$ between $x = 0$ and $x = 3$.
4. $f(x) = (\frac{2}{3}x + 1)^{3/2}$ between $x = 0$ and $x = 9$.
5. $f(x) = (4 - x^{2/3})^{3/2}$ between $x = 0$ and $x = 8$.
6. $f(x) = \frac{1}{6}x^3 + \frac{1}{2}x^{-1}$ between $x = 1$ and $x = 3$.
7. $f(x) = \frac{2}{3}x^{3/2} - \frac{1}{2}x^{1/2}$ between $x = 1$ and $x = 4$.
8. $f(x) = x^4/4 + x^{-2}/8$ between $x = 1$ and $x = 2$.
9. $f(x) = \int_1^x \sqrt{t^2 - 1}\, dt$ between $x = 1$ and $x = 3$.
10. $f(x)$ between $x = 1$ and $x = a$ $(a > 1)$, where $y = f(x)$ is any solution of the differential equation $dy/dx = (x^4 - 1)^{1/2}$.

3.2   Circular sectors. Circumference[‡]

We use the length formula to establish two theorems of elementary geometry.

The area of a circular sector is half the product of its radius and the length of its arc.

The length of a circle is $2\pi r$, where r is the radius and π the number defined by the definite integral

$$\frac{\pi}{4} = \int_0^1 \sqrt{1 - x^2}\, dx;\tag{2}$$

see Chapter 8, §3.5.

To simplify writing, we consider the case of unit radius. We must show that (i) *the area of a circular sector of radius 1 equals half the length of its arc*, and (ii) *the circumference of a circle of radius 1 is 2π.*

Let the center of the circle be chosen as the origin O, let a and b be two numbers such that $-1 < a < b < 1$, and let P and Q be points on the upper semicircle with first coordinates a and b. Then $P = (a, \sqrt{1 - a^2})$, $Q = (b, \sqrt{1 - b^2})$; see Figure 9.38. The upper semicircle is the graph of the function $y = \sqrt{1 - x^2}$. In order to apply the length formula (1), we compute

$$\frac{dy}{dx} = -\frac{x}{\sqrt{1 - x^2}}, \qquad 1 + \left(\frac{dy}{dx}\right)^2 = 1 + \frac{x^2}{1 - x^2} = \frac{1}{1 - x^2},$$

$$\sqrt{1 + \left(\frac{dy}{dx}\right)^2} = \frac{1}{\sqrt{1 - x^2}}.$$

Therefore

$$\text{length of circular arc } PQ = \int_a^b \frac{dx}{\sqrt{1 - x^2}}.\tag{3}$$

[‡]Optional subsection.

FIGURE 9.38

FIGURE 9.39

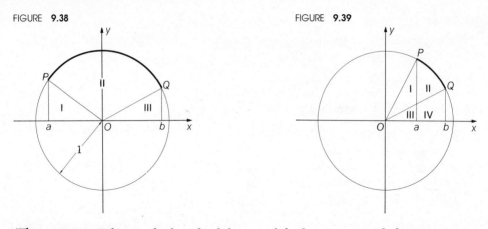

(The same integral gives the length of the arc of the lower semicircle between $x = a$ and $x = b$.)

Does formula (3) remain valid if $a = -1$ and/or $b = 1$? For $a = -1$ or for $b = 1$, the definition of the integral given in §1 of Chapter 8 is not applicable, since the integrand $(1 - x^2)^{-1/2}$ is not bounded in the interval $(-1,1)$; it becomes arbitrarily large as x approaches -1 or 1. But it is geometrically obvious that, for a fixed a, the length of the circular arc PQ is a continuous function of b defined also for $b = 1$, and that, for a fixed b, the length PQ is a continuous function of a defined also for $a = -1$. (We shall soon verify this analytically.) Therefore (3) holds also for $a = -1$ or for $b = 1$ provided we interpret the integral $\int_a^1 (1 - x^2)^{-1/2}\, dx$ as the limit of $\int_a^b (1 - x^2)^{-1/2}\, dx$ as $b \to 1$ from the left. This is an improper integral; see Chapter 8, §4.)

Next we compute the area of the sector POQ. Consider first the case $-1 < a < 0 < b < 1$; see Figure 9.38. Let I, II, III be the areas indicated in the figure. Then I + II + III is the area under the curve $y = \sqrt{1 - x^2}$ from $x = a$ to $x = b$; hence

$$\text{I + II + III} = \int_a^b \sqrt{1 - x^2}\, dx.$$

Also,

$$\text{I} = -\tfrac{1}{2}a\sqrt{1 - a^2}, \qquad \text{III} = \tfrac{1}{2}b\sqrt{1 - b^2}$$

(since $a < 0$ and $b > 0$). The area of the sector is II. Thus

$$\text{area of sector } POQ = \int_a^b \sqrt{1 - x^2}\, dx - \tfrac{1}{2}b\sqrt{1 - b^2} + \tfrac{1}{2}a\sqrt{1 - a^2}. \qquad (4)$$

The same formula holds in the case $0 < a < b < 1$ shown in Figure 9.39. Now the integral in (4) equals II + IV, and $\tfrac{1}{2}a\sqrt{1 - a^2} = \text{I + III}$, $\tfrac{1}{2}b\sqrt{1 - b^2} = \text{III + IV}$. The right side of (4) equals $(\text{II + IV}) - (\text{III + IV}) + (\text{I + III}) = \text{II + I}$; this is the area of the sector. The reader should convince himself that (4) holds also if $a = -1$ or $a = 0$ and if $b = 0$ or $b = 1$.

The geometric statement (i) that we want to establish reads

$$\int_a^b \sqrt{1 - x^2}\, dx - \frac{1}{2}b\sqrt{1 - b^2} + \frac{1}{2}a\sqrt{1 - a^2} = \frac{1}{2}\int_a^b \frac{dx}{\sqrt{1 - x^2}}. \qquad (5)$$

[We note that the left-hand side depends continuously on a and on b for $|a| \le 1$, $|b| \le 1$. Thus once we prove (5), we shall have verified the statement we made above about the limits of (3) for $a \to -1^+$ and $b \to 1^-$.]

We rewrite (5) in the form

$$\int_a^b \sqrt{1 - x^2}\, dx - \frac{1}{2} b\sqrt{1 - b^2} + \frac{1}{2} a\sqrt{1 - a^2} - \frac{1}{2} \int_a^b \frac{dx}{\sqrt{1 - x^2}} = 0;$$

keep a fixed, and consider the left-hand side as a function of b. The derivative of this function is, by the fundamental theorem,

$$\frac{d}{db} \left(\int_a^b \sqrt{1 - x^2}\, dx - \frac{1}{2} b\sqrt{1 - b^2} + \frac{1}{2} a\sqrt{1 - a^2} - \frac{1}{2} \int_a^b \frac{dx}{\sqrt{1 - x^2}} \right)$$

$$= \sqrt{1 - b^2} - \frac{1}{2}\sqrt{1 - b^2} + \frac{1}{2}\frac{b^2}{\sqrt{1 - b^2}} + 0 - \frac{1}{2}\frac{1}{\sqrt{1 - b^2}} = 0.$$

The function is constant. But, for $b = a$, it equals 0. Hence (5) holds for all b.

For $a = 0$, $b = 1$, Equation (5) becomes

$$\int_0^1 \frac{dx}{\sqrt{1 - x^2}} = 2 \int_0^1 \sqrt{1 - x^2}\, dx \qquad (6)$$

or, recalling the definition of π,

$$\int_0^1 \frac{dx}{\sqrt{1 - x^2}} = \frac{\pi}{2}. \qquad (7)$$

Similarly, we obtain from (5), for $a = -1$, $b = 1$,

$$\int_{-1}^1 \frac{dx}{\sqrt{1 - x^2}} = \pi. \qquad (8)$$

Since the integral above is half the unit circumference, this proves the geometric statement (ii): the unit circumference has length 2π.

REMARK   In the next chapter we will need the following theorem: *for two circular sectors with the same opening, the ratios of length of arc to radius are the same.* In the notations of Figure 9.40, $l/r = L/R$.

FIGURE **9.40**

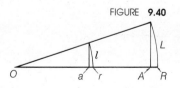

To prove this, we compute the length l by the length formula and obtain

$$\frac{l}{r} = \int_a^r (r^2 - x^2)^{-1/2}\, dx = \int_{a/r}^1 (1 - t^2)^{-1/2}\, dt$$

(where we used the substitution $x = rt$). Similarly, $L/R = \int_{A/R}^1 (1 - t^2)^{-1/2}\, dt$. But $a/r = A/R$, by similar triangles. Hence $l/r = L/R$.

PROBLEMS

Problems 1 to 6 refer to a circle of radius r with center O.

1. Let $L(a,b)$ denote the length of the arc of the upper semicircle from the point $P = (a, \sqrt{r^2 - a^2})$ to the point $Q = (b, \sqrt{r^2 - b^2})$, with $-r \leq a < b \leq r$. Let $A(a,b)$ denote the area of the sector POQ. Write the two theorems stated in the beginning of this subsection as equations involving L and A.
2. Write $L(a,b)$ as an integral.
3. Find a formula for $A(a,b)$, analogous to relation (4). Assume $a < 0 < b$.
4. Prove that $A(a,b) = \frac{1}{2}rL(a,b)$.
5. Show that $L(0,r) = \frac{1}{2}\pi r$.
6. Compute $L(-r,r)$.

§4   Moments‡

In this section we consider integrals of the form

$$\int_a^b (x - \alpha)f(x)\, dx, \qquad \int_a^b (x - \alpha)^2 f(x)\, dx.$$

Such integrals arise in two quite different connections: in mechanics and in statistics.

4.1   Mass center of a rod

Consider a system of n particles situated along a horizontal line, say, the x-axis. Let their masses be $m_1, m_2, \ldots, m_n$ and their positions $x_1, x_2, \ldots, x_n$. If α is any point on the real axis, then the number $(x_i - \alpha)m_i$ is called the **moment** of the ith particle, with respect to α, and the sum

$$(x_1 - \alpha)m_1 + (x_2 - \alpha)m_2 + \cdots + (x_n - \alpha)m_n \tag{1}$$

is called the moment of the system, with respect to α. The physical significance of moments is disclosed by the **law of the lever,** discovered by Archimedes: if the particles are attached to a weightless rod, pivoted at α, the rod will be in equilibrium (will not move under the influence of gravity) if and only if the moment with respect to α is 0. For example, the system of four particles shown in Figure 9.41(a) is in equilibrium, while the one in Figure 9.41(b) is not.

FIGURE **9.41**

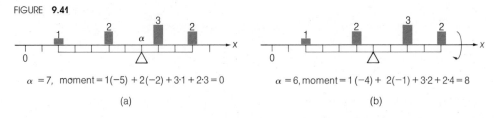

$\alpha = 7$, moment $= 1(-5) + 2(-2) + 3\cdot1 + 2\cdot3 = 0$

(a)

$\alpha = 6$, moment $= 1(-4) + 2(-1) + 3\cdot2 + 2\cdot4 = 8$

(b)

The point α with respect to which the system of particles has moment 0 is called the **mass center** (or **center of mass**) of the system. To find the mass center, set the

‡Optional section.

sum (1) equal to 0, and solve for α. This gives

$$\alpha = \frac{x_1 m_1 + x_2 m_2 + \cdots + x_n m_n}{m_1 + m_2 + \cdots + m_n}. \tag{2}$$

The denominator in (2) is the total mass.

Consider next a thin rod of nonuniform density. We may think of the rod as lying over the interval $[a,b]$ of the x-axis. The density of the rod is a nonnegative function $f(x)$, $a \leq x \leq b$, such that the mass of the part of the rod between $x = \xi$ and $x = \eta$ is $\int_\xi^\eta f(x)\,dx$; see Figure 9.42. The integral $\int_a^b f(x)\,dx$ is then the total mass of the rod. How shall we define the moment of the rod with respect to a point α?

FIGURE **9.42**

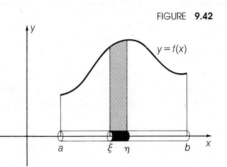

We divide the rod into n small subintervals: $a = x_0 < x_1 < x_2 < \cdots < x_n = b$. The part of the rod between $x = x_{j-1}$ and $x = x_j$ has mass $\int_{x_{j-1}}^{x_j} f(x)\,dx$, and since we may consider f as nearly constant in the very small interval $[x_{j-1}, x_j]$, this mass is nearly $m_j = f(\xi_j)(x_j - x_{j-1})$; here ξ_j is some point between x_{j-1} and x_j. Let us regard the rod as a system of n particles, of masses $f(\xi_1)(x_1 - x_0), f(\xi_2)(x_2 - x_0), \ldots, f(\xi_n)(x_n - x_{n-1})$, located at $\xi_1, \xi_2, \ldots, \xi_n$. The moment, about α, of this system is

$$(\xi_1 - \alpha)f(\xi_1)(x_1 - x_0) + (\xi_2 - \alpha)f(\xi_2)(x_2 - x_1) + \cdots + (\xi_n - \alpha)f(\xi_n)(x_n - x_{n-1}).$$

But this is a Riemann sum for the integral

$$\int_a^b (x - \alpha)f(x)\,dx. \tag{3}$$

Therefore we define (3) to be the **moment** of our rod about α.

The value of α for which the moment is 0 is called the **mass center**. To find the mass center, set the integral in (3) equal to 0, and solve for α. Since $\int_a^b (x - \alpha)f(x)\,dx = \int_a^b xf(x)\,dx - \alpha\int_a^b f(x)\,dx$, we find that $\int_a^b xf(x)\,dx - \alpha\int_a^b f(x)\,dx = 0$ or

$$\alpha = \frac{\int_a^b xf(x)\,dx}{\int_a^b f(x)\,dx}. \tag{4}$$

As in Equation (2), the total mass appears in the denominator.

EXAMPLE   The density function of a rod is $f(x) = 2x$ for $0 \le x \le 1$ and $f(x) = \frac{5}{2} - \frac{1}{2}x$ for $1 \le x \le 5$. What is the total mass? Where is the mass center?

ANSWER   The total mass is

$$\int_0^5 f(x)\,dx = \int_0^1 2x\,dx + \int_1^5 \left(\frac{5}{2} - \frac{1}{2}x\right) dx = x^2 \Big|_0^1 + \left(\frac{5}{2}x - \frac{1}{4}x^2\right) \Big|_1^5 = 5.$$

The numerator in (4) is

$$\int_0^5 xf(x)\,dx = \int_0^1 2x^2\,dx + \int_1^5 \left(\frac{5}{2}x - \frac{1}{2}x^2\right) dx = \frac{2x^3}{3} \Big|_0^1 + \left(\frac{5}{4}x^2 - \frac{1}{6}x^3\right) \Big|_1^5 = 10.$$

The coordinate of the mass center is, therefore, $\alpha = \frac{10}{5} = 2$.

PROBLEMS

1. Suppose four particles, with masses of 2, 4, 3, and 2 grams, are situated along the x-axis at the points -4, -2, 2, and 5, respectively. What is the mass center of the system?
2. Suppose six particles, with masses of 20, 10, 8, 6, 4, and 2 grams, are situated along the x-axis at the points 1, 2, 3, 4, 8, and 10, respectively. What is the mass center of the system?
3. Six particles of equal mass are attached to a weightless rod of length 100 cm at distances of 10, 20, 40, 60, 80, and 90 cm from the left end. If the rod is pivoted at the center, is the rod in equilibrium?
4. Four particles of mass 25, 40, 80, and 100 grams are attached to a weightless rod of length 100 cm at distances of 10, 20, 40, and 80 cm from the left end. If the rod is pivoted at the center, is the rod in equilibrium?
5. The density function of a rod is given by $f(x) = 3x^2$ from $0 \le x \le 1$. Find the total mass and mass center of the rod.
6. The density function of a rod is given by $f(x) = 2 + 4x - x^2$ for $0 \le x \le 4$. Find the total mass and mass center of the rod.

4.2   Frequency densities and means

Mathematical statistics deals with performing an experiment or measurement, whose outcome is a number, and performing it *many times*. The precise nature of the experiment need not concern us here. We may think, for instance, of measuring the height of babies at birth, reading off the temperature at noon on different days, counting the number of calls made from different pay telephones, and so on. Suppose that the possible outcomes of our experiment are numbers x satisfying $a \le x \le b$. To record the outcome of our series of experiments, we could divide the interval $[a,b]$ into n subintervals, say, of equal length, $a = x_0 < x_1 < x_2 < \cdots < x_n = b$, and note the number m_i of times the outcome of our experiment gave a number between x_{i-1} and x_i. This can be represented by a step function whose graph has height m_i over the ith subinterval. The resulting diagram is called a **histogram.**

From the histogram in Figure 9.43, for instance, we can read off that the result of our experiment was 400 times between 1.5 and 2.5, 500 times between 2.5 and 3.5, 600 times between 3.5 and 4.5, 550 times between 4.5 and 5.5, 500 times between 5.5 and 6.5, and 350 times between 6.5 and 7.5. All in all, we performed $400 + 500 + 600 + 550 + 500 + 350 = 2900$ experiments. The sum of all outcomes is

FIGURE 9.43

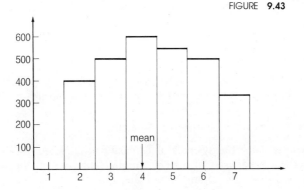

nearly $2 \cdot 500 + 3 \cdot 500 + 4 \cdot 600 + 5 \cdot 550 + 6 \cdot 500 + 7 \cdot 550 = 14{,}500$. The average outcome, also called the mean, is therefore nearly $14{,}500/2900 = 4.1$.

Assume now that the graph in Figure 9.43 represents the density function of a horizontal rod. To compute the mass center of this rod, we carry out the same calculation as before, and we find that the mass center is located at 4.1.

If we use for our histogram more and more smaller and smaller subintervals, we expect the graph to approach a graph of a continuous (or piecewise continuous) function $f(x)$. We may choose the unit of length along the vertical axis so that the total area under the curve is 1. Then we obtain a function $y = f(x)$, called the **frequency density**, which has the following properties: (i) $f(x) \geq 0$; (ii) if all results x of our experiment lie between a and b, then

$$\int_a^b f(x)\,dx = 1; \tag{5}$$

(iii) for $a \leq \xi < \eta \leq b$, the integral

$$\int_\xi^\eta f(x)\,dx \tag{6}$$

gives the ratio of times the experiment yielded a value between ξ and η to the total number of experiments performed. We also call (6) the **probability** of a random experiment to have had an outcome between ξ and η. We may, of course, interpret $f(x)$ as the density of a material rod of total mass 1. In view of (4) and (5), the mass center of the rod is at the point

$$\alpha = \int_a^b x f(x)\,dx. \tag{7}$$

This value is called the **mean** of the frequency distribution, or the **expected value** of our experiment.

PROBLEMS

In Problems 1 to 7 determine whether the given expression represents a frequency distribution function. If it does, then find its mean. If it does not, state why.

1. $f(x) = \tfrac{1}{4}(x + 1), \quad 0 \leq x \leq 2.$ 2. $f(x) = \dfrac{x^2}{16}, \quad 0 \leq x \leq 4.$

3. $f(x) = \frac{1}{2}(2x + 1)$, $-1 \le x \le 1$.

4. $f(x) = \begin{cases} \frac{1}{16}(x + 4), & -4 \le x \le 0 \\ \frac{1}{16}(4 - x), & 0 \le x \le 4. \end{cases}$

5. $f(x) = \frac{3}{8}(x - 1)^2$, $|x| \le 1$.

6. $f(x) = \frac{3}{64}x^2$, $0 \le x \le 4$.

7. $f(x) = \frac{3}{8}\left(1 - \frac{x^2}{4}\right)$, $|x| \le 2$.

8. If the frequency distribution function $f(x)$ is an even function of x over the interval $|x| \le a$, where a is any positive number, what is the value of the mean?

4.3 Moment of inertia and standard deviation

Return, for an instant, to the systems of n masses $m_1, \ldots, m_n$ located at the points $x_1, \ldots, x_n$ along the x-axis (see §4.1). The number $m_i(x_i - \alpha)^2$ is called the **moment of inertia** of the ith particle, about α, and the sum

$$m_1(x_1 - \alpha)^2 + m_2(x_2 - \alpha)^2 + \cdots + m_n(x_n - \alpha)^2 \tag{8}$$

is called the moment of inertia of the system, about α. For the two systems depicted in Figure 9.39, the moments of inertia are 54 and 62, respectively. If $f(x)$, $a \le x \le b$, is the density function of a rod, the moment of inertia about α is defined as

$$\int_a^b (x - \alpha)^2 f(x) \, dx. \tag{9}$$

It determines the resistance of the rod pivoted at α against a torque. The moment of inertia would be 0 if all mass were concentrated at α; it would be large if there were masses, even small ones, far away from α.

Assume now that $f(x)$ is a frequency density, so that (5) holds, and let α be its mean, so that (7) holds. Then the number (9) is called the **variance,** and its square root

$$\sigma = \sqrt{\int_a^b (x - \alpha)^2 f(x) \, dx} \tag{10}$$

is called the **standard deviation.** If σ is small, the results of our experiments bunch around the mean (expected value) α; if σ is large, a significant portion of the experiments have had outcomes removed from α.

EXAMPLE   Find the standard deviation for the frequency function $f(x) = \frac{1}{2}x$, $0 \le x \le 2$.
ANSWER   First we check whether condition (5) holds; it does. Next we use (7) to compute the mean α:

$$\alpha = \int_0^2 \frac{1}{2}x^2 \, dx = \frac{4}{3}.$$

Now, by (10),

$$\sigma^2 = \int_0^2 (x - \frac{4}{3})^2 \frac{1}{2}x \, dx = \int_0^2 (\frac{1}{2}x^3 - \frac{4}{3}x^2 + \frac{8}{9}x) \, dx = \frac{2}{9}.$$

Hence $\sigma = \sqrt{2}/3$.

REMARK   It is customary, in mathematical statistics, to consider frequency densities $f(x)$ defined for *all* x, that is, not to impose any limitations on the possible outcome of the experiments or measurements considered. A frequency density $f(x)$ must always satisfy the conditions

$$f(x) \ge 0 \quad \text{for all } x, \quad \text{and} \quad \int_{-\infty}^{+\infty} f(x) \, dx = 1. \tag{11}$$

The mean α and the standard deviation σ are defined by

$$\alpha = \int_{-\infty}^{+\infty} xf(x)\,dx, \qquad \sigma^2 = \int_{-\infty}^{+\infty} (x - \alpha)^2 f(x)\,dx. \tag{12}$$

(All integrals above are improper integrals; see Chapter 8, §4.)

PROBLEMS

1. Suppose four particles, with masses of 8, 4, 2, and 6 grams, are situated along the x-axis at the points -4, 1, 2, and 4, respectively. Find the moment of inertia of the system about the mass center.

2. Suppose six particles, with masses of 10, 20, 20, 20, 20, and 10 grams, are situated along the x-axis at the points -6, -4, -2, 0, 2, and 4, respectively. What is the moment of inertia of the system about the mass center?

3. Six particles of masses 50, 50, 50, 50, 100, and 100 grams are attached to a weightless rod of length 100 cm at distances of 10, 20, 30, 60, 80, and 100 cm from the left end. Find the moment of inertia about the center of the rod.

4. Two weightless rods, each of length 100 cm, are pivoted about their centers. The first rod has four masses of 50 grams each located at distances of 20, 30, 40, and 80 cm from the left end. The second rod has five masses of 40 grams each located at distances of 10, 20, 30, 60, and 80 cm from the left end. Which rod has the greater tendency to rotate about the pivot?

In Problems 5 to 8 find the variance and standard deviation for the given frequency function.

5. $f(x) = 2(1 - x), \quad 0 \le x \le 1.$

6. $f(x) = \dfrac{1}{4}\left(1 - \dfrac{x^2}{9}\right), \quad |x| \le 3.$

7. $f(x) = \frac{3}{8}x^2, \quad 0 \le x \le 2.$

8. $f(x) = \begin{cases} \dfrac{3}{8}, & |x| < 1 \\[2mm] \dfrac{3}{8x^4}, & |x| \ge 1. \end{cases}$

9. Find the standard deviation for the frequency distribution $f(x) = 1/2a$, $|x| \le a$. Here a is any positive number. As a increases, what happens to the standard deviation? Is this consistent with the notion that the standard deviation is a measure of spread about the mean?

10. Find the standard deviation for the frequency distribution $f(x) = \dfrac{3}{4a}\left(1 - \dfrac{x^2}{a^2}\right)$, $|x| \le a$. Here a is any positive number.

§5 Energy[‡]

In this section, integrals (and the chain rule) are applied to mechanics. This leads to the all-important concept of energy and to the principle of conservation of energy.

5.1 Forces depending on position

We consider a particle moving on a straight line, so that its position is described by a single number. We assume that the force F acting on the particle (in the direction of the line along which the particle moves) depends only on its location. This means that $s \mapsto F(s)$ is a given function.

[‡]Optional section.

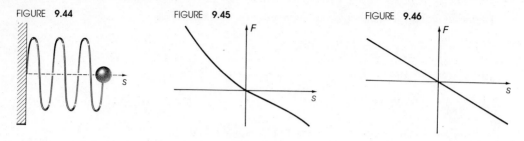

FIGURE 9.44 FIGURE 9.45 FIGURE 9.46

An example is the weight of a vertically moving particle of mass m. In this case $F(s) = -mg$, where g is the acceleration of gravity and the upward direction is considered as positive. This force is a constant.

Another example is a particle attached to a **spring** (Figure 9.44). We choose $s = 0$ as the point at which the spring is relaxed and exerts no force on the particle. Thus $F(0) = 0$. If the particle is at a point $s \neq 0$, the spring will exert a **restoring force** on the particle, since it will tend to contract if extended and to expand if compressed. Thus $F(s)$ will be negative for $s > 0$ and positive for $s < 0$ (compare Figure 9.45). The precise shape of the curve $s \mapsto F(s)$ depends on the spring. The simplest case of a **linear restoring force**

$$F = -ks$$

(see Figure 9.46) is of particular importance. Most springs will exhibit approximately such a restoring force for small deflections. Indeed, if the curve $s \mapsto F(s)$ has a tangent at $s = 0$, and we consider only small values of s, we can replace the curve by its tangent without committing a significant error. A spring with a linear restoring force is called **elastic,** and the value k is called the **spring constant.** Since F is measured in units of force (say, gram cm/sec²) and s in units of length (say, cm), k is measured in units of mass per unit of time per unit of time (say, gram/sec²).

PROBLEMS

1. An elastic spring has a spring constant of 3 grams/sec². What restoring force will the spring exert on a body attached to one end if the spring is stretched 2 cm from its relaxed position?
2. When a $\frac{1}{2}$-gram mass is hung from a vertical elastic spring, the spring is stretched 5 cm from its relaxed position. What is the spring constant? (Neglect the weight of the spring.)
3. A 2-gram mass is hung from a vertical elastic spring whose spring constant is 100 grams/sec². How far is the spring stretched from its relaxed position? (Neglect the weight of the spring.)
4. Suppose that a spring has a restoring force given by $F(s) = 5s - 16s^3 + s^5$. For small deflections, we may treat it as an elastic spring with constant k. What is k?

5.2 Work

We call the quantity

$$W = -\int_{s_0}^{s} F(x)\, dx \tag{1}$$

the **work against the force** F needed to move the particle from s_0 to s. This is a *definition;* it cannot be proved, but it can be motivated. Let us do so.

It is a common experience that in moving a particle against a force acting on it, say, in lifting a heavy stone, effort is expended. Even before we define work, we have a feeling that it takes twice as much work to lift two equal weights as to lift one, and that the work needed to lift a stone to the height, say, of 3 ft, is three times that needed to lift it 1 ft. We therefore define work against a constant force as the product of the *force times distance*. The work is positive if the object is moved in the direction opposite to that of the force, negative otherwise.

Consider now the work done against a nonconstant force in moving the particle from s_0 to, say, x: denote it by $W(x)$. We still have to define $W(x)$, but we certainly want to have $W(s_0) = 0$, since no work is involved in not moving the particle at all. It is also natural to require that

$$W(x + h) - W(x) = w$$

be the work done in moving the particle from the position x to the position $x + h$. Assume now that h is small. Over the small interval $(x, x + h)$, the force F will differ only a little from its value at x. We treat it as a constant and see that $w \approx -F(x)h$. [The minus sign is correct. If $F(x)$ is negative (say, the force pushes to the left) and h is positive (the particle is moved to the right), we want w to be positive.] From the preceding equations, we conclude that $[W(x + h) - W(x)]/h \approx -F(x)$. This suggests the requirement $W'(x) = -F(x)$, and, by the condition $W(s_0) = 0$ and the fundamental theorem of calculus, we obtain (1).

Work is measured in units of force times distance, say, gram $\cdot$ cm^2/sec^2.

5.3   Potential energy

Let us fix s_0 ("the zero level") arbitrarily. Then $W(s)$ is called the **potential energy** of the particle located at s. In the case of a body subject to weight $F = -mg$, it is convenient to choose for $s_0 = 0$ the ground level. Then the potential energy of a particle at height s is

$$W(s) = -\int_0^s (-mg)\, dx = mgs, \tag{2}$$

which is weight times height.

In the case of a spring, we also take $s_0 = 0$ as the zero level. If the tip of the spring is at the point s, the potential energy is

$$W(s) = -\int_0^s F(x)\, dx.$$

The reader should check that $W(s) > 0$ for $s < 0$ and for $s > 0$. For the linear restoring force $F = -ks$, we have

$$W(s) = \tfrac{1}{2}ks^2. \tag{3}$$

One of the everyday meanings of the word "energy" is "the ability to do work." This explains the name "potential energy" for $W(s)$. A stone lifted to the height s will do work (fall) if released. So will a stretched spring.

PROBLEMS

1. Two positively charged particles with charge 1 statcoulomb repel each other with a force equal to $1/r^2$, where r is the distance between the particles measured in centimeters. One particle is held fixed, and the second is moved directly toward the first from a point 100 cm away to a point 10 cm away. How much work is done?
2. How much work is done in lifting a body whose mass is 10 grams from a height of 100 cm to a height of 200 cm?
3. How much work is done in stretching the end of an elastic spring 3 cm from its relaxed position if the spring constant is 15 grams/sec²?
4. What is the potential energy of a stone of weight 10 lb at a height of 50 ft?
5. What is the potential energy of an elastic spring if it is compressed 1 cm from its relaxed position and the spring constant is 4.5 grams/sec²?
6. How far is it necessary to compress an elastic spring whose spring constant is 3 grams/sec² in order that the compressed spring have a potential energy of 13.5 grams · cm/sec²?

5.4 Kinetic energy. Conservation of energy

A moving particle can do work. For instance, a stone that falls into water produces waves. Therefore we want to define the energy of motion or **kinetic energy** K of a moving particle. Since K should be measured in units of work, that is, in gram · cm²/sec², mass is measured in grams and velocity in cm/sec, the proper definition is $K = \alpha m v^2$, where α is a numerical constant. It turns out that we should choose $\alpha = \frac{1}{2}$ and set

$$K = \tfrac{1}{2}mv^2. \tag{4}$$

Why $\frac{1}{2}$ is the right constant will be seen presently.

If a particle moves along a straight line under the influence of a force that depends only on its position, then the sum

$$E = W + K$$

of its potential and kinetic energies remains constant during the motion.

We call E the total energy of the particle. The fact that E is a constant is a special case of the **law of conservation of energy**. It is a mathematical consequence of Newton's law of motion and the definitions of W and K. In following the proof, the reader should observe that it would have failed had we used a number different from $\frac{1}{2}$ in defining the kinetic energy K.

We have that E, W, and K are functions of time t. In particular, $K = \frac{1}{2}mv^2$ and $v = ds/dt$, so that, by the chain rule,

$$\frac{dK}{dt} = \frac{1}{2}m\frac{dv^2}{dt} = \frac{1}{2}m\frac{dv^2}{dv}\frac{dv}{dt} = mv\frac{d^2s}{dt^2} = mva,$$

where a is the acceleration. On the other hand, the potential energy at time t is the potential energy at the point $s(t)$, that is,

$$W(s(t)) = -\int_{s_0}^{s(t)} F(x)\, dx.$$

By the fundamental theorem of calculus, $dW/ds = -F(s)$, so that, by the chain rule,

$$\frac{dW}{dt} = \frac{dW}{ds}\frac{ds}{dt} = -Fv.$$

Thus

$$\frac{dE}{dt} = \frac{dW}{dt} + \frac{dK}{dt} = mva - Fv = (ma - F)v = 0,$$

since $ma = F$, by Newton's law of motion. Since $dE/dt = 0$, E is constant as asserted.

5.5 Free fall

Let us apply the energy principle just proved to free fall. Here W is given by (2) so that the total energy is

$$E = \tfrac{1}{2}mv^2 + mgs.$$

This quantity is constant during motion. If a particle is released from height s_0, its velocity, at the instant of release, is 0. Hence

$$\tfrac{1}{2}mv^2 + mgs = mgs_0$$

during the motion. When the particle hits the ground, $s = 0$. Thus the terminal velocity, call it v_1, is computed from the equation $\tfrac{1}{2}mv_1^2 = mgs_0$; so

$$v_1 = \sqrt{2gs_0}. \tag{5}$$

If a particle is thrown upward from the ground with speed v_0, its potential energy at the initial instant is 0, and the total energy is therefore $E = \tfrac{1}{2}mv_0^2$. When the particle reaches ground level again, its kinetic energy is $\tfrac{1}{2}mv_1^2$, v_1 being the terminal velocity, and its potential energy is again 0. Thus $\tfrac{1}{2}mv_0^2 = \tfrac{1}{2}mv_1^2$, and since $v_0 > 0$ and $v_1 < 0$,

$$v_1 = -v_0, \tag{6}$$

as we already observed in Chapter 7, §1.3.

The results (5) and (6) can also be derived from the general formula describing the position s at the time t of a freely falling particle which had the position s_0 and the velocity v_0 at the time $t = 0$ [Equation (4) in Chapter 7, §1.3].

PROBLEMS

1. What velocity must a $\tfrac{1}{2}$-lb snowball have in order to have the same kinetic energy as a 3000-lb car traveling at 10 mph?
2. From what height must a 2-lb flower pot be dropped in order to have the same kinetic energy as a 3000-lb car traveling at 10 mph?
3. A particle of mass m is moving in a straight line. If distance s from some fixed point on the line is measured in centimeters and time t is measured in seconds, then the position of the particle is given by $s = t^{3/2} + t + 1$ for $t \geq 0$. What is the kinetic energy of the particle when $t = 49$?
4. Let $K(t)$ denote the kinetic energy at time t of a body of mass m falling freely under the influence of gravity. Show that $K''(t)$ is constant, and find its value.

5. A particle of unit mass is moving in a straight line under the influence of a force whose action at each point of the line is independent of position. Suppose the force acting on the moving particle at time t is found to be $6t$. Suppose also that the particle is at rest at time $t = 0$.

    (a) Find the velocity of the particle.

    (b) Find the kinetic energy of the particle directly from the definition.

    (c) Find the position of the particle (take the position at $t = 0$ as a reference point).

    (d) Find the force at each point s of the line.

    (e) Find the potential energy of the particle (take the position at $t = 0$ as the zero level) directly from the definition.

    (f) Verify that the law of conservation of energy holds.

°6. A particle of unit mass is moving in a straight line under the influence of a force whose action at each point of the line is independent of time. The distance of the particle from some fixed point on the line is given by $s = t^{5/2}$, $t \geq 0$, where t denotes time.

    (a) Find the kinetic energy of the particle directly from the definition.

    (b) Find the force at each point s of the line.

    (c) Find the potential energy of the particle (take the base point as zero level) directly from the definition.

    (d) Verify that the law of conservation of energy holds.

5.6 Gravitation

Newton's law of universal gravitation states that, between any two particles with masses M and m and situated at a distance r from each other, there is an attracting force of magnitude

$$\frac{\gamma M m}{r^2}.$$

Here γ is a constant (**gravitational constant**), which is the same for all bodies. Since force is measured in units of mass times length per unit of time squared (say, gram $\cdot$ cm/sec²), and the expression Mm/r^2 is measured in units of mass squared per length squared (say, gram/cm²), γ must be measured in units of length cubed per unit of mass per unit of time squared (say, cm³/gram $\cdot$ sec²). In applying Newton's law to a spherical body, such as the sun or a planet, we may consider the total mass concentrated at the center. (This last statement is a mathematical theorem, which we shall not prove here.)

    We shall show later (see Chapter 17, §5) how the law of gravitation follows from observed facts about planetary motions. Let us now apply it to the case where one "particle" is the earth and the other some body moving along a straight line passing through the center of the earth. We denote this center by O and the distance of the moving particle from O by s. Let M be the mass of the earth, m be the mass of the moving particle. Also, let R denote the radius of the earth, and let s denote the distance of the particle from the center of the earth.

    By the law of gravity, the force F exerted by the earth on the body is

$$F = -\frac{\gamma M m}{s^2}; \tag{7}$$

it is negative, since it pulls the body toward the earth. In everyday life we deal with bodies whose distance x from the surface of the earth is very small compared to the radius R of the earth. For such a body, we have $s = R + x$ and

$$F = -\frac{\gamma Mm}{(R + x)^2}.$$

If x/R is small, the number $1/(R + x)^2$ differs very little from $1/R^2$. In fact, for $x > 0$,

$$\frac{1}{R^2} - \frac{1}{(R + x)^2} = \frac{2Rx + x^2}{(R + x)^2 R^2} < \frac{2x}{R^3} + \frac{x^2}{R^4}. \tag{8}$$

An approximate value of R, determined by geodesic measurements, is

$$R = 6.37 \cdot 10^8 \text{ cm} = 6370 \text{ km} = 3959 \text{ miles}.$$

If x is less than $\frac{1}{1000}$ of the earth's radius, that is, less than 3.9 miles, the difference (8) is less than $1/10^{19}$ cm^2. We commit, therefore, an exceedingly small error if, for bodies close to the surface of the earth, we treat F as a constant and write

$$F = -\frac{\gamma Mm}{R^2}.$$

Setting

$$\frac{\gamma M}{R^2} = g, \tag{9}$$

we obtain, from the law of motion,

$$-mg = ma \qquad (a = \text{acceleration}).$$

This is Galileo's law of constant acceleration $a = -g$.

(As a matter of fact, the value of g, the acceleration of gravity at the surface of the earth, is not the same everywhere. The dependence of g on geographic latitude was discovered in 1672. Newton concluded from this that the earth is not a perfect sphere; the equator is farther from the earth's center than are the poles. Very precise measurements disclose variations in the value of g due to the nonuniformity in the crust of the earth. This fact is used in scientific prospecting to locate the probable whereabouts of oil deposits, minerals, and the like.)

5.7 Escape velocity

We return to a body whose distance from the earth's surface is large enough to warrant the use of Equation (7). Its law of motion is $ma = -\gamma Mm/s^2$ or

$$\frac{d^2s}{dt^2} = -\frac{\gamma M}{s^2}. \tag{10}$$

We do not discuss here how to solve this differential equation, that is, to compute from it the function $s(t)$. However, since F depends only on s, the energy principle is applicable. We compute the potential energy of the body at the distance s from the center O, choosing $s = R$ (the earth's surface) as the zero level. By (1) and (7), we have

$$W = -\int_R^s \frac{-\gamma Mm}{x^2}\, dx = \gamma Mm \int_R^s \frac{dx}{x^2} = \gamma Mm \left(\frac{1}{R} - \frac{1}{s}\right). \tag{11}$$

Let us use this to compute the initial velocity v_0 which must be given to a body, say, a rocket, in order that it reach the distance A from O before beginning to fall back. At the initial instance, we have $K = \frac{1}{2}mv_0^2$, $W = 0$. Assume that when the body reaches height A, its velocity becomes zero, so that it begins to fall back toward the earth. At this instant we have $s = A$, $v = 0$, and hence

$$K = 0, \qquad W = \gamma Mm \left(\frac{1}{R} - \frac{1}{A} \right).$$

But the sum $K + W$ remains the same throughout the motion. Hence

$$\frac{1}{2}mv_0^2 = \gamma Mm \left(\frac{1}{R} - \frac{1}{A} \right) \qquad \text{or} \qquad v_0 = \sqrt{2\gamma M \left(\frac{1}{R} - \frac{1}{A} \right)}.$$

In view of (9), we have $\gamma M = gR^2$. Thus

$$v_0 = \sqrt{2g \left(R - \frac{R^2}{A} \right)}. \tag{12}$$

Note that v_0 does not depend on m.

As A grows without bound, v_0 approaches the value

$$V = \sqrt{2gR}. \tag{13}$$

A body projected with a speed greater than, or equal to, V, will *never* return; we therefore call V the *escape velocity*. Substitution of the value of R given above and of the value $g = 980$ cm/sec^2 yields $V = 1.17 \cdot 10^6$ cm/sec $\approx 26,100$ miles per hour. Of course, our calculation neglected the effect of air resistance and the gravitational effect of other heavenly bodies. But the value we obtained for V is not too far from the actual escape velocity from earth.

PROBLEMS

In Problems 1 to 5 the value of the gravitational constant γ should be taken as 6.67×10^{-8} cm^3 gram $\cdot$ sec^2.

1. The radius of the Moon is approximately 1.74×10^8 cm; its mass is approximately 7.35×10^{25} grams. Compute the escape velocity of a particle from the Moon.
2. The radius of the planet Jupiter is approximately 70.96×10^8 cm; its mass is approximately 1.88×10^{30} grams. Compute the escape velocity of a particle from Jupiter.
3. The radius of the sun is approximately 692.32×10^8 cm; its mass is approximately 1.97×10^{33} grams. Compute the escape velocity of a particle from the sun.
4. Let v_0 denote the escape velocity of a particle from a large spherical body of mass M and radius R. Suppose R remains constant while M varies; for example, if meteors continually fall from space and are embedded in the sphere, the mass will increase while the radius remains effectively unchanged. Find dv_0/dM.
5. In Problem 4 suppose M remains constant while R varies. (An example of a spherical astronomical body with changing radius is a pulsating star, although in the case of a star the mass is not constant either.) Find dv_0/dR.

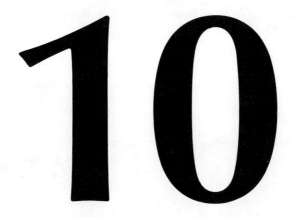

10

TRIGONOMETRIC FUNCTIONS AND THEIR INVERSES

A function $y = f(x)$ is called **algebraic** if, given the number x, the number y can be computed by performing a finite number of additions, multiplications, subtractions, and divisions and by solving a finite number of algebraic equations, that is, finding roots of polynomials. (This definition is equivalent to the one in Chapter 6, §3.) All polynomials, rational functions, and radical functions are algebraic.

A function that is not algebraic is called **transcendental.** The most important transcendental functions are the logarithmic functions, the power function $[f(x) = a^x]$, and the trigonometric functions and their inverses. These functions have been named and used before calculus was invented, and they are studied today in high school without using calculus. But a true mastery of these functions requires calculus, as we shall see in this and the following two chapters.

§1 Review of basic properties

Trigonometric functions are used in surveying, astronomy, and navigation, and they play a part in almost all applications of mathematics to science and technology. The astronomical applications predate all others. Tables of the sine function were compiled as early as the second century B.C. The oldest preserved sine tables are contained in Ptolomy's work, known under its Arabic name *Almagest*. The word sine, *sinus* in Latin, is a corruption of an Arabic word meaning "chord."

This section is an account of basic trigonometry and may be reviewed rapidly by readers familiar with the material.

1.1   Angles and their measures

A point O on a line divides the line into two **rays;** O is called the origin of the two rays. An **angle** is a pair of rays with a common origin. The origin is called the **vertex** of the angle, and the rays are called the **legs.** To describe an angle, it suffices to name the vertex O and to name two points, say, P and Q, distinct from O, one on each leg (see Figure 10.1). We call the angle so described $\angle POQ$ or $\angle QOP$. At this point, we deal only with undirected angles, that is, we do not prescribe the order of the legs.

Given an angle with distinct legs, say, $\angle POQ$, draw a circle of some radius r about O, and note the points, say, A and B, at which the legs intersect the circle. These points divide the circle into two arcs; let s be the length of the shorter arc (see Figure 10.1). The ratio

$$\alpha = \frac{s}{r} \tag{1}$$

(length of subtended arc divided by the radius) is called the **measure** of the angle or, more precisely, the **radian measure.**

To justify this definition we must know that the ratio s/r does not depend on the choice of the radius r. This is indeed so (in Chapter 9 we proved this using calculus; see the Remark in §3.2).

If the legs of the angle divide the circle of radius r into two equal arcs, then each arc has length $s = \frac{1}{2}2\pi r = \pi r$. The ratio s/r is π (for both arcs). In this case, the two legs are two rays of one line, and the measure of the angle is π.

If the two legs of an angle coincide, the radian measure is 0, by definition.

One should remember that the radian measure of an angle is a number (being the ratio of two lengths). *No units are involved.*

The **degree measure** of an angle is the radian measure multiplied by $180/\pi$. The tradition of dividing the full circle into 360 parts, called **degrees,** each degree into 60 **minutes** $(1° = 60')$, and each minute into 60 **seconds** $(1' = 60'')$ goes back to the sexagesimal system of the Babylonians. In everyday life and in most technological and scientific applications, angles are measured by degrees, and degrees are used in practically all trigonometric tables.

In mathematics it is almost imperative to use radian measure (see the Remark in §2.2). The simplest way of reconciling the two methods is to agree, once and for all, that 360° (360 degrees) is another name for the number 2π,

$$360° = 2\pi,$$

1° (1 degree) is another name for $\pi/180$, and in general $k°$ (k degrees) is another name for $k\pi/180$. Thus

$$k° = \frac{k\pi}{180} \quad \text{and} \quad \theta = \left(\frac{180\,\theta}{\pi}\right)°. \tag{2}$$

In particular, $90° = 90\pi/180 = \pi/2$, $1 = (180/\pi)° \approx 57°\,17'45''$. An angle of measure $\frac{1}{2}\pi = 90°$ is called a **right angle.** An angle of measure 1 is called a **radian** (see Figure 10.2).

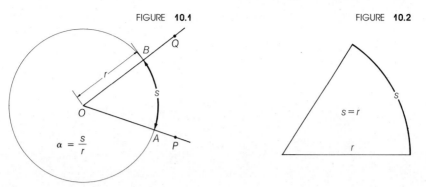

FIGURE **10.1**   FIGURE **10.2**

Two angles with the same measure are called **congruent.** Often we call two angles "equal" instead of "congruent," and we often say "the angle POQ equals θ, or is θ" instead of saying that "the measure of $\sphericalangle POQ$ is θ." Such abuses of language are harmless if they do not lead to confusion.

EXAMPLES 1. What is the radian measure of an angle of 30°?
ANSWER   Use the first relation (2), with $k = 30$: $30° = 30\pi/180 = \pi/6$. The desired radian measure is $\pi/6$.

2. What is the degree measure of an angle of radian measure $\pi/5$?
ANSWER   Use the second relation (2), with $\theta = \pi/5$:

$$\frac{\pi}{5} = \left(\frac{180 \cdot \pi/5}{\pi}\right)^\circ = \left(\frac{180}{5}\right)^\circ = 36°.$$

3. What is the radian measure of an angle of 1′?
ANSWER   Since $1' = (1/60)°$, we have, by (2),

$$1' = \left(\frac{1}{60}\right)^\circ = \frac{(1/60)\pi}{180} = \frac{\pi}{10{,}800}.$$

PROBLEMS

In Problems 1 to 10 the measure of an angle is given in degrees. Convert these to the equivalent radian measure.

1. 10°.   3. 24°.   5. 45°.   7. 90°.   9. 150°.
2. 15°.   4. 36°.   6. 60°.   8. 120°.   10. 270°.

In Problems 11 to 20 the measure of an angle is given in radians. Convert these to the equivalent degree measure.

11. $\dfrac{\pi}{8}$.   14. $\dfrac{\pi}{3}$.   17. $\dfrac{7\pi}{6}$.   19. $\dfrac{7\pi}{4}$.

12. $\dfrac{\pi}{6}$.   15. $\dfrac{2\pi}{3}$.   18. $\dfrac{5\pi}{3}$.   20. $\dfrac{11\pi}{6}$.

13. $\dfrac{\pi}{4}$.   16. $\dfrac{3\pi}{4}$.

21. Express 1′15″ in radian measure.
22. Express 1.5 radians in degrees, minutes, and seconds.

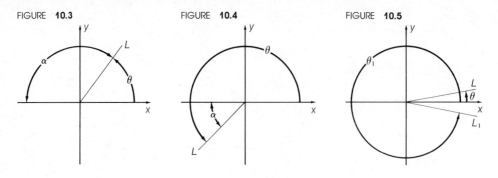

FIGURE **10.3**   FIGURE **10.4**   FIGURE **10.5**

1.2 Directed angles

Thus far we considered undirected angles; the measure of such an angle is a number between $0 = 0°$ and $\pi = 180°$. Now we define the measure θ of the directed angle between two rays: the positive x-axis and another ray, call it L, with vertex at the origin. Let α be the undirected angle between L and the negative x-axis; then the directed angle from the positive x-axis to L has measure $\theta = \pi - \alpha$ if L lies above the x-axis, $\theta = \pi + \alpha$ if L lies below the x-axis (see Figures 10.3 and 10.4). If L is the negative x-axis, we set $\theta = \pi = 180°$.

How shall we define θ if L is the positive x-axis? In the case depicted in Figure 10.5, L is close to the positive x-axis, and θ is close to $0 = 0°$; L_1 is close to the positive x-axis, and θ_1 is close to $2\pi = 360°$. This suggests that we define *both* 0 and 2π to be measures of the angle from the positive x-axis to itself.

In the situation depicted in Figures 10.3 and 10.4, we say that L is obtained by rotating the positive x-axis, in the positive direction, by the angle θ. If we rotate by $2\pi = 360°$, the positive x-axis returns to its initial position. This suggests that the measure of every directed angle be defined only up to a multiple of $2\pi = 360°$. In Figure 10.6, for instance, L can be thought of as obtained by rotating the positive x-axis by $60° = \frac{1}{3}\pi$, by $420° = 2\pi + \frac{1}{3}\pi$, by $780° = 4\pi + \frac{1}{3}\pi$, and so forth. The same L may be thought of as obtained by rotating the positive x-axis by $-300° = -360° + 60° = -\frac{5}{3}\pi$, by $-660° = -720° + 60° = -\frac{11}{3}\pi$, and so forth.

The positive y-axis is obtained by rotating the positive x-axis by $90° = \frac{1}{2}\pi$, or by $450° = \frac{5}{2}\pi$, or by $810° = \frac{9}{2}\pi$, or by $-270° = -\frac{3}{2}\pi$, and so forth.

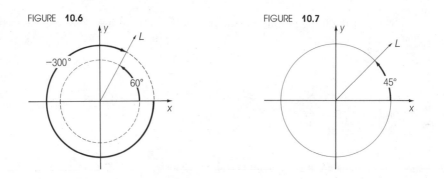

FIGURE **10.6**   FIGURE **10.7**

EXAMPLE Draw the rays L obtained by rotating the positive x-axis, in the positive direction by 405°, by $(33/4)\pi$ and by $-315°$.

ANSWER We want to write each of the given numbers as a number between 0 and $2\pi = 360°$ plus a multiple of 2π. Now,

$$405° = 360° + 45°, \qquad \frac{33}{4}\pi = 8\pi + \frac{\pi}{4} = 4 \cdot 360° + 45°, \qquad -315° = -360° + 45°.$$

All three rays are therefore the same; see Figure 10.7.

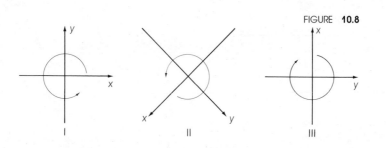

FIGURE **10.8**

REMARK If we are given a coordinate system, we can define a positive direction of rotation; this is the direction of the 90° swing which takes the positive x-axis into the positive y-axis. Two different coordinate systems may define the same positive rotations, as systems I and II in Figure 10.8, or they may lead to different positive rotations, as systems I and III in the same figure. If we call the coordinate system I right-handed, we must call system II right-handed also and system III left-handed.

(The statement just made can be proved analytically, but we shall not do so.)

While the difference between right-handed and left-handed coordinate systems can be defined mathematically, we cannot define a right-handed system; we can only exhibit one. Sometimes one says that in a right-handed system positive rotations are "counterclockwise." But not all clocks look alike. (A famous medieval clock on the so-called Jewish City Hall in Prague is marked by Hebrew letters rather than numerals. Its hands move "counterclockwise." The author's teacher, C. Loewner, liked to ask whether a man who knew only this clock and read only mathematics books without pictures would ever discover that other clocks move differently.)

When we choose a coordinate system, which will be called right-handed, we **orient** the plane. In an **oriented plane,** we can talk about a "counterclockwise" direction. We assume from now on that the plane has been oriented.

PROBLEMS

Draw the rays obtained by rotating the positive x-axis, in the positive direction, by the angles given below.

1. 480°.
2. $-270°$.
3. $-150°$.

4. $-330°$.
5. 1380°.
6. 5π.

7. $\dfrac{7\pi}{3}$.

8. $-\dfrac{\pi}{4}$.

9. $-\dfrac{5\pi}{4}$.

10. $-\dfrac{8\pi}{3}$.

CHARLES LOEWNER (1895–1968) came to the United States after his native Czechoslovakia was overrun by the Nazis. He made important discoveries in several fields of mathematics.

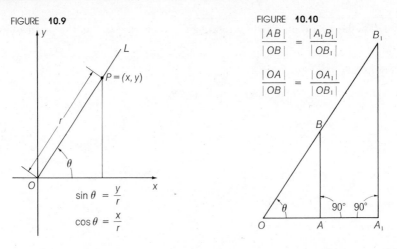

FIGURE **10.9**

FIGURE **10.10**

1.3 Sines and cosines

We now define the two basic trigonometric functions, the sine and the cosine. Let θ be a number, and let L be the ray through the origin obtained by rotating the positive x-axis by θ. (In other words, θ is the radian measure of the directed angle from the positive x-axis to L.) Let P be some point on L (distinct from O), x and y the coordinates of P, and r the distance from P to O, so that

$$r = \sqrt{x^2 + y^2} \tag{2}$$

(see Figure 10.9). Then the **sine** and **cosine** of θ are the ratios

$$\sin \theta = \frac{y}{r}, \qquad \cos \theta = \frac{x}{r}. \tag{3}$$

These ratios do not depend on the position of P, only on θ. This is seen by similar triangles (see Figure 10.10). If we choose P at distance 1 from the origin, formulas (3) simplify to

$$\sin \theta = y, \qquad \cos \theta = x \qquad \text{(for } r = 1\text{);} \tag{3'}$$

see the four cases shown in Figure 10.11.

(The trigonometric functions and their inverses are sometimes called *circular functions*. Figure 10-11 explains this name.)

FIGURE **10.11**

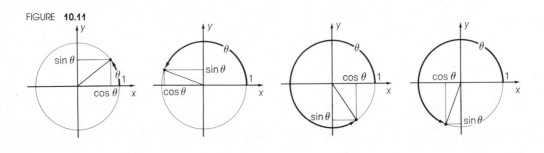

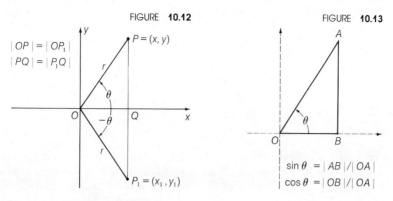

FIGURE 10.12

$|OP| = |OP_1|$
$|PQ| = |P_1Q|$

$P = (x, y)$

$P_1 = (x_1, y_1)$

FIGURE 10.13

$\sin\theta = |AB|/|OA|$
$\cos\theta = |OB|/|OA|$

We proceed to derive some properties of the sine and cosine functions. First of all, since L is not changed if we add 2π to θ, we have

$$\sin(\theta + 2\pi) = \sin\theta, \qquad \cos(\theta + 2\pi) = \cos\theta. \tag{4}$$

Also

$$\sin(\theta + 2\pi n) = \sin\theta, \qquad \cos(\theta + 2\pi n) = \cos\theta, \qquad n = 0, \pm 1, \pm 2, \ldots. \tag{5}$$

Next,

$$\sin^2\theta + \cos^2\theta = 1. \tag{6}$$

Here $\sin^2\theta$, $\cos^2\theta$ is the traditional (and unfortunate) way of denoting $(\sin\theta)^2$ and $(\cos\theta)^2$; the above equation is the same as

$$(\sin\theta)^2 + (\cos\theta)^2 = 1.$$

This is verified by replacing $\sin\theta$, $\cos\theta$ by their values (3) and noting (2). It follows from (6) that $0 \leq \sin^2\theta \leq 1$, $0 \leq \cos^2\theta \leq 1$. Hence

$$-1 \leq \sin\theta \leq 1, \qquad -1 \leq \cos\theta \leq 1, \qquad \text{for all } \theta. \tag{7}$$

We show next that *the sine function is odd* and *the cosine function is even*. Let L_1 be the ray obtained by rotating the positive x-axis by $-\theta$, let P_1 be a point on L_1 at the distance r from O, and let x_1, y_1 be the coordinates of P_1 (see Figure 10.12). Then $\sin(-\theta) = y_1/r$, $\cos(-\theta) = x_1/r$. Since P and P_1 are situated symmetrically with respect to the x-axis, $x_1 = x$ and $y_1 = -y$. Hence $x_1/r = -x/r$, $y_1/r = y/r$, and

$$\sin(-\theta) = -\sin\theta, \qquad \cos(-\theta) = \cos\theta. \tag{8}$$

(In Figure 10.12, L is in the first quadrant. The reader should make similar drawings for the case when L lies in other quadrants.)

If θ lies between $0 = 0°$ and $\pi/2 = 90°$, $\sin\theta$ and $\cos\theta$ have the following geometric meaning. In a right triangle OAB, let $\angle OBA$ be the right angle, and let θ be the measure of $\angle AOB$. Then $\sin\theta = |AB|/|OA|$ (= "ratio of the leg, opposite to the angle, to the hypotenuse") and $\cos\theta = |OB|/|OA|$ (= "ratio of the adjacent leg to the hypotenuse"). This high school definition of sine and cosine is shown in Figure 10.13.

We note some values of the functions $\sin\theta$ and $\cos\theta$ in the following table. (See Examples 1, 2 and Problems 1, 2, 3 for the proofs.)

θ	$0 = 0°$	$\frac{\pi}{6} = 30°$	$\frac{\pi}{4} = 45°$	$\frac{\pi}{3} = 60°$	$\frac{\pi}{2} = 90°$	$\pi = 180°$	$\frac{3\pi}{2} = 270°$	$2\pi = 360°$
$\sin\theta$	0	$\frac{1}{2}$	$\frac{1}{\sqrt{2}}$	$\frac{\sqrt{3}}{2}$	1	0	-1	0
$\cos\theta$	1	$\frac{\sqrt{3}}{2}$	$\frac{1}{\sqrt{2}}$	$\frac{1}{2}$	0	-1	0	1

It is geometrically obvious that a small change in θ produces a small change in the position of L and hence a small change in $\sin\theta$ and $\cos\theta$. We conclude that $\cos\theta$ *and* $\sin\theta$ *are continuous functions of* θ. (This can also be proved analytically ▶.) The graphs of $\sin\theta$ and $\cos\theta$, for $0 \le \theta \le \pi/2$, are shown in Figures 10.14 and 10.15. Note that in this interval $\sin\theta$ increases and $\cos\theta$ decreases. (This will be verified analytically once we compute, in §2.2, the derivatives of these functions.)

A less obvious but very important property of the sine and cosine functions is the **addition theorem:**

$$\sin(\phi + \psi) = \sin\phi\cos\psi + \cos\phi\sin\psi, \tag{9}$$

$$\cos(\phi + \psi) = \cos\phi\cos\psi - \sin\phi\sin\psi \tag{10}$$

for all numbers ϕ and ψ. We postpone the proof of this theorem until §1.6.

If we set $\psi = \pi/2$ in (9), (10), and note that $\cos(\pi/2) = 0$, $\sin(\pi/2) = 1$, we obtain the identities

$$\sin\left(\phi + \frac{\pi}{2}\right) = \cos\phi, \qquad \cos\left(\phi + \frac{\pi}{2}\right) = -\sin\phi. \tag{11}$$

FIGURE **10.14**
sin θ

FIGURE **10.15**
cos θ

FIGURE **10.16**

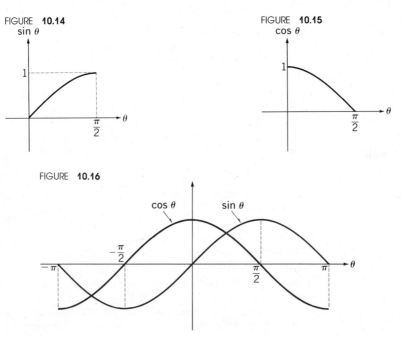

This implies that

$$\sin\left(\frac{\pi}{2} - \phi\right) = \cos\phi, \qquad \cos\left(\frac{\pi}{2} - \phi\right) = \sin\phi. \tag{12}$$

Indeed, noting (8) and (11), we have $\sin(\frac{1}{2}\pi - \phi) = \sin[(-\phi) + \frac{1}{2}\pi] = \cos(-\phi) = \cos\phi$. This is the first relation (12). The second says the same thing, since $\frac{1}{2}\pi - (\frac{1}{2}\pi - \phi) = \phi$.

Setting $\psi = \pi$ in (9), (10), and noting that $\sin\pi = 0$, $\cos\pi = -1$ (by the definition of sine and cosine), we see that

$$\sin(\phi + \pi) = -\sin\phi, \qquad \cos(\phi + \pi) = -\cos\phi. \tag{13}$$

Using relations (4), (8), (11), and (13), we can extend the graphs in Figures 10.14 and 10.15 to larger intervals; for example, see Figures 10.16 and 10.17. The shape of these graphs is quite important and should be remembered.

FIGURE **10.17**

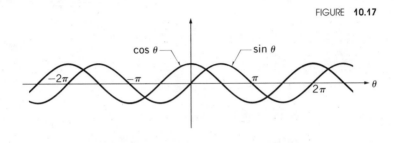

EXAMPLES 1. Verify the third column in the table on p. 246.

ANSWER In Figure 10.18, COB is an equilateral triangle with $|OC| = |OB| = |CB| = 1$, and A is the midpoint of CB. Hence $\angle COB = \frac{1}{3}180° = 60°$, $\angle OAB = 90°$, and $|AB| = \frac{1}{2}$. Also, $\angle AOB = \frac{1}{2}60° = 30°$. Thus $\sin 30° = |AB|/|OB| = \frac{1}{2}$. By the Pythagorean theorem, $|OB|^2 = \sqrt{1 - (\frac{1}{2})^2} = \sqrt{3}/2$. Thus $\cos 30° = \sqrt{3}/2$.

FIGURE **10.18**

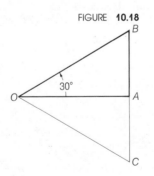

2. Verify the sixth column in the table on p. 246.

ANSWER Draw a circle of radius $r = 1$ (any radius will do), and let L be the ray through the origin obtained by rotating the positive x-axis by 90°. The point P, where L intersects the circle, has coordinates (0,1). From (3), we obtain $\sin 90° = 1$, $\cos 90° = 0$.

3. What is $\cos(-\pi/6)$?

ANSWER Using (8) and Example 1, we have $\cos(-\pi/6) = \cos(\pi/6) = \cos 30° = \sqrt{3}/2$.

4. What is $\sin 3645°$?

ANSWER $\sin 3645° = \sin(3600° + 45°) = \sin 45° = 1/\sqrt{2}$, by (5) and the table.

5. What is $\cos 95°$? (Use the tables at the end of the book.)

ANSWER   By (11), $\cos 95° = \cos(90° + 5°) = -\sin 5° = -.0872$, approximately. (Function values found in tables are almost always approximate.)

The following examples are applications of the addition theorem.

6. Express $\sin 2\phi$ and $\cos 2\phi$ in terms of $\sin \phi$, $\cos \phi$.

ANSWER   Set $\psi = \phi$ in (9), (10). This yields

$$\sin 2\phi = 2 \sin \phi \cos \phi, \qquad \cos 2\phi = \cos^2 \phi - \sin^2 \phi. \tag{14}$$

Since $\sin^2 \phi = 1 - \cos^2 \phi$, the second relation (14) can be written as

$$\cos 2\phi = 2 \cos^2 \phi - 1 = 1 - 2 \sin^2 \phi. \tag{15}$$

7. Express $\sin(\theta/2)$ and $\cos(\theta/2)$ in terms of $\sin \theta$ and $\cos \theta$, assuming that $0 \le \theta \le \pi$.

ANSWER   In (14), set $\phi = \theta/2$; we see that

$$\cos \theta = \cos^2 \frac{\theta}{2} - \sin^2 \frac{\theta}{2}.$$

Also

$$1 = \cos^2 \frac{\theta}{2} + \sin^2 \frac{\theta}{2}.$$

If we add and subtract these two equations, we obtain

$$1 - \cos \theta = 2 \sin^2 \frac{\theta}{2}, \qquad 1 + \cos \theta = 2 \cos^2 \frac{\theta}{2}.$$

Dividing by 2, and noting that $\sin(\theta/2)$ and $\cos(\theta/2)$ must be nonnegative for $0 \le \theta \le \pi$, we get

$$\sin \frac{\theta}{2} = \sqrt{\frac{1 - \cos \theta}{2}}, \qquad \cos \frac{\theta}{2} = \sqrt{\frac{1 + \cos \theta}{2}}. \tag{16}$$

8. Express $\cos 3x$ and $\sin 3x$ in terms of $\cos x$ and $\sin x$.

SOLUTION   We use (10), with $\phi = 2x$, $\psi = x$, and then (14) with $\phi = x$:

$$\begin{aligned}
\cos 3x &= \cos 2x \cos x - \sin 2x \sin x \\
&= (\cos^2 x - \sin^2 x) \cos x - (2 \sin x \cos x) \cos x \\
&= \cos^3 x - \cos x \sin^2 x - 2 \sin^2 x \cos x = \cos^3 x - 3 \cos x \sin^2 x.
\end{aligned}$$

We show similarly that $\sin 3x = 3 \cos^2 x \sin x - \sin^3 x$.

9. Prove the formulas

$$\cos x + \cos y = 2 \cos \frac{x + y}{2} \cos \frac{x - y}{2},$$

$$\sin x + \sin y = 2 \sin \frac{x + y}{2} \sin \frac{x - y}{2}.$$

ANSWER   We prove only the first relation; the second is established similarly. By the addition theorem,

$$\cos(\alpha + \beta) = \cos \alpha \cos \beta - \sin \alpha \sin \beta$$

and, since $\alpha - \beta = \alpha + (-\beta)$,

$$\cos(\alpha - \beta) = \cos \alpha \cos \beta + \sin \alpha \sin \beta.$$

Adding the two displayed equations, we obtain

$$\cos(\alpha + \beta) + \cos(\alpha - \beta) = 2 \cos \alpha \cos \beta.$$

Setting $\alpha + \beta = x$, $\alpha - \beta = y$, we have $\alpha = (x + y)/2$ and $\beta = (x - y)/2$. The result follows.

PROBLEMS

1. Verify the fourth column in the table on p. 246.
2. Verify the fifth column in the table.
3. Verify the second, seventh, eighth, and ninth columns in the table.
4. Using the definitions of sine and cosine, find their algebraic signs in each of the four quadrants.

In Problems 5 to 16 find the indicated values.

5. $\cos(-30°)$.
6. $\sin(-60°)$.
7. $\cos\left(-\dfrac{\pi}{4}\right)$.

8. $\cos 120°$.
9. $\sin 135°$.
10. $\sin \dfrac{5\pi}{4}$.

11. $\cos \dfrac{4\pi}{3}$.
12. $\sin(-225°)$.
13. $\cos\left(-\dfrac{5\pi}{6}\right)$.

14. $\cos 300°$.
15. $\sin(-570°)$.
16. $\sin \dfrac{25\pi}{4}$.

Using tables, find the following values:

17. $\sin(-72°)$.
18. $\cos 124°$.

19. $\sin 157°$.
20. $\cos 214°$.

21. $\cos 311°$.
22. $\sin(-142°)$.

23. $\cos(-249°)$.
24. $\sin 1336°$.

1.4 Other trigonometric functions

There are four more trigonometric functions: the tangent, the cotangent, the secant, and the cosecant. They can be all expressed in terms of the sine and the cosine:

$$\tan \theta = \frac{\sin \theta}{\cos \theta}, \tag{17}$$

$$\cot \theta = \frac{1}{\tan \theta} = \frac{\cos \theta}{\sin \theta}, \qquad \sec \theta = \frac{1}{\cos \theta}, \qquad \csc \theta = \frac{1}{\sin \theta}. \tag{18}$$

The functions $\tan \theta$ and $\sec \theta$ are not defined for those θ for which $\cos \theta = 0$, that is, for $\theta = \pm\pi/2$, $\pm 3\pi/2$, $\pm 5\pi/2$, and so forth. The functions $\cot \theta$ and $\csc \theta$ are not defined for $\theta = 0$, $\pm\pi$, $\pm 2\pi, \ldots$; for these values of θ we have $\sin \theta = 0$. In the situation depicted in Figure 10.9 in §1.3, we have

$$\tan \theta = \frac{y}{x}, \qquad \cot \theta = \frac{x}{y}, \qquad \sec \theta = \frac{r}{x}, \qquad \csc \theta = \frac{r}{y}. \tag{19}$$

From (2), (3), and (18) [or from (6), (17), and (18)], we obtain the identities

$$1 + \tan^2 \theta = \sec^2 \theta, \qquad 1 + \cot^2 \theta = \csc^2 \theta. \tag{20}$$

$$\tan(-\theta) = -\tan \theta, \tag{21}$$

$$\cot(-\theta) = -\cot \theta, \qquad \sec(-\theta) = \sec \theta, \qquad \csc(-\theta) = -\csc \theta \tag{22}$$

FIGURE 10.19

FIGURE 10.20

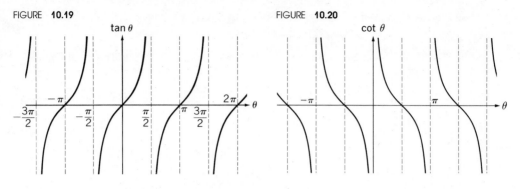

tan θ

cot θ

(see Example 1 and Problems 3, 4, 5 for proofs), and

$$\tan\left(\frac{\pi}{2} - \theta\right) = \cot\theta, \qquad \sec\left(\frac{\pi}{2} - \theta\right) = \csc\theta, \tag{23}$$

$$\tan(\theta + \pi) = \tan\theta, \qquad \cot(\theta + \pi) = \cot\theta, \tag{24}$$

$$\sec(\theta + 2\pi) = \sec\theta, \qquad \csc(\theta + 2\pi) = \csc\theta \tag{25}$$

(see Example 2 and Problems 6–10).

The graphs of the four functions considered are shown in Figures 10.19–10.22. At the points at which the functions are not defined (odd multiples of $\pi/2$ for the tangent and the secant, multiples of π for the cotangent and the cosecant), there are infinite limits from the left and from the right. For instance,

$$\lim_{\theta \to (\pi/2)^-} \tan\theta = +\infty. \tag{26}$$

Indeed, for θ near $\pi/2$, but smaller than $\pi/2$, $\tan\theta = \sin\theta/\cos\theta$ is positive, while $1/\tan\theta = \cot\theta = \cos\theta/\sin\theta$ is continuous at $\theta = \pi/2$, and equals 0 at $\theta = \pi/2$. Hence $\tan\theta$ becomes arbitrarily large as θ approaches $\pi/2$ from the left. We see similarly that

$$\lim_{\theta \to (-\pi/2)^+} \tan\theta = -\infty. \tag{27}$$

(Of the four functions just introduced, the tangent is the most important. It is

FIGURE 10.21

FIGURE 10.22

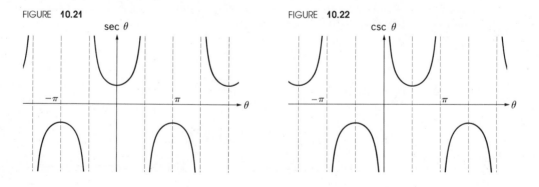

sec θ

csc θ

regrettable that the word "tangent" is used to denote two different things, the line that "just touches" a curve and the function $\tan \theta$. But both names are so traditional that they cannot be changed.)

EXAMPLES 1. Show that the tangent is an odd function.
ANSWER By (8), (17) we have

$$\tan(-\theta) = \frac{\sin(-\theta)}{\cos(-\theta)} = \frac{-\sin \theta}{\cos \theta} = -\tan \theta.$$

2. Verify the first relation (24).
ANSWER By (17), (13) we have

$$\tan(\theta + \pi) = \frac{\sin(\theta + \pi)}{\cos(\theta + \pi)} = \frac{-\sin \theta}{-\cos \theta} = \frac{\sin \theta}{\cos \theta} = \tan \theta.$$

3. Derive the addition theorem for the tangent function:

$$\tan(\phi + \psi) = \frac{\tan \phi + \tan \psi}{1 - \tan \phi \tan \psi}. \tag{28}$$

SOLUTION By (17), (9), and (10),

$$\tan(\phi + \psi) = \frac{\sin(\phi + \psi)}{\cos(\phi + \psi)} = \frac{\sin \phi \cos \psi + \cos \phi \sin \psi}{\cos \phi \cos \psi - \sin \phi \sin \psi}.$$

In the last fraction, divide the numerator and denominator by $\cos \phi \cos \psi$. This yields

$$\tan(\phi + \psi) = \frac{\dfrac{\sin \phi}{\cos \phi} + \dfrac{\sin \psi}{\cos \psi}}{1 - \dfrac{\sin \phi}{\cos \phi} \dfrac{\sin \psi}{\cos \psi}},$$

which is the same as (28).

4. Show that

$$\sin 2\phi = \frac{2 \tan \phi}{1 + \tan^2 \phi}, \qquad \cos 2\phi = \frac{1 - \tan^2 \phi}{1 + \tan^2 \phi}. \tag{29}$$

ANSWER We prove only the first relation; the second is proved similarly. Using (17), (14), and (6), we have

$$\frac{2 \tan \phi}{1 + \tan^2 \phi} = \frac{2 \dfrac{\sin \phi}{\cos \phi}}{1 + \dfrac{\sin^2 \phi}{\cos^2 \phi}} = \frac{2 \sin \phi \cos \phi}{\cos^2 \phi + \sin^2 \phi} = \frac{\sin 2\phi}{1} = \sin 2\phi.$$

PROBLEMS

1. Make a table of values of $\tan \theta$ and $\cot \theta$ for $\theta = 0°$, $30°$, $45°$, $60°$, $90°$, $180°$, $270°$, and $360°$.
2. Make a table of values of $\sec \theta$ and $\csc \theta$ for $\theta = 0°$, $30°$, $45°$, $60°$, $90°$, $180°$, $270°$, and $360°$.
3. Prove that the cotangent is an odd function, that is, show that $\cot(-\theta) = -\cot \theta$.
4. Prove that the secant is an even function, that is, show that $\sec(-\theta) = \sec \theta$.
5. Prove that the cosecant is an odd function, that is, show that $\csc(-\theta) = -\csc \theta$.

In Problems 6 to 10 prove the given identities.

6. $\tan\left(\dfrac{\pi}{2} - \theta\right) = \cot\theta.$

7. $\sec\left(\dfrac{\pi}{2} - \theta\right) = \csc\theta.$

8. $\cot(\theta + \pi) = \cot\theta.$
9. $\sec(\theta + 2\pi) = \sec\theta.$
10. $\csc(\theta + 2\pi) = \csc\theta.$

In Problems 11 to 20 find the indicated values.

11. $\tan(-45°).$
12. $\cot 225°.$

13. $\sec\left(-\dfrac{\pi}{3}\right).$

14. $\csc 150°.$

15. $\tan\dfrac{2\pi}{3}.$

16. $\sec\left(-\dfrac{5\pi}{6}\right).$

17. $\cot 330°.$

18. $\tan\left(-\dfrac{5\pi}{4}\right).$

19. $\csc 1500°.$

20. $\cot\dfrac{13\pi}{4}.$

1.5 Periodicity

Many natural and artificial phenomena are periodic, that is, they repeat themselves in time. This is true of planetary motions, tides, various kinds of wave motions (sound, light, radio signals), breathing, the movements of clocks and other mechanisms. Periodic phenomena are described by periodic functions.

Let $t \mapsto f(t)$ be a function and $p \neq 0$ a number. The function f is called **periodic** with **period** p if

$$f(t + p) = f(t) \qquad \text{for all } t$$

(for which f is defined). A graph of a periodic function is shown in Figure 10.23.

FIGURE **10.23**

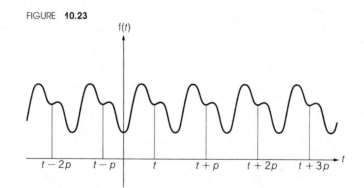

The functions $\sin\theta$, $\cos\theta$, $\sec\theta$, and $\csc\theta$ have period 2π; the functions $\tan\theta$ and $\cot\theta$ have period π [see (4), (24), and (25)].

If f has periods p and q, then $p + q$ is also a period, for $f(t + p + q) = f(t + p) = f(t)$. If f has period p, then $(-p)$ is also a period, for $f(t - p) = f(t - p + p) = f(t)$. The smallest positive period of f, if there is such a number, is called the **primitive period** of f. If f has primitive period p, then it has periods p, $-p$, $2p$, $-2p$, $3p$, $-3p$, ..., and no others.

The graphs of trigonometric functions show that 2π is the primitive period of the sine, cosine, secant, and cosecant functions, and π is the primitive period of the tangent and cotangent.

1.6 Proof of the addition theorem‡

We now prove the addition theorem stated in §1.3 without proof.

Let ϕ and ψ be given, let L be the ray obtained by turning the positive x-axis by the angle $\phi + \psi$, and let P be the point on L at the distance 1 from the origin. By the definition of sines and cosines, P has the coordinates $x = \cos(\phi + \psi)$, $y = \sin(\phi + \psi)$ [to see this, use Equation (3) with $r = 1$, $\theta = \phi + \psi$]. Let Q be the point, on the positive x-axis, with coordinates $x = 1$, $y = 0$. By the distance formula,

$$
\begin{aligned}
|PQ|^2 &= (\cos(\phi + \psi) - 1)^2 + (\sin(\phi + \psi) - 0)^2 \\
&= \cos^2(\phi + \psi) - 2\cos(\phi + \psi) + 1 + \sin^2(\phi + \psi) \qquad (30) \\
&= 2 - 2\cos(\phi + \psi)
\end{aligned}
$$

[here we used (6)].

Now we introduce a "new" coordinate system (ξ, η) obtained by rotating the "old" (x,y)-system by the angle ϕ; see Figure 10.24. We see from the figure that L is obtained by rotating the positive ξ-axis by the angle ϕ, so that the point P has the "new" coordinates $\xi = \cos\phi$, $\eta = \sin\phi$. The positive x-axis is obtained by rotating the positive ξ-axis by the angle $(-\phi)$; therefore the "new" coordinates of Q are $\xi = \cos(-\phi) = \cos\phi$, $\eta = \sin(-\phi) = -\sin\phi$. Thus, by the distance formula,

$$
\begin{aligned}
|PQ|^2 &= (\cos\psi - \cos\phi)^2 + (\sin\psi + \sin\phi)^2 \\
&= \cos^2\psi - 2\cos\phi\cos\psi + \cos^2\phi + \sin^2\psi + 2\sin\phi\sin\psi + \sin^2\phi \qquad (31) \\
&= 2 - 2(\cos\phi\cos\psi - \sin\phi\sin\psi).
\end{aligned}
$$

FIGURE **10.24**

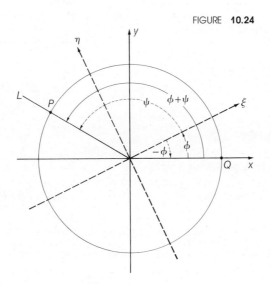

Comparing (30) and (31) we see that

$$\cos(\phi + \psi) = \cos\phi\,\cos\psi - \sin\phi\,\sin\psi. \tag{32}$$

(The argument is valid for all ϕ and ψ, not only for the situation depicted in Figure 10.24.)

In (32) set first $\phi = \frac{1}{2}\pi$, $\psi = -\alpha$, and then $\phi = \frac{1}{2}\pi - \beta$, $\psi = \frac{1}{2}\pi$. Noting that $\cos\frac{1}{2}\pi = 0$, $\sin\frac{1}{2}\pi = 1$, and that the cosine function is even and the sine function odd, we obtain

$$\cos\left(\frac{\pi}{2} - \alpha\right) = \sin\alpha, \qquad \sin\left(\frac{\pi}{2} - \beta\right) = \cos\beta. \tag{33}$$

Using (32) and (33), we may write

$$\sin(\phi + \psi) = \cos\left[\frac{\pi}{2} - (\phi + \psi)\right] = \cos\left[\left(\frac{\pi}{2} - \phi\right) + (-\psi)\right]$$

$$= \cos\left(\frac{\pi}{2} - \phi\right)\cos(-\psi) - \sin\left(\frac{\pi}{2} - \phi\right)\sin(-\psi)$$

or

$$\sin(\phi + \psi) = \sin\phi\,\cos\psi + \cos\phi\,\sin\psi. \tag{34}$$

Relations (32) and (34) constitute the addition theorem.

We can also prove the addition theorem analytically ▶.

§2 Differentiation of trigonometric functions

The rules for differentiating trigonometric functions follow from the addition theorem and from two limit relations described below.

2.1 Two limits

Examining the graphs of the functions cos and sin, or better, tables for these functions, we note that the sine of a small number h (that is, of an angle of radian measure h) is nearly h. For example, $5° \approx .08727$, $\sin 5° \approx .08716$, and $1° \approx .01745$, $\sin 1° \approx .01745$ (correct to five decimal places). This suggests that

$$\lim_{h \to 0} \frac{\sin h}{h} = 1. \tag{1}$$

In order to prove (1), let h first be a number such that $0 < h < \pi/2$, and consider Figure 10.25. The area of the circular sector OCD is $\frac{1}{2}(\cos h)^2 h$ (by Chapter 9, §3.2), that of the sector OAB is $\frac{1}{2}(1)^2 h = \frac{1}{2}h$. The area of the triangle OBC equals $\frac{1}{2}\cos h\,\sin h$. It is larger than the area of the smaller sector OCD and smaller than that of the larger sector OAB. Therefore

$$\tfrac{1}{2}(\cos^2 h)h \leq \tfrac{1}{2}\sin h\,\cos h \leq \tfrac{1}{2}h.$$

Dividing this inequality by the positive number $(\cos h)h$, we see that

$$\cos h \leq \frac{\sin h}{h} \leq \frac{1}{\cos h}. \tag{2}$$

FIGURE **10.25**

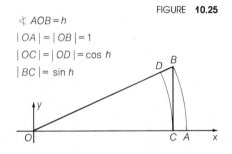

$\sphericalangle\, AOB = h$

$|OA| = |OB| = 1$

$|OC| = |OD| = \cos h$

$|BC| = \sin h$

Since $\cos h$, $\sin h/h$, and $1/\cos h$ are even functions of h, inequality (2) is true also if $-\pi/2 < h < 0$. Now the functions $\cos h$ and $1/\cos h$ are continuous at $h = 0$, and both have the value 1 at $h = 0$. For $h \neq 0$ and $|h|$ small enough, $\cos h$ and $1/\cos h$ are both as close to 1 as we want. The same is therefore true of the number $(\sin h)/h$, which lies between $\cos h$ and $1/\cos h$. Hence (1) is true.

From (1) we derive another important limit relation:

$$\lim_{h \to 0} \frac{\cos h - 1}{h} = 0. \tag{3}$$

Proof.   For $0 < |h| < \pi/2$,

$$\frac{\cos h - 1}{h} = \frac{(\cos h - 1)(\cos h + 1)}{h(\cos h + 1)} = \frac{\cos^2 h - 1}{h(\cos h + 1)} = \frac{-\sin^2 h}{h(\cos h + 1)},$$

so that by (1) and by the rules for computing limits (Chapter 3, §4.3),

$$\lim_{h \to 0} \frac{\cos h - 1}{h} = \lim_{h \to 0} \frac{\sin h}{h} \cdot \frac{-\sin h}{\cos h + 1}$$

$$= \lim_{h \to 0} \frac{\sin h}{h} \lim_{h \to 0} \frac{-\sin h}{\cos h + 1} = 1 \cdot \frac{0}{2} = 0.$$

2.2 Derivatives of the sine and cosine functions

Theorem 1. *The sine and cosine functions are differentiable and*

$$\frac{d \sin \theta}{d\theta} = \cos \theta, \qquad \frac{d \cos \theta}{d\theta} = -\sin \theta. \tag{4}$$

Proof.   First we compute the difference quotients, using the addition theorems:

$$\frac{\sin(\theta + h) - \sin \theta}{h} = \frac{\sin \theta \cos h + \cos \theta \sin h - \sin \theta}{h}$$

$$= \sin \theta \frac{\cos h - 1}{h} + \cos \theta \frac{\sin h}{h},$$

$$\frac{\cos(\theta + h) - \cos \theta}{h} = \frac{\cos \theta \cos h - \sin \theta \sin h - \cos \theta}{h}$$

$$= \cos \theta \frac{\cos h - 1}{h} - \sin \theta \frac{\sin h}{h}.$$

Using the rules for computing limits (Chapter 3, §4.3) and relations (1), (3), we have

$$\frac{d \sin \theta}{d\theta} = \lim_{h \to 0} \frac{\sin(\theta + h) - \sin \theta}{h}$$

$$= \lim_{h \to 0} \left(\sin \theta \, \frac{\cos h - 1}{h} + \cos \theta \, \frac{\sin h}{h} \right)$$

$$= \sin \theta \lim_{h \to 0} \frac{\cos h - 1}{h} + \cos \theta \lim_{h \to 0} \frac{\sin h}{h}$$

$$= \sin \theta \cdot 0 + \cos \theta \cdot 1 = \cos \theta.$$

Similarly,

$$\frac{d \cos \theta}{d\theta} = \lim_{h \to 0} \frac{\cos(\theta + h) - \cos \theta}{h}$$

$$= \lim_{h \to 0} \left(\cos \theta \, \frac{\cos h - 1}{h} - \sin \theta \, \frac{\sin h}{h} \right)$$

$$= \cos \theta \lim_{h \to 0} \frac{\cos h - 1}{h} - \sin \theta \lim_{h \to 0} \frac{\sin h}{h}$$

$$= \cos \theta \cdot 0 - \sin \theta \cdot 1 = -\sin \theta.$$

It is worthwhile to rewrite Theorem 1 in the language of differentials:

$$d \sin u = \cos u \, du, \qquad d \cos u = -\sin u \, du. \tag{5}$$

Here u may be considered as an independent variable, in which case (5) means that $d \sin u/du = \cos u$, $d \cos u/du = -\sin u$ [this is the same as (4)]. Or, we may think of u as a differentiable function of some other variable, say, x. In this case, (5) means that

$$\frac{d \sin u(x)}{dx} = \cos u(x) \, \frac{du(x)}{dx}, \qquad \frac{d \cos u(x)}{dx} = -\sin u(x) \, \frac{du(x)}{dx}.$$

Of course, the same thing follows directly from (4) and the chain rule.

Every differentiation rule may be rewritten as an integration rule. In our case, we obtain

$$\int \cos u \, du = \sin u + C, \qquad \int \sin u \, du = -\cos u + C. \tag{6}$$

Here, again, u may be the name of an independent variable or u may denote a continuously differentiable function of another variable, say, x, in which case $du = u'(x) \, dx$.

REMARK The differentiation rule (4) is so simple only because we use radian measure for angles.

EXAMPLES 1. What is the derivative of $f(x) = x^2 \sin x$?

ANSWER By the product rule,

$$\frac{d(x^2 \sin x)}{dx} = \frac{dx^2}{dx} \sin x + x^2 \frac{d \sin x}{dx} = 2x \sin x + x^2 \cos x.$$

2. Differentiate $f(x) = \sin(1 + x^2)$.
ANSWER   Set $u = 1 + x^2$. Then $f(x) = \sin u$ and $df(x) = \cos u \, du = (\cos u)(du/dx) \, dx = (\cos(1 + x^2))2x \, dx$. Hence

$$\frac{df(x)}{dx} = 2x \cos(1 + x^2).$$

Or, using the chain rule directly,

$$f'(x) = \frac{d \sin u}{du} \frac{du}{dx} = (\cos u)2x = 2x \cos(1 + x^2).$$

3. Investigate the local maxima and minima of the function $f(t) = 2 \sin t + \cos 2t$.
ANSWER We note that $f(t)$ has period 2π. Hence it suffices to consider the interval $0 \leq t \leq 2\pi$. We have

$$f'(t) = 2 \cos t - 2 \sin 2t, \qquad f''(t) = -2 \sin t - 4 \cos 2t.$$

To find the zeros of $f'(t)$ use the formula (14) of §1.3 for $\sin 2t$. Thus

$$f'(t) = 2 \cos t - 4 \sin t \cos t = 2(\cos t)(1 - 2 \sin t).$$

The zeros of $f'(t)$ for $0 \leq t \leq 2\pi$ are: the zeros of $\cos t$ ($t = 90° = \pi/2$, $t = 270° = 3\pi/2$) and the points where $\sin t = \frac{1}{2}$ (that is, $t = 30° = \pi/6$, $t = 150° = 5\pi/6$). Now, $f''(t) = 4 > 0$ for $t = 90°, 270°$, and $f''(t) = -3 < 0$ for $t = 30°, 150°$. Hence f has local minima at $t = 90°$, $270°$, and local maxima at $t = 30°, 150°$.

4. What is the area between the graph of $y = \sin x$ and the x-axis, from $x = 0$ to $x = \pi$?
ANSWER The desired area is

$$\int_0^\pi \sin x \, dx = -\cos x \bigg|_0^\pi = -\cos \pi + \cos 0 = 2.$$

5. Find a primitive function of $f(t) = t \cos(1 + t^2)$.
ANSWER We note that, setting $u = 1 + t^2$, we have $du = 2t \, dt$. Hence

$$\int t \cos(1 + t^2) \, dt = \frac{1}{2} \int (\cos(1 + t^2))2t \, dt$$

$$= \frac{1}{2} \int \cos u \, du = \frac{1}{2} \sin u + C = \frac{1}{2} \sin(1 + t^2) + C.$$

That $\frac{1}{2} \sin(1 + t^2)$ is indeed a primitive function of $t \cos(1 + t^2)$ can (and should) be checked by differentiation.

PROBLEMS

In Problems 1 to 16 find the derivatives of the given functions.

1. $f(x) = \sin(2x + 1)$.
2. $f(x) = 2 \cos(3x + 1)$.
3. $f(x) = \cos(x^2 + 4x + 1)$.
4. $x \mapsto \cos^2 4x$.
5. $x \mapsto \sin \sqrt{1 + x^2}$.
6. $x \mapsto (\cos 2x)^{3/2}$.
7. $f(\theta) = \sqrt{1 - \cos \theta}$.
8. $f(\theta) = (\theta + 2 \cos 4\theta)^3$.

9. $y = x^2 \sin(4x^2 - 1)$.

10. $y = 2 \sin \dfrac{x}{2} \cos \dfrac{x}{2}$.

11. $f(\theta) = \dfrac{1 - \cos \theta}{1 + \cos \theta}$.

12. $f(\theta) = \dfrac{\sin^2 2\theta}{\cos 2\theta}$.

13. $g(t) = \cos^2(\sin 4t)$.
14. $g(t) = \cos(t^2 \sqrt{1 + t^2})$.
15. $g(t) = \sin^3 \left(\dfrac{1 - t}{1 + t} \right)$.
16. $g(t) = \sin^4 \left(\dfrac{\cos t}{1 + \cos^2 t} \right)$.

In Problems 17 to 21 find all maxima, minima, and inflection points over the period indicated, and sketch the curves.

17. $y = \frac{1}{2}\cos 4x, \quad 0 \leq x \leq \frac{\pi}{2}$.

18. $y = 4 \sin \frac{x}{3}, \quad 0 \leq x \leq 6\pi$.

19. $y = 3 \sin \left(\frac{x}{2} + \frac{3\pi}{4} \right), \quad 0 \leq x \leq 4\pi$.

20. $f(x) = \cos \left(2x - \frac{\pi}{4} \right), \quad 0 \leq x \leq \pi$.

21. $f(x) = \sin x + \sqrt{3} \cos x, \quad 0 \leq x \leq 2\pi$.

Evaluate the following indefinite integrals.

22. $\int \cos 4\theta \; d\theta$.

23. $\int \sin \frac{\theta}{2} \; d\theta$.

24. $\int \cos(1 - 2x) \; dx$.

25. $\int x^2 \sin(4x^3 + 1) \; dx$.

26. $\int \sin^2 3x \cos 3x \; dx$.

27. $\int \cos^3 \frac{t}{2} \sin \frac{t}{2} \; dt$.

28. $\int \frac{\cos 2t}{\sin^4 2t} \; dt$.

29. $\int \frac{\cos 4x}{\sqrt{1 + \sin 4x}} \; dx$.

[*Hint:* Let $u = 1 + \sin 4x$.]

30. $\int \frac{x + \cos 2x}{(x^2 + \sin 2x)^2} \; dx$.

31. $\int \sin 2x \cos 2x \; dx$.

[*Hint:* Use a trig identity.]

32. $\int \cos^3 4\theta \; d\theta$.

33. $\int \sqrt{1 - \cos 6\theta} \; d\theta$.

Evaluate the following definite integrals.

34. $\int_{-\pi/2}^{\pi/2} \cos \frac{\theta}{2} \; d\theta$.

35. $\int_0^{\pi/6} \sin 3\theta \; d\theta$.

36. $\int_0^{\pi/6} \cos \frac{\theta}{2} \sin \frac{\theta}{2} \; d\theta$.

37. $\int_0^{\pi/3} \sqrt{1 + \cos 2t} \; dt$.

38. $\int_0^{\pi/3} \frac{\sin 4t \; dt}{\cos^3 4t}$.

39. $\int_0^{\pi/12} \sqrt{1 - \sin 2t} \cos 2t \; dt$.

40. $\int_0^{\pi/3} \sin \left(x + \frac{\pi}{6} \right) dx$.

41. $\int_0^{\sqrt{\pi/2}} x \cos \left(\frac{\pi}{2} - x^2 \right) dx$.

42. $\int_{-\pi/4}^{\pi/4} \sin^2 \left(\theta + \frac{\pi}{4} \right) \cos \left(\theta + \frac{\pi}{4} \right) d\theta$.

43. $\int_0^{\pi} |\sin \theta - \cos \theta| \; d\theta$.

*44. Show that the function $y = x \sin(1/x)$ for $x \neq 0$, $y = 0$ for $x = 0$, is (i) continuous for all x, (ii) differentiable for $x \neq 0$, and (iii) not differentiable at $x = 0$. [*Hint:* For (i) and (iii), use the definitions of continuity and differentiability.]

°45. Show that the function $y = x^2 \sin(1/x)$ for $x \neq 0$, $y = 0$ for $x = 0$, has a derivative for all x, but this derivative is discontinuous at $x = 0$. [*Hint:* First show that $(dy/dx)_{x=0} = 0$ using the definition of derivative.]

2.3   Derivatives of other trigonometric functions

From Theorem 1 and the definitions of the tangent and the other trigonometric functions [see (17) and (18) in §1.4], we obtain the formulas

$$\frac{d \tan \theta}{d\theta} = \frac{1}{\cos^2 \theta} = \sec^2 \theta, \tag{7}$$

which can be also written as

$$\frac{d \tan \theta}{d\theta} = 1 + \tan^2 \theta \tag{8}$$

and

$$\frac{d \sec \theta}{d\theta} = \frac{\sin \theta}{\cos^2 \theta} = \tan \theta \sec \theta, \tag{9}$$

$$\frac{d \csc \theta}{d\theta} = -\frac{\cos \theta}{\sin^2 \theta} = -\cot \theta \csc \theta, \tag{10}$$

$$\frac{d \cot \theta}{d\theta} = -\frac{1}{\sin^2 \theta} = -\csc^2 \theta = -1 - \cot^2 \theta. \tag{11}$$

(See Example 1 and Problems 1, 2, 3 for the proofs.) We rewrite these formulas as differentiation and integration rules, using the language of differentials:

$$d \tan u = \sec^2 u \, du, \qquad \int \sec^2 u \, du = \tan u + C, \tag{12}$$

$$d \sec u = \tan u \sec u \, du, \qquad \int \tan u \sec u \, du = \sec u + C, \tag{13}$$

$$d \csc u = -\cot u \csc u \, du, \qquad \int \cot u \csc u \, du = -\csc u + C, \tag{14}$$

$$d \cot u = -\csc^2 u \, du, \qquad \int \csc^2 u \, du = -\cot u + C. \tag{15}$$

Of course, all these formulas become meaningless at points at which the functions are not defined. Equation (7), for instance, has no meaning if $\theta = \pi/2$.

EXAMPLES 1. Verify relations (7) and (8).
ANSWER Both relations mean the same thing [see identity (20) in §1.4]. To prove (7) we use the definition of the tangent, the quotient rule for derivative, and Theorem 1. This yields

$$(\tan \theta)' = \left(\frac{\sin \theta}{\cos \theta}\right)' = \frac{(\sin \theta)'(\cos \theta) - (\sin \theta)(\cos \theta)'}{\cos^2 \theta}$$

$$= \frac{\cos^2 \theta + \sin^2 \theta}{\cos^2 \theta} = \frac{1}{\cos^2 \theta} = \sec^2 \theta,$$

by the definition of the secant.

2. Find the derivative of $f(x) = \tan^3 2x$.
ANSWER It is necessary to use the chain rule directly. Set $z = \tan 2x$, $u = 2x$. We have $f(x) = z^3$, $z = \tan u$. Thus

$$\frac{df(x)}{dx} = \frac{dz^3}{dz}\frac{dz}{du}\frac{du}{dx} = \frac{dz^3}{dz}\frac{d \tan u}{du}\frac{d2x}{dx}$$

$$= (3z^2)(\sec^2 u)(2x) = (3 \tan^2 2x)(\sec^2 2x)2x = 6x(\tan^2 2x)(\sec^2 2x).$$

This can also be written as $f'(x) = \dfrac{6x \sin^2 2x}{\cos^4 2x}$.

3. Find the area under the curve $y = \sec^2 x$ from $x = \pi/6$ to $x = \pi/4$.
ANSWER Observe that $\sec^2 x$ is defined in the interval $\pi/6 \le x \le \pi/4$. The desired area is

$$\int_{\pi/6}^{\pi/4} \sec^2 x \, dx = \tan x \Big|_{\pi/6}^{\pi/4} = 1 - \frac{1}{\sqrt{3}} \approx .423.$$

PROBLEMS

1. Prove that $\dfrac{d}{d\theta}(\sec\theta) = \tan\theta\,\sec\theta$.

2. Prove that $\dfrac{d}{d\theta}(\csc\theta) = -\cot\theta\,\csc\theta$.

3. Prove that $\dfrac{d}{d\theta}(\cot\theta) = -\csc^2\theta = -1 - \cot^2\theta$.

In Problems 4 to 17 find the derivatives of the given functions.

4. $f(x) = \frac{1}{2}\tan 4x$.
5. $f(x) = 3\sec 2x$.
6. $f(x) = \cot\sqrt{2x^2 + 1}$.
7. $x \mapsto \csc^2(3x + 2)$.
8. $x \mapsto (1 + \sec^2 x)^{1/2}$.

9. $x \mapsto x\tan\dfrac{1}{x}$.

10. $g(t) = \cot 2t\,\csc^2 2t$.
11. $g(t) = (\sec 4t + \cot 4t)^3$.

12. $g(t) = \sec^3\left(\dfrac{1}{2t}\right)$.

13. $f(x) = \tan^2(2x^2 + 4x + 1)$.
14. $f(x) = \cot(\sec 2x)$.

15. $y = \csc(x\sqrt{1 - x})$.

16. $y = \tan\left(\dfrac{1 - x^2}{1 + x^2}\right)$.

17. $y = \tan\left(\dfrac{1 - \cos x}{1 + \cos x}\right)$.

Evaluate the following indefinite integrals.

18. $\displaystyle\int \tan 2\theta\,\sec 2\theta\,d\theta$.

19. $\displaystyle\int \sec^2 \tfrac{1}{2}\theta\,d\theta$.

20. $\displaystyle\int \csc^2 4\theta\,d\theta$.

21. $\displaystyle\int \cot(3\theta + 1)\,\csc(3\theta + 1)\,d\theta$.

22. $\displaystyle\int x^2\sec^2(x^3 + 2)\,dx$.

23. $\displaystyle\int \tan^2 x\,dx$.

[Hint: Use a trig identity.]

24. $\displaystyle\int \sec^3 x\,\tan x\,dx$.

25. $\displaystyle\int \tan^2\dfrac{x}{2}\,\sec^2\dfrac{x}{2}\,dx$.

26. $\displaystyle\int \dfrac{\tan 2t}{\sec^4 2t}\,dt$.

27. $\displaystyle\int \csc^4 6t\,\cot 6t\,dt$.

28. $\displaystyle\int \dfrac{\sec^2 t}{\sqrt{\tan t + 1}}\,dt$.

29. $\displaystyle\int \dfrac{\csc^2 2t}{(\cot 2t + 1)^4}\,dt$.

Evaluate the following definite integrals.

30. $\displaystyle\int_0^{\pi/4} \sec^2 x\,dx$.

31. $\displaystyle\int_{\pi/3}^{\pi/2} \cot\dfrac{x}{2}\,\csc\dfrac{x}{2}\,dx$.

32. $\displaystyle\int_0^{\pi/6} \sec 2x\,\tan 2x\,dx$.

33. $\displaystyle\int_0^{\pi/2} \sec^2\dfrac{x}{2}\,\tan\dfrac{x}{2}\,dx$.

34. $\displaystyle\int_0^{\pi/6} \tan^5 2x\,\sec^2 2x\,dx$.

35. $\displaystyle\int_{\pi/6}^{\pi/2} \csc^3\theta\,\cot\theta\,d\theta$.

36. $\displaystyle\int_{\pi/6}^{\pi/3} \dfrac{\cot\theta}{\csc^2\theta}\,d\theta$.

37. $\displaystyle\int_{\pi/4}^{\pi/2} \dfrac{\csc^2\theta}{(\cot\theta + 1)^3}\,d\theta$.

38. $\displaystyle\int_0^{\pi/6} \dfrac{\tan\theta}{1 + \tan^2\theta}\,d\theta$.

39. $\displaystyle\int_0^{\pi/4} \dfrac{\tan\theta\,\sec^2\theta}{(2 + \tan^2\theta)^2}\,d\theta$.

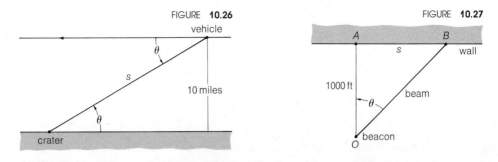

FIGURE **10.26**

vehicle

10 miles

crater

FIGURE **10.27**

A   B

wall

1000 ft

beam

beacon

2.4 Some rate of change problems‡

We illustrate the use of trigonometric functions by some problems. In solving these, we should begin by making a drawing and labeling all relevant quantities.

EXAMPLES 1. A space vehicle is orbiting the moon at a speed of 1000 mph at an elevation of 10 miles above the lunar surface. How fast must an automatic camera aboard be turning to photograph a crater directly ahead if the angle between the path of the space vehicle and the line of sight to the crater is 30°? Give your answer in revolutions per hour.

ANSWER The situation is depicted in Figure 10.26. We must find $d\theta/dt$ (t being time) for $\theta = 30°$. From the figure we see that $\tan\theta = 10/s$, so that $s = 10\cot\theta$. Hence

$$\frac{ds}{dt} = 10\frac{d\cot\theta}{dt} = -10\csc^2\theta\,\frac{d\theta}{dt}\,(\text{mph}).$$

We are told that $ds/dt = -1000$ (minus because of the way we drew the figure). For $\theta = 30°$, $\csc\theta = 2$. Thus, at the instant when $\theta = 30°$,

$$-1000 = -10(2^2)\frac{d\theta}{dt}$$

or $d\theta/dt = 25$. The camera should be turning at the rate of 25 radians per hour or $25(180/\pi)$ degrees per hour or $25(360/2\pi)$ degrees per hour. Since a full revolution is one of 360°, the camera must be turning at the rate of $25/2\pi \approx 3.98$ revolutions per hour.

2. A revolving beacon is located 1000 ft from a straight sea wall and rotates at the constant rate of 2 rpm. How fast does the beam of light cast by the beacon sweep along the sea wall at the nearest point to the beacon? How fast is the beam moving at a point 500 ft from this point?

ANSWER The situation is depicted in Figure 10.27. The angle θ is a function of time t with a constant derivative. In 1 min θ increases by 4π (two full revolutions). Hence $d\theta/dt = 4\pi$ [the unit in which $d\theta/dt$ is measured is $(\text{minute})^{-1}$ since θ is a pure number and t is measured in minutes]. Let A be the point on the wall closest to the beacon O, B some other point on the wall, $s = |AB|$. Then $|OA| = 1000$ (feet), and $|AB|/|OA| = \tan\theta$, so that $s = 1000\tan\theta$. The speed with which the beam sweeps along the sea wall is

$$\frac{ds}{dt} = \frac{d(1000\tan\theta)}{dt} = \frac{d(1000\tan\theta)}{d\theta}\frac{d\theta}{dt}$$

$$= (1000\sec^2\theta)4\pi = 4000\pi\sec^2\theta\,(\text{ft/min}).$$

In order to express everything in terms of s, we note that $\sec^2\theta = 1 + \tan^2\theta = 1 + (s/1000)^2$.

‡This subsection may be omitted without loss of continuity.

Thus

$$\frac{ds}{dt} = 4000\pi \left[1 + \left(\frac{s}{1000} \right)^2 \right].$$

For $B = A$, $s = 0$ and $ds/dt = 4000\pi$ ft/min $= 4000\pi/60$ ft/sec ≈ 210 ft/sec. For $s = 500$, $ds/dt = 4000\pi(1 + \frac{1}{4}) = 5000\pi$ ft/min ≈ 262 ft/sec.

PROBLEMS

1. A rocket leaves the ground 2000 ft from an observer and rises vertically at the constant rate of 100 ft/sec. How fast is the angle between the observer's line of sight and the ground increasing after 20 sec? Give your answer in degrees per second.

2. An airplane flying at an altitude of 20,000 ft and on a horizontal course passes directly over an observer on the ground below. The observer notes that when the angle between the ground and his line of sight is 60°, the angle is decreasing at the rate of 2° per second. What is the speed of the airplane?

3. A vertical pole 30 ft high is located 30 ft east of a tall building. If the sun is rising at the rate of 18° per hour, how fast is the shadow of the pole on the building shortening when the elevation of the sun is 30°?

4. A television tower 200 ft high casts a shadow on the flat ground below. If the angle the sun makes with the horizontal (angle of elevation) decreases at the rate of $\pi/10$ radians per hour, how fast is the shadow lengthening when the elevation of the sun is 30°?

5. A ladder 30 ft long leans against a wall. Suppose the bottom of the ladder slides away from the wall at the rate of 3 ft/sec. How fast is the angle between the ladder and the ground changing when the bottom of the ladder is 15 ft from the wall?

6. A yacht is pulled toward a dock by means of a winch. If the deck of the yacht is 15 ft below the level of the dock, and the yacht moves at the rate of 5 ft/min, how fast is the angle between the rope and the horizontal changing when the rope is 30 ft long?

7. A kite at an altitude of 200 ft moves horizontally at the rate of 20 ft/sec. At what rate is the angle between the line and the ground changing when 400 ft of line is out?

8. A searchlight on a beach is trained on a boat moving in a direction parallel to the beach at the rate of 20 miles per hour. At what rate is the searchlight revolving when the boat is 1 mile offshore at a point nearest the light? Give your answer in degrees per second.

9. A weather balloon is released on the ground 1500 ft from an observer and rises vertically at the constant rate of 250 ft/min. How fast is the angle between the observer's line of sight and the ground increasing when the balloon is at an altitude of 2000 ft? Give your answer in degrees per minute.

10. Two ships steam from the same island at the same time. One ship steams north at 15 miles per hour. The other steams west at 20 miles per hour for 3 hours and then turns south. What is the rate of rotation of the line joining them 4 hours after leaving the island?

11. A weather balloon is released on the ground 1000 ft from an observer. Due to the wind, the balloon is carried away from the point of release at an angle of 60° with the horizontal. How fast is the angle between the observer's line of sight and the ground increasing when the balloon is at an altitude of 1732 ft? Assume the balloon rises vertically at the constant rate of 280 ft/min.

12. A rocket is launched 1000 ft from an observer. Due to a malfunction in the navigation mechanism, the rocket leaves the launching site at an angle of 45° with the horizontal. How fast is the angle of elevation of the observer's line of sight increasing 6 sec after the launching if the horizontal velocity of the rocket is 500 ft/sec?

13. A billboard 10 ft high is located on the edge of a building 45 ft tall. A girl 5 ft in height approaches the building at the rate of 3.4 ft/sec. How fast is the angle subtended at her eye by the billboard changing when she is 30 ft from the billboard?

14. A girl skates at the rate of 10 ft/sec along a diameter in a circular skating rink. A light located at one end of a diameter perpendicular to her path casts a shadow of the girl on the side of the rink. How fast is her shadow moving when she reaches the center of the rink?

15. A helicopter, hovering at an altitude of 4600 ft, drops a lighted flare that falls toward the ground below. An observer on the ground measures the angle subtended by the helicopter and flare. How fast is the angle changing 10 sec after the flare is dropped if the observer is stationed 4000 ft from the point on the ground directly below the helicopter? [*Hint:* Assume free fall motion. Then the distance s (feet) the flare falls in time t (seconds) is $s = 16t^2$.]

§3 Inverse trigonometric functions

We now apply to trigonometric functions the concept of inverse functions (see Chapter 6, §1). We recall that a continuous function f has a continuous inverse g in an interval if and only if f is strictly monotone in that interval, and that if f has a derivative $f' \neq 0$, then $g' = 1/f'$.

3.1 Inverse sine and inverse cosine

The function $\sin\theta$ is not monotone; in order to define an inverse we must select an interval on which $\sin\theta$ is either increasing or decreasing. We choose the interval $[-\pi/2, \pi/2]$. In this interval $\sin\theta$ increases; this is seen from the graph and also by noting that the derivative $d\sin\theta/d\theta = \cos\theta$ is positive for $-\pi/2 < \theta < \pi/2$. The function inverse to $\sin\theta$ in this interval is called the **arc sine** (or inverse sine) and is denoted by arc sin. Thus

$$\theta = \text{arc sin } x \qquad \text{means that} \qquad x = \sin\theta, \qquad -\frac{\pi}{2} \le \theta \le \frac{\pi}{2}, \tag{1}$$

or

$$\text{arc sin}(\sin\theta) = \theta \qquad \text{for } -\frac{\pi}{2} \le \theta \le \frac{\pi}{2}, \quad \sin(\text{arc sin } x) = x. \tag{2}$$

The function $\theta = \text{arc sin } x$ is defined only for $-1 \le x \le 1$, since $\sin\theta$ takes on no values less than -1 or greater than 1. The graph of the function arc sin x is shown in Figure 10.28; it is obtained by "flipping" a part of the graph of $\sin\theta$.

By the theorems on inverse functions, arc sin x is continuous for $-1 \le x \le 1$ and differentiable for $-1 < x < 1$. Its derivative is computed as follows. Set $\theta = \text{arc sin } x$; then $x = \sin\theta$ and (since $|\theta| \le \pi/2$) $\cos\theta \ge 0$ so that $\cos\theta = \sqrt{1 - \sin^2\theta}$. Now, by Theorem 2 in Chapter 6, §1.2,

$$\frac{d \text{ arc sin } x}{dx} = \frac{d\theta}{dx} = \frac{1}{dx/d\theta} = \frac{1}{d\sin\theta/d\theta} = \frac{1}{\cos\theta} = \frac{1}{\sqrt{1 - \sin^2\theta}} = \frac{1}{\sqrt{1 - x^2}}$$

or

$$\frac{d \text{ arc sin } x}{dx} = \frac{1}{\sqrt{1 - x^2}}, \qquad \text{for } -1 < x < 1. \tag{3}$$

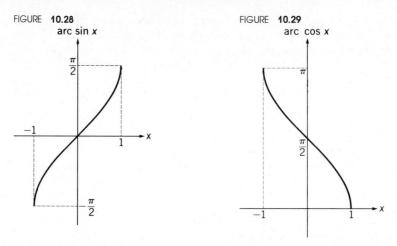

FIGURE 10.28
arc sin x

FIGURE 10.29
arc cos x

It is remarkable that the derivative of the transcendental function arc sin x is a radical function.

The function arc cos x (called the **arc cosine** or the inverse cosine of x) is defined as the function inverse to the function cos θ in the interval $0 \leq \theta \leq \pi$. (In this interval cos θ is decreasing since its derivative, $-\sin \theta$, is negative for $0 < \theta < \pi$.) Thus

$$\theta = \text{arc cos } x \qquad \text{means that} \qquad x = \cos \theta, \quad 0 \leq \theta \leq \pi, \tag{4}$$

or

$$\text{arc cos}(\cos \theta) = \theta \qquad \text{for } 0 \leq \theta \leq \pi, \quad \cos(\text{arc cos } x) = x. \tag{5}$$

The function arc cos x is defined only for $-1 \leq x \leq 1$, since $|\cos \theta|$ is never greater than 1. The graph of the inverse cosine is shown in Figure 10.29; it is obtained by "flipping" the graph of cos θ in the interval $0 \leq \theta \leq \pi$. We have

$$\frac{d \text{ arc cos } x}{dx} = -\frac{1}{\sqrt{1 - x^2}}, \qquad \text{for } -1 < x < 1. \tag{6}$$

To prove this, set $\theta = \text{arc cos } x$, so that $\cos \theta = x$, and (since $0 \leq \theta \leq \pi$) $\sin \theta = \sqrt{1 - \cos^2 \theta}$. Then

$$\frac{d \text{ arc cos } x}{dx} = \frac{d\theta}{dx} = \frac{1}{dx/d\theta} = \frac{1}{d \cos \theta/d\theta} = \frac{1}{-\sin \theta}$$

$$= -\frac{1}{\sqrt{1 - \cos^2 \theta}} = -\frac{1}{\sqrt{1 - x^2}}.$$

Note that

$$\text{arc sin } 0 = 0, \qquad \text{arc cos } 0 = \frac{\pi}{2} \tag{7}$$

(since $\sin 0 = 0$, and 0 lies in the interval $[-\pi/2, \pi/2]$; $\cos \pi/2 = 0$ and $\pi/2$ lies in the interval $[0, \pi]$).

Relations (3), (6) show that the function arc sin x + arc cos x has the derivative 0. It is thus a constant; but at 0 it equals $\pi/2$. Hence

$$\text{arc sin } x + \text{arc cos } x = \frac{\pi}{2}. \tag{8}$$

This also follows from relations (12) in §1.3.

Using (3), (6), (8), and the fundamental theorem of calculus, we can write the arc sine and the arc cosine of a number x as integrals:

$$\text{arc sin } x = \int_0^x \frac{dt}{\sqrt{1 - t^2}}, \qquad \text{arc cos } x = \frac{\pi}{2} + \int_x^0 \frac{dt}{\sqrt{1 - t^2}}. \qquad (9)$$

We proved this for $-1 < x < 1$. But we know that $\lim_{x \to 1^-}$ arc sin $x = \pi/2$. Hence

$$\lim_{x \to 1^-} \int_0^x \frac{dt}{\sqrt{1 - t^2}} = \frac{\pi}{2} \qquad \text{or} \qquad \int_0^1 \frac{dt}{\sqrt{1 - t^2}} = \frac{\pi}{2}. \qquad (10)$$

(We already evaluated this improper integral in Chapter 9, §3.2.) Using (10), we write the second relation (9) as

$$\text{arc cos } x = \int_x^1 \frac{dt}{\sqrt{1 - t^2}}. \qquad (11)$$

We rewrite once more the differentiation rules (3), (6), and the corresponding integration rule.

$$d \text{ arc sin } u = \frac{du}{\sqrt{1 - u^2}} = -d \text{ arc cos } u, \qquad (12)$$

$$\int \frac{du}{\sqrt{1 - u^2}} = \text{arc sin } u + C. \qquad (13)$$

REMARK   One sometimes calls every number θ, such that $\sin \theta = x$, an arc sine of x, and denotes the θ satisfying both $\sin \theta = x$ and $-\pi/2 \le \theta \le \pi/2$ the principal branch of the arc sine. A similar remark applies to other inverse trigonometric functions.

EXAMPLES   1.   What is arc cos $(-\frac{1}{2})$?
ANSWER   We look for a θ such that $0 < \theta < \pi$ and cos $\theta = -\frac{1}{2}$. This θ must satisfy $\frac{1}{2}\pi < \theta < \pi$ (refer to the graph of the cosine). Set $\theta = \frac{1}{2}\pi + \alpha$, $0 < \alpha < \frac{1}{2}\pi$. We need $-\frac{1}{2} = \cos \theta = \cos(\frac{1}{2}\pi + \alpha) = -\sin \alpha$. Hence $\sin \alpha = \frac{1}{2}$, $\alpha = 30° = \pi/6$, $\theta = 120° = 2\pi/3$.

2.   What is arc cos$(-.6)$?
ANSWER   Proceeding as in Example 1, we see that arc cos$(-.6) = \frac{1}{2}\pi + \alpha$ where $0 < \alpha < \frac{1}{2}\pi$ and sin $\alpha = .6$. From the tables in the back of the book, we see that $\alpha = 37°$, approximately. Hence arc cos$(-.6) = 127° = 2.22$, approximately.

3.   Differentiate $f(t) = $ arc sin$(1 - 2t)$.
ANSWER   Set $u = 1 - 2t$. Then $f = $ arc sin u, and

$$f'(t) = \frac{d \text{ arc sin } u}{du} \frac{du}{dt} = \frac{-2}{\sqrt{1 - u^2}} = -\frac{1}{\sqrt{t(1 - t)}}.$$

4.   Evaluate $\int (1 - t^4)^{-1/2} t \, dt$.
ANSWER   Set $u = t^2$. Then $du = 2t \, dt$, and

$$\int (1 - t^4)^{-1/2} t \, dt = \frac{1}{2} \int (1 - u^2)^{-1/2} \, du = \frac{1}{2} \text{ arc sin } u + C = \frac{1}{2} \text{ arc sin}(t^2) + C.$$

5.   Evaluate $\int_0^{1/\sqrt{2}} (1 - t^4)^{-1/2} \, dt$.
ANSWER   Using the result of the preceding example, we have

$$\int_0^{1/\sqrt{2}} (1 - t^4)^{-1/2} t \, dt = \frac{1}{2} \text{ arc sin}(t^2) \Big|_0^{1/\sqrt{2}} = \frac{1}{2} \text{ arc sin } \frac{1}{2} - \frac{1}{2} \text{ arc sin } 0 = \frac{\pi}{12}.$$

PROBLEMS

In Problems 1 to 10 find the indicated numbers.

1. $\arc \sin(\frac{1}{2})$.

2. $\arc \cos\left(\dfrac{\sqrt{3}}{2}\right)$.

3. $\arc \sin\left(-\dfrac{\sqrt{2}}{2}\right)$.

4. $\arc \cos(-\frac{1}{2})$.

5. $\arc \sin\left(-\dfrac{\sqrt{3}}{2}\right)$.

6. $\arc \cos\left(\cos\dfrac{\pi}{3}\right)$.

7. $\arc \sin\left(\cos\dfrac{5\pi}{6}\right)$.

8. $\arc \sin(\sin 30°)$.

9. $\cos\left(\arc \cos\left(-\dfrac{\sqrt{3}}{2}\right)\right)$.

10. $\sin(\arc \cos \frac{1}{2})$.

In Problems 11 to 22 find the derivatives of the given functions.

11. $x \mapsto \arc \cos 2x$.

12. $x \mapsto \arc \sin \sqrt{x}$.

13. $x \mapsto \arc \cos(x^2 - 1)$.

14. $f(t) = \arc \sin(2t^2 + 1)^4$.

15. $f(t) = (\arc \cos t)^3$.

16. $f(t) = \arc \sin(\sin 2t)$.

17. $g(\theta) = \sin(\arc \sin 2\theta)$.

18. $g(\theta) = (\arc \cos \theta)(\arc \sin \theta)$.

19. $g(\theta) = (1 - \theta^2)^{1/2} \arc \cos \theta$.

20. $y = \arc \sin \dfrac{1}{\sqrt{x + 1}}$.

21. $y = \arc \cos \dfrac{2 - x}{2 + x}$.

22. $y = \arc \sin \left(\dfrac{1}{x^2 + 1}\right)$.

Evaluate the following indefinite integrals.

23. $\displaystyle\int \dfrac{dt}{\sqrt{1 - 16t^2}}$.

24. $\displaystyle\int (9 - 4t^2)^{-1/2}\, dt$.

25. $\displaystyle\int \dfrac{t\, dt}{\sqrt{1 - 16t^4}}$.

26. $\displaystyle\int x^2(1 - x^6)^{-1/2}\, dx$.

27. $\displaystyle\int \dfrac{dx}{\sqrt{9 - (x + 3)^2}}$.

28. $\displaystyle\int \dfrac{(1 - 8x)}{\sqrt{1 - 4x^2}}\, dx$.

29. $\displaystyle\int \dfrac{8z^{3/2}}{\sqrt{1 - z^5}}\, dz$.

30. $\displaystyle\int \left(\dfrac{\arc \cos z}{1 - z^2}\right)^{1/2}\, dz$.

Evaluate the following definite integrals.

31. $\displaystyle\int_0^{1/4} \dfrac{dt}{\sqrt{1 - 4t^2}}$.

32. $\displaystyle\int_0^2 \dfrac{dt}{\sqrt{16 - t^2}}$.

33. $\displaystyle\int_0^{\sqrt{2}} \dfrac{t}{\sqrt{4 - t^4}}\, dt$.

34. $\displaystyle\int_{-1}^1 \dfrac{1 - \theta}{\sqrt{1 - \theta^2}}\, d\theta$.

35. $\displaystyle\int_{1/4}^{1/2} \dfrac{d\theta}{\sqrt{\theta - \theta^2}}$.   [*Hint:* Let $\theta = u^2$.]

36. $\displaystyle\int_{\sqrt{2}/2}^{\sqrt{3}/2} \dfrac{\arc \cos \theta}{\sqrt{1 - \theta^2}}\, d\theta$.

37. Prove that $\arc \sin x + \arc \sin y = \arc \sin(x\sqrt{1 - y^2} + y\sqrt{1 - x^2})$. [*Hint:* Use the addition formula for the sine.]

38. Prove that $\arc \cos x + \arc \cos y = \arc \cos(xy - \sqrt{1 - x^2}\,\sqrt{1 - y^2})$.

3.2 The arc tangent

Since the function $x = \tan \theta$ is strictly increasing and has a positive derivative in the interval $-\pi/2 < \theta < \pi/2$, it has a differentiable inverse function in this interval. We call this function the **arc tangent** or inverse tangent of x, and we denote it by arc $\tan x$. In view of relations (26) and (27) in §1.4, the equation $\tan \theta = x$ has a solution

θ in the interval $-\pi/2 < \theta < \pi/2$ for every number x. The function $x \mapsto \operatorname{arc\,tan} x$ is therefore defined for all values of x. The graph of the arc tan function (see Figure 10.30) is obtained by "flipping" the central branch of the graph in Figure 10.19.

FIGURE **10.30**

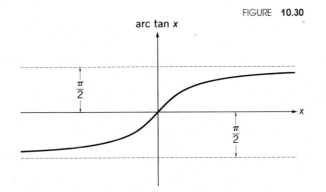

We have

$$\tan(\operatorname{arc\,tan} x) = x, \quad -\infty < x < \infty; \qquad \operatorname{arc\,tan}(\tan \theta) = \theta, \quad -\frac{\pi}{2} < \theta < \frac{\pi}{2}. \quad (14)$$

Since $\tan 0 = 0$,

$$\operatorname{arc\,tan} 0 = 0. \tag{15}$$

Next, we compute the derivative of the arc tan. Set $\theta = \operatorname{arc\,tan} x$. Then $x = \tan \theta$, and, by the theorem on derivatives of inverse functions,

$$\frac{d \operatorname{arc\,tan} x}{dx} = \frac{d\theta}{dx} = \frac{1}{dx/d\theta} = \frac{1}{d \tan \theta/d\theta} = \frac{1}{1 + \tan^2 \theta} = \frac{1}{1 + x^2}$$

or

$$\frac{d \operatorname{arc\,tan} x}{dx} = \frac{1}{1 + x^2}. \tag{16}$$

The derivative of the arc tangent is a rational function! From (15), (16), and the fundamental theorem of calculus, we obtain

$$\operatorname{arc\,tan} x = \int_0^x \frac{dt}{1 + t^2}. \tag{17}$$

The differentiation and integration rules

$$d \operatorname{arc\,tan} u = \frac{du}{1 + u^2}, \qquad \int \frac{du}{1 + u^2} = \operatorname{arc\,tan} u + C \tag{18}$$

are among the most important formulas in calculus.

The arc tan function satisfies the relation

$$\operatorname{arc\,tan} x + \operatorname{arc\,tan} y = \operatorname{arc\,tan} \frac{x + y}{1 - xy}. \tag{19}$$

This follows from the addition theorem for the tangent (see Problem 21).

EXAMPLES 1. We have arc tan $1 = 45° = \pi/4$ [since $\pi/4$ lies in the interval $(-\pi/2, \pi/2)$ and tan $45° = 1$]. By (17),

$$\int_0^1 \frac{dt}{1 + t^2} = \frac{\pi}{4}.$$

Similarly,

$$\int_{-1}^0 \frac{dt}{1 + t^2} = \frac{\pi}{4}.$$

2. Prove that

$$\int_1^{+\infty} \frac{dt}{1 + t^2} = \frac{\pi}{4}, \qquad \int_{-\infty}^{-1} \frac{dt}{1 + t^2} = \frac{\pi}{4}.$$

ANSWER For $A > 1$, we have

$$\int_1^A \frac{dt}{1 + t^2} = \text{arc tan } A - \text{arc tan } 1 = \text{arc tan } A - \frac{\pi}{4}.$$

We must show that this has limit $\pi/4$ as $A \to +\infty$, that is, $\lim_{A \to +\infty} \text{arc tan } A = \pi/2$. This is so, in view of relation (26) in §1.4; see also the graph in Figure 10.33. The second relation is proved similarly.

3. Show that

$$\int_{-\infty}^{+\infty} \frac{dt}{1 + t^2} = \pi. \tag{20}$$

ANSWER The integral (20) is the sum of the four integrals computed in Examples 1 and 2. The result is illustrated in Figure 10.31.

FIGURE **10.31**

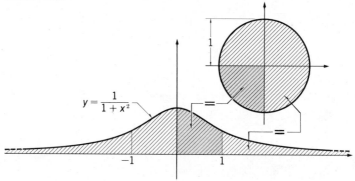

4. Compute $\int_2^3 \frac{x \, dx}{1 + x^4}$.

SOLUTION Recalling that $(1 + u^2)^{-1}$ is the primitive function of arc tan u, we set $u = x^2$. Then $du = 2x \, dx$, $u = 4$ for $x = 2$, $u = 9$ for $x = 3$. Thus

$$\int_2^3 \frac{x \, dx}{1 + x^4} = \frac{1}{2} \int_4^9 \frac{du}{1 + u^2} = \frac{1}{2} \text{ arc tan } u \, \Big|_4^9 \approx .0672.$$

PROBLEMS

In Problems 1 to 8 find the derivatives of the given functions.

1. $f(x) = \text{arc tan}(4x + 1)$.

2. $f(x) = \text{arc tan}\dfrac{1}{x}$.

3. $g(s) = \text{arc tan}(s^2 + 1)^{1/2}$.

4. $g(s) = \text{arc tan}\dfrac{s}{2s + 1}$.

5. $y = x\,\text{arc tan}\dfrac{1}{\sqrt{x}}$.

6. $y = (\text{arc tan } 4x)^3$.

7. $h(z) = \frac{1}{2}\,\text{arc tan}(\text{arc sin } 2z)$.

8. $h(z) = \text{arc tan}(\sqrt{\text{arc tan } z})$.

Evaluate the following integrals.

9. $\displaystyle\int \dfrac{dt}{1 + 9t^2}$.

10. $\displaystyle\int \dfrac{dt}{16 + 25t^2}$.

11. $\displaystyle\int \dfrac{dt}{1 + (2t - 1)^2}$.

12. $\displaystyle\int \dfrac{dx}{x^2 + 2x + 2}$.

   [Hint: Complete the square.]

13. $\displaystyle\int \dfrac{dx}{4x^2 - 12x + 13}$.

14. $\displaystyle\int \dfrac{x^2}{1 + 4x^6}\,dx$.

15. $\displaystyle\int \dfrac{\sqrt{z}}{1 + z^3}\,dz$.

16. $\displaystyle\int \dfrac{\text{arc tan } z}{1 + z^2}\,dz$.

   [Hint: Let $u = \text{arc tan } z$.]

17. $\displaystyle\int_0^{1/4} \dfrac{1}{1 + 16s^2}\,ds$.

18. $\displaystyle\int_{-\sqrt{3}}^{\sqrt{3}} \dfrac{1}{s^2 + 9}\,ds$.

19. $\displaystyle\int_{-1}^{1} \dfrac{s\,ds}{1 + s^4}$.

20. $\displaystyle\int_0^{\pi/2} \dfrac{\sin\theta\,\cos\theta}{1 + \sin^4\theta}\,d\theta$.

21. Prove that $\text{arc tan } x + \text{arc tan } y = \text{arc tan}\dfrac{x + y}{1 - xy}$.

22. Prove that $\text{arc tan } x + \text{arc tan}\dfrac{1}{x} = \dfrac{\pi}{2}$. [Hint: Differentiate both sides.]

23. Prove that $\text{arc sin } x = \text{arc tan}\dfrac{x}{\sqrt{1 - x^2}}$.

24. Prove that $\text{arc cos } x = \text{arc tan}\dfrac{\sqrt{1 - x^2}}{x}$.

25. Show that $\sin(2\,\text{arc tan } u) = \dfrac{2u}{1 + u^2}$. $\left[\text{Hint: Use the identity } \sin 2\theta = \dfrac{2\tan\theta}{1 + \tan^2\theta}.\right]$

26. Show that $\cos(2\,\text{arc tan } u) = \dfrac{1 - u^2}{1 + u^2}$.

27. Show that $\sin(\text{arc tan } u) = \dfrac{u}{\sqrt{1 + u^2}}$. [Hint: Use the result in Problem 25.]

28. Show that $\cos(\text{arc tan } u) = \dfrac{1}{\sqrt{1 + u^2}}$.

29. Show that $\text{arc tan}(x + y) = \text{arc tan } x + \text{arc tan}\dfrac{y}{1 + xy + x^2}$.

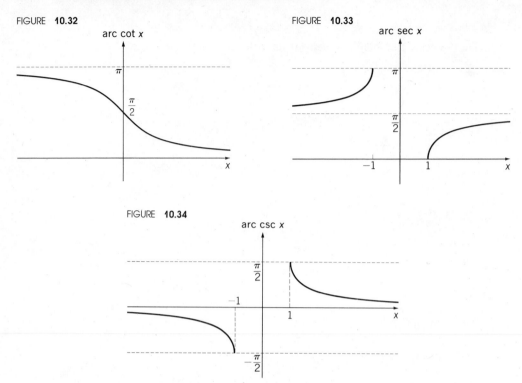

FIGURE **10.32**

arc cot x

FIGURE **10.33**

arc sec x

FIGURE **10.34**

arc csc x

3.3   Other inverse trigonometric functions

In order to define functions inverse to the cotangent, secant, and cosecant, we must choose for each of these functions an interval on which it is monotone. This involves a certain amount of arbitrariness and is sometimes done differently by different authors.

We define the arc cotangent, arc secant, and arc cosecant as the functions inverse to the cotangent, secant, and cosecant, respectively, by the formulas:

$$\cot(\text{arc cot } x) = x, \quad -\infty < x < +\infty; \qquad \text{arc cot}(\cot \theta) = \theta, \quad 0 < \theta < \pi; \quad (21)$$

$$\sec(\text{arc sec } x) = x, \quad |x| \geq 1; \qquad \text{arc sec}(\sec \theta) = \theta, \quad 0 \leq \theta \leq \pi, \theta \neq \frac{\pi}{2}; \qquad (22)$$

$$\csc(\text{arc csc } x) = x, \quad |x| \geq 1; \qquad \text{arc csc}(\csc \theta) = \theta, \quad -\frac{\pi}{2} \leq \theta \leq \frac{\pi}{2}, \theta \neq 0. \quad (23)$$

The functions arc cot x, arc sec x, and arc csc x are graphed in Figures 10.32–10.34. They have the derivatives

$$\frac{d \text{ arc cot } x}{dx} = -\frac{1}{1 + x^2}, \qquad (24)$$

$$\frac{d \text{ arc sec } x}{dx} = \frac{1}{|x| \sqrt{x^2 - 1}}, \qquad \frac{d \text{ arc csc } x}{dx} = -\frac{1}{|x| \sqrt{x^2 - 1}} \qquad (25)$$

(see Example 1 and Problems 1, 2 for the proofs). Thus in the language of differentials,

$$d \text{ arc cot } u = -\frac{du}{1 + u^2}, \tag{26}$$

$$d \text{ arc sec } u = \frac{1}{|u| \sqrt{u^2 - 1}} = -d \text{ arc csc } u, \tag{27}$$

and

$$\int \frac{du}{|u| \sqrt{u^2 - 1}} = \text{arc sec } u + C, \tag{28}$$

or

$$\int \frac{du}{u \sqrt{u^2 - 1}} = \text{arc sec } |u| + C. \tag{28'}$$

Our definitions are such that

$$\text{arc sec } x = \text{arc cos} \left(\frac{1}{x}\right), \qquad \text{arc csc } x = \text{arc sin} \left(\frac{1}{x}\right), \tag{29}$$

for $|x| \geq 1$. Indeed, if $0 < \theta < \pi$ and $\sec \theta = x$, then $\cos \theta = 1/x$, which proves the first relation. The second is proved similarly.

REMARK   Of the six trigonometric functions, it is advisable to use only $\sin x$, $\cos x$, and $\tan x$. Of the six inverse trigonometric functions, it is advisable to use only arc tan x, arc sin x, and arc sec x, the primitive functions of

$$\frac{1}{1 + x^2}, \qquad \frac{1}{\sqrt{1 - x^2}}, \qquad \frac{1}{|x| \sqrt{x^2 - 1}},$$

respectively.

In many books, the inverse trigonometric functions are denoted as follows: $\sin^{-1} x$ for arc sin$x$, $\cos^{-1} x$ for arc cos x, $\tan^{-1} x$ for arc tan x, and so on. This is inconsistent with writing $\sin^2 x$ for $(\sin x)^2$, $\tan^3 x$ for $(\tan x)^3$, and so on. Therefore we do not use this notation.

EXAMPLES 1. Prove the first relation (25).
ANSWER   Let $\theta = $ arc sec x. Then $x = \sec \theta$ and $0 < \theta < \pi$, and, by Equation (9) in §2.3 and by the theorem on differentiating inverse functions,

$$\frac{d \text{ arc sec } x}{dx} = \frac{d\theta}{dx} = \frac{1}{dx/d\theta} = \frac{1}{d \sec \theta / d\theta} = \frac{1}{\tan \theta \sec \theta}.$$

Now, $\tan \theta \sec \theta = \sin \theta / \cos^2 \theta$ is positive for $0 < \theta < \pi$. Since $1 + \tan^2 \theta = \sec^2 \theta$, we can write $\tan \theta \sec \theta = |\sec \theta| \sqrt{\sec^2 \theta - 1}$, or

$$\frac{d \text{ arc sec } x}{dx} = \frac{1}{|x| \sqrt{x^2 - 1}}.$$

2. Prove that arc cot x + arc tan $x = \pi/2$.
ANSWER   By (16) and (24), the function $x \mapsto$ arc tan x + arc cot x has the derivative 0 for all x; hence it is constant. But arc cot $0 = \pi/2$ and arc tan $0 = 0$.

PROBLEMS

1. Prove that $\dfrac{d}{dx}$ arc cot $x = -\dfrac{1}{1 + x^2}$.

2. Prove that $\dfrac{d}{dx}$ arc csc $x = -\dfrac{1}{|x|\sqrt{x^2 - 1}}$.

3. Prove that arc cot $x = $ arc tan $\left(\dfrac{1}{x}\right)$.

4. Prove that arc sec $x = $ arc tan $\sqrt{x^2 - 1}$.

5. Prove that arc csc $x = \dfrac{\pi}{2} - $ arc tan $\sqrt{x^2 - 1}$.

6. Prove that arc csc $x = $ arc sin $\left(\dfrac{1}{x}\right)$.

7. Prove that arc sec $x = \dfrac{\pi}{2} - $ arc sin $\left(\dfrac{1}{x}\right)$.

In Problems 8 to 19 find the derivatives of the given functions.

8. $x \mapsto$ arc cot $\dfrac{x}{2}$.

9. $x \mapsto$ arc sec $\sqrt{x}$.

10. $y = $ arc csc$(2x + 1)$.

11. $y = $ arc cot x^2.

12. $t \mapsto$ arc csc $\dfrac{1}{t}$.

13. $t \mapsto$ arc sec $\sqrt{t^2 + 1}$.

14. $f(z) = (\text{arc sec } 3z)^2$.

15. $f(z) = (\text{arc cot } \sqrt{z})^{3/2}$.

16. $f(z) = \dfrac{\text{arc cot } z}{1 + z^2}$.

17. $y = $ arc cot $\dfrac{1 - x}{1 + x}$.

18. $y = x$ arc csc $\dfrac{1}{x} + \sqrt{1 - x^2}$.

19. $y = \frac{1}{2}$ arc tan$(2 \tan x)$.

Evaluate the following integrals.

20. $\displaystyle\int \dfrac{dx}{x\sqrt{16x^2 - 1}}$.

21. $\displaystyle\int \dfrac{dx}{x\sqrt{9x^2 - 25}}$.

22. $\displaystyle\int \dfrac{dx}{x\sqrt{x^4 - 1}}$.   [Hint: Let $x^2 = u$.]

23. $\displaystyle\int_{1/2}^{1} \dfrac{dt}{t\sqrt{4t^2 - 1}}$.

24. $\displaystyle\int_{-2\sqrt{3}}^{-3} \dfrac{dt}{t\sqrt{t^2 - 9}}$.

25. $\displaystyle\int_{2}^{4} \dfrac{dt}{t\sqrt{t - 1}}$.   [Hint: Let $t = u^2$.]

11

LOGARITHMS AND EXPONENTIALS

§1 Natural logarithms

Logarithms were invented by Napier, who published his tables in 1614. (Another mathematician, Bürgi, developed logarithms independently of Napier and at about the same time. The simultaneous discovery of important ideas by more than one person occurs frequently.) The purpose of logarithmic tables is to facilitate computations, or, more precisely, to reduce "difficult" operations (like multiplications) to "easy" ones (like additions).

Logarithmic tables enable one to perform not only multiplication but also division, raising to powers, and extraction of roots with comparative ease. Soon after the appearance of Napier's first tables, a better set was computed by Napier jointly with Briggs. Much more elaborate tables became available later. For over three centuries

---

JOHN NAPIER (1550–1617), a Scots nobleman, lived and worked on his ancestral estate near Edinburgh. He was an advocate of the Protestant cause and the author of a theological treatise. His book on logarithms was the result of 20 years work.

  Beside inventing logarithms, Napier also constructed another device to facilitate calculations—the Napier rods.

JOST BÜRGI (1552–1632) was a Swiss watchmaker, mathematician, and inventor. He published his tables in 1620, but the promised book of instructions and explanations never appeared.

HENRY BRIGGS (1561–1631) was a professor of mathematics at London University and later at Oxford.

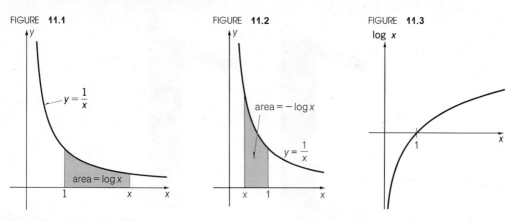

FIGURE **11.1**

$y = \dfrac{1}{x}$

area = log x

1   x   x

FIGURE **11.2**

area = − log x

$y = \dfrac{1}{x}$

x   1   x

FIGURE **11.3**

log x

1

x

all extensive calculations were performed by logarithms. (In the nineteenth century engineers learned to rely on slide rules for rapid calculations. A slide rule is simply a table of logarithms in which certain numbers are represented by lengths.) There is hardly a scientific discovery or a technological advance that did not use Napier's invention, either directly or indirectly.

Only with the advent of the modern automatic computer have logarithmic (and other) tables become somewhat obsolete. An electronic computer performs multiplications and divisions directly and with great speed. And when a computer needs a logarithm, or a cosine of a number, it does not look up a table but finds it by performing the necessary calculation. (Currently pocket computers are replacing slide rules, contrary to a prediction made in the first edition of this book.)

The lasting value, however, of Napier's work lies in the important functions to which it led.

1.1　The logarithmic function

Thus far we have not encountered a function with derivative $1/x$. But $1/x$ is continuous for $x > 0$, and we know, by the fundamental theorem of calculus, that every continuous function is the derivative of another function. The primitive function of $1/x$, which equals 0 at 1, is called the **natural logarithmic function** and is denoted by $\log x$. Thus

$$\log 1 = 0, \qquad \frac{d \log x}{dx} = \frac{1}{x}, \quad \text{for } x > 0, \tag{1}$$

and

$$\log x = \int_{1}^{x} \frac{dt}{t}. \tag{2}$$

The **natural logarithm** $\log a$ of a positive number a is, for $a > 1$, the area under the graph of the equilateral hyperbola $y = 1/x$ from $x = 1$ to $x = a$ (see Figure 11.1). If $a = 1$, then $\log a = 0$, and if $0 < a < 1$, then $\log a$ is minus the area under the hyperbola from $x = a$ to $x = 1$ (see Figure 11.2).

The values of the logarithmic function $\log x$ can be computed using the geometric interpretation of relation (2) just given. In a later chapter we shall learn more efficient methods. A table of $\log x$ will be found at the end of this book, and the graph of the function is shown in Figure 11.3.

The logarithmic function is increasing (since its derivative $1/x$ is positive) and convex (the second derivative $-1/x^2$ is negative). In the next subsection we shall show that

$$\lim_{x \to +\infty} \log x = +\infty \tag{3}$$

($\log x$ becomes arbitrarily large for large x) and

$$\lim_{x \to 0^+} \log x = -\infty \tag{4}$$

($\log x$ is negative and becomes arbitrarily large, in absolute value, for x close to 0). We also note that

$$\frac{d \log |x|}{dx} = \frac{1}{x}, \qquad \text{for } x \neq 0. \tag{5}$$

Indeed, for $x > 0$, (5) is the same as the second relation in (1). If $x < 0$, then $|x| = -x$, and, setting $\xi = -x$, we have $\xi > 0$ and

$$\frac{d \log |x|}{dx} = \frac{d \log \xi}{dx} = \frac{d \log \xi}{d\xi} \frac{d\xi}{dx} = \frac{1}{\xi}(-1) = -\frac{1}{\xi} = \frac{1}{x}.$$

If $x \mapsto u(x)$ is any function (which is differentiable and not equal to 0 in the interval considered), then by the chain rule and (5),

$$\frac{d \log |u(x)|}{dx} = \frac{d \log |u|}{du} \frac{du}{dx} = \frac{1}{u(x)} u'(x) = \frac{u'(x)}{u(x)}. \tag{5a}$$

In other words, the function $\log |u(x)|$ is a primitive function for $u'(x)/u(x)$:

$$\int \frac{u'(x)}{u(x)} dx = \log |u(x)| + C. \tag{5b}$$

Using the abbreviation $du = u'(x) dx$ (see Chapter 8, §3.3), we rewrite (5a) and (5b) as

$$d \log |u| = \frac{du}{u}, \qquad \int \frac{du}{u} = \log |u| + C. \tag{6}$$

EXAMPLES 1. What is $\int_2^3 \frac{dx}{x}$? Use tables.

ANSWER   By (6), $\int_2^3 \frac{dx}{x} = \log x \Big|_2^3 = \log 3 - \log 2 \approx 1.0986 - .6931 = .4055$ (approximately, since the values of $\log 2$ and $\log 3$ in the table are only approximate).

2. Differentiate $\log(1 + 2x + 3x^2)$.
ANSWER   Set $u = 1 + 2x + 3x^2$. Then

$$\frac{d \log(1 + 2x + 3x^2)}{dx} = \frac{d \log u}{du} \frac{du}{dx} = \frac{1}{u}(2 + 6x) = \frac{2 + 6x}{1 + 2x + 3x^2}.$$

3. Evaluate $\int \dfrac{1 + 2x}{-1 + x + x^2}\, dx.$

ANSWER   Set $u = -1 + x + x^2$. Then $du = (1 + 2x)\, dx$. By (6),

$$\int \frac{1 + 2x}{-1 + x + x^2}\, dx = \int \frac{du}{u} = \log |u| + C = \log |-1 + x + x^2| + C.$$

4. Evaluate $\int \dfrac{dx}{x \log x}.$

ANSWER   Set $u = \log x$. Then $du = dx/x$ and

$$\int \frac{dx}{x \log x} = \int \frac{du}{u} = \log |u| + C = \log |\log x| + C.$$

5. Evaluate $\int \tan x\, dx$ and $\int \cot x\, dx.$

ANSWER   Recall that $\tan x = \dfrac{\sin x}{\cos x}$, $\cot x = \dfrac{\cos x}{\sin x}$, and $\dfrac{d \sin x}{dx} = \cos x, \dfrac{d \cos x}{dx} = -\sin x.$ If we

put $u = \cos x$, then $du = -\sin x\, dx$, so that $\tan x\, dx = -du/u$, and the first integral to be evaluated becomes

$$\int \tan x\, dx = -\log |\cos x| + C.$$

Setting $u = \sin x$ we obtain, similarly,

$$\int \cot x\, dx = \log |\sin x| + C.$$

REMARKS   1.   We do *not* define $\log x$ for $x < 0$. (This would require so-called complex numbers.)
2.   It can be shown that $\log x$ is not an algebraic function.
3.   In some books, especially engineering texts, the natural logarithm of x is denoted by $\ln x$.

PROBLEMS

Differentiate the following functions.

1. $y = \log(3 + 2x).$
2. $y = \log |5 - 3x|.$
3. $f(x) = \log(1 + 2x^2).$
4. $f(x) = \log(1 + 3x^2 + 5x^4).$
5. $f(x) = \log |\log x|.$
6. $f(x) = \log(x + \sqrt{x^2 + 4}).$
7. $y = (\log 2x)^4.$

8. $y = \dfrac{\log 2x}{(x^2 + 1)}.$

9. $g(t) = \log \sin 3t.$
10. $g(t) = \log \cos(t^2 + 1).$
11. $g(t) = \log(t + \cos 2t).$
12. $g(t) = \log |\sec 4t + \tan 4t|.$

Evaluate the following integrals.

13. $\int \dfrac{dx}{2x + 5}.$

14. $\int_0^1 \dfrac{4x}{1 + x^2}\, dx.$

15. $\int \dfrac{(x + 2)\, dx}{x^2 + 4x + 1}.$

16. $\int_1^2 \dfrac{x^2\, dx}{9 - x^3}.$

17. $\int \dfrac{(x^2 - 2x + 2)\, dx}{1 - 6x + 3x^2 - x^3}.$

18. $\int \dfrac{x^{1/2}}{1 + x^{3/2}}\, dx.$

19. $\int \dfrac{\log x}{x}\, dx.$

[*Hint:* Set $u = \log x$.]

20. $\int \dfrac{dx}{\sqrt{x}(1 + \sqrt{x})}.$

21. $\int \dfrac{\sin x\, dx}{1 + 2\cos x}.$

22. $\int \dfrac{\sin x \cos x\, dx}{2 + \sin^2 x}.$

23. $\int \sec x\, dx.$

$\left[\textit{Hint:} \text{ Write } \sec x \text{ as } \sec x \dfrac{(\sec x + \tan x)}{(\sec x + \tan x)}. \right]$

24. $\int \csc x\, dx.$

1.2 The functional equation and its consequences

The most important property of the logarithmic function is the **functional equation**:

$$\log(xy) = \log x + \log y. \tag{7}$$

This remarkable relation was published, as a property of areas under the hyperbola, by Gregory of St. Vincent in 1647. The connection between the area under the hyperbola and Napier's work was not perceived at that time.

To prove (7), note that

$$\log(xy) = \int_1^{xy} \frac{dt}{t} = \int_1^x \frac{dt}{t} + \int_x^{xy} \frac{dt}{t}.$$

In the second integral, we make the substitution $t = xs$, so that $dt = x\, ds$, $s = 1$ for $t = x$, $s = y$ for $t = xy$; this yields

$$\int_1^x \frac{dt}{t} + \int_x^{xy} \frac{dt}{t} = \int_1^x \frac{dt}{t} + \int_1^y \frac{x\, ds}{xs} = \int_1^x \frac{dt}{t} + \int_1^y \frac{ds}{s} = \log x + \log y,$$

as asserted.

The functional equation has important consequences. For $y = (1/x)$ we obtain that $\log(x \cdot 1/x) = \log 1 = 0 = \log x + \log(1/x)$. Thus,

$$\log \frac{1}{x} = -\log x. \tag{8}$$

Also $\log(x/y) = \log(x \cdot 1/y) = \log x + \log(1/y) = \log x - \log y$, by (8). Thus

$$\log \frac{x}{y} = \log x - \log y. \tag{9}$$

Next, if n is a positive integer, then by repeated application of (7),

$$\log x^n = \log \overbrace{(x \cdot x \ldots x)}^{n \text{ times}} = \overbrace{\log x + \log x + \cdots + \log x}^{n \text{ times}}$$

or

$$\log x^n = n \log x. \tag{10}$$

The same is true if $n = 0$, since $x^0 = 1$ and $\log 1 = 0$.

GREGORY OF ST. VINCENT (1584–1667), a Jesuit priest, was born in Ghent and taught in Rome,

Prague, and Madrid, where he was a tutor at the royal court.

If n is a negative integer, then $n = -m$ where n is positive and, by (8) and (10),

$$\log x^n = \log x^{-m} = \log \frac{1}{x^m} = -\log x^m = -m \log x = n \log x. \tag{11}$$

Since $(\sqrt[n]{x})^n = x$, we have, by (10),

$$\log x = \log(\sqrt[n]{x})^n = n \log \sqrt[n]{x}$$

or

$$\log \sqrt[n]{x} = \frac{1}{n} \log x. \tag{12}$$

We conclude that, for p a positive integer and q any integer,

$$\log x^{q/p} = \log(x^{1/p})^q = q \log x^{1/p} = q \log \sqrt[p]{x} = q \left(\frac{1}{p} \log x \right) = \frac{q}{p} \log x.$$

Thus we have proved that

$$\log x^r = r \log x, \tag{13}$$

for every rational number r.

We show next that $\lim_{x \to +\infty} \log x = +\infty$, $\lim_{x \to 0^+} \log x = -\infty$. Since $\log 1 = 0$ and $\log x$ is increasing, we have

$$\log x > 0 \qquad \text{if and only if} \qquad x > 1. \tag{14}$$

Now let A be any given number. We shall show that $\log x > A$ if x is large enough. Let n be an integer such that $n > A/\log 2$. If $x > 2^n$, then $\log x > \log 2^n = n \log 2 > A$. Thus $\log x$ will become as large as we want if x becomes sufficiently large. Also, since $\log x = -\log(1/x)$, $\log x$ will be negative and as large as we want in absolute value if the positive number x is sufficiently close to 0. This proves relations (3) and (4) in §1.1.

Equation (7) and its consequences are useful in simplifying calculations that involve logarithms. We illustrate their use by means of examples.

EXAMPLES 1. Given that $\log 2 \approx .6931$ and $\log 3 \approx 1.0986$, compute $\log 6$, $\log \frac{1}{3}$, $\log 1.5$, $\log 27$, and $\log \sqrt{2}$.

ANSWER $\log 6 = \log(2 \cdot 3) = \log 2 + \log 3 \approx 1.7917$; $\log \frac{1}{3} = \log 3^{-1} = -\log 3 \approx -1.0986$; $\log 1.5 = \log \frac{3}{2} = \log 3 - \log 2 \approx .4055$; $\log 27 = \log 3^3 = 3 \log 3 \approx 3.2958$; $\log \sqrt{2} = \log 2^{1/2} = \frac{1}{2} \log 2 \approx .3465$.

2. Find the derivative of $f(x) = \log \sqrt{\dfrac{4 + x^2}{4 - x^2}}$, $|x| < 2$.

ANSWER We first rewrite the given expression as

$$f(x) = \tfrac{1}{2} \log \frac{4 + x^2}{4 - x^2} = \tfrac{1}{2} \log(4 + x^2) - \tfrac{1}{2} \log(4 - x^2).$$

Then

$$f'(x) = \frac{1}{2} \frac{1}{(4 + x^2)} (2x) - \frac{1}{2} \frac{1}{(4 - x^2)} (-2x)$$

or

$$f'(x) = \frac{8x}{(4 + x^2)(4 - x^2)} = \frac{8x}{16 - x^4}.$$

PROBLEMS

Given that $\log 2 = .6391$, $\log 3 = 1.0986$, and $\log 5 = 1.6094$ (approximately), compute the following numbers.

1. $\log 10$.
2. $\log 30$.
3. $\log \frac{6}{5}$.
4. $\log \frac{2}{3}$.
5. $\log \frac{1}{12}$.
6. $\log \sqrt{\frac{5}{3}}$.
7. $\log \sqrt[3]{4}$.
8. $\log \dfrac{1}{\sqrt{15}}$.

Find the derivatives of the following functions.

9. $y = \log \dfrac{1}{(2x + 1)}$.

10. $y = \log \sqrt{9 - x^2}$.

11. $y = \log \dfrac{x^2 - 1}{x^2 + 1}$.

12. $y = \log x^2 \sqrt{x^2 + 3}$.

13. $f(x) = \log \sqrt{\dfrac{x + 1}{x - 1}}$.

14. $f(x) = \log(x^2 + 1)^5 (1 - x)^{10}$.

15. $f(x) = \log \dfrac{(x^2 + 4)^8}{(4x^2 + 1)^6}$.

16. $f(x) = \log \dfrac{\sqrt{x + 1}}{\sqrt[3]{x^2 + x + 1}}$.

17. $g(t) = \log(\sin t \cos 2t)$.

18. $g(t) = \log(\sin^4 t \cos^2 t)$.

19. $g(t) = \log \dfrac{\tan t}{1 + \tan t}$.

20. $g(t) = \log \sqrt{\dfrac{1 + \cos t}{1 - \cos t}}$.

*21. Prove that $\log(xy) = \log x + \log y$ by studying the function $F(x) = \log(xy) - \log x - \log y$ where y is some fixed positive number (that is, a constant). [*Hint:* Show that $F'(x) = 0$ for all $x > 0$ and $F(1) = 0$.]

*22. Prove that, for every rational r, $\log x^r = r \log x$ by studying the function $f(x) = \log x^r - r \log x$.

1.3 The number e

In view of (3) and (4), and by the intermediate value theorem (Chapter 3, §3.4), the function $\log x$ takes on all values in the interval $(-\infty, +\infty)$. In other words, *for every number α there is a positive number x such that $\log x = \alpha$*; this x is unique, since the log is an increasing function.

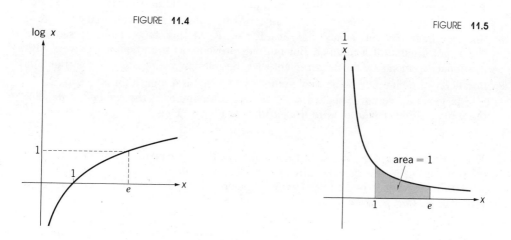

FIGURE 11.4

FIGURE 11.5

In particular, there is a number e such that

$$\log e = 1. \tag{15}$$

This means: the area under the hyperbola $xy = 1$ from 1 to e is 1 (see Figures 11.4 and 11.5). The number e is irrational, as we will prove in Chapter 14, §2.4; its decimal expansion begins thus:

$$e = 2.718282\ldots.$$

The importance of the number e, and its surprising connection with the number π, will become apparent later.

EXAMPLE   What is $\log e^7$, $\log \sqrt{e}$, $\log(1/e)$?

ANSWER   By (13) and (15) we have: $\log e^7 = 7 \log e = 7$, $\log \sqrt{e} = \frac{1}{2}$, $\log(1/e) = \log e^{-1} = -1$.

PROBLEM

1. Draw the curve $y = 1/x$ on a piece of graph paper as carefully as you can. Then determine the area under the curve (by counting spaces) from $x = 1$ to $x = 2.6$, and from $x = 1$ to $x = 2.8$. (This should convince you that there exists a number e between 2.6 and 2.8 for which the area is 1.)

§2   Irrational exponents. Logarithms to different bases

The natural logarithmic function has been defined, in §1.1, by an integral. Using this function we shall assign a meaning to the symbol a^x for all $a > 0$ and all rational and irrational x, and we shall recapture the usual definition of logarithms.

2.1   Irrational exponents

The relation

$$\log a^r = r \log a \tag{1}$$

has been proved in §1.2 for rational numbers r. Is it valid also for irrational r? For instance, is it true that $\log 2^{\sqrt{2}} = \sqrt{2} \log 2$? As it stands, this question makes no sense since we have not yet defined the meaning of a^r. At this stage of our presentation $2^{\sqrt{2}}$ is not a name of a number. But nothing prevents us from defining powers with irrational exponents by the requirement that the relation analogous to (1) be true. For instance, we define $2^{\sqrt{2}}$ to be that number whose natural logarithm is $\sqrt{2} \log 2$.

In general, we agree that, for $a > 0$ and any number x, *the symbol a^x denotes the unique number whose natural logarithm is $x \log a$.* Thus

$$\log a^x = x \log a. \tag{2}$$

This is a **definition** of a^x. If x is a rational number, then a^x can, of course, be computed differently; but in this case we already know that (2) holds.

It turns out that the familiar laws of exponents hold also for irrational exponents: *for all x, and all positive a and b,*

$$1^x = 1, \tag{3}$$

$$a^{x+y} = a^x a^y, \qquad a^{x-y} = a^x/a^y, \qquad (a^x)^y = a^{xy}, \qquad (4)$$

and

$$(ab)^x = a^x b^x. \qquad (5)$$

To prove these relations we must verify that in each case both sides of the equation have the same natural logarithm. For example, since $\log 1 = 0$ we have, by (2), that $\log 1^x = x \log 1 = x \cdot 0 = 0$. Hence $\log 1^x = \log 1$ so that (3) is true. To verify (5) we use (2) and the functional equation of the logarithmic function [see §1.2, Equation (7)]; we have

$$\log(ab)^x = x \log(ab) = x(\log a + \log b) = x \log a + x \log b$$
$$= \log a^x + \log b^x = \log(a^x b^x);$$

therefore $(ab)^x = a^x b^x$. The reader is asked to verify the three relations (4).

PROBLEMS

Verify the following relations using the properties of the natural logarithmic function. Assume $a > 0$.

1. $a^{x+y} = a^x a^y$.   2. $a^{x-y} = \dfrac{a^x}{a^y}$.   3. $(a^x)^y = a^{xy}$.

2.2 Logarithms to different bases

We recall now the usual "elementary" definition of logarithms. Let a and b be positive numbers, $b \neq 1$. The **logarithm of a to the base b** is the number y such that $a = b^y$; we write $y = \log_b a$. For example, $\log_2 8 = 3$ since $8 = 2^3$; $\log_2 16 = 4$ since $16 = 2^4$; $\log_{10} 100 = 2$ since $100 = 10^2$; $\log_3 \frac{1}{3} = -1$ since $\frac{1}{3} = 3^{-1}$. Thus

$$y = \log_b a \qquad \text{means that} \qquad a = b^y. \qquad (6)$$

But $a = b^y$ means that $\log a = y \log b$ (see §2.1). Hence $y = \log a / \log b$, or

$$\log_b a = \frac{\log a}{\log b}. \qquad (7)$$

If $b = e$, then $\log b = 1$ (see §1.3); we conclude that

$$\log_e a = \log a = \int_1^a \frac{dt}{t}; \qquad (8)$$

natural logarithms are logarithms to the base e. Thus

$$\log e^y = y \qquad \text{and} \qquad e^{\log a} = a \qquad (9)$$

for all y and all $a > 0$.

The properties of logarithms, to any base, are similar to those of natural logarithms:

$$\log_b 1 = 0, \qquad (10)$$

$$\log_b xy = \log_b x + \log_b y, \qquad \log_b \frac{x}{y} = \log_b x - \log_b y, \qquad (11)$$

$$\log_b a^x = x \log_b a, \qquad \log_a a^x = x. \qquad (12)$$

The above results follow from (7). For example, to obtain the first relation (11), use (7) above and Equation (7) in §1.2. We have

$$\log_b xy = \frac{\log xy}{\log b} = \frac{\log x + \log y}{\log b} = \frac{\log x}{\log b} + \frac{\log y}{\log b} = \log_b x + \log_b y.$$

(The remaining relations are given as problems; see 15, 16, 17, and 18.) Relations (9), (10), and (11) can also be obtained from the definition (6) and the laws of exponents. The function $x \mapsto \log_b x$ has the derivative

$$\frac{d \log_b x}{dx} = \frac{1}{x \log b}. \qquad (13)$$

To prove this, use (7) and relation (1) in §1.1. We have

$$\frac{d \log_b x}{dx} = \frac{d(\log x/\log b)}{dx} = \frac{1}{\log b} \frac{d \log x}{dx} = \frac{1}{x \log b},$$

since b is constant.

REMARK   The usual definition (7) of logarithms: the logarithm of a to base b is the power to which b must be raised to obtain a, is due to Euler. Why did we not begin our discussion with this definition? Because we would have had to define the meaning of a^x for x irrational, and we would have had to prove that the equation $a^x = u$ has, for given a and u, a unique solution. Both steps could be accomplished without calculus, but it would take much longer.

EXAMPLE   Find the value of the derivative of

$$f(x) = \log_{10}(1 + x^2) \qquad \text{at } x = 1.$$

ANSWER   By (12) and the chain rule (setting $u = 1 + x^2$)

$$f'(x) = \frac{d \log_{10}(1 + x^2)}{dx} = \frac{d \log_{10} u}{du} \frac{du}{dx} = \frac{1}{\log 10} \frac{1}{u} \frac{du}{dx} = \frac{1}{\log 10} \frac{1}{1 + x^2}.$$

For $x = 1$, $f'(1) = 1/\log 10 = 1/2.3026 \approx .4343$.

PROBLEMS

In Problems 1 to 6 evaluate the given logarithms.

1. $\log_2 16$.
2. $\log_4 64$.
3. $\log_{16} 4$.
4. $\log_2 \frac{1}{32}$.
5. $\log_4 \frac{1}{16}$.
6. $\log_{27} 3$.

7. What is $2^{4 \log_2 3}$?
8. What is $\log_{z^2} z^{12}$? Assume $z > 0$.
9. What is $\log_{v^2+1}(v^4 + 2v^2 + 1)$?
10. For what values of x is $\log_x(x + 6) = 2$?

LEONHARD EULER (1707–1789) was born in Basel, Switzerland, and studied there under Johann Bernoulli. He spent most of his life as a member of the Academies in Berlin and in St. Petersburg (now Leningrad). Euler was not only the most productive mathematician of all time but one of the best; he wrote on all mathematical subjects and on some nonmathematical ones as well. During the last 13 years of his life, he worked in total blindness.

Euler also wrote textbooks of calculus of which all subsequent texts are derivatives. Euler's collected works consist of over 70 large volumes.

In Problems 11 to 14 find the values of the derivatives of the given functions at the indicated points.

11. $f(x) = \log_{10}(x^2 + 6x + 1)$, $x = 1$.

12. $f(x) = \log_{10} \sqrt{9 - x^2}$, $x = 2$.

13. $f(x) = \log_2(x^2 + x + 2)^4$, $x = 2$.

14. $f(x) = \log_2 \sqrt{\dfrac{2x - 1}{2x + 1}}$, $x = 1$.

Prove the following relations.

15. $\log_b 1 = 0$.

16. $\log_b \dfrac{x}{y} = \log_b x - \log_b y$.

17. $\log_b a^x = x \log_b a$.

18. $\log_a a^x = x$.

2.3 Arithmetic calculations with logarithms

A table of logarithms to some base b can be used to compute products, quotients, and fractional powers. To find xy (or x/y, or x^r), we find $\alpha = \log_b x$, $\beta = \log_b y$, and then the number whose logarithm to base b is $\alpha + \beta$ (or $\alpha - \beta$, or $r\alpha$).

Of course, if we work with tables, the result will be only approximate. For instance, given a number γ, we shall, in general, not find in the tables a number z with $\log_b z = \gamma$, but only a number z such that $\log_b z$ is close to γ. The more extensive the tables, though, the more accurate the results. Accuracy can be also improved by using interpolation.

The number $\log_{10} x$ is called the **common logarithm** of x. Most tables, beginning with those computed by Napier and Briggs, list common logarithms. (Authors who denote natural logarithms by ln usually reserve the notation log for common logarithms.) Common logarithms are convenient to use, since $1 = 10^0$, $10 = 10^1$, $100 = 10^2$, $1000 = 10^3$, $.1 = 10^{-1}$, $.01 = 10^{-2}$, and so forth. Hence we know, without any tables, that $\log_{10} 1 = 0$, $\log_{10} 10 = 1$, $\log_{10} 100 = 2$, $\log_{10} 1000 = 3$, $\log_{10}(.1) = -1$, $\log_{10}(.01) = -2$, and so on. Thus, from the table at the end of the book, we can read off that $\log_{10} 224 = 2.3502$, $\log_{10} 22.4 = 1.3502$, $\log_{10} 2.24 = .3502$, $\log_{10} .224 = .3502 - 1$, and so on.

REMARK   Before logarithms had been invented, mathematicians performed multiplications using trigonometric tables and the identity

$$2 \cos \alpha \cos \beta = \cos(\alpha + \beta) + \cos(\alpha - \beta),$$

which follows from the addition theorem for the cosine.

Suppose we want, using this identity, to multiply two numbers a and b; we assume that $0 < a < 1$ and $0 < b < 1$. From trigonometric tables, we find numbers (or "angles") α and β such that $a = \cos \alpha$ and $b = \cos \beta$. Then we compute $\alpha + \beta$ and $\alpha - \beta$, look up in the tables the values of $\cos(\alpha + \beta)$ and $\cos(\alpha - \beta)$, add these values, and take one-half of the sum. This is the desired product $ab = \cos \alpha \cos \beta$.

PROBLEMS‡

In Problems 1 to 10 perform the indicated calculations by using logarithms to base 10.

1. 10.8×2.91.

2. $168 \times .0157$.

3. $1.24 \times .637 \times .895$.

4. $24.8 \div .0366$.

5. $\dfrac{.143 \times 76.2}{2380}$.

‡Optional problems.

6. $\sqrt{117}$.
7. $\sqrt[3]{22500}$.

8. $(.326)^5$.
9. $(1.04)^{10}$.

10. $\sqrt[4]{\dfrac{.0463}{.516}}$.

11. Multiply .90475 by .75836 using cosines.

12. Find the square of .82534 using cosines.

§3 Exponential functions

The function $f(x) = e^x$ is one of the most important functions in mathematics. We shall describe here its properties, and in the next chapter we shall discuss some of its applications.

3.1 The function e^x and its derivative

Having defined powers with arbitrary (rational or irrational) exponents, we can study functions such as $x \mapsto a^x$, where a is any fixed positive number. First we investigate the so-called **exponential function** $x \mapsto e^x$. Since [see §2.2, Equations (9)]

$$\log e^x = x \qquad \text{and} \qquad e^{\log y} = y, \tag{1}$$

for all x and all $y > 0$, the exponential function is *inverse* to the logarithm function $y \mapsto \log y$ (see Chapter 6, §1). The graph of $x \mapsto e^x$ (obtained by "flipping" the graph in Figure 11.4) is shown in Figure 11.6. We note that

$$e^x > 0, \qquad \text{for all } x. \tag{2}$$

FIGURE **11.6**

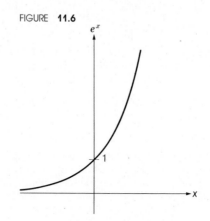

To differentiate $y = e^x$ we use Theorem 2 in Chapter 6, §1.2, and obtain

$$\frac{de^x}{dx} = \frac{dy}{dx} = 1 \bigg/ \frac{dx}{dy} = 1 \bigg/ \frac{d\log y}{dy} = \frac{1}{1/y} = y = e^x$$

or

$$\frac{de^x}{dx} = e^x; \tag{3}$$

the exponential function is its own derivative! If $u(x)$ is a differentiable function, the chain rule and (3) imply that

$$\frac{de^{u(x)}}{dx} = \frac{de^u}{du}\frac{du}{dx} = e^{u(x)}u'(x). \tag{3a}$$

Thus $e^{u(x)}$ is a primitive function of $e^{u(x)}u'(x)$:

$$\int e^{u(x)}u'(x)\,dx = e^{u(x)} + C. \tag{3b}$$

Using the language of differentials (see Chapter 8, §3.3), we write (3a) and (3b) as

$$de^u = e^u\,du, \qquad \int e^u\,du = e^u + C. \tag{4}$$

A particular case is worth noting: if $u = \alpha x$, α a constant, then $du = \alpha\,dx$ and

$$\frac{de^{\alpha x}}{dx} = \alpha e^{\alpha x}. \tag{5}$$

EXAMPLES   1.   Differentiate e^{x^2}.
ANSWER Set $u = x^2$. Then $du = 2x\,dx$ and, by (4),

$$\frac{de^{x^2}}{dx} = \frac{de^u}{du}\frac{du}{dx} = e^u 2x = 2xe^{x^2}.$$

2. Evaluate $\displaystyle\int_0^2 xe^{x^2}\,dx$.
ANSWER With $u = x^2$ we find that $x\,dx = \frac{1}{2}\,du$, $u = 0$ for $x = 0$ and $u = 4$ for $x = 2$. Thus by (4)

$$\int_0^2 xe^{x^2}\,dx = \frac{1}{2}\int_0^2 2xe^{x^2}\,dx = \frac{1}{2}\int_0^4 e^u\,du = \frac{1}{2}e^u\,\bigg|_0^4$$

$$= \frac{1}{2}(e^4 - 1) \approx 26.8.$$

3. Evaluate $\displaystyle\int \frac{e^{-2x}\,dx}{(1 + 4e^{-2x})}$.
ANSWER Set $u = 1 + 4e^{-2x}$. Then $du = -8e^{-2x}\,dx$ and

$$\int \frac{e^{-2x}\,dx}{(1 + 4e^{-2x})} = -\frac{1}{8}\int \frac{du}{u} = -\frac{1}{8}\log u + C = -\frac{1}{8}\log(1 + 4e^{-2x}) + C.$$

PROBLEMS

Find the derivatives of the following functions.

1. $y = e^{3x}$.

2. $y = 2e^{-4x}$.

3. $y = \frac{1}{2}e^{(x^2 - 4x + 3)}$.

4. $f(x) = e^{\sqrt{x^2 - 4}}$.

5. $f(x) = x^2 e^{2x}$.

6. $f(x) = \dfrac{e^{-2x}}{x^2}$.

7. $g(t) = e^{\sin 2t}$.

8. $g(t) = e^{-2\tan t}$.

9. $g(t) = e^{\arctan 4t}$.

10. $g(t) = 4\sin e^{t/2}$.

11. $g(t) = \arctan e^t$.

12. $y = \log(e^x + e^{-x})$.

13. $y = \log\sqrt{e^x + 1}$.

14. $y = e^{\sqrt{\log x}}$.

15. $f(x) = e^{\cos^2(2x+1)}$.

16. $f(x) = \dfrac{e^x - e^{-x}}{e^x + e^{-x}}$.

17. $f(x) = \dfrac{e^{2x}}{\sqrt{e^{2x} - 1}}$.

18. $f(x) = \log\sqrt{\dfrac{e^x + 1}{e^x - 1}}$.

Find the following integrals.

19. $\displaystyle\int_0^{1/2} e^{-2x}\,dx.$

20. $\displaystyle\int \frac{(e^x - e^{-x})}{2}\,dx.$

21. $\displaystyle\int \frac{e^x\,dx}{(e^x + 1)}.$

22. $\displaystyle\int \frac{x\,dx}{e^{x^2}}.$

23. $\displaystyle\int_1^2 (2x^3 - 5x)e^{(x^4 - 5x^2 + 4)}\,dx.$

24. $\displaystyle\int \frac{e^{4x}\,dx}{(2e^{4x} + 1)}.$

25. $\displaystyle\int e^x(1 + 3e^x)^6\,dx.$

26. $\displaystyle\int \frac{e^{-x}\,dx}{(e^{-x} + 1)^4}.$

27. $\displaystyle\int_0^1 \frac{e^{2t} - 1}{e^{2t} + 1}\,dt.$ $\left[\begin{array}{l}\textit{Hint: Multiply numerator} \\ \text{and denominator by } e^{-t}.\end{array}\right]$

28. $\displaystyle\int e^{-\sin 2t} \cos 2t\,dt.$

29. $\displaystyle\int \sec^2 t\, e^{4\tan t}\,dt.$

30. $\displaystyle\int_0^{\pi/4} \frac{e^{\arctan t}}{(1 + t^2)}\,dt.$

31. $\displaystyle\int e^t \cos e^t\,dt.$

32. $\displaystyle\int \frac{e^t \log(1 + e^t)}{(1 + e^t)}\,dt.$

33. $\displaystyle\int \sqrt{e^t}\,dt.$

34. $\displaystyle\int_1^4 \frac{e^{\sqrt{t}}}{\sqrt{t}}\,dt.$

In Problems 35 to 42 find all maximum, minimum, and inflection points of the given functions, and sketch their graphs.

35. $y = e^{-x}.$
36. $y = 1 - e^{x/2}.$
37. $y = \frac{1}{2}(e^x + e^{-x}).$
38. $y = \frac{1}{2}(e^x - e^{-x}).$
39. $y = xe^{-x}.$
40. $y = x^2 e^{-2x}.$
41. $y = e^{-x^2}.$
42. $y = e^{1/x}, \quad x \neq 0.$

43. Verify that $y = (a + bx)e^x$, where a and b are arbitrary numbers, satisfies the differential equation $y'' - 2y' + y = 0$.
44. Verify that $y = e^x(a \sin x + b \cos x)$, where a and b are arbitrary numbers, satisfies the differential equation $y'' - 2y' + 2y = 0$.
45. Verify that $y = e^{1/x}$ is a solution of the differential equation $x^3 y'' + xy' - 2y = 0$.

3.2 The function a^x

Let a be a positive number. In order to investigate the function a^x, we set $\alpha = \log a$. Then $a = e^\alpha$ and $a^x = (e^\alpha)^x = e^{\alpha x}$; thus

$$a^x = e^{(\log a)x}. \tag{6}$$

By (5), with $\alpha = \log a$, we obtain

$$\frac{da^x}{dx} = (\log a)a^x. \tag{7}$$

Just as (3) led to (4), (7) leads to the relations

$$da^u = (\log a)a^u\,du, \qquad \int a^u\,du = \frac{1}{\log a}a^u + C. \tag{8}$$

But there is no great need to remember these formulas. It is preferable to use (6) and always replace a^x by $e^{(\log a)x}$.

EXAMPLES 1. Find the derivative of $f(x) = 10^x$ at $x = 2$.
ANSWER   By (7), $d(10^x)/dx = (\log 10)10^x$. For $x = 2$, $f'(2) = (\log 10)100 \approx (2.3026)100 \approx 230$.

2. Differentiate $f(x) = 37^{\cos x}$.
ANSWER   Since $f(x) = e^{(\log 37)\cos x}$,

$$f'(x) = e^{(\log 37)\cos x}(\log 37)(-\sin x) = -37^{\cos x}(\log 37)\sin x.$$

PROBLEMS

Find the derivatives of the following functions.

1. $y = 4^{2x}$.

2. $y = 2^{\sqrt{x}}$.

3. $y = 3^{1/x}$.

4. $y = 10^{(2x^2 - x + 1)}$.

5. $f(x) = 10^{\sqrt{x^2 - 4}}$.

6. $f(x) = 6^x 4^{-x}$.

7. $f(x) = 2^{\sin x/2}$.

8. $f(x) = 2^{\arcsin 4x}$.

Evaluate the following integrals.

9. $\displaystyle\int 16^x \, dx$.

10. $\displaystyle\int_0^1 4^{2x} \, dx$.

11. $\displaystyle\int (x^2 + 1)10^{(x^3 + 3x + 1)} \, dx$.

12. $\displaystyle\int_{-2}^{-1} 10^{-x} \, dx$.

13. $\displaystyle\int 2^x e^x \, dx$.

14. $\displaystyle\int 5^x 2^{-2x} \, dx$.

15. $\displaystyle\int 4^{\sin x} \cos x \, dx$.

16. $\displaystyle\int_0^{\pi/4} \frac{4^{\arctan x}}{1 + x^2} \, dx$.

3.3 Logarithmic differentiation

If $f(x)$ is a positive differentiable function, then, by the chain rule [and setting $u = f(x)$],

$$\frac{d \log f(x)}{dx} = \frac{d \log u}{du}\frac{du}{dx} = \frac{1}{u}\frac{du}{dx} = \frac{1}{f(x)}\frac{df(x)}{dx}.$$

Thus

$$f'(x) = f(x)\frac{d \log f(x)}{dx} \tag{8a}$$

or

$$df = f\, d \log f. \tag{8b}$$

The use of this formula in finding derivatives is called logarithmic differentiation.

EXAMPLES 1. Differentiate x^α, where α is any number (rational or irrational).
SOLUTION   Note that $\log x^\alpha = \alpha \log x$; hence $d \log x^\alpha = \alpha\, d \log x = (\alpha/x)\, dx$. By (8), $dx^\alpha = x^\alpha(\alpha/x)\, dx = \alpha x^{\alpha - 1}\, dx$. Thus

$$\frac{dx^\alpha}{dx} = \alpha x^{\alpha - 1}. \tag{9}$$

Of course, we already know this for rational α, but now we have proved it for irrational exponents as well!

2. Differentiate the function $f(x) = x^x (x > 0)$.
SOLUTION   Since $\log(x^x) = x \log x$, (8) yields $dx^x = x^x d \log x^x = x^x d(x \log x) = x^x(\log x\, dx + x\, d \log x) = x^x(\log x\, dx + dx) = (x^x \log x + x^x)\, dx$. Hence

$$\frac{dx^x}{dx} = x^x(1 + \log x). \tag{10}$$

3. Find $\dfrac{dy}{dx}$ if $y = \dfrac{(x^2 + 2)^5(x^3 + 1)^{1/2}}{(2x + 1)^{4/3}}$.

SOLUTION We can find the derivative by using the product formula and the chain rule; however, the algebra involved is simpler if we use logarithmic differentiation. Taking logs,

$$\log y = 5\log(x^2 + 2) + \tfrac{1}{2}\log(x^3 + 1) - \tfrac{4}{3}\log(2x + 1).$$

Hence

$$\frac{1}{y}\frac{dy}{dx} = 5\frac{2x}{(x^2 + 2)} + \frac{3x^2}{2(x^3 + 1)} - \frac{4(2)}{3(2x + 1)}$$

or

$$\frac{dy}{dx} = \frac{(x^2 + 2)^5(x^3 + 1)^{1/2}}{(2x + 1)^{4/3}}\left[\frac{10x}{(x^2 + 2)} + \frac{3x^2}{2(x^3 + 1)} - \frac{8}{3(2x + 1)}\right].$$

PROBLEMS

Differentiate the following functions.

1. $y = x^{-2x}$.

2. $y = x^{1/x}$.

3. $y = x^{\sqrt{x}}$.

4. $y = (1 + x^2)^{2x}$.

5. $y = \dfrac{1}{(1 + e^x)^x}$.

6. $y = (x^2 + 4)^6(x^2 - 1)^4$.

7. $y = \dfrac{(2x + 1)^8}{(x + 4)^2(x + 1)^4}$.

8. $y = \sqrt{\dfrac{x^2 + 2x + 4}{x^2 - 4x + 1}}$.

9. $y = \sqrt{(x - 1)(x - 2)(x - 3)}$.

10. $y = \dfrac{(x^2 + 1)^{2/3}(2x - 1)^{4/3}}{(x^2 + 4)^{1/3}}$.

3.4 The growth of the exponential function

As x gets large the exponential function $y = e^x$ increases very rapidly. For instance, e^2 is slightly larger than 7, and e^{10} exceeds 22,000. If $a > 0$ and we consider a sequence of values for x,

$$x = 0, \qquad x = a, \qquad x = 2a, \qquad x = 3a, \ldots$$

(we call such a sequence an arithmetic progression), the corresponding values of $y = e^x$ are

$$y = 1, \qquad y = q, \qquad y = q^2, \qquad y = q^3, \ldots \qquad \text{with } q = e^a > 1$$

(such a sequence is called a geometric progression). People long ago noted how fast the terms of a geometric progression with $q > 1$ increase. (Recall the oriental legend about the reward requested by the inventor of chess: 1 grain of wheat on the first square of the board, 2 on the second, 4 on the third, 8 on the fourth, and so on, up to 2^{63} on the last square.)

Theorem 1. *If α is any number, then*

$$\lim_{x \to +\infty} \frac{e^x}{x^\alpha} = +\infty. \tag{11}$$

We express this by saying that "e^x goes to infinity faster than any power of x, as x approaches $+\infty$." The theorem is of interest only for $\alpha > 0$.

If $\alpha > 0$, then $\lim_{x \to +\infty} x^\alpha = +\infty$. Hence the denominator in (11) will be very large for large positive x. The theorem asserts that e^x will eventually be even larger. The word "eventually" is essential. For instance, if we choose $\alpha = 10^{10}/\log 10$, then $x^\alpha > e^x$ for $x = 10^{10}$. Indeed, we have $(10^{10})^\alpha = 10^{10\alpha} = e^{10\alpha \log 10} = e^{10^{11}} > e^{10^{10}}$. But even for this tremendous α, the number e^x will eventually, as x grows, become much greater than x^α.

Proof of Theorem 1.   Consider the function $\phi(t) = e^t - 1 - t$. Since $\phi'(t) = e^t - 1$ and therefore $\phi'(t) > 0$ for $t > 0$, we have that $\phi(t)$ is an increasing function of t for $t > 0$. Therefore $\phi(t) > \phi(0) = 0$ so that $e^t > 1 + t$, and

$$\frac{e^t}{t} > 1 + \frac{1}{t} > 1, \qquad \text{for } t > 0. \tag{12}$$

Let a number $\alpha > 0$ be given. We choose an integer $N > \alpha$ and apply (12) to $t = x/N$. We obtain, for $x > 0$,

$$\frac{e^{x/N}}{x/N} > 1 \qquad \text{or} \qquad \frac{e^{x/N}}{x} > \frac{1}{N}$$

so that (raising each side of the last inequality to the Nth power)

$$\frac{e^x}{x^N} > \frac{1}{N^N} \qquad \text{and} \qquad \frac{e^x}{x^\alpha} = \frac{e^x}{x^N} x^{N-\alpha} > \frac{x^{N-\alpha}}{N^N}.$$

Since $\lim_{x \to \infty} x^{N-\alpha} = +\infty$, assertion (11) follows.

We record a consequence of Theorem 1.

$$\lim_{x \to +\infty} \frac{x^\alpha}{\log x} = +\infty, \qquad \text{for every } \alpha > 0. \tag{13}$$

We express this by saying: "the function $\log x$ goes to infinity slower than any power of x, as x approaches $+\infty$."

Proof.   Set $\log(x^\alpha) = \alpha \log x = y$. Then $y \to +\infty$ for $x \to +\infty$. Also,

$$\frac{x^\alpha}{\log x} = \frac{x^\alpha}{(1/\alpha)\log(x^\alpha)} = \alpha \frac{e^y}{y} \to +\infty, \qquad \text{for } y \to +\infty$$

by (11), since $\alpha > 0$.

EXAMPLES 1.   What is $\lim_{x \to +\infty} x^\alpha e^{-x}$?
ANSWER   Since $x^\alpha e^{-x} = (e^x/x^\alpha)^{-1}$, and (e^x/x^α) becomes as large as we want for $x \to +\infty$, the desired limit is 0. Hence

$$\lim_{x \to +\infty} x^\alpha e^{-x} = 0. \tag{14}$$

2.   What is $\lim_{x \to +\infty} \log x/x^\alpha$, for $\alpha > 0$?
ANSWER   Since $\log x/x^\alpha = (x^\alpha/\log x)^{-1}$, and $x^\alpha/\log x$ becomes as large as we want for $x \to +\infty$ provided $\alpha > 0$, the desired limit is 0. Therefore

$$\lim_{x \to +\infty} \frac{\log x}{x^\alpha} = 0, \qquad \alpha > 0. \tag{15}$$

3.   What is $\lim_{x \to +\infty} e^{1/x}$?
ANSWER   When x is large, $1/x$ is small, as small as we like for large enough x. Hence the limit is $e^0 = 1$.

PROBLEMS

Evaluate the following limits.

1. $\lim\limits_{x \to +\infty} \dfrac{e^x}{\sqrt{x}}$.

2. $\lim\limits_{x \to +\infty} \dfrac{x^2}{\log x}$.

3. $\lim\limits_{x \to +\infty} x^{3/2} e^{-x}$.

4. $\lim\limits_{x \to +\infty} x^4 e^{-x^2}$. [*Hint:* Set $u = x^2$.]

5. $\lim\limits_{x \to +\infty} x^3 e^{-\sqrt{x}}$.

6. $\lim\limits_{x \to +\infty} \dfrac{\log x}{\sqrt{x}}$.

7. $\lim\limits_{x \to +\infty} e^{1/x^2}$.

8. $\lim\limits_{x \to 0^+} e^{1/x}$.

9. $\lim\limits_{x \to 0^-} e^{1/x}$.

10. $\lim\limits_{x \to 0^+} x^\alpha \log x$. $\left[\text{*Hint:* Set } u = \dfrac{1}{x}. \right]$

11. $\lim\limits_{x \to 0^+} x^x$. [*Hint:* Write $x^x = e^{x \log x}$.]

12. $\lim\limits_{x \to +\infty} x^{1/x}$.

3.5 Hyperbolic functions‡

Starting with the exponential functions e^x and e^{-x}, we can form certain functions, called hyperbolic, which exhibit surprising similarities with trigonometric functions.

The basic hyperbolic functions are the **hyperbolic sine** and the **hyperbolic cosine** defined by

$$\sinh x = \frac{e^x - e^{-x}}{2}, \qquad \cosh x = \frac{e^x + e^{-x}}{2}, \tag{16}$$

FIGURE **11.7**

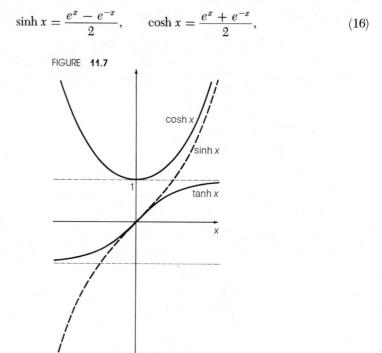

‡This subsection may be omitted without loss of continuity.

and the **hyperbolic tangent** defined by

$$\tanh x = \frac{\sinh x}{\cosh x} = \frac{e^x - e^{-x}}{e^x + e^{-x}}. \tag{17}$$

These functions are graphed in Figure 11.7. We also set $\operatorname{sech} x = 1/\cosh x$, $\operatorname{csch} x = 1/\sinh x$, $\coth x = 1/\tanh x$.

Note that

$$\cosh^2 x - \sinh^2 x = 1, \tag{18}$$

$$d \sinh u = \cosh u \, du, \qquad d \cosh u = \sinh u \, du, \tag{19}$$

$$\int \sinh u \, du = \cosh u + C, \qquad \int \cosh u \, du = \sinh u + C. \tag{20}$$

These (and other) formulas for hyperbolic functions follow from the definitions (16), (17) and the properties of exponential functions.

PROBLEMS

1. Show that $\sinh 0 = 0$, $\cosh 0 = 1$, and $\tanh 0 = 0$.

   In Problems 2 to 10 establish the given identities.

2. $\sinh(-x) = -\sinh x$.
3. $\cosh(-x) = \cosh x$.
4. $\cosh^2 x - \sinh^2 x = 1$.
5. $\tanh^2 x + \operatorname{sech}^2 x = 1$.
6. $\coth^2 x - \operatorname{csch}^2 x = 1$.
7. $\sinh(x + y) = \sinh x \cosh y + \cosh x \sinh y$.
8. $\cosh(x + y) = \cosh x \cosh y + \sinh x \sinh y$.

9. $\tanh(x + y) = \dfrac{\tanh x + \tanh y}{1 + \tanh x \tanh y}$.

10. $(\cosh x + \sinh x)^\alpha = \cosh \alpha x + \sinh \alpha x$.

    In Problems 11 to 20 evaluate the given limits.

11. $\lim\limits_{x \to +\infty} \dfrac{\sinh x}{\frac{1}{2}e^x}$.

12. $\lim\limits_{x \to -\infty} \dfrac{\sinh x}{\frac{1}{2}e^{-x}}$.

13. $\lim\limits_{x \to +\infty} \dfrac{\cosh x}{\frac{1}{2}e^x}$.

14. $\lim\limits_{x \to -\infty} \dfrac{\cosh x}{\frac{1}{2}e^{-x}}$.

15. $\lim\limits_{x \to +\infty} \tanh x$.

16. $\lim\limits_{x \to -\infty} \tanh x$.

17. $\lim\limits_{x \to \pm\infty} \operatorname{sech} x$.

18. $\lim\limits_{x \to 0^-} \operatorname{csch} x$.

19. $\lim\limits_{x \to 0^-} \coth x$.

20. $\lim\limits_{x \to 0^+} \coth x$.

21. Sketch the graph of $y = \operatorname{sech} x$.
22. Sketch the graph of $y = \operatorname{csch} x$.
23. Sketch the graph of $y = \coth x$.

In Problems 24 to 29 establish the given differentiation rules.

24. $\dfrac{d}{dx} \sinh x = \cosh x$.

25. $\dfrac{d}{dx} \cosh x = \sinh x$.

26. $\dfrac{d}{dx} \tanh x = \operatorname{sech}^2 x$.

27. $\dfrac{d}{dx} \coth x = -\text{csch}^2 x.$ 29. $\dfrac{d}{dx} \text{csch } x = -\text{csch } x \coth x.$

28. $\dfrac{d}{dx} \text{sech } x = -\text{sech } x \tanh x.$

30. Write out the corresponding integration formulas for the differentiation rules given in Problems 26 to 29.

Differentiate the functions given in Problems 31 to 40.

31. $y = \cosh 4x.$ 34. $y = \cosh \sqrt{x^2 + 4}.$ 38. $f(x) = \coth \dfrac{1}{x}.$

32. $y = \sinh(2x + 1).$ 35. $f(x) = \text{sech}^4 2x.$

36. $f(x) = \sinh^4(x^2 + 1).$ 39. $g(t) = \log(\text{sech } 4t).$

33. $y = \tanh \dfrac{x}{2}.$ 37. $f(x) = \text{csch } \sqrt{x}.$

40. $g(t) = \dfrac{\tanh t}{1 + \tanh t}.$

Evaluate the following integrals.

41. $\displaystyle\int \sinh(3x + 2)\, dx.$ 47. $\displaystyle\int \cosh^2 2z \sinh 2z\, dz.$

42. $\displaystyle\int \text{sech } 4x \tanh 4x\, dx.$ 48. $\displaystyle\int \dfrac{\cosh 4z}{\sinh^3 4z}\, dz.$

43. $\displaystyle\int \text{csch}^2(1 - 6x)\, dx.$ 49. $\displaystyle\int \tanh z\, dz.$

44. $\displaystyle\int x \cosh(4x^2 + 1)\, dx.$ $\left[\text{\textit{Hint}: Set } \tanh z = \dfrac{\sinh z}{\cosh z}.\right]$

45. $\displaystyle\int \dfrac{1}{x^2} \text{sech}^2 \dfrac{1}{x}\, dx.$ 50. $\displaystyle\int \sinh^2 z\, dz.$

46. $\displaystyle\int \dfrac{\sinh \sqrt{z}}{\sqrt{z}}\, dz.$ [*Hint*: Use the identities in Problems 4 and 8.]

APPLICATIONS OF
TRANSCENDENTAL FUNCTIONS

§1 Exponential growth and decay

In this chapter (which may be omitted without loss of continuity) we study various applications of the transcendental functions introduced in Chapters 10 and 11. Most of these involve the differential equations for the functions in question.

1.1 A first-order differential equation

The exponential function e^x is equal to its own derivative. This is its most essential property. Most applications of the exponential function are based on the following proposition:

Theorem 1. *Let A and α be numbers. The function*

$$f(x) = Ae^{\alpha x} \tag{1}$$

is a solution of the differential equation

$$f'(x) = \alpha f(x) \qquad \text{(for all } x) \tag{2}$$

satisfying the "initial condition"

$$f(0) = A; \tag{3}$$

it is the only function satisfying (2) and (3).

Proof.   The first statement is verified by a direct calculation. Assume now that f satisfies (2) and (3). Define $g(x) = f(x)e^{-\alpha x}$. Then $g'(x) = f'(x)e^{-\alpha x} - \alpha f(x)e^{-\alpha x} = e^{-\alpha x}(f'(x) - \alpha f(x)) = 0$ for all x. Hence g is constant. Since $g(x) = g(0) = f(0) = A$, this constant is A. Relation (1) follows.

For $A = \alpha = 1$ we obtain the statement: the function $x \mapsto e^x$ is the only function which is equal to its own derivative and takes on the value 1 at 0.

REMARK   The differential equation (2) is said to be of first order, since it involves the first derivative of the function f, and no higher derivatives.

EXAMPLE   If $f'(t) = -2f(t)$ for all t and $f(3) = e$, what is $f(0)$? Find a formula for $f(t)$.
ANSWER   By Theorem 1, $f(t) = Ae^{-2t}$ for some A. Now $f(3) = Ae^{-6} = e$. Hence $A = e^7$ and $f(t) = e^{7-2t}$ for all t.

PROBLEMS

1. Find a function $g(t)$ such that $g'(t) = g(t)$ and $g(0) = -2$.
2. Find a function $f(x)$ such that $f'(x) = \frac{1}{2}f(x)$ and $f(0) = 4$.
3. Find a function $g'(t)$ such that $g'(t) + \frac{1}{2}g(t) = 0$ and $g(0) = 6$. What is $g(2)$?
4. Find a function $f(x)$ such that $f'(x) = 4f(x)$ and $f(1) = e$.
5. Find a function $f(x)$ such that $3f'(x) = -2f(x)$ and $f(3) = e$. What is $f(0)$?
6. Find a function $f(x)$ such that $f'(x)/f(x) = \frac{2}{5}$ and $f'(1) = e$.

In Problems 7 to 10 assume that a solution of the form $f(x) = e^{g(x)}$ exists, and try to determine $g(x)$. Then verify that $e^{g(x)}$ actually is a solution. Be sure to include a constant of integration in your computations.

7. Find a function $f(x)$ such that $f'(x) = 2xf(x)$ and $f(0) = 1$.
8. Find a function $f(x)$ such that $f'(x) = 3x^2f(x)$ and $f(0) = 2$.
9. Find a function $f(x)$ such that $2\sqrt{x}f'(x) = f(x)$ and $f(4) = 1$.
10. Find a function $f(x)$ such that $x^2f'(x) = -f(x)$ and $\lim_{x\to\infty} f(x) = e$.

1.2   Continuously compounded interest

The discovery of the exponential function was anticipated in a question asked by Jacob Bernoulli: suppose a borrower pays a lender a certain annual interest, but compounds the interest *continuously*; what will he owe after one year for each unit lent?

Assume that the interest rate is k percent, and set $\alpha = k/100$. If the interest is computed at the end of the year, the borrower owes $1 + \alpha$ times the original amount. We assume, for the sake of simplicity, that this amount was unity. If the interest is compounded semiannually, then the borrower owes $1 + \frac{1}{2}\alpha$ after half a year, and therefore

---

BERNOULLI. This Basel (Switzerland) family produced several outstanding mathematicians in three generations. The most famous were Jacob (1654–1705), his brother, Johann (1667–1748), and Johann's son Daniel (1700–1784). The two older Bernoullis were the first scholars to understand, use, and advance Leibniz's calculus. Jacob Bernoulli is also the author of the first treatise on the mathematical theory of probability. Daniel is remembered primarily for his work in fluid mechanics. Daniel's two brothers were also mathematicians.

The Bernoullis maintained close scientific contacts with each other and occasionally quarrelled bitterly about priorities.

$$\left(1 + \frac{\alpha}{2}\right)\left(1 + \frac{\alpha}{2}\right) = \left(1 + \frac{\alpha}{2}\right)^2$$

after one year. If the interest is compounded quarterly, the amount owed increases by a factor of $1 + \frac{1}{4}\alpha$ each quarter, and it increases by the factor $(1 + \frac{1}{4}\alpha)^4$ after a year. Compounding monthly yields $(1 + \frac{1}{12}\alpha)^{12}$. In general, if we divide the year into n equal periods and compound the interest at the end of each period, a debt of 1 will grow during the year into

$$\left(1 + \frac{\alpha}{n}\right)^n. \tag{4}$$

We have the feeling that, for a very large n, the number (4) will be close to the true answer, that is, to the result of the continual compounding of interest.

We now look at the problem from the point of view of calculus. If the interest is compounded continuously, the amount owed will be an increasing function of the time t elapsed; call this function $f(t)$. (Assume that we measure time in years.) Since the debt was 1 at $t = 0$, we have

$$f(0) = 1. \tag{5}$$

We assume f to be differentiable and try to compute $f'(t)$. Let h be a small positive number. Consider some instant t_0 and a very short time interval from t_0 to $t_0 + h$. During this short interval $f(t)$ will change very little. If we neglect this change, we may say that during this short time interval the amount $f(t_0)$, which would have earned $\alpha f(t_0)$ interest in a year, increased by $h\alpha f(t_0)$. Hence $f(t_0 + h) \approx f(t_0) + h\alpha f(t_0)$ and

$$\frac{f(t_0 + h) - f(t_0)}{h} = \alpha f(t_0) + \text{a small error};$$

we guess the error to be arbitrarily small for h sufficiently small. If so, then

$$f'(t) = \alpha f(t). \tag{6}$$

By Theorem 1, we see from (5) and (6) that

$$f(t) = e^{\alpha t}, \tag{7}$$

and, for $t = 1$, $f(1) = e^\alpha$. We conclude that the answer to Bernoulli's question is e^α.

REMARK   For an interest rate of 5 percent we have $\alpha = 5/100 = .05$; hence $e^\alpha \approx 1.0513$, while advertisements of savings banks promise that the effective annual interest rate for 5 percent compounded "daily" or "continually" is .052. This discrepancy is explained as follows. The banks first define the daily interest rate as $.05/360$ and then compound it for 365 days.

The procedure by which we arrived at the "solution" to Bernoulli's question is typical for applications of calculus. It consisted of a "fuzzy" part and a "sharp" part. In getting to the differential equation (6), we freely used our intuition. We certainly did not "prove" that $f(t)$ satisfies (6). How could we? We do not even possess a clear definition of "continuous compounding of interest." Rather, we are inclined to consider the differential equation as a substitute for such a definition.

Yet once we arrived at the differential equation, the fuzzy part of the argument was completed. The next step, that is, obtaining (7), was accomplished by a purely mathematical and completely rigorous method, the application of Theorem 1.

If we dealt with a natural process, we could next check how successful our "fuzzy" guessing was by comparing the solution of the differential equation with observations. Here we cannot do it, but we can perform an indirect check. If our reasoning was sound, then the number $(1 + \alpha/n)^n$ should be, for very large n, very close to e^α. This is indeed so, as we prove below.

1.3 Exponentials as limits

We want to show that $(1 + \alpha/n)^n$ is as close as we like to e^α if n is large enough in absolute value. (We do not require that n or α be positive or that n take on only integral values.) The mathematical shorthand for the statement just made is

$$\lim_{n \to \infty} \left(1 + \frac{\alpha}{n}\right)^n = e^\alpha. \tag{8}$$

Set $h = 1/n$; then $|h|$ becomes small as $|n|$ becomes large and the statement to be proved can be written as

$$\lim_{h \to 0} (1 + \alpha h)^{1/h} = e^\alpha. \tag{9}$$

The proof involves a trick. Set $f(x) = \log(1 + \alpha x)$ and apply to the function f the linear approximation theorem (Theorem 3 in Chapter 4, §1.6) with $x_0 = 0$ and $x = h$. Since $f(0) = \log 1 = 0$, $f'(x) = \alpha/(1 + \alpha x)$, we have $m = f'(0) = \alpha$ and

$$\log(1 + \alpha h) = 0 + \alpha h + hr(h),$$

where $r(h)$ is continuous at 0, and $r(0) = 0$. Hence

$$\log(1 + \alpha h)^{1/h} = \frac{1}{h}\log(1 + \alpha h) = \alpha + r(h)$$

and

$$(1 + \alpha h)^{1/h} = e^{\alpha + r(h)}. \tag{10}$$

This function of h is continuous at $h = 0$ (by the theorem on composite functions, Theorem 1 in Chapter 6, §2.1) and its value at $h = 0$ is $e^{\alpha + r(0)} = e^\alpha$. This establishes (9).

We note the special case $\alpha = 1$:

$$\lim_{n \to +\infty} \left(1 + \frac{1}{n}\right)^n = e. \tag{11}$$

This is sometimes used to define e (then we must prove that the limit exists). We also can take either (8) or (9) as a definition of the exponential function. There is a similar definition of the natural logarithm. It reads

$$\lim_{h \to 0} \frac{a^h - 1}{h} = \log a, \qquad \text{for every } a > 0. \tag{12}$$

The proof follows by noting that

$$\lim_{h \to 0} \frac{a^h - 1}{h} = \lim_{h \to 0} \frac{a^h - a^0}{h} = \left(\frac{da^x}{dx}\right)_{x=0} = (\log a)a^0 = \log a,$$

as asserted.

PROBLEMS

Evaluate the following limits.

1. $\displaystyle\lim_{x\to\infty}\left(1+\frac{1}{2x}\right)^x$.

2. $\displaystyle\lim_{x\to\infty}\left(1-\frac{2}{x}\right)^x$.

3. $\displaystyle\lim_{x\to\infty}\left(1+\frac{1}{x}\right)^{2x}$.

4. $\displaystyle\lim_{x\to\infty}\left(1+\frac{3}{x}\right)^{x+2}$.

5. $\displaystyle\lim_{x\to\infty}\left(1+\frac{1}{x+1}\right)^x$.

   [*Hint:* Set $u = x + 1$.]

6. $\displaystyle\lim_{x\to0}\left(1-\frac{x}{4}\right)^{2/x}$.

7. $\displaystyle\lim_{x\to0}(1-x)^{(x+1)/x}$.

8. $\displaystyle\lim_{x\to0}\frac{e^{2x}-1}{x}$.

1.4 Population explosion

There are natural phenomena that imitate Bernoulli's notion that capital continuously increases at a rate proportional to itself. An example is the **growth of a bacteria population.** Bacteria are virtually indestructible unless killed, and a given population of N bacteria will, after a very short time interval of length h, increase approximately by αNh, where α is some positive constant. Thus $[N(t + h) - N(t)]/h \approx \alpha N(t)$. We assume therefore that $N(t)$, the number of bacteria at the time t, satisfies the differential equation

$$N'(t) = \alpha N(t), \tag{13}$$

and is therefore (by Theorem 1) given by the formula

$$N(t) = N_0 e^{\alpha t}, \qquad N_0 = N(0).$$

The rapid growth of $e^{\alpha t}$ as $t \to +\infty$ describes a "population explosion."

[The example just considered illustrates an important point. The number of bacteria is an integer. Also, $N(t)$ is a step function which cannot, strictly speaking, be regarded as a continuous function. But N is a very large integer, and the appearance of a few new bacteria change it by a relatively insignificant amount. We therefore obtain a sufficiently accurate picture of reality by treating $N(t)$ *as if* it were a continuous and even differentiable function.]

A differential equation describing the growth of a human population would have to be more complicated than (13). At the very least, it would have to contain two constants, the birth rate α and the death rate β. If we assume that in a population of N there occur, during a short time interval h, αhN births and βhN deaths, we obtain the differential equation

$$\frac{dN}{dt} = \alpha N - \beta N$$

[since, at the time $t + h$, the population is approximately $N(t) + \alpha hN(t) - \beta hN(t)$]. This can be written as

$$N'(t) = (\alpha - \beta)N(t) \tag{14}$$

and has the solution

$$N = N_0 e^{(\alpha-\beta)t}, \qquad N_0 = N(0). \tag{15}$$

The shape of the population curve described by (15) depends on the sign of $\alpha - \beta$. If $\alpha = \beta$, the population is constant. If $\alpha < \beta$, the population decreases and ultimately becomes extinct, since $\lim_{t \to +\infty} N(t) = 0$. If the birth rate α exceeds the death rate β, the population will increase, maybe slowly at first, but then very fast.

While a human population explosion is a serious danger, the assumptions underlying the differential equation (14) are too simplified to give anything approaching a true picture. For instance, the numbers of deaths and births depend on both the size of the population and the age distribution within it.

PROBLEMS

1. Suppose that, in an experiment, a culture of 100 bacteria is allowed to reproduce under favorable conditions. Twelve hours later the culture is found to have 500 bacteria. How many bacteria will there be two days after the start of the experiment?
2. Suppose an isolated community had a population of 10,000 people in the year 1800. Assuming an exponential growth law, what is the population in the year 1900 if the population was 12,000 people in the year 1810?

1.5 Radioactive disintegration

As an example of exponential decay, we consider a radioactive element—for instance, radium. Given a sample containing N atoms of such an element, we find that, after a short time interval h, approximately $\alpha N h$ of the atoms will disintegrate. Therefore, the amount $F(t)$ of the radioactive material present at time t obeys the differential equation $dF/dt = -\alpha F$, so that

$$F(t) = F_0 e^{-\alpha t}, \qquad F_0 = F(0). \tag{16}$$

The time t after which the original amount of a radioactive substance will be reduced by $\frac{1}{2}$ is called the **half-life** of the substance. The half-life is found from the equation $e^{-\alpha t} = \frac{1}{2}$. Solving this equation for t, we obtain

$$t = \frac{\log 2}{\alpha}. \tag{17}$$

The half-life t depends only on the substance considered. For radium, it is 1656 years; for the isotopes radium A and radium C', it is $t = 2.75$ min, and $t = 1.4 \times 10^{-6}$ sec. The various isotopes of uranium have half-lives of the order of 10^9 years.

The universality of the decay rule (16) shows that the laws of atomic and nuclear physics are quite different from the more familiar laws describing objects of ordinary size. Given N atoms of polonium, we find that about half of them will disintegrate in 140 days (for this element $t = 140$ days, approximately). Suppose we divide the N atoms into several piles and observe them separately. Half of *each* pile will disintegrate in 140 days. There is no way to predict when any *individual* atom will decay.

PROBLEMS

1. Suppose a radioactive material has a half-life of one year. How long will it take for 10 grams of this material to decay to 1 gram?
2. If 10 percent of a certain radioactive material decays in five days, what is the half-life of this material?

§2 Harmonic motion‡

While the applications of trigonometric functions to geometry continue to be very significant, the importance of the sine and cosine functions in mathematics and science derives primarily from the differential equations they satisfy.

2.1 The differential equation of sine and cosine

Recall that $(\cos x)' = -\sin x$ and $(\sin x)' = \cos x$. Therefore, $(\cos x)'' = -\cos x$. Similarly, $(\sin x)'' = -\sin x$, for $(\sin x)'' = (\cos x)' = -\sin x$. Thus both functions $\cos x$ and $\sin x$ satisfy the differential equation $f''(x) + f(x) = 0$. We proceed to find all solutions of this second-order differential equation.

(The equation is said to be of second order, since it involves a second derivative. We now denote the independent variable by x rather than by θ. We chose this notation, since the geometric origin of the functions sine and cosine now recedes into the background.)

Theorem 1. *Let a and b be two numbers. If*

$$f(x) = a \cos x + b \sin x, \tag{1}$$

then

$$f''(x) + f(x) = 0 \tag{2}$$

and

$$f(0) = a, \qquad f'(0) = b. \tag{3}$$

Conversely, if the function $f(x)$ satisfies Equation (2) and the conditions (3), then (1) holds.

In the converse, it is assumed that $f(x)$ is defined for all x or, at least, in an interval containing $x = 0$. We note three important special cases: If $f'' + f = 0$ and $f(0) = f'(0) = 0$, then $f(x) = 0$. If $f'' + f = 0$ and $f(0) = 1, f'(0) = 0$, then $f(x) = \cos x$. If $f'' + f = 0$ and $f(0) = 0$, $f'(0) = 1$, then $f(x) = \sin x$.

Proof of Theorem 1. If $f(x) = a \cos x + b \sin x$, then we compute directly that $f''(x) + f(x) = 0$ and $f(0) = a$, $f'(0) = b$.

We establish first a special case of the converse. Assume that $f(x)$ satisfies $f'' + f = 0$ and that $f(0) = f'(0) = 0$. We want to show that $f(x) = 0$. To do this, consider the function $\phi(x) = f(x)^2 + f'(x)^2$. Then $\phi(0) = 0$ and

$$\phi'(x) = 2f(x)f'(x) + 2f'(x)f''(x) = 2f'(x)(f''(x) + f(x)) = 0.$$

Hence ϕ is constant, and since $\phi(0) = 0$, we have $\phi(x) = 0$. Since $f(x)^2$ and $f'(x)^2$ cannot be negative, we have $f(x)^2 = 0$ and $f(x) = 0$ as asserted.

Now let $f(x)$ be a given function satisfying $f'' + f = 0$ with $f(0) = a$ and $f'(0) = b$. We set $g(x) = f(x) - a \cos x - b \sin x$, and we must show that $g(x) = 0$. But we have $g(0) = f(0) - a - 0 = 0$, and

$$g'(x) = f'(x) + a \sin x - b \cos x, \qquad g'(0) = f'(0) + 0 - b = 0,$$
$$g''(x) = f''(x) + a \cos x + b \sin x = -f(x) + a \cos x + b \sin x = -g(x).$$

‡Optional section.

Thus $g''(x) + g(x) = 0$ and $g(0) = g'(0) = 0$. By what was proved before, $g(x) = 0$. This completes the argument.

The content of Theorem 1 is sometimes expressed by saying that $f(x) = a \cos x + b \sin x$ is the *general solution* of the differential equation $f''(x) + f(x) = 0$. This means that, for every choice of constants a and b, the function $a \cos x + b \sin x$ satisfies the equation, and every function satisfying the equation can be written as $a \cos x + b \sin x$ for some choice of a and b.

The theorem also asserts that a solution $f(x)$ of $f'' + f = 0$ is uniquely determined by the so-called **initial conditions,** that is, by the values of $f(0)$ and $f'(0)$, and these values may be prescribed. We can also assign the values of f and its derivative f' at any point x_0. For if $f(x)$ is a solution satisfying the conditions $f(0) = a$, $f'(0) = b$, then $f_1(x) = f(x - x_0)$ is a solution satisfying $f_1(x_0) = a$, $f_1'(x_0) = b$. In other words, the solution of $f_1'' + f_1 = 0$ satisfying the conditions $f_1(x_0) = a$, $f_1'(x_0) = b$ is $f_1(x) = a \cos(x - x_0) + b \sin(x - x_0)$.

EXAMPLES 1. If $f''(x) + f(x) = 0$, $f(0) = f'(0) = 1$, what is $f(x)$?
ANSWER By the theorem, $f(x) = \cos x + \sin x$.

2. Using Theorem 1, prove the addition theorem for the sine function.
ANSWER We must show that

$$\sin(x + y) = \sin x \cos y + \cos x \sin y.$$

For a fixed y, set

$$\phi(x) = \sin(x + y) - \sin x \cos y - \cos x \sin y;$$

it will suffice to show that $\phi(x) = 0$ for all x. Now, remembering that y is a constant, we have

$$\phi'(x) = \cos(x + y) - \cos x \cos y + \sin x \sin y,$$
$$\phi''(x) = -\sin(x + y) + \sin x \cos y + \cos x \sin y,$$

so that $\phi''(x) + \phi(x) = 0$. Also $\phi(0) = \sin y - 0 \cdot \cos y - 1 \cdot \sin y = 0$, $\phi'(0) = \cos y - 1 \cdot \cos y + 0 \cdot \sin y = 0$. Hence $\phi(x) = 0$.

PROBLEMS

Solve the following differential equations subject to the given initial conditions.

1. $f''(x) + f(x) = 0$, $f(0) = -1$, $f'(0) = 1$.
2. $f''(x) + f(x) = 0$, $f(0) = 2$, $f'(0) = -4$.

3. $\dfrac{d^2y}{dx^2} + y = 0$, $y = 8$ when $x = 0$, $\dfrac{dy}{dx} = 4$ when $x = 0$.

4. $f''(x) + f(x) = 0$, $f\left(\dfrac{\pi}{3}\right) = 1$, $f'\left(\dfrac{\pi}{3}\right) = -1$.

5. $\dfrac{d^2y}{dx^2} + y = 0$, $y = 0$ when $x = \dfrac{\pi}{2}$, $\dfrac{dy}{dx} = 2$ when $x = \dfrac{\pi}{2}$.

6. $f''(x) + f(x) = 0$, $f\left(-\dfrac{\pi}{4}\right) = 1$, $f'\left(-\dfrac{\pi}{4}\right) = -3$.

*7. Using Theorem 1, prove the addition theorem for the cosine.

2.2   A second-order differential equation

Theorem 1 has a simple but important generalization.

Theorem 2.   *Let $\omega > 0$ be a given number. The general solution of the differential equation*

$$f''(x) + \omega^2 f(x) = 0 \tag{4}$$

is

$$f(x) = a \cos \omega x + b \sin \omega x. \tag{5}$$

We can prove this in the same way as we proved Theorem 1. It is perhaps simpler to reduce Theorem 2 to Theorem 1. Set $F(t) = f(t/\omega)$, and let $t = \omega x$. Then $f(x) = F(\omega x)$, $f'(x) = \omega F'(\omega x)$, and $f''(x) = \omega^2 F''(\omega x)$. Hence the equation $f'' + \omega^2 f = 0$ is equivalent to $F'' + F = 0$. We know that $F'' + F = 0$ if and only if $F = a \cos t + b \sin t$. Therefore $f'' + \omega^2 f = 0$ if and only if $f = a \cos \omega x + b \sin \omega x$.

The constants a and b in (5) can be determined if we know the values of f and f' at some point. For instance, if $f(0) = A$ and $f'(0) = B$, then $a = A$ and $b = B/\omega$.

We remark that *the general solution of* (4) *is periodic with period* $2\pi/\omega$. Indeed

$$\cos \omega \left[x + \frac{2\pi}{\omega} \right] = \cos(\omega x + 2\pi) = \cos \omega x. \text{ Similarly, } \sin \omega \left[x + \frac{2\pi}{\omega} \right] = \sin \omega x.$$

EXAMPLE   If $f''(x) + 4f(x) = 0$, $f(\pi/4) = 2$, and $f'(\pi/4) = 1$, what is $f(x)$?
ANSWER   By Theorem 2, the general solution is given by $f(x) = a \cos 2x + b \sin 2x$. Hence $f'(x) = -2a \sin 2x + 2b \cos 2x$. Therefore $f(\pi/4) = a \cos(\pi/2) + b \sin(\pi/2) = 2$, that is, $a = 2$. Similarly, $f'(\pi/4) = -2a \sin(\pi/2) + 2b \cos(\pi/2) = 1$, that is, $b = \frac{1}{2}$. The answer is $f(x) = 2 \cos 2x + \frac{1}{2} \sin 2x$.

PROBLEMS

In Problems 1 to 6 find solutions of the differential equations satisfying the indicated initial conditions.

1. $f''(x) + 16f(x) = 0$, $f(0) = f'(0) = 1$.
2. $f''(x) + 4f(x) = 0$, $f\left(\dfrac{\pi}{2}\right) = 1$, $f'\left(\dfrac{\pi}{2}\right) = 2$.
3. $\dfrac{d^2 y}{dx^2} + 9y = 0$, $y = \dfrac{dy}{dx} = -1$ when $x = \dfrac{\pi}{6}$.
4. $\dfrac{d^2 y}{dx^2} + \dfrac{1}{4} y = 0$, $y = \dfrac{dy}{dx} = \sqrt{2}$ when $x = \dfrac{\pi}{2}$.
5. $f''(x) + \dfrac{25}{16} f(x) = 0$, $f\left(\dfrac{\pi}{5}\right) = f'\left(\dfrac{\pi}{5}\right) = 1$.
6. $f''(x) + \frac{1}{9}f(x) = 0$, $f(\pi) = -3$, $f'(\pi) = 1$.

2.3   Motion of an elastic spring

We apply Theorem 2 to the motion of a particle attached to an elastic spring. Such a spring is shown in Figure 9.44 (p. 232). Recall that the spring has an equilibrium position, and that if it is displaced from that position by an amount s, there is a restoring force F acting on the particle that tends to restore the spring to its original position. (We may assume that $s > 0$ if the spring is stretched, and $s < 0$ if the spring is compressed.) The spring is called elastic if the restoring force is a linear function of the displacement

$$F = -ks,$$

where k is a positive constant (**spring constant**). Practically all springs may be considered as elastic, provided we consider only small displacements. (This has already been discussed in Chapter 9, §5.1.)

Let m be the mass of the particle attached to the spring. If we neglect the mass of the spring itself, and if we observe that ds/dt (the time derivative of s), is the velocity and d^2s/dt^2 is the acceleration, Newton's law of motion reads

$$m\frac{d^2s}{dt^2} = -ks. \tag{6}$$

We rewrite this as

$$\frac{d^2s}{dt^2} + \omega^2 s = 0 \qquad \text{where } \omega = \sqrt{\frac{k}{m}}.$$

Equation (6) is identical to Equation (4), except for the names of the quantities involved. Using Theorem 2, we conclude that the function $t \mapsto s(t)$ describing the motion is of the form

$$s(t) = a \cos \omega t + b \sin \omega t. \tag{7}$$

In particular, the motion is periodic with period T, where

$$T = \frac{2\pi}{\omega} = 2\pi\sqrt{\frac{m}{k}}. \tag{8}$$

Such a perpetual motion is possible, because we neglect friction.

In order to know the motion completely, we must know the numbers a and b in (7). This means that we must know the position $s(t)$ and velocity $v(t) = s'(t)$ of our spring at some fixed time t_0, say, at the initial time $t = 0$. Once these *initial conditions* are given, we can compute the constants a and b, and hence also the position $s(t)$ and the velocity $s'(t)$ of the spring for any time t.

We claim that $s(t)$ can also be written in the form

$$s(t) = A \cos[\omega(t - t_0)]. \tag{9}$$

where $A > 0$ and t_0 are constants depending on the initial conditions. To see this, start with the expression (7) for $s(t)$. If $a = b = 0$, set $A = 0$. If $a^2 + b^2 > 0$, let $A = \sqrt{a^2 + b^2}$. There is a number α such that $a/A = \cos \alpha$, $b/A = \sin \alpha$. Set $t_0 = \alpha/\omega$. Then we have

$$s(t) = a \cos \omega t + b \sin \omega t = A \cos \omega t_0 \cos \omega t + A \sin \omega t_0 \sin \omega t$$
$$= A \cos(\omega t - \omega t_0) = A \cos[\omega(t - t_0)]$$

by the addition theorem.

Since the cosine oscillates between -1 and 1, the number A in (9) gives the maximal deviation of the spring from the equilibrium position; A is called the **amplitude** of the oscillation. The number t_0 describes the time at which this maximal deviation takes place; it is called the **phase difference.** Between the times t_0 and $t_0 + T$ the spring carries out a complete cycle.

Amplitude and phase difference depend on how the motion was started at $t = 0$; the period T, however, depends only on the spring.

PROBLEMS

In Problems 1 to 6 a particle of mass m is attached to one end of a horizontal elastic spring whose spring constant is k. The other end of the spring is held fixed, and the particle is set vibrating along the axis of the spring. The displacement of the particle from the rest position is denoted by $s = s(t)$ with $s > 0$ when the spring is stretched.

1. Suppose that $m = 5$ grams and $s = 2\cos(t/3)$. Find the spring constant.
2. Suppose that $m = 2$ grams and $k = 8$ grams/sec². Suppose also that, at time $t = 0$ sec, the spring is stretched 1 cm from its rest position, and that the speed of the particle is 4 cm/sec and is increasing. Find $s(t)$. Find the amplitude.
3. Suppose that $m = 1.5$ grams and $k = 4.5$ grams/sec². Suppose also that at times $t = 0$ sec and $t = \sqrt{3}\pi/6$ sec, the spring is compressed 1 cm from its rest position. Find an expression for the velocity of the particle.
4. Suppose that $m = .2$ gram and $k = 1.8$ grams/sec². Suppose also that at time $t = 0$ sec, the spring is compressed 2 cm from its rest position. Find the velocity of the particle at time $t = \pi/6$ sec.
5. Suppose that $m = 1$ gram and $k = .25$ gram/sec². Suppose also that at time $t = 0$ sec, the spring is stretched 1 cm from its rest position, and at time $t = \pi/2$ sec the velocity of the particle is $(\sqrt{2}/2)$ cm/sec. Find $s(t)$. Find the amplitude.
6. Suppose that the period of the motion is $(2\pi/15)$ sec. Suppose also that at time $t = 0$ sec, the velocity of the particle is -15 cm/sec, and at time $t = \pi/6$ sec the velocity of the particle is 3 cm/sec. Find $s(t)$. Find the amplitude.

2.4 Harmonic oscillations

When we strike a tuning fork, we always hear the same tone no matter how hard or how lightly we strike it. Similarly, hitting the same piano key once lightly and once full force will produce sounds of different intensities but of the same pitch. These and many similar phenomena can be understood by considering a vibrating body as an elastic spring. This may be done because the differential equations describing the motion of a vibrating tuning fork, a vibrating piano string, and so on are approximately the same as the equation of motion of an elastic spring; that is, Equation (6). In other words, there is a number s describing the deviation of the vibrating body from the position of equilibrium, a constant k that measures the **stiffness** of the body, and a constant m that measures the **inertia** of the body. These three quantities and the time t are connected by Equation (6). Therefore s as a function of time is given by (7) or by (9).

The amplitude and the phase difference depend on the initial conditions; the period is given by formula (8) and depends only on the stiffness and the inertia of the vibrating body. So does the number

$$\nu = \frac{1}{T},$$

the **frequency** of the oscillation; it is the number of cycles per unit time. In sound (audible oscillations of air) the frequency determines the **pitch.** By (8),

$$\nu = \frac{1}{2\pi}\sqrt{\frac{k}{m}}. \tag{10}$$

The frequency is always the same. It does not depend on the amplitude A (which in sound determines the intensity).

A motion governed by the same differential equation as the elastic spring is called a **harmonic oscillation.** Harmonic oscillations occur not only in elastic bodies but in many other situations, for instance, in a radio transmitter or inside an atom. The constancy of pitch is their most important property. It explains why a wrist watch keeps correct time and why an atom emits light of a definite frequency. The surprising fact that the number π appears in the formula for the frequency suggests a connection between circles and harmonic oscillations. This connection will be pursued in Chapter 17, §3.2.

§3 The normal distribution‡

One of the most important and surprising discoveries in mathematical statistics is the Error Law due to Laplace. Suppose we perform many times an experiment or measurement whose outcome depends on many independent random events. (For example, we measure the weight of babies at birth in some hospital.) The graph that describes how the results of the measurements are distributed around the average outcome always has nearly *the same* basic shape, and this shape can be described by a function involving the two numbers e and π!

3.1 The function $y = e^{-x^2}$. Some improper integrals

To formulate the Error Law we must study the function

$$y = e^{-x^2} \tag{1}$$

graphed in Figure 12.1. The function is nonnegative and even, and the graph is

FIGURE **12.1**

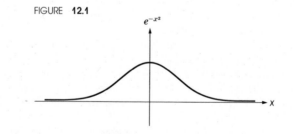

e^{-x^2}

x

‡Optional section.

bell-shaped. By this we mean that there is a single maximum (at $x = 0$) and two symmetrically situated inflection points (at $x = \pm\frac{1}{2}\sqrt{2}$; the reader should verify this). For large values of $|x|$, e^{-x^2} is very small;

$$\lim_{x \to \pm\infty} e^{-x^2} = 0. \tag{2}$$

(This is so, since $1/e^{-x^2} = e^{x^2}$ has limits $+\infty$ as $x \to \pm\infty$; see Chapter 11, §3.4.) Also

$$\lim_{x \to \pm\infty} xe^{-x^2} = 0. \tag{3}$$

(This is so since, for $|x| > 1$, we have that $|xe^{-x^2}| = |x|e^{-x^2} < x^2 e^{-x^2}$, and we know that $\lim_{x \to \pm\infty} x^2 e^{-x^2} = \lim_{u \to +\infty} ue^{-u} = 0$ by Chapter 11, §3.4.)

The area between the graph of $y = e^{-x^2}$ and the x-axis is finite, and

$$\int_{-\infty}^{+\infty} e^{-x^2}\, dx = \sqrt{\pi}. \tag{4}$$

This surprising relation between e and π is best proved using calculus of two variables; we postpone the proof until Chapter 20, §3.6. Since e^{-x^2} is even, the integral in (4) is twice the integral $\int_0^{+\infty} e^{-x^2}\, dx$ (see Chapter 8, §3.6). Thus

$$\int_0^{+\infty} e^{-x^2}\, dx = \frac{1}{2}\sqrt{\pi}. \tag{5}$$

Next, xe^{-x^2} is the derivative of $-\frac{1}{2}e^{-x^2}$, so that

$$\int_0^{+\infty} xe^{-x^2}\, dx = \lim_{A \to +\infty} \int_0^A xe^{-x^2}\, dx = \lim_{A \to +\infty} \left(-\frac{1}{2}e^{-x^2} \right)\Big|_0^A$$

$$= \lim_{A \to +\infty} \left(\frac{1}{2} - e^{-A^2} \right) = \frac{1}{2}.$$

Here we used (2). In the same way we show that $\int_{-\infty}^0 xe^{-x^2}\, dx = -\frac{1}{2}$. Hence

$$\int_{-\infty}^{+\infty} xe^{-x^2}\, dx = 0. \tag{6}$$

Finally, we note that

$$\frac{d}{dx}\left(-\frac{1}{2}xe^{-x^2} \right) = -\frac{1}{2}e^{-x^2} + x^2 e^{-x^2}.$$

Therefore

$$\int_0^A \frac{d}{dx}\left(-\frac{1}{2}xe^{-x^2} \right) dx = -\frac{1}{2}\int_0^A e^{-x^2}\, dx + \int_0^A x^2 e^{-x^2}\, dx$$

and, since the left side equals $-\frac{1}{2}Ae^{-A^2}$,

$$\int_0^A x^2 e^{-x^2}\, dx = -\frac{1}{2}Ae^{-A^2} + \frac{1}{2}\int_0^A e^{-x^2}\, dx.$$

Noting (3) and (5), we conclude that

$$\int_0^{+\infty} x^2 e^{-x^2}\, dx = \lim_{A\to+\infty}\left(-\frac{1}{2} Ae^{-A^2}\right) + \lim_{A\to+\infty}\frac{1}{2}\int_0^A e^{-x^2}\, dx$$

$$= 0 + \frac{1}{2}\int_0^{+\infty} e^{-x^2}\, dx = \frac{1}{4}\sqrt{\pi}.$$

By the same method, or by noting that $x^2 e^{-x^2}$ is an even function of x (see Chapter 8, §3.6), we conclude that $\int_{-\infty}^0 x^2 e^{-x^2}\, dx = \frac{1}{4}\sqrt{\pi}$. Thus

$$\int_{-\infty}^{+\infty} x^2 e^{-x^2}\, dx = \frac{1}{2}\sqrt{\pi}. \tag{7}$$

PROBLEMS

1. Show that the function $f(x) = e^{-x^2}$ has a maximum at $x = 0$ and two inflection points at $x = \pm\frac{1}{2}\sqrt{2}$.
2. Show that $\lim_{x\to+\infty} x^\alpha e^{-x^2} = 0$, where α is any integer.

3.2 The normal curve

Let m be any number, and σ a positive number. We construct the function

$$f(x) = \frac{1}{\sigma\sqrt{2\pi}}\, e^{-(x-m)^2/2\sigma^2}. \tag{8}$$

It has the following properties:

$$f(x) > 0, \qquad \text{for all } x, \tag{9}$$

$$\int_{-\infty}^{+\infty} f(x)\, dx = 1, \tag{10}$$

$$\int_{-\infty}^{+\infty} xf(x)\, dx = m, \tag{11}$$

$$\int_{-\infty}^{+\infty} (x - m)^2 f(x)\, dx = \sigma^2. \tag{12}$$

The first follows from (8), the others will be proved below. Referring to the Remark at the end of Chapter 9, §4.3, we conclude from (9) and (10) that $f(x)$ is a *frequency distribution*, from (11) that m is the *mean* (or expected value), and from (12) that σ is the *standard deviation*.

To prove (10), (11), and (12) we use the substitution $t = (x - m)/\sqrt{2}\sigma$. Then $t = -\infty$ for $x = -\infty$ and $t = +\infty$ for $x = +\infty$,

$$x = m + \sqrt{2}\sigma t, \qquad dx = \sqrt{2}\sigma\, dt, \qquad f(x) = e^{-t^2}/\sqrt{2\pi}\,\sigma.$$

Using (5), (6), and (7), we obtain

$$\int_{-\infty}^{+\infty} f(x)\, dx = \frac{1}{\sigma\sqrt{2\pi}}\int_{-\infty}^{+\infty} e^{-t^2}\sqrt{2}\sigma\, dt = \frac{1}{\sqrt{\pi}}\int_{-\infty}^{+\infty} e^{-t^2}\, dt = 1,$$

$$\int_{-\infty}^{+\infty} xf(x)\, dx = \frac{1}{\sigma\sqrt{2\pi}}\int_{-\infty}^{+\infty} (m + \sqrt{2}\sigma t)e^{-t^2}\sqrt{2}\sigma\, dt$$

$$= \frac{m}{\sqrt{\pi}} \int_{-\infty}^{+\infty} e^{-t^2} \, dt + \sigma \sqrt{\frac{2}{\pi}} \int_{-\infty}^{+\infty} t e^{-t^2} \, dt = m,$$

$$\int_{-\infty}^{+\infty} (x-m)^2 f(x) \, dx = \frac{1}{\sigma\sqrt{2\pi}} \int_{-\infty}^{+\infty} 2\sigma^2 t^2 e^{-t^2} \sqrt{2}\sigma \, dt$$

$$= \frac{2\sigma^2}{\sqrt{\pi}} \int_{-\infty}^{+\infty} t^2 e^{-t^2} \, dt = \sigma^2.$$

The graph of (8) is called a **normal curve** with mean m and standard deviation σ. It is (a) symmetric with respect to the line $x = m$, and (b) bell-shaped, with two inflection points at $x = m + \sigma$ and $x = m - \sigma$.

The proof of (a) is by direct calculation. If $x_1 = m + \xi$ and $x_2 = m - \xi$, Equation (8) assigns the same values to x_1 and x_2.

To prove (b), we compute the first and second derivatives of (8):

$$f'(x) = -\frac{x-m}{\sigma^3\sqrt{2\pi}} e^{-(x-m)^2/2\sigma^2},$$

$$f''(x) = -\frac{1}{\sigma^3\sqrt{2\pi}} e^{-(x-m)^2/2\sigma^2} + \frac{(x-m)^2}{\sigma^5\sqrt{2\pi}} e^{-(x-m)^2/2\sigma^2}$$

$$= \frac{1}{\sigma^3\sqrt{2\pi}} \left\{ \left(\frac{x-m}{\sigma}\right)^2 - 1 \right\} e^{-(x-m)^2/2\sigma^2}.$$

This second derivative is positive for $|x - m| > \sigma$, and negative for $|x - m| < \sigma$.

Normal curves with $m = 0$ and $\sigma = 1, \frac{1}{2}, \frac{1}{4}$ are shown in Figure 12.2. Note that the curve is peaked for small σ, flat for large σ.

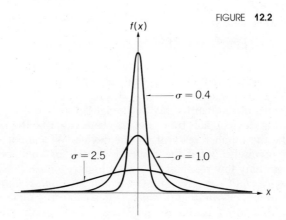

FIGURE **12.2**

f(x)

σ = 0.4

σ = 2.5   σ = 1.0

x

3.3   The Error Law

The normal distribution function with mean m and standard deviation σ is the function

$$F(x) = \frac{1}{\sigma\sqrt{2\pi}} \int_{-\infty}^{x} e^{-(t-m)^2/2\sigma^2} \, dt. \qquad (13)$$

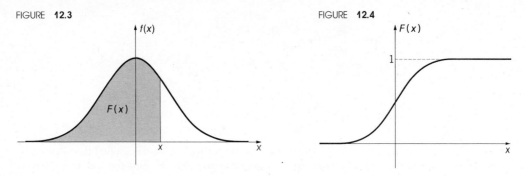

FIGURE **12.3**

FIGURE **12.4**

Thus $F(x)$ is the area under the normal curve from $-\infty$ to x (see Figure 12.3). The function $F(x)$, whose graph for $m = 0$, $\sigma = 1$ appears in Figure 12.4, is "nonelementary"; we know no way of writing it in terms of known functions without using an integral sign or its equivalent. We have $\lim_{x\to-\infty} F(x) = 0$, $\lim_{x\to+\infty} F(x) = 1$. Also,

$$F(m) = \tfrac{1}{2}, \tag{14}$$

since half of the total area lies to the left of line $x = m$. Using the change of variables $\xi = (t - m)/\sigma$, we obtain

$$F(x) = \frac{1}{\sqrt{2\pi}} \int_{-\infty}^{(x-m)/\sigma} e^{-\xi^2/2} \, d\xi.$$

Hence

$$F(m - \sigma) = \frac{1}{\sqrt{2\pi}} \int_{-\infty}^{-1} e^{-\xi^2/2} \, d\xi, \qquad F(m + \sigma) = \frac{1}{\sqrt{2\pi}} \int_{-\infty}^{1} e^{-\xi^2/2} \, d\xi; \tag{15}$$

numerical calculation shows that

$$F(m - \sigma) \approx .159, \qquad F(m + \sigma) \approx .841.$$

Thus the area to the left of the line $x = m - \sigma$ and to the right of $x = m + \sigma$ are each about 16 percent of the total area, while about 68 percent of the total area lies between the lines $x = m - \sigma$ and $x = m + \sigma$.

Suppose we measure a quantity, say, the weight of some object, many times, using the same measuring device each time. Let the true value of the quantity measured be a, and let ξ be the number read off on our device. We cannot expect, of course, to obtain $\xi = a$, precisely. But if the device is "good" or "unbiased," we shall get, on the average, as many low readings ($\xi < a$) as high readings ($\xi > a$). Moreover, *the fraction of times we shall get a reading $\xi < x$ will be, approximately, equal to $F(x)$, where $F(x)$ is the normal distribution function with mean a and some standard deviation σ*. The smaller σ, the more sensitive is our device, for a larger fraction of all measurements ξ is close to a.

This statement about the distribution of the results of measurements is known as the **Error Law** or as **Laplace's Law** as we already mentioned; actually its validity is much wider than may appear at first glance. Whenever we make many measurements, the results of which are determined by many independent random events, we may expect to encounter a normal distribution. This applies, for instance, to the distribution of IQ's in a large population, to the distribution of heights among a large number of men of the same age, and so forth. We formulated the Error Law as a "fuzzy" empirical rule. In the mathematical theory of probability, there is a corresponding "sharp" theorem. We cannot discuss this here.

§4   Other applications of the exponential function‡

In this section we consider several differential equations related to the differential equation of the exponential function and apply one of these equations to atomic fission.

4.1   The general solution of a nonhomogeneous equation

Consider the differential equation

$$f'(x) - \alpha f(x) = \phi(x), \tag{1}$$

where ϕ is a given function. [This equation is called nonhomogeneous since we do not have 0 on the right; $f(x)$ is therefore not a solution.]

We can solve this equation, that is, find all functions f satisfying (1) by a device that is also useful for other differential equations. If $\phi = 0$, we know that the solution is $Ae^{\alpha x}$, where A is a constant (see Theorem 1 in §1.1). We try to solve (1) by writing the desired solution f in the form

$$f(x) \doteq A(x)e^{\alpha x},$$

where $A(x)$ is a function to be determined. (If there is a solution, it certainly can be so written.) Substituting this expression for f into (1), we obtain

$$f'(x) - \alpha f(x) = A'(x)e^{\alpha x} + A(x)\alpha e^{\alpha x} - \alpha A(x)e^{\alpha x} = \phi(x)$$

or

$$A'(x) = e^{-\alpha x}\phi(x).$$

Hence, by the fundamental theorem of calculus,

$$A(x) = \int_0^x e^{-\alpha \xi}\phi(\xi)\, d\xi + A,$$

where A is a constant, the value of $A(x)$ at $x = 0$. Thus

$$f(x) = Ae^{\alpha x} + e^{\alpha x}\int_0^x e^{-\alpha \xi}\phi(\xi)\, d\xi. \tag{2}$$

We compute that this f indeed satisfies (1). Then we have

Theorem 1.   *The (only) solution of the differential equation (1) subject to the initial condition $f(0) = A$ is given by (2).*

Note that Theorem 2 in §1.1 is a special case, obtained by setting $\phi(x) = 0$.

‡Optional section.

PROBLEMS

1. Find a function $f(x)$ such that $f'(x) - f(x) = 4$ and $f(0) = 1$.
2. Find a function $f(x)$ such that $f'(x) - 2f(x) = 1$ and $f(0) = \frac{1}{2}$.
3. Find a function $f(x)$ such that $f'(x) - \frac{1}{2}f(x) = e^{x/2}$ and $f(0) = 2$.
4. Find a function $f(x)$ such that $f'(x) + f(x) = e^{2x}$ and $f(0) = -1$.
5. Find a function $f(x)$ such that $f'(x) + 3f(x) = (e^{3x} + 1)^{-2}$ and $f(0) = -\frac{1}{6}$.

4.2   Atomic fission

Atomic **fission,** the breakup of the nucleus into several parts, can occur in certain elements, such as U-235 (the uranium isotope of atomic weight 235), when a nucleus of an atom is hit by a neutron, an uncharged subatomic particle. Stray neutrons are always present in a material. Some will hit atomic nuclei, producing fission and thus new neutrons; some of the newly produced neutrons will in turn cause fission of other atoms. At the same time, some neutrons will escape from the material. (In an atomic reactor there is also an independent source of neutrons.) When neutrons are produced by fission, large amounts of energy are liberated, mainly in the form of heat. We shall derive a differential equation describing what happens. It is, of course, very simplified. (Among other things that we neglect in what follows is the use of "moderators," materials that scatter neutrons and slow down the process.)

Consider a ball of radius r and of uniform density ρ, made of a fissionable material. If N is the number of neutrons at a time t, then, after a very short time interval of length h, a certain number of new neutrons will be produced by fission. This number is proportional to N and to h and therefore equals αNh. The positive proportionality factor α depends on the material used and on the density but not on the shape or size of the piece of material. This is so since each individual process of atomic fission takes place in a volume whose dimensions are insignificant compared to the "ordinary" dimensions of a piece of metal.

During the same very short time interval h, a certain number of neutrons will escape. This number is βNh, since it is proportional to the number N of neutrons present and to the time elapsed. It is reasonable to assume that the proportionality factor β depends on the radius r, for only neutrons already close to the surface of the ball will have a chance to escape during the short interval h. Therefore β will be proportional to the ratio of the surface area of the ball to its volume. This ratio is $(4\pi r^2)/(\frac{4}{3}\pi r^3) = 3/r$. Thus $\beta = \beta_0/r$, where the positive constant β_0 depends on the material and the density.

The number of neutrons after the time interval h is

$$N(t + h) \approx N(t) + \alpha N(t)h - \beta N(t)h.$$

The differential equation of N as a function of time t reads

$$\frac{dN}{dt} = (\alpha - \beta)N$$

and has the solution

$$N = N_0 e^{(\alpha - \beta)t}, \tag{3}$$

where N_0 is the number of stray neutrons present at time $t = 0$. If t is measured in seconds, then the dimension of the constant rates α and β is 1/sec; the dimension of β_0 is therefore that of a velocity, namely, cm/sec if the radius of the ball is given in centimeters.

The value of the radius r for which $\alpha = \beta = \beta_0/r$ is called the **critical radius**; it is given by $r_{cr} = \beta_0/\alpha$. The corresponding mass of fissionable material, $m_{cr} = (\frac{4}{3})\pi r_{cr}{}^3 \rho$, is known as the **critical mass**.

The amount of energy liberated during fission is proportional to the number of neutrons produced. If this number increases very fast, enormous amounts of energy are liberated in a very short time. This will occur whenever the mass exceeds the critical mass. For if $r > r_{cr}$, then $\alpha - \beta > 0$ and N increases "exponentially." This is how an **atomic explosion** takes place.

4.3 Atomic reactors

Consider next the case in which there is also a steady neutron source that introduces into the ball q neutrons per unit time. Now, if there are N neutrons at time t, the number at time $t + h$, h small, will be

$$N + \alpha N h - \beta N h + q h,$$

approximately, and we obtain the differential equation

$$\frac{dN}{dt} = (\alpha - \beta)N + q. \tag{4}$$

This is a special case of Equation (1), the "given function" ϕ being the constant q. By Theorem 1, we have

$$N = N_0 e^{(\alpha-\beta)t} + e^{(\alpha-\beta)t} \int_0^t e^{-(\alpha-\beta)\xi} q \, d\xi. \tag{5}$$

Since

$$\int_0^t e^{-(\alpha-\beta)\xi} q \, d\xi = -\frac{q e^{-(\alpha-\beta)\xi}}{\alpha - \beta} \Big|_0^t = -\frac{q}{\alpha - \beta}\{e^{-(\alpha-\beta)t} - 1\},$$

we obtain

$$N = N_0 e^{(\alpha-\beta)t} + \frac{q}{\beta - \alpha}\{1 - e^{(\alpha-\beta)t}\}. \tag{6}$$

We assumed here that $\alpha - \beta \neq 0$, that is, $r \neq r_{cr}$. If $\alpha - \beta = 0$, we obtain directly from (4) that $N = N_0 + qt$.

If $\alpha > \beta$ (supercritical mass), we see from (6) that N grows exponentially; in this case, we have an explosion. If $\alpha < \beta$ (subcritical mass), then the quantity $e^{(\alpha-\beta)t}$ in (6) rapidly becomes very close to zero, and we obtain

$$N \approx \frac{q}{\beta - \alpha}. \tag{7}$$

This means that the number of neutrons is kept steady and we have a source of heat. This is the principle of an **atomic reactor**. (Note that in order for N to be large for q small, $\beta - \alpha$ must be small. Thus we need a subcritical mass that is close to the critical one. This requires delicate controls.)

4.4   The differential equation of hyperbolic sine and cosine

The following statement is analogous to Theorem 1 in §2.1. *The general solution of the equation*

$$f''(t) - f(t) = 0 \tag{8}$$

is

$$f(t) = a \cosh t + b \sinh t. \tag{9}$$

Proof. (Note how the proof differs from that of Theorem 1 in §2.1.) Using the definitions in Chapter 11, §3.5, we can verify that, for any choice of numbers a and b, the function (9) satisfies (8). Now let $f(t)$ be a given solution of (8). We set $a = f(0)$, $b = f'(0)$, and

$$g(t) = f(t) - a \cosh t - b \sinh t.$$

We must show that $g(t) = 0$ for all t. We compute that $g''(t) = g(t)$ for all t and $g(0) = 0$, $g'(0) = 0$. Set $\phi(t) = g(t) - g'(t)$. Then $\phi(0) = g(0) - g'(0) = 0$ and $\phi'(t) = g'(t) - g''(t) = g'(t) - g(t) = -\phi(t)$. Applying Theorem 1 of §1.1, we conclude that $\phi(t) = 0$ for all t. But this means that $g(t) = g'(t)$. Since $g(0) = 0$, we have $g(t) = 0$, for all t, again by Theorem 1 in §1.1.

Let $\omega > 0$ be a given number. The general solution of the equation $f''(t) - \omega^2 f(t) = 0$ is

$$f(t) = a \cosh \omega t + b \sinh \omega t.$$

This follows from (8) and (9) in the same way as Theorem 2 in §2.2 follows from Theorem 1 in §2.1. The details are left to the reader.

PROBLEMS

1. Find the solution of $f''(t) - 9f(t) = 0$ such that $f(0) = f'(0) = 2$.
2. Find the solution of $f''(t) - 4f(t) = 0$ such that $f(0) = 8$ and $f'(0) = -4$.
3. Find the solution of $f''(t) - f(t) = 0$ such that $f(0) = 1$ and $f(1) = e$.
4. Find the solution of $f'''(t) - f'(t) = 0$ such that $f(0) = 1, f'(0) = 1, f''(0) = 2$. [*Hint:* First solve for $f'(t)$.]
5. Show that the general solution of $f''(t) - \omega^2 f(t) = 0$, $\omega > 0$, is given by $f(t) = a \cosh \omega t + b \sinh \omega t$.

TECHNIQUES
OF INTEGRATION

§1 Numerical integration

In this chapter we study methods for integrating a given function. We first discuss numerical integration, a method which in principle always works, and then formal integration, a method which works only sometimes but is effective and elegant when it does.

1.1 Quadrature formulas

A function $x \mapsto f(x)$ may be "given" in various ways. It can be defined by an explicit formula involving a finite number of rational operations, extractions of roots, and substitutions into certain "known" or "elementary" functions. We mean by these the elementary transcendental functions which we studied in the preceding chapters. Finding the derivative of such a function is a purely mechanical process, provided, of course, that we know the derivatives of the "elementary" functions. No matter how complicated the formula, if we systematically apply the differentiation rules we have learned, after finitely many steps we shall obtain an explicit formula for the derivative $f'(x)$. It is even possible to program a computer to perform differentiations.

It is quite different with integration. We know, by the fundamental theorem of calculus, that every continuous function $f(x)$ has a primitive function $F(x)$. But if $f(x)$ is defined by an explicit formula, there is no reason at all why $F(x)$ should be defined by such a formula. In general, it will not. Forming primitive functions of simple

313

"known" functions can lead to interesting "new" functions. Recall, in this connection, how log and arc tan can be defined as integrals of rational functions.

In applying mathematics we also encounter functions $x \mapsto f(x)$ defined in ways other than by an explicit formula. A function may be "given" numerically, that is, we may be presented with a table of certain values of the function. This is so, for instance, when $f(x)$ is the result of an experiment of measurement. And when $f(x)$ is defined by an explicit formula, what is this formula but the description of a method for computing the number $f(x)$ for a given number x? Integration of numerically given functions is therefore a basic technique that is used quite often. Even if a function $f(x)$ is given by an explicit formula, the only way of computing

$$\int_a^b f(x)\,dx \tag{1}$$

may be to treat f as a numerically given function and to use **numerical integration.** The only way to compute a primitive function $F(x)$ may consist in tabulating F by computing $\int_a^x f(t)\,dt$ for various values of x.

We assume in what follows that $f(x)$ is continuous for $a \le x \le b$. Divide the interval into n equal subintervals, each of length

$$h = \frac{b-a}{n}.$$

The endpoints of these small intervals are

$$x_0 = a, \qquad x_1 = a + h, \qquad x_2 = a + 2h, \ldots, \qquad x_n = a + nh = b$$

(see Figure 13.1, where $n = 6$). We assume that the values

$$y_0 = f(x_0), \qquad y_1 = f(x_1), \qquad y_2 = f(x_2), \ldots, \qquad y_n = f(x_n)$$

are known, and we want to find a formula for an approximate value of the integral (1) in terms of the numbers $x_0, x_1, \ldots, x_n, y_0, y_1, y_2, \ldots, y_n$.

FIGURE 13.1

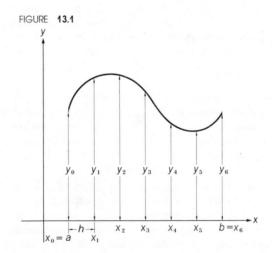

Such a formula is called a **quadrature formula.**

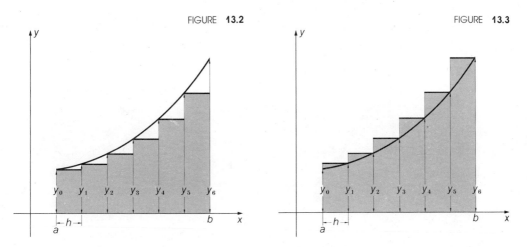

FIGURE **13.2**   FIGURE **13.3**

1.2 Trapezoidal rule

In Chapter 8, §1, we have already discussed one approximate integration formula for a monotone function $f(x)$. It amounts to replacing the area under the curve $x \mapsto f(x)$ by the sum of areas of n rectangles of width h and heights $y_0 = f(x_0)$, $y_1 = f(x_1), \ldots, y_{n-1} = f(x_{n-1})$; see Figure 13.2. Each rectangle has area hy_j (which is negative if $y_j < 0$). Hence the approximate formula reads

$$\int_a^b f(x) \, dx \approx hy_0 + hy_1 + \cdots + hy_{n-1} = \frac{b-a}{n}(y_0 + y_1 + y_2 + \cdots + y_{n-1}). \quad (2)$$

We could also use the rectangles with heights $y_1, y_2, \ldots, y_n$. Then we get (see Figure 13.3)

$$\int_a^b f(x) \, dx \approx hy_1 + hy_2 + \cdots + hy_n = \frac{b-a}{n}(y_1 + y_2 + \cdots + y_n). \quad (3)$$

For a monotone function, the true value of the integral lies between the two approximate values (2) and (3). This suggests that a better quadrature formula may be obtained by using the arithmetic mean of (2) and (3). Adding (2) and (3) and dividing by 2, we obtain the approximate formula

$$\int_a^b f(x) \, dx \approx \frac{b-a}{2n}(y_0 + 2y_1 + 2y_2 + \cdots + 2y_{n-1} + y_n)$$

$$= \frac{h}{2}(y_0 + 2y_1 + 2y_2 + \cdots + 2y_{n-1} + y_n) \quad (4)$$

$$= \frac{h}{2}y_0 + hy_1 + hy_2 + \cdots + hy_{n-1} + \frac{h}{2}y_n.$$

This is known as the **trapezoidal rule.** The numbers multiplying $y_0, y_1, y_2, \ldots, y_n$ are called weight factors.

There is another way of arriving at this formula, which also explains its name. Let $\phi(x)$ denote a continuous function which coincides with $f(x)$ for $x = x_0, x_1, x_2, \ldots, x_n$ but is linear between two consecutive values x_j and x_{j+1}; this function is graphed in

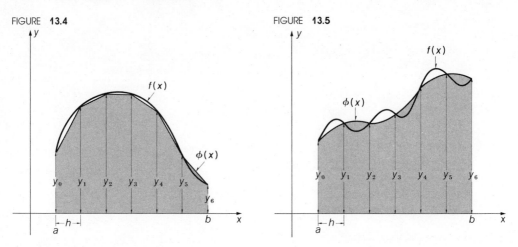

FIGURE 13.4

FIGURE 13.5

Figure 13.4 [in this figure $f(x)$ is not monotone]. We want to use $\int_a^b \phi(x)\,dx$ as an approximation to $\int_a^b f(x)\,dx$. But $\int_a^b \phi(x)\,dx$ is the sum of n integrals:

$$\int_a^b \phi(x)\,dx = \int_{x_0}^{x_1} \phi(x)\,dx + \int_{x_1}^{x_2} \phi(x)\,dx + \cdots + \int_{x_{n-1}}^{x_n} \phi(x)\,dx.$$

Since $\int_{x_0}^{x_1} \phi(x)\,dx$ is the area of a trapezoid,

$$\int_{x_0}^{x_1} \phi(x)\,dx = \frac{h}{2}(y_0 + y_1).$$

By the same token,

$$\int_{x_1}^{x_2} \phi(x)\,dx = \frac{h}{2}(y_1 + y_2),$$
$$\vdots$$
$$\int_{x_{n-1}}^{x_n} \phi(x)\,dx = \frac{h}{2}(y_{n-1} + y_n).$$

Adding these equations, we obtain

$$\int_a^b \phi(x)\,dx = \frac{h}{2}(y_0 + 2y_1 + 2y_2 + \cdots + 2y_{n-1} + y_n).$$

The right-hand side here is the same as in (4).

1.3 Simpson's rule

There is another quadrature formula which, in most cases, gives a better approximation. It is the **parabolic rule,** or **Simpson's rule.** We assume now that n is even, so that $n = 2m$, and agree to take as an approximation to $\int_a^b f(x)\,dx$ the number $\int_a^b \phi(x)\,dx$,

THOMAS SIMPSON (1710–1761) was a silkweaver by trade. His mathematical books earned him a professorship at the military college in Woolwich.

where $\phi(x)$ coincides with $f(x)$ for $x = x_0, x_1, x_2, \ldots, x_n$, and $\phi(x)$ equals a quadratic polynomial in each of the m intervals of length $2h$:$[x_0,x_2], [x_2,x_4], \ldots, [x_{2m-2},x_{2m}]$. The graph of $\phi(x)$, shown in Figure 13.5, consists of parabolic arcs.

Let us compute $\int_{x_0}^{x_2} \phi(x)\,dx$. In the interval $[x_0,x_2]$, the function $\phi(x)$ is a quadratic polynomial, $\phi(x) = Ax^2 + Bx + C$. We can substitute for x the expression $(x - x_1) + x_1$, collect terms, and write

$$\phi(x) = \alpha(x - x_1)^2 + \beta(x - x_1) + \gamma.$$

The constants α, β, and γ are found from the conditions

$$\phi(x_0) = f(x_1 - h) = f(x_0) = y_0, \qquad \phi(x_1) = f(x_1) = y_1,$$
$$\phi(x_2) = f(x_1 + h) = f(x_2) = y_2.$$

The second relation gives $\gamma = y_1$; the first and third can now be written as

$$\alpha h^2 - \beta h = y_0 - y_1, \qquad \alpha h^2 + \beta h = y_2 - y_1.$$

Adding and subtracting the two equations, we get

$$\alpha = \frac{y_0 + y_2 - 2y_1}{2h^2}, \qquad \beta = \frac{y_2 - y_0}{2h}.$$

Thus

$$\phi(x) = \frac{y_0 + y_2 - 2y_1}{2h^2}(x - x_1)^2 + \frac{y_2 - y_0}{2h}(x - x_1) + y_1$$

for $x_0 \le x \le x_2$. This polynomial has the primitive function

$$\frac{y_0 + y_2 - 2y_1}{6h^2}(x - x_1)^3 + \frac{y_2 - y_0}{4h}(x - x_1)^2 + y_1x + C.$$

Substituting $x = x_2 = x_1 + h$ and $x = x_0 = x_1 - h$ and subtracting, we obtain

$$\int_{x_0}^{x_2} \phi(x)\,dx = \frac{y_0 + y_2 - 2y_1}{6h^2}[h^3 - (-h)^3] + \frac{y_2 - y_1}{4h}[h^2 - (-h)^2] + y_1[h - (-h)]$$

$$= \frac{y_0 + y_2 - 2y_1}{3}h + 2y_1 h = \left(\frac{1}{3}y_0 + \frac{4}{3}y_1 + \frac{1}{3}y_2\right)h.$$

Thus

$$\int_{x_0}^{x_2} \phi(x)\,dx = \left(\frac{1}{3}y_0 + \frac{4}{3}y_1 + \frac{1}{3}y_2\right)h,$$

and, by the same token,

$$\int_{x_2}^{x_4} \phi(x)\,dx = \left(\frac{1}{3}y_2 + \frac{4}{3}y_3 + \frac{1}{3}y_4\right)h,$$

$$\int_{x_4}^{x_6} \phi(x)\,dx = \left(\frac{1}{3}y_4 + \frac{4}{3}y_5 + \frac{1}{3}y_6\right)h,$$

$$\vdots$$

$$\int_{x_{2m-2}}^{x_{2m}} \phi(x)\,dx = \left(\frac{1}{3}y_{2m-2} + \frac{4}{3}y_{2m-1} + \frac{1}{3}y_{2m}\right)h.$$

If we add all these equations, we obtain $\int_a^b \phi(x)\, dx$. We want to use this as an approximation to $\int_a^b f(x)\, dx$. Thus

$$\int_a^b f(x)\, dx \approx \frac{b-a}{n}\left(\frac{1}{3}y_0 + \frac{4}{3}y_1 + \frac{2}{3}y_2 + \frac{4}{3}y_3 + \frac{2}{3}y_4 + \cdots\right.$$

$$\left. + \frac{2}{3}y_{n-2} + \frac{4}{3}y_{n-1} + \frac{1}{3}y_n\right) \tag{5}$$

$$= \frac{h}{3}(y_0 + 4y_1 + 2y_2 + 4y_3 + 2y_4 + \cdots + 2y_{n-2} + 4y_{n-1} + y_n).$$

This quadrature formula is Simpson's rule.

1.4 Error estimates

Let E denote the absolute value of the error committed in evaluating the integral (1) by the above rules:

$$E = \left| \int_a^b f(x)\, dx - \frac{h}{2}(y_0 + 2y_1 + 2y_2 + \cdots + 2y_{n-1} + y_n) \right|,$$

for the trapezoidal rule;

$$E = \left| \int_a^b f(x)\, dx - \frac{h}{3}(y_0 + 4y_1 + 2y_2 + \cdots + 4y_{n-1} + y_n) \right|, \quad \text{for Simpson's rule.}$$

The precise value of E depends, of course, on the function f, on a and b, and on n. It turns out, however, that if we have certain additional knowledge about f, we can give an estimate (inequality) for E. One can show ▶ that

$$E \leq \frac{1}{12}(b-a)M_2 h^2 \quad \text{if } |f''(x)| \leq M_2, \qquad \text{for the trapezoidal rule;} \tag{6}$$

$$E \leq \frac{1}{180}(b-a)M_4 h^4 \quad \text{if } |f^{(4)}(x)| \leq M_4, \qquad \text{for Simpson's rule.} \tag{7}$$

In order to apply (6) or (7) we must know that the indicated derivatives exist and satisfy the stated inequalities in the whole interval (a,b). If h is small, h^4 is much smaller than h^2; hence we expect the error in Simpson's rule to be smaller than that for the trapezoidal rule.

Simpson's rule gives the precise value of the integral if f is a polynomial of degree at most 3; then $f^{(4)} = 0$, $M_4 = 0$, and $E = 0$. Similarly, $E = 0$ for the trapezoidal rule if f is linear.

EXAMPLES 1. Compute

$$\int_0^6 x^2\, dx \qquad \left(\int_0^6 x^2\, dx = \frac{x^3}{3}\Big|_0^6 = 72\right),$$

using the trapezoidal rule and also Simpson's rule, with 6 subintervals. Compare the actual error with the error estimates given above.

The calculations are shown in the following table. Here j denotes the subscript of the point considered and v_j denotes the "weight factor." Thus for the trapezoidal rule $v_j = 1$, if $j = 0$ or n, $v_j = 2$ for all other j. For Simpson's rule $v_j = 1$, if $j = 1$ and n, and for other j we have $v_j = 2$ or $v_j = 4$, according to whether j is odd or even.

j	x_j	$y_j = x_j^2$	Trapezoidal Rule		Simpson's Rule	
			v_j	$v_j y_j$	v_j	$v_j y_j$
0	0	0	1	0	1	0
1	1	1	2	2	4	4
2	2	4	2	8	2	8
3	3	9	2	18	4	36
4	4	16	2	32	2	32
5	5	25	2	50	4	100
6	6	36	1	36	1	36
				146		216

The result of addition must be multiplied by $h/2$ (for the trapezoidal rule) or by $h/3$ (for Simpson's rule). In our case, $h = 1$. Hence

$$\int_0^6 x^2 \, dx \approx \begin{cases} \frac{1}{2} \cdot 146 = 73 & \text{trapezoidal rule,} \\ \frac{1}{3} \cdot 216 = 72 & \text{Simpson's rule.} \end{cases}$$

For the trapezoidal rule the absolute error E is 1. Estimate (6), with $a = 0$, $b = 6$, and $h = 1$, gives $E \le \frac{1}{12}(6)M_2(1)^2 = \frac{1}{2}M_2$. Since $f''(x) = 2$, $|f''(x)| \le 2$ for all x and so $M_2 = 2$. Hence $E \le 1$. For Simpson's rule we have $E = 0$ since $f(x)$ is a quadratic function.

2. Compute

$$\log 3 = \int_1^3 \frac{dx}{x}$$

by the trapezoidal rule and by Simpson's rule, using $n = 4$. Compare the error with the error estimates.

The calculations are:

j	x_j	$y_j = \dfrac{1}{x_j}$	Trapezoidal Rule		Simpson's Rule	
			v_j	$v_j y_j$	v_j	$v_j y_j$
0	1.0	1.0000	1	1.0000	1	1.0000
1	1.5	.6667	2	1.3334	4	2.6668
2	2.0	.5000	2	1.0000	2	1.0000
3	2.5	.4000	2	.8000	4	1.6000
4	3.0	.3333	1	.3333	1	.3333
				4.4667		6.6001

$$\log 3 \approx \begin{cases} \frac{1}{2} \cdot \frac{1}{2} \cdot (4.4667) = 1.1167 & \text{trapezoidal rule,} \\ \frac{1}{3} \cdot \frac{1}{2} \cdot (6.6001) = 1.1000 & \text{Simpson's rule.} \end{cases}$$

Now for $f(x) = 1/x$ we have $f''(x) = 2/x^3$, $f^{(4)}(x) = 24/x^5$. For $1 \le x \le 3$ we may apply (6) with $M_2 = 2$ and (7) with $M_4 = 24$. This gives (for $b - a = 2$, $h = \frac{1}{2}$)

$$E \le \begin{cases} \frac{1}{12} = .0833 \ldots & \text{trapezoidal rule,} \\ \frac{1}{60} = .0166 \ldots & \text{Simpson's rule.} \end{cases}$$

Actually, $\log 3 = 1.0986\ldots$, and therefore

$$E = \begin{cases} .0181\ldots & \text{trapezoidal rule,} \\ .0014\ldots & \text{Simpson's rule.} \end{cases}$$

Both errors are well below the theoretical bounds, and Simpson's rule gives a more accurate result.

PROBLEMS

Find approximate values for the following integrals using both the trapezoidal rule and Simpson's rule for the given value of n. Estimate the errors in both cases. Use the tables in the back of the book when necessary.

1. $\displaystyle\int_0^2 (1 + x^2)\,dx$ with $n = 4$.

2. $\displaystyle\int_1^2 \frac{dx}{x}$ with $n = 4$.

3. $\displaystyle\int_0^3 \frac{x}{x+1}\,dx$ with $n = 6$.

4. $\displaystyle\int_0^1 \frac{dx}{x^2+1}$ with $n = 4$.

5. $\displaystyle\int_0^{.8} xe^x\,dx$ with $n = 4$.

6. $\displaystyle\int_0^{1.2} e^{-x^2}\,dx$ with $n = 6$.

7. $\displaystyle\int_0^3 x\sqrt{1+x}\,dx$ with $n = 6$.

1.5   Truncation error and round-off error

We have discussed only two quadrature formulas and have estimated only the error due to replacing the integral by a weighted sum. This error is called the **truncation error.** In actual calculations, there are other sources of error. Often the values y_j may be the result of an experiment and therefore cannot be measured precisely. And even with precise values, when arithmetic operations are performed, it is necessary to round-off, that is, throw away digits. In practice, especially with many subdivisions, there will be a **round-off error,** and we must be able to control it.

These matters, however, as well as more sophisticated quadrature formulas, are best learned together with the technique of automatic computing.

§2   Integration by parts

The exceptional cases in which a given function $f(x)$ can be **integrated in closed form,** that is, has a primitive function (indefinite integral) $F(x)$ given by an explicit formula, are quite important. The process of finding a formula for F is called **formal integration.** In almost all cases, it involves trial and error; this task cannot be delegated to a computer. Formal integration will be discussed in this and the following sections of the present chapter.

2.1   The method

The rule for differentiating products

$$\frac{d[f(x)g(x)]}{dx} = f'(x)g(x) + f(x)g'(x) \tag{1}$$

leads to a method of evaluating integrals known as integration by parts. From (1), we have

$$\int_a^b \frac{d[f(x)g(x)]}{dx}\,dx = \int_a^b f'(x)g(x)\,dx + \int_a^b f(x)g'(x)\,dx$$

and, by the fundamental theorem of calculus,

$$f(x)g(x)\Big|_a^b = \int_a^b f'(x)g(x)\,dx + \int_a^b f(x)g'(x)\,dx,$$

that is,

$$\int_a^b f(x)g'(x)\,dx = f(x)g(x)\Big|_a^b - \int_a^b f'(x)g(x)\,dx. \tag{2}$$

In order to apply this rule to computing an integral $\int_a^b \phi(x)\,dx$, we should be able first to write $\phi(x)$ in the form $f(x)g'(x)$, and we must know a primitive function for $f'(x)g(x)$. Here is an example.

We want to compute

$$\int_0^1 x(x-1)^8\,dx.$$

We could first expand $x(x-1)^8$ and then integrate this polynomial term by term; this would be quite time consuming. Instead, we note that $(x-1)^8$ is the derivative of $g(x) = \frac{1}{9}(x-1)^9$. If we also set $f(x) = x$, then our integrand $x(x-1)^8$ can be written as $f(x)g'(x)$. On the other hand, $f'(x) = 1$ so that $f'(x)g(x) = \frac{1}{9}(x-1)^9$; we know how to integrate this. Thus (2) yields

$$\int_0^1 \underbrace{x(x-1)^8}_{f(x)g'(x)}\,dx = \underbrace{x\frac{1}{9}(x-1)^9\Big|_0^1}_{f(x)g(x)} - \int_0^1 \underbrace{\frac{1}{9}(x-1)^9}_{f'(x)g(x)}\,dx$$

$$= x\frac{1}{9}(x-1)^9\Big|_0^1 - \frac{1}{9}\frac{1}{10}(x-1)^{10}\Big|_0^1 = \frac{1}{90}.$$

If we omit the limits of integration in (2), we obtain the formula for indefinite integrals:

$$\int f(x)g'(x)\,dx = f(x)g(x) - \int f'(x)g(x)\,dx. \tag{3}$$

For instance,

$$\int x(x-1)^8\,dx = x\frac{1}{9}(x-1)^9 - \int \frac{1}{9}(x-1)^9\,dx = \frac{1}{9}x(x-1)^9 - \frac{1}{90}(x-1)^{10} + C.$$

The rule for integration by parts is best remembered, and applied, using the language of differentials (see Chapter 8, §3.3). In this language the product rule reads

$$d(fg) = g\,df + f\,dg,$$

and the rule for integrating by parts becomes

$$\int f\,dg = fg - \int g\,df. \tag{3'}$$

[We recall that for a function $f(x)$, df denotes $f'(x)\,dx$.] In applying (3') we try to choose f and g in such a way that the integral on the right, $\int g\,df$, is simpler than the one to the left, $\int f\,dg$.

We treat the example considered above [$\int x(x-1)^8\,dx$] once more, using the differential notation. We set $f(x) = x$ and $g(x) = \frac{1}{9}(x-1)^9$, so that $df = dx$ and $dg = (x-1)^8\,dx$. Then

$$\int x(x-1)^8\,dx = \underbrace{\int x\,d\,\frac{1}{9}(x-1)^9}_{f\,dg} = \underbrace{x\,\frac{1}{9}(x-1)^9}_{fg} - \underbrace{\int \frac{1}{9}(x-1)^9\,dx}_{g\,df}$$

$$= \frac{1}{9}x(x-1)^9 - \frac{1}{90}(x-1)^{10} + C.$$

EXAMPLE Evaluate $\int x^2 e^x\,dx$.

ANSWER Set $f = x^2$, $dg = e^x\,dx$. Then $df = 2x\,dx$, $g = \int e^x\,dx = e^x$, and

$$\int x^2 e^x\,dx = x^2 e^x - \int e^x(2x)\,dx = x^2 e^x - 2\int xe^x\,dx.$$

Next consider $\int xe^x\,dx$. We must integrate by parts once more. Let $f = x$, $dg = e^x\,dx$. Then $df = dx$, $g = \int e^x\,dx = e^x$, and

$$\int xe^x\,dx = xe^x - \int e^x\,dx = xe^x - e^x + C.$$

Putting the pieces together, we obtain

$$\int x^2 e^x\,dx = x^2 e^x - 2xe^x + 2e^x + C.$$

We should always check an integration by differentiation. In our case, we have

$$\frac{d}{dx}(x^2 e^x - 2xe^x + 2e^x + C) = 2xe^x + x^2 e^x - 2e^x - 2xe^x + 2e^x = x^2 e^x,$$

as expected.

PROBLEMS

Evaluate the following integrals using the method of integration by parts.

1. $\int x(1+2x)^6\,dx$.

2. $\int xe^{-2x}\,dx$.

3. $\int_0^1 x(2x-1)^5\,dx$.

4. $\int_0^\infty x^2 e^{-4x}\,dx$.

5. $\int x\cos\frac{x}{2}\,dx$.

6. $\int x(2-3x)^8\,dx$.

7. $\int_0^{\pi/6} x\sin 3x\,dx$.

8. $\int_0^1 \frac{x\,dx}{(1+4x)^3}$.

9. $\int x\sqrt{1+x}\,dx$.

10. $\int_0^3 \frac{x\,dx}{\sqrt{1+5x}}$.

11. $\int_{-1}^3 \frac{3x+1}{(x+2)^3}\,dx$.

12. $\int \frac{2x-3}{\sqrt{x-3}}\,dx$.

13. $\int_0^{\pi/2} x^2 \sin x\,dx$.

14. $\int x^2 \cos 4x\,dx$.

15. $\int x^3 \sin x^2\,dx$.

[*Hint:* Let $u = x^2$, then integrate by parts.]

16. $\int \sqrt{x} \cos \sqrt{x} \; dx.$ **17.** $\int x^2 \sqrt{5 - 2x} \; dx.$ **18.** $\int_{-1}^{2} \frac{x^2 \; dx}{\sqrt{3 - x}}.$

2.2 Integrals of certain elementary functions

We shall illustrate the power of integration by parts by several examples. At the same time we shall find primitive functions for some of the elementary functions which we did not yet integrate.

EXAMPLES 1. Evaluate $\int \log x \; dx.$
ANSWER Set $f = \log x$, $dg = dx$. Then $df = (1/x) \; dx$, $g = \int dx = x$, and

$$\int \log x \; dx = (\log x)x - \int x \left(\frac{1}{x} dx \right) = x \log x - x + C.$$

2. Find a primitive function for $x \mapsto x \log x$.
ANSWER Set $f = \log x$, $dg = x \; dx$. Then $df = (1/x) \; dx$, $g = \int x \; dx = x^2/2$, and

$$\int x \log x \; dx = (\log x)\frac{x^2}{2} - \int \frac{x^2}{2}\left(\frac{1}{x} dx\right) = \frac{x^2}{2}\log x - \frac{1}{2}\int x \; dx = \frac{x^2}{2}\log x - \frac{x^2}{4} + C.$$

3. Find a primitive function for $x \mapsto$ arc tan x.
ANSWER Set $f = $ arc tan x, $dg = dx$. Then $df = [1/(1 + x^2)] \; dx$, $g = x$, and

$$\int \text{arc tan } x \; dx = (\text{arc tan } x)x - \int x \left(\frac{1}{1 + x^2} dx \right).$$

The last integral can be evaluated by the substitution $1 + x^2 = t$, $2x \; dx = dt$; we obtain

$$\int \text{arc tan } x \; dx = x \text{ arc tan } x - \frac{1}{2}\log(1 + x^2) + C.$$

4. Evaluate $\int$ arc cot $x \; dx$.
ANSWER Use the same derivation as above.

$$\int \text{arc cot } x \; dx = x \text{ arc cot } x + \frac{1}{2}\log(1 + x^2) + C.$$

5. Find primitive functions for $x \mapsto$ arc sin x and $x \mapsto$ arc cos x.
ANSWER

$$\int \text{arc sin } x \; dx = x \text{ arc sin } x - \int x \; d \text{ arc sin } x$$

$$= x \text{ arc sin } x - \int \frac{x}{\sqrt{1 - x^2}} dx = x \text{ arc sin } x + \sqrt{1 - x^2} + C$$

(where we used the substitution $1 - x^2 = t$, $-2x \; dx = dt$). We obtain similarly that

$$\int \text{arc cos } x \; dx = x \text{ arc cos } x - \sqrt{1 - x^2} + C.$$

REMARK We postpone until §4.6 finding primitive functions for arc sec x and arc csc x.

6. Evaluate $\int e^{ax} \sin bx \; dx$ (a and b fixed numbers).
ANSWER We integrate by parts twice. Set $f = \sin bx$, $dg = e^{ax} \; dx$. Then $df = b \cos bx \; dx$, $g = (1/a)e^{ax}$, and

$$\int e^{ax} \sin bx \; dx = \frac{1}{a}e^{ax} \sin bx - \frac{b}{a}\int e^{ax} \cos bx \; dx.$$

Next, set $f = \cos bx$, $dg = e^{ax}\, dx$ in the integral on the right-hand side. Then $df = -b\sin bx$, $g = (1/a)e^{ax}$, and

$$\int e^{ax} \sin bx\, dx = \frac{1}{a} e^{ax} \sin bx - \frac{b}{a}\left\{\frac{1}{a} e^{ax} \cos bx + \frac{b}{a}\int e^{ax} \sin bx\, dx\right\}$$

or

$$\int e^{ax} \sin bx\, dx = \frac{1}{a} e^{ax} \sin bx - \frac{b}{a^2} e^{ax} \cos bx - \frac{b^2}{a^2}\int e^{ax} \sin bx\, dx.$$

Transposing terms, we see that

$$\left(1 + \frac{b^2}{a^2}\right)\int e^{ax} \sin bx\, dx = \frac{1}{a^2} e^{ax}(a\sin bx - b\cos bx)$$

or finally

$$\int e^{ax} \sin bx\, dx = \frac{e^{ax}(a\sin bx - b\cos bx)}{a^2 + b^2} + C.$$

(We tacitly assumed that $a \neq 0$; the final formula is valid whenever $a^2 + b^2 \neq 0$.) We obtain similarly that

$$\int e^{ax} \cos bx\, dx = \frac{e^{ax}(a\cos bx + b\sin bx)}{a^2 + b^2} + C,$$

as the reader should verify.

PROBLEMS

In Problems 1 to 10 evaluate the given integrals using the results of this subsection.

1. $\int \log(2x + 1)\, dx$.

2. $\int_0^{\sqrt{3}} \arcsin \frac{x}{2}\, dx$.

3. $\int x \arctan x^2\, dx$.

4. $\int_0^{1/2} \frac{1}{\sqrt{x}} \arccos \sqrt{x}\, dx$.

5. $\int x^2 \arcsin(x^3 - 8)\, dx$.

6. $\int \frac{x}{\sqrt{x^2 + 1}} \operatorname{arc\,cot} \sqrt{x^2 + 1}\, dx$.

7. $\int e^x \sin 5x\, dx$.

8. $\int_0^{\infty} e^{-x} \cos 2x\, dx$.

9. $\int e^{2x} \cos 4x\, dx$.

10. $\int_0^{\infty} e^{-2x} \sin \frac{1}{2} x\, dx$.

Evaluate the following integrals using the method of integration by parts and the results of this subsection.

11. $\int x \arcsin x\, dx$.

12. $\int x \arctan \frac{x}{3}\, dx$.

13. $\int_0^{\sqrt{2}} x \arccos \frac{x}{2}\, dx$.

14. $\int x^2 \arcsin \frac{x}{6}\, dx$.

15. $\int x\, e^x \cos ax\, dx$, $a \neq 0$.

2.3 Reduction formulas

A typical reduction formula for evaluating integrals is

$$\int x^n e^x\, dx = x^n e^x - n \int x^{n-1} e^x\, dx. \tag{4}$$

This is true for all n (as we shall see soon), and since we know what the integral is for $n = 0$, we can use (4) to find $\int x^n e^x \, dx$ for every integer $n > 0$. For example, to find a primitive function for $x \mapsto x^3 e^x$, we write down (4) for $n = 3$, $n = 2$, $n = 1$:

$$\int x^3 e^x \, dx = x^3 e^x - 3 \int x^2 e^x \, dx, \qquad \int x^2 e^x \, dx = x^2 e^x - 2 \int x e^x \, dx,$$

$$\int x e^x \, dx = x e^x - \int e^x \, dx = x e^x - e^x + C,$$

and therefore

$$\int x^3 e^x \, dx = x^3 e^x - 3x^2 e^x + 3 \cdot 2x e^x - 3 \cdot 2 e^x + 3 \cdot 2C$$
$$= (x^3 - 3x^2 + 6x - 6)e^x + C'.$$

(Here C' is another constant of integration.)

In order to verify (4), it is enough to verify that the two sides have the same derivative; they do: $x^n e^x = n x^{n-1} e^x + x^n e^x - n x^{n-1} e^x$. But how was (4) discovered? By integration by parts,

$$\int x^n e^x \, dx = \int x^n \, d(e^x) = x^n e^x - \int e^x \, d(x^n) = x^n e^x - n \int x^{n-1} e^x \, dx.$$

[We also can proceed differently:

$$\int x^m e^x \, dx = \frac{1}{m+1} \int e^x \, d(x^{m+1}) = \frac{1}{m+1} x^{m+1} e^x - \frac{1}{m+1} \int x^{m+1} e^x \, dx.$$

This is the same as (4), with $m = n - 1$.]

The following reduction formulas are also consequences of integration by parts:

$$\int x^n \cos x \, dx = x^n \sin x - n \int x^{n-1} \sin x \, dx, \qquad (5)$$

$$\int x^n \sin x \, dx = -x^n \cos x + n \int x^{n-1} \cos x \, dx, \qquad (6)$$

$$\int x^\alpha (\log x)^n \, dx = \frac{x^{\alpha+1}(\log x)^n}{\alpha + 1} - \frac{n}{\alpha + 1} \int x^\alpha (\log x)^{n-1} \, dx, \qquad \alpha \neq -1. \quad (7)$$

In order to obtain (7) by integration by parts, note that $x^\alpha \, dx = d(x^{\alpha+1})/(\alpha + 1)$. The case $\alpha = -1$ can be handled by the substitution $t = \log x$, $dt = dx/x$. This yields

$$\int \frac{(\log x)^n}{x} \, dx = \frac{(\log x)^{n+1}}{n + 1} + C. \qquad (8)$$

EXAMPLE Compute $\displaystyle\int_0^{\pi/2} x^3 \sin x \, dx$.

ANSWER   Using (6) and (5), we have

$$\int_0^{\pi/2} x^3 \sin x \, dx = -x^3 \cos x \Big|_0^{\pi/2} + 3 \int_0^{\pi/2} x^2 \cos x \, dx$$

$$= -x^3 \cos x \Big|_0^{\pi/2} + 3x^2 \sin x \Big|_0^{\pi/2} - 3 \cdot 2 \int_0^{\pi/2} x \sin x \, dx$$

$$= (-x^3 \cos x + 3x^2 \sin x + 6x \cos x) \Big|_0^{\pi/2} - 6 \int_0^{\pi/2} \cos x \, dx$$

$$= (-x^3 \cos x + 3x^2 \sin x + 6x \cos x - 6 \sin x) \Big|_0^{\pi/2} = \frac{3\pi^2}{4} - 6.$$

REMARK We note a more general form of (4), (5), and (6), involving a fixed number $a \neq 0$:

$$\int x^n e^{ax} \, dx = \frac{x^n e^{ax}}{a} - \frac{n}{a} \int x^{n-1} e^{ax} \, dx, \tag{4'}$$

$$\int x^n \cos ax \, dx = \frac{1}{a} \left[x^n \sin ax - n \int x^{n-1} \sin ax \, dx \right], \tag{5'}$$

$$\int x^n \sin ax \, dx = \frac{1}{a} \left[-x^n \cos ax + n \int x^{n-1} \cos ax \, dx \right]. \tag{6'}$$

These reduction formulas can be either (i) verified by differentiating both sides, (ii) reduced to (4), (5), (6) by an appropriate change of variables [set $x = ay$, $dx = a \, dy$ in (4), (5), (6)], or (iii) proved by integration by parts.

PROBLEMS

Perform the following integrations.

1. $\int x^2 e^{-x} \, dx$.

2. $\int x^2 e^{3x} \, dx$.

3. $\int_0^2 x^2 e^{-\frac{1}{2}x} \, dx$.

4. $\int x^3 e^{\frac{1}{2}x} \, dx$.

5. $\int_0^1 x^3 e^{-2x} \, dx$.

6. $\int x^4 e^x \, dx$.

7. $\int x^2 \cos x \, dx$.

8. $\int_0^{\pi/2} x^2 \sin x \, dx$.

9. $\int x^2 \sin \frac{x}{2} \, dx$.

10. $\int_0^{\pi} x^2 \cos \frac{x}{4} \, dx$.

11. $\int x^3 \sin x \, dx$.

12. $\int x^3 \cos 2x \, dx$.

13. $\int_1^e \frac{(\log x)^4}{x} \, dx$.

14. $\int x^2 \log x \, dx$.

15. $\int (\log x)^3 \, dx$.

16. $\int x^2 (\log x)^2 \, dx$.

17. $\int \frac{(\log x)^2}{x^3} \, dx$.

18. $\int \frac{(\log x)^2}{\sqrt{x}} \, dx$.

[*Hint:* Use Equation (7) and then integrate by parts.]

2.4   More reduction formulas

We list now some more complicated reduction formulas, the first of which will turn out to be very useful.

$$\int \frac{dx}{(1 + x^2)^n} = \frac{1}{(2n - 2)} \frac{x}{(1 + x^2)^{n-1}} + \frac{2n - 3}{2n - 2} \int \frac{dx}{(1 + x^2)^{n-1}}, \qquad n \neq 1, \tag{9}$$

$$\int \sin^m x \cos^n x \, dx = - \frac{1}{m + n} \sin^{m-1} x \cos^{n+1} x$$
$$+ \frac{m - 1}{m + n} \int \sin^{m-2} x \cos^n x \, dx, \qquad m + n \neq 0, \tag{10}$$

$$\int \cos^n x \, dx = \frac{1}{n} \cos^{n-1} x \sin x + \frac{n - 1}{n} \int \cos^{n-2} x \, dx, \qquad n \neq 0. \tag{11}$$

The reader *should* verify these formulas by checking that both sides of each formula have identical derivatives. He *may* want to find by himself how these formulas were discovered, by integration by parts. (One should not memorize reduction formulas, but it pays to remember which kind of reduction formulas are available.) Note that the reduction formula (9) permits us to compute the integral $\int (1 + x^2)^{-n-1} \, dx$ for every integer $n > 0$, since repeated application of this formula reduces the integral to $\int (1 + x^2)^{-1}\, dx = \arctan x + C$. Similarly, (10) enables us to compute $\int \cos^n x\, dx$ for all integers $m > 0$. Now we see that the reduction formula (10) is sufficient for calculating the integral $\int \sin^m x \cos^n x\, dx$ for all integers $m \geq 0$, $n \geq 0$. Indeed, by repeated application of (10), we can reduce this integral either to $\int \cos^n x\, dx$ (if m is even) or to $\int \cos^n x \sin x\, dx = -(\cos^{n+1} x)/(n + 1) + C$ (if m is odd).

EXAMPLES 1. Compute $\displaystyle\int_0^1 (1 + x^2)^{-2}\, dx$.

ANSWER Using (9) with $n = 2$, we obtain

$$\int_0^1 \frac{dx}{(1 + x^2)^2} = \frac{1}{2}\frac{x}{1 + x^2}\bigg|_0^1 + \frac{1}{2}\int_0^1 \frac{dx}{1 + x^2} = \left(\frac{1}{2}\frac{x}{1 + x^2} + \frac{1}{2}\arctan x\right)\bigg|_0^1 = \frac{1}{4} + \frac{\pi}{8}.$$

2. Verify that, for $m \geq 2$,

$$\int_0^{\pi/2} \sin^m x\, dx = \frac{m - 1}{m}\int_0^{\pi/2} \sin^{m-2} x\, dx. \tag{12}$$

ANSWER This follows from (10) by setting $n = 0$ and noting that $\cos x \sin^{m-1} x\, |_0^{\pi/2} = 0$.

3. What is $\displaystyle\int_0^{\pi/2} \sin^6 x\, dx$?

ANSWER Applying (12) we get

$$\int_0^{\pi/2} \sin^6 x\, dx = \frac{5}{6}\int_0^{\pi/2} \sin^4 x\, dx = \frac{5}{6}\cdot\frac{3}{4}\int_0^{\pi/2} \sin^2 x\, dx$$

$$= \frac{5}{6}\cdot\frac{3}{4}\cdot\frac{1}{2}\int_0^{\pi/2} dx = \frac{5}{6}\cdot\frac{3}{4}\cdot\frac{1}{2}\cdot\frac{\pi}{2}.$$

4. What is $\displaystyle\int_0^{\pi/2} \sin^7 x\, dx$?

ANSWER Applying (12) one obtains

$$\int_0^{\pi/2} \sin^7 x\, dx = \frac{6}{7}\int_0^{\pi/2} \sin^5 x\, dx = \frac{6}{7}\cdot\frac{4}{5}\int_0^{\pi/2} \sin^3 x\, dx$$

$$= \frac{6}{7}\cdot\frac{4}{5}\cdot\frac{2}{3}\int_0^{\pi/2} \sin x\, dx = \frac{6}{7}\cdot\frac{4}{5}\cdot\frac{2}{3}.$$

5. Compute $\displaystyle\int_0^{\pi/2} \sin^{2k} x\, dx$, $k \geq 0$ an integer.

ANSWER Relation (12) with $m = 2k$, $k \geq 1$ gives

$$\int_0^{\pi/2} \sin^{2k} x\, dx = \frac{2k - 1}{2k}\int_0^{\pi/2} \sin^{2k-2} x\, dx. \tag{*}$$

On the other hand, for $k = 0$ the desired integral is $\pi/2$. Repeated application of the reduction formula (*) therefore gives

$$\int_0^{\pi/2} \sin^{2k} x\, dx = \frac{2k - 1}{2k}\cdot\frac{2k - 3}{2k - 2}\cdot\frac{2k - 5}{2k - 4}\cdots\frac{1}{2}\cdot\frac{\pi}{2}. \tag{13}$$

6. Compute $\displaystyle\int_0^{\pi/2} \sin^{2k+1} x\, dx$, $k \geq 0$ an integer.

ANSWER First of all, we have for $k = 0$, $\int_0^{\pi/2} \sin x\, dx = \cos x\, |_0^{\pi/2} = 1$. Also, Relation (12) yields, for $m = 2k + 1$, $k \geq 1$, the reduction formula

$$\int_0^{\pi/2} \sin^{2k+1} x\, dx = \frac{2k}{2k + 1}\int_0^{\pi/2} \sin^{2k-1} x\, dx. \tag{14}$$

Repeated application of this reduction formula gives

$$\int_0^{\pi/2} \sin^{2k+1} x \, dx = \frac{2k}{2k+1} \cdot \frac{2k-2}{2k-1} \cdot \frac{2k-4}{2k-3} \cdots \frac{2}{3}. \qquad (15)$$

PROBLEMS

Perform the following integrations using the appropriate reduction formula.

1. $\int \dfrac{dx}{(1+9x^2)^2}$.

2. $\int_0^2 \dfrac{dx}{(4+x^2)^2}$.

3. $\int \dfrac{dx}{(1+x^2)^3}$.

4. $\int \dfrac{dx}{(9x^2+4)^3}$.

5. $\int \cos^4 x \sin^3 x \, dx$.

6. $\int_0^{\pi/2} \sin^2 x \cos^2 x \, dx$.

7. $\int \cos^5 2x \, dx$.

8. $\int \sin^4 \dfrac{x}{2} \, dx$.

9. $\int \cos^5 4x \sin^3 4x \, dx$.

10. $\int_0^{\pi/2} \sin^5 x \, dx$.

11. $\int_0^{\pi/2} \sin^6 x \, dx$.

12. $\int \cos^4 x \sin^4 x \, dx$.

§3 Integration of rational functions

It turns out that, having defined the transcendental functions $x \mapsto \log |x|$ and $x \mapsto \arctan x$, which have derivatives $1/x$ and $1/(1+x^2)$, respectively, we are in a position to write down *an explicit formula for the indefinite integral of every rational function.*

This is so since there are certain special rational functions, called **partial fractions,** that can all be integrated in closed form, and since every rational function can be written as a sum of a polynomial and of partial fractions.

3.1 Partial fractions

A partial fraction is a rational function either of the form constant/(linear function)^n, or of the form (linear function)/(quadratic function without real roots)^n, where n is a positive integer. For instance,

$$\frac{2}{x-3}, \qquad \frac{6}{(x+100)^{100}}, \qquad \frac{7+3x}{2x^2+2x+10}, \qquad \frac{3}{(x^2+2x+10)^3}$$

are partial fractions. We do *not* call $\dfrac{1}{(x^2-3x+2)^2}$ a partial fraction, since the denominator has (real) roots $x = 2$ and $x = 1$. Our first task is to learn how to integrate partial fractions.

A partial fraction with the power of a linear function in the denominator can always be integrated by the method of substitution.

EXAMPLES 1. What is $\int \dfrac{2 \, dx}{3x-4}$?

ANSWER Set $t = 3x - 4$; then $dt = 3 \, dx$, so that

$$\int \frac{2 \, dx}{3x-4} = \int \frac{2\frac{1}{3} \, dt}{t} = \frac{2}{3} \log |t| + C = \frac{2}{3} \log |3x - 4| + C.$$

2. What is $\int \dfrac{dx}{(2x - 5)^2}$?

ANSWER   Set $t = 2x - 5$, so that $dt = 2\,dx$. Thus

$$\int \frac{dx}{(2x - 5)^2} = \int \frac{\frac{1}{2}\,dt}{t^2} = \frac{1}{2}\int t^{-2}\,dt = -\frac{1}{2}t^{-1} + C = -\frac{1}{2}(2x - 5)^{-1} + C.$$

The integral of a partial fraction of the form (quadratic polynomial without real roots)$^{-n}$ can be reduced, by substitution, to an integral of the form

$$\int \frac{dx}{(1 + x^2)^n}$$

which can be computed using the reduction formula (9) of §2.3. The procedure involves completing the square. We explain it by several examples.

EXAMPLES   3. Compute $\int \dfrac{dx}{26 - 4x + 4x^2}$.

ANSWER   We check that $26 - 4x + 4x^2$ has no real roots and proceed to write it in the form (constant) $(1 + t^2)$. We complete the square and obtain

$$26 - 4x + 4x^2 = 25 + (2x - 1)^2 = 25\left[1 + \frac{1}{25}(2x - 1)^2\right]$$

$$= 25\left[1 + \left(\frac{2x - 1}{5}\right)^2\right].$$

This suggests the substitution $t = (2x - 1)/5$, so that $dt = \frac{2}{5}\,dx$, $26 - 4x + 4x^2 = 25(1 + t^2)$. Thus

$$\int \frac{dx}{26 - 4x + 4x^2} = \int \frac{\frac{5}{2}\,dt}{25(1 + t^2)} = \frac{1}{10}\int \frac{dt}{1 + t^2}$$

$$= \frac{1}{10}\text{ arc tan } t + C = \frac{1}{10}\text{ arc tan }\frac{2x - 1}{5} + C.$$

The reader should check by differentiation.

4. What is $\int \dfrac{dx}{(9x^2 + 12x + 13)^2}$?

ANSWER   The denominator has no real roots, so we rewrite the denominator as a constant times $(1 + t^2)^2$. Completing the square, we obtain

$$9x^2 + 12x + 13 = (3x + 2)^2 + 9 = 9\left[1 + \left(\frac{3x + 2}{3}\right)^2\right] = 9\left[1 + \left(x + \frac{2}{3}\right)^2\right].$$

Hence, with $t = x + \frac{2}{3}$, $dt = dx$, we get [using the reduction formula (9) in §2.3]

$$\int \frac{dx}{(9x^2 + 12x + 13)^2} = \int \frac{dt}{81(1 + t^2)^2} = \frac{1}{81}\int \frac{dt}{(1 + t^2)^2}$$

$$= \frac{1}{81}\cdot\frac{1}{2}\frac{t}{1 + t^2} + \frac{1}{81}\cdot\frac{1}{2}\int \frac{dt}{1 + t^2}$$

$$= \frac{1}{162}\cdot\frac{t}{1 + t^2} + \frac{1}{162}\text{ arc tan } t + C$$

$$= \frac{1}{162}\cdot\frac{x + \frac{2}{3}}{1 + (x + \frac{2}{3})^2} + \frac{1}{162}\text{ arc tan }\left(x + \frac{2}{3}\right) + C$$

$$= \frac{1}{54}\cdot\frac{3x + 2}{9x^2 + 12x + 13} + \frac{1}{162}\text{ arc tan }\left(x + \frac{2}{3}\right) + C.$$

We are now in a position to integrate every partial fraction of the form (linear function)/(quadratic function)^n. We shall show, by examples, how every such partial fraction can be written as a fraction whose integral can be reduced to $\int (1 + t^2)^{-2} dt$ plus a fraction which can be integrated by the method of substitution.

EXAMPLES 5. Evaluate $\int \dfrac{x + 1}{1 + x^2} dx$.

ANSWER   The derivative of $1 + x^2$ is $2x$. We write the numerator in the integral in the form $x + 1 = \frac{1}{2} 2x + 1$. Thus

$$\int \frac{x + 1}{1 + x^2} dx = \int \frac{\frac{1}{2} 2x + 1}{1 + x^2} dx = \frac{1}{2} \int \frac{2x \, dx}{1 + x^2} + \int \frac{dx}{1 + x^2}.$$

The first integral on the right can be computed by the substitution $t = 1 + x^2$, $dt = 2x \, dx$.

$$\int \frac{x + 1}{1 + x^2} dx = \frac{1}{2} \int \frac{dt}{t} + \int \frac{dx}{1 + x^2} = \frac{1}{2} \log |t| + \arctan x + C$$

$$= \frac{1}{2} \log(1 + x^2) + \arctan x + C.$$

6. Evaluate $\int \dfrac{3 + 16x}{26 - 4x + 4x^2} dx$.

ANSWER   The denominator has no real roots. We compute the derivative of $26 - 4x + 4x^2$; it is $-4 + 8x$. The numerator in our integral is $3 + 16x = 2(-4 + 8x) + 11$. Thus

$$\int \frac{3 + 16x}{26 - 4x + 4x^2} dx = \int \frac{2(-4 + 8x) + 11}{26 - 4x + 4x^2} dx$$

$$= 2 \int \frac{-4 + 8x}{26 - 4x + 4x^2} dx + 11 \int \frac{dx}{26 - 4x + 4x^2}.$$

The first integral on the right can be integrated by the substitution $t = 26 - 4x + 4x^2$, $dt = -4 + 8x$:

$$2 \int \frac{-4 + 8x}{26 - 4x + 4x^2} dx = 2 \int \frac{dt}{t} = 2 \log |t| + C = 2 \log(26 - 4x + 4x^2) + C.$$

(We need no absolute value sign since $26 - 4x + 4x^2 > 0$ for all x!) Using the result of Example 3, we conclude that

$$\int \frac{3 + 16x}{26 - 4x + 4x^2} dx = \frac{11}{10} \arctan \frac{2x - 1}{5} + 2 \log(26 - 4x + 4x^2) + C.$$

7. Evaluate $\int \dfrac{x \, dx}{(9x^2 + 12x + 13)^2}$.

ANSWER   The derivative of $9x^2 + 12x + 13$ is $18x + 12$. We write the numerator of the integrand as $x = \frac{1}{18}(18x + 12) - \frac{2}{3}$ and obtain

$$\int \frac{x \, dx}{(9x^2 + 12x + 13)^2} = \int \frac{\frac{1}{18}(18x + 12) - \frac{2}{3}}{(9x^2 + 12x + 13)^2} dx$$

$$= \frac{1}{18} \int \frac{18x + 12}{(9x^2 + 12x + 13)^2} dx - \frac{2}{3} \int \frac{dx}{(9x^2 + 12x + 13)^2}.$$

The first integral on the right can be integrated by the substitution $t = 9x^2 + 12x + 13$, $dt = (18x + 12) \, dx$:

$$\int \frac{(18x + 12) \, dx}{(9x^2 + 12x + 13)^2} = \int \frac{dt}{t^2} = - \frac{1}{t} + C = - \frac{1}{9x^2 + 12x + 13} + C.$$

The second integral has been considered in Example 4 above. Using this result we see that

$$\int \frac{x\,dx}{(9x^2 + 12x + 13)^2}$$

$$= -\frac{1}{18} \cdot \frac{1}{9x^2 + 12x + 13} - \frac{2}{3} \cdot \frac{1}{54} \cdot \frac{3x + 2}{9x^2 + 12x + 13} - \frac{2}{3} \cdot \frac{1}{162} \text{arc tan}\left(x + \frac{2}{3}\right) + C$$

$$= -\frac{6x + 13}{162(9x^2 + 12x + 13)} - \frac{1}{243}\text{arc tan}\left(x + \frac{2}{3}\right) + C.$$

The examples in this section are typical. They show that *a primitive function of any partial fraction is the sum of a rational function, a logarithmic function, and an arc tangent function.*

Of course, in any given case, not all three types need be present. [By a logarithmic function we mean, in this connection, a function of the form $x \mapsto \alpha \log|ax + b|$ or $x \mapsto \alpha \log|ax^2 + bx + c|$. By an arc tangent function we mean a function of the form $x \mapsto \alpha \text{ arc tan}(ax + b)$.]

PROBLEMS

Perform the following integrations. When an indefinite integral is asked for, check your answer by differentiation.

1. $\int_1^3 \frac{1}{(2x - 1)^2}\,dx.$

2. $\int_0^1 \frac{1}{(2 - x)^3}\,dx.$

3. $\int \frac{1}{8 - 7x}\,dx.$

4. $\int \frac{1}{(3x - 7)^5}\,dx.$

5. $\int_2^3 \frac{1}{3x - 5}\,dx.$

6. $\int_1^{6/5} \frac{1}{(5x - 4)^6}\,dx.$

7. $\int \frac{dx}{x^2 + 4x + 5}.$

8. $\int_3^5 \frac{dx}{x^2 - 6x + 13}.$

9. $\int \frac{dx}{2x^2 + 2x + 1}.$

10. $\int \frac{dx}{5x^2 - 2x + 1}.$

11. $\int \frac{dx}{2x^2 - 6x + 17}.$

12. $\int_0^1 \frac{dx}{x^2 + x + 1}.$

13. $\int \frac{dx}{(x^2 + 6x + 10)^2}.$

14. $\int_1^2 \frac{dx}{(x^2 - 2x + 2)^2}.$

15. $\int \frac{dx}{(2x^2 + 2x + 1)^2}.$

16. $\int \frac{dx}{(5x^2 - 2x + 1)^2}.$

17. $\int \frac{dx}{(x^2 + 8x + 17)^3}.$

18. $\int_0^1 \frac{dx}{(x^2 - 2x + 2)^3}.$

19. $\int_0^{1/2} \frac{3x + 2}{4x^2 + 1}\,dx.$

20. $\int \frac{4x - 1}{x^2 + 9}\,dx.$

21. $\int \frac{3x - 1}{x^2 + 4x + 5}\,dx.$

22. $\int \frac{1 - 2x}{4x^2 + 4x + 5}\,dx.$

23. $\int_0^2 \frac{x\,dx}{x^2 - 4x + 8}.$

24. $\int \frac{(3x + 5)\,dx}{9x^2 - 6x + 10}.$

25. $\int \frac{x - 2}{(2x^2 + 2x + 1)^2}\,dx.$ (See Problem 15.)

26. $\int \frac{(3 - 5x)\,dx}{(5x^2 - 2x + 1)^2}.$ (See Problem 16.)

27. $\int \frac{(x + 1)\,dx}{(x^2 + 8x + 17)^3}.$ (See Problem 17.)

28. $\int_0^1 \frac{(3 - x)\,dx}{(x^2 - 2x + 2)^3}.$ (See Problem 18.)

3.2   Decomposition into partial fractions. Theory

The key to integrating all rational functions in closed form is the so-called **fundamental theorem of algebra,** which says that every nonconstant polynomial has roots, provided that one has extended the real number system and also admits so-called complex numbers. A form of the theorem which involves no complex numbers asserts that *every polynomial (of positive degree, with real coefficients) can be written as a product of linear and quadratic polynomials.*

There are many proofs of the fundamental theorem of algebra; most of them are due to Gauss, the greatest mathematician of the nineteenth century. The truth of the theorem had been suspected, however, long before Gauss.

Unfortunately there is no simple way of actually finding the roots or the linear and quadratic factors of a given polynomial. During the sixteenth century, Italian mathematicians found explicit formulas (associated with the name of Cardano) for the roots of third- and fourth-degree equations, thus generalizing the age-old formula for solving quadratic equations. But the search for an explicit formula for solving all fifth-degree equations was unsuccessful, and finally Abel (at the age of 22) showed that there can be no such formula. An even more remarkable discovery was made by Galois (who died before reaching the age of 21). Galois found that, given any specific polynomial of degree $n \geq 5$, one cannot, except under very special circumstances, obtain its roots from the coefficients by performing rational operations (additions, subtractions, multiplications, divisions) and by extracting roots.

The fundamental theorem of algebra will not be proved in this book. A consequence of this theorem is the following proposition.

Every rational function that is not a polynomial can be written, uniquely, as the sum of a polynomial and one or more partial fractions.

Assuming this result, and noting the results of §3.1, we obtain

HIERONIMO CARDANO (1501–1576), a colorful Renaissance character and a many-sided scientist, published in 1545 a method for solving cubic equations. He acknowledged that a special case of this method had been communicated to him, under pledge of secrecy, by Tartaglia. A bitter polemic followed.

CARL FRIEDRICH GAUSS (1777–1855) was the son of a poor laborer. He recalled, half jokingly, that he was able to do sums before he could talk. A perceptive grammar school teacher noticed Gauss' genius. He secured the patronage that enabled Gauss to continue his education. The young Gauss hesitated between philology and mathematics; he made his choice after his first big discovery—the ruler and compass construction of a regular polygon with 17 sides.

While he was still alive, Gauss was recognized as an equal of Archimedes and Newton. The fame of Goettingen as a mathematical center dates from Gauss, who was professor there and director of the Goettingen observatory. Gauss' interests embraced all of mathematics, astronomy, physics, and geodesy. This aloof, awe-inspiring scholar published somewhat reluctantly, and his notebooks revealed that he anticipated many discoveries made later by other mathematicians.

NIELS HENRIK ABEL (1802–1829). His work on fifth-degree equations is only one of his great achievements. Abel contributed to the rigorous theory of infinite series, and his discovery of "elliptic" and other transcendental functions initiated a new era in mathematical analysis.

At 16, Abel began, under the influence of a perceptive teacher, to read the works of Newton, Euler, and Lagrange; after a few years he started to make his discoveries. Meanwhile his father died, and Abel became responsible for a large family; he was poor for the rest of his life. A Norwegian government subsidy enabled him to visit Germany and France, but the leading mathematicians there failed to recognize his genius. Abel died of tuberculosis at the age of 27.

ÉVARISTE GALOIS (1811–1832) was persecuted by his government for radicalism and by some of his teachers for impertinence. He was twice denied admission to the prestigious École Polytechnique, expelled from the École Normale, and in jail for several months. On May 30, 1832, Galois fought a duel and received a fatal wound. The night before, he composed a farewell letter to a friend, describing his discoveries. This letter contains the "Galois theory," one of the foundation stones of modern algebra.

Theorem 1. *A primitive function of a rational function is always a sum of a rational function, logarithmic functions, and arc tangent functions.*

For example, since

$$\frac{x^3 - x^2 + x + 2}{x^2(x^2 + 1)} = \frac{2}{x^2} + \frac{1}{x} - \frac{3}{x^2 + 1}$$

(as the reader should check), we have

$$\int \frac{x^3 - x^2 + x + 2}{x^2(x^2 + 1)} \, dx = 2 \int \frac{dx}{x^2} + \int \frac{dx}{x} - 3 \int \frac{dx}{x^2 + 1}$$

$$= -\frac{2}{x} + \log|x| - 3 \arctan x + C.$$

This result can (and should) be verified by differentiation.

3.3   Decomposition into partial fractions. Practice

In order to find an explicit primitive function for a rational function, we must decompose this function into partial fractions. This can be done effectively only for relatively simple cases. Thus it may be necessary to use numerical integration even for rational functions, and the main significance of Theorem 1 is theoretical.

We shall explain how the decomposition can be accomplished in some cases. Let $x \mapsto f(x)$ be a given rational function. Then $f(x) = \phi(x)/\psi(x)$, where ϕ and ψ are polynomials, and ϕ is not divisible by ψ.

If the degree of ϕ is not less than that of ψ, we apply long division and obtain $\phi(x) = q(x)\psi(x) + r(x)$, where q and r are polynomials and the degree of r is less than that of ψ. Thus

$$f(x) = \frac{\phi(x)}{\psi(x)} = q(x) + \frac{r(x)}{\psi(x)}.$$

In order to integrate f, it suffices to integrate r/ψ, since we know how to integrate q. Thus we may assume, to begin with, that the degree of ϕ is less than that of ψ, and that the two polynomials have no common linear or quadratic factors.

The first and most difficult step in representing ϕ/ψ as a sum of partial fractions is to represent the denominator ψ as a product of a constant A and factors, each of which is either a power of a linear function $x - \alpha$, or a power of a quadratic function $x^2 + \beta x + \gamma$ with no (real) roots. Thus

$$\psi(x) = A(x - \alpha_1)^{\nu_1} \cdots (x - \alpha_k)^{\nu_k}(x^2 + \beta_1 x + \gamma_1)^{\mu_1} \cdots (x^2 + \beta_l x + \gamma_l)^{\mu_l}, \quad (1)$$

where the ν's and μ's are positive integers, the α's are distinct numbers, and the quadratic functions $x^2 + \beta x + \gamma$ are all distinct.

Let us assume that the denominator ψ is already given in the form (1). We may also assume that $A = 1$, since we can divide both numerator and denominator by A. Now we must find partial fractions whose sum is $\phi(x)/\psi(x)$. Only powers of the linear and quadratic factors, with exponents not greater than those in (1), can occur as denominators. Once this is realized, finding the actual representation of ϕ/ψ as a sum of partial fractions is reduced to solving a system of simultaneous linear equations.

This is best explained by examples. Suppose we want to decompose

$$\frac{x^2 - 2x + 4}{x(x - 1)^2}$$

into partial fractions. The factors x and $x - 1$ occur with exponents 1 and 2, respectively. Therefore, if we can write the given function as a sum of partial fractions, we *may* encounter the denominators x, $x - 1$, $(x - 1)^2$, and no others. Since the factors x and $x - 1$ are linear (and not quadratic), the numerators of the partial fractions will be constants (and not linear functions). Thus we look for numbers A, B, C such that

$$\frac{x^2 - 2x + 4}{x(x - 1)^2} = \frac{A}{x} + \frac{B}{x - 1} + \frac{C}{(x - 1)^2}.$$

(We do not yet know what the numbers A, B, C are; hence the method we use is called the *method of indeterminate coefficients*.) If we add the three partial fractions above, we obtain

$$\frac{A}{x} + \frac{B}{x - 1} + \frac{C}{(x - 1)^2} = \frac{A(x - 1)^2 + Bx(x - 1) + Cx}{x(x - 1)^2}$$

$$= \frac{Ax^2 - 2Ax + A + Bx^2 - Bx + Cx}{x(x - 1)^2}$$

$$= \frac{(A + B)x^2 + (-2A - B + C)x + A}{x(x - 1)^2}.$$

We want the numerator to be $x^2 - 2x + 4$, that is, we want to have

$$A + B = 1,$$
$$-2A - B + C = -2,$$
$$A = 4.$$

Solving this system of simultaneous equations, we obtain $A = 4, B = -3, C = 3$. Thus

$$\frac{x^2 - 2x + 4}{x(x - 1)^2} = \frac{4}{x} - \frac{3}{x - 1} + \frac{3}{(x - 1)^2}. \qquad (*)$$

Consider next

$$\frac{2x^3}{x^4 - 1}.$$

The denominator can be written as

$$x^4 - 1 = (x^2 - 1)(x^2 + 1) = (x - 1)(x + 1)(x^2 + 1).$$

Each factor in the denominator occurs in the first power; the last factor is a quadratic function without real roots. Therefore, we need numbers A, B, C, and D such that

$$\frac{2x^3}{x^4 - 1} = \frac{2x^3}{(x - 1)(x + 1)(x^2 + 1)} = \frac{A}{x - 1} + \frac{B}{x + 1} + \frac{Cx + D}{x^2 + 1}$$

$$= \frac{A(x + 1)(x^2 + 1) + B(x - 1)(x^2 + 1) + (Cx + D)(x - 1)(x + 1)}{x^4 - 1}.$$

The numerator is

$$A(x^3 + x^2 + x + 1) + B(x^3 - x^2 + x - 1) + (Cx + D)(x^2 - 1)$$
$$= (A + B + C)x^3 + (A - B + D)x^2 + (A + B - C)x + A - B - D.$$

We want the numerator to be $2x^3$, so that we need

$$A + B + C = 2,$$
$$A - B + D = 0,$$
$$A + B - C = 0,$$
$$A - B - D = 0.$$

Solving this system of equations, we obtain $A = B = \frac{1}{2}$, $C = 1$, $D = 0$. Thus

$$\frac{2x^3}{x^4 - 1} = \frac{\frac{1}{2}}{x - 1} + \frac{\frac{1}{2}}{x + 1} + \frac{x}{x^2 + 1}.$$

As a final example, consider

$$\frac{x^5}{(1 + x^2)^3}.$$

We want to find numbers A, B, C, D, E, F such that

$$\frac{x^5}{(1 + x^2)^3} = \frac{A + Bx}{1 + x^2} + \frac{C + Dx}{(1 + x^2)^2} + \frac{E + Fx}{(1 + x^2)^3}.$$

Adding the partial fractions, we obtain

$$\frac{(A + Bx)(1 + x^2)^2 + (C + Dx)(1 + x^2) + E + Fx}{(1 + x^2)^3}$$

$$= \frac{A + 2Ax^2 + Ax^4 + Bx + 2Bx^3 + Bx^5 + C + Cx^2 + Dx + Dx^3 + E + Fx}{(1 + x^2)^3}$$

$$= \frac{(A + C + E) + (B + D + F)x + (2A + C)x^2 + (2B + D)x^3 + Ax^4 + Bx^5}{(1 + x^2)^3}.$$

The numerator must be 1, so that we obtain the equations

$$A + C + E = 0, \qquad B + D + F = 0,$$
$$2A + C = 0, \qquad 2B + D = 0,$$
$$A = 0, \qquad B = 1.$$

Solving these we see that $A = C = E = 0$, $B = 1$, $D = -2$, $F = 1$. Thus

$$\frac{x^5}{(1 + x^2)^3} = \frac{x}{1 + x^2} - \frac{2x}{(1 + x^2)^2} + \frac{x}{(1 + x^2)^3}.$$

EXAMPLES 1. Find a primitive function of $x \mapsto \dfrac{x^2 - 2x + 4}{x(x - 1)^2}$.

ANSWER Using the partial fraction decomposition (*) above, we have

$$\int \frac{x^2 - 2x + 4}{x(x - 1)^2}\, dx = 4 \int \frac{dx}{x} - 3 \int \frac{dx}{x - 1} + 3 \int \frac{dx}{(x - 1)^2}$$

$$= 4 \log |x| - 3 \log |x - 1| - \frac{3}{x - 1} + C.$$

2. Find $\int \dfrac{x^3\, dx}{x^2 + x - 12}$.

ANSWER   Since the numerator of the integrand has degree 3, and the denominator degree 2, we first divide x^3 by $x^2 + x - 12$ and obtain the quotient $x - 1$ and the remainder $13x - 12$. Thus

$$x^3 = (x - 1)(x^2 + x - 12) + (13x - 12)$$

or

$$\frac{x^3}{x^2 + x - 12} = (x - 1) + \frac{13x - 12}{x^2 + x - 12}.$$

Now, $x^2 + x - 12$ factors into $(x + 4)(x - 3)$. The method of indeterminate coefficients yields

$$\frac{13x - 12}{x^2 + x - 12} = \frac{13x - 12}{(x + 4)(x - 3)} = \frac{A}{x + 4} + \frac{B}{x - 3}$$

$$= \frac{A(x - 3) + B(x + 4)}{(x + 4)(x - 3)} = \frac{(A + B)x + (-3A + 4B)}{(x + 4)(x - 3)}.$$

We need $A + B = 13$, $-3A + 4B = -12$. Thus $A = 64/7$, $B = 27/7$. Hence

$$\frac{13x - 12}{x^2 + x - 12} = \frac{64}{7} \frac{1}{x + 4} + \frac{27}{7} \frac{1}{x - 3}$$

and

$$\int \frac{x^3}{x^2 + x - 12}\, dx = \int (x - 1)\, dx + \frac{64}{7} \int \frac{dx}{x + 4} + \frac{27}{7} \int \frac{dx}{x - 3}$$

$$= \frac{1}{2}x^2 - x + \frac{64}{7} \log |x + 4| + \frac{27}{7} \log |x - 3| + C.$$

PROBLEMS

Perform the following integrations.

1. $\int \dfrac{(3x + 2)}{(x + 2)(x - 4)}\, dx.$

2. $\int \dfrac{(1 - x)}{(2x + 1)(x + 2)}\, dx.$

3. $\int \dfrac{(2x + 4)}{x(x - 1)(x - 2)}\, dx.$

4. $\int \dfrac{x^2 - 2}{x^3 - 4x}\, dx.$

5. $\int \dfrac{x^2 + x - 3}{x^3 - 2x^2 - x + 2}\, dx.$

6. $\int \dfrac{3x^2 - 6x + 2}{x^3 - 2x^2 + x}\, dx.$

7. $\int \dfrac{3x^2 - 4}{(x + 2)^3}\, dx.$

8. $\int \dfrac{(2 + x)}{x(4x^2 + 1)}\, dx.$

9. $\int \dfrac{x^2 - 1}{x(x^2 + x + 1)}\, dx.$

10. $\int \dfrac{2x^2 + 4x - 1}{(x^2 + 2x + 2)(x - 1)}\, dx.$

11. $\int \dfrac{x^4 + 3x^2 + x + 1}{x^3 + x}\, dx.$

12. $\int \dfrac{x^4 + 3x^3 + 8x^2 + 13x + 18}{(x^2 + 4)(x + 2)}\, dx.$

13. $\int \dfrac{2x^3 - 8x - 5}{x^2(x + 1)^2}\, dx.$

14. $\int \dfrac{8x^3\, dx}{(9x^2 + 1)(x^2 + 1)}\, dx.$

15. $\int \dfrac{(x^2 + 2x)}{(x^2 + 4)^2}\, dx.$

16. $\int \dfrac{(x^2 + 8)}{x^2(x^2 + 2x + 4)}\, dx.$

17. $\int \dfrac{7 - x - 2x^2}{(x - 1)^2(x^2 + x + 2)}\, dx.$

18. $\int \dfrac{x^2 + x + 2}{x(x^2 + 1)^2}\, dx.$

§4 Rationalizable integrals

In this section we exhibit various classes of integrals that can be reduced, by an appropriate substitution, to an integral of a rational function and hence evaluated in closed form. In each case, the proof will provide a method of carrying out this evaluation. But often it is simpler to compute the integral by some special method rather than by proceeding with the rationalization process. We shall illustrate this by several examples.

In studying this section, the reader should strive to develop the ability to recognize integrals that can be rationalized. Once it is known that a given integral can be computed in closed form (that is, by a formula involving elementary functions), we can perform the integration in various ways. In particular, we can consult a table of integrals (see §5).

4.1 Notations

In order to describe rationalizable integrals, we use the following conventions. The symbol $R(u)$ will always denote a rational function of the variable u. The symbol $R(u,v)$ will always denote a rational function of two variables, u and v. This means that $R(u,v)$ is a quotient of two polynomials in u and v. Note that if $x \mapsto f(x)$ and $x \mapsto g(x)$ are rational functions, then $x \mapsto R(f(x),g(x))$ is a rational function of the variable x. For instance, suppose that

$$R(u,v) = \frac{u^2 - v^2}{2uv}.$$

Then

$$R(2x^3, 4x^2) = \frac{(2x^3)^2 - (4x^2)^2}{2(2x^3)(4x^2)} = \frac{4x^6 - 16x^4}{16x^5} = \frac{x}{4} - \frac{1}{x}.$$

4.2 Rational functions of the exponential function

We begin with a particularly simple case of a rationalizable integral.

(I) *An integral of the form $\int R(e^x)\,dx$ can be rationalized by the substitution* $y = e^x$.

Proof.   We have $x = \log y$; therefore $dx = dy/y$ and

$$\int R(e^x)\,dx = \int R(y)\frac{1}{y}\,dy.$$

This is an integral of a rational function.

EXAMPLE $\displaystyle \int (1 + e^x)^{-1}\,dx = \int (1 + y)^{-1} y^{-1}\,dy = \int \frac{dy}{y(1 + y)}$

$$= \int \left(\frac{1}{y} - \frac{1}{1 + y}\right) dy = \log|y| - \log|1 + y| + C$$

$$= \log\left|\frac{y}{1 + y}\right| + C = \log\left|\frac{e^x}{1 + e^x}\right| + C.$$

PROBLEMS

Perform the following integrations.

1. $\int \dfrac{e^{2x}}{e^{2x}-1}\,dx.$

6. $\int \dfrac{e^{2x}\,dx}{(e^{2x}+2e^x+2)^2}.$

2. $\int_0^\infty \dfrac{e^x}{e^{2x}+1}\,dx.$

7. $\int \dfrac{dx}{1+\cosh x}.$

3. $\int \dfrac{e^x\,dx}{(e^x+1)^2}.$

8. $\int \dfrac{\cosh x}{3\cosh x + \sinh x}\,dx.$

4. $\int \dfrac{2e^{2x}+e^x+1}{e^{2x}+1}\,dx.$

9. $\int \dfrac{1+\sinh x}{1+\cosh x}\,dx.$

5. $\int \dfrac{dx}{(e^{2x}+4e^x+5)}.$

10. $\int \dfrac{2\cosh x + 3\sinh x}{3\cosh x + 5\sinh x}\,dx.$

4.3   Roots of fractional linear functions

Here is another class of rationalizable integral.

(II) *An integral* $\int R\left(x, \sqrt[q]{\dfrac{\alpha x + \beta}{\gamma x + \delta}}\right) dx$, $q>1$ *an integer, can be rationalized by the substitution*

$$y = \sqrt[q]{\dfrac{\alpha x + \beta}{\gamma x + \delta}}. \tag{1}$$

A special case is worth mentioning.

(II′) *An integral* $\int R(x, \sqrt[q]{\alpha x + \beta})\,dx$, $q>1$ *an integer, can be rationalized by the substitution*

$$y = \sqrt[q]{\alpha x + \beta}. \tag{1′}$$

Before proving (II), we consider some examples.

EXAMPLES 1. Evaluate $\displaystyle\int_{3/2}^2 x\sqrt[3]{2x-3}\,dx.$

SOLUTION   We set

$$y = \sqrt[3]{2x-3}, \qquad y^3 = 2x-3, \qquad x = \tfrac12 y^3 + \tfrac32,$$

so that

$$dx = \tfrac32 y^2\,dy$$

and

$$\int_{3/2}^2 x\sqrt[3]{2x-3}\,dx = \int_0^1 \left(\tfrac12 y^3 + \tfrac32\right) y\,\tfrac32 y^2\,dy$$

$$= \int_0^1 \left(\tfrac34 y^6 + \tfrac94 y^3\right) dy = \left(\tfrac{3}{28} y^7 + \tfrac{9}{16} y^4\right)\Big|_0^1 = \tfrac{75}{112}.$$

2. Evaluate $\displaystyle\int \dfrac{dx}{1+\sqrt x}.$

SOLUTION   We set $y = \sqrt{x}$, $y^2 = x$, $dx = 2y\,dy$ and obtain

$$\int \frac{dx}{1 + \sqrt{x}} = \int \frac{2y\,dy}{1 + y} = \int \left(2 + \frac{-2}{1 + y}\right) dy = 2y - 2\log(1 + y) + C$$
$$= 2\sqrt{x} - 2\log(1 + \sqrt{x}) + C.$$

We check by differentiating:

$$\frac{d}{dx}[2x^{1/2} - 2\log(1 + x^{1/2})] = x^{-1/2} - \frac{x^{-1/2}}{1 + x^{1/2}} = \frac{1}{1 + x^{1/2}},$$

as expected.

3. Evaluate $\int \sqrt{\dfrac{1 - x}{1 + x}}\,dx$.

ANSWER   We set $y = \sqrt{\dfrac{1 - x}{1 + x}}$, that is,

$$y^2 = \frac{1 - x}{1 + x} \quad \text{so that} \quad y^2 + y^2 x = 1 - x, \quad x = \frac{1 - y^2}{1 + y^2}, \quad dx = \frac{-4y}{(1 + y^2)^2}\,dy,$$

and obtain

$$\int \sqrt{\frac{1 - x}{1 + x}}\,dx = \int y\,\frac{-4y}{(1 + y^2)^2}\,dy = \int \left(\frac{-4}{(1 + y^2)} + \frac{4}{(1 + y^2)^2}\right) dy.$$

In order to evaluate this integral, we apply the recursion formula (9) derived in §2.4. We obtain

$$\int \left(\frac{-4}{1 + y^2}\right) dy + 4\left\{\frac{1}{2}\frac{y}{1 + y^2} + \frac{1}{2}\int \frac{dy}{1 + y^2}\right\}$$

$$= -2\int \frac{dy}{1 + y^2} + 2\frac{y}{1 + y^2} = -2\arctan y + 2\frac{y}{1 + y^2} + C$$

$$= -2\arctan \sqrt{\frac{1 - x}{1 + x}} + (1 + x)\sqrt{\frac{1 - x}{1 + x}} + C$$

$$= -2\arctan \sqrt{\frac{1 - x}{1 + x}} + \sqrt{1 - x^2} + C.$$

The reader should check by differentiation.

Proof of (II).   If y is defined by (1), we have

$$y^q = \frac{\alpha x + \beta}{\gamma x + \delta}.$$

After multiplying both sides by $\gamma x + \delta$, solve for x. This yields

$$x = \frac{-\delta y^q + \beta}{\gamma y^q - \alpha}, \quad dx = \frac{\alpha\delta - \beta\gamma}{(\gamma y^q - \alpha)^2}\,qy^{q-1}\,dy$$

and

$$\int R\left(x, \sqrt[q]{\frac{\alpha x + \beta}{\gamma x + \delta}}\right) dx = \int R\left(\frac{-\delta y^q + \beta}{\gamma y^q - \alpha}, y\right)\frac{(\alpha\delta - \beta\gamma)qy^{q-1}}{(\alpha y^q - \alpha)^2}\,dy;$$

this is an integral of a rational function. (We assume that $\alpha\delta - \beta\gamma \neq 0$, otherwise the radical is a constant.)

PROBLEMS

Perform the following integrations.

1. $\int \sqrt[3]{4x + 1}\, dx.$

5. $\int x^2(1 - 2x)^{1/4}\, dx.$

9. $\int \sqrt{\dfrac{x}{1 - x}}\, dx.$

2. $\int x\sqrt{3 - x}\, dx.$

6. $\int \dfrac{\sqrt{x + 3}}{x + 4}\, dx.$

10. $\int \sqrt{\dfrac{1 - x}{4 + x}}\, dx.$

3. $\int \dfrac{x^2}{\sqrt{x - 3}}\, dx.$

7. $\int \dfrac{2 - \sqrt{x}}{2 + \sqrt{x}}\, dx.$

11. $\int \left(\dfrac{1 + x}{1 - x}\right)^{3/2} dx.$

4. $\int \dfrac{x^2}{(x + 4)^{3/2}}\, dx.$

8. $\int \dfrac{x + \sqrt{x + 1}}{x + 2}\, dx.$

12. $\int \dfrac{1}{x^2}\sqrt{\dfrac{2 - x}{2 + x}}\, dx.$

4.4 Rational functions of trigonometric functions

In Chapter 10 [see §1.4, Equation (29)] we derived two important trigonometric identities that we now use. Setting $x = 2\phi$ in these equations, we obtain

$$\sin x = \frac{2 \tan \dfrac{x}{2}}{1 + \tan^2 \dfrac{x}{2}}, \qquad \cos x = \frac{1 - \tan^2 \dfrac{x}{2}}{1 + \tan^2 \dfrac{x}{2}}. \tag{2}$$

The above identities lead to the following basic result.

(III) *An integral of the form $\int R(\cos x, \sin x)\, dx$ can be rationalized by the substitution*

$$y = \tan \frac{x}{2}. \tag{3}$$

Proof.   We use (2) and obtain

$$\sin x = \frac{2y}{1 + y^2}, \qquad \cos x = \frac{1 - y^2}{1 + y^2}.$$

Also, by (3), we have $x/2 = \operatorname{arc} \tan y$, $x = 2 \operatorname{arc} \tan y$. Hence $dx = \dfrac{2}{1 + y^2}\, dy$, and

$$\int R(\cos x, \sin x)\, dx = \int R\left(\frac{1 - y^2}{1 + y^2}, \frac{2y}{1 + y^2}\right) \frac{2\, dy}{1 + y^2};$$

this is an integral of a rational function.

This method permits us to calculate integrals of functions formed rationally from $\cos x, \sin x, \sec x = 1/\cos x, \csc x = 1/\sin x, \tan x = \sin x/\cos x, \cot x = \cos x/\sin x.$ But it often pays to use some special device, rather than the general method.

EXAMPLES 1. Compute $\int \sec x\, dx.$
ANSWER   We set $y = \tan \tfrac{1}{2}x$ and obtain

$$\int \sec x\, dx = \int \frac{dx}{\cos x} = \int \frac{1 + y^2}{1 - y^2} \frac{2\, dy}{1 + y^2} = \int \frac{2}{1 - y^2}\, dy$$

$$= \int \left(\frac{1}{1+y} + \frac{1}{1-y} \right) dy = \log|1 + y| - \log|1 - y| + C$$

$$= \log \frac{|1+y|}{|1-y|} dy = \log \left| \frac{1 + \tan \frac{x}{2}}{1 - \tan \frac{x}{2}} \right| + C.$$

Since

$$\frac{1 + \tan \frac{x}{2}}{1 - \tan \frac{x}{2}} = \frac{\cos \frac{x}{2} + \sin \frac{x}{2}}{\cos \frac{x}{2} - \sin \frac{x}{2}} = \frac{\cos \frac{x}{2} + \sin \frac{x}{2}}{\cos \frac{x}{2} - \sin \frac{x}{2}} \cdot \frac{\cos \frac{x}{2} + \sin \frac{x}{2}}{\cos \frac{x}{2} + \sin \frac{x}{2}}$$

$$= \frac{1 + 2 \cos \frac{x}{2} \sin \frac{x}{2}}{\cos^2 \frac{x}{2} - \sin^2 \frac{x}{2}} = \frac{1 + \sin x}{\cos x} = \sec x + \tan x,$$

we have

$$\int \sec x \, dx = \log|\sec x + \tan x| + C. \tag{4}$$

This is the formula that appears in most tables of integrals.

2. Compute $\int \csc x \, dx$.
ANSWER We could obtain this integral in closed form using the same method as above. It is simpler, however, to use (4) and write

$$\int \csc x \, dx = \int \sec \left(\frac{\pi}{2} - x \right) dx = - \int \sec \left(\frac{\pi}{2} - x \right) d \left(\frac{\pi}{2} - x \right)$$

$$= -\log \left| \sec \left(\frac{\pi}{2} - x \right) + \tan \left(\frac{\pi}{2} - x \right) \right| + C = -\log|\csc x + \cot x| + C.$$

This can also be written in the form

$$\int \csc x \, dx = \log|\csc x - \cot x| + C, \tag{5}$$

as the reader should verify.

3. Evaluate $\int \sin^2 x \, dx$ and $\int \cos^2 x \, dx$.
ANSWER Instead of using the general method, we note that

$$\int \cos^2 x \, dx + \int \sin^2 x \, dx = \int 1 \, dx = x + C_1,$$

$$\int \cos^2 x \, dx - \int \sin^2 x \, dx = \int \cos 2x \, dx = \frac{1}{2} \sin 2x + C_2.$$

Adding and subtracting the two equations, we obtain

$$\int \cos^2 x \, dx = \frac{1}{2} x + \frac{1}{4} \sin 2x + C, \tag{6}$$

$$\int \sin^2 x \, dx = \frac{1}{2} x - \frac{1}{4} \sin 2x + C. \tag{7}$$

We could also have obtained this result as follows:

$$\int \sin^2 x \, dx = \int \frac{1 - \cos 2x}{2} \, dx = \frac{1}{2} x - \frac{1}{4} \sin 2x + C.$$

[We can also compute $\int \sin^2 x \, dx$ and $\int \cos^2 x \, dx$ using the reduction formulas (10) and (11) in §2.4; set $n = 0$, $m = 2$ in (10) and $n = 2$ in (11).]

4. Compute $\displaystyle\int \frac{dx}{\sin x + \cos x}$.

ANSWER   Instead of using the general method, note that

$$\sin x + \cos x = \sqrt{2}\left(\frac{1}{\sqrt{2}}\sin x + \frac{1}{\sqrt{2}}\cos x\right)$$

$$= \sqrt{2}\left(\sin x \cos \frac{\pi}{4} + \cos x \sin \frac{\pi}{4}\right) = \sqrt{2}\sin\left(x + \frac{\pi}{4}\right).$$

Therefore, using Example 2,

$$\int \frac{dx}{\sin x + \cos x} = \frac{1}{\sqrt{2}}\int \frac{dx}{\sin\left(x + \dfrac{\pi}{4}\right)}$$

$$= -\frac{1}{\sqrt{2}}\log\left|\csc\left(x + \frac{\pi}{4}\right) + \cot\left(x + \frac{\pi}{4}\right)\right| + C.$$

5. Evaluate $\displaystyle\int \frac{dx}{1 - \sin x}$.

ANSWER   We set $y = \tan \frac{1}{2}x$ and obtain

$$\int \frac{dx}{1 - \sin x} = \int \frac{1 + y^2}{(y - 1)^2}\cdot\frac{2}{1 + y^2}\,dy = 2\int \frac{dy}{(y - 1)^2}$$

$$= -\frac{2}{y - 1} + C = \frac{2}{1 - \tan\dfrac{x}{2}} + C.$$

A more elegant form, the one often given in tables of integrals, can be obtained using the addition formula for the tangent. Note that

$$\tan\left(\frac{\pi}{4} + \frac{x}{2}\right) = \frac{\tan\dfrac{\pi}{4} + \tan\dfrac{x}{2}}{1 - \tan\dfrac{\pi}{4}\tan\dfrac{x}{2}} = \frac{1 + \tan\dfrac{x}{2}}{1 - \tan\dfrac{x}{2}}$$

$$= \frac{1 + \tan\dfrac{x}{2}}{1 - \tan\dfrac{x}{2}} + \frac{1 - \tan\dfrac{x}{2}}{1 - \tan\dfrac{x}{2}} - 1 = \frac{2}{1 - \tan\dfrac{x}{2}} - 1.$$

Hence

$$\int \frac{dx}{1 - \sin x} = \tan\left(\frac{\pi}{4} + \frac{x}{2}\right) + 1 + C = \tan\left(\frac{\pi}{4} + \frac{x}{2}\right) + C',$$

where C' is a new constant of integration.

PROBLEMS

Perform the following integrations.

1. $\displaystyle\int \frac{dx}{1 + \cos x}$.

2. $\displaystyle\int \frac{dx}{2 + \sin x}$.

3. $\displaystyle\int \frac{dx}{5 + 3\cos x}$.

4. $\displaystyle\int \frac{dx}{(1 - \cos x)^2}$.

5. $\int \dfrac{\sin x \, dx}{1 + \sin x}.$

6. $\int \dfrac{dx}{\sin x(1 + \sin x)}.$

7. $\int \dfrac{\sec x \, dx}{(1 + \sec x)^2}.$

8. $\int \dfrac{dx}{4 + 5 \sin x}.$

9. $\int \dfrac{dx}{5 + 3 \sec x}.$

10. $\int \dfrac{\sin x \, dx}{5 + 4 \sin x}.$

Use the method of Example 4 to compute the following integrals.

11. $\int \dfrac{dx}{a \cos x + b \sin x},$ $a^2 + b^2 \neq 0.$

12. $\int \dfrac{a + \tan x}{1 - a \tan x} \, dx,$ a any number.

Evaluate the following integrals using the substitution $y = \tan x.$

13. $\int \dfrac{dx}{a^2 \cos^2 x + b^2 \sin^2 x},$ $a^2 + b^2 \neq 0.$

14. $\int \dfrac{dx}{a^2 \cos^2 x - b^2 \sin^2 x},$ $a^2 + b^2 \neq 0.$

4.5 Trigonometric substitutions

We consider next integrals of the form $\int R(u,x) \, dx$ where u is either $\sqrt{1 - x^2}$, or $\sqrt{x^2 - 1}$, or $\sqrt{1 + x^2}$. These integrals can be reduced, as we shall see presently, to integrals of rational functions of sine and cosine and can be computed, either by a special device or by the general method of §4.4.

(IV) *The following integrals can be rationalized:*

$$\int R(x, \sqrt{1 - x^2}) \, dx \qquad (set \ x = \sin y), \qquad (a)$$

$$\int R(x, \sqrt{x^2 - 1}) \, dx \qquad (set \ x = \sec y), \qquad (b)$$

$$\int R(x, \sqrt{1 + x^2}) \, dx \qquad (set \ x = \tan y). \qquad (c)$$

Proof. In case (a), x may take on all values between -1 and $+1$, so that it makes sense to set

$$x = \sin y, \qquad dx = \cos y \, dy, \qquad \sqrt{1 - x^2} = \cos y \qquad (a')$$

with y restricted to $-\pi/2 \leq y \leq \pi/2$. We obtain

$$\int R(x, \sqrt{1 - x^2}) \, dx = \int R(\sin y, \cos y) \cos y \, dy$$

and the right integral is rationalizable, by (III). Similarly, in case (b) we have

$$x = \sec y, \qquad dx = \tan y \sec y \, dy, \qquad \sqrt{x^2 - 1} = \tan y \qquad (b')$$

(we assume that $x > 1$ and $-\pi/2 < y < \pi/2$). The integral considered becomes $\int R \,(\sec y, \tan y) \tan y \sec y \, dy$ and is rationalizable, by (III). In case (c), finally,

$$x = \tan y, \qquad dx = \sec^2 y \, dy, \qquad \sqrt{1 + x^2} = \sec y \qquad (c')$$

(with $-\pi/2 < y < \pi/2$); the integral is transformed into the rationalizable integral $\int R(\tan y, \sec y) \sec^2 y \, dy$.

FIGURE **13.6**

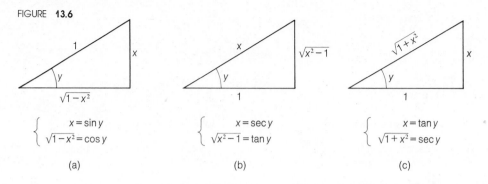

$$\begin{cases} x = \sin y \\ \sqrt{1-x^2} = \cos y \end{cases}$$

$$\begin{cases} x = \sec y \\ \sqrt{x^2-1} = \tan y \end{cases}$$

$$\begin{cases} x = \tan y \\ \sqrt{1+x^2} = \sec y \end{cases}$$

(a)   (b)   (c)

REMARK   Figure 13.6 may be helpful to remember the trigonometric substitutions (a'), (b'), and (c'). In each case, x and 1 are taken as two sides of a right triangle, and y is an acute angle. The third side is found using the Pythagorean theorem.

EXAMPLES 1. Compute $\int \sqrt{1-x^2}\,dx$.
ANSWER We use (a') and obtain

$$\int \sqrt{1-x^2}\,dx = \int \cos y\,(\cos y\,dy) = \int \cos^2 y\,dy.$$

Now, using relation (6), we have

$$\int \sqrt{1-x^2}\,dx = \frac{1}{2}y + \frac{1}{4}\sin 2y + C$$

$$= +\frac{1}{2}\arcsin x + \frac{1}{2}\sin y \cos y + C = +\frac{1}{2}\arcsin x + \frac{1}{2}x\sqrt{1-x^2} + C.$$

2. Compute $\int \dfrac{dx}{x^2\sqrt{x^2-1}}$.

ANSWER We use (b') and obtain

$$\int \frac{dx}{x^2\sqrt{x^2-1}} = \int \frac{\tan y \sec y\,dy}{\sec^2 y \tan y} = \int \cos y\,dy = \sin y + C.$$

But what is $\sin y$? To find out, we use the identity $\sin^2 y + \cos^2 y = 1$. Then

$$\sin y = \pm\sqrt{1-\cos^2 y} = \pm\sqrt{1-\frac{1}{\sec^2 y}} = \pm\frac{\sqrt{x^2-1}}{x}.$$

To resolve the ambiguity in sign, we require that $\sin y$ be consistent with the original substitutions $x = \sec y$, $\sqrt{x^2-1} = \tan y$. Since $\sin y$ and $\tan y$ have the same sign for $-\pi/2 < y < \pi/2$, we must take the positive sign. Hence

$$\int \frac{dx}{x^2\sqrt{x^2-1}} = \frac{\sqrt{x^2-1}}{x} + C.$$

3. Compute $\int \dfrac{dx}{\sqrt{x^2+4}}$.

ANSWER Set $t = x/2$, $dt = \frac{1}{2}\,dx$, and rewrite the given integral as

$$\int \frac{dx}{\sqrt{x^2+4}} = \int \frac{dt}{\sqrt{t^2+1}}.$$

Now use (c'), that is, let $t = \tan y$. Then

$$\int \frac{dt}{\sqrt{t^2 + 1}} = \int \frac{\sec^2 y\, dy}{\sec y} = \int \sec y\, dy$$

$$= \log|\sec y + \tan y| + C = \log|\sqrt{t^2 + 1} + t| + C.$$

Hence

$$\int \frac{dx}{\sqrt{x^2 + 4}} = \log\left|\sqrt{\left(\frac{x}{2}\right)^2 + 1} + \frac{x}{2}\right| + C = \log|\sqrt{x^2 + 4} + x| + C',$$

where $C' = C - \log 2$ is a new constant of integration.

4. Evaluate $\int \operatorname{arc\,sec} x\, dx$ and $\int \operatorname{arc\,csc} x\, dx$.

ANSWER   We shall use first an integration by parts, then substitution (b'), and then Example 1 in §4.4:

$$\int \operatorname{arc\,sec} x\, dx = x \operatorname{arc\,sec} x - \int x\, d(\operatorname{arc\,sec} x)$$

$$= x \operatorname{arc\,sec} x - \int \frac{dx}{\sqrt{x^2 - 1}} = x \operatorname{arc\,sec} x - \int \sec y\, dy$$

$$= x \operatorname{arc\,sec} x - \log|\sec y + \tan y| + C.$$

Thus

$$\int \operatorname{arc\,sec} x\, dx = x \operatorname{arc\,sec} x - \log|x + \sqrt{x^2 - 1}| + C.$$

We obtain, in a similar way,

$$\int \operatorname{arc\,csc} x\, dx = x \operatorname{arc\,csc} x + \log|x + \sqrt{x^2 - 1}| + C.$$

PROBLEMS

In Problems 1 to 8 compute the given integrals using substitution (a') of this subsection.

1. $\displaystyle\int \frac{dx}{\sqrt{1 - x^2}}.$

2. $\displaystyle\int \frac{dx}{(4 - x^2)^{3/2}}.$

3. $\displaystyle\int \frac{dx}{\sqrt{1 - 4x^2}}.$

4. $\displaystyle\int \frac{dx}{x\sqrt{25 - 16x^2}}.$ [Hint: Set $u = \frac{4}{5}x$.]

5. $\displaystyle\int \frac{\sqrt{1 - x^2}}{x^2}\, dx.$

6. $\displaystyle\int \frac{x^2\, dx}{(1 - x^2)^{3/2}}.$

7. $\displaystyle\int \frac{x^2\, dx}{\sqrt{36 - x^2}}.$

8. $\displaystyle\int \frac{dx}{x\sqrt{4 - x^2}}.$

In Problems 9 to 16 compute the given integrals using substitution (b') of this subsection.

9. $\displaystyle\int \frac{dx}{\sqrt{x^2 - 16}}.$

10. $\displaystyle\int \frac{dx}{(x^2 - 1)^{3/2}}.$

11. $\displaystyle\int \frac{dx}{\sqrt{9x^2 - 1}}.$

12. $\displaystyle\int \frac{dx}{x^2\sqrt{x^2 - 4}}.$

13. $\displaystyle\int \frac{dx}{x\sqrt{x^2 - 25}}.$

14. $\displaystyle\int \frac{dx}{x^2(x^2 - 1)^{3/2}}.$

15. $\displaystyle\int \frac{\sqrt{x^2 - 9}}{x}\, dx.$

16. $\displaystyle\int \frac{dx}{x(x^2 - 4)^{3/2}}.$

In Problems 17 to 24 compute the given integrals using substitution (c′) of this subsection.

17. $\int \dfrac{dx}{\sqrt{x^2 + 16}}$.

18. $\int \dfrac{dx}{x\sqrt{x^2 + 1}}$.

19. $\int \dfrac{dx}{x^2\sqrt{x^2 + 1}}$.

20. $\int \dfrac{x^3\,dx}{\sqrt{x^2 + 4}}$.

21. $\int \sqrt{x^2 + 9}\,dx$.

22. $\int \dfrac{x^2\,dx}{\sqrt{x^2 + 1}}$.

23. $\int \dfrac{x^2\,dx}{(x^2 + 1)^{3/2}}$.

24. $\int \dfrac{dx}{x^2(x^2 + 1)^{3/2}}$.

4.6 Square roots of quadratic functions

We are now able to make a rather general statement.

(V) *Any integral of the form* $\int R(x, \sqrt{\alpha x^2 + 2\beta x + \gamma})\,dx$ *can be rationalized.*

Indeed, if $\alpha = 0$, we have case (II′) of §4.3. If $\alpha x^2 + 2\beta x + \gamma$ is the square of a linear function, there is nothing to prove. In all other cases, we can complete the square and make a substitution which will reduce our integrals to one of the three forms considered in §4.5.

EXAMPLE Compute $\int \dfrac{dx}{\sqrt{-4x^2 + 8x - 3}}$.

ANSWER We have

$$-4x^2 + 8x - 3 = -4(x - 1)^2 + 1 = 1 - (2x - 2)^2.$$

Therefore we set $\xi = 2x - 2$, $d\xi = 2\,dx$, and obtain

$$\int \frac{dx}{\sqrt{-4x^2 + 8x - 3}} = \frac{1}{2}\int \frac{d\xi}{\sqrt{1 - \xi^2}} = \frac{1}{2}\,\text{arc sin }\xi + C$$

$$= \frac{1}{2}\,\text{arc sin}(2x - 2) + C.$$

PROBLEMS

Evaluate the following integrals.

1. $\int \dfrac{dx}{\sqrt{8x - x^2}}$.

2. $\int \dfrac{dx}{\sqrt{-9x^2 + 24x - 15}}$.

3. $\int \dfrac{dx}{(x^2 - 4x + 5)^{3/2}}$.

4. $\int \dfrac{dx}{\sqrt{x^2 - 2x - 8}}$.

5. $\int \dfrac{x\,dx}{\sqrt{x^2 - 6x + 10}}$.

6. $\int \dfrac{x\,dx}{(5 - 4x - x^2)^{3/2}}$.

7. $\int \dfrac{dx}{x\sqrt{7x^2 - 6x - 1}}$.

$\left[\text{Hint: Let } x = \dfrac{1}{z}.\right]$

8. $\int \dfrac{dx}{x^2\sqrt{2x^2 + 2x + 1}}$.

$\left[\text{Hint: Let } x = \dfrac{1}{z}.\right]$

4.7   Products of sines and cosines

We conclude this section by mentioning a type of trigonometric integral that can be evaluated in closed form, but by a somewhat different method from the ones used up to now. A typical example is the integral $\int \cos mx \cos nx \, dx$, where m and n are two distinct numbers. To evaluate it, we make use of the trigonometric identity

$$\cos \alpha \cos \beta = \tfrac{1}{2}[\cos(\alpha + \beta) + \cos(\alpha - \beta)],$$

which follows from the addition theorem (see Example 9 in Chapter 10, §1.3). This yields $\cos mx \cos nx = \tfrac{1}{2}[\cos(m + n)x + \cos(m - n)x]$. Therefore

$$\int \cos mx \cos nx \, dx = \frac{\sin(m + n)x}{2(m + n)} + \frac{\sin(m - n)x}{2(m - n)} + C.$$

We observe that, if m and n are integers, then we could, by repeated application of the addition theorems, write $\cos mx \cos nx$ as a rational function of $\cos x$ and $\sin x$. In this case, we could conclude, from statement (III), that the integral under consideration could be evaluated in closed form. A similar reasoning would hold if the ratio m/n were rational. At any rate, the method by which we arrived at the result is more convenient. It is also the only method that works when m/n is irrational.

Integrals of the form

$$\int \sin mx \sin nx \, dx, \qquad \int \cos mx \sin nx \, dx,$$

can be evaluated similarly by making use of the trigonometric identities

$$\sin \alpha \sin \beta = \tfrac{1}{2}[\cos(\alpha - \beta) - \cos(\alpha + \beta)],$$
$$\cos \alpha \sin \beta = \tfrac{1}{2}[\sin(\alpha + \beta) - \sin(\alpha - \beta)].$$

The reader is asked to derive these identities from the addition theorems and to verify the following integration formulas:

$$\int \sin mx \sin nx \, dx = -\frac{\sin(m + n)x}{2(m + n)} + \frac{\sin(m - n)x}{2(m - n)} + C, \tag{9}$$

$$\int \cos mx \sin nx \, dx = \frac{\cos(m + n)x}{2(m + n)} + \frac{\cos(m - n)x}{2(m - n)} + C. \tag{10}$$

It is clear that the method just described permits us to evaluate in closed form a primitive function of a product of more than two functions, each of which is a sine or a cosine of a multiple of x. We show this by an example.

EXAMPLE   Evaluate $\int \cos x \cos 2x \cos 3x \, dx$.
ANSWER   We have

$$
\begin{aligned}
\cos x \cos 2x \cos 3x &= \tfrac{1}{2}[\cos 3x + \cos x] \cos 3x \\
&= \tfrac{1}{2} \cos^2 3x + \tfrac{1}{2} \cos x \cos 3x \\
&= \tfrac{1}{2} \cdot \tfrac{1}{2}(\cos 6x + 1) + \tfrac{1}{2} \cdot \tfrac{1}{2}(\cos 4x + \cos 2x) \\
&= \tfrac{1}{4} + \tfrac{1}{4} \cos 2x + \tfrac{1}{4} \cos 4x + \tfrac{1}{4} \cos 6x.
\end{aligned}
$$

Hence

$$\int \cos x \cos 2x \cos 3x \, dx = \frac{1}{4}x + \frac{1}{8} \sin 2x + \frac{1}{16} \sin 4x + \frac{1}{24} \sin 6x + C.$$

PROBLEMS

Perform the following integrations.

1. $\int \cos 2x \cos 4x \, dx.$

6. $\int_0^{\pi/8} \sin 3x \cos 5x \, dx.$

2. $\int_0^{\pi/8} \cos x \cos 3x \, dx.$

7. $\int \sin x \sin 2x \sin 4x \, dx.$

3. $\int \sin 2x \sin 6x \, dx.$

8. $\int \cos x \cos 3x \cos 5x \, dx.$

4. $\int_0^{\pi/4} \sin 3x \sin 5x \, dx.$

9. $\int_0^{\pi/2} \sin x \cos 2x \sin 3x \, dx.$

5. $\int \cos 4x \sin 5x \, dx.$

10. $\int \cos x \cos 2x \sin 3x \sin 4x \, dx.$

§5 Evaluation of integrals using a table

The evaluation of integrals is often a tedious and time-consuming process. For this reason, it is desirable to have available a list of useful integral formulas arranged in some convenient order. Such a list is called a **table of integrals.** In this section we describe its use.

A short table of integrals will be found at the end of this book. The table is arranged according to the basic form of the integrand to facilitate table reference. The following forms are grouped together: $(a + bu)$, $\sqrt{a + bu}$, $(a^2 + u^2)$, $(a^2 - u^2)$, $\sqrt{a^2 + u^2}$, $\sqrt{a^2 - u^2}$, $\sqrt{u^2 - a^2}$, trigonometric forms, inverse trigonometric forms, exponential and logarithmic forms, and hyperbolic forms. It is to be understood that all symbols appearing in the formulas are to be interpreted in such a way that both sides of the equation are well defined.

The following examples illustrate the use of the table.

EXAMPLES 1. Evaluate $\int \dfrac{\sqrt{2 + 3x}}{x} \, dx.$

ANSWER The radical appearing in the integrand is of the form $\sqrt{a + bu}$ with $a = 2$, $b = 3$, and $u = x$. We search the table and find that formula 17a "fits." Hence

$$\int \frac{\sqrt{2 + 3x}}{x} \, dx = 2\sqrt{2 + 3x} + \sqrt{2} \log \left| \frac{\sqrt{2 + 3x} - \sqrt{2}}{\sqrt{2 + 3x} + \sqrt{2}} \right| + C.$$

The reader should check by differentiation.

2. Evaluate $\int \dfrac{dx}{x^2(16 + 9x^2)}.$

ANSWER The integrand fits formula 27, except for a constant, if we set $a = 4$ and $u = 3x$. We have

$$\int \frac{dx}{x^2(16 + 9x^2)} = \int \frac{\frac{1}{3} du}{\left(\dfrac{u}{3}\right)^2 (a^2 + u^2)} = 3 \int \frac{du}{u^2(a^2 + u^2)}.$$

Hence

$$\int \frac{dx}{x^2(16 + 9x^2)} = 3 \left\{ -\frac{1}{a^2 u} - \frac{1}{a^3} \arctan \frac{u}{a} + C \right\}.$$

Substituting for a and u, we obtain the desired integral in terms of x. It is

$$\int \frac{dx}{x^2(16 + 9x^2)} = -\frac{1}{16x} - \frac{3}{64} \arctan \frac{3x}{4} + C'.$$

3. Evaluate $\int t^2 \sqrt{4 + t^2}\, dt$.

ANSWER The above integral does not appear explicitly in the table; however, a suitable reduction formula does. Set $n = 2$, $a = 2$, and $u = t$ in formula 41. Then

$$\int t^2 \sqrt{4 + t^2}\, dt = \frac{1}{4} t(4 + t^2)^{3/2} - \int \sqrt{4 + t^2}\, dt.$$

The integral on the right is now just formula 40 with $a = 2$, $u = t$. Hence

$$\int t^2 \sqrt{4 + t^2}\, dt = \frac{1}{4} t(4 + t^2)^{3/2} - \frac{1}{2} t\sqrt{4 + t^2} - 2 \log|t + \sqrt{4 + t^2}| + C.$$

4. Evaluate $\int \tan^5 2t\, dt$.

ANSWER The appropriate reduction formula is given by formula 107 with $n = 5$ and $u = 2t$. We have

$$\int \tan^5 2t\, dt = \frac{1}{2} \int \tan^5 u\, du = \frac{1}{2} \left\{ \frac{1}{4} \tan^4 u - \int \tan^3 u\, du \right\}.$$

Applying the same reduction formula again, this time with $n = 3$, we obtain

$$\int \tan^5 2t\, dt = \frac{1}{8} \tan^4 u - \frac{1}{2} \left\{ \frac{1}{2} \tan^2 u - \int \tan u\, du \right\}.$$

The integral on the right is given by formula 105. Substituting, and writing u in terms of t, we obtain

$$\int \tan^5 2t\, dt = \frac{1}{8} \tan^4 2t - \frac{1}{4} \tan^2 2t + \frac{1}{2} \log|\sec 2t| + C.$$

PROBLEMS

Evaluate the following integrals using the table at the end of the book.

1. $\int \dfrac{x\, dx}{(2 + 3x)}$.

2. $\int \dfrac{x\, dx}{(1 - 3x)^4}$.

3. $\int \dfrac{dx}{x(2 - x)^2}$.

4. $\int \dfrac{dx}{x^2 \sqrt{1 + x}}$.

5. $\int \dfrac{\sqrt{x - 9}}{x^2}\, dx$.

6. $\int \dfrac{dt}{(4 + t^2)^2}$.

7. $\int \dfrac{t^5\, dt}{(1 + t^4)}$.

[Hint: Set $u = t^2$.]

8. $\int \dfrac{dt}{(9 - t^2)^2}$.

9. $\int \dfrac{dt}{t^{3/2}(1 - t)}$.

[Hint: Set $u = \sqrt{t}$.]

10. $\int \dfrac{t^2\, dt}{\sqrt{9 + 4t^2}}$.

11. $\int \dfrac{dx}{x^2(9 + x^2)^{3/2}}$.

12. $\int \dfrac{(1 + x^2)^{3/2}}{x^3}\, dx$.

13. $\int \dfrac{dx}{x^2 \sqrt{9 - 16x^2}}$.

14. $\int \dfrac{\sqrt{16 - x^2}}{x^3}\, dx$.

15. $\int x^2(1 - x^2)^{3/2}\, dx$.

16. $\int \dfrac{\sqrt{4s^2 - 1}}{s^2}\, ds$.

17. $\int \dfrac{s^4\, ds}{(s^2 - 9)^{3/2}}$.

18. $\int \dfrac{(s^3 - 8)^{3/2}}{s}\, ds$.

[Hint: Set $u = s^{3/2}$.]

19. $\int \dfrac{ds}{(5 - 3\sin 2s)}$.

20. $\int s^3 \sin s\, ds$.

21. $\int \cos^4 3x\, dx$.

22. $\int \sin^4 x \cos^2 x \, dx.$

23. $\int \sin^3 x \cos^3 x \, dx.$

24. $\int \sin 10x \cos 6x \, dx.$

25. $\int \sec^6 4x \, dx.$

26. $\int \cot^5 (2t + 1) \, dt.$

27. $\int \dfrac{\sin^3 t}{\cos^6 t} \, dt.$

    [*Hint:* Use sec and tan.]

28. $\int \dfrac{\cos^4 t}{\sin^7 t} \, dt.$

29. $\int t \arc \sin \dfrac{t^2}{2} \, dt.$

    $\left[\textit{Hint: Set } u = \dfrac{t^2}{2}. \right]$

30. $\int \dfrac{1}{\sqrt{t}} \arc \cot \sqrt{t} \, dt.$

31. $\int x^3 e^{2x} \, dx.$

32. $\int e^{-x} \sin^2 4x \, dx.$

33. $\int e^{4x} \cos^3 8x \, dx.$

34. $\int x^4 (\log x)^2 \, dx.$

35. $\int x^2 \left(\log \dfrac{x}{2} \right)^3 \, dx.$

36. $\int \dfrac{ds}{s (\log 2s)^4}.$

37. $\int \sech(1 - 4s) \, ds.$

38. $\int \dfrac{1}{\sqrt{s}} \coth \sqrt{s} \, ds.$

14

MEAN VALUE THEOREM
AND TAYLOR'S FORMULA

§1 Mean value theorems

At the beginning of our discussion of calculus, soon after we had defined the derivative, we arrived at an approximate method for computing the value of a function $f(x)$ close to a point x_0 at which we knew the value of f, namely, $f(x_0)$, and the value of its derivative, namely, $f'(x_0)$. The formula reads

$$f(x) \approx f(x_0) + f'(x_0)(x - x_0), \tag{1}$$

where "$\approx$" stands for "approximately equal." [This is formula (2) in Chapter 7, §2.1, where we set $x = x_0 + h$.] Now we take up this subject once more. We shall give a precise meaning to the words "approximately equal" by estimating the error committed when using (1). Calculation of this error has far-reaching consequences.

1.1 The mean value theorem and Rolle's theorem

Consider a continuous function $x \mapsto f(x)$ defined for $a \le x \le b$, and assume that $f'(x)$ exists for $a < x < b$. The graph is a smooth curve joining the points $P = (a, f(a))$ and $Q = (b, f(b))$. It seems geometrically obvious that there must be points on this graph, between P and Q, at which the tangent to the curve is parallel to the chord PQ. (See Figure 14.1.)

FIGURE **14.1**

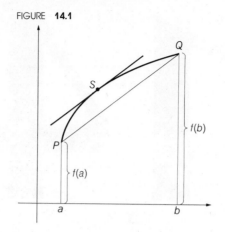

Let us formulate the statement "there is a tangent parallel to the chord" analytically. The slope of the chord is

$$\frac{f(b) - f(a)}{b - a};$$

the slope of a tangent at a point $(\xi, f(\xi))$ is $f'(\xi)$. The tangent and the chord are parallel if their slopes are equal. Thus we expect the following to hold.

Theorem 1 (mean value theorem). *If the function $x \mapsto f(x)$ is continuous for $a \leq x \leq b$, and has a derivative for $a < x < b$, then there is a number ξ such that $a < \xi < b$ and*

$$\frac{f(b) - f(a)}{b - a} = f'(\xi). \tag{2}$$

Let us set $a = x_0$, $b = x$; then we can rewrite (2) as $f(x) - f(x_0) = f'(\xi)(x - x_0)$ or

$$f(x) = f(x_0) + f'(\xi)(x - x_0). \tag{3}$$

This looks very much like (1), except that we have an $=$ sign instead of the $\approx$ sign and $f'(\xi)$ instead of $f'(x_0)$, where ξ is a point between x_0 and x.

There may be, of course, more than one point on the arc $y = f(x)$, $a \leq x \leq b$, at which the tangent is parallel to the chord. In Figure 14.1 there is one and only one such point, namely, S; in Figure 14.2 there are three points, S_1, S_2, and S_3, with the desired property; in Figure 14.3 all points between P_1 and Q_1 will do. [We note that, if the derivative $f'(x)$ fails to exist at a single point, there need be no tangent parallel to the chord; such a case is considered in the Example 2.]

MICHEL ROLLE (1652–1719) was a French mathematician. He stated the theorem that bears his name as a rule for locating roots of polynomials.

FIGURE **14.2**   FIGURE **14.3**

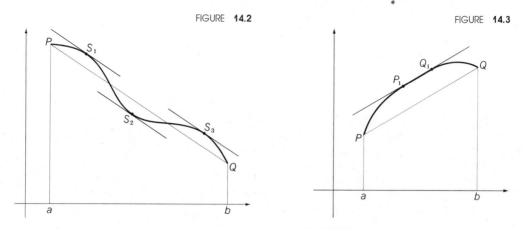

An important special case of the mean value theorem is

Theorem 1a (Rolle's theorem). *If $x \mapsto f(x)$ is continuous for $a \le x \le b$ and differentiable for $a < x < b$, and if $f(a) = f(b) = 0$, then there is a number ξ between a and b such that $f'(\xi) = 0$.*

Rolle's theorem is sometimes stated as follows:

Between two zeros of a function, there is a zero of the derivative.

We prove Rolle's theorem first. By Theorem 1 in Chapter 5, §1.1, the continuous function $f(x)$ defined on the closed finite interval $[a,b]$ assumes a maximum and a minimum. Let f assume its maximum at $x = \alpha$ and its minimum at $x = \beta$. If $a < \alpha < b$, then f has a local maximum at α and hence (see Chapter 5, §2.5) $f'(\alpha) = 0$. If $a < \beta < b$, then f has a local minimum at β and $f'(\beta) = 0$. If, on the other hand, both α and β are endpoints of $[a,b]$, then the maximum and minimum values of f are 0. Hence f is the constant 0 and $f'(\xi) = 0$ for $a < \xi < b$.

The general form of Theorem 1 follows from Rolle's theorem. Define the function $\phi(x)$ by

$$\phi(x) = \frac{f(b) - f(a)}{b - a}(x - a) - [f(x) - f(a)]. \tag{4}$$

Then ϕ is continuous for $a \le x \le b$, $\phi(a) = \phi(b) = 0$, and ϕ has the derivative

$$\phi'(x) = \frac{f(b) - f(a)}{b - a} - f'(x)$$

for $a < x < b$. Therefore Rolle's theorem applies: there is a number ξ between a and b with $\phi'(\xi) = 0$. But the relation $\phi'(\xi) = 0$ is the same as (2).

[We note the geometric meaning of the function $\phi(x)$ used in the proof. The equation of the chord joining the points $P = (a,f(a))$ and $Q = (b,f(b))$ (see Figure 14.1) is $y = l(x)$ where $l(x) = f(a) + \frac{f(b) - f(a)}{b - a}(x - a)$. Since $\phi(x) = l(x) - f(x)$, ϕ measures the vertical distance between the curve $y = f(x)$ and the chord.]

FIGURE **14.4**   FIGURE **14.5**

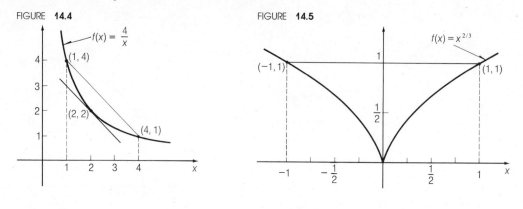

EXAMPLES   1.   If $f(x) = 4/x$, find all numbers ξ in the open interval $(1,4)$ such that

$$f'(\xi) = \frac{f(4) - f(1)}{4 - 1} = -1.$$

ANSWER $f'(\xi) = -4/\xi^2$, hence ξ must satisfy $-4/\xi^2 = -1$, that is, $\xi = \pm 2$. We reject the value $\xi = -2$ since it lies outside the interval $(1,4)$. The answer is $\xi = 2$. See Figure 14.4.

2.   If $f(x) = x^{2/3}$, find all numbers ξ in the open interval $(-1,1)$ for which the mean value theorem is satisfied.

ANSWER   If we proceed in a formal way, we find that

$$f'(\xi) = \tfrac{2}{3}\xi^{-1/3} = \frac{(1)^{2/3} - (-1)^{2/3}}{1 - (-1)} = 0.$$

But there is no number ξ for which $\tfrac{2}{3}\xi^{-1/3} = 0$. Hence there is no number in the interval $(-1,1)$ which satisfies the mean value theorem. The reason the theorem need not hold in this case is that $f(x)$ does not have a derivative for *all* x in the interval $(-1,1)$. Thus one of the hypotheses of the theorem is not met. See Figure 14.5.

PROBLEMS

In Problems 1 to 10 determine all numbers ξ in the given intervals (a,b) for which $[f(b) - f(a)]/(b - a) = f'(\xi)$.

1. $f(x) = 1 - 3x$, $(1,4)$.
2. $f(x) = x^2 + 1$, $(1,2)$.
3. $f(x) = \log x$, (e^2, e^3).
4. $f(x) = x^3 - x$, $(0,1)$.
5. $f(x) = e^x$, $(1,2)$.

6. $f(x) = x^3$, $(-2,2)$.
7. $f(x) = 1 - x^2 + x^4$, $(-2,2)$.
8. $f(x) = (x - 1)/x$, $(1,3)$.
9. $f(x) = \sqrt{x^2 + 9}$, $(0,4)$.
10. $f(x) = \tan x$, $(0,\pi/4)$.

The mean value theorem does not hold for the following functions and intervals. Give the reason why in each case.

11. $f(x) = |x|$, $(-1,3)$.
12. $f(x) = x + 1/x$, $(-1,2)$.

13. $f(x) = x^2/(x - 1)$, $(0,2)$.
14. $f(x) = \sqrt[3]{x}$, $(-8,8)$.

1.2   Generalized mean value theorem

There is a useful generalization of the mean value theorem.

Theorem 2 (generalized mean value theorem).   *Let $F(x)$ and $G(x)$ be continuous functions defined for $a \leq x \leq b$, and assume that the derivatives $F'(x)$ and $G'(x)$ exist and $G'(x) \neq 0$, for $a < x < b$. Then there is a number ξ such that $a < \xi < b$ and*

$$\frac{F(b) - F(a)}{G(b) - G(a)} = \frac{F'(\xi)}{G'(\xi)}. \tag{5}$$

[We can dispense with the condition $G' \neq 0$, if we agree to interpret (5) to mean that $[F(b) - F(a)]G'(\xi) = [G(b) - G(a)]F'(\xi)$.]

   Proof.   Define the new function

$$f(x) = [F(b) - F(a)][G(x) - G(a)] - [G(b) - G(a)][F(x) - F(a)].$$

It is continuous for $a \leq x \leq b$, and if $a < x < b$, then

$$f'(x) = [F(b) - F(a)]G'(x) - [G(b) - G(a)]F'(x).$$

Also $f(a) = f(b) = 0$. By Rolle's theorem, there is a number ξ between a and b with $f'(\xi) = 0$. Thus $[F(b) - F(a)]G'(\xi) = [G(b) - G(a)]F'(\xi)$. This is the same as (5), provided $G'(\xi) \neq 0$.

Corollary.   *If also $F(a) = G(a) = 0$, then for every $b \neq a$ there is a ξ between a and b with*

$$\frac{F(b)}{G(b)} = \frac{F'(\xi)}{G'(\xi)}. \tag{6}$$

(In the corollary we may have $b < a$.)

EXAMPLE If $F(x) = x^2 + 3x - 4$ and $G(x) = x^2 - 5x + 1$, find all numbers ξ in the open interval $(0,1)$ such that

$$\frac{F(1) - F(0)}{G(1) - G(0)} = \frac{F'(\xi)}{G'(\xi)}.$$

ANSWER $F'(\xi) = 2\xi + 3$, $G'(\xi) = 2\xi - 5$, and so ξ must satisfy

$$\frac{2\xi + 3}{2\xi - 5} = \frac{F(1) - F(0)}{G(1) - G(0)} = -1.$$

Hence $\xi = \frac{1}{2}$.

PROBLEMS

   In the following problems, find all numbers ξ in the given intervals (a,b) for which
$$\frac{F(b) - F(a)}{G(b) - G(a)} = \frac{F'(\xi)}{G'(\xi)}.$$

1. $F(x) = 2x + 1$, $G(x) = 3x - 4$, $(1,3)$.
2. $F(x) = x^3$, $G(x) = 2 - x$, $(0,3)$.
3. $F(x) = 1/x$, $G(x) = x^2$, $(1,2)$.
4. $F(x) = \log x$, $G(x) = 1/x$, $(1,e)$.
5. $F(x) = \sin x$, $G(x) = \cos x$, $(\pi/4, 3\pi/4)$.
6. $F(x) = \frac{1}{4}x^4 - x^3 + x^2$, $G(x) = x^2$, $(0,2)$.

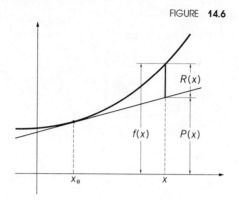

FIGURE **14.6**

1.3   The error in linear approximation

We now have the tools for estimating the error in the approximate formula (1). The use of this formula amounts to replacing the graph of f by its tangent at $(x_0, f(x_0))$. This tangent is the graph of the linear function $x \mapsto P(x) = f(x_0) + f'(x_0)(x - x_0)$, which has at x_0 the value and slope of $f(x)$; see Figure 14.6. Let $R(x)$ denote the error committed if we compute the value of $f(x)$ using P instead of f; that is, let $R(x) = f(x) - P(x)$. Then we have

$$f(x) = f(x_0) + f'(x_0)(x - x_0) + R(x). \tag{7}$$

Therefore

$$\frac{R(x)}{x - x_0} = \frac{f(x) - f(x_0)}{x - x_0} - f'(x_0),$$

and since

$$\lim_{x \to x_0} \frac{f(x) - f(x_0)}{x - x_0} = f'(x_0)$$

we have that

$$\lim_{x \to x_0} \frac{R(x)}{x - x_0} = 0. \tag{8}$$

This means that $|R(x)|$ is small if x is close to x_0, and it even becomes arbitrarily small compared to $|x - x_0|$ (compare Theorem 1 in Chapter 4, §1.6). But statement (8) does not tell us *how* small $R(x)$ is for a given $x \neq x_0$.

In order to answer this question, we need additional information on the function $f(x)$. We assume that f has a *second* derivative, and consider the remainder $R(x)$ for a fixed x_0 and a variable x. Clearly

$$R(x_0) = 0$$

[this is seen by substituting $x = x_0$ in (7)]. Also, differentiating both sides of (7), and noting that x_0 is fixed, we get $f'(x) = f'(x_0) + R'(x)$ so that

$$R'(x_0) = 0.$$

Another differentiation gives

$$f''(x) = R''(x).$$

Now we apply the corollary to Theorem 2 with $F(t) = R(t)$, $G(t) = (t - x_0)^2$, $a = x_0$, $b = x$. There exists a number τ between x_0 and x such that

$$\frac{R(x)}{(x - x_0)^2} = \frac{R'(\tau)}{2(\tau - x_0)}.$$

Applying the corollary to Theorem 2 once more, this time to $F(x) = R'(x)$ and $G(x) = 2(x - x_0)$, we conclude that there exists a ξ between x_0 and τ with

$$\frac{R'(\tau)}{2(\tau - x_0)} = \frac{R''(\xi)}{2} = \frac{f''(\xi)}{2}.$$

Thus

$$\frac{R(x)}{(x - x_0)^2} = \frac{f''(\xi)}{2}$$

or

$$R(x) = \frac{f''(\xi)(x - x_0)^2}{2}. \tag{9}$$

In particular

$$|R(x)| \le \frac{M}{2}(x - x_0)^2 \quad \text{if } |f''(t)| \le M \quad \text{for } all \text{ } t \text{ between } x_0 \text{ and } x. \tag{10}$$

Substituting (9) into (7), we see that

$$f(x) = f(x_0) + f'(x_0)(x - x_0) + \tfrac{1}{2}f''(\xi)(x - x_0)^2 \tag{11}$$

for *some* ξ between x_0 and x.

EXAMPLES 1. Use the linear approximation formula (11) to compute $\sqrt{1 + x}$ for small $|x|$, and estimate the error.

ANSWER We use (11) with $f(x) = \sqrt{1 + x}$, $x_0 = 0$. Since $f(0) = 1$, $f'(x) = \tfrac{1}{2}(1 + x)^{-1/2}$, $f'(0) = \tfrac{1}{2}$, $f''(x) = -\tfrac{1}{4}(1 + x)^{-3/2}$, the linear approximation formula reads

$$\sqrt{1 + x} = 1 + \frac{x}{2} + R(x), \qquad R(x) = -\frac{1}{8}(1 + \xi)^{-3/2}x^2 \quad \text{for some } \xi \text{ between 0 and } x.$$

To estimate $R(x)$ we must estimate $(1 + \xi)^{-3/2}$ for ξ between 0 and x. For $\xi > 0$, $1 + \xi > 1$, $(1 + \xi)^{3/2} > 1$, and $(1 + \xi)^{-3/2} < 1$. Thus $|R(x)| < \tfrac{1}{8}x^2$ for $x > 0$. If $x < 0$, then $\xi < 0$. Since $(1 + \xi)^{-3/2}$ will be very large for ξ close to -1, we restrict ourselves to values of x between $-\tfrac{3}{4}$ and 0. Then $-\tfrac{3}{4} < \xi < 0$; hence $1 + \xi > \tfrac{1}{4}$, $(1 + \xi)^{3/2} > \tfrac{1}{8}$, $(1 + \xi)^{-3/2} < 8$, and $|R(x)| < x^2$. Conclusion: the error in replacing $\sqrt{1 + x}$ by $1 + x/2$ is at most $\tfrac{1}{8}x^2$ for $x > 0$, at most x^2 for $x > -\tfrac{3}{4}$.

For instance, $\sqrt{1.02} \approx 1.01000$ with an error at most $\tfrac{1}{8}(.02)^2 = .00005$ (here $x = .02$); $\sqrt{.98} \approx .9900$ with an error at most $.0004$ (here $x = -.02$).

2. Using the linear approximation formula (11), find an approximate value for $e^{-.1}$ and estimate the error.

ANSWER Take $f(x) = e^{-x}$, and set $x_0 = 0$ in formula (11). Then

$$e^{-x} = 1 - x + R(x), \qquad R(x) = \tfrac{1}{2}e^{-\xi}x^2,$$

for some ξ between $x_0 = 0$ and $x = .1$. Since $e^{-\xi}$ is decreasing for $0 \le \xi \le .1$ and $e^{-0} = 1$, $0 < e^{-\xi} \le 1$ and $0 < R(x) \le \tfrac{1}{2}(1)(.1)^2 = .005$. Therefore, $e^{-.1} \approx .9$ with an error not greater than $.005$.

PROBLEMS

In Problems 1 to 5 find an approximate value for the given expression, and estimate the error. Use the results derived in Example 1.

1. $\sqrt{5}$.
   [*Hint:* Write $\sqrt{5} = 2\sqrt{1 + \frac{1}{4}}$.]
2. $\sqrt{50}$.

3. $(.044)^{1/2}$.
4. $\sqrt{78}$.
5. $\sqrt{.0062}$.

In Problems 6 to 12 find an approximate value for the given expression, and estimate the error. Use the linear approximation of this subsection.

6. $e^{-.01}$.
7. $e^{-.04}$.
8. $\sqrt[3]{9}$.
9. $(.03)^{1/3}$.

10. $\sqrt[3]{130}$.
11. $\log(1.04)$.
12. $\arctan(.1)$.

1.4 Some consequences of the mean value theorem‡

Using Theorem 1, we can supply proofs of some basic theorems stated in Chapters 5 and 6 without demonstrations. Suppose we know that f is continuous for $a \le x \le b$ and that $f'(x) > 0$ for all x, $a < x < b$. We conclude from the mean value theorem

$$f(b) - f(a) = f'(\xi)(b - a) \tag{12}$$

(where ξ is some number between a and b), that

$$f(b) - f(a) > 0, \qquad \text{that is,} \qquad f(a) < f(b).$$

Thus: *a function with a positive derivative increases* (this is Theorem 1 in Chapter 5, §2.2). Also, if we know only that $f'(x) \ge 0$ for $a < x < b$, then (12) shows that $f(b) \ge f(a)$. *A function with a nonnegative derivative is nondecreasing* (this is Theorem 2 in Chapter 5, §2.2). Finally, if $f'(x) = 0$ for all x, $a < x < b$, then (12) shows that $f(b) = f(a)$. *If the derivative is zero everywhere, the function is constant* (this is Theorem 1 in Chapter 6, §4.1).

Here is another application. The ξ in (12) depends on b. If we let b approach a, ξ also approaches a. Hence if $\lim_{\xi \to a^+} f'(\xi)$ exists, then

$$f'(a^+) = \lim_{b \to a^+} \frac{f(b) - f(a)}{b - a} = \lim_{\xi \to a^+} f'(\xi).$$

This is so also if the limit is infinite. Similarly, if there is a limit $\lim_{\xi \to a^-} f'(\xi)$, then

$$f'(a^-) = \lim_{c \to a^-} \frac{f(a) - f(c)}{a - c} = \lim_{\xi \to a^-} f'(\xi).$$

This establishes Theorem 1 in Chapter 6, §5.2.

§2 Taylor's theorem

Taylor's theorem permits us to represent any function $f(x)$, defined near a point x_0 (and having derivatives up to order $n + 1$) as a polynomial of degree n plus a "re-

‡This subsection may be omitted without loss of continuity.

mainder." For x close to x_0, the remainder is small of the order of $|x - x_0|^{n+1}$. The linear approximation formula (11) in §1.3 is Taylor's theorem for $n = 1$. Before discussing the general case, we look at $n = 2$.

2.1 Parabolic approximation

We consider a thrice differentiable function $f(x)$ for which we know the values of $f(x_0)$, $f'(x_0)$, and $f''(x_0)$, and a quadratic polynomial $P(x)$ such that

$$P(x_0) = f(x_0), \qquad P'(x_0) = f'(x_0), \qquad P''(x_0) = f''(x_0). \tag{1}$$

FIGURE **14.7**

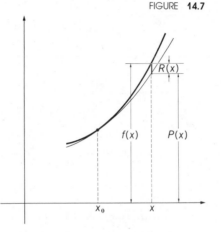

In other words: the parabola $y = P(x)$ should pass through the point $(x_0, f(x_0))$ and should, at this point, have the same slope and the same derivative of the slope as the graph of $y = f(x)$; see Figure 14.7. We expect that the "remainder"

$$R(x) = f(x) - P(x) \tag{2}$$

will be, for x close to x_0, significantly smaller than the error in the linear approximation considered before.

First we write down $P(x)$:

$$P(x) = f(x_0) + f'(x_0)(x - x_0) + \tfrac{1}{2}f''(x_0)(x - x_0)^2. \tag{3}$$

Indeed, this is a quadratic polynomial in x (x_0 is kept fixed), and we compute that (1) holds. In order to determine the remainder $R(x)$, we note that $R'(x) = f'(x) - P'(x)$, $R''(x) = f''(x) - P''(x)$, and $R'''(x) = f'''(x) - P'''(x)$. Therefore

$$R(x_0) = R'(x_0) = R''(x_0) = 0,$$

since P has been chosen to satisfy condition (1), and

$$R'''(x) = f'''(x), \tag{4}$$

BROOKS TAYLOR (1685–1731) published the series which bears his name in a book that appeared in 1715. Taylor for a time was secretary of the Royal Society.

since $P'''(x) = 0$ for all x. Now we shall apply several times the corollary to the generalized mean value theorem (see §1.2). We obtain

$$\frac{R(x)}{(x - x_0)^3} = \frac{R'(\tau_1)}{3(\tau_1 - x_0)^2}, \qquad \text{for some } \tau_1 \text{ between } x_0 \text{ and } x,$$

$$\frac{R'(\tau_1)}{3(\tau_1 - x_0)^2} = \frac{R''(\tau_2)}{3 \cdot 2(\tau_2 - x_0)}, \qquad \text{for some } \tau_2 \text{ between } x_0 \text{ and } \tau_1,$$

$$\frac{R''(\tau_2)}{3 \cdot 2(\tau_2 - x_0)} = \frac{R'''(\xi)}{3 \cdot 2}, \qquad \text{for some } \xi \text{ between } x_0 \text{ and } \tau_2.$$

Noting (4), we combine these results into

$$\frac{R(x)}{(x - x_0)^3} = \frac{f'''(\xi)}{3 \cdot 2}, \qquad \text{for some } \xi \text{ between } x \text{ and } x_0.$$

This may be written as

$$R(x) = \frac{f'''(\xi)}{1 \cdot 2 \cdot 3} (x - x_0)^3. \tag{5}$$

Thus we obtain, by (2) and (3),

$$f(x) = f(x_0) + f'(x_0)(x - x_0) + \frac{1}{1 \cdot 2} f''(x_0)(x - x_0)^2 + R(x), \tag{6}$$

where R is given by (5), for a suitable point ξ between x_0 and x. We note that

$$|R| \leq \frac{M|x - x_0|^3}{6}, \qquad \text{if } |f'''(t)| \leq M, \text{ for all } t \text{ between } x_0 \text{ and } x.$$

EXAMPLES 1. Use the quadratic approximation formula (6) to compute $\sqrt{1 + x}$ for small $|x|$, and estimate the error. In particular, compute $\sqrt{1.02}$ and $\sqrt{.98}$. See Example 1 in §1.3.

ANSWER   For $f(x) = \sqrt{1 + x}$, we have $f'(x) = \frac{1}{2}(1 + x)^{-1/2}$, $f''(x) = -\frac{1}{4}(1 + x)^{-3/2}$, $f'''(x) = \frac{3}{8}(1 + x)^{-5/2}$. Equations (5) and (6) yield, for $x_0 = 0$,

$$\sqrt{1 + x} = 1 + \tfrac{1}{2}x - \tfrac{1}{8}x^2 + R(x), \qquad R(x) = \tfrac{1}{16}(1 + \xi)^{-5/2}x^3,$$

where ξ is some point between 0 and x. We have

$$|R(x)| \leq \tfrac{1}{16}x^3, \quad \text{for } x \geq 0, \qquad |R(x)| \leq |2x^3|, \quad \text{for } -\tfrac{3}{4} \leq x \leq 0$$

[since $(1 + \xi)^{5/2} \geq 1$ if $\xi \geq 0$ and $(1 + \xi)^{5/2} \geq 32$ if $-\tfrac{3}{4} \leq \xi \leq 0$]. In particular,

$$\sqrt{1.02} \approx 1 + \tfrac{1}{2}(.02) - \tfrac{1}{8}(.02)^2 = 1.0099500$$

with an error not exceeding $\tfrac{1}{16}(.02)^3 = .0000005$, and

$$\sqrt{.98} \approx 1 - \tfrac{1}{2}(.02) + \tfrac{1}{8}(.02)^2 = .990050$$

with an error not exceeding $2(.02)^3 = .000016$.

2. Using the quadratic approximation formula (6), find an approximate value for $e^{-.1}$, and estimate the error. See Example 2 in §1.3.

ANSWER   Take $f(x) = e^{-x}$ and set $x_0 = 0$ in formula (6). Then

$$e^{-x} = 1 - x + \tfrac{1}{2}x^2 + R(x), \qquad R(x) = \tfrac{1}{6}e^{-\xi}x^3,$$

for some ξ between $x_0 = 0$ and $x = .1$. Since $e^{-\xi}$ is monotone decreasing for $0 \leq \xi \leq .1$, $e^{-\xi} \leq 1$ and $0 < R(x) \leq \tfrac{1}{6}(1)(.1)^3 \approx .000167$. Thus $e^{-.1} \approx .9050$ with an error less than .0002.

PROBLEMS

In Problems 1 to 5 find an approximate value for the given expression, and estimate the error. Use the results derived in Example 1.

1. $\sqrt{110}$.

[*Hint:* Write $\sqrt{110} = 10\sqrt{1 + \frac{1}{10}}$.]

2. $\sqrt{18}$.

3. $(.164)^{1/2}$.

4. $\sqrt{216}$.

5. $\sqrt{.00038}$.

In Problems 6 to 12 find an approximate value for the given expression, and estimate the error. Use the quadratic approximation of this subsection.

6. $e^{-.02}$.

7. $e^{-.05}$.

8. $\sin(.06)$.

9. $\log(1.2)$.

10. $\sqrt[3]{1050}$.

11. $\sqrt[3]{30}$.

12. $(1.1)^{2/3}$.

2.2 Taylor's theorem: statement and proof

The linear approximation formula (11) in §1.3 and the quadratic approximation formula (6) are answers to special cases of the following question. Let $x \mapsto f(x)$ be a function; suppose that we know the values of f and its first n derivatives f', f'', ..., $f^{(n)}$ at a point x_0. Now can we compute, approximately, the value of $f(x)$ at a point x close to x_0? At first glance, this question seems somewhat contrived. It turns out, however, that it leads to some of the most important topics in mathematics.

The answer to the question is given by

Theorem 1 (Taylor's formula). *Let $f(x)$ be defined and continuous in an interval containing x and x_0, and let it have a derivative of order n at x and a derivative of order $n + 1$ at all inner points of the interval. Then*

$$f(x) = f(x_0) + \frac{f'(x_0)}{1}(x - x_0) + \frac{f''(x_0)}{1 \cdot 2}(x - x_0)^2 + \frac{f'''(x_0)}{1 \cdot 2 \cdot 3}(x - x_0)^3$$
$$+ \cdots + \frac{f^{(n)}(x_0)}{1 \cdot 2 \cdot 3 \ldots n}(x - x_0)^n + \frac{f^{(n+1)}(\xi)(x - x_0)^{n+1}}{1 \cdot 2 \cdot 3 \ldots (n + 1)}, \tag{7}$$

where ξ is some number between x_0 and x.

The last term

$$R(x) = \frac{f^{(n+1)}(\xi)(x - x_0)^{n+1}}{1 \cdot 2 \ldots (n + 1)} \tag{8}$$

is called the **remainder.** Note that

$$|R(x)| \leq \frac{M|x - x_0|^{n+1}}{1 \cdot 2 \ldots (n + 1)} \qquad \text{if } |f^{(n+1)}(t)| \leq M \text{ for all } t \text{ between } x_0 \text{ and } x. \tag{9}$$

For $x_0 = 0$, Taylor's formula (7) reads

$$f(x) = f(0) + \frac{f'(0)}{1}x + \frac{f''(0)}{1 \cdot 2}x^2 + \cdots + \frac{f^{(n)}(0)}{1 \cdot 2 \ldots n}x^n$$
$$+ \frac{f^{(n+1)}(\xi)}{1 \cdot 2 \ldots n(n + 1)}x^{n+1}. \tag{10}$$

Here ξ is some point between 0 and x. This special case is sometimes called **Maclaurin's formula.**

Proof of Taylor's theorem.   Set

$$P(x) = f(x_0) + \frac{f'(x_0)}{1}(x - x_0) + \frac{f''(x_0)}{1 \cdot 2}(x - x_0)^2$$
$$+ \frac{f'''(x_0)}{1 \cdot 2 \cdot 3}(x - x_0)^3 + \cdots + \frac{f^{(n)}(x_0)}{1 \cdot 2 \cdot 3 \ldots n}(x - x_0)^n. \tag{11}$$

This is a polynomial of degree n in x (note that x_0 is a fixed number). It is called the nth **Taylor polynomial** of f at x_0. All terms in $P(x)$, except for the first, are 0 when $x = x_0$. Hence $P(x_0) = f(x_0)$. Differentiating (11), we obtain

$$P'(x) = f'(x_0) + \frac{f''(x_0)}{1}(x - x_0) + \frac{f'''(x_0)}{1 \cdot 2}(x - x_0)^2$$
$$+ \cdots + \frac{f^{(n)}(x_0)}{1 \cdot 2 \ldots (n - 1)}(x - x_0)^{n-1}.$$

Again, all terms on the right, except for the first, are 0 at $x = x_0$; thus $P'(x_0) = f'(x_0)$. Differentiating once more and setting $x = x_0$, we obtain $P''(x_0) = f''(x_0)$. In general,

$$P(x_0) = f(x_0), \, P'(x_0) = f'(x_0), \ldots, P^{(n)}(x_0) = f^{(n)}(x_0) \tag{12}$$

and, since P is an nth-degree polynomial,

$$P^{(n+1)}(x) = 0, \qquad \text{for all } x. \tag{13}$$

Now set

$$R(x) = f(x) - P(x). \tag{14}$$

By (12) and (14), $R(x_0) = R'(x_0) = \cdots = R^{(n)}(x_0) = 0$, and repeated application of the corollary to the generalized mean value theorem (see §1.2) gives

$$\frac{R(x)}{(x - x_0)^{n+1}} = \frac{R'(\tau_1)}{(n + 1)(\tau_1 - x_0)^n}, \qquad \text{for some } \tau_1 \text{ between } x_0 \text{ and } x,$$

$$= \frac{R''(\tau_2)}{(n + 1)n(\tau_2 - x_0)^{n-1}}, \qquad \text{for some } \tau_2 \text{ between } x_0 \text{ and } \tau_1,$$

$$= \frac{R'''(\tau_3)}{(n + 1)n(n - 1)(\tau_3 - x_0)^{n-2}}, \qquad \text{for some } \tau_3 \text{ between } x_0 \text{ and } \tau_2,$$

$$= \cdots,$$

$$= \frac{R^{(n+1)}(\xi)}{(n + 1)n(n - 1) \ldots 2 \cdot 1}, \qquad \text{for some } \xi \text{ between } x_0 \text{ and } \tau_n, \text{ hence between } x_0 \text{ and } x.$$

COLIN MACLAURIN (1698–1746), a professor at Edinburgh, was a former pupil of Newton.

Since $R^{(n+1)}(x) = f^{(n+1)}(x)$, by (13) and (14), we have

$$\frac{R(x)}{(x-x_0)^{n+1}} = \frac{f^{(n+1)}(\xi)}{1 \cdot 2 \cdot 3 \ldots (n+1)},$$

and this, together with (11) and (14), gives (7).

EXAMPLES **1.** Find the fourth Taylor polynomial, at $x_0 = 0$, for

$$f(x) = \sqrt{1+x}.$$

ANSWER   We have $f'(x) = \frac{1}{2}(1+x)^{-1/2}$, $f''(x) = -\frac{1}{4}(1+x)^{-3/2}$, $f'''(x) = \frac{3}{8}(1+x)^{-5/2}$, $f^{(4)}(x) = -\frac{15}{16}(1+x)^{-7/2}$. Therefore

$$f(0) = 1, \, f'(0) = \frac{1}{2}, \, f''(0) = -\frac{1}{4}, \, f'''(0) = \frac{3}{8}, \, f^{(4)}(0) = -\frac{15}{16}.$$

The desired polynomial is

$$P(x) = 1 + \frac{1}{2}x - \frac{1}{1 \cdot 2} \cdot \frac{1}{4}x^2 + \frac{1}{1 \cdot 2 \cdot 3} \cdot \frac{3}{8}x^3 - \frac{1}{1 \cdot 2 \cdot 3 \cdot 4} \cdot \frac{15}{16}x^4$$

or

$$P(x) = 1 + \frac{x}{2} - \frac{x^2}{8} + \frac{x^3}{16} - \frac{5x^4}{128}.$$

2. Find the third Taylor polynomial of the function $f(x) = 1 + x^2 + 2x^3$ at $x_0 = 1$.
ANSWER   Differentiating $f(x)$ three times, we have $f'(x) = 2x + 6x^2$, $f''(x) = 2 + 12x$, and $f'''(x) = 12$. Equation (11) gives

$$P(x) = 4 + \frac{8}{1}(x-1) + \frac{14}{1 \cdot 2}(x-1)^2 + \frac{12}{1 \cdot 2 \cdot 3}(x-1)^3$$

or

$$P(x) = 4 + 8(x-1) + 7(x-1)^2 + 2(x-1)^3.$$

Note that the remainder R in this case is zero since $f^{(n)}(x) = 0$ for all $n \geq 4$. Hence $f(x) = P(x)$. Thus we have rearranged the given function $f(x) = 1 + x^2 + 2x^3$ into one involving powers of $(x-1)$.

3. The function $f(x) = x^{11/3}$ has at $x = 0$ the third Taylor polynomial $P(x) = 0$, since $f(0) = f'(0) = f''(0) = f'''(0) = 0$. There is *no* fourth Taylor polynomial, since the function has no fourth derivative at $x = 0$.

4. Find the third Taylor polynomial $P(x)$ for $f(x) = (1+x)^{3/2}$ at $x_0 = 0$, and find an estimate (upper bound) for the absolute value of the remainder $R(x) = f(x) - P(x)$ in the interval $[0,1]$.
ANSWER   Differentiating $f(x)$ four times, we have $f'(x) = \frac{3}{2}(1+x)^{1/2}$, $f''(x) = \frac{3}{4}(1+x)^{-1/2}$, $f'''(x) = -\frac{3}{8}(1+x)^{-3/2}$, and $f^{(4)}(x) = +\frac{9}{16}(1+x)^{-5/2}$. Using Equation (11), we find that

$$P(x) = 1 + \tfrac{3}{2}x + \tfrac{3}{8}x^2 - \tfrac{1}{16}x^3.$$

From $f^{(4)}(x)$ we have that $|f^{(4)}(t)| \leq \frac{9}{16}$ for all t between 0 and 1. Hence, from Equation (9),

$$|R(x)| \leq \frac{9}{16} \frac{|1-0|^4}{1 \cdot 2 \cdot 3 \cdot 4} = \frac{3}{128} \approx .0234.$$

$|R(x)|$ is the maximum error committed in using $P(x)$ as an approximation for $f(x)$ in the interval 0 to 1. This error does not exceed 24/1000.

PROBLEMS

In Problems 1 to 12 find the Taylor polynomial for the given function f at the indicated point x_0 for the specified value of n. [*Note: n is the exponent in the last term $(x - x_0)^n$ in the Taylor polynomial.*]

1. $f(x) = \dfrac{1}{x + 1}$, $x_0 = 0, n = 3$.

   [*Answer:* $P(x) = 1 - x + x^2 - x^3$.]

2. $f(x) = \dfrac{1}{\sqrt{x + 1}}$, $x_0 = 0, n = 4$.

3. $f(x) = \cos x$, $x_0 = 0, n = 4$.

4. $f(x) = \sin x$, $x_0 = 0, n = 5$.

5. $f(x) = e^{-x}$, $x_0 = 0, n = 4$.

6. $f(x) = \log(1 + x)$, $x_0 = 0, n = 4$.

7. $f(x) = x^4$, $x_0 = 1, n = 4$.

8. $f(x) = x^3 + 2x^2 + 4x + 3$,
   $x_0 = -1, n = 3$.

9. $f(x) = \sin x$, $x_0 = \pi/6, n = 3$.

10. $f(x) = e^x$, $x_0 = 1, n = 4$.

11. $f(x) = \log x$, $x_0 = 1, n = 4$.

12. $f(x) = \arctan x$, $x_0 = 1, n = 3$.

In Problems 13 to 18 find the Taylor polynomial for the given function f at the indicated point x_0 for the specified value n. Then find an estimate for the absolute value of the remainder term in the given interval.

13. $f(x) = e^x$, $x_0 = 0, n = 4$, $[0,1]$.

14. $f(x) = \sin x$, $x_0 = 0, n = 4$, $\left[-\dfrac{\pi}{3}, \dfrac{\pi}{3}\right]$.

15. $f(x) = \log x$, $x_0 = 1$, $n = 4$, $[1, \frac{3}{2}]$.

16. $f(x) = \dfrac{1}{x + 1}$, $x_0 = 1, n = 3$, $[\frac{1}{2}, \frac{3}{2}]$.

17. $f(x) = \dfrac{1}{(x + 1)^{3/2}}$, $x_0 = 0, n = 3$, $[0, \frac{1}{3}]$.

18. $f(x) = \log \cos x$, $x_0 = 0, n = 5$, $\left[0, \dfrac{\pi}{3}\right]$.

2.3   Binomial formula

As the first application we consider the function

$$f(x) = (1 + x)^\alpha,$$

where $\alpha \neq 0$ is any real number. For $x \neq -1$, this function has derivatives of all orders:

$$f'(x) = \alpha(1 + x)^{\alpha-1}, \qquad f''(x) = \alpha(\alpha - 1)(1 + x)^{\alpha-2},$$
$$f'''(x) = \alpha(\alpha - 1)(\alpha - 2)(1 + x)^{\alpha-3}, \ldots,$$

and, in general, for $j = 1, 2, 3, \ldots,$

$$f^{(j)}(x) = \alpha(\alpha - 1)(\alpha - 2) \ldots (\alpha - j + 1)(1 + x)^{\alpha-j},$$

so that

$$f^{(j)}(0) = \alpha(\alpha - 1) \ldots (\alpha - j + 1).$$

Substitution into the Maclaurin formula (10) yields the relation

$$(1 + x)^\alpha = 1 + \frac{\alpha}{1}x + \frac{\alpha(\alpha - 1)}{1 \cdot 2}x^2 + \frac{\alpha(\alpha - 1)(\alpha - 2)}{1 \cdot 2 \cdot 3}x^3$$

$$+ \cdots + \frac{\alpha(\alpha - 1)\ldots(\alpha - n + 1)}{1 \cdot 2 \ldots n}x^n + R, \tag{15}$$

with

$$R = \frac{\alpha(\alpha - 1)\ldots(\alpha - n)}{1 \cdot 2 \ldots(n + 1)}(1 + \xi)^{\alpha - n - 1}x^{n+1},$$

where ξ is a number between 0 and x. We define the **binomial coefficients** $\binom{\alpha}{j}$, j a positive integer, α *any* number, as

$$\binom{\alpha}{0} = 1, \qquad \binom{\alpha}{j} = \frac{\alpha(\alpha - 1)(\alpha - 2)\ldots(\alpha - j + 1)}{1 \cdot 2 \cdot 3 \ldots j}, \qquad j = 1, 2, 3, \ldots . \tag{16}$$

Then we can write (15) as

$$(1 + x)^\alpha = 1 + \binom{\alpha}{1}x + \binom{\alpha}{2}x^2 + \cdots + \binom{\alpha}{n}x^n + R, \qquad \text{for } x \neq -1, \tag{17}$$

with

$$R = \binom{\alpha}{n + 1}(1 + \xi)^{\alpha - n - 1}x^{n+1}. \tag{18}$$

If α is a positive integer n, the restriction $x \neq -1$ is unnecessary. For $\alpha = n$, $\binom{\alpha}{n} = $

$\binom{n}{n} = 1, \binom{\alpha}{n + 1} = \binom{n}{n + 1} = 0$, all $\binom{\alpha}{m} = \binom{n}{m}$ with $m > n$ are 0 and (17) becomes

$$(1 + x)^n = 1 + \binom{n}{1}x + \binom{n}{2}x^2 + \binom{n}{3}x^3 + \cdots + \binom{n}{n}x^n. \tag{19}$$

This is an algebraic identity which can be proved directly.

EXAMPLES 1.   Compute $\sqrt{1.06}$ using formula (17) with $n = 2$, and then estimate the error using (18).
ANSWER   We want to write $\sqrt{1.06}$ in the form $(1 + x)^\alpha$, so we set $\alpha = \frac{1}{2}$ and $x = .06$. For $n = 2$ we have, by (17),

$$\sqrt{1.06} = (1 + .06)^{1/2} = 1 + \frac{\frac{1}{2}}{1}(.06) + \frac{\frac{1}{2}(\frac{1}{2} - 1)}{1 \cdot 2}(.06)^2 + R,$$

where

$$R = \frac{\frac{1}{2}(\frac{1}{2} - 1)(\frac{1}{2} - 2)}{1 \cdot 2 \cdot 3}(1 + \xi)^{(1/2) - 3}(.06)^3.$$

Simplifying,

$$\sqrt{1.06} = 1.02955 + R, \qquad R = .0000135(1 + \xi)^{-5/2}.$$

Here $0 < \xi < .06$, so that $0 < (1 + \xi)^{-5/2} < 1$. Hence $R < (.0000135)(1) < .00002$. We conclude that $\sqrt{1.06} \approx 1.02955$ with an error not greater than .00002.

2. Compute $\sqrt{.84}$ using formula (17) with $n = 2$, and then estimate the error using (18).
ANSWER   Proceeding in the same way as in Example 1, we obtain

$$\sqrt{.84} = .9168 + R, \qquad R = .000256(1 + \xi)^{-5/2}.$$

Here we have $-.16 < \xi < 0$, so that $0 < (1 + \xi)^{-5/2} < (.84)^{-5/2} < (.81)^{-5/2} = (.9)^{-5}$. Hence $R < .000256/.59049 < .0005$. Therefore $\sqrt{.84} \approx .9168$ with an error not greater than .0005.

PROBLEMS

Compute the following numbers using the binomial formula (17) for the given value of n, and estimate the error using Equation (18).

1. $\sqrt{1.2}, \quad n = 2.$
2. $\sqrt{.9}, \quad n = 2.$
3. $\sqrt{1.4}, \quad n = 3.$
4. $\sqrt{2}, \quad n = 3.$

5. $\sqrt{2}, \quad n = 4.$
6. $\sqrt[3]{1.03}, \quad n = 2.$
7. $\sqrt[3]{.9}, \quad n = 2.$

8. $\sqrt[3]{1.1}, \quad n = 3.$
9. $(1.1)^{.2}, \quad n = 3.$
10. $(1.2)^{-.4}, \quad n = 3.$

11. Show that $\dfrac{n}{j}\dbinom{n-1}{j-1} = \dbinom{n}{j}$ for $j = 1, \ldots, n$. Use Equation (16).

12. Show that $\dbinom{n}{j} = \dbinom{n}{n-j}$ for $j = 0, 1, \ldots, n$. Use Equation (16) and note that it suffices to consider the case $j \le n - j$.

13. Show that $\dbinom{n}{j-1} + \dbinom{n}{j} = \dbinom{n+1}{j}$ for $j \ge 1$. Use Equation (16).

14. Show that $\dbinom{n}{0} + \dbinom{n}{1} + \dbinom{n}{2} + \cdots + \dbinom{n}{n} = 2^n$. Use the binomial theorem with $x = 1$.

15. Show that $\dbinom{n}{0} - \dbinom{n}{1} + \dbinom{n}{2} - \cdots \dbinom{n}{n} = 0$ for $n \ge 1$. Use the binomial theorem.

16. Show that $\dbinom{m}{1} + 2\dbinom{m}{2} + 3\dbinom{m}{3} + \cdots + m\dbinom{m}{m} = m2^{m-1}$ for $m \ge 2$. Use the result of Problem 14 with $n = m - 1$ and the result of Problem 11.

2.4   Exponentials, sines, and cosines

Let us apply Maclaurin's formula (10) to $f(x) = e^x$. Since all derivatives of e^x are e^x, and therefore equal to 1 at $x = 0$, we obtain

$$e^x = 1 + \frac{x}{1} + \frac{x^2}{1 \cdot 2} + \frac{x^3}{1 \cdot 2 \cdot 3} + \cdots + \frac{x^n}{1 \cdot 2 \cdot 3 \ldots n} + \frac{e^\xi x^{n+1}}{1 \cdot 2 \cdot 3 \ldots n(n + 1)} \qquad (20)$$

where ξ is some number between 0 and x.
In particular, for $x = 1$, we have

$$e = 1 + 1 + \frac{1}{1 \cdot 2} + \frac{1}{1 \cdot 2 \cdot 3} + \frac{1}{1 \cdot 2 \cdot 3 \cdot 4} + \cdots + \frac{1}{1 \cdot 2 \cdot 3 \ldots n}$$
$$+ \frac{e^\xi}{1 \cdot 2 \cdot 3 \ldots (n + 1)}, \qquad \text{where } 0 < \xi < 1. \qquad (21)$$

This can be used to show that e *is irrational*. Indeed, assume that e is rational. Then, since $2 < e < 3$, we must have $e = p/q$, where p and q are integers, $q \ge 2$. Let n be chosen so that $n > q$. We multiply both sides of (21) by $1 \cdot 2 \cdot 3 \ldots n$ and observe that, since q is one of the numbers $2, 3, \ldots, n$, all terms in (21), except perhaps the last, become integers. We conclude that $e^{\xi}/(n+1)$ is an integer for some ξ between 0 and 1. This is impossible; for such ξ, $0 < e^{\xi} < e < 3$ and $n + 1 \geq 3$.

As another example, we consider Maclaurin's formula for the functions $x \mapsto \cos x$ and $x \mapsto \sin x$. We note the table:

		Value at $x = 0$		Value at $x = 0$
Function	$\cos x$	1	$\sin x$	0
First derivative	$-\sin x$	0	$\cos x$	1
Second derivative	$-\cos x$	-1	$-\sin x$	0
Third derivative	$\sin x$	0	$-\cos x$	-1
Fourth derivative	$\cos x$	1	$\sin x$	0
Fifth derivative	$-\sin x$	0	$\cos x$	1
Sixth derivative	$-\cos x$	-1	$-\sin x$	0

and obtain

$$\cos x = 1 - \frac{x^2}{1 \cdot 2} + \frac{x^4}{1 \cdot 2 \cdot 3 \cdot 4} - \cdots + (-1)^n \frac{x^{2n}}{1 \cdot 2 \cdot 3 \ldots (2n)}$$

$$+ (-1)^{n+1} \frac{(\cos \xi)x^{2n+2}}{1 \cdot 2 \cdot 3 \ldots (2n+2)},$$

where ξ is between 0 and x. Since $|\cos \xi| \leq 1$, we have

$$\cos x = 1 - \frac{x^2}{1 \cdot 2} + \frac{x^4}{1 \cdot 2 \cdot 3 \cdot 4} - \cdots + (-1)^n \frac{x^{2n}}{1 \cdot 2 \cdot 3 \ldots (2n)} + R,$$

$$|R| \leq \frac{|x|^{2n+2}}{1 \cdot 2 \cdot 3 \ldots (2n+2)}. \tag{22}$$

In the same way, we get

$$\sin x = x - \frac{x^3}{1 \cdot 2 \cdot 3} + \frac{x^5}{1 \cdot 2 \cdot 3 \cdot 4 \cdot 5} - \cdots + (-1)^n \frac{x^{2n+1}}{1 \cdot 2 \cdot 3 \ldots (2n+1)} + R,$$

$$|R| \leq \frac{|x|^{2n+3}}{1 \cdot 2 \cdot 3 \ldots (2n+3)}. \tag{23}$$

EXAMPLES 1. Compute $\cos(.2)$ correctly to four decimal places.
ANSWER Using (22), we have

$$\cos(.2) = 1 - \frac{(.2)^2}{1 \cdot 2} + \frac{(.2)^4}{1 \cdot 2 \cdot 3 \cdot 4} + R = 1 - .02 + .000067 + R \approx .9801 + R,$$

where we have rounded off the sum of the three terms on the right to four decimal places. It appears that we have taken a sufficient number of terms in the series. But we can be sure only by checking that the value of R does not affect the fourth decimal place. Indeed, $|R| \leq$

$$\left| \frac{(.2)^6}{1 \cdot 2 \cdot 3 \cdot 4 \cdot 5 \cdot 6} \right| = .000000\overline{08} < .0000001,$$ that is, $|R|$ cannot possibly affect the fourth decimal place. Hence $\cos(.2) \approx .9801$, correct to four decimal places.

2. Compute $e^{.1}$ correctly to four decimal places.
ANSWER   Using (20), we have

$$e^{.1} = 1 + \frac{(.1)}{1} + \frac{(.1)^2}{1 \cdot 2} + \frac{(.1)^3}{1 \cdot 2 \cdot 3} + R$$

$$= 1 + .1 + .005 + .000167 + R \approx 1.1052 + R.$$

We now verify that R does not affect the fourth decimal place. The remainder is $R = \dfrac{e^{\xi}(.1)^4}{1 \cdot 2 \cdot 3 \cdot 4}$, where $0 < \xi < .1$. Since $1 < e^{.1} < e < 3$, $R < .0003/24 = .0000125 < .00002$. Hence $e^{.1} \approx 1.1052$, correct to four decimal places.

PROBLEMS

Compute the following quantities to the degree of accuracy specified.

1. e^2,   three decimal places.
2. $e^{-.1}$,   four decimal places.
3. e^4,   three decimal places.
4. $e^{-.2}$,   four decimal places.

5. $\sin(.2)$,   four decimal places.
6. $\sin(-.4)$,   three decimal places.
7. $\cos(-.1)$,   four decimal places.
8. $\cos(.3)$,   four decimal places.

2.5   Logarithms

Some important special cases of Taylor's polynomials can be obtained by "elementary" methods without using Theorem 1. We begin with the identity

$$\frac{1}{1 - x} = 1 + x + x^2 + \cdots + x^n + \frac{x^{n+1}}{1 - x}, \tag{24}$$

which can be proved by multiplying both sides by $1 - x$. Replacing x by $(-x)$, we obtain

$$\frac{1}{1 + x} = 1 - x + x^2 - \cdots + (-1)^n x^n + \frac{(-1)^{n+1} x^{n+1}}{1 + x}. \tag{25}$$

Now we replace x in (25) by x^2 and obtain the identity

$$\frac{1}{1 + x^2} = 1 - x^2 + x^4 - \cdots + (-1)^n x^{2n} + \frac{(-1)^{n+1} x^{2n+2}}{1 + x^2}. \tag{26}$$

The three formulas just written are essentially Taylor formulas for the functions $(1 - x)^{-1}$, $(1 + x)^{-1}$, and $(1 + x^2)^{-1}$, with different expressions for the remainders.
    We obtain interesting cases of Taylor's polynomials by integrating the identities just obtained from $x = 0$. First we rewrite (25), replacing x by t and n by $n - 1$. We obtain

$$\frac{1}{1 + t} = 1 - t + t^2 - t^3 + \cdots + (-1)^{n-1} t^{n-1} + \frac{(-1)^n t^n}{1 + t}.$$

Integrate both sides from 0 to x, and note that $\int_0^x (1 + t)^{-1} \, dt = \int_0^{1+x} s^{-1} \, ds = \log(1 + x)$. We obtain

$$\log(1 + x) = x - \frac{x^2}{2} + \frac{x^3}{3} - \frac{x^4}{4} + \cdots + (-1)^{n-1} \frac{x^n}{n} + (-1)^n \int_0^x \frac{t^n \, dt}{1 + t}. \tag{27}$$

To estimate the remainder, assume first that $x > 0$. For $0 < t < x$, we have $1/(1 + t) < 1$, and therefore

$$0 < \int_0^x \frac{t^n \, dt}{1 + t} \le \int_0^x t^n \, dt = \frac{x^{n+1}}{n + 1}, \qquad \text{for } x > 0.$$

If $x < 0$, we must assume that $x > -1$. For $x < t < 0$, we have

$$\frac{1}{1 + t} < \frac{1}{1 + x} = \frac{1}{1 - |x|}$$

and

$$\left| \int_0^x \frac{t^n \, dt}{1 + t} \right| \le \frac{1}{1 - |x|} \left| \int_0^x t^n \, dt \right| = \frac{|x|^{n+1}}{(1 - |x|)(n + 1)}.$$

We conclude that

$$\log(1 + x) = x - \frac{x^2}{2} + \frac{x^3}{3} - \cdots + (-1)^{n-1} \frac{x^n}{n} + R, \tag{28}$$

where

$$|R| \le \begin{cases} \dfrac{x^{n+1}}{n + 1}, & \text{for } x \ge 0, \\[3mm] \dfrac{1}{1 - |x|} \dfrac{|x|^{n+1}}{n + 1}, & \text{for } -1 < x < 0. \end{cases}$$

Starting with relation (24) we obtain, in a similar way,

$$\frac{1}{1 - t} = 1 + t + t^2 + \cdots + t^{n-1} + \frac{t^n}{1 - t},$$

and upon integration from $t = 0$ to $t = x$,

$$\log \frac{1}{1 - x} = x + \frac{x^2}{2} + \frac{x^3}{3} + \cdots + \frac{x^n}{n} + \int_0^x \frac{t^n \, dt}{1 - t}. \tag{29}$$

The estimate of the remainder is given as Problem 10.

If we write the relations (27) and (29) for n even, that is, for $n = 2m + 2$, and add, we see that

$$\log \frac{1 + x}{1 - x} = 2x + \frac{2x^3}{3} + \cdots + \frac{2x^{2m+1}}{2m + 1} + R, \tag{30}$$

where

$$R = 2 \int_0^x \frac{t^{2m+2} \, dt}{1 - t^2}.$$

Assume that $x > 0$. Then $1 - t^2 \ge 1 - x^2$ for $0 \le t \le x$ and

$$\int_0^x \frac{t^{2m+2} \, dt}{1 - t^2} \le \int_0^x \frac{t^{2m+2} \, dt}{1 - x^2} = \frac{1}{1 - x^2} \frac{x^{2m+3}}{2m + 3},$$

so that

$$0 < R \le \frac{2}{1 - x^2} \frac{x^{2m+3}}{2m + 3}, \qquad \text{for } 0 < x < 1. \tag{31}$$

Now we observe that every number $a > 0$ can be written as

$$a = \frac{1 + x}{1 - x} \qquad \text{with} \qquad x = \frac{a - 1}{a + 1}.$$

This x satisfies $0 < x < 1$ if $a > 1$. Substituting into (30) and (31), we obtain

$$\log a = 2 \frac{a - 1}{a + 1} + \frac{2}{3} \left(\frac{a - 1}{a + 1} \right)^3 + \cdots + \frac{2}{2m + 1} \left(\frac{a - 1}{a + 1} \right)^{2m+1} + R \qquad (32)$$

with

$$0 < R \le \frac{(a + 1)^2}{2a} \frac{1}{2m + 3} \left(\frac{a - 1}{a + 1} \right)^{2m+3}, \qquad \text{for } a > 1. \qquad (33)$$

The error can be made arbitrarily small by choosing m large, but if $a - 1$ is small, even a small m gives good accuracy.

For instance, setting $a = 2$ and $m = 1$ in (32), we get

$$\log 2 = \frac{2}{3} + \frac{2}{3} \frac{1}{3^3} + R, \qquad 0 < R \le \frac{9}{4} \frac{1}{5} \frac{1}{3^5},$$

so that

$$.6914 < \log 2 < .6932$$

(actually $\log 2 = .6931 \ldots$). On the other hand, (28) applied to $x = 1$ gives the beautiful relation

$$\log 2 = 1 - \frac{1}{2} + \frac{1}{3} - \frac{1}{4} + \cdots + (-1)^n \frac{1}{n} + R, \qquad |R| \le \frac{1}{n + 1}. \qquad (34)$$

But to compute $\log 2$ with an accuracy of .002 by this formula, we need about 50 terms.

PROBLEMS

In Problems 1 to 4 find approximate values for the given quantities, and estimate the errors. Use Equation (28) with $n = 3$.

1. $\log 1.1$. 2. $\log .8$. 3. $\log \frac{3}{2}$. 4. $\log \frac{5}{4}$.

In Problems 5 to 8 find approximate values for the given quantities, and estimate the errors. Use Equations (32) and (33) with $m = 1$.

5. $\log 3$. 6. $\log 4$. 7. $\log \frac{3}{2}$. 8. $\log \frac{5}{4}$.

9. Find the nth Taylor polynomial for the function $f(x) = 1/(1 + x)$ at $x_0 = 0$, and then find the remainder term. Compare these results with Equation (25). Why are the remainder terms different?

10. Estimate the remainder term $\int_0^x \frac{t^n \, dt}{1 - t}$ in Equation (29). [*Hint:* Consider the cases $0 < t < x < 1$ and $x < t < 0$ separately.]

2.6 The arc tangent and π

Now we rewrite (26), replacing x by t and n by $n - 1$:

$$\frac{1}{1 + t^2} = 1 - t^2 + t^4 - \cdots + (-1)^{n-1} t^{2n-2} + (-1)^n \frac{t^{2n}}{1 + t^2}.$$

Integration yields

$$\text{arc tan } x = x - \frac{x^3}{3} + \frac{x^5}{5} - \frac{x^7}{7} + \cdots + (-1)^{n-1}\frac{x^{2n-1}}{2n-1} + R, \tag{35}$$

where

$$R = (-1)^n \int_0^x \frac{t^{2n}\, dt}{1+t^2}$$

and, since $1/(1+t^2) \le 1$, we see that

$$|R| \le \frac{|x|^{2n+1}}{2n+1}. \tag{36}$$

For $x = 1$, arc tan $x = \pi/4$ and we obtain the interesting relation

$$\frac{\pi}{4} = 1 - \frac{1}{3} + \frac{1}{5} - \frac{1}{7} + \cdots + \frac{(-1)^{n-1}}{2n-1} + R, \qquad |R| \le \frac{1}{2n+1}. \tag{37}$$

This is, however, not an efficient way to compute $\pi/4$; we would need about $n/2$ terms to obtain an accuracy of $1/n$.

The formula (35) for arc tan x gives good results with relatively few terms if x is close to 0. Now if u and v are small, we can use this formula to compute arc tan u and arc tan v, and we obtain, from the addition theorem for the arc tangent (see Chapter 10, §3.2),

$$\text{arc tan } \frac{u+v}{1-uv} = \text{arc tan } u + \text{arc tan } v. \tag{38}$$

For $u = \frac{1}{2}$, $v = \frac{1}{3}$, we have $(u+v)/(1-uv) = 1$, and Equation (38) gives

$$\frac{\pi}{4} = \text{arc tan } \frac{1}{2} + \text{arc tan } \frac{1}{3}. \tag{39}$$

For $u = \frac{1}{3}$, $v = \frac{1}{7}$ and for $u = \frac{1}{5}$, $v = \frac{1}{8}$, we obtain

$$\text{arc tan } \frac{1}{2} = \text{arc tan } \frac{1}{3} + \text{arc tan } \frac{1}{7}, \qquad \text{arc tan } \frac{1}{3} = \text{arc tan } \frac{1}{5} + \text{arc tan } \frac{1}{8}.$$

Combining these formulas we obtain

$$\frac{\pi}{4} = 2 \text{ arc tan } \frac{1}{5} + \text{arc tan } \frac{1}{7} + 2 \text{ arc tan } \frac{1}{8}. \tag{40}$$

This, together with (35), gives a good basis for computing π.

EXAMPLE Compute π, correctly to two decimal places.
ANSWER From (40), we have the formula

$$\pi = 8 \text{ arc tan } \frac{1}{5} + 4 \text{ arc tan } \frac{1}{7} + 8 \text{ arc tan } \frac{1}{8}.$$

It is sufficient to consider the remainder for arc tan $\frac{1}{5}$ since it will be larger than the remainders for arc tan $\frac{1}{7}$ and arc tan $\frac{1}{8}$ for the same number of terms. According to (36), the error in the calculation of arc tan $\frac{1}{5}$ will not exceed $\dfrac{1}{2n+1}\left|\dfrac{1}{5}\right|^{2n+1}$. Therefore, the error in the calculation of π will not exceed $\dfrac{20}{2n+1}\left|\dfrac{1}{5}\right|^{2n+1}$. For $n = 2$, this gives $\dfrac{20}{5}\left|\dfrac{1}{5}\right|^5 = \dfrac{20}{15625} = .00128$. Since $n = 2$ corresponds to the third term in (35), only two terms are needed in the Taylor formulas.

Thus

$$\arctan\frac{1}{5} \approx \frac{1}{5} - \frac{1}{3}\left(\frac{1}{5}\right)^3 \approx .1973, \qquad \arctan\frac{1}{7} \approx \frac{1}{7} - \frac{1}{3}\left(\frac{1}{7}\right)^3 \approx .1419,$$

$$\arctan\frac{1}{8} \approx \frac{1}{8} - \frac{1}{3}\left(\frac{1}{8}\right)^3 \approx .1243.$$

Hence $\pi \approx (8)(.1973) + (4)(.1419) + (8)(.1243) \approx 3.1404$, that is, $\pi = 3.14$, correct to two decimal places.

PROBLEMS

In Problems 1 to 3 find approximate values for the given quantities, and estimate the errors. Use Equations (35) and (36), as indicated.

1. $\arctan\frac{1}{10}$ with $n = 2$.
2. $\arctan(-\frac{1}{4})$ with $n = 2$.
3. $\arctan\frac{1}{2}$ with $n = 3$.
4. Using Equation (40), compute π correctly to four decimal places.

2.7 The sigma notation

From now on we shall use an abbreviated notation for sums using the letter Σ (Greek capital "sigma"). The sum

$$a_1 + a_2 + a_3 + a_4 + a_5 + a_6$$

is abbreviated by

$$\sum_{j=1}^{6} a_j$$

(read: "summation, j from 1 to 6, of a sub j"). For any integer $n > 1$, the sum

$$a_1 + a_2 + a_3 + \cdots + a_n \tag{41}$$

is abbreviated by the symbol

$$\sum_{j=1}^{n} a_j \tag{42}$$

(read: "summation, j from 1 to n, of a sub j"). The symbol j is called the **index of summation**; it is a "dummy subscript" and can be replaced by any other letter (not used in another capacity). Thus the sum (42) can be denoted also by any of the following symbols:

$$\sum_{k=1}^{n} a_k, \qquad \sum_{r=1}^{n} a_r, \qquad \sum_{\mu=1}^{n} a_\mu, \qquad \text{and so forth.}$$

(See the remark about the "variable of integration" in Chapter 8, §1.1.) For typographical convenience, we sometimes write $\sum_{j=1}^{n} a_j$ instead of $\displaystyle\sum_{j=1}^{n} a_j$.

The expression

$$\sum_{j=0}^{n} b_j \tag{43}$$

denotes the sum

$$b_0 + b_1 + b_2 + \cdots + b_n. \tag{44}$$

Similarly, the sum $a_2 + a_3 + a_4 + a_5$ is abbreviated by $\Sigma_{j=2}^{5} a_j$, the sum $c_4 + c_5 + c_6 + \cdots + c_{100}$ by $\Sigma_{k=4}^{100} c_k$, and so forth.

EXAMPLES 1. What is $\Sigma_{j=1}^{n} 1$?
ANSWER The desired sum is $\underbrace{1 + 1 + 1 + \cdots + 1}_{n \text{ times}}$. Thus $\Sigma_{j=1}^{n} 1 = n$.

2. Is it true that $\Sigma_{j=1}^{n} (a_j + b_j) = \Sigma_{j=1}^{n} a_j + \Sigma_{j=1}^{n} b_j$?
ANSWER Yes. The assertion is that

$$(a_1 + b_1) + (a_2 + b_2) + \cdots + (a_n + b_n) = (a_1 + a_2 + \cdots + a_n) + (b_1 + b_2 + \cdots + b_n).$$

This is so.

3. Compute $\Sigma_{j=0}^{4} (1 + j^2)$.
ANSWER We have

$$\sum_{j=0}^{4} (1 + j^2) = (1 + 0^2) + (1 + 1^2) + (1 + 2^2) + (1 + 3^2) + (1 + 4^2)$$
$$= 1 + 2 + 5 + 10 + 17 = 35.$$

4. What is $\displaystyle\sum_{k=1}^{4} \frac{1}{k}$?

ANSWER $\dfrac{1}{1} + \dfrac{1}{2} + \dfrac{1}{3} + \dfrac{1}{4} = \dfrac{25}{12}.$

We now write down, in the sigma notation, two important algebraic identities:

$$\sum_{j=1}^{n} j = \frac{1}{2} n(n + 1), \tag{45}$$

$$\sum_{j=1}^{n} q^j = \frac{1 - q^{n+1}}{1 - q}, \qquad \text{if } q \neq 1. \tag{46}$$

Relation (45) gives a way to compute the sum of the first n positive integers:

$$\sum_{j=1}^{n} j = 1 + 2 + 3 + 4 + \cdots + n.$$

We may also write

$$\sum_{j=1}^{n} j = n + (n - 1) + (n - 2) + (n - 3) + (n - 4) + \cdots + 1$$

(this is the same sum in reverse order). Adding the two expressions, we see that

$$2 \sum_{j=1}^{n} j = [1 + n] + [2 + (n - 1)] + [3 + (n - 2)] + \cdots + [n + 1]$$

$$= \underbrace{(n + 1) + (n + 1) + (n + 1) + \cdots + (n + 1)}_{n \text{ terms}} = n(n + 1).$$

Equation (45) follows. (This method of proof was discovered by Gauss at the age of 10.)

Equation (46) is called the formula for the sum of a **geometric progression**. The proof is as follows:

$$\sum_{j=0}^{n} q^j = 1 + q + q^2 + q^3 + \cdots + q^n,$$

hence

$$q \sum_{j=0}^{n} q^j = q + q^2 + q^3 + \cdots + q^n + q^{n+1}.$$

Subtraction yields

$$(1 - q) \sum_{j=0}^{n} q^j = 1 + q - q + q^2 - q^2 + \cdots + q^n - q^n - q^{n+1} = 1 - q^{n+1},$$

and (46) is obtained by dividing both sides by $1 - q \neq 0$.

Equation (46) is equivalent to Equation (24) in §2.5.

PROBLEMS

Evaluate the following sums.

1. $\displaystyle\sum_{k=1}^{4} (2k + 1).$

2. $\displaystyle\sum_{k=0}^{4} (k - 1)^2.$

3. $\displaystyle\sum_{m=1}^{4} \frac{1}{m(m + 1)}.$

4. $\displaystyle\sum_{m=0}^{5} \frac{m - 1}{m + 1}.$

5. $\displaystyle\sum_{\alpha=0}^{3} \frac{(-1)^{\alpha+1}}{(\alpha + 1)^2}.$

6. $\displaystyle\sum_{\alpha=10}^{12} (\alpha - 1)(\alpha + 1).$

7. $\displaystyle\sum_{j=-2}^{2} 2^{j+1}.$

8. $\displaystyle\sum_{j=2}^{6} (-1)^j \cos j\pi.$

9. $\displaystyle\sum_{p=1}^{4} \frac{1 \cdot 3 \cdot 5 \ldots (2p - 1)}{2 \cdot 4 \cdot 6 \ldots (2p)}.$

10. $\displaystyle\sum_{p=1}^{5} \binom{p + 1}{p - 1}.$

Write the following sums in sigma notation.

11. $2 + 4 + 6 + 8 + 10 + 12.$ $\left[Answer: \displaystyle\sum_{k=1}^{6} 2k. \right]$

12. $\sqrt{1} + \sqrt{3} + \sqrt{5} + \sqrt{7} + \sqrt{9}.$

13. $\dfrac{1}{2} + \dfrac{1}{6} + \dfrac{1}{10} + \dfrac{1}{14} + \dfrac{1}{18} + \dfrac{1}{22}.$

14. $1 - 8 + 27 - 64 + 125 - 216$.

15. $\dfrac{1}{3} + \dfrac{3}{6} + \dfrac{5}{9} + \dfrac{7}{12} + \dfrac{9}{15}$.

16. $\dfrac{1 \cdot 2}{3 \cdot 4} - \dfrac{2 \cdot 3}{4 \cdot 5} + \dfrac{3 \cdot 4}{5 \cdot 6} - \dfrac{4 \cdot 5}{6 \cdot 7} + \dfrac{5 \cdot 6}{7 \cdot 8}$.

17. $\dfrac{1 \cdot 4}{3 \cdot 4} + \dfrac{9 \cdot 16}{5 \cdot 6} + \dfrac{25 \cdot 36}{7 \cdot 8} + \dfrac{49 \cdot 64}{9 \cdot 10} + \dfrac{81 \cdot 100}{11 \cdot 12}$.

18. $\dfrac{1}{1} + \dfrac{1 \cdot 3}{1 \cdot 4} + \dfrac{1 \cdot 3 \cdot 5}{1 \cdot 4 \cdot 7} + \dfrac{1 \cdot 3 \cdot 5 \cdot 7}{1 \cdot 4 \cdot 7 \cdot 10}$.

19. Show that $\displaystyle\sum_{j=1}^{n} c = nc$, where c is a constant.

20. Show that $\displaystyle\sum_{j=1}^{n} ca_j = c \sum_{j=1}^{n} a_j$, where c is a constant.

21. Show that $\displaystyle\sum_{j=1}^{n} (a_j + cb_j) = \sum_{j=1}^{n} a_j + c \sum_{j=1}^{n} b_j$, where c is a constant.

22. Evaluate $\displaystyle\sum_{k=1}^{2n} \left[\cos\left(\dfrac{k\pi}{n}\right) - \cos\left(\dfrac{(k-1)\pi}{n}\right) \right]$ for any positive integer n. This is an example of what is called a telescoping sum.

23. Show that $\displaystyle\sum_{j=0}^{n+1} (-1)^j \binom{n+1}{j} = (-1)^{n+1} \binom{n}{n+1}$ for any nonnegative integer n. [*Hint:* Use the result of Problem 13, §2.3 and the principle of the telescoping sum.]

24. Show that $\displaystyle\sum_{\lambda=0}^{n} \binom{n-\lambda}{j-1} = \binom{n+1}{j} - \binom{0}{j}$ for any nonnegative integer n and positive integer j. See hint above.

25. Show that $\displaystyle\sum_{\lambda=0}^{n} \binom{\lambda+k}{k} = \binom{n+k+1}{k+1}$ for any nonnegative integers n and k. See hint above.

26. Show that

$$\sum_{\lambda=0}^{\left[\frac{n}{2}\right]} \binom{n}{2\lambda} = \sum_{\lambda=0}^{\left[\frac{n-1}{2}\right]} \binom{n}{2\lambda+1} = 2^{n-1}$$

for any positive integer n where $[\beta]$, as usual, denotes the largest integer not exceeding β. First prove the left-hand equality using the result of Problem 15, §2.3. (You may find it helpful to consider the cases n even and n odd separately.) Then use the left-hand equality and the result of Problem 14, §2.3, to prove the right-hand equality.

2.8 Taylor's formula in the sigma notation

To make full use of the sigma notation, we recall the notation

$$0! = 1, \quad 1! = 1, \quad 2! = 1 \cdot 2, \quad 3! = 1 \cdot 2 \cdot 3, \dots, \quad j! = 1 \cdot 2 \cdot 3 \dots j$$

($j!$ is read j **factorial**). The binomial coefficients [see Equation (16)] may be written

$$\binom{\alpha}{j} = \frac{\alpha(\alpha - 1)(\alpha - 2) \ldots (\alpha - j + 1)}{j!}.$$

If n is a positive integer, then

$$n(n - 1)(n - 2) \ldots (n - j + 1) = \frac{n!}{(n - j)!}$$

and

$$\binom{n}{j} = \frac{n!}{(n - j)!j!}.$$

This explains the symmetry

$$\binom{n}{j} = \binom{n}{n - j}.$$

The binomial formula can be written as

$$(1 + x)^n = \sum_{j=0}^{n} \binom{n}{j} x^j.$$

Setting $x = 1$ and $x = -1$, we see that

$$\sum_{j=0}^{n} \binom{n}{j} = 2^n, \qquad \sum_{j=0}^{n} \binom{n}{j} (-1)^j = 0.$$

Now we rewrite Taylor's general formula in the sigma notation:

$$f(x) = \sum_{j=0}^{n} \frac{f^{(j)}(x_0)}{j!} (x - x_0)^j + \frac{f^{(n+1)}(\xi)}{(n + 1)!} (x - x_0)^{n+1}, \qquad \text{for some } \xi \text{ between } x_0 \text{ and } x.$$

It is assumed here that $f^{(0)}(x) = f(x)$. We adopt this convention in what follows.
Some particular Taylor formulas follow.

$$(1 + x)^\alpha = \sum_{j=0}^{n} \binom{\alpha}{j} x^j + R, \tag{47}$$

where $R = \binom{\alpha}{n + 1}(1 + \xi)^{\alpha-n-1}x^{n+1}$ with ξ between 0 and x,

$$\log(1 + x) = \sum_{j=1}^{n} (-1)^{j+1} \frac{x^j}{j} + R, \tag{48}$$

where

$$|R| \leq \begin{cases} \dfrac{x^{n+1}}{(n + 1)}, & \text{if } x \geq 0 \\[2ex] \dfrac{|x|^{n+1}}{(1 - |x|)(n + 1)}, & \text{if } x < 0 \end{cases}$$

and, in particular,

$$\log 2 = \sum_{j=1}^{n} \frac{(-1)^{j+1}}{j} + R, \qquad |R| \leq \frac{1}{n + 1}. \tag{49}$$

Also,

$$\arctan x = \sum_{j=1}^{n} (-1)^{j+1} \frac{x^{2j-1}}{2j-1} + R, \qquad |R| \leq \frac{|x|^{2n+1}}{2n+1} \tag{50}$$

and, in particular,

$$\frac{\pi}{4} = \sum_{j=1}^{n} \frac{(-1)^{j+1}}{2j-1} + R, \qquad |R| \leq \frac{1}{2n+1}. \tag{51}$$

Also

$$e^x = \sum_{j=0}^{n} \frac{x^j}{j!} + R, \qquad |R| \leq \frac{e^x |x|^{n+1}}{(n+1)!}, \tag{52}$$

$$\cos x = \sum_{j=0}^{n} (-1)^j \frac{x^{2j}}{(2j)!} + R, \qquad |R| \leq \frac{|x|^{2n+2}}{(2n+2)!}, \tag{53}$$

$$\sin x = \sum_{j=0}^{n} (-1)^j \frac{x^{2j+1}}{(2j+1)!} + R, \qquad |R| \leq \frac{|x|^{2n+3}}{(2n+3)!}. \tag{54}$$

The reader is asked to check that these are indeed the same expansions we obtained before.

§3 L'Hospital's rule‡

There is a method for evaluating certain limits which is known as l'Hospital's rule. It is based on the generalized mean value theorem, stated in §1.2 as Theorem 2, and on its corollary. The method is almost as old as calculus, and, its name notwithstanding, is due to Johann Bernoulli.

3.1 Indeterminate forms 0/0

The basic prescription is the l'Hospital rule for evaluating the limit of a fraction $F(x)/G(x)$ at a point a when $F(a) = G(a) = 0$. It is applicable if F and G are continuous at a, if F and G have continuous derivatives, and $G' \neq 0$ near a, except perhaps for $x = a$. The rule reads:

If $\quad \lim_{x \to a} F(x) = \lim_{x \to a} G(x) = 0 \quad$ *and* $\quad \lim_{x \to a} \frac{F'(x)}{G'(x)} \quad$ *exists,*

then

$$\lim_{x \to a} \frac{F(x)}{G(x)} = \lim_{x \to a} \frac{F'(x)}{G'(x)}. \tag{1}$$

GUILLAUME FRANÇOIS MARQUIS DE L'HOSPITAL (1661–1704), a member of the high French nobility, was an amateur mathematician and a friend and patron of mathematicians. He studied under Johann Bernoulli; this resulted in a textbook of calculus which contains the l'Hospital rule.

---

‡This section may be omitted without loss of continuity.

Indeed, by the corollary to Theorem 2 in §1.2 there is, for every $x \neq a$, a ξ between a and x such that $F(x)/G(x) = F'(\xi)/G'(\xi)$. If x is close to a, then ξ is close to a, and $F'(\xi)/G'(\xi)$ is close to its limit.

L'Hospital's rule is called the rule for evaluating "indeterminate forms 0/0." It may happen, of course, that $F'(a) = G'(a) = 0$. In this case, we can try applying l'Hospital's rule once more. If $\lim_{x \to a} [F''(x)/G''(x)]$ exists, we have

$$\lim_{x \to a} \frac{F(x)}{G(x)} = \lim_{x \to a} \frac{F''(x)}{G''(x)}.$$

We may have to do this several times.

L'Hospital's rule holds also for one-sided limits; no new proof is needed. The rule holds also if $\lim_{x \to a} F'/G'$ is $+\infty$ or $-\infty$; no new proof is needed. The rule applies also if $a = +\infty$ or if $a = -\infty$. Indeed, if $\lim_{x \to +\infty} F(x) = 0$, $\lim_{x \to +\infty} G(x) = 0$, we have by the definition of limits at infinity and by l'Hospital's rule:

$$\lim_{x \to +\infty} \frac{F(x)}{G(x)} = \lim_{x \to 0^+} \frac{F(1/x)}{G(1/x)} = \lim_{x \to 0^+} \frac{dF(1/x)/dx}{dG(1/x)/dx}$$

$$= \lim_{x \to 0^+} \frac{F'(1/x)(-1/x^2)}{G'(1/x)(-1/x^2)} = \lim_{x \to 0^+} \frac{F'(1/x)}{G'(1/x)} = \lim_{x \to +\infty} \frac{F'(x)}{G'(x)},$$

provided that the last limit exists.

EXAMPLES 1. Compute $\lim_{x \to 0} \dfrac{\tan x}{x}$.

ANSWER Numerator and denominator are 0 at $x = 0$. L'Hospital's rule gives

$$\lim_{x \to 0} \frac{\tan x}{x} = \lim_{x \to 0} \frac{\sec^2 x}{1} = \frac{1}{1} = 1.$$

2. Compute $\lim_{x \to 1} \dfrac{x^2 - 1}{x^2 - 5x + 4}$.

ANSWER Numerator and denominator are 0 at $x = 1$. L'Hospital's rule gives

$$\lim_{x \to 1} \frac{x^2 - 1}{x^2 - 5x + 4} = \lim_{x \to 1} \frac{2x}{2x - 5} = -\frac{2}{3}.$$

3. Find $\lim_{x \to 0} \dfrac{\sin x - e^x + 1}{x^2}$.

ANSWER Numerator and denominator are 0 at $x = 0$. Applying l'Hospital's rule twice, we have

$$\lim_{x \to 0} \frac{\sin x - e^x + 1}{x^2} = \lim_{x \to 0} \frac{\cos x - e^x}{2x}$$

$$= \lim_{x \to 0} \frac{-\sin x - e^x}{2} = -\frac{1}{2}.$$

4. What is wrong with the following evaluation based on l'Hospital's rule?

$$\lim_{x \to 1} \frac{x^3 - 3x + 2}{x^3 - 2x^2 + x} = \lim_{x \to 1} \frac{3x^2 - 3}{3x^2 - 4x + 1}$$

$$= \lim_{x \to 1} \frac{6x}{6x - 4} = \lim_{x \to 1} \frac{6}{6} = 1.$$

ANSWER We have applied l'Hospital's rule too many times. After using l'Hospital's rule twice, we have $\lim_{x \to 1} 6x/(6x - 4)$. But here $6x$ and $6x - 4$ are equal to 6 and 2 at $x = 1$ (and not to 0 and 0). Hence further use of l'Hospital's rule is incorrect. The correct answer is $\lim_{x\to 1} 6x/(6x-4) = 3$.

5. Find $\lim_{x\to 0^+} \dfrac{\sin x}{\sqrt{x}}$.

ANSWER Numerator and denominator are 0 at $x = 0$. Applying l'Hospital's rule, we have

$$\lim_{x\to 0^+} \frac{\sin x}{\sqrt{x}} = \lim_{x\to 0^+} \frac{\cos x}{(-1/2\sqrt{x})} = \lim_{x\to 0^+}(\cos x)(-2\sqrt{x}) = (1)(0) = 0.$$

6. Find $\lim_{x\to 0^+} \dfrac{e^x - 1}{x^2}$.

ANSWER Numerator and denominator are 0 at $x = 0$. Applying l'Hospital's rule, we have

$$\lim_{x\to 0^+} \frac{e^x - 1}{x^2} = \lim_{x\to 0^+} \frac{e^x}{2x} = +\infty$$

(since $2x/e^x$ is positive for $x > 0$ and has limit 0 at $x = 0$).

PROBLEMS

Evaluate the following limits using l'Hospital's rule.

1. $\lim_{x\to 2} \dfrac{3x^2 - 5x - 2}{3x^2 - 7x + 2}$.

2. $\lim_{x\to 1} \dfrac{x^3 + x^2 - 2x}{x^3 - x^2 + x - 1}$.

3. $\lim_{x\to -2} \dfrac{x^3 + 3x^2 - 4}{x^3 + 5x^2 + 8x + 4}$.

4. $\lim_{x\to 1} \dfrac{x^3 - x^2 - x + 1}{x^3 + x^2 - 5x + 3}$.

5. $\lim_{x\to \pi/2} \dfrac{1 - \sin x}{\cos x}$.

6. $\lim_{x\to 0} \dfrac{x - \sin x}{1 - \cos x}$.

7. $\lim_{x\to 0} \dfrac{x - \sin x}{x^3}$.

8. $\lim_{x\to \pi/4} \dfrac{\cos x - \sin x}{\log \tan x}$.

9. $\lim_{x\to 0} \dfrac{\arctan x}{\arcsin x}$.

10. $\lim_{x\to 0} \dfrac{(\arctan x)^2}{\log(1 + x^2)}$.

11. $\lim_{x\to +\infty} \dfrac{\sin(4/x)}{(1/x)}$.

12. $\lim_{x\to +\infty} \dfrac{\arctan(1/x)}{\sin(1/x)}$.

3.2 Indeterminate forms ∞/∞

There is another case of l'Hospital's rule. It refers to "indeterminate forms ∞/∞" and reads:

If $\lim_{x\to a} F(x) = +\infty,$ $\lim_{x\to a} G(x) = +\infty,$ *and* $\lim_{x\to a} \dfrac{F'(x)}{G'(x)}$ exists,

then

$$\lim_{x\to a} \frac{F(x)}{G(x)} = \lim_{x\to a} \frac{F'(x)}{G'(x)}. \qquad (2)$$

The proof is more complicated; we shall only sketch it. Set

$$\lim_{x\to a} \frac{F'(x)}{G'(x)} = L.$$

We pick a number $x_1 \neq a$ and consider a number x between a and x_1. By the generalized mean value theorem (Theorem 2 in §1.2), there is a number x_0 between x and x_1 such that

$$\frac{F(x) - F(x_1)}{G(x) - G(x_1)} = \frac{F'(x_0)}{G'(x_0)}. \tag{3}$$

On the other hand,

$$\frac{F(x) - F(x_1)}{G(x) - G(x_1)} = \frac{F(x)}{G(x)} \frac{1 - (F(x_1)/F(x))}{1 - (G(x_1)/G(x))},$$

so that we can write (3) as

$$\frac{F(x)}{G(x)} = \frac{1 - (G(x_1)/G(x))}{1 - (F(x_1)/F(x))} \frac{F'(x_0)}{G'(x_0)}.$$

Now x_0 lies between x_1 and a; hence x_0 is closer to a than x_1 is. Therefore, if x_1 is sufficiently close to a, the fraction $F'(x_0)/G'(x_0)$ will be as close to L as we want. On the other hand, if we keep x_1 fixed and then choose x sufficiently close to a, then $|F(x)|$ and $|G(x)|$ will be as large as we want, the fractions $G(x_1)/G(x)$ and $F(x_1)/F(x)$ will be as close to 0 as we want, and the factor

$$\frac{1 - (G(x_1)/G(x))}{1 - (F(x_1)/F(x))}$$

will be as close to 1 as we want. Hence $F(x)/G(x)$ will be as close to L as we want, provided $x \neq a$ is sufficiently close to a. That is what we set out to prove.

Rule (2) holds also with $+\infty$ replaced by $-\infty$, and for one-sided limits. The rule applies also if $a = +\infty$ or if $a = -\infty$, and if $\lim_{x \to a} F'/G'$ is $+\infty$ or $-\infty$.

EXAMPLES 1. Find $\lim\limits_{x \to +\infty} \dfrac{x + \log x}{x \log x}$.

ANSWER   Numerator and denominator have limits $+\infty$ at $x = +\infty$. Applying l'Hospital's rule, we have

$$\lim_{x \to +\infty} \frac{x + \log x}{x \log x} = \lim_{x \to \infty} \frac{1 + 1/x}{(1)\log x + x(1/x)}$$

$$= \lim_{x \to \infty} \left(1 + \frac{1}{x}\right)\left(\frac{1}{1 + \log x}\right) = (1)(0) = 0.$$

2. Evaluate $\lim\limits_{x \to +\infty} \dfrac{x^2 + 3x + 1}{4x^2 - 5x + 1}$.

ANSWER   Numerator and denominator have limits $+\infty$ at $x = +\infty$. Applying l'Hospital's rule twice, we have

$$\lim_{x \to +\infty} \frac{x^2 + 3x + 1}{4x^2 - 5x + 1} = \lim_{x \to +\infty} \frac{2x + 3}{8x - 5} = \lim_{x \to +\infty} \frac{2}{8} = \frac{1}{4}.$$

3. Find $\lim\limits_{x \to 0^+} \dfrac{\log \sin x}{\log \tan x}$.

ANSWER   Numerator and denominator have limits $-\infty$ at $x = 0$. L'Hospital's rule gives

$$\lim_{x \to 0^+} \frac{\log \sin x}{\log \tan x} = \lim_{x \to 0^+} \frac{\dfrac{1}{\sin x}(\cos x)}{\dfrac{1}{\tan x}(\sec^2 x)} = \lim_{x \to 0^+} \cos^2 x = 1.$$

4. Find $\lim\limits_{x \to 0} \dfrac{\log \sin^2 x}{\cot x}$.

ANSWER Numerator and denominator have limits ∞ at $x = 0$ (one-sided limits $+\infty$ and $-\infty$ in the case of the denominator, but this does not matter). L'Hospital's rule gives

$$\lim_{x \to 0} \frac{\log \sin^2 x}{\cot x} = \lim_{x \to 0} \frac{(1/\sin^2 x)2 \sin x \cos x}{-1/\sin^2 x}$$

$$= \lim_{x \to 0} (-2 \sin x \cos x) = 0.$$

PROBLEMS

Evaluate the following limits using l'Hospital's rule.

1. $\lim\limits_{x \to +\infty} \dfrac{x^3 + 2x + 2}{2x^3 + x^2 - 3x + 1}$.

2. $\lim\limits_{x \to +\infty} \dfrac{\log x}{x}$.

3. $\lim\limits_{x \to \pi^-/2} \dfrac{\tan x}{\log \cos x}$.

4. $\lim\limits_{x \to \pi^-/2} \dfrac{\log \tan x}{\tan x}$.

5. $\lim\limits_{x \to \frac{1}{2}^+} \dfrac{\sqrt{2x - 1}}{\log 2x}$.

6. $\lim\limits_{x \to 1^+} \dfrac{\log(x - 1)}{\log(x^2 - 1)}$.

7. $\lim\limits_{x \to +\infty} \dfrac{x \log x}{(x + 1)^2}$.

8. $\lim\limits_{x \to +\infty} \dfrac{\log(x + e^x)}{x}$.

9. $\lim\limits_{x \to +\infty} \dfrac{e^x + \log x}{e^x + x}$.

10. $\lim\limits_{x \to 0^+} \dfrac{\log x}{e^{1/x}}$.

3.3   Other indeterminate forms

Other "indeterminate forms," that is, forms of the type $0 \cdot \infty$, $\infty - \infty$, ∞^0, 1^∞, and 0^0, can sometimes be evaluated by l'Hospital's rule. The following "hints" may be helpful: (1) reduce the forms $0 \cdot \infty$ and $\infty - \infty$ to $0/0$ or ∞/∞ by algebraic manipulation, and evaluate using l'Hospital's rule; (2) rewrite the forms ∞^0, 1^∞, and 0^0 using the identity $[f(x)]^{g(x)} = e^{g(x) \log f(x)}$, and reduce to the forms $0/0$ or ∞/∞. Then evaluate using l'Hospital's rule. The "hints" are best explained by examples.

EXAMPLES **1.** A case of $0 \cdot \infty$ reduced to ∞/∞.
Find $\lim_{x \to 0^+} x \log x$. We have

$$\lim_{x \to 0^+} x \log x = \lim_{x \to 0^+} \frac{\log x}{1/x} = \lim_{x \to 0^+} \frac{1/x}{-1/x^2} = \lim_{x \to 0^+} (-x) = 0.$$

2. A case of $\infty - \infty$ reduced to $0/0$.
What is $\lim_{x \to 0} (\cot x - 1/x)$? Since

$$\cot x - \frac{1}{x} = \frac{\cot x}{x}\left(x - \frac{1}{\cot x}\right) = \frac{x - \tan x}{x \tan x},$$

we have

$$\lim_{x\to0}\left(\cot x - \frac{1}{x}\right) = \lim_{x\to0}\frac{x - \tan x}{x \tan x} = \lim_{x\to0}\frac{1 - \dfrac{1}{\cos^2 x}}{\tan x + \dfrac{x}{\cos^2 x}} = \lim_{x\to0}\frac{\cos^2 x - 1}{\sin x \cos x + x}$$

$$= \lim_{x\to0}\frac{-\sin^2 x}{\frac{1}{2}\sin 2x + x} = \lim_{x\to0}\frac{-2\sin x \cos x}{\cos 2x + 1} = \frac{0}{2} = 0.$$

3. A case of ∞^0 reduced to $0/0$.

Suppose we want to find the limit, for $x \to +\infty$, of $x^{e^{-x}}$. We first consider the limit of the function $\log(x^{e^{-x}})$, and obtain

$$\lim_{x\to+\infty}\log x^{e^{-x}} = \lim_{x\to+\infty} e^{-x}\log x$$

$$= \lim_{x\to+\infty}\frac{\log x}{e^x} = \lim_{x\to+\infty}\frac{1/x}{e^x} = \lim_{x\to+\infty}\frac{1}{xe^x} = 0.$$

Now we write

$$\lim_{x\to+\infty} x^{e^{-x}} = \lim_{x\to+\infty} e^{e^{-x}\log x}$$

$$= e^{\lim_{x\to+\infty}e^{-x}\log x} = e^0 = 1.$$

Here we used Theorem 2 in Chapter 6, §2.1, and Equation (14) in Chapter 11, §3.4.

PROBLEMS

Evaluate the following indeterminate forms using l'Hospital's rule.

1. $\lim\limits_{x\to0^+} (e^x - 1)\cot x$.

2. $\lim\limits_{x\to0^+} (\sin x)(\log x)$.

3. $\lim\limits_{x\to0^+} x^3 e^{1/x}$.

4. $\lim\limits_{x\to0^+}\left(\dfrac{1}{e^x - 1} - \dfrac{1}{x}\right)$.

5. $\lim\limits_{x\to0^+}\left(\dfrac{1}{x} - \dfrac{1}{\sin x}\right)$.

6. $\lim\limits_{x\to1}\left(\dfrac{1}{\log x} - \dfrac{x}{x - 1}\right)$.

7. $\lim\limits_{x\to+\infty} x^{1/x}$.

8. $\lim\limits_{x\to0}\left(1 + \dfrac{1}{x^2}\right)^{x^2}$.

9. $\lim\limits_{x\to0^+} (\cot x)^x$.

10. $\lim\limits_{x\to0^+} (\cos x)^{1/x}$.

11. $\lim\limits_{x\to1^+} x^{1/(x-1)}$.

12. $\lim\limits_{x\to0^+} (e^x + x)^{1/x}$.

13. $\lim\limits_{x\to0^+} x^x$.

14. $\lim\limits_{x\to0^+} (\sin x)^x$.

15. $\lim\limits_{x\to0^+} (\log(1 + x))^x$.

INFINITE SERIES

§1 Infinite sequences

Taylor's theorem leads to the concept of infinite series. For example, we have seen in Chapter 14, §2.4, that $e = 1 + \dfrac{1}{1!} + \dfrac{1}{2!} + \cdots + \dfrac{1}{n!}$ plus a remainder of absolute value not greater than $3/n!$. Perhaps we can dispense with the remainder and write e as an infinite sum $1 + \dfrac{1}{1!} + \dfrac{1}{2!} + \dfrac{1}{3!} + \cdots$? The theory of such infinite series is the subject of the present chapter.

1.1 Infinite sequences of numbers

In order to develop a theory of infinite series, we first consider **infinite sequences** of numbers. ("Series" and "sequences" are closely related but distinct terms; care should be taken not to confuse them.) An **infinite sequence** is a rule that associates a number to each integer $1, 2, 3, \ldots$; the number associated with the integer k is called the kth term of the sequence.

Sequences are often denoted by writing down the first few terms; in this case we assume that the rule of forming the "general" term, that is, the kth term for any k, is clear. For instance, consider the sequences

$$1, 3, 5, 7, \ldots \tag{1}$$

$$1, 0, 1, 0, 1, 0, \ldots \tag{2}$$

$$1, 1, 2, 3, 5, 8, 13, \ldots . \tag{3}$$

In writing $\ldots$, we indicate that the sequence is infinite. Sequence (1) is the sequence of odd integers. The fifth term is 9. The nth term is $2n - 1$. In sequence (2) the nth term is 0 or 1 according to whether n is even or odd. We may write this nth term as $[1 - (-1)^n]/2$. Sequence (3) is called the **Fibonacci sequence;** each term of this sequence is computed by adding the preceding two, the first two terms being 1. Therefore the eighth term in (3) is $8 + 13 = 21$, the ninth is $13 + 21 = 34$, and so forth. The Fibonacci numbers have been a source of delight to professional and amateur mathematicians for seven centuries.

In talking about sequences, we use variables; the term "the sequence $a_1, a_2, a_3, \ldots$" or "the sequence a_n, $n = 1, 2, 3, \ldots$" or "the sequence $\{a_j\}$" or simply "the sequence a_j" denotes a definite but unspecified rule $k \mapsto a_k$ which assigns to every positive integer k the number a_k. If we want to specify the sequence, we write out the rule. For instance,

$$a_n = 2n - 1, \qquad n = 1, 2, \ldots$$

denotes the sequence (1) of odd numbers, whereas

$$a_1 = a_2 = 1, \qquad a_n = a_{n-1} + a_{n-2}, \qquad n = 3, 4, \ldots$$

denotes the Fibonacci sequence. Of course, in the above formulas a and n are "dummy variables." The sequence of odd integers can be just as well described by writing $\alpha_s = 2s - 1$, $s = 1, 2, \ldots$.

Finally, we note that labeling the terms in a sequence by the integers $1, 2, 3, \ldots$ is only a matter of convenience. We may begin at 0 or at -1 or at 5.

EXAMPLE If $a_n = 3n - 5$, $n = 0, 1, \ldots$, what are the first five terms of this sequence?
ANSWER $a_0 = 3 \cdot 0 - 5 = -5$, $a_1 = 3 \cdot 1 - 5 = -2$, $a_2 = 3 \cdot 2 - 5 = 1$, $a_3 = 3 \cdot 3 - 5 = 4$, $a_4 = 3 \cdot 4 - 5 = 7$.

PROBLEMS

In Problems 1 to 8 write the first five terms of the given sequence:

1. $a_n = n^2 - n$, $n = 1, 2, \ldots$.

2. $a_n = n^3 - 1$, $n = 1, 2, \ldots$.

3. $a_n = \dfrac{(-1)^n}{n+1}$, $n = 0, 1, \ldots$.

4. $a_n = \dfrac{n+1}{n^2+1}$, $n = 0, 1, \ldots$.

5. $a_j = 2j^2 + j + 1$, $j = -1, 0, \ldots$.

6. $a_j = \dfrac{(-1)^j}{1+j^2}$, $j = 0, 1, \ldots$.

7. $a_j = (-1)^j(j-1)^3$, $j = 1, 2, \ldots$.

8. $a_j = \dfrac{(j-1)(2j-1)}{(2j+1)}$, $j = 2, 3, \ldots$.

LEONARDO of Pisa, also known as FIBONACCI (13th century), wrote two mathematical books—one on algebra, the other on geometry. They contain mate-

rial that Leonardo acquired during his travels in the Orient and, probably, also include some of his original investigations.

9. If $x_1 = 1$, $x_2 = 2$, and $x_n = 2x_{n-1} - x_{n-2}$ for $n = 3, 4, \ldots$, find x_3, x_4, and x_5.

10. If $x_0 = 2$, $x_1 = 4$, and $x_n = \left(\dfrac{x_{n-1}}{x_{n-2}}\right)$ for $n = 2, 3, \ldots$, find x_2, x_3, and x_4.

11. If $b_{-1} = 1$, $b_0 = 4$, $b_1 = 9$, $b_2 = 16$, $b_3 = 25$, $b_4 = 36, \ldots$, what is b_n?

12. If $b_2 = 1$, $b_3 = 2$, $b_4 = 4$, $b_5 = 8$, $b_6 = 16$, $b_7 = 32$, what is b_k?

13. What is the "general" term of the sequence 0, 4, 8, 12, 16, 20, ...? Indicate from where you begin labeling the terms.

14. What is the "general" term of the sequence 1, 4, 7, 10, 13, 16, ...? Indicate from where you begin labeling the terms.

1.2   Convergent and divergent sequences. Limits

Consider the four sequences

$$1, \frac{1}{2}, \frac{1}{3}, \frac{1}{4}, \frac{1}{5}, \ldots$$

$$1, -\frac{1}{2}, \frac{1}{3}, -\frac{1}{4}, \frac{1}{5}, \ldots$$

$$1, 0, \frac{1}{10}, 0, \frac{1}{100}, 0, \frac{1}{1000}, \ldots$$

$$\frac{1}{\sqrt{2}}, \frac{1}{\sqrt{3}}, \frac{1}{\sqrt{4}}, \frac{1}{\sqrt{5}}, \frac{1}{\sqrt{6}}, \ldots$$

All four have one property in common. If we go far enough in the sequence, the terms become as small as we like in absolute value. In the first sequence, for instance, all terms after the fifth lie between 0 and $\frac{1}{5}$, all terms after the tenth between 0 and $\frac{1}{10}$, and so on. We describe this property by saying that the four sequences considered have the limit 0, or that they converge to 0, or that they approach 0.

If we go far enough in the sequence

$$5.1, 5.01, 5.001, 5.0001, \ldots,$$

the terms become as close as we want to 5. We say that this sequence has the limit 5 (or converges to 5, or approaches 5).

We now define the terms **convergence** and **limit of a sequence** precisely. It is convenient to use the expression **nearly all** to mean "all but a finite number." (For instance, in the sequence $-3, -2, -1, 0, 1, 2, \ldots$ nearly all terms are positive.)

The statement

$$\lim_{n \to \infty} a_n = \alpha$$

means that, for every positive number ϵ, nearly all terms satisfy $|a_n - \alpha| < \epsilon$. In other words, for every positive number ϵ, there is a number N such that $|a_n - \alpha| < \epsilon$ for $n > N$.

Let us use the definition to prove that

$$\lim_{n \to \infty} \frac{1}{n} = 0. \tag{4}$$

Let $\epsilon > 0$ be given. For which n is it true that $\left|\dfrac{1}{n} - 0\right| = \dfrac{1}{n} < \epsilon$? For $n > 1/\epsilon$, that is, for nearly all n. $\left(\text{In other words, for every positive number } \epsilon, \text{ there is a number}\right.$

$N = 1/\epsilon$ such that $\left|\dfrac{1}{n} - 0\right| < \epsilon$ for $n > N.\left.\right)$

Let us use the same method to prove that

$$\lim_{n \to \infty} \frac{1}{n^{\alpha}} = 0, \qquad \alpha > 0. \tag{4'}$$

Let an $\epsilon > 0$ be given. For which n is it true that $\left|\dfrac{1}{n^{\alpha}} - 0\right| = \dfrac{1}{n^{\alpha}} < \epsilon$? For $n >$ $(1/\epsilon)^{1/\alpha}$, that is, for nearly all n. $\left[\text{In other words, for every positive number } \epsilon, \text{ there}\right.$ is a number $N = (1/\epsilon)^{1/\alpha}$ such that $\left|\dfrac{1}{n^{\alpha}} - 0\right| < \epsilon$ for $n > N.\left.\right]$

Not every sequence has a limit; a sequence without a limit is called **divergent.** A sequence $a_1, a_2, a_3, \ldots$ is said to *diverge to* $+\infty$; in symbols, $\lim_{n \to \infty} a_n = +\infty$ if, for *every* number A, nearly all terms a_n are $>A$. An example is the sequence 1, 2, 3, 4, 5, 6, $\ldots$. Similarly, $\lim_{n \to \infty} a_n = -\infty$ means that for *every* number A, nearly all terms a_n are $<A$.

A sequence may diverge (that is, fail to converge) without having even an infinite limit. For instance, the sequence 1, 2, 3, 1, 2, 3, 1, 2, 3, 1, 2, $\ldots$ diverges.

EXAMPLES 1. Evaluate $\lim_{n \to \infty} 1/\sqrt{n}$ and $\lim_{n \to \infty} 1/n^2$.
ANSWER Both of these limits are special cases of the result given in Equation (4'). We have $\lim_{n \to \infty} 1/\sqrt{n} = 0$, $\lim_{n \to \infty} 1/n^2 = 0$.

2. Does the sequence $\frac{2}{2}, \frac{3}{4}, \frac{4}{6}, \frac{5}{8}, \frac{6}{10}, \frac{7}{12}, \ldots$ converge?
ANSWER To find out, we first determine the general term. We note that the numerators are $2 = 1 + 1$, $3 = 2 + 1$, $4 = 3 + 1, \ldots$ while the denominators are $2 = 2 \cdot 1$, $4 = 2 \cdot 2$, $6 = 2 \cdot 3, \ldots$. Therefore, the general term is $a_n = \dfrac{n + 1}{2n} = \dfrac{1}{2} + \dfrac{1}{2n}$, where $n = 1, 2, 3, \ldots$. The term $1/2n$ is small for large n, so we "guess" the limit of the sequence is $\alpha = \frac{1}{2}$. To prove we have the correct limit, let $\epsilon > 0$ be given. Then $|a_n - \alpha| = \left|\dfrac{n + 1}{2n} - \dfrac{1}{2}\right| = \dfrac{1}{2n}$. Now, $1/2n < \epsilon$ if $n > 1/2\epsilon$. Hence there is a number $N = 1/2\epsilon$ such that $|a_n - \alpha| < \epsilon$ for $n > N$.

3. Find the limit of the sequence whose general term is $a_n = 4n^2/(n^2 + 1)$, where $n = 1, 2, 3, \ldots$.
ANSWER We use a trick. Divide numerator and denominator by n^2. Then $a_n = 4/(1 + (1/n^2))$. For large n, the denominator approaches 1 and a_n approaches 4. We guess that $\lim_{n \to \infty} 4n^2/(n^2 + 1) = 4$. To prove this, let $\epsilon > 0$ be given. Then

$$|a_n - \alpha| = \left|\frac{4n^2}{n^2 + 1} - 4\right| = \left|\frac{-4}{n^2 + 1}\right| < \frac{4}{n^2} < \epsilon$$

if $n > 2/\sqrt{\epsilon}$. Hence there is a number $N = 2/\sqrt{\epsilon}$ such that $|a_n - \alpha| < \epsilon$ for $n > N$.

PROBLEMS

The application of the definition of convergence of a sequence directly to a concrete problem consists of two stages. First, we must decide if we are trying to prove that the sequence converges or that it does not converge, and, in the former case, we must also decide on the value to which we believe it converges. Only after this can we pass to the second stage, namely, the actual formal proof; of course, if our guess in the first stage is incorrect, it will be impossible to complete the second stage. Remember that the process is not complete without the formal proof, but the *main thing here is to guess at the correct answer*.

In each of Problems 1 to 16 decide if the given sequence converges or not, and, in the former case, decide on what the limit should be.

1. $\dfrac{1}{1}, \dfrac{1}{8}, \dfrac{1}{27}, \dfrac{1}{64}, \dfrac{1}{125}, \ldots$

2. $1, -\dfrac{1}{\sqrt{2}}, \dfrac{1}{\sqrt{3}}, -\dfrac{1}{\sqrt{4}}, \dfrac{1}{\sqrt{5}}, \ldots$

3. $a_n = n^2 - 2n + 6, \quad n = 1, 2, \ldots$

4. $a_n = \dfrac{(-1)^n}{2n}, \quad n = 0, 1, \ldots$

5. $\dfrac{1}{2}, -\dfrac{1}{2}, \dfrac{1}{2}, -\dfrac{1}{2}, \dfrac{1}{2}, -\dfrac{1}{2}, \ldots$

6. $0, \dfrac{1}{2}, \dfrac{2}{3}, \dfrac{3}{4}, \dfrac{4}{5}, \ldots$

7. $a_n = \dfrac{3n + 1}{2n}, \quad n = 1, 2, \ldots$

8. $a_n = \dfrac{n + 2}{n^2}, \quad n = 1, 2, \ldots$

9. $\dfrac{3}{2}, \dfrac{5}{3}, \dfrac{7}{4}, \dfrac{9}{5}, \dfrac{11}{6}, \ldots$

10. $\dfrac{1 \cdot 3}{2}, \dfrac{2 \cdot 6}{5}, \dfrac{3 \cdot 9}{10}, \dfrac{4 \cdot 12}{17}, \dfrac{5 \cdot 15}{26}, \ldots$

11. $a_n = 0$ for n even, $a_n = \dfrac{2}{n}$ for n odd.

12. $a_n = 1$ for n even, $a_n = \dfrac{1}{n}$ for n odd.

13. $\dfrac{2}{3}, \dfrac{5}{6}, \dfrac{10}{11}, \dfrac{17}{18}, \dfrac{26}{27}, \ldots$

14. $\dfrac{1}{2}, \dfrac{1}{4}, \dfrac{1}{8}, \dfrac{1}{16}, \dfrac{1}{32}, \ldots$

15. $1.1, 1.01, 1.001, 1.0001, 1.00001, \ldots$

16. $1.1, -1.01, 1.001, -1.0001, 1.00001, \ldots$

1.3 Properties of limits

(i) A sequence can have *only one* limit. (For if $\lim_{j \to \infty} a_j = \alpha$ and also $\lim_{j \to \infty} a_j = \beta > \alpha$, then we would have $|a_j - \alpha| < \frac{1}{2}|\beta - \alpha|$ for nearly all j and $|a_j - \beta| < \frac{1}{2}|\beta - \alpha|$ for nearly all j. Hence $\beta - \alpha = |\beta - \alpha| = |\beta - a_j + a_j - \alpha| \le |\beta - a_j| + |a_j - \alpha| < \beta - \alpha$ or $\beta - \alpha < \beta - \alpha$, which is absurd.)

(ii) If $\lim_{j \to \infty} a_j = \alpha$, and if the sequence $b_1, b_2, b_3, \ldots$ is obtained from $a_1, a_2, a_3, \ldots$ by changing finitely many terms, then $\lim_{j \to \infty} b_j = \alpha$. Also, if $c_1, c_2, c_3, \ldots$ is a **subsequence** of $a_1, a_2, a_3, \ldots$, that is, a sequence obtained from $a_1, a_2, a_3, \ldots$ by dropping some, perhaps infinitely many, terms, then $\lim_{j \to \infty} c_j = \alpha$. (For anything that holds for nearly all a_j is true for nearly all b_j and nearly all c_j.)

(iii) We also note that the statements "$\lim_{j \to \infty} a_j = 0$" and "$\lim_{j \to \infty} |a_j| = 0$" are equivalent. So are the statements "$\lim_{j \to \infty} a_j = \alpha$" and "$\lim_{j \to \infty} a_{j+1} = \alpha$."

Evaluating limits is often possible by using the following theorems.

Theorem 1. *Assume that* $\lim_{n \to \infty} a_n = \alpha$, $\lim_{n \to \infty} b_n = \beta$. *Then* $\lim_{n \to \infty} (a_n + b_n) = \alpha + \beta$, $\lim_{n \to \infty} (c a_n) = c\alpha$, $\lim_{n \to \infty} (a_n b_n) = \alpha\beta$, *and* $\lim_{n \to \infty} a_n / b_n = \alpha/\beta$ *(if* $\beta \neq 0$*).*

Theorem 2. *If* $x \mapsto f(x)$ *is a function continuous at* $x = \alpha$, *and if* $\lim_{n \to \infty} a_n = \alpha$, *then* $\lim_{n \to \infty} f(a_n) = f(\alpha)$, *that is,* $\lim_{n \to \infty} f(a_n) = f(\lim_{n \to \infty} a_n)$.

These theorems are very similar to the corresponding theorems about limits of functions (see Chapter 3, §4.3, and Chapter 6, §2.1) and can be proved ▶ in the same way. Before applying Theorems 1 and 2 to examples, we must derive two important limits:

$$\lim_{n \to \infty} x^n = 0, \qquad \text{for } |x| < 1, \tag{5}$$

$$\lim_{n \to \infty} \frac{x^n}{n!} = 0, \qquad \text{for all } x. \tag{6}$$

We prove (5) first. If $x = 0$, then $x^n = 0$ for all n and hence $\lim_{n \to \infty} x^n = 0$. Assume that $x \neq 0$. Let $\epsilon > 0$ be given. We must show that $|x^n - 0| = |x^n| = |x|^n < \epsilon$ for nearly all n. Since $|x| < 1$, we know that $1/|x| > 1$. Therefore $1/|x| = 1 + a$ where $a > 0$. Hence $1/|x|^n = (1 + a)^n > 1 + na > na$ and $|x|^n < 1/na$. This is less than ϵ if $1/na < \epsilon$; that is, if $na > 1/\epsilon$; that is, if $n > 1/(a\epsilon)$; that is, for nearly all n.

The proof of (6) will use (5) and a trick. Since (6) is obvious for $x = 0$, we may assume that $x \neq 0$. Let m be a *fixed* number, such that $m > |x|$. For $n > m$, we have

$$\left| \frac{x^n}{n!} \right| = \left| \frac{x}{1} \cdot \frac{x}{2} \cdot \frac{x}{3} \cdots \frac{x}{m} \cdot \frac{x}{m+1} \cdot \frac{x}{m+2} \cdots \frac{x}{n} \right|$$

$$= \left| \frac{x^m}{m!} \cdot \underbrace{\frac{x}{m+1} \cdot \frac{x}{m+2} \cdots \frac{x}{n}}_{n - m \text{ terms}} \right| \le \left| \frac{x^m}{m!} \cdot \underbrace{\frac{x}{m+1} \cdot \frac{x}{m+1} \cdot \frac{x}{m+1}}_{n - m \text{ terms}} \right|$$

$$= \left| \frac{x^m}{m!} \right| \left| \frac{x}{m+1} \right|^{n-m} = \left| \frac{x^m}{m!} \left(\frac{x}{m+1} \right)^{-m} \right| \left| \frac{x}{m+1} \right|^n$$

$$= \frac{(m+1)^m}{m!} \left| \frac{x}{m+1} \right|^n = cq^n,$$

where

$$c = \frac{(m+1)^m}{m!}, \quad q = \left| \frac{x}{m+1} \right|.$$

For large n, q^n is as small as we like [by (5), since $|q| < 1$]. So is cq^n (since c does not depend on n), and so is $|x^n/n!|$ (since $|x^n/n!| \le cq^n$).

EXAMPLES 1. Find the limit of the sequence $a_n = 3 \dfrac{1 + n^{-1}}{2 - n^{-2}}$, $n = 1, 2, \dots$.

SOLUTION   By Theorem 1,

$$\lim_{n \to \infty} a_n = \lim_{n \to \infty} 3 \frac{1 + \dfrac{1}{n}}{2 - \dfrac{1}{n^2}} = 3 \lim_{n \to \infty} \frac{1 + \dfrac{1}{n}}{2 - \dfrac{1}{n^2}}$$

$$= 3 \frac{\lim_{n \to \infty} \left(1 + \dfrac{1}{n} \right)}{\lim_{n \to \infty} \left(2 - \dfrac{1}{n^2} \right)} = 3 \frac{\lim_{n \to \infty} 1 + \lim_{n \to \infty} \dfrac{1}{n}}{\lim_{n \to \infty} 2 - \lim_{n \to \infty} \dfrac{1}{n^2}} = 3 \frac{1 + 0}{2 - 0} = \frac{3}{2}.$$

We used the fact that the limit of a sequence $c, c, c, \dots$ is c and relation (4') for $\alpha = 1$ and $\alpha = 2$.

2. Compute $\lim\limits_{n\to\infty} \dfrac{3n^2 + 2}{5n^2 - 2n + 1}$.

SOLUTION By a trick used in Example 3 in §1.2, and by repeated application of Theorem 1,

$$\lim_{n\to\infty} \frac{3n^2 + 2}{5n^2 - 2n + 1} = \lim_{n\to\infty} \frac{3 + \dfrac{2}{n^2}}{5 - \dfrac{2}{n} + \dfrac{1}{n^2}} = \frac{\lim\limits_{n\to\infty}\left(3 + \dfrac{2}{n^2}\right)}{\lim\limits_{n\to\infty}\left(5 - \dfrac{2}{n} + \dfrac{1}{n^2}\right)}$$

$$= \frac{\lim\limits_{n\to\infty} 3 + \lim\limits_{n\to\infty} \dfrac{2}{n^2}}{\lim\limits_{n\to\infty} 5 - \lim\limits_{n\to\infty} \dfrac{2}{n} + \lim\limits_{n\to\infty} \dfrac{1}{n^2}} = \frac{3 + 0}{5 - 0 + 0} = \frac{3}{5}.$$

We used Equation (4′) and the fact that $\lim_{n\to\infty} c = c$, where c is a constant.

3. Compute $\lim\limits_{n\to\infty} \sqrt[3]{8 + \dfrac{1}{n^2}}$.

SOLUTION Use Theorem 2 with $f(x) = \sqrt[3]{x}$, $a_n = 8 + \dfrac{1}{n^2}$. Then

$$\lim_{n\to\infty} \sqrt[3]{8 + \frac{1}{n^2}} = \sqrt[3]{\lim_{n\to\infty}\left(8 + \frac{1}{n^2}\right)}.$$

But $\lim\limits_{n\to\infty}\left(8 + \dfrac{1}{n^2}\right) = \lim\limits_{n\to\infty} 8 + \lim\limits_{n\to\infty} 1/n^2 = 8$ by Theorem 1 in §1.3 and Equation (4′) in §1.2. Hence the desired limit is 2.

4. Show that for every $a > 1$, we have $\lim_{n\to\infty} a^n/n = +\infty$.

Proof. Since $a > 1$, we may write $a = 1 + b$, $b > 0$. By the binomial formula [see Chapter 14, §2.3, Equation (19)]

$$a^n = (1 + b)^n = \sum_{j=0}^{n} \binom{n}{j} b^j > \binom{n}{2} b^2 = \frac{n(n-1)}{2} b^2.$$

Hence $a^n/n > (b^2/2)(n - 1)$. This becomes as large as we want.

5. Does the sequence $a_n = \dfrac{3 + 2n^2}{1 - n}$, $n = 2, 3, \ldots$, have a finite limit? An infinite limit?

ANSWER The sequence diverges to $-\infty$. Indeed,

$$a_n = (-n) \frac{2 + \dfrac{3}{n^2}}{1 - \dfrac{1}{n}};$$

hence a_n is $(-n)$ times a number that is, for large n, very close to 2. Therefore a_n is, for large n, nearly $-2n$, therefore $\lim_{n\to\infty} a_n = -\infty$.

6. Show that for every x,

$$\lim_{n\to\infty} \frac{x^{2n}}{(2n)!} = 0, \qquad \lim_{n\to\infty} \frac{x^{2n+1}}{(2n+1)!} = 0.$$

Proof. Note that the sequences $\dfrac{x^2}{2!}, \dfrac{x^4}{4!}, \dfrac{x^6}{6!}, \dfrac{x^8}{8!}, \ldots$ and $\dfrac{x^3}{3!}, \dfrac{x^5}{5!}, \dfrac{x^7}{7!}, \dfrac{x^9}{9!}, \ldots$ are subsequences of $\dfrac{x}{1!}, \dfrac{x^2}{2!}, \dfrac{x^3}{3!}, \dfrac{x^4}{4!}, \dfrac{x^5}{5!}, \ldots$ and apply (6), together with the observation (ii).

PROBLEMS

In Problems 1 to 20 find which of the given sequences converge, which diverge to $+\infty$ or $-\infty$, and which diverge with no infinite limit. In the first case, find the limit. (Assume all subscripts begin at 1.)

1. $a_n = 1 + \dfrac{2}{n}$.

2. $a_n = n^2 - 2n + 3$.

3. $a_n = 1 + 4n - n^2$.

4. $a_n = \dfrac{(-1)^n}{n^2}$.

5. $a_k = \dfrac{3k^2 + 2k + 1}{3k^2}$.

6. $a_k = \dfrac{(-1)^k k^2 + 1}{3k^2}$.

7. $a_k = (\tfrac{1}{2})^k + 4$.

8. $a_k = \dfrac{x^{k+1}}{k!}$, x any number.

9. $b_i = \sin \dfrac{i\pi}{2} + \cos \dfrac{i\pi}{2}$.

10. $b_i = \dfrac{2^{i+1} - 1}{2^i}$.

11. $b_i = \dfrac{(i-1)^2 - (i+1)^2}{i}$.

12. $b_i = \sqrt{9 + 1/i}$.

13. $c_m = \dfrac{1 - 2m^2}{1 + m + m^2}$.

14. $c_m = \dfrac{2m^3 - m}{3m^2 + n^2 + 1}$.

15. $c_m = \dfrac{m^4 - 3m + 4}{m^2 + m}$.

16. $c_m = \sqrt{\dfrac{m^2 + 2m + 2}{4m^2 + 2m + 1}}$.

17. $c_j = \dfrac{(\tfrac{1}{4})^j + 1}{(\tfrac{1}{4})^j - 1}$.

18. $c_j = \dfrac{2j + \sin j}{5j + 1}$.

19. $c_j = \dfrac{(j+2)! - j!}{(j+3)!}$.

20. $c_j = \sqrt{j+1} - \sqrt{j}$.

*21. Prove that $\lim_{n \to \infty} q^{1/n} = 1$ for $q > 0$.
*22. Prove that $\lim_{n \to \infty} q^{n/(n+1)} = q$ for $q > 0$.

1.4   Monotone sequences

We conclude this section by stating a basic convergence theorem which we will need later.

Theorem 3.   *Let a_n be a sequence such that $a_1 \le a_2 \le a_3 \le \cdots$. If there is a number A such that $a_n \le A$ for all n, then the series converges (and $\alpha = \lim_{n \to \infty} a_n$ satisfies $\alpha \le A$). If there is no number A such that $a_n \le A$ for all n, then the series diverges (and $\lim_{n \to \infty} a_n = +\infty$).*

The theorem is rather obvious geometrically (see Figure 15.1), but a proof ▶ depends on a deeper discussion of real numbers.

FIGURE   **15.1**

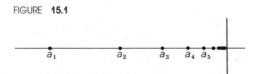

Corollary. *If $b_1 \geq b_2 \geq b_3 \geq \cdots$ and there is a number B such that $b_n \geq B$ for all n, the sequence b_n converges.*

Proof.   Apply Theorem 3 to the sequence $a_n = -b_n$.

A sequence $a_1, a_2, a_3, \ldots$ is called **increasing** if $a_1 < a_2 < a_3 < \ldots$, **nondecreasing** if $a_1 \leq a_2 \leq a_3 \leq \ldots$. Thus, in an increasing sequence, every term is greater than its predecessor, while in a nondecreasing sequence every term is not less than its predecessor. Similarly, $\{a_i\}$ is called **decreasing** if $a_i > a_{i+1}$ for all i, **nonincreasing** if $a_i \geq a_{i+1}$ for all i. A sequence that is either nondecreasing or nonincreasing is called **monotone**.

A sequence $\{a_i\}$ is called **bounded** if there is a number M such that $|a_i| \leq M$ for all i.

Theorem 3'.   *A bounded monotone sequence converges.*

EXAMPLE   Is the sequence $a_n = \dfrac{2n+3}{n+1}$, $n = 1, 2, \ldots$ monotone? Is it convergent?

ANSWER   To check monotonicity, compute

$$a_{n+1} - a_n = \frac{2(n+1)+3}{(n+1)+1} - \frac{2n+3}{n+1}$$

$$= \frac{2n+4}{n+2} - \frac{2n+3}{n+1} = \frac{(2n+4)(n+1) - (2n+3)(n+2)}{(n+2)(n+1)}$$

$$= \frac{2n^2 + 4n + 2n + 4 - (2n^2 + 3n + 4n + 6)}{(n+2)(n+1)}$$

$$= \frac{-n-2}{(n+2)(n+1)}$$

This is negative for all $n = 1, 2, \ldots$. Thus $a_{n+1} - a_n < 0$ or $a_{n+1} < a_n$. The sequence is decreasing.

Instead of checking whether the sequence is bounded, compute the limit:

$$\lim_{n \to \infty} \frac{2n+3}{n+1} = \lim_{n \to \infty} \frac{2 + \dfrac{3}{n}}{1 + \dfrac{1}{n}} = 2.$$

The sequence is bounded (otherwise it would diverge to $-\infty$).

PROBLEMS

Determine which of the following sequences are nondecreasing or nonincreasing (and hence monotone). Among these, determine which are bounded (and so converge) and which are unbounded (and so diverge). In the first case, find limits.

1. $a_n = \dfrac{2n-1}{3n+4}$, $n = 1, 2, \ldots$.

2. $c_k = \dfrac{k^2+1}{k^2+k}$, $k = 1, 2, \ldots$.

3. $a_k = \dfrac{k}{k^2+4}$, $k = 0, 1, \ldots$.

4. $c_n = \dfrac{n^2}{2n^2-1}$, $n = 1, 2, \ldots$.

5. $a_n = n^2 - 1000n$, $n = 0, 1, \ldots$.

6. $b_k = \dfrac{k^2+20k}{k+10}$, $k = 0, 1, \ldots$.

7. $u_k = 1 - \dfrac{1}{k^2}$, $k = 1, 2, \ldots$.

9. $c_k = \tan \dfrac{\pi}{4k}$, $k = 1, 2, \ldots$.

8. $a_n = n \sin \dfrac{\pi n}{2}$, $n = 0, 1, \ldots$.

10. $u_n = \dfrac{10^n}{n!}$, $n = 0, 1, \ldots$.

11. $a_n = \dfrac{1 \cdot 3 \cdot 5 \ldots (2n - 1)}{2 \cdot 4 \cdot 6 \ldots (2n)}$, $n = 1, 2, \ldots$.

12. $x_n = \left(\sin \dfrac{1}{1} \right) \left(\sin \dfrac{1}{2} \right) \ldots \left(\sin \dfrac{1}{n} \right)$, $n = 1, 2, \ldots$.

13. $a_n = \dfrac{n}{\log(n!)}$, $n = 2, 3, \ldots$.

14. $a_n = 1 + 3^{-1} + 3^{-2} + \cdots + 3^{-n}$, $n = 0, 1, 2, \ldots$.

1.5   Euler's constant‡

As an application of Theorem 3, we shall prove that there exists a number γ such that

$$\lim_{n \to \infty} \left(1 + \frac{1}{2} + \frac{1}{3} + \cdots + \frac{1}{n} - \log n \right) = \gamma. \tag{7}$$

This means that $\log n$ is approximately $1 + \dfrac{1}{2} + \dfrac{1}{3} + \cdots + \dfrac{1}{n} - \gamma$ for large integers n. The number γ is called **Euler's constant**; its decimal expansion begins thus:

$$\gamma = .5772156649 \ldots .$$

Nobody knows whether γ is rational.

In order to prove the convergence of the sequence

$$a_n = 1 + \frac{1}{2} + \frac{1}{3} + \cdots + \frac{1}{n} - \log n, \qquad n = 1, 2, \ldots,$$

we start with the inequality

$$\frac{1}{2} + \frac{1}{3} + \cdots + \frac{1}{n-1} + \frac{1}{n} < \log n < 1 + \frac{1}{2} + \frac{1}{3} + \cdots + \frac{1}{n-1}.$$

This inequality is obvious from Figure 15.2, where $n = 4$. It expresses the fact that the sum of the areas of the "short" rectangles is less than the area (from $x = 1$ to $x = n$) under the function $1/x$, which in turn is less than the sum of the areas of the "tall" rectangles. Set

$$b_n = \log n - \left(\frac{1}{2} + \frac{1}{3} + \cdots + \frac{1}{n} \right).$$

Subtracting $\dfrac{1}{2} + \dfrac{1}{3} + \cdots + \dfrac{1}{n}$ from the second inequality above, we see that

$$0 < b_n < 1 - \frac{1}{n} < 1,$$

‡This subsection may be omitted without loss of continuity.

FIGURE **15.2**

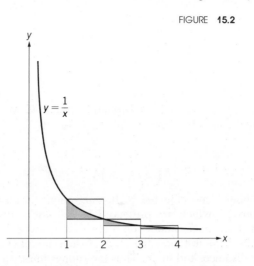

so that the sequence $\{b_n\}$ is bounded. Since b_n is the sum of the shaded areas in Figure 15.2, the sequence $\{b_n\}$ is increasing. Hence, by Theorem 3′, the finite limit, $\lim_{n\to\infty} b_n$, exists; so does $\lim_{n\to\infty} a_n$, for $a_n = 1 - b_n$.

§2 Infinite series

An **infinite series** is a sequence of numbers with plus signs between these numbers; for instance, $1 + \frac{1}{2} + \frac{1}{4} + \frac{1}{8} + \cdots$ or $1 + (-\frac{1}{3}) + (\frac{1}{9}) + (-\frac{1}{27}) + \cdots$. The numbers, called the **terms** of the series, may be labeled by integers $0, 1, 2, \ldots$, or $1, 2, 3, \ldots$, or in any other way. If some term is a negative number, say, $-\frac{1}{3}$, we agree to write $-\frac{1}{3}$, instead of $+(-\frac{1}{3})$. When we say "the infinite series $a_0 + a_1 + a_2 + \ldots$," we mean that the terms a_j are definite but unspecified numbers. We also use for infinite series the sigma notation introduced in Chapter 14, §2.7. Thus

$$\sum_{j=0}^{\infty} a_j = a_0 + a_1 + a_2 + \cdots.$$

For example,

$$\sum_{j=0}^{\infty} 2^{-j} = 1 + \frac{1}{2} + \frac{1}{4} + \frac{1}{8} + \cdots.$$

Our task is to decide when an infinite series has a "sum."

2.1 Adding infinitely many numbers

How can infinitely many numbers "add up" to a number? Perhaps the simplest way of visualizing this is by looking at the following example.

Consider a segment of length 2 (see Figure 15.3). Divide it into two equal segments (of length 1 each). Leave the left segment alone, and divide the right one into two equal

FIGURE **15.3**

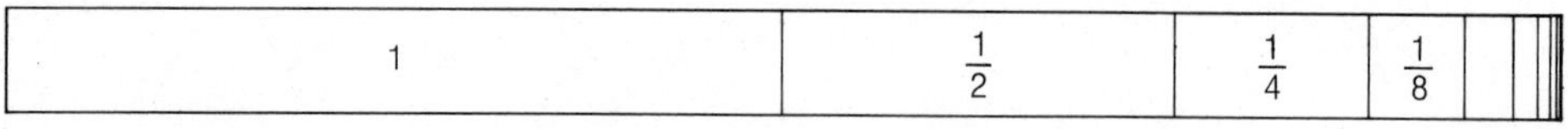

segments (of length $\frac{1}{2}$ each). Divide the right segment of length $\frac{1}{2}$ into two equal segments (of length $\frac{1}{4}$ each). Continue this process indefinitely. We obtain a decomposition of the segment of length 2 into segments of length $1, \frac{1}{2}, \frac{1}{4}, \frac{1}{8}, \frac{1}{16}$, and so forth. "Therefore,"

$$1 + \frac{1\cdot}{2} + \frac{1}{4} + \frac{1}{8} + \frac{1}{16} + \cdots = 2. \tag{1}$$

The Greek philosopher Zeno objected to this argument. Zeno is known to us only through his "paradoxes," which are preserved in the works of others. One of these paradoxes asserts that a runner cannot complete a race course, because he would first have to traverse half of the distance, then half of the remaining distance, then half of the then remaining distance, and so on. Thus the runner must traverse infinitely many distances, and this will take forever.

Zeno certainly saw runners arriving at the finishing line, and precisely what he meant by this and other paradoxes is a matter of controversy. But if he wanted to say that one cannot talk about adding infinitely many numbers as if this were a procedure equivalent to adding finitely many numbers, he was certainly right. If we were to compute the sum in (1) by carrying out all infinitely many additions, this would indeed take forever.

Yet we feel that Equation (1) is correct. What is its precise meaning?

We can compute, for any positive integer n, the sum of the first n terms on the left-hand side. We obtain

$$1 + \frac{1}{2} = \frac{3}{2}, \quad 1 + \frac{1}{2} + \frac{1}{4} = \frac{7}{4}, \quad 1 + \frac{1}{2} + \frac{1}{4} + \frac{1}{8} = \frac{15}{8}, \quad \cdots,$$

and, in general [see Equation (46) in Chapter 14, §2.7]

$$\underbrace{1 + \frac{1}{2} + \frac{1}{4} + \cdots + \frac{1}{2^{n-1}}}_{n \text{ terms}} = \frac{1 - (\frac{1}{2})^n}{1 - \frac{1}{2}} = 2 - \frac{1}{2^{n-1}}.$$

Thus *the difference between the sum of the first n terms in the series* (1) *and the number 2 will be as small as we like if n is sufficiently large.* We interpret (1) to be an abbreviation for the statement just made.

A reader may object that the preceding discussion amounts to hairsplitting. "It is true, is it not, that $1 + \frac{1}{2} + \frac{1}{4} + \frac{1}{8} + \cdots = 2$? What difference does it make how you choose to interpret this statement?" To counter such an attitude, we present an example which shows that we cannot always operate with infinite sums as we do with finite ones. In Chapter 14, §2.5, we showed [see (34)] that the difference between the finite sum

---

ZENO of Elea (ca. 490 B.C.) was a follower of the Eleatic philosopher Parmenides, who taught that ultimate reality is unchangeable. Zeno's paradoxes may be interpreted as an attempt to show that motion is illusory. To refute Zeno's paradoxes one needs the concept of limits.

$1 - \frac{1}{2} + \frac{1}{3} - \frac{1}{4} + \frac{1}{5} - \cdots + (-1)^{n-1}(1/n)$ and $\log 2$ is at most $1/(n+1)$ in absolute value, that is, as small as we like, provided n is large enough. Therefore

$$1 - \frac{1}{2} + \frac{1}{3} - \frac{1}{4} + \frac{1}{5} - \frac{1}{6} + \frac{1}{7} - \frac{1}{8} + \frac{1}{9} - \frac{1}{10} + \cdots = \log 2. \qquad (2)$$

Consider now the infinite sum

$$1 + \frac{1}{3} - \frac{1}{2} + \frac{1}{5} + \frac{1}{7} - \frac{1}{4} + \frac{1}{9} + \frac{1}{11} - \frac{1}{6} + \cdots . \qquad (3)$$

It consists of the same terms as the sum in (2) arranged in a different order. We expect therefore that this sum is again $\log 2$. On the other hand, we conclude from (2), on dividing both sides by 2, that

$$\frac{1}{2} - \frac{1}{4} + \frac{1}{6} - \frac{1}{8} + \frac{1}{10} - \cdots = \frac{1}{2}\log 2.$$

We write (2) and the relation just obtained one above the other:

$$1 - \frac{1}{2} + \frac{1}{3} - \frac{1}{4} + \frac{1}{5} - \frac{1}{6} + \frac{1}{7} - \frac{1}{8} + \frac{1}{9} - \frac{1}{10} + \frac{1}{11} - \cdots = \log 2$$

$$\frac{1}{2} \qquad - \frac{1}{4} \qquad + \frac{1}{6} \qquad - \frac{1}{8} \qquad + \frac{1}{10} \qquad - \cdots = \frac{1}{2}\log 2$$

and add. This yields

$$1 + \frac{1}{3} - \frac{1}{2} + \frac{1}{5} + \frac{1}{7} - \frac{1}{4} + \frac{1}{9} + \frac{1}{11} - \frac{1}{6} + \cdots = \frac{3}{2}\log 2. \qquad (4)$$

So what is the true value of the infinite sum (3), $\log 2$ or $\frac{3}{2}\log 2$? We can answer this only on the basis of a precise definition. When we give this definition, it will turn out that (4) is a correct equation (see §2.3 below).

2.2 Convergent and divergent series

The nth **partial sum** of the series

$$\sum_{n=0}^{\infty} a_n = a_0 + a_1 + a_2 + \cdots$$

is, by definition, the (finite) sum of the terms up to, and including, the nth, that is, the sum

$$a_0 + a_1 + \cdots + a_n = \sum_{j=0}^{n} a_j. \qquad (5)$$

For example, the partial sums of the series $1 + \frac{1}{2} + \frac{1}{4} + \cdots = \sum_{j=0}^{\infty} 2^{-j}$ are

$$S_0 = 1, \qquad S_1 = 1 + \frac{1}{2} = \frac{3}{2}, \qquad S_2 = 1 + \frac{1}{2} + \frac{1}{4} = \frac{7}{4},$$

$$S_3 = 1 + \frac{1}{2} + \frac{1}{4} + \frac{1}{8} = \frac{15}{16}, \qquad \cdots.$$

It is convenient to think of every infinite series as representing the sequence of its partial sums. We note that given any sequence $S_0, S_1, S_2, S_3, \ldots$ there is a series for which these numbers are the partial sums, namely, the series $\Sigma_{j=0}^{\infty} a_j$ with $a_0 = S_0$, $a_j = S_j - S_{j-1}$ for $j > 0$, that is, the series

$$S_0 + (S_1 - S_0) + (S_2 - S_1) + (S_3 - S_2) + \cdots.$$

Example (1) considered in §2.1 suggests the following definition.

An infinite series $a_0 + a_1 + a_2 + \cdots = \Sigma_{j=0}^{\infty} a_j$ is said to **converge** if the sequence of partial sums converges, that is, if there is a finite limit

$$S = \lim_{n \to \infty} \sum_{j=0}^{n} a_j = \lim_{n \to \infty} (a_0 + a_1 + \cdots + a_n). \tag{6}$$

In this case, S is called the **sum of the infinite series,** and we write

$$a_0 + a_1 + a_2 + \cdots = S \quad \text{or} \quad \sum_{j=0}^{\infty} a_j = S.$$

For instance, relation (1) asserts that

$$\sum_{j=0}^{\infty} \frac{1}{2^j} = 1 + \frac{1}{2} + \frac{1}{4} + \cdots = 2. \tag{7}$$

This is a special case $(x = \frac{1}{2})$ of the important relation

$$\sum_{j=0}^{\infty} x^j = 1 + x + x^2 + \cdots = \frac{1}{1 - x} \qquad \text{if } |x| < 1. \tag{8}$$

The series in (8) is called the **geometric series.** If $|x| < 1$, this series converges to $1/(1 - x)$. This follows from the identity

$$\sum_{j=0}^{n} x^j = 1 + x + \cdots + x^n = \frac{1}{1 - x} - \frac{x^{n+1}}{1 - x}$$

[see Equation (24) in Chapter 14, §2.5]. The identity shows that the limit of the partial sum $1 + x + \cdots + x^n$ is $1/(1 - x)$, provided that $|x| < 1$ [see relation (5) in §1.3].

We shall see soon that the geometric series $1 + x + x^2 + \cdots$ does *not* converge if $|x| \geq 1$.

Here is another example of a convergent series:

$$1 - \frac{1}{3} + \frac{1}{5} - \frac{1}{7} + \frac{1}{9} - + \cdots = \frac{\pi}{4}. \tag{9}$$

This follows from relation (37) in Chapter 14, §2.6. (The reader should supply the details.)

A series that does not converge is called **divergent.**

If $\lim_{n \to \infty} \Sigma_{j=0}^{n} a_j = +\infty$, we say that the series $\Sigma_{j=0}^{\infty} a_j$ *diverges to* $+\infty$, and we write

$$\sum_{j=0}^{\infty} a_j = +\infty.$$

The meaning of the symbol $\Sigma_{j=0}^{\infty} a_j = -\infty$ needs no separate explanation. For instance, $1 + 1 + 1 + \cdots = +\infty, \quad 1 + 2 + 3 + 4 + \cdots = +\infty, \quad$ and $\quad -1 - 3 - 5 - 7 - \cdots = -\infty.$

The so-called **harmonic series**

$$1 + \frac{1}{2} + \frac{1}{3} + \frac{1}{4} + \frac{1}{5} + \frac{1}{6} + \cdots$$

also diverges to $+\infty$. For according to §1.4, the nth partial sum of this series $1 + \frac{1}{2} + \cdots + \frac{1}{n}$, is, for large n, very close to $\gamma + \log n$, γ being Euler's constant. That

$$1 + \frac{1}{2} + \frac{1}{3} + \frac{1}{4} + \frac{1}{5} + \cdots = +\infty \tag{10}$$

also follows by noting that, for every integer $k > 0$, we have

$$\underbrace{\frac{1}{k+1} + \frac{1}{k+2} + \cdots + \frac{1}{2k}}_{k \text{ terms}} > k\frac{1}{2k} = \frac{1}{2}.$$

Hence, if we take n large enough, the partial sum $1 + \frac{1}{2} + \cdots + \frac{1}{n}$ will contain as many segments each of which adds up to more than $\frac{1}{2}$ as we wish.

This beautiful argument is due to the medieval scholar Oresme.

The series

$$\sum_{j=0}^{\infty} (-1)^j = 1 - 1 + 1 - 1 + 1 - 1 + \cdots$$

diverges, since $a_0 + a_1 + \cdots + a_n = 1$ or 0, depending on whether n is even or odd. But the series does not diverge to either $+\infty$ or $-\infty$.

PROBLEMS

In Problems 1 to 6 evaluate the first four partial sums of the given series.

1. $\displaystyle\sum_{j=0}^{\infty} \frac{j}{j+1}.$

2. $\displaystyle\sum_{j=0}^{\infty} \frac{j-1}{2j+1}.$

3. $\displaystyle\sum_{j=0}^{\infty} \frac{1}{1+j^2}.$

4. $\displaystyle\sum_{k=0}^{\infty} (-1)^k k!.$

5. $\displaystyle\sum_{k=1}^{\infty} \frac{1 \cdot 3 \cdot 5 \ldots (2k-1)}{2 \cdot 4 \cdot 6 \ldots (2k)}.$

6. $\displaystyle\sum_{k=1}^{\infty} \frac{1 \cdot 4 \cdot 7 \ldots (3k-2)}{3^k}.$

7. Find an infinite series whose nth partial sum is given by $n/(n+2)$ for $n = 0, 1, 2, \ldots$.

8. Find an infinite series whose nth partial sum is given by $2^n/n!$ for $n = 0, 1, 2, \ldots$.

NICOLE ORESME (1323–1382), a Parisian scholar who became Bishop of Liseux, was probably the most outstanding European mathematician between antiquity and the Renaissance. He invented graphs of functions and fractional powers and may have anticipated Galileo's law of uniform acceleration.

In Problems 9 to 12 evaluate the given infinite sums. [*Hint:* Use (8).]

9. $1 + \dfrac{1}{3} + \dfrac{1}{9} + \dfrac{1}{27} + \dfrac{1}{81} + \cdots$.

11. $4 - 2 + 1 - \dfrac{1}{2} + \dfrac{1}{4} - \dfrac{1}{8} + \cdots$.

10. $1 - \dfrac{2}{3} + \dfrac{4}{9} - \dfrac{8}{27} + \dfrac{16}{81} - \cdots$.

12. $12 + .12 + .0012 + .000012 + \cdots$.

2.3   Properties of convergent series

If a series converges, the partial sums $(a_0 + a_1 + \cdots + a_n)$ and $(a_0 + a_1 + \cdots + a_{n-1})$ are, for large n, very close to the sum of the series. Hence a_n, the nth term, must be of small absolute value when n is large. More precisely:

Theorem 1. *If $\Sigma_{n=0}^{\infty} a_n$ converges, then $\lim_{n \to \infty} a_n = 0$.*

Proof.   Let $S = \Sigma_{n=0}^{\infty} a_n$. Then $\lim_{n \to \infty} (a_0 + a_1 + \cdots + a_n) = S$ and (see §1.3) also $\lim_{n \to \infty} (a_0 + a_1 + \cdots + a_{n-1}) = S$. Hence, by Theorem 1, §1.3,

$$\lim_{n \to \infty} a_n = \lim_{n \to \infty} [(a_0 + a_1 + \cdots + a_n) - (a_0 + a_1 + \cdots + a_{n-1})] = S - S = 0.$$

Corollary. *If the sequence a_n does not converge to 0, the series $a_0 + a_1 + a_2 + \cdots$ cannot converge.*

This shows that the geometric series $1 + x + x^2 + x^3 + \cdots$ diverges for $|x| \geq 1$. Indeed, if $|x| \geq 1$, then the sequence $1, x, x^2, \ldots$ does not have the limit 0.

We should *not* think, however, that every series with $\lim_{n \to \infty} a_n = 0$ converges. For instance, the harmonic series (10) does not converge.

Theorem 1 is, therefore, only a necessary condition for the convergence of an infinite series.

EXAMPLES 1.   Show that the series $\Sigma_{n=1}^{\infty} n/(n + 2)$ diverges.
ANSWER   The general term of the series is $a_n = n/(n + 2)$. But $\lim_{n \to \infty} n/(n + 2) = \lim_{n \to \infty} 1/(1 + (2/n)) = 1$. Hence the series diverges.

2.   Does the series $\Sigma_{n=1}^{\infty} n/(n^2 + 1)$ converge?
ANSWER   The general term of the series is $a_n = n/(n^2 + 1)$ and $\lim_{n \to \infty} n/(n^2 + 1) = 0$. However, we cannot conclude from this result alone that the series converges. All we know, at this stage, is that the series *may* converge. (Actually the series diverges.)

The following is a consequence of our definition and of Theorem 1 in §1.3.

Theorem 2.   *If $\Sigma_{n=0}^{\infty} a_n = A$, and $\Sigma_{n=0}^{\infty} b_n = B$, then $\Sigma_{n=0}^{\infty} c a_n = cA$, and*
$\Sigma_{n=0}^{\infty} (a_n + b_n) = A + B$.

In words: a convergent infinite series can be multiplied term by term by a number, and two convergent infinite series can be added term by term.

In applying this theorem, it is well to remember that every infinite series in which nearly all terms are 0 is convergent, and that every finite sum can be considered as an infinite series with nearly all terms 0. Also, we can insert as many zeros as we want between the terms of a convergent series; this does not change its sum.

We can now verify that the manipulations by which we arrived at Equation (4) in §2.1 are legitimate. On the other hand, we may not, in general, interchange infinitely many terms in a convergent series (see §2.8, below).

Theorem 3.   *If an infinite series converges, so does the series obtained by adding, dropping, or changing finitely many terms.*

Proof.   It suffices to consider what happens if we change *one* term. Suppose $a_0 + a_1 + a_2 + \cdots = S$. This means that $\lim_{n\to\infty} (a_0 + a_1 + \cdots + a_n) = S$. Consider now the series $b_0 + a_1 + a_2 + \cdots$. Its partial sums are

$$b_0 + a_1 + a_2 + \cdots + a_n = a_0 + a_1 + a_2 + \cdots + a_n + (b_0 - a_0).$$

Hence, by Theorem 1 of §1.3, $\lim_{n\to\infty} (b_0 + a_1 + \cdots + a_n) = S + (b_0 - a_0)$. The series $b_0 + a_1 + a_2 + \cdots$ converges.

PROBLEMS

Determine which of the following series diverge and which *may* converge.

1. $\displaystyle\sum_{n=1}^{\infty} \frac{1}{4n + 1}$.

2. $\displaystyle\sum_{n=1}^{\infty} \frac{n + 1}{2n - 3}$.

3. $\displaystyle\sum_{n=1}^{\infty} \frac{(-1)^n}{n(n + 1)}$.

4. $\displaystyle\sum_{n=1}^{\infty} \frac{n^2 + 1}{n + 1}$.

5. $\displaystyle\sum_{n=1}^{\infty} \frac{n(n + 1)}{(n + 2)(n + 3)}$.

6. $\displaystyle\sum_{k=1}^{\infty} \frac{\sin k}{k}$.

7. $\displaystyle\sum_{k=1}^{\infty} \cos \frac{1}{k}$.

8. $\displaystyle\sum_{k=1}^{\infty} \frac{10^k}{k!}$.

9. $\displaystyle\sum_{k=1}^{\infty} (-1)^k \log\left(\frac{k}{k + 2}\right)$.

10. $\displaystyle\sum_{n=1}^{\infty} \frac{1}{\sqrt{n + 1} + \sqrt{n}}$.

2.4   Series with positive terms

Consider now an infinite series $a_0 + a_1 + \cdots$ with only positive terms. Then the sequence of partial sums is increasing, for the $(k + 1)$st partial is obtained from the kth by adding to it the positive number a_{k+1}. Hence Theorem 3 of §1.4 is applicable to the sequence of partial sums. This yields

Theorem 4.   *If all terms in $\sum_{i=0}^{\infty} a_i$ are positive, then either $\sum_{i=0}^{\infty} a_i = +\infty$, or the sequence of partial sums is bounded and the series converges.*

This theorem has many applications. As a first application, we derive a test for determining whether a series converges or not.

Theorem 5 (the comparison test).   *Assume that*

$$0 \le a_n \le b_n, \qquad \text{for } n = 0, 1, 2, \ldots. \tag{11}$$

If the series $b_0 + b_1 + b_2 + \cdots$ converges, so does the series $a_0 + a_1 + a_2 + \cdots$, and $a_0 + a_1 + a_2 + \cdots \le b_0 + b_1 + b_2 + \cdots$.

Proof. Assume that $b_0 + b_1 + b_2 + \cdots$ converges and set $B = b_0 + b_1 + b_2 + \cdots$. We claim that $b_0 + b_1 + \cdots + b_n \leq B$ for all n. Indeed, if there would be an m with $b_0 + b_1 + \cdots + b_m > B$, then, for every $n > m + 1$, we would have $b_0 + b_1 + \cdots + b_n = b_0 + b_1 + \cdots + b_m + b_{m+1} + \cdots + b_n > B + b_{m+1}$. This would contradict the statement that $\lim_{n \to \infty} (b_0 + b_1 + \cdots + b_n) = B$. Now, by (11), we know that $a_0 + \cdots + a_n \leq b_0 + \cdots + b_n \leq B$ for all n. By Theorem 4, the series $a_0 + a_1 + a_2 + \cdots$ converges. We have that $a_0 + a_1 + a_2 + \cdots = \lim_{n \to \infty} (a_0 + a_1 + \cdots + a_n) \leq B = b_0 + b_1 + b_2 + \cdots$ follows from Theorem 3 in §1.4.

Corollary. *If (11) holds, and the series* $a_0 + a_1 + a_2 + \cdots$ *diverges, the series* $b_0 + b_1 + b_2 + \cdots$ *also diverges.*

REMARK   If (11) holds, we say that the series $a_0 + a_1 + \cdots$ is **dominated** by the series $b_0 + b_1 + \cdots$.

EXAMPLES 1. Does the series $\sum_{n=0}^{\infty} \cos^2 n / 2^n$ converge or diverge?
ANSWER   The series converges since it is dominated by the geometric series $\sum_{n=0}^{\infty} 2^{-n}$.

2. Investigate the convergence of the series

$$\sum_{n=0}^{\infty} \frac{1 + \cos^2 n}{n + 1}.$$

ANSWER   The given series dominates the harmonic series $1 + \frac{1}{2} + \frac{1}{3} + \cdots$. The harmonic series diverges. So does the given series, by the corollary.

PROBLEMS

Determine which of the following series converge and which diverge by using the comparison test.

1. $\displaystyle\sum_{n=1}^{\infty} \frac{4}{2n - 1}.$

2. $\displaystyle\sum_{n=1}^{\infty} \frac{1}{\sqrt{n}}.$

3. $\displaystyle\sum_{n=1}^{\infty} \frac{1}{100n}.$

4. $\displaystyle\sum_{n=1}^{\infty} \frac{1}{n + 10}.$

5. $\displaystyle\sum_{n=2}^{\infty} \frac{1}{\log n}.$

6. $\displaystyle\sum_{k=1}^{\infty} \frac{k}{2k + 1}.$

7. $\displaystyle\sum_{k=1}^{\infty} \frac{k + 4}{k(k + 1)}.$

8. $\displaystyle\sum_{k=1}^{\infty} \frac{\sqrt{k}}{k + 3}.$

9. $\displaystyle\sum_{k=1}^{\infty} \frac{1}{2^k + 1}.$

10. $\displaystyle\sum_{k=1}^{\infty} \frac{k^2 + 1}{k^3 + 1}.$

11. $\displaystyle\sum_{n=3}^{\infty} \frac{1}{\sqrt{n^2 - 4}}.$

12. $\displaystyle\sum_{n=0}^{\infty} \frac{\cos^2 n}{2^n}.$

13. $\displaystyle\sum_{n=1}^{\infty} \frac{1}{n3^{n-1}}.$

14. $\displaystyle\sum_{n=1}^{\infty} \frac{1}{n^n}.$

15. $\displaystyle\sum_{k=1}^{\infty} \frac{2 + \cos k}{k}.$

16. $\displaystyle\sum_{k=1}^{\infty} \frac{\log k}{2^{k-1}}.$

17. $\displaystyle\sum_{k=1}^{\infty} \frac{1}{k!}.$

18. $\displaystyle\sum_{k=1}^{\infty} \frac{1}{3^k - \sin k}.$

2.5 Decimals as series‡

The concept of convergent infinite series is already present, implicitly, in the use of infinite decimals. Consider the decimal fraction

$$\alpha_0.\alpha_1\alpha_2\alpha_3 \ldots$$

(α_0 being a positive integer, and each other α_j being one of the digits $0, 1, 2, \ldots, 9$). This fraction can be written as an infinite series with nonnegative terms

$$\alpha_0 + \frac{\alpha_1}{10} + \frac{\alpha_2}{10^2} + \frac{\alpha_3}{10^3} + \cdots. \tag{12}$$

The terms of this series are not greater than those of the convergent series

$$\alpha_0 + \frac{9}{10} + \frac{9}{10^2} + \frac{9}{10^3} + \cdots = \alpha_0 + \frac{9}{10}\left(1 + \frac{1}{10} + \frac{1}{10^2} + \cdots\right).$$

Hence (12) converges. The sum of the series is, of course, the number $\alpha_0.\alpha_1\alpha_2\alpha_3\ldots$

The formula (8) for summing a geometric series shows that *a repeating decimal represents a rational number* (as was asserted in Chapter 1, §2.3).

EXAMPLES

$$.\overline{3} = .333\ldots = \frac{3}{10} + \frac{3}{10^2} + \frac{3}{10^3} + \cdots = \frac{3}{10}\left(1 + \frac{1}{10} + \frac{1}{10^2} + \cdots\right)$$

$$= \frac{3}{10}\frac{1}{1 - \frac{1}{10}} = \frac{3}{9} = \frac{1}{3};$$

$$.\overline{32} = .323232\ldots = \frac{32}{100} + \frac{32}{100^2} + \frac{32}{100^3} + \cdots$$

$$= \frac{32}{100}\left(1 + \frac{1}{100} + \frac{1}{100^2} + \cdots\right) = \frac{32}{100}\frac{1}{1 - \frac{1}{100}} = \frac{32}{99};$$

$$1.23\overline{5} = 1.23555\ldots = 1.23 + .00555\ldots = 1.23 + \frac{5}{10^3} + \frac{5}{10^4} + \frac{5}{10^5} + \cdots$$

$$= 1.23 + \frac{5}{10^3}\left(1 + \frac{1}{10} + \frac{1}{10^2} + \cdots\right) = 1.23 + \frac{5}{1000}\frac{1}{1 - \frac{1}{10}}$$

$$= 1.23 + \frac{5}{900} = \frac{123}{100} + \frac{5}{900} = \frac{1112}{900}.$$

PROBLEMS

Express the following repeating decimals as the quotient of two integers.

1. $.\overline{4}$.	4. $.\overline{39}$.	7. $2.01\overline{3}$.	10. $6.45\overline{15}$.
2. $.2\overline{7}$.	5. $1.\overline{29}$.	8. $4.01\overline{7}$.	11. $.21\overline{80}$.
3. $.\overline{18}$.	6. $.06\overline{3}$.	9. $.21\overline{69}$.	12. $.32\overline{117}$.

‡This subsection may be omitted without loss of continuity.

2.6 The integral test

Here is a useful convergence criterion based on the concept of improper integrals (see Chapter 8, §4).

Theorem 6 (the integral test). *Let* $t \mapsto f(t)$ *be a continuous nonnegative decreasing function defined for* $t \geq 1$. *If the improper integral*

$$\int_1^{+\infty} f(t)\, dt \tag{13}$$

exists, the series

$$\sum_{n=1}^{\infty} f(n) = f(1) + f(2) + f(3) + \cdots \tag{14}$$

converges. If the integral does not exist, the series diverges.

Before proving the theorem, we note a consequence.

Corollary. *The series*

$$1 + \frac{1}{2^s} + \frac{1}{3^s} + \frac{1}{4^s} + \frac{1}{5^s} + \cdots = \sum_{n=1}^{\infty} n^{-s}$$

converges for $s > 1$, *diverges for* $s \leq 1$.

Indeed, if $s > 0$, then $t \mapsto t^{-s}$ is a decreasing function of t and we can apply the integral test. For $s \neq 1$, we have

$$\int_1^{+\infty} t^{-s}\, dt = \lim_{A \to +\infty} \int_1^A t^{-s}\, dt = \lim_{A \to +\infty} \frac{t^{1-s}}{1-s} \Big|_1^A$$

$$= \lim_{A \to +\infty} \frac{1}{s-1}(1 - A^{1-s}) = \begin{cases} \dfrac{1}{s-1}, & \text{if } s > 1, \\ +\infty, & \text{if } s < 1. \end{cases}$$

For $s = 1$, we get the harmonic series, which we know diverges. The integral test also works; the improper integral $\int_1^{+\infty} t^{-1}\, dt$ does not exist. For $s < 0$, we have $\lim_{n \to \infty} n^{-s} = +\infty$; the series certainly diverges.

Proof of Theorem 6. The graph of $t \mapsto f(t)$ descends. Therefore

$$f(2) + f(3) + \cdots + f(n) \leq \int_1^n f(t)\, dt \leq f(1) + f(2) + \cdots + f(n-1). \tag{15}$$

Indeed (see Figure 15.4, where $n = 6$), the left sum is the area under a step function below the curve $t \mapsto f(t)$ from 1 to n, whereas the right sum is the area from 1 to n under a step function above the curve.

Assume that the (finite limit) $\lim_{x \to +\infty} \int_1^x f(t)\, dt = \int_1^{+\infty} f(t)\, dt$ exists. Set $A = \int_1^{+\infty} f(t)\, dt$. Then $A = \int_1^n f(t)\, dt + \int_n^{+\infty} f(t)\, dt$, hence $\int_1^n f(t)\, dt \leq A$, hence by (15),

$$f(2) + f(3) + \cdots + f(n) \leq A, \qquad \text{for all } n = 2, 3, \ldots.$$

We conclude, by Theorem 4, that the series $f(2) + f(3) + \cdots$ converges; so does the series (14) which differs from it by one term.

FIGURE **15.4**

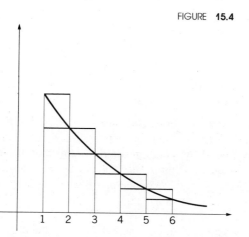

$$1 \quad 2 \quad 3 \quad 4 \quad 5 \quad 6$$

To prove the second statement (if the integral does not exist, the series diverges), we shall show that if the series (14) converges, the integral (13) exists.

Assume that $f(1) + f(2) + f(3) + \cdots = A$, where A is some number. Then $f(1) + f(2) + \cdots + f(n - 1) \leq A$ and, by (15)

$$\int_1^2 f(t)\, dt + \int_2^3 f(t)\, dt + \int_3^4 f(t)\, dt + \cdots + \int_{n-1}^n f(t)\, dt = \int_1^n f(t)\, dt \leq A.$$

We conclude, by Theorem 4, that the series $\int_1^2 f\, dt + \int_2^3 f\, dt + \int_3^4 f\, dt + \cdots$ converges. Let B be its sum. Then $\lim_{n \to \infty} \int_1^n f\, dt = B$. Here n denotes positive integers. But if x is a number between n and $n + 1$, then $\int_1^n f\, dt \leq \int_1^x f\, dt \leq \int_1^{n+1} f\, dt$, so that $\lim_{x \to +\infty} \int_1^x f\, dt = B$, that is, $\int_1^{+\infty} f\, dt = B$.

PROBLEMS

Determine which of the following series converge and which diverge by using the integral test.

1. $\displaystyle\sum_{n=1}^{\infty} \frac{1}{n^2}.$

2. $\displaystyle\sum_{n=1}^{\infty} \frac{1}{3n + 1}.$

3. $\displaystyle\sum_{n=1}^{\infty} \frac{1}{(2n + 1)^2}.$

4. $\displaystyle\sum_{n=1}^{\infty} \frac{2n}{n^2 + 1}.$

5. $\displaystyle\sum_{n=1}^{\infty} \frac{1}{(n + 3)^{3/2}}.$

6. $\displaystyle\sum_{n=1}^{\infty} \frac{2n + 1}{n^2 + n + 4}.$

7. $\displaystyle\sum_{n=1}^{\infty} \frac{e^{1/n}}{n^2}.$

8. $\displaystyle\sum_{n=1}^{\infty} \frac{1}{n^2} \sin \frac{\pi}{n}.$

9. $\displaystyle\sum_{k=1}^{\infty} \frac{\arctan k}{1 + k^2}.$

10. $\displaystyle\sum_{k=1}^{\infty} \frac{e^{\arctan k}}{1 + k^2}.$

11. $\displaystyle\sum_{k=1}^{\infty} \frac{1}{\sqrt{k^2 + 1}}.$

12. $\displaystyle\sum_{k=1}^{\infty} \frac{k}{e^k}.$

13. $\displaystyle\sum_{k=1}^{\infty} \log\left(\frac{k + 5}{k}\right).$

14. $\displaystyle\sum_{k=1}^{\infty} \frac{1}{(k + 1)(\log(k + 1))}.$

15. $\displaystyle\sum_{k=1}^{\infty} \frac{1}{(k + 1)\log^\alpha (k + 1)},$
    where $\alpha > 1$.

16. $\displaystyle\sum_{k=1}^{\infty} \frac{1}{\cosh^2 k}.$

2.7 The ratio test

This convergence criterion is based on Theorem 5 and on what we know about geometric series.

Theorem 7 (the ratio test). *Let the terms in the series $\sum_{n=0}^{\infty} a_n$ all be positive. If $\lim_{n\to\infty} a_{n+1}/a_n < 1$, the series converges. If $\lim_{n\to\infty} a_{n+1}/a_n > 1$, the series diverges.*

Proof. Assume first that $\lim_{n\to\infty} a_{n+1}/a_n = L < 1$. Let θ be a number such that $L < \theta < 1$. Then $0 < a_{n+1}/a_n < \theta$ for nearly all n. Since convergence or divergence of a series is not affected by changing or dropping finitely many terms, we assume that $a_{n+1}/a_n < \theta$ for all n. Then $a_1/a_0 < \theta$; hence $a_1 < \theta a_0$. Also $a_2/a_1 < \theta$; hence $a_2 < \theta a_1 < \theta^2 a_0$. Also $a_3/a_2 < \theta$; hence $a_3 < \theta a_2 < \theta^2 a_1 < \theta^3 a_0$. In general,

$$0 < a_n < a_0 \theta^n,$$

and since the series $a_0 + a_0\theta + a_0\theta^2 + a_0\theta^3 + \cdots$ converges (we have $0 < \theta < 1$), the series $\sum a_n$ converges, by Theorem 5.

Assume next that $\lim_{n\to\infty} a_{n+1}/a_n = L > 1$. Let θ be a number such that $1 < \theta < L$. Then $a_{n+1}/a_n > \theta$ for nearly all n. Since convergence or divergence of a series is not affected by dropping finitely many terms, we may assume that $a_{n+1}/a_n > \theta$ for all n. Then $a_1 > a_0\theta$, $a_2 > a_1\theta > a_0\theta^2$, and so forth. In general, $a_n > a_0\theta^n$, with $\theta > 1$. Hence it is not true that $\lim_{n\to\infty} a_n = 0$, and $\sum a_n$ diverges.

EXAMPLES 1. The series $1 + 1 + \dfrac{1}{\sqrt{2!}} + \dfrac{1}{\sqrt{3!}} + \dfrac{1}{\sqrt{4!}} + \cdots = \sum_{n=0}^{\infty} \dfrac{1}{\sqrt{n!}}$ converges.

Proof. We have

$$\lim_{n\to\infty} \frac{a_{n+1}}{a_n} = \lim_{n\to\infty} \frac{\sqrt{n!}}{\sqrt{(n+1)!}} = \lim_{n\to\infty} \frac{1}{\sqrt{n+1}} = 0 < 1.$$

2. The series $3 + \frac{1}{2} + \frac{3}{4} + \frac{1}{4} + \frac{3}{8} + \frac{1}{8} + \frac{3}{16} + \cdots$ converges (to 7) as the reader is asked to verify. (Note that $3 + \frac{3}{2} + \frac{3}{4} + \frac{3}{8} + \cdots = 6$ and $\frac{1}{2} + \frac{1}{4} + \frac{1}{8} + \cdots = 1$.) The ratio test is not applicable, however, since $\lim_{n\to\infty} a_{n+1}/a_n$ fails to exist. Indeed, the sequence a_1/a_0, a_2/a_1, a_3/a_2, ... is the sequence $\frac{1}{6}$, 3, $\frac{1}{6}$, 3, $\frac{1}{6}$, 3, ...

3. We learned in §2.6 that the series $\sum_{n=1}^{\infty} n^{-s}$ diverges for $s \le 1$, converges for $s > 1$. But $\lim_{n\to\infty} a_{n+1}/a_n = \lim_{n\to\infty} (n/(n+1))^s = 1$ for all s. The ratio test is not applicable.

PROBLEMS

Use the ratio test to determine the convergence or divergence of the following series. If the test fails, say so.

1. $\displaystyle\sum_{n=0}^{\infty} \frac{n}{2^{n+1}}.$

2. $\displaystyle\sum_{n=0}^{\infty} \frac{n(n+1)}{(n+2)^2}.$

3. $\displaystyle\sum_{n=1}^{\infty} \frac{(n+3)}{n(n+1)(n+2)}.$

4. $\displaystyle\sum_{n=1}^{\infty} \frac{n(n+2)}{(n+1)!}.$

5. $\displaystyle\sum_{n=1}^{\infty} \frac{1}{9+n^2}.$

6. $\displaystyle\sum_{n=1}^{\infty} \frac{4^n}{n(n+2)}.$

7. $\displaystyle\sum_{n=0}^{\infty} \frac{e^n}{n!}.$

8. $\displaystyle\sum_{n=0}^{\infty} \frac{n!}{10^n}.$

9. $\displaystyle\sum_{k=1}^{\infty} \frac{k^3}{e^k}.$

10. $\sum_{k=1}^{\infty} \frac{2 + \sqrt{k}}{k!}.$

11. $\sum_{k=1}^{\infty} \frac{2^k}{(k!)^\alpha},$ where $\alpha > 0.$

12. $\sum_{k=1}^{\infty} \frac{k!}{1 \cdot 3 \cdot 5 \ldots (2k - 1)}.$

13. $\sum_{k=1}^{\infty} \frac{1 \cdot 3 \cdot 5 \ldots (2k - 1)}{2 \cdot 4 \cdot 6 \ldots (2k)}.$

14. $\sum_{k=1}^{\infty} \frac{1 \cdot 3 \cdot 5 \ldots (2k - 1)}{1 \cdot 4 \cdot 7 \ldots (3k - 2)}.$

15. $\sum_{k=1}^{\infty} \frac{(2k)!}{k! \, k!}.$

16. $\sum_{k=1}^{\infty} \frac{k!}{k^k}.$

2.8   Alternating series

An alternating series is one whose terms are alternately positive and negative. Examples are the famous series [Equations (2) and (9)]

$$1 - \frac{1}{2} + \frac{1}{3} - \frac{1}{4} + \frac{1}{5} - \frac{1}{6} + \cdots = \log 2, \qquad 1 - \frac{1}{3} + \frac{1}{5} - \frac{1}{7} + \frac{1}{9} - \cdots = \frac{\pi}{4}.$$

The following theorem is known as Leibniz's convergence criterion.

Theorem 8 (on alternating series). If $p_0 > p_1 > p_2 > p_3 > \cdots > 0$ and $\lim_{n \to \infty} p_n = 0$, then the series

$$p_0 - p_1 + p_2 - p_3 + p_4 - p_5 + \cdots$$

converges.

Proof. The odd partial sums form an increasing sequence, since the $(2n + 1)$st sum can be written as

$$S_{2n+1} = (p_0 - p_1) + (p_2 - p_3) + (p_4 - p_5) + \cdots + (p_{2n} - p_{2n+1})$$

and all terms $(p_0 - p_1), (p_2 - p_3), \ldots$ are positive. Since also

$$S_{2n+1} = p_0 - (p_1 - p_2) - (p_3 - p_4) - \cdots - (p_{2n-1} - p_{2n}) - p_{2n+1}$$

and the terms $(p_1 - p_2), (p_3 - p_4), \ldots$ are positive, we have $S_{2n+1} < p_0$. Hence $A = \lim_{n \to \infty} S_{2n+1}$ exists, by Theorem 3 in §1.4. We see similarly that $B = \lim_{n \to \infty} S_{2n}$ exists. (The reader should supply the details.) Since $\lim_{n \to \infty} (S_{2n+1} - S_{2n}) = \lim_{n \to \infty} p_{2n+1} = 0$, we have that $A = B$. Clearly A is the sum of our series.

EXAMPLES   1. Test the alternating series $\sum_{n=0}^{\infty} (-1)^n/(n^2 + n + 1)$ for convergence.
SOLUTION   Let $p_n = 1/(n^2 + n + 1)$. Then $\lim_{n \to \infty} p_n = 0$, so the first requirement of Theorem 8 is satisfied. To show that $p_0 > p_1 > p_2 > \cdots > 0$, it is sufficient to prove that $p_n > p_{n+1}$ for all $n \geq 0$. Substituting for p_n and p_{n+1}, we must have

$$\frac{1}{n^2 + n + 1} > \frac{1}{(n + 1)^2 + (n + 1) + 1}$$

or $n^2 + 3n + 3 > n^2 + n + 1$ or $(n + 1) > 0$. This is certainly true for $n \geq 0$. Hence the series converges.

2. Does the alternating series $\sum_{n=1}^{\infty} (-1)^{n+1} n/e^n$ converge?
ANSWER   Let $p_n = n/e^n$. Then $\lim_{n \to \infty} p_n = 0$. To show that p_n is a monotone decreasing sequence, it is sufficient to prove that $p_n > p_{n+1}$ for all $n \geq 1$. This is so: $n/e^n > (n + 1)/e^{n+1}$, that is, $e > (n + 1)/n$ for all $n \geq 1$. The series converges.

PROBLEMS

Examine the following alternating series for convergence using the Leibniz convergence criterion. If this test fails, use another method to determine convergence or divergence.

1. $\displaystyle\sum_{n=1}^{\infty} \frac{(-1)^{n-1}}{n(n+1)}$.

5. $\displaystyle\sum_{n=0}^{\infty} \frac{(-1)^{n}}{n!}$.

9. $\displaystyle\sum_{n=1}^{\infty} \frac{(-1)^{n+1}}{n2^{n}}$.

2. $\displaystyle\sum_{n=0}^{\infty} \frac{(-1)^{n}}{n^{2}+4}$.

6. $\displaystyle\sum_{n=1}^{\infty} \frac{(-1)^{n+1}}{n\sqrt{n}}$.

10. $\displaystyle\sum_{n=1}^{\infty} \frac{(-1)^{n-1}}{(n!)^{4}}$.

3. $\displaystyle\sum_{n=1}^{\infty} (-1)^{n+1} \frac{n}{n+1}$.

7. $\displaystyle\sum_{n=1}^{\infty} (-1)^{n+1} \frac{n}{2n-1}$.

11. $\displaystyle\sum_{n=2}^{\infty} (-1)^{n} \frac{n^{2}}{n^{4}-1}$.

4. $\displaystyle\sum_{n=1}^{\infty} (-1)^{n+1} \frac{(n+1)^{2}}{e^{n}}$.

8. $\displaystyle\sum_{n=1}^{\infty} \frac{(-1)^{n+1}}{n(n+1)}$.

12. $\displaystyle\sum_{n=1}^{\infty} \frac{(-1)^{n-1}}{2^{n}+1}$.

2.9 Absolute convergence

Given a series $a_0 + a_1 + a_2 + \cdots$, it is of interest to consider also the series of absolute values $|a_0| + |a_1| + |a_2| + \cdots$.

Theorem 9. *If the series $\sum_{j=0}^{\infty} |a_j|$ converges, then the series $\sum_{j=0}^{\infty} a_j$ also converges, and we have*

$$\sum_{j=0}^{\infty} a_j \leq \sum_{j=0}^{\infty} |a_j|. \tag{16}$$

Proof. We assume that $\sum_{j=0}^{\infty} |a_j|$ converges and set $A = \sum_{j=0}^{\infty} |a_j|$,

$$\alpha_j = \begin{cases} a_j, & \text{if } a_j \geq 0, \\ 0, & \text{if } a_j \leq 0; \end{cases} \qquad \beta_j = \begin{cases} -a_j, & \text{if } a_j \leq 0, \\ 0, & \text{if } a_j \geq 0. \end{cases}$$

We have $0 \leq \alpha_j \leq |a_j|$, $0 \leq \beta_j \leq |\beta_j|$. Hence $\sum_{j=0}^{n} a_j \leq \sum_{j=0}^{n} |a_j| \leq A$, $\sum_{j=0}^{n} \beta_j \leq \sum_{j=0}^{n} |\beta_j| \leq A$. Applying Theorem 4 to the series $\sum_{j=0}^{\infty} \alpha_j$, $\sum_{j=0}^{\infty} \beta_j$, we conclude that these series converge. Set

$$P = \sum_{j=0}^{\infty} \alpha_j, \qquad N = \sum_{j=0}^{\infty} \beta_j.$$

We have $|a_j| = \alpha_j + \beta_j$. Hence, by Theorem 2, $P + N = A$. Also, $a_j = \alpha_j - \beta_j$. Hence, again by Theorem 2, $\sum_{j=0}^{\infty} a_j$ converges and

$$\sum_{j=0}^{\infty} a_j = P - N.$$

Since $P \geq 0$, $N \geq 0$, we have $|P - N| \leq P + N = A$, which proves (16).

A series $a_0 + a_1 + a_2 + \cdots$ such that the series $|a_1| + |a_2| + |a_3| + \cdots$ converges, is called **absolutely convergent**. To find whether the series $\Sigma |a_j|$ converges, we may apply to this series the tests given above as Theorems 4, 5, 6, and 7.

There are convergent series that do not converge absolutely. Such series are called **conditionally convergent**. An example is the Leibniz series $1 - \frac{1}{2} + \frac{1}{3} - \frac{1}{4} + \frac{1}{5} - \cdots$. It turns out that the "paradoxical" behavior encountered in §2.1 *cannot* happen in the case of absolute convergence and *must* happen in the case of a convergent series for which the series of absolute values diverges. We state, without proof ▶,

Theorem 10. *Assume that $\Sigma_{j=0}^{\infty} a_j$ converges. If $\Sigma_{j=0}^{\infty} |a_j|$ converges, then every series obtained from $\Sigma_{j=0}^{\infty} a_j$ by rearranging terms converges, and to the same sum. If $\Sigma_{j=0}^{\infty} |a_j|$ diverges, then the series $\Sigma_{j=0}^{\infty} a_j$ can be rearranged so as to diverge, and also so as to converge to any given sum.*

EXAMPLES   1.   Show that the series $\Sigma_{n=1}^{\infty} (-1)^{n+1}/n^2$ converges absolutely.

SOLUTION   Consider the series $\Sigma_{n=1}^{\infty} 1/n^2$ containing positive terms only. Using the integral test with $f(x) = 1/x^2$, we obtain

$$\int_1^{\infty} \frac{1}{x^2}\, dx = \left(-\frac{1}{x}\right)\Big|_1^{\infty} = 1.$$

Hence the series of positive terms converges and so the given series is absolutely convergent.

2.   Determine if the series $\Sigma_{n=1}^{\infty} (-1)^{n-1}/\sqrt{n}$ converges absolutely, converges conditionally, or diverges.

SOLUTION   Consider the series $\Sigma_{n=1}^{\infty} 1/\sqrt{n}$ containing positive terms only. Using the comparison test, we see that the terms of this series dominate the corresponding terms of the harmonic series $\Sigma_{n=1}^{\infty} 1/n$, which is known to diverge. Hence the given series is not absolutely convergent. To test for conditional convergence, we use Theorem 8 with $p_n = 1/\sqrt{n}$. Clearly $\lim_{n\to\infty} p_n = 0$. Also, $p_n > p_{n+1}$ for all $n \geq 1$, so the given series converges. The given series, therefore, is conditionally convergent.

PROBLEMS

Determine if the following series converge absolutely, converge conditionally, or diverge.

1. $\displaystyle\sum_{n=0}^{\infty} \frac{(-1)^n}{3^n}$.

2. $\displaystyle\sum_{n=1}^{\infty} \frac{(-1)^{n-1}}{2n-1}$.

3. $\displaystyle\sum_{n=1}^{\infty} \frac{(-1)^{n-1}}{\log(n+1)}$.

4. $\displaystyle\sum_{n=0}^{\infty} \frac{(-1)^n}{\sqrt{n+4}}$.

5. $\displaystyle\sum_{n=1}^{\infty} (-1)^{n+1} \frac{n}{1+n^2}$.

6. $\displaystyle\sum_{n=1}^{\infty} (-1)^{n+1} \frac{2n}{(n+1)!}$.

7. $\displaystyle\sum_{n=1}^{\infty} (-1)^{n-1} \frac{\sqrt{2n+1}}{n}$.

8. $\displaystyle\sum_{n=1}^{\infty} \frac{(-1)^{n+1}}{n\sqrt{n}}$.

9. $\displaystyle\sum_{n=0}^{\infty} (-1)^n \frac{n}{2^n}$.

10. $\displaystyle\sum_{n=1}^{\infty} \frac{\sin n}{n^2}$.

11. $\displaystyle\sum_{n=1}^{\infty} \frac{\cos n}{2^n}$.

12. $\displaystyle\sum_{n=2}^{\infty} \frac{(-1)^n}{n\log n}$.

13. $\displaystyle\sum_{n=1}^{\infty} \frac{(-1)^{n+1}}{\log\left(1+\dfrac{1}{n}\right)}$.

14. $\displaystyle\sum_{n=1}^{\infty} \frac{(-1)^{n+1}}{n^s}$, $0 < s \leq 1$.

§3   Power series

A power series is an infinite series of the form

$$a_0 + a_1(x - x_0) + a_2(x - x_0)^2 + a_3(x - x_0)^3 + \cdots = \sum_{n=0}^{\infty} a_n(x - x_0)^n. \qquad (1)$$

Here we think of $x_0, a_0, a_1, a_2, \ldots$ as fixed numbers, whereas x will be ranging over an interval. More precisely, we shall call (1) a power series with center at x_0, or a power series **about** x_0 or a power series **in** $(x - x_0)$. The numbers $a_0, a_1, a_2, \ldots$ are called the **coefficients** of the power series. If nearly all coefficients are zero, the power series is a polynomial.

If $x_0 = 0$, the power series becomes

$$a_0 + a_1 x + a_2 x^2 + a_3 x^3 + \cdots = \sum_{n=0}^{\infty} a_n x^n. \qquad (1')$$

In order to simplify writing, we often consider such a series; we can recapture the general case (1) by replacing x by $x - x_0$.

3.1   Radius of convergence

If we choose a specific number for x, the series (1) is an infinite series of numbers, and it either converges or diverges. For $x = x_0$, we obtain $a_0 + 0 + 0 + \cdots$; hence the series always converges at least at this one point.

Before stating a general theorem, let us note some examples of power series.

First of all, any polynomial may be considered a power series, with all but finitely many coefficients equal to 0. For instance,

$$1 + 3x - 5x^3 = \sum_{n=0}^{\infty} a_n x^n \qquad \begin{array}{l} \text{with } a_0 = 1, a_1 = 3, a_2 = 0, a_3 = -5, \\ \text{and } a_n = 0 \text{ for } n > 3. \end{array} \qquad (2)$$

Next,

$$1 + x + x^2 + x^3 + \cdots = \sum_{n=0}^{\infty} x^n = \frac{1}{1 - x} \qquad \text{(for } -1 < x < 1) \qquad (3)$$

as we observed in §2.1, Equation (8). The power series in (3) converges for $|x| < 1$, diverges for all other x.

Taylor's theorem (Chapter 14, §2.2) gives many examples of convergent power series. For instance, since (see Chapter 14, §2.4)

$$e^x = 1 + x + \frac{x^2}{2!} + \frac{x^3}{3!} + \cdots + \frac{x^n}{n!} + R_n(x) \qquad \text{with } |R_n(x)| \leq \frac{e^{|x|}|x|^{n+1}}{(n + 1)!}$$

and $\lim_{n \to \infty} R_n(x) = 0$ for all x, by relation (6) in §1.3 we conclude that

$$\lim_{n \to \infty}\left(1 + x + \frac{x^2}{2!} + \cdots + \frac{x^n}{n!}\right) = e^x$$

so that

$$1 + x + \frac{x^2}{2!} + \frac{x^3}{3!} + \cdots = \sum_{n=0}^{\infty} \frac{x^n}{n!} = e^x \qquad \text{(for all } x). \qquad (4)$$

Similarly, we obtain from relations (53) and (54) in Chapter 14, §2.8, that

$$\cos x = 1 - \frac{x^2}{2!} + \frac{x^4}{4!} - \frac{x^6}{6!} + - \cdots \qquad \text{(for all } x\text{)} \tag{5}$$

$$\sin x = x - \frac{x^3}{3!} + \frac{x^5}{5!} - \frac{x^7}{7!} + - \cdots \qquad \text{(for all } x\text{).} \tag{6}$$

Next we look at relations (48) and (50) in Chapter 14, §2.8. For the function $\log(1 + x)$ the remainder R has the limit 0 as $n \to 0$ provided $-1 < x \le 1$; for the function arc tan x the same is true for $-1 \le x \le 1$. We conclude that

$$\log(1 + x) = x - \frac{x^2}{2} + \frac{x^3}{3} - \frac{x^4}{4} + - \cdots \qquad \text{(for } -1 < x \le 1\text{)} \tag{7}$$

$$\text{arc tan } x = x - \frac{x^3}{3} + \frac{x^5}{5} - \frac{x^7}{7} + - \cdots \qquad \text{(for } -1 \le x \le 1\text{).} \tag{8}$$

The series in (7) diverges for $x = -1$ (it becomes the harmonic series $1 + 2^{-1} + 3^{-1} + \cdots$; see §2.2). The series in (7) and in (8) diverge for $|x| > 1$ since $\lim_{n \to \infty} |x^n|/n = +\infty$ for $|x| > 1$.

The series (7), known as **Mercator's series**, and (8), called **Gregory's series**, predate calculus; the geometric series can be traced back to antiquity.

We state now without proof ▶ the basic theorem on convergence of power series.

Theorem 1. *Given a power series $\sum_{n=0}^{\infty} a_n(x - x_0)^n$, there is a number $R \ge 0$, which may take on the "value" $R = +\infty$, such that the series converges absolutely for $|x - x_0| < R$, and diverges for $|x - x_0| > R$. (At $x = x_0 + R$, and at $x = x_0 - R$, the series may either converge or diverge.)*

The examples above illustrate the validity of the result. We call R the **radius of convergence** of the series considered and the interval $|x - x_0| < R$ the **interval of convergence,** provided, of course, that $R > 0$.

For the series (4), (5), (6) the radius of convergence is $R = \infty$; for the series (3), (7), (8) the radius of convergence is $R = 1$.

For the series

$$1 + x + 2!x^2 + 3!x^3 + 4!x^4 + \cdots,$$

$R = 0$. The ratio test shows that this series diverges for every $x \ne 0$.

To find the radius of convergence of a given power series, we may try the ratio test (Theorem 7 in §2.7) applied to the series of absolute values.

EXAMPLES 1. Find the radius of convergence R of the power series $1 + x + x^2/2! + x^3/3! + \cdots$, without using previous knowledge of this series.

ANSWER The nth term of the series is $x^{n-1}/(n-1)!$, the $(n + 1)$st term is $x^n/n!$. We apply the ratio test. In view of the relation

---

NICOLAUS MERCATOR (1620–1687) was born in Denmark (he latinized his original name of Kaufmann) but was active in England, where he was one of the first members of the Royal Society, and later in France, where he designed the Versailles Fountains.

He is not to be confused with the sixteenth-century mapmaker Gerardus Mercator.

JAMES GREGORY (1638–1675), a brilliant Scottish mathematician, studied in Aberdeen and later in Italy. During his short life he discovered many results of what was to become calculus, for instance, the fact that the area under the graph of $y = \sec x$ from $x = 0$ to $x = t$ is log(sec t + tan t). The term "convergence" is due to Gregory.

$$\lim_{n\to\infty} \frac{\left|\dfrac{x^n}{n!}\right|}{\left|\dfrac{x^{n-1}}{(n-1)!}\right|} = \lim_{n\to\infty} \frac{|x|}{n} = 0 < 1,$$

the series converges for all x. Hence $R = \infty$. The interval of convergence is $-\infty < x < +\infty$.

2. Find the radius of convergence R of the power series $\Sigma_{n=0}^{\infty} (2n + 1)(x - 1)^n$.
ANSWER   We apply the ratio test:

$$\lim_{n\to\infty} \frac{|[2(n + 1) + 1](x - 1)^{n+1}|}{|(2n + 1)(x - 1)^n|} = \lim_{n\to\infty} \frac{2n + 3}{2n + 1}|x - 1| = |x - 1| \lim_{n\to\infty} \frac{2 + \dfrac{3}{n}}{2 + \dfrac{1}{n}} = |x - 1|.$$

This is < 1 for $|x - 1| < 1$, > 1 for $|x - 1| > 1$. The series of absolute values converges for $|x - 1| < 1$, diverges for $|x - 1| > 1$. Hence $R = 1$. The interval of convergence is $0 < x < 2$.

PROBLEMS

In Problems 1 to 20 find the interval of convergence of the given power series.

1. $\displaystyle\sum_{n=0}^{\infty} nx^n$.

2. $\displaystyle\sum_{n=1}^{\infty} \frac{x^n}{\sqrt{n}}$.

3. $\displaystyle\sum_{n=0}^{\infty} (-1)^n \frac{x^{2n+1}}{(2n + 1)!}$.

4. $\displaystyle\sum_{n=1}^{\infty} (-1)^{n-1} \frac{x^n}{n4^n}$.

5. $\displaystyle\sum_{n=0}^{\infty} \frac{x^{2n}}{(n + 1)(n + 2)(n + 3)}$.

6. $\displaystyle\sum_{n=0}^{\infty} \frac{(n + 1)x^n}{3^n}$.

7. $\displaystyle\sum_{n=1}^{\infty} \frac{x^n}{4 + n^2}$.

8. $\displaystyle\sum_{n=1}^{\infty} \frac{x^n}{\log(n + 1)}$.

9. $\displaystyle\sum_{n=0}^{\infty} \frac{n!}{2^n} x^n$.

10. $\displaystyle\sum_{n=2}^{\infty} \frac{x^n}{(\log n)^n}$.

11. $\displaystyle\sum_{n=1}^{\infty} (-1)^{n-1} n(x - 4)^n$.

12. $\displaystyle\sum_{n=1}^{\infty} \frac{(x - 6)^4}{n^2}$.

13. $\displaystyle\sum_{n=1}^{\infty} (-1)^{n-1} \frac{(x + 2)^n}{n}$.

14. $\displaystyle\sum_{n=1}^{\infty} \frac{(x - 10)^n}{n10^n}$.

15. $\displaystyle\sum_{n=1}^{\infty} \frac{n(x - 1)^{n-1}}{3^n}$.

16. $\displaystyle\sum_{n=1}^{\infty} \frac{(x + 2)^n}{2^n \sqrt{n + 1}}$.

17. $\displaystyle\sum_{n=0}^{\infty} (\log 2)^n \frac{(x + 1)^n}{n!}$.

18. $\displaystyle\sum_{n=1}^{\infty} \frac{1 \cdot 3 \cdot 5 \ldots (2n - 1)}{2 \cdot 5 \cdot 8 \ldots (3n - 1)}(x - 1)^n$.

19. $\displaystyle\sum_{n=1}^{\infty} \frac{1 \cdot 5 \cdot 9 \ldots (4n - 3)}{2 \cdot 4 \cdot 6 \ldots (2n)}(x + 2)^n$.

20. $\displaystyle\sum_{n=1}^{\infty} \frac{n^n}{n!}\left(x + \frac{1}{2}\right)^n$.

21. Suppose that $\lim\limits_{n \to \infty} \left| \dfrac{a_{n+1}}{a_n} \right| = L$. Show that the power series $\Sigma_{n=0}^{\infty} a_n(x - x_0)^n$ has radius of convergence $1/L$ if $L \neq 0$, ∞ if $L = 0$.

22. If the power series $\Sigma_{n=0}^{\infty} a_n x^n$ has radius of convergence R, what is the radius of convergence of $\Sigma_{n=0}^{\infty} a_n p^n x^n$, where p is any constant?

23. If the power series $\Sigma_{n=0}^{\infty} a_n x^n$ has radius of convergence R, what is the radius of convergence of $\Sigma_{n=0}^{\infty} a_n x^{nk}$, where k is a positive integer?

24. If the power series $\Sigma_{n=0}^{\infty} a_n(x - x_0)^n$ has a radius of convergence R, what is the radius of convergence of $\Sigma_{n=0}^{\infty} a_n(x - x_0)^{n+k}$, where k is a positive integer?

3.2 Operations on power series

A power series may be thought of as a "polynomial of infinite degree," and, when functions are represented by power series, we can work with them almost as easily as if they were polynomials. For instance, if we set, in the expansion [see (7) in §3.1]

$$\log(1 + x) = x - \frac{x^2}{2} + \frac{x^3}{3} - \frac{x^4}{4} + \cdots \qquad \text{(for } |x| < 1) \qquad (9)$$

$x = -y$, we obtain

$$\log(1 - y) = -y - \frac{y^2}{2} - \frac{y^3}{3} - \frac{y^4}{4} - \cdots \qquad (|y| < 1),$$

and if we then set $y = x$ (the name of a number does not matter!), we see that

$$\log(1 - x) = -x - \frac{x^2}{2} - \frac{x^3}{3} - \frac{x^4}{4} - \cdots \qquad (|x| < 1). \qquad (10)$$

We say: "(10) is obtained from (9) by replacing x by $-x$." If we replace, in (9), x by x^2, we see that

$$\log(1 + x^2) = x^2 - \frac{x^4}{2} + \frac{x^6}{3} - + \cdots \qquad (|x| < 1),$$

and if we replace, in (10), x by $2x$, we obtain the expansion

$$\log(1 - 2x) = -2x - 2x^2 - \frac{8x^3}{3} - 4x^4 - \cdots \qquad \left(|x| < \frac{1}{2}\right).$$

Recall now that convergent series may be added or subtracted term by term; see Theorem 2 in §2.3. If we subtract (10) from (9), we obtain

$$\log(1 + x) - \log(1 - x) = \left(x - \frac{x^2}{2} + \frac{x^3}{3} - \frac{x^4}{4} + \cdots \right)$$
$$- \left(-x - \frac{x^2}{2} - \frac{x^3}{3} - \frac{x^4}{4} - \cdots \right)$$

or

$$\log \frac{1 + x}{1 - x} = 2x + \frac{2x^3}{3} + \frac{2x^5}{3} + \cdots \qquad (|x| < 1). \qquad (11)$$

According to Theorem 2 in §2.3, a convergent series may be multiplied by a number term by term. Multiplying the series [see (6) in §3.1]

$$\sin x = x - \frac{x^3}{3!} + \frac{x^5}{5!} - \frac{x^7}{7!} + \cdots \tag{12}$$

by $1/x$ we obtain

$$\frac{\sin x}{x} = 1 - \frac{x^2}{3!} + \frac{x^4}{5!} - \frac{x^6}{7!} + - \cdots. \tag{13}$$

Since (12) holds for all x, (13) holds for all $x \neq 0$. Can we conclude from (13) that $\lim_{x \to \infty} (\sin x/x) = 0$? (See Chapter 10, §2.1.) We could, if we knew that the right-hand side in (13) represents a continuous function of x, also for $x = 0$. This is so. Indeed, one can show ▶ that the following general theorem holds.

Theorem 2. *Let* $f(x) = \sum_{n=0}^{\infty} a_n(x - x_0)^n$ *for* $|x - x_0| < R$, $R > 0$ *being the radius of convergence of the power series. Then the function* $x \mapsto f(x)$ *is continuous and differentiable for* $|x - x_0| < R$, *and*

$$f'(x) = \sum_{n=1}^{\infty} na_n(x - x_0)^{n-1}, \qquad \int_{x_0}^{x} f(t)\, dt = \sum_{n=0}^{\infty} \frac{a_n}{n+1}(x - x_0)^{n+1}.$$

Furthermore, the series in the above formulas have radius of convergence R.

The theorem asserts that it is legitimate to differentiate or integrate a convergent power series term by term, as if it were an ordinary polynomial.

For instance, if we differentiate (12) term by term, we obtain

$$\frac{d \sin x}{dx} = 1 - \frac{3x^2}{3!} + \frac{5x^5}{5!} - + \cdots = 1 - \frac{x^2}{2!} + \frac{x^4}{4!} - + \cdots = \cos x,$$

in view of relation (5) in §3.1.

Integration of the geometric series

$$\frac{1}{1 + x} = 1 - x + x^2 - x^3 + \cdots \tag{14}$$

yields the expansion (7) for the function $\log(1 + x) = \int_0^x [dt/(1 + t)]$. Replacing, in (14), x by x^2 we obtain

$$\frac{1}{1 + x^2} = 1 - x^2 + x^4 - x^6 + \cdots. \tag{15}$$

Term by term integration of this series yields the arc tangent series (8).

Differentiating (13), we see that

$$\frac{x \cos x - \sin x}{x^2} = -\frac{2x}{3!} + \frac{4x^3}{5!} - \frac{6x^5}{7!} + \cdots \tag{16}$$

(which can also be obtained in other ways). Integration of (13) yields

$$\int_0^x \frac{\sin t}{t}\, dt = x - \frac{x^3}{2! \, 3^2} + \frac{x^5}{4! \, 5^2} - \frac{x^7}{6! \, 7^2} + \cdots, \tag{17}$$

as the reader should check. This is interesting, since we know of no way to write $\int_0^x [(\sin t)/t]\, dt$ in terms of elementary functions, without using the integral sign.

A similar situation arises if we replace x by $(-x^2)$ in (4) and then integrate the resulting series term by term. We obtain

$$e^{-x^2} = 1 - \frac{x^2}{1!} + \frac{x^4}{2!} - \frac{x^6}{3!} + \cdots \tag{18}$$

and

$$\int_0^x e^{-t^2}\, dt = x - \frac{x^3}{1!\,3} + \frac{x^5}{2!\,5} - \frac{x^7}{3!\,7} + \cdots. \tag{19}$$

For small values of x, we can use this expansion to compute the "error integral" $\int_0^x e^{-t^2}\, dt$ (see Chapter 12, §3.3).

REMARK   The following observation may be useful in calculating power series. If $f(x)$ is even, and $f(x) = A_0 + A_1 x + A_2 x^2 + A_3 x^3 + \cdots$, then

$$A_1 = A_3 = A_5 = \cdots = 0.$$

Also, if $g(x)$ is odd, and $g(x) = A_0 + A_1 x + A_2 x^2 + A_3 x^3 + \cdots$, then

$$A_2 = A_4 = A_6 = \cdots = 0.$$

(For the proof, see Problem 23 in §3.4.)

PROBLEMS

In Problems 1 to 20 find the power series of the given functions about the point $x_0 = 0$. Use only the power series expansions derived in this subsection and the theorems on manipulating series discussed so far. In all cases, find the general term of the series, the interval of convergence, and write your answer in Σ notation (you may have to write out the first few terms separately).

1. $f(x) = \dfrac{1}{1-x}$.

2. $f(x) = \dfrac{x^4}{1-x}$.

3. $f(x) = \dfrac{1}{(1-x)^2}$.

$\left[Hint:\ \dfrac{d}{dx}\left(\dfrac{1}{1-x}\right) = \dfrac{1}{(1-x)^2}. \right]$

4. $f(x) = \dfrac{1}{(2-x)^2}$.

5. $f(x) = \sin \dfrac{x}{2}$.

6. $f(x) = \cos 4x^2$.

7. $f(x) = \log\left(1 + \dfrac{x}{2}\right)$.

8. $f(x) = \log(2 + x)$.

9. $f(x) = \sin^2 x$.
   [Hint: Write $\sin^2 x = \frac{1}{2}(1 - \cos 2x)$.]

10. $f(x) = \cos^2 x$.
11. $f(x) = e^{-x/2}$.

12. $f(x) = \dfrac{e^{2x} - 1}{2x}$.

13. $f(x) = 4^x$.
    [Hint: $4^x = e^{x \log 4}$.]

14. $f(x) = a^x$, $a > 0$.
15. $f(x) = \arctan 2x$.

16. $f(x) = \dfrac{\sin x - \arctan x}{x^3}$.

17. $f(x) = \displaystyle\int_0^x \dfrac{dt}{1 + t^3}$.

18. $f(x) = \displaystyle\int_0^x \log(1 + t^2)\, dt$.

19. $f(x) = \displaystyle\int_0^x \dfrac{\log(1 + t)}{t}\, dt$.

20. $f(x) = \displaystyle\int_0^x \dfrac{\arctan t}{t}\, dt$.

3.3 Binomial series

An important power series is the **binomial series**

$$(1 + x)^\alpha = 1 + \alpha x + \frac{\alpha(\alpha - 1)}{2!} x^2 + \frac{\alpha(\alpha - 1)(\alpha - 2)}{3!} x^3 + \cdots$$

$$= 1 + \sum_{n=1}^{\infty} \frac{\alpha(\alpha - 1) \ldots (\alpha - n + 1)}{n!} x^n = \sum_{n=0}^{\infty} \binom{\alpha}{n} x^n, \qquad |x| < 1. \quad (20)$$

(This result is due to Newton himself!) First we shall show that the series in (20) converges for $|x| < 1$, then that its sum is $(1 + x)^\alpha$.

Apply the ratio test to the series of absolute values $\sum_{n=0}^{\infty} \left| \binom{\alpha}{n} x^n \right|$:

$$\lim_{n \to \infty} \frac{\left| \binom{\alpha}{n+1} x^{n+1} \right|}{\left| \binom{\alpha}{n} x^n \right|} = \lim_{n \to \infty} \left| \frac{\alpha(\alpha - 1) \ldots (\alpha - n + 1)(\alpha - n)n!\, x^{n+1}}{\alpha(\alpha - 1) \ldots (\alpha - n + 1)(n + 1)!\, x^n} \right|$$

$$= \lim_{n \to \infty} \left| \frac{(\alpha - n)x}{n + 1} \right| = \lim_{n \to \infty} \left| \frac{\dfrac{\alpha}{n} - 1}{1 + \dfrac{1}{n}} x \right| = |x|$$

so that the series converges for $|x| < 1$, diverges for $|x| > 1$. For $|x| < 1$, set

$$f(x) = \sum_{n=0}^{\infty} \binom{\alpha}{n} x^n = 1 + \binom{\alpha}{1} x + \binom{\alpha}{2} x^2 + \binom{\alpha}{3} x^3 + \cdots.$$

Then

$$f'(x) = \sum_{n=1}^{\infty} \binom{\alpha}{n} n x^{n-1} = \sum_{n=0}^{\infty} \binom{\alpha}{n + 1} (n + 1)x^n$$

(by changing dummy variables), and

$$xf'(x) = \sum_{n=0}^{\infty} \binom{\alpha}{n} n x^n,$$

so that

$$(1 + x)f'(x) = \sum_{n=0}^{\infty} \left\{ \binom{\alpha}{n + 1} (n + 1) + \binom{\alpha}{n} n \right\} x^n.$$

But

$$\binom{\alpha}{n + 1} (n + 1) + \binom{\alpha}{n} n$$

$$= \frac{\alpha(\alpha - 1)(\alpha - 2) \ldots (\alpha - n)}{(n + 1)!} (n + 1) + \frac{\alpha(\alpha - 1) \ldots (\alpha - n + 1)}{n!} n$$

$$= \frac{\alpha(\alpha - 1) \ldots (\alpha - n + 1)}{n!} (\alpha - n + n) = \alpha \binom{\alpha}{n}.$$

Therefore $(1 + x)f'(x) = \alpha f(x)$ or

$$\frac{f'(x)}{f(x)} = \frac{\alpha}{1 + x},$$

which is the same as

$$\frac{d \log f(x)}{dx} = \alpha \frac{d \log(1 + x)}{dx} = \frac{d(\alpha \log(1 + x))}{dx} = \frac{d \log(1 + x)^\alpha}{dx}.$$

Therefore $\log f(x) - \log(1 + x)^\alpha$ is a constant; call it c. Thus $\log f(x) = \log(1 + x)^\alpha + c$ or $f(x) = e^c(1 + x)^\alpha$. But $f(0) = 1$ and $(1 + 0)^\alpha = 1$. Therefore $e^c = 1$ and $f(x) = (1 + x)^\alpha$, as asserted.

EXAMPLES 1. For $\alpha = -\frac{1}{2}$, we obtain

$$\frac{1}{\sqrt{1 + x}} = 1 + \frac{-\frac{1}{2}}{1}x + \frac{(-\frac{1}{2})(-\frac{3}{2})}{1 \cdot 2}x^2 + \frac{(-\frac{1}{2})(-\frac{3}{2})(-\frac{5}{2})}{1 \cdot 2 \cdot 3}x^3 + \frac{(-\frac{1}{2})(-\frac{3}{2})(-\frac{5}{2})(-\frac{7}{2})}{1 \cdot 2 \cdot 3 \cdot 4}x^4 + \cdots$$

$$= 1 - \frac{x}{2} + \frac{1 \cdot 3}{2!}\left(\frac{x}{2}\right)^2 - \frac{1 \cdot 3 \cdot 5}{3!}\left(\frac{x}{2}\right)^3 + \frac{1 \cdot 3 \cdot 5 \cdot 7}{4!}\left(\frac{x}{2}\right)^4 - \cdots.$$

2. Replacing x by $-x^2$ in Example 1, we get

$$\frac{1}{\sqrt{1 - x^2}} = 1 + \frac{x^2}{2} + \frac{1 \cdot 3x^4}{2! \, 2^2} + \frac{1 \cdot 3 \cdot 5x^6}{3! \, 2^3} + \frac{1 \cdot 3 \cdot 5 \cdot 7x^8}{4! \, 2^4} + \cdots.$$

Integration yields

$$\text{arc sin } x = \int_0^x \frac{dt}{\sqrt{1 - t^2}}$$

$$= x + \frac{x^3}{2 \cdot 3} + \frac{1 \cdot 3x^5}{2! \, 2^2 5} + \frac{1 \cdot 3 \cdot 5x^7}{3! \, 2^3 7} + \frac{1 \cdot 3 \cdot 5 \cdot 7x^9}{4! \, 2^4 9} + \cdots.$$

The expansions are valid for $|x| < 1$.

PROBLEMS

Use the binomial expansion to determine the power series of $f(x)$ at $x_0 = 0$, and determine where the series represents $f(x)$.

1. $f(x) = (x + 1)^{-1/3}$.

2. $f(x) = (1 + x^2)^{-1/2}$.

3. $f(x) = (x^2 + 1)^{2/5}$.

4. $f(x) = (2x + 1)^{1/2}$.

5. $f(x) = (x + 4)^{-3/2}$.

6. $f(x) = x^2(16 + x^2)^{-3/4}$.

7. $f(x) = \int_0^x (1 + u^2)^{3/2} \, du$.

8. $f(x) = \int_0^x (1 + u^3)^{-5/2} \, du$.

9. $f(x) = \frac{2}{x} \text{ arc sin } \frac{x}{2}$.

10. $f(x) = \log(\sqrt{1 + x^2} + x)$.

$$\left[\textit{Hint: } \frac{d}{dx} \log(\sqrt{1 + x^2} + x) = \frac{1}{\sqrt{1 + x^2}}. \right]$$

3.4   Taylor series. Analytic functions

Let $x \mapsto f(x)$ be a function represented by a convergent power series

$$f(x) = \sum_{n=0}^{\infty} a_n(x - x_0)^n. \tag{21}$$

Applying Theorem 2 repeatedly, we conclude that the function $f(x)$ has derivatives of all orders. Also,

$$f'(x) = \sum_{n=1}^{\infty} na_n(x - x_0)^{n-1}, \qquad f''(x) = \sum_{n=2}^{\infty} n(n - 1)a_n(x - x_0)^{n-2},$$

$$f'''(x) = \sum_{n=3}^{\infty} n(n - 1)(n - 2)a_n(x - x_0)^{n-3},$$

and so forth, and, in general,

$$f^{(j)}(x) = \sum_{n=j}^{\infty} n(n - 1)(n - 2) \ldots (n - j + 1)a_n(x - x_0)^{n-j}.$$

Substituting into the series for $f(x)$, $f'(x)$, $f''(x)$, $f'''(x)$, and so forth, $x = x_0$, we obtain $f(x_0) = a_0$, $f'(x_0) = a_1$, $f''(x_0) = 2a_2$, $f'''(x_0) = 3 \cdot 2 \cdot a_3$, and, in general, $f^{(j)}(x_0) = j!a_j$. Thus

$$a_j = \frac{f^{(j)}(x_0)}{j!}, \qquad j = 0, 1, 2, \ldots. \tag{22}$$

These formulas show that a given function can be represented by a power series about x_0 *in only one way* (if at all). The coefficients must be given by the formulas (22).

The series in (21) is called the **Taylor series** of f at x_0. The partial sums of the Taylor series are the polynomials appearing in Taylor's formulas for f at x_0. (If $x_0 = 0$, one sometimes calls the Taylor series the **Maclaurin series.**)

We have already given many examples of Taylor series in the preceding subsections.

Let $f(x)$ be a given function, defined near $x = x_0$, and having derivatives of all orders. We can compute the numbers $a_j = f^{(j)}(x_0)/j!$ and form the power series

$$\sum_{j=0}^{\infty} a_j(x - x_0)^j.$$

Let R be the radius of convergence of this series. Now three cases are possible.

   (i) $R > 0$ and $f(x) = \sum_{j=0}^{\infty} a_j(x - x_0)^j$ for $|x - x_0| < R$. Then $f(x)$ is represented by its Taylor series. This is the case considered in all the examples above.

   (ii) $R = 0$. Then the series does not converge for any $x \neq x_0$.

   (iii) $R > 0$ but the sum $\sum_{j=0}^{\infty} a_j(x - x_0)^j$ *does not equal* $f(x)$. (See Example 3 below.)

A function $x \mapsto f(x)$ defined near x_0 is called **analytic** at x_0 if it can be represented as the sum of a convergent power series about x_0. It follows from Theorem 2 that an analytic function has derivatives of all orders.

We mention an important fact: the sum of a convergent power series is analytic at every point in the interval of convergence. Also, the sum, difference, product of two analytic functions is analytic, so is the quotient, at points at which the denominator is not 0, and the composition of two analytic functions. The proofs of these results belong to the general theory of analytic functions and will not be given here.

Analytic functions are in many ways the most "well-behaved" functions imaginable. Fortunately, most functions encountered in applications of calculus are analytic, except at certain points.

EXAMPLES   1.   The function $x \mapsto f(x) = 1/x$ is analytic near every point $x_0 \neq 0$, for we can write

$$\frac{1}{x} = \frac{1}{(x - x_0) + x_0} = \frac{1}{x_0} \frac{1}{1 + ((x - x_0)/x_0)}$$

$$= \frac{1}{x_0}\left(1 - \frac{x - x_0}{x_0} + \left(\frac{x - x_0}{x_0}\right)^2 - \cdots\right) = \sum_{j=0}^{\infty} \frac{(-1)^j}{x_0^{j+1}}(x - x_0)^j,$$

the series converging for $|(x - x_0)/x_0| < 1$, that is, for $|x - x_0| < x_0$.

The function $x \mapsto f(x) = 1/x$ is not analytic at $x_0 = 0$, for no matter how we define $f(0)$, the function will not be continuous at this point.

2.   The exponential function $x \mapsto e^x$ can be expanded in a power series about every point x_0; all these power series have radius of convergence ∞. We may write

$$e^x = e^{x_0}e^{x-x_0} = e^{x_0}\left(1 + \frac{x - x_0}{1!} + \frac{(x - x_0)^2}{2!} + \cdots\right) = \sum_{j=0}^{\infty} \frac{e^{x_0}}{j!}(x - x_0)^j.$$

3.   The function $x \mapsto f(x)$ defined by $f(x) = e^{-1/x^2}$ for $x \neq 0$, $f(0) = 0$, has derivatives of all orders but is not analytic near $x = 0$. We have that $f(0) = f'(0) = f''(0) = \cdots = 0$, all Taylor coefficients at $x_0 = 0$ are 0, and the Taylor series converges to 0, but $f(x) > 0$ for $x \neq 0$. (See Problems 21, 22 below.) The graph of the function is shown in Figure 15.5.

FIGURE **15.5**

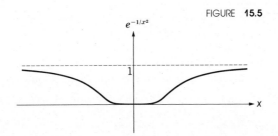

e^{-1/x^2}

The same function is analytic near every point $x_0 \neq 0$.

4.   Expand the function $f(x) = \sec x$ in a Taylor series about the point $x_0 = 0$, assuming the expansion is valid.

SOLUTION   We have

$$
\begin{aligned}
f(x) &= \sec x, & f(0) &= 1, \\
f'(x) &= \sec x \tan x, & f'(0) &= 0, \\
f''(x) &= \sec x(\tan^2 x + \sec^2 x), & f''(0) &= 1, \\
f'''(x) &= \sec x \tan x(\tan^2 x + 5 \sec^2 x), & f'''(0) &= 0, \\
f^{(4)}(x) &= \sec x(\tan^4 x + 5 \sec^4 x) + 18 \sec^3 x \tan^2 x, & f^{(4)}(0) &= 5.
\end{aligned}
$$

Hence the first three nonzero terms are given by

$$f(x) = \sec x = 1 + \frac{1}{2}x^2 + \frac{5}{24}x^4 + \cdots.$$

The calculation of the general nth-order derivative $f^{(n)}(x)$ is difficult; nevertheless, the above expansion for $\sec x$ is indeed the Taylor series about $x_0 = 0$ with radius of convergence $R < \pi/2$. We shall not prove it here.

PROBLEMS

In Problems 1 to 20 assume the given functions can be expanded in Taylor series about the indicated points $x = x_0$. Find the first three nonzero terms in each case. Use Equations (21) and (22).

1. $f(x) = \log x$ at $x_0 = 1$.
2. $f(x) = \log x$ at $x_0 = a > 0$.
3. $f(x) = \sqrt{x}$ at $x_0 = 9$.
4. $f(x) = x^{1/3}$ at $x_0 = 8$.
5. $f(x) = x^{3/2}$ at $x_0 = 4$.
6. $f(x) = 1/\sqrt{x}$ at $x_0 = 1$.
7. $f(x) = e^{-2x}$ at $x_0 = 1$.
8. $f(x) = e^{x^2+2x-1}$ at $x_0 = 0$.
9. $f(x) = \sin x$ at $x_0 = \pi/6$.
10. $f(x) = \cos x$ at $x_0 = \pi/3$.
11. $f(x) = \tan x$ at $x_0 = 0$.
12. $f(x) = e^{\tan x}$ at $x_0 = 0$.

13. $f(x) = \log(1 + e^x)$ at $x_0 = 0$.
14. $f(x) = \log \cos x$ at $x_0 = 0$.
15. $f(x) = \dfrac{\sin x}{1 + x}$ at $x_0 = 0$.
16. $f(x) = \dfrac{\cos x}{1 + x}$ at $x_0 = 0$.
17. $f(x) = \dfrac{\arctan x}{1 + x}$ at $x_0 = 0$.
18. $f(x) = e^{-x} \sin x$ at $x_0 = 0$.
19. $f(x) = e^{-x} \tan x$ at $x_0 = 0$.
20. $f(x) = \dfrac{e^{-x}}{\sqrt{1 + x^2}}$ at $x_0 = 0$.

*21. Suppose $f(x) = e^{-1/x^2}$ for $x \neq 0$, 0 for $x = 0$. If $x \neq 0$, prove that $f^{(n)}(x) = P_{3n}(1/x)e^{-1/x^2}$, where $n = 1, 2, 3, \ldots$ and P_{3n} is a polynomial of degree $3n$. [*Hint:* Show this first for $n = 1, 2, 3$. Then use mathematical induction.]

*22. For the function given in Problem 21, prove that $f^{(n)}(0) = 0$, where $n = 1, 2, 3, \ldots$. [*Hint:* Show that $\lim_{x \to 0} f^{(n)}(x) = 0$, and refer to Chapter 6, §5.2.]

*23. Prove the statements in the Remark in §3.3. [*Hint:* Assume that $f(x) = A_0 + A_1 x + A_2 x^2 + \cdots$. If $f(x)$ is even, write down the Maclaurin series for $f(x) - f(-x)$. If $f(x)$ is odd, write down the Maclaurin series for $f(x) + f(-x)$. Now use Equation (22).]

16

POLAR COORDINATES
AND PARAMETRIC
REPRESENTATIONS

§1 Polar coordinates

Thus far we have represented points in the plane by their Cartesian coordinates, and we considered curves in the plane as graphs of functions. In this chapter we discuss another way of representing points by numbers and another way of representing curves by functions. We also resume the study of conics begun in Chapter 2, §4.

1.1 Polar coordinates of a point

The Cartesian coordinates (defined in Chapter 2, §1.1) are not the only way of describing the position of a point in the plane by two numbers. There are others, and among them polar coordinates are particularly useful.

Let P be a point in the plane with Cartesian coordinates (x,y). Assume that P is not the origin O, so that the distance $r = \sqrt{x^2 + y^2}$ from P to O is positive. Then $(x/r)^2 + (y/r)^2 = 1$. Thus $(x/r, y/r)$ are the coordinates of a point on the unit circle, and according to the definition of the function sine and cosine (see Chapter 10, §1), there is a number θ such that $x/r = \cos\theta$, $y/r = \sin\theta$. (This θ is determined only up to a multiple of 2π.) The numbers (r,θ) are called the **polar coordinates** of P. The connection between polar coordinates (r,θ) and Cartesian coordinates (x,y) is given by the formulas

$$x = r\cos\theta, \qquad y = r\sin\theta, \tag{1}$$

$$r = \sqrt{x^2 + y^2}, \qquad \cos\theta = \frac{x}{r}, \qquad \sin\theta = \frac{y}{r}. \tag{2}$$

These can also be seen in Figure 16.1.

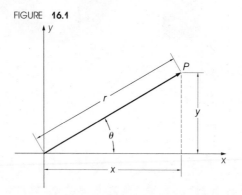

FIGURE **16.1**

The curves consisting of points with Cartesian coordinates $x = $ const. and $y = $ const. are lines parallel to coordinate axes (Figure 16.2). They form a "rectangular grid" (not unlike the avenues and streets in Manhattan). The "grid" formed by curves with polar coordinates $r = $ const. (circles with center at 0) and the curves with polar coordinates $\theta = $ const. (rays emanating from 0) are shown in Figure 16.3. The ray $\theta = 0°$ is called the **polar axis.**

It is convenient to extend the definition and to consider *every* pair of numbers (r,θ) satisfying (1) as the polar coordinates of the point with Cartesian coordinates (x,y).

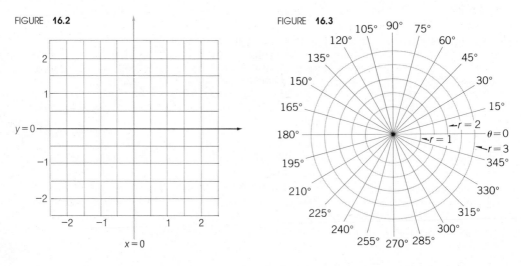

FIGURE **16.2**

FIGURE **16.3**

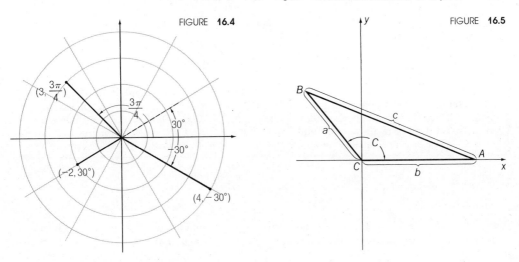

FIGURE 16.4

FIGURE 16.5

This **convention**, which we shall follow, means that

  I. *if (r,θ) are polar coordinates of P, so are $(r,\theta \pm 2\pi)$;*
 II. *if (r,θ) are polar coordinates of P, so are $(-r,\theta \pm \pi)$;*
III. *for every θ, the numbers $(0,\theta)$ are polar coordinates of the origin.*

EXAMPLES 1.   Find the Cartesian coordinates of the points with the polar coordinates (a) $(3,3\pi/4)$, (b) $(4,-30°)$, and (c) $(-2,30°)$.

ANSWER   (a) Use Equation (1) with $r = 3$, $\theta = 3\pi/4$.   Then $x = 3 \cos(3\pi/4) = 3(-\sqrt{2}/2) \approx -2.121$, $y = 3 \sin(3\pi/4) = 3(\sqrt{2}/2) \approx 2.121$.

(b) Use Equation (1) with $r = 4$, $\theta = -30°$. Then $x = 4 \cos(-30°) = 4(\sqrt{3}/2) \approx 3.464$, $y = 4 \sin(-30°) = 4(-1/2) = -2$.

(c) Use Equation (1) with $r = -2$, $\theta = 30°$. Then $x = -2 \cos 30° = -2(\sqrt{3}/2) \approx -1.732$, $y = -2 \sin 30° = -2(1/2) = -1$.

The polar coordinates for the given points are plotted in Figure 16.4.

2.   Find polar coordinates for the point with Cartesian coordinates $(3,4)$.

ANSWER   Use Equation (2) with $x = 3$, $y = 4$. Then $r = \sqrt{3^2 + 4^2} = \sqrt{25} = 5$ and $\theta = $ arc tan $\frac{4}{3} = $ arc tan $1.333 \cdots \approx 53°$.

In view of our convention, the point considered has also polar coordinates $(5, 53° + 360°) = (5, 413°)$, $(-5, 53° + 180°) = (-5, 233°)$, and so forth.

3.   The **distance formula** in polar coordinates. If two points have polar coordinates (r_1,θ_1) and (r_2,θ_2), respectively, then the distance d between them is

$$d = \sqrt{r_1^2 - 2r_1 r_2 \cos(\theta_2 - \theta_1) + r_2^2}.$$

Proof.   The Cartesian coordinates of the two points are $(r_1 \cos\theta_1,\ r_1 \sin\theta_1)$ and $(r_2 \cos\theta_2,\ r_2 \sin\theta_2)$. The distance d is found using the Cartesian distance formula:

$$\begin{aligned}
d^2 &= (x_1 - x_2)^2 + (y_1 - y_2)^2 = (r_1 \cos\theta_1 - r_2 \cos\theta_2)^2 + (r_1 \sin\theta_1 - r_2 \sin\theta_2)^2 \\
&= r_1^2 \cos^2\theta_1 - 2r_1 r_2 \cos\theta_1 \cos\theta_2 + r_2^2 \cos^2\theta_2 + r_1^2 \sin^2\theta_1 \\
&\quad - 2r_1 r_2 \sin\theta_1 \sin\theta_2 + r_2^2 \sin^2\theta_2 \\
&= r_1^2 - 2r_1 r_2(\cos\theta_1 \cos\theta_2 + \sin\theta_1 \sin\theta_2) + r_2^2 \\
&= r_1^2 - 2r_1 r_2 \cos(\theta_2 - \theta_1) + r_2^2.
\end{aligned}$$

4. The **cosine law** in trigonometry. In a triangle with vertices A, B, C, we have

$$c^2 = a^2 + b^2 - 2ab \cos C.$$

(Here, as usual, C denotes the measure of the angle ACB, c denotes the length $|AB|$, and so forth; see Figure 16.5.)

Proof. Choose a Cartesian coordinate system with C as the origin, A on the positive x-axis, and B in the upper half-plane. Then A has polar coordinates $(b,0)$ and B has polar coordinates (a,C). Now apply Example 3 with $r_1 = b, \theta_1 = 0, r_2 = a, \theta_2 = C, d = c$. [Or, apply directly the Cartesian distance formula, noting that A has Cartesian coordinates $(b,0)$ and B has Cartesian coordinates $(a \cos C, a \sin C)$.]

PROBLEMS

In each of Problems 1 to 6 the polar coordinates of a point are given. Find the coordinates of the point in the corresponding Cartesian coordinate system. Draw a figure.

1. $(2,\pi/2)$. 3. $(3,-\pi/2)$. 5. $(\sqrt{3},5\pi/6)$.

2. $(2,3\pi/4)$. 4. $(\tfrac{1}{2},\pi/4)$. 6. $(\sqrt{2},4\pi/3)$.

In each of Problems 7 to 12 the coordinates of a point in a Cartesian coordinate system are given. Find the corresponding polar coordinates. Draw a figure.

7. $(0,\tfrac{1}{2})$. 9. $(-\sqrt{3},1)$. 11. $(1,-\sqrt{3})$.

8. $(1,1)$. 10. $(-3,3)$. 12. $(-2 \sin \tfrac{1}{9}\pi, 2 \cos \tfrac{1}{9}\pi)$.

13. If P has polar coordinates $(1,\pi/2)$ and Q has polar coordinates $(2,\pi/6)$, find the distance between P and Q. [*Hint:* Use the formula in Example 3.]

14. If P has polar coordinates $(\sqrt{2},\pi/3)$ and Q has polar coordinates $(1,\pi/12)$, find the distance between P and Q.

15. If P has polar coordinates $(\sqrt{3},7\pi/6)$ and Q has polar coordinates $(2,\pi/3)$, find the distance between P and Q.

16. For a triangle, like the one in Figure 16.5, write a formula expressing a^2 in terms of b, c, and $\cos A$.

17. In a triangle, $a = 2$, $b = 3$, $c = \sqrt{6}$, find the angle C. (Use tables of the function cos.)

18. In a triangle, $a = 2.3$, $b = 4$, and $C = \pi/3$, find c.

19. In a triangle, $b = 1$, $c = 2$, and $A = 45°$, find a.

1.2 Equations in polar coordinates

Like Cartesian coordinates, polar coordinates can be used to represent equations and functions by geometric figures, as we shall show by examples. Since the equations considered may involve negative values of r, the convention stated above should be remembered.

EXAMPLES 1. Find the equation in polar coordinates for a circle of radius 4 with center at the origin.
ANSWER   The given circle is a circular grid of the polar coordinate system. Hence $r = 4$.

2. Describe the curve given by $\theta = \pi/4$, $r \neq 0$.
ANSWER   A ray, emanating from the origin, making an angle of $45°$ with the positive x-axis.

3. Find an equation, in polar coordinates, of a circle with center at the point P [polar coordinates: (r_0, θ_0)] and radius R.

FIRST SOLUTION   We write down the condition: the distance from a point with polar coordinates (r, θ) to P is R. By Example 3 in §1.1, this reads

$$r^2 - 2r_0 r \cos(\theta - \theta_0) + r_0^2 - R^2 = 0. \tag{3}$$

SECOND SOLUTION   The Cartesian coordinates of the center are $r_0 \cos \theta_0$ and $r_0 \sin \theta_0$ [by Equation (1)]. Therefore the equation of our circle in Cartesian coordinates reads (see Chapter 2, §3.1)

$$(x - r_0 \cos \theta_0)^2 + (y - r_0 \sin \theta_0)^2 = R^2.$$

Set $x = r \cos \theta$, $y = r \sin \theta$, and simplify. This yields the same equation as before.

4. What is the equation, in polar coordinates, of a line?

SOLUTION   The general equation of a line in Cartesian coordinates reads $Ax + By + C = 0$. Using (1), we obtain the equation

$$r(A \cos \theta + B \sin \theta) + C = 0. \tag{4}$$

For a vertical line, $A = 0$ and (4) becomes $rB \sin \theta + C = 0$ or $r \sin \theta = a$, where a is a new constant. For a horizontal line, $B = 0$ and (4) becomes $rA \cos \theta + C = 0$ or $r \cos \theta = b$, where b is a new constant.

5. What is the equation, in polar coordinates, of the line through the origin of slope m?

SOLUTION   The equation of this line in Cartesian coordinates reads $y = mx$. Using (1), we obtain the equation $r \sin \theta = mr \cos \theta$. If $r \neq 0$, we obtain $\sin \theta = m \cos \theta$ or $\tan \theta = m$. Thus we have the answer: $\tan \theta = m$ or $r = 0$. (The same result follows from Example 4 with $C = 0$, $B \neq 0$, and $m = -A/B$.)

6. The equation of a parabola, in Cartesian coordinates, reads $x^2 = 4y$. Find the equation of this curve in polar coordinates.

ANSWER   Using (1), we obtain the equation $r^2 \cos^2 \theta = 4r \sin \theta$. Dividing by $r \cos^2 \theta$, we find that $r = 4 \sec \theta \tan \theta$. This equation does not include the origin (since we divided by $r \cos^2 \theta$ and hence assumed that $r \neq 0$). To include the origin we must write: $r = 4 \sec \theta \tan \theta$ or $r = 0$.

PROBLEMS

1. Describe the curve in polar coordinates given by $r = 2$. By $r = \frac{1}{2}$.
2. Describe the curve in polar coordinates given by $\theta = \pi/3$, $r \neq 0$. By $\theta = -\pi/3$, $r \neq 0$.
3. Describe the curve in polar coordinates given by $\theta = 2\pi/3$, $r \neq 0$. By $\theta = 4\pi/3$, $r \neq 0$.
4. Find the equation, in polar coordinates, of a circle with center at $(2, -\pi/4)$ and radius 4.
5. Find the equation, in polar coordinates, of a circle with center at $(2, \pi/2)$ and radius 1.
6. Find the equation, in polar coordinates, of a line whose Cartesian equation reads $2x + 3y - 1 = 0$.
7. Find the equation, in polar coordinates, of a line whose Cartesian equation reads $4x - y + 2 = 0$.
8. The equation of a line in polar coordinates is $r(3 \cos \theta + 5 \sin \theta) = 1$. What is the equation in Cartesian coordinates?
9. The equation of a line in polar coordinates is $r \cos \theta - 2r \sin \theta + 4 = 0$. What is the equation in Cartesian coordinates?
10. Find the equation, in polar coordinates, of the parabola whose Cartesian equation reads $x = \frac{1}{2}y^2$.

In each of Problems 11 to 16 the equation of a curve is given in polar coordinates (r,θ). Find an equation for the curve in Cartesian coordinates (x,y). Identify the curve.

11. $r^2 - 2r \cos \theta - 3 = 0$.

12. $r = 2 \sin \theta$.

13. $r = \cos \theta + \sin \theta$.

14. $r^2 \cos \theta \sin \theta = 1$.

15. $r = \dfrac{1}{1 - \cos \theta}$.

16. $r = \dfrac{3}{2 - \cos \theta}$.

1.3 Curves in polar coordinates

A curve may be defined by a function f such that

$$r = f(\theta), \tag{5}$$

where r and θ are polar coordinates of a point on the curve. The Cartesian coordinates of this point are [see (1) in §1.1]

$$x = f(\theta) \cos \theta, \qquad y = f(\theta) \sin \theta. \tag{6}$$

Relation (6) is a special case of a so-called "parametric representation" of a curve, which we shall study in §2; the Cartesian coordinates x and y of a point on the curve are given as functions of the "polar angle" θ.

The simplest example is the equation

$$r = 1, \tag{7}$$

which defines the unit circle. For this curve, relations (6) read

$$x = \cos \theta, \qquad y = \sin \theta. \tag{8}$$

Other examples will be given below.

Given (5), how can we compute the slope m of the curve at a point corresponding to some value θ? Let us try to answer this question by trusting the Leibniz notation. If we knew y as a function of x, m would be found as the derivative $m = dy/dx$. From (6), using the language of differentials, we have

$$dx = \frac{d[f(\theta) \cos \theta]}{d\theta} \, d\theta = [f'(\theta) \cos \theta - f(\theta) \sin \theta] \, d\theta,$$

$$dy = \frac{d[f(\theta) \sin \theta]}{d\theta} \, d\theta = [f'(\theta) \sin \theta + f(\theta) \cos \theta] \, d\theta.$$

Substituting into the expression $m = dy/dx$, we obtain

$$m = \frac{f'(\theta) \sin \theta + f(\theta) \cos \theta}{f'(\theta) \cos \theta - f(\theta) \sin \theta} = \frac{\dfrac{dr}{d\theta} \sin \theta + r \cos \theta}{\dfrac{dr}{d\theta} \cos \theta - r \sin \theta}. \tag{9}$$

In §2.2 we shall see that this formula is indeed valid, assuming, of course, that f is continuously differentiable.

If Θ denotes the angle from the x-axis to the tangent line (measured counter-

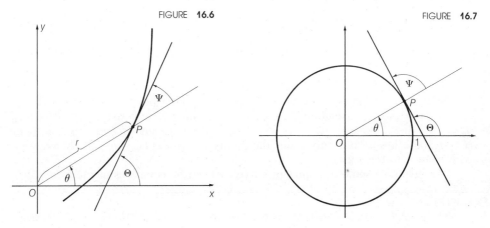

FIGURE **16.6**

FIGURE **16.7**

clockwise), then $m = \tan \Theta$ and (9) may be written as

$$\tan \Theta = \frac{dy}{dx} = \frac{f'(\theta) \sin \theta + f(\theta) \cos \theta}{f'(\theta) \cos \theta - f(\theta) \sin \theta} = \frac{\dfrac{dr}{d\theta} \sin \theta + r \cos \theta}{\dfrac{dr}{d\theta} \cos \theta - r \sin \theta}. \qquad (9')$$

Let P be a point on the curve, and let Ψ be the angle from the line OP to the tangent line to the curve through P, measured counterclockwise, see Figure 16.6. Then $\Psi = \Theta - \theta$. By (9') and the addition theorem for the function $\tan$ (compare Chapter 10, §1.4), we have

$$\tan \Psi = \frac{\tan \Theta - \tan \theta}{1 + \tan \Theta \tan \theta} = \frac{\dfrac{f' \sin \theta + f \cos \theta}{f' \cos \theta - f \sin \theta} - \dfrac{\sin \theta}{\cos \theta}}{1 + \left(\dfrac{f' \sin \theta + f \cos \theta}{f' \cos \theta - f \sin \theta}\right)\left(\dfrac{\sin \theta}{\cos \theta}\right)} = \frac{f}{f'}.$$

This may be written as

$$\tan \Psi = \frac{r}{dr/d\theta} \qquad \text{or} \qquad \cot \Psi = \frac{1}{r} \frac{dr}{d\theta}. \qquad (10)$$

Let us check formulas (9) and (10) for the case of the circle defined by (7) and (8). Since $r = 1$, $dr/d\theta = 0$, we obtain $\tan \Theta = -\cot \theta = \tan(\theta + \frac{1}{2}\pi)$, $\cot \Psi = 0$. This is correct (see Figure 16.7) since $\Theta = \theta + \frac{1}{2}\pi$, $\Psi = \frac{1}{2}\pi$.

In plotting a curve $r = f(\theta)$ in polar coordinates, it is convenient to make a table of $f(\theta)$ for various values of θ. It is also convenient to obtain or prepare a "polar graph paper" with a grid like the one in Figure 16.3.

Relation (9) may be used in order to compute the slope of the curve. If there is a number θ such that $f(\theta) = 0$, then the curve passes through the origin. *In this case, it has at the origin a tangent with slope* $\tan \theta$ *(a vertical tangent if θ is an odd multiple of $\pi/2$).* This is seen from (9): setting $f(\theta) = 0$, we obtain $\tan \Theta = \tan \theta$.

One should also remember the following points.

If $f(\theta)$ is an even function, $[f(-\theta) = f(\theta)]$, the curve is symmetric with respect to the x-axis. If $f(\theta)$ is an odd function, $[f(-\theta) = -f(\theta)]$, the curve is symmetric with respect to the y-axis. If $f(\theta)$ satisfies the relation $f(\theta + \pi) = f(\theta)$, then the curve is symmetric with respect to the origin.

We prove only the first statement. The other two are proved similarly.

Suppose that f is an even function. Let (x_0, y_0) be a point on the curve, x_0 and y_0 being Cartesian coordinates. Then there is a number θ_0 such that $x_0 = f(\theta_0) \cos \theta_0$, $y_0 = f(\theta_0) \sin \theta_0$. Since $f(-\theta_0) \cos(-\theta_0) = x_0$ and $f(-\theta_0) \sin(-\theta_0) = -y_0$, the point $(x_0, -y_0)$ also lies on our curve. But the points (x_0, y_0) and $(x_0, -y_0)$ are symmetrical with respect to the x-axis.

We give now some examples of curves defined by equations of the form (5).

EXAMPLES 1. $r = \cos \theta$.

The graph is a circle of radius $\frac{1}{2}$ with center at $x = \frac{1}{2}$, $y = 0$. This is seen, for instance, by going over to Cartesian coordinates: if $r = \cos \theta$, then $r^2 = r \cos \theta$ or $x^2 + y^2 = x$. The latter equation can be written as $(x - \frac{1}{2})^2 + y^2 = \frac{1}{4}$.

(We also note that the equation in Example 1, §1.2, becomes $r = \cos \theta$ if we set $\theta_0 = 0$, $R = r_0 = \frac{1}{2}$.)

The graph is shown in Figure 16.8. Note that, for $\theta = 90°$, we have $r = 0$. The tangent at this point is vertical. Note also that every point of the circle corresponds to two values of θ such that $-180° \leq \theta \leq 180°$. For instance, if $\theta = -45°$, then $r = 1/\sqrt{2}$, and if $\theta = 135°$, then $r = -1/\sqrt{2}$. These two values of θ yield the same point, the point with Cartesian coordinates $x = (1/\sqrt{2}) \cos(-45°) = (-1/\sqrt{2}) \cos 135° = \frac{1}{2}$, $y = (1/\sqrt{2}) \sin(-45°) = (-1/\sqrt{2}) \sin 135° = \frac{1}{2}$.

2. $r = \cos 2\theta$.

The graph, shown in Figure 16.9, is called a **four-leaved rose**. Since $\cos 2\theta$ is even, the graph is symmetric about the x-axis. Since $\cos 2(\theta + \pi) = \cos(2\theta + 2\pi) = \cos 2\theta$, the graph is symmetric about the origin. Hence the graph is also symmetric about the y-axis. [The function $\cos 2\theta$ is not odd, but there is no contradiction. We said above that if $f(\theta)$ is odd, then the graph $r = f(\theta)$ is symmetric about the y-axis; we did *not* say that if the graph is symmetric, then $f(\theta)$ is odd.]

A table of easily computable values of r and θ is given to aid in sketching the graph.

θ	2θ	$r = \cos 2\theta$	θ	2θ	$r = \cos 2\theta$
0°	0°	1.000	±90°	±180°	−1.000
±15°	±30°	.866	±105°	±210°	−.866
±30°	±60°	.500	±120°	±240°	−.500
±45°	±90°	.000	±135°	±270°	.000
±60°	±120°	−.500	±150°	±300°	.500
±75°	±150°	−.866	±165°	±330°	.866
±90°	±180°	−1.000	±180°	±360°	1.000

3. $r = \cos 3\theta$.

The graph is called a **three-leaved rose**; see Figure 16.10. (The reader is asked to supply a table of easily computable values of r and θ.)

4. $r = Ae^{a\theta}$, A and a positive constants.

The graph is shown in Figure 16.11. The curve is called a **logarithmic spiral**, since its equation may be written in the form

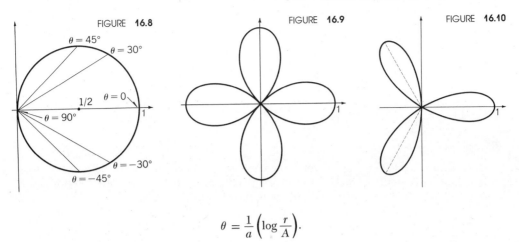

FIGURE **16.8**

$\theta = 45°$
$\theta = 30°$
$\theta = 0$
1/2
1
$\theta = 90°$
$\theta = -30°$
$\theta = -45°$

FIGURE **16.9**

FIGURE **16.10**

$$\theta = \frac{1}{a}\left(\log\frac{r}{A}\right).$$

5. $r = k\theta$, k a positive constant.

This curve (shown in Figure 16.12 for a positive value of k) is called the **spiral of Archimedes.** It enters the origin but spirals infinitely often as it recedes from the origin. The dotted curve corresponds to negative values of θ.

6. Find the slope of the four-leaved rose (Example 2) as a function of θ. In particular, find the slopes of the four-leaved rose at the origin.

ANSWER   We use Equation (9) with $f(\theta) = \cos 2\theta$, $f'(\theta) = -2\sin 2\theta$, and obtain

$$\tan \Theta = \frac{-2\sin 2\theta \sin \theta + \cos 2\theta \cos \theta}{-2\sin 2\theta \cos \theta - \cos 2\theta \sin \theta}.$$

Next, $f(\theta) = \cos 2\theta = 0$ for $\theta = \pi/4$, $3\pi/4$, $5\pi/4$, $7\pi/4$ (we need only values of θ between 0 and 2π). For each of these four values of θ, we have either $\tan \Theta = 1$ or $\tan \Theta = -1$.

7. Find the slope of Archimedes' spiral (Example 5) as a function of θ.

ANSWER   Using (9), with $f(\theta) = k\theta$, we obtain

$$\tan \Theta = \frac{\sin \theta + \theta \cos \theta}{\cos \theta - \theta \sin \theta}.$$

Note that $\tan \Theta$ does not depend on k, is 0 for $r = \theta = 0$, and is nearly $-\cot \theta$ for large θ.

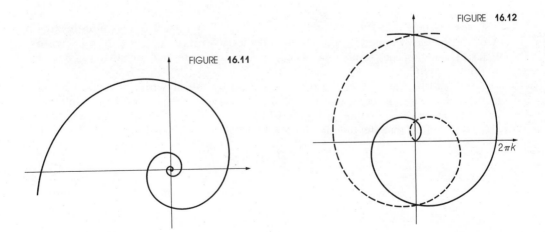

FIGURE **16.11**

FIGURE **16.12**

$2\pi k$

REMARK Equations of the form $r = a \sin n\theta$ and $r = a \cos n\theta$ are called **rose** (or petal) curves. The number of leaves (or petals) is equal to n if n is an odd integer, $2n$ if n is even. The special case $n = 1$ is a circle.

Equations of the form $r = a + b \cos \theta$ and $r = a + b \sin \theta$ are called **limacons**. When $|b| > |a|$, the limacon has an inner loop; when $|b| < |a|$, it does not. In the special case $|b| = |a|$, the curve has a cusp at the pole and is called a **cardioid**.

Equations of the form $r^2 = a^2 \cos 2\theta$ and $r^2 = a^2 \sin 2\theta$ are called **lemniscates** and resemble a figure eight.

PROBLEMS

In each of Problems 1 to 20 sketch the curve whose equation is given with respect to the polar coordinates (r,θ). Make a table listing easily computable values of r and θ to aid in your sketch. Try to exploit any symmetry in the equation.

1. $r = 2 \sin \theta$.
2. $r = \sin 2\theta$.
3. $r = \sin 3\theta$.
4. $r = \cos 4\theta$.
5. $r = 1 + \cos \theta$.
6. $r = 2(1 - \sin \theta)$.
7. $r = 2 - \sin \theta$.
8. $r = 1 - 2 \sin \theta$.
9. $r = 2 + 4 \cos \theta$.
10. $r = 3 + 2 \cos \theta$.
11. $r = \pi/\theta$.
12. $r = 2/\cos \theta$.
13. $r = 1/(\sin \theta - \cos \theta)$.
14. $r^2 = 4 \cos 2\theta$.
15. $r^2 = -9 \sin 2\theta$.
16. $r = 3 - \cos 2\theta$.
17. $r = \cos(\theta/2)$.
18. $r = \cos \theta \cos 2\theta$.
19. $r = 1 + 2 \sin 2\theta$.
20. $r = \cos(2\theta/3)$.

In each of Problems 21 to 26 find $\tan \Psi$ and $\tan \Theta$ for the given curves at the indicated points. Sketch the curve, draw the tangent line, and show the angles Ψ and Θ.

21. $r = 2 \cos \theta, \quad \left(\sqrt{3}, \dfrac{\pi}{6}\right)$.

22. $r = \sin \theta, \quad \left(\dfrac{\sqrt{3}}{2}, \dfrac{2\pi}{3}\right)$.

23. $r = 1 - \cos \theta, \quad \left(1, \dfrac{\pi}{2}\right)$.

24. $r^2 = 2 \cos 2\theta, \quad \left(1, \dfrac{7\pi}{6}\right)$.

25. $r = 2 \sin 3\theta, \quad \left(\sqrt{2}, \dfrac{\pi}{4}\right)$.

26. $r = 3 - 2 \cos \theta, \quad \left(2, -\dfrac{\pi}{3}\right)$.

1.4 Areas in polar coordinates

We consider again a curve defined (in polar coordinates) by the equation

$$r = f(\theta), \qquad \alpha \le \theta \le \beta, \tag{11}$$

where we assume that $f(\theta) \ge 0$ and $0 \le \alpha < \beta \le 2\pi$. We want to compute the area A of the region bounded by the curve and the two rays $\theta = \alpha$, $\theta = \beta$; see Figure 16.13. [The region consists of all points with polar coordinates (r,θ) such that $\alpha < \theta < \beta$ and $0 < r < f(\theta)$.]

Let $A(\theta)$ be the area of the region bounded by the rays from 0 corresponding to the polar angles α and θ and the arc of the curve corresponding to the interval $[\alpha,\theta]$; this is the shaded region in Figure 16.14. We set $A(\alpha) = 0$. Our intuitive notions about area tell us that the function $\theta \mapsto A(\theta)$ is continuous. Since $A(\beta) = A$, we have

$$A = \int_\alpha^\beta A'(\theta)\, d\theta,$$

assuming that $A'(\theta)$ exists and is piecewise continuous. We proceed to find or rather guess the value of the derivative $A'(\theta)$.

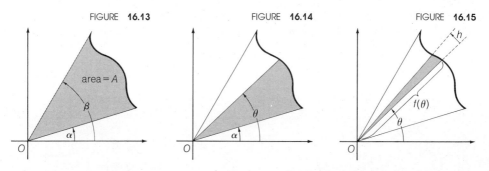

FIGURE **16.13**   FIGURE **16.14**   FIGURE **16.15**

Let h be a small positive number. The number $A(\theta + h) - A(\theta)$ is the area of the narrow sector shown in Figure 16.15. This area differs very little from the area of the circular sector shown in the same figure; the latter area is $\frac{1}{2}$ of the radius squared times the central angle, that is, $\frac{1}{2}f(\theta)^2\, h$. Hence $\dfrac{A(\theta + h) - A(\theta)}{h} \approx \dfrac{1}{2}f(\theta)^2$. This suggests that

$$A'(\theta) = \tfrac{1}{2}f(\theta)^2 \tag{12}$$

and

$$A = \frac{1}{2}\int_\alpha^\beta f(\theta)^2\, d\theta, \tag{13}$$

which may also be written as

$$A = \frac{1}{2}\int_\alpha^\beta r^2\, d\theta. \tag{13'}$$

(In the preceding reasoning, we used our intuitive notions about area to derive a formula for it. One can ▶ verify that this formula leads to the same number as the area formula in Chapter 9, §1.)

EXAMPLES 1. Find the area A bounded by the arc of the Archimedes spiral $r = \theta$ (see Example 5 in §1.3) from $\theta = 0$ to $\theta = 2\pi$ and the x-axis.
ANSWER See Figure 16.16. We use (13) with $\alpha = 0$, $\beta = 2\pi$, $f(\theta) = \theta$, and obtain

$$A = \frac{1}{2}\int_0^{2\pi} \theta^2\, d\theta = \frac{\theta^3}{6}\bigg|_0^{2\pi} = \frac{4}{3}\pi^3.$$

2. Find the area enclosed by the four-leaved rose, $r = 2\cos 2\theta$.
SOLUTION We plot the curve (see §1.3, Example 2) and note that, by symmetry, the desired area is eight times the area of the shaded half-petal in Figure 16.17. The curved boundary of this half-petal is described as the polar angle goes from 0 to $\pi/4$. The desired area is therefore

$$A = 8 \cdot \frac{1}{2}\int_0^{\pi/4} (2\cos 2\theta)^2\, d\theta = 16\int_0^{\pi/4} \cos^2 2\theta\, d\theta = 8\int_0^{\pi/4} (1 + \cos 4\theta)\, d\theta$$

$$= 8\left(\theta + \frac{1}{4}\sin 4\theta\right)\bigg|_0^{\pi/4} = 2\pi.$$

Here we used the identity $2\cos^2 x = 1 + \cos 2x$.
    Note that the area of the rose is half of the area of the circle circumscribed around it.

FIGURE **16.16**   FIGURE **16.17**

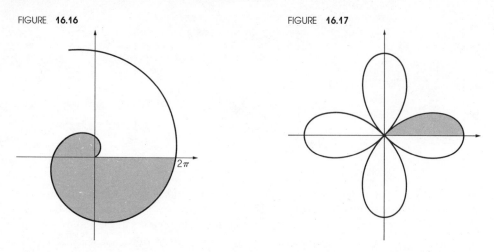

3. Find the area of the inner loop of the limacon $r = 1 + 2 \sin \theta$.
SOLUTION We first sketch the curve (see Figure 16.18) and observe that the inner loop is described as the polar angle goes from $7\pi/6$ to $11\pi/6$. By symmetry, the desired area is

$$A = 2 \cdot \frac{1}{2} \int_{7\pi/6}^{3\pi/2} (1 + 2 \sin \theta)^2 \, d\theta = \int_{7\pi/6}^{3\pi/2} (1 + 4 \sin \theta + 4 \sin^2 \theta) \, d\theta$$

$$= \int_{7\pi/6}^{3\pi/2} (3 + 4 \sin \theta - 2 \cos 2\theta) \, d\theta = (3\theta - 4 \cos \theta - \sin 2\theta) \Big|_{7\pi/6}^{3\pi/2}$$

$$= \left(\frac{9\pi}{2} + 4 - 0 \right) - \left(\frac{21\pi}{6} + 2\sqrt{3} - \frac{\sqrt{3}}{2} \right) = \pi + 4 - \frac{3\sqrt{3}}{2}.$$

4. Find the area of a right triangle. Let the triangle be bounded by the x-axis, the line $x = 1$, and the line $y = mx$, with $m > 0$; see Figure 16.19. The equation of the line $x = 1$ in polar coordinates is $r \cos \theta = 1$, and our triangle is obtained for $0 \leq \theta \leq \arctan m$. Using (13), we obtain the desired area A:

$$\frac{1}{2} \int_0^{\arctan m} \frac{d\theta}{\cos^2 \theta} = \frac{1}{2} \tan \theta \Big|_0^{\arctan m} = \frac{1}{2} m,$$

as expected.

FIGURE **16.18**   FIGURE **16.19**   FIGURE **16.20**

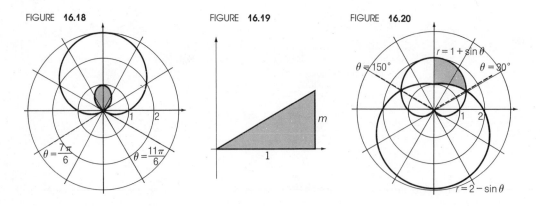

5. Find the area outside the limacon $r = 2 - \sin\theta$ and inside the cardioid $r = 1 + \sin\theta$.

SOLUTION   The curves are plotted in Figure 16.20. The intersection points, which determine the limits of integration, are found by solving the given equations simultaneously: we have $2 - \sin\theta = 1 + \sin\theta$, that is, $\sin\theta = \frac{1}{2}$, hence $\theta = \pi/6$ and $\theta = 7\pi/6$, and the intersection points are $(3/2, \pi/6)$ and $(3/2, 7\pi/6)$. By symmetry, the desired area is twice the shaded part shown in the figure. Therefore

$$A = 2\left\{\left(\begin{array}{c}\text{area of cardioid}\\\text{from } \pi/6 \text{ to } \pi/2\end{array}\right) - \left(\begin{array}{c}\text{area of limacon}\\\text{from } \pi/6 \text{ to } \pi/2\end{array}\right)\right\}$$

$$= 2\left\{\frac{1}{2}\int_{\pi/6}^{\pi/2}(1 + \sin\theta)^2 \, d\theta - \frac{1}{2}\int_{\pi/6}^{\pi/2}(2 - \sin\theta)^2 \, d\theta\right\}$$

$$= \int_{\pi/6}^{\pi/2}(1 + 2\sin\theta + \sin^2\theta) \, d\theta - \int_{\pi/6}^{\pi/2}(4 - 4\sin\theta + \sin^2\theta) \, d\theta$$

$$= \int_{\pi/6}^{\pi/2}(-3 + 6\sin\theta) \, d\theta = (-3\theta - 6\cos\theta)\Big|_{\pi/6}^{\pi/2} = 3\sqrt{3} - \pi.$$

PROBLEMS

In each of the following problems find the area enclosed by the given curves using polar coordinates. It is advisable to make a sketch before computing the area.

1. Find the area enclosed by the circle $r = 2\cos\theta$.
2. Find the area bounded by the arc of the logarithmic spiral $r = e^\theta$ from $\theta = 0$ to $\theta = 2\pi$ and the x-axis.
3. Find the area enclosed by the lemniscate $r^2 = \cos 2\theta$.
4. Find the area enclosed by the cardioid $r = 1 + \cos\theta$.
5. Find the area enclosed by *all* the leaves of the three-leaved rose $r = 4\cos 3\theta$.
6. Find the area inside *one* leaf of the four-leaved rose $r = 8\sin 2\theta$.
7. Find the area enclosed by the limacon $r = 4 + 2\cos\theta$.
8. Find the area inside the inner loop of the limacon $r = 2 + 4\cos\theta$.
9. Find the area of one of the smaller (inner) loops of the curve $r^2 = \cos\theta/2$.
10. Find the area of one of the smaller (left-hand) loops of the curve $r = \cos\theta\cos 2\theta$.
11. Find the area common to the circles $r = 2\sin\theta$ and $r = 2\cos\theta$.
12. Find the area inside the circle $r = 4\sin\theta$ and outside the cardioid $r = 4 + 4\cos\theta$.
13. Find the area outside the circle $r = \sqrt{2}$ and inside the lemniscate $r^2 = 4\cos 2\theta$.
14. Find the area inside the circle $r = 5\sin\theta$ and outside the limacon $r = 2 + \sin\theta$.
15. Find the area outside the cardioid $r = 2 - 2\cos\theta$ and inside the circle $r = 1$.

§2   Parametric representation

The parabola is, as we know, the graph of a function $y = f(x) = x^2$. The circle $x^2 + y^2 = 1$, on the other hand, is not a graph of a function since for every x, $-1 < x < 1$, there are two points (x,y) on the circle, one with $y = \sqrt{1 - x^2}$, the other with $y = -\sqrt{1 - x^2}$. In §1.3, however, we learned how to represent all points on the circle by two functions: $x = \cos\theta$, $y = \sin\theta$. This is a special case of a parametric representation of curves to be studied in this section.

2.1   Examples

Suppose we are given two continuous functions, $f(t)$ and $g(t)$, defined on the same interval, say, for $a \leq t \leq b$. The set of points (x,y), which can be written as

$$x = f(t), \qquad y = g(t), \qquad a \leq t \leq b, \tag{1}$$

is called a **curve represented parametrically** by the function f and g. The point with the coordinates $(f(a),g(a))$ is called the **initial point** of the curve, the point with the coordinates $(f(b),g(b))$ is called the **terminal point.** Occasionally, it is convenient to interpret the **parameter** t as time. Then we say that the formulas (1) describe a motion of a point in the plane; at the time t the point is at $(f(t),g(t))$. Since a curve represented parametrically has an initial point and a terminal point, we say that it is an **oriented** curve.

EXAMPLES 1.   The circle of radius R, $x^2 + y^2 = R^2$, has a standard parametric representation

$$x = R \cos t, \qquad y = R \sin t, \qquad 0 \leq t \leq 2\pi. \tag{2}$$

Here t is also the polar angle of the point (x,y). The initial point is the point $x = R \cos 0 = R$, $y = R \sin 0 = 0$, that is, the point $(R,0)$. The terminal point is also $(R,0)$. (A curve for which the two endpoints coincide is called **closed.**)

2.   The relations

$$x = R \cos t, \qquad y = R \sin t, \qquad 0 \leq t \leq \frac{\pi}{2} \tag{3}$$

represent an arc of the circle $x^2 + y^2 = R^2$, the arc lying in the first quadrant. Here the initial point is $(R,0)$, the endpoint is $(R \cos \pi/2, R \sin \pi/2) = (0,R)$.

Equation (3) is the same as (2) except that the parameter t is defined over a subinterval of $[0,2\pi]$. The curve given by (3) is called a **subarc** of (2).

3.   The formulas

$$x = R \cos t, \qquad y = -R \sin t, \qquad 0 \leq t \leq 2\pi \tag{4}$$

represent the same circle as do (2). But in (2) the circle is traversed counterclockwise, in (4) clockwise. Indeed, if t is small and positive, then the point (2) lies above the x-axis, the point (4) below.

The oriented curve (4) is said to be the **opposite** of the oriented curve (2). The endpoint of (2) is the initial point of (4), and conversely.

4.   Consider now the parametric representation

$$x = R \cos t, \qquad y = R \sin t, \qquad 0 \leq t \leq 4\pi. \tag{5}$$

This is again the same circle, traversed in the counterclockwise direction, but traversed *twice.* Indeed, for every point (x,y) on the circle, distinct from $(R,0)$, there is *one* value of t such that (2) holds, and *two* values of t such that (5) holds. For instance, the point $(\frac{1}{2},\frac{1}{2}\sqrt{3})$ corresponds to the value $t = 30° = \frac{1}{6}\pi$ in the interval $[0,2\pi]$ and to the values $t = \frac{1}{6}\pi$, $t = \frac{13}{6}\pi$ in the interval $[0,4\pi]$.

5.   A parametric representation may be defined also in an open and in an infinite interval. For instance,

$$x = t, \qquad y = t^2, \qquad -\infty < t < +\infty \tag{6}$$

represents the parabola $y = x^2$.

6. Set

$$x = t^2, \qquad y = t^4, \qquad -\infty < t < +\infty. \tag{7}$$

This represents all points of the parabola $y = x^2$ with $x \geq 0$. Here every such point, with $x > 0$, corresponds to two values of t, $t = \sqrt{x}$, and $t = -\sqrt{x}$.

7. Set

$$x = 2 + t, \qquad y = 1 - 2t, \qquad -\infty < t < +\infty. \tag{8}$$

This is a parametric representation of the straight line $2x + y - 5 = 0$.

To see this, solve the first Equation (8) for t to obtain $t = x - 2$, and substitute this into the second equation. We get $y = 1 - 2(x - 2)$ or $y = -2x + 5$.

8. To obtain a parametric representation of the ellipse

$$\frac{x^2}{a^2} + \frac{y^2}{b^2} = 1,$$

we note that (x,y) lies on this ellipse if and only if the point $(x/a, y/b)$ lies on the unit circle, that is, if and only if there is a number t, $0 \leq t \leq 2\pi$, such that $x/a = \cos t$, $y/b = \sin t$. Hence

$$x = a \cos t, \qquad y = b \sin t, \qquad 0 \leq t \leq 2\pi$$

is a parametric representation of the ellipse.

9. Consider two parametric representations of curves,

$$x = 1 + 2t, \qquad y = 3 - t, \qquad 0 \leq t \leq 27 \tag{9}$$

and

$$x = 1 + 2\tau^3, \qquad y = 3 - \tau^3, \qquad 0 \leq \tau \leq 3. \tag{10}$$

The first represents a straight segment joining the point $(1,3)$ to the point $(55, -24)$. Indeed, we conclude from (9) that $t = \frac{1}{2}(x - 1)$, $y = 3 - \frac{1}{2}(x - 1) = -\frac{1}{2}x + \frac{7}{2}$, which is the equation of a line. Also $x = 1$, $y = 3$ for $t = 0$ and $x = 55$, $y = -24$ for $t = 27$. Now (10) represents the *same* straight segment. To see this, note that (9) becomes (10) if we set $t = \tau^3$, and that τ^3 goes from 0 to 27 as τ goes from 0 to 3.

There is no reason to be surprised that a curve has more than one parametric representation. There are various ways to walk along the same path.

PROBLEMS

In Problems 1 to 10 sketch the curves given by the parametric representations. Indicate the initial and terminal points on the sketch.

1. $x = \frac{1}{2}t - 1$, $\quad y = 2 - t \quad$ for $-2 \leq t \leq 6$.
2. $x = |t - 2|$, $\quad y = 2t \quad$ for $0 \leq t \leq 8$.
3. $x = t^2 - 1$, $\quad y = t + 1 \quad$ for $-3 \leq t \leq 3$.
4. $x = 2t$, $\quad y = (t - 2)^2 \quad$ for $0 \leq t < +\infty$.
5. $x = 2t$, $\quad y = t^3 + 4 \quad$ for $-2 \leq t \leq 2$.

6. $x = 1 - t$, $\quad y = \dfrac{8}{t^2 + 1} \quad$ for $1 \leq t < +\infty$.

7. $x = 4 \sin t$, $\quad y = 4 \cos t \quad$ for $0 \leq t \leq \pi$.
8. $x = 1 + 4 \cos t$, $\quad y = 2 - 4 \sin t \quad$ for $0 \leq t \leq 2\pi$.

9. $x = \sin t, \quad y = \cos^2 t, \quad 0 \le t \le \dfrac{3\pi}{2}.$

10. $x = -2 \sin t, \quad y = 3 \cos t, \quad 0 \le t \le 3\pi.$

In each of Problems 11 to 14 show that the two parametric representations determine the same curve by finding the relation between the two parameters. (See Example 9.).

11. $x = \cos \theta, \quad y = \sin \theta, \quad 0 \le \theta \le \pi/2;$ $x = \cos 2t, \quad y = \sin 2t, \quad 0 \le t \le \pi/4.$

12. $x = t, \quad y = 0, \quad 0 \le t \le 1;$ $x = \sin \theta, \quad y = 0, \quad 0 \le \theta \le \pi/2.$

13. $x = e^t, \quad y = e^{-t}, \quad -\infty < t < +\infty;$ $x = u^2, \quad y = u^{-2}, \quad 0 < u < +\infty.$

*14. $x = \cos \theta, \quad y = \sin \theta, \quad 0 \le \theta \le \pi/2;$ $x = (1 - u^2)/(1 + u^2),$
    $y = 2u/(1 + u^2), \quad 0 \le u \le 1.$

2.2   Smooth curves. Tangent lines

We consider now parametric representations such that the functions $x = f(t),\ y = g(t)$ have continuous derivatives and

$$\left(\frac{dx}{dt}\right)^2 + \left(\frac{dy}{dt}\right)^2 > 0, \tag{11}$$

for all t considered. (At the endpoints, think of the derivative $dx/dt,\ dy/dt$ as one-sided.) Such a representation is called smooth, and a curve that can be so represented is called a **smooth curve.** Curves occurring in applications are usually smooth, but there may be exceptional points at which this condition is violated.

All parametric representations given in the examples above, with one single exception, are smooth. Let us verify this for the ellipse

$$x = a \cos t, \qquad y = b \sin t.$$

We have

$$\left(\frac{dx}{dt}\right)^2 + \left(\frac{dy}{dt}\right)^2 = a^2 \cos^2 t + b^2 \sin^2 t.$$

Since $a \neq 0,\ b \neq 0$, the above number can be 0 only if $\cos^2 t = 0$ *and* $\sin^2 t = 0$. But this is impossible, since $\cos^2 t + \sin^2 t = 1$.

The parabola $y = x^2$ is also a smooth curve, since it can be represented parametrically by $x = t,\ y = t^2$, so that $(dx/dt)^2 + (dy/dt)^2 = 1 + 4t^2 > 0$.

FIGURE 16.21   FIGURE 16.22

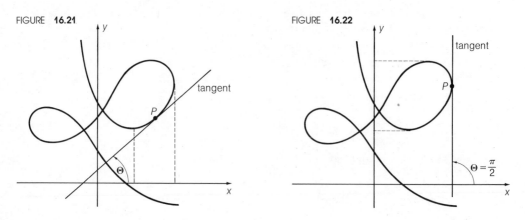

(The representation in Example 6 in §2.1 is not smooth for one value of t, namely, for $t = 0$.)

We show next how to compute the *slope of a tangent* to a smooth curve (1) at some point P. Let P be the point (x_0, y_0) with $x_0 = f(t_0)$, $y_0 = g(t_0)$, and let us assume that $f'(t_0) \neq 0$. There is an interval $[t_1, t_2]$ about t_0 such that $f'(t) \neq 0$ in this interval. and, therefore, $f'(t)$ is either always positive or always negative for $t_1 \leq t \leq t_2$. [This is so since $f'(t)$ is a continuous function.] Hence $f(t)$ is monotone and has an inverse function (see Chapter 6, §1.1 and §1.2); call this inverse function $t = w(x)$, it is defined and differentiable near x_0. Now, for x near x_0, we have $y = g(t) = g(w(x))$. Thus the arc of the curve near P is the graph of the function $y = g(w(x))$; see Figure 16.21. The slope m of the tangent line to the curve is the derivative

$$m = \frac{dy}{dx} = \frac{dg(w(x))}{dx} = g'(w(x))w'(x) \quad \text{by the chain rule}$$

$$= g'(w(x)) \frac{1}{f'(t)} \quad \begin{array}{l} \text{by the rule for differentiating} \\ \text{inverse functions} \end{array} \tag{12}$$

$$= \frac{g'(t)}{f'(t)}.$$

Let Θ denote the angle between the tangent line and the x-axis defined in §1.3. We have just proved that

$$m = \tan \Theta \frac{dy/dt}{dx/dt} \tag{12'}$$

(as we could have guessed from the Leibniz notation).

What if $dx/dt = 0$ at $t = t_0$? Then $dy/dt \neq 0$, by (11), and we can repeat the previous argument by interchanging x and y. We conclude that, at P, the tangent line is vertical so that $\Theta = \pi/2$; see Figure 16.22.

REMARK   If our curve is represented by an equation in polar coordinates $r = f(\theta)$, it admits the parametric representation

$$x = f(\theta) \cos \theta, \qquad y = f(\theta) \sin \theta,$$

as we noted in §1.3. Applying relation (12), with t replaced by θ, we obtain formula (7) in §1.3.

EXAMPLE   Find the slope of the tangent line to the curve $x = 2 + 4 \cos t$, $y = 1 + 2 \sin t$ for $0 \leq t \leq 2\pi$. In particular, find those points where the tangent is vertical or horizontal. Sketch the curve.

SOLUTION   We first check that the curve is smooth. Since $dx/dt = -4 \sin t$, $dy/dt = 2 \cos t$, we have

$$\left(\frac{dx}{dt}\right)^2 + \left(\frac{dy}{dt}\right)^2 = 16 \sin^2 t + 4 \cos^2 t = 4 + 12 \sin^2 t > 0$$

and so the curve is smooth for all t. The curve is also closed. Next, $dx/dt = 0$ when $t = 0$, π, or 2π, so the tangent line is vertical at the points $(6,1)$ and $(-2,1)$. Since $dy/dt = 0$ when $t = \pi/2$ or $3\pi/2$, the tangent line is horizontal at the points $(2,3)$ and $(2,-1)$. The slope at all other points is given by

$$m = \tan \Theta = \frac{dy/dt}{dx/dt} = \frac{2 \cos t}{-4 \sin t} = -\frac{1}{2} \cot t.$$

FIGURE **16.23**

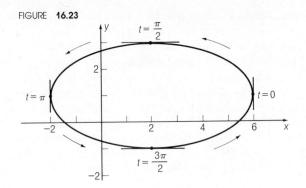

The curve, an ellipse, is shown in Figure 16.23. (The parameter t is *not* the polar angle.)

PROBLEMS

In each of Problems 1 to 6 sketch the curve whose parametric representation is given. To aid in your sketch, you may want to make a table of easily computable values of x and y and determine all points where the tangent is vertical or horizontal. (The parameter t is unrestricted, that is, t may take on all values for which x and y are both defined.)

1. $x = 2 + 2 \cos t, \quad y = 4 + 2 \sin t.$
2. $x = -4 + 3 \cos t, \quad y = 2 + 6 \sin t.$
3. $x = 4t - t^2, \quad y = t^2.$
4. $x = t^2 - 2t, \quad y = t^2 + 2t.$
5. $x = t^3 - 3t, \quad y = 2t^2.$
6. $x = t^2 - 3, \quad y = t^3 - 3t.$

*7. If $x = f(t)$ and $y = g(t)$ have continuous second derivatives for all t considered, show that

$$\frac{d^2y}{dx^2} = \frac{\dfrac{d}{dt}(dy/dx)}{(dx/dt)} = \frac{f'(t)g''(t) - f''(t)g'(t)}{[f'(t)]^3}$$

provided $f'(t) \neq 0$.

* 8. Find dy/dx and d^2y/dx^2 if $x = 3t + 1$, $y = t^2 - 2$.
* 9. Find dy/dx and d^2y/dx^2 if $x = t^2 - t$, $y = t^3 + 1$.
* 10. Find dy/dx and d^2y/dx^2 if $x = \frac{1}{3}\cos^3 t$, $y = \frac{1}{3}\sin^3 t$.

2.3   Arc length

We now derive a formula for the length of a smooth curve defined parametrically by

$$x = f(t), \qquad y = g(t), \qquad a \leq t \leq b. \tag{1'}$$

Let us assume first that $f'(t) > 0$ for $a \leq t \leq b$. Then f is an increasing function; it has an inverse function $t = w(x)$, so that $y = g(t) = g(w(x))$, and our curve is the graph of the function $y = g(w(x))$, $f(a) \leq x \leq f(b)$. By Chapter 9, §3.1, the length of the curve is

$$L = \int_{x=f(a)}^{x=f(b)} \sqrt{1 + \left(\frac{dy}{dx}\right)^2}\, dx.$$

In the preceding subsection we obtained relation (12) for $dy/dx = dg(w(x))/dx$. Using it, and the substitution rule for integrals, we obtain

$$L = \int_{x=f(a)}^{x=f(b)} \sqrt{1 + \frac{g'(w(x))^2}{f'(w(x))^2}}\, dx = \int_{t=a}^{t=b} \sqrt{1 + \frac{g'(t)^2}{f'(t)^2}}\, f'(t)\, dt$$

$$= \int_a^b \sqrt{f'(t)^2 + g'(t)^2}\, dt,$$

which can also be written as

$$L = \int_a^b \sqrt{\left(\frac{dx}{dt}\right)^2 + \left(\frac{dy}{dt}\right)^2}\, dt. \tag{13}$$

If we assume that $f'(t) < 0$ for $a \le t \le b$, a similar argument gives again formula (13). Also, if we assume that $g'(t) \ne 0$ for $a \le t \le b$ so that g has an inverse function, call it ρ, and our curve is the graph of $x = f(\rho(y))$, we are again led to Equation (13). Finally, since every smooth arc (1') can be divided into subarcs along each of which either $f'(t) \ne 0$ or $g'(t) \ne 0$, the formula (13) holds *always!*

For a curve defined by an equation in polar coordinates, $r = f(\theta)$, $\alpha \le \theta \le \beta$, we have the parametric representation

$$x = f(\theta) \cos \theta, \qquad y = f(\theta) \sin \theta, \qquad \alpha \le \theta \le \beta.$$

Hence

$$\left(\frac{dx}{d\theta}\right)^2 + \left(\frac{dy}{d\theta}\right)^2 = (f'(\theta) \cos \theta - f(\theta) \sin \theta)^2 + (f'(\theta) \sin \theta + f(\theta) \cos \theta)^2$$

$$= f'(\theta)^2 + f(\theta)^2 = r^2 + \left(\frac{dr}{d\theta}\right)^2,$$

and, by (13) with θ replacing t,

$$L = \int_\alpha^\beta \sqrt{r^2 + \left(\frac{dr}{d\theta}\right)^2}\, d\theta. \tag{14}$$

This is the length formula in polar coordinates.

EXAMPLES 1. Find the length of the curve

$$x = \tfrac{1}{2} t \sqrt{1 + t^2} + \tfrac{1}{2} \log(t + \sqrt{1 + t^2}), \qquad y = \tfrac{4}{3} t^{3/2}, \qquad 1 \le t \le 2.$$

ANSWER We use (13). Since

$$\frac{dx}{dt} = \sqrt{1 + t^2}, \qquad \frac{dy}{dt} = 2\sqrt{t},$$

we have

$$L = \int_1^2 \sqrt{1 + t^2 + 4t}\, dt = \int_1^2 (1 + 2t)\, dt = (t + t^2) \Big|_1^2 = 4.$$

2. Find the length of the circular arc

$$x = R \cos t, \qquad y = R \sin t, \qquad 0 \le t \le \alpha.$$

ANSWER Since

$$\left(\frac{dx}{dt}\right)^2 + \left(\frac{dy}{dt}\right)^2 = R^2 \sin^2 t + R^2 \cos^2 t = R^2,$$

(13) yields

$$L = \int_0^\alpha R\, dt = R\alpha$$

as expected.

3. How long is the arc of the logarithmic spiral $r = e^\theta$ from $\theta = 0$ to $\theta = \pi/2$? (Here r and θ are polar coordinates.)
ANSWER We use (14):

$$L = \int_0^{\pi/2} \sqrt{(e^\theta)^2 + (e^\theta)^2}\; d\theta = \int_0^{\pi/2} \sqrt{2e^{2\theta}}\; d\theta$$

$$= \sqrt{2} \int_0^{\pi/2} e^\theta\, d\theta = \sqrt{2}(e^{\pi/2} - 1).$$

PROBLEMS

In each of Problems 1 to 6 find the length of the curve determined by the given parametric representation between the indicated values of the parameter t.

1. $x = \tfrac{1}{2}t^2$, $\;y = \tfrac{1}{3}(2t + 1)^{3/2}$, $\;0 \le t \le 1$.
2. $x = 1 + 5\cos t$, $\;y = 3 - 5\sin t$, $\;0 \le t \le 2\pi$.
3. $x = 2t^2$, $\;y = t^3$, $\;-1 \le t \le 1$.
4. $x = \tfrac{1}{2}t^2$, $\;y = t + 1$, $\;0 \le t \le 1$.
5. $x = \cos^3 t$, $\;y = \sin^3 t$, $\;0 \le t \le \pi/2$.
6. $x = \tfrac{1}{3}\cos^3 t$, $\;y = \sin t - \tfrac{1}{3}\sin^3 t$, $\;\pi/6 \le t \le \pi/3$.

In each of Problems 7 to 12 find the length of the curve given in polar coordinates between the indicated values of the parameter θ.

7. $r = 4\cos\theta$, $\;0 \le \theta \le 2\pi$.   10. $r = 2(1 - \cos\theta)$, $\;0 \le \theta \le \pi$.
8. $r = e^{\tfrac{1}{2}\theta}$, $\;0 \le \theta \le 2\pi$.   11. $r = \sin^3(\theta/3)$, $\;0 \le \theta \le 3\pi$.
9. $r = 3\sec\theta$, $\;-\pi/4 \le \theta \le \pi/4$.   12. $r = 2\theta$, $\;0 \le \theta \le 2\pi$.

*13. Show that the total arc length of the ellipse $x = a\cos t$, $y = b\sin t$, $a > b > 0$, is given by

$$L = 4a \int_0^{\pi/2} \sqrt{1 - \epsilon^2 \sin^2 \psi}\; d\psi,$$

where

$$\epsilon = \frac{\sqrt{a^2 - b^2}}{a} < 1$$

is the eccentricity. (This result is an example of an elliptic integral; it cannot be integrated in terms of elementary functions.) [*Hint:* Use the substitution $t = \pi/2 - \psi$.]
*14. Show that the total arc length of the lemniscate $r^2 = 2a^2 \cos 2\theta$ is given by the elliptic integral

$$L = 4a \int_0^{\pi/2} \frac{d\psi}{\sqrt{1 - \tfrac{1}{2}\sin^2 \psi}}.$$

[*Hint:* Use the substitution $\sqrt{2}\sin\theta = \sin\psi$.]

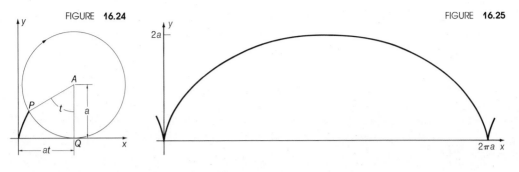

FIGURE 16.24

FIGURE 16.25

2.4 The cycloid‡

We conclude this section by defining an interesting curve, the **cycloid,** for which it is easier to find a parametric representation than a nonparametric one.

Consider a circle of radius a which rolls along a straight line without slipping or sliding. The cycloid is the trajectory described by a fixed point P on this circle.

To derive the parametric representation of the cycloid, we assume that the circle rolls along the x-axis and that the motion begins with P at the origin (see Figure 16.24). Assume that the circle traveled a certain distance at to the right, $0 \le t \le 2\pi$, and let Q be the point on the circle which is now on the "ground," that is, on the x-axis. Then the length of the circular arc between P and Q is precisely at; this is what is meant by saying that there is no slipping or sliding. (If $t > 2\pi$, the length of the arc is at minus a multiple of $2\pi a$.) Let A be the present position of the center. Clearly A has the coordinates (at,a) and $t = \angle QAP$. Thus the coordinates (x,y) of P are $x = at - a \sin t$ and $y = a - a \cos t$. Then we have the representation

$$x = a(t - \sin t), \qquad y = a(1 - \cos t). \tag{21}$$

The curve is shown in Figure 16.25. The points at which the cycloid reaches the x-axis are called *cusps.* They correspond to the parameter values $t = 0, \pm 2\pi, \pm 4\pi, \ldots$.

The cycloid was a favorite subject of mathematicians during the period when calculus was being created.

PROBLEMS

1. For every point Q on the upper unit semicircle (centered at the origin), a line is drawn from the origin, through Q, and meeting the line l whose equation is $y = 2$ at the point P. Let R be the midpoint of the line segment PQ. Find a parametric representation for the curve described by the point R.

2. For every point Q on the upper unit semicircle (centered at the origin), a line is drawn from the point P with Cartesian coordinates $(-1,0)$, through Q, and ending in the point R whose distance from Q is 1. Find a parametric representation for the curve described by the point R.

3. Consider a disk of radius a which rolls along a straight line without slipping or sliding. Suppose that this disk is glued face to face and concentrically to a second disk of radius b. Fix a point R on the circumference of the second disk. Find a parametric representation for the curve described by the point R. This curve is called a **trochoid.**

‡This subsection may be omitted without loss of continuity.

4. Suppose a string of length $2\pi a$ is unwound from around the circumference of a disk of radius a whose center is kept fixed. Find a parametric equation for the curve described by the free end of the string. This curve is called the **involute** of the circle.
5. Suppose a disk of radius b rolls along the outside of a fixed circle of radius a without slipping or sliding. Find the parametric representation for the curve described by a fixed point R on the circumference of the second disk. This curve is called an **epicycloid** and the special case $a = b$ is called a **cardioid.**
6. In Problem 5 suppose $b < a$ and the moving disk rolls along the inside of the circle. This curve is called a **hypocycloid.** Find its parametric representation.

§3 Conics, continued

In this section we return to the theory of conics begun in Chapter 2, §4. Our aim is twofold: to give a unified geometric definition for parabolas, ellipses, and hyperbolas and to write down a convenient equation for these curves in polar coordinates.

3.1 Eccentricity

In Chapter 2, §4, we derived for parabolas, ellipses, and hyperbolas the three **standard equations**

$$y^2 = 4ax, \qquad a \neq 0, \tag{P}$$

$$\frac{x^2}{a^2} + \frac{y^2}{b^2} = 1, \qquad a \geq b > 0, \tag{E}$$

$$\frac{x^2}{a^2} - \frac{y^2}{b^2} = 1, \qquad a > 0, b > 0. \tag{H}$$

(In the case of the parabola, we write a instead of p for convenience.) We recall that the ellipse (E) and the hyperbola (H) have two foci each, a left focus F_- and a right focus F_+:

$$F_- = (-c,0), \qquad F_+ = (c,0), \tag{1}$$

where the focal distance c is

$$c = \sqrt{a^2 - b^2}, \qquad \text{for the ellipse,} \tag{2}$$

$$c = \sqrt{a^2 + b^2}, \qquad \text{for the hyperbola.} \tag{3}$$

(The two foci of the ellipse coincide with the center if $a = b$, that is, if the ellipse is a circle.) We recall also that the parabola (P) has one focus,

$$F = (a,0); \tag{4}$$

it will be convenient to call it a left focus.

The eccentricity ϵ is the number defined by the relation

$$\epsilon = \frac{c}{a}. \tag{5}$$

Thus, for an ellipse, $\epsilon = \sqrt{1 - (b/a)^2}$. Hence

$$0 \leq \epsilon < 1, \qquad \text{for an ellipse}$$

FIGURE **16.26**

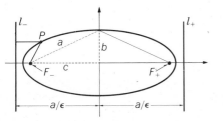

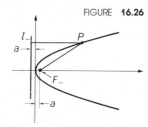

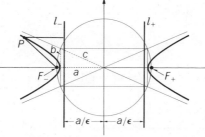

(and $\epsilon = 0$ for a circle). Similarly, for a hyperbola, $\epsilon = \sqrt{1 + (b/a)^2}$. Hence

$$1 < \epsilon, \qquad \text{for a hyperbola}$$

(with $\epsilon = \sqrt{2}$ for an equilateral hyperbola). The relation

$$\epsilon = 1, \qquad \text{for a parabola}$$

holds, by definition.
    The lines

$$l_-: x = -\frac{a}{\epsilon}, \qquad l_+: x = \frac{a}{\epsilon} \tag{6}$$

are called the left and right **directrices** of the ellipse (E) or hyperbola (H). The circle has no directrices. We recall that the line

$$l: x = -a \tag{7}$$

is called the directrix of the parabola (P). It will be convenient to call it the left directrix. The three kinds of conics, with their foci and directrices are shown in Figure 16.26.

Theorem 1. *For a conic of eccentricity $\epsilon > 0$ (that is, not a circle), the distance from a point P on the curve to a focus is ϵ times the distance from P to a directrix (right focus and right directrix or left focus and left directrix).*

    Proof. Let P be a point on the conic. We assume that the conic is given by one of the equations (P), (E), and (H), and that P has coordinates (x,y). From Figure 16.26 we see that

$$\text{distance from } P \text{ to } l_- = \left| \frac{a}{\epsilon} + x \right|, \qquad \text{distance from } P \text{ to } l_+ = \left| \frac{a}{\epsilon} - x \right|. \tag{8}$$

To prove the theorem we must show that

$$|F_- P| = \epsilon \left| \frac{a}{\epsilon} + x \right| = |a + \epsilon x|, \qquad |F_+ P| = \epsilon \left| \frac{a}{\epsilon} - x \right| = |a - \epsilon x|. \tag{9}$$

We prove the first relation (9) only for an ellipse. The proof for a hyperbola is similar, that for a parabola even simpler. These proofs, as well as the analogous proof of the second relation (9), are left to the reader.

The proof for an ellipse proceeds as follows. If $P = (x,y)$ satisfies (E), then $y^2 = b^2 - (b^2/a^2)x^2$, and therefore

$$|F_- P|^2 = (x + c)^2 + y^2 = x^2 + 2cx + c^2 + b^2 - \frac{b^2}{a^2}x^2$$

$$= x^2 \left(1 - \frac{b^2}{a^2}\right) + 2cx + a^2 \qquad \text{since } c^2 + b^2 = a^2$$

$$= x^2 \frac{a^2 - b^2}{a^2} + 2cx + a^2 = a^2 + 2cx + \frac{c^2}{a^2}x^2$$

$$= \left|a + \frac{c}{a}x\right|^2 = |a + \epsilon x|^2 \qquad \text{since } c = a\epsilon.$$

This is the assertion to be established.

EXAMPLE   Find the foci, eccentricity, and directrices for the hyperbola $16x^2 - 25y^2 = 400$.
ANSWER   Divide both sides of the given equation by 400. We obtain $x^2/9 - y^2/16 = 1$, which is in the standard form (H) with $a = 3$, $b = 4$. Using Equation (2), $c = \sqrt{a^2 + b^2} = \sqrt{9 + 16} = 5$, and so the foci are at $(\pm 5,0)$. Using Equations (5) and (6), the eccentricity and directrices are given by $\epsilon = c/a = 5/3$ and $x = \pm a/\epsilon = \pm 9/5$.

REMARK   If we interchange x and y in (P), (E), and (H), we obtain the equations

$$x^2 = 4ay, \qquad a \neq 0, \tag{P'}$$

$$\frac{x^2}{b^2} + \frac{y^2}{a^2} = 1, \qquad a \geq b > 0, \tag{E'}$$

$$\frac{y^2}{a^2} - \frac{x^2}{b^2} = 1, \qquad a > 0, b > 0. \tag{H'}$$

These represent a parabola, ellipse, and hyperbola, respectively, with foci lying on the y-axis. The parabola has a focus at $(0,a)$, directrix $y = -a$, and eccentricity $\epsilon = 1$. The ellipse and hyperbola have foci at $(0,\pm c)$, directrices $y = \pm a/\epsilon$, and eccentricity $\epsilon = c/a$. The focal distance c for an ellipse is given by Equation (2), for a hyperbola by Equation (3).

PROBLEMS

In each of Problems 1 to 6 find the foci, eccentricity, and directrices for the given conic.

1. $9x^2 + 25y^2 - 225 = 0$.

2. $\dfrac{x^2}{9} + \dfrac{y^2}{4} = 1$.

3. $16x^2 - 9y^2 = 144$.

4. $\dfrac{x^2}{16} - \dfrac{y^2}{4} = 1$.

5. $\dfrac{x^2}{25} + \dfrac{y^2}{169} = 1$.

6. $\dfrac{y^2}{25} - \dfrac{x^2}{144} = 1$.

*7. An ellipse with foci along the x-axis has eccentricity ϵ and vertices located at $(\pm 1,0)$. Write the equation of this ellipse.

*8. An ellipse has foci located at $(\pm 1,0)$ and eccentricity ϵ. Write the equation of this ellipse.

3.2   A unified definition of conics

Theorem 1 has a valid converse.

Theorem 2. *Let l be a line, F a point not on l, and ϵ a positive number. The set of all points P such that*

$$|PF| = \epsilon \text{ times the distance from } P \text{ to } l$$

is a conic of eccentricity ϵ, with focus F and directrix l.

We assume $\epsilon > 1$. The proof for $\epsilon < 1$ is similar, the one for $\epsilon = 1$ even simpler. We choose a coordinate system such that F is the origin and l is vertical and intersects the positive ray of the x-axis at a point $x = -\alpha$, $\alpha > 0$ (see Figure 16.27).

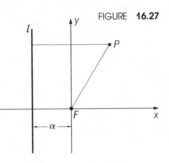

FIGURE **16.27**

If $P = (x,y)$, then the distance from P to l is $|x + \alpha|$ and $|FP| = \sqrt{x^2 + y^2}$. If $\sqrt{x^2 + y^2} = \epsilon|x + \alpha|$, then

$$x^2 + y^2 = \epsilon^2(x + \alpha)^2 = \epsilon^2 x^2 + 2\alpha\epsilon^2 x + \epsilon^2\alpha^2$$

or

$$(\epsilon^2 - 1)x^2 + 2\alpha\epsilon^2 x - y^2 + \epsilon^2\alpha^2 = 0;$$

that is,

$$(\epsilon^2 - 1)\left(x^2 + \frac{2\alpha\epsilon^2}{\epsilon^2 - 1}x + \frac{\alpha^2\epsilon^4}{(\epsilon^2 - 1)^2}\right) - y^2 - \frac{\alpha^2\epsilon^4}{\epsilon^2 - 1} + \epsilon^2\alpha^2 = 0,$$

or

$$(\epsilon^2 - 1)\left(x + \frac{\alpha\epsilon^2}{\epsilon^2 - 1}\right)^2 - y^2 = \frac{\alpha^2\epsilon^2}{\epsilon^2 - 1}.$$

We translate the coordinates and set

$$\xi = x + \frac{\alpha\epsilon^2}{\epsilon^2 - 1}, \qquad \eta = y. \tag{10}$$

Our equation becomes (in the new "translated" coordinates ξ,η)

$$(\epsilon^2 - 1)^2\xi^2 - \eta^2 = \frac{\alpha^2\epsilon^2}{\epsilon^2 - 1}$$

or

$$\frac{(\epsilon^2 - 1)^2}{\alpha^2\epsilon^2}\xi^2 - \frac{\epsilon^2 - 1}{\alpha^2\epsilon^2}\eta^2 = 1. \tag{11}$$

This is a standard equation of a hyperbola. To show that all points of this hyperbola have the desired property, we must verify that the eccentricity is ϵ and that F and l are the focus and directrix, respectively.

From (11) we see that the semiaxes are

$$a = \frac{\alpha\epsilon}{\epsilon^2 - 1}, \qquad b = \frac{\alpha\epsilon}{\sqrt{\epsilon^2 - 1}}.$$

Therefore

$$c^2 = a^2 + b^2 = \alpha^2\epsilon^2\left(\frac{1}{(\epsilon^2 - 1)^2} + \frac{1}{\epsilon^2 - 1}\right) = \frac{\alpha^2\epsilon^4}{(\epsilon^2 - 1)^2}.$$

Hence $c = \alpha\epsilon^2/(\epsilon^2 - 1) = \alpha\epsilon$, so that ϵ is indeed the eccentricity. The foci lie on the ξ-axis, which is also the x-axis. They have the ξ-coordinates $\xi = \pm c = \pm\alpha\epsilon^2/(\epsilon^2 - 1)$. Hence the x-coordinates of the foci are [compare (10)] $x = \pm\alpha\epsilon^2/(\epsilon^2 - 1) - \alpha\epsilon^2/(\epsilon^2 - 1)$. Hence $F = (0,0)$ is the right focus. Finally, the right directrix of (11) is the line

$$\xi = \frac{a}{\epsilon} = \frac{\alpha}{\epsilon^2 - 1},$$

that is,

$$x = \frac{\alpha}{\epsilon^2 - 1} - \frac{\alpha\epsilon^2}{\epsilon^2 - 1} = -\alpha.$$

But this is the line l with which we started. This completes the argument.

PROBLEMS

1. Suppose the distance from the point $P = (x,y)$ to the directrix $x = 6$ is twice the distance from the point P to the focus $(3,4)$. What is the equation of the conic?
2. Suppose the distance from the point $P = (x,y)$ to the directrix $x = 2$ is half the distance from the point P to the focus $(5,-3)$. What is the equation of the conic?

3.3 Conics in polar coordinates

The theory of conics as developed by the Greeks was a branch of "pure mathematics," although it was probably motivated by observing sun dials. Two thousand years later mathematicians discovered that conics had many applications in natural sciences and engineering. The most spectacular of these is to planetary motion, which we shall consider in Chapter 17, §5. As a preparation, we derive here the equations of the conics in polar coordinates.

Consider first a conic of eccentricity $\epsilon > 0$, that is, not a circle. Let O be a focus of the conic and l the corresponding directrix. We chose a Cartesian coordinate system with O as the origin, with the axis of the conic (the major axis in the case of an ellipse) as the x-axis, and with the positive x-direction chosen so that the equation of l reads $x = -p$, where p is the distance from O to l. See Figure 16.28. A point P with Cartesian coordinates (x,y) has distance $\delta = |x - (-p)| = |x + p|$ from the directrix l. If P has polar coordinates (r,θ), then $x = r \cos\theta$ and

$$\delta = |r \cos\theta + p|.$$

FIGURE **16.28**

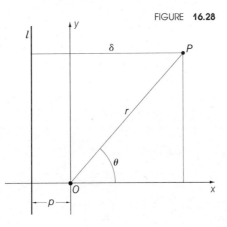

The distance from P to O is r. By Theorems 1 and 2, P lies on our conic if and only if $r = \epsilon\delta$. If $r\cos\theta + p > 0$, then $\delta = r\cos\theta + p$ and the condition $r = \epsilon\delta$ becomes

$$r = \epsilon(r\cos\theta + p). \tag{12}$$

If $r\cos\theta + p < 0$, then $\delta = -r\cos\theta - p$ and the condition $r = \epsilon\delta$ becomes

$$r = -\epsilon(r\cos\theta + p). \tag{13}$$

Solving (12) and (13) for r, we obtain

$$r = \frac{\epsilon p}{1 - \epsilon\cos\theta} \tag{14}$$

or

$$r = \frac{-\epsilon p}{1 + \epsilon\cos\theta}. \tag{15}$$

Now we recall the convention made in §1.1. According to this convention, we can replace, in (15), r by $(-r)$ and θ by $(\theta + \pi)$. This amounts to replacing the numerator $(-\epsilon p)$ by (ϵp) and the denominator $1 + \epsilon\cos\theta$ by $1 - \epsilon\cos\theta$. Then (15) becomes identical with (14).

Theorem 3.   *The equation, in suitable polar coordinates, of a conic of eccentricity ϵ, with a focus at O, reads*

$$r = \frac{A}{1 - \epsilon\cos\theta}, \tag{16}$$

with a positive constant A. Conversely, (16) *is always the equation of a conic.*

We have just proved this for a noncircle [set $A = \epsilon p$ in (14)]. For $\epsilon = 0$, Equation (16) reads $r = A$. This is, of course, the equation of a circle.

REMARK 1   For an ellipse $(0 < \epsilon < 1)$, the denominator in (16) is never 0.

For a parabola $(\epsilon = 1)$, the denominator is never negative. It is 0 for $\cos\theta = 1$. For $\theta = 0$ is the polar angle of points on the axis, far away from the focus, and there are no points on the parabola with this polar angle.

For the hyperbola ($\epsilon > 1$), the denominator is sometimes negative, sometimes positive; this corresponds to the two branches of the hyperbola. The denominator is 0 for $\cos \theta = 1/\epsilon$; the two rays with $\theta = \pm \arccos 1/\epsilon$ are parallel to the asymptotes and do not meet the curve.

REMARK 2   By changing the unit of length, we can replace the numerator A in (16) by 1. We conclude that *the shape of a conic depends only on the eccentricity* ϵ. Two conics with the same eccentricity are similar.

EXAMPLE   Identify the conic whose equation in polar coordinates is $r = \dfrac{8}{2 - 6 \cos \theta}$.

ANSWER   Divide the numerator and denominator by 2. Then

$$r = \frac{4}{1 - 3 \cos \theta},$$

which is in the form of Equation (14) with $\epsilon = 3$, $\epsilon p = 4$. Since $\epsilon > 1$, the conic is a hyperbola. The equation of the directrix is $x = -\frac{4}{3}$, that is, the directrix is $\frac{4}{3}$ units to the left of the pole (focus) and perpendicular to the polar axis.

PROBLEMS

In each of Problems 1 to 6 find the eccentricity and identify the conic as an ellipse, hyperbola, or parabola. Sketch the curve.

1. $r = \dfrac{1}{1 - \cos \theta}$.

2. $r = \dfrac{3}{2 - 2 \cos \theta}$.

3. $r = \dfrac{5}{3 - 2 \cos \theta}$.

4. $r = \dfrac{8}{3 - \cos \theta}$.

5. $r = \dfrac{3}{1 - 2 \cos \theta}$.

6. $r = \dfrac{8}{3 - 5 \cos \theta}$.

7. Show that the asymptotes of the hyperbola $r = \dfrac{\epsilon p}{1 - \epsilon \cos \theta}$, $\epsilon > 1$, pass through the point $\left(\dfrac{\epsilon^2 p}{\epsilon^2 - 1}, \pi \right)$ and have slopes given by $\pm \sqrt{\epsilon^2 - 1}$.

8. If the focus of a conic is at the origin, and the directrix is perpendicular to the x-axis and a distance p to the *right* of the origin, show that the equation of the conic is $r = \dfrac{\epsilon p}{1 + \epsilon \cos \theta}$.

9. Show that the equation of a conic, with focus at the origin and directrix parallel to the x-axis and distance p *below* it, is $r = \dfrac{\epsilon p}{1 - \epsilon \sin \theta}$.

10. Show that the equation of a conic, with focus at the origin and directrix parallel to the x-axis and distance p *above* it, is $r = \dfrac{\epsilon p}{1 + \epsilon \sin \theta}$.

§4   Change of coordinates‡

A Cartesian coordinate system in the plane is determined by choosing a point O and two mutually perpendicular directions: the x-direction and the y-direction. In this

‡Optional section.

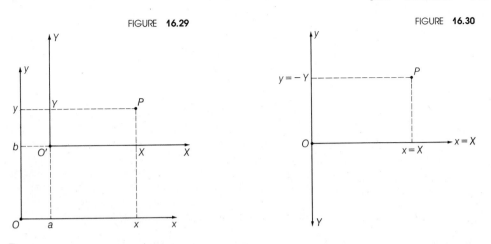

FIGURE **16.29**   FIGURE **16.30**

section we discuss what happens when we change the coordinate system. The result of this discussion will be applied to conics.

4.1 Translations and reflections

In Chapter 2, §1.2, we described what happens when we **translate** the axes, that is, when we go over from a system with origin O to a "new" system with origin O' and the *same* direction of axes; see Figure 16.29. We recall that the old coordinates (x,y) of a point P and the new (X,Y) coordinates of the same point are connected by the formulas

$$x = a + X, \qquad y = b + Y, \tag{1}$$

where (a,b) are the coordinates of the new origin O' in the old system.

Another way of changing a coordinate system is to leave the origin O where it is, not to change the direction of the x-axis but to reverse the y-direction; see Figure 16.30. In this case we say that the new system (coordinates X and Y) has been obtained from the old (coordinates x and y) by **reflection.** The formula connecting the old and new coordinates are

$$x = X, \qquad y = -Y, \tag{2}$$

as is seen from the figure.

(Let us call the coordinate system with which we start right-handed; see Chapter 10, §1.2. Then a new coordinate system obtained by a translation is again right-handed, whereas a new coordinate system obtained by reflection, or by reflection followed by a translation, is left-handed.)

4.2 Rotations

Recall first that to every Cartesian coordinate system there belongs a polar coordinate system (see §1.1 and Figure 16.1). Consider next two Cartesian coordinate systems that have the *same* origin but different directions of the axes (Figure 16.31). Let (x,y) be the Cartesian coordinates of a point P in the old system, (r,θ), the polar coordinates of P in the old system. Also, let (X,Y) be the new Cartesian coordinates and (R,Θ)

FIGURE 16.31

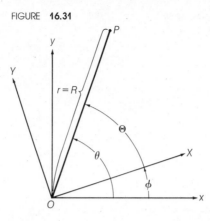

the new polar coordinates of P. We say that the new system has been obtained from the old by a **rotation** by the angle ϕ if

$$r = R, \qquad \theta = \Theta + \phi. \tag{3}$$

Let us derive the formulas connecting the old and the new polar coordinates. We have (see §1.1)

$$x = r \cos \theta, \qquad y = r \sin \theta \tag{4}$$

and analogously

$$X = R \cos \Theta, \qquad Y = R \sin \Theta. \tag{5}$$

Therefore, by (3), (4), and (5),

$$x = R \cos(\Theta + \phi) = R \cos \Theta \cos \phi - R \sin \Theta \sin \phi = X \cos \phi - Y \sin \phi,$$
$$y = R \sin(\Theta + \phi) = R \sin \Theta \cos \phi + R \cos \Theta \sin \phi = X \cos \phi + Y \sin \phi.$$

Thus

$$x = X \cos \phi - Y \sin \phi, \qquad y = X \sin \phi + Y \cos \phi. \tag{6}$$

(A rotation of a coordinate system takes a right-handed system into a right-handed one. A rotation by $0° = 0$ or by $360° = 2\pi$ leaves the coordinate system unchanged. A rotation by α followed by a rotation by β amounts to a rotation by $\alpha + \beta$. These statements are geometrically obvious and could be proved analytically. In Figure 16.32 we show coordinate systems obtained from a given one by various rotations.)

FIGURE 16.32

$\phi = 0 \qquad \phi = 45° \qquad \phi = 135° \qquad \phi = 180° \qquad \phi = 225° \qquad \phi = 315°$

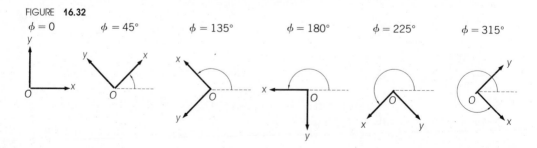

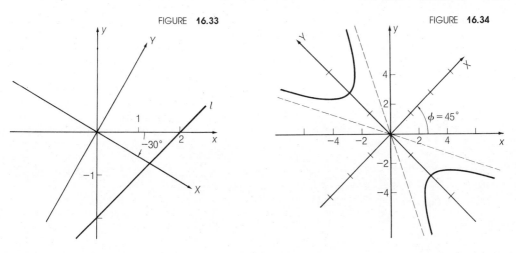

FIGURE **16.33**

FIGURE **16.34**

EXAMPLES 1. A line l has the equation $x - y - 2 = 0$. What is the equation of l in a coordinate system obtained by rotating the x,y system by $(-30°)$?

ANSWER   Call the new coordinate X, Y. They are connected with the old coordinates by Equations (6) with $\phi = -30°$. Since $\cos(-30°) = \frac{1}{2}\sqrt{3}$, $\sin(-30°) = -\frac{1}{2}$, we obtain

$$x = \tfrac{1}{2}\sqrt{3}\,X + \tfrac{1}{2}Y, \qquad y = -\tfrac{1}{2}X + \tfrac{1}{2}\sqrt{3}\,Y.$$

Substituting this into the equation $x - y - 2 = 0$, we obtain

$$(\tfrac{1}{2}\sqrt{3}X + \tfrac{1}{2}Y) - (-\tfrac{1}{2}X + \tfrac{1}{2}\sqrt{3}Y) - 2 = 0.$$

Multiplying, collecting terms, and simplifying, we obtain the "new" equation for l:

$$(1 + \sqrt{3})X + (1 - \sqrt{3})Y - 4 = 0.$$

The coordinate systems and l are shown in Figure 16.33.

2. Find the equation of the curve $3x^2 + 10xy + 3y^2 + 32 = 0$ in a coordinate system obtained by rotating the old system by $45°$.

ANSWER   Call the new coordinates X, Y. Using Equations (6) with $\phi = 45°$, we have

$$x = X \cos 45° - Y \sin 45° = \frac{1}{\sqrt{2}}(X - Y), \qquad y = X \sin 45° + Y \cos 45° = \frac{1}{\sqrt{2}}(X + Y).$$

Substituting the above into the given equation, we obtain

$$3\tfrac{1}{2}(X - Y)^2 + 10\tfrac{1}{2}(X - Y)(X + Y) + 3\tfrac{1}{2}(X + Y)^2 + 32 = 0,$$

or

$$3(X^2 - 2XY + Y^2) + 10(X^2 - Y^2) + 3(X^2 + 2XY + Y^2) + 64 = 0,$$

or

$$16X^2 - 4Y^2 + 64 = 0,$$

or

$$\frac{Y^2}{16} - \frac{X^2}{4} = 1.$$

In the new system we recognize the equation as a hyperbola with focus along the Y-axis. The vertices and asymptotes in the new system are $(0, \pm 4)$ and $Y = \pm 2X$, respectively. The coordinate systems and hyperbola are shown in Figure 16.34.

PROBLEMS

In each of Problems 1 to 12 the equation of a curve is given in terms of the (old) Cartesian coordinate system (x,y). Find the equation of the curve in the (new) Cartesian coordinate system (X,Y) obtained by rotating the old system through the indicated angle ϕ. Draw the old and new coordinate axes, and sketch the curve.

1. $x - y - 4 = 0$, $\phi = 90°$.
2. $x + 2y + 4 = 0$, $\phi = 180°$.
3. $4x^2 + 9y^2 = 36$, $\phi = -90°$.
4. $4x^2 - 9y^2 = 36$, $\phi = 180°$.
5. $4x^2 + y = 0$, $\phi = 270°$.
6. $3x + 5y - 4\sqrt{2} = 0$, $\phi = 45°$.
7. $x + \sqrt{3}y - 4 = 0$, $\phi = -30°$.
8. $13x^2 - 10xy + 13y^2 - 72 = 0$, $\phi = 45°$.
9. $6x^2 + 4\sqrt{3}xy + 2y^2 + x - \sqrt{3}y = 0$, $\phi = 30°$.
10. $47x^2 - 34\sqrt{3}xy + 13y^2 + 64 = 0$, $\phi = -30°$.
11. $x^2 + 2xy + y^2 - \sqrt{2}x + \sqrt{2}y = 0$, $\phi = -45°$.
12. $13x^2 + 6\sqrt{3}xy + 7y - 64 = 0$, $\phi = 30°$.

4.3   Second-degree curves

A quadratic equation in two variables, x and y, is an equation of the form

$$Ax^2 + 2Bxy + Cy^2 + Dx + Ey + F = 0, \tag{7}$$

where $A, B, \ldots, F$ are some numbers. In such an equation, we call the terms Ax^2, $2Bxy$ (the mixed term), and Cy^2 quadratic terms, the terms Dx and Ey linear terms, and F the constant term. The coefficient of the mixed term is called $2B$ for the sake of convenience. We assume that A, B, and C are not all three equal to 0.

A second-degree curve is the set of points whose Cartesian coordinates (x,y) satisfy a quadratic equation in x and y. Conics are second-degree curves, since the standard equations, (P), (E), and (H) in §2.1, are special cases of (7).

A second-degree curve may also be empty (for instance, the equation $x^2 + 1 = 0$ defines the empty set), or it may consist of a single point (for instance, $x^2 + y^2 = 0$), a single line (for instance, $x^2 = 0$), two parallel lines (for instance, $x^2 - 1 = 0$), or two intersecting lines (for instance, $xy = 0$). We call such curves **degenerate.**

It is a remarkable fact that a nondegenerate second-degree curve is always a conic. As a first step in proving this we establish

Lemma 1.   *Consider a quadratic equation in x and y without a mixed term:*

$$Ax^2 + Cy^2 + Dx + Ey + F = 0. \tag{8}$$

If $AC > 0$, the equation represents an ellipse (which may be a circle), or a point, or the empty set.

If $AC < 0$, the equation represents a hyperbola or two intersecting lines.

If $AC = 0$, the equation represents a parabola, or two parallel lines, or a line, or the empty set.

To prove the lemma we shall simplify Equation (8) by completing squares and translating the coordinate system.

Assume first that $AC \neq 0$. Then $A \neq 0$, $C \neq 0$, and we may write

$$Ax^2 + Dx = A\left(x^2 + \frac{D}{A}x\right) = A\left(x + \frac{D}{2A}\right)^2 - \frac{D^2}{4A^2}, \tag{9}$$

$$Cy^2 + Ey = C\left(y^2 + \frac{E}{C}y\right) = C\left(y + \frac{E}{2C}\right)^2 - \frac{E^2}{4C^2}. \tag{10}$$

Equation (8) may be rewritten as

$$A\left(x + \frac{D}{2A}\right)^2 + C\left(y + \frac{E}{2C}\right)^2 = \frac{D^2}{4A^2} + \frac{E^2}{4C^2} - F. \tag{11}$$

We make a translation of the coordinate system by introducing new variables X, Y such that

$$x + \frac{D}{2A} = X, \qquad y + \frac{E}{2C} = Y.$$

We also set

$$\frac{D}{4A^2} + \frac{E}{4C^2} - F = G.$$

Now (11) becomes

$$AX^2 + CY^2 = G. \tag{11'}$$

If $G = 0$ and $AC > 0$, then (11') represents the point $X = Y = 0$.

If $G = 0$ and $AC < 0$, then (11') may be rewritten as $Y^2 = (-C/A)X^2$; this represents two lines: $Y = \sqrt{-C/A}\,X$ and $Y = -\sqrt{-C/A}\,Y$.

If $G \neq 0$, we divide Equation (11') by G or by $-G$ according to whether $A > 0$ or $A < 0$ and obtain an equation

$$\alpha X^2 + \beta Y^2 = 1, \tag{11''}$$

where $\alpha > 0$, and $\beta > 0$ for $AC > 0$, $\beta < 0$ for $AC < 0$. This may be written as either

$$\frac{X^2}{a^2} + \frac{Y^2}{b^2} = 1 \quad \left(a = \frac{1}{\sqrt{\alpha}}, b = \frac{1}{\sqrt{\beta}}\right) \qquad \text{if } AC > 0,$$

or as

$$\frac{X^2}{a^2} - \frac{Y^2}{b^2} = 1 \quad \left(a = \frac{1}{\sqrt{\alpha}}, b = \frac{1}{\sqrt{|\beta|}}\right) \qquad \text{if } AC < 0,$$

and represents either an ellipse or a hyperbola.

It remains to consider the case $AC = 0$. Then either $A = 0$ and $C \neq 0$ or $A \neq 0$ and $C = 0$ (A and C are not both 0, by hypothesis). We consider only the first case and leave the second to the reader.

If $A = 0$, $C \neq 0$, we use (10) and rewrite (8) as

$$C\left(y + \frac{E}{C}\right)^2 + Dx = \frac{E^2}{4C^2} - F \tag{12}$$

or, dividing the equation by C and setting

$$-\frac{D}{C} = H, \qquad \frac{E^2}{4C^3} - \frac{F}{C} = K,$$

as

$$\left(y + \frac{E}{C}\right)^2 = Hx + K. \tag{12'}$$

If $H = 0$, (12') represents the empty set (for $K < 0$), the line $y = -E/C$ (for $K = 0$), the two parallel lines $y = -E/C + \sqrt{K}$, $y = -E/C - \sqrt{K}$ (for $K > 0$). If $H \neq 0$, we translate the coordinates and introduce the new variables

$$X = x + \frac{K}{H}, \qquad Y = y + \frac{E}{C}.$$

Equation (12') becomes

$$Y^2 = HX;$$

it represents a parabola.

This proves the lemma. One should remember the method, not the formulas!

EXAMPLE   What curve is described by the equation

$$4x^2 - 9y^2 - 40x - 72y - 44 = 0?$$

ANSWER   We rewrite the equation as $4(x^2 - 10x) - 9(y^2 + 8y) - 44 = 0$,

that is,

$$4(x - 5)^2 - 9(y + 4)^2 = 0.$$

Set

$$x - 5 = X, \qquad y + 4 = Y$$

(this amounts to a translation of coordinates). Our equation becomes

$$4X^2 - 9Y^2 = 0 \qquad \text{or} \qquad (2X - 3Y)(2X + 3Y) = 0.$$

It represents the two lines $2X - 3Y = 0$, $2X + 3Y = 0$, that is, the two lines $2x - 3y - 22 = 0$ and $2x + 3y + 2 = 0$. See Figure 16.35.

FIGURE **16.35**

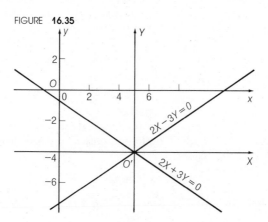

PROBLEMS

In each of the following problems, identify the given conic, and make a sketch. (Two main steps are involved: first, complete the square; second, introduce an appropriate translation of axes that reduces the conic to standard form.)

1. $25x^2 - 4y^2 + 50x - 8y + 21 = 0.$
2. $3x^2 - 24x - y + 50 = 0.$
3. $x^2 + 4y^2 + 8x - 16y + 16 = 0.$
4. $4x^2 - 16y^2 + 8x - 8y + 3 = 0.$
5. $2y^2 + x - 12y + 21 = 0.$
6. $x^2 - 4y^2 + 4x + 8y - 4 = 0.$
7. $4x^2 - 9y^2 - 16x + 36y - 56 = 0.$
8. $9x^2 + 4y^2 - 18x + 16y - 11 = 0.$
9. $4x^2 + y^2 + 16x - 8y + 36 = 0.$
10. $4x^2 + y^2 - 24x - 10y + 61 = 0.$

4.4 The main theorem on second-degree curves

We return to the general quadratic equation (7) $Ax^2 + 2Bxy + Cy^2 + Dx + Ey + F = 0$ and investigate how the equation changes when we rotate the coordinate system by an angle ϕ. According to §3.2, this amounts to introducing new coordinates (X,Y) by the formulas

$$x = X \cos \phi - Y \sin \phi, \qquad y = X \sin \phi + Y \cos \phi.$$

Therefore

$$x^2 = X^2 \cos^2 \phi - 2XY \cos \phi \sin \phi + Y^2 \sin^2 \phi,$$
$$xy = X^2 \cos \phi \sin \phi + XY(\cos^2 \phi - \sin^2 \phi) - Y^2 \sin \phi \cos \phi,$$
$$y^2 = X^2 \sin^2 \phi + 2XY \sin \phi \cos \phi + Y^2 \cos^2 \phi.$$

Substituting into (7), we obtain the equation in the form

$$\widehat{A}X^2 + 2\widehat{B}XY + \widehat{C}Y^2 + \widehat{D}X + \widehat{E}Y + \widehat{F} = 0, \tag{13}$$

where

$$\begin{aligned}
\widehat{A} &= A \cos^2 \phi + 2B \sin \phi \cos \phi + C \sin^2 \phi, \\
\widehat{B} &= (C - A) \sin \phi \cos \phi + B(\cos^2 \phi - \sin^2 \phi), \\
\widehat{C} &= A \sin^2 \phi - 2B \sin \phi \cos \phi + C \cos^2 \phi, \\
\widehat{D} &= D \cos \phi + E \sin \phi, \\
\widehat{E} &= -D \sin \phi + E \cos \phi, \\
\widehat{F} &= F.
\end{aligned} \tag{14}$$

We assume that $B \neq 0$ and rewrite the formula (14) for $\widehat{B}$ (pronounced "B hat") as

$$2\widehat{B} = (C - A) \sin 2\phi + 2B \cos 2\phi.$$

We want to choose ϕ so as to have $\widehat{B} = 0$. If $A \neq C$, then we must have

$$\tan 2\phi = \frac{2B}{A - C}. \tag{15}$$

In this case, there will be four values ϕ with $\widehat{B} = 0$, namely,

$$\phi_0 = \frac{1}{2} \arctan \frac{2B}{A - C}, \qquad \phi_0 + 90°, \qquad \phi_0 + 180°, \qquad \text{and} \qquad \phi_0 + 270°. \tag{16}$$

If $A = C$ and $B \neq 0$, then $\widehat{B} = 0$ if $\cos 2\phi = 0$, that is, if ϕ has one of the values $45°$, $135°$, $225°$, or $315°$.

Thus we have proved

Lemma 2. *By a rotation of the coordinate system, a second-order equation (7) can be brought into a form with no mixed term (no term with xy).*

Combining this with Lemma 1 we obtain

Theorem 1. *The set of points (x,y) satisfying a second-degree equation (7) is either empty, or a point, or consists of one or two lines, or it is a parabola, an ellipse, or a hyperbola.*

The procedure described in the proofs of Lemmas 1 and 2 is called "reducing a second-degree equation to standard form." The first step is rotating the coordinate system so as to get rid of the mixed term; this is the most important step. Then we remove as many linear and constant terms as possible by completing the square and translating the coordinate systems. Finally, we simplify the equation by transposing terms and multiplying the equation by a constant.

REMARK   To find the sine and cosine of the rotation angle ϕ which gets rid of the mixed term, we don't need any tables. If $A = C$, then $\phi = 45°$ will do. If $A \neq C$, we find $\cos \phi$ and $\sin \phi$ as follows: set

$$t = \frac{2B}{A - C}, \qquad s = \sqrt{1 + t^2}; \tag{17}$$

then

$$\cos \phi = \sqrt{\frac{s + 1}{2s}}, \qquad \sin \phi = \sqrt{\frac{s - 1}{2s}} \text{ if } t > 0, \qquad \sin \phi = -\sqrt{\frac{s - 1}{2s}} \text{ if } t < 0. \tag{18}$$

Proof.   We rewrite (18) as

$$\cos \phi = \sqrt{\frac{s + 1}{2s}}, \qquad \sin \phi = \frac{t}{|t|} \sqrt{\frac{s - 1}{2s}}. \tag{19}$$

There is a number ϕ such that (19) holds, since

$$\left(\sqrt{\frac{s + 1}{2s}}\right)^2 + \left(\sqrt{\frac{s - 1}{2s}}\right)^2 = 1.$$

Now, by (19) and (17),

$$\sin 2\phi = 2 \sin \phi \cos \phi = \frac{t}{|t|} \frac{\sqrt{s^2 - 1}}{s} = \frac{t}{|t|} \frac{\sqrt{t^2}}{s} = \frac{t}{s},$$

$$\cos 2\phi = \cos^2 \phi - \sin^2 \phi = \frac{1}{s},$$

so that $\tan 2\phi = t$, as required by (15).

[This way of finding ϕ amounts to choosing that value of the possible four values (16) for which $-45° < \phi < 45°$.]

EXAMPLES   1.   What curve is defined by the equation

$$x^2 - 2xy + y^2 - \frac{\sqrt{2}}{4}x - \frac{\sqrt{2}}{4}y = 0? \tag{*}$$

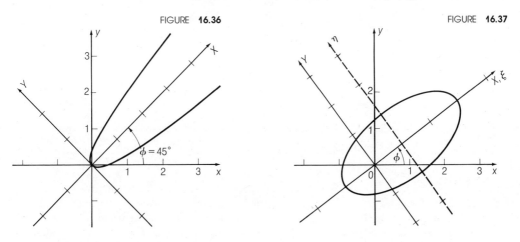

FIGURE **16.36**   FIGURE **16.37**

ANSWER   We shall reduce the equation to standard form. First we determine which rotation will eliminate the mixed term. We try to use (17) and (18) with $A = 1$, $2B = -2$, and $C = 1$ and find that t cannot be defined. We recall that for $A = C$, $\phi = 45°$ will do. Thus we set

$$x = X \cos 45° - Y \sin 45° = \frac{1}{\sqrt{2}}(X - Y),$$

$$y = X \sin 45° + Y \cos 45° = \frac{1}{\sqrt{2}}(X + Y).$$

Substitution into (*) yields

$$\tfrac{1}{2}(X - Y)^2 - 2\tfrac{1}{2}(X - Y)(X + Y) + \tfrac{1}{2}(X + Y)^2 - \tfrac{1}{4}(X - Y) - \tfrac{1}{4}(X + Y) = 0$$

or

$$\tfrac{1}{2}(X^2 - 2XY + Y^2) - (X^2 - Y^2) + \tfrac{1}{2}(X^2 + 2XY + Y^2) - \tfrac{1}{2}X = 0,$$

that is,

$$4Y^2 - X = 0.$$

The curve is a parabola. See Figure 16.36.

2.   What curve is defined by the equation

$$52x^2 - 72xy + 73y^2 - 50x - 30y - 75 = 0?$$ (**)

ANSWER   We shall reduce the equation to standard form. First we determine the required rotation, that is, $\sin \phi$ and $\cos \phi$. Since $A = 52$, $2B = -72$, $C = 73$, we have, by (17) and (18), $t = \tfrac{24}{7}$, $s = \tfrac{25}{7}$, $\cos \phi = \tfrac{4}{5}$, $\sin \phi = \tfrac{3}{5}$. (This means that $\phi = 36.9°$, approximately). We introduce new variables X, Y by setting [see (6)]

$$x = \tfrac{4}{5}X - \tfrac{3}{5}Y, \qquad y = \tfrac{3}{5}X + \tfrac{4}{5}Y.$$

We substitute into (**) [this is preferable to using formulas (14)] and obtain

$$52(\tfrac{4}{5}X - \tfrac{3}{5}Y)^2 - 72(\tfrac{4}{5}X - \tfrac{3}{5}Y)(\tfrac{3}{5}X + \tfrac{4}{5}Y)$$

$$+ 73(\tfrac{3}{5}X + \tfrac{4}{5}Y)^2 - 40(\tfrac{4}{5}X - \tfrac{3}{5}Y) - 30(\tfrac{3}{5}X + \tfrac{4}{5}Y) - 75 = 0,$$

$$25X^2 + 100Y^2 - 50X - 75 = 0,$$

or

$$X^2 - 2X + 4Y^2 - 3 = 0.$$

Completing the square, we write this as

$$(X - 1)^2 + 4Y^2 - 4 = 0.$$

Set

$$X - 1 = \xi, \qquad Y = \eta$$

(this amounts to a translation of the coordinate system) and write our equation in the form $\xi^2 - 4\eta^2 = 4$ or $(\xi^2/2^2) + \eta^2 = 1$.

Thus (**) represents an ellipse with semiaxes 2 and 1. The center of the ellipse is at $\xi = \eta = 0$; that is, at $X = 1$, $Y = 0$; that is, at $x = \frac{4}{5}$, $y = \frac{3}{5}$. See Figure 16.37.

PROBLEMS

In each of Problems 1 to 8 find a rotation of the given (x,y) Cartesian coordinate system that eliminates the mixed term of the given second-degree equation in x and y, and write down the equivalent equation in the new (X,Y) coordinate system. Then find a translation of the (X,Y) Cartesian coordinate system that eliminates the linear terms, and write down the equivalent equation in the final (ξ,η) coordinate system. Identify the curve, and make a sketch.

1. $7x^2 - 6\sqrt{3}xy + 13y^2 - 16\sqrt{3}x - 16y = 0$.
2. $10x^2 + 16xy + 10y^2 + 32\sqrt{2}x + 40\sqrt{2}y + 62 = 0$.
3. $11x^2 - 10\sqrt{3}xy + y^2 + 32\sqrt{3}x - 32y + 80 = 0$.
4. $5x^2 - 26xy + 5y^2 - 36\sqrt{2}x + 36\sqrt{2}y + 144 = 0$.
5. $x^2 + 2xy + y^2 - 3\sqrt{2}x - 5\sqrt{2}y + 12 = 0$.
6. $73x^2 + 72xy + 52y^2 + 30x - 40y - 75 = 0$.
7. $18x^2 - 48xy + 32y^2 + 80x - 65y = 0$.
8. $7x^2 - 48xy - 7y^2 - 20x + 140y - 100 = 0$.

4.5   The discriminant

Given a quadratic equation (7), the number

$$AC - B^2$$

is called the **discriminant.**

The discriminant is not changed by rotating the coordinate system. Indeed, if we rotate by an angle ϕ, we obtain the equation (13) with the coefficient given by (14). The "new" discriminant is

$$
\begin{aligned}
\widehat{AC} - \widehat{B}^2 &= A^2 \sin^2 \phi \cos^2 \phi - 2AB \sin \phi \cos^3 \phi + AC \cos^4 \phi \\
&\quad + 2AB \sin^3 \phi \cos \phi - 4B^2 \sin^2 \phi \cos^2 \phi + 2BC \sin \phi \cos^3 \phi \\
&\quad + AC \sin^4 \phi - 2BC \sin^3 \phi \cos \phi + C^2 \sin^2 \phi \cos^2 \phi \\
&\quad - (C^2 - 2AC + A^2) \sin^2 \phi \cos^2 \phi \\
&\quad - 2(BC - AB)(\sin \phi \cos^3 \phi - \sin^3 \phi \cos \phi) \\
&\quad - B^2(\cos^4 \phi - 2 \sin^2 \phi \cos^2 \phi + \sin^4 \phi) \\
&= AC(\cos^4 \phi + \sin^4 \phi + 2 \sin^2 \phi \cos^2 \phi) \\
&\quad - B^2(4 \sin^2 \phi \cos^2 \phi + \cos^4 \phi - 2 \sin^2 \phi \cos^2 \phi + \sin^4 \phi) \\
&= (AC - B^2)(\cos^2 \phi + \sin^2 \phi)^2 = AC - B^2.
\end{aligned}
$$

Using this remark and Lemmas 1 and 2, we can state a more precise form of Theorem 1.

Theorem 2. *A quadratic equation* (7) *represents an ellipse, a point, or the empty set if $AC - B^2 > 0$, a hyperbola or two intersecting lines if $AC - B^2 < 0$, a parabola, or two parallel lines, or one line, or the empty set if $AC - B^2 = 0$.*

EXAMPLES 1. What curve is represented by

$$11x^2 - 2xy + 2y^2 - 2 = 0?$$

ANSWER The discriminant is $11 \cdot 2 - 1 = 21 > 0$. We can find two points on the curve: $(0,1)$ and $(0,-1)$. Hence, by Theorem 2, the curve is an ellipse.

2. Identify the curve represented by $4x^2 + 5xy + y^2 + 3x - 1 = 0$.
ANSWER The discriminant is $(4)(1) - (\frac{5}{2})^2 = -\frac{9}{4} < 0$. By Theorem 2, the curve is either a hyperbola or two intersecting lines. To decide, we rewrite the given equation as a quadratic in y and solve. Thus

$$y^2 + (5x)y + (4x^2 + 3x - 1) = 0.$$

Solving,

$$y = \frac{-5x \pm \sqrt{(5x)^2 - 4(4x^2 + 3x - 1)}}{2} = \frac{-5x \pm \sqrt{9x^2 - 12x + 4}}{2}$$

$$= \frac{-5x \pm \sqrt{(3x - 2)^2}}{2} = \frac{-5x \pm (3x - 2)}{2},$$

that is, $y = -x - 1$ and $y = -4x + 1$. Hence the curve represents two intersecting lines.

PROBLEMS

In each of the following problems find the discriminant of the given second-degree equation. Identify the given equation as an ellipse, a hyperbola, a parabola, a point, one line, two lines, or the empty set. Do *not* sketch the curve.

1. $9x^2 + 6xy + y^2 - 24x - 8y + 16 = 0$.
2. $2x^2 - xy - y^2 - 2x + 5y - 4 = 0$.
3. $16x^2 - 8xy + y^2 - 4x + y - 2 = 0$.
4. $5x^2 + 6xy + 5y^2 - 20x - 12y - 12 = 0$.
5. $9x^2 - 6xy + y^2 + 12x - 4y + 6 = 0$.
6. $16x^2 + 24xy + 9y^2 - 65x - 80y + 105 = 0$.
7. $x^2 + 2\sqrt{3}xy - y^2 + \sqrt{3}x - y + 4 = 0$.
8. $5x^2 - 6xy + 5y^2 - 8x + 24y + 32 = 0$.

VECTORS AND CURVES
IN THE PLANE

§1 Algebra of vectors

This chapter introduces the language of vectors (in the plane) and applies vectors to geometry and mechanics. Space vectors will be considered in Chapter 18.

Vector analysis was invented in the nineteenth century. Its foundations were laid by Hamilton; the present form is due largely to the American mathematician Gibbs. At a first glance, vectors seem merely a special way of writing the formulas of analytic geometry. Yet the development of vector analysis had a far-reaching effect on science and mathematics.

Vectors are used to describe quantities "which have a magnitude and a direction." The prime example is a force acting on a particle (see §3.3 for this example).

1.1   Directed segments and vectors

Let P and Q be two points. They determine a **directed segment** leading from P to Q. We denote such a directed segment by the symbol $\overrightarrow{PQ}$, and we usually draw it as an arrow (Figure 17.1). The first point is called the **initial point,** the second the **terminal point.** The distance $|PQ|$ between the points P and Q is called the **length** or **magnitude** of $\overrightarrow{PQ}$. The length may be 0, since P and Q may coincide.

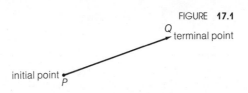

FIGURE **17.1**

Q terminal point

initial point P

The meaning of the statement "two directed segments have the same magnitude and direction, although perhaps different initial points" is illustrated in Figure 17.2. The four directed segments 2, 3, 4, 5 have the same magnitude and direction as 1, segment 6 has the same direction as 1 but a different magnitude, and 7 has the same magnitude as 1 but a different direction. A precise definition follows.

We say that the directed segments $\overrightarrow{PQ}$ and $\overrightarrow{RS}$ have the same magnitude and direction, and we write

$$\overrightarrow{PQ} \sim \overrightarrow{RS} \qquad (\text{read: } \overrightarrow{PQ} \text{ is } \textbf{equivalent to } \overrightarrow{RS}) \tag{1}$$

if the following conditions are satisfied: (i) $|PQ| = |RS|$, (ii) if $P \neq Q$, then the line through P and Q is parallel to or coincides with the line through R and S, (iii) $|PR| = |QS|$, and (iv) if $P \neq R$, then the line through P and R is parallel to or coincides with the line through Q and S.

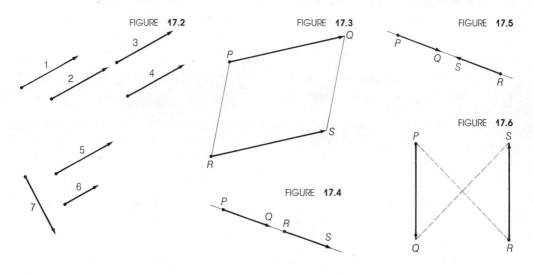

FIGURE **17.2**

FIGURE **17.3**

FIGURE **17.5**

FIGURE **17.6**

FIGURE **17.4**

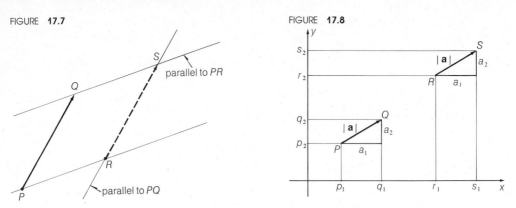

FIGURE 17.7

parallel to *PR*

parallel to *PQ*

FIGURE 17.8

[Conditions (i) and (ii) are illustrated in Figures 17.3 and 17.4. Condition (iii) is needed so that the two directed segments in Figure 17.5 are *not* considered as having the same magnitude and direction. Similarly, condition (iv) is needed so that the directed segments in Figure 17.6 are *not* considered as having the same magnitude and direction.]

Our definition is such that, given a directed segment $\overrightarrow{PQ}$ and a point R, there is precisely one point S such that $\overrightarrow{PQ} \sim \overrightarrow{RS}$. (See Figure 17.7 where we show how to find S, assuming that P, Q, and R do not lie on one line.)

Consider now a directed segment $\overrightarrow{PQ}$. The set (collection) **a** of *all* directed segments having the same magnitude and direction as $\overrightarrow{PQ}$ is called a **vector.** We call **a** the vector represented by $\overrightarrow{PQ}$. If $\overrightarrow{RS} \sim \overrightarrow{PQ}$, then $\overrightarrow{RS}$ belongs to **a**, and $\overrightarrow{RS}$ represents **a.** (In Figure 17.2, for example, the directed segments 1, 2, 3, 4, 5 all represent the same vector.) It is legitimate and standard to visualize vectors as directed segments or arrows, keeping in mind that the initial point is not important, and that any two arrows with the same magnitude and direction are "equivalent."

Vectors will be denoted by boldface letters.

The **length** of a vector **a,** denoted by $|\mathbf{a}|$, is defined as the length of any directed segment representing **a.** A vector of length 1 is called a **unit vector.**

The vector represented by $\overrightarrow{PP}$ is called the **null-vector** and is denoted by **0;** its length is 0.

In order to compute with vectors, we must express vectors by numbers. The key is the following theorem (see Figure 17.8):

Theorem 1. *Let the points P, Q, R, S have Cartesian coordinates (p_1,p_2), (q_1,q_2), (r_1,r_2) and (s_1,s_2), respectively. Then $\overrightarrow{PQ} \sim \overrightarrow{RS}$, that is, $\overrightarrow{PQ}$ and $\overrightarrow{RS}$ represent the same vector, if and only if*

$$q_1 - p_1 = s_1 - r_1, \qquad q_2 - p_2 = s_2 - r_2.$$

To prove the theorem, we show that if P, Q, R are given, the point S with the coordinates

$$s_1 = (q_1 - p_1) + r_1, \qquad s_2 = (q_2 - p_2) + r_2$$

satisfies conditions (i)–(iv). We check (i). $|RS| = [(r_1 - s_1)^2 + (r_2 - s_2)^2]^{1/2} =$

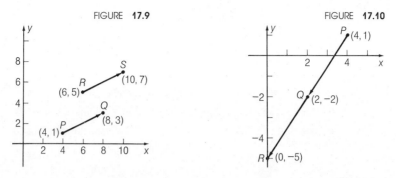

FIGURE **17.9**   FIGURE **17.10**

$[(q_1 - p_1)^2 + (q_2 - p_2)^2]^{1/2} = |PQ|$. We check (ii). Let $P \neq Q$. If the line through P and Q is vertical, $p_1 = q_1$; then $r_1 = s_1$ and the line through R and S is also vertical. If $p_1 \neq q_1$, both lines have the same slope:

$$\frac{p_2 - q_2}{p_1 - q_1} = \frac{s_2 - r_2}{s_1 - r_1}.$$

Conditions (iii) and (iv) are checked similarly. (A similar argument shows that no point distinct from S satisfies $\overrightarrow{PQ} \sim \overrightarrow{RS}$.)

REMARK   *Given a vector* **a** *and a point P, there is exactly one point Q such that $\overrightarrow{PQ}$ represents* **a**.

EXAMPLES 1.   Suppose P, Q, R, and S have coordinates $(4,1)$, $(8,3)$, $(6,5)$, and $(10,7)$, respectively. Is $\overrightarrow{PQ} \sim \overrightarrow{RS}$?

ANSWER   We use Theorem 1 and verify that

$$q_1 - p_1 = s_1 - r_1 \quad \text{or} \quad 8 - 4 = 10 - 6 \quad \text{or} \quad 4 = 4$$

and

$$q_2 - p_2 = s_2 - r_2 \quad \text{or} \quad 3 - 1 = 7 - 5 \quad \text{or} \quad 2 = 2.$$

Hence $\overrightarrow{PQ} \sim \overrightarrow{RS}$. See Figure 17.9.

2.   If $P = (4,1)$, $Q = (2,-2)$, what coordinates must T have if $\overrightarrow{PQ} \sim \overrightarrow{QT}$?

ANSWER   Denote the coordinates of T by (t_1, t_2). By Theorem 1, $\overrightarrow{PQ} \sim \overrightarrow{QT}$ if and only if $q_1 - p_1 = t_1 - q_1$ and $q_2 - p_2 = t_2 - q_2$. Hence $2 - 4 = t_1 - 2$ and $-2 - 1 = t_2 - (-2)$; that is, $t_1 = 0$ and $t_2 = -5$. Therefore the point T has coordinates $(0,-5)$. See Figure 17.10.

PROBLEMS

In Problems 1 to 10 assume a fixed Cartesian coordinate system with origin at O. For each problem, make a sketch.

1. Suppose P, Q, R, and S have coordinates $(-2,2)$, $(1,4)$, $(2,1)$, and $(5,3)$, respectively. Is $\overrightarrow{PQ} \sim \overrightarrow{RS}$?

2. Suppose P, Q, R, and S have coordinates $(5,-1)$, $(1,3)$, $(-2,2)$, and $(2,-2)$, respectively. Is $\overrightarrow{PQ} \sim \overrightarrow{RS}$?

3. Suppose A, B, C, and D have coordinates $(2,3)$, $(-2,-1)$, $(6,2)$, and $(2,-2)$, respectively. Is $\overrightarrow{AB} \sim \overrightarrow{CD}$?

4. If $A = (-4,-1)$, $B = (-1,1)$, $C = (2,-3)$, and $D = (5,-1)$, is $\overrightarrow{AC} \sim \overrightarrow{BD}$?

5. If $P = (-1,-2)$, $Q = (1,1)$, what coordinates must T have if $\overrightarrow{PQ} \sim \overrightarrow{QT}$?

6. If $A = (7,-1)$, $B = (2,1)$, what coordinates must C have if $\overrightarrow{AB} \sim \overrightarrow{BC}$?

7. Suppose $A = (1,2)$ and $B = (5,3)$. What coordinates must C have if $\overrightarrow{AB} \sim \overrightarrow{CA}$?

8. If $P = (2,2)$, $Q = (6,-1)$, and $R = (1,-1)$, what coordinates must S have if $\overrightarrow{PQ} \sim \overrightarrow{RS}$?

9. If $P = (4,3)$, $Q = (2,-1)$, and $S = (-3,-2)$, what coordinates must R have if $\overrightarrow{PQ} \sim \overrightarrow{RS}$?

10. Suppose $A = (-1,3)$, $B = (2,-1)$, and $C = (-5,1)$. What coordinates must D have if $\overrightarrow{AC} \sim \overrightarrow{BD}$?

1.2 Components

Let us choose a Cartesian coordinate system with origin at O. Given a vector **a**, there is a point Q, such that **a** is represented by $\overrightarrow{OQ}$. Let Q have the Cartesian coordinates (a_1,a_2). The numbers a_1, a_2 determine the point Q and also the vector **a**. We call a_1, a_2 the **components** of **a** (with respect to the coordinate system used), and we write

$$\mathbf{a} = \langle a_1,a_2 \rangle. \tag{2}$$

The components of a vector with respect to two different coordinate systems are, in general, different. In Figure 17.11, for instance, the vector represented by $\overrightarrow{OQ}$ has components $\langle 4,3 \rangle$ in the xy-coordinate system and components $\langle 5,0 \rangle$ in the XY-system.

FIGURE **17.11**

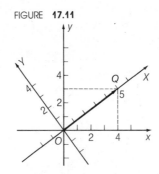

There are two exceptions to the statement just made. (1) If we translate the coordinate system, the components of a vector are unchanged; this follows from Theorem 1 in §1.1. (2) The null-vector **0** has components $\langle 0,0 \rangle$ in every system.

Note that (once a coordinate system is chosen) the vector equation

$$\langle a_1,a_2 \rangle = \langle b_1,b_2 \rangle$$

is equivalent to two equations for numbers

$$a_1 = b_1 \quad \text{and} \quad a_2 = b_2.$$

Note also that the length $|\mathbf{a}|$ of the vector (2) is given by the formula

$$|\mathbf{a}| = \sqrt{a_1^2 + a_2^2}, \tag{3}$$

as can be seen in Figure 17.8. Thus the sum of the squares of the components is the same in all coordinate systems (see the case in Figure 17.11).

Theorem 2. *If the point P has Cartesian coordinates* (p_1,p_2) *and the point Q has Cartesian coordinates* (q_1,q_2), *then the vector represented by* $\overrightarrow{PQ}$ *has the components* $\langle q_1 - p_1, q_2 - p_2 \rangle$.

This follows from the discussion in this and the preceding section.

REMARK If the vector **a** is represented by the directed segment $\overrightarrow{PQ}$, the components of **a** are also called components of $\overrightarrow{PQ}$.

EXAMPLES 1. If P has coordinates $(2,-3)$ and Q has coordinates $(5,1)$, what are the components and length of $\overrightarrow{PQ}$?

ANSWER By Theorem 2, $\overrightarrow{PQ}$ has components $\langle q_1 - p_1, q_2 - p_2 \rangle = \langle 5 - 2, 1 - (-3) \rangle = \langle 3,4 \rangle$. Using Equation (3), the length of $\overrightarrow{PQ}$ is $|PQ| = \sqrt{3^2 + 4^2} = 5$.

2. If $P = (1,3)$, $Q = (4,-2)$, and $R = (-2,4)$, what are the components of $\overrightarrow{RS}$ if $\overrightarrow{PQ} \sim \overrightarrow{RS}$? What are the coordinates of S?

ANSWER The vector $\overrightarrow{PQ}$ has components $\langle 4 - 1, -2 - 3 \rangle = \langle 3,-5 \rangle$. But $\overrightarrow{PQ}$ and $\overrightarrow{RS}$ have the same components since $\overrightarrow{PQ} \sim \overrightarrow{RS}$. Hence $\overrightarrow{RS} = \langle 3,-5 \rangle$. Denote the coordinates of S by (s_1,s_2). Then the components of $\overrightarrow{RS}$ are $\langle s_1 - (-2), s_2 - 4 \rangle = \langle s_1 + 2, s_2 - 4 \rangle$. Therefore $\overrightarrow{RS} = \langle s_1 + 2, s_2 - 4 \rangle = \langle 3,-5 \rangle$, that is, $s_1 + 2 = 3$ and $s_2 - 4 = -5$. Hence $s_1 = 1$, $s_2 = -1$, and the coordinates of S are $(1,-1)$.

PROBLEMS

In each of Problems 1 to 8 find the components and length of the vector represented by $\overrightarrow{PQ}$ for the given points P and Q.

1. $P = (2,1)$, $Q = (3,4)$.
2. $P = (1,2)$, $Q = (-3,-1)$.
3. $P = (-4,-2)$, $Q = (0,2)$.
4. $P = (1,3)$, $Q = (3,-1)$.

5. $P = (-3,-3)$, $Q = (2,-1)$.
6. $P = (-1,-2)$, $Q = (-4,2)$.
7. $P = (-3,1)$, $Q = (3,-3)$.
8. $P = (6,2)$, $Q = (-4,-2)$.

9. If $P = (-3,1)$, $Q = (2,3)$, and $R = (1,-3)$, what are the components of $\overrightarrow{RS}$ if $\overrightarrow{PQ} \sim \overrightarrow{RS}$? What are the coordinates of S?

10. If $P = (-1,3)$, $Q = (-3,-1)$, and $S = (2,-2)$, what are the components of $\overrightarrow{RS}$ if $\overrightarrow{PQ} \sim \overrightarrow{RS}$? What are the coordinates of R?

1.3   Sum of two vectors. Product of a number and a vector

We proceed to define the sum of two vectors, **a** and **b**. The sum, denoted by **a** + **b**, will again be a vector. The definition reads:

$$\text{if } \mathbf{a} = \langle a_1,a_2 \rangle \text{ and } \mathbf{b} = \langle b_1,b_2 \rangle, \text{ then } \mathbf{a} + \mathbf{b} = \langle a_1 + b_1, a_2 + b_2 \rangle \qquad (4)$$

or, which is the same,

$$\langle a_1,a_2 \rangle + \langle b_1,b_2 \rangle = \langle a_1 + a_2, b_1 + b_2 \rangle. \qquad (4')$$

In words: *we add vectors by adding their components.*

The definition just given uses components and therefore seems to depend on the coordinate system. This is not so, since there is an equivalent, purely geometric

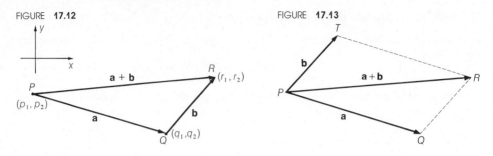

FIGURE 17.12

FIGURE 17.13

definition: *let* **a** *be represented by* $\overrightarrow{PQ}$, *and choose a point* R *such that* $\overrightarrow{QR}$ *represents* **b**; *then* $\overrightarrow{PR}$ *represents* **a** + **b**.

This is indeed equivalent to (4). Let P have coordinates (p_1, p_2), Q coordinates (q_1, q_2), R coordinates (r_1, r_2). See Figure 17.12. The vectors represented by $\overrightarrow{PQ}$, $\overrightarrow{QR}$, and $\overrightarrow{PR}$ have components $\langle q_1 - p_1, q_2 - p_2 \rangle$, $\langle r_1 - q_1, r_2 - q_2 \rangle$, and $\langle r_1 - p_1, r_2 - p_2 \rangle$, respectively, by Theorem 2 in §1.2. But by the definition (4) of vector addition,

$$\langle q_1 - p_1, q_2 - p_2 \rangle + \langle r_1 - q_1, r_2 - q_2 \rangle$$
$$= \langle (q_1 - p_1) + (r_1 - q_1), (q_2 - p_2) + (r_2 - q_2) \rangle$$
$$= \langle r_1 - p_1, r_2 - p_2 \rangle.$$

Thus the sum of the vectors represented by $\overrightarrow{PQ}$ and $\overrightarrow{QR}$ is the vector represented by $\overrightarrow{PR}$.

Let us represent **a** and **b** by directed segments $\overrightarrow{PQ}$ and $\overrightarrow{PT}$, with the same initial point P, and assume that P, Q, and T do not lie on the same line. Then, as is seen from Figure 17.13, the sum **a** + **b** is represented by $\overrightarrow{PR}$, where P, Q, T, and R are the vertices of a parallelogram. The definition of vector addition is therefore called the **Parallelogram Law**.

[The way of adding vectors has been suggested by mechanics. If **a** and **b** represent forces acting on a particle, **a** + **b** is the resultant force: the particle moves as if subject to the force **a** + **b**; see §3.3.]

We define next the **product** of a number α and a vector **a**. The product, denoted by α**a**, is again a vector:

$$\text{if } \mathbf{a} = \langle a_1, a_2 \rangle, \qquad \alpha\mathbf{a} = \langle \alpha a_1, \alpha a_2 \rangle, \tag{5}$$

or

$$\alpha\langle a_1, a_2 \rangle = \langle \alpha a_1, \alpha a_2 \rangle. \tag{5'}$$

In words: *we multiply a vector by a number by multiplying each component by the number*.

The definition is again independent of the coordinate system used, since it has a geometric meaning: if $\alpha \neq 0$, $\mathbf{a} \neq \mathbf{0}$, then $\alpha\mathbf{a}$ is obtained from **a** by preserving the direction if $\alpha > 0$ (and reversing it if $\alpha < 0$) and multiplying the length by $|\alpha|$. See Figure 17.14. We must add two statements to the geometric definition:

$$0\mathbf{a} = \mathbf{0} \quad \text{for all vectors} \quad \mathbf{a}, \qquad \alpha\mathbf{0} = \mathbf{0} \quad \text{for all numbers} \quad \alpha. \tag{6}$$

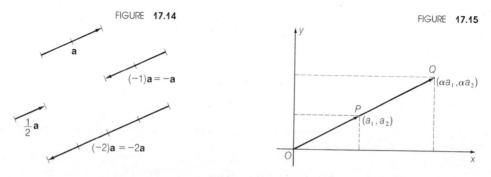

FIGURE **17.14**

$\mathbf{a}$

$(-1)\mathbf{a} = -\mathbf{a}$

$\frac{1}{2}\mathbf{a}$

$(-2)\mathbf{a} = -2\mathbf{a}$

FIGURE **17.15**

y

Q

$(\alpha a_1, \alpha a_2)$

P

(a_1, a_2)

O

x

Let us verify that this geometric definition is indeed equivalent to (5). Let P have the coordinate (a_1, a_2) so that $\overrightarrow{OP}$ represents $\mathbf{a}$. Let $\alpha > 0$. By (5), $\alpha \mathbf{a}$ is represented by $\overrightarrow{OQ}$ where Q has coordinates $(\alpha a_1, \alpha a_2)$. Now the points O, P, and Q lie on one line since the line OQ has slope $\alpha a_2 / \alpha a_1 = a_2/a_1$ (or is vertical if $a_1 = 0$), and the same is true of the line OP. Also, P and Q lie on the same side of O since the coordinate αa_1 is negative, zero, or positive if a_1 is, and αa_2 is negative, zero, or positive if a_2 is. Hence $\overrightarrow{OP}$ and $\overrightarrow{OQ}$ have the same direction. Finally, $|OQ| = \sqrt{(\alpha a_1)^2 + (\alpha a_2)^2} = |\alpha| \sqrt{a_1^2 + a_2^2} = \alpha |OP|$. (See Figure 17.15 where $\alpha > 1$.) The case $\alpha < 0$ is treated similarly.

In talking about vectors, numbers are often called **scalars**. The vector $\alpha \mathbf{a}$ is called a **scalar multiple** of the vector $\mathbf{a}$. Instead of $(-1)\mathbf{a}$, we write $-\mathbf{a}$; thus

$$-\langle a_1, a_2 \rangle = (-1)\langle a_1, a_2 \rangle = \langle -a_1, -a_2 \rangle. \tag{7}$$

Instead of $\mathbf{b} + (-1)\mathbf{a}$ we write $\mathbf{b} - \mathbf{a}$. Thus

$$\langle b_1, b_2 \rangle - \langle a_1, a_2 \rangle = \langle b_1 - a_1, b_2 - a_2 \rangle. \tag{8}$$

Comparing this with the definition (4) of addition, we see that

$$\mathbf{a} + \mathbf{x} = \mathbf{b} \quad \text{if and only if} \quad \mathbf{x} = \mathbf{b} - \mathbf{a}, \tag{9}$$

just as it is for numbers. The geometric construction of the **difference** $\mathbf{b} - \mathbf{a}$ is shown in Figure 17.16. Note that one diagonal of a parallelogram spanned by two vectors represents their sum, and the other diagonal represents their difference.

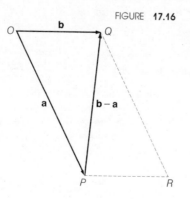

FIGURE **17.16**

O $\mathbf{b}$ Q

$\mathbf{a}$ $\mathbf{b} - \mathbf{a}$

P R

EXAMPLE If **a** has components $\langle 3,1 \rangle$ and **b** has components $\langle -6,2 \rangle$, what are the components of $\frac{1}{3}(\mathbf{a} + \mathbf{b})$?

ANSWER Using Equation (4), we have $\mathbf{a} + \mathbf{b} = \langle 3,1 \rangle + \langle -6,2 \rangle = \langle 3 - 6, 1 + 2 \rangle = \langle -3,3 \rangle$. Using Equation (5'), we have $\frac{1}{3}(\mathbf{a} + \mathbf{b}) = \frac{1}{3}\langle -3,3 \rangle = \langle \frac{1}{3}(-3), \frac{1}{3}(3) \rangle = \langle -1,1 \rangle$.

PROBLEMS

1. If **a** has components $\langle 1, -2 \rangle$ and **b** has components $\langle 3, -2 \rangle$, what are the components of $\mathbf{a} + \mathbf{b}$? Of $4\mathbf{a}$? Of $(-5)\mathbf{b}$?
2. If $\mathbf{a} = \langle 1,4 \rangle$, $\mathbf{b} = \langle -3, -4 \rangle$, what is $\mathbf{a} + \mathbf{b}$? What is $\frac{1}{2}(\mathbf{a} + \mathbf{b})$? What is $(-\frac{1}{2})(\mathbf{a} + \mathbf{b})$?
3. If $\mathbf{a} = \langle -2,4 \rangle$, $\mathbf{b} = \langle 0,6 \rangle$, what is $\mathbf{a} - \mathbf{b}$? What is $\frac{1}{4}(\mathbf{a} - \mathbf{b})$? What is $-2(\mathbf{a} - \mathbf{b})$?
4. If $\mathbf{a} = \langle 12,7 \rangle$, $\mathbf{b} = \langle -4,5 \rangle$, what is $-\frac{1}{2}(\mathbf{a} - \mathbf{b})$? What is $\frac{1}{8}(\mathbf{a} - \mathbf{b})$?
5. If $\mathbf{a} = \langle 2,4 \rangle$, $\mathbf{b} = \langle 5,8 \rangle$, find $2\mathbf{a} - \mathbf{b}$.
6. If $\mathbf{a} = \langle x^2,4 \rangle$, $\mathbf{b} = \langle x,2 \rangle$, $\mathbf{c} = \langle 1, -3 \rangle$, find $\mathbf{v} = \mathbf{a} + 4(\mathbf{b} + \mathbf{c})$. For which x is $\mathbf{v} = \mathbf{0}$?

1.4 Calculating with vectors

In calculating with vectors, we use certain formal rules; it will be convenient to state them all at once.

Theorem 3. *Vector addition and multiplication of vectors by numbers obey the following rules:*

1. *For any two vectors **a** and **b**, there is a unique vector called the **sum** of **a** and **b**, denoted by $\mathbf{a} + \mathbf{b}$.*
2. *$\mathbf{a} + \mathbf{b} = \mathbf{b} + \mathbf{a}$ for all **a** and **b**.*
3. *$(\mathbf{a} + \mathbf{b}) + \mathbf{c} = \mathbf{a} + (\mathbf{b} + \mathbf{c})$ for all **a**, **b**, and **c**.*
4. *There is a unique vector called the **null-vector**, denoted by **0**, such that $\mathbf{0} + \mathbf{a} = \mathbf{a}$ for all **a**.*
5. *For every vector **a**, there is a unique vector called its **negative** and denoted by $(-\mathbf{a})$ such that $\mathbf{a} + (-\mathbf{a}) = \mathbf{0}$.*
6. *For any number α and any vector **a**, there is a unique vector called the **product** of α and **a**. It is denoted by $\alpha\mathbf{a}$.*
7. *$1\mathbf{a} = \mathbf{a}$ for all **a**.*
8. *$\alpha(\beta\mathbf{a}) = (\alpha\beta)\mathbf{a}$ for all α, β, and **a**.*
9. *$(\alpha + \beta)\mathbf{a} = \alpha\mathbf{a} + \beta\mathbf{a}$ for all α, β, and **a**.*
10. *$\alpha(\mathbf{a} + \mathbf{b}) = \alpha\mathbf{a} + \alpha\mathbf{b}$ for all α, **a**, and **b**.*

We could have added

$$-\mathbf{0} = \mathbf{0}, \quad -(-\mathbf{a}) = \mathbf{a}, \quad 0\mathbf{a} = \mathbf{0}, \quad \alpha\mathbf{0} = \mathbf{0}, \tag{10}$$

$$0\mathbf{a} = \mathbf{0}, \quad \alpha\mathbf{0} = \mathbf{0}, \quad (-1)\mathbf{a} = -\mathbf{a}. \tag{11}$$

But these relations are consequences of the basic rules **1–10**.

The proof of Theorem 3 follows from Equations (4) and (5) and properties of numbers. We carry it out for **4** only; the reader is asked to prove the rest.

Statement **4** says: there are two numbers, x_1 and x_2, such that $\langle x_1,x_2 \rangle + \langle a_1,a_2 \rangle = \langle a_1,a_2 \rangle$ for any two numbers a_1, a_2. But, by our definition, $\langle x_1,x_2 \rangle + \langle a_1,a_2 \rangle = \langle x_1 + a_1, x_2 + a_2 \rangle$. For this to equal $\langle a_1,a_2 \rangle$, we need $a_1 = x_1 + a_1$, $a_2 = x_2 + a_2$. This means that $x_1 = 0$, $x_2 = 0$. Theorem 3 can also be proved geometrically. The proof of the "associative law" **3** is shown in Figure 17.17.

FIGURE **17.17**

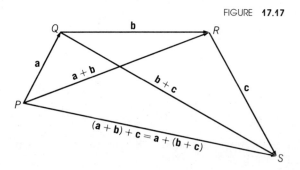

In view of **3** we may write **a** + **b** + **c**, instead of **a** + (**b** + **c**), and similarly for sums of more than three vectors. We recall (see §1.3) that **b** − **a** = **b** + (−**a**) by definition.

Statements (9) in §1.3 and statements (10) and (11) can be proved either from the definitions (4), (5), or from Rules **1–10**. For instance, the proof of (9) goes as follows. Assume that **x** = **b** − **a** = **b** + (−1)**a**. Then **a** + **x** = **a** + (**b** + (−**a**)) = **a** + ((−**a**) + **b**) = (**a** + (−**a**)) + **b** = **0** + **b** = **b** by **2, 3, 5,** and **4.** Now assume that **a** + **x** = **b**. Then **x** = **0** + **x** = (**a** + (−**a**)) + **x** = ((−**a**) + **a**) + **x** = (−**a**) + (**a** + **x**) = (−**a**) + **b** = **b** + (−**a**) by **4, 5, 2, 3, 2.**

Theorem 3 enables us to compute with vectors.

EXAMPLES **1.** If **a** = ⟨3,−2⟩, **b** = ⟨0,4⟩, and **c** = ⟨5,10⟩, what are the components of 2(**a** − 3**b**) + **c**?
ANSWER Using Theorem 3 we have 2(**a** − 3**b**) + **c** = 2**a** − 6**b** + **c** = 2⟨3,−2⟩ − 6⟨0,4⟩ + ⟨5,10⟩ = ⟨6,−4⟩ + ⟨0,−24⟩ + ⟨5,10⟩ = ⟨6,−28⟩ + ⟨5,10⟩ = ⟨11,−18⟩.

2. Let **a, b, c** be vectors such that 3**a** − 4(**a** + **b** − **c**) = **0**. Assuming **b** and **c** to be known, find **a**.
SOLUTION We "solve" this equation for **a** as follows:

$$3\mathbf{a} - 4\mathbf{a} - 4\mathbf{b} + 4\mathbf{c} = \mathbf{0},$$
$$(3 - 4)\mathbf{a} - 4\mathbf{b} + 4\mathbf{c} = \mathbf{0},$$
$$-\mathbf{a} - 4\mathbf{b} + 4\mathbf{c} = \mathbf{0},$$
$$\mathbf{a} + 4\mathbf{b} - 4\mathbf{c} = \mathbf{0},$$
$$\mathbf{a} = -4\mathbf{b} + 4\mathbf{c}.$$

(The reader may want to justify each step by Rules **1–10**.)

3. Let **a, b, c** be vectors such that **a** − 3**b** = 5**c**, 2**a** + **b** = 3**c**. Assuming **c** is known, find **a** and **b**.
SOLUTION Solving the first equation for **a** we obtain **a** = 3**b** + 5**c**. Substituting this into the second equation we have 2(3**b** + 5**c**) + **b** = 3**c**, 6**b** + 10**c** + **b** = 3**c**, 7**b** + 10**c** = 3**c**, 7**b** = −7**c**, that is, **b** = −**c**. Using **a** = 3**b** + 5**c**, we find that **a** = 3(−**c**) + 5**c** = 2**c**. Hence **a** = 2**c**, **b** = −**c**.

PROBLEMS

In Problems 1 to 10 assume a fixed Cartesian coordinate system with origin at *O*.

1. If **a** = ⟨2,4⟩, **b** = ⟨4,−3⟩, and **c** = ⟨1,5⟩, what are the components of 3**a** − 2**b** + 4**c**?
2. If **a** = ⟨1,3⟩, **b** = ⟨2,2⟩, and **c** = ⟨2,−1⟩, what are the components of 4(**a** − 3**b**) − 2**c**?
3. If **a** = ⟨1,6⟩, **b** = ⟨−4,2⟩, and **c** = ⟨3,−4⟩, what are the components of 2(**a** − 4**b**) − 3(**a** − **b** + 2**c**)?

4. Find the components of $\langle 3,0 \rangle + 4\langle 6,1 \rangle - 3\langle 0,2 \rangle$.

5. Find the components of $2\langle 0,1 \rangle - 5\langle 3,2 \rangle + 3\langle 3,1 \rangle$.

6. Find the components of $4\langle 2,\frac{1}{2} \rangle - \langle 1,2 \rangle + 2\langle -1,-\frac{1}{2} \rangle$.

7. Let $\mathbf{a}$, $\mathbf{b}$, $\mathbf{c}$ be vectors such that $-2(\mathbf{a} + 3\mathbf{c}) = \mathbf{a} + 3\mathbf{b}$. Assuming $\mathbf{b}$ and $\mathbf{c}$ to be known, find $\mathbf{a}$.

8. If $\mathbf{a}$, $\mathbf{b}$, and $\mathbf{c}$ are vectors and $3(\mathbf{a} - 2\mathbf{b}) + 2(\mathbf{a} + 4\mathbf{c}) = 6\mathbf{a} - 2\mathbf{b} + \mathbf{c}$, solve for $\mathbf{a}$. What are the components of $\mathbf{a}$ if $\mathbf{b} = \langle 1,-2 \rangle$ and $\mathbf{c} = \langle 2,1 \rangle$?

9. Let $\mathbf{a}$, $\mathbf{b}$, $\mathbf{c}$ be vectors such that $2\mathbf{a} + 4\mathbf{b} = \mathbf{c}$, $\mathbf{a} - 3\mathbf{b} = -2\mathbf{c}$. Assuming $\mathbf{c}$ is known, find $\mathbf{a}$ and $\mathbf{b}$.

10. Suppose $\mathbf{a}$, $\mathbf{b}$, $\mathbf{c}$ are vectors such that $\mathbf{a} - 5\mathbf{b} = 8\mathbf{c}$, $2\mathbf{a} + 3\mathbf{b} = 3\mathbf{c}$. What are the components of $\mathbf{a}$ and $\mathbf{b}$ if $\mathbf{c} = \langle 4,2 \rangle$?

11. Prove Statements **2** and **3** of Theorem **3**.

12. Prove Statements **8** and **9** of Theorem **3**.

13. Prove Statement **5** of Theorem **3**.

1.5 The unit vectors **i** and **j**

We associate with every Cartesian coordinate system two unit vectors $\mathbf{i}$ and $\mathbf{j}$; these are vectors which in the given system have the components

$$\mathbf{i} = \langle 1,0 \rangle, \qquad \mathbf{j} = \langle 0,1 \rangle. \tag{12}$$

The vector $\mathbf{i}$ is represented by a segment of length 1 in the x-direction; $\mathbf{j}$ is represented by a directed segment of length 1 in the y-direction. If $\mathbf{a}$ has components $\langle a_1,a_2 \rangle$,

$$\mathbf{a} = \langle a_1,a_2 \rangle = a_1\mathbf{i} + a_2\mathbf{j}. \tag{13}$$

Indeed, by (4), (5), and (12),

$$a_1\mathbf{i} + a_2\mathbf{j} = a_1\langle 1,0 \rangle + a_2\langle 0,1 \rangle = \langle a_1,0 \rangle + \langle 0,a_2 \rangle = \langle a_1,a_2 \rangle = \mathbf{a}.$$

The geometric meaning of our statement is shown in Figure 17.18.

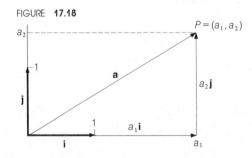

FIGURE **17.18**

EXAMPLES 1. Express the vector $\mathbf{a} = \langle 4,-3 \rangle$ in terms of $\mathbf{i}$ and $\mathbf{j}$.

ANSWER   Using Equation (13), we have at once $\mathbf{a} = 4\mathbf{i} + (-3)\mathbf{j}$ or $\mathbf{a} = 4\mathbf{i} - 3\mathbf{j}$.

2. If $\mathbf{a} = 3\mathbf{i} + 4\mathbf{j}$ and $\mathbf{b} = \mathbf{i} + 2\mathbf{j}$, what are the components of $\mathbf{a} + 2\mathbf{b}$?

FIRST SOLUTION   We have $\mathbf{a} = \langle 3,4 \rangle$, $\mathbf{b} = \langle 1,2 \rangle$, $2\mathbf{b} = \langle 2,4 \rangle$. Therefore $\mathbf{a} + 2\mathbf{b} = \langle 3,4 \rangle + \langle 2,4 \rangle = \langle 5,8 \rangle$.

SECOND SOLUTION   This time we work directly with $\mathbf{i}$ and $\mathbf{j}$ and use Theorem 3. We have

$$\mathbf{a} + 2\mathbf{b} = 3\mathbf{i} + 4\mathbf{j} + 2(\mathbf{i} + 2\mathbf{j}) = 3\mathbf{i} + 4\mathbf{j} + 2\mathbf{i} + 4\mathbf{j} = 3\mathbf{i} + 2\mathbf{i} + 4\mathbf{j} + 4\mathbf{j} = (3 + 2)\mathbf{i} + (4 + 4)\mathbf{j} = 5\mathbf{i} + 8\mathbf{j}.$$

PROBLEMS

In Problems 1 to 12 assume a fixed Cartesian coordinate system with origin at O. $\mathbf{i}$ and $\mathbf{j}$ are unit vectors with components $\langle 1,0 \rangle$ and $\langle 0,1 \rangle$, respectively.

1. Express the vector $\mathbf{a} = \langle -1,5 \rangle$ in terms of $\mathbf{i}$ and $\mathbf{j}$.
2. Express the vector $\mathbf{a} = \langle \frac{1}{2}, \frac{1}{3} \rangle$ in terms of $\mathbf{i}$ and $\mathbf{j}$.
3. If $\mathbf{a} = 2\mathbf{i} + 3\mathbf{j}$ and $\mathbf{b} = 4\mathbf{i} + \mathbf{j}$, what are the components of $\mathbf{a} + 4\mathbf{b}$? Of $\mathbf{b} - 3\mathbf{a}$?
4. If $\mathbf{a} = 4\mathbf{i} - 2\mathbf{j}$ and $\mathbf{b} = -\frac{1}{2}\mathbf{i} + \frac{3}{2}\mathbf{j}$, what are the components of $2\mathbf{a} + 4\mathbf{b}$? Of $\frac{1}{2}(\mathbf{a} - 2\mathbf{b})$?
5. If $\mathbf{a}$ has components $\langle 2,3 \rangle$ and $\mathbf{b} = \mathbf{i} - \mathbf{j} + \mathbf{a}$, what are the components of $3\mathbf{a} - 2\mathbf{b}$?
6. If $\mathbf{a}$ has components $\langle 4,-2 \rangle$ and $\mathbf{b} = 2\mathbf{i} - 4\mathbf{j} - 3\mathbf{a}$, what are the components of $\frac{1}{2}(5\mathbf{a} + 3\mathbf{b})$?
7. Suppose $\mathbf{a} = \mathbf{i} + 4\mathbf{j}$, $\mathbf{b} = 3\mathbf{i} - 2\mathbf{j}$, $\mathbf{c} = 2\mathbf{i} - 6\mathbf{j}$. Find the components of $\mathbf{a} - 3\mathbf{b} + 6\mathbf{c}$.
8. Suppose $\mathbf{a} = 2\mathbf{i} - \mathbf{j}$, $\mathbf{b} = \mathbf{i} + 3\mathbf{j}$, $\mathbf{c} = -3\mathbf{i} + 4\mathbf{j}$. Find the components of $4\mathbf{a} - 2\mathbf{b} - \mathbf{c}$.
9. Suppose $\mathbf{a} + \mathbf{b} = \mathbf{i}$ and $\mathbf{a} - \mathbf{b} = \mathbf{j}$. What are the components of $\mathbf{a}$ and $\mathbf{b}$? [*Hint:* Solve for $\mathbf{a}$ and $\mathbf{b}$ simultaneously.]
10. Suppose $2\mathbf{a} + 4\mathbf{b} = \mathbf{i} + 3\mathbf{j}$ and $\mathbf{a} - 2\mathbf{b} = -2\mathbf{i} + \mathbf{j}$. What are the components of $\mathbf{a}$ and $\mathbf{b}$?
11. Show that $|a_1\mathbf{i} + a_2\mathbf{j}| = \sqrt{a_1^2 + a_2^2}$.
12. Prove that $a_1\mathbf{i} + a_2\mathbf{j} = \mathbf{0}$ if and only if $a_1 = a_2 = 0$.

1.6 Position vectors. Colinear points

Let O be a point, which shall be kept fixed. For every point Q in the plane, the directed segment $\overrightarrow{OQ}$ represents a vector $\mathbf{a}$. We call $\mathbf{a}$ the **position vector** of Q with respect to O. By the Remark in §1.1, every vector is the position vector of some point. For example, the null-vector $\mathbf{0}$ is the position vector of O.

Let O be the origin of a Cartesian coordinate system. The position vector of a point P, with respect to O, that is, the vector represented by $\overrightarrow{OP}$, will be denoted by $\mathbf{P}$. We have

$$\mathbf{P} = x\mathbf{i} + y\mathbf{j},$$

where $\mathbf{i}$ and $\mathbf{j}$ are the vectors (12); that is, the position vectors of the points $(1,0)$ and $(0,1)$, and x, y are the Cartesian coordinates of P. If $\mathbf{P} \neq \mathbf{0}$, let (r,θ) be the polar coordinates of P. Then $r = |\mathbf{P}|$ is the length of $\mathbf{P}$, and $\theta = \arctan(y/x)$ is called the **polar angle** of $\mathbf{P}$. (See Figure 17.19.)

FIGURE **17.19**

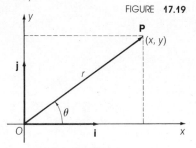

EXAMPLE 1. Find the length and polar angle of $\mathbf{P} = \mathbf{i} + \mathbf{j}$.

ANSWER P has coordinates $(1,1)$. The length is $|\mathbf{P}| = \sqrt{1^2 + 1^2} = \sqrt{2}$. The polar angle is $\theta = \arctan(1/1) = 45°$.

FIGURE 17.20

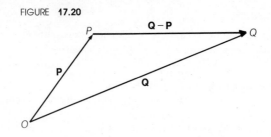

REMARK Let P and Q be two points; the vector determined by $\overrightarrow{PQ}$ is $\mathbf{Q} - \mathbf{P}$ (so that $|PQ| = |\mathbf{Q} - \mathbf{P}|$). The geometric proof is illustrated in Figure 17.20.

The following theorem is the basis of many geometric applications of vectors.

Theorem 4. Let P and Q be two distinct points, and let l be the line through P and Q. A point R lies on l if and only if the position vectors $\mathbf{P}, \mathbf{Q}, \mathbf{R},$ *satisfy*

$$\mathbf{R} = \mathbf{P} + t(\mathbf{Q} - \mathbf{P}) \tag{14}$$

for some number t.

EXAMPLES 2. Does the point $R = (1,2)$ lie on the line through $P = (0,2)$ and $Q = (3,4)$?
ANSWER We ask: is there a number t such that

$$\langle 1,2 \rangle = \langle 0,2 \rangle + t\{\langle 3,4 \rangle - \langle 0,2 \rangle\}$$

or

$$\langle 1,2 \rangle = \langle 0,2 \rangle + t\langle 3,2 \rangle = \langle 0 + 3t, 2 + 2t \rangle,$$

or

$$1 = 3t, \qquad 2 = 2 + 2t?$$

The first equation shows that $t = \frac{1}{3}$. Substituting into the second equation, we obtain the wrong statement $2 = 2 + \frac{2}{3}$. There is no such t; $(1,3)$ does *not* lie on the line considered.

3. Do the points $(1,1), (2,2), (4,4)$ lie on one line?
ANSWER We ask: is there a t such that $\langle 4,4 \rangle = \langle 1,1 \rangle + t\{\langle 2,2 \rangle - \langle 1,1 \rangle\}$ or $\langle 4,4 \rangle = \langle 1,1 \rangle + \langle t,t \rangle = \langle 1 + t, 1 + t \rangle$? There is such a t, $t = 3$. The three points lie on one line.

Proof of Theorem 4. In view of the definition of multiplication of vectors by numbers, the condition "R lies on the line l through P and Q" means that "the vector represented by $\overrightarrow{PR}$ is some number t times the vector represented by $\overrightarrow{PQ}$." By the Remark, this can be written as $\mathbf{R} - \mathbf{P} = t(\mathbf{Q} - \mathbf{P})$. This is the same as (14) and can also be written as

$$\mathbf{R} = (1 - t)\mathbf{P} + t\mathbf{Q}. \tag{14'}$$

In Figure 17.21 we show the case $0 < t < 1$; here R lies between P and Q. If $t > 1$, Q lies between P and R (see Figure 17.22). If $t = 0$, then $R = P$, and if $t < 0$, then P lies between R and Q. In all cases $|PR| = |\mathbf{R} - \mathbf{P}| = |\mathbf{P} - t\mathbf{P} + t\mathbf{Q} - \mathbf{P}| = |t(\mathbf{Q} - \mathbf{P})| = |t|\,|\mathbf{Q} - \mathbf{P}| = |t|\,|PQ|$. We see similarly that $|QR| = |1 - t|\,|QP|$.

Corollary 1. Let P and Q be distinct points, and let R be the point on the segment PQ, such that $|PR|/|RQ| = m/n$. *Then*

$$\mathbf{R} = \frac{n}{n + m}\mathbf{P} + \frac{m}{n + m}\mathbf{Q}. \tag{15}$$

FIGURE **17.21**

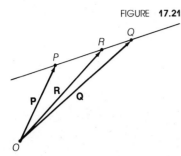

FIGURE **17.22**

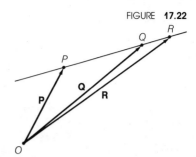

Proof. By what was said above, **R** can be written in the form (14′) with $0 < t < 1$ and $t/(1 - t) = m/n$. Solving this equation for t, we obtain $t = m/(n + m)$, $1 - t = n/(n + m)$.

Corollary 2. *The position vector of the midpoint R between two points P and Q is*

$$\mathbf{R} = \tfrac{1}{2}\mathbf{P} + \tfrac{1}{2}\mathbf{Q}.$$

This is a special case of Corollary 1 (set $m = n = 1$).

All results above can be rewritten in terms of coordinates (components of position vectors). For instance, Corollary 2 says that the midpoint of a segment with endpoints (p_1,p_2), (r_1,r_2) has coordinates

$$\frac{p_1 + r_1}{2}, \frac{p_2 + r_2}{2}.$$

EXAMPLE 4. If P and Q have coordinates $(-3,4)$ and $(6,7)$, respectively, find the coordinates of R lying on the segment joining P and Q if $|PR| = 9|RQ|$.

ANSWER   We use Corollary 1 with $\mathbf{P} = \langle -3,4 \rangle$, $\mathbf{Q} = \langle 6,7 \rangle$, $m = 9$, $n = 1$ and obtain

$$\mathbf{R} = \frac{1}{10}\langle -3,4 \rangle + \frac{9}{10}\langle 6,7 \rangle = \langle -.3,.4 \rangle + \langle 5.4,6.3 \rangle = \langle 5.1,6.7 \rangle.$$

Hence R has coordinates $(5.1,6.7)$.

PROBLEMS

In Problems 1 to 10 assume a fixed Cartesian coordinate system with origin O. **P, Q,** and **R** denote the position vectors of the points P, Q, and R, respectively.

1. Find the length and polar angle of $\mathbf{P} = (\sqrt{3}/2)\mathbf{i} + \tfrac{1}{2}\mathbf{j}$.
2. Find the length and polar angle of $\mathbf{Q} = -(\sqrt{2}/2)\mathbf{i} + (\sqrt{2}/2)\mathbf{j}$.
3. Find the length and polar angle of $\mathbf{R} = -\mathbf{i} + \sqrt{3}\mathbf{j}$.
4. Find the length and polar angle of $\mathbf{P} = 3\mathbf{i} - 4\mathbf{j}$.
5. Does the point $R = (2,-1)$ lie on the line through $P = (-1,1)$ and $Q = (5,-3)$?
6. Does the point $R = (3,2)$ lie on the line through $P = (1,\tfrac{4}{3})$ and $Q = (\tfrac{1}{2},\tfrac{7}{6})$?
7. Do the points $(\tfrac{1}{2},\tfrac{1}{2})$, $(1,\tfrac{3}{2})$, $(\tfrac{5}{2},4)$ lie on one line?
8. Do the points $(-1,-2)$, $(\tfrac{1}{2},\tfrac{7}{4})$, $(1,3)$ lie on one line?
9. If P has coordinates $(1,2)$ and Q has coordinates $(-4,5)$, find the midpoint R of the segment joining P and Q.
10. If $\mathbf{P} = \langle \tfrac{1}{2},3 \rangle$, $\mathbf{Q} = \langle \tfrac{3}{2},-1 \rangle$, find the position vector **R** such that R is the midpoint of the segment joining P and Q.

11. P has coordinates $(2,3)$, Q has coordinates $(-1,9)$, and R lies on the segment joining P and Q. If $|PR| = 2|RQ|$, find R.

12. If $\mathbf{P} = \langle 3,-3 \rangle$, $\mathbf{Q} = \langle -1,5 \rangle$, and R lies on the segment joining P and Q, find $\mathbf{R}$ if $|PR| = \frac{1}{3}|RQ|$.

13. P has coordinates $(-1,-6)$, Q has coordinates $(3,2)$, and R lies on the line through P and Q. If P lies between Q and R, and $|RP| = \frac{1}{4}|PQ|$, find $\mathbf{R}$. [*Hint:* Use Corollary 1 but interchange $\mathbf{P}$ and $\mathbf{R}$, that is, regard P as the point on the segment RQ.]

14. Suppose $\mathbf{P} = \langle -1,2 \rangle$, $\mathbf{Q} = \langle 2,4 \rangle$, and R lies on the line through P and Q. If Q lies between P and R, and $|PR| = 3|PQ|$, find $\mathbf{R}$.

1.7 Parametric representations of lines

We conclude this section with the most important application of Theorem 4.

Corollary 3 to Theorem 4. *Let* $\mathbf{a} \neq \mathbf{0}$ *and* $\mathbf{b}$ *be fixed vectors. Consider all points with position vectors*

$$\mathbf{R} = t\mathbf{a} + \mathbf{b}. \tag{16}$$

These points form a line. Every line can be so represented.

Before proving (16), let us rewrite this vector equation in components. Let $\mathbf{a} = \langle a_1,a_2 \rangle$, $\mathbf{b} = \langle b_1,b_2 \rangle$, and let $\mathbf{R}$ have coordinates (x,y). Then

$$\langle x,y \rangle = t\langle a_1,a_2 \rangle + \langle b_1,b_2 \rangle = \langle a_1 t + b_1, a_2 t + b_2 \rangle$$

so that

$$x = a_1 t + b_1, \qquad y = a_2 t + b_2. \tag{17}$$

Equation (16) and the equivalent system (17) are called a **parametric representation** of a line (see Chapter 16, §2). The number t is called the parameter; different values of t give different points on the line.

Proof of the corollary. Indeed, if a line l is given, let P and Q be two distinct points on l, and set $\mathbf{a} = \mathbf{Q} - \mathbf{P}$, $\mathbf{b} = \mathbf{P}$. Now (16) becomes (14') in §1.6. On the other hand, if $\mathbf{a}$ and $\mathbf{b}$ are given, let P be the point with position vector $\mathbf{b}$ and Q the point with position vector $\mathbf{a} + \mathbf{b}$. Then (16) represents the points on the line l through P and Q.

EXAMPLES 1. Find a parametric representation of the line through the points $(2,1)$ and $(7,6)$.
ANSWER Set $\mathbf{P} = \langle 2,1 \rangle$, $\mathbf{Q} = \langle 7,6 \rangle$. By Theorem 4 all points on this line have position vectors of the form

$$\mathbf{R} = \langle 2,1 \rangle + t(\langle 7,6 \rangle - \langle 2,1 \rangle) = \langle 2,1 \rangle + t\langle 5,5 \rangle = \langle 2 + 5t, 1 + 5t \rangle.$$

If we set $\mathbf{R} = \langle x,y \rangle$, we obtain

$$x = 5t + 2, \qquad y = 5t + 1. \tag{*}$$

2. Find the equation of the line in Example 1.
FIRST SOLUTION Solve the first equation (*) for t. This yields $t = \frac{1}{5}x - \frac{2}{5}$. Substitute this into the second equation. This yields $y = 5(\frac{1}{5}x - \frac{2}{5}) + 1$ or $y = x - 1$.
SECOND SOLUTION The two-point form of the equation of a line (Chapter 2, §2.2) gives

$$\frac{y - 1}{x - 2} = \frac{6 - 1}{7 - 2} \qquad \text{or} \qquad y = x - 1.$$

3. Find a parametric representation for the line $2x - 3y + 1 = 0$.

SOLUTION   We locate two points on the line, say, $(0,\frac{1}{3})$ and $(-\frac{1}{2},0)$, and proceed as in Example 1:

$$\langle x,y \rangle = \langle 0,\tfrac{1}{3} \rangle + t(\langle -\tfrac{1}{2},0 \rangle - \langle 0,\tfrac{1}{3} \rangle) = \langle 0,\tfrac{1}{3} \rangle + t\langle -\tfrac{1}{2},-\tfrac{1}{3} \rangle = \langle -\tfrac{1}{2}t, -\tfrac{1}{3}t + \tfrac{1}{3} \rangle,$$

or

$$x = -\tfrac{1}{2}t, \qquad y = -\tfrac{1}{3}t + \tfrac{1}{3}.$$

4. Find a parametric representation of the line $x = 3$.

ANSWER $x = 3$, $y = t$.

PROBLEMS

In Problems 1 to 6 find a parametric representation for each of the lines passing through the indicated points. [Remember that parametric representations are not unique.]

1. $(1,2)$ and $(4,3)$.
2. $(-1,2)$ and $(1,-1)$.
3. $(-2,-2)$ and $(1,4)$.

4. $(-2,3)$ and $(4,6)$.
5. $(5,0)$ and $(2,-3)$.
6. $(-2,1)$ and $(4,-2)$.

In Problems 7 to 12 find a parametric representation for each of the given lines.

7. $y = 0$ (x-axis).
8. $x = 0$ (y-axis).
9. $x + 2y + 2 = 0$.

10. $x - 4y + 8 = 0$.
11. $2x + 3y - 6 = 0$.
12. $3x - 2y - 1 = 0$.

§2 Vector calculus

In this section we combine calculus with vector algebra, as a preparation for the applications to be discussed later, and give some geometric applications.

2.1 Vector-valued functions

A **vector-valued function**

$$t \mapsto \mathbf{f}(t) \tag{1}$$

is a rule that associates a vector $\mathbf{f}(t)$ with every real number t in some interval or collection of intervals.

It is often convenient to think of vectors as position vectors of points, with respect to some fixed point O. We can represent a vector-valued function (1) geometrically by drawing all points with position vector $\mathbf{f}(t)$.

We choose a Cartesian coordinate system with origin O and unit vectors $\mathbf{i}$, $\mathbf{j}$. A vector-valued function $t \mapsto \mathbf{f}(t)$ can be written as

$$\mathbf{f}(t) = x(t)\mathbf{i} + y(t)\mathbf{j}. \tag{2}$$

Thus a vector-valued function is represented by a pair of ordinary (number-valued) functions

$$t \mapsto x(t), \qquad t \mapsto y(t). \tag{3}$$

We call these functions the **components** of $\mathbf{f}(t)$.

We see now that a vector-valued function of t may be thought of as a *parametric representation* of a curve (see Chapter 16, §2.1). Using vectors has an important advantage: no explicit reference must be made to a coordinate system.

An example of a vector-valued function is a nonconstant linear function

$$\mathbf{f}(t) = t\mathbf{a} + \mathbf{b}, \qquad -\infty < t + \infty, \tag{4}$$

where $\mathbf{a} \neq \mathbf{0}$ and $\mathbf{b}$ are fixed vectors. It represents a straight line as we saw in §1.7. Another example is the vector-valued function which represents the unit circle:

$$\mathbf{f}(t) = (\cos t)\mathbf{i} + (\sin t)\mathbf{j}, \qquad 0 \leq t < 2\pi. \tag{5}$$

It is often convenient to think of a vector-valued function $\mathbf{f}(t)$ as describing the motion of a point in the plane; namely, the point with the position vector $\mathbf{f}(t)$. In this case, the independent variable t (also called the **parameter**) is interpreted as time, and the components $x(t)$, $y(t)$ are interpreted as the horizontal and vertical displacements along the x- and y-axes, respectively.

Equation (4) describes the motion of a point along a straight line. At $t = 0$, the moving point is at the point P with position vector $\mathbf{b}$; at time $t = 1$, the moving point is at the point Q with position vector $\mathbf{a} + \mathbf{b}$. More generally, at any time t the point divides the segment $\overrightarrow{PQ}$ in the ratio $t/(1 - t)$.

The function (5) describes a motion of a point along the unit circle, starting at the point (0,1) at $t = 0$, proceeding in the counterclockwise direction, and traversing the whole circle in 2π time units.

EXAMPLES 1. Describe in words the motion of a point described by the function $\mathbf{f}(t) = 2t\mathbf{i}$.
ANSWER The point moves along the x-axis, moves to the right, starts at the origin at $t = 0$, and has distance $2t$ from the origin at time t.

2. A point moves on the curve $y = x^2$, starts at the origin at $t = 0$, and moves to the right. The distance of the point from the y-axis is proportional to the time t. At $t = 1$, the point is at (2,4). What vector-valued function describes the motion?
ANSWER We need a function $\mathbf{f}(t) = x(t)\mathbf{i} + y(t)\mathbf{j}$ with $y(t) = (x(t))^2$. Hence $f(t) = x(t)\mathbf{i} + x(t)^2\mathbf{j}$. The distance from the y-axis is $x(t)$. Hence $x(t) = kt$ (since this distance is proportional to time). Thus $\mathbf{f}(t) = kt\mathbf{i} + k^2t^2\mathbf{j}$. For $t = 1$ we must have $\mathbf{f}(1) = 2\mathbf{i} + 4\mathbf{j}$. Hence $k = 2$. Conclusion: $\mathbf{f}(t) = 2t\mathbf{i} + 4t^2\mathbf{j}$.

3. A point moves on the curve $y = \sqrt{x^2 + 1}$, starting at (0,1) at time $t = \frac{1}{4}$ and moving to the right. If the distance of the point from the origin is proportional to t, find the vector-valued function describing the motion.
SOLUTION We must express $\mathbf{f}(t)$ as $\mathbf{f}(t) = x(t)\mathbf{i} + y(t)\mathbf{j}$, where the functions x and y satisfy the equation of the curve; that is, $y(t) = \sqrt{(x(t))^2 + 1}$ for each $t \geq \frac{1}{4}$. Since the distance of the point from the origin is proportional to t, we also have $\sqrt{(x(t))^2 + (y(t))^2} = kt$. Thus $\sqrt{(x(\frac{1}{4}))^2 + (y(\frac{1}{4}))^2} = \frac{1}{4}k$ or $\sqrt{0 + 1} = \frac{1}{4}k$, that is, $k = 4$. Therefore

$$\sqrt{(x(t))^2 + (y(t))^2} = 4t.$$

Replacing $y(t)$ by $\sqrt{(x(t))^2 + 1}$ in the above equation, and squaring both sides, we obtain $2(x(t))^2 + 1 = 16t^2$ or $x(t) = \pm\sqrt{8t^2 - \frac{1}{2}}$. We may drop the minus sign, since the point is always in the first quadrant. Therefore

$$x(t) = \sqrt{8t^2 - \frac{1}{2}} \quad \text{and} \quad y(t) = \sqrt{(x(t))^2 + 1} = \sqrt{8t^2 + \frac{1}{2}}.$$

Hence $\mathbf{f}(t) = \sqrt{8t^2 - \frac{1}{2}}\,\mathbf{i} + \sqrt{8t^2 + \frac{1}{2}}\,\mathbf{j}.$

PROBLEMS

In Problems 1 to 7 find the vector-valued function $\mathbf{f}(t)$ describing the given motion. Assume that (x,y) denotes the coordinates with respect to a fixed Cartesian coordinate system.

1. A point moves on the parabola $y = x^2$, starting at $(1,1)$ at time $t = 0$, moving to the right, and reaching $(2,4)$ at time $t = 3$. Suppose the horizontal displacement of the point from the start of the motion is proportional to the square of the elapsed time from the start of the motion.

2. A point moves on the curve $y = x^3 + 1$, starting at $(-1,0)$ at time $t = 1$, moving to the right, and reaching $(1,2)$ at time $t = 5$. Suppose the horizontal displacement of the point from the start of the motion is proportional to the square root of the elapsed time from the start of the motion.

3. A point moves on the curve $y = x^3$, starting at the origin at time $t = 2$, moving to the right, and reaching $(2,8)$ at time $t = 4$. Suppose the slope of the line through the particle tangent to the path of the particle is proportional to the cube of the elapsed time from the start of the motion.

4. A point moves on the parabola $y = x^2 - x$, starting at the origin at time $t = -1$, moving to the left, and reaching $(-6,42)$ at time $t = 1$. Suppose the sum of the horizontal and vertical displacements of the point from the start of the motion is proportional to the square of the elapsed time from the start of the motion.

5. A point moves on the curve $y = \log x$, starting at $(1,0)$ at time $t = -3$, moving to the right, and reaching $(10,\log 10)$ at time $t = 0$. Suppose the rate of change of the horizontal displacement of the point is proportional to the elapsed time from the start of the motion.

6. A point moves on the line $y = 8x$, starting at the origin at time $t = 1$, moving to the right, and reaching $(1,8)$ at time $t = 2$. Suppose that the area of the right triangle formed by the origin, the point, and the projection of the point on the x-axis is proportional to the cube of the elapsed time from the start of the motion.

7. A point moves on the curve $3y - 2x^{3/2} = 0$, starting at the origin at time $t = 2$, moving to the right, and reaching $x = 3$ at time $t = 16$. Suppose the total distance traveled by the point (measured along the curve) is proportional to the elapsed time from the start of the motion.

2.2   Calculus of vector-valued functions

Using the representation (2), we can transfer to vector-valued functions most concepts of calculus. The rule of thumb is: *calculate in the usual way, treating the unit vectors* $\mathbf{i}$ *and* $\mathbf{j}$ *as constants.* Let us make this explicit.

The function $\mathbf{f}(t)$ is said to be **bounded** in an interval if both components are bounded. It is said to be **continuous** at a point if both components $x(t)$ and $y(t)$ are continuous at the point. The relation

$$\lim_{t \to t_0} \mathbf{f}(t) = a\mathbf{i} + b\mathbf{j} = \mathbf{c} \tag{6}$$

is defined to mean that

$$\lim_{t \to t_0} x(t) = a, \qquad \lim_{t \to t_0} y(t) = b. \tag{7}$$

This means: for t sufficiently close to t_0, $|\mathbf{f}(t) - \mathbf{c}|^2 = |x(t) - a|^2 + |y(t) - b|^2$ is arbitrarily small. Hence (6) can be written as

$$\lim_{t \to t_0} |\mathbf{f}(t) - \mathbf{c}| = 0. \tag{6'}$$

The derivative is defined by the relation

$$\frac{d\mathbf{f}}{dt} = \mathbf{f}'(t) = \frac{dx}{dt}\mathbf{i} + \frac{dy}{dt}\mathbf{j} \tag{8}$$

(if x' and y' exist at the point considered). We can check that this is equivalent to

$$\mathbf{f}'(t) = \lim_{h \to 0} \frac{1}{h}(\mathbf{f}(t + h) - \mathbf{f}(t)). \tag{8'}$$

For a fixed t, the derivative $\mathbf{f}'(t)$ is a vector; if we consider it for all (relevant) values of t, we have a vector-valued function. We can differentiate (8) again.

$$\frac{d^2\mathbf{f}}{dt^2} = \frac{d^2x}{dt^2}\mathbf{i} + \frac{d^2y}{dt^2}\mathbf{j}, \tag{9}$$

and so forth. We often denote differentiation with respect to the parameter t by a dot: $d\mathbf{f}/dt = \dot{\mathbf{f}}$, $dx/dt = \dot{x}$ (this tradition goes back to Newton). The preceding relations can be rewritten as

$$\dot{\mathbf{f}} = \dot{x}\mathbf{i} + \dot{y}\mathbf{j}, \qquad \ddot{\mathbf{f}} = \ddot{x}\mathbf{i} + \ddot{y}\mathbf{j}.$$

We also write

$$\mathbf{f} = \langle x,y \rangle, \qquad \dot{\mathbf{f}} = \langle \dot{x},\dot{y} \rangle, \qquad \ddot{\mathbf{f}} = \langle \ddot{x},\ddot{y} \rangle.$$

The integral of a vector-valued function is the vector defined by

$$\int_a^b \mathbf{f}(t)\, dt = \left(\int_a^b x(t)\, dt \right)\mathbf{i} + \left(\int_a^b y(t)\, dt \right)\mathbf{j}.$$

REMARK   The components of a vector depend upon the choice of the coordinate system. It is possible to verify that the concepts introduced above (boundedness, continuity, limit, derivative, and so forth) do not depend upon this choice. We can accomplish this by giving descriptions of these concepts not involving components. For limits and derivatives this is done by Equations (6') and (8').

EXAMPLES 1.   Find the derivative $(d/dt)(\mathbf{i} + 2t^2\mathbf{j})$.
ANSWER   The function $\mathbf{i} + 2t^2\mathbf{j}$ can be written as $1\mathbf{i} + 2t^2\mathbf{j}$. We have

$$\frac{d}{dt}(\mathbf{i} + 2t^2\mathbf{j}) = \frac{d1}{dt}\mathbf{i} + \frac{d2t^2}{dt}\mathbf{j} = 0\mathbf{i} + 4t\mathbf{j} = 4t\mathbf{j}.$$

2.   Compute $\int_0^{2\pi} (t^3\mathbf{a} + \sin t\mathbf{b})\, dt$, where $\mathbf{a}$ and $\mathbf{b}$ are constant vectors.
SOLUTION   The integral equals $(\frac{1}{4}t^4\mathbf{a} - \cos t\mathbf{b})|_0^{2\pi} = 4\pi^4\mathbf{a}$.

PROBLEMS

In Problems 1 to 20 $\mathbf{i}$ and $\mathbf{j}$ are constant unit vectors. Evaluate each of the following derivatives.

1. $\dfrac{d}{dt}[(\cos t)^2\mathbf{i} + (\sin t)\mathbf{j}]$.

2. $\dfrac{d}{dt}[t\mathbf{i} - (\log t)^3\mathbf{j}]$.

3. $\dfrac{d}{dt}[t(\sqrt{t}\,\mathbf{i} - (\sec t)\mathbf{j})]$.

4. $\dfrac{d}{dt}\left\{\dfrac{1}{t}(t\mathbf{i} - \mathbf{j}) + (\cos t)(t^2\mathbf{i} + \mathbf{j})\right\}$.

5. The second derivative of $t \mapsto t^{5/2}\mathbf{i} + (t-1)\mathbf{j}$.

6. $\mathbf{f}''(t)$ if $\mathbf{f}(t) = (\cos t)\mathbf{i} + (\tan t)\mathbf{j}$.

7. $\dfrac{d^2}{dt^2}(\sqrt{t}\,\mathbf{i} + e^t\mathbf{j})$.

8. $\dfrac{d}{dt}\left(\dfrac{t\mathbf{i} + \mathbf{j}}{|t\mathbf{i} + \mathbf{j}|}\right)$.

9. $\mathbf{f}'(0)$ if $\mathbf{f}(t) = (a(t) + t)(t\mathbf{i} - a(t)\mathbf{j})$ and $a(0) = 1$, $a'(0) = -2$.

10. $\mathbf{f}''(0)$ if $\mathbf{f}(t) = a(t)(t^2\mathbf{i} + (\cos t)\mathbf{j})$ and $a(0) = 1$, $a'(0) = -2$, $a''(0) = 4$.

Evaluate each of the following integrals.

11. $\displaystyle\int_0^{1/2} \left(t^2\mathbf{i} - \frac{4t}{1+t^2}\mathbf{j}\right) dt$.

12. $\displaystyle\int_0^{\pi/3} ((\cos 2t)\mathbf{i} + (\sin 3t)\mathbf{j})\, dt$.

13. $\displaystyle\int_1^2 \left(\frac{1}{t}\mathbf{i} + \frac{1}{t+1}\mathbf{j}\right) dt$.

14. $\displaystyle\int_1^3 \left(2^t\mathbf{i} - \frac{1}{\sqrt{t+1}}\mathbf{j}\right) dt$.

15. $\displaystyle\int_0^{\sqrt{\pi}} t(\sqrt{t^2+1}\,\mathbf{i} + (\cos t^2)\mathbf{j})\, dt$.

16. $\displaystyle\int_0^{\sqrt{3}} \frac{t\mathbf{i} + 4\mathbf{j}}{t^2+1}\, dt$.

17. $\displaystyle\int_0^{\log 2} e^t(\mathbf{i} + t\mathbf{j})\, dt$.

18. Find a vector-valued function $\mathbf{f}(t)$ such that $\mathbf{f}'(t) = t^{1/3}\mathbf{i} - t^2\mathbf{j}$ and $\mathbf{f}(0) = 2\mathbf{i} + \mathbf{j}$.

19. Find a vector-valued function $\mathbf{f}(t)$ such that $\mathbf{f}''(t) = t^2\mathbf{i}$ and $\mathbf{f}(0) = \mathbf{i} + \mathbf{j}$, $\mathbf{f}(1) = 2\mathbf{i} - \mathbf{j}$.

20. Find a vector-valued function $\mathbf{f}(t)$ such that $\mathbf{f}''(t) = (\cos t)\,\mathbf{i} + \mathbf{j}$ and $\mathbf{f}(0) = 3\mathbf{i} - \mathbf{j}$, $\mathbf{f}'(0) = \frac{3}{2}\mathbf{i} - \mathbf{j}$.

21. Prove that $(d/dt)\displaystyle\int_a^\tau \mathbf{f}(t)\, dt = \mathbf{f}(\tau)$ if $\mathbf{f}(t)$ is continuous at $t = \tau$.

22. Prove that $\displaystyle\int_a^b \mathbf{f}(t)\, dt = \mathbf{F}(b) - \mathbf{F}(a)$ if $\mathbf{f}(t)$ is continuous and bounded for $a < t < b$ and $\mathbf{F}(t)$ is a primitive function for $\mathbf{f}(t)$.

2.3   The unit tangent and normal vectors to a curve

A vector-valued function $\mathbf{f}(t) = x(t)\mathbf{i} + y(t)\mathbf{j}$ defines a smooth curve (see Chapter 16, §2.2) if

$$\mathbf{f}'(t) = x'(t)\mathbf{i} + y'(t)\mathbf{j} \tag{10}$$

exists and is continuous, and if $\mathbf{f}'(t) \neq 0$. This last condition is equivalent to relation (11) in Chapter 16, §2.2.

The derivative $\mathbf{f}'(t)$ has a geometric meaning. To guess at it, let h be a small positive number. Then the vector $(1/h)(\mathbf{f}(t+h) - \mathbf{f}(t))$ is close to the vector $\mathbf{f}'(t)$. But (see Figure 17.23) the vector $\mathbf{f}(t+h) - \mathbf{f}(t)$ is represented by the directed segment $\overrightarrow{PP_h}$, where P_h has the position vector $\mathbf{f}(t+h)$. Thus P_h is a point close to P and located in the direction of increasing t. We conclude that $\mathbf{f}'(t)$ has the direction of the "tangent line" to our curve and points in the direction of increasing parameter values, or, as we say, in the direction of the curve. The same result can be obtained analytically.

FIGURE **17.23**

FIGURE **17.24**

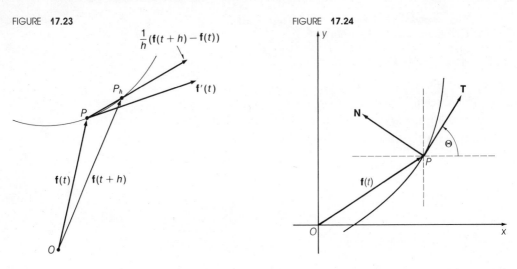

The polar angle of $\mathbf{f}'(t)$ is, by §1.6 and (10),

$$\Theta = \arctan \frac{y'(t)}{x'(t)}, \tag{11}$$

and this is, as we know from Chapter 16, §2.2, the slope of the tangent line to the curve.

A vector of length 1 is called, as we said before, a **unit vector**. If $\mathbf{a} \neq \mathbf{0}$ is any vector, $(1/|\mathbf{a}|)\mathbf{a}$ is a unit vector in the direction of $\mathbf{a}$. We apply this remark to $\mathbf{f}'(t)$ and obtain the unit vector

$$\mathbf{T}(t) = \frac{1}{|\mathbf{f}'(t)|} \mathbf{f}'(t), \tag{12}$$

called the **unit tangent vector** to the curve, at the point corresponding to the parameter value t. We compute that

$$\mathbf{T}(t) = \frac{x'(t)}{\sqrt{x'(t)^2 + y'(t)^2}} \mathbf{i} + \frac{y'(t)}{\sqrt{x'(t)^2 + y'(t)^2}} \mathbf{j}, \tag{13}$$

and that

$$\cos \Theta = \frac{x'(t)}{\sqrt{x'(t)^2 + y'(t)^2}}, \qquad \sin \Theta = \frac{y'(t)}{\sqrt{x'(t)^2 + y'(t)^2}}. \tag{14}$$

The **unit normal vector** $\mathbf{N}(t)$ to our curve is defined as the vector of length 1 obtained by rotating $\mathbf{T}(t)$ in the counterclockwise direction by 90°. Thus the polar angle of $\mathbf{N}(t)$ is $\Theta + 90°$, and its components are $\cos(\Theta + 90°) = -\sin \Theta$, $\sin(\Theta + 90°) = \cos \Theta$. Hence

$$\mathbf{N}(t) = -\frac{y'(t)}{\sqrt{x'(t)^2 + y'(t)^2}} \mathbf{i} + \frac{x'(t)}{\sqrt{x'(t)^2 + y'(t)^2}} \mathbf{j}. \tag{15}$$

Using Equation (14), Equations (13) and (15) can be written as

$$\mathbf{T}(t) = \cos \Theta \, \mathbf{i} + \sin \Theta \, \mathbf{j} \tag{16}$$

and

$$N(t) = -\sin\Theta\,\mathbf{i} + \cos\Theta\,\mathbf{j}. \tag{17}$$

See Figure 17.24.

EXAMPLE Find the unit tangent and normal vectors to the curve represented by $\mathbf{f}(t) = 2t\mathbf{i} + t^2\mathbf{j}$.
SOLUTION Differentiating $\mathbf{f}(t)$, we have $\mathbf{f}'(t) = 2\mathbf{i} + 2t\mathbf{j}$, and so $|\mathbf{f}'(t)| = \sqrt{4 + 4t^2} = 2\sqrt{1 + t^2}$.
Using Equation (12), the tangent vector is just

$$\mathbf{T}(t) = \frac{\mathbf{f}'(t)}{|\mathbf{f}'(t)|} = \frac{1}{\sqrt{1 + t^2}}\,\mathbf{i} + \frac{t}{\sqrt{1 + t^2}}\,\mathbf{j}.$$

From Equation (16) we have $\cos\Theta = 1/\sqrt{1 + t^2}$, $\sin\Theta = t/\sqrt{1 + t^2}$. Hence, by Equation (17), the normal vector is

$$N(t) = -\frac{t}{\sqrt{1 + t^2}}\,\mathbf{i} + \frac{1}{\sqrt{1 + t^2}}\,\mathbf{j}.$$

PROBLEMS

In each of Problems 1 to 5 find $\mathbf{T}(t)$ and $N(t)$ for the given vector-valued functions $\mathbf{f}(t)$.

1. $\mathbf{f}(t) = (\cos t)\mathbf{i} + t\mathbf{j}$.
2. $\mathbf{f}(t) = \sqrt{t^2 + 1}\mathbf{i} + t\mathbf{j}$.
3. $\mathbf{f}(t) = (t^3 - 1)\mathbf{i} - t^2\mathbf{j}$.
4. $\mathbf{f}(t) = e^{2t}\mathbf{i} + e^{-2t}\mathbf{j}$.
5. $\mathbf{f}(t) = (\sin^3 t)\mathbf{i} + (\cos^3 t)\mathbf{j}$.

6. Find $\mathbf{T}(t)$ and $N(t)$ for the graph of $y = \log(x^2 + 1)$. [*Hint:* Let $x = t$; then $y = \log(t^2 + 1)$.]
7. Find $\mathbf{T}(t)$ and $N(t)$ for the graph of $y = x^3 + 2$.
8. Find $\mathbf{T}(t)$ and $N(t)$ for the graph of $y^3 = x^2 + 4$ at the point with Cartesian coordinates (2,2).

2.4 Arc length

In Chapter 16, §2.3, we obtained the formula (13) for the length of a smooth curve defined parametrically. We rewrite this formula in vector language: the length of the curve defined by

$$t \mapsto \mathbf{f}(t), \qquad a \le t \le b \tag{18}$$

is

$$L = \int_a^b |\mathbf{f}'(t)|\,dt. \tag{19}$$

Note that the length formula (19) is given in terms of vectors. It presupposes no special choice of a coordinate system. Thus the length of a curve does not depend on the position of the Cartesian coordinates used to calculate it. This is, of course, as it should be. We have now settled a point left open in Chapter 9, §3.1.

If we denote the length of an arc of our curve from a (fixed) point $\mathbf{f}(t_0)$ to a variable point $\mathbf{f}(t)$ with $t > t_0$ by $s = L(t)$, we have

$$L(t) = \int_{t_0}^t |\mathbf{f}'(\tau)|\,d\tau. \tag{20}$$

This is the same formula as (19), except that we replaced a by t_0, b by t, and used a new name for the "dummy" variable of integration. Also, $L(t_0) = 0$, $L(b) = L$.

Assume that the parameter representation is such that $|\mathbf{f}'(\tau)| = 1$ for all τ. Then, by (20), $L(t) = t - t_0$. The length of an arc is the difference of the parameter values at the endpoints! We say that the parameter is the arc length. (We assume here, for the sake of simplicity, that the curve has no self-intersections: different values of the parameter correspond to different points.)

On any smooth curve we can introduce the arc length, given by (20), as a new parameter, as we show in an example.

EXAMPLE   Rewrite the curve given by $y = \frac{3}{2}(x - 1)$, $x \geq 1$, with the arc length s as parameter. Assume $s = 0$ for $x = 1$, $y = 0$.

ANSWER   The curve (a straight line) has the parametric representation

$$t \mapsto t\mathbf{i} + \tfrac{3}{2}(t - 1)\mathbf{j} = \mathbf{f}(t),$$

with $t = 1$ for $x = 1$, $y = 0$. We have, by (20),

$$s = L(t) = \int_{t_0}^{t} |\mathbf{f}'(\tau)| \, d\tau = \int_{1}^{t} |\mathbf{i} + \tfrac{3}{2}\mathbf{j}| \, d\tau$$

$$= \int_{1}^{t} \tfrac{1}{2}\sqrt{13} \, d\tau = \tfrac{1}{2}\sqrt{13}(t - 1)$$

so that $t = 1 + (2/\sqrt{13})s$. Hence the desired parametric representation is

$$s \mapsto \left(1 + \frac{2}{\sqrt{13}} s\right)\mathbf{i} + \frac{3}{\sqrt{13}} s\mathbf{j}.$$

PROBLEMS

In each of Problems 1 to 5 rewrite the given curve as a vector-valued function using the arc length s as a parameter.

1. $y = -\frac{1}{2}x + 1$ for $x \geq 0$. Assume $s = 0$ for $x = 0$, $y = 1$.
2. $y = 2x + 4$ for $x \geq -2$. Assume $s = 0$ for $x = -2$, $y = 0$.
3. $y = \frac{2}{3}x^{3/2}$ for $x \geq 0$. Assume $s = 0$ for $x = 0$, $y = 0$.

4. $y = \displaystyle\int_{1}^{x} \sqrt{u^2 - 1} \, du$ for $x \geq 1$. Assume $s = 0$ for $x = 1$, $y = 0$.

5. $y = \displaystyle\int_{0}^{x} \sqrt{3 + 4u} \, du$ for $x \geq 0$. Assume $s = 0$ for $x = 0$, $y = 0$.

§3   Plane motion

The most important applications of vectors occur in physics. In this section we discuss the motion of a particle in a plane, a more complicated problem than rectilinear motion studied in Chapter 4, §3.1, and Chapter 7, §1.3. The considerations of the present section extend to motion in space, once we define (in Chapter 18, §2) vectors in space.

3.1   Velocity and acceleration

Motion in a plane is described by a vector-valued function

$$t \mapsto \mathbf{r}(t) = x(t)\mathbf{i} + y(t)\mathbf{j}$$

which indicates the position vector $\mathbf{r}$ of the particle at the time t. (We assume that this function has a continuous second derivative. The position vector is denoted by the letter $\mathbf{r}$ in order to reserve $\mathbf{f}$ for force.) The derivative $\mathbf{v}(t) = d\mathbf{r}(t)/dt$ is called **velocity.** It is again a vector-valued function. Throughout this and the following section, we shall denote differentiation with respect to time by a dot. Thus we have

$$\mathbf{v} = \dot{\mathbf{r}} \qquad \text{or, in components,} \qquad \mathbf{v} = \dot{x}\mathbf{i} + \dot{y}\mathbf{j};$$

$\dot{x}$ and $\dot{y}$ are called the horizontal and vertical components of velocity. The length of the velocity vector, $|\mathbf{v}| = \sqrt{\dot{x}^2 + \dot{y}^2}$, is called **speed.** The time derivative of the velocity vector

$$\mathbf{a}(t) = \frac{d\mathbf{v}(t)}{dt} = \frac{d^2\mathbf{r}(t)}{dt^2}, \qquad \text{that is,} \qquad \mathbf{a} = \dot{\mathbf{v}} = \ddot{\mathbf{r}} = \ddot{x}\mathbf{i} + \ddot{y}\mathbf{j},$$

is called **acceleration;** $\ddot{x}$ and $\ddot{y}$ are called the horizontal and vertical components of acceleration.

The function $t \mapsto \mathbf{r}(t)$ describing the motion of our particle defines an oriented curve, the **trajectory** of the motion. Now the parameter t has a physical meaning: it is the time at which the particle is at the point $\mathbf{r}(t)$.

The scientific use of the terms velocity, speed, and acceleration corresponds closely to the everyday meaning of these words. Indeed, consider a moving particle, and assume that it is at the point P with position vector $\mathbf{r}(t)$ at the time t, and at the point Q with position vector $\mathbf{r}(t + h)$ after a short time interval h (see Figure 17.25). The

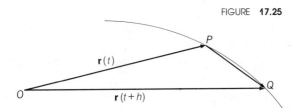

FIGURE **17.25**

vector $(1/h)[\mathbf{r}(t + h) - \mathbf{r}(t)]$ is nearly the velocity vector $\mathbf{v}(t)$. Its direction is that from P to Q, which is nearly the direction of motion. The magnitude of this vector is $|PQ|/h$, which is nearly the distance traveled, divided by the time elapsed. Thus the velocity vector $\mathbf{v}(t)$ may be said to describe the instantaneous change of position. We see similarly that the acceleration vector describes the instantaneous change in velocity.

Assume that at the time t the velocity vector $\mathbf{v}(t)$ is not the null-vector. Then the velocity vector $\mathbf{v}(t)$ is tangent to the trajectory at the point $\mathbf{r}(t)$. Indeed, as we saw in §2.3, the unit tangent vector $\mathbf{T}(t)$ is

$$\mathbf{T}(t) = \frac{1}{|\dot{\mathbf{r}}(t)|}\dot{\mathbf{r}}(t) = \frac{1}{|\mathbf{v}(t)|}\mathbf{v}(t),$$

so that $\mathbf{T}$ and $\mathbf{v}$ have the same direction.

Let $s(t)$ denote the length on the trajectory, measured from some fixed time t_0, in the direction of the motion. Then, as we saw in §2.4 (where we wrote $\mathbf{f}$ instead of $\mathbf{r}$),

$$s(t) = \int_{t_0}^{t} |\dot{\mathbf{r}}(\tau)|\, d\tau = \int_{t_0}^{t} |\mathbf{v}(\tau)|\, d\tau.$$

By the fundamental theorem of calculus,

$$\frac{ds}{dt} = |\mathbf{v}(t)|;$$

thus the speed is the time rate of change of the distance traveled. Since $\mathbf{v}(t)$ and $\mathbf{T}(t)$ have the same direction, and $\mathbf{T}(t)$ is a unit vector,

$$\mathbf{v}(t) = \frac{ds}{dt}\,\mathbf{T}(t). \tag{1}$$

EXAMPLES 1.   The path of a point (or "particle") is described by the position vector $\mathbf{r}(t) = (3t^2 + 1)\mathbf{i} + t^3\mathbf{j}$, where $\mathbf{r}(t)$ is measured in feet and t is measured in seconds. Find the velocity, speed, and acceleration when $t = 2$ seconds.
SOLUTION   We have

$$\mathbf{v}(t) = \dot{\mathbf{r}}(t) = (6t)\mathbf{i} + (3t^2)\mathbf{j}, \qquad \mathbf{a}(t) = \dot{\mathbf{v}}(t) = 6\mathbf{i} + (6t)\mathbf{j}.$$

Hence $\mathbf{v}(2) = (12\mathbf{i} + 12\mathbf{j})$ ft/sec, $|\mathbf{v}(2)| = \sqrt{12^2 + 12^2} = 12\sqrt{2}$ ft/sec, and $\mathbf{a}(2) = (6\mathbf{i} + 12\mathbf{j})$ ft/sec^2.

2.   A particle moves along the curve $y = x^2 - 6x + 8$ in such a way that the x component of velocity is always 2 ft/sec. Find the y component of velocity and acceleration when the particle passes through the point $(4,0)$.
ANSWER   Differentiating $y = x^2 - 6x + 8$ with respect to t, and noting that the x component of the velocity is $\dot{x} = 2$, we obtain

$$\dot{y} = 2x\dot{x} - 6\dot{x}, \qquad \ddot{y} = 2\dot{x}^2 + 2x\ddot{x} - 6\ddot{x} = 8$$

since $\ddot{x} = 0$. For $x = 4$, the y component of the velocity is $\dot{y} = 4$ ft/sec, and the y component of the acceleration is $\ddot{y} = 8$ ft/sec^2.

PROBLEMS

In Problems 1 to 6 the path of a point (particle) is described by the position vector $\mathbf{r}(t) = x(t)\mathbf{i} + y(t)\mathbf{j}$. For each of the given motions, find $\mathbf{v}(t)$, $|\mathbf{v}(t)|$, and $\mathbf{a}(t)$. Assume that $t \geq 0$.

1. $\mathbf{r}(t) = t\mathbf{i} + \frac{1}{2}t^2\mathbf{j}$.
2. $\mathbf{r}(t) = (t^3 + t)\mathbf{i} + (t^3 - t)\mathbf{j}$.
3. $\mathbf{r}(t) = (2\cos t)\mathbf{i} + (3\cos t)\mathbf{j}$.
4. $\mathbf{r}(t) = (\cos^2 t)\mathbf{i} + (\sin^2 t)\mathbf{j}$.

5. $\mathbf{r}(t) = \sqrt{t^2 + 1}\,\mathbf{i} + \dfrac{1}{\sqrt{t^2 + 1}}\mathbf{j}$.

6. $\mathbf{r}(t) = (e^{-t}\cos t)\mathbf{i} + (e^{-t}\sin t)\mathbf{j}$.

7. A particle moves along the curve $y = \frac{1}{16}x^2$ in such a way that the y component of velocity is always 8 ft/sec. Find $\dot{x}$ and $\mathbf{a}$ when the particle passes through the point $(4,1)$.
8. A particle moves along the curve $y = e^{-\frac{1}{2}x}$ in such a way that the x component of velocity is always 2 ft/sec. Find $\dot{y}$ and $\mathbf{a}$ when the particle passes through the point $(2, 1/e)$.
9. A particle moves up the curve $y = \frac{1}{4}x^2$ with a constant speed of 5 ft/sec. Find $\mathbf{v}$ and $\mathbf{a}$ when the particle passes through the point $(2,1)$.
10. If the path of a particle is described by $\mathbf{r}(t) = 3t^2\mathbf{i} + (9t - t^3)\mathbf{j}$, find the minimum value of the speed, and the Cartesian coordinates where this value occurs.

3.2   Circular motion

An important example is a plane motion given by

$$x = R \cos 2\pi\nu t, \qquad y = R \sin 2\pi\nu t,$$

where R and ν are positive constants. The trajectory is a circle of radius R. The velocity has the components

$$\dot{x} = -2\pi\nu R \sin 2\pi\nu t, \qquad \dot{y} = 2\pi\nu R \cos 2\pi\nu t;$$

the speed $|\mathbf{v}(t)| = (\dot{x}^2 + \dot{y}^2)^{1/2} = 2\pi\nu R$ is constant; per unit time the moving particle covers the distance $2\pi\nu R$, that is, ν times the length of the full circle. For this reason, ν is called the **frequency**. The **period** of the motion, that is, the time required for covering the circle once, is

$$T = \frac{1}{\nu}.$$

We note the interesting connection between a circular motion and harmonic oscillation (see Chapter 12, §2.4). When a particle moves along a circle with constant speed, each of its components (in a Cartesian system) executes a harmonic oscillation.

The acceleration of our motion has the components $\ddot{x} = -4\pi^2\nu^2 R \cos 2\pi\nu t$, $\ddot{y} = -4\pi^2\nu^2 R \sin 2\pi\nu t$ or $\ddot{x} = -4\pi^2\nu^2 x$, $\ddot{y} = -4\pi^2\nu^2 y$. This shows that acceleration vector $\mathbf{a} = -4\pi^2\nu^2(x\mathbf{i} + y\mathbf{j}) = -4\pi^2\nu^2\mathbf{r}$ has the direction opposite to that of the position vector $\mathbf{r}$. The acceleration is not null, although the speed is constant. There is no paradox: the *direction* of the velocity vector changes in time. The magnitude of the acceleration, however, is constant; we have

$$|\mathbf{a}| = \sqrt{\ddot{x}^2 + \ddot{y}^2} = 4\pi^2\nu^2 R = \frac{4\pi^2 R}{T^2}. \tag{2}$$

EXAMPLE   The motion of an **artificial satellite** may be considered, in first approximation, as a circular motion with radius R equal to that of the earth. The satellite experiences, therefore, an acceleration directed toward the center of the earth of magnitude $4\pi^2 R/T^2$. But the satellite in orbit is subject only to gravity (or, as we say, is in free fall). Thus $4\pi^2 R/T^2 = g = 980$ cm/sec^2. Hence the period $T = 2\pi\sqrt{R/g}$ can be computed by knowing R. Since $R \approx 4000$ miles $\approx 6.44 \cdot 10^8$ cm, we obtain $T = 5.1 \cdot 10^3$ sec $= 85$ minutes, approximately. This explains why most artificial satellites have periods of this order of magnitude.

3.3   Law of motion

We formulate now **Newton's Law of Motion:** at any moment the acceleration of a particle of mass m equals $1/m$ times the force acting on the particle. Here m is a number; since the acceleration $\mathbf{a}$ is a vector, the force must be also a vector. If we denote it by $\mathbf{f}$, the mathematical expression of Newton's Law reads

$$m\mathbf{a} = \mathbf{f}. \tag{3}$$

Of course, this becomes a law of nature that can be confirmed or disproved by experiments only if something is said about finding $\mathbf{f}$.

One statement about forces is implicit in the usual formulation of Newton's Law but should be made explicit: if two forces $\mathbf{F}_1$ and $\mathbf{F}_2$ act on a particle, the particle moves as if it were acted upon by the sum $\mathbf{f} = \mathbf{F}_1 + \mathbf{F}_2$.

The reader will observe that the "Parallelogram Law" for vector addition (see §1.3) is a mathematical definition, but the statement that the effect of two forces can be computed by forming their sum according to this definition is a physical law.

If no force acts on the particle, Newton's Law gives $\mathbf{a} = \mathbf{0}$; that is, $\dot{\mathbf{v}}(t) = \mathbf{0}$ and $\mathbf{v} = \mathbf{a}$ constant vector $\mathbf{c}$, and therefore $\mathbf{r}(t) = t\mathbf{v} + \mathbf{c}$. A particle subject to no force moves along a straight line and with constant speed. This statement is the **Law of Inertia.**

3.4   Motion of a projectile

As an application of Newton's Law, consider plane motion under the influence of gravity. We choose the x, y plane as the plane of motion with the y-axis pointing "up," so that the weight of the particle is $-mg\mathbf{j}$, where g is the acceleration of gravity (compare Chapter 7, §1.3, where we considered only vertical motion). If weight is the only force present, then the equation of motion reads

$$m\ddot{\mathbf{r}} = m\mathbf{a} = -mg\mathbf{j}$$

or, canceling m and using components,

$$\ddot{x} = 0, \qquad \ddot{y} = -g.$$

The two differential equations are independent of each other and can be solved independently. We have

$$\dot{x}(t) = \alpha, \qquad \dot{y}(t) = -gt + \beta, \tag{4}$$

where α and β are constants, and therefore

$$x(t) = \alpha t + \gamma, \qquad y(t) = -\tfrac{1}{2}gt^2 + \beta t + \delta, \tag{5}$$

with some new constants γ, δ. It is seen from (5) that (γ,δ) are the coordinates of our particle at time $t = 0$, whereas (4) shows that (α,β) are the components of the velocity vector at time $t = 0$. The same formula shows that the horizontal velocity $\dot{x}$ is constant during the motion. (It must be since there is no horizontal force.) The whole motion is determined if we know the **initial position** (γ,δ) and the **initial velocity** (α,β).

Formulas (5) are a parametric representation of the trajectory. To find a non-parametric representation of this curve, we "eliminate t." The first equation yields $t = (x - \gamma)/\alpha$, provided that $\alpha \neq 0$. Substituting this value of t in the second equation, we obtain

$$y = -\frac{1}{2}g\left(\frac{x-\gamma}{\alpha}\right)^2 + \beta\frac{x-\gamma}{\alpha} + \delta$$

or

$$y = -\frac{g}{2\alpha^2}x^2 + \frac{g\gamma + \alpha\beta}{\alpha^2}x - \frac{g\gamma^2 + 2\alpha\beta\gamma - 2\alpha^2\delta}{2\alpha^2}.$$

Hence the trajectory is a **parabola,** with a vertical axis, that is convex upward (see Figure 17.26). If $\alpha = 0$, then x is constant, and we have rectilinear vertical motion as considered in Chapter 7, §1.3.

It should be remembered, of course, that in deriving (5) we neglected air resistance, wind, the influence of the rotation of the earth, and so forth.

FIGURE **17.26**   FIGURE **17.27**   FIGURE **17.28**

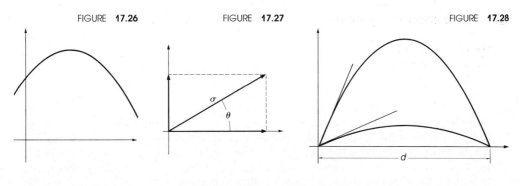

EXAMPLE The initial speed of a projectile is σ cm/sec. It is shot from ground level at an angle θ, $0 < \theta < 90°$, and hits the ground again at a distance d cm. Find θ. [We assume, of course, that the projectile moves according to the law (5).]

SOLUTION We have $\delta = 0$ and may assume $\gamma = 0$. The initial velocity has components $\alpha = \sigma \cos \theta$ and $\beta = \sigma \sin \theta$ (see Figure 17.27). Therefore $y(t) = -\frac{1}{2}gt^2 + \beta t$; this equals 0 for $t = 0$, the initial moment, and for $t = 2\beta/g = 2\sigma \sin \theta/g$. For this value of t, we have

$$x(t) = \alpha t = \frac{2\sigma^2 \sin \theta \cos \theta}{g} = \frac{\sigma^2}{g} \sin 2\theta,$$

and this should be d. Thus θ is found from the relation $\sin 2\theta = gd/\sigma^2$. This gives two possible values for θ: $\theta = \theta_1$ and $\theta = 90° - \theta_1$ (see Figure 17.28).

PROBLEMS

1. A projectile is shot from ground level with an initial speed σ_0 and strikes the ground again with a terminal speed σ_1. Show that $\sigma_0 = \sigma_1$.

2. A projectile is shot from ground level at time $t = 0$ and strikes the ground again at time $t = t_1$. Show that the projectile attains its maximum height at time $t = t_1/2$.

3. A projectile is shot from point O at ground level with an initial velocity $\alpha \mathbf{i} + \beta \mathbf{j}$. Let P be the highest point on the trajectory and let Q be the point where the particle strikes the ground again. Let h be the height of P above ground level, let r ($=$ range) be the distance from O to Q, and let θ ($=$ angle of inclination) be the polar angle of the initial velocity. Let ϕ be the angle $\angle QOP$. (a) Compute r and h. (b) Show that $2 \tan \phi = \tan \theta$.

In Problems 4 to 8 assume $g = 32$ ft/sec for computational convenience.

4. A projectile is shot from ground level with an initial velocity of $640\mathbf{i} + 320\mathbf{j}$ ft/sec. How far away will the projectile again hit the ground?

5. A projectile is shot from ground level with an initial velocity of $250\mathbf{i} + 48\mathbf{j}$ ft/sec. How far above ground level will the projectile strike a sheer vertical cliff which is 500 ft away?

6. A projectile is shot from ground level with an initial velocity of $500\mathbf{i} + 32\mathbf{j}$ ft/sec. The projectile strikes a sheer vertical cliff 12 ft above ground level. How far away is the cliff? Explain why there are two possible answers.

7. A projectile shot from ground level attains a maximum height of 100 ft above ground level and a range (distance to the point where the projectile strikes the ground again) of 2000 ft. Find the initial velocity.

8. A projectile shot from a 16-ft-high platform strikes the ground after 8 sec. If the tangent of the angle of the initial velocity vector is $\frac{1}{4}$, find the horizontal distance traveled.

§4   Plane motion, continued[‡]

We discuss next some more sophisticated cases of plane motion. In this connection we introduce the concept of curvature.

4.1   Curvature of a plane curve

Consider a curve defined by a vector-valued function $t \mapsto \mathbf{f}(t) = \mathbf{r}(t) = x(t)\mathbf{i} + y(t)\mathbf{j}$, and assume that this function has a continuous second derivative. Two such curves are shown in Figure 17.29. It is evident that the first is "more curved" than the second and that the first curve is "more curved" at the point P than at the point Q. We want to find a way to *measure* the property of "being curved."

To this end, let $\Theta(t)$ be the polar angle of the unit tangent vector $\mathbf{T}(t)$ to the curve (see §2.3). If the curve is a straight segment, the angle Θ is constant. It is natural to take the rate of change of Θ as we move along the curve as the measure of how much the curve is curved. But the rate of change with respect to what? There is only one "natural parameter" on a curve; this is the arc length defined by

$$s(t) = \int_{t_0}^{t} |\mathbf{r}'(\tau)| \, d\tau$$

(see §2.4). We therefore define the **curvature** κ of the curve at a point P as the value, at this point, of the derivative of the polar angle of the unit tangent vector with respect to arc length.

We may consider the polar angle Θ and also t as functions of s. By the chain rule,

$$\kappa = \frac{d\Theta}{dt} \frac{dt}{ds}, \tag{1}$$

and, since $ds/dt = |\mathbf{r}'(t)|$, we have $dt/ds = 1/|\mathbf{r}'(t)|$, and therefore

$$\kappa(t) = \frac{d\Theta}{ds} = \frac{\Theta'(t)}{|\mathbf{r}'(t)|}. \tag{2}$$

Since [by Equation (11) in §2.3] $\tan \Theta(t) = y'(t)/x'(t)$, we also have

$$\Theta(t) = \arctan \frac{y'(t)}{x'(t)}, \tag{3}$$

and therefore

$$\Theta'(t) = \frac{1}{1 + y'(t)^2/x'(t)^2} \frac{x'(t)y''(t) - x''(t)y'(t)}{x'(t)^2}$$

$$= \frac{x'(t)y''(t) - x''(t)y'(t)}{x'(t)^2 + y'(t)^2}.$$

At points at which $x'(t) = 0$, we arrive at the same result starting with the formula $\cot \Theta(t) = x'(t)/y'(t)$; we assume that $x'(t)$ and $y'(t)$ cannot both be zero.

[‡]Optional section.

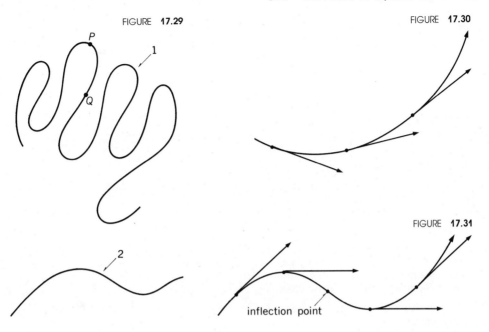

FIGURE **17.29**

FIGURE **17.30**

FIGURE **17.31**

inflection point

Substituting into (2) and noting that $|\mathbf{r}'| = (x'(t)^2 + y'(t)^2)^{1/2}$, we obtain the final formula:

$$\kappa(t) = \frac{x'(t)y''(t) - x''(t)y'(t)}{(x'(t)^2 + y'(t)^2)^{3/2}}. \qquad (4)$$

It may appear that the value of the curvature depends on the position of the coordinate axes, since the polar angle of a vector depends on this position, but this is not so. For if we rotate the coordinates, we must add a constant to $\Theta(t)$, and this does not affect the derivative of Θ. On the other hand, the sign of the curvature depends on the orientation of the plane. (For this reason, some authors call the absolute value $|\kappa(t)|$ the curvature.)

If $\kappa = d\Theta/ds > 0$, then the angle Θ increases as we proceed along the curve in the direction of increasing parameter values; that means the unit tangent vector is rotated counterclockwise (see Figure 17.30). If $\kappa < 0$, then Θ decreases; that is, $\mathbf{T}$ is rotated in the clockwise direction. A point at which the curvature changes sign is called an **inflection point** (see Figure 17.31).

The graph of a number-valued function $y = f(x)$, $a \le x \le b$ has the parametric representation $x(t) = t$, $y(t) = f(t)$ so that

$$x'(t) = 1, \qquad x''(t) = 0, \qquad y'(t) = f'(t), \qquad y''(t) = f''(t),$$

and we obtain from (4) the following formula for the curvature of the graph of a function f at point $(x, f(x))$:

$$\kappa(x) = \frac{f''(x)}{(1 + f'(x)^2)^{3/2}}. \qquad (5)$$

EXAMPLES **1.**   **Circle** of radius R. A parametric representation is

$$x = R \cos t, \qquad y = R \sin t, \qquad 0 \le t < 2\pi; \tag{6}$$

hence $x' = -R \sin t$, $x'' = -R \cos t$, $y' = R \cos t$, $y'' = -R \sin t$, and, by (4), $\kappa = 1/R$. The curvature at all points of the circle is $1/R$. Of course, we could have obtained this directly by noting that $\Theta(t) = t + \pi/2$ so that $d\Theta/dt = 1$, and $s(t) = Rt$ so that $dt/ds = 1/R$, and hence $d\Theta/ds = 1/R$.

The upper half of our circle is the graph of the function

$$y = f(x) = \sqrt{R^2 - x^2}. \tag{7}$$

We have

$$f'(x) = \frac{-x}{\sqrt{R^2 - x^2}}, \qquad f''(x) = \frac{-R^2}{(R^2 - x^2)\sqrt{R^2 - x^2}},$$

$$[1 + f'(x)^2]^{3/2} = \frac{R^3}{(R^2 - x^2)\sqrt{R^2 - x^2}},$$

and, by (5), $\kappa = -1/R$. The apparent discrepancy is easily explained. In the parametric representation (6), the parameter t increases as we proceed along the circle in the counterclockwise direction. In the representation (7), the parameter is x; here x increases as we proceed along the circle in the clockwise direction.

2.   The **parabola** $y = x^2$. Here $\kappa(x) = 2(1 + 4x^2)^{-3/2}$. The curvature is positive for all x. There are no inflection points.

3.   The **cubic parabola.** This is the graph of $y = x^3$. We have $\kappa(x) = 6x(1 + 9x^4)^{-3/2}$. The curvature is positive for $x > 0$, negative for $x < 0$. There is an inflection point at $x = y = 0$.

PROBLEMS

In Problems 1 to 8 find the curvature and the sign of the curvature of the given curve. Find any inflection points on the curve.

1. $y = x^2 - 2x - 8$ for $-\infty < x < +\infty$.   3. $y = 1/x$ for $0 < x < +\infty$.
2. $y = xe^x$ for $-\infty < x < +\infty$.   4. $y = \log \sin x$ for $0 < x < \pi$.

5. $\mathbf{f}(t) = \frac{1}{3}t^3\mathbf{i} + (2t + 1)\mathbf{j}$ for $-\infty < t < +\infty$.
6. $\mathbf{f}(t) = (\cos t)\mathbf{i} + (2 \sin t)\mathbf{j}$ for $0 \le t \le 2\pi$.
7. $\mathbf{f}(t) = e^t \sin t\,\mathbf{i} + e^t \cos t\,\mathbf{j}$ for $-\infty < t < +\infty$.
8. $\mathbf{f}(t) = e^t\mathbf{i} + \cos t\,\mathbf{j}$ for $-\infty < t < +\infty$.

9. If $y = -\log \cos x$ for $-\pi/2 < x < \pi/2$, find any points on the curve where the curvature attains a local maximum or minimum.
10. If $y = \log x$ for $0 < x < +\infty$, find $\lim_{x \to 0^+} \kappa(x)$ and $\lim_{x \to \infty} \kappa(x)$.
11. If $2x^2 - xy - y^2 = 0$, find the curvature at $(1,1)$. [*Hint:* Use implicit differentiation.]
12. If $xy^2 + x^3 - y^3 + x + 2 = 0$, find the curvature at the point $(1,2)$.

4.2   Tangential and normal accelerations

Let t again denote time, and let $\mathbf{r}(t)$ be the position vector of a moving particle at time t. We saw in §3.1 [see Equation (1)] that

$$\mathbf{v} = \dot{s}\mathbf{T}, \tag{8}$$

where $\mathbf{T}$ is the unit tangent vector to the path, $s = s(t)$ is the arc length measured along the path from the initial position of the particle to its position at time t, and $\dot{s}(t)$ is the speed. Assume that $s(t)$ increases with t. Then $\dot{s}(t)$ is positive and $\dot{s} = |\mathbf{v}|$. This follows from (8) since $|\mathbf{T}| = 1$.

The acceleration vector $\mathbf{a} = \dot{\mathbf{v}}$ is, by (8),

$$\mathbf{a} = \ddot{s}\mathbf{T} + \dot{s}\dot{\mathbf{T}}. \tag{9}$$

(Here we used the rule for differentiating products.) Recall [see §2.3, Equations (16) and (17)] that $\mathbf{T} = \cos\Theta\mathbf{i} + \sin\Theta\mathbf{j}$, $\mathbf{N} = -\sin\Theta\mathbf{i} + \cos\Theta\mathbf{j}$. Hence

$$\dot{\mathbf{T}} = \frac{d\cos\Theta}{dt}\mathbf{i} + \frac{d\sin\Theta}{dt}\mathbf{j}.$$

But, by the chain rule,

$$\frac{d\cos\Theta}{dt} = \frac{d\cos\Theta}{d\Theta}\frac{d\Theta}{ds}\frac{ds}{dt} = (-\sin\Theta)\kappa\dot{s}, \qquad \frac{d\sin\Theta}{dt} = \frac{d\sin\Theta}{d\Theta}\frac{d\Theta}{ds}\frac{ds}{dt} = (\cos\Theta)\kappa\dot{s}.$$

Thus $\dot{\mathbf{T}} = \kappa\dot{s}(-\sin\Theta\mathbf{i} + \cos\Theta\mathbf{j}) = \kappa\dot{s}\mathbf{N}$, and substitution into (9) yields the formula

$$\mathbf{a} = \ddot{s}\mathbf{T} + \kappa\dot{s}^2\mathbf{N}. \tag{10}$$

Recall that $\mathbf{T}$ and $\mathbf{N}$ are mutually perpendicular unit vectors. Equation (10) shows that the acceleration vector $\mathbf{a}$ is the sum of two vectors. The **tangential acceleration** $\ddot{s}\mathbf{T}$ is in the direction of motion; its magnitude $|\ddot{s}|$ is the absolute value of the rate of change of the speed $\dot{s}$ with respect to time. The **normal acceleration** $\kappa\dot{s}^2\mathbf{N}$ is perpendicular to the direction of motion; its magnitude is the absolute curvature $|\kappa|$ times the square of the speed. It is this acceleration you feel when negotiating a curve in a car. In order that the normal acceleration be not too large, you must slow down (make $|\dot{s}|$ small) before entering a curve (the more so, the bigger the curvature $|\kappa|$).

EXAMPLE Find the tangential and normal accelerations for a particle whose motion is described by the position vector $\mathbf{r} = (t^2 + 1)\mathbf{i} + t^3\mathbf{j}$. Assume that the arc length s is measured from some initial point along the path and s increases with t.

SOLUTION We have

$$\mathbf{v}(t) = (2t)\mathbf{i} + (3t^2)\mathbf{j} \qquad \text{and} \qquad \mathbf{a}(t) = 2\mathbf{i} + (6t)\mathbf{j}.$$

Hence $\dot{x} = 2t$, $\dot{y} = 3t^2$, $\ddot{x} = 2$, $\ddot{y} = 6t$, and so

$$\dot{s} = |\mathbf{v}| = (\dot{x}^2 + \dot{y}^2)^{1/2} = (4t^2 + 9t^4)^{1/2} = t(4 + 9t^2)^{1/2},$$

$$\ddot{s} = \frac{d}{dt}\dot{s} = (4 + 9t^2)^{1/2} + \frac{9t^2}{(4 + 9t^2)^{1/2}} = \frac{4 + 18t^2}{(4 + 9t^2)^{1/2}},$$

and

$$\kappa = \frac{\dot{x}\ddot{y} - \ddot{x}\dot{y}}{(\dot{x}^2 + \dot{y}^2)^{3/2}} = \frac{(2t)(6t) - (2)(3t^2)}{(4t^2 + 9t^4)^{3/2}} = \frac{6}{t(4 + 9t^2)^{3/2}}.$$

By Equation (10), the tangential and normal accelerations are

$$\ddot{s}\mathbf{T} = \frac{4 + 18t^2}{(4 + 9t^2)^{1/2}}\mathbf{T} \qquad \text{and} \qquad \kappa\dot{s}^2\mathbf{N} = \frac{6t}{(4 + 9t^2)^{1/2}}\mathbf{N}.$$

PROBLEMS

In Problems 1 to 6 the path of a point (particle) is described by the position vector $\mathbf{r}(t) = x(t)\mathbf{i} + y(t)\mathbf{j}$. For each of the given motions, find a_T and a_N, that is, the tangential and normal accelerations.

1. $\mathbf{r}(t) = t\mathbf{i} + (1 + t^2)\mathbf{j}$.
2. $\mathbf{r}(t) = \frac{1}{3}t^3\mathbf{i} + \frac{1}{2}t^2\mathbf{j}$.
3. $\mathbf{r}(t) = t\mathbf{i} + (\log t)\mathbf{j}$ for $t > 0$.

4. $\mathbf{r}(t) = (4\cos t)\mathbf{i} + (2\sin t)\mathbf{j}$.
5. $\mathbf{r}(t) = e^t\mathbf{i} + e^{-t}\mathbf{j}$.
6. $\mathbf{r}(t) = (e^{2t}\cos 2t)\mathbf{i} + (e^{2t}\sin 2t)\mathbf{j}$.

4.3 Motion on a rigid path

Let a heavy particle be constrained to move along a perfectly smooth rigid path (see Figure 17.32). The forces acting are the weight $= -mg\mathbf{j}$ and an unknown force $\mathbf{G}$, the constraint exerted by the path on the particle. We assume that this force is always perpendicular to the path; this is what is meant by saying that the path is perfectly smooth. Thus $\mathbf{G} = \gamma\mathbf{N}$, where $\mathbf{N}$ is the unit normal vector (see §2.3) to the path and γ some function of t.

Let the path be represented parametrically by $\mathbf{r} = x(s)\mathbf{i} + y(s)\mathbf{j}$ where s is arc length. Then $|d\mathbf{r}/ds|^2 = 1$, that is, $x'(s)^2 + y'(s)^2 = 1$. In view of relations (13) and (15) in §2.3, we have $\mathbf{T}(s) = x'(s)\mathbf{i} + y'(s)\mathbf{j}$, $\mathbf{N}(s) = -y'(s)\mathbf{i} + x'(s)\mathbf{j}$. Multiply the first equation by $y'(s)$, the second by $x'(s)$, and add. We obtain $\mathbf{j} = y'(s)\mathbf{T}(s) + x'(s)\mathbf{N}(s)$. Hence the weight of the particle is $-mg\mathbf{j} = -mgy'\mathbf{T} - mgx'\mathbf{N}$, and the total force is $\mathbf{f} = -mg\mathbf{j} + \gamma\mathbf{N} = -mgy'\mathbf{T} + (\gamma - mgx')\mathbf{N}$. Noting the expression (10) for the acceleration $\mathbf{a}$, we write Newton's Law $m\mathbf{a} = \mathbf{f}$ as

$$m\ddot{s}\mathbf{T} + m\kappa\dot{s}^2\mathbf{N} = -mgy'\mathbf{T} + (\gamma - mgx')\mathbf{N}$$

or

$$m(\ddot{s} + gy')\mathbf{T} = (\gamma - mgx' - m\kappa\dot{s}^2)\mathbf{N}.$$

Since $\mathbf{T}$ and $\mathbf{N}$ are mutually perpendicular unit vectors, the above equation means that

$$\ddot{s} + gy' = 0, \qquad \gamma - mgx' - m\kappa\dot{s}^2 = 0. \tag{11}$$

The first equation (11), with all details filled in, reads

$$s''(t) + gy'(s(t)) = 0. \tag{12}$$

This is a second-order differential equation for the unknown function $s(t)$ describing the motion.

We shall solve Equation (12) in two important special cases.

4.4 Motion on an inclined plane

We think of a particle sliding down an inclined straight path; the path is given by $x = s\cos\theta$, $y = s\sin\theta$ (see Figure 17.33). Here θ is the constant angle of inclination, and s is the length along the path (measured in the upward direction). Since $y' = dy/ds = \sin\theta$, Equation (12) becomes

$$\frac{d^2s}{dt^2} = -g\sin\theta.$$

Hence $ds/dt = -(g\sin\theta)t + A$, and $s(t) = (-\frac{1}{2}g\sin\theta)t^2 + At + B$ (where A and B

FIGURE **17.32**

FIGURE **17.33**

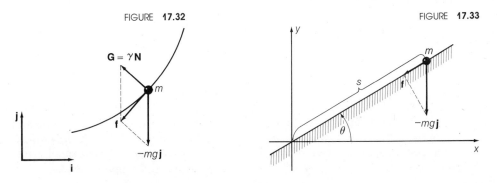

are constants). Assuming that the particle is released at $t = 0$ from the point $s = 0$ [so that $s(0) = B = 0$] with initial speed 0 [so that $\dot{s}(0) = A = 0$], we obtain $s = -\frac{1}{2}(g \sin \theta)t^2$. The distance traveled is proportional to the square of the time elapsed times the sine of the angle of inclination. This was first stated by Galileo.

PROBLEMS

1. The second equation (11) permits us to compute the magnitude γ of the force exerted by the path on the particle, once we know the function $s(t)$. Show that for the case considered in this subsection, this magnitude does not depend on the motion but only on θ.

2. Let a particle slide down a line joining any point on the circle $x^2 + (y - 1)^2 = 1$ to the origin. If the initial speed is 0, the time of descent is the same for all points. This is Galileo's "law of chords." Prove it.

4.5 Mathematical pendulum

We consider next a particle attached to a weightless rod of length l, which hangs on a frictionless pivot (see Figure 17.34). It is constrained to move on the semicircle $x = l \sin \theta$, $y = -l \cos \theta$, where the angle θ measures the deviation from the vertical position. The particle moves as if it were sliding along a smooth circular arc.

FIGURE **17.34**

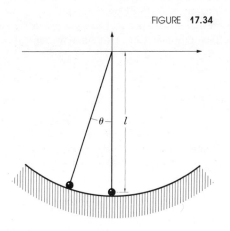

The arc length along the semicircle is $s = l\theta$. Hence $\theta = s/l$, and the parametric representation of the path in terms of s is $x = l\sin(s/l)$, $y = -l\cos(s/l)$. (Note that $s = 0$ corresponds to the pendulum being vertical.) Now $dy/ds = \sin(s/l)$, and Equation (12) reads

$$\frac{d^2s}{dt^2} + g\sin\frac{s}{l} = 0. \tag{13}$$

This is a complicated differential equation, which we shall not discuss. Instead, we restrict ourselves to motions, during which the pendulum remains *nearly vertical* (as is the case in a clock). Then $\theta = s/l$ is small, and we commit a *very* small error if we replace $\sin(s/l) = \sin\theta$ by $s/l = \theta$. Thus we obtain the simplified equation

$$\frac{d^2s}{dt^2} + \frac{g}{l}s = 0 \quad\text{or}\quad \frac{d^2\theta}{dt^2} + \frac{g}{l}\theta = 0. \tag{14}$$

This equation we know well (see Chapter 12, §2.1). Its general solution is

$$\theta = A\cos\sqrt{\frac{g}{l}}\,t + B\sin\sqrt{\frac{g}{l}}\,t.$$

The pendulum executes (within the error committed when we replaced $\sin\theta$ by θ) a simple harmonic motion with period

$$T = 2\pi\sqrt{\frac{l}{g}}; \tag{15}$$

the period does *not* depend on the amplitude, only on the length l and the value of g. Formula (15) explains how a pendulum stabilizes the movement of a clock and how a pendulum is used to measure the acceleration of gravity g.

Galileo discovered experimentally that the period of a pendulum does not depend on the amplitude. But he never succeeded, although he tried, to prove it mathematically. Moreover, Galileo did not realize that the independence of the period of the amplitude holds only for small oscillations of the pendulum.

PROBLEMS

1. Find the rate of change of the period of a mathematical pendulum with respect to the length l. Is this rate bigger on earth or on the moon?
2. Compute the force exerted on the particle (considered in this subsection) by the rod to which it is attached. (See Problem 1 in §4.4.) Since we set $\sin\theta = \theta$, you may set $\cos\theta = 1 - \frac{1}{2}\theta^2$.

4.6 Remarks on the cycloid

The cycloid (see Chapter 16, §2.4) has a remarkable property discovered by Huygens. Consider a cycloidal smooth rigid path, situated as in Figure 17.35, and a particle released from a point P and allowed to slide down to the lowest point Q. The time the descent will take is precisely the same for all release points P. The cycloid is an *isochrone* (curve of equal descent time) and is the only curve with this property.

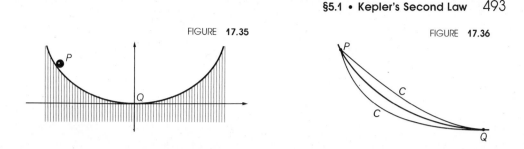

FIGURE **17.35**   FIGURE **17.36**

The cycloid is also a *brachistochrone* (curve of fastest descent). If we join a point P on the cycloid to the lowest point Q by *any* curve C other than the cycloid (see Figure 17.36), then the descent from P to Q along this other curve will take *longer* than the descent along the cycloid. Several famous mathematicians established this result independently of each other, among them two Bernoulli brothers (Jacob and Johann) and Pascal. (Pascal solved the problem at night, while suffering from toothache, *in order* not to think about the pain.)

§5 Planetary motion[‡]

We come now to one of the most dramatic and most important applications of calculus—the derivation of the Law of Universal Gravitation from astronomical observations. This application was made by Newton himself, but when he published it, he did not mention calculus.

5.1 Kepler's Second Law

The first step toward this momentous discovery was purely conceptual. This was the recognition, obvious to us but against all traditions of antiquity, that the laws of physics abstracted from observations on ordinary (terrestrial) objects are applicable also to heavenly (celestial) bodies. For instance, the same force that makes an apple fall to the ground keeps the moon in orbit around the earth.

CHRISTIAN HUYGENS (1629–1695), a mathematician, astronomer, and physicist, developed the wave theory of light. (This theory was the only one accepted by physicists until the advent of quantum theory, which reconciled Huygens' wave theory with the particle theory of light advanced by Newton.)

A Dutchman of independent means, Huygens lived for many years in Paris. There he met the 26-year-old Leibniz and tutored him in the then-modern mathematics.

BLAISE PASCAL (1623–1662) was perhaps the most amazing prodigy in the history of mathematics. As a child he is supposed to have proved, entirely on his own, that the sum of the angles in a triangle is 180° At 16 he discovered a fundamental theorem about conics and then wrote a long treatise about these curves, now lost. Pascal's most notable achievements are the creation of the theory of probability (with Fermat) and his work on the pressure of gases and liquids. He also built the first computing machine.

Later, however, mystical and religious interests became predominant in Pascal's life. His *Provincial Letters* (a brilliant theological polemic) and his posthumous *Pensées* are considered masterpieces of French literature.

[‡]Optional section.

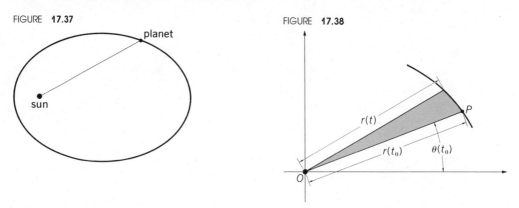

FIGURE **17.37**

planet

sun

FIGURE **17.38**

$r(t)$

$r(t_0)$ $\theta(t_0)$

O

P

The second step was mathematical. Kepler analyzed the astronomical observation of Tycho Brahe and extracted from them his three laws of planetary motion. The first two state that

(I) *Planets move in planes containing the sun, and their orbits are ellipses with the sun at a focus* (see Figure 17.37).

(II) *A segment joining the sun and a planet sweeps out equal areas in equal time intervals.*

Newton calculated (presumably at the age of 23) that Kepler's laws imply that the acceleration of a planet is directed toward the sun and has a magnitude inversely proportional to the square of the distance from the sun.

The statement that the planet's acceleration is directed along the line joining it to the sun follows from the Second Law alone. To see this, we introduce a coordinate system with the sun at the origin O. Let r and θ be the polar coordinates of the planet P. The motion of the planet is then described by two functions

$$r = r(t), \qquad \theta = \theta(t),$$

where t denotes time. The area swept out by the segment $\overrightarrow{OP}$ from the time t_0 to the time t equals [see Figure 17.38 and Equation (11′), §1.4 in Chapter 16]

$$\frac{1}{2} \int_{\theta(t_0)}^{\theta(t)} r(\theta(t))^2 \, d\theta = \frac{1}{2} \int_{t_0}^{t} r(t)^2 \dot{\theta}(t) \, dt.$$

(Here and hereafter a dot indicates differentiation with respect to time.) Law II says

JOHANNES KEPLER (1571–1630) devoted years of laborious calculations to uncovering the hidden mathematical harmonies which, he was sure, underlay the motion of the planets. Not all of his guesses were right, but in 1609 he published his first two laws, and 10 years later he announced the third. Kepler also made pioneering contributions to optics; his treatise on measuring the volumes of wine barrels anticipates calculus.

Ancient superstition influenced the life of this creator of modern astronomy. His official duties as mathematician to Emperor Rudolph II included the preparation of horoscopes, and his mother was tried for witchcraft (and acquitted).

TYCHO BRAHE (1546–1607), a Danish nobleman, perfected the art of astronomical observations without telescopes and built a big observatory on the island of Ven. He also had the insight to entrust the records of his observations to the man who could make the best use of them, Kepler.

that the time derivative of this area,

$$\frac{d}{dt} \frac{1}{2} \int_{t_0}^{t} r(t)^2 \dot{\theta}(t) \, dt = \frac{1}{2} r(t)^2 \dot{\theta}(t),$$

is a constant; call it α. Thus

$$r^2 \dot{\theta} = 2\alpha. \tag{1}$$

Note that α is the "areal velocity." In a time interval τ the segment $\overrightarrow{OP}$ sweeps out an area $\alpha\tau$.

Differentiating (1), we obtain

$$2r\dot{r}\dot{\theta} + r^2\ddot{\theta} = 0,$$

so that

$$2\dot{r}\dot{\theta} + r\ddot{\theta} = 0. \tag{2}$$

The Cartesian coordinates of P are $x = r \cos\theta$, $y = r \sin\theta$, so that the velocity vector $\mathbf{v}$ has the components

$$\dot{x} = \dot{r} \cos\theta - r\dot{\theta} \sin\theta, \qquad \dot{y} = \dot{r} \sin\theta + r\dot{\theta} \cos\theta.$$

The components of the acceleration vector $\mathbf{a}$ are therefore

$$\ddot{x} = \ddot{r} \cos\theta - 2\dot{r}\dot{\theta} \sin\theta - r\dot{\theta}^2 \cos\theta - r\ddot{\theta} \sin\theta,$$

$$\ddot{y} = \ddot{r} \sin\theta + 2\dot{r}\dot{\theta} \cos\theta - r\dot{\theta}^2 \sin\theta + r\ddot{\theta} \cos\theta.$$

Using (2), we obtain

$$\ddot{x} = [\ddot{r} - r\dot{\theta}^2] \cos\theta, \qquad \ddot{y} = [\ddot{r} - r\dot{\theta}^2] \sin\theta,$$

that is,

$$\ddot{x} = \left[\frac{\ddot{r}}{r} - \dot{\theta}^2 \right] x, \qquad \ddot{y} = \left[\frac{\ddot{r}}{r} - \dot{\theta}^2 \right] y.$$

This shows that the acceleration vector $(\ddot{x}, \ddot{y})$ is a scalar multiple of the position vector (x,y). Thus the acceleration of the planet is directed toward the sun.

The magnitude of the acceleration vector is $|\mathbf{a}| = |\ddot{r} - r\dot{\theta}^2|$. To complete the argument, we must show that $|\mathbf{a}|$ is inversely proportional to the square of the distance from the sun. This will follow if we can show that

$$\ddot{r} - r\dot{\theta}^2 = -\frac{\mu}{r^2}, \tag{3}$$

where μ is a positive constant. Now we must use the First Law.

5.2 Application of Kepler's First Law

The trajectory of the planet is an ellipse with focus O; let ϵ be its eccentricity. We recall the formulas for ellipses in polar coordinates (see §3.3 in Chapter 16) and conclude that, if the position of the axes is chosen suitably, then

$$r(t) = \frac{A}{1 - \epsilon \cos\theta(t)},$$

where A is a positive constant. Thus

$$r(1 - \epsilon \cos \theta) = A. \tag{4}$$

Differentiating with respect to time, we obtain

$$\dot{r}(1 - \epsilon \cos \theta) + \epsilon r \dot{\theta} \sin \theta = 0.$$

We multiply both sides of this equation by r and obtain

$$\dot{r}r(1 - \epsilon \cos \theta) + \epsilon r^2 \dot{\theta} \sin \theta = 0,$$

which, in view of (1) and (4), may be written as $A\dot{r} + 2\alpha\epsilon \sin \theta = 0$. Another differentiation gives

$$A\ddot{r} + 2\alpha\epsilon\dot{\theta} \cos \theta = 0. \tag{5}$$

Since, by virtue of (1),

$$\dot{\theta} = \frac{2\alpha}{r^2}, \tag{6}$$

we have from (5)

$$\ddot{r} = -\frac{4\alpha^2}{r^2} \frac{\epsilon \cos \theta}{A}.$$

Since, by (4), $-\epsilon \cos \theta = (A/r) - 1$, we obtain

$$r = \frac{4\alpha^2}{r^2} \left(\frac{1}{r} - \frac{1}{A} \right).$$

Now $r\dot{\theta}^2 = 4\alpha^2/r^3$ by (6), so that

$$\ddot{r} - r\dot{\theta}^2 = -\frac{4\alpha^2}{A} \frac{1}{r^2}. \tag{7}$$

This is the same as relation (3) which we wanted to prove, with

$$\mu = \frac{4\alpha^2}{A}. \tag{8}$$

5.3　Application of Kepler's Third Law

We have seen that Kepler's first two laws show that a planet P moving around the sun O experiences an acceleration of magnitude

$$\frac{4\alpha^2}{A} \frac{1}{r^2}.$$

Here $r = |OP|$ is the distance from the sun and $r = A/(1 - \epsilon \cos \theta)$ is the equation of the orbit in polar coordinates. The constant α is the area swept out by the segment $\overrightarrow{OP}$ per unit time. Let T be the period of the planet. Then αT is the area of the ellipse. If the semiaxes of the ellipse are a and b, its area is πab (see the example in Chapter 8, §3.5). Therefore

$$\alpha = \frac{\pi ab}{T}. \tag{9}$$

FIGURE **17.39**

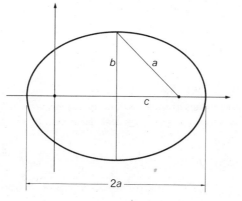

From (4), we conclude that the largest and smallest distances of P from O are $A/(1 - \epsilon)$ and $A/(1 + \epsilon)$. Therefore $A/(1 - \epsilon) + A/(1 + \epsilon) = 2a$ (see Figure 17.39) or $A = a(1 - \epsilon^2)$. Also, $b = a\sqrt{1 - \epsilon^2}$ so that, by (9),

$$\alpha = \frac{\pi a^2 \sqrt{1 - \epsilon^2}}{T}$$

and

$$\mu = \frac{4\alpha^2}{A} = \frac{4\pi^2 a^3}{T^2}. \tag{10}$$

The magnitude of the attraction exerted by the sun on the planet is equal to mass times acceleration, that is,

$$\frac{4\pi^2 a^3 m}{T^2} \frac{1}{r^2}. \tag{11}$$

At this point, we use Kepler's Third Law:

(III) *The ratio a^3/T^2 is the same for all planets.*

Newton concluded from this that the attractive force of the sun per unit mass of a planet is "universal," the same for all planets, and dependent only on the distance from the sun. Furthermore, Kepler's laws hold also for the moons of planets, in particular for the moons of Jupiter. The value of a^3/T^2 for the moons of Jupiter, however, is different from the value of this ratio for planets.

These considerations led Newton to postulate that an attractive force exists between *any* two masses, M and m, and has the magnitude

$$\frac{\gamma M m}{r^2}, \tag{12}$$

r being the distance between the masses and γ a universal constant.

Comparing (11) and (12), we conclude that the mass of the sun is

$$M = \frac{4\pi^2 a^3}{\gamma T^2}.$$

If M_1 is the mass of a planet (or of another sun) and a_1 and T_1 the major semiaxis and period of a moon (or planet) rotating about it, we have $M_1 = 4\pi^2 a_1^3/\gamma T_1^2$, and therefore

$$\frac{M}{M_1} = \frac{a^3 T_1^2}{a_1^3 T^2}.$$

Thus astronomical observations permit us to compare masses of two heavenly bodies. To determine the actual values of these masses, however, we need a direct measurement of the gravitational constant γ. This was first done accurately by Cavendish in 1798.

5.4   Celestial mechanics

The reader will note that, in deriving Newton's Gravitation Law from Kepler's laws, we never used the fact that the eccentricity of the orbit was $\epsilon < 1$. The same argument would hold if the orbit were a parabola ($\epsilon = 1$) or a hyperbola ($\epsilon > 1$). For a circular orbit ($\epsilon = 0$), the argument would give little information, since then r would be a constant.

We can reverse the argument and show that a particle subject to an attractive force directed toward a fixed center 0, and varying as the inverse square of the distance from this center, *always* moves on a conic with a focus at O and obeys Kepler's Second Law of constant "areal velocity." We shall not carry out this calculation.

Actual planets satisfy Kepler's laws only approximately. For each planet is also subject to the attraction of all other planets, and so is the sun. Beginning with Newton himself, generations of mathematicians have developed methods, based on calculus, for taking these effects into account and for solving the problems of "celestial mechanics." An important triumph of this science was the discovery of the planet Neptune in 1846. Before anyone saw this planet through a telescope, Adams and Leverrier had calculated its mass and its orbit from observed perturbations in the motion of the planet Uranus. Today space exploration poses new challenges to celestial mechanics, and modern computers permit us to meet these challenges.

There is one planet, however, whose motion defies Newtonian mechanics. A certain peculiarity in the motion of Mercury can be explained only on the basis of Einstein's general theory of relativity.

HENRY CAVENDISH (1731–1810) was a very rich, very eccentric English aristocrat who lived almost as a recluse and performed some of the most brilliant experiments in the history of science.

JOHN COUCH ADAMS (1819–1892) was educated at Cambridge University and spent the rest of his life there. As a student, he became interested in the unexplained irregularities in the motion of Uranus, and within a few years he calculated the motion of a more distant planet (later named Neptune) whose

gravitational pull would account for these irregularities.

U. J. J. LEVERRIER (1811–1877) was director of the Paris Observatory when he solved the same problem as Adams. The two men worked without knowledge of each other. A German astronomer directed his telescope at the point of the sky indicated by Leverrier and saw the "new" planet. It seems that Leverrier himself never cared to take a look.

PROBLEMS

1. Suppose, in place of Kepler's First Law, we assume that the orbits of planets are spirals that have the form $r(t) = \beta e^{\theta(t)}$ in a polar coordinate system with origin at the sun, and we also assume the Second Law.

   (a) Show that $r(t) = (4\alpha t + r_0^2)^{1/2}$, where $r_0 = r(0)$.

   (b) Show that the acceleration is inversely proportional to the cube of the distance from the sun.

2. In Problem 1 suppose that the spirals have the form $r(t) = \beta\theta(t)$.

   (a) Show that $r(t) = (6\alpha\beta t + r_0^3)^{1/3}$, where $r_0 = r(0)$.

   (b) Show that the acceleration is not proportional to any single power (positive or negative) of the distance from the sun.

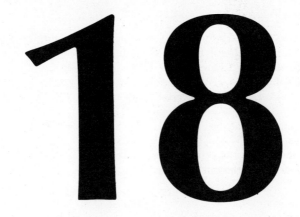

18

VECTORS, CURVES, AND SURFACES IN SPACE

§1 Coordinate systems in space

In this chapter we develop analytic geometry in space by generalizing the concepts and techniques developed in Chapters 2, 16, and 17 from two to three dimensions.

Cartesian coordinates can be used not only for the plane but also in space (this was recognized already by Fermat). Polar coordinates have two distinct spatial analogs—cylindrical and spherical coordinates. We shall discuss these coordinate systems in the present section. From now on "point" and "line" will mean "point in space" and "line in space."

1.1 Cartesian coordinates in space

We select a unit of length and a point O in space called the origin. Through this point we draw three mutually perpendicular lines and give each of them a direction. The lines are called the x-**axis**, the y-**axis**, and the z-**axis**, respectively, and they are usually drawn as in Figure 18.1. Each pair of axes determines a **coordinate plane** containing them.

A point P in space is described by the numbers (a,b,c), its coordinates. P is the point reached by going from the origin a units in the x-direction, b units in the y-direction, and c units in the z-direction (see Figure 18.1).

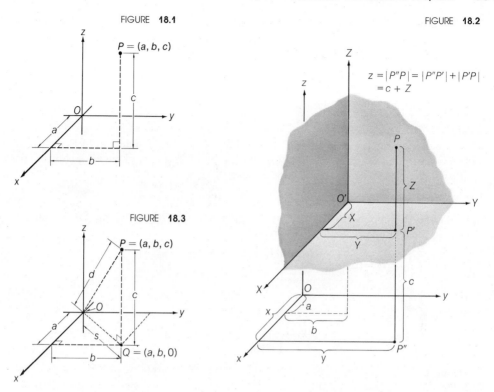

FIGURE **18.1**

$P = (a, b, c)$

FIGURE **18.2**

$z = |P''P| = |P''P'| + |P'P|$
$= c + Z$

FIGURE **18.3**

$P = (a, b, c)$

$Q = (a, b, 0)$

The set $x = a$, that is, the set of all points with x-coordinate (first coordinate) equal to a, is a plane parallel to the yz plane and thus perpendicular to the x-axis. It coincides with the yz plane if $a = 0$; otherwise it has distance $|a|$ from that plane. The plane $x = a$ lies on the side of the yz plane toward which the x-axis is pointing if $a > 0$ and on the opposite side if $a < 0$. Similarly, the sets $y = b$ (points with y-coordinate b) and $z = c$ (points with z-coordinate c) are planes perpendicular to the y-axis and z-axis, respectively. The point (a,b,c) is the intersection of the three planes $x = a$, $y = b$, and $z = c$. (Equations of planes not parallel to the coordinate planes will be discussed later; see §4.1.)

The three coordinate planes divide the space into eight **octants.** The one containing points with all three coordinates positive is called the first octant.

If we choose a "new" origin O' and do not change the direction of the coordinate axes, we obtain a new coordinate system, said to be obtained by a translation of the original one; see Figure 18.2. Let a point P have coordinates (x,y,z) in the old system and coordinates (X,Y,Z) in the new one, and let (a,b,c) be the old coordinates of the new origin O'. Then

$$x = a + X, \qquad y = b + Y, \qquad z = c + Z. \tag{1}$$

These formulas are analogous to the corresponding relations in the plane (Chapter 2, §1.2). In Figure 18.2 we indicate the proof for the z-coordinate and for the case when O' lies in the old first octant and P in the new first octant. Other cases can be treated similarly.

The **distance** d between two points, (a_1,b_1,c_1) and (a_2,b_2,c_2), is given by

$$d = \sqrt{(a_1 - a_2)^2 + (b_1 - b_2)^2 + (c_1 - c_2)^2}. \tag{2}$$

By translating the coordinate system, we can put one point at the origin. Therefore (2) will be verified if we show that the distance d of a point P with coordinates (a,b,c) from the origin is

$$d = \sqrt{a^2 + b^2 + c^2}. \tag{3}$$

The proof, for the case when P lies in the first octant, is indicated in Figure 18.3. Let s denote the distance from O to Q; Q is the point in the xy plane with coordinates (a,b). By the distance formula for the plane,

$$s^2 = a^2 + b^2.$$

Since OP is the hypotenuse of a right triangle with legs OQ and QP and the length of QP is c, the Pythagorean theorem gives

$$d^2 = s^2 + c^2.$$

Hence $d^2 = a^2 + b^2 + c^2$ and (3) follows.

PROBLEMS

In Problems 1 to 12 draw a three-dimensional coordinate system and locate the points with the given coordinates.

1. $(1,0,0)$.   4. $(1,1,0)$.   7. $(1,2,4)$.   10. $(1,-2,-4)$.
2. $(0,1,0)$.   5. $(1,0,1)$.   8. $(1,-2,4)$.   11. $(-1,2,4)$.
3. $(0,0,1)$.   6. $(0,1,1)$.   9. $(1,2,-4)$.   12. $(-1,-2,-4)$.

13. What is the equation of the plane through the point $(-4,2,7)$ that is perpendicular to the x-axis? Parallel to the xz plane? Parallel to the xy plane?
14. What is the equation of the plane through the point $(3,-4,-6)$ that is parallel to both the x- and y-axes? Perpendicular to the y-axis? Parallel to the yz plane?
15. If the coordinates of the new origin in the old system are $(0,4,2)$ and the coordinates of a point P in the old system are $(1,-2,3)$, what are the coordinates of P in the new system?
16. If the coordinates of a point P in the old system are $(4,2,3)$ and in the new system are $(1,-3,1)$, what are the coordinates of the new origin in the old system?
17. If the coordinates of a new origin in the old system are $(2,-4,-6)$ and the coordinates of P in the new system are $(-1,2,-4)$, what are the coordinates of P in the old system?
18. Find the distance between the points whose coordinates are given by $(0,1,1)$ and $(-2,1,5)$. By $(1,-3,2)$ and $(3,1,-4)$.

1.2 Spheres

In space, the set of all points that have the same distance $r > 0$ from a fixed point (center) is called a **sphere**; the number r is called the **radius**. The sphere is the set of all points (x,y,z) such that

$$(x - a)^2 + (y - b)^2 + (z - c)^2 = r^2,$$

where (a,b,c) is the center. This follows from the distance formula (2).

The solution set of any equation of the form

$$x^2 + y^2 + z^2 + Ax + By + Cz + C = 0$$

is a sphere, or a point, or the empty set. This can be proved by "completing the square," just as we proved the corresponding statement in the plane (Theorem 1 in Chapter 2, §3.1).

EXAMPLES 1. What is the solution set of

$$x^2 + 2x + y^2 - 2y + z^2 - 4z + 6 = 0?$$

ANSWER   Completing the squares, we rewrite the equation in the form

$$(x + 1)^2 - 1 + (y - 1)^2 - 1 + (z - 2)^2 - 4 + 6 = 0$$

or

$$(x + 1)^2 + (y - 1)^2 + (z - 2)^2 = 0.$$

The solution set consists of one point, the point $(-1,1,2)$.

2. Find the intersection of the plane $z = 13$ and the sphere

$$(x - 2)^2 + (y - 3)^2 + (z - 4)^2 = 100.$$

SOLUTION   On the plane, z has the value 13. We set $z = 13$ in the equation of the sphere and obtain the equation

$$(x - 2)^2 + (y - 3)^2 = 19.$$

This is true for every point of the intersection and for no other points. In the plane $z = 13$, we can use x and y as Cartesian coordinates; hence the intersection is a circle. Its center is $(2,3,13)$; its radius is $\sqrt{19}$.

We can use the method of this example for an analytic proof of the theorem: *if a plane intersects a sphere, the intersection is a point or a circle.*

PROBLEMS

1. What is the solution set of $x^2 + y^2 + z^2 - 4x + 2y - 8z + 5 = 0$?
2. What is the solution set of $x^2 + y^2 + z^2 + 2x - 2y - 4z + 10 = 0$?
3. Find the center and radius of the sphere whose equation is $x^2 + y^2 + z^2 - 4x + 6y - z + 49/4 = 0$.
4. Find the center and radius of the sphere whose equation is $4x^2 + 4y^2 + 4z^2 + 40x - 32y - 4z = 31$.
5. The plane $z = 1$ intersects the sphere $x^2 + y^2 + z^2 = 10$ in a circle. Find the radius and center of this circle.
6. The plane $y = 3$ intersects the sphere $x^2 + y^2 + z^2 - 2x + 2z = 16$ in a circle. Find the radius and center of this circle.
7. Find the equation of the sphere whose center is at $(1,-2,2)$ and which passes through the point $(2,0,3)$.
8. The spheres $x^2 + y^2 + z^2 = 19$ and $x^2 + y^2 + z^2 - 10z = -21$ intersect in a circle. Find the radius and center of this circle.
9. Find all planes σ such that σ is parallel to the xy plane, and σ intersects the sphere $x^2 + y^2 + z^2 = 100$ in a circle of radius 8.
10. Find the equation of a sphere of radius $\sqrt{114}$ which passes through the points $(2,6,-3)$, $(8,3,-2)$, and $(9,4,0)$.

FIGURE **18.4**   FIGURE **18.5**

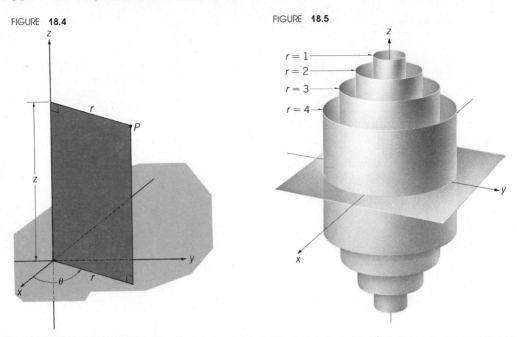

1.3 Cylindrical coordinates

If a point P has Cartesian coordinates (x,y,z), its **cylindrical coordinates** are, by definition, the three numbers (r,θ,z), where (r,θ) are the polar coordinates of the point (x,y) in
the xy plane. Note that $r = \sqrt{x^2 + y^2}$ is the distance from the point $P = (x,y,z)$ to the
point $(0,0,z)$, that is, the distance from P to the z-axis (see Figure 18.4).
    The surfaces corresponding to fixed values of r are circular cylinders (see Figure
18.5), hence the name of this coordinate system. The surfaces corresponding to fixed
values of θ are planes containing the z-axis (see Figure 18.6); the surfaces corresponding
to fixed values of z are planes parallel to the xy plane.
    The connection between the cylindrical coordinates (r,θ,z) and Cartesian coordi-
nates (x,y,z) of a point is given by the formulas

$$r^2 = x^2 + y^2, \qquad \cos\theta = \frac{x}{r}, \qquad \sin\theta = \frac{y}{r}, \qquad z = z, \tag{4}$$

and

$$x = r\cos\theta, \qquad y = r\sin\theta, \qquad z = z. \tag{5}$$

(As in the case of polar coordinates, it is convenient to permit negative values of r.)

PROBLEMS

    In Problems 1 to 8 the cylindrical coordinates (r,θ,z) of a point are given. Find the
corresponding Cartesian coordinates of the point.

1. $\left(2, \frac{\pi}{4}, 0\right).$ 2. $\left(-2, \frac{\pi}{4}, 0\right).$ 3. $(2,0,2).$ 4. $\left(1, -\frac{\pi}{3}, 2\right).$

5. $\left(1, \frac{\pi}{6}, -4\right)$.   6. $\left(2, \frac{5\pi}{3}, -1\right)$.   7. $\left(-2, -\frac{3\pi}{2}, 3\right)$.   8. $\left(-5, \frac{3\pi}{4}, 4\right)$.

In Problems 9 to 16 the Cartesian coordinates of a point in space are given. Find the corresponding cylindrical coordinates of the point.

9. $(2, -2, -3)$.

10. $(-\sqrt{2}, \sqrt{2}, 1)$.

11. $(2, -2\sqrt{3}, 2)$.

12. $\left(-\frac{\sqrt{3}}{2}, \frac{1}{2}, \frac{1}{4}\right)$.

13. $(1, 2, 3)$.

14. $(3, 4, -\frac{1}{2})$.

15. $(4 \cos 15°, -4 \sin 15°, 1)$.

16. $\left(\frac{1}{2} \sin \frac{\pi}{8}, \frac{1}{2} \cos \frac{\pi}{8}, \frac{\sqrt{3}}{2}\right)$.

1.4 Surfaces of revolution

We now present an application of cylindrical coordinates.

Assume that we are given a curve C in the xz plane, that is, in the plane $y = 0$, located in the half-plane $x \geq 0$. Suppose we rotate it about the z-axis to obtain a **surface of revolution** S. Then consider a point P on S, with the distance r from the z-axis. Let P_0 be the point on C with the same z coordinate as P (see Figure 18.7). Then $r =$ the distance x from P_0 to the z-axis. We conclude the following.

The equation of S in cylindrical coordinates is obtained from that of C by replacing x by r.

Once we have the equation of S in cylindrical coordinates, we can replace r^2 by $x^2 + y^2$ and obtain the equation of S in Cartesian coordinates.

Instead of assuming the curve C to lie in the half-plane $x \geq 0$, we may assume it to be symmetric about the z-axis.

For instance, if, in the xz plane, C is the line $x = p$, p a positive number, then the surface S obtained by rotating C about the z-axis is the same surface we would obtain by

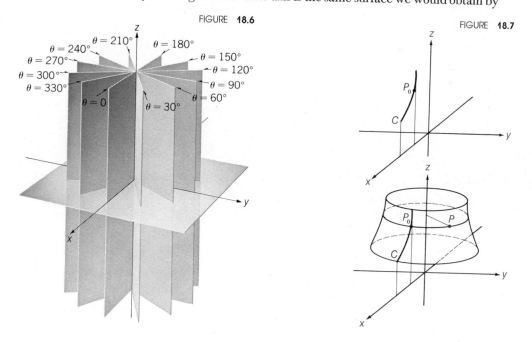

FIGURE **18.6**

$\theta = 210°$ $\theta = 180°$
$\theta = 240°$
$\theta = 270°$
$\theta = 300°$
$\theta = 330°$
$\theta = 0$ $\theta = 30°$
$\theta = 150°$
$\theta = 120°$
$\theta = 90°$
$\theta = 60°$

FIGURE **18.7**

FIGURE 18.8   FIGURE 18.9

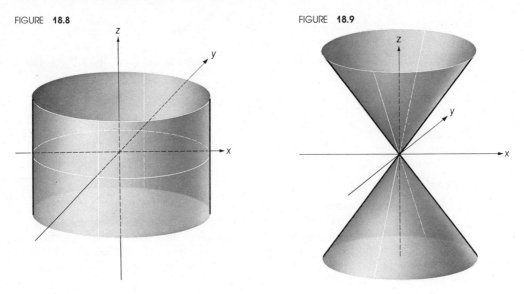

rotating the pair consisting of the lines $x = p$ and $x = -p$ (see Figure 18.8). The equation of this pair is $x^2 = p^2$. The equation of S is therefore $r^2 = p^2$ or

$$x^2 + y^2 = p^2.$$

The surface S is called a **circular cylinder,** with the z-axis as axis.

EXAMPLE Let m be a positive number. We rotate about the z-axis the line $z = mx$ or (which is the same) the pair of lines $z = mx$ and $z = -mx$ (see Figure 18.9). The equation of the pair is

$$(z + mx)(z - mx) = 0 \qquad \text{or} \qquad z^2 - m^2x^2 = 0. \tag{6}$$

The equation of the surface of revolution is therefore $z^2 - m^2r^2 = 0$, or, setting $a = 1/m$,

$$\frac{x^2}{a^2} + \frac{y^2}{a^2} - z^2 = 0. \tag{7}$$

The surface is called a **circular cone,** with the z-axis as axis.

PROBLEMS

In Problems 1 to 8 a surface S is obtained by rotating a curve C in the xz plane about the z-axis. Express the equation of the surface first in cylindrical coordinates, then in three-dimensional Cartesian coordinates.

1. $z^2 = 4x^2$.

2. $x^2 + z^2 = 16$.

3. $z = \frac{1}{2}x^2$.

4. $z = \dfrac{1}{1 + x^2}$.

5. $4x^2 + z^2 = 100$.

6. $z = |x|$.

7. $z = e^{-|x|}$.

8. $(x - 2)^2 + z^2 = 1$.

1.5 Spherical coordinates

The spherical coordinates of a point P with Cartesian coordinates (x,y,z) are three numbers, ρ, θ, and ϕ, defined as follows (see Figure 18.10): ρ is the distance from the origin O to P, θ has the same significance as in cylindrical coordinates, ϕ is the angle between the positive z-axis and the ray from O to P; θ is not defined if $x^2 + y^2 = 0$; ϕ is not defined if $\rho = 0$.

Denote by Q the point $(x,y,0)$ and by T the point $(0,0,z)$. We see from Figure 18.10 that $z = \rho \cos \phi$. On the other hand, $x^2 + y^2 = |OQ|^2 = |TP|^2 = \rho^2 \sin^2 \phi$. Hence the length $|OQ|$ is $\rho \sin \phi$. Thus the polar coordinates of Q are $\rho \sin \phi$ and θ, so that $x = \rho \sin \phi \cos \theta$ and $y = \rho \sin \phi \sin \theta$. We summarize the formulas connecting the spherical coordinates of a point with the Cartesian coordinates:

$$\rho^2 = x^2 + y^2 + z^2, \qquad \cos \theta = \frac{x}{\sqrt{x^2 + y^2}}, \qquad \sin \theta = \frac{y}{\sqrt{x^2 + y^2}}, \tag{8}$$

$$\cos \phi = \frac{z}{\rho}, \qquad \sin \phi = \frac{\sqrt{x^2 + y^2}}{\rho}$$

and

$$x = \rho \sin \phi \cos \theta, \qquad y = \rho \sin \phi \sin \theta, \qquad z = \rho \cos \phi. \tag{9}$$

By construction, we have $\rho \geq 0$ and $0 \leq \phi < \pi$. But it is convenient to call any set of numbers (ρ,θ,ϕ) for which (9) holds spherical coordinates of (x,y,z).

All points with $\phi = 0$ or $\phi = 180°$ lie on the z-axis. The surfaces corresponding to a fixed value of ϕ between 0 and $\pi/2$ or between $\pi/2$ and π are circular cones, with the z-axis as axis (see Figure 18.11). This is geometrically obvious and can be seen analytically as follows.

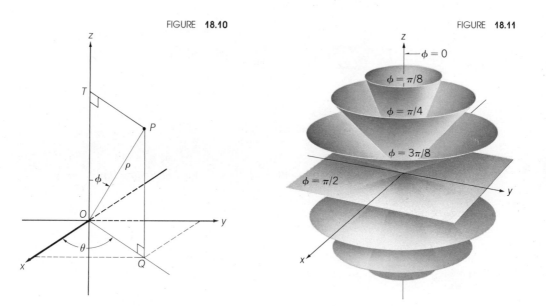

FIGURE **18.10**

FIGURE **18.11**

Let a be a given positive number; by (9), we have

$$x^2 + y^2 = \rho^2 \sin^2 \phi \cos^2 \theta + \rho^2 \sin^2 \phi \sin^2 \theta = \rho^2 \sin^2 \phi,$$

$$\frac{x^2}{a^2} + \frac{y^2}{a^2} - z^2 = \frac{\rho^2}{a^2} \sin^2 \phi - \rho^2 \cos^2 \phi = \rho^2 \cos^2 \phi \left(\frac{\tan^2 \phi}{a^2} - 1 \right),$$

or

$$\frac{x^2}{a^2} + \frac{y^2}{a^2} - z^2 = z^2 \left(\frac{\tan^2 \phi}{a^2} - 1 \right).$$

Thus the condition $|\tan \phi| = a$ is equivalent to Equation (7) of a cone (see §1.4). For $\phi = 90°$, we obtain the xy plane.

The surfaces corresponding to a fixed value of θ are the same as in the case of cylindrical coordinates. The surfaces corresponding to fixed values of r are spheres about the origin, hence the name of this coordinate system.

Consider now a point P on a fixed sphere about the origin. We can specify the position of P by specifying the numbers ϕ and θ. We restrict ϕ and θ by the inequalities $0 \le \phi \le 180°$, $-180° < \theta \le 180°$. In this case, we call θ the *longitude* of P and $90° - \phi$ the *latitude* of P. This corresponds to the use of these terms in geography where $\phi = 0$, $\phi = 90°$, and $\phi = 180°$ correspond to the North Pole, the equator, and the South Pole, respectively. The curves on the sphere corresponding to fixed values of

FIGURE **18.12**

ϕ are called *parallels*, those corresponding to fixed values of θ *meridians*; see Figure 18.12.

PROBLEMS

In Problems 1 to 8 the spherical coordinates (ρ, θ, ϕ) of a point are given. Find the corresponding Cartesian coordinates of the point.

1. $\left(2, 0, \frac{\pi}{4} \right)$.   2. $\left(2, \frac{\pi}{4}, \frac{\pi}{2} \right)$.   3. $\left(\sqrt{2}, \frac{\pi}{6}, 0 \right)$.   4. $\left(1, \frac{\pi}{4}, \frac{\pi}{4} \right)$.

5. $\left(\sqrt{2}, \frac{3\pi}{2}, \frac{\pi}{2}\right)$.   6. $\left(1, \frac{\pi}{3}, \pi\right)$.   7. $\left(2, \frac{5\pi}{3}, \frac{\pi}{3}\right)$.   8. $\left(4, \frac{3\pi}{4}, \frac{5\pi}{6}\right)$.

In Problems 9 to 16 the Cartesian coordinates of a point in space are given. Find the corresponding spherical coordinates of the point.

9. $(1,0,0)$.

10. $(0,1,0)$.

11. $(1,1,0)$.

12. $(1,0,-1)$.

13. $(3,\sqrt{3},2)$.

14. $(-1,1,\sqrt{2})$.

15. $(\sqrt{3},-1,2\sqrt{3})$.

16. $(-\frac{1}{2}\sqrt{3},-\frac{3}{2},1)$

§2 Vectors and curves in space

The concepts of vector algebra and vector calculus (see Chapter 17, §§1, 2, and 3) will now be extended to space. This can be done briefly, since no essentially new ideas are involved.

From now on (until §5.6) we assume that a fixed Cartesian coordinate system has been chosen.

2.1 Vectors and their components

The definitions of directed segments, and when two directed segments are equivalent (have the same magnitude and direction) given in §1.1 of Chapter 17 apply without any changes to points in space. There is a valid analog of Theorem 1 in Chapter 17, §1.1: *given four points* $P = (p_1,p_2,p_3)$, $Q = (q_1,q_2,q_3)$, $R = (r_1,r_2,r_3)$, *and* $S = (s_1,s_2,s_3)$, *the directed segments* $\overrightarrow{PQ}$ *and* $\overrightarrow{RS}$ *have the same magnitude and direction; in symbols,*

$$\overrightarrow{PQ} \sim \overrightarrow{RS}, \tag{1}$$

if and only if

$$q_1 - p_1 = s_1 - r_1, \qquad q_2 - p_2 = s_2 - r_2, \qquad q_3 - p_3 = s_3 - r_3. \tag{2}$$

The proof is similar to, but more complicated than, the one given for the case of segments in the plane in Chapter 17, §1.1. It will not be carried out here. Indeed, it may be more convenient to take an "analytic" rather than a "geometric" point of view and to use (2) as a definition of the relation (1).

A directed segment $\overrightarrow{PQ}$ determines or represents a vector (in space) **a**, the set of all directed segments $\overrightarrow{RS}$ satisfying (1). The vector **a** is uniquely determined by the three numbers (2), that is, by the numbers

$$a_1 = q_1 - p_1, \qquad a_2 = q_2 - p_2, \qquad a_3 = q_3 - p_3.$$

These numbers are called the **components** of **a**; instead of saying that **a** is the vector with the components $\langle a_1,a_2,a_3 \rangle$, we sometimes write

$$\mathbf{a} = \langle a_1,a_2,a_3 \rangle. \tag{3}$$

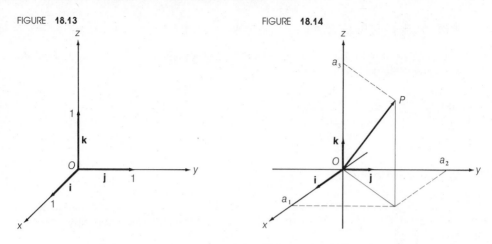

FIGURE 18.13

FIGURE 18.14

The length $|\mathbf{a}|$ of $\mathbf{a}$ is the length of any directed segment representing $\mathbf{a}$; for instance, the segment from the origin $O = (0,0,0)$ to the point (a_1,a_2,a_3). By the distance formula [Equation (3) in §1.1],

$$|\mathbf{a}| = \sqrt{a_1^2 + a_2^2 + a_3^2}. \tag{4}$$

The sum of two vectors and the scalar multiple of a vector (product of a number and a vector) are defined exactly as in Chapter 17, §1.3. Equations (4) and (5) from that section have a valid generalization. *If*

$$\mathbf{a} = \langle a_1,a_2,a_3 \rangle, \qquad \mathbf{b} = \langle b_1,b_2,b_3 \rangle, \tag{5}$$

then

$$\mathbf{a} + \mathbf{b} = \langle a_1 + b_1, a_2 + b_2, a_3 + b_3 \rangle, \qquad \lambda\mathbf{a} = \langle \lambda a_1, \lambda a_2, \lambda a_3 \rangle. \tag{6}$$

The algebraic rules stated in Theorem 3, Chapter 17, §1.4, remain valid for the same reasons as before.

We introduce three mutually perpendicular unit vectors (vectors of length 1) $\mathbf{i}, \mathbf{j}, \mathbf{k}$ by the formulas

$$\mathbf{i} = \langle 1,0,0 \rangle, \qquad \mathbf{j} = \langle 0,1,0 \rangle, \qquad \mathbf{k} = \langle 0,0,1 \rangle; \tag{7}$$

they can be represented by directed segments of length 1 extending from the origin in the x-, y-, and z-directions, respectively (see Figure 18.13). If $\mathbf{a}$ is a vector with components $\langle a_1,a_2,a_3 \rangle$, then

$$\mathbf{a} = a_1\mathbf{i} + a_2\mathbf{j} + a_3\mathbf{k}. \tag{8}$$

This can be seen geometrically, from Figure 18.14, or algebraically, using the rules (6) and the definition (7):

$$a_1\mathbf{i} + a_2\mathbf{j} + a_3\mathbf{k} = a_1\langle 1,0,0 \rangle + a_2\langle 0,1,0 \rangle + a_3\langle 0,0,1 \rangle$$

$$= \langle a_1,0,0 \rangle + \langle 0,a_2,0 \rangle + \langle 0,0,a_3 \rangle = \langle a_1,a_2,a_3 \rangle.$$

We should remember that every algebraic relation between vectors (in space) is equivalent to three relations between numbers (the components of the vectors). For instance, $\mathbf{a} = \langle a_1,a_2,a_3 \rangle = \mathbf{0}$ (the null-vector) means that $a_1 = 0$, $a_2 = 0$, $a_3 = 0$.

EXAMPLES 1. Given that $\mathbf{u} = \langle 1,2,3 \rangle$, $\mathbf{v} = \langle 1,1,0 \rangle$, $\mathbf{w} = \langle 0,1,2 \rangle$, what is $2\mathbf{u} - 3\mathbf{v} + 4\mathbf{w}$?
ANSWER $2\mathbf{u} - 3\mathbf{v} + 4\mathbf{w} = 2\langle 1,2,3 \rangle - 3\langle 1,1,0 \rangle + 4\langle 0,1,2 \rangle = \langle 2,4,6 \rangle - \langle 3,3,0 \rangle + \langle 0,4,8 \rangle = \langle 2-3+0,4-3+4,6-0+8 \rangle = \langle -1,5,14 \rangle$.

2. Let $\mathbf{u}$, $\mathbf{v}$, $\mathbf{w}$ be as in Example 1. Let $\mathbf{a} = \langle 7,-3,5 \rangle$. Find numbers x, y, z such that $\mathbf{a} = x\mathbf{u} + y\mathbf{v} + z\mathbf{w}$.
ANSWER   The vector equation above reads

$$\langle 7,-3,5 \rangle = x\langle 1,2,3 \rangle + y\langle 1,1,0 \rangle + z\langle 0,1,2 \rangle \tag{*}$$

and is equivalent to three numerical equations

$$7 = x + y,$$
$$-3 = 2x + y + z,$$
$$5 = 3x + 2z.$$

The first equation shows that $x = 7 - y$. Substitution into the other two equations gives

$$-y + z = -17,$$
$$-3y + 2z = -16.$$

This shows that $y = -18$, $z = -35$. Thus $x = 25$. The reader should check that $x = 25$, $y = -18$, $z = -35$ satisfy the vector equation (*).

PROBLEMS

In Problems 1 to 6 make a sketch illustrating the situation.

1. If P has coordinates $(1,4,2)$ and Q has coordinates $(4,4,2)$, find the coordinates of R such that $\overrightarrow{PQ} \sim \overrightarrow{QR}$.
2. If P has coordinates $(1,2,0)$, Q has coordinates $(3,6,0)$, and R has coordinates $(3,6,4)$, find the coordinates of the point S such that $\overrightarrow{PQ} \sim \overrightarrow{RS}$.
3. If P has coordinates $(2,4,4)$, Q has coordinates $(2,4,8)$, and R has coordinates $(4,6,2)$, find the coordinates of the point S such that $\overrightarrow{PQ} \sim \overrightarrow{RS}$.
4. If P has coordinates $(2,-4,-2)$, Q has coordinates $(2,-2,2)$, and S has coordinates $(-4,4,6)$, find the coordinates of the point R such that $\overrightarrow{PQ} \sim \overrightarrow{RS}$.
5. If P has coordinates $(1,1,-1)$, and Q has coordinates $(3,-3,5)$, what are the components and length of $\overrightarrow{PQ}$?
6. If R and S have coordinates $(-2,-1,-1)$ and $(2,-4,-6)$, respectively, what are the components and length of $\overrightarrow{RS}$?
7. If $\mathbf{a}$ has components $\langle -1,2,3 \rangle$ and $\mathbf{b}$ has components $\langle 3,-3,0 \rangle$, what are the components of $\frac{1}{2}(\mathbf{a} + \mathbf{b})$? Of $-3(2\mathbf{a} - \mathbf{b})$?
8. If $\mathbf{a}$ has components $\langle 2,0,1 \rangle$ and $\mathbf{b}$ has components $\langle -1,4,-1 \rangle$, what are the components of $2\mathbf{a} - 3(\mathbf{b} + \mathbf{i})$? Of $-\mathbf{a} + 2(\mathbf{b} - \mathbf{k})$?
9. Suppose that $\mathbf{u} = \langle 2,-3,7 \rangle$, $\mathbf{v} = \langle -1,4,-2 \rangle$, $\mathbf{w} = \langle 4,3,5 \rangle$. What is $2(\mathbf{u} - \mathbf{v}) + 3\mathbf{w}$? Express this result in the form $a_1\mathbf{i} + a_2\mathbf{j} + a_3\mathbf{k}$.
10. Suppose that $\mathbf{u} = \langle 1,1,1 \rangle$, $\mathbf{v} = \langle -1,3,2 \rangle$, $\mathbf{w} = \langle 2,-1,-1 \rangle$. If $\mathbf{a} = \langle 3,-4,9 \rangle$, find numbers x, y, z such that $\mathbf{a} = x\mathbf{u} + y\mathbf{v} + z\mathbf{w}$.
11. Suppose $\mathbf{a} + \mathbf{b} - 2\mathbf{c} = 2\mathbf{j} - \mathbf{k}$, $2\mathbf{a} - \mathbf{b} + \mathbf{c} = 2\mathbf{i} + \mathbf{j}$, and $\mathbf{a} - \mathbf{b} - \mathbf{c} = -\mathbf{i} - 2\mathbf{k}$. What are the components of $\mathbf{a}$, $\mathbf{b}$, and $\mathbf{c}$?
12. Suppose $\mathbf{u} + \mathbf{v} + \mathbf{w} = \mathbf{i}$, $\mathbf{u} - \mathbf{v} + \mathbf{w} = \mathbf{j}$, and $3\mathbf{u} - \mathbf{v} - 2\mathbf{w} = \mathbf{k}$. What are the components of $\mathbf{u}$, $\mathbf{v}$, and $\mathbf{w}$?

2.2   Position vectors and lines

If a point R has Cartesian coordinates (x,y,z), the vector $\mathbf{R}$ determined by the directed segment $\overrightarrow{OR}$, that is, the vector

$$\mathbf{R} = x\mathbf{i} + y\mathbf{j} + z\mathbf{k}, \tag{9}$$

is called the **position vector** of R (with respect to O). This is the same definition as the one given in Chapter 17, §1.6, for the points in the plane. As before, *the vector determined by* $\overrightarrow{PQ}$ *is* $\mathbf{Q} - \mathbf{P}$ (see Figure 17.20).

Theorem 4 of Chapter 17, §1.6, and its corollaries (including the important Corollary 3 in Chapter 17, §1.7) remain valid, without any changes. *A point R lies on the line through P and Q ($Q \neq P$) if and only if the position vectors of these points satisfy*

$$\mathbf{R} = (1 - t)\mathbf{P} + t\mathbf{Q} \tag{10}$$

for some number t.

For every vector $\mathbf{a} \neq \mathbf{0}$ and every vector $\mathbf{b}$, all points with position vectors

$$\mathbf{R} = t\mathbf{a} + \mathbf{b}, \qquad -\infty < t < \infty \tag{11}$$

form a line. Every line can be so represented.

[If we want to take a strictly analytic point of view, we may think of (11) as the *definition* of a line.]

It is convenient to rewrite the above formulas in coordinates. Let (p_1,p_2,p_3) and (q_1,q_2,q_3) be the coordinates of P and Q (that is, the components of the position vectors $\mathbf{P}$ and $\mathbf{Q}$). Noting (9), we rewrite (10) as

$$x = (1 - t)p_1 + tq_1, \qquad y = (1 - t)p_2 + tq_2, \qquad z = (1 - t)p_3 + tq_3, \tag{10'}$$

or

$$x = p_1 + (q_1 - p_1)t, \qquad y = p_2 + (q_2 - p_2)t, \qquad z = p_3 + (q_3 - p_3)t. \tag{10''}$$

The existence of a number t satisfying these three conditions is the condition (necessary and sufficient!) that the points P, Q, R lie on one line.

If $\langle a_1,a_2,a_3 \rangle$ are the components of $\mathbf{a}$ and $\langle b_1,b_2,b_3 \rangle$ those of $\mathbf{b}$, then the parametric representation (11) may be rewritten as

$$x = a_1 t + b_1, \qquad y = a_2 t + b_2, \qquad z = a_3 t + b_3, \qquad -\infty < t < \infty. \tag{11'}$$

Note that (10'') is a parametric representation of a line through P and Q. Note also that in the plane we may represent a line either by a single linear equation or parametrically. In space we do not have the first option.

EXAMPLES 1. Do the points $(1,2,3)$, $(2,3,4)$, and $(4,9,6)$ lie on one line?
SOLUTION Set $P = (p_1,p_2,p_3) = (1,2,3)$, $Q(q_1,q_2,q_3) = (2,3,4)$. The point $(x,y,z) = (4,9,6)$ lies on the same line as P and Q if (and only if) Equations (10') hold for some t. Thus in view of (10'), we must determine whether there is a number t such that

$$4 = (1 - t)1 + t2, \qquad 9 = (1 - t)2 + t3, \qquad 6 = (1 - t)3 + t4;$$

that is,

$$4 = 1 + t, \qquad 9 = 2 + t, \qquad 6 = 3 + t.$$

The first and last equations assert that $t = 3$, the second that $t = 7$. Hence there is no number t satisfying the desired condition. The answer is thus *no*.

2. Find a parametric representation of the line through the points $(2, -1, 4)$ and $(3, 2, 6)$.
SOLUTION   Using $(10'')$ with $(p_1, p_2, p_3) = (2, -1, 4)$ and $(q_1, q_2, q_3) = (3, 2, 6)$, we obtain the representation

$$x = 2 + t, \qquad y = -1 + 3t, \qquad z = 4 + 2t. \tag{*}$$

3. At which points does the line in Example 2 intersect the coordinate planes $x = 0$, $y = 0$, and $z = 0$?
ANSWER   We see from $(*)$ that $x = 0$ for $t = -2$. But for $t = -2$ we have, again by $(*)$, $y = -7$, $z = 0$. Hence our line intersects the plane $x = 0$ at the point $(0, -7, 0)$. We note that this is also the intersection of the line $(*)$ with the plane $z = 0$. In the same way, we see that $y = 0$ for $t = \frac{1}{3}$. For this value of t, relations $(*)$ yield $x = \frac{7}{3}$, $z = \frac{14}{3}$. Hence our line intersects the xz plane (the plane $y = 0$) at $(\frac{7}{3}, 0, \frac{14}{3})$.

4. Does the line given by

$$x = 1 + 2t, \qquad y = -4 + 6t, \qquad z = 2 + 4t \tag{**}$$

coincide with the line in Example 2?
ANSWER   First of all, in order to prevent misunderstandings, it is much better to denote the parameters in two representations by different letters. Therefore we rewrite $(**)$ in the form

$$x = 1 + 2s, \qquad y = -4 + 6s, \qquad z = 2 + 4s. \tag{***}$$

In order to obtain two points of $(***)$, we choose two values of the parameter s, for instance, $s = 0$ and $s = 1$. This gives us the points $(1, -4, 2)$ and $(3, 2, 6)$. We ask whether these points lie on the first line, that is, whether their coordinates can be represented in the form $(*)$.
Consider first the point $(1, -4, 2)$. Does there exist a t such that

$$1 = 2 + t, \qquad -4 = -1 + 3t, \qquad 2 = 4 + 2t?$$

The first equation gives $t = -1$; for this value of t the other two equations also hold. Hence $(1, -4, 2)$ lies on $(*)$. In the same way, we see that $(3, 2, 6)$ is a point of $(*)$, corresponding to the parameter value $t = 1$. Since the two lines have two points in common, they coincide.

PROBLEMS

In Problems 1 to 6 find a parametric representation for the line passing through the indicated points.

1. $(3, 4, 5)$ and $(5, 4, 3)$.   3. $(1, 4, 2)$ and $(-2, 8, 4)$.   5. $(-3, 2, 6)$ and $(12, -4, 2)$.
2. $(-1, 2, 3)$ and $(0, 4, 5)$.   4. $(0, 12, 7)$ and $(-5, 0, 3)$.   6. $(-2, 0, 1)$ and $(1, 6, 7)$.

7. At which points does the line in Problem 1 meet the coordinate planes?
8. At which points does the line in Problem 4 meet the coordinate planes?
9. At which point does the line $x = 1 - t$, $y = 1 + t$, $z = 2 - 2t$ meet the plane $x = 2$?
10. Do the lines in Problems 3 and 4 coincide?
11. Do the lines in Problems 2 and 6 coincide?
12. Find numbers α and β such that the line $x = 1 + \alpha t$, $y = \beta + \frac{1}{2}t$, $z = 2 - t$ coincides with the line in Problem 9.
13. Let the points P and Q have coordinates $(1, -3, 0)$ and $(2, 5, -7)$, respectively. Find the midpoint of the segment $\overrightarrow{PQ}$.
14. Let P and Q be as in Problem 1. If R lies on the segment joining P and Q, find R if $|PR| = \frac{4}{5}|RQ|$.

15. Find a point P on the coordinate plane $y = 0$ and a point Q on the coordinate plane $z = 0$ such that the point $(1,2,3)$ is the midpoint of the segment $\overrightarrow{PQ}$. [*Note:* P and Q are not unique.]

16. Find numbers x and y such that $(x,y,0)$ lies on the straight line passing through $(4,3,-1)$ and $(2,8,2)$.

2.3 Vector-valued functions and curves

A **space curve** is defined by a vector-valued function $t \mapsto \mathbf{f}(t)$, $\alpha \leq t \leq \beta$, where we interpret $\mathbf{f}$ as the position vector of a point in space. If we denote the components of $\mathbf{f}(t)$ by $x(t)$, $y(t)$, $z(t)$, we can write the vector-valued function as

$$t \mapsto x(t)\mathbf{i} + y(t)\mathbf{j} + z(t)\mathbf{k}, \qquad \alpha \leq t \leq \beta \tag{12}$$

or, which amounts to the same, as a triple of number-valued functions

$$x = x(t), \qquad y = y(t), \qquad z = z(t), \qquad \alpha \leq t \leq \beta. \tag{12'}$$

The curve is smooth if the derivative

$$\mathbf{f}'(t) = x'(t)\mathbf{i} + y'(t)\mathbf{j} + z'(t)\mathbf{k} \tag{13}$$

exists, is continuous, and is not the null-vector. Everything said in Chapter 17, §2, about differentiation and integration and parameter changes applies to vector-valued functions in space and space curves. In particular,

$$\mathbf{T}(t) = \frac{1}{|\mathbf{f}'(t)|} \mathbf{f}'(t) \tag{14}$$

is the unit tangent vector to the curve. Its components are

$$\frac{x'(t)}{\sqrt{x'(t)^2 + y'(t)^2 + z'(t)^2}}, \qquad \frac{y'(t)}{\sqrt{x'(t)^2 + y'(t)^2 + z'(t)^2}},$$

$$\frac{z'(t)}{\sqrt{x'(t)^2 + y'(t)^2 + z'(t)^2}}.$$

The length L of (12) is given by the formula

$$L = \int_\alpha^\beta |\mathbf{f}'(t)|\, dt = \int_\alpha^\beta \sqrt{x'(t)^2 + y'(t)^2 + z'(t)^2}\, dt. \tag{15}$$

The definitions of velocity and acceleration and the statement of Newton's Law of Motion in Chapter 17, §3.1 and §3.3, go over, without any changes, to motion in space.

EXAMPLES 1. Let a and b be positive numbers. The curve defined by

$$x = a \cos t, \qquad y = a \sin t, \qquad z = bt \qquad (-\infty < t < \infty) \tag{*}$$

is called a **right-handed helix** (see Figure 18.15). This is the shape of the cutting edge of an ordinary ("right-handed") screw. The left-handed helix is defined by

$$x = a \cos t, \qquad y = a \sin t, \qquad z = -bt \qquad (-\infty < t < +\infty);$$

see Figure 18.16.

FIGURE **18.15**   FIGURE **18.16**

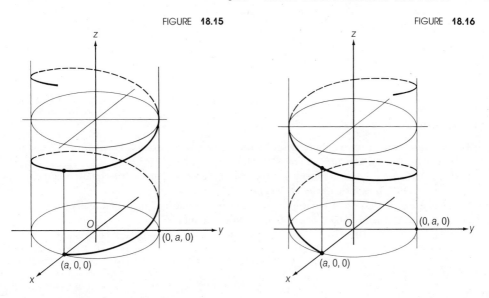

2. Find the length of the arc on the helix (*) for $t_0 \le t \le t_1$.

ANSWER   By (15), the desired length is

$$L = \int_{t_0}^{t_1} \sqrt{\left(\frac{d\,a\cos t}{dt}\right)^2 + \left(\frac{d\,a\sin t}{dt}\right)^2 + \left(\frac{d\,bt}{dt}\right)^2}\, dt$$

$$= \int_{t_0}^{t_1} \sqrt{a^2\sin^2 t + a^2\cos^2 t + b^2}\, dt = \sqrt{a^2 + b^2}\,(t_1 - t_0).$$

3. What is the velocity, speed, and acceleration of a particle whose trajectory is described by the position vector $\mathbf{r}(t) = t\mathbf{i} + \frac{1}{2}t^2\mathbf{j} + \frac{1}{3}t^3\mathbf{k}$, $t \ge 0$?

ANSWER   The velocity and acceleration vectors are given by $\mathbf{v}(t) = \dot{\mathbf{r}}(t)$ and $\mathbf{a} = \dot{\mathbf{v}}(t) = \ddot{\mathbf{r}}(t)$, respectively. Hence $\mathbf{v}(t) = \mathbf{i} + t\mathbf{j} + t^2\mathbf{k}$, and $\mathbf{a}(t) = \mathbf{j} + 2t\mathbf{k}$. The speed is the magnitude of the velocity, that is, $|\mathbf{v}(t)| = \sqrt{1 + t^2 + t^4}$.

PROBLEMS

In Problems 1 to 4 find the unit tangent vector to the given space curve at the indicated value of t.

1. $\mathbf{f}(t) = \sqrt{2}t\mathbf{i} + \frac{1}{2}t^2\mathbf{j} + \frac{1}{3}t^3\mathbf{k}$, $t = 1$.

2. $\mathbf{f}(t) = (t^2 + 1)\mathbf{i} + (\cos t)\mathbf{j} + e^t\mathbf{k}$, $t = 0$.

3. $\mathbf{f}(t) = (t\cos t)\mathbf{i} + (t\sin t)\mathbf{j} + t\mathbf{k}$, $t = \dfrac{\pi}{2}$.

4. $\mathbf{f}(t) = (\cos t)\mathbf{i} + (2\sin t)\mathbf{j} + (\sin t\cos t)\mathbf{k}$, $t = \dfrac{\pi}{3}$.

In Problems 5 to 8 find the length of the given space curve between the indicated limits.

5. $\mathbf{f}(t) = e^t\mathbf{i} + 2\sqrt{2}e^{t/2}\mathbf{j} + t\mathbf{k}$,   for $0 \le t \le 1$.

6. $\mathbf{f}(t) = \left(\dfrac{t^3}{3}\right)\mathbf{i} + \left(\dfrac{t^2}{\sqrt{2}}\right)\mathbf{j} + t\mathbf{k}$,   for $0 \le t \le 1$.

7. $\mathbf{f}(t) = (\sin t)\mathbf{i} + (\cos t)\mathbf{j} + \frac{2}{3}t^{3/2}\mathbf{k}$,   for $0 \leq t \leq 8$.
8. $\mathbf{f}(t) = \frac{1}{2}t^2\mathbf{i} + \frac{1}{3}(t^2 - 1)^{3/2}\mathbf{j} + \frac{1}{2}t^2\mathbf{k}$,   for $1 \leq t \leq \sqrt{3}$.

In Problems 9 to 12 the trajectory of a particle is described by the position vector $\mathbf{r}(t)$. Find the velocity and acceleration in Cartesian coordinates and determine the speed. Assume $t \geq 0$.

9. $\mathbf{r}(t) = 2t\mathbf{i} + (t^2 + 1)\mathbf{j} + 2t\mathbf{k}$.   12. $\mathbf{r}(t) = \dfrac{1}{\sqrt{1 + t}}\mathbf{i} + \dfrac{1}{\sqrt{1 + t}}\mathbf{j} + \sqrt{1 + t}\,\mathbf{k}$.
10. $\mathbf{r}(t) = (\cos t)\mathbf{i} + (\sin t)\mathbf{j} + te^{-t}\mathbf{k}$.
11. $\mathbf{r}(t) = (e^t \cos t)\mathbf{i} + (e^t \sin t)\mathbf{j} + e^t\mathbf{k}$.

13. If a and b are positive numbers, describe the curve whose parametric equations are given by $x = at$, $y = at$, $z = bt^2$, where $-\infty < t < +\infty$.
14. If a, b, and c are positive numbers, describe the curve whose parametric equations are given by $x = a \cos t$, $y = b \sin t$, $z = ct$, where $-\infty < t < +\infty$.
15. What curve is described by the parametric representation, in cylindrical coordinates, $r = a$, $\theta = t$, $z = bt$, where a and b are positive numbers?
16. What curve is described by the parametric representation, in spherical coordinates, $\rho = 2$, $\phi = \pi/6$, $\theta = t$?

§3   Inner product

We shall now describe a way of multiplying two vectors; the product of two vectors will not be a vector but a number. In the next section we shall apply vector multiplication to geometry.

3.1   Definition of inner product

Let $\mathbf{a}$ and $\mathbf{b}$ be two vectors, with $\mathbf{a} \neq \mathbf{0}, \mathbf{b} \neq \mathbf{0}$. Choose a point O and represent $\mathbf{a}$ and $\mathbf{b}$ by directed segments $\overrightarrow{OP}$ and $\overrightarrow{OQ}$ (see Figure 18.17). The angle $\sphericalangle POQ$ is called the angle between the vectors $\mathbf{a}$ and $\mathbf{b}$ and is denoted by $\sphericalangle(\mathbf{a},\mathbf{b})$.

We associate with any two vectors $\mathbf{a}$ and $\mathbf{b}$ a number called their **inner product** and denoted by $(\mathbf{a},\mathbf{b})$ or by $\mathbf{a} \cdot \mathbf{b}$. The definition reads:

$$(\mathbf{a},\mathbf{b}) = |\mathbf{a}|\,|\mathbf{b}|\cos\theta, \qquad \theta = \text{angle between } \mathbf{a} \text{ and } \mathbf{b}, \tag{1}$$
$$(\mathbf{a},\mathbf{b}) = 0 \qquad \text{if either } \mathbf{a} = \mathbf{0} \text{ or } \mathbf{b} = \mathbf{0}.$$

(The inner product is also called the "scalar product" and the "dot product.") The definition just given applies to plane vectors and vectors in space.

It follows from (1) that

$$\text{if } (\mathbf{a},\mathbf{b}) = 0, \qquad \text{then either } \ |\mathbf{a}| = 0, \ \text{ or } \ |\mathbf{b}| = 0, \ \text{ or } \ \theta = 90°. \tag{2}$$

Vectors $\mathbf{a}$, $\mathbf{b}$ with $(\mathbf{a},\mathbf{b}) = 0$ are called **orthogonal** or perpendicular.

For $\mathbf{a} = \mathbf{b}$, we have $\theta = 0$, $\cos\theta = 1$. Therefore

$$\mathbf{a} \cdot \mathbf{a} = (\mathbf{a},\mathbf{a}) = |\mathbf{a}|^2, \qquad |\mathbf{a}| = \sqrt{(\mathbf{a},\mathbf{a})}. \tag{3}$$

Theorem 1. *Let the vectors $\mathbf{a}$ and $\mathbf{b}$ have the components a_1, a_2, a_3 and b_1, b_2, b_3, respectively. Then*

$$\mathbf{a} \cdot \mathbf{b} = (\mathbf{a},\mathbf{b}) = a_1 b_1 + a_2 b_2 + a_3 b_3. \tag{4}$$

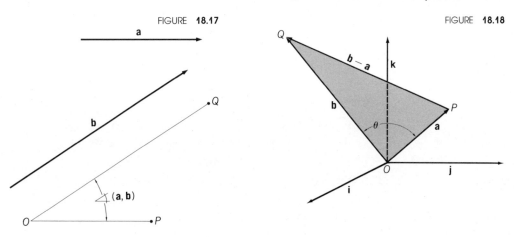

FIGURE **18.17**   FIGURE **18.18**

Proof.   Relation (4) is certainly true if either $\mathbf{a} = \mathbf{0}$ or $\mathbf{b} = \mathbf{0}$, for then both sides of (4) are 0. Assume that $\mathbf{a} \neq \mathbf{0}$, $\mathbf{b} \neq \mathbf{0}$, and represent $\mathbf{a}$ by $\overrightarrow{OP}$ and $\mathbf{b}$ by $\overrightarrow{OQ}$, where O is the origin. Then P has Cartesian coordinates (a_1,a_2,a_3), and Q has coordinates (b_1,b_2,b_3); see Figure 18.18. By the distance formula,

$$|\mathbf{a}|^2 = |OP|^2 = a_1^2 + a_2^2 + a_3^2, \qquad |\mathbf{b}|^2 = |OQ|^2 = b_1^2 + b_2^2 + b_3^2,$$

and

$$|\mathbf{b} - \mathbf{a}|^2 = |PQ|^2 = (b_1 - a_1)^2 + (b_2 - a_2)^2 + (b_3 - a_3)^2$$

$$= b_1^2 + b_2^2 + b_3^2 + a_1^2 + a_2^2 + a_3^2 - 2(a_1 b_1 + a_2 b_2 + a_3 b_3).$$

Since $\theta = \measuredangle(\mathbf{a},\mathbf{b}) = \measuredangle(POQ)$, we can write the cosine law (see Chapter 16, §1.1, Example 4) in the form

$$|\mathbf{b} - \mathbf{a}|^2 = |\mathbf{a}|^2 + |\mathbf{b}|^2 - 2|\mathbf{a}| \, |\mathbf{b}| \cos \theta = |\mathbf{a}|^2 + |\mathbf{b}|^2 - 2(\mathbf{a},\mathbf{b}).$$

Thus

$$(\mathbf{a},\mathbf{b}) = \tfrac{1}{2}(|\mathbf{a}|^2 + |\mathbf{b}|^2 - |\mathbf{b} - \mathbf{a}|^2).$$

Substituting for $|\mathbf{a}|^2$, $|\mathbf{b}|^2$, and $|\mathbf{b} - \mathbf{a}|^2$ their expressions in terms of components, we obtain (4).

Corollary 1.   *If* $\mathbf{a}$ *and* $\mathbf{b}$ *are plane vectors with components* a_1, a_2 *and* b_1, b_2, *then*

$$\mathbf{a} \cdot \mathbf{b} = (\mathbf{a},\mathbf{b}) = a_1 b_1 + a_2 b_2. \tag{5}$$

Indeed, $\mathbf{a}$ and $\mathbf{b}$ may be considered as space vectors with components a_1, a_2, 0 and b_1, b_2, 0. Hence (5) follows from (4).

Theorem 1 permits us to calculate the angle θ between two vectors $\mathbf{a}$ and $\mathbf{b}$ (of positive length) or between the directed segments representing these vectors. Indeed, by (1) and (3), we have

$$\theta = \measuredangle(\mathbf{a},\mathbf{b}) = \text{arc cos} \frac{(\mathbf{a},\mathbf{b})}{|\mathbf{a}| \, |\mathbf{b}|} = \text{arc cos} \frac{(\mathbf{a},\mathbf{b})}{\sqrt{(\mathbf{a},\mathbf{a})} \, \sqrt{(\mathbf{b},\mathbf{b})}}, \tag{6}$$

and, by (4), we are able to compute the numbers $(\mathbf{a},\mathbf{b})$, $(\mathbf{a},\mathbf{a})$, and $(\mathbf{b},\mathbf{b})$.

EXAMPLES 1. We have

$$(\mathbf{i,i}) = (\mathbf{j,j}) = (\mathbf{k,k}) = 1, \qquad (\mathbf{i,j}) = (\mathbf{j,k}) = (\mathbf{k,i}) = 0.$$

Indeed, $|\mathbf{i}| = |\mathbf{j}| = |\mathbf{k}| = 1, \measuredangle(\mathbf{i,i}) = \measuredangle(\mathbf{j,j}) = \measuredangle(\mathbf{k,k}) = 0°, \measuredangle(\mathbf{i,j}) = \measuredangle(\mathbf{j,k}) = \measuredangle(\mathbf{k,i}) = 90°.$

2. Let the points P, Q, R have the Cartesian coordinates $(1,2,3)$, $(4,-1,7)$, and $(6,-1,4)$, respectively. Compute $\theta = \measuredangle PQR$.

SOLUTION   We have $\theta = \measuredangle(\mathbf{P - Q, R - Q})$; see Figure 18.19. Since $\mathbf{P - Q}$ has components $\langle 1 - 4, 2 + 1, 3 - 7\rangle = \langle -3,3,-4\rangle$, and $\mathbf{R - Q}$ has components $\langle 2,0,-3\rangle$, we have

$$|\mathbf{P - Q}|^2 = (-3)^2 + 3^2 + 4^2 = 34, \qquad |\mathbf{R - Q}|^2 = 2^2 + 0^2 + (-3)^2 = 13,$$

$$(\mathbf{P - Q, R - Q}) = (-3)2 + 3 \cdot 0 + (-4)(-3) = 6.$$

FIGURE **18.19**

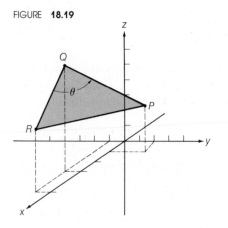

By (6), we have $\cos \theta = 6/(\sqrt{34}\sqrt{13}) \approx .28$ and $\theta \approx 74°$.

PROBLEMS

In Problems 1 to 4 find the number $(\mathbf{a,b})$ for the given vectors $\mathbf{a}$ and $\mathbf{b}$.

1. $\mathbf{a} = \langle 2,4,0\rangle, \quad \mathbf{b} = \langle 6,-1,0\rangle.$
2. $\mathbf{a} = \langle -1,4,2\rangle, \quad \mathbf{b} = \langle 5,1,\frac{1}{2}\rangle.$
3. $\mathbf{a} = \langle 2,3,1\rangle, \quad \mathbf{b} = \langle -20,10,30\rangle.$
4. $\mathbf{a} = \langle \frac{1}{10},\frac{3}{10},\frac{5}{10}\rangle, \quad \mathbf{b} = \langle \frac{1}{2},-1\frac{1}{8}\rangle.$

In Problems 5 to 8 find the angle between the two given vectors.

5. $\mathbf{a} = \langle \frac{1}{2},1,\frac{3}{2}\rangle, \quad \mathbf{b} = \langle 1,-\frac{3}{2},-\frac{1}{2}\rangle.$
6. $\mathbf{a} = \mathbf{i} - \mathbf{j}, \quad \mathbf{b} = \mathbf{j} - \mathbf{k}.$
7. $\mathbf{u} = \langle 1,-2,4\rangle, \quad \mathbf{v} = \langle 1,2,4\rangle.$
8. $\mathbf{u} = \mathbf{i} + \mathbf{j} + \mathbf{k}, \quad \mathbf{v} = 2\mathbf{i} + \mathbf{j} + 2\mathbf{k}.$

9. Find all numbers α such that the vectors $\mathbf{u} = \mathbf{i} + 5\alpha\mathbf{j} + 6\alpha\mathbf{k}$ and $\mathbf{v} = -\mathbf{i} - \mathbf{j} + \alpha\mathbf{k}$ are orthogonal.
10. Find all numbers α such that $\cos \measuredangle (\mathbf{i} + 2\mathbf{j} + \alpha\mathbf{k}, 2\mathbf{i} + \mathbf{j} + 2\mathbf{k}) = \frac{2}{3}.$

3.2   Rules for inner products

We note the formal properties of inner products.

Theorem 2.   *Inner products obey the following rules:*

> **I.** *For any two vectors* **a** *and* **b**, *there is a unique number called their* **inner product** *and denoted by* (**a**,**b**).
> **II.** (**a**,**b**) = (**b**,**a**) *for all* **a** *and* **b**.
> **III.** (λ**a**,**b**) = λ(**a**,**b**) *for all* **a**, **b** *and all numbers* λ.
> **IV.** (**a**,**b** + **c**) = (**a**,**b**) + (**a**,**c**) *for all* **a**, **b**, **c**.
> **V.** (**a**,**a**) > 0 *unless* **a** = **0**.

Proof.   Rule **I** needs no proof. The other rules can be proved either geometrically, from the definition (1), or analytically, using (4) and the properties of numbers. We carry out the second approach.

Let **a**, **b**, and **c** be vectors with components $\langle a_1,a_2,a_3 \rangle$, $\langle b_1,b_2,b_3 \rangle$, and $\langle c_1,c_2,c_3 \rangle$, respectively. The commutative law **II** follows from (4) by noting that each term on the right-hand side of (4) is unchanged if we interchange the a's and b's. To prove the associative law **III** note that λ**a** has components $\langle \lambda a_1, \lambda a_2, \lambda a_3 \rangle$. Hence **III** says that

$$(\lambda a_1)b_1 + (\lambda a_2)b_2 + (\lambda a_3)b_3 = \lambda(a_1 b_1 + a_2 b_2 + a_3 b_3).$$

This is true. Since **b** + **c** has the components $\langle b_1 + c_1, b_2 + c_2, b_3 + c_3 \rangle$, the distributive law **IV** asserts that

$$a_1(b_1 + c_1) + a_2(b_2 + c_2) + a_3(b_3 + c_3)$$
$$= (a_1 b_1 + a_2 b_2 + a_3 b_3) + (a_1 c_1 + a_2 c_2 + a_3 c_3).$$

This is true. Finally, $a_1^2 + a_2^2 + a_3^2$ is 0 only if $a_1 = a_2 = a_3 = 0$. This proves **V**.

We note that

$$(\mathbf{a},\mathbf{0}) = 0, \qquad \text{for all } \mathbf{a}, \tag{7}$$

$$(\mathbf{b} + \mathbf{c},\mathbf{a}) = (\mathbf{b},\mathbf{a}) + (\mathbf{c},\mathbf{a}). \tag{8}$$

The proofs are left to the reader.

Here is an application of these rules.

Theorem 3.   *If* $\mathbf{a} = a_1\mathbf{i} + a_2\mathbf{j} + a_3\mathbf{k} = \langle a_1,a_2,a_3 \rangle$, *then*

$$a_1 = (\mathbf{a},\mathbf{i}), \qquad a_2 = (\mathbf{a},\mathbf{j}), \qquad a_3 = (\mathbf{a},\mathbf{k}). \tag{9}$$

Proof.   Using the rules of Theorem 2 and the relations in §3.1, Example 1, we have

$$(\mathbf{a},\mathbf{i}) = (a_1\mathbf{i} + a_2\mathbf{j} + a_3\mathbf{k},\mathbf{i}) = (a_1\mathbf{i},\mathbf{i}) + (a_2\mathbf{j},\mathbf{i}) + (a_3\mathbf{k},\mathbf{i})$$
$$= a_1(\mathbf{i},\mathbf{i}) + a_2(\mathbf{j},\mathbf{i}) + a_3(\mathbf{k},\mathbf{i}) = a_1 \cdot 1 + a_2 \cdot 0 + a_3 \cdot 0 = a_1.$$

This proves the first relation (9). The others are proved similarly.

EXAMPLES 1.   What is the angle between the diagonal of a cube and an edge?
ANSWER   We must find the angle α between $\mathbf{i} + \mathbf{j} + \mathbf{k}$ and $\mathbf{i}$ (this is seen from Figure 18.20).
We have

$$\cos \alpha = \frac{(\mathbf{i} + \mathbf{j} + \mathbf{k},\mathbf{i})}{|\mathbf{i} + \mathbf{j} + \mathbf{k}| \, |\mathbf{i}|} = \frac{1}{\sqrt{3}}$$

so that $\alpha \approx 54°40'$.

FIGURE **18.20**   FIGURE **18.21**

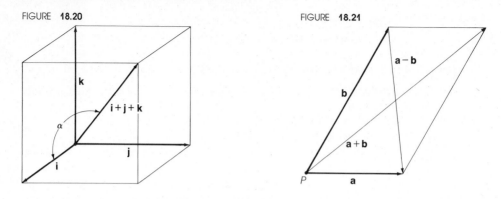

2. *In a parallelogram, the two diagonals are equal (have the same length) if and only if the parallelogram is a rectangle.*

Proof.   Let the two sides issuing from a vertex, P, of the parallelogram represent the vectors **a** and **b** (see Figure 18.21). The two diagonals represent the vectors $\mathbf{a} + \mathbf{b}$ and $\mathbf{a} - \mathbf{b}$. The equality of the diagonals means that $|\mathbf{a} + \mathbf{b}|^2 = |\mathbf{a} - \mathbf{b}|^2$ or

$$0 = (\mathbf{a} + \mathbf{b},\mathbf{a} + \mathbf{b})^2 - (\mathbf{a} - \mathbf{b},\mathbf{a} - \mathbf{b})$$
$$= (\mathbf{a},\mathbf{a}) + 2(\mathbf{a},\mathbf{b}) + (\mathbf{b},\mathbf{b}) - [(\mathbf{a},\mathbf{a}) - 2(\mathbf{a},\mathbf{b}) + (\mathbf{b},\mathbf{b})] = 4(\mathbf{a},\mathbf{b}).$$

But $(\mathbf{a},\mathbf{b}) = 0$ means that **a** is perpendicular to **b**; that is, the parallelogram is a rectangle.

PROBLEMS

1. Prove the identity $(\mathbf{a} + \mathbf{b},\mathbf{a} + \mathbf{b}) = |\mathbf{a}|^2 + 2(\mathbf{a},\mathbf{b}) + |\mathbf{b}|^2$.
2. Prove the identity $(\mathbf{a} + \mathbf{b} + \mathbf{c},\mathbf{a} + \mathbf{b} + \mathbf{c}) = |\mathbf{a}|^2 + |\mathbf{b}|^2 + |\mathbf{c}|^2 + 2(\mathbf{a},\mathbf{b}) + 2(\mathbf{a},\mathbf{c}) + 2(\mathbf{b},\mathbf{c})$.
3. Find a vector **a** with $(\mathbf{a},\mathbf{i}) = (\mathbf{a},\mathbf{j}) = (\mathbf{a},\mathbf{k})$ and $|\mathbf{a}| = 2\sqrt{3}$.
4. Suppose that $(\mathbf{a},\mathbf{i}) = (\mathbf{a},\mathbf{j}) = 2$ and $|\mathbf{a}| = \sqrt{17}$. Find $(\mathbf{a},\mathbf{k})$.
5. If $(\mathbf{a},\mathbf{i}) = (\mathbf{a},\mathbf{j}) = 5$, can we have $|\mathbf{a}| \leq 7$?
6. Find a vector **a** such that $(\mathbf{a},\mathbf{i} + \mathbf{j}) = 2$, $(\mathbf{a},\mathbf{i} - \mathbf{j}) = 3$, and $(\mathbf{a},\mathbf{k}) = 0$.
7. Suppose that $(\mathbf{a},\mathbf{i}) = 1$, $(\mathbf{a},\mathbf{j}) = 2$, $(\mathbf{a},\mathbf{k}) = 3$, $(\mathbf{b},\mathbf{i}) = 3$, $(\mathbf{b},\mathbf{j}) = 2$, $(\mathbf{b},\mathbf{k}) = 1$. Find $\sphericalangle(\mathbf{a},\mathbf{b})$.
8. Show that $(\mathbf{a},\mathbf{i})^2 \leq (\mathbf{a},\mathbf{a})$ for all **a**. When is $(\mathbf{a},\mathbf{i})^2 = (\mathbf{a},\mathbf{a})$?
9. In a cube, find the angle between an edge and a diagonal of a face. [*Hint:* Compute $\sphericalangle(\mathbf{i},\mathbf{i} + \mathbf{j})$.]
10. In a cube, find the angle between two diagonals emanating from the same vertex if the diagonals lie in different faces of the cube.
11. What is the angle between the diagonal of a cube and the diagonal of a face?
12. Prove that in a parallelogram the two diagonals are perpendicular if and only if the parallelogram is a rhombus.

3.3   Direction cosines

Let $\mathbf{a} \neq \mathbf{0}$ be a vector, and let α_1, α_2, α_3 be the angles between **a** and **i**, **j**, **k** (see Figure 18.22). These angles are called the **direction angles** of **a**, and the numbers $\cos \alpha_1$, $\cos \alpha_2$, $\cos \alpha_3$ are called the **direction cosines** of **a**.

FIGURE **18.22**

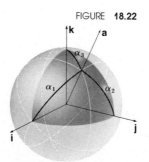

The direction cosines of a vector in space have properties similar to those of the cosine and sine of the polar angle of a vector in the plane.

If $\mathbf{a} = a_1\mathbf{i} + a_2\mathbf{j} + a_3\mathbf{k}$, then the direction cosines of $\mathbf{a}$ are

$$\cos \alpha_1 = \frac{a_1}{|\mathbf{a}|}, \qquad \cos \alpha_2 = \frac{a_2}{|\mathbf{a}|}, \qquad \cos \alpha_3 = \frac{a_3}{|\mathbf{a}|}. \tag{10}$$

For we have, by (6) and Theorem 3,

$$\cos \alpha_1 = \cos \sphericalangle(\mathbf{a},\mathbf{i}) = \frac{(\mathbf{a},\mathbf{i})}{|\mathbf{a}| \, |\mathbf{i}|} = \frac{a_1}{|\mathbf{a}|},$$

since $|\mathbf{i}| = 1$. The other two relations (10) are proved in the same way. Knowing the length $|\mathbf{a}|$ of a vector $\mathbf{a}$ and its direction angles α_1, α_2, α_3, we can write the vector in the form

$$\mathbf{a} = |\mathbf{a}| \, (\cos \alpha_1)\mathbf{i} + |\mathbf{a}| \, (\cos \alpha_2)\mathbf{j} + |\mathbf{a}| \, (\cos \alpha_3)\mathbf{k}. \tag{11}$$

Formulas (10) and (11) should be compared with the corresponding formulas for plane vectors:

$$\cos \theta = \frac{a_1}{|\mathbf{a}|}, \qquad \sin \theta = \frac{a_2}{|\mathbf{a}|}, \qquad \mathbf{a} = |\mathbf{a}| \, (\cos \theta)\mathbf{i} + |\mathbf{a}| \, (\sin \theta)\mathbf{j}. \tag{12}$$

The relation $\cos^2 \theta + \sin^2 \theta = 1$ also has an analog. Since $|\mathbf{a}|^2 = (\mathbf{a},\mathbf{a}) = a_1^2 + a_2^2 + a_3^2$, we conclude from (10) that

$$\cos^2 \alpha_1 + \cos^2 \alpha_2 + \cos^2 \alpha_3 = 1. \tag{13}$$

If α_1, α_2, α_3 are any numbers between 0 and π, inclusive, such that (13) holds, then they are the direction angles of a unique **unit vector** (vector of length 1). This vector is

$$\mathbf{a}_0 = (\cos \alpha_1)\mathbf{i} + (\cos \alpha_2)\mathbf{j} + (\cos \alpha_3)\mathbf{k}.$$

REMARK 1 Let $\overrightarrow{PQ}$ and $\overrightarrow{RS}$ be parallel directed segments (neither of zero length), representing vectors $\mathbf{a}$ and $\mathbf{b}$. Then $\mathbf{b} = t\mathbf{a}$, where t is a number, $t \neq 0$, and the direction cosines of $\mathbf{b}$ are either the same as those of $\mathbf{a}$ (if $t > 0$) or the negative of those of $\mathbf{a}$ (if $t < 0$).

Proof.   Represent **a** and **b** by directed segments $\overrightarrow{OA}$ and $\overrightarrow{OB}$. Then $\overrightarrow{OA}$ is parallel to $\overrightarrow{PQ}$ and $\overrightarrow{OB}$ is parallel to $\overrightarrow{RS}$ so that O, A, and B lie on one line. If O lies between A and B, then $\mathbf{b} = t\mathbf{a}$ where $t = -|OB|/|OA|$. If not, $\mathbf{b} = t\mathbf{a}$ when $t = |OB|/|OA|$. The statement about direction cosines follows.

REMARK 2   Let **a** be a vector orthogonal to **k** so that $\alpha_3 = 90°$, $\cos \alpha_3 = 0$. In this case, we may think of **a** as a vector in the xy plane. Let θ be the polar angle of **a**. Then $\cos \alpha_1 = \cos \theta$ and $\cos \alpha_2 = \sin \theta$, as is seen by comparing (10) with (12). Note, however, that direction angles must satisfy the inequality $0 \leq \alpha \leq \pi$, while the polar angle can take any value.

EXAMPLES 1.   Find the direction cosines and direction angles of the vector $\mathbf{a} = 3\mathbf{i} + 2\mathbf{j} + 6\mathbf{k}$.
ANSWER   We have $\mathbf{a} = \langle 3,2,6 \rangle$. Hence $|\mathbf{a}|^2 = 3^2 + 2^2 + 6^2 = 49$, $|\mathbf{a}| = 7$, $\cos \alpha_1 = \frac{3}{7} \approx .429$, $\cos \alpha_2 = \frac{2}{7} \approx .286$, $\cos \alpha_3 = \frac{6}{7} \approx .857$, $\alpha_1 \approx 65°$, $\alpha_2 \approx 73°$, $\alpha_3 \approx 31°$.

2.   *Let θ be the angle between the vectors **a** and **b** with direction angles α_1, α_2, α_3 and β_1, β_2, β_3, respectively. Then*

$$\cos \theta = \cos \alpha_1 \cos \beta_1 + \cos \alpha_2 \cos \beta_2 + \cos \alpha_3 \cos \beta_3. \tag{14}$$

Proof.   We note Equation (11) and the analogous equation for **b**,

$$\mathbf{b} = |\mathbf{b}| (\cos \beta_1)\mathbf{i} + |\mathbf{b}| (\cos \beta_2)\mathbf{j} + |\mathbf{b}| (\cos \beta_3)\mathbf{k}.$$

Substituting (11) and the equation just written into (6), we obtain (14).
In particular, **a** and **b** are orthogonal if and only if

$$\cos \alpha_1 \cos \beta_1 + \cos \alpha_2 \cos \beta_2 + \cos \alpha_3 \cos \beta_3 = 0.$$

PROBLEMS

1. Find the direction cosines of the vector $\mathbf{a} = 2\mathbf{i} - 3\mathbf{j} + 4\mathbf{k}$.
2. Find the direction cosines of the vector $\mathbf{a} = -\mathbf{i} + 2\mathbf{j} - \sqrt{11}\mathbf{k}$.
3. Find the direction cosines of the vector **a** with components $\langle -3,4,0 \rangle$.
4. Find the direction cosines of the vector **a** with components $\langle 7,-1,-5 \rangle$.
5. If P has coordinates $(2,-4,8)$, find the direction cosines of the vector represented by $\overrightarrow{OP}$.
6. P has coordinates $(3 \cos 13°, -3 \sin 13°, 0)$. Find the direction angles of the vector represented by $\overrightarrow{OP}$.
7. Find the direction angles of the vector **a** with components $\langle 0, 20 \cos 20°, 20 \sin 20° \rangle$.
8. Find direction cosines of the line represented by $x = 2 - 2t$, $z = 3 + t$, $y = 4 - 5t$.
9. **a** has direction angles satisfying $\alpha_1 = \alpha_2 = \alpha_3$, $0 < \alpha_1 < 90°$. Find α_1.
10. **a** has direction angles satisfying $\cos \alpha_1 = \cos \alpha_2 = \frac{1}{2}$, $0 < \alpha_3 < 90°$. Find α_3.
11. Find the angle between the vectors $\mathbf{a} = \langle 3,4,0 \rangle$ and $\mathbf{b} = \langle 4,0,3 \rangle$.
12. Find the angle between the vectors $\mathbf{a} = \langle 1,-2,2 \rangle$ and $\mathbf{b} = \langle 4,4,2 \rangle$.
13. If **a** has direction cosines $2/\sqrt{30}$, $-1/\sqrt{30}$, $5/\sqrt{30}$, and **b** has direction cosines $1/\sqrt{5}$, $2/\sqrt{5}$, 0, show that **a** and **b** are orthogonal.
14. Show that the line passing through the points $(-1,3,2)$ and $(0,5,3)$ is perpendicular to the line passing through the points $(2,1,5)$ and $(5,1,2)$.
15. A point P has spherical coordinates (ρ, θ, ϕ) with $\rho > 0$. Find the direction cosines of the vector represented by $\overrightarrow{OP}$.
16. A point P has cylindrical coordinates (r, θ, z) with $r > 0$. Find the direction cosines of the vector represented by $\overrightarrow{OP}$.

§4   Planes and lines

In this section we consider planes in space and also the relative position of planes and lines.

4.1   Planes and linear equations

In the plane, a linear equation $(Ax + By + C = 0)$ defines a line. In space, a linear equation defines a plane, as we shall see below. By a linear equation in three variables, we mean an equation of the form

$$Ax + By + Cz + D = 0, \tag{1}$$

where A, B, C, D are given numbers and A, B, C are not all three 0. The last condition can be written as

$$A^2 + B^2 + C^2 \neq 0. \tag{2}$$

In order to prove that planes are defined by linear equations, consider a plane σ and the line l through the origin O which is perpendicular to σ; it intersects σ at a point O'. The number $\delta = |\overline{OO'}|$ is the distance from O to σ (see Figure 18.23).

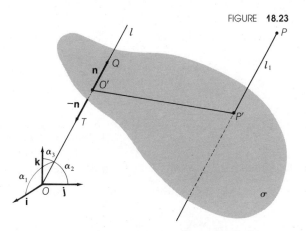

FIGURE   **18.23**

There are two points, Q and T, on l whose distance from O' is 1. The directed segments $\overrightarrow{O'Q}$ and $\overrightarrow{O'T}$ define unit vectors $\mathbf{n}$ and $-\mathbf{n}$; we call them the **unit normals** to σ. (If σ does not pass through O, we choose Q so that O and Q lie on different sides of σ, and we call $\mathbf{n}$ the unit normal pointing from O to σ. If σ passes through O, then $O' = O$ and the choice of Q is arbitrary.) The position vector of O' is $\mathbf{O'} = \delta\mathbf{n}$.

Now let P be a point with Cartesian coordinates (x,y,z), that is, with position vector

$$\mathbf{P} = x\mathbf{i} + y\mathbf{j} + z\mathbf{k}. \tag{3}$$

We draw a line l_1 through P perpendicular to σ; it meets the plane σ at a point P'. (If P lies on σ, $P' = P$. If P lies on l, $P' = O'$.) The distance $|P'P|$ is the distance from P to σ. Let $\mathbf{P'}$ be the position vector of P'. The directed segment $\overrightarrow{P'P}$ represents the vector $\mathbf{P} - \mathbf{P'}$. Since l_1 is parallel to l, we conclude (see Remark 1 in §3.3) that there is a number t such that

$$\mathbf{P} - \mathbf{P}' = t\mathbf{n}. \tag{4}$$

Note that

$$\text{distance from } P \text{ to } \sigma = |P'P| = |\mathbf{P} - \mathbf{P}'| = |t| \tag{5}$$

since $|\mathbf{n}| = 1$.

Since P' lies on σ, the segment $\overrightarrow{O'P'}$ is perpendicular to l. This segment represents the vector $\mathbf{P}' - \mathbf{O}' = \mathbf{P}' - \delta\mathbf{n}$. Therefore

$$(\mathbf{P}' - \delta\mathbf{n}, \mathbf{n}) = 0. \tag{6}$$

(This holds also if $P' = O'$; then $\mathbf{P}' - \delta\mathbf{n} = \mathbf{O}$.) Using (4), we conclude that $\mathbf{P}' = \mathbf{P} - t\mathbf{n}$, so that we may rewrite (6) as

$$0 = (\mathbf{P}' - \delta\mathbf{n}, \mathbf{n}) = (\mathbf{P} - t\mathbf{n} - \delta\mathbf{n}, \mathbf{n}) = (\mathbf{P}, \mathbf{n}) - (t\mathbf{n}, \mathbf{n}) - (\delta\mathbf{n}, \mathbf{n})$$
$$= (\mathbf{P}, \mathbf{n}) - t(\mathbf{n}, \mathbf{n}) - \delta(\mathbf{n}, \mathbf{n}) = (\mathbf{P}, \mathbf{n}) - t - \delta.$$

Hence $t = (\mathbf{P}, \mathbf{n}) - \delta$ and, by (5),

$$\text{distance from } P \text{ to } \sigma = |(\mathbf{P}, \mathbf{n}) - \delta|. \tag{7}$$

Let the direction angles of $\mathbf{n}$ be α_1, α_2, α_3. Then

$$\mathbf{n} = \cos\alpha_1 \mathbf{i} + \cos\alpha_2 \mathbf{j} + \cos\alpha_3 \mathbf{k} \tag{8}$$

and, noting (3) and (7), we conclude that

$$\text{distance from } P \text{ to } \sigma = |(\cos\alpha_1)x + (\cos\alpha_2)y + (\cos\alpha_3)z - \delta|. \tag{9}$$

When will P lie on σ? If and only if the distance (9) is 0, that is,

$$x\cos\alpha_1 + y\cos\alpha_2 + z\cos\alpha_3 - \delta = 0. \tag{10}$$

This is an equation of the form (1), with $A = \cos\alpha_1$, $B = \cos\alpha_2$, $C = \cos\alpha_3$, $D = -\delta$, and $A^2 + B^2 + C^2 = 1$ [see Equation (13) in §3.3]. Thus we have proved

Theorem 1. *Every plane is the solution set of a linear equation.*

The converse is also true. To see this, consider a given linear equation (1), and set

$$\rho = \begin{cases} \sqrt{A^2 + B^2 + C^2}, & \text{if } D \leq 0 \\ -\sqrt{A^2 + B^2 + C^2}, & \text{if } D \geq 0. \end{cases} \tag{11}$$

(If $D = 0$, either sign may be used.) By (2), $\rho \neq 0$. Now multiply both sides of (1) by $1/\rho$; this does not change the solution set. We obtain the equation

$$\left(\frac{A}{\rho}\right)x + \left(\frac{B}{\rho}\right)y + \left(\frac{C}{\rho}\right)z + \left(\frac{D}{\rho}\right) = 0 \tag{1'}$$

with $D/\rho \leq 0$ and $(A/\rho)^2 + (B/\rho)^2 + (C/\rho)^2 = 1$. There are angles α_1, α_2, α_3, and a nonnegative number δ, such that

$$\cos\alpha_1 = \frac{A}{\rho}, \qquad \cos\alpha_2 = \frac{B}{\rho}, \qquad \cos\alpha_3 = \frac{C}{\rho}, \qquad \delta = -\frac{D}{\rho} \tag{12}$$

(see §3.3). Now we can write (1') in the form (10). We have proved

Theorem 2. *The solution set of a linear equation* (1) *is a plane. The direction angles* α_1, α_2, α_3 *of the unit normal to this plane, and the distance* δ *from the plane to the origin, are given by* (11), (12).

Let us now substitute the values $\cos \alpha_1$, $\cos \alpha_2$, $\cos \alpha_3$, and δ given by (11), (12), into (5). We obtain the following.

Theorem 3. *The distance from a point* (x_0, y_0, z_0) *to the plane* $Ax + By + Cz + D = 0$ *is* $|Ax_0 + By_0 + Cz_0 + D| / \sqrt{A^2 + B^2 + C^2}$.

Theorem 3 has an analog in plane analytic geometry. If a line l in the xy plane is given by the equation $Ax + By + C = 0$, then the distance between a point (x_0, y_0) in the plane and the line l is $|Ax_0 + By_0 + C| / \sqrt{A^2 + B^2}$.

We can prove this by imitating the proof of Theorem 3 above. It is simpler to observe that the desired distance is also the distance from the point $(x_0, y_0, 0)$ in space to the plane σ with the equation $Ax + By + C = 0$.

[The form (10) of a linear equation is called its **normal form**.]

EXAMPLES **1.** Find the direction angles of the normal to the plane

$$x + y + z + 2 = 0 \qquad (*)$$

and its distance from the origin.

ANSWER   Since $A = B = C = 1$ and $D = 2 > 0$, we have $\rho = -\sqrt{3}$. Now multiply (*) by $1/\rho = -1/\sqrt{3}$ and obtain (*) in normal form:

$$-\frac{1}{\sqrt{3}} x - \frac{1}{\sqrt{3}} y - \frac{1}{\sqrt{3}} z - \frac{2}{\sqrt{3}} = 0.$$

Hence the direction cosines are $\cos \alpha_1 = \cos \alpha_2 = \cos \alpha_3 = -1/\sqrt{3}$ so that $\alpha_1 = \alpha_2 = \alpha_3 \approx 147°$. The distance from the origin is $\delta = 2/\sqrt{3} \approx 1.15$.

2. Find the equation of a plane that passes through the points $(5,4,1)$, $(4,-2,-3)$, and $(0,6,5)$.
SOLUTION   The three points do not lie on one line, so that the plane will be uniquely determined. Its equation must be of the form (1) and, since the three points must lie on the plane, we should have

$$5A + 4B + C + D = 0,$$
$$4A - 2B - 3C + D = 0,$$
$$6B + 5C + D = 0.$$

There are three linear equations for four unknowns; A, B, C, and D. [This is a reflection of the fact that we can multiply all coefficients in (1) by a number $\alpha \neq 0$ and obtain an equivalent equation.] We try to express three unknowns in terms of the fourth. The last equation shows that $D = -6B - 5C$. Substituting this expression in the first two equations, we obtain

$$5A + 4B + C - 6B - 5C = 0, \qquad 4A - 2B - 3C - 6B - 5C = 0,$$

or

$$5A - 2B - 4C = 0, \qquad\qquad\qquad 4A - 8B - 8C = 0.$$

The second equation here says that

$$A = 2B + 2C;$$

we substitute this in the first equation above and obtain $5(2B + 2C) - 2B - 4C = 0$ or

$$B = -\tfrac{3}{4}C.$$

Hence

$$A = 2(-\tfrac{3}{4}C) + 2C = \tfrac{1}{2}C$$

and

$$D = -6(-\tfrac{3}{4}C) - 5C = -\tfrac{1}{2}C.$$

The value of C is up to us; we choose $C = 4$ and obtain $A = 2$, $B = -3$, and $D = -2$. The desired equation is

$$2x - 3y + 4z - 2 = 0.$$

We can easily check that the given points actually satisfy this equation.

3. If $a \neq 0$, $b \neq 0$, and $c \neq 0$, then the plane through the points $(a,0,0)$, $(0,b,0)$, $(0,0,c)$ has the equation $x/a + y/b + z/c = 1$.

   Proof. This is an equation of the form (1) satisfied by the coordinates of the points considered. The result should be compared with the intercept form of the equation of a line in the xy plane (Chapter 2, §2.2).

PROBLEMS

   In Problems 1 to 5 find the direction angles of the normal to the given plane and the distance of the plane from the origin.

1. $4x + 2y + 4z - 3 = 0$.
2. $x - 2y + 2z + 4 = 0$.
3. $x + 4y - 8z + 9 = 0$.

4. $x + 4y - z - 3 = 0$.
5. $3x - 4y - 5z = 0$.

   In Problems 6 to 10 find the equation of the plane that passes through the given points.

6. $(-1,0,0)$, $(0,3,0)$, $(0,0,2)$.
7. $(1,-4,5)$, $(6,7,0)$, $(3,-2,5)$.
8. $(2,-1,3)$, $(4,0,5)$, $(2,1,7)$.

9. $(1,-2,-1)$, $(1,1,5)$, $(-1,3,3)$.
10. $(-2,1,3)$, $(3,2,16)$, $(1,-3,-3)$.

11. What is the distance from the point $(-1,2,-4)$ to the plane $x + 4y - 5z + 15 = 0$?
12. What is the distance from the point $(1,2,3)$ to the plane $2x - 3y + 4z - 8 = 0$?
13. Find the equation of the plane passing through the point $(4,-1,0)$ and the line l: $x = t$, $y = 2t$, $z = 3t$. [*Hint:* Find any two points on l.]
14. Find the equation of the plane passing through the point $(1,2,3)$ and the line l: $x = 2 + t$, $y = 2 + 2t$, $z = -3t$.

4.2   Parallel planes. Intersection of two planes

Theorem 4. *Two distinct planes, defined by*

$$Ax + By + Cz + D = 0 \quad and \quad A_1x + B_1y + C_1z + D_1 = 0,$$

are parallel if and only if there is a number $\alpha \neq 0$ with

$$A_1 = \alpha A, \qquad B_1 = \alpha B, \qquad C_1 = \alpha C. \tag{13}$$

   Proof. By Theorem 2, the unit normals to the two planes have direction cosines A/ρ, B/ρ, C/ρ and A_1/ρ_1, B_1/ρ_1, C_1/ρ_1, where $\rho = \pm(A^2 + B^2 + C^2)^{1/2}$, $\rho_1 = \pm(A_1^2 + B_1^2 + C_1^2)^{1/2}$. The two planes are parallel if and only if the direction cosines of the two normals are the same, or differ only in sign (see Remark 1 in §3.3). In the first case, (13) holds with $\alpha = \rho_1/\rho$; in the second case, (13) holds with $\alpha = -\rho_1/\rho$.

Two nonparallel planes, say,

$$2x - 3y + 4z - 5 = 0 \quad \text{and} \quad 6x - 9y + 5z - 1 = 0, \qquad (14)$$

intersect in a line l. We can therefore describe the line l by the two equations (14): the line consists of all points (x,y,z) satisfying *both* equations.

In order to get a parametric representation of this line, we consider one of the three variables x, y, z as given and solve for the other two. We must decide which variable may be prescribed; this depends upon the equation considered. In our case, we can solve Equations (14) for y and z in terms of x, and for x and z in terms of y. But we *cannot* solve for x and y in terms of z (as the reader is asked to verify). We give to x an arbitrary value, call it t, and consider (14) as a system of two equations for y and z:

$$\begin{aligned} -3y + 4z &= 5 - 2t, \\ -9y + 5z &= 1 - 6t. \end{aligned} \qquad (15)$$

Multiply the first equation by 5, the second by 4, and subtract. This yields $[(-3)5 - 4(-9)]y = 5(5 - 2t) - 4(1 - 6t)$ so that $21y = 21 + 14t$ or

$$y = 1 + \tfrac{2}{3}t.$$

Substituting this into the first equation (15) and solving for z, we obtain $-3(1 + \tfrac{2}{3}t) + 4z = 5 - 2t$ or $z = 2$. Thus the intersection of the planes (14) has a parametric representation

$$x = t, \quad y = 1 + \tfrac{2}{3}t, \quad z = 2.$$

If a line l is given by a parametric representation, it is possible to represent it by two linear equations, that is, as the intersection of two planes. This can be done in infinitely many ways, since there are infinitely many planes through a given line.

For instance, if a line l is given by

$$x = 2 - 2t, \quad y = -3 + 4t, \quad z = 6 - t, \qquad (16)$$

we can obtain linear equations satisfied by all points of l as follows. We "solve" each Equation (16) for t, that is, we rewrite (16) in the form

$$t = \frac{2 - x}{2}, \quad t = \frac{3 + y}{4}, \quad t = 6 - z.$$

Now we "eliminate" t: if (x,y,z) is a point on the line l, we must have

$$\frac{2 - x}{2} = \frac{3 + y}{4} = 6 - z$$

or, multiplying all terms by 4,

$$4 - 2x = 3 + y = 24 - 4z. \qquad (17)$$

These are called equations of the line (16) in *symmetric form*. We can now consider our line to be the intersection of the two planes $4 - 2x = 3 + y$ and $3 + y = 24 - 4z$, that is,

$$2x + y - 1 = 0 \quad \text{and} \quad y + 4z - 21 = 0.$$

We can also consider our line to be the intersection of the planes $4 - 2x = 24 - 4z$ and $3 + y = 24 - 4z$, that is,

$$x - 2z + 10 = 0 \quad \text{and} \quad y + 4z - 21 = 0,$$

and so forth.

REMARK   In general, if a, b, c are three numbers, none of them zero, the equations

$$\frac{x - x_0}{a} = \frac{y - y_0}{b} = \frac{z - z_0}{c} \tag{L}$$

represent a line passing through the point (x_0, y_0, z_0). A parametric representation of this line reads

$$x = x_0 + at, \quad y = y_0 + bt, \quad z = z_0 + ct. \tag{L$'$}$$

If (x_0, y_0, z_0) and (x_1, y_1, z_1) are two points such that $x_1 - x_0 \neq 0$, $y_1 - y_0 \neq 0$, $z_1 - z_0 \neq 0$, then

$$\frac{x - x_0}{x_1 - x_0} = \frac{y - y_0}{y_1 - y_0} = \frac{z - z_0}{z_1 - z_0} \tag{L$''$}$$

are the equations of the line through these two points. The reader is asked to verify these statements.

EXAMPLES **1.** Find a plane through the point $(2, -1, 4)$ that is parallel to the plane

$$5x - 7y + 8z - 6 = 0.$$

SOLUTION   The coefficients of x, y, z in the equation of the desired plane must be of the form 5α, -7α, 8α, in view of Theorem 4. We may choose $\alpha = 1$ and write the desired equation as

$$5x - 7y + 8z + D = 0,$$

where D is to be determined. Substituting $x = 2$, $y = -1$, and $z = 4$, we see that $(5)(2) + (-7)(-1) + (8)(4) + D = 0$. Hence $D = -49$ and the desired plane has the equation

$$5x - 7y + 8z - 49 = 0.$$

2. Find a parametric representation for the intersection line of the planes

$$x + y + z = 1, \quad 2x + y + z = 1.$$

ANSWER   If we consider x as given and try to solve for y and z we fail. Consider z as given and set, to simplify writing, $t = 1 - z$. Then we have two equations:

$$x + y = t, \quad 2x + y = t.$$

Solving them simultaneously, we obtain $x = 0$, $y = t$. Thus the desired line has a parametric representation

$$x = 0, \ y = t, \ z = 1 - t.$$

3. Represent the line $x = 2 + t$, $y = 2 - t$, $z = 2$ as the intersection of two planes.
ANSWER   The parameter t does not occur in the expression for z. We eliminate t from the expressions for x and y and conclude that the line considered is the intersection of the planes $x + y = 4$ and $z = 2$.

PROBLEMS

1. Find the equation of the plane passing through the point $(7, 4, 5)$ and parallel to the plane $2x - 3y - 6 = 0$.

2. Find the equation of the plane passing through the point $(4,0,6)$ and parallel to the plane $x + y + z = 0$.

3. Find the equation of the plane passing through the point $(1,1,1)$ and parallel to the plane $x - 20y + 7z - 5 = 0$.

4. Find a number a such that the plane $2x - ay + 6z + 7 = 0$ is parallel to the plane $x + y + 3z + 1 = 0$.

5. Find a number a such that the plane $ax + 2ay + 10z - 2 = 0$ is parallel to the plane $x + 2y + 5z - 7 = 0$.

In Problems 6 to 10 determine whether the given planes are parallel. If they are not, find a parametric representation for the intersection line.

6. $x - y = 1$, $y - z = 1$.

7. $2x - 3y - 5 = 0$, $-4x + 6y - 4 = 0$.

8. $7x - 5y + 6z = 0$, $21x - 15y + 18z - 62 = 0$.

9. $x + y + 2z - 3 = 0$, $x + y - 2z + 3 = 0$.

10. $2x - 3y + 2z - 2 = 0$, $6x + y + z - 1 = 0$.

In Problems 11 to 15 find two linear equations representing the given line.

11. $x = 2 - t$, $y = 2 - 2t$, $z = 6t$.

12. $x = -1 + 2t$, $y = -2 + 3t$, $z = 6 - 7t$.

13. $x = 7 - 6t$, $y = 5t$, $z = 6 - t$.

14. The line through $(0,0,0)$ and $(7,6,3)$.

15. The line through $(2,-1,4)$ and $(6,2,-3)$.

In Problems 16 to 20 the equations of a line are given in symmetric form. Find a parametric representation of the line.

16. $x = y = 0$.

17. $x = y = z$.

18. $2x - y = x + z = z - 4y$.

19. $x - y + z = 2x - 3y + z = x - 10y + 5z$.

20. $2x + y - 4 = x - 3y + z + 5 = x + 2y - z - 10$.

21. What points (x,y,z) satisfy the equation $(x - y + 1)^2 + (x + y - z + 4)^2 = 0$?

22. For which values of q does the equation

$$(2x + 3y + 4z - 1)^2 + (qx + 3y + 4z - 1)^2 = 0$$

represent a line? What does it represent if it does not represent a line?

4.3 Relative position of lines and planes. Angles

A vector $\mathbf{a}$ is said to be parallel to a line l if it can be represented by a directed segment of l; if so, then $(-\mathbf{a})$ is also a vector parallel to l. Note that if l admits the parametric representation $t\mathbf{a} + \mathbf{b}$, then $\mathbf{a}$ is a vector parallel to l.

Let l and k be two lines, let $\mathbf{a}$ and $\mathbf{b}$ be vectors parallel to these lines, and set $\theta = \measuredangle(\mathbf{a},\mathbf{b})$. If $0 < \theta \leq 90°$, we call $\phi = \theta$ the **acute angle** between l and k. If $90° < \theta \leq 180°$, then we set $\phi = 180° - \theta$ and call ϕ the acute angle between the two lines. Thus $\cos \phi = |\cos \theta|$ in all cases. Recalling Equation (6) in §3.1, we obtain

$$\cos \phi = \frac{|(\mathbf{a},\mathbf{b})|}{|\mathbf{a}|\,|\mathbf{b}|}. \tag{18}$$

FIGURE 18.24

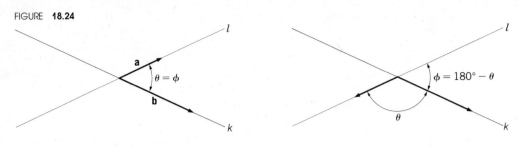

Figure 18.24 shows that, if l and k intersect, our definition makes geometric sense.

If l and k are parallel, then $\phi = 0$. If $\phi = 0$, then l and k either coincide (if they have a point in common) or are parallel (if they do not). If $\phi \neq 0$, then l and k either have precisely one point in common, or none. In the second case, the lines are called **skew.** Parallel lines lie in one plane, skew lines do not.

The angle ϕ between two planes is, by definition, the acute angle between their unit normal vectors. In view of Theorem 2 in §4.1, the angle ϕ between the planes

$$Ax + By + Cz + D = 0 \qquad \text{and} \qquad A_1x + B_1y + C_1z + D_1 = 0 \qquad (19)$$

is given by

$$\cos \phi = \frac{|AA_1 + BB_1 + CC_1|}{\sqrt{A^2 + B^2 + C^2} \, \sqrt{A_1^2 + B_1^2 + C_1^2}}. \qquad (20)$$

If $\phi = 0$, the two planes either coincide (if they have a point in common) or are parallel. If $0 < \phi \leq 90°$, the planes intersect in a line.

Consider, finally, the line

$$x = x_0 + at, \qquad y = y_0 + bt, \qquad z = z_0 + ct \qquad (21)$$

and the plane

$$Ax + By + Cz + D = 0. \qquad (22)$$

The acute angle ϕ between the line (21) and the normal to the plane (22) is given by

$$\cos \phi = \frac{|Aa + Bb + Cc|}{\sqrt{A^2 + B^2 + C^2} \, \sqrt{a^2 + b^2 + c^2}}. \qquad (23)$$

If $\phi = 90°$, that is, if $Aa + Bb + Cc = 0$, the line either lies in the plane, or is parallel to it. If $0 \leq \phi < 90°$, the line meets the plane at precisely one point.

EXAMPLES 1. Determine the relative position of the lines

$$x = 2 - 3t, \qquad y = 6 + t, \qquad z = -1 - 2t, \qquad (24)$$

and

$$x = -1 + 6s, \qquad y = -2s, \qquad z = 4 + 4s. \qquad (25)$$

(Note that we use different letters for the two parameters, as we should!)

SOLUTION   We determine first whether the two lines have a point in common. If there is such a point, then it corresponds to a value of t and a value of s such that both give the same values to x, y, z. Thus we must have

$$2 - 3t = -1 + 6s, \qquad 6 + t = -2s, \qquad -1 - 2t = 4 + 4s.$$

These equations are not satisfied for any values of s and t. (Indeed, the first equation says that $3t + 6s = 3$, that is, $t + 2s = 1$. But the second equation asserts that $t + 2s = -6$.) Hence the two lines do not meet. They are either parallel or skew.

To determine what takes place, note that the vectors $\langle -3,1,-2 \rangle$ and $\langle 6,-2,4 \rangle$ are parallel to the given lines. Since $\langle 6,-2,4 \rangle = -2\langle -3,1,-2 \rangle$, we have $\phi = 0$; this can be also seen by computing ϕ by formula (18). Thus the two lines are parallel.

2. Determine the relative position of the line (24) and the plane

$$3x - y + 2z - 5 = 0. \tag{26}$$

SOLUTION   We ask which points (24) lie on (26). To find out, replace x, y, z in (26) by their expression in (24). We obtain

$$3(2 - 3t) - (6 + t) + 2(-1 - 2t) - 5 = 0.$$

This is true if $t = -\frac{1}{2}$, and for no other value of t. Thus the line intersects the plane in one point, the point with the coordinates $x = 2 - 3(-\frac{1}{2})$, $y = 6 + (-\frac{1}{2})$, $z = -1 - 2(-\frac{1}{2})$, that is, the point $(\frac{7}{2},\frac{11}{2},0)$.

PROBLEMS

1. What is the acute angle between the planes $x + y + z - 2 = 0$ and $x - y + z + 4 = 0$?
2. What is the acute angle between the planes $2x - 4y + 2z + 5 = 0$ and $x + 2y - z + 1 = 0$?
3. Find the acute angle between the normal to the plane $x + 2y + z + 1 = 0$ and the line $l: x = 1 + t, y = 1 - 2t, z = 1 - t$.
4. Find the acute angle between the normal to the plane $3x + y - z + 3 = 0$ and the line $l: x = 2 - t, y = 2 + 3t, z = 1 + t$.

In Problems 5 to 12 determine the relative position of the given lines; that is, determine whether the lines l_1 and l_2 are skew or parallel or whether they coincide or meet at precisely one point. In the last case, determine the point.

5. $l_1: x = 1 - t, y = 2 + t, z = 2t;$ $l_2: x = 1 - s, y = 1 + 2s, z = 4 - s.$
6. $l_1: x = 1 - t, y = 2 + t, z = 2t;$ $l_2: x = 3 - 2s, y = 4 + 2s, z = 6 + 4s.$
7. $l_1: x - 1 = y + 2 = 2z - 3;$ $l_2: x = s, y = s - 3, z = \frac{1}{3}s - 1.$
8. $l_1: 2x - 3y + 4z - 1 = x - y + 2z = 0;$ $l_2: 3x = 3y = z - 2x.$
9. $l_1: x = 2 + 3t, y = -2 + 3t, z = 4;$ $l_2: x + y - 2 = z - 4 = 0.$
10. $l_1: x - 2y = z - 4 = 3x + z - 5;$ $l_2: x + y - z = x + y - 2z = 1.$
11. l_1 passes through $(2,3,4)$ and $(1,6,7);$ $l_2: x = 1 + t, y = z = t.$
12. l_1 passes through $(1,-1,0)$ and $(0,1,0);$ $l_2: x + y + z - 2 = x - y + z - 2 = x + y - z - 2.$

In Problems 13 to 20 determine the relative position of the given line and plane; that is, determine whether the line l lies in the plane σ, is parallel to it, or intersects it at one point. In the latter case, find the point of intersection.

13. $l: x = 2 - 2t, y = 3 + 2t, z = 4 + t;$ $\sigma: x + 2y - 4z + 8 = 0.$
14. $l: x = 2 + t, y = 6 - t, z = 3 - 4t;$ $\sigma: 3x - y + z - 3 = 0.$
15. $l: x = -1 + t, y = 2 + 3t, z = 4 + 7t;$ $\sigma: 2x - 3y + z + 5 = 0.$
16. $l: x - 2y = 3$ and $x + z = 5;$ $\sigma: x + y + z = 1.$
17. $l: x - 2y + z = 7$ and $x - y - 2z = 0;$ $\sigma: 2x - 4y = 0.$
18. $l: x - 1 = y - 2 = 1 - z;$ $\sigma: x + 2y + 3z + 4 = 0.$
19. $l: x - 2 = y - 3 = 2z - 5;$ $\sigma: 2x + 3y - 4z + 5 = 0.$
20. l passes through $(0,2,1)$ and $(4,-1,0);$ σ passes through $(0,0,0)$, $(2,-1,6)$, and $(-2,1,7)$.

4.4   Spaces of n dimensions‡

The arithmetic interpretation of geometry, by Cartesian coordinates, leads to the concept of n-dimensional space, where n is a positive integer. A point on a line can be described by one number, a point in the plane by two, a point in (ordinary) space by three. A point in n-dimensional space or n-**space** is, *by definition*, an n-tuple of numbers $(x_1, x_2, \ldots, x_n)$. The **distance** between two points $(x_1, x_2, \ldots, x_n)$ and $(y_1, y_2, \ldots, y_n)$ is *by definition*

$$d = \sqrt{(x_1 - y_1)^2 + \cdots + (x_n - y_n)^2}.$$

A **line** in n-space is, *by definition*, the set of all points (vectors) $(x_1, x_2, \ldots, x_n)$ that can be written in the form

$$x_j = a_j + tb_j, \qquad j = 1, 2, \ldots, n, \qquad -\infty < t < \infty,$$

where the a's and b's are fixed numbers and at least one b is not 0.

A **hyperplane** in n-space is, *by definition*, the set of all points $(x_1, x_2, \ldots, x_n)$ which satisfy a linear equation, that is, an equation of the form

$$A_1 x_1 + A_2 x_2 + \cdots + A_n x_n + D = 0,$$

where $A_1, A_2, \ldots, A_n$ and D are numbers and $A_1^2 + A_2^2 + \cdots + A_n^2 \neq 0$. (If $n = 1$, a hyperplane is a point; if $n = 2$, a hyperplane is a line; if $n = 3$, a hyperplane is a plane.)

Familiar geometric theorems still hold. For instance, there is exactly one line through two distinct points. And if two distinct points on a line lie in a hyperplane, so does the whole line.

Proceeding in this way, we can develop an n-dimensional geometry. It contains none of the mystery with which science-fiction writers sometimes surround the "fourth dimension." Every statement in n-dimensional geometry is simply a statement about numbers. The geometric language helps, by the analogy with the "visible" cases $n = 1, 2, 3$, to ask interesting questions and guess at the correct answers.

The primary motivation for developing n-dimensional geometry was the desire of mathematicians to generalize and develop concepts in their natural setting. But n-dimensional spaces, as well as more complicated spaces of infinitely many dimensions, are not at all mere playthings of pure mathematicians; they are also indispensable working tools of modern physicists, engineers, statisticians, and economists.

§5   Quadric surfaces

A **quadric surface** is the set of points with Cartesian coordinates (x, y, z) satisfying an equation of the form

$$Ax^2 + By^2 + Cz^2 + 2Dxy + 2Eyz + 2Fxz + Gx + Hy + Iz + K = 0, \qquad (1)$$

where not all numbers A, B, C, D, E, F are 0. Such an equation is called a second-degree equation in three variables.

‡This subsection may be omitted at first reading.

5.1   Degenerated cases. Cylinders. Circular cones

First we look at some degenerate cases. An equation of the form (1) may represent the empty set $(x^2 = -1)$, a point $(x^2 + y^2 + z^2 = 0)$, a line $(x^2 + y^2 = 0)$, a plane $(z^2 = 0)$, two parallel planes $(z^2 = 1)$, or two intersecting planes $(xy = 0)$.

If no term with z appears in (1), the surface is called a quadric **cylinder.** To construct such a cylinder, consider a second-degree curve γ in the xy plane and "draw" all points (x,y,z) in space with (x,y) on γ. Some examples (**parabolic cylinder, elliptic cylinder,** and **hyperbolic cylinder**) are shown in Figure 18.25.

FIGURE **18.25**

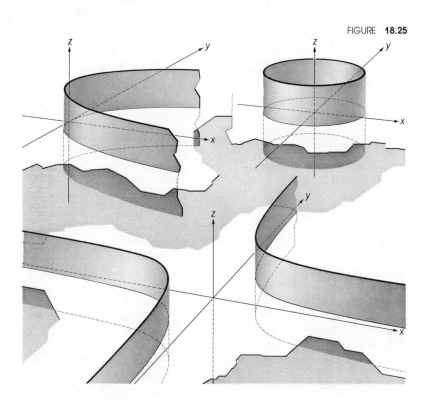

In the example in §1.4, we noted that, by rotating the line $z = mx$ in the xz plane about the z-axis, we obtain the surface

$$\frac{x^2}{a^2} + \frac{y^2}{a^2} - z^2 = 0. \tag{2}$$

The quadric surface is called a **circular cone.**

In order to study this surface, we compute its intersections with planes parallel to the coordinate planes. A plane parallel to the coordinate plane $z = 0$ (that is, the xy plane) has the equation $z = \gamma$ (γ some constant). In this plane, x and y are Cartesian coordinates. Similarly, (y,z) are Cartesian coordinates in a plane $x = \alpha$, and (x,z) are Cartesian coordinates in a plane $y = \beta$.

FIGURE **18.26**   FIGURE **18.27**

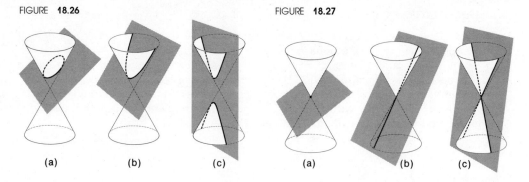

(a)   (b)   (c)   (a)   (b)   (c)

The equation of the intersection of the cone (2) with a plane $z = \gamma$ is obtained by replacing z in (2) by γ. This equation reads

$$x^2 + y^2 = a^2\gamma^2.$$

It represents a point if $\gamma = 0$ and a circle if $\gamma \neq 0$.

The intersection with a plane $x = \alpha$ has the equation

$$z^2 - \frac{y^2}{a^2} = \frac{\alpha^2}{a^2}.$$

This is, for $\alpha \neq 0$, a hyperbola with semiaxes α/a and α. For $\alpha = 0$, we obtain a pair of lines. The intersection with a plane $y = \beta$ is also either a hyperbola or a pair of lines.

It turns out (see §5.6 below) that *the intersection of a circular cone with any plane is a quadric curve.* If the plane does not pass through the vertex ("tip") of the cone, then the intersection is either (a) an ellipse, (b) a parabola, or (c) a hyperbola; see Figure 18.26. In cases (a) and (b), the plane meets only one "nappe" of the cones; in case (b), it meets both nappes. If the plane passes through the vertex of the cone, the intersection is either (a) a point, (b) a line, or (c) a pair of lines; see Figure 18.27. This is how the ancient Greek mathematicians defined ellipses, parabolas, and hyperbolas, and this explains the name "conic sections" or "conics."

PROBLEMS

In Problems 1 to 12 identify the given quadric surface as a parabolic cylinder, elliptic cylinder, hyperbolic cylinder, or circular cone.

1. $x^2 - y = 0$.
2. $4x - y^2 = 0$.
3. $x^2 + 4y^2 - 16 = 0$.
4. $4x^2 + 9y^2 - 36 = 0$.
5. $x^2 - y^2 - 1 = 0$.
6. $4x^2 - 25y^2 - 100 = 0$.
7. $x^2 + y^2 - z^2 = 0$.

8. $x^2 + y^2 - 4z^2 = 0$.
9. $x - y^2 + 2y - 1 = 0$.
   [*Hint:* Factor and translate coordinates; see §1.1.]
10. $x^2 + 4y^2 - 8x - 16y + 16 = 0$.
11. $x^2 - 4y^2 + 16y - 20 = 0$.
12. $x^2 + y^2 - z^2 - 6x - 6y + 18 = 0$.

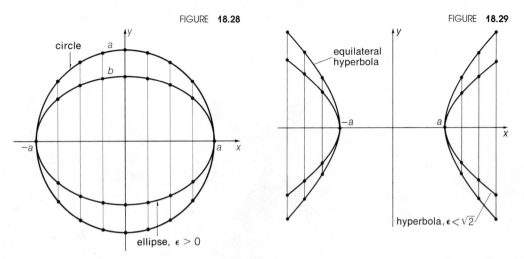

FIGURE **18.28**   FIGURE **18.29**

5.2 Contractions. Elliptic cones

Consider once again an ellipse and a hyperbola defined by the equations

$$\frac{x^2}{a^2} + \frac{y^2}{b^2} = 1 \quad (a \geq b), \qquad \frac{x^2}{a^2} - \frac{y^2}{b^2} = 1.$$

For $b = a$, we obtain a circle and an equilateral hyperbola defined by

$$x^2 + y^2 = a^2 \qquad \text{and} \qquad x^2 - y^2 = a^2.$$

We rewrite these equations in the form

ellipse: $y = \pm \dfrac{b}{a} \sqrt{a^2 - x^2}$,   hyperbola: $y = \pm \dfrac{b}{a} \sqrt{x^2 - a^2}$,

circle: $y = \pm \sqrt{a^2 - x^2}$,   equilateral hyperbola: $y = \pm \sqrt{x^2 - a^2}$.

In Figures 18.28 (and 18.29) we indicate points on their curves corresponding to the same values of x. Comparing the formulas, we see that *the ellipse is obtained from the circle by "contracting in the y-direction in the ratio b/a,"* and that *the hyperbola is obtained from the equilateral hyperbola by "contracting (or expanding) in the y-direction in the ratio b/a."*

   Contraction and expansion in one direction can be also applied to quadric surfaces in space. In particular, we obtain the **elliptic cone** by "contracting (or expanding) the circular cone in the y-direction." If the contraction ratio is b/a, the new surface has the equation

$$\frac{x^2}{a^2} + \frac{y^2}{b^2} - z^2 = 0; \tag{3}$$

it is shown in Figure 18.30.

   The reader should verify that the intersection of the cone (3) with a plane $z = \gamma$ is an ellipse or a point and with a plane $x = \alpha$ or $y = \beta$ a hyperbola or a pair of lines.

FIGURE   18.30

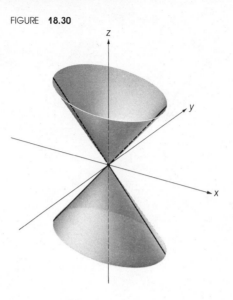

PROBLEMS

1. Show that the intersection of the cone (3) with a plane $z = \gamma$ is an ellipse or a point.
2. Show that the intersection of the cone (3) with a plane $x = \alpha$ or $y = \beta$ is a hyperbola or a pair of lines.
3. Identify and sketch the curve $x^2 + 4y^2 - 4z^2 = 0$.
4. Identify and sketch the curve $x^2 + 16y^2 - 16z^2 - 4x + 4 = 0$.

5.3   Ellipsoids and hyperboloids

Rotating a conic about an axis, we also obtain a quadric surface. The rotation of an ellipse

$$\frac{x^2}{a^2} + \frac{z^2}{c^2} = 1$$

about the z-axis yields an ellipsoid of revolution. The equation of this surface in cylindrical coordinates (see §1.4) is

$$\frac{r^2}{a^2} + \frac{z^2}{c^2} = 1.$$

The equation in Cartesian coordinates is therefore (since $r^2 = x^2 + y^2$)

$$\frac{x^2}{a^2} + \frac{y^2}{a^2} + \frac{z^2}{c^2} = 1. \tag{4}$$

The surface is called an **ellipsoid of revolution**, a **prolate spheroid** if $c > a$ (Figure 18.31), an **oblate spheroid** if $c < a$ (Figure 18.32), and, of course, a sphere if $a = c$. In other words, a prolate spheroid is obtained by rotating an ellipse about its major axis; an oblate one by rotating an ellipse about its minor axis. The surface of the earth is, approximately, an oblate spheroid.

FIGURE **18.32**

FIGURE **18.31**

FIGURE **18.33**

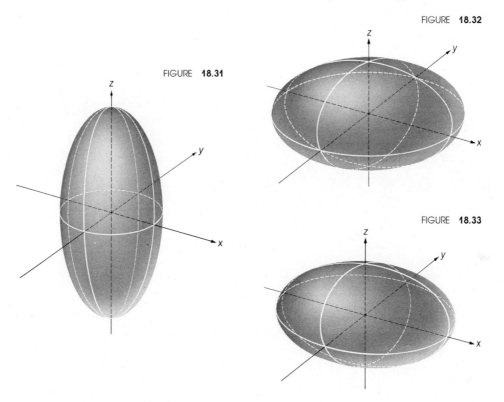

A **general ellipsoid** is obtained from (4) by a contraction in the y-direction; its equation reads

$$\frac{x^2}{a^2} + \frac{y^2}{b^2} + \frac{z^2}{c^2} = 1 \qquad (5)$$

(see Figure 18.33, where $a > b > c$). The intersection of a plane $x = \alpha$, $y = \beta$, or $z = \gamma$ with (5) is either empty, a point, or an ellipse, as the reader may verify.

Rotating a hyperbola *around its axis* gives a **hyperboloid of revolution of two sheets.** If the equation of the hyperbola is

$$\frac{z^2}{c^2} - \frac{x^2}{z^2} = 1,$$

the equation of the hyperboloid reads

$$-\frac{x^2}{a^2} - \frac{y^2}{a^2} + \frac{z^2}{c^2} = 1.$$

A contraction in the y-direction gives the **general hyperboloid of two sheets:**

$$-\frac{x^2}{a^2} - \frac{y^2}{b^2} + \frac{z^2}{c^2} = 1 \qquad (6)$$

(Figure 18.34). The intersection of this surface with a plane $z = \gamma$ is empty, a point, or an ellipse, while the intersection with either $x = \alpha$ or $y = \beta$ is always a hyperbola.

FIGURE **18.34**   FIGURE **18.35**

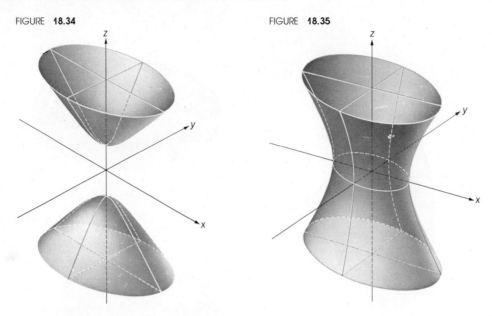

A **hyperboloid of revolution of one sheet** is obtained by rotating a hyperbola *about its conjugate axis* (see Chapter 2, §4.6). Starting with

$$\frac{x^2}{a^2} - \frac{z^2}{c^2} = 1,$$

we obtain

$$\frac{x^2}{a^2} + \frac{y^2}{a^2} - \frac{z^2}{c^2} = 1.$$

A contraction in the y-direction gives us the **general hyperboloid of one sheet** (see Figure 18.35).

$$\frac{x^2}{a^2} + \frac{y^2}{b^2} - \frac{z^2}{c^2} = 1. \tag{7}$$

The intersections with planes $z = \gamma$ are ellipses, those with planes $x = \alpha$ or $y = \beta$ are hyperbolas which for $\alpha = a$ or $\beta = b$ degenerate into pairs of straight lines.

PROBLEMS

In Problems 1 to 12 identify the given quadric surface as a general ellipsoid, prolate spheroid, oblate spheroid, hyperboloid of revolution of one or two sheets, or a general hyperboloid of one or two sheets.

1. $4x^2 + 4y^2 + z^2 - 16 = 0$.
2. $x^2 + y^2 + 4z^2 - 4 = 0$.
3. $36x^2 + 16y^2 + 9z^2 - 144 = 0$.
4. $9x^2 + 9y^2 - z^2 + 9 = 0$.
5. $36x^2 + 9y^2 - 4z^2 + 36 = 0$.
6. $4x^2 + 4y^2 - z^2 - 4 = 0$.

7. $9x^2 + 36y^2 - 4z^2 - 36 = 0$.
8. $16x^2 + 16y^2 + z^2 - 8z = 0$.
9. $4x^2 + 4y^2 + z^2 - 16x - 16y + 28 = 0$.
10. $x^2 + y^2 + 9z^2 - 54z + 72 = 0$.
11. $9x^2 + 9y^2 - z^2 - 36x - 72y + 189 = 0$.
12. $9x^2 + 36y^2 - 4z^2 + 24z - 72 = 0$.

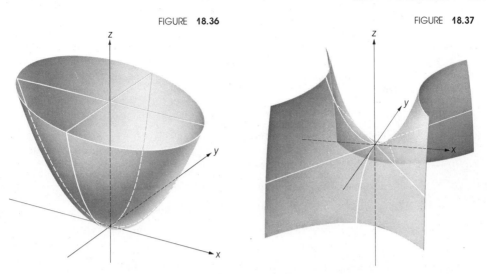

FIGURE **18.36**

FIGURE **18.37**

5.4 Paraboloids

There are two kinds of quadric surfaces called paraboloids. An **elliptic paraboloid** has the equation

$$\frac{x^2}{a^2} + \frac{y^2}{b^2} = z. \tag{8}$$

The surface lies in the half-space $z \geq 0$ (see Figure 18.36). For $\gamma > 0$, the plane $z = \gamma$ intersects the surface in an ellipse. The intersections with the planes $x = \alpha$ and $y = \beta$ are parabolas. We may think of (8) as having been obtained by a contraction in the y-direction of a **paraboloid of revolution**

$$x^2 + y^2 = a^2 z,$$

which is the result of rotating the parabola

$$x^2 - a^2 z = 0$$

about its axis.

The **hyperbolic paraboloid** (Figure 18.37) is *not* the result of contracting a surface of revolution. Its equation is of the form

$$\frac{x^2}{a^2} - \frac{y^2}{b^2} = z. \tag{9}$$

The intersections with the planes $x = \alpha$ or $y = \beta$ are parabolas, those with planes $z = \gamma$ are hyperbolas, with the exception that for $z = 0$ we get two lines.

PROBLEMS

In Problems 1 to 6 determine if the given paraboloid is elliptic or hyperbolic.

1. $x^2 + 16y^2 - 16z = 0$.
2. $4x^2 + 9y^2 - 36z = 0$.
3. $x^2 - y^2 - z = 0$.

4. $x^2 - 4y^2 - 16z = 0$.
5. $x^2 + 4y^2 - 4x - 32y - 4z + 68 = 0$.
6. $x^2 - 4y^2 - 4x - 4z + 4 = 0$.

5.5   Ruled surfaces‡

A surface S is called **ruled** if for every point P on S there is a line l that passes through P and lies on S (that is, such that every point on the line is a point on the surface). Planes and cylinders are ruled surfaces, and so are cones. It is somewhat surprising that there are also other ruled quadric surfaces.

Hyperboloids of one sheet and hyperbolic paraboloids are ruled surfaces.

More precisely, the following statements are true:
(I) There are two families of straight lines lying on the hyperboloid $(x/a)^2 + (y/b)^2 - (z/c)^2 = 1$. These lines are given by

$$x = a \cos \alpha + (a \sin \alpha)t, \qquad y = b \sin \alpha - (b \cos \alpha)t, \qquad z = ct, \qquad (10)$$

$$x = a \cos \beta - (a \sin \beta)t, \qquad y = b \sin \beta + (b \cos \beta)t, \qquad z = ct. \qquad (11)$$

Similarly, there are two families of straight lines lying on the paraboloid $z = (x/a)^2 - (y/b)^2$:

$$x = a\alpha + at, \qquad y = b\alpha - bt, \qquad z = 4\alpha t, \qquad (12)$$

$$x = a\beta + at, \qquad y = -b\beta + bt, \qquad z = 4\beta t. \qquad (13)$$

(Here α and β are arbitrary fixed numbers, and t is the parameter.)
(II) Through every point of the hyperboloid (7), there passes a line of the family (10) and one of the family (11), and through every point of the paraboloid (9), there passes a line of the family (12) and one of the family (13).
Statement (I) is proved by substituting the values x, y, z given by (10) or (11) [or by (12) or (13)] into (7) [or into (9)] and checking that we obtain a correct statement. The reader is urged to do so.
We do not prove statement (II) in all generality but only in some special cases; see the following problems. The general proof would proceed along the same lines.

EXAMPLES 1.   Find two lines on the hyperboloid $x^2 + 4y^2 - 9z^2 = 1$, passing through the point (1,3,2).
ANSWER   First we check that (1,3,2) lies on the surface: $1^2 + 4 \cdot 3^2 - 9 \cdot 2^2 = 1 + 36 - 36 = 1$.
Next we try to find α so that the line (10), with $a = 1$, $b = \frac{1}{2}$, $c = \frac{1}{3}$, passes through (1,3,2). We must have, for some t,

$$1 = \cos \alpha + (\sin \alpha)t, \qquad 3 = \tfrac{1}{2} \sin \alpha - (\tfrac{1}{2} \cos \alpha)t, \qquad 2 = \tfrac{1}{3}t.$$

The last relation shows that $t = 6$. Substitute this into the expressions for x and y to obtain

$$1 = \cos \alpha + 6 \sin \alpha, \qquad 3 = \tfrac{1}{2} \sin \alpha - 3 \cos \alpha.$$

Solving for $\cos \alpha$ and $\sin \alpha$, we find that $\sin \alpha = 12/37$, $\cos \alpha = -35/37$. [There is such an α, since $(12/37)^2 + (-35/37)^2 = 1$.] The desired line is the line (10) with this α and the a, b, c mentioned before, that is, the line

$$x = -\frac{35}{37} + \frac{12}{37}t, \qquad y = \frac{6}{37} + \frac{35}{74}t, \qquad z = \frac{1}{3}t.$$

Starting with (11), we find, in the same way, the line

$$x = 1, \qquad y = \tfrac{1}{2}t, \qquad z = \tfrac{1}{3}t.$$

---

‡This subsection may be omitted without loss of continuity.

FIGURE 18.38   FIGURE 18.39

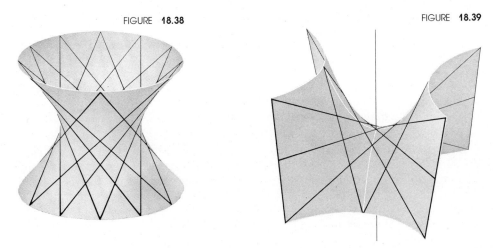

2. Find two lines on the paraboloid $z = x^2 - y^2$ which pass through the point $(2,1,3)$.
ANSWER   The point lies on the paraboloid: $2^2 - 1^2 = 3$.
In order for the line (12), with $a = b = 1$, to pass through the point $(2,1,3)$, we must have, for some t,

$$2 = \alpha + t, \qquad 1 = \alpha - t, \qquad 3 = 4\alpha t.$$

The first two equations show that $\alpha = \frac{3}{2}$, $t = \frac{1}{2}$. Then $4\alpha t = 3$, as we need. The desired line is

$$x = \tfrac{3}{2} + t, \qquad y = \tfrac{3}{2} - t, \qquad z = 6t.$$

Starting with (13) we obtain, in the same way, the line

$$x = \tfrac{1}{2}t, \qquad y = -\tfrac{1}{2} + t, \qquad z = 6t.$$

Since the hyperboloid of one sheet and the hyperbolic paraboloid are ruled surfaces, we can make models of these surfaces out of straight rods. Such models are shown in Figures 18.38 and 18.39.

PROBLEMS

1. Find two lines through the point $(\frac{1}{2}\sqrt{3},\frac{1}{2},0)$ that lie on the hyperboloid of one sheet $x^2 + y^2 - 100z^2 = 1$.
2. Find two lines through the point $(2,2,\sqrt{3})$ that lie on the hyperboloid of one sheet $x^2 + y^2 - z^2 = 1$.
3. Find two lines through the point $(0,0,0)$ that lie on the hyperbolic paraboloid $4x^2 - 5y^2 = z$.
4. Find two lines through the point $(4,5,-9)$ that lie on the hyperbolic paraboloid $x^2 - y^2 = z$.
5. Show that two lines of the form (10), corresponding to two different values of α from the range $0 \le \alpha \le 2\pi$, are skew to each other.
6. Show that no line in the family (10) coincides with a line in the family (11).
7. Show that two lines in the family (12), corresponding to different values of α, are skew to each other.
8. Show that no line in the family (12) coincides with a line in the family (13).

5.6   Changing coordinates[‡]

It turns out that the examples of quadric surfaces given thus far *exhaust all possibilities.* This important result is analogous to Theorem 1 in Chapter 16, §4.4.

Theorem 1.   *A quadratic equation in three variables represents either the empty set, a point, a line, two planes, an elliptic, parabolic, or hyperbolic cylinder, an elliptic cone, or an ellipsoid, hyperboloid, or paraboloid.*

We do not carry out the proof in this book. It consists of three steps. First, we show that a quadric surface remains one if a new Cartesian coordinate system is used. Then we show that by choosing appropriate new directions of the coordinate axes, we can transform a given quadratic equation into one without mixed terms, that is, without terms with xy, xz, yz. Finally, we show that an equation without mixed terms can be transformed, by a translation of the coordinate system, either into an equation of the kind considered in the beginning of §5.1 or into one of the "standard" forms (3), (4), (5), (6), (7), (8), (9).

Theorem 2.   *The intersection of a quadric surface with a plane (not contained in it) is a second-degree curve, a line, a point, or empty.*

Proof.   We may assume that the coordinate system is chosen so that the plane σ is the plane $z = 0$. The surface S is represented by an equation of the form (1); see §5.1. If we set $z = 0$ in (1), we obtain for the intersection the equation

$$Ax^2 + By^2 + 2Dxy + Gx + Hy + K = 0.$$

Not all numbers A, B, D, G, H, K are 0; otherwise σ would be part of S. Hence the assertion.

REMARK   A circular cone is a quadric surface. The intersection of a plane and a circular cone is never empty. Hence this intersection is a parabola, an ellipse, a hyperbola, a pair of lines, a line, or a point. This is the proposition with which the Greek mathematicians started their investigations.

§6   Cross product[†]

In this section we describe a way of multiplying two space vectors which yields a new vector. This kind of multiplication, which has many applications in physics, depends on orientation in space.

6.1   Orientation in space

Space is **oriented** by choosing one Cartesian coordinate system and calling it **right-handed.** Then every other coordinate system is either right-handed or left-handed, but not both. In Figure 18.40 (a) is chosen to be right-handed; (b) and (c) are also right-handed, while (d) and (e) are left-handed.

Intuitively, if we imagine the "frame" consisting of the unit vectors **i, j, k** to be made of a rigid material, then any right-handed frame can be moved into the position of

[‡]Optional subsection.
[†]Optional section.

FIGURE   **18.40**

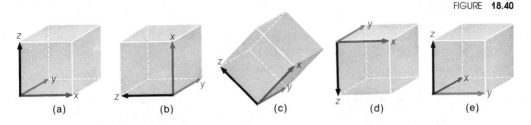

(a)   (b)   (c)   (d)   (e)

any other right-handed frame and similarly for left-handed frames. But it is impossible to turn a right-handed frame into a left-handed one, just as it is impossible to fit a right glove over a left hand.

The statements just made must be reinforced by precise definitions and by an analytic proof. We do not do it here, since we shall make no use of determinants.

One sometimes "defines" a right-handed coordinate system as follows: a right-threaded screw, pointing in the z-direction, advances when its head is rotated by 90° from the x-direction to the y-direction. This, however, is not a mathematical definition, since we cannot define a right-threaded screw, we can only show one. (See the Remark on the orientation in the plane in Chapter 10, §1.2.) Rather, this statement is a definition of a right-threaded screw *in an oriented space.*

We assume from now on that the space has been oriented.

6.2   Cross product of two vectors

Let **a** and **b** be vectors. We assume that $\mathbf{a} \neq \mathbf{0}, \mathbf{b} \neq \mathbf{0},$ and we represent the vectors by directed segments $\overrightarrow{OP}$ and $\overrightarrow{OQ}$, emanating from the same point O. We also assume that O, P, Q do not lie on the same line (that is, the directions of **a** and **b** are neither the same nor opposite to each other). Let θ be the measure of the angle $\angle POQ$; note that $0 < \theta < \pi$. There is a unique point R such that O, P, R, Q are the vertices of a parallelogram (so that $\overrightarrow{OR}$ represents the vector $\mathbf{a} + \mathbf{b}$). Let A be the area of this parallelogram.

FIGURE   **18.41**

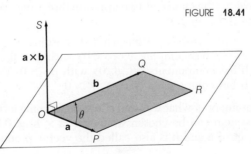

Let $\overrightarrow{OS}$ be a segment of length A, perpendicular to both $\overrightarrow{OP}$ and $\overrightarrow{OQ}$ (see Figure 18.41). We have two choices for the direction of $\overrightarrow{OS}$, and we choose S so that a right-threaded screw, pointing in the direction of $\overrightarrow{OS}$, advances when its head is rotated by the angle θ from the direction $\overrightarrow{OP}$ to the direction $\overrightarrow{OQ}$.

FIGURE 18.42

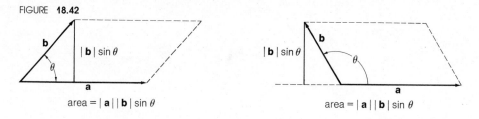

$$\text{area} = |\mathbf{a}| |\mathbf{b}| \sin \theta \qquad\qquad \text{area} = |\mathbf{a}| |\mathbf{b}| \sin \theta$$

The vector determined by $\overrightarrow{OS}$ is called the **cross product** of **a** and **b** (in this order!) and is denoted by

$$\mathbf{a} \times \mathbf{b}.$$

Our construction fails if either **a** or **b** is a null vector, or if **a** and **b** point in the same or in opposite directions (that is, when $\theta = 0$ or $\theta = \pi$). We *define* $\mathbf{a} \times \mathbf{b}$ to be $\mathbf{0}$ in these cases.

We note three properties of the cross product.

$$|\mathbf{a} \times \mathbf{b}| = |\mathbf{a}| |\mathbf{b}| \sin \theta, \qquad \theta = \measuredangle(\mathbf{a},\mathbf{b}). \tag{1}$$

Indeed, the right side of (1) is the area A; see Figure 18.42. The right side is 0, as it should be, if $\mathbf{a} = \mathbf{0}$ or $\mathbf{b} = \mathbf{0}$ or $\theta = 0$ or $\theta = \pi$, and in no other case. Since $|\mathbf{a}|^2|\mathbf{b}|^2 \sin^2 \theta = |\mathbf{a}|^2|\mathbf{b}|^2 (1 - \cos^2 \theta) = |\mathbf{a}|^2|\mathbf{b}|^2 - [|\mathbf{a}||\mathbf{b}| \cos \theta]^2 = |\mathbf{a}|^2|\mathbf{b}|^2 - (\mathbf{a},\mathbf{b})^2$, we can rewrite (1) as

$$|\mathbf{a} \times \mathbf{b}|^2 = |\mathbf{a}|^2|\mathbf{b}|^2 - (\mathbf{a},\mathbf{b})^2. \tag{1'}$$

The second property of the cross product is

$$(\mathbf{a} \times \mathbf{b},\mathbf{a}) = 0, \qquad (\mathbf{a} \times \mathbf{b},\mathbf{b}) = 0. \tag{2}$$

This tells us nothing new if $|\mathbf{a} \times \mathbf{b}| = 0$. If $|\mathbf{a} \times \mathbf{b}| \neq 0$, (2) asserts that the vector $\mathbf{a} \times \mathbf{b}$ is perpendicular to **a** and **b**, hence to the plane spanned by two directed segments representing **a** and **b**, as it should be. This almost determines the direction of $\mathbf{a} \times \mathbf{b}$; more precisely, it gives us the choice of two opposite directions. Which is to be chosen is determined by the third property:

Choose (if possible) a right-handed coordinate system
in which **a** has components $\langle \alpha,0,0 \rangle$ with $\alpha > 0$,
and **b** has components $\langle \beta_1,\beta_2,0 \rangle$ with $\beta_2 > 0$. Then
$\mathbf{a} \times \mathbf{b}$ has components $\langle 0,0,\gamma \rangle$, $\gamma > 0$. $\tag{3}$

This is merely a restatement of what was said above concerning a right-threaded screw. (Such a coordinate system can be chosen whenever $|\mathbf{a} \times \mathbf{b}| \neq 0$.)

The cross product of **a** and **b** is also called the vector product and is sometimes denoted by $[\mathbf{a},\mathbf{b}]$.

PROBLEMS

1. If P has coordinates $(1,0,0)$ and Q has coordinates $(0,1,0)$, find the coordinates of S such that the segment $\overrightarrow{OS}$ is the cross product of $\overrightarrow{OP}$ and $\overrightarrow{OQ}$.
2. If P has coordinates $(1,0,0)$ and Q has coordinates $(1,1,0)$, find the coordinates of S such that the segment $\overrightarrow{OS}$ is the cross product of $\overrightarrow{OP}$ and $\overrightarrow{OQ}$.

3. If $\mathbf{a} = \langle 0,1,0 \rangle$ and $\mathbf{b} = \langle 0,0,1 \rangle$, what is $\mathbf{a} \times \mathbf{b}$?
4. If $\mathbf{a} = \langle 0,0,1 \rangle$ and $\mathbf{b} = \langle 1,0,0 \rangle$, what is $\mathbf{a} \times \mathbf{b}$?
5. If $\mathbf{a} = \langle 1,2,0 \rangle$ and $\mathbf{b} = \langle -4,6,0 \rangle$, what is $\mathbf{a} \times \mathbf{b}$? What is $\mathbf{b} \times \mathbf{a}$?
6. If $\mathbf{a} = \langle 1,0,1 \rangle$ and $\mathbf{b} = \langle -1,0,2 \rangle$, what is $\mathbf{a} \times \mathbf{b}$? What is $\mathbf{b} \times \mathbf{a}$?
7. Find the area of the parallelogram spanned by the vectors $\mathbf{a} = \langle 2,0,0 \rangle$, $\mathbf{b} = \langle 2,2,0 \rangle$.
8. Find the area of the parallelogram spanned by the vectors $\mathbf{a} = \langle 1,0,0 \rangle$, $\mathbf{b} = \langle 1,0,\sqrt{3} \rangle$.

6.3   Rules for cross products

Theorem 1. *If* $\mathbf{a} = a_1\mathbf{i} + a_2\mathbf{j} + a_3\mathbf{k}$ *and* $\mathbf{b} = b_1\mathbf{i} + b_2\mathbf{j} + b_3\mathbf{k}$, *then* $\mathbf{a} \times \mathbf{b}$ *is given by*

$$\mathbf{a} \times \mathbf{b} = (a_2 b_3 - a_3 b_2)\mathbf{i} + (a_3 b_1 - a_1 b_3)\mathbf{j} + (a_1 b_2 - a_2 b_1)\mathbf{k}. \tag{4}$$

Proof.   We verify that the vector defined by (4) has properties (1), (2), (3). We have

$$|\mathbf{a} \times \mathbf{b}|^2 = (a_2 b_3 - a_3 b_2)^2 + (a_3 b_1 - a_1 b_3)^2 + (a_1 b_2 - a_2 b_1)^2$$
$$= a_2^2 b_3^2 + a_3^2 b_2^2 - 2a_2 a_3 b_2 b_3 + a_3^2 b_1^2 + a_1^2 b_3^2 - 2a_3 a_1 b_3 b_1$$
$$+ a_1^2 b_2^2 + a_2^2 b_1^2 - 2a_1 a_2 b_1 b_2.$$

Also,

$$|\mathbf{a}|^2 |\mathbf{b}|^2 - (\mathbf{a},\mathbf{b})^2 = (a_1^2 + a_2^2 + a_3^2)(b_1^2 + b_2^2 + b_3^2) - (a_1 b_1 + a_2 b_2 + a_3 b_3)^2$$
$$= a_1^2 b_1^2 + a_1^2 b_2^2 + a_1^2 b_3^2 + a_2^2 b_1^2 + a_2^2 b_2^2 + a_2^2 b_3^2 + a_3^2 b_1^2 + a_3^2 b_2^2$$
$$- a_1^2 b_1^2 - a_2^2 b_2^2 - a_3^2 b_3^2 - 2a_1 a_2 b_1 b_2 - 2a_1 a_3 b_1 b_3 - 2a_2 a_3 b_2 b_3.$$

Simplifying the above, we see that $|\mathbf{a} \times \mathbf{b}|^2 = |\mathbf{a}|^2 |\mathbf{b}|^2 - (\mathbf{a},\mathbf{b})^2$, so that (1′) holds.
   Next,

$$(\mathbf{a} \times \mathbf{b},\mathbf{a}) = (a_2 b_3 - a_3 b_2)a_1 + (a_3 b_1 - a_1 b_3)a_2 + (a_1 b_2 - a_2 b_1)a_3$$
$$= a_1 a_2 b_3 - a_1 a_3 b_2 + a_2 a_3 b_1 - a_1 a_2 b_3 + a_1 a_3 b_2 - a_2 a_3 b_1 = 0.$$

One sees in the same way that $(\mathbf{a} \times \mathbf{b},\mathbf{b}) = 0$. Hence (2) holds.
   Finally, if $a_1 > 0$, $a_2 = a_3 = 0$, $b_2 > 0$, $b_3 = 0$, then $\mathbf{a} \times \mathbf{b} = 0\mathbf{i} + 0\mathbf{j} + a_1 b_2 \mathbf{k}$, so that property (3) holds. The theorem is proved.

Theorem 2. *The cross product obeys the following rules:*

$$\mathbf{a} \times \mathbf{b} = \mathbf{0} \text{ if and only if either } \mathbf{a} = \mathbf{0} \text{ or } \mathbf{b} = \mathbf{0} \text{ or} \tag{5}$$
there is a number t *such that* $\mathbf{a} = t\mathbf{b}$.

$$\mathbf{a} \times \mathbf{b} = -\mathbf{b} \times \mathbf{a}, \tag{6}$$

$$(s\mathbf{a}) \times \mathbf{b} = s(\mathbf{a} \times \mathbf{b}), \qquad \mathbf{a} \times (s\mathbf{b}) = s(\mathbf{a} \times \mathbf{b}), \tag{7}$$

$$\mathbf{a} \times (\mathbf{b} + \mathbf{c}) = (\mathbf{a} \times \mathbf{b}) + (\mathbf{a} \times \mathbf{c}), \qquad (\mathbf{a} + \mathbf{b}) \times \mathbf{c} = (\mathbf{a} \times \mathbf{c}) + (\mathbf{b} \times \mathbf{c}). \tag{8}$$

Here $\mathbf{a}$, $\mathbf{b}$, $\mathbf{c}$ *are any three vectors,* s *any number.*

   Proof.   Statement (5) follows from (1).
   The anticommutative law (6), the associative laws (7), and the distributive laws (8) follow from Theorem 1 and properties of numbers by direct calculations.

REMARK (for readers familiar with determinants). Formula (4) can be remembered in the form

$$\mathbf{a} \times \mathbf{b} = \begin{vmatrix} \mathbf{i} & \mathbf{j} & \mathbf{k} \\ a_1 & a_2 & a_3 \\ b_1 & b_2 & b_3 \end{vmatrix};$$ (4′)

this also explains the anticommutative law.

EXAMPLES 1. For every vector $\mathbf{a}$, $\mathbf{a} \times \mathbf{a} = \mathbf{0}$.

This follows from (1′), since $|\mathbf{a}|^2 = (\mathbf{a},\mathbf{a})$.

2. $\mathbf{i} \times \mathbf{j} = \mathbf{k}, \mathbf{j} \times \mathbf{k} = \mathbf{i}, \mathbf{k} \times \mathbf{i} = \mathbf{j}$.

First proof. Use Theorem 1.
Second proof. Use the geometric definition in §6.2.

3. $\mathbf{j} \times \mathbf{i} = -\mathbf{k}, \mathbf{k} \times \mathbf{j} = -\mathbf{i}, \mathbf{i} \times \mathbf{k} = -\mathbf{j}$.

Proof. Use Example 2 and the anticommutative law (6).

PROBLEMS

1. Prove that $\mathbf{i} \times \mathbf{i} = \mathbf{0}, \mathbf{j} \times \mathbf{j} = \mathbf{0}, \mathbf{k} \times \mathbf{k} = \mathbf{0}$.
2. Prove the anticommutative law (6).
3. Prove the associative laws (7).
4. Prove the distributive laws (8).
5. Find $\mathbf{a} \times \mathbf{b}$ if $\mathbf{a} = 2\mathbf{i} + 3\mathbf{j} - \mathbf{k}$ and $\mathbf{b} = \mathbf{i} + 2\mathbf{j} - 4\mathbf{k}$.
6. Find $\mathbf{a} \times \mathbf{b}$ if $\mathbf{a} = -\mathbf{i} + 2\mathbf{j} + \mathbf{k}$ and $\mathbf{b} = \mathbf{i} - \mathbf{j} + 2\mathbf{k}$.
7. Find $\mathbf{a} \times \mathbf{b}$ if $\mathbf{a} = 3\mathbf{i} + 5\mathbf{j} - \mathbf{k}$ and $\mathbf{b} = 2\mathbf{i} + 3\mathbf{j} - \mathbf{k}$.
8. If $\mathbf{a} = \langle 3,-2,6 \rangle$, $\mathbf{b} = \langle 1,-1,0 \rangle$, what is $\mathbf{a} \times \mathbf{b}$?
9. If $\mathbf{a} = \langle 0,7,1 \rangle$, $\mathbf{b} = \langle -1,-2,-4 \rangle$, what is $\mathbf{a} \times \mathbf{b}$?
10. If $\mathbf{a} = \langle 2,0,-5 \rangle$, $\mathbf{b} = \langle -3,1,0 \rangle$, what is $\mathbf{a} \times \mathbf{b}$?
11. Find the area of the parallelogram spanned by the vectors $\mathbf{a} = \langle 0,2,4 \rangle$, $\mathbf{b} = \langle 2,4,2 \rangle$.
12. Find the area of the parallelogram spanned by the vectors $\mathbf{a} = \langle -1,3,1 \rangle$, $\mathbf{b} = \langle 2,-4,1 \rangle$.
13. Find a unit vector perpendicular to the vectors $\mathbf{a} = \langle 0,1,4 \rangle$, $\mathbf{b} = \langle 2,0,-2 \rangle$.
14. Find a unit vector perpendicular to the vectors $\mathbf{a} = \langle 1,-3,2 \rangle$, $\mathbf{b} = \langle 2,-1,3 \rangle$.
15. Find the area of the triangle with vertices $(2,4,0)$, $(4,1,0)$ and $(6,5,0)$.
16. Find the area of the triangle whose vertices are the terminal points of the position vectors $\mathbf{a}, \mathbf{b},$ and $\mathbf{c}$.

6.4 Triple products

Given three vectors $\mathbf{a}, \mathbf{b}, \mathbf{c}$, we can form, in different ways, their products. The product $(\mathbf{a},\mathbf{b})\mathbf{c}$ is of little interest; it is the vector $\mathbf{c}$ multiplied by the number $(\mathbf{a},\mathbf{b})$. The **triple vector product** $\mathbf{a} \times (\mathbf{b} \times \mathbf{c})$ can be computed by the following formula:

$$\mathbf{a} \times (\mathbf{b} \times \mathbf{c}) = (\mathbf{a},\mathbf{c})\mathbf{b} - (\mathbf{a},\mathbf{b})\mathbf{c}.$$ (9)

The proof is a lengthy calculation based on (4) which the reader is asked to carry out.

In particular, (9) implies that $\mathbf{a} \times (\mathbf{a} \times \mathbf{b}) = (\mathbf{a},\mathbf{b})\mathbf{a} - (\mathbf{a},\mathbf{a})\mathbf{b}$ while $(\mathbf{a} \times \mathbf{a}) \times \mathbf{b} = \mathbf{0} \times \mathbf{b} = \mathbf{0}$. Hence the triple vector product is not associative: $\mathbf{a} \times (\mathbf{b} \times \mathbf{c})$ is, in general, distinct from $(\mathbf{a} \times \mathbf{b}) \times \mathbf{c}$.

FIGURE **18.43**

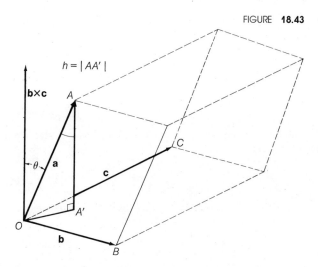

Of particular interest is the **mixed triple product**

$$\mathbf{a} \cdot (\mathbf{b} \times \mathbf{c}) = (\mathbf{a}, \mathbf{b} \times \mathbf{c}); \tag{10}$$

note that this is a number.

Theorem 3. *Represent the vectors* **a, b, c** *by directed segments* $\overrightarrow{OA}$, $\overrightarrow{OB}$, $\overrightarrow{OC}$. *If O, A, B, C lie in one plane,* $\mathbf{a} \cdot (\mathbf{b} \times \mathbf{c}) = 0$. *If not,* $|\mathbf{a} \cdot (\mathbf{b} \times \mathbf{c})|$ *is the volume of the parallelopiped spanned by* $\overrightarrow{OA}$, $\overrightarrow{OB}$, $\overrightarrow{OC}$.

   Proof.   If O, B, C lie on one line, $\mathbf{b} \times \mathbf{c} = \mathbf{0}$ (by Theorem 2) and $\mathbf{a} \cdot (\mathbf{b} \times \mathbf{c}) = 0$. Assume that O, A, B do not lie on one line; then $F = |\mathbf{b} \times \mathbf{c}|$ is the area of the parallelogram spanned by $\overrightarrow{OB}$ and $\overrightarrow{OC}$ [see (1)]. If $A = 0$, then $\mathbf{a} = \mathbf{0}$ and $\mathbf{a} \cdot (\mathbf{b} \times \mathbf{c}) = 0$. Assume that $\mathbf{a} \neq \mathbf{0}$. Then

$$|\mathbf{a} \cdot (\mathbf{b} \times \mathbf{c})| = |(\mathbf{a}, \mathbf{b} \times \mathbf{c})| = |\mathbf{a}| F |\cos \theta|,$$

where θ is the angle between $\mathbf{a}$ and $\mathbf{b} \times \mathbf{c}$. Assume A lies in the plane OBC. Since $\mathbf{a} \times \mathbf{b}$ is perpendicular to this plane, $\theta = 90°$, $\cos \theta = 0$, and $\mathbf{a} \cdot (\mathbf{b} \times \mathbf{c}) = 0$. If A does not lie in this plane, $|\mathbf{a}| |\cos \theta|$ is the distance from A to the plane OBC, that is, the height h of the parallelopiped spanned by $\overrightarrow{OA}$, $\overrightarrow{OB}$, $\overrightarrow{OC}$; see Figure 18.43. Since F is the area of the base, $|\mathbf{a} \cdot (\mathbf{b} \times \mathbf{c})| = Fh$ is the volume.

Theorem 4. *Let* **a, b, c** *be vectors, t a number. Then*

$$(\mathbf{a}, \mathbf{b} \times \mathbf{c}) = (\mathbf{a} + \mathbf{b}, \mathbf{b} \times \mathbf{c}), \tag{11}$$

$$(\mathbf{a}, \mathbf{b} \times \mathbf{c}) = -(\mathbf{a}, \mathbf{c} \times \mathbf{b}), \tag{12}$$

$$(t\mathbf{a}, \mathbf{b} \times \mathbf{c}) = t(\mathbf{a}, \mathbf{b} \times \mathbf{c}). \tag{13}$$

   Proof of (11). $(\mathbf{a} + \mathbf{b}, \mathbf{b} \times \mathbf{c}) = (\mathbf{a}, \mathbf{b} \times \mathbf{c}) + (\mathbf{b}, \mathbf{b} \times \mathbf{c}) = (\mathbf{a}, \mathbf{b} \times \mathbf{c}) + 0$ since $\mathbf{b} \times \mathbf{c}$ is orthogonal to $\mathbf{b}$.
   Proof of (12). $-(\mathbf{a}, \mathbf{c} \times \mathbf{b}) = -(\mathbf{a}, -\mathbf{b} \times \mathbf{c}) = (\mathbf{a}, \mathbf{b} \times \mathbf{c}).$

The reader is asked to justify the steps above by referring to specific rules for inner products and cross products. Statement (13) is clear.

REMARK   Readers familiar with determinants should verify the formula

$$\mathbf{a} \cdot (\mathbf{b} \times \mathbf{c}) = \begin{vmatrix} a_1 & a_2 & a_3 \\ b_1 & b_2 & b_3 \\ c_1 & c_2 & c_3 \end{vmatrix}$$

(the a's, b's, and c's being the components of $\mathbf{a}$, $\mathbf{b}$, $\mathbf{c}$) and recognize in statements (11), (12), (13) known properties of determinants.

PROBLEMS

1. Find the volume of the parallelopiped spanned by the directed segments $\overrightarrow{OA}$, $\overrightarrow{OB}$, and $\overrightarrow{OC}$ if the coordinates of A, B, and C are (1,0,0), (1,1,0) and (0,0,4), respectively.
2. Find the volume of the parallelopiped spanned by the position vectors $\mathbf{a} = \langle 8,4,0 \rangle$, $\mathbf{b} = \langle 2,6,0 \rangle$, $\mathbf{c} = \langle 0,4,6 \rangle$.
3. Find the volume of the parallelopiped spanned by the position vectors $\mathbf{a} = \langle 2,2,2 \rangle$, $\mathbf{b} = \langle 0,2,2 \rangle$, $\mathbf{c} = \langle 2,0,8 \rangle$.
4. Prove that $\mathbf{a} \times (\mathbf{b} \times \mathbf{c}) = (\mathbf{a},\mathbf{c})\mathbf{b} - (\mathbf{a},\mathbf{b})\mathbf{c}$.
5. Prove that $\mathbf{a} \cdot (\mathbf{b} \times \mathbf{c}) = \mathbf{b} \cdot (\mathbf{c} \times \mathbf{a}) = \mathbf{c} \cdot (\mathbf{a} \times \mathbf{b})$.
6. Prove that $\mathbf{a} \times (\mathbf{b} \times \mathbf{c}) + \mathbf{b} \times (\mathbf{c} \times \mathbf{a}) + \mathbf{c} \times (\mathbf{a} \times \mathbf{b}) = \mathbf{0}$.
7. Prove that $(\mathbf{a} \times \mathbf{b}) \cdot (\mathbf{c} \times \mathbf{d}) = (\mathbf{a} \cdot \mathbf{c})(\mathbf{b} \cdot \mathbf{d}) - (\mathbf{a} \cdot \mathbf{d})(\mathbf{b} \cdot \mathbf{c})$.
8. Prove that $(\mathbf{a} \times \mathbf{b}) \times (\mathbf{c} \times \mathbf{d}) = ((\mathbf{a} \times \mathbf{b}) \cdot \mathbf{d})\mathbf{c} - ((\mathbf{a} \times \mathbf{b}) \cdot \mathbf{c})\mathbf{d}$.

In Problems 9 to 12 $\mathbf{a}$, $\mathbf{b}$, $\mathbf{c}$, and λ are differentiable functions of the variable t.

9. Show that $\dfrac{d}{dt}(\lambda \mathbf{a}) = \dfrac{d\lambda}{dt}\mathbf{a} + \lambda \dfrac{d\mathbf{a}}{dt}$.

10. Show that $\dfrac{d}{dt}(\mathbf{a} \cdot \mathbf{b}) = \dfrac{d\mathbf{a}}{dt} \cdot \mathbf{b} + \mathbf{a} \cdot \dfrac{d\mathbf{b}}{dt}$.

11. Show that $\dfrac{d}{dt}(\mathbf{a} \times \mathbf{b}) = \dfrac{d\mathbf{a}}{dt} \times \mathbf{b} + \mathbf{a} \times \dfrac{d\mathbf{b}}{dt}$.

12. Show that $\dfrac{d}{dt}(\mathbf{a} \cdot (\mathbf{b} \times \mathbf{c})) = \dfrac{d\mathbf{a}}{dt} \cdot (\mathbf{b} \times \mathbf{c}) + \mathbf{a} \cdot \left(\dfrac{d\mathbf{b}}{dt} \times \mathbf{c}\right) + \mathbf{a} \cdot \left(\mathbf{b} \times \dfrac{d\mathbf{c}}{dt}\right)$.

PARTIAL DERIVATIVES

§1 Function of several variables

In this and the following chapters we extend the basic concepts of calculus to functions of several variables.

There are significant differences between functions of one and functions of several variables. But in most cases, the transition from two to more than two variables is easy; for this reason, we shall often consider in detail only the case of two variables.

1.1 Functions of two variables

A typical function of two variables is the rule

$$(x,y) \mapsto z = f(x,y) = x^2y^3$$

which assigns to any two ordered numbers the product of the square of the first number and the cube of the second. In general, a **function of two variables** is a rule that assigns to every ordered pair of numbers (from some set of such pairs) another number. In working with functions of two variables, and later with functions of more than two variables, we shall use variables, as well as the arrow $\mapsto$, just as we did for functions of one variable.

In dealing with functions of two variables, $(x,y) \mapsto z$, it is usually illuminating to consider the two numbers x and y as the coordinates of a point in the plane, with respect to some Cartesian coordinate system. We may think of a function of two variables as a rule by which we assign a number to every point in the plane or to every point in some part of the plane.

The definitions of sums, differences, products, and quotients of functions, as well as of composite functions, are similar to the corresponding definitions for functions of one variable; there is no need to state them explicitly.

EXAMPLES 1. The function $f(x,y) = x/y$ is defined for all x and y such that $y \neq 0$. We have $f(4,2) = 2$, $f(2,4) = .5$, $f(0,y) = 0$ for all $y \neq 0$, and so forth.

2. The function $f(x,y) = \sqrt{x - y}$ is defined for all (x,y) such that $x \geq y$.

3. Set $f(x,y) = x^2 + y^2$, $g(x,y) = x^2 - y^2$, $\phi(x) = \cos x$, $\psi(x) = \sin x$. Then

$$f(x,y) + g(x,y) = (x^2 + y^2) + (x^2 - y^2) = 2x^2,$$
$$f(g(x,y),y^2) = g(x,y)^2 + (y^2)^2 = (x^2 - y^2)^2 + (y^2)^2 = x^4 - 2x^2y^2 + 2y^4,$$
$$f(\phi(x),\psi(x)) = \phi(x)^2 + \psi(x)^2 = \cos^2 x + \sin^2 x = 1,$$
$$g(\phi(x),\psi(x)) = \phi(x)^2 - \psi(x)^2 = \cos^2 x - \sin^2 x = \cos 2x,$$
$$\phi(g(x,y)) = \cos g(x,y) = \cos(x^2 - y^2),$$
$$g(f(x,y),g(x,y)) = f(x,y)^2 - g(x,y)^2 = (x^2 + y^2)^2 - (x^2 - y^2)^2 = 4x^2y^2.$$

All these functions are defined everywhere, that is, for all x and y.

4. The function

$$(x,y) \mapsto z = \begin{cases} x^2 + y, & \text{if } y \leq 0 \\ \dfrac{x^3 + xy}{y}, & \text{if } x > 0, y > 0 \\ 315, & \text{if } x \leq 0, y > 0 \end{cases}$$

is defined everywhere.

PROBLEMS

In Problems 1 to 4 find the values of x and y for which the given functions are defined.

1. $g(x,y) = \sqrt{2x - y + 3}$.

2. $g(x,y) = \log(x^2 + 4y^2 - 16)$.

3. $g(x,y) = \dfrac{xy}{\sqrt{x^2 - y^2 - 9}}$.

4. $g(x,y) = \dfrac{\sqrt{xy - 1}}{\sqrt{16 - x^2 - y^2}}$.

5. Suppose $f(x,y) = x^3 + 4x^2y + y^3$. Find $f(2,1)$ and $f(-3,0)$.

6. Suppose $f(x,y) = \dfrac{2xy + 1}{x^2 + 4xy + y^2}$. Find $f(1,-1)$ and $f(-2,4)$.

7. If $F(x,y) = x + y^2$, what is $F(2x,y^2)$? What is $F(x^2, \sqrt{x + y})$?
8. If $F(x,y) = \sin(2x + y)$, what is $F(3x^2,y^3)$? What is $F(2x,x + y)$?
9. If $F(x,y) = x^3y^3/(x^3 + y^3)$, what is $F(e^x,e^{-x})$? What is $F(e^x,2e^x)$?
10. If $f(x,y) = e^{-2xy} \tan(x^2 + y^2)$, what is $f(-s,r/2)$? What is $f(u + v,u - v)$?
11. If $f(x,y) = x/(x + y)$, what is $f(x,f(x,y))$? What is $f(f(x,y),f(x,y))$?
12. If $f(x,y) = \log xy$, $\phi(x) = e^{2x}$, $\psi(x) = 2/x$, what is $f(\phi(2),\psi(5))$?
13. If $u(x,y) = \arctan x/y$, $v(x,y) = x^2 + y^2$, show that $\sqrt{v} \sin u = x$ and $\sqrt{v} \cos u = y$.
14. If $r(m,n) = e^m \cos n$, $s(m,n) = e^m \sin n$, show that $[r(m,n)]^2 - [s(m,n)]^2 = r(2m,2n)$ and $2r(m,n)s(m,n) = s(2m,2n)$.

FIGURE **19.1**   FIGURE **19.2**

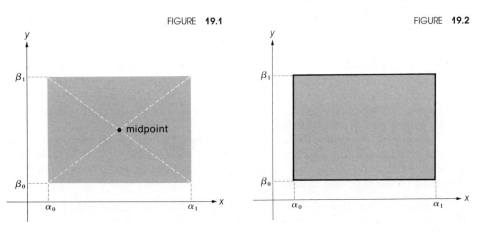

1.2   Intervals. Continuity

The set of points inside a rectangle, with sides parallel to the coordinate axes, is called a (two-dimensional) interval, or **open interval.** Such a set is shown in Figure 19.1. It can be described as the set of all points (x,y) satisfying the inequalities

$$\alpha_0 < x < \alpha_1, \qquad \beta_0 < y < \beta_1, \tag{1}$$

where α_0, α_1, β_0, β_1 are numbers such that $\alpha_0 < \alpha_1$, $\beta_0 < \beta_1$. The point with the coordinates $x = (\alpha_1 + \alpha_0)/2$, $y = (\beta_1 + \beta_0)/2$ is called the **midpoint** of the interval. This is the intersection point of the diagonals of the rectangle. The **boundary** of the interval consists of all points on the rectangle proper.

The set of all points in the open interval *and* on the boundary is called a **closed interval** (see Figure 19.2). The closed interval corresponding to (1) is the set of all points (x,y) satisfying the inequalities

$$\alpha_0 \leq x \leq \alpha_1, \qquad \beta_0 \leq x \leq \beta_1. \tag{2}$$

We agree to use the phrase "**near a point** P" (in the plane) to mean "in some interval with P as midpoint." Since every interval with midpoint P contains a circular disk with center P, and conversely (see Figure 19.3), we conclude that "near P" also means "in some circular disk with center P" or "at all points Q whose distance from P is less than some fixed positive number."

FIGURE **19.3**

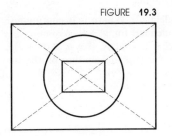

The following terminology will be useful. A set D of points in the plane is **open** if whenever it contains a point P, it also contains all points near P. For example, the interval $0 < x < 1$, $0 < y < 1$ is an open set. On the other hand, the segment

$0 \leq x \leq 1$, $y = 0$ is not open. For instance, in every interval with midpoint $(\frac{1}{2},0)$ there are points, say, $(1/2,1/n)$ with n large, which do not lie on the segment. The closed interval $0 \leq x \leq 1$, $0 \leq y \leq 1$ is also not open.

A nonempty open set D is a **domain** if any two points in D can be joined by a smooth curve lying in D. The **boundary** of any set D consists of all points Q such that in every interval with midpoint Q there are points of D and also points not in D. We shall deal mostly with functions of two variables defined in some domain and perhaps on its boundary or a part of the boundary.

The concept of **continuity** for functions of two variables is an extension of the concept of continuity of functions of one variable (see Chapter 3, §3.2). Roughly speaking, a function $f(x,y)$ is continuous if a small change in x and y results in a small change in $f(x,y)$. A precise definition follows.

A function $f(x,y)$ defined for $x = x_0$, $y = y_0$ is called **continuous** *at the point (x_0,y_0) if, given any positive number ϵ, we can find a positive δ such that*

$$|f(x,y) - f(x_0,y_0)| < \epsilon$$

for all (x,y) for which $f(x,y)$ is defined and

$$\sqrt{(x - x_0)^2 + (y - y_0)^2} < \delta.$$

Finally, a function is called **continuous on a set** S if it is defined and continuous at every point of S.

Sums, products, and differences of continuous functions are continuous; so is the quotient of two continuous functions, except where the denominator is 0. Composing continuous functions, we again obtain continuous functions. All this is proved exactly as for functions of one variable. Also, if $f(x)$ is a continuous function of one variable, the functions of two variables $F(x,y) = f(x)$ and $G(x,y) = f(y)$ are continuous. Every monomial $(x,y) \mapsto cx^n y^m$, c a number, n and m nonnegative integers, is a continuous function. Hence, every polynomial in two variables, that is, a finite sum of monomials, is continuous everywhere. A rational function of two variables, that is, a quotient of two polynomials, is continuous where the denominator is not 0. A radical function of two variables (the definition of such functions parallels the one given in Chapter 6, §2.2) is continuous wherever defined.

EXAMPLES 1. Where is the function $(x,y) \mapsto e^{\sin(x^2+2y)} + \cos(xy)$ continuous?
ANSWER It is continuous everywhere, for it is the sum of two functions, each of which is a continuous function of a polynomial.

2. Where is $f(x,y) = \tan(x/y)$ defined? Where is it continuous?
ANSWER The quotient x/y is defined for $y \neq 0$. Assuming $y \neq 0$, we see that f is defined if x/y is not an odd multiple of $\pi/2$, that is, if $x/y \neq (n + \frac{1}{2})\pi$, $n = 0, \pm 1, \pm 2, \ldots$. In other words, $f(x,y)$ is defined whenever (x,y) lies neither on the line $y = 0$ nor on one of the lines $x = \frac{1}{2}\pi y$, $x = -\frac{1}{2}\pi y$, $x = \frac{3}{2}\pi y$, $x = -\frac{3}{2}\pi y, \ldots$. Also, f is continuous wherever it is defined, being a composite of continuous functions.

3. The function

$$g(x,y) = \begin{cases} \dfrac{1}{x^2 + y^2}, & \text{if } (x,y) \neq (0,0) \\ \dfrac{1}{134}, & \text{if } x = y = 0 \end{cases}$$

is defined everywhere; it is continuous everywhere except at $x = y = 0$. Indeed, if $x^2 + y^2 \neq 0$, g is the quotient of two continuous functions, 1 and $x^2 + y^2$, different from 0. But for (x,y) close to and distinct from $(0,0)$, $x^2 + y^2$ is very small, $g(x,y)$ very large, hence not close to $g(0,0) = 1/134$.

4. Consider the function

$$\phi(x,y) = \begin{cases} \dfrac{xy}{(x^2 + y^2)^2}, & \text{if } x^2 + y^2 \neq 0 \\ 0, & \text{if } x = y = 0 \end{cases}$$

which is defined everywhere. If we choose a fixed value of x and consider $\phi(x,y)$ as a function of y alone, we always obtain an everywhere continuous function of one variable. Indeed, for a fixed $x = a \neq 0$, we have $\phi(x,y) = ay/(a^2 + y^2)^2$ for all y, and for $x = 0$ we have $\phi(x,y) = 0$ for all y. Similarly, for every fixed value of y, the function $x \mapsto \phi(x,y)$ is an everywhere continuous function of x. But ϕ considered as a function of (x,y) is not continuous at the origin. Indeed, $\phi(0,0) = 0 < 1$, and in every interval with midpoint $(0,0)$, there are points (x,y) with $\phi(x,y) > 1$; for instance, the points $x = y = t$ with $0 < t < \tfrac{1}{2}$.

This example shows that continuity in each variable x and y separately is a weaker condition than continuity in (x,y).

PROBLEMS

In Problems 1 to 5 state whether the set of points in the plane is open or not open.

1. The set of all points the sum of whose coordinates is negative.
2. The set of all points *at least one* of whose coordinates is negative.
3. The set of all points *both* of whose coordinates are positive.
4. The set of all points *neither* of whose coordinates is negative.
5. The set of points whose x-coordinate satisfies $\sin x = 0$.

6. If A and B are open sets, let C be the set of points that belong to either A or B, and let D be the set of points belonging to both A and B. Must C be an open set? Must D be an open set?
7. Find the boundary of the set given in Problem 1.
8. Find the boundary of the set given in Problem 2.
9. Find the boundary of the set given in Problem 5.
10. If A is the set of all points in the x, y plane such that both x and y are rational numbers, what is the boundary of A?
11. Is the set of points (x,y) with $x^2 + y^2 \neq 1$ open? What is the boundary of this set?
12. There are exactly two sets of points (x,y) which have the property that they possess no boundary points at all. Find them.

In each of Problems 13 to 25 determine if the given function is continuous. If not, state where it is discontinuous.

13. $f(x,y) = \log(1 + x^2y^2 + y^4)$.

14. $f(x,y) = \dfrac{e^{xy} \sin(x + y)}{1 + x^2 + y^2}$.

15. $f(x,y) = \log|1 + xy|$.

16. $g(x,y) = |x|^y$.

17. $g(x,y) = |x|^{1/|y|}$.

18. $h(x,y) = \dfrac{x^2 + y^2}{x^2 - y^2}$.

19. $h(x,y) = \sin(x \tan y)$.

20. $h(x,y) = \arctan \dfrac{y}{x}$.

21. $F(x,y) = \begin{cases} \dfrac{x^2y^2}{x^4 + y^4}, & x \neq 0, y \neq 0 \\ 0, & x = y = 0. \end{cases}$

24. $F(x,y) = \begin{cases} \sin \dfrac{x}{y}, & x \neq 0, y \neq 0 \\ 1, & x = y = 0. \end{cases}$

22. $F(x,y) = \begin{cases} \dfrac{xy}{x^2 + y^2}, & x \neq 0, y \neq 0 \\ 1, & x = y = 0. \end{cases}$

25. $F(x,y) = \begin{cases} \dfrac{\sin(x + y)}{x + y}, & x + y \neq 0 \\ 1, & x + y = 0. \end{cases}$

23. $F(x,y) = \begin{cases} \dfrac{x^3 + y^3}{x^2 + y^2}, & x \neq 0, y \neq 0 \\ 0, & x = y = 0. \end{cases}$

1.3   Functions of three or more variables

A function of three variables is a rule $(x,y,z) \mapsto u = f(x,y,z)$ which assigns, to some ordered triples of numbers, a number. Similarly, for any positive integer n, a **function of n variables** is a rule

$$(x_1, x_2, \ldots, x_n) \mapsto u = f(x_1, x_2, \ldots, x_n) \tag{3}$$

which assigns numbers u to certain ordered n-tuples of numbers. It is useful to think of the ordered n-tuples of numbers as the coordinates of a point.

The definition of continuity and the theorems on continuity stated in §1.2 extend to functions of three or more variables. So do the definitions of "interval," "open set," "domain," and "boundary."

REMARK   A continuous function defined on a set S remains continuous if we consider it on a subset of S. In particular, a continuous function of several variables remains continuous if we keep one or several variables fixed.

PROBLEMS

1. Suppose $f(x,y,z) = xy + yz + xz$. Find $f(1,-2,3)$ and $f(-1,0,4)$.

2. Suppose $f(x,y,z) = \dfrac{xyz}{\sqrt{x + y + z}}$. Find $f(1,1,2)$ and $f(-1,7,5)$.

3. Suppose $f(x,y,z) = xy - 2yz + 4xz$. Find $f(a,-2a,a^2)$ and $f(a + b, b^2, a - b)$.

4. Suppose $u(x,y,z,t) = \sqrt{x^2 + y^2 + z^2 - t^2}$. Find $u(1,-2,3,2)$ and $u(\tfrac{3}{2},\tfrac{1}{4},\tfrac{1}{4},\tfrac{1}{2})$.

5. Suppose $u(x,y,z,t) = e^{xy} \cos(z - t) + e^{yz} \cos(x - t) + e^{xz} \cos(y - t)$. Find $u(\pi,\tfrac{1}{2}\pi,0,\tfrac{1}{4}\pi)$.

6. If $F(x,y,z) = \sin(2x + 4y - z)$, what is $F(2x,-y,x + y)$? What is $F(x + 1,-2y + x,x - 8y + 2)$?

7. If $F(x,y,z) = \dfrac{2xyz}{x + 2y + 3z}$, what is $F\left(\dfrac{3}{x^2 + 1}, \dfrac{2}{x^2 + 1}, \dfrac{1}{x^2 + 1}\right)$?

8. If $G(x,y,z,t) = e^{xyz} \sin t + e^{-xyz} \cos t$, what is $G\left(\log x, 2x, \dfrac{1}{x}, \dfrac{\pi}{4}\right)$?

For each of the functions in Problems 9 to 14 state where it is defined and where it is continuous.

9. $g(x,y,z) = \sin(x + y + z) + e^{-xyz}$.

10. $g(x,y,z) = xyz \log(x^2 + y^2 + z^2 + 1)$.

11. $g(x,y,z) = \dfrac{y}{x} + \dfrac{x}{z} + \dfrac{z}{y}.$

12. $g(x,y,z) = xy \sin\dfrac{1}{z} + yz \sin\dfrac{1}{x} + xz \sin\dfrac{1}{y}.$

13. $G(x,y,z,t) = \dfrac{xyzt}{x^2 + y^2 + z^2 - t^2}.$

14. $G(x,y,z,t) = \dfrac{\sin(x + y)}{(x + y)} + \dfrac{\sin(z + t)}{(z + t)}.$

1.4 Geometric and physical interpretations

A function $z = f(x,y)$ defined on some domain can be represented by its **graph,** the set of all points (x,y,z) in space such that $f(x,y)$ is defined and $z = f(x,y)$. Unfortunately, it is rather difficult to draw such graphs on paper. If $f(x,y)$ is a continuous function defined on a domain, the graph of f is called a **surface.**

EXAMPLES   1.   A function of several variables is called **linear** if it is a polynomial of degree 1. *The graph of a linear function of two variables is a plane.* Indeed, such a function can be written as

$$(x,y) \mapsto z = ax + by + c,$$

where a, b, c are fixed numbers, and we know that the set of points (x,y,z) satisfying $ax + by + c - z = 0$ is a plane (see Chapter 18, §4.1).

2.   The graph of $z = \sqrt{1 - x^2 - y^2}$ is the upper unit hemisphere. Indeed, if $z = \sqrt{1 - x^2 - y^2}$, then $z \geq 0$ and $x^2 + y^2 + z^2 = 1$.

3.   The graph of the function $z = x^2$ considered as a function of two variables, that is, the graph of $(x,y) \mapsto x^2$ is a parabolic cylinder (compare Chapter 18, §5.1).

4.   The graph of $f(x,y) = x^2 - y^2$ is a hyperbolic paraboloid. We discussed this surface in Chapter 18, §5.4.

For a function of three (or more) variables, we can also define a graph, but this is now a set of points in a space of four (or more) dimensions. It cannot be drawn or modeled.

It is sometimes useful to think of a function of three variables, $(x,y,z) \mapsto f(x,y,z) = \phi$, as describing the temperature ϕ at a point with coordinates (x,y,z). A function of four variables, $(x,y,z,t) \mapsto \phi$, can be thought of as describing the temperature at the point (x,y,z) at the time t. There are many similar interpretations of functions of three or more variables.

§2 Derivatives of functions of several variables

In this section we extend the first basic process of calculus, differentiation, to functions of several variables.

2.1 Partial derivatives

We consider a function $(x,y) \mapsto f(x,y)$ of two variables and fix the value of the second variable at y_0. In other words, we consider the function of *one* variable $x \mapsto f(x,y_0)$.

Assume that this function has, at $x = x_0$, a derivative; this means that the (finite) limit

$$\lim_{h \to 0} \frac{f(x_0 + h, y_0) - f(x_0, y_0)}{h}$$

exists. This number is called the **partial derivative with respect to** x of $f(x,y)$ at (x_0, y_0) and is denoted by

$$\left(\frac{\partial f}{\partial x} \right)_{x=x_0, \, y=y_0} \qquad \text{or} \qquad \left(\frac{\partial f}{\partial x} \right)_{(x_0, \, y_0)}.$$

(The symbol ∂ is called the "round d.") If we compute the partial derivative with respect to x at all points of some set, we obtain a new function of (x,y):

$$\frac{\partial f}{\partial x} = \frac{\partial f(x,y)}{\partial x} = \lim_{h \to 0} \frac{f(x + h, y) - f(x,y)}{h}. \tag{1}$$

The partial derivative with respect to y is defined similarly:

$$\frac{\partial f}{\partial y} = \lim_{h \to 0} \frac{f(x, y + h) - f(x,y)}{h}; \tag{1'}$$

it is the derivative of the function of one variable $y \mapsto f(x,y)$, where x is a fixed number.

Since partial derivatives are, essentially, derivative of functions of one variable, we compute them by the differentiation rules learned earlier, taking care to remember in each case which variable is "kept fixed" and is to be treated like a constant and which is the "variable of differentiation."

EXAMPLES 1. Find the partial derivatives (at $x = 2$, $y = 3$) of

$$f(x,y) = 3x^3 y + 4xy^2 - 2x + 4y - 5.$$

SOLUTION   We have

$$f(x,3) = 9x^3 + 36x - 2x + 12 - 5 = 9x^3 + 34x + 7.$$

The derivative of this function of x is $27x^2 + 34$; for $x = 2$, we obtain 142. Thus

$$\left(\frac{\partial f}{\partial x} \right)_{x=2, \, y=3} = 142.$$

Next,

$$f(2,y) = 24y + 8y^2 - 4 + 4y - 5 = 8y^2 + 28y - 9.$$

This function of y has the derivative $16y + 28$. For $y = 3$, we obtain 76; hence

$$\left(\frac{\partial f}{\partial y} \right)_{x=2, \, y=3} = 76.$$

2. Find general formulas for the partial derivatives of the function $f(x,y)$ of Example 1.

ANSWER   We treat y as a constant and obtain, by the usual differentiation rules,

$$\frac{\partial f}{\partial x} = \frac{\partial(3x^3 y + 4xy^2 - 2x + 4y - 5)}{\partial x} = 9x^2 y + 4y^2 - 2.$$

Similarly, treating x as a fixed number, we have

$$\frac{\partial f}{\partial y} = \frac{\partial(3x^3 y + 4xy^2 - 2x + 4y - 5)}{\partial y} = 3x^3 + 8xy + 4.$$

Substituting into these formulas $x = 2$, $y = 3$, we have

$$\left(\frac{\partial f}{\partial x}\right)_{(2,3)} = 142, \qquad \left(\frac{\partial f}{\partial y}\right)_{(2,3)} = 76,$$

as before.

3. Find the partial derivatives of $f(x,y) = \sin(x^2 + y)$ at $x = 0$, $y = \pi$.
FIRST SOLUTION   We follow the method of Example 1 and write

$$\left(\frac{\partial \sin(x^2 + y)}{\partial x}\right)_{x=0,y=\pi} = \left(\frac{d \sin(x^2 + \pi)}{dx}\right)_{x=0} = (2x) \cos(x^2 + \pi) \bigg|_{x=0} = 0,$$

$$\left(\frac{\partial \sin(x^2 + y)}{\partial y}\right)_{x=0,y=\pi} = \left(\frac{d \sin y}{dy}\right)_{y=\pi} = (\cos y)_{y=\pi} = -1.$$

SECOND SOLUTION   We follow the method of Example 2 and obtain first the general formulas

$$\frac{\partial \sin(x^2 + y)}{\partial x} = 2x \cos(x^2 + y), \qquad \frac{\partial \sin(x^2 + y)}{\partial y} = \cos(x^2 + y).$$

Setting $x = 0$, $y = \pi$, we obtain for the two partial derivatives the values 0 and -1, respectively.

4. What are the partial derivatives of the function $f(x,y) = |x|$ at $x = y = 0$?
ANSWER   Since $f(x,0) = |x|$, and the function $x \mapsto |x|$ has no derivative at $x = 0$, there is no partial derivative with respect to x. Since $f(0,y) = 0$, we have $\partial f/\partial y = 0$ at $x = y = 0$.

5. We exhibit a function of two variables that has partial derivatives at a point, without being continuous at this point. Set

$$f(x,y) = 0 \quad \text{if } xy = 0, \qquad 1 \quad \text{if } xy \neq 0.$$

Then $\partial f/\partial x = \partial f/\partial y = 0$ at $(0,0)$, since $f(x,0) = f(0,y) = 0$. We observe that $f(0,0) = 0$ and that, in every interval around $(0,0)$, there are points (x,y) with $f(x,y) = 1$. Thus f is not continuous at $(0,0)$.

REMARK   Many symbols other than ∂ are used to denote partial derivatives. For instance, instead of $\partial f/\partial x$, we write f_x and $D_x f$. Similarly, $\partial f/\partial y$ is sometimes denoted by f_y or by $D_y f$. Thus, if $f(x,y) = 2x^3 y - xy^2$, we have

$$f_x(x,y) = 6x^2 y - y^2, \qquad f_x(1,2) = 12 - 4 = 8,$$
$$D_y f(x,y) = 2x^3 - 2xy, \qquad D_y f(1,2) = 2 - 4 = -2.$$

We shall use only the ∂ notation and the subscript notation, f_x and f_y.

PROBLEMS

In Problems 1 to 12 find the partial derivatives $\partial f/\partial x$ and $\partial f/\partial y$ of the given function $f(x,y)$.

1. $f(x,y) = x + 2y$.
2. $f(x,y) = x^2 y^5$.
3. $f(x,y) = x^2 y^3 - 2xy^2$.
4. $f(x,y) = \sqrt{1 + x^2 + y^2}$.
5. $f(x,y) = \sin(x^2 + y)$.
6. $f(x,y) = xy \cos(x + y)$.

7. $f(x,y) = (xy)^{-1} - \log y$.
8. $f(x,y) = \sin(x/y)$.
9. $f(x,y) = (1/x)e^{x^2+y^2}$.
10. $f(x,y) = \arctan(y/x)$.
11. $f(x,y) = \log(x \tan y)$.
12. $f(x,y) = (x^2 - y^2)/(x^2 + y^2)$.

13. Find $f_x(-1,2)$ and $f_y(-1,2)$ for the function in Problem 2.
14. Find $f_x(3,1)$ and $f_y(3,1)$ for the function in Problem 3.
15. Find $f_x(1,0)$ and $f_y(1,0)$ for the function in Problem 4.
16. For the function $f(x,y)$ defined in Problem 10, let $g(x) = f(x,3)$. Compute $g'(2)$ and compare with the value of $\partial f/\partial x$ at $x = 2$, $y = 3$.
17. Repeat Problem 16 for the function in Problem 11.
18. Repeat Problem 16 for the function in Problem 12.
19. If $z = \sqrt{x^2 + y^2}$, show that $xz_x + yz_y = z$.
20. If $\phi = \log \sqrt{a^2 + b^2}$, show that $a\phi_a + b\phi_b = 1$.

2.2 Continuously differentiable functions. Tangent plane and surface normal

Suppose that the function $f(x,y)$ is defined and has partial derivatives [say, near a point (x_0,y_0)], and that these partial derivatives, $\partial f/\partial x$ and $\partial f/\partial y$, are continuous functions of (x,y). Such an f is called **continuously differentiable.** Most functions of two variables occurring in applications of calculus are continuously differentiable. For instance, the function $f(x,y) = \sin(e^{x \cos y} + 2x^3 y)$ is continuously differentiable everywhere, since $\partial f/\partial x$ and $\partial f/\partial y$, which can be computed by the usual rules, are obtained by composition and addition and multiplication from functions of one variable that have derivatives of all orders.

Theorem 1   (linear approximation theorem for two variables).   *Let $f(x,y)$ be continuously differentiable near (x_0,y_0), and let z_0, a, b be the values of f and its partial derivatives at (x_0,y_0):*

$$z_0 = f(x_0,y_0), \qquad a = f_x(x_0,y_0), \qquad b = f_y(x_0,y_0). \tag{2}$$

Then

$$f(x,y) = z_0 + a(x - x_0) + b(y - y_0) + \sqrt{(x - x_0)^2 + (y - y_0)^2}\, r(x,y), \tag{3}$$

where

$$r(x,y) \text{ is continuous at } (x_0,y_0) \text{ and } r(x_0,y_0) = 0. \tag{4}$$

To appreciate the meaning of this theorem, note that

$$z = z_0 + a(x - x_0) + b(y - y_0) \tag{5}$$

is a linear function. The term $\sqrt{(x - x_0)^2 + (y - y_0)^2}\, r(x,y)$ in (3) is the error committed if, in computing the value of $f(x,y)$, we replace the given function f by the linear function (5). If the distance $\sqrt{(x - x_0)^2 + (y - y_0)^2}$ is small, then this error is *very* small, much smaller than the distance.

The proof of Theorem 1 will be found in §5.1.

We note a consequence of the theorem: *a function with continuous partial derivatives near a point is continuous at this point.* Indeed, the right side of (3) is a sum of continuous functions and hence continuous.

The graph of the linear function (5) is a plane, called the **tangent plane** to the surface $z = f(x,y)$ at the point $P = (x_0,y_0,z_0)$, with z_0 given by (2). The geometric meaning of Theorem 1 is shown in Figure 19.4. The plane and the surface pass through

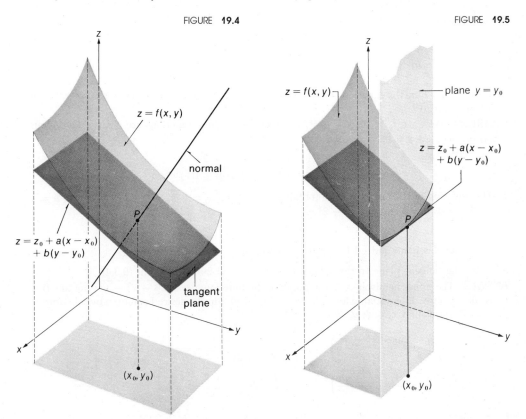

FIGURE **19.4**

FIGURE **19.5**

the same point P, and at this point the plane attaches itself to the surface as closely as possible—it "just touches" the surface, like the tangent to a curve "just touches" the curve.

Let us intersect the surface $z = f(x,y)$ and its tangent plane at (x_0,y_0) by the plane $y = y_0$; see Figure 19.5. The intersection of this plane with the surface is a curve—the graph of the function $z = f(x,y_0)$ of one variable x. The intersection of the tangent plane with $y = y_0$ is a line—the tangent to the curve. The slope of this tangent is a, the partial derivative $\partial f/\partial x$ at (x_0,y_0). The partial derivative $\partial f/\partial y$ has a similar geometric meaning.

The line through P perpendicular to the tangent plane is called the **surface normal** to the graph (see Figure 19.4).

The surface normal admits the parametric representation

$$x = x_0 + at, \qquad y = y_0 + bt, \qquad z = z_0 - t, \tag{6}$$

where z_0, a, and b are given by (2).

To verify this, note that the line (6) contains the point (x_0,y_0,z_0); this point corresponds to the value $t = 0$. Equation (5) of the tangent plane is of the form $Ax + By + Cz + D = 0$ with $A = a$, $B = b$, $C = -1$. Hence the acute angle ϕ between the line (6) and the normal to the tangent plane is computed from Equation (23) in Chapter 18, §4.3, by setting $A = a$, $B = b$, $C = c = -1$. We obtain $\cos \phi = 1$, $\phi = 0$, as asserted. The result just obtained may be stated as follows: *the vector with components*

$$\langle f_x, f_y, -1 \rangle \tag{7}$$

is perpendicular to the surface $z = f(x,y)$, *at the point* (x,y,z).

EXAMPLES 1. Find the equation of the tangent plane to the surface $z = f(x,y) = \sin x + e^{xy} + y$ at the point $(0,2,3)$.

SOLUTION   We have $f_x = \cos x + ye^{xy}$, $f_y = xe^{xy} + 1$. These functions are clearly continuous everywhere. At $x = 0$, $y = 2$, we find

$$f(0,2) = z_0 = \sin 0 + e^0 + 2 = 3,$$
$$f_x(0,2) = \cos 0 + 2e^0 = 3, \qquad f_y(0,2) = 0 \cdot e^0 + 1 = 1.$$

The equation of the tangent plane [obtained by setting $z_0 = 3$, $x_0 = 0$, $y_0 = 2$, $a = 3$, $b = 1$ in (5)] reads $z = 3 + 3x + (y - 2)$ or

$$z = 3x + y + 1.$$

Note that this plane indeed passes through $(0,2,3)$.

2. Find the tangent plane to the sphere $x^2 + y^2 + z^2 = 29$ at the point $(2,3,4)$.

ANSWER   The upper hemisphere is the graph of the function $z = \sqrt{29 - x^2 - y^2}$; the function is continuously differentiable for $x^2 + y^2 < 29$. We obtain the desired equation by substituting into (5) the values $x_0 = 2$, $y_0 = 3$, $z_0 = 4$, and

$$a = \left(\frac{\partial \sqrt{29 - x^2 - y^2}}{\partial x} \right)_{x=2, y=3} = \left(-\frac{x}{\sqrt{29 - x^2 - y^2}} \right)_{x=2, y=3} = -\frac{1}{2},$$

$$b = \left(\frac{\partial \sqrt{29 - x^2 - y^2}}{\partial y} \right)_{x=2, y=3} = \left(-\frac{y}{\sqrt{29 - x^2 - y^2}} \right)_{x=2, y=3} = -\frac{3}{4}.$$

The desired equation reads $z = 4 - \frac{1}{2}(x - 2) - \frac{3}{4}(y - 3)$ or $2x + 3y + 4z - 29 = 0$.

3. Find the line normal to the surface $z = e^x \cos y$ through the point with the coordinates $x = 1$, $y = \pi$, $z = 1/e$.

ANSWER   Using the formulas (6) with $x_0 = 1$, $y_0 = \pi$, $z_0 = 1/e$, and noting that $\partial z/\partial x = \cos y\, e^{x \cos y}$, $\partial z/\partial y = -x \sin y\, e^{x \cos y}$, we obtain $a = -1/e$, $b = 0$, so that the normal has a parametric representation

$$x = 1 - \frac{t}{e}, \qquad y = \pi, \qquad z = \frac{1}{e} - t.$$

4. Find the intersection of the normal to the surface $z = x^4y - 5xy^6$ at the point $(2,1,6)$ and the plane $x = 0$.

SOLUTION   We have $\partial z/\partial x = 4x^3y - 5y^6$, $\partial z/\partial y = x^4 - 30xy^5$. Hence, at $x_0 = 2$, $y_0 = 1$, $z_0 = 6$, we obtain $a = 27$, $b = -44$. By (6), the normal has a parametric representation

$$x = 2 + 27t, \qquad y = 1 - 44t, \qquad z = 6 - t.$$

If $x = 0$, then $t = -2/27$; hence $y = 115/27$, $z = 164/27$. The desired intersection point is $(0,115/27,164/27)$.

REMARK   In the case of functions of one variable, the existence of a derivative tells us a great deal about the function. If $f'(x_0)$ exists, then the function $x \mapsto f(x)$ is continuous at x_0, and close to x_0 it can be approximated, with only a very small error, by a linear function (compare the linear approximation theorem, Theorem 1, in Chapter 4, §1.6). In the case of functions of two variables, we cannot expect to learn so much about a function $(x,y) \mapsto f(x,y)$ from the mere fact that f has partial derivatives at a point (x_0,y_0). Indeed, in computing the partial derivatives at (x_0,y_0), we make use only of the values of the function on the vertical and the horizontal segments through this point. Thus it is not surprising that the mere existence of partial derivatives does not even imply continuity. (See Example 5 in §2.1.)

Because of this, the linear approximation theorem for functions of two variables requires a stronger hypothesis: existence and continuity of partial derivatives.

PROBLEMS

In Problems 1 to 8 find the equation of the tangent plane and the equation of the normal line to the given surface at the given point P.

1. $z = x^2 + y^2$, $P = (1,1,2)$.

2. $z = 5x - y^2$, $P = (4,2,16)$.

3. $z = x^2 + 2y^2 - 16$, $P = (3,2,1)$.

4. $z = xy$, $P = (3,-2,-6)$.

5. $z = \dfrac{3}{1 + x^2 + y^2}$, $P = (1,1,1)$.

6. $z = x^3 + y^3 + 3xy$, $P = (1,-1,-3)$.

7. $z = \sqrt{1 - x^2 - y^2}$, $P = (\tfrac{2}{3},\tfrac{2}{3},\tfrac{1}{3})$.

8. $z = e^{x^2+y^2}$, $P = (0,0,1)$.

9. Show that the surfaces $z = xy/(4x - y)$ and $z = \sqrt{(5x - y)/3}$ intersect at right angles at the point $(1,2,1)$. (This means that both surfaces pass through this point and their normals are perpendicular.)

10. Show that the tangent plane to the sphere $x^2 + y^2 + z^2 = 1$, at a point (x_0,y_0,z_0) on this sphere, has the equation $xx_0 + yy_0 + zz_0 = 1$. (Assume that $z_0 > 0$.)

11. Show that the tangent plane to the ellipsoid $(x/a)^2 + (y/b)^2 + (z/c)^2 = 1$, at a point (x_0,y_0,z_0) on it, has the equation $xx_0/a^2 + yy_0/b^2 + zz_0/c^2 = 1$. (Assume that $z_0 > 0$.)

12. Find the equation of the tangent plane to the hyperboloid $(x/a)^2 + (y/b)^2 - (z/c)^2 = 1$ at the point (x_0,y_0,z_0). Do the same for the hyperboloid $(x/a)^2 - (y/b)^2 - (z/c)^2 = 1$. (Assume $z_0 > 0$.)

2.3 The chain rule for two variables

The rule for finding partial derivatives of composite functions is similar to, although somewhat more complicated than, the corresponding rule for functions of one variable. Given three (continuously differentiable) functions

$$z = F(x,y), \qquad x = \phi(u,v), \qquad y = \psi(u,v),$$

we form the composed function

$$z = F(\phi(u,v),\psi(u,v)). \tag{8}$$

[For example: $z = xy$, $x = u + v$, $y = u - v$. Substituting the expressions for x and for y into that for z, we obtain the composed function $z = (u + v)(u - v) = u^2 - v^2$.] We claim that the partial derivative of the composed function (8) with respect to the new variable u is given by the formula

$$\frac{\partial z}{\partial u} = \frac{\partial F}{\partial x}\frac{\partial \phi}{\partial u} + \frac{\partial F}{\partial y}\frac{\partial \psi}{\partial u}, \tag{9'}$$

or, using the subscript notation for derivatives,

$$z_u = F_x\phi_u + F_y\psi_u.$$

[In our example, $F_x\phi_u + F_y\psi_u = (y)(1) + (x)(1) = (u - v) + (u + v) = 2u$; this is indeed the partial derivative of $z = u^2 - v^2$ with respect to u.] Similarly,

$$\frac{\partial z}{\partial v} = \frac{\partial F}{\partial x}\frac{\partial \phi}{\partial v} + \frac{\partial F}{\partial y}\frac{\partial \psi}{\partial v} \tag{9''}$$

or

$$z_v = F_x\phi_v + F_y\psi_v.$$

Note the structure of the formulas (9') and (9''). The partial derivative of z with respect to a "new" variable (u or v) is the sum of products: the partial derivatives of z with respect to both "old" variables (x and y), each multiplied by the partial derivative of an old variable [$x = \phi(u,v)$ and $y = \psi(u,v)$] with respect to the new variable (u or v). We now state our claim formally.

Theorem 2 (chain rule). *Let $F(x,y)$ be continuously differentiable at (x_0,y_0), let $x = \phi(u,v)$ and $y = \psi(u,v)$ be continuously differentiable at (u_0,v_0), and assume that $\phi(u_0,v_0) = x_0$, $\psi(u_0,v_0) = y_0$. Then the composed function $z = F(\phi(u,v),\psi(u,v))$ is continuously differentiable at (u_0,v_0) and*

$$\frac{\partial z}{\partial u} = \frac{\partial F}{\partial x}\frac{\partial \phi}{\partial u} + \frac{\partial F}{\partial y}\frac{\partial \psi}{\partial u}, \qquad \frac{\partial z}{\partial v} = \frac{\partial F}{\partial x}\frac{\partial \phi}{\partial v} + \frac{\partial F}{\partial y}\frac{\partial \psi}{\partial v}, \tag{9}$$

where the partial derivatives with respect to u, v are computed at (u_0,v_0), those with respect to x, y at (x_0,y_0).

We can also write the chain rule in the form

$$z_u = F_x x_u + F_y y_u, \qquad z_v = F_x x_v + F_y y_v. \tag{9'''}$$

Theorem 2 follows from Theorem 1 in the same way as the chain rule for functions of one variable follows from Theorem 1 in Chapter 4, §1.6 (see the proof in Chapter 6, §2.3). We omit the details.

EXAMPLES   1.   Suppose that $F(x,y) = 3x^2y$, $x = \phi(u,v) = u + v$ and $y = \psi(u,v) = uv$. Set $z = F(\phi(u,v), \psi(u,v))$ and find the partial derivatives $\partial z/\partial u$, $\partial z/\partial v$ at $u = 2$, $v = 3$.
FIRST SOLUTION   We have

$$z = 3x^2y = 3(u + v)^2uv = 3u^3v + 6u^2v^2 + 3uv^3$$

so that

$$z_u = 9u^2v + 12uv^2 + 3v^3, \qquad z_v = 3u^3 + 12u^2v + 9uv^2$$

and $z_u(2,3) = 405$, $z_v(2,3) = 330$.
SECOND SOLUTION   We have

$$F_x = 6xy, \qquad F_y = 3x^2, \qquad \phi_u = 1, \qquad \phi_v = 1, \qquad \psi_u = v, \qquad \psi_v = u.$$

For $u = 2$, $v = 3$, we obtain

$$\phi_u = 1, \qquad \phi_v = 1, \qquad \psi_u = 3, \qquad \psi_v = 2$$

and, setting $x = \phi(u,v) = u + v$, $y = \psi(u,v) = uv$,

$$x = 5, \qquad y = 6, \qquad F_x = 180, \qquad F_y = 75.$$

Thus, by the chain rule,

$$z_u = F_x\phi_u + F_y\psi_u = 180 \cdot 1 + 75 \cdot 3 = 405,$$
$$z_v = F_x\phi_v + F_y\psi_v = 180 \cdot 1 + 75 \cdot 2 = 330,$$

as before.

2. Find the partial derivatives with respect to u and v of $z = e^{xy}$ where $x = u^2$ and $y = uv$.
ANSWER   Direct substitution gives $z = e^{u^3v}$ so that $z_u = 3u^2ve^{u^3v}$ and $z_v = u^3e^{u^3v}$. If we apply the chain rule, we must note that a function of u alone may be considered as a function of (u,v) with a partial derivative 0 with respect to v. In our case, we have

$$z_x = ye^{xy}, \qquad z_y = xe^{xy}, \qquad x_u = 2u, \qquad x_v = 0, \qquad y_u = v, \qquad y_v = u$$

so that

$$z_u = z_x x_u + z_y y_u = ye^{xy}2u + xe^{xy}v = 3u^2ve^{u^3v},$$
$$z_v = z_x x_v + z_y y_v = ye^{xy}0 + xe^{xy}u = u^3e^{u^3v}$$

as before.

From now on we consider only continuously differentiable functions, unless stated otherwise.

PROBLEMS

For each of Problems 1 to 6 find $\partial z/\partial u$ and $\partial z/\partial v$. Do each problem twice, once by direct substitution and once by use of the chain rule.

1. $z = x^2 + xy + y^2, \quad x = 2u - v, \quad y = u - 2v.$
2. $z = x^2 - y^2, \quad x = u^2 + v^2, \quad y = 2uv.$
3. $z = x^2 + y^2, \quad x = u\cos v, \quad y = u\sin v.$
4. $z = xe^y - ye^x, \quad x = uv, \quad y = u^2 - v^2.$
5. $z = x/(x^2 + y^2), \quad x = u\cos v, \quad y = u\sin v.$
6. $z = \log(x^2 + y^2), \quad x = ue^{-v}, \quad y = ue^v.$

7. If $u = \dfrac{1}{x+y}, x = \sin(s+t), y = \cos(s+t)$, find $\partial u/\partial s$ and $\partial u/\partial t$ at $s = \dfrac{\pi}{4}, t = \dfrac{\pi}{4}$.

8. If $u = e^{x/y}, x = s\cos t, y = t\cos s$, find u_s and u_t at $s = 0, t = \dfrac{\pi}{2}$.

9. If $u = \dfrac{x-y}{1+xy}, x = \tan s, y = \tan t$, find u_s and u_t at $s = 0, t = \dfrac{\pi}{4}$.

10. If $w = \arcsin uv, u = \log x, v = \log y$, find $\partial w/\partial x$ and $\partial w/\partial y$ at $x = 1, y = 1$.

11. If $w = u^2 + v^2, u = \dfrac{x+1}{y}, v = \dfrac{y+1}{x}$, find w_x and w_y at $x = 1, y = 2$.

12. If $w = \arctan\dfrac{v}{u}, u = x + \sin y, v = x + \cos y$, find w_x and w_y at $x = 0, y = \dfrac{\pi}{2}$.

13. If $f(t)$ is a differentiable function, and $F(x,y) = f(x - 2y)$, show that $2F_x(x,y) + F_y(x,y) = 0$. [Hint: Let $t = x - 2y$.]

14. If $f(t)$ is a differentiable function, and $F(x,y) = f(x^2 + y^2)$, show that $xF_y(x,y) - yF_x(x,y) = 0$.

15. If $f(t)$ is a differentiable function and $f'(t) > 0$, and if $\phi(u,v) = f(u + v)$, what can you say about $\phi_u + \phi_v$?

*16. Carry out the proof of Theorem 2.

2.4   Directional derivatives and gradients in the plane

The chain rule is applicable also when we form a function of one variable, $s \mapsto f(s)$, by substituting into a function $F(x,y)$ of two variables, two functions $x = \phi(s), y = \psi(s)$ of one variable. The derivative of the function

$$z = F(\phi(s),\psi(s)) \tag{12}$$

is

$$\frac{dz}{ds} = \frac{\partial F}{\partial x}\frac{dx}{ds} + \frac{\partial F}{\partial y}\frac{dy}{ds} \tag{13}$$

or

$$z'(s) = F_x((\phi(s),\psi(s))\phi'(s) + F_y(\phi(s),\psi(s))\psi'(s); \tag{13'}$$

here we used the usual notations (d instead of ∂, and the prime $'$) for functions of one variable.

Now let $F(x,y)$ be a continuously differentiable function. We want to compute the rate of change of the function as the point (x,y) moves, from (x_0,y_0) in some direction. The direction is best described by a unit vector

$$\mathbf{u} = \cos\theta\mathbf{i} + \sin\theta\mathbf{j} \tag{14}$$

(see Figure 19.6). A point reached by moving from (x_0,y_0) in the direction of $\mathbf{u}$, by a distance s, has Cartesian coordinates $x = x_0 + s\cos\theta$, $y = y_0 + s\sin\theta$; the value of F at this point is

$$F(x_0 + s\cos\theta, y_0 + s\sin\theta). \tag{15}$$

FIGURE **19.6**

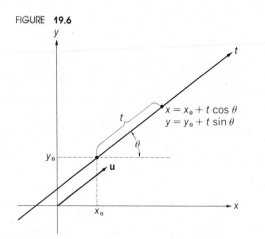

The derivative of (15) with respect to s, at $s = 0$, is called the **directional derivative** of F in the direction $\mathbf{u}$, at (x_0,y_0). It equals, by the chain rule (13),

$$\left.\frac{dF(x_0 + s\cos\theta, y_0 + s\sin\theta)}{ds}\right|_{s=0} = F_x(x_0,y_0)\cos\theta + F_y(x_0,y_0)\sin\theta;$$

or, more simply,

$$F_x\cos\theta + F_y\sin\theta. \tag{16}$$

Setting $\theta = 0$, we see that the directional derivative in the x-direction is F_x. Similarly, the directional derivative in the y-direction is F_y (set $\theta = 90°$), the one in the negative x-direction is $-F_x$ (set $\theta = 180°$), and so forth.

Now we define, at every relevant point (x,y), the vector with components F_x,F_y;

this vector is called the **gradient** of F and is denoted by grad F. Thus grad $F = \langle F_x, F_y \rangle$ or

$$\operatorname{grad} F = \frac{\partial F}{\partial x}\mathbf{i} + \frac{\partial F}{\partial y}\mathbf{j}. \tag{17}$$

Note that grad F is a rule that assigns a vector to every point in the domain of definition of F; it is an example of a **vector-valued function of two variables.** Recalling the definition of the inner product (see Chapter 18, §3.1), we write the directional derivative (16) as

$$(\operatorname{grad} F, \mathbf{u}) \tag{16'}$$

This is equal to $|\operatorname{grad} F| \, |\mathbf{u}| \cos \alpha = |\operatorname{grad} F| \cos \alpha$, where α is the angle between $\mathbf{u}$ and grad F. The directional derivative, at a given point, is largest when $\cos \alpha = 1$, that is, when $\mathbf{u}$ has the direction of grad F. Thus *the gradient points in the direction of steepest increase (or, as we say, steepest ascent). The magnitude of the gradient* $|\operatorname{grad} F| = \sqrt{F_x^2 + F_y^2}$ *is the value of the directional derivative in the direction of steepest ascent.*

Here we have tacitly assumed that, at the point considered, F_x and F_y are not both zero. Points with grad $F = 0$, that is, with $F_x = F_y = 0$, are called **critical points.** At such points, the directional derivatives in all directions are 0.

EXAMPLE   For the function $F = x^2y - 2x - y$ find (a) the gradient, (b) the critical points, (c) the unit vector pointing in the direction of steepest increase at $(3,1)$, and (d) the directional derivative at $(3,1)$ in the direction $\mathbf{u} = \mathbf{i}\cos 45° + \mathbf{j}\sin 45°$.

ANSWER   (a) Since $F_x = 2xy - 2$ and $F_y = x^2 - 1$, grad $F = 2(xy - 1)\mathbf{i} + (x^2 - 1)\mathbf{j}$.

(b) Note that $F_y = 0$ if $x = 1$ or $x = -1$. If $x = 1$, then $F_x = 2(y - 1) = 0$ for $y = 1$, and if $x = -1$, $F_x = 2(-y - 1) = 0$ for $y = -1$. Thus there are two critical points, $(1,1)$ and $(-1,-1)$.

(c) At $(3,1)$ we have that $F_x = 4$, $F_y = 8$, grad $F = 4\mathbf{i} + 8\mathbf{j} = 4(\mathbf{i} + 2\mathbf{j})$. The unit vector in the direction of grad F is

$$\frac{1}{\sqrt{1 + 4}}(\mathbf{i} + 2\mathbf{j}) = \frac{1}{\sqrt{5}}\mathbf{i} + \frac{2}{\sqrt{5}}\mathbf{j}.$$

This vector points in the direction of steepest increase. It can be written as $\cos \alpha \mathbf{i} + \sin \alpha \mathbf{j}$, where $\cos \alpha = 1/\sqrt{5}$, $\sin \alpha = 2/\sqrt{5}$, that is, $\alpha \approx 63°30'$.

(d) The desired directional derivative is, by (16), $F_x \cos 45° + F_y \sin 45° = 4(1/\sqrt{2}) + 8(1/\sqrt{2}) = 12/\sqrt{2} \approx 8.484$.

PROBLEMS

For each of the functions z given in Problems 1 to 5 find dz/ds.

1. $z = x^2 + 3xy - y^2$, $\quad x = \cos 2s$, $\quad y = \sin 2s$.
2. $z = x^2 + y^{-2}$, $\quad x = e^s \cos s$, $\quad y = 1 + \log s$.
3. $z = \log(x^2 + y^2)$, $\quad x = e^{-s}$, $\quad y = e^s$.
4. $z = \sin(xy + 1)$, $\quad x = \dfrac{1}{1 + s}$, $\quad y = \dfrac{s}{1 + s}$.
5. $z = \log \dfrac{x}{y}$, $\quad x = \tan s$, $\quad y = \sec^2 s$.

For each of the functions $F(x,y)$ given in Problems 6 to 10 find (a) the gradient, (b) the critical points, (c) the unit vector pointing in the direction of steepest increase at the given point P, and (d) the directional derivative at P in the given direction θ.

6. $F = 3x^2 - 6y^2$, $P = (8,2)$, $\theta = 120°$.

7. $F = ye^{2x}$, $P = (0,4)$, $\theta = 135°$.

8. $F = x^2 + y^2 - 6x + 4y + 25$, $P = (1,-4)$, $\theta = \pi/4$.

9. $F = x^2 - 2xy + 4y^2 + 6x$, $P = (2,1)$, $\theta = 4\pi/3$.

10. $F = e^x \cos y$, $P = (0,0)$, $\theta = \pi/3$.

2.5   Level curves

Let us think of a continuously differentiable function F of two variables as describing a mountainous landscape: $F(x,y)$ is the height (elevation) of the point above (x,y); the graph $z = F(x,y)$ is the surface of the mountain. A **level curve** is a smooth curve (compare Chapter 16, §2.2) $x = \phi(t)$, $y = \psi(t)$ such that $F(\phi(t),\psi(t))$ is a constant, say, c. In other words, a level curve is the projection on the x, y plane of a trail of constant elevation. We assume that *through every point (x_0,y_0) that is not a critical point of F there passes a unique level curve.* (If the partial derivatives F_x and F_y are themselves continuously differentiable, this need not be assumed but can be proved. We shall not do so here.)

Let (x_0,y_0) be a noncritical point of F. The level curve through this point, $x = \phi(t)$, $y = \psi(t)$, with $t = t_0$ corresponding to (x_0,y_0), has the property

$$F(\phi(t),\psi(t)) = c$$

(this property makes it into a level curve). We differentiate both sides of this equation with respect to t and obtain, for $t = t_0$,

$$F_x(x_0,y_0)\phi'(t_0) + F_y(x_0,y_0)\psi'(t_0) = 0. \tag{18}$$

Now the vector $\mathbf{v} = \phi'(t_0)\mathbf{i} + \psi'(t_0)\mathbf{j}$ is a tangent vector to the level curve at (x_0,y_0); it may be written as $\mathbf{v} = \sqrt{\phi'(t_0)^2 + \psi'(t_0)^2}\,\mathbf{T}$, where $\mathbf{T}$ is the unit tangent vector (see Chapter 17, §2.3). Relation (18) says that $(\mathbf{v},\text{grad } F) = 0$, so that $\mathbf{v}$ is perpendicular

FIGURE **19.7**

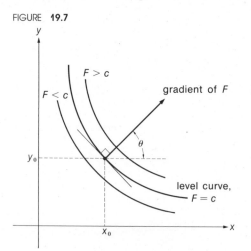

FIGURE **19.8**   FIGURE **19.9**

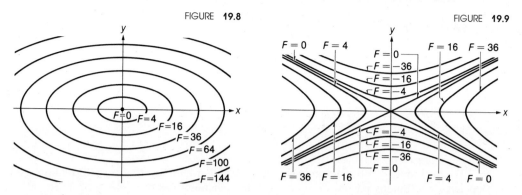

to grad F. We conclude that *the level curve through a (noncritical) point is perpendicular to the gradient at this point* (see Figure 19.7). This is not surprising: the direction of the gradient is that of steepest ascent; the direction perpendicular to it is the level direction; the derivative in this direction is 0.

EXAMPLES 1. The level curves of the function $F(x,y) = x^2 + 4y^2$ are ellipses $x^2 + 4y^2 = c$. We have $F_x = 2x$, $F_y = 8y$, so that there is one critical point, $x = y = 0$. There is no level curve through this point. Some level curves are shown in Figure 19.8.

2. The level curve of the function $F(x,y) = x^2 + 4y^2$ passing through the point $(2, \sqrt{3})$ is the ellipse $x^2 + 4y^2 = 16$. To find a tangent vector $\mathbf{v}$ through the point $(2, \sqrt{3})$, let $\mathbf{v} = v_1\mathbf{i} + v_2\mathbf{j}$ and substitute into $(\mathbf{v}, \text{grad } F) = 0$. Since grad $F = 4\mathbf{i} + 8\sqrt{3}\mathbf{j}$ at the point $(2, \sqrt{3})$, $(\mathbf{v}, \text{grad } F) = (v_1\mathbf{i} + v_2\mathbf{j}) \cdot (4\mathbf{i} + 8\sqrt{3}\mathbf{j}) = 4v_1 + 8\sqrt{3}v_2 = 0$. Hence $v_1 = -2\sqrt{3}v_2$. For $v_2 = 1$, $v_1 = -2\sqrt{3}$ and a tangent vector is $\mathbf{v} = -2\sqrt{3}\mathbf{i} + \mathbf{j}$.

3. The level curves of the function $F(x,y) = x^2 - 4y^2$ are hyperbolas $x^2 - 4y^2 = c$; the coordinate axes are also level curves. We have $F_x = 2x$, $F_y = -8y$ and $(0,0)$ is the only critical point. Two level curves intersect at this point. Some level curves are shown in Figure 19.9.

PROBLEMS

1. For the function F and point P given in Problem 6, §2.4, write the equation of the level curve, and find a tangent vector through the point.
2. Repeat Problem 1 above for the function in Problem 7, §2.4.
3. Repeat Problem 1 above for the function in Problem 8, §2.4.
4. Repeat Problem 1 above for the function in Problem 9, §2.4.
5. Repeat Problem 1 above for the function in Problem 10, §2.4.

2.6 Differentiation of functions of three or more variables

The definition of partial derivatives extends to functions of three and more variables.
For instance, for the function $f(x,y,z)$, the partial derivative with respect to z at (x,y,z) is the derivative of the function of one variable $z \mapsto f(x,y,z)$. This partial derivative is denoted by any of the symbols

$$\frac{\partial f}{\partial z} \quad \text{or} \quad f_z(x,y,z)$$

and also by $D_z f(x,y,z)$. Thus

$$f_z(x,y,z) = \lim_{h \to 0} \frac{f(x,y,z+h) - f(x,y,z)}{h}.$$

To specify the value of the partial derivatives at some point (x_0, y_0, z_0), we write

$$\left(\frac{\partial f}{\partial x} \right)_{(x_0, y_0, z_0)} \qquad \text{or} \qquad f_x(x_0, y_0, z_0).$$

For functions of n variables, the definition is analogous.

We deal mostly with continuously differentiable functions, that is, with functions that have continuous partial derivatives. For such functions the analog of Theorem 1 is valid. So is an analog of Theorem 2, the chain rule. The derivatives of the composed function

$$z = F(\phi_1(u_1, \ldots, u_m), \ldots, \phi_n(u_1, \ldots, u_m)) \tag{19}$$

are given by the formulas

$$\frac{\partial z}{\partial u_j} = \frac{\partial F}{\partial x_1} \frac{\partial x_1}{\partial u_j} + \frac{\partial F}{\partial x_2} \frac{\partial x_2}{\partial u_j} + \cdots + \frac{\partial F}{\partial x_n} \frac{\partial x_n}{\partial u_j}. \tag{20}$$

The same rule applies also if some of the ϕ are functions of only some of the variables u, since a function of less than m variables may be considered as a function of all m variables, with partial derivatives 0 with respect to the "missing" variables.

As an application of the chain rule, suppose $z = F(x_1, x_2, x_3)$ where $x_1 = \phi(u_1, u_2)$, $x_2 = \phi_2(u_1, u_2)$, and $x_3 = \phi_3(u_1, u_2)$. Then, by Equation (20), we have the partial derivatives

$$\frac{\partial z}{\partial u_1} = \frac{\partial F}{\partial x_1} \frac{\partial x_1}{\partial u_1} + \frac{\partial F}{\partial x_2} \frac{\partial x_2}{\partial u_1} + \frac{\partial F}{\partial x_3} \frac{\partial x_3}{\partial u_1}, \qquad \frac{\partial z}{\partial u_2} = \frac{\partial F}{\partial x_1} \frac{\partial x_1}{\partial u_2} + \frac{\partial F}{\partial x_2} \frac{\partial x_2}{\partial u_2} + \frac{\partial F}{\partial x_3} \frac{\partial x_3}{\partial u_2}.$$

As another application, suppose $z = F(x_1, x_2)$ where $x = \phi_1(u_1, u_2, u_3)$ and $x_2 = \phi_2(u_1, u_2, u_3)$. Then, by Equation (20), we have

$$\frac{\partial z}{\partial u_1} = \frac{\partial F}{\partial x_1} \frac{\partial x_1}{\partial u_1} + \frac{\partial F}{\partial x_2} \frac{\partial x_2}{\partial u_1},$$

$$\frac{\partial z}{\partial u_2} = \frac{\partial F}{\partial x_1} \frac{\partial x_1}{\partial u_2} + \frac{\partial F}{\partial x_2} \frac{\partial x_2}{\partial u_2},$$

$$\frac{\partial z}{\partial u_3} = \frac{\partial F}{\partial x_1} \frac{\partial x_1}{\partial u_3} + \frac{\partial F}{\partial x_2} \frac{\partial x_2}{\partial u_3}.$$

Other cases follow the same pattern.

EXAMPLES   1.   Find the partial derivatives for the function $f(x,y,z) = x^2 y^2 z^2$.
ANSWER   Treating y and z as constants and differentiating, we obtain $f_x(x,y,z) = 2xy^2 z^2$. Proceeding in a similar way, we obtain $f_y(x,y,z) = 2x^2 yz^2$ and $f_z(x,y,z) = 2x^2 y^2 z$.

2.   Find the partial derivative $\partial G / \partial t$ for the function $G(x,y,z,t) = (x^2 + y^2 + z^2 - c^2 t^2)^{1/2}$, where c is a constant.
ANSWER   Treating x, y, and z as constants and differentiating with respect to t, we find that $\partial G / \partial t = -c^2 t (x^2 + y^2 + z^2 - c^2 t^2)^{-1/2}$.

3. Suppose $w = x^2 - y^2$, where $x = rs \cos t$ and $y = rs \sin t$. What is $\partial w/\partial r$ in terms of r, s, and t?

ANSWER   Let $w = F(x,y)$, $x = \phi_1(r,s,t) = rs \cos t$, and $y = \phi_2(r,s,t) = rs \sin t$. By the chain rule,

$$\frac{\partial w}{\partial r} = \frac{\partial w}{\partial x}\frac{\partial x}{\partial r} + \frac{\partial w}{\partial y}\frac{\partial y}{\partial r}.$$

Hence

$$\frac{\partial w}{\partial r} = (2x)(s \cos t) + (-2y)(s \sin t) = 2s(x \cos t - y \sin t).$$

Substituting for x and y, we find that

$$\frac{\partial w}{\partial r} = 2s(rs \cos^2 t - rs \sin^2 t) = 2rs^2 \cos 2t.$$

Calculating partial derivatives of composite functions can be facilitated by using the language of differentials discussed next.

PROBLEMS

In Problems 1 to 8 find the partial derivatives $f_x(x,y,z)$, $f_y(x,y,z)$, and $f_z(x,y,z)$ of the given functions.

1. $f = xyz$.

2. $f = 2xz^2 + yz^3 - xyz$.

3. $f = x^2 \tan yz$.

4. $f = e^{x + \sin yz}$.

5. $f = (\cos xy)(\cos yz)$.

6. $f = \dfrac{1}{x^2 + 2y^2 + 4z^2}$.

7. $f = \dfrac{x^2 - y^2}{y^2 + z^2}$.

8. $f = \arctan(1 + xyz)$.

9. If $f(x,y,z) = e^{xy + yz + zx}$, find f_x, f_y, and f_z at $x = 1$, $y = 2$, $z = -1$.

10. If $f(x,y,z) = \cos x \cos y \cos z$, find f_x, f_y, and f_z at $x = 0$, $y = \dfrac{\pi}{4}$, $z = \dfrac{\pi}{2}$.

In Problems 11 to 14 find the partial derivatives of $G_x(x,y,z,t)$, $G_y(x,y,z,t)$, $G_z(x,y,z,t)$, and $G_t(x,y,z,t)$ of the given functions.

11. $G = xyzt^2$.

12. $G = \sin(x + y) \cos(z - t)$.

13. $G = \dfrac{1}{\sqrt{x^2 + y^2 + z^2 - t^2}}$.

14. $G = \dfrac{z^2 - t^2}{1 + x^2 + y^2}$.

In Problems 15 to 20 find $\partial w/\partial r$, $\partial w/\partial s$, and $\partial w/\partial t$ using the chain rule.

15. $w = \frac{1}{2}(x^2 + y^2)$, $x = r + e^{st}$, $y = r - e^{st}$.

16. $w = x \log y$, $x = rst^2$, $y = rst$.

17. $w = \dfrac{xy}{x^2 + y^2}$, $x = r + 2s + t$, $y = r - 2s + t$.

18. $w = x^2 - 2y^2 + z^2$, $x = r^2 - s^2$, $y = s^2 - t^2$, $z = t^2 - r^2$.

19. $w = xy + yz + xz$, $x = e^{rs} \cos t$, $y = e^{rs} \sin t$, $z = e^{rs}$.

20. $w = x^2 + y^2 + z^2$, $x = r \sin s \cos t$, $y = r \sin s \sin t$, $z = r \cos s$.

2.7   Differentials for functions of several variables

If $\phi(t)$ is a function of one variable, $d\phi$ is an abbreviation for the expression $\phi'(t)\,dt$; we saw in Chapter 8, §3.3, how useful this notation is. We now extend the language to differentials to functions of several variables. If $f(x_1,x_2,\ldots,x_n)$ is such a function, we define, formally, its differential as

$$df = \frac{\partial f}{\partial x_1}\,dx_1 + \frac{\partial f}{\partial x_2}\,dx_2 + \cdots + \frac{\partial f}{\partial x_n}\,dx_n, \tag{21}$$

and we calculate as if the symbols dx_1, dx_2, $\ldots$ were names of numbers. "Formally" means: in calculating, we use the left-hand side of (21) as an abbreviation for the right-hand side.

The chain rule (compare §2.6) is "built into" this language. If the x_i's are functions of m new variables $u_1, u_2, \ldots, u_m$, that is, if $x_i = \phi_i(u_1,\ldots,u_m)$, $i = 1, 2, \ldots, n$, then

$$dx_i = \frac{\partial \phi_i}{\partial u_1}\,du_1 + \frac{\partial \phi_i}{\partial u_2}\,du_2 + \cdots + \frac{\partial \phi_i}{\partial u_m}\,du_m, \qquad i = 1, 2, \ldots, n.$$

Let us substitute these expressions into (21):

$$df = \frac{\partial f}{\partial x_1}\left(\frac{\partial x_1}{\partial u_1}\,du_1 + \frac{\partial x_1}{\partial u_2}\,du_2 + \cdots + \frac{\partial x_1}{\partial u_m}\,dx_m\right)$$

$$+ \frac{\partial f}{\partial x_2}\left(\frac{\partial x_2}{\partial u_1}\,du_1 + \frac{\partial x_2}{\partial u_2}\,du_2 + \cdots + \frac{\partial x_2}{\partial u_m}\,du_m\right)$$

$$+ \cdots$$

$$+ \frac{\partial f}{\partial x_n}\left(\frac{\partial x_n}{\partial u_1}\,du_1 + \frac{\partial x_n}{\partial u_2}\,du_2 + \cdots + \frac{\partial x_n}{\partial u_m}\,du_m\right).$$

If we carry out the multiplications and collect terms, we see that

$$df = \left(\frac{\partial f}{\partial x_1}\frac{\partial x_1}{\partial u_1} + \frac{\partial f}{\partial x_2}\frac{\partial x_2}{\partial u_1} + \cdots + \frac{\partial f}{\partial x_n}\frac{\partial x_n}{\partial u_1}\right)du_1$$

$$+ \left(\frac{\partial f}{\partial x_1}\frac{\partial x_1}{\partial u_2} + \frac{\partial f}{\partial x_2}\frac{\partial x_2}{\partial u_2} + \cdots + \frac{\partial f}{\partial x_n}\frac{\partial x_n}{\partial u_2}\right)du_2$$

$$+ \cdots$$

$$+ \left(\frac{\partial f}{\partial x_1}\frac{\partial x_1}{\partial u_m} + \frac{\partial f}{\partial x_2}\frac{\partial x_2}{\partial u_m} + \cdots + \frac{\partial f}{\partial x_n}\frac{\partial x_n}{\partial u_m}\right)du_m$$

$$= A_1\,du_1 + A_2\,du_2 + \cdots + A_m\,du_m,$$

where

$$A_j = \frac{\partial f}{\partial x_1}\frac{\partial x_1}{\partial u_j} + \frac{\partial f}{\partial x_2}\frac{\partial x_2}{\partial u_j} + \cdots + \frac{\partial f}{\partial x_n}\frac{\partial x_n}{\partial u_j}, \qquad j = 1, \ldots, m.$$

Since

$$df = \frac{\partial f}{\partial u_1}\,du_1 + \frac{\partial f}{\partial u_2}\,du_2 + \cdots + \frac{\partial f}{\partial u_m}\,du_m,$$

A_j is the coefficient of du_j in the expression for df. Thus

$$A_j = \frac{\partial f}{\partial u_j},$$

or

$$\frac{\partial f}{\partial u_j} = \frac{\partial f}{\partial x_1}\frac{\partial x_1}{\partial u_j} + \frac{\partial f}{\partial x_2}\frac{\partial x_2}{\partial u_j} + \cdots + \frac{\partial f}{\partial x_n}\frac{\partial x_n}{\partial u_j}.$$

This is the chain rule.

REMARK   The formula just written differs from Equation (20) only in notation. In Equations (19) and (20) we distinguish between the function of n variables $F(x_1, \ldots, x_n)$ and the composed function z obtained by replacing each x_i by $\phi_i(u_1, \ldots, u_m)$. In this subsection we use the same letter, f, for both functions. Both ways of denoting composed functions are used in technical literature.

EXAMPLES   1.   If $f(x,y) = x^2 e^{xy}$, what is df?
ANSWER   Since $f_x = \partial f/\partial x = 2xe^{xy} + x^2 y e^{xy}$ and $f_y = \partial f/\partial y = x^3 e^{xy}$,

$$df = f_x\,dx + f_y\,dy = (2xe^{xy} + x^2 y e^{xy})\,dx + x^3 e^{xy}\,dy$$
$$= xe^{xy}[(2 + xy)\,dx + x^2\,dy].$$

2.   If $F(u,v,w) = u^2 - w^2$, what is dF?
ANSWER   We have

$$dF = F_u\,du + F_v\,dv + F_w\,dw = 2u\,du + 0 \cdot dv + (-2w)\,dw$$
$$= 2u\,du - 2w\,dw.$$

3.   Let $f(x,y,z) = x^2 e^{yz} + z^2$, and $x = u + v$, $y = uv$, $z = uvw$. Using differentials, compute the partial derivatives of f with respect to u, v, w, at $u = 1$, $v = 0$, $w = 2$.
SOLUTION   We have

$$df = 2xe^{yz}\,dx + x^2 z e^{yz}\,dy + (x^2 y e^{yz} + 2z)\,dz, \qquad (*)$$
$$dx = du + dv, \qquad dy = v\,du + u\,dv,$$
$$dz = vw\,du + uw\,dv + uv\,dw.$$

For $u = 1$, $v = 0$, $w = 2$, we have

$$dx = du + dv, \qquad dy = dv, \qquad dz = 2\,dv, \qquad x = 1, \qquad y = z = 0.$$

Substitution into (*) yields

$$df = 2(du + dv) \qquad \text{at} \qquad (u,v,w) = (1,0,2).$$

Hence

$$\left(\frac{\partial f}{\partial u}\right)_{(1,0,2)} = \left(\frac{\partial f}{\partial v}\right)_{(1,0,2)} = 2, \qquad \left(\frac{\partial f}{\partial w}\right)_{(1,0,2)} = 0.$$

PROBLEMS

In Problems 1 to 6 compute df for the given function $f = f(x,y)$.

1. $f = x^3 + 2x^2 y + 2xy^2 + y^3$.
2. $f = 4 \sin 2x \cos 2y$.
3. $f = \log\left(\dfrac{x + 1}{y + 1}\right)$.

4. $f = e^x \tan y$.
5. $f = \frac{1}{2}\arccos(x + y)$.
6. $f = (x^4 + y^4)^{10}$.

7. If $g(x,y) = x^3 + y^3 + x^2y$, $x = rs$ and $y = r - s$, find g_r and g_s at $r = s = 1$ using differentials.

8. If $g(x,y) = \log(x^2 + xy + y^2)$, $x = r^2 + s^2$, and $y = r^2 - s^2$, find g_r and g_s at $r = 2$, $s = 1$ using differentials.

In Problems 9 to 14 compute dF for the given function $F = F(u,v,w)$.

9. $F = u^2vw + 4uv^2w + uvw^2$.

10. $F = \sin u \sin v \sin w$.

11. $F = \log(u^2v + v^2w + uw^2)$.

12. $F = \dfrac{u + v + w}{1 + uvw}$.

13. $F = $ arc tan uvw.

14. $F = e^{-w} \sin(u + 2v + 3w)$.

15. If $G(x,y,z) = \cos(x + y + z)$, $x = re^{s+t}$, $y = se^{r+t}$, and $z = te^{r+s}$, find G_r, G_s, and G_t at $r = s = t = 0$ using differentials.

16. If $G(x,y,z) = x^3 + y^3 + z^3$, $x = \sin(r + s)$, $y = \cos(s + t)$, and $z = \sin(r + t)$, find G_r, G_s, and G_t at $r = 0$, $s = \pi/4$, $t = \pi/2$ using differentials.

17. If $u = u(x_1,x_2, \ldots, x_n)$ and $v = v(x_1,x_2, \ldots, x_n)$, is it true that $d(u + v) = du + dv$? Is it true that $d(uv) = (du)v + u(dv)$?

2.8   Gradients, directional derivatives, and level surfaces in space

In §2.4 we introduced, for functions of two variables, the concepts of directional derivatives and gradients. Now we do the same for functions of three variables.

Let $F(x,y,z)$ be a continuously differentiable function. The vector whose components are the partial derivatives of F, at some point P, is called the **gradient** of F at P and is denoted by grad F. Thus

$$\text{grad } F = \frac{\partial F}{\partial x}\mathbf{i} + \frac{\partial F}{\partial y}\mathbf{j} + \frac{\partial F}{\partial z}\mathbf{k}.$$

The points where grad $F = 0$ are called **critical points** of F.

Now let P have the position vector $\mathbf{P}$ with components (coordinates of P) x_0, y_0, and z_0. Let $\mathbf{n}$ be a unit vector. We recall (see Chapter 18, §4.1) that the components of $\mathbf{n}$ are $\cos \alpha_1$, $\cos \alpha_2$, $\cos \alpha_3$, where α_1, α_2, and α_3 are the angles between $\mathbf{n}$ and the coordinate axes:

$$\mathbf{n} = (\cos \alpha_1)\mathbf{i} + (\cos \alpha_2)\mathbf{j} + (\cos \alpha_3)\mathbf{k}.$$

The point with position vector $\mathbf{P} + t\mathbf{n}$ lies on the line through P in the direction of $\mathbf{n}$, at the distance s from P (see Figure 19.10); its coordinates are $x = x_0 + s \cos \alpha_1$, $y = y_0 + s \cos \alpha_2$, $z = z_0 + s \cos \alpha_3$. The value of F at this point, that is,

$$F(x_0 + s \cos \alpha_1, \ y_0 + s \cos \alpha_2, \ z_0 + s \cos \alpha_3)$$

is a function of s. The derivative of this function at $s = 0$ is called the **directional derivative** of F at P in the direction $\mathbf{n}$. It measures the rate of change in F as we move from P in the direction $\mathbf{n}$. By the chain rule, this derivative is equal to

$$F_x(x_0,y_0,z_0) \cos \alpha_1 + F_y(x_0,y_0,z_0) \cos \alpha_2 + F_z(x_0,y_0,z_0) \cos \alpha_3. \tag{22}$$

We can also write the directional derivative as

$$(\text{grad } F, \mathbf{n}) = (\text{grad } F) \cdot \mathbf{n} = |\text{grad } F| \cos \theta, \tag{23}$$

FIGURE **19.10**

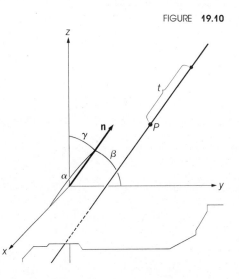

where θ is the angle between **n** and the gradient. This is so since $|\mathbf{n}| = 1$ and since $F_x/\sqrt{F_x^2 + F_y^2 + F_z^2}, F_y/\sqrt{F_x^2 + F_y^2 + F_z^2}, F_z/\sqrt{F_x^2 + F_y^2 + F_z^2}$ are the direction cosines of grad F. (See Chapter 18, §3.3.)

We conclude from (23) that, at a noncritical point, *the gradient points in the direction of steepest increase of the function and the length of the gradient is the rate of this increase.* Indeed, the directional derivative (23) is largest if $\cos \theta = 1$, that is, if $\theta = 0$ and **n** has the direction of grad F; in this case the directional derivative is $|\text{grad } F|$.

Suppose now that $z = \phi(x,y)$ is the equation of a surface along which F is constant, that is,

$$F(x,y,\phi(x,y)) = c. \tag{24}$$

Such a surface is called a **level surface.** Differentiating both sides of (24) with respect to x and to y, we obtain

$$F_x + F_z\phi_x = 0, . \quad F_y + F_z\phi_y = 0. \tag{25}$$

According to Equation (7) of §2.2, the numbers $\langle \phi_x, \phi_y, -1 \rangle$ are the components of a vector perpendicular to the surface $z = \phi(x,y)$ at a point (x,y,z). By (25), we have $\phi_x = -F_x/F_z$, $\phi_y = -F_y/F_z$, so that the vector considered has components

$$\left\langle -\frac{F_x}{F_z}, -\frac{F_y}{F_z}, -1 \right\rangle$$

and thus is parallel to the gradient $\langle F_x, F_y, F_z \rangle$. Hence *the gradient at a point is perpendicular to the level surface through this point.*

The same result can be proved if the level surface is represented not by $z = \phi(x,y)$ but by a function $x = \psi(y,z)$ or $y = \chi(z,x)$. We can also prove, although we shall not do this here, that if the derivatives F_x, F_y, F_z are themselves continuously differentiable, then there is a unique level surface through every noncritical point.

(If the function represents a temperature distribution, level surfaces are called isothermal surfaces. If the function represents an electric potential, the level surfaces are called equipotentials.)

Everything said in this subsection applies, with obvious changes, to functions of $n > 3$ variables.

EXAMPLE   For the function $F = 2x^2 - y^2 + z^2 + xy + 9x$, find (a) the gradient, (b) the critical points, (c) the unit vector pointing in the direction of steepest increase at the point $P = (-1,1,3)$, (d) the magnitude of the steepest increase at P, and (e) the directional derivative at P in the direction of the unit vector $\mathbf{n} = \frac{1}{3}\mathbf{i} - \frac{2}{3}\mathbf{j} + \frac{2}{3}\mathbf{k}$.

ANSWER   (a) Since $F_x = 4x + y + 9$, $F_y = -2y + x$, and $F_z = 2z$, we have grad $F = (4x + y + 9)\mathbf{i} + (x - 2y)\mathbf{j} + 2z\mathbf{k}$.

(b) The critical points occur when $4x + y + 9 = 0$, $-2y + x = 0$, and $2z = 0$. Solving the first two equations simultaneously, we obtain $x = -2$, $y = -1$. The third equation gives $z = 0$. Thus there is one critical point at $(2, -1, 0)$.

(c) At $(-1,1,3)$ we have $F_x = 6$, $F_y = -3$, $F_z = 6$, grad $F = 6\mathbf{i} - 3\mathbf{j} + 6\mathbf{k}$. The unit vector in the direction of grad F is

$$\frac{1}{\sqrt{6^2 + 3^2 + 6^2}}(6\mathbf{i} - 3\mathbf{j} + 6\mathbf{k}) = \frac{2}{3}\mathbf{i} - \frac{1}{3}\mathbf{j} + \frac{2}{3}\mathbf{k}.$$

This vector points in the direction of steepest increase and has direction cosines $\frac{2}{3}$, $-\frac{1}{3}$, $\frac{2}{3}$.

(d) At $(-1,1,3)$ we have $|\text{grad } F| = \sqrt{6^2 + 3^2 + 6^2} = 9$, and this is the magnitude of the steepest increase.

(e) The directional derivative at $(-1,1,3)$ in the direction $\mathbf{n} = \frac{1}{3}\mathbf{i} - \frac{2}{3}\mathbf{j} + \frac{2}{3}\mathbf{k}$ is, by (23),

$$(\text{grad } F) \cdot \mathbf{n} = (6\mathbf{i} - 3\mathbf{j} + 6\mathbf{k}) \cdot (\tfrac{1}{3}\mathbf{i} - \tfrac{2}{3}\mathbf{j} + \tfrac{2}{3}\mathbf{k}) = 2 + 2 + 4 = 8.$$

PROBLEMS

For each of the functions $F(x,y,z)$ given in Problems 1 to 5 find (a) the gradient, (b) the critical points, (c) the unit vector pointing in the direction of steepest increase at the given point P, (d) the magnitude of the steepest increase at P, and (e) the directional derivative at P in the direction of the given unit vector $\mathbf{n}$.

1. $F = x^2 - 2y^2 + z^2$, $P = (2,1,2)$, $\mathbf{n} = \frac{1}{4}(\sqrt{2}\mathbf{i} + \mathbf{j} + \mathbf{k})$.

2. $F = xy + z^2 + x + y$, $P = (1,1,1)$, $\mathbf{n} = \frac{1}{\sqrt{6}}(\mathbf{i} - \mathbf{j} + 2\mathbf{k})$.

3. $F = xy + yz - xz + x$, $P = (1,2,1)$, $\mathbf{n} = \frac{1}{\sqrt{3}}(\mathbf{i} + \mathbf{j} - \mathbf{k})$.

4. $F = x^2 + xy + yz + 4z$, $P = (1,0,3)$, $\mathbf{n} = \frac{1}{3}(2\mathbf{i} - \mathbf{j} + 2\mathbf{k})$.

5. $F = x^3 - y^2 + yz - 12x$, $P = (2,1,0)$, $\mathbf{n} = \frac{1}{3}(\mathbf{i} + 2\mathbf{j} + 2\mathbf{k})$.

6. If $F(x,y,z) = \sqrt{1 + x^2 + xy + z^2}$, find the rate of change of F at the point $(1,3,2)$ in the direction pointing toward the point $(2,5,4)$.

7. If $F(x,y,z) = \dfrac{xyz}{1 + x^2 + y^2 + z^2}$, find the rate of change of F at the point $(1,2,2)$ in the direction pointing toward the origin.

8. If $F(x,y,z) = \dfrac{x^2}{a^2} + \dfrac{y^2}{b^2} + \dfrac{z^2}{c^2}$, write the equation of the level surface through the point $(-1,3,4)$. Find the normal to this surface at this point, and verify that it is parallel to grad F.

2.9 Euler's theorem on homogeneous functions‡

A function $f(x,y)$ is called homogeneous of degree α if

$$f(tx,ty) = t^\alpha f(x,y), \tag{26}$$

for all $t \neq 0$; here α is some number. For example, $f(x,y) = x/y$ is homogeneous of degree 0, $f(x,y) = x^3y + x^2y^2$ is homogeneous of degree 4, and $f(x,y) = x^2y$ is homogeneous of degree 3, while the functions $f(x,y) = x + y^2$ and $f(xy) = \sin(xy)$ are not homogeneous.

Euler's theorem asserts that *a (continuously differentiable) function which is homogeneous of degree α satisfies the relation*

$$xf_x(x,y) + yf_y(x,y) = \alpha f(x,y). \tag{27}$$

To prove this, differentiate both sides of (26) with respect to t, using the chain rule for the left side. This yields

$$xf_x(tx,ty) + yf_y(tx,ty) = \alpha t^{\alpha-1}f(x,y).$$

Setting $t = 1$, we obtain (27).

The definition and the theorem hold for functions of more than two variables. If $f(x_1, \ldots, x_n)$ is homogeneous of degree α, that is, if $f(tx_1, \ldots, tx_n) = t^\alpha f(x_1, \ldots, x_n)$, for all $t \neq 0$, then $x_1 f_{x_1} + x_2 f_{x_2} + \cdots + x_n f_{x_n} = \alpha f$.

The proof is left to the reader.

PROBLEMS

In each of Problems 1 to 8 determine which of the given functions $f(x,y)$ is homogeneous. If the function is homogeneous, determine the degree.

1. $f = x^3 + 4x^2y + xy^2$.
2. $f = x^3 + 3xy$.
3. $f = e^{2xy}$.

4. $f = \arctan\dfrac{x}{y}$.

5. $f = \dfrac{2xy}{1 + x^2 + y^2}$.

6. $f = \sqrt{x^5 + y^5}$.
7. $f = (x^4 + x^2y^2 + y^4)^{1/3}$.
8. $f = (x^2 + 3y^2)^{-2/3}$.

9. Prove that a continuously differentiable function $f(x_1, x_2, \ldots, x_n)$ of n variables which is homogeneous of degree α satisfies the relation $\displaystyle\sum_{j=1}^{n} x_i \dfrac{\partial f}{\partial x_i} = \alpha f$.

‡This subsection may be omitted without loss of continuity.

§3   Partial derivatives of higher order

3.1   Second partial derivatives. Equality of mixed derivatives

If the function $f(x,y)$ has partial derivatives

$$f_x(x,y) = \frac{\partial f(x,y)}{\partial x}, \qquad f_y(x,y) = \frac{\partial f(x,y)}{\partial y},$$

these are themselves functions of two variables and may have partial derivatives. For instance, if $f(x,y) = x^2y + xy^3$, then

$$f_x = 2xy + y^3, \qquad f_y = x^2 + 3xy^2,$$

and therefore

$$\frac{\partial f_x}{\partial x} = \frac{\partial}{\partial x}\frac{\partial f}{\partial x} = 2y, \qquad \frac{\partial f_x}{\partial y} = \frac{\partial}{\partial y}\frac{\partial f}{\partial x} = 2x + 3y^2,$$

$$\frac{\partial f_y}{\partial x} = \frac{\partial}{\partial x}\frac{\partial f}{\partial y} = 2x + 3y^2, \qquad \frac{\partial f_y}{\partial y} = \frac{\partial}{\partial y}\frac{\partial f}{\partial y} = 6xy.$$

Partial derivatives of partial derivatives are called **second partial derivatives,** or partial derivatives of second order.

It may seem that a function of two variables, say, x and y, can have four distinct second partials: the x derivative of the x derivative, the y derivative of the x derivative, the x derivative of the y derivative, and the y derivative of the y derivative. But in the example just computed $(f = x^2y + xy^3)$, the two "mixed" derivatives

$$\frac{\partial}{\partial y}\frac{\partial f}{\partial x} \qquad \text{and} \qquad \frac{\partial}{\partial x}\frac{\partial f}{\partial y}$$

turned out to be the same function (namely, $2x + 3y^2$). This was no accident. In fact, the following is true.

Theorem 1 (equality of mixed derivatives).   *If a function $f(x,y)$ (defined in some domain) has all second-order partial derivatives, and if these derivatives are continuous functions, then*

$$\frac{\partial}{\partial y}\frac{\partial f}{\partial x} = \frac{\partial}{\partial x}\frac{\partial f}{\partial y}.$$

In other words, given a twice continuously differentiable function f, that is, a function whose partial derivatives are themselves continuously differentiable, we obtain the same result if we differentiate f first with respect to x and then with respect to y, or first with respect to y and then with respect to x. We also express this by saying: differentiations with respect to two different variables may be interchanged.

Let us verify Theorem 1 for monomials $f(x,y) = Cx^my^n$, where C is some number and m and n are nonnegative integers. In this case, $f_x = mCx^{m-1}y^n$, $(f_x)_y = nmCx^{m-1}y^{n-1}$, $f_y = nCx^my^{n-1}$, $(f_y)_x = mnCx^{m-1}y^{n-1}$, so that $(f_x)_y = (f_y)_x$ as asserted. It follows that Theorem 1 is true for polynomials and hence also for rational functions. The proof for the general case will be found in §5.2.

The following notations will be used for second partial derivatives:

$$\frac{\partial}{\partial x}\frac{\partial f}{\partial x} = \frac{\partial^2 f}{\partial x^2} = f_{xx}$$

(read: "∂ squared f with respect to x squared" or "second partial of f with respect to x" or "f sub x, x"). Similarly,

$$\frac{\partial}{\partial y}\frac{\partial f}{\partial x} = \frac{\partial^2 f}{\partial y\,\partial x} = f_{xy}, \qquad \frac{\partial}{\partial y}\frac{\partial f}{\partial y} = \frac{\partial^2 f}{\partial y^2} = f_{yy},$$

and so forth. Theorem 1 asserts that $f_{xy} = f_{yx}$.

PROBLEMS

In each of Problems 1 to 10 verify Theorem 1 for the given functions $f(x,y)$.

1. $f = 4x^2 + 3xy - y^2$.
2. $f = \sqrt{2x - y}$.
3. $f = \log \sqrt{x^2 + y^2}$.
4. $f = \sin(2x + 3y)$.

5. $f = \dfrac{x - y}{x + y}$.

6. $f = (x^2 + y^2)e^{xy}$.
7. $f = \tan 2x \sec 4y$.

8. $f = \arcsin \dfrac{y}{x}$.

9. $f = \sec(x^2 + y^2)$.
10. $f = \sin(x \cos y)$.

11. Prove Theorem 1 for functions of the form $f(x,y) = \sum_{j=1}^{n} a_j(x)b_j(y)$, that is, finite sums of products of differentiable functions of one variable.
12. Use Theorem 1 to prove that, if f_x, f_y have second partial derivatives, and these derivatives are all continuous functions, then $(f_{xx})_y = (f_{xy})_x = (f_{yx})_x$.

3.2 Differentiation under the integral sign

Theorem 1 above says that differentiations with respect to two different variables may be interchanged. There is a similar theorem which asserts that, under suitable continuity conditions, differentiation with respect to one variable and integration with respect to another can be interchanged. We call the variable of integration y and the variable of differentiation x, but it should be clear that the conclusion is independent of the notations used.

Theorem 2. *If $f(x,y)$ is continuously differentiable, the function of x, $\int_{\gamma}^{\delta} f(x,y)\,dy$, has a continuous derivative:*

$$\frac{d}{dx}\int_{\gamma}^{\delta} f(x,y)\,dy = \int_{\gamma}^{\delta} f_x(x,y)\,dy.$$

(It is assumed that f is defined and continuously differentiable in some interval $A < x < B$, $C < y < D$ and that $C < \gamma < \delta < D$.)

A similar conclusion holds if f depends on the variable of integration y and on *several* other variables (see Example 2 below).

The proof of Theorem 2 is not elementary ▶.

Corollary.   *Under the same hypothesis, if $\gamma(x)$ and $\delta(x)$ are continuously differentiable functions, then*

$$\frac{d}{dx}\int_{\gamma(x)}^{\delta(x)} f(x,y)\,dy = \int_{\gamma(x)}^{\delta(x)} f_x(x,y)\,dy - f[x,\gamma(x)]\gamma'(x) + f[x,\delta(x)]\,\delta'(x).$$

Proof.   Consider the function of three variables $G(x,\gamma,\delta) = \int_\gamma^\delta f(x,y)\,dy$. It is continuously differentiable, with

$$G_x(x,\gamma,\delta) = \int_\gamma^\delta f_x(x,y)\,dy, \qquad G_\gamma(x,\gamma,\delta) = -f(x,\gamma), \qquad G_\delta(x,\gamma,\delta) = f(x,\delta),$$

by Theorem 2 and the fundamental theorem of calculus. By the chain rule,

$$\frac{d}{dx}\int_{\gamma(x)}^{\delta(x)} f(x,y)\,dy = \frac{d}{dx}\,G[x,\gamma(x),\delta(x)] = G_x + G_\gamma\gamma'(x) + G_\delta\,\delta'(x)$$

$$= \int_{\gamma(x)}^{\delta(x)} f_x(x,y)\,dy - f[x,\gamma(x)]\gamma'(x) + f[x,\delta(x)]\,\delta'(x),$$

as asserted.

EXAMPLES 1.   Compute $F'(x)$ if $F(x) = \int_0^2 (x^2y + xy^3)\,dy$.

FIRST SOLUTION   We use Theorem 2 with $f(x,y) = x^2y + xy^3$. We have $f_x(x,y) = 2xy + y^3$, and

$$F'(x) = \frac{d}{dx}\int_0^2 f(x,y)\,dy = \int_0^2 f_x(x,y)\,dy$$

$$= \int_0^2 (2xy + y^3)\,dy = \left[xy^2 + \frac{1}{4}y^4\right]_{y=0}^{y=2} = 4x + 4.$$

SECOND SOLUTION   We compute directly:

$$F(x) = \int_0^2 (x^2y + xy^3)\,dy = \left[\frac{1}{2}x^2y^2 + \frac{1}{4}xy^4\right]_{y=0}^{y=2} = 2x^2 + 4x,$$

so that $F'(x) = 4x + 4$, as expected.

2.   Set $u(p,q) = \int_0^1 [e^{x^2} + x\sin(pq^2)]\,dx$. Find u_p and u_q.

ANSWER   Differentiation under the integral sign is legitimate [by Theorem 2, since the integrand $e^{x^2} + x\sin(pq^2)$ is a continuously differentiable function of the three variables x, p, q]. We have

$$u_p(p,q) = \int_0^1 \frac{\partial}{\partial p}[e^{x^2} + x\sin(pq^2)]\,dx$$

$$= \int_0^1 xq^2\cos(pq^2)\,dx = q^2\cos(pq^2)\int_0^1 x\,dx = \frac{1}{2}q^2\cos(pq^2).$$

In the same way, we see that $u_q(p,q) = pq\cos(pq^2)$.

3.   Find $g'(x)$ if

$$g(x) = \int_0^{4x} (x + y)^2\,dy.$$

FIRST SOLUTION   We apply the corollary with $f(x,y) = (x + y)^2$, $\gamma(x) = 0$, $\delta(x) = 4x$, $f_x(x,y) = 2(x + y)$, $\gamma'(x) = 0$, $\delta'(x) = 4$, and obtain

$$g'(x) = \int_0^{4x} 2(x + y)\,dy + (x + 4x)^2 4 = (x + y)^2\Big|_{y=0}^{y=4x} + 100x^2 = 124x^2.$$

SECOND SOLUTION   We integrate first

$$g(x) = \frac{1}{3}(x+y)^3 \Big|_{y=0}^{y=4x} = \frac{124x^3}{3} \qquad \text{so that} \qquad g'(x) = 124x^2.$$

PROBLEMS

In Problems 1 to 5 find $F'(x)$ in two different ways: by integrating first and then differentiating the result; and by differentiating under the integral sign, and then integrating the result.

1. $F(x) = \displaystyle\int_0^1 (x^3 + 3xy^2 - 2y^3)\,dy.$

4. $F(x) = \displaystyle\int_6^8 \log(xy)\,dy.$

2. $F(x) = \displaystyle\int_2^3 e^{xy}\,dy.$

5. $F(x) = \displaystyle\int_0^{\pi/2} (y\cos x + x\cos y)\,dy.$

3. $F(x) = \displaystyle\int_1^2 \frac{dy}{(x+y)}.$

In Problems 6 to 12 use any method you like.

6. If $F(x) = \displaystyle\int_1^2 \frac{e^{-xy}}{y}\,dy$, find $F'(2)$.

7. If $F(x) = \displaystyle\int_0^1 \cos(xe^y)\,dy$, find $F'(x)$.

8. If $F(x) = \displaystyle\int_1^2 \frac{\log(1+x^2y^2)}{y}\,dy$, find $F'(1)$.

9. If $F(x) = \displaystyle\int_0^x \frac{\sin xy}{y}\,dy$, find $F'(x)$.

10. If $F(x) = \displaystyle\int_0^x \frac{1-e^{-xy}}{y}\,dy$, find $F'(x)$.

11. If $F(x) = \displaystyle\int_x^{x^2} xy\,dy$, find $F'(x)$.

12. If $F(x) = \displaystyle\int_{\sin x}^{\cos x} x^2y\,dy$, find $F'(x)$.

3.3   Functions with prescribed partial derivatives

A function of one variable $f(x)$, with $f'(x) = 0$ for all x, is a constant. There is a similar result for functions of two variables.

Theorem 3.   *If $f(x,y)$ satisfies the condition $f_x(x,y) = f_y(x,y) = 0$ at all points of a domain, f is constant in this domain.*

Proof.   Let (x_0,y_0) and (x_1,y_1) be two points in a domain. Then there is a smooth curve joining them:

$$x = x(t), \quad y = y(t), \quad 0 \le t \le 1,$$

with

$$x(0) = x_0, \quad y(0) = y_0, \quad x(1) = x_1, \quad y(1) = y_1$$

(see §1.2). Set $F(t) = f[x(t),y(t)]$. Then, by the chain rule,

$$F'(t) = f_x[x(t),y(t)]x'(t) + f_y[x(t),y(t)]y'(t) = 0,$$

for all t. Hence $F(0) = F(1)$. But $f(x_0,y_0) = F(0)$ and $f(x_1,y_1) = F(1)$. We have shown that f takes on the same value at any two points of the domain; thus f is constant.

Corollary. *If two functions, f and g, have the same partial derivatives in a domain, then there is a constant c such that $f(x,y) = g(x,y) + c$.*

Proof. Apply Theorem 2 to $\phi = (f - g)$. Since $\phi_x = (f - g)_x = f_x - g_x = 0$ and, similarly, $\phi_y = 0$, ϕ is a constant.

Now let $u(x,y)$ and $v(x,y)$ be continuously differentiable functions defined in some domain. Does there exist a function $f(x,y)$ with

$$f_x = u, \qquad f_y = v \tag{1}$$

in this domain? Theorem 1 tells us that the answer cannot be yes in all cases, for if (1) holds, then we must also have $(f_x)_y = (f_y)_x$, that is,

$$u_y = v_x. \tag{2}$$

We call this equation the **integrability condition.** If it is satisfied, and if the domain is an interval, the answer to our question is affirmative. The proof of this, given below, involves Theorem 2.

Theorem 4. *If $u(x,y)$, $v(x,y)$ are continuously differentiable functions defined in an open interval I, and $u_y = v_x$ in I, then there is a function $f(x,y)$ in I such that $f_x = u$, $f_y = v$. [The function can be chosen so as to equal a given value, say, A, at a given point, say, (x_0,y_0), in I and is then uniquely determined.]*

Proof. Let f be a solution with

$$f(x_0,y_0) = A. \tag{3}$$

If (x,y) is some other point of I, then the point (x,y_0) is also in I; see Figure 19.11.

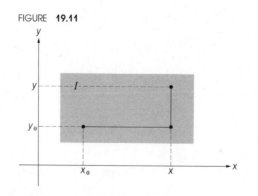

FIGURE **19.11**

By (1), (3), and the fundamental theorem of calculus,

$$f(x,y_0) - A = f(x,y_0) - f(x_0,y_0) = \int_{x_0}^{x} f_x(t,y_0)\, dt = \int_{x_0}^{x} u(t,y_0)\, dt,$$

$$f(x,y) - f(x,y_0) = \int_{y_0}^{y} f_y(x,s)\, ds = \int_{y_0}^{y} v(x,s)\, ds$$

and therefore

$$f(x,y) = A + \int_{x_0}^{x} u(t,y_0)\, dt + \int_{y_0}^{y} v(x,s)\, ds. \tag{4}$$

This shows that, given the value $f(x_0,y_0) = A$, our problem has a unique solution, if any. (This statement also follows from the preceding corollary.)

We show next that, if the integrability condition (2) holds, then the function f defined by (4) has the required partial derivatives. Indeed, by the fundamental theorem of calculus and by the theorem about differentiating under the integral sign (Theorem 2), we have, assuming that $u_y = v_x$,

$$f_x(x,y) = \frac{\partial A}{\partial x} + \frac{\partial}{\partial x} \int_{x_0}^{x} u(t,y_0)\, dt + \frac{\partial}{\partial x} \int_{y_0}^{y} v(x,s)\, ds$$

$$= 0 + u(x,y_0) + \int_{y_0}^{y} v_x(x,s)\, ds = u(x,y_0) + \int_{y_0}^{y} u_s(x,s)\, ds$$

$$= u(x,y_0) + [u(x,y) - u(x,y_0)] = u(x,y).$$

Also,

$$f_y(x,y) = \frac{\partial A}{\partial y} + \frac{\partial}{\partial y} \int_{x_0}^{x} u(t,y_0)\, dt + \frac{\partial}{\partial y} \int_{y_0}^{y} v(x,s)\, ds$$

$$= 0 + 0 + v(x,y) = v(x,y).$$

Finally, (4) implies (3). Thus the theorem is proved.

REMARK The proof of Theorem 4 gives a method of computing a function with derivatives u and v, namely, formula (4). It may be easier not to use the formula explicitly but to proceed as follows. Treat y as a constant, and find a "primitive function" ϕ for u treated as a function of x. This gives $f = \phi + c$, the "constant of integration" $c = c(y)$ being a function of y. Now determine $c(y)$ from the conditions: $\phi_y(x,y) + c'(y) = v(x,y)$, $\phi(x_0,y_0) + c(y_0) = A$. We can also interchange the roles of x and y. We illustrate this by Examples 2 and 3.

EXAMPLES 1. Is there a function $f(x,y)$ such that $f_x = 5y$, $f_y = 2x$?
ANSWER No. For setting $u = 5y$, $v = 2x$, we have $u_y = 5$, $v_x = 2$; hence $u_y \neq v_x$. Suppose we try to use formula (4) with, say, $x_0 = y_0 = 1$, $A = 3$. This gives

$$f(x,y) = 3 + \int_{1}^{x} 5\, dt + \int_{1}^{y} 2x\, ds = 3 + 5(x - 1) + 2x(y - 1) = 2xy + 3x - 2.$$

Then $f(1,1) = 3$, $f_y = 2x$, but $f_x = 2y + 3 \neq 5y$.

2. Find a function $f(x,y)$ with

$$f_x = 3x^2 y + 2y^2, \qquad f_y = x^3 + 4xy - 1, \qquad f(1,1) = 4.$$

FIRST SOLUTION We check the integrability condition:

$$\frac{\partial(3x^2 y + 2y^2)}{\partial y} = 3x^2 + 4y, \qquad \frac{\partial(x^3 + 4xy - 1)}{\partial x} = 3x^2 + 4y.$$

Now we apply (4) with $x_0 = y_0 = 1$, $A = 4$, $u(x,y) = 3x^2y + 2y^2$, $v(x,y) = x^3 + 4xy - 1$, and obtain

$$f(x,y) = 4 + \int_1^x (3t^2 + 2)\, dt + \int_1^y (x^3 + 4xs - 1)\, ds$$

$$= 4 + (x^3 - 1) + 2(x - 1) + x^3(y - 1) + (2xy^2 - 2x) - (y - 1)$$

$$= x^3y + 2xy^2 - y + 2.$$

It is advisable always to check the answer by differentiation:

$$(x^3y + 2xy^2 - y + 2)_x = 3x^2y + 2y^2, \qquad (x^3y + 2xy^2 - y + 2)_y = x^3 + 4xy - 1,$$

as required.

SECOND SOLUTION   After verifying the integrability condition, we use the equation $f_x = 3x^2y + 2y^2$ to conclude that

$$f(x,y) = x^3y + 2xy^2 + c(y).$$

In order to have $f_y = x^3 + 4xy + c'(y) = x^3 + 4xy - 1$, we need $c(y) = -y + C$, C a constant. Thus $f = x^3y + 2xy^2 - y + C$, and the condition $f(1,1) = 4$ gives $C = 2$.

3.   Find all functions $\phi(x,y)$ with $\phi_x = e^{xy} + xye^{xy} + \cos x + 1$, $\phi_y = x^2e^{xy}$.

ANSWER   The integrability condition reads

$$(e^{xy} + xye^{xy} + \cos x + 1)_y = (x^2e^{xy})_x.$$

It is satisfied; both sides equal $2xe^{xy} + x^2ye^{xy}$. The second equation ($\phi_y = x^2e^{xy}$) yields

$$\phi(x,y) = xe^{xy} + c(x),$$

where $c(x)$ is to be determined. In order to satisfy the condition $\phi_x = e^{xy} + xye^{xy} + \cos x + 1$, we need

$$e^{xy} + xye^{xy} + c'(x) = e^{xy} + xye^{xy} + \cos x + 1.$$

Hence $c'(x) = \cos x + 1$, or $c(x) = \sin x + x + C$, C a constant. The final answer

$$\phi(x,y) = xe^{xy} + \sin x + x + C$$

should be checked by differentiation.

PROBLEMS

1. Find a function $f(x,y)$ such that $f_x = 2x - 4y$, $f_y = -4x + 6y + 1$, and $f(2,1) = 0$.

2. Find a function $f(x,y)$ such that $f_x = 3x^2 + 2xy + 3y^2$, $f_y = x^2 + 6xy - 3y^2$, and $f(-1,1) = -2$.

3. Find a function $f(x,y)$ such that $f_x = e^x \cos y + e^y \cos x$, $f_y = -e^x \sin y + e^y \sin x$, and $f(0,0) = 0$.

4. Find a function $f(x,y)$ such that $f_x = \dfrac{x + 2y - 1}{(x + 2y)^2}$, $f_y = \dfrac{2(x + 2y - 1)}{(x + 2y)^2}$, and $f(0,\tfrac{1}{2}) = 1$.

5. Find all functions $\phi(x,y)$ with $\phi_x = 3x^2y + 6xy^2 - y^3 + 1$, $\phi_y = x^3 + 6x^2y - 3xy^2$.

6. Find all functions $\phi(x,y)$ with $\phi_x = \sin 2y - 2y \sin 2x$, $\phi_y = 2x \cos 2y + \cos 2x$.

7. Find all functions $\phi(x,y)$ with $\phi_x = x \cos(x + y) + 1$, $\phi_y = x \cos(x + y) - \sin(x + y)$.

8. Find all functions $\phi(x,y)$ with $\phi_x = e^{x^2y^2} + 2x^2y^2e^{x^2y^2} + x$, $\phi_y = 2x^3ye^{x^2y^2} + y$.

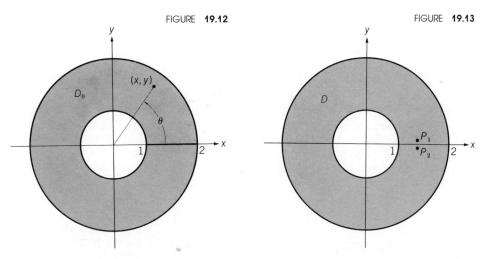

FIGURE **19.12**

FIGURE **19.13**

3.4 Functions with prescribed partial derivatives in general domains‡

Does Theorem 4 remain valid if we replace the interval I by some other domain? Not necessarily!

To see this, consider the domain D bounded by the circles $x^2 + y^2 = 1$ and $x^2 + y^2 = 4$, that is, the set of all points (x,y) with $1 < x^2 + y^2 < 4$. Let D_0 denote the set of all points (x,y) with $1 < x^2 + y^2 < 4$ *except* the points on the segment from $(1,0)$ to $(2,0)$; see Figure 19.12. This is also a domain. Every point (x,y) of D_0 has a unique polar angle θ, if we require the inequality $0 < \theta < 2\pi$. Hence $\theta = \theta(x,y)$ is a function in D_0. This function has the partial derivatives

$$\theta_x = -\frac{y}{x^2 + y^2}, \qquad \theta_y = \frac{x}{x^2 + y^2}. \tag{5}$$

This can be verified by starting with the expressions

$$x = r \cos \theta, \qquad y = r \sin \theta \tag{6}$$

(here $r = \sqrt{x^2 + y^2}$). Take the partial derivatives with respect to x to obtain

$$\frac{\partial x}{\partial x} = 1 = \frac{\partial r}{\partial x} \cos \theta - r \sin \theta \frac{\partial \theta}{\partial x}, \qquad \frac{\partial y}{\partial x} = 0 = \frac{\partial r}{\partial x} \sin \theta + r \cos \theta \frac{\partial \theta}{\partial x}.$$

Multiply the first equation by $(-\sin \theta)$, the second by $\cos \theta$, and add. This yields $-\sin \theta = r(\partial \theta / \partial x)$ or

$$\frac{\partial \theta}{\partial x} = -\frac{\sin \theta}{r} = -\frac{r \sin \theta}{r^2} = -\frac{y}{x^2 + y^2}.$$

‡This subsection may be omitted without loss of continuity.

The second equation (5) is obtained similarly.
Now set

$$u(x,y) = -\frac{y}{x^2 + y^2}, \qquad v(x,y) = \frac{x}{x^2 + y^2}.$$

These functions have continuous partial derivatives of all orders in D and satisfy the integrability condition $u_y = v_x$. But in D there is no function $f(x,y)$ with $f_x = u, f_y = v$. For if there were such a function, then in D_0 it would have to be of the form $\theta + c$, c a constant (by the corollary to Theorem 3). Hence, at a point like P_1 in Figure 19.13, it would have to be very close to c, and at a point like P_2, it would be very close to $2\pi + c$. Such an f could not be continuous in D.

In general, given a domain D and two continuously differentiable functions u and v in D satisfying $u_y = v_x$, we can always find a continuous function f in D with $f_x = u, f_y = v$ if D is what we call **simply connected**, intuitively speaking: without holes. (An interval is simply connected. So is the domain D_0 in Figure 19.12.) In a domain that is not simply connected (like the domain D in Figure 19.13), there are functions u, v with $u_y = v_x$, which are not the partial derivatives of a function. We touch here upon an area in which calculus needs the help of a modern branch of geometry called topology.

3.5 Higher derivatives

The process of partial differentiation need not stop with second derivatives. We may be able to find partial derivatives of the second partial derivatives, called third partial derivatives, then fourth partial derivatives, and so forth.

If $f(x,y)$ has all partial derivatives up to, and including those of some order N, and these derivatives are continuous, the "mixed" partial derivatives of f depend only on the number of times we differentiate with respect to each variable, *not* on the order in which we perform these differentiations. For instance,

$$\frac{\partial}{\partial x}\frac{\partial}{\partial y}\frac{\partial f}{\partial y} = \frac{\partial}{\partial y}\frac{\partial}{\partial x}\frac{\partial f}{\partial y} = \frac{\partial}{\partial y}\frac{\partial}{\partial y}\frac{\partial f}{\partial x},$$

which can also be written as $f_{yyx} = f_{yxy} = f_{xyy}$, and this third-order partial derivative is denoted simply by

$$\frac{\partial^3 f}{\partial x\, \partial y^2}$$

(read: "third partial of f, once with respect to x and twice with respect to y"). This follows from Theorem 1 in §3.1. For instance, the equality

$$\frac{\partial}{\partial x}\frac{\partial}{\partial y}\frac{\partial f}{\partial y} = \frac{\partial}{\partial y}\frac{\partial}{\partial x}\frac{\partial f}{\partial y}$$

is simply Theorem 1 applied to the function f_y.

A function $f(x,y)$ has four distinct third partial derivatives:

$$\frac{\partial^3 f}{\partial x^3}, \qquad \frac{\partial^3 f}{\partial x^2\, \partial y}, \qquad \frac{\partial^3 f}{\partial x\, \partial y^2}, \qquad \frac{\partial^3 f}{\partial y^3},$$

five distinct fourth partials:

$$\frac{\partial^4 f}{\partial x^4}, \quad \frac{\partial^4 f}{\partial x^3 \, \partial y}, \quad \frac{\partial^4 f}{\partial x^2 \, \partial y^2}, \quad \frac{\partial^4 f}{\partial x \, \partial y^3}, \quad \frac{\partial^4 f}{\partial y^4},$$

and so forth.

EXAMPLE Find the values of all second partials of $f(x,y) = e^{x^2 y}$ at $x = 0$, $y = 1$.
ANSWER   We have $f_x = 2xye^{x^2 y}$, $f_y = x^2 e^{x^2 y}$. Hence $f_{xx} = (f_x)_x = 2ye^{x^2 y} + 4x^2 y^2 e^{x^2 y}$,
$f_{xy} = (f_y)_x = 2xe^{x^2 y} + 2x^3 ye^{x^2 y}$, $f_{yy} = (f_y)_y = x^4 e^{x^2 y}$.
Thus $f_{xx}(0,1) = 2$, $f_{xy}(0,1) = f_{yy}(0,1) = 0$.

PROBLEMS

1. Find all third partials of $f(x,y) = x^4 + x^2 y^2 + y^4$ at $x = 1$, $y = 2$.
2. Find all third partials of $f(x,y) = \log(x^2 + y^2)$.
3. Find all third partials of $f(x,y) = \sin(xy)$.
4. Find all fourth partials of $f(x,y) = x^6 + 3x^3 y^3 + y^6$ at $x = 1$, $y = 1$.
5. Find all fourth partials of $f(x,y) = e^{2xy}$.
6. Find all fourth partials of $f(x,y) = e^x \cos 2y$.

3.6   Higher derivatives for more than two variables

The definition of, and notation for, derivatives of higher order for functions of more than two variables are self-evident. If a function of several variables is differentiated several times, the result depends only on how many times we differentiated with respect to each variable, not on the order in which we performed the differentiations. For instance, if $u = f(x,y,z)$, then

$$\frac{\partial}{\partial x}\frac{\partial}{\partial y}\frac{\partial f}{\partial z} = \frac{\partial}{\partial x}\frac{\partial}{\partial z}\frac{\partial f}{\partial y} = \frac{\partial}{\partial y}\frac{\partial}{\partial x}\frac{\partial f}{\partial z} = \frac{\partial}{\partial y}\frac{\partial}{\partial z}\frac{\partial f}{\partial x} = \frac{\partial}{\partial z}\frac{\partial}{\partial x}\frac{\partial f}{\partial y} = \frac{\partial}{\partial z}\frac{\partial}{\partial y}\frac{\partial f}{\partial x},$$

and the common value of these six derivatives is denoted by

$$f_{xyz} = \frac{\partial^3 f}{\partial x \, \partial y \, \partial z}.$$

No new proof is needed; the assertion follows by repeated application of Theorem 1 in §3.1. Indeed, if we perform the differentiations with respect to two variables, the others are kept fixed, so that we deal, in effect, with functions of two variables.

EXAMPLE If $v = xe^{yz} \cos u^2$, find the value $\partial^4 v / \partial x \, \partial y \, \partial u^2$ for $x = y = 0$, $z = 1$, $u = \sqrt{\pi}$.
ANSWER We have

$$\frac{\partial v}{\partial x} = e^{yz} \cos u^2, \qquad \frac{\partial^2 v}{\partial x \, \partial y} = \frac{\partial}{\partial y}\frac{\partial v}{\partial x} = ze^{yz} \cos u^2,$$

$$\frac{\partial^3 v}{\partial x \, \partial y \, \partial u} = \frac{\partial}{\partial u}\frac{\partial^2 v}{\partial x \, \partial y} = -ze^{yz} \sin u^2 \, 2u,$$

$$\frac{\partial^4 v}{\partial x \, \partial y \, \partial u^2} = \frac{\partial}{\partial u}\frac{\partial^3 v}{\partial x \, \partial y \, \partial u} = -2ze^{yz} \sin u^2 - 4u^2 ze^{yz} \cos u^2.$$

The value of this derivative at $x = y = 0$, $z = 1$, $u = \sqrt{\pi}$ is 4π.

If $f(x_1, x_2, \ldots, x_n)$ is a (twice continuously differentiable) function of n variables, with the partial derivatives

$$u_i(x_1, \ldots, x_n) = \frac{\partial f(x_1, \ldots, x_n)}{\partial x_i}, \qquad i = 1, 2, \ldots, n, \qquad (7)$$

then these derivatives satisfy $\frac{1}{2}n(n-1)$ **integrability conditions:**

$$\frac{\partial u_i}{\partial x_j} = \frac{\partial u_j}{\partial x_i}, \qquad i = 1, \ldots, n; \qquad j = 1, \ldots, n; \qquad i \neq j. \qquad (8)$$

Indeed, both sides of (8) are equal to $\partial^2 u / \partial x_i\, \partial x_j$.

Conversely, if we are given, in some interval I, n continuously differentiable functions u_i satisfying the integrability conditions, then there is a function f in I such that $\partial f / \partial x_i = u_i$ for all i. The solution of (7) is uniquely determined if we prescribe the value of f at some point of I.

All this is proved exactly as for two variables (see §3.3). For instance, in the case $n = 3$, if u, v, w are functions of (x,y,z) defined in I, and we want to find a function f in I with

$$f_x = u, \qquad f_y = v, \qquad f_z = w,$$

then we need the integrability conditions

$$u_y = v_x, \qquad u_z = w_x, \qquad v_z = w_y.$$

If these are satisfied, and (x_0, y_0, z_0) is a point in I, the function

$$f(x,y,z) = A + \int_{x_0}^{x} u(t, y_0, z_0)\, dt + \int_{y_0}^{y} v(x, s, z_0)\, ds + \int_{z_0}^{z} w(x, y, r)\, dr$$

takes on the value A at (x_0, y_0, z_0) and has the required partial derivatives. (The reader may want to prove this, imitating the proof of Theorem 4 in §3.3.)

We mention without proof that we can find a function with prescribed partial derivatives in a domain D provided (1) the prescribed partials satisfy the integrability conditions and (2) D is what we call **simply connected.** (This means that any closed curve in D can be deformed into a point without leaving D. In the plane, this condition is equivalent to D having "no holes.")

EXAMPLES 1. Find a function $f(x,y,z)$ such that

$$f_x = 2x, \qquad f_y = 2y, \qquad f_z = 2z, \qquad \text{and} \qquad f(0,0,0) = 1.$$

ANSWER The integrability conditions are satisfied. Since $f_x = 2x$ we have $f = x^2 + c_1(y,z)$ where the "constant of integration with respect to x" is a function of y and z. The condition $f_y = 2y$ becomes $(c_1)_y = 2y$. Hence, $c_1(y,z) = y^2 + c_2(z)$ and $f = x^2 + y^2 + c_2(z)$. The condition $f_z = 2z$ becomes $c_2'(z) = 2z$. Hence $c_2(z) = z^2 + c_3$; c_3 is a constant. Thus $f = x^2 + y^2 + z^2 + c_3$. Since $f = 1$ at $x = y = z = 0$, we have $c_3 = 1$. So we get $f = x^2 + y^2 + z^2 + 1$.

2. Find a function $f(x,y,z)$ such that

$$df = x\, dx - 2y\, dy \qquad \text{and} \qquad f(0,0,0) = 0.$$

ANSWER The relation "$df = x\, dx - 2y\, dy$" means that $f_x = x$, $f_y = -2y$, $f_z = 0$. The integrability conditions are satisfied. Since $f_x = x$, $f = \frac{1}{2}x^2 + c(y,z)$. The condition $f_y = -2y$ yields: $c_y = -2y$. Hence $c(y,z) = -y^2 + c(z)$. Since $f_z = 0$, we must have $c(z) = c$, a constant. Thus $f = \frac{1}{2}x^2 - y^2 + c$, and since $f(0,0,0) = 0$, $c = 0$ and $f(x,y,z) = \frac{1}{2}x^2 - y^2$.

3. Find a function $g(x,y,z)$ such that

$$dg = 2 \, dx - 3z \, dy + y \, dz.$$

ANSWER   There is *no* such function. Indeed, the condition above means that $g_x = 2$, $g_y = -3z$, $g_z = y$. Of the three integrability conditions

$$(g_x)_y = (g_y)_x, \qquad (g_x)_z = (g_z)_x, \qquad (g_y)_z = (g_z)_y$$

or

$$\partial 2/\partial y = \partial(-3z)/\partial x, \qquad \partial 2/\partial z = \partial y/\partial x, \qquad \partial(-3z)/\partial z = \partial y/\partial y,$$

only the first two are satisfied.

PROBLEMS

1. Find all second partials of $f(x,y,z) = xy^2z^3$.
2. Find all second partials of $f(x,y,z) = \log(1 + xyz)$ at $x = 0$, $y = 1$, $z = -1$.
3. Find all third partials of $f(x,y,z) = x^2y^2 + y^2z^2 + x^2z^2$ at $x = y = z = 1$.
4. If $f(x,y,z) = \cos(xyz)$, find f_{xxy} and f_{yyz}.
5. Find all third partials of $f(x,y,z) = x \cos y \sin z$.
6. If $G(x,y,z,t) = \cos(x + 2y + 3z + 4t)$, find G_{xyzt}, G_{xyzz}, and G_{xyyz} at $x = y = z = t = 0$.
7. If $F(x_1,x_2,\ldots,x_n) = (\cos x_1)(\sin x_1)(\cos x_2)(\sin x_2) \ldots (\cos x_n)(\sin x_n)$, find $\partial F^n/\partial x_1 \partial x_2 \ldots \partial x_n$ at $x_1 = x_2 = \cdots = x_n = \pi/6$.

In Problems 8 to 15 determine whether there exists a function $f(x,y,z)$ satisfying the prescribed conditions; if one exists, find it.

8. $f_x = 4x^3 + y$, $f_y = z^3 + x$, $f_z = 3yz^2 + 3$, and $f(1,3,-1) = -2$.
9. $f_x = x^2 + yz$, $f_y = y^2 + xz$, $f_z = z^2 + xy$, and $f(1,-1,1) = 0$.
10. $f_x = 2xy + z$, $f_y = x^2 + y^2$, $f_z = 2yz + 1$, and $f(1,2,1) = 6$.
11. $f_x = e^{yz} + yze^{xz}$, $f_y = xze^{yz} + e^{xz}$, $f_z = xye^{yz} + xye^{xz}$, and $f(1,0,0) = 1$.
12. $df = (y + 2z) \, dx + (x - z) \, dy + (2x - y) \, dz$.
13. $df = \left(\dfrac{1}{x} + yz\right) dx + \left(\dfrac{1}{y} + xz\right) dy + \left(\dfrac{1}{z} + xy\right) dz$ and $f(1,1,1) = 1$.
14. $df = (2ye^{xy} + z) \, dx + (2xe^{xy} + 4) \, dy + (2e^{xy} + x) \, dz$.
15. $df = (\sin yz) \, dx + (xz \cos yz) \, dy + (xy \cos yz) \, dz$ and $f(0,0,0) = 1$.

---

§4   Maxima and minima for functions of several variables

In this section we extend to functions of several variables the methods discussed in Chapter 5, §2. We shall consider primarily functions of two variables.

4.1   Critical points

Let $f(x,y)$ be a continuously differentiable function of two variables. We recall (see §2.4) that a point at which f_x and f_y are both 0 is called a critical point. For instance, if f has a **local maximum** (or a **local minimum**) at (x_0,y_0), that is, if $f(x,y) \leq f(x_0,y_0)$ for all (x,y) near (x_0,y_0) [or if $f(x,y) \geq f(x_0,y_0)$ for all (x,y) near (x_0,y_0)], then (x_0,y_0) is a critical point. Indeed, in this case, the two functions of one variable

$$x \mapsto f(x,y_0), \qquad y \mapsto f(x_0,y)$$

have local maxima (or local minima) at $x = x_0$ and $y = y_0$, and therefore their derivatives are zero at these points. But this means that $f_x(x_0,y_0) = 0$ and $f_y(x_0,y_0) = 0$.

The converse need not be true; at a critical point the function need not have a maximum or a minimum. Whether it does can sometimes be determined by looking at the second partials of f, more precisely, at the signs of $f_{xx}f_{yy} - f_{xy}^2$ and f_{xx} (compare the corresponding statement about functions of one variable in Chapter 5, §2.4).

Before stating the rule, we introduce some terminology. Let $f(x,y)$ be defined near (x_0,y_0). We say that (x_0,y_0) is a **strict minimum point** if $f(x_0,y_0) < f(x,y)$ for all (x,y) near to, and distinct from, (x_0,y_0). If $f(x_0,y_0) > f(x,y)$ for all (x,y) near to, and distinct from, (x_0,y_0) we say that (x_0,y_0) is a **strict maximum point**. Finally, if $f_x = f_y = 0$ at (x_0,y_0) and in *every* interval with midpoint (x_0,y_0) there are points (x_1,y_1) with $f(x_1,y_1) < f(x_0,y_0)$ and also points (x_2,y_2) with $f(x_2,y_2) > f(x_0,y_0)$, then (x_0,y_0) is called a **saddle point**.

For instance, if $f(x,y) = xy$, then the origin is a saddle point. Indeed, $f = f_x = f_y = 0$ at the origin, $f(x,y) > 0$ in the first and third quadrants, and $f(x,y) < 0$ in the second and fourth quadrants.

Theorem 1. *If $f(x,y)$ is twice continuously differentiable at (x_0,y_0), and if at this point $f_x = f_y = 0$ and*

if at this point	then the point is
$f_{xx}f_{yy} - f_{xy}^2 < 0$	a saddle point
$f_{xx}f_{yy} - f_{xy}^2 > 0,$ $f_{xx} > 0$	a strict minimum point
$f_{xx}f_{yy} - f_{xy}^2 > 0,$ $f_{xx} < 0$	a strict maximum point

The proof will be given in §4.3.

EXAMPLES 1.   The function $z = f(x,y) = x^2 + 4y^2$ has a strict minimum point at the origin. This is seen directly; the graph is an elliptic paraboloid (see Chapter 18, §5.4), the level lines are shown in Figure 19.8. Let us verify Theorem 1: we have $f_x = 2x, f_y = 8y, f_{xx} = 2, f_{yy} = 8, f_{xy} = 0$. At $(0,0)$ we have $f_x = f_y = 0, f_{xx} = 2 > 0$, and $f_{xx}f_{yy} - f_{xy}^2 = 16 > 0$.

2.   The function $z = f(x,y) = x^2 - 4y^2$ has the origin as the only critical point. Since $f_{xx}f_{yy} - f_{xy}^2 = -16 < 0$, this is a saddle point. The level lines are shown in Figure 19.9. The surface $z = f(x,y)$ is a hyperbolic paraboloid (see Chapter 18, §5.4).

3.   The function $z = (x^2 + y^2)^2$ has a strict minimum point at $x = y = 0$, but this could not have been detected by Theorem 1, since $f_{xx} = f_{xy} = f_{yy} = 0$ at the origin.

4.   The function $z = f(x,y) = x^2$ has a critical point at every point $x = 0$, and no other critical points. All critical points are local minima, but none is strict. Theorem 1 is not applicable.

PROBLEMS

Find and classify all critical points of the following functions $f(x,y)$.

1. $f = 2x^2 + y^2 - 2x + y.$
2. $f = x^2 - y^2 + 4x + 2y.$

3. $f = xy.$
4. $f = -x^2 - y^2 + xy + 6x.$

5. $f = x^2 + 2y^2 - xy + 3x + 2y.$

6. $f = xy^2 + 4x^2 + 2y.$

7. $f = x^2y + y^2 + 8x.$

8. $f = x^3 + y^3 + 3xy.$

9. $f = x^3 + y^3 - 3x - 3y.$

10. $f = x^4 + y^4 + 4x - 4y.$

4.2 Finding largest and smallest values

Suppose we are given a continuous function $(x,y) \mapsto f(x,y)$ defined in a bounded domain and on its boundary. (A domain is called **bounded** if it lies in some finite interval.) We want to find the points where the function takes on its largest and its smallest values—its **maximum** and **minimum** in the domain. (We take the existence of these points for granted at this stage ▶.) The maximum and minimum *may* be achieved on the boundary. If the maximum (or minimum) of f is attained at a point P not on the boundary of the domain considered, it must be a local maximum (or a local minimum) point and hence a critical point.

This suggests the following procedure for finding the absolute maxima and minima. First, find all critical points inside the domain by solving the equations $f_x(x,y) = 0$, $f_y(x,y) = 0$. Second, determine which of these critical points are local maxima and local minima; Theorem 1 may be helpful. Third, find the points on the boundary of the region that may be maxima or minima points. If the boundary consists of smooth curves, this is reduced to finding maxima and minima of functions of one variable. Finally, compare the values of the function at the points considered.

EXAMPLE What is the largest value (maximum) and smallest value (minimum) of the function

$$f(x,y) = 2x^2 - xy + y^2 + 7x$$

in the closed interval

$$-3 \le x \le 3, \qquad -3 \le y \le 3,$$

and where is it achieved?

ANSWER We have

$$f_x = 4x - y + 7, \qquad f_y = -x + 2y.$$

Solving the simultaneous equations $f_x = f_y = 0$, we see that the only critical point is at $x = -2, y = -1$; it lies in the interval considered. Since $f_{xx} = 4, f_{xy} = -1, f_{yy} = 2$, we have $f_{xx}f_{yy} - f_{xy}^2 = 7$. Hence (see Theorem 1 in §4.1) the critical point is a local minimum point.

Next we consider the function $f(x,y)$ for $y = -3$, that is, we look at the function

$$x \mapsto f(x,-3) = 2x^2 + 10x + 9$$

in the interval $-3 \le x \le 3$. The derivative is $4x + 10$. This is seen either by differentiation, or by setting $y = -3$ in the expression for f_x. The derivative is 0 at $x = -5/2$. The values of x at which $f(x,-3)$ can achieve extreme values are, therefore, the endpoints of the interval $x = -3$, $x = +3$, and the root $x = -5/2$ of the derivative.

In a similar way we investigate the other three edges of our interval; on the edge $y = 3$, the function $x \mapsto f(x,3)$ has a critical point at $x = -1$. On the two vertical edges, the functions $y \mapsto f(-3,y)$ and $y \mapsto f(3,y)$ have no critical points. Thus the only points at which f may take an extreme value are the critical point inside the interval, the four vertices, and the two points on the horizontal edges at which $f_x = 0$. Now we compute the values of f at the seven "interesting" points and obtain the table

x	-3	-1	3	-2	-3	$-\frac{5}{2}$	3
y	3	3	3	-1	-3	-3	-3
f	15	7	39	-7	-3	$-\frac{7}{2}$	57

Thus the maximum of $f(x,y)$ in the closed interval $|x| \leq 3$, $|y| \leq 3$ is 57; it is achieved at $x = 3$, $y = -3$. The minimum is -7 and is achieved at $x = -2$, $y = -1$.

PROBLEMS

In Problems 1 to 4 find the largest value (maximum) and smallest value (minimum) of the given functions $f(x,y)$ in the indicated closed intervals.

1. $f = x^2 + y^2 - 8x - 2y$; $0 \leq x \leq 2$, $0 \leq y \leq 4$.
2. $f = -x^2 - 2y^2 + 2x - 4y + 1$; $0 \leq x \leq 4$, $|y| \leq 2$.
3. $f = x^2 + y^2 - xy + 3y$; $|x| \leq 2$, $|y| \leq 3$.
4. $f = -x^2 - 2y^2 + 2xy + 6x - 4y$; $0 \leq x \leq 8$, $|y| \leq 2$.

5. If $f(x,y) = x^3 + y^3 - 3xy$, find the largest and smallest values of f in the triangle with vertices $(0,0)$, $(0,1)$, $(1,0)$.
6. If $f(x,y) = e^{x^2 + y^2 + y}$, find the largest and smallest values of f in the square with corners at $(1,0)$, $(0,1)$, $(-1,0)$, $(0,-1)$.
7. Find three positive numbers whose sum is 300 such that the sums of squares is a minimum.
8. Find three positive numbers whose sum is 240 such that the sums of products taken two at a time is a maximum.
9. What dimensions should a closed rectangular box have if its volume is V and its surface area is to be as small as possible?
10. What dimensions should an open rectangular box have if its volume is V and its surface area is to be as small as possible?
11. A rectangular storage bin is to be constructed to hold 288 cubic feet of grain. If the top and bottom cost 20 cents per square foot, and the sides cost 15 cents per square foot, what dimensions should the bin have to minimize the cost?
12. Find the shortest distance from the origin to the plane $ax + by + cz + d = 0$.
13. Find the point on the surface $z = xy - 1$ nearest the origin.
14. Find the minimum distance between the lines with parametric equations $x = 2t - 1$, $y = t$, $z = t + 2$ and $x = s$, $y = s - 2$, $z = s + 4$.
15. Find the volume of the largest rectangular box that can be inscribed in the ellipsoid $(x/a)^2 + (y/b)^2 + (z/c)^2 = 1$.

4.3 Taylor's theorem and the classification of critical points‡

We show first how Taylor's theorem can be extended to functions of two variables. Actually, we shall deal only with one special case of Taylor's theorem, but the method is general (and applicable also to functions of more than two variables).

Theorem 2. *Let $f(x,y)$ be twice continuously differentiable near (x_0,y_0). For (x,y) near (x_0,y_0),*

$$f(x,y) = f(x_0,y_0) + f_x(x_0,y_0)(x - x_0) + f_y(x_0,y_0)(y - y_0)$$
$$+ \tfrac{1}{2}\{ f_{xx}(\hat{x},\hat{y})(x - x_0)^2 + 2f_{xy}(\hat{x},\hat{y})(x - x_0)(y - y_0) + f_{yy}(\hat{x},\hat{y})(y - y_0)^2 \}, \quad (1)$$

where $(\hat{x},\hat{y})$ is a (suitably chosen) point on the straight segment from (x_0,y_0) to (x,y), not an endpoint.

‡This subsection may be omitted without loss of continuity.

FIGURE **19.14**

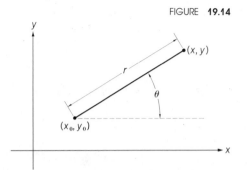

Proof. Set

$$x = x_0 + r \cos \theta, \qquad y = y_0 + r \sin \theta$$

(see Figure 19.14) and consider, for a fixed θ, the value of f at (x,y) as a function of r. In other words, consider the auxiliary function

$$F(r) = f(x_0 + r \cos \theta, y_0 + r \sin \theta).$$

Using the chain rule we compute the derivatives of F:

$$F'(r) = f_x(\ldots) \cos \theta + f_y(\ldots) \sin \theta, \tag{2}$$
$$F''(r) = f_{xx}(\ldots) \cos^2 \theta + 2f_{xy}(\ldots) \cos \theta \sin \theta + f_{yy}(\ldots) \sin^2 \theta. \tag{3}$$

Here the symbol $\ldots$ stands for $x_0 + r \cos \theta, y_0 + r \sin \theta$. Now write down the first case of Taylor's theorem for the function F:

$$F(r) = F(0) + F'(0)r + \frac{1}{2!} F''(\rho)r^2; \tag{4}$$

here ρ is a number such that $0 < \rho < r$. If we substitute the expressions obtained above for F, F', F'', note that $r \cos \theta = x' - x_0$, $r \sin \theta = y - y_0$, and set $\hat{x} = x_0 + \rho \cos \theta$, $\hat{y} = y_0 + \rho \sin \theta$, we obtain (1). Note that we do not know in advance where the point $(\hat{x},\hat{y})$ lies, we know only that there *is* such a point on the line from (x_0,y_0) to (x,y).

Now we prove Theorem 1 in §4.1. Let $\widehat{A}$, $\widehat{B}$, $\widehat{C}$ denote the values of f_{xx}, f_{xy}, f_{yy} at the point $(\hat{x},\hat{y})$ in (1), and let A, B, C denote the values of these derivatives at (x_0,y_0). If $f_x = f_y = 0$ at (x_0,y_0), as is required by the conditions of Theorem 1, (1) may be rewritten as

$$f(x,y) - f(x_0,y_0) = \tfrac{1}{2}(\widehat{A}\xi^2 + 2\widehat{B}\xi\eta + \widehat{C}\eta^2), \tag{5}$$

where we set $\xi = x - x_0$, $\eta = y - y_0$.

Assume that $AC - B^2 < 0$. Then the quadratic polynomial $At^2 + 2Bt + C = 0$ has 2 distinct roots. Hence there are numbers p and q such that $Ap^2 + 2Bp + C > 0$, $Aq^2 + 2Bq + C < 0$.

For a positive ϵ, set $\xi = p\epsilon$, $\eta = \epsilon$ (that is, $x = x_0 + \epsilon p$, $y = y_0 + \epsilon$). If ϵ is small, (x,y) is close to (x_0,y_0) and so is $(\hat{x},\hat{y})$. Hence $\widehat{A}p^2 + 2\widehat{B}p + \widehat{C} > 0$ and, by (5),

$$f(x,y) - f(x_0,y_0) = \frac{1}{2}(\widehat{A}\xi^2 + 2\widehat{B}\xi\eta + \widehat{C}\eta^2) = \frac{\epsilon^2}{2}(\widehat{A}p^2 + 2\widehat{B}p + \widehat{C}) > 0.$$

If we set $\xi = q\epsilon$, $\eta = \epsilon$, we obtain a point (x,y), as close as we want to (x_0,y_0), with $f(x,y) - f(x_0,y_0) < 0$. Hence (x_0,y_0) is a saddle point for f.

Assume next that $AC - B^2 > 0$, $A > 0$. Then $C > 0$. For (x,y) close to (x_0,y_0), $\widehat{A}\widehat{C} - \widehat{B}^2 > 0$, $\widehat{A} > 0$, $\widehat{C} > 0$. By (1),

$$\phi(x,y) - \phi(x_0,y_0) = \frac{1}{2} \{\widehat{A}\xi^2 + 2\widehat{B}\xi\eta + \widehat{C}\eta^2\}$$

$$= \frac{1}{2} (\widehat{A}^{1/2}\xi + \widehat{B}\widehat{A}^{-1/2}\eta)^2 - \widehat{B}^2\widehat{A}^{-1}\eta^2 + \widehat{C}\eta^2$$

$$= \frac{1}{2} (\widehat{A}^{1/2}\xi + \widehat{B}\widehat{A}^{-1/2}\eta)^2 + \frac{\widehat{A}\widehat{C} - \widehat{B}^2}{\widehat{A}} \eta^2 > 0,$$

provided $\eta \neq 0$. We see similarly that $\phi(x,y) - \phi(x_0,y_0) > 0$ if $\xi \neq 0$. Since ξ and η are not both 0 if $(x,y) \neq (x_0,y_0)$, we conclude that (x_0,y_0) is a strict minimum point for f.

The case $AC - B^2 > 0$, $A < 0$ is treated similarly.

§5 Two theorems on partial derivatives[‡]

In this section we prove two important theorems on partial derivatives, stated in §2.2 and in §3.1. Both proofs rely on the mean value theorem (Theorem 1 in Chapter 14, §1.1).

5.1 Linear approximation

We begin by proving Theorem 1 in §2.2. To simplify writing, assume that $(x_0,y_0) = (0,0)$, and set $g(x,y) = f(x,y) - z_0 - ax - by$. Then $g(x,y)$ is continuously differentiable, and

$$g(0,0) = g_x(0,0) = g_y(0,0) = 0.$$

Next, set $r(x,y) = g(x,y)(x^2 + y^2)^{-1/2}$ [for (x,y) close to $(0,0)$ but $(x,y) \neq (0,0)$]. We must show that, for every given $\epsilon > 0$, $|r(x,y)| < \epsilon$ if $x^2 + y^2$ is small enough. Let $\delta > 0$ be chosen so that $|g_x(x,y)| < |g_x(x,y)| + |g_y(x,y)| < \epsilon$ for $(x^2 + y^2)^{1/2} < \delta$. (There is such a δ since g_x and g_y are continuous.) By the mean value theorem,

$$g(x,y) = g(0,y) + g_x(\xi,y)x \qquad \text{(for some ξ between 0 and x)}$$
$$= g_y(0,\eta)y + g_x(\xi,y)x \qquad \text{(for some η between 0 and y).}$$

If $x^2 + y^2 < \delta^2$, then $\xi^2 + y^2 < \delta^2$ and $y^2 < \delta^2$; hence $|g_y(0,\eta)| + |g_x(\xi,y)| < \epsilon$ and $|r(x,y)| < \epsilon$.

5.2 Mixed derivatives

We prove next Theorem 1 of §3.1 which asserts that if $f(x,y)$ is defined and has continuous partial derivatives of the first and second order in an open set, then $f_{xy} = f_{yx}$.

We choose a point (x_0,y_0) and consider the expression

[‡]Optional section.

FIGURE **19.15**

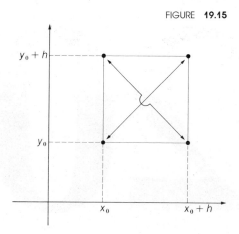

$$\delta(h) = f(x_0 + h, y_0 + h) - f(x_0 + h, y_0) - f(x_0, y_0 + h) + f(x_0, y_0), \qquad (1)$$

where $h \neq 0$ and $|h|$ is small. (This is the sum of the values of f at two opposite corners of the square shown in Figure 19.15 minus the sum of the values at the other corners. In the figure, we have $h > 0$.) We shall show that

$$\lim_{h \to 0} \frac{\delta(h)}{h} = f_{xy}(x_0, y_0) \qquad \text{and also} \qquad \lim_{h \to 0} \frac{\delta(h)}{h} = f_{yx}(x_0, y_0), \qquad (2)$$

which establishes that the two mixed partials are equal.

Consider the functions

$$F(x) = f(x, y_0 + h) - f(x, y_0), \qquad F'(x) = f_x(x, y_0 + h) - f_x(x, y_0). \qquad (3)$$

Now

$$
\begin{aligned}
\delta(h) &= F(x_0 + h) - F(x_0) & &\text{[by (1) and (3)]} \\
&= hF'(\xi) & &\text{(where ξ is between 0 and h, by the mean value theorem)} \\
&= h[f_x(\xi, y_0 + h) - f_x(\xi, y_0)] & &\text{[by (3)]} \\
&= h^2 f_{xy}(\xi, \eta) & &\text{(by the mean value theorem; η is between y_0 and $y_0 + h$).}
\end{aligned}
$$

Thus

$$\frac{\delta(h)}{h^2} = f_{xy}(\xi, \eta).$$

The numbers ξ and η depend on h, but if $|h|$ is small enough, the point (ξ, η) is as close as we want to (x_0, y_0). Since f_{xy} is a continuous function, the first assertion (2) follows.

Next we consider the function $G(y) = f(x_0 + h, y) - f(x_0, y)$ and, using the mean value theorem twice, obtain

$$
\begin{aligned}
\delta(h) &= G(y_0 + h) - G(y_0) = hG'(\widehat{\eta}) \\
&= hf_y(x_0 + h, \widehat{\eta}) - f_y(x_0, \widehat{\eta}) = h^2 f_{yx}(\widehat{\xi}, \widehat{\eta}),
\end{aligned}
$$

where $\widehat{\xi}$ and $\widehat{\eta}$ are numbers, depending on h, which lie between x_0 and $x_0 + h$ and between y_0 and $y_0 + h$, respectively. Hence $\delta(h)/h^2 = f_{yx}(\widehat{\xi}, \widehat{\eta})$, and the second assertion (2) follows.

MULTIPLE INTEGRALS

§1 Double integrals as volumes

We now extend the second basic procedure of calculus, integration, to functions of two (or more) variables. We proceed as we did in Chapter 8 where we introduced integrals for functions of one variable. First we base the concept of a double integral on geometric intuition; only then do we give an analytic definition.

1.1 The double integral of a nonnegative function over an interval

Consider first a continuous nonnegative function of two variables, $z = f(x,y)$, defined in an interval I: $a < x < b$, $c < y < d$ (and also on its boundary). The graph of this function, that is, the surface $z = f(x,y)$, together with the five planes, $x = a$, $x = b$, $y = c$, $y = d$, and $z = 0$, bounds a solid. Our geometric intuition tells us that this solid has a **volume** V.

The number V is called the **double integral** of $f(x,y)$ over the interval I; the Leibniz notation for this is

$$V = \iint_I f(x,y)\, dx\, dy \qquad \text{or} \qquad V = \iint_{\substack{a < x < b \\ c < y < d}} f(x,y)\, dx\, dy. \tag{1}$$

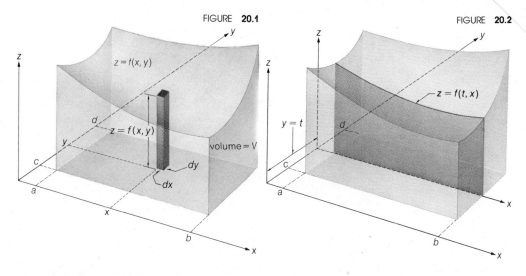

FIGURE 20.1   FIGURE 20.2

The notation can be justified by a mathematical myth. The volume under the surface is composed of infinitely many infinitely thin three-dimensional boxes. Each of these is erected over a point (x,y) in I: it has the infinitely small width and breadth, dx and dy, and the height $f(x,y)$. Its volume is therefore $f(x,y)\,dx\,dy$ (see Figure 20.1). The total volume is the sum of all these infinitely small volumes.

To compute (1) we apply the volume formula in Chapter 9, §2.2, choosing as the line L the y-axis, and obtain

$$V = \int_c^d A(t)\,dt,$$

where $A(t)$ is the area of the intersection of our solid with the plane $y = t$ (see Figure 20.2). But what is $A(t)$? In the plane $y = t$, we can use x and z as Cartesian coordinates. The intersection of the plane $y = t$ and our solid is the region under the curve $z = f(x,t)$ from $x = a$ to $x = b$, as is seen from Figure 20.2. The area $A(t)$ of this intersection is therefore $A(t) = \int_a^b f(x,t)\,dx$, and the desired volume is $V = \int_c^d \{\int_a^b f(x,t)\,dx\}\,dt$. The name of a dummy variable is of no importance, and we obtain a better looking formula if we replace t by y; noting (1), we write

$$\iint_I f(x,y)\,dx\,dy = \int_c^d \left\{ \int_a^b f(x,y)\,dx \right\} dy. \tag{2}$$

Agreeing once and for all that the *inner integral sign goes with the inner differential and the outer integral sign goes with the outer differential*, we omit the braces and write

$$\iint_I f(x,y)\,dx\,dy = \int_c^d \int_a^b f(x,y)\,dx\,dy. \tag{2'}$$

We could have reversed the parts of x and y. In this case, an entirely analogous argument would have given

$$\iint_I f(x,y)\,dx\,dy = \int_a^b \left\{ \int_c^d f(x,y)\,dy \right\} dx = \int_a^b \int_c^d f(x,y)\,dy\,dx. \tag{3}$$

Formulas (2) and (3) reduce the calculation of double integrals to two ordinary integrations (we also say: to **iterated integrals,** or repeated integrals).

REMARK An important special case of formulas (2) and (3) is the statement: *the double integral (over an interval) of a product of functions of one variable is a product of single integrals:*

$$\iint\limits_{\substack{a<x<b \\ c<y<d}} \phi(x)\psi(y)\,dx\,dy = \int_a^b \phi(x)\,dx \int_c^d \psi(y)\,dy.$$

Indeed, by (2),

$$\iint\limits_{\substack{a<x<b \\ c<y<d}} \phi(x)\psi(y)\,dx\,dy = \int_c^d \left\{ \int_a^b \phi(x)\psi(y)\,dx \right\} dy$$

$$= \int_c^d \psi(y) \left\{ \int_a^b \phi(x)\,dx \right\} dy = \int_a^b \phi(x)\,dx \int_c^d \psi(y)\,dy,$$

since $\int_a^b \phi(x)\,dx$ is a constant.

EXAMPLES 1. Compute the volume under the graph of $z = f(x,y) = xy + 1$ over the interval $0 < x < 2,\, 0 < y < 4$.
ANSWER The desired volume is the double integral

$$\iint\limits_{\substack{0<x<2 \\ 0<y<4}} (xy + 1)\,dx\,dy.$$

Using (2), and remembering that the inner integral sign goes with the inner differential and the outer integral sign goes with the outer differential, we may write this double integral as an iterated integral:

$$\int_0^4 \int_0^2 (xy + 1)\,dx\,dy = \int_0^4 \left\{ \int_0^2 (xy + 1)\,dx \right\} dy.$$

But $xy + 1$, considered as a function of x (with y treated as a constant) has $\frac{1}{2}x^2y + x$ as a primitive function. Hence

$$\int_0^2 (xy + 1)\,dx = (\tfrac{1}{2}x^2y + x) \,\Big|_{x=0}^{x=2} = (2y + 2) - (0 + 0) = 2y + 2.$$

The double integral equals

$$\int_0^4 (2y + 2)\,dy = (y^2 + 2y) \,\Big|_0^4 = 24.$$

The desired volume is 24.

2. Compute the double integral $\int_0^{\pi/2} \int_0^{\pi} \sin x \cos y\,dx\,dy$.
ANSWER Using the Remark, we rewrite the given integral as

$$\left(\int_0^{\pi} \sin x\,dx \right)\left(\int_0^{\pi/2} \cos y\,dy \right).$$

Evaluating each single integral separately, we obtain

$$(-\cos x) \,\Big|_{x=0}^{x=\pi} \cdot (\sin y) \,\Big|_{y=0}^{y=\pi/2} = (2)(1) = 2.$$

3. Find the integral $\iint_I(6x^2y + 8xy^3)\,dx\,dy$, where I is the interval $1 < x < 2$, $3 < y < 4$.
ANSWER   The desired integral equals

$$\int_3^4 \int_1^2 (6x^2y + 8xy^3)\,dx\,dy = \int_3^4 \left\{ \int_1^2 (6x^2y + 8xy^3)\,dx \right\} dy.$$

But $6x^2y + 8xy^3$, considered as a function of x, with y treated as a constant, has $2x^3y + 4x^2y^3$ as a primitive function. Hence

$$\int_1^2 (6x^2y + 8xy^3)\,dx = (2x^3y + 4x^2y^3) \, \Big|_{x=1}^{x=2} = (16y + 16y^3) - (2y + 4y^3)$$

$$= 14y + 12y^3,$$

and the double integral equals

$$\int_3^4 (14y + 12y^3)\,dy = (7y^2 + 3y^4) \, \Big|_3^4 = 574.$$

We integrated first with respect to x and then with respect to y; that is, we used Equation (2). But we can also integrate first with respect to y, then with respect to x; that is, we can use Equation (3). In this case, we obtain

$$\int_1^2 \int_3^4 (6x^2y + 8xy^3)\,dy\,dx = \int_1^2 \left\{ \int_3^4 (6x^2y + 8xy^3)\,dy \right\} dx = \int_1^2 \left\{ (3x^2y^2 + 2xy^4) \, \Big|_{y=3}^{y=4} \right\} dx$$

$$= \int_1^2 \{(48x^2 + 512x) - (27x^2 + 162x)\}\,dx = \int_1^2 (21x^2 + 350x)\,dx$$

$$= (7x^3 + 175x^2) \, \Big|_1^2 = 574,$$

as expected.

PROBLEMS

In each of Problems 1 to 6 compute the volume under the graph of $z = f(x,y)$ over the given interval I. Do each problem twice, once by integrating first with respect to x and once by integrating first with respect to y.

1. $f(x,y) = x + 4y$; $I: 0 < x < 2, 1 < y < 2$.
2. $f(x,y) = 3x^2y + 1$; $I: 1 < x < 2, 1 < y < 3$.
3. $f(x,y) = \sqrt{xy} + 2x + y$; $I: 0 < x < 1, 0 < y < 4$.
4. $f(x,y) = \cos x \cos 2y$; $I: 0 < x < \pi/2, 0 < y < \pi/4$.
5. $f(x,y) = e^{2x+y}$; $I: 0 < x < \tfrac{1}{2}, 0 < y < 1$.
6. $f(x,y) = y^2e^{-x}$; $I: 0 < x < 1, 0 < y < 9$.

In each of Problems 7 to 12 evaluate the given integrals.

7. $\displaystyle\int_0^9 \int_0^3 (3x^2 + y^2)\,dx\,dy.$

8. $\displaystyle\int_0^2 \int_0^1 (4x^2y^3 + x^4y)\,dx\,dy.$

9. $\displaystyle\int_0^{\pi/4} \int_0^1 e^{-x} \sin 2y\,dx\,dy.$

10. $\displaystyle\int_0^1 \int_0^1 x\sqrt{x^2 + y}\,dx\,dy.$

11. $\displaystyle\int_0^1 \int_0^{1/2} ye^{2xy}\,dx\,dy.$

12. $\displaystyle\int_0^{\pi/2} \int_0^1 x \cos xy\,dx\,dy.$

1.2   Integrals of functions that take on negative values

If the function $f(x,y)$ also takes on negative values, then the double integral

$$\iint\limits_{\substack{a<x<b \\ c<y<d}} f(x,y)\, dx\, dy$$

is interpreted as the *difference:* the volume above the xy plane and below the surface $z = f(x,y)$ minus the volume below the xy plane and above the surface $z = f(x,y)$.

The argument given in §1.1 can be repeated, with some modifications, to show that formulas (2) and (3) still hold. (The Remark also holds.)

PROBLEMS

Evaluate the following integrals.

1. $\displaystyle\int_0^2 \int_{-1}^0 (2x - y)\, dx\, dy.$

2. $\displaystyle\int_{-2}^1 \int_0^1 x^2 y^3\, dx\, dy.$

3. $\displaystyle\int_{-2}^2 \int_{-3}^3 (3x^2 y - xy^2)\, dx\, dy.$

4. $\displaystyle\int_0^{3\pi/2} \int_0^{\pi/2} \cos x \sin 2y\, dx\, dy.$

5. $\displaystyle\int_0^{\pi/3} \int_0^{\pi/2} \cos(2x + y)\, dx\, dy.$

6. $\displaystyle\int_0^1 \int_{-1}^0 xe^{-xy}\, dx\, dy.$

1.3   Double integrals over general regions

The description of the double integral as a volume (or, in the case of a function that takes on negative values, as a difference of volumes) is not restricted to integration over intervals or to continuous functions. We will admit as region of integration D any "reasonable" domain. (What is meant by "reasonable" will be defined in §2.1 and §2.3 below.)

If $f(x,y)$ is a "reasonable," nonnegative function defined in D, the number

$$\iint\limits_{D} f(x,y)\, dx\, dy \tag{4}$$

is the volume of the solid above D and below the surface $z = f(x,y)$. This solid consists of all points (x,y,z) with (x,y) in D and $0 < z < f(x,y)$. If the function f also takes on negative values, (4) is the difference of the volume above D and below $z = f(x,y)$ and the volume below D and above $z = f(x,y)$.

Integrals like (4) can be computed by a method which we explain in an example.

Let D denote the unit disk, the set of all points (x,y) with $x^2 + y^2 < 1$. We want to compute the volume V enclosed between the disk D, the plane $z = 3 + x$, and the cylinder $x^2 + y^2 = 1$ (see Figure 20.3), that is, the double integral

$$V = \iint\limits_{D} (3 + x)\, dx\, dy. \tag{5}$$

Define a new function

$$f(x,y) = \begin{cases} 3 + x, & \text{if } x^2 + y^2 < 1, \\ 0, & \text{if } x^2 + y^2 \geq 1. \end{cases} \tag{6}$$

FIGURE **20.3**   FIGURE **20.4**

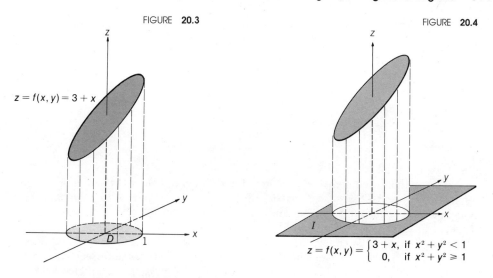

$z = f(x, y) = 3 + x$

$$z = f(x, y) = \begin{cases} 3 + x, & \text{if } x^2 + y^2 < 1 \\ 0, & \text{if } x^2 + y^2 \geq 1 \end{cases}$$

If I is any interval containing the disk D, then

$$\iint_I f(x,y)\, dx\, dy = \iint_D (3 + x)\, dx\, dy. \tag{7}$$

This means that the volume V may be thought of as the volume enclosed between the interval I in the plane $z = 0$ and the graph of the function $z = f(x,y)$; the graph over points outside the disk lies on the plane $z = 0$ and contributes nothing to the volume (compare Figure 20.4). [We can also verify (7) using the volume formula in Chapter 9, §2.2.]

Using (7) and (6), we can compute V by an iterated integration. It does not matter how large I is, as long as it encloses D; we may assume that I is the interval $-1 < x < 1$, $-1 < y < 1$; see Figure 20.5. We have

$$V = \iint_D (3 + x)\, dx\, dy = \int_{-1}^{1} \int_{-1}^{1} f(x,y)\, dx\, dy.$$

FIGURE **20.5**

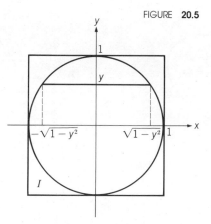

But, for a fixed y, we have $f(x,y) = 0$ for $x^2 + y^2 > 1$, that is, for $x^2 > 1 - y^2$, that is, for x outside the interval $|x| < |\sqrt{1 - y^2}|$. Hence

$$\int_{-1}^{1} f(x,y)\, dx = \int_{-\sqrt{1-y^2}}^{\sqrt{1-y^2}} f(x,y)\, dx = \int_{-\sqrt{1-y^2}}^{\sqrt{1-y^2}} (3 + x)\, dx$$

and

$$V = \int_{-1}^{1} \int_{-\sqrt{1-y^2}}^{\sqrt{1-y^2}} (3 + x)\, dx\, dy = \int_{-1}^{1} 6\sqrt{1 - y^2}\, dy = 3\pi$$

(here we used the result of Chapter 8, §3.5).

The method used above is quite general. *In order to compute the double integral of f over a region D, set $f = 0$ outside D, and integrate the resulting function over some interval containing D.* The double integral over an interval is computed as an iterated integral (see §1.1 and §1.2).

We say that D is convex in the x-direction if there are numbers γ and δ such that for $y_0 < \gamma$ or $y_0 > \delta$ the horizontal line $y = y_0$ does not intersect D, and for $\gamma < y_0 < \delta$ the intersection of D and the line $y = y_0$ is a single segment, $\alpha(y_0) < x < \beta(y_0)$; see Figure 20.6. In this case, we have

$$\iint_D f(x,y)\, dx\, dy = \int_{\gamma}^{\delta} \left\{ \int_{\alpha(y)}^{\beta(y)} f(x,y)\, dx \right\} dy$$

or, recalling the convention that "the outer integral sign belongs with the outer differential,"

$$\iint_D f(x,y)\, dx\, dy = \int_{\gamma}^{\delta} \int_{\alpha(y)}^{\beta(y)} f(x,y)\, dx\, dy. \qquad (8)$$

If D is convex in the y-direction (in the sense indicated in Figure 20.7), we can first integrate with respect to y and then with respect to x:

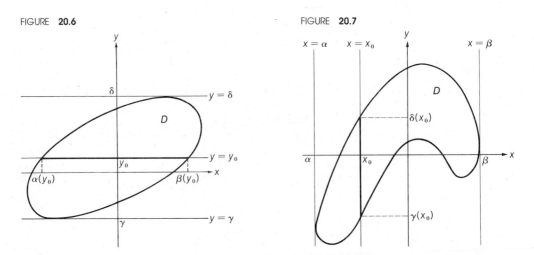

FIGURE **20.6**

FIGURE **20.7**

$$\iint_D f(x,y)\,dx\,dy = \int_\alpha^\beta \int_{\gamma(x)}^{\delta(x)} f(x,y)\,dy\,dx; \tag{9}$$

the meaning of α, β, γ, and δ is clear from the figure. In practice, we encounter only regions of integration D that are either convex in one direction or can be decomposed into several *disjoint* regions $D_1, D_2, \ldots$ with this property. In the latter case the integral (4) equals the sum of the integrals over $D_1, D_2, \ldots$. (This follows from the interpretation of the double integral as a volume.)

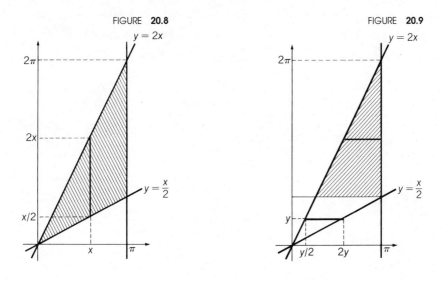

FIGURE **20.8**   FIGURE **20.9**

EXAMPLES **1.**   Let D be the triangular region bounded by the lines $2y = x$, $y = 2x$, and $x = \pi$. Compute $\iint_D \sin y\,dx\,dy$.

ANSWER   In doing such problems, it is essential to make a sketch of the region of integration; see Figure 20.8. The sketch shows that D is convex in both directions, x and y. We shall first integrate with respect to y. For every point (x,y) in D, x lies between 0 and π. If we draw a vertical line at the distance x from the y-axis, its intersection with D has endpoints with $y = x/2$ and $y = 2x$. Hence we have

$$\iint_D \sin y\,dx\,dy = \int_0^\pi \int_{x/2}^{2x} \sin y\,dy\,dx = \int_0^\pi [-\cos y]_{y=x/2}^{y=2x}\,dx$$

$$= \int_0^\pi \left(\cos\frac{x}{2} - \cos 2x\right)dx = \left(2\sin\frac{x}{2} - \frac{1}{2}\sin 2x\right)\Big|_0^\pi = 2\sin 90° = 2.$$

If we want to integrate first with respect to x, we note that, for (x,y) in D, we have $0 < y < 2\pi$, and (see Figure 20.9) that a horizontal line (a line with fixed y) intersects D in a segment with endpoints at

$$x = \frac{y}{2} \quad\text{and}\quad x = 2y \quad\text{if } 0 < y < \frac{\pi}{2}, \qquad x = \frac{y}{2} \quad\text{and}\quad x = \pi \quad\text{if } \frac{\pi}{2} < y < 2\pi.$$

Therefore we think of D as decomposed into two regions and write

$$\iint_D \sin y \, dx \, dy = \int_0^{\pi/2} \left\{ \int_{y/2}^{2y} \sin y \, dx \right\} dy + \int_{\pi/2}^{2\pi} \left\{ \int_{y/2}^{\pi} \sin y \, dx \right\} dy$$

$$= \int_0^{\pi/2} \left(2y - \frac{y}{2} \right) \sin y \, dy + \int_{\pi/2}^{2\pi} \left(\pi - \frac{y}{2} \right) \sin y \, dy$$

$$= \frac{3}{2} \int_0^{\pi/2} y \sin y \, dy - \frac{1}{2} \int_{\pi/2}^{2\pi} y \sin y \, dy + \pi \int_{\pi/2}^{2\pi} \sin y \, dy.$$

Since $y \sin y$ has a primitive function $\sin y - y \cos y$ [as can be found by integration by parts (compare Chapter 13, §2.1) and checked by direct differentiation], the above equals

$$\tfrac{3}{2}[\sin y - y \cos y]_0^{\pi/2} - \tfrac{1}{2}[\sin y - y \cos y]_{\pi/2}^{2\pi} - \pi[\cos y]_{\pi/2}^{2\pi} = \tfrac{3}{2} + \tfrac{1}{2} + \pi - \pi = 2,$$

at it must.

2. Let D be the set of points (x,y) with $x > 0$, $y > x^2$, $y < 2 - x^2$. Compute $\iint_D \sqrt{xy} \, dx \, dy$.

ANSWER   The set of points (x,y) with $x > 0$ is the right half-plane. The set of points (x,y) with $y > x^2$ is the set above the parabola $y = x^2$. The set of points (x,y) with $y < 2 - x^2$ is the set below the parabola $y = 2 - x^2$. The set D is shown in Figure 20.10; it is the *intersection*

FIGURE **20.10**

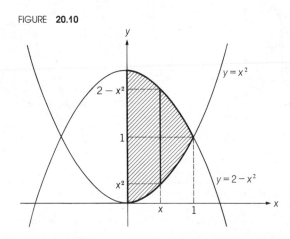

of the three sets; that is, it consists of points belonging to all of them. The two parabolas intersect when $2 - x^2 = x^2$, that is, when $2x^2 = 2$; the intersection point in the right half-plane has the coordinates $(1,1)$. In D the coordinate x varies from $x = 0$ to $x = 1$. A vertical line with a fixed x intersects D in a segment with endpoints $y = x^2$ and $y = 2 - x^2$. Therefore

$$\iint_D \sqrt{xy} \, dx \, dy = \int_0^1 \int_{x^2}^{2-x^2} \sqrt{xy} \, dy \, dx = \int_0^1 \sqrt{x} \left[\frac{y^2}{2} \right]_{y=x^2}^{y=2-x^2} dx$$

$$= \int_0^1 \sqrt{x} \, \frac{1}{2} [(2 - x^2)^2 - x^4] \, dx = \int_0^1 (2x^{1/2} - 2x^{5/2}) \, dx$$

$$= \left[\frac{4}{3} x^{3/2} - \frac{4}{7} x^{7/2} \right]_0^1 = \frac{16}{21}.$$

Our region of integration is also convex in the x-direction. The reader should compute the integral, integrating first in the x-direction.

PROBLEMS

In Problems 1 to 9 we are given a set D and a function $f(x,y)$. Compute $\iint_D f(x,y)\,dx\,dy$.

1. D is the interior of the triangle with vertices $(0,0)$, $(0,1)$, $(1,1)$; $f = x^2 y$.
2. D is the interior of the triangle with vertices at $(0,0)$, $(0,1)$, $(1,0)$; $f = x^2 y$.
3. D is the interior of the square with vertices $(1,0)$, $(0,1)$, $(-1,0)$, $(0,-1)$; $f = xy$.
4. D is the interior of the unit circle; $f = 3x + 4y$.
5. D is the region bounded by the curve $y = \sin x$ and the segment $0 \le x \le \pi$; $f = xy$.
6. D is the bounded region enclosed by the two parabolas $y = x^2$ and $y = -x^2 + 1$; $f(x,y) = x\sqrt{y}$.
7. D is the interior of the ellipse $(x/a)^2 + (y/b)^2 = 1$; $f = c\sqrt{1 - (x/a)^2 - (y/b)^2}$.
8. D is the interior of the circle $x^2 + y^2 = R^2$; $f = \sqrt{R^2 - x^2 - y^2}$. (Give a geometric interpretation of your result.)
9. D is the region cut off by the x- and y-axes between the two parallel lines of slope 2 that pass through $(1,0)$ and $(2,0)$; $f = \cos(x + y)$.

1.4 Properties of double integrals

We describe now some properties of double integrals analogous to properties of integrals of functions of one variable, stated in Chapter 8, §1.3 and §3.2. These properties are:

$$\iint_D dx\,dy = \iint_D 1\,dx\,dy = A = \text{area of } D, \tag{10}$$

$$\iint_D \alpha f\,dx\,dy = \alpha \iint_D f\,dx\,dy, \qquad \alpha \text{ a constant}, \tag{11}$$

$$\iint_D (f + g)\,dx\,dy = \iint_D f\,dx\,dy + \iint_D g\,dx\,dy, \tag{12}$$

$$\iint_D f\,dx\,dy \le \iint_D g\,dx\,dy, \qquad \text{if } f \le g \text{ in } D, \tag{13}$$

$$\iint_D \alpha\,dx\,dy = \alpha \cdot \text{area of } D \qquad (\alpha \text{ a constant}), \tag{14}$$

and

$$\iint_D f\,dx\,dy = \iint_{D_1} f\,dx\,dy + \iint_{D_2} f\,dx\,dy, \tag{15}$$

provided that D_1 and D_2 have no points in common and D consists of all points belonging to either D_1 or D_2.

This corresponds to the additivity of the single integral for functions of one variable; see Chapter 8, §1.3.

The above properties are geometrically rather obvious. For instance, the integral in (10) is, by the geometric interpretation of the integral, the volume of a cylinder of height 1 and base area A; this volume is $A \cdot 1 = A$.

Relations (10)–(15) can also be proved from properties of integrals of functions of one variable, using Equations (2) and (3) in §1.1.

We begin with (10), which we rewrite as

$$\iint_D dx\, dy = \iint_I \chi(x,y)\, dx\, dy = \text{area of } D, \tag{16}$$

where $\chi(x,y)$ is a function equal to 1 in D and to 0 outside D, and I is some interval containing D, say, the interval $a < x < b$, $c < y < d$. We shall show that this is the same as the area formula in Chapter 9, §1.1.

The integral in (16) equals

$$\iint_{\substack{a<x<b \\ c<y<d}} \chi(x,y)\, dx\, dy = \int_a^b \left\{ \int_c^d \chi(x,y)\, dy \right\} dx.$$

If, for a fixed x_0, the intersection of the line $x = x_0$ with D consists of one or several segments, then $\int_c^d \chi(x_0,y)\, dy = l(x_0)$ is the sum of the length of these segments. Hence (10) may be rewritten as

$$\text{area of } D = \int_a^b l(x)\, dx.$$

This is the area formula in Chapter 9, §1.1 (with L the x-axis).

Next we prove (12). We may assume that D is an interval $a < x < b$, $c < y < d$, since we may set f and g equal to 0 outside D and then integrate over some interval containing D. Now, using (2) and the rule (1) in Chapter 8, §3.2,

$$\iint_{\substack{a<x<b \\ c<y<d}} [f(x,y) + g(x,y)]\, dx\, dy = \int_c^d \left\{ \int_a^b [f(x,y) + g(x,y)]\, dx \right\} dy$$

$$= \int_c^d \left\{ \int_a^b f(x,y)\, dx + \int_a^b g(x,y)\, dx \right\} dy$$

$$= \int_c^d \left\{ \int_a^b f(x,y)\, dx \right\} dy + \int_c^d \left\{ \int_a^b g(x,y)\, dx \right\} dy$$

$$= \iint_{\substack{a<x<b \\ c<y<d}} f(x,y)\, dx\, dy + \iint_{\substack{a<x<b \\ c<y<d}} g(x,y)\, dx\, dy.$$

Formulas (11) and (13) are proved similarly. From (10) and (11) we obtain (14).

In order to prove Property (15), we define a function f_1 which is equal to f in D_1 and to 0 outside D_1 and a function f_2 equal to f in D_2 and to 0 outside D_2. Since D_1 and D_2 have no common points, $f(x,y) = f_1(x,y) + f_2(x,y)$ for all (x,y) in D. Also $\iint_D f_1\, dx\, dy = \iint_{D_1} f\, dx\, dy$, $\iint_D f_2\, dx\, dy = \iint_{D_2} f\, dx\, dy$. Now, by Property (12),

$$\iint_D f\, dx\, dy = \iint_D (f_1 + f_2)\, dx\, dy = \iint_D f_1\, dx\, dy + \iint_D f_2\, dx\, dy$$

$$= \iint_{D_1} f_1\, dx\, dy + \iint_{D_2} f_2\, dx\, dy = \iint_{D_1} f\, dx\, dy + \iint_{D_2} f\, dx\, dy,$$

as we wanted to prove.

PROBLEMS

*1. Prove Equation (11).   *2. Prove Equation (13).

§2 Analytic definition of the double integral‡

We proceed to define the double integral analytically (from properties of numbers).
The definition is patterned after the one given in Chapter 8, §1.6, for simple integrals.

2.1 Piecewise continuous functions

Recall that in Chapter 8, §1.4, we called a function of one variable piecewise continuous if it was continuous at all points of a (finite) interval, except perhaps at finitely
many points.

A function of two variables will be called **piecewise continuous** in a two-dimensional interval if it is defined and continuous at all points of this interval, except perhaps
at finitely many points and along finitely many curves, each of which is either a straight
line or a graph of a continuous strictly monotone function.

A function $f(x,y)$ of two variables is called **bounded** if there is a number M such
that $|f(x,y)| \leq M$. This is the same definition as for functions of one variable.

REMARK Bounded, piecewise continuous functions are the "reasonable" functions in §1.3.

EXAMPLES 1. Let $x \mapsto f(x)$, $0 < x < 1$ be a piecewise continuous function [say, the function
$f(x) = 0$ for $0 < x < \frac{1}{2}$, $f(x) = -1$ for $\frac{1}{2} < x < \frac{3}{4}$, $f(x) = 5$ for $\frac{3}{4} < x < 1$]. Then $(x,y) \mapsto f(x)$
is a piecewise continuous function for $0 < x < 1$, $0 < y < 1$. Indeed, it is discontinuous only
on finitely many vertical segments (the segments $x = \frac{1}{2}$, $x = \frac{3}{4}$, $0 < y < 1$, in our case).

2. Let $f(x,y)$ be continuous for $-2 \leq x \leq 2$, $-2 \leq y \leq 2$, except for points lying on the circle
$x^2 + y^2 = 1$. Then f is piecewise continuous, since the circle can be decomposed into four
arcs, each of which satisfies the condition of the definition: the arc from $(1,0)$ to $(0,1)$, the
arc from $(1,1)$ to $(-1,0)$, the arc from $(-1,0)$ to $(0,-1)$, and the arc from $(0,-1)$ to $(1,0)$.

3. The function $f(x,y) = 1$ if $x^2 + y^2 \leq 1$, $f(x,y) = 0$ if $x^2 + y^2 > 1$, is piecewise continuous.
The reader should supply the reasons.

4. The function $f(x,y) = 1/(x^2 + y^2)$ is piecewise continuous. The only discontinuity is at one
point, the origin.

PROBLEMS

1. Suppose $f(x,y) = (x^2 + y^4 + 6)/xy$ if $xy \neq 0$, $f(x,y) = 77$ if $xy = 0$. Is $f(x,y)$ piecewise
   continuous in every finite interval?
2. Is the function $f(x,y)$ of Problem 1 bounded?
3. If $f(x,y)$ and $g(x,y)$ are piecewise continuous in an interval I, are the functions $f + g$, $f - g$
   and fg piecewise continuous? Give reasons for your answers.
4. Under the condition of Problem 3, under what *additional* condition is f/g certainly
   piecewise continuous?
5. If $f(x,y)$ is piecewise continuous in an interval I, is the function $\sin f$ piecewise continuous?
   Give a reason for your answer.

‡This section may be omitted at first reading.

2.2 Step functions

If we subdivide the interval $a < x < b$ into several, say, p, subintervals and draw vertical lines through each of the dividing lines, and if we subdivide the interval $c < y < d$ into several, say, q, subintervals and draw horizontal lines through the division points, then the two-dimensional interval I: $a < x < b$, $c < y < d$ gets subdivided into pq *two-dimensional subintervals.* In Figure 20.11, for instance, the

FIGURE **20.11**

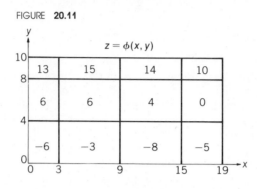

interval $0 < x < 19$ is subdivided into 4 subintervals, the interval $0 < y < 10$ is subdivided into 3 subintervals, and the two-dimensional interval $0 < x < 19$, $0 < y < 10$ is subdivided into $4 \cdot 3 = 12$ two-dimensional subintervals.

A **step function** $(x,y) \mapsto \phi(x,y)$ in I is a function which, for a suitable subdivision of I, is constant in each subinterval. For a given subdivision we can define a step function $z = \phi(x,y)$ by giving its values, $z_1, z_2, z_3, \ldots$ in the various subintervals. We did so in Figure 20.11; the graph of this function is shown in Figure 20.12. A step function is always piecewise continuous (and bounded). Its values along the division lines are of no interest.

The integral of a step function $z = \phi(x,y)$ which takes on the constant values z_1, $z_2, z_3, \ldots$ in the subinterval $I_1, I_2, I_3, \ldots$ is *defined* as the finite sum

$$\iint_I \phi(x,y) \, dx \, dy = z_1 \cdot \text{area of } I_1 + z_2 \cdot \text{area of } I_2 + \cdots.$$

This corresponds to the geometric definition of §1.1 and §1.2.

For instance, in the case of the function defined by Figure 20.11 we have

$$\iint_{\substack{0 < x < 19 \\ 0 < y < 10}} \phi(x,y) \, dx \, dy = 13 \cdot 6 + 15 \cdot 12 + 14 \cdot 12 + 10 \cdot 8 + 6 \cdot 12 + 6 \cdot 24 + 4 \cdot 24$$
$$+ 0 \cdot 16 + (-6) \cdot 12 + (-3) \cdot 24 + (-8) \cdot 24 + (-5) \cdot 16$$
$$= 818 - 416 = 402.$$

This is the sum of the volumes of the boxes in Figure 20.12 above the plane $z = 0$ minus the sum of the volumes of the boxes below that plane.

Let $f(x,y)$ be a bounded piecewise continuous function defined in I, and let $\phi(x,y)$, $\psi(x,y)$ be step functions such that $\phi \leq f \leq \psi$ in I. By the geometric definition of double integrals, we expect that

$$\iint_I \phi(x,y) \, dx \, dy \leq \iint_I f(x,y) \, dx \, dy \leq \iint_I \psi(x,y) \, dx \, dy.$$

FIGURE 20.12

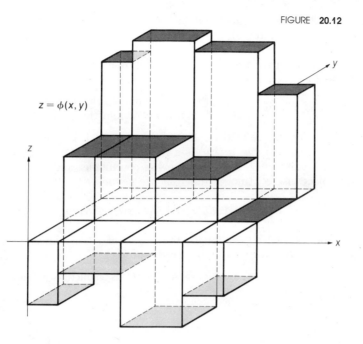

$z = \phi(x, y)$

We shall make this observation into the basis for the analytic definition.

PROBLEMS

*1. If ϕ and ψ are step functions, so are $\phi + \psi$ and $\phi\psi$. Prove this statement.
*2. If $\phi(x,y)$ and $\psi(x,y)$ are step functions, so is $M(x,y) = \max[\phi(x,y),\psi(x,y)] =$ the larger of the numbers $\phi(x,y)$, $\psi(x,y)$. Explain why this is so.

2.3 The analytic definition

Theorem 1 and Definition. *Let $f(x,y)$ be a bounded piecewise continuous function defined in an interval I. Then there is a unique number V with the property: for all step functions $\phi(x,y)$ and $\psi(x,y)$ such that $\phi(x,y) \leq f(x,y) \leq \psi(x,y)$, we have*

$$\iint_I \phi(x,y)\, dx\, dy \leq V \leq \iint_I \psi(x,y)\, dx\, dy. \tag{1}$$

The number V *is called the* double integral *of f over I and is denoted by* $\iint_I f(x,y)\, dx\, dy$.

The next theorem justifies formulas (2) and (3) in §1.1.

Theorem 2. *If $f(x,y)$ is bounded and piecewise continuous for $a < x < b$, $c < x < d$, then*

$$\iint_{\substack{a<x<b\\c<y<d}} f(x,y)\, dx\, dy = \int_c^d \int_a^b f(x,y)\, dx\, dy = \int_a^b \int_c^d f(x,y)\, dy\, dx. \tag{2}$$

The proofs ▶ of Theorems 1 and 2 are rather difficult. [It is part of the assertion of the theorem that the integrals of functions of one variable occurring in (2) are well defined.]

Now let D be a bounded set in the plane, that is, a set contained in some interval. The **characteristic function** of D, denoted by $\chi_D(x,y)$, is defined as follows:

$$\chi_D(x,y) = \begin{cases} 1, & \text{if } (x,y) \text{ is a point in the set } D, \\ 0, & \text{if } (x,y) \text{ is } not \text{ a point in } D. \end{cases} \tag{3}$$

Let $f(x,y)$ be a function defined on D and perhaps on a bigger set. The function $\chi_D(x,y)f(x,y)$ is equal to $f(x,y)$ if (x,y) is a point in D and is (by definition) equal to 0 if (x,y) is not a point in D. If χ_D is piecewise continuous, and if f is such that $\chi_D f$ is a bounded piecewise continuous function, we *define*

$$\iint_D f(x,y)\,dx\,dy = \iint_I \chi_D(x,y)f(x,y)\,dx\,dy, \tag{4}$$

where I is some interval containing D; the choice of I is irrelevant. The definition just given corresponds exactly to what we did in §1.3.

REMARK   Bounded domains D with a piecewise continuous characteristic function χ_D are the "reasonable domains" mentioned in §1.3.

2.4   Riemann sums

Let $f(x,y)$ be a (bounded, piecewise continuous) function defined in the open interval I. Suppose we subdivide I, as in §2.2, choose a point (x_j,y_j) in each subinterval I_j or on its boundary, and form the sum

$$S = f(x_1,y_1) \cdot \text{area of } I_1 + f(x_2,y_2) \cdot \text{area of } I_2 + \cdots. \tag{5}$$

The number S is called a **Riemann sum** for the integral

$$V = \iint_I f(x,y)\,dx\,dy. \tag{6}$$

We can define a step function, call it $\omega(x,y)$, which equals $f(x_j,y_j)$ in I_j; S is then the double integral of this function over I. If ϕ and ψ are two step functions such that $\phi \leq f \leq \psi$, then also $\phi \leq \omega \leq \psi$. By the monotonicity property of integrals [Property (13) in §1.4], we have

$$\iint_I \phi\,dx\,dy \leq S \leq \iint_I \psi\,dx\,dy, \qquad \iint_I \phi\,dx\,dy \leq V \leq \iint_I \psi\,dx\,dy. \tag{7}$$

It follows ▶ from Theorem 1 that we can find step functions ϕ and ψ with $\phi \leq f \leq \psi$, such that the difference between the two integrals in (7) is smaller than any given number $\epsilon > 0$. Then $|S - V| < \epsilon$. Thus, *we can approximate a double integral by a Riemann sum as closely as we want.*

EXAMPLE   Consider the integral

$$V = \iint_{\substack{0<x<6 \\ 0<y<4}} (x^2y + xy^2)\,dx\,dy = \int_0^4 \int_0^6 (x^2y + xy^2)\,dx\,dy$$

$$= \int_0^4 (72y + 18y^2)\,dy = 576 + 384 = 960.$$

FIGURE   **20.13**

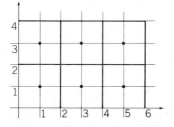

We subdivide the interval $0 < x < 6$, $0 < y < 4$ into six subintervals and choose one point in each subinterval, as shown in Figure 20.13. Each subinterval has area 4. The values of $x^2y + xy^2$ at the points $(1,1)$, $(3,1)$, $(5,1)$, $(1,3)$, $(3,3)$, and $(5,3)$ are 2, 12, 30, 12, 54, and 120. The corresponding Riemann sum is therefore

$$(2 + 12 + 30 + 12 + 54 + 120)4 = 230 \cdot 4 = 920,$$

which is about 4 percent less than the integral.

PROBLEMS

In Problems 1 to 6 you are given an integrand $f(x,y)$ and an interval $I = \{x_0 < x < x_0 + n, y_0 < y < y_0 + m\}$. In each case, define nm subintervals by dividing the x- and y-axes at integer values. Then compute a Riemann sum using the value of the function taken at the midpoint of each subinterval. Compare this Riemann sum with the exact value of $\iint_I f(x,y)\, dx\, dy$.

1. $f = 3x + 2y$; $0 < x < 4$, $0 < y < 2$.
2. $f = 2x - 5y$; $-2 < x < 2$, $-2 < y < 2$.
3. $f = x^2y$; $1 < x < 4$, $1 < y < 4$.
4. $f = x^2 - 2y^2$; $-2 < x < 2$, $0 < y < 4$.
5. $f = x^2 + xy + 2y^2$; $0 < x < 4$, $-2 < y < 2$.
6. $f = \sin\dfrac{\pi x}{8}\cos\dfrac{\pi y}{8}$; $0 < x < 4$, $0 < y < 4$.

§3   Applications of double integrals

3.1   Surface area

Our next aim is to compute areas of regions on curved surfaces. As a preparation, let us look at a nonvertical plane and a reasonable set Δ on this plane (Figure 20.14). Through each point P of Δ, we drop a perpendicular on the plane $z = 0$; let it meet the plane $z = 0$ at the point Q. Then Q is called the **projection** of P on the plane $z = 0$. [In other words, if P has coordinates (x,y,z), its projection Q has coordinates $(x,y,0)$.] The set D of projections of all points of Δ is called the projection of Δ.

The area of an inclined plane set Δ equals the area of its projection D divided by $\cos \theta$, where θ is the angle between the planes of Δ and D.

To verify this statement, we may assume that the plane of Δ passes through the origin (we may achieve this by translating the coordinate system in the z-direction),

FIGURE **20.14**   FIGURE **20.15**

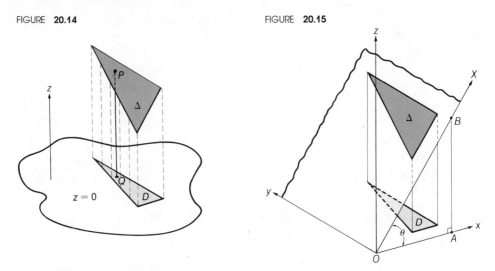

and that this plane passes through the y-axis (this can be achieved by rotating the coordinate system about the z-axis). We then have the situation depicted in Figure 20.15. In the plane of Δ, we introduce the Cartesian coordinate system (X,y). A point in the plane of Δ with coordinates (X,y) has the projection (x,y) with $x = X \cos \theta$, as is seen from the right triangle OAB. (Note that B is the projection of A, $|OA| = x$, and $|OB| = X$.) Thus we obtain Δ from D by keeping all distances in the y-direction unchanged and multiplying all distances in the x-direction by $1/\cos \theta$. This implies our statement. For a triangle such as the one shown in the figure, this follows by elementary geometry; for every other shape it follows, for instance, by using the area formula of Chapter 9, §1.1, with the y-axis as the line L.

Consider now a continuously differentiable function $z = f(x,y)$ defined in some reasonable bounded domain D; we assume that the partial derivatives f_x, f_y are bounded. We want to compute the area A of the graph of $z = f(x,y)$, trusting, for the time being, our intuitive ideas about areas.

Assume, for the sake of simplicity, that D is an interval (as in Figure 20.16). We subdivide D into small subintervals. Then our intuitive notion of area tells us that the total area A is the sum of all small areas that project on the small subintervals. Each of these small areas is approximately the area of a plane figure—for a small piece of a surface may be replaced by a piece of its tangent plane. We just learned how to compute areas of inclined plane sets; we conclude that we commit a small error if we replace the area of the portion of our surface lying above a subinterval I_j by the number: (area of I_j)/$\cos \theta_j$, θ_j being the angle between the plane $z = 0$ and the tangent plane to our surface at some point (x_j, y_j) in I_j. Let us compute θ_j. It is the acute angle between a vector normal to the surface at the point with coordinates (x_j, y_j) and $z_j = f(x_j, y_j)$ and a vector in the z-direction. We know (see Chapter 19, §2.2) that the vector with the components $\langle -f_x(x_j,y_j), -f_y(x_j,y_j), 1 \rangle$ is normal to the surface at (x_j, y_j, z_j). The unit vector in the z-direction has the components $\langle 0,0,1 \rangle$. Thus (see Chapter 18, §3.1)

$$\cos \theta_j = \frac{1}{\sqrt{1 + f_x^2 + f_y^2}} \qquad \text{computed at } (x_j, y_j).$$

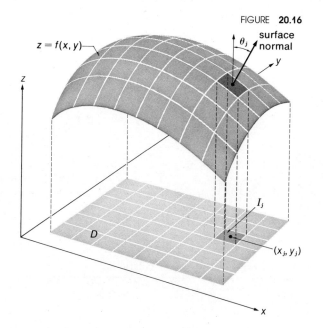

FIGURE **20.16**

An approximate value of the total area A is

$$\frac{(\text{area of } I_1)}{\cos \theta_1} + \frac{(\text{area of } I_2)}{\cos \theta_2} + \frac{(\text{area of } I_3)}{\cos \theta_3} + \cdots$$

$$= \sqrt{1 + f_x(x_1,y_1)^2 + f_y(x_1,y_1)^2} \cdot \text{area of } I_1 + \sqrt{1 + f_x(x_2,y_2)^2 + f_y(x_2,y_2)^2} \cdot \text{area of } I_2$$
$$+ \sqrt{1 + f_x(x_3,y_3)^2 + f_y(x_3,y_3)^2} \cdot \text{area of } I_3 + \cdots.$$

But this is a Riemann sum for a double integral (see §2.4), namely, the integral of the function $\sqrt{1 + f_x(x,y)^2 + f_y(x,y)^2}$. We are therefore led to suspect that

$$A = \iint_D \sqrt{1 + f_x(x,y)^2 + f_y(x,y)^2} \, dx \, dy. \tag{1}$$

This formula gives the right result if $f(x,y) = 0$ (then the integrand is 1) or if $f(x,y)$ is a linear function $f = ax + by + c$ [then the integrand is $\sqrt{1 + a^2 + b^2} = 1/\cos \theta$, θ being the angle between the plane $z = f(x,y)$ and the plane $z = 0$].

Now we change our point of view and adopt formula (1) as the *definition* of the area of the surface $z = f(x,y)$, (x,y) in D.

EXAMPLES **1.** Find the area of the surface $z = \frac{2}{3}x^{3/2}$, $0 < x < 1$, $1 < y < 2$.
ANSWER   We have $z_x = x^{1/2}$, $z_y = 0$. By the area formula (1),

$$A = \int_1^2 \int_0^1 \sqrt{1 + z_x^2 + z_y^2} \, dx \, dy = \int_1^2 \int_0^1 \sqrt{1 + x} \, dx \, dy$$

$$= \int_1^2 \left[\tfrac{2}{3}(1 + x)^{3/2} \right]_{x=0}^{x=1} dy = \tfrac{2}{3}(2\sqrt{2} - 1) \approx 1.219.$$

FIGURE **20.17**

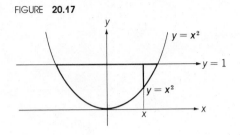

2. Find the area of the surface $z = x + \sqrt{2}y + 1$, $y > x^2$, $y < 1$.

ANSWER   We have $z_x = 1$, $z_y = \sqrt{2}$, hence $\sqrt{1 + z_x^2 + z_y^2} = 2$. The inequalities $x^2 < y < 1$ determine a domain (see Figure 20.17) bounded by a parabola and a horizontal line. The desired area is

$$\int_{-1}^{1} \int_{x^2}^{1} 2 \, dy \, dx = \int_{-1}^{1} 2(1 - x^2) \, dx = \tfrac{8}{3}.$$

In general, we should not expect such easy calculations.

PROBLEMS

1. Find the area of the surface $z = \tfrac{4}{3}x^{3/2}$, $0 < x < 2$, $-1 < y < 0$.
2. Find the area of the surface $z = 2x - y + 1$, $0 < x < 2$, $0 < y < 2$.
3. Find the area of the surface $z = x^2 + \sqrt{3}y + 1$, $0 < x < 1$, $0 < y < 1$.
4. Find the area of the surface $z = \sqrt{x^2 + y^2}$, $y > \tfrac{1}{4}x^2$, $y < 1$.
5. Find the area of the surface of the cylinder $x^2 + z^2 = 9$ lying in the first octant and cut off by the planes $y = 0$ and $y = x$.
6. Find the area of the surface of the cylinder $x^2 + z^2 = 4$ lying in the first octant and cut off by the cylinder $x^2 + y^2 = 4$.
7. Find the area of the surface of the cylinder $y^2 + z^2 - 4z = 0$ cut off by the sphere $x^2 + y^2 + z^2 = 16$.
8. Find the area of the spherical surface $x^2 + y^2 + z^2 - 4z = 0$ lying inside the cone $z^2 = x^2 + y^2$.
9. Find the area of the spherical surface $x^2 + y^2 + z^2 - 4z = 0$ lying inside the paraboloid $z = x^2 + y^2$.
10. Find the area above the xy plane cut off from the cone $z^2 = x^2 + y^2$ by the cylinder $x^2 + y^2 - 2x = 0$.

3.2   Double integrals in polar coordinates

In some cases it is convenient to compute double integrals using polar coordinates.
    Let $f(x,y)$ be a (bounded, continuous) function defined in the disk $x^2 + y^2 < R^2$. Then

$$\iint_{x^2 + y^2 < R^2} f(x,y) \, dx \, dy = \int_{0}^{2\pi} \int_{0}^{R} f(r \cos \theta, r \sin \theta) r \, dr \, d\theta$$

$$= \int_{0}^{R} \int_{0}^{2\pi} f(r \cos \theta, r \sin \theta) r \, d\theta \, dr.$$

(2)

FIGURE 20.18   FIGURE 20.19

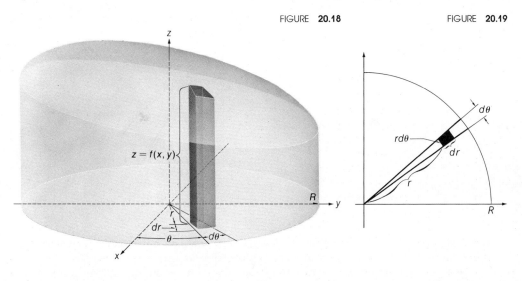

(Remember the convention: outer integral sign goes with outer differential. Note also that $dx\,dy$ is replaced by $r\,dr\,d\theta$.) Where does relation (2) come from? Let us assume that f is a positive function, so that

$$V = \iint\limits_{x^2+y^2<R^2} f(x,y)\,dx\,dy \tag{3}$$

is the volume of the solid bounded by the plane $z = 0$, the cylinder $x^2 + y^2 = R^2$, and the graph of our function.

Let (r,θ) denote the polar coordinates of (x,y). We now tell one more mathematical myth: the volume in question consists of infinitely many infinitely thin boxes. Each box is erected at some point in the disk with coordinates $x = r \cos \theta$, $y = r \sin \theta$ (see Figure 20.18). Its base is an infinitely small figure bounded by two segments of infinitely small length dr, lying on rays through the origin, and two circular arcs with center at the origin joining the two rays. The two rays form the infinitely small angle $d\theta$ (see Figure 20.19). The height of the box is $z = f(x,y) = f(r \cos \theta, r \sin \theta)$. The base of the box is nearly a rectangle, the lengths of the sides are dr and $r\,d\theta$ (by the formula for the length of a circular arc). Hence the area of the base is $(dr)\cdot(r\,d\theta) = r\,dr\,d\theta$ and the infinitely small volume of the box is height times area of the base, that is,

$$f(r \cos \theta, r \sin \theta)r\,dr\,d\theta.$$

The whole volume is the sum of all these infinitely small volumes. We write

$$V = \iint\limits_{\text{disk}} f(r \cos \theta, r \sin \theta)r\,dr\,d\theta$$

and, guided by the Leibniz notation, we interpret this to mean that

$$V = \int_0^{2\pi} \int_0^R f(r \cos \theta, r \sin \theta)r\,dr\,d\theta = \int_0^R \int_0^{2\pi} f(r \cos \theta, r \sin \theta)r\,d\theta\,dr. \tag{4}$$

From (3) and (4), relation (2) follows.

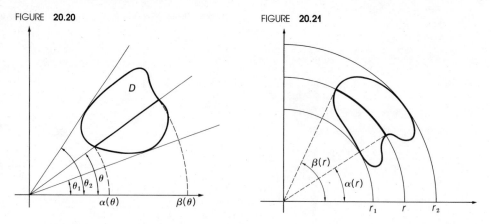

FIGURE **20.20**

D

θ_1 θ_2 θ

$\alpha(\theta)$ $\beta(\theta)$

FIGURE **20.21**

$\beta(r)$

$\alpha(r)$

r_1 r r_2

A rigorous proof ▶ of relation (2) shows that we may permit f to take on negative values, and we may replace continuity by piecewise continuity.

In applying relation (2), we sometimes deal with a function that is 0 outside some set D, so that the integral to the left may be replaced by an integral over D. It may happen that the region of integration D is convex in the radial direction; that is, there are numbers θ_1 and θ_2 such that the ray $x = r\cos\theta$, $y = r\sin\theta$, $r > 0$ intersects D only if $\theta_1 < \theta < \theta_2$. If so, the intersection is a segment between $r = \alpha(\theta)$ and $r = \beta(\theta) > \alpha(\theta)$. In such a case, shown in Figure 20.20, (2) may be written as

$$\iint_D f(x,y)\,dx\,dy = \int_{\theta_1}^{\theta_2} \int_{\alpha(\theta)}^{\beta(\theta)} f(r\cos\theta, r\sin\theta)r\,dr\,d\theta. \tag{5}$$

It may also happen that D is convex with respect to θ, so that we may write

$$\iint_D f(x,y)\,dx\,dy = \int_{r_1}^{r_2} \int_{\alpha(r)}^{\beta(r)} f(r\cos\theta, r\sin\theta)r\,d\theta\,dr. \tag{6}$$

The meaning of the term convex and the numbers r_1, r_2, $\alpha(r)$, $\beta(r)$ is seen from Figure 20.21.

EXAMPLES **1.** Compute the volume of the hemisphere of radius R, that is, the integral

$$V = \iint_{x^2+y^2<R^2} \sqrt{R^2 - x^2 - y^2}\,dx\,dy$$

using polar coordinates.

SOLUTION   The inequality $x^2 + y^2 < R^2$ determines a circular domain with center at the origin and radius R. Hence r varies from 0 to R, and θ varies from 0 to 2π. Using Equation (5), the volume becomes

$$V = \int_0^{2\pi} \int_0^R \sqrt{R^2 - r^2}\,r\,dr\,d\theta = \int_0^{2\pi} d\theta \int_0^R (R^2 - r^2)^{1/2}r\,dr$$

$$= 2\pi \int_0^R -\frac{1}{3}\frac{d(R^2-r^2)^{3/2}}{dr}\,dr = -\frac{2}{3}\pi(R^2-r^2)^{3/2}\Big|_0^R = \frac{2}{3}\pi R^3.$$

2. Let D be the region shown in Figure 20.22. Compute

$$\iint_D e^{x^2+y^2}\,dx\,dy.$$

FIGURE **20.22**

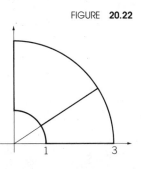

ANSWER Here θ varies from 0 to $\pi/2$, and r varies from 1 to 3 in the domain considered. Also, $x^2 + y^2 = r^2$. Therefore

$$\iint_D e^{x^2+y^2}\, dx\, dy = \int_0^{\pi/2} \int_1^3 e^{r^2} r\, dr\, d\theta$$

$$= \int_0^{\pi/2} \left[\frac{1}{2} e^{r^2} \right]_{r=1}^{r=3} d\theta = \int_0^{\pi/2} \frac{e^9 - e}{2}\, d\theta = \frac{\pi e(e^8 - 1)}{4}.$$

Without using polar coordinates, we can compute this integral only by numerical integration.

3. Assume that D consists of all points with polar coordinates (r,θ) such that $\theta_1 < \theta < \theta_2$, $0 < r < \phi(\theta)$, where $\phi(\theta)$ is a positive continuous function. We apply (5) with $f(x,y) = 1$. Since $\alpha(\theta) = 0$, $\beta(\theta) = \phi(\theta)$, and $\int_0^{\phi(\theta)} r\, dr = \frac{1}{2}\phi(\theta)^2$, we obtain

$$\text{area of } D = \frac{1}{2} \int_{\theta_1}^{\theta_2} \phi(\theta)^2\, d\theta.$$

This is the same formula that we obtained in Chapter 16, §1.4, except for notations.

PROBLEMS

In Problems 1 to 8 a function $f(x,y)$ and a set D are given. Compute $\iint_D f(x,y)\, dx\, dy$ using polar coordinates.

1. $f(x,y) = \sqrt{x^2 + y^2}$; D is the region in the first quadrant bounded by the unit circle and the coordinate axes.
2. $f(x,y) = (x^2 + y^2)^{3/2}$; D is the region in the first quadrant bounded by the circles $x^2 + y^2 = 1$, $x^2 + y^2 = 4$, and the coordinate axes.
3. $f(x,y) = x/\sqrt{x^2 + y^2}$; D is the region in the upper half-plane bounded by the circle $x^2 + y^2 = 16$ and the x-axis.
4. $f(x,y) = x^2 y^2/(x^2 + y^2)^2$; D is the region enclosed by the ring $1 < x^2 + y^2 < 2$.
5. $f(x,y) = Ax^2 + Bxy + Cy^2$; D is the region between two concentric circles with centers at the origin and radii R_1 and R_2.
6. $f(x,y) = 1$; D is the region bounded by the segments $0 < x < 1$, $|y| < 2$.
7. $f(x,y) = \arctan y/x$; D is the region bounded by the spiral $\theta = r$ and the segment $0 \le x \le 2\pi$.
8. $f(x,y) = \sqrt{x^2 + y^2}$; D is the interior of one leaf of the four-leaved rose $r = \cos 2\theta$.
9. Find the surface area of the paraboloid $z = \frac{1}{6}(x^2 + y^2)$ which lies directly above the lemniscate $r^2 = 9 \cos 2\theta$.
10. Find the surface area of the ellipsoid of revolution $(x/a)^2 + (y/a)^2 + (z/c)^2 = 1$.

FIGURE **20.23**   FIGURE **20.24**

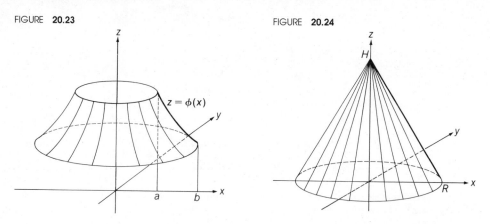

3.3   Surfaces of revolution

We recall that a surface of revolution S is obtained by rotating a plane curve C about a straight line l in the plane of the curve (compare Chapter 18, §1.4). Assume that l is the z-axis, that C lies in the (x,z) plane, and that C is the graph of the function

$$z = \phi(x), \qquad 0 \le a < x < b$$

(see Figure 20.23). If C is rotated about the z-axis by $360°$, it sweeps out a surface S, the graph of

$$(x,y) \mapsto z = \phi(\sqrt{x^2 + y^2}) = \phi(r), \qquad a < r < b,$$

where r, θ are the polar coordinates of (x,y). By the chain rule,

$$z_x = \phi'(r)\frac{\partial r}{\partial x} = \phi'(r)\frac{x}{r}, \qquad z_y = \phi'(r)\frac{\partial r}{\partial y} = \phi'(r)\frac{y}{r}$$

so that

$$1 + z_x^2 + z_y^2 = 1 + \phi'(r)^2.$$

By (1), the area of S is

$$A = \iint_{a^2 < x^2 + y^2 < b^2} \sqrt{1 + \phi'(\sqrt{x^2 + y^2})^2}\, dx\, dy$$

$$= \int_0^{2\pi} \int_a^b \sqrt{1 + \phi'(r)^2}\, r\, dr\, d\theta = 2\pi \int_a^b \sqrt{1 + \phi'(r)^2}\, r\, dr.$$

(7)

We may call a dummy variable whatever we please, and we may write

$$A = 2\pi \int_a^b \sqrt{1 + \phi'(x)^2}\ x\, dx.$$

(8)

We often write this in another form:

$$A = 2\pi \int_a^b x\, ds \qquad [\text{where } ds = \sqrt{1 + (dz/dx)^2}\, dx].$$

(8′)

We call ds the differential of arc length along the curve $z = \phi(x)$ in the xz plane. Indeed, $\int_a^b ds = \int_a^b \sqrt{1 + (dz/dx)^2}\, dx$ is the length of the curve $z = \phi(x)$, $a \leq x \leq b$. If the curve $y = \phi(x)$, $a \leq x \leq b$, is rotated about the y-axis by $360°$, the resulting surface has the area

$$A = 2\pi \int_a^b x\, ds \qquad [\text{where } ds = \sqrt{1 + (dy/dx)^2}\, dx]. \qquad (9)$$

If the curve $x = \phi(y)$, $a \leq x \leq b$, is rotated about the x-axis by $360°$, we obtain a surface of area

$$A = 2\pi \int_a^b y\, ds \qquad [\text{where } ds = \sqrt{1 + (dx/dy)^2}\, dy]. \qquad (10)$$

The three formulas (8′), (9), (10) differ only in the names of the coordinate axes.

REMARK   If we rotate the curve C not by 2π but only by the angle ω, $0 < \omega < 2\pi$, then we must replace the factor 2π in (8) by ω. For in that case the region of integration (in the x, y plane) in (7) is given by $a < r < b$, $0 < \theta < \omega$. The same applies to the area formulas (9) and (10).

EXAMPLES   1.   Find the mantle surface of a circular cone of height H and radius R.
ANSWER   The surface is obtained by rotating about the z-axis the segment $z = H - (H/R)x$, $0 < x < R$, called the generator of the cone (see Figure 20.24). Using (8) with $a = 0$, $b = R$, $\phi(x) = H - (H/R)x$, we obtain

$$A = 2\pi \int_0^R \sqrt{1 + H^2/R^2}\; x\, dx = \frac{2\pi \sqrt{1 + H^2/R^2}\, R^2}{2} = \pi R \sqrt{R^2 + H^2}.$$

Since the length of the generator is $L = \sqrt{R^2 + H^2}$, we have $A = \pi RL$, a well-known formula from elementary geometry.

2.   Find the surface area of a sphere of radius R.
ANSWER   The desired area is twice the area obtained by rotating the circular arc $z = \sqrt{R^2 - x^2}$, $0 < x < R$ about the z-axis, or

$$2 \cdot 2\pi \int_0^R \left[1 + \left(\frac{d\sqrt{R^2 - x^2}}{dx}\right)^2\right]^{1/2} x\, dx = 4\pi \int_0^R \frac{Rx\, dx}{\sqrt{R^2 - x^2}} = 4\pi R^2,$$

as we are taught in elementary geometry.

3.   What is the area of the part of the sphere $x^2 + y^2 + z^2 = R^2$ which lies over the sector $0 < \theta < \omega < 2\pi$ in the (x,y) plane?
ANSWER $2\omega R^2$, by the Remark made above and by Example 2.

4.   Compute the area of the surface obtained by rotating the curve

$$x = \phi(y) = \tfrac{1}{2}y\sqrt{y^2 - 1} - \tfrac{1}{2}\log\left(y + \sqrt{y^2 - 1}\right), \qquad 1 \leq y \leq 2$$

about the x-axis by $180°$.
ANSWER   By the Remark above, and by (10), the desired area is

$$A = \pi \int_1^2 y\, ds = \pi \int_1^2 y\sqrt{1 + (dx/dy)^2}\, dy.$$

But an elementary although lengthy calculation yields $dx/dy = \phi'(y) = \sqrt{y^2 - 1}$, so that $ds = \sqrt{1 + y^2 - 1}\, dy = y\, dy$ and $A = \pi\int_1^2 y^2\, dy = \pi\tfrac{8}{3} \approx 8.38$.

PROBLEMS

1. Find a formula for the area of the surface of revolution generated by revolving the curve $z = \phi(x)$ from (x_0, z_0) to (x_1, z_1) about the x-axis.
2. Repeat Problem 1 for the curve $y = \phi(x)$ from (x_0, y_0) to (x_1, y_1) about the x-axis.
3. Repeat Problem 1 for the curve $x = \phi(y)$ from (x_0, y_0) to (x_1, y_1) about the y-axis.

In each of Problems 4 to 14 find the area of the surface generated by rotating the given curve C about the indicated axis.

4. C: $z = \sqrt{3}x$ from $x = 1$ to $x = 2$;   z-axis.
5. C: $z = x^2$ from $x = 0$ to $x = \sqrt{2}$;   z-axis.
6. C: $x = \frac{1}{4}y^2$ from $y = 0$ to $y = 2$;   x-axis.
7. C: $y = \frac{1}{3}x^3$ from $x = 0$ to $x = 2$;   x-axis.
8. C: $y = \frac{1}{2}(e^x + e^{-x})$ from $x = 0$ to $x = 1$;   x-axis.
9. C: $y = \frac{1}{3}x^3 + 1/4x$ from $x = 1$ to $x = 2$;   y-axis.
10. C: $z = \sin x$ from $x = 0$ to $x = \pi$;   x-axis.
11. C: $z = \log x$ from $x = 1$ to $x = 2$;   z-axis.
12. C: $y = \frac{1}{2}x^2$ from $x = 0$ to $x = 1$;   y-axis.
13. C: $y = x^3$ from $x = 0$ to $x = 1$;   y-axis.
14. C: $x = \frac{1}{16}y^2$ from $y = 0$ to $y = 4$;   y-axis.

3.4 Improper double integrals‡

A double integral is called improper if either the region of integration or the integrand (the function to be integrated) or both are unbounded. We shall consider such integrals only for the case of nonnegative integrands, as we did for simple integrals (see Chapter 8, §4).

Suppose that $f(x,y)$ is a nonnegative, piecewise continuous but unbounded function defined in some interval I. What meaning can we ascribe to the expression

$$\iint_I f(x,y) \, dx \, dy? \tag{11}$$

The geometric meaning of the double integral, the volume under the graph of $z = f(x,y)$, suggests the following definition. Let A be a positive number. We denote by f_A the function equal to f whenever $f < A$ and equal to 0 otherwise, that is,

$$f_A(x,y) = \begin{cases} f(x,y), & \text{if } f(x,y) < A, \\ 0, & \text{if } f(x,y) \geq A. \end{cases} \tag{12}$$

The function f_A is bounded, since $0 \leq f_A(x,y) < A$ in I, so we can form

$$\iint_I f_A(x,y) \, dx \, dy. \tag{13}$$

This is the volume under the part of the graph of $z = f(x,y)$ where $f \leq A$. [Since $f_A \leq f_B$ for $A \leq B$, the number (13) is an increasing function of A.] We say that f is **integrable** over I [or that the integral (11) **converges**] if there is a finite limit

$$\iint_I f(x,y) \, dx \, dy = \lim_{A \to +\infty} \iint f_A(x,y) \, dx \, dy. \tag{14}$$

‡This subsection, as well as the two subsections that follow, may be omitted at first reading.

If the limit in (14) is $+\infty$, we say that the integral (11) **diverges**, and we write $\iint_I f(x,y)\, dx\, dy = +\infty$.

It can be shown that if f is integrable over I, then

$$\iint_I f(x,y)\, dx\, dy = \int_a^b \left\{ \int_c^d f(x,y)\, dx \right\} dy = \int_c^d \left\{ \int_a^b f(x,y)\, dy \right\} dx,$$

where a, b, c, d have the usual meaning, provided the single integrals occurring above exist, as improper integrals.

EXAMPLES   1.   Evaluate the improper integral

$$\iint_{x^2+y^2<1} \frac{dx\, dy}{x^2 + y^2}.$$

FIRST SOLUTION   The integrand $(x^2 + y^2)^{-1}$ is unbounded near the origin. For large $A > 0$, the integrand is less than A for $x^2 + y^2 > 1/A$. Hence the integral to be computed is

$$\lim_{A \to +\infty} \iint_{D_A} \frac{dx\, dy}{x^2 + y^2},$$

where D_A is the set of (x,y) with $1/A < x^2 + y^2 < 1$. It is convenient to set $1/A = \epsilon^2$ and use polar coordinates:

$$\iint_{D_A} \frac{dx\, dy}{x^2 + y^2} = \int_0^{2\pi} \int_\epsilon^1 \frac{1}{r}\, r\, dr\, d\theta = \int_0^{2\pi} \int_1^\epsilon dr\, d\theta = 2\pi(1 - \epsilon).$$

Hence

$$\iint_{x^2+y^2<1} \frac{dx\, dy}{x^2 + y^2} = 2\pi.$$

SECOND SOLUTION   We use polar coordinates from the beginning:

$$\iint_{x^2+y^2<1} \frac{dx\, dy}{x^2 + y^2} = \int_0^{2\pi} \int_0^1 \frac{1}{r}\, r\, dr\, d\theta = \int_0^{2\pi} \int_0^1 dr\, d\theta = 2\pi.$$

2.   Compute

$$\iint_I (xy)^{-1/2}\, dx\, dy,$$

where I is the interval $0 < x < 1$, $0 < y < 1$.

ANSWER   In this case, it is convenient to write the integral as an iterated integral:

$$\iint_I (xy)^{-1/2}\, dx\, dy = \int_0^1 \int_0^1 (xy)^{-1/2}\, dx\, dy = \int_0^1 x^{-1/2}\, dx \int_0^1 y^{-1/2}\, dy$$

$$= \left(\int_0^1 x^{-1/2}\, dx \right)^2 ;$$

the name of a dummy variable is of no importance. Since

$$\int_0^1 x^{-1/2} = 2x^{1/2} \Big|_0^1 = 2$$

(this is an improper integral), the original integral equals 4.

PROBLEMS

1. Consider $\iint_D (x^2 + y^2)^{-\alpha/2} \, dx \, dy$ where D is the unit circle. Determine for which α the integral is finite and compute its value for such α.

In Problems 2 to 6 determine whether the given integral converges or diverges. If the integral converges, find its value.

2. $\displaystyle\iint_I \frac{\sqrt{x} + \sqrt{y}}{\sqrt{xy}} \, dx \, dy$, where I is the interval $0 < x < 1$, $0 < y < 1$.

3. $\displaystyle\iint_I \frac{x + y}{xy} \, dx \, dy$, where I is the interval $0 < x < 1$, $0 < y < 1$.

4. $\displaystyle\iint_D \frac{xy}{(x^2 + y^2)^{3/2}} \, dx \, dy$, where D is the unit circle.

5. $\displaystyle\iint_D \frac{x + y}{(x^2 + y^2)^{3/2}} \, dx \, dy$, where D is the unit circle.

6. $\displaystyle\iint_D \log \sqrt{x^2 + y^2} \, dx \, dy$, where D is the unit circle.

3.5   Improper double integrals, continued

Suppose we want to integrate a nonnegative function $f(x,y)$ over an unbounded region D. We might as well set $f(x,y) = 0$ for (x,y) not in D and integrate over the whole plane. We write such an integral as

$$\int\int_{-\infty}^{+\infty} f(x,y) \, dx \, dy \tag{15}$$

and interpret it as the volume between the whole (x,y) plane and the surface $z = f(x,y)$.

In order to assign a numerical value to (15), we choose a sequence of bounded regions of integration, $D_1, D_2, D_3, \ldots$ such that (a) each D_j is contained in the next region D_{j+1} and (b) each disk $x^2 + y^2 < R^2$ is contained in some D_j and hence also in all following ones. For instance, we can choose for D_j the interval $-j < x < j$, $-j < y < j$.

We assume that f is piecewise continuous and the integrals

$$\iint_{D_j} f(x,y) \, dx \, dy \tag{16}$$

converge. (They may be improper integrals.) Since $f > 0$ and condition (a) holds, the numbers (16) form an increasing sequence (see Chapter 15, §1.4). We write

$$\lim_{j \to +\infty} \iint_{D_j} f(x,y) \, dx \, dy = \int\int_{-\infty}^{+\infty} f(x,y) \, dx \, dy. \tag{17}$$

If this is a finite limit, we say that (15) **converges** or that f is **integrable** over the plane. Otherwise (15) is called **divergent.**

[It may seem that the value of (15) just defined depends on the choice of the regions $D_1, D_2, D_3, \ldots$. This is not so. Indeed, let $\widehat{D}_1, \widehat{D}_2, \widehat{D}_3, \ldots$ be another sequence satisfying conditions (a) and (b). Choose a fixed k; the region $\widehat{D}_k$ lies in some disk about the origin, hence also in all D_j, for j sufficiently large. Since $f \geq 0$, the integral $\iint_{\widehat{D}_k} f \, dx \, dy$ is not greater than (16), for large j, and hence not greater than the limit (17). The value of (15) obtained by using the sequence $\widehat{D}_1, \widehat{D}_2, \ldots$ is therefore not greater than that obtained through the sequence $D_1, D_2, \ldots$. An analogous argument shows that the value obtained by using the D_j's is not greater than the value obtained through the $\widehat{D}_k$'s. Hence the two values are equal.]

Integrals over the whole plane can be computed by iterated integrals and by polar coordinates; for instance,

$$\int_{-\infty}^{+\infty} \int f(x,y) \, dx \, dy = \int_{-\infty}^{+\infty} \left\{ \int_{-\infty}^{+\infty} f(x,y) \, dx \right\} dy = \int_{-\infty}^{+\infty} \int_{-\infty}^{+\infty} f(x,y) \, dx \, dy. \qquad (18)$$

EXAMPLE Compute

$$\int_{-\infty}^{+\infty} \int \frac{dx \, dy}{(1 + x^2 + y^2)^2}.$$

ANSWER   We choose as D_j the disk $|x^2 + y^2| < j^2$. Using polar coordinates, we have

$$\int_{-\infty}^{+\infty} \int \frac{dx \, dy}{(1 + x^2 + y^2)^2} = \lim_{j \to +\infty} \iint_{D_j} \frac{dx \, dy}{(1 + x^2 + y^2)^2} = \lim_{j \to +\infty} \int_0^{2\pi} \int_0^j \frac{r \, dr \, d\theta}{(1 + r^2)^2}$$

$$= \lim_{j \to +\infty} \int_0^{2\pi} d\theta \int_0^j \frac{r \, dr}{(1 + r^2)^2} = \lim_{j \to +\infty} 2\pi \left[-\frac{1}{2} \frac{1}{1 + r^2} \right]_{r=0}^{r=j}$$

$$= \lim_{j \to +\infty} 2\pi \left(\frac{1}{2} - \frac{1}{2(1 + j^2)} \right) = \pi.$$

It is simpler to use polar coordinates at once:

$$\int_{-\infty}^{+\infty} \int \frac{dx \, dy}{(1 + x^2 + y^2)^2} = \int_0^{2\pi} \int_0^{+\infty} \frac{r \, dr \, d\theta}{(1 + r^2)^2} = 2\pi \int_0^{+\infty} \frac{r \, dr}{(1 + r^2)^2}$$

$$= 2\pi \left[-\frac{1}{2} \frac{1}{1 + r^2} \right]_0^{+\infty} = \pi.$$

The reader should try to compute the same integral using iterated integration in Cartesian coordinates. This will take longer.

PROBLEMS

1. The area under the curve $xy = 1$ for $x \geq 1$ is rotated about the x-axis. Show that the resulting solid of revolution has a finite volume but an infinite area. (If we were to build a paint can in the shape of the figure thus constructed, it would hold only a finite amount of paint; yet to paint it would require an infinite amount of paint!)

2. If D is given by $x^2 + y^2 > 1$, for which values of α does $\iint_D (x^2 + y^2)^{-\alpha/2} \, dx \, dy$ converge? What is its value?

In Problems 3 to 6 determine which integrals converge and which diverge. If the integral converges, find its value. D is always the region *exterior* of the unit circle.

3. $\displaystyle \iint_D \frac{dx \, dy}{1 + x^2 + y^2}.$

4. $\displaystyle \iint_D \frac{x^2 y^2}{(x^2 + y^2)^{5/2}} \, dx \, dy.$

5. $\displaystyle \iint_D \frac{x^2}{(x^2 + y^2)^{5/2}} \, dx \, dy.$

6. $\displaystyle \iint_D \frac{\log \sqrt{x^2 + y^2}}{(x^2 + y^2)} \, dx \, dy.$

3.6 An important example

We prove the remarkable relation

$$\int_{-\infty}^{+\infty} e^{-x^2}\, dx = \sqrt{\pi}\,.$$

(19)

This is important in mathematical statistics (see Chapter 12, §3.1).
    The proof is based on the double integral

$$T = \iint_{-\infty}^{+\infty} e^{-x^2-y^2}\, dx\, dy.$$

We have, using polar coordinates,

$$T = \int_0^{2\pi}\int_0^{+\infty} e^{-r^2} r\, dr\, d\theta = 2\pi \int_0^{+\infty} \frac{d}{dr}\left(-\frac{e^{-r^2}}{2}\right) dr = \pi[-e^{-r^2}]_0^{+\infty} = \pi.$$

On the other hand, by (18) and the addition theorem for e^x,

$$T = \iint_{-\infty}^{+\infty} e^{-x^2} e^{-y^2}\, dx\, dy = \int_{-\infty}^{+\infty}\int_{-\infty}^{+\infty} e^{-x^2} e^{-y^2}\, dx\, dy$$

$$= \int_{-\infty}^{+\infty} e^{-x^2}\, dx \int_{-\infty}^{+\infty} e^{-y^2}\, dy = \left(\int_{-\infty}^{+\infty} e^{-x^2}\, dx\right)^2,$$

the name of a dummy variable being irrelevant. Now (19) follows.

PROBLEMS

1. Prove that $\displaystyle\int_0^{+\infty} (e^{-u}/\sqrt{u})\, du = \sqrt{\pi}\,.$

2. Prove that $\displaystyle\int_0^{+\infty} \sqrt{u}e^{-u}\, du = \sqrt{\pi}/2.$

§4 Triple integrals

Now that we have discussed double integrals, we can rapidly extend the theory to functions of three (and more) variables. No essentially new phenomena appear.

4.1 Computing triple integrals

A function $f(x,y,z)$ defined in a three-dimensional interval $x_0 < x < x_1,\ y_0 < y < y_1,\ z_0 < z < z_1$ will be called **piecewise continuous** if it is either continuous in the whole interval or the points at which f fails to be continuous lie on finitely many surfaces. Each of these surfaces must be either the intersection of the interval with a plane parallel to the coordinate axes or representable as a graph of a function $x = \phi_1(y,z)$ *and* as a graph of a function $y = \phi_2(z,x)$ *and* as a graph of a function $z = \phi_3(x,y)$, all three functions being defined and continuous in some closed two-dimensional intervals.

If $f(x,y,z)$ is a bounded and piecewise continuous function defined in the interval I: $x_0 < x < x_1$, $y_0 < y < y_1$, $z_0 < z < z_1$, then the number

$$\iiint_I f(x,y,z)\, dx\, dy\, dz \tag{1}$$

(the integral of f over I) can be computed by integrating $f(x,y,z)$ with respect to x from x_0 to x_1, with respect to y from y_0 to y_1, and with respect to z from z_0 to z_1 *in any order*. For instance, if we decide to integrate in the order x, y, z, then we have

$$\iiint_I f(x,y,z)\, dx\, dy\, dz = \int_{z_0}^{z_1} \int_{y_0}^{y_1} \int_{x_0}^{x_1} f(x,y,z)\, dx\, dy\, dz$$

$$= \int_{z_0}^{z_1} \left\{ \int_{y_0}^{y_1} \left[\int_{x_0}^{x_1} f(x,y,z)\, dx \right] dy \right\} dz. \tag{2}$$

We again follow the *convention:* the outer integral sign corresponds to the outer differential.

The definition of the integral based on step functions, the concept of Riemann sums (see §2.2 and §2.3), and Theorem 1 of §2.3 extend to functions of three variables. Also, triple integrals have the properties (11), (12), (13) stated in §1.4 for double integrals.

Note that we can compute (1) by one single and one double integration. Indeed, since

$$\int_{y_0}^{y_1} \int_{x_0}^{x_1} f(x,y,z)\, dx\, dy = \iint_{\substack{y_0 < y < y_1 \\ x_0 < x < x_1}} f(x,y,z)\, dx\, dy,$$

we can write instead of (2)

$$\iiint_I f(x,y,z)\, dx\, dy\, dz = \int_{z_0}^{z_1} \left[\iint_{\substack{y_0 < y < y_1 \\ x_0 < x < x_1}} f(x,y,z)\, dx\, dy \right] dz. \tag{3}$$

All orders of integration lead to the same value of (1).

To compute the triple integral

$$\iiint_D f(x,y,z)\, dx\, dy\, dz,$$

where D is a region of integration, other than an interval, we set $f = 0$ outside D and then integrate over some three-dimensional interval containing D.

EXAMPLES **1.** Let I be the cube $0 < x < 1$, $0 < y < 1$, $0 < z < 1$. Compute $\iiint_I xy^2 z^3\, dx\, dy\, dz$.

ANSWER (Compare the Remark in §1.1.) We have

$$\iiint_I xy^2 z^3\, dx\, dy\, dz = \int_0^1 \int_0^1 \int_0^1 xy^2 z^3\, dx\, dy\, dz$$

$$= \left(\int_0^1 x\, dx \right) \left(\int_0^1 y^2\, dy \right) \left(\int_0^1 z^3\, dz \right) = \frac{1}{2} \cdot \frac{1}{3} \cdot \frac{1}{4} = \frac{1}{24}.$$

2. Compute $\iiint_I yz^3 \cos(xyz)\, dx\, dy\, dz$, where I is as in Example 1.
ANSWER   The integral equals

$$\int_0^1 \int_0^1 \int_0^1 yz^3 \cos(xyz)\, dx\, dy\, dz = \int_0^1 \int_0^1 [z^2 \sin(xyz)]_{x=0}^{x=1}\, dy\, dz$$

$$= \int_0^1 \int_0^1 z^2 \sin(yz)\, dy\, dz = \int_0^1 [-z \cos(yz)]_{y=0}^{y=1}\, dz$$

$$= \int_0^1 (z - z \cos z)\, dz = \tfrac{1}{2} - \int_0^1 z\, d\sin z$$

$$= \tfrac{1}{2} - [z \sin z]_{z=0}^{z=1} + \int_0^1 \sin z\, dz = \tfrac{1}{2} - \sin 1 - [\cos z]_{z=0}^{z=1}$$

$$= \tfrac{3}{2} - \sin 1 - \cos 1 \approx .1182.$$

In the last step we used integration by parts.

3. Let D be the region between the planes $x = 0$, $y = 0$, $z = 4$ and the paraboloid $z = x^2 + y^2$ (see Figure 20.25). Compute $\iiint_D 2x\, dy\, dy\, dz$.

FIGURE **20.25**

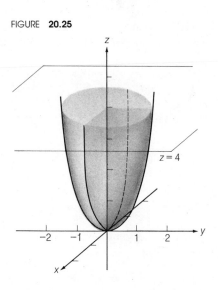

SOLUTION   In region D, z varies between 0 and 4. For a fixed z, x and y are positive and vary so that $x^2 + y^2 < z$. This means that y varies from 0 to $\sqrt{z}$ and x varies from 0 to $\sqrt{z - y^2}$. Hence the integral equals

$$\int_0^4 \int_0^{\sqrt{z}} \int_0^{\sqrt{z-y^2}} 2x\, dx\, dy\, dz = \int_0^4 \int_0^{\sqrt{z}} [x^2]_{x=0}^{x=\sqrt{z-y^2}}\, dy\, dz = \int_0^4 \int_0^{\sqrt{z}} (z - y^2)\, dy\, dz$$

$$= \int_0^4 \left[zy - \frac{y^3}{3} \right]_{y=0}^{y=\sqrt{z}}\, dz = \int_0^4 \frac{2}{3} z^{3/2}\, dz = \left[\frac{4}{15} z^{5/2} \right]_{z=0}^{z=4} = \frac{128}{15}.$$

REMARK   Everything said above can be repeated, with suitable modification, for functions of more than three variables.

PROBLEMS

In Problems 1 to 8 integrate the given function $f(x,y,z)$ over the indicated interval I.

1. $f = x^2 + y^2 + z^2$; $I: 0 < x < 1, \, 0 < y < 1, \, 0 < z < 1$.
2. $f = x^2 y^4 z$; $I: -1 < x < 0, \, 0 < y < 2, \, 0 < z < 1$.
3. $f = yz^2 e^{xyz}$; $I: 0 < x < 1, \, 0 < y < 1, \, 0 < z < 1$.
4. $f = xyz e^{x+y+z}$; $I: 1 < x < 2, \, 1 < y < 2, \, 1 < z < 2$.
5. $f = (x + y)/z$; $I: -1 < x < 1, \, -2 < y < 2, \, 1 < z < 2$.
6. $f = \dfrac{x}{y} + \dfrac{y}{z} + \dfrac{z}{x}$; $I: 1 < x < 2, \, 1 < y < 2, \, 1 < z < 2$.
7. $f = (x + y + z)^4$; $I: 0 < x < 1, \, 0 < y < 1, \, 0 < z < 1$.
8. $f = \cos(x + 2y + 3z)$; $I: 0 < x < \dfrac{\pi}{2}, \, 0 < y < \dfrac{\pi}{4}, \, 0 < z < \dfrac{\pi}{6}$.

In Problems 9 to 14 evaluate the given integrals.

9. $\displaystyle \int_0^2 \int_0^{4z} \int_{y-z}^{y+z} x \, dx \, dy \, dz.$

12. $\displaystyle \int_0^2 \int_0^{1+2x} \int_0^{\sqrt{4-y^2}} \sqrt{4 - y^2} \, dz \, dy \, dz.$

10. $\displaystyle \int_0^4 \int_0^{1+y^2} \int_{-2x}^{2x} xyz \, dz \, dx \, dy.$

13. $\displaystyle \int_{-\pi/4}^{\pi/4} \int_0^{\sin x} \int_0^{1+z \cos x} z \, dy \, dz \, dx.$

11. $\displaystyle \int_0^1 \int_0^{\sqrt{4-x^2}} \int_{2x-y}^{2x+y} z \, dz \, dy \, dx.$

14. $\displaystyle \int_0^{\pi/6} \int_0^{2z} \int_0^{\sqrt{\sin x}} y^3 \cos x \, dy \, dx \, dz.$

4.2   Volume

If D is a reasonable region in (x,y,z) space, then

$$\iiint_D dx \, dy \, dz$$

is the **volume** of D. This definition corresponds to the volume formula given in Chapter 9, §2.2, and to our common sense idea of volume.

EXAMPLES 1. Find the volume inside the region between the planes $x = 0$, $y = 0$, $z = 4$ and the paraboloid $z = x^2 + y^2$.

SOLUTION   The region is the same as that considered in Example 3, §4.1. See Figure 20.25. Hence the volume is given by

$$V = \int_0^4 \int_0^{\sqrt{z}} \int_0^{\sqrt{z-y^2}} dx \, dy \, dz = \int_0^4 \int_0^{\sqrt{z}} [x]_{x=0}^{x=\sqrt{z-y^2}} \, dy \, dz$$

$$= \int_0^4 \int_0^{\sqrt{z}} \sqrt{z - y^2} \, dy \, dz = \int_0^4 \left[\frac{1}{2} y \sqrt{z - y^2} + \frac{z}{2} \arcsin \frac{y}{\sqrt{z}} \right]_{y=0}^{y=\sqrt{z}} \, dz$$

$$= \int_0^4 \frac{\pi}{4} z \, dz = 2\pi.$$

2.   Find the volume inside the surface $x^2 + y^2 = 9$ bounded above by $x + z = 4$ and below by $z = 0$.

SOLUTION   We first sketch the volume to be determined. The surface $x^2 + y^2 = 9$ is a cylinder of radius 3 with axis along the z-axis, $z = 0$ is the xy plane, and $x + z = 4$ is a tilted plane parallel to the y-axis and passing through the points $(4,0,0)$, $(0,0,4)$. See Figure 20.26. The

FIGURE 20.26

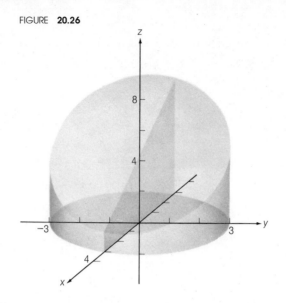

limits of integration for y are from -3 to 3. Using $x^2 + y^2 = 9$, the corresponding limits for x are $-\sqrt{9 - y^2}$ to $\sqrt{9 - y^2}$. Finally, the limits for z are just 0 to $4 - x$, since z must lie above the plane $z = 0$ and below the plane $z = 4 - x$. If we evaluate the integral for the volume in the order z, x, and y, we have

$$\dot{V} = \int_{-3}^{3} \int_{-\sqrt{9-y^2}}^{\sqrt{9-y^2}} \int_{0}^{4-x} dz \, dx \, dy = \int_{-3}^{3} \left(4x - \frac{x^2}{2}\right) \Big|_{-\sqrt{9-y^2}}^{\sqrt{9-y^2}} dy$$

$$= 8 \int_{-3}^{3} \sqrt{9 - y^2} \, dy = 4 \left(y\sqrt{9 - y^2} + 9 \arcsin \frac{y}{3}\right) \Big|_{-3}^{3} = 36\pi.$$

PROBLEMS

1. Find the volume interior to the tetrahedron formed by the planes $x = 0$, $y = 0$, $z = 0$, $6x + 4y + 3z = 12$.
2. Find the volume of the tetrahedron bounded by the planes $x = 0$, $y = 0$, $z = 0$, and $(x/a) + (y/b) + (z/c) - 1 = 0$.
3. Find the volume of the sphere $x^2 + y^2 + z^2 = a^2$.
4. Find the volume inside the cylinder $x^2 + y^2 = 9$, above $z = 0$, and below $x + z = 4$.
5. Find the volume bounded by the paraboloid $z = x^2 + y^2$ and the cylinder $x^2 + y^2 = a^2$.
6. Find the volume common to the two cylinders $x^2 + y^2 = 9$, $x^2 + z^2 = 9$.
7. Find the volume of the ellipsoid $(x/a)^2 + (y/b)^2 + (z/c)^2 = 1$.
8. Find the volume bounded by the two paraboloids $z = x^2 + 4y^2$ and $z = 8 - x^2 - 4y^2$.

4.3 Integration in cylindrical coordinates

In §3.2 we saw that when we want to compute a double integral using polar coordinates, it is not enough to rewrite the integrand $f(x,y)$ as a function of polar coordinates (r,θ), but we also must multiply the integrand by a factor, namely, r

FIGURE **20.27**

FIGURE **20.28**

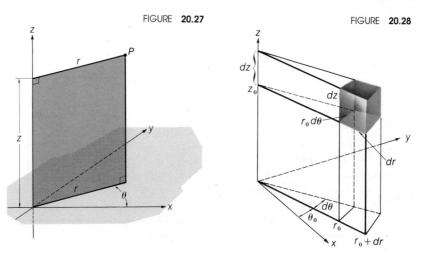

[compare Equation (2) in §3.2]. This phenomenon occurs whenever we use "new" coordinates for integration—the integrand is to be multiplied by a factor, depending on the coordinates used. This factor is called the **Jacobian,** in honor of the mathematician Jacobi.

Without going into the general theory, we shall show how to compute triple integrals in cylindrical coordinates. These coordinates, usually denoted by (r,θ,z), have been defined in Chapter 18, §1.3; they are shown in Figure 20.27. We recall that the connection with rectangular coordinates is given by the formulas

$$x = r \cos \theta, \qquad y = r \sin \theta, \qquad z = z. \tag{4}$$

Assume now that $f(x,y,z)$ is a bounded piecewise continuous function which is 0 outside the cylinder $x^2 + y^2 < R^2$, $|z| < H$ and that the function $(r,\theta,z) \mapsto f(r \cos \theta, r \sin \theta, z)$ is also piecewise continuous. Under these circumstances, we have

$$\iiint_{\substack{x^2+y^2<R^2 \\ |z|<H}} f(x,y,z)\, dx\, dy\, dz = \int_{-H}^{H} \int_{0}^{2\pi} \int_{0}^{R} f(r \cos \theta, r \sin \theta, z) r\, dr\, d\theta\, dz. \tag{5}$$

The Jacobian is again r; this is not surprising since cylindrical coordinates differ from polar coordinates only by the presence of the third coordinate z.

To prove (5), we note that by Equation (3) in §4.1, and since f is 0 outside the cylinder, we have

$$\iiint_{\substack{x^2+y^2<R^2 \\ |z|<H}} f(x,y,z)\, dx\, dy\, dz = \int_{-H}^{H} \left\{ \iint_{x^2+y^2<R^2} f(x,y,z)\, dx\, dy \right\} dz. \tag{6}$$

G. G. J. JACOBI (1804–1857). Among his greatest achievements are the theory of so-called "elliptic" functions and a new mathematical approach to problems of mechanics that continued the work of Hamilton. The Hamilton-Jacobi theory played an un-

expected part in the creation of quantum theory. Jacobi established the tradition of combining university teaching with research apprenticeship. It seems he was the first mathematician to lecture to his students on his own discoveries.

But by Equation (2) in §3.2 we have, for a fixed z,

$$\iint_{x^2+y^2<R^2} f(x,y,z)\, dx\, dy = \int_0^{2\pi} \int_0^R f(r\cos\theta, r\sin\theta, z)r\, dr\, d\theta.$$

Substituting this into (6), we obtain (5). Of course, we can rewrite (5) in various ways by changing the order of integration in the right-hand side.

We could also have guessed at (5) by a mathematical myth as depicted in Figure 20.28. We consider a point with cylindrical coordinates (r_0, θ_0, z_0), the two cylinders $r = r_0$, $r = r_0 + dr$, the two planes $\theta = \theta_0$, $\theta = \theta_0 + d\theta$, and the two planes $z = z_0$, $z = z_0 + dz$. Here dr, $d\theta$, dz are infinitely small numbers (we may say so since we are telling a myth). The six surfaces bound an infinitely small box whose volume equals the product of the lengths of the sides, that is, $(dr)(r_0\, d\theta)(dz) = r_0\, dr\, d\theta\, dz$. The triple integral of a function f is the sum of infinitely many terms: value of f at a point times the volume of the small box at this point.

EXAMPLE Let D be the region defined by the inequalities

$$x^2 + y^2 < 1, \qquad 0 < z < x^2 + y^2. \tag{7}$$

Find $\iiint_D x^2 y^2\, dx\, dy\, dz$.

ANSWER We use cylindrical coordinates; the inequalities (7) can be rewritten as $0 < r < 1$, $0 < z < r^2$. Hence the triple integral considered equals

$$\int_0^{2\pi} \int_0^1 \int_0^{r^2} (r\cos\theta)^2 (r\sin\theta)^2 r\, dz\, dr\, d\theta$$

$$= \int_0^{2\pi} \int_0^1 \int_0^{r^2} \frac{1}{4} r^5 \sin^2 2\theta\, dz\, dr\, d\theta = \int_0^{2\pi} \int_0^1 \frac{1}{4} r^7 \sin^2 2\theta\, dr\, d\theta$$

$$= \frac{1}{4} \int_0^{2\pi} \sin^2 2\theta\, d\theta \int_0^1 r^7\, dr = \frac{1}{4}\pi\frac{1}{8} = \frac{\pi}{32}.$$

PROBLEMS

1. Find the volume cut from the sphere $x^2 + y^2 + z^2 = 25$ by the cylinder $x^2 + y^2 = 9$.
2. Two circular cylinders have the same radius R, and their axes intersect at right angles. Find the volume of their intersection.
3. Find the volume cut off the cylinder $r = 4\cos\theta$ by the sphere $x^2 + y^2 + z^2 = 16$ and the xy plane.
4. Find the volume above the paraboloid $z = x^2 + y^2$ and below the plane $z = 2y$.
5. Find the volume bounded by the xy plane, the circular cylinder $x^2 + y^2 = 1$, and the parabolic cylinder $z = x^2 - 1$.
6. A round hole of radius 1 is bored through a sphere of radius 2. The axis of the hole is a diameter of the sphere. What volume remains?
7. Compute $\iiint_D [z/(y^2 + z^2)]\, dx\, dy\, dz$ where D is the region $1 < y^2 + z^2 < 2$, $3 < x < 4$.
8. If D is the cylindrical shell bounded by the planes $x = 1$, $x = 2$, and by the circular cylinders $y^2 + z^2 = 4$, $y^2 + z^2 = 9$, compute $\iiint_D e^x \sqrt{y^2 + z^2}\, dx\, dy\, dz$.

4.4 Bodies of revolution

Let D be a region and l a line, both lying in one plane, D on one side of l. If we rotate D about l by $360°$, it sweeps out a body of revolution B. To compute the volume

FIGURE 20.29

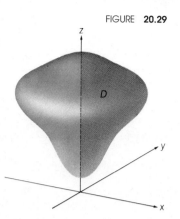

of B, choose l as the z-axis, and let D lie in the right half of the (x,z) plane; see Figure 20.29. Let $\chi_D(x,z)$ be the characteristic function of D. Then the characteristic function of B is $\chi_D(\sqrt{x^2 + y^2},z)$. Using cylindrical coordinates, we compute the volume V of B to be (T is some large number)

$$V = \iiint_{x^2+y^2+z^2<T^2} \chi_D(\sqrt{x^2 + y^2},z)\, dx\, dy\, dz$$

$$= \int_0^{2\pi} \iint_{r^2+z^2<T^2} \chi_D(r,z) r\, dr\, dz\, d\theta = 2\pi \iint_D r\, dr\, dz$$

or, since it does not matter what we call dummy variables,

$$V = 2\pi \iint_D x\, dx\, dz. \tag{8}$$

Similarly, if we rotate a region D in the right half of the (x,y) plane about the y-axis, the volume is

$$V = 2\pi \iint_D x\, dx\, dy, \tag{8'}$$

and if D is a region in the upper half of the (x,y) plane, and we rotate it about the x-axis, we obtain the volume

$$V = 2\pi \iint_D y\, dx\, dy. \tag{8''}$$

In the latter case, assume that D is the region under the graph of the positive function $y = f(x)$, $a < x < b$. Then

$$V = 2\pi \int_a^b \int_0^{f(x)} y\, dy\, dx = 2\pi \int_a^b [\tfrac{1}{2}y]_{y=0}^{y=f(x)}\, dx = \pi \int_a^b f(x)^2\, dx.$$

We have already obtained this formula before; see Chapter 9, §2.3.

REMARK If D is rotated not by 2π but by an angle ω, $0 < \omega < 2\pi$, we should replace, in the formula (8), 2π by ω. The proof is left to the reader.

FIGURE **20.30**

FIGURE **20.31**

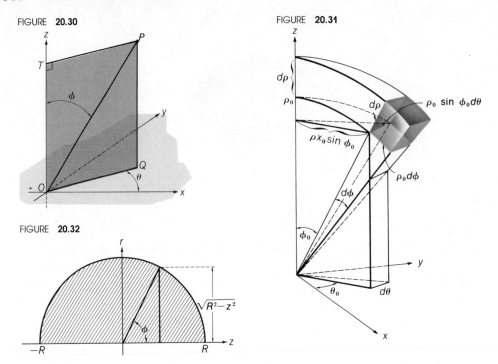

FIGURE **20.32**

4.5 Integration in spherical coordinates

We recall (see Chapter 18, §1.5, and Figure 20.30) that spherical coordinates (ρ,θ,ϕ) are connected with Cartesian coordinates by the relations

$$x = \rho \sin \phi \cos \theta, \qquad y = \rho \sin \phi \sin \theta, \qquad z = \rho \cos \phi \tag{9}$$

and with the cylindrical coordinates by the relations

$$r = \rho \sin \phi, \qquad \theta = \theta, \qquad z = \rho \cos \phi. \tag{9'}$$

The integration formula (valid under appropriate conditions) reads

$$\iiint_{x^2+y^2+z^2<R^2} f(x,y,z)\, dx\, dy\, dz$$
$$= \int_0^\pi \int_0^{2\pi} \int_0^R f(\rho \sin \phi \cos \theta, \rho \sin \phi \sin \theta, \rho \cos \phi)\rho^2 \sin \phi\, d\rho\, d\theta\, d\phi. \tag{10}$$

Thus the Jacobian is $\rho^2 \sin \phi$.

Before proving this, we note that the result could have been guessed from a mathematical myth similar to the one told above and has been depicted in Figure 20.31. The two infinitely close spheres $\rho = \rho_0$, $\rho = \rho_0 + d\rho$, the two infinitely close planes $\theta = \theta_0$, $\theta = \theta_0 + d\theta$, and the two infinitely close cones $\phi = \phi_0$, $\phi = \phi_0 + d\phi$ bound a "box" of volume $(d\rho)(\rho_0 \sin \phi_0\, d\theta)(\rho_0\, d\phi) = \rho_0^2 \sin \phi_0\, d\rho\, d\theta\, d\phi$.

Now we prove (10); we begin with an identity that follows from (5) by assuming that f vanishes outside the sphere of radius R about the origin. Since the inequality

$x^2 + y^2 + z^2 < R^2$ can be written in cylindrical coordinates as $z^2 + r^2 < R^2$, we have

$$\iiint_{x^2+y^2+z^2<R^2} f(x,y,z)\, dx\, dy\, dz = \int_{-R}^{R} \int_{0}^{2\pi} \int_{0}^{R^2-z^2} f(r\cos\theta, r\sin\theta, z) r\, dr\, d\theta\, dz$$

$$= \int_{0}^{2\pi} \int_{-R}^{R} \int_{0}^{R^2-z^2} f(r\cos\theta, r\sin\theta, z) r\, dr\, dz\, d\theta. \tag{11}$$

The last step is justified since it is legitimate to interchange the order of integration in a multiple integral over an interval. For a fixed θ, consider

$$\int_{-R}^{R} \int_{0}^{R^2-z^2} f(r\cos\theta, r\sin\theta, z) r\, dr\, dz. \tag{12}$$

We treat z and r as Cartesian coordinates in a (z,r) plane. Then (12) is a double integral over an upper semidisk (see Figure 20.32). Noting (9′), we interpret (ρ,ϕ) as polar coordinates in the (z,r) plane so that

$$\int_{-R}^{R} \int_{0}^{R^2-z^2} f(r\cos\theta, r\sin\theta, z) r\, dr\, dz$$

$$= \int_{0}^{\pi} \int_{0}^{R} f(\rho\sin\phi\cos\theta, \rho\sin\phi\sin\theta, \rho\cos\phi)(\rho\sin\phi)\rho\, d\rho\, d\phi.$$

If we substitute this into (11), we obtain an identity that differs from (10) only by the order of integrations.

EXAMPLES 1. We compute the volume bounded by the sphere of radius R; the problem is tailor-made for spherical coordinates. Thus

$$\iiint_{x^2+y^2+z^2<R^2} dx\, dy\, dz = \int_{0}^{\pi} \int_{0}^{2\pi} \int_{0}^{R} \rho^2 \sin\phi\, d\rho\, d\theta\, d\phi$$

$$= \left(\int_{0}^{\pi} \sin\phi\, d\phi \right) \left(\int_{0}^{2\pi} d\theta \right) \left(\int_{0}^{R} \rho^2\, d\rho \right) = 2 \cdot 2\pi \frac{R^3}{3} = \frac{4\pi R^3}{3},$$

as expected.

2. Evaluate the improper triple integral

$$\iiint_{D} (x^2 + y^2 + z^2)^{-\alpha/2}\, dx\, dy\, dz,$$

where D is the unit ball $(x^2 + y^2 + z^2 < 1)$ and α a positive number.
ANSWER We use spherical coordinates. The integral equals

$$\int_{0}^{\pi} \int_{0}^{2\pi} \int_{0}^{1} \rho^{-\alpha}\rho^2 \sin\phi\, d\rho\, d\theta\, d\phi = \int_{0}^{\pi} \sin\phi\, d\phi \int_{0}^{2\pi} d\theta \int_{0}^{1} \rho^{2-\alpha}\, d\rho$$

$$= 4\pi \int_{0}^{1} \rho^{2-\alpha}\, d\rho = \begin{cases} 4\pi \left. \dfrac{\rho^{3-\alpha}}{3-\alpha} \right|_{0}^{1} = \dfrac{4\pi}{3-\alpha}, & \text{if } \alpha < 3, \\[2ex] +\infty, & \text{if } \alpha \geq 3. \end{cases}$$

PROBLEMS

1. Compute $\iiint_{D} z^2\, dx\, dy\, dz$ where D is the volume bounded by the sphere $x^2 + y^2 + z^2 = 4$.

2. Compute $\iiint_D xyz \, dx \, dy \, dz$, where D is the portion of the unit sphere in the "first octant," $x > 0$, $y > 0$, $z > 0$.

3. Compute $\iiint_D (x^2 + y^2 + z^2)^3 \, dx \, dy \, dz$, where D is the region $x > 0$, $y > 0$, $x^2 + y^2 + z^2 < \sqrt{3}$.

4. Compute $\iiint_D xy^2 \, dx \, dy \, dz$, where D is the intersection of the unit sphere, the half-space $z > 0$, and the cone $x^2 + y^2 = z^2$.

5. If D is the region defined by $x^2 + y^2 + z^2 > 1$, for which value of α does the integral $\iiint_D (x^2 + y^2 + z^2)^{-\alpha/2} \, dx \, dy \, dz$ exist?

6. Find the volume bounded by the sphere $x^2 + y^2 + z^2 = 25$ and the planes $z = 3$, $z = 4$.

7. Find the volume remaining when the cone $3(x^2 + y^2) = z^2$ is removed from the hemisphere $z > 0$, $x^2 + y^2 + z^2 = 9$.

8. Find the volume bounded by the spheres $x^2 + y^2 + z^2 = 1$, $x^2 + y^2 + z^2 = 4$, and the cone $x^2 + y^2 = z^2$.

9. Find the volume bounded above by the sphere $x^2 + y^2 + z^2 = 2z$ and below by the cone $z^2 = x^2 + y^2$.

§5   Mass centers and centroids‡

5.1   Density

In applications the integrand in a triple integral is often interpreted as **density**. This means that we think of the region of integration, D, as a material body and we assume that there is a positive continuous function $f(x,y,z)$ such that the mass of every part D_0 of D is given by $\iiint_{D_0} f(x,y,z) \, dx \, dy \, dz$. The total mass is the integral over all of D. Since mass cannot be negative, the density function f must be positive or zero. (Negative densities may occur in describing the distribution of electric charges.) The density is measured in units of mass divided by length cubed.

(The image of continuously distributed matter, given by a density function, is very useful in many branches of physics. Yet we should not forget that this is a radical idealization of reality. It neglects the fact that matter consists of molecules, and molecules of atoms, and that the atoms themselves have a complicated internal structure.)

Occasionally we also consider a density function of two variables (measured in units of mass divided by length squared) defined in some region D of the plane $z = 0$. Then we think of D as a material **lamina** which is so thin that its width may be set equal to 0. The mass of the part of the lamina corresponding to a part D_0 of D is given by $\iint_{D_0} f(x,y) \, dx \, dy$.

We can also discuss a continuous distribution of matter along a curved surface or along a curve. In the latter case, we speak of a material **wire,** and the density is measured in units of mass divided by length. If the wire is the curve $x = x(t)$, $y = y(t)$, $z = z(t)$, $a \leq t \leq b$, the density function is a function $f(x,y,z)$ such that the mass of a subarc corresponding to $t_0 \leq t \leq t_1$ is given by

$$\int_{t_0}^{t_1} f(x(t),y(t),z(t)) \sqrt{x'(t)^2 + y'(t)^2 + z'(t)^2} \, dt.$$

The density of a wire is measured by units of mass divided by units of length.

‡Optional section.

EXAMPLES 1. If the density is $f(x,y,z) = x^2y^4z^6$, the mass contained in the cube $0 < x < 1$, $0 < y < 1, 0 < z < 1$ is

$$\int_0^1 \int_0^1 \int_0^1 x^2y^4z^6 \, dx \, dy \, dz = \frac{1}{105}.$$

2. Suppose the density of a distribution of matter in the plane $z = 0$ is $f(x,y) = 1$. Then the mass contained in a (reasonable) set D in that plane is equal to the area of D.

PROBLEMS

1. A hemisphere of radius R has density equal to the cube of the distance from the center. What is its mass? [Hint: Use spherical coordinates.]

2. A right circular cylinder of unit height and unit radius has density proportional to the square of the distance from its axis. What is the mass if the density is K on the lateral surface?

3. A right circular cone has base area π and height h. The density at each point is equal to the distance from the apex of the cone. What is the mass?

4. A plane lamina has the form of a circle of radius a. Find the mass if the planar density is equal to the distance from the center.

5. A plane lamina has the shape of an isosceles right triangle. The density at each point equals the square of the distance from the right angle. The area of the lamina is 1. What is its mass?

6. A plane lamina has the form of an ellipse with semiaxes a and b, $a > b$. It has planar density equal to the positive distance from the major axis. Compute its mass.

7. A straight wire of length a has linear density equal to the square of the distance measured from its midpoint. What is the mass?

8. A circular wire of radius 1 has linear density equal to $(1 + \sin^2 \phi)$, where ϕ is the angular measure around the wire from a given fixed point on the wire. Compute the mass of the wire.

5.2 Mass center of a continuous distribution of matter. Centroids

The **mass center** of N particles with position vectors $\mathbf{x}_1, \mathbf{x}_2, \ldots, \mathbf{x}_N$ and masses $m_1, m_2, \ldots, m_N$ is defined as the point with position vector

$$\mathbf{X} = \frac{1}{M} \sum_{j=1}^{N} m_j \mathbf{x}_j, \qquad \text{where } M = \sum_{j=1}^{N} m_j.$$

(Compare the definition in Chapter 9, §4.1.) This suggests that the reasonable way of *defining* the position of the mass center of a solid D with variable density $f(x,y,z) = f(\mathbf{x})$ is

$$\mathbf{X} = \frac{1}{M} \iiint_D f(\mathbf{x})\mathbf{x} \, dx \, dy \, dz, \qquad \text{where } M = \iiint_D f(\mathbf{x}) \, dx \, dy \, dz$$

or

$$\mathbf{X} = \frac{1}{M} \iiint_D f(x,y,z)x \, dx \, dy \, dz \, \mathbf{i} + \frac{1}{M} \iiint_D f(x,y,z)y \, dx \, dy \, dz \, \mathbf{j}$$
$$+ \frac{1}{M} \iiint_D f(x,y,z)z \, dx \, dy \, dz \, \mathbf{k}.$$

Needless to say, there is a myth justifying this definition: $\mathbf{X}$ is the mass center of infinitely many infinitely small boxes with position vectors $\mathbf{x}$, volumes $dx\,dy\,dz$, and masses $f(\mathbf{x})\,dx\,dy\,dz$.

If D is a region in the plane $z = 0$, with a density function $f(x,y)$ defined in D, the mass center of the corresponding material lamina is, by definition, located at

$$\mathbf{X} = \frac{1}{M} \iint_D f(x,y)(x\mathbf{i} + y\mathbf{j})\,dx\,dy, \qquad \text{where } M = \iint_D f(x,y)\,dx\,dy.$$

If we are given a wire $\mathbf{x} = \mathbf{x}(t)$, $a \le t \le b$, and if $f(t) = f(x(t),y(t),z(t))$ is the density function, the mass center of the wire is at

$$\mathbf{X} = \frac{1}{M} \int_a^b f(t)|\mathbf{x}'(t)|\mathbf{x}(t)\,dt, \qquad \text{where } M = \int_a^b f(t)|\mathbf{x}'(t)|\,dt.$$

These definitions can be justified in the same way as for a solid. In all cases M is the total mass.

REMARK   In order that the "wires" should have physical meaning, we require from now on that the curve $\mathbf{x} = \mathbf{x}(t)$ be **simple**; this means that different values of t, $a \le t \le b$ should correspond to different points $\mathbf{x}(t)$, except that we may have $\mathbf{x}(b) = \mathbf{x}(a)$ if the curve is closed.

The **centroid** of a reasonable set D in space is the mass center of D for a constant density function. Centroids of plane sets and curves are defined similarly. The formulas for centroids are obtained from those for mass centers by setting $f = 1$. They read

$$\mathbf{X} = \frac{1}{V} \iiint_D \mathbf{x}\,dx\,dy\,dz \qquad \left(V = \iiint_D dx\,dy\,dz = \text{volume}\right), \tag{1}$$

$$\mathbf{X} = \frac{1}{A} \iint_D (x\mathbf{i} + y\mathbf{j})\,dx\,dy \qquad \left(A = \iint_D dx\,dy = \text{area}\right), \tag{2}$$

$$\mathbf{X} = \frac{1}{L} \int_a^b \mathbf{x}(t)|\mathbf{x}'(t)|\,dt \qquad \left(L = \int_a^b |\mathbf{x}'(t)|\,dt = \text{length}\right). \tag{3}$$

In other words,

$$X = \frac{1}{V} \iiint_D x\,dx\,dy\,dz, \qquad Y = \frac{1}{V} \iiint_D y\,dx\,dy\,dz, \qquad Z = \frac{1}{V} \iiint_D z\,dx\,dy\,dz \tag{1'}$$

are the coordinates of the centroid of the three-dimensional set D with volume V,

$$X = \frac{1}{A} \iint_D x\,dx\,dy, \qquad Y = \frac{1}{A} \iint_D y\,dx\,dy, \qquad Z = 0 \tag{2'}$$

are the coordinates of the centroid of a set D in the plane $z = 0$, with area A, and

$$X = \frac{1}{L} \int_a^b x(t)\sqrt{x'(t)^2 + y'(t)^2 + z'(t)^2}\,dt,$$

$$Y = \frac{1}{L} \int_a^b y(t)\sqrt{x'(t)^2 + y'(t)^2 + z'(t)^2}\,dt, \tag{3'}$$

$$Z = \frac{1}{L} \int_a^b z(t)\sqrt{x'(t)^2 + y'(t)^2 + z'(t)^2}\,dt$$

are the coordinates of the centroid of the curve $\mathbf{x} = \mathbf{x}(t)$, $a \leq t \leq b$, of length L.

The mass centers have an important **additivity property.** Let us, for instance, divide the solid D into two disjoint regions D_1 and D_2. Let M_1 and M_2 be the masses of D_1 and D_2, and let $\mathbf{X}_1$ and $\mathbf{X}_2$ be the corresponding mass centers. Then

$$\mathbf{X} = \frac{1}{M_1 + M_2}(M_1\mathbf{X}_1 + M_2\mathbf{X}_2).$$

In other words, $\mathbf{X}$ is the mass center of two particles of mass M_1 and M_2, located at $\mathbf{X}_1$ and $\mathbf{X}_2$, respectively.

To prove this, we note that by the additivity property of integrals (discussed in §1.4 for double integrals),

$$M_1\mathbf{X}_1 + M_2\mathbf{X}_2 = \iiint_{D_1} f(\mathbf{x})\mathbf{x}\, dx\, dy\, dz + \iiint_{D_2} f(\mathbf{x})\mathbf{x}\, dx\, dy\, dz$$

$$= \iiint_{D} f(\mathbf{x})\mathbf{x}\, dx\, dy\, dz = M\mathbf{X}.$$

A similar argument establishes the additivity property for laminae and wires.

The additivity property can be extended to the case where the solid (or the lamina or the wire) is divided into *several* parts.

Next we note the **symmetry properties.** If the density $f(x,y,z)$ and the region D are symmetric with respect to a plane, then the mass center lies in that plane. If the density $f(x,y)$ and the region D in the plane $z = 0$ are symmetric with respect to a line l, the mass center lies on this line. The reader should state himself the symmetric property of mass centers of wires.

We prove only one symmetry property. Suppose the plane region D and the density $f(x,y)$ are symmetric with respect to some line l. Choose the coordinate system so that l is the x-axis, D is an interval of the form $-a < x < a$, $-b < y < b$ (for we may set $f = 0$ outside the original D), and $f(x,y)$ is an even function of y for every fixed x. Hence the function $y \mapsto yf(x,y)$ is odd and

$$\int_{-b}^{b} yf(x,y)\, dy = 0$$

so that the y-component of the mass center is

$$Y = \frac{1}{M}\int_{-a}^{a}\int_{-b}^{b} yf(x,y)\, dy\, dx = 0,$$

as asserted.

The additivity and symmetry properties often simplify the calculation of centroids.

EXAMPLES 1. The centroid of a rectangle is the intersection point of the diagonals.

Proof. Draw two lines, l_1 and l_2, through the intersection point P of the diagonals, parallel to the sides. The rectangle is symmetric with respect to l_1 and to l_2; hence the centroid lies on both lines.

The reader should confirm this by a direct computation.

2. The centroid of a three-dimensional interval (parallelopiped, box) is at the intersection of the main diagonals. In other words: the centroid is the midpoint.

This is proved as the corresponding statement in the example above.

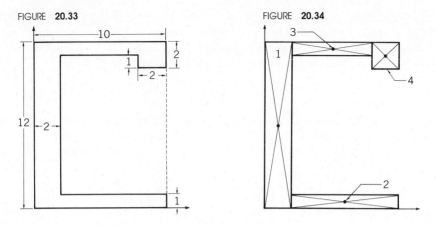

FIGURE **20.33**   FIGURE **20.34**

3. Find the centroid of the region shown in Figure 20.33.

SOLUTION.   We divide the region into four rectangles as shown in Figure 20.34. They have areas 24, 8, 6, and 4, respectively, and their centroids are at $(1,6)$, $(6,\frac{1}{2})$, $(5,\frac{23}{2})$, and $(9,11)$. The coordinates of the centroid are therefore

$$X = \frac{24 \cdot 1 + 8 \cdot 6 + 6 \cdot 5 + 4 \cdot 9}{24 + 8 + 6 + 4} = \frac{23}{7}, \qquad Y = \frac{24 \cdot 6 + 8 \cdot \frac{1}{2} + 6 \cdot \frac{23}{2} + 4 \cdot 11}{24 + 8 + 6 + 4} = \frac{261}{42}.$$

Since we deal with a figure in the plane, there is no need to state explicitly that $Z = 0$.

PROBLEMS

1. Find the center of mass of a right circular cylinder of radius R and height h if the density equals twice the distance from the base. [Hint: Use cylindrical coordinates.]
2. Find the center of mass of a right circular cone of radius R and height h if the density equals the square of the distance from the base.
3. Find the center of mass of a plane lamina in the shape of a square of side a whose density equals the square of the distance from one corner.
4. Find the center of mass of a plane lamina in the shape of an equilateral triangle of side a, whose density equals the distance from one vertex.
5. A helical wire lies on a curve $x = \cos t$, $y = \sin t$, $z = t$, $0 \le t \le 6\pi$. The density at each point equals $4t$. Where is the center of mass?
6. Find the center of mass of a bent wire of length 3. The wire travels east for one unit, then north for one unit, then straight up for one unit. Density equals the distance *along* the wire from the starting point (westernmost end).

   In each of Problems 7 to 14 find the centroid of the indicated plane region. (Assume the region lies in the xy plane.)

7. The region below the line $y = 2x + 3$ and above the x-axis from $x = 0$ to $x = 2$.
8. The region below the parabola $y = x^2$ and above the x-axis from $x = 0$ to $x = 4$.
9. The region below the curve $y = 1/x$ and above the x-axis from $x = 1$ to $x = 3$.
10. The region below the curve $y = \sin x$ and above the x-axis from $x = 0$ to $x = \pi/2$.
11. The region between the two parabolas $y = 4 - x^2$ and $y = x^2 - 4$.
12. The region determined by connecting the points $(0,0)$, $(2,0)$, $(2,4)$, $(1,4)$, $(1,1)$, $(0,1)$, $(0,0)$ in the given order by line segments.

13. The region determined by connecting the points (0,0), (8,0), (8,2), (6,2), (6,8), (2,8), (2,2), (0,2), (0,0) in the given order by line segments.

14. The region determined by connecting the points (0,0), (8,0), (8,8), (2,8), (2,4), (4,4), (4,6), (6,6), (6,2), (0,2), (0,0) in the given order by line segments.

In each of Problems 15 to 21 find the centroid of the given volume.

15. A hemisphere of radius R.
16. A right circular cone of radius r and height h.
17. The volume cut from one nappe of a cone of vertex angle 60° by a sphere of radius 2 whose center is the vertex of the cone.
18. The tetrahedron with vertices (0,0,0), (1,0,0), (0,1,0), (0,0,1).
19. The portion of the sphere $x^2 + y^2 + z^2 = R^2$ contained between the half-planes $\theta = -\pi/4$ and $\theta = \pi/4$.
20. The cube $0 < x < 1, 0 < y < 1, 0 < z < 1$, in which a vertical hole of radius $\frac{1}{4}$ is drilled through the point $x = \frac{1}{4}$, $y = \frac{1}{4}$. [Hint: Treat the hole as a negative mass point.]
21. The drilled cube of Problem 20 with an additional hole of radius $\frac{1}{8}$ parallel to the x-axis through the point $y = \frac{3}{4}$, $z = \frac{1}{2}$.

In Problems 22 to 28 find the centroid of the indicated curve.

22. The semicircular arc $y = \sqrt{1 - x^2}$, $|x| \leq 1$.
23. The quarter-circular arc $y = \sqrt{1 - x^2}$, $0 \leq x \leq 1$.
24. The first quadrant arc of the hypocycloid $x^{2/3} + y^{2/3} = 1$, $0 \leq x \leq 1$.
25. The arc of the catenary $y = \frac{1}{2}(e^x + e^{-x})$ lying between $x = -1$ and $x = 1$.
26. The line segments joining the points (0,0), (4,3), (7,7). Assume the curve lies in the xy plane.
27. The bent wire of Problem 6.
28. A wire shaped like the letter A, if the vertex angle is 30° and the crossbar is at the midpoints of the two sides.

5.3 Pappus' theorems

Centroids were studied already in ancient Greece. The following two propositions are due to Pappus, the last of the great Greek mathematicians of antiquity.

Theorem 1. *The volume V produced by rotating a plane area A about an axis l in the same plane is*

$$V = 2\pi\, \delta A, \tag{4}$$

where δ is the distance from the centroid of the area to the axis.

Theorem 2. *The area A produced by rotating a plane curve of length L about an axis l in the same plane is*

$$A = 2\pi\, \delta L, \tag{5}$$

where δ is the distance from the centroid of the curve to the axis.

PAPPUS of Alexandria (ca. 340 A.D.) is the author of a historical commentary on Greek geometrical litera-ture known as *Synagoge* (that is, Collection).

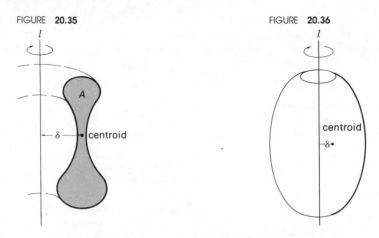

FIGURE 20.35

FIGURE 20.36

In Theorem 1 (see Figure 20.35), it is assumed that A is the area of a domain D lying *on one side* of l. Let D lie in the (x,y) plane, and let l be chosen as the y-axis. The x-coordinate of the centroid of D, that is, the distance δ, is

$$\delta = X = \frac{1}{A} \iint_D x \, dx \, dy,$$

as follows from (2'). By Equation (8') in §4.4, the volume V of the solid B, obtained by rotating D about the y-axis, is

$$V = 2\pi \iint_D x \, dx \, dy.$$

The two displayed formulas imply (4).

In Theorem 2 it is assumed that the curve, call it C, lies *on one side* of l. See Figure 20.36. Let C lie in the (x,y) plane with l chosen as the y-axis. We assume that C is the graph of $y = \phi(x)$, $0 \le a < x < b$. The x-coordinate of the centroid of C, that is, the distance δ, is

$$\delta = X = \frac{1}{L} \int_a^b x\sqrt{1 + \phi'(x)^2} \, dx.$$

This is seen by setting $x = t$, $y = \phi(t)$, $z = 0$ in (3'). By Equation (9) in §3.3, the area A, of the surface S obtained by rotating D about the y-axis, is

$$A = 2\pi \int_a^b x \, ds = 2\pi \int_a^b x\sqrt{1 + \phi'(x)^2} \, dx.$$

The two displayed formulas imply (5). It is even simpler to verify (5) when C is a segment $x = a$, $\alpha < y < \beta$. It follows that (5) holds whenever C is composed of vertical segments and graphs of continuously differentiable functions $x = \phi(y)$.

We give here some typical applications of Pappus' theorems. In some of them the volumes and areas are known from other considerations, and we use this knowledge to find centroids. In others, the centroids are known from symmetry considerations, and we use this to find volumes and areas.

EXAMPLES 1. Find the centroid of a semidisk.
ANSWER   We represent the semidisk as the set D of points (x,y) with $x^2 + y^2 < 1$, $x > 0$. Its area is $A = \pi/2$. The centroid lies on the x-axis, by symmetry. Let X be its x-coordinate. Rotating D about the y-axis, we obtain the unit sphere; its volume is $V = 4\pi/3$. Hence $4\pi/3 = 2\pi X(\pi/2)$, by Pappus' first theorem, and $X = 4/(3\pi)$.

2. Find the centroid of a semicircle.
ANSWER   Represent the semicircle as the curve C: $x^2 + y^2 = 1$, $x > 0$. Its length is π. The centroid lies at a point $(X,0)$, by symmetry. Rotating C about the y-axis, we obtain the sphere of area 4π. By Pappus' second theorem, $4\pi = 2\pi X\pi$. Hence $X = 2/\pi$.

3. If we rotate a circular disk of radius r about a line at the distance $R > r$ from its center, we obtain a body known as the **solid torus** of radii R and r. The surface obtained by rotating the circumference of radius r about a line at the distance $R > r$ from its center is called the **torus** of radii R and r. (The torus looks like an inner tube; it is the boundary surface of the solid torus.)

Pappus' theorems permit us to determine the volume of the solid torus, and the area of the torus, practically without computing. In both cases $\delta = R$. The volume in question is $2\pi R \cdot \pi r^2 = 2\pi^2 R r^2$. The area is $2\pi R \cdot 2\pi r = 4\pi R r$.

PROBLEMS

In Problems 1 to 3 use Pappus' first theorem to find the volumes of the indicated bodies of revolution.

1. A circular cylinder of radius r and height H.
2. The body obtained by rotating a square of side length 1, with one vertex at the origin and the other at $(0, \sqrt{2})$, about the x-axis.
3. The body obtained by rotating the rectangle bounded by the lines $y = a > 0$, $y = b > a$, $x = 0$, and $x = c$, about the x-axis.

In Problems 4 to 6 use Pappus' first theorem to find the centroid of the indicated plane region.

4. The region bounded by two concentric circles and a line through their center.
5. An equilateral triangle.
6. A right triangle.

7. Use Pappus' second theorem to compute the mantle surface of a straight circular cone.
8. Let C be the circle passing through the corners of a given rectangle R. Prove that for all lines l tangent to C, the area generated by revolving R around l is the same.
9. Suppose, in Pappus' second theorem, the plane curve is a line segment perpendicular to l. What does the theorem reduce to in this case?
10. Use a symmetry argument together with Pappus' second theorem to find the centroid of a quarter circle. (What surface of known area can be obtained by revolving a quarter circle about an axis?)

LINE INTEGRALS

§1 Line integrals in the plane

There are two ways of integrating functions of several variables: multiple integrals, discussed in Chapter 20, and integrals over curves and curved surfaces. Here we discuss only the simplest and most important case: line integrals.

1.1 Definition and notations

Let C be a smooth curve in the plane defined, with respect to a Cartesian coordinate system, by two functions $x = x(t)$, $y = y(t)$, $a \le t \le b$, and let $P(x,y,z)$, $Q(x,y,z)$ be two continuous functions defined in some set containing C. Under these conditions, we define a number called "the **line integral** of $P\,dx + Q\,dy$ over C" and denoted by

$$\int_C P\,dx + Q\,dy. \tag{1}$$

This number is

$$\int_a^b \{P[x(t),y(t)]x'(t) + Q[x(t),y(t)]y'(t)\}\,dt. \tag{2}$$

Note that we could have guessed at the meaning of (1) by trusting the Leibniz notation. What could dx and dy mean besides $x'(t)\,dt$ and $y'(t)\,dt$?

EXAMPLES 1. Let C be the curve $x = \sin t$, $y = \cos t$, $0 \leq t \leq 1$. Compute $\int_C y\,dx - x\,dy$.
ANSWER   The line integral is of the form (1) with $P = y$, $Q = x$. On the curve C, $x = \sin t$, $y = \cos t$, and therefore $dx = \cos t\,dt$, $dy = -\sin t\,dt$. Hence the line integral equals

$$\int_0^1 \{(\cos t)(\cos t) - (\sin t)(-\sin t)\}\,dt = \int_0^1 dt = 1.$$

2. Let C be the curve $x = t$, $y = 0$, $a \leq t \leq b$. Compute $\int_C f(x)\,dx + g(y)\,dy$.
ANSWER   Since $dx/dt = 1$, $dy/dt = 0$, the line integral is $\int_a^b f(t)\,dt$. Thus the ordinary integral is a special case of the line integral.

The value of a line integral over a curve C does *not* depend on the parameter used to represent the curve.

We verify this for a line integral of the form

$$\int_C P\,dx, \tag{3}$$

that is, for the case when $Q = 0$. But the argument is general. Suppose C is represented by $x = x(t)$, $y = y(t)$, $a \leq t \leq b$. Let $t = \phi(\tau)$, $\alpha \leq \tau \leq \beta$, be a smooth parameter change. This means that $\phi(\alpha) = a$, $\phi(\beta) = b$, $\phi'(\tau) > 0$, and is continuous. The curve C can be represented by the two functions $x(\phi(\tau))$, $y(\phi(\tau))$, $\alpha \leq \tau \leq \beta$. If we compute the line integral using the t-representation, we obtain

$$\int_a^b P[x(t),y(t)]x'(t)\,dt. \tag{4}$$

Computing (3) using τ, we get

$$\int_\alpha^\beta P[x(\phi(\tau)),y(\phi(\tau))]x'(\phi(\tau))\phi'(\tau)\,d\tau;$$

this is the same as (4), by the substitution rule for ordinary integrals (Chapter 8, §3.4).

PROBLEMS

In each of Problems 1 to 8 evaluate the given line integral over the indicated path C.

1. $\int_C y\,dx + x\,dy$; C: $x = t + 1, y = 2t - 1, 0 \leq t \leq 2$.

2. $\int_C (x^2 - y^2)\,dx + (x^2 + y^2)\,dy$; C: $x = 1 - t, y = t, 0 \leq t \leq 1$.

3. $\int_C (x + y)\,dx + (x - y)\,dy$; C: $x = t^2 + 1, y = t^2 - 1, 1 \leq t \leq 2$.

4. $\int_C \sqrt{xy}\,dx + (1/\sqrt{xy})\,dy$; C: $x = t, y = t^2, 1 \leq t \leq 2$.

5. $\int_C y \cos x\,dx - x \cos y\,dy$; C: $x = t, y = 2t, 0 \leq t \leq \pi/4$.

6. $\int_C ye^x\,dx + xe^y\,dy$; C: $x = t, y = 4t, 0 \leq t \leq 1$.

7. $\int_C (x^2 + y)\,dx - (x + y^2)\,dy$; C: $x = \sin t, y = \cos t, 0 \leq t \leq \pi/2$.

8. $\int_C y^2\,dx + x^2\,dy$; C: $x = \sin t, y = \cos t, 0 \leq t \leq \pi/2$.

1.2 Properties of line integrals

We have

$$\int_C (P + \hat{P}) \, dx + (Q + \hat{Q}) \, dy = \int_C P \, dx + Q \, dy + \int_C \hat{P} \, dx + \hat{Q} \, dy, \qquad (5)$$

$$\int_C cP \, dx + cQ \, dy = c \int_C P \, dx + Q \, dy, \qquad c \text{ a constant}, \qquad (6)$$

$$\int_{-C} P \, dx + Q \, dy = - \int_C P \, dx + Q \, dy, . \qquad (7)$$

where $-C$ denotes the curve C traversed in the opposite direction, and

$$\int_{C_1 + C_2} P \, dx + Q \, dy = \int_{C_1} P \, dx + Q \, dy + \int_{C_2} P \, dx + Q \, dy, \qquad (8)$$

where $C_1 + C_2$ denotes the curve C_1 followed by C_2.

All four relations are proved by direction calculations. We give only the proof of (7); to save writing, we consider the integral (3). If C is represented by $t \mapsto x(t)$, $t \mapsto y(t)$, $a \le t \le b$, then $-C$ can be represented by $s \mapsto x(b - s)$, $s \mapsto y(b - s)$, $0 \le s \le b - a$. Hence

$$\int_{-C} P \, dx = \int_0^{b-a} P[x(b - s), y(b - s)][-x'(b - s)] \, ds.$$

Setting $\tau = b - s$, we have $d\tau = -ds$, $\tau = b$ for $s = 0$, $\tau = a$ for $s = b - a$. Hence

$$\int_{-C} P \, dx = \int_b^a P[x(\tau), y(\tau)] x'(\tau) \, d\tau = - \int_a^b P[x(\tau), y(\tau)] x'(\tau) \, d\tau = - \int_C P \, dx.$$

Property (8) permits us to extend the definition of the line integral to the case when the **path of integration** C is only **piecewise smooth,** that is, consists of several smooth arcs (see the following example).

EXAMPLE Let C be the curve shown in Figure 21.1, traversed once starting at (0,0) and ending at (0,1). Compute

$$\int_C (2 + y) \, dx + x \, dy.$$

FIGURE **21.1**

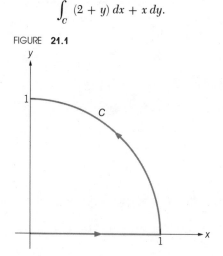

ANSWER   We have $C = C_1 + C_2$ where C_1 is defined by $x = t$, $y = 0$, $0 \leq t \leq 1$, and C_2 by $x = \cos \theta$, $y = \sin \theta$, $0 \leq \theta \leq \pi/2$. On C_1: $dx = dt$, $dy = 0$. On C_2: $dx = -\sin \theta \, d\theta$, $dy = \cos \theta \, d\theta$. Hence, by (5), (6), and (8),

$$\int_C (2 + y) \, dx + x \, dy = 2 \int_C dx + \int_C y \, dx + x \, dy$$

$$= 2 \int_{C_1} dx + 2 \int_{C_2} dx + \int_{C_1} y \, dx + x \, dy + \int_{C_2} y \, dx + x \, dy$$

$$= 2 \int_0^1 dt + 2 \int_0^{\pi/2} (-\cos \theta) \, d\theta + \int_0^1 0 \cdot dt$$

$$+ \int_0^{\pi/2} (-\sin^2 \theta + \cos^2 \theta) \, d\theta$$

$$= 2 - 2 \sin \theta \, \Big|_0^{\pi/2} + \tfrac{1}{2} \sin 2\theta \, \Big|_0^{\pi/2} = 0.$$

PROBLEMS

In each of Problems 1 to 6 evaluate the given line integral over the indicated path C.

1. $\int_C xy \, dx + (x + y) \, dy$; C is the line segment from $(0,0)$ to $(1,1)$.
2. $\int_C y^2 \, dx + x^2 \, dy$; C is the line segment from $(0,0)$ to $(1,0)$ followed by the line segment from $(1,0)$ to $(1,2)$.
3. $\int_C x\sqrt{y} \, dx + y\sqrt{x} \, dy$; C is the curve $y = \tfrac{1}{4}x^2$ from $(0,0)$ to $(2,1)$.
4. $\int_C y \, dx + (1 + x) \, dy$; C is the semicircle $y = \sqrt{1 - x^2}$ from $(1,0)$ to $(-1,0)$ followed by the line segment from $(-1,0)$ to $(0,0)$.
5. $\int_C (x^2 + y^2) \, dx + 2xy \, dy$; C is the shortest path by which a bug crawling in the xy plane can get from $(-2,0)$ to $(1,0)$ if it is not permitted to enter the unit circle. (There are two such paths, but both give the same value to the integral.)
6. $\int_C (x + y) \, dx + (x - y) \, dy$; C is the ellipse $(x^2/a^2) + (y^2/b^2) = 1$, traversed once in the counterclockwise direction starting at $(0,b)$.

1.3 Path independence

In this and the next subsection, D denotes some domain in the plane.

Theorem 1.   *Let $P(x,y), Q(x,y)$ be continuous functions defined in D. If there is a function $f(x,y)$ defined in D with*

$$df = P \, dx + Q \, dy, \tag{9}$$

that is, with

$$P = \frac{\partial f}{\partial x}, \qquad Q = \frac{\partial f}{\partial y}, \tag{9'}$$

then, for every piecewise smooth curve C in D leading from a point A to a point B, we have

$$\int_C P \, dx + Q \, dy = f(B) - f(A). \tag{10}$$

In particular,

$$\int_C P\, dx + Q\, dy = 0 \qquad \text{for every closed curve } C \text{ in } D. \tag{11}$$

Conversely, if (11) holds, then there is a function f defined in D and satisfying (9).

The Leibniz notation could have helped us to guess at statement (10), since, if (9) holds, we can rewrite (10) in the form

$$\int_A^B df = f(B) - f(A),$$

which looks like the fundamental theorem of calculus. Indeed, the present theorem follows from the fundamental theorem.

Proof of Theorem 1.   Assume that (9) holds, and let C be defined by $x = x(t)$, $y = y(t)$, $a \le t \le b$. If C leads from $A = (a_1,a_2)$ to $B = (b_1,b_2)$, then $x(a) = a_1$, $y(a) = a_2$, $x(b) = b_1$, $x(b) = b_2$. Set $F(t) = f[x(t),y(t)]$. Then $F(a) = f(a_1,a_2) = f(A)$, $F(b) = f(b_1,b_2) = F(B)$. By (9′) and the chain rule,

$$F'(t) = P[x(t),y(t)]x'(t) + Q[x(t),y(t)]y'(t).$$

Now, by the definition of line integrals,

$$\int_C P\, dx + Q\, dy = \int_a^b \{P[x(t),y(t)]x'(t) + Q[x(t),y(t)]y'(t)\}\, dt$$

$$= \int_a^b F'(t)\, dt = F(b) - F(a) = f(B) - f(A).$$

This proves (10). For $A = B$, we obtain (11).

Assume next that (11) holds. If C_1 and C_2 are two curves in D with the same initial points and the same terminal points, then $C_1 + (-C_2)$ is a closed curve (see Figure 21.2) so that

$$0 = \int_{C_1+(-C_2)} P\, dx + Q\, dy = \int_{C_1} P\, dx + Q\, dy + \int_{-C_2} P\, dx + Q\, dy$$

$$= \int_{C_1} P\, dx + Q\, dy - \int_{C_2} P\, dx + Q\, dy.$$

In other words: condition (11) means that *the line integral $\int_C P\, dx + Q\, dy$ depends only on the endpoints of C;* we say in this case that this integral is **path-independent.**

If so, we can define unambiguously a function $f(x,y)$ by choosing a point $(\hat{x},\hat{y})$ in D and setting

$$f(x,y) = \int_{(\hat{x},\hat{y})}^{(x,y)} P\, dx + Q\, dy, \tag{12}$$

where the integration is performed along *some* curve in D leading from $(\hat{x},\hat{y})$ to (x,y); since the integral is path-independent, the choice of the curve does not affect the value of $f(x,y)$. It turns out that this f satisfies (9).

We must show that (9′) holds in D, where $f(x,y)$ is defined by the (path-independent) line integral (12). Let (x_1,y_1) be some point in D; we show that (9′) holds

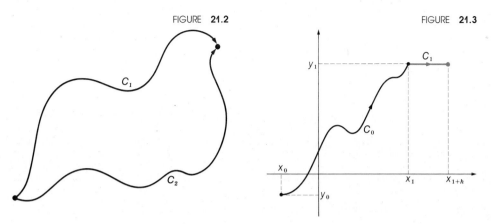

FIGURE 21.2   FIGURE 21.3

at this point. Let C_0 be a curve leading from $(\hat{x},\hat{y})$ to (x_1,y_1) so that

$$f(x_1,y_1) = \int_{C_0} P\,dx + Q\,dy.$$

Let C_1 be the horizontal segment joining (x_1,y_1) to $(x_1 + h, y_1)$ where $|h|$ is small. Then (see Figure 21.3)

$$f(x_1 + h, y_1) = \int_{C_0 + C_1} P\,dx + Q\,dy = \int_{C_0} P\,dx + Q\,dy + \int_{C_1} P\,dx + Q\,dy$$

$$= f(x_1, y_1) + \int_{C_1} P\,dx + Q\,dy.$$

Now C_1 admits the parametric representation $x = x_1 + t$, $y = y_1$, $0 \le t \le h$, if $h \ge 0$ and $x = x_1 - s$, $y = y_1$, $0 \le s \le h$, if $h < 0$. Hence if $h \ge 0$, then

$$f(x_1 + h, y_1) = f(x_1, y_1) + \int_0^h P(x_1 + t, y_1)\,dt. \tag{13}$$

If $h < 0$,

$$f(x_1 + h, y_1) = f(x_1, y_1) - \int_0^h P(x_1 - s, y_1)\,ds.$$

If we set $s = -t$ here, we see that (13) holds in all cases. We differentiate both sides of (13) with respect to h, using the fundamental theorem of calculus, and set $h = 0$. This gives $f_x(x_1, y_1) = P(x_1, y_1)$. The second relation (9') is proved similarly, using for C_1 a vertical segment.

EXAMPLES **1.** Let C be a curve leading from $(1,2)$ to $(10,20)$. What is $\int_C x\,dx + y\,dy$?
ANSWER Since, as one can guess almost at once, $x\,dx + y\,dy = d(\frac{1}{2}x^2 + \frac{1}{2}y^2)$, the integral equals
$(\frac{1}{2}10^2 + \frac{1}{2}20^2) - (\frac{1}{2} + \frac{1}{2}2^2) = 247.5$.

2. We compute once more the example in §1.2. We note that

$$2\,dx = d(2x), \qquad y\,dx + x\,dy = d(xy) \qquad \text{so that} \qquad (2 + y)\,dx + x\,dy = d(2x + xy),$$

and that C leads from $(0,0)$ to $(0,1)$. At these endpoints $2x + xy = 0$. Therefore $\int_C (2 + y)\,dx + x\,dy = 0 - 0 = 0$.

PROBLEMS

In each of Problems 1 to 8 compute the given line integral for the curve C from the indicated points P and Q.

1. $\displaystyle\int_C A\,dx + B\,dy;\quad P = (x_0, y_0),\; Q = (x_1, y_1).$

2. $\displaystyle\int_C x\,dx - y\,dy;\quad P = (-1, -1),\; Q = (1, 1).$

3. $\displaystyle\int_C (2x + y)\,dx + (x + 2y)\,dy;\quad P = (1, 1),\; Q = (2, 4).$

4. $\displaystyle\int_C xy^2\,dx + x^2 y\,dy;\quad P = (0, 1),\; Q = (1, 2).$

5. $\displaystyle\int_C (y^2 - 2xy)\,dx + (2xy - x^2)\,dy;\quad P = (1, 1),\; Q = (-1, -2).$

6. $\displaystyle\int_C y\cos(xy)\,dx + x\cos(xy)\,dy;\quad P = (0, 0),\; Q = (1, \pi/2).$

7. $\displaystyle\int_C y^2 e^x\,dx + 2ye^x\,dy;\quad P = (0, 1),\; Q = (1, 4).$

8. $\displaystyle\int_C x(x^2 + y^2)^{-1/2}\,dx + y(x^2 + y^2)^{-1/2}\,dy;\quad P = (3, 4),\; Q = (5, 12).$

1.4   Integrability condition. Simply connected domains‡

We continue to study the line integral $\int_C P\,dx + Q\,dy$, assuming now that the functions $P(x,y)$, $Q(x,y)$ are continuously differentiable. We noted in Chapter 19, §2.7, that if $P\,dx + Q\,dy = df$, then P and Q satisfy the integrability condition

$$P_y = Q_x. \tag{14}$$

If the domain we deal with is an interval, the converse is also true, that is, if the integrability condition holds, then P and Q are the partial derivatives of a function. This is not so for every domain.

Consider, for instance, the two functions

$$P = -\frac{y}{x^2 + y^2}, \qquad Q = \frac{x}{x^2 + y^2},$$

which are continuously differentiable in the domain $D = $ the whole plane with the origin removed. Let C be a circle with center at the origin, traversed once in the positive direction. This is a closed curve, and

$$\int_C P\,dx + Q\,dy = 2\pi. \tag{15}$$

Indeed, C admits the parametric representation $x = R\cos\theta,\, y = R\sin\theta,\, 0 \le \theta \le 2\pi$. Along C we have $dx = -R\sin\theta\,d\theta,\, dy = R\cos\theta\,d\theta,\, P = -\sin\theta/R,\, Q = \cos\theta/R$.

---

‡Optional subsection.

FIGURE **21.4**   FIGURE **21.5**

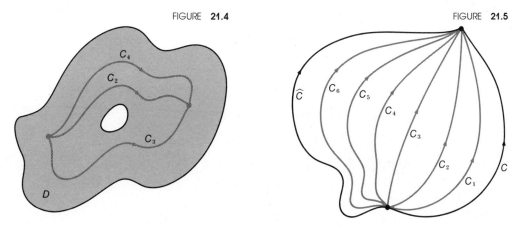

Hence $P\,dx + Q\,dy = d\theta$, and (15) follows. But P and Q satisfy the integrability condition, since

$$\frac{\partial P}{\partial y} = \frac{\partial Q}{\partial x} = \frac{y^2 - x^2}{(x^2 + y^2)^2}.$$

(The reader will note that the above considerations shed light on the discussion in §3.4 in Chapter 19.)

In order to state a result valid in *any* domain, we need a geometric concept for which we give only an intuitive definition. Two curves, C_1 and C_2, in a domain D, with the same endpoints, are called **homotopic** in D if one can be continuously deformed into another, keeping the endpoints fixed, and without leaving the domain D. If we think of the two curves as strings made of a flexible material, the meaning of the term "continuously deformed" becomes clear. The curve C_1 in Figure 21.4, for instance, can be deformed into the curve C_2 but not into the curve C_3; a deformation of C_1 into C_3 would be obstructed by the hole in D. A precise analytic definition of homotopy may be found in any introductory topology text.

Theorem 2. *If the functions $P(x,y)$, $Q(x,y)$ are continuously differentiable in D and satisfy the integrability condition (14), then*

$$\int_C P\,dx + Q\,dy = \int_{\widehat{C}} P\,dx + Q\,dy \tag{16}$$

for any two curves C and $\widehat{C}$ that are homotopic in D.

We do not give a formal proof but an intuitive argument which shows why Theorem 2 is true. If C can be deformed into $\widehat{C}$, then we can find curves C_1, C_2, $C_3, \ldots, C_N$ such that not only are all curves $C, C_1, C_2, \ldots, C_N, \widehat{C}$ homotopic in D, but also C is very close to C_1, C_1 very close to C_2, and so forth (see Figure 21.5). It suffices to show that the integral of $P\,dx + Q\,dy$ over C is equal to the integral over C_1, the integral over C_1 is equal to the integral over C_2, and so forth. It will suffice to consider C and C_1.

If two curves are sufficiently close to each other, as we assume C and C_1 to be, we can go from C to C_1 by passing through a succession of several homotopic curves,

FIGURE 21.6   FIGURE 21.7

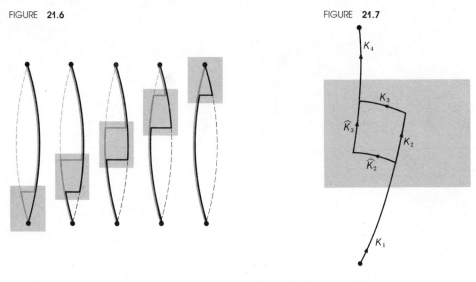

any two of which run together except inside some small interval that belongs to D (see Figure 21.6). It suffices to show that the integrals of $P\,dx + Q\,dy$ over any two such curves are equal.

A typical situation is depicted in Figure 21.7. We must show that

$$\int_{K_1+K_2+K_3+K_4} P\,dx + Q\,dy = \int_{K_1+\widehat{K}_2+\widehat{K}_3+K_4} P\,dx + Q\,dy.$$

In view of the rules in §1.2, this amounts to showing that

$$\int_{K_2+K_3+(-\widehat{K}_2)+(-\widehat{K}_3)} P\,dx + Q\,dy = 0. \tag{17}$$

But this is so, for $K_2 + K_3 + (-\widehat{K}_2) + (-\widehat{K}_3)$ is a closed curve lying in an *interval* in which the functions P and Q satisfy (14), so that in this interval there is, as we noted before, a function f such that $df = P\,dx + Q\,dy$. Hence (17) holds, by Theorem 1.

We call D **simply connected** if any two curves in D with the same endpoints are homotopic. Every interval is simply connected. A domain D (in the plane) is simply connected if and only if D "has no holes."

Theorem 3. *If D is simply connected, and if two continuously differentiable functions P and Q defined in D satisfy the integrability condition, then there is a function f defined in D such that $df = P\,dx + Q\,dy$.*

This follows by combining Theorems 1 and 2.

§2 Green's theorem

We now exhibit, without complete proof, an identity that connects line integrals and double integrals. This identity, known as Green's theorem, is actually the simplest and most important special case of a general result which we shall not discuss here.

FIGURE **21.8**

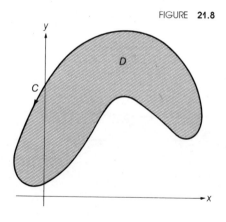

2.1 Statement of the theorem

Let D be a domain in the plane, and let C be an oriented, piecewise smooth, **simple** closed curve $x = x(t)$, $y = y(t)$, $0 \le t \le T$. [By "simple" we mean that for no two different parameter values t, other than 0 and T, do the points $(x(t),y(t))$ coincide.] We call C the **boundary curve** of D if every point $(x(t),y(t))$ is a boundary point of D, and every boundary point of D can be written as $(x(t),y(t))$ for some t. A typical situation is shown in Figure 21.8.

We say that the boundary curve C of D is **properly oriented** if, "as we go along C in its direction, that is, in the direction of increasing parameter values, the region D is to the left." Here we accept this concept as intuitively clear; a precise definition could be given. In Figure 21.8 the boundary curve is properly oriented.

Theorem 4 (Green's theorem). *Let C be a properly oriented (piecewise smooth) boundary curve of the open set D. Let $P(x,y)$, $Q(x,y)$ be continuously differentiable functions (defined in some open set containing C and D). Then*

$$\iint_D \left(\frac{\partial Q}{\partial x} - \frac{\partial P}{\partial y} \right) dx\, dy = \int_C P\, dx + Q\, dy. \tag{1}$$

We can replace this by *two* statements

$$\int_C P\, dx = -\iint_D \frac{\partial P}{\partial y}\, dx\, dy, \tag{2}$$

$$\int_C Q\, dy = \iint_D \frac{\partial Q}{\partial x}\, dx\, dy. \tag{3}$$

Indeed, if we set $P = 0$ or $Q = 0$ in (1), we obtain (2) or (3), and if we add (2) and (3), we obtain (1).

GEORGE GREEN (1793–1841) published the theorem that bears his name in a book on the mathematical theory of electricity and magnetism. Green was an essentially self-taught mathematician. He entered Cambridge University at the age of 40, several years after his pioneering work appeared in print.

EXAMPLE   Verify Green's theorem for $P = y$, $Q = 0$, if D is the unit disk $x^2 + y^2 < 1$.
SOLUTION   Since D is the unit disk, the properly oriented boundary curve C of D is $x = \cos t$,
$y = \sin t$, $0 \le t \le 2\pi$. Hence

and

$$\iint_D \left(\frac{\partial Q}{\partial x} - \frac{\partial P}{\partial y} \right) dx\, dy = \iint_D (-1)\, dx\, dy = - \iint_{x^2+y^2<1} dx\, dy = -\pi$$

$$\int_C P\, dx + Q\, dy = \int_C y\, dx = \int_0^{2\pi} (\sin t)(-\sin t\, dt) = - \int_0^{2\pi} \sin^2 t\, dt = -\pi.$$

Green's theorem (1) implies that

$$\int_C P\, dx + Q\, dy = 0 \qquad \text{if} \qquad \frac{\partial P}{\partial y} = \frac{\partial Q}{\partial x}.$$

This is in accordance with the results in §1.4. Since D is simply connected and
$\partial P/\partial y = \partial Q/\partial x$, there is a function f with $P\, dx + Q\, dy = df$, so that $\int_C P\, dx + Q\, dy = \int_C df = 0$.

PROBLEMS

In each of Problems 1 to 6 verify Green's theorem for the given functions $P(x,y)$, $Q(x,y)$ and
the indicated domain D.

1. $P = y$, $Q = x$; D is the unit disk.
2. $P = x + y$, $Q = x^2 - y^2$; D is the unit disk.
3. $P = xy$, $Q = \sqrt{x} + \sqrt{y}$; D is the unit square $0 < x < 1$, $0 < y < 1$.
4. $P = x^2 + y^2$, $Q = 2xy$; D is the triangle $x > 0$, $y > 0$, $x + y < 1$.
5. $P = y \cos x$, $Q = x \sin y$; D is the square with corners at $(1,0)$, $(0,1)$, $(1,2)$, $(2,1)$.
6. $P = xy^2$, $Q = x^2 y$; D is the portion of the first quadrant where $1 < x^2 + y^2 < 4$.

2.2   Area formulas

If we apply (2) to $P = -y$ and (3) to $Q = x$, we obtain interesting formulas for the
area $A = \iint_D dx\, dy$, namely,

$$A = - \int_C y\, dx, \tag{4}$$

$$A = \int_C x\, dy, \tag{5}$$

so that also

$$A = \frac{1}{2} \int_C x\, dy - y\, dx, \tag{6}$$

as is seen by adding (4) and (5).
   Suppose now that C can be described in polar coordinates as $r = \phi(\theta)$, $0 \le \theta \le 2\pi$.
A parametric representation for C, with the proper orientation, is

$$x = \phi(\theta) \cos \theta, \qquad y = \phi(\theta) \sin \theta, \qquad 0 \le \theta \le 2\pi$$

so that

$$dx = [\phi'(\theta) \cos \theta - \phi(\theta) \sin \theta] \, d\theta, \qquad dy = [\phi'(\theta) \sin \theta + \phi(\theta) \cos \theta] \, d\theta,$$

and by (6) the area A equals

$$\frac{1}{2} \int_0^{2\pi} [-\phi(\theta)\phi'(\theta) \sin \theta \cos \theta + \phi(\theta)^2 \sin^2 \theta + \phi(\theta)\phi'(\theta) \cos \theta \sin \theta + \phi(\theta)^2 \cos^2 \theta] \, d\theta$$

$$= \frac{1}{2} \int_0^{2\pi} \phi(\theta)^2 \, d\theta.$$

This proves formula (13) in Chapter 16, §1.4, for a simple closed curve, that is, for $\alpha = 0$, $\beta = 2\pi$. The reader should try to treat, in the same way, the case $\alpha = 0$, $0 < \beta < 2\pi$.

EXAMPLE   Sketch the graphs of, and calculate the area bounded by, the curves $y = x^2$, $x = y^2$.
SOLUTION   The graphs of the curves and the domain D are shown in Figure 21.9. The properly

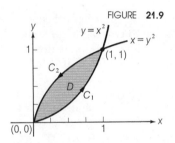

FIGURE **21.9**

oriented boundary curve $C = C_1 + C_2$ is traversed in the direction indicated by the arrows in the figure (the area is "to the left"). Using Equation (6), we have

$$A = \frac{1}{2} \int_C x \, dy - y \, dx = \frac{1}{2} \int_{C_1} x \, dy - y \, dx + \frac{1}{2} \int_{C_2} x \, dy - y \, dx$$

$$= \frac{1}{2} \int_0^1 x(2x \, dx) - x^2 \, dx + \frac{1}{2} \int_1^0 y^2 \, dy - y(2y \, dy)$$

$$= \frac{1}{2} \int_0^1 x^2 \, dx - \frac{1}{2} \int_1^0 y^2 \, dy = \frac{1}{6} + \frac{1}{6} = \frac{1}{3}.$$

PROBLEMS

1. Use Green's theorem to compute the area of the ellipse whose boundary is given by $x = a \cos t$, $y = b \sin t$, $0 \le t \le 2\pi$.
2. Calculate the area bounded by the *hypocycloid* $x^{2/3} + y^{2/3} = R^{2/3}$. (A sketch may help.)
3. Calculate the area bounded by the *cardioid*, defined by $x = 2 \cos \theta(1 - \cos \theta) + 1$, $y = 2 \sin \theta(1 - \cos \theta)$. (A sketch may help.)
4. Calculate the area bounded by $r = \theta$, $0 \le \theta < 2\pi$, and the coordinate axis.
5. Calculate the area bounded by the semicubical parabola $y = x^{3/2}$ and the lines $y = 0$, $x = 1$, $x = 4$.
6. Calculate the area under one arch of the cycloid $x = t - \sin t$, $y = 1 - \cos t$.

FIGURE **21.10**   FIGURE **21.11**

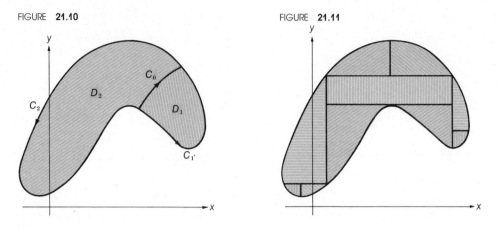

2.3 Remarks on the proof

We sketch the main ideas in proving Green's theorem.

Suppose that we divide D into two regions, D_1 and D_2 (as shown in Figure 21.10) by a smooth simple curve C_0, whose endpoints divide C into two curves C_1 and C_2. We claim that Green's theorem holds for D if it holds for D_1 and for D_2.

The (properly oriented) boundary curves of D_1 and D_2 are $C_1 + C_0$ (that is, C_1 followed by C_0) and $(-C_0) + C_2$ (C_0 traversed in the opposite direction followed by C_2). We *assume* Green's theorem for D_1 and for D_2. Then

$$- \iint_{D_1} \frac{\partial P}{\partial y}\, dx\, dy = \int_{C_1} P\, dx + \int_{C_0} P\, dx,$$

$$- \iint_{D_2} \frac{\partial P}{\partial y}\, dx\, dy = \int_{-C_0} P\, dx + \int_{C_2} P\, dx = - \int_{C_0} P\, dx + \int_{C_2} P\, dx.$$

Adding the two equations, we get

$$- \iint_{D} \frac{\partial P}{\partial y}\, dx\, dy = - \iint_{D_1} \frac{\partial P}{\partial y}\, dx\, dy - \iint_{D_2} \frac{\partial P}{\partial y}\, dx\, dy = \int_{C_1} P\, dx + \int_{C_2} P\, dx$$

$$= \int_{C} P\, dx,$$

since $C = C_1 + C_2$. This proves (2). A similar argument holds for (3).

The above argument can be made precise (it is not, since we depended on looking at the diagram). Also, the argument can be continued, subdividing D_1 and D_2, and so forth. We conclude that in order to establish Green's theorem, it suffices to subdivide D into "nice" pieces and prove Green's theorem for each nice piece.

A subdivision of D is shown in Figure 21.11. Each piece is either a rectangle or a region bounded by two segments parallel to the coordinate axes and by a graph of a monotone smooth function.

We show next how to prove part (2) of Green's theorem for a typical piece, say, for the domain D shown in Figure 21.12. Its properly oriented boundary is the curve $C = C_1 + C_2 + C_3$ where

FIGURE 21.12

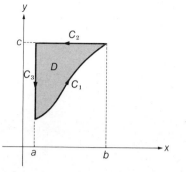

$$C_1: x = t, \qquad y = f(t), \qquad a \leq t \leq b, \qquad [f'(t) \geq 0 \text{ for } a < t < b],$$

$$C_2: x = b - t, y = c, \qquad 0 \leq t \leq b - a,$$

$$C_3: x = a, \qquad y = c - t, \qquad 0 \leq t \leq c - f(a).$$

Now, by the rule for computing double integrals and by the fundamental theorem of calculus,

$$-\iint_D \frac{\partial P}{\partial y} \, dx \, dy = -\int_a^b \int_{f(x)}^c \frac{\partial P(x,y)}{\partial y} \, dy \, dx = -\int_a^b \{ P(x,c) - P[x,f(x)] \} \, dx$$

$$= \int_a^b P[x,f(x)] \, dx - \int_a^b P(x,c) \, dx. \tag{7}$$

Also, since $dx = 0$ on C_3, we have

$$\int_C P \, dx = \int_{C_1} P \, dx + \int_{C_2} P \, dx = \int_a^b P[t,f(t)] \, dt - \int_0^{b-a} f(b - t,c) \, dt.$$

We make the substitutions $t = x$ and $b - t = x$, respectively, and obtain

$$\int_C P \, dx = \int_a^b P[x,f(x)] \, dx - \int_a^b P(x,c) \, dx,$$

which is equal to (7); this proves (2). Relation (3) is proved similarly; we must only write C_1 in the form $C_1: x = g(t), y = t, f(a) \leq t \leq c$, where g is the function inverse to f.

The proof of Green's theorem for an interval (rectangle with sides parallel to the coordinate axes) is relatively simple and is left to the reader.

PROBLEMS

1. Prove relation (3) for the domain D considered above.
2. Prove Green's theorem for a rectangular region with sides parallel to the coordinate axes.

§3   Line integrals in space

3.1   Definition and main properties

The line integral

$$\int_C P\,dx + Q\,dy + R\,dz, \tag{1}$$

where C is a smooth, or piecewise smooth, curve in space defined by $x = x(t)$, $y = y(t)$, $z = z(t)$, $a \le t \le b$, and $P(x,y,z)$, $Q(x,y,z)$, $R(x,y,z)$ are continuous functions defined in some domain containing C, is defined as

$$\int_a^b \{P[x(t),y(t),z(t)]x'(t) + Q[x(t),y(t),z(t)]y'(t) + R[x(t),y(t),z(t)]z'(t)\}\,dt.$$

Rules similar to those stated in §1 are valid (and are proved in the same way). In particular, if P, Q, and R are the partial derivatives of a function $f(x,y,z)$:

$$P = f_x, \quad Q = f_y, \quad R = f_z \qquad \text{or} \qquad P\,dx + Q\,dy + R\,dz = df, \tag{2}$$

then

$$\int_C P\,dx + Q\,dy + R\,dz = f(\text{terminal point}) - f(\text{initial point}), \tag{3}$$

that is, the integral is path-independent. Conversely, if (in the domain considered)

$$\int_C P\,dx + Q\,dy + R\,dz = 0 \qquad \text{for every closed curve } C, \tag{4}$$

then there is a function $f(x,y,z)$ such that (2) holds.

If P, Q, and R are continuously differentiable, *necessary* conditions for path-independence of the integral (1) are the three integrability conditions

$$P_y = Q_x, \qquad P_z = R_x, \qquad Q_z = R_y. \tag{5}$$

These conditions are *sufficient* if the domain considered is simply connected, that is, if any two curves in this domain, with the same endpoints, are homotopic. [In space a simply connected domain may have "holes." For instance, the domain consisting of all points (x,y,z) with $x^2 + y^2 + z^2 > 1$ is simply connected.]

EXAMPLES   1.   Let C be the curve $x = t$, $y = t^2$, $z = t^3$, $0 \le t \le 1$. Compute $\int_C xz\,dx + xz\,dy + xy\,dz$.
ANSWER   The line integral equals

$$\int_0^1 (t)(t^3)(dt) + \int_0^1 (t)(t^3)(2t\,dt) + \int_0^1 (t)(t^2)(3t^2\,dt) = \int_0^1 (t^4 + 5t^5)\,dt = \frac{31}{30}.$$

2.   Let C be a curve leading from $(1,1,1)$ to $(2,4,6)$. What is $\int_C yz\,dx + xz\,dy + xy\,dz$?
ANSWER   Since $d(xyz) = yz\,dx + xz\,dy + xy\,dz$, the integral equals $(2)(4)(6) - (1)(1)(1) = 47$.

3.   Find a function $f(x,y,z)$ such that

$$df = y\,dx + (x + z^2)\,dy + 2yz\,dz,$$

and specify the domain in which the solution is defined.

SOLUTION We first check the integrability conditions (5). Setting $P = f_x = y$, $Q = f_y = x + z^2$, $R = f_z = 2yz$, we easily verify that $P_y = Q_x$, $P_z = R_x$, $Q_z = R_y$. Hence the integrability conditions are satisfied.

Knowing f_x, we conclude that $f(x,y,z) = xy + g(y,z)$. The condition on f_y gives $x + z^2 = x + g_y$, that is, $g_y = z^2$. Thus $g(y,z) = yz^2 + h(z)$ and $f(x,y,z) = xy + yz^2 + h(z)$. The condition on f_z gives $2yz = 0 + 2yz + h_z$, that is, $h_z = 0$. Thus $h(z) = c$, c a constant. Therefore

$$f(x,y,z) = xy + yz^2 + c.$$

The solution is valid for all x, y, z.

PROBLEMS

In each of Problems 1 to 8 evaluate the given line integral over the indicated path C.

1. $\int_C yz\, dx + zx\, dy + xy\, dz$; $C: x = t$, $y = t - 1$, $z = t + 1$, $0 \le t \le 1$.

2. $\int_C (y + z)\, dx + (z + x)\, dy + (x + y)\, dz$; $C: x = e^t$, $y = e^{-t}$, $z = e^{2t}$, $0 \le t \le 1$.

3. $\int_C x \cos yz\, dx + y \cos zx\, dy + z \cos xy\, dz$; $C: x = t$, $y = 2t$, $z = 3t$, $0 \le t \le \sqrt{\pi/4}$.

4. $\int_C y^2\, dx + x^2\, dy + (x^2 + y^2)\, dz$; $C: x = \sin t$, $y = \cos t$, $z = t$, $0 \le t \le \pi/2$.

5. $\int_C y\, dx + x\, dy + xyz\, dz$; C is the line segment from $(0,0,0)$ to $(1,2,4)$.

6. $\int_C \cos y\, dx + \sin x\, dy + xy\, dz$; C is the line segment from $(0,0,1)$ to $(\pi/2, \pi/2, 4)$.

7. $\int_C (x^2 + y^2)\, dx + xyz\, dy + (x + y + z)\, dz$; C is the rectangle with vertices $(0,0,1)$, $(1,0,1)$, $(1,1,1)$, $(0,1,1)$ and is traversed in this order.

8. $\int_C yz\, dx + xz\, dy + xy\, dz$; C is the intersection of the sphere $x^2 + y^2 + z^2 = 4$ with the plane $z = 1$. C is traversed once in the counterclockwise direction starting at $(0, \sqrt{3}, 1)$.

In each of Problems 9 to 14 compute the given line integral for the curve C from the point P to the point Q.

9. $\int_C A\, dx + B\, dy + C\, dz$; $P = (x_0, y_0, z_0)$, $Q = (x_1, y_1, z_1)$.

10. $\int_C xy^2z^2\, dx + x^2yz^2\, dy + x^2y^2z\, dz$; $P = (1,1,1)$, $Q = (2,2,2)$.

11. $\int_C y^2z^3\, dx + 2xyz^3\, dy + 3xy^2z^2\, dz$; $P = (1,0,1)$, $Q = (2,1,2)$.

12. $\int_C yze^{xyz}\, dx + xze^{xyz}\, dy + xye^{xyz}\, dz$; $P = (0,0,0)$, $Q = (1,1,1)$.

13. $\int_C \cos y \sin z\, dx - x \sin y \sin z\, dy + x \cos y \cos z\, dz$; $P = (1,0,\pi/2)$, $Q = (2,\pi,3\pi/2)$.

14. $\int_C (2xy + z^2)\, dx + (x^2 + 2yz)\, dy + (y^2 + 2zx)\, dz$; $P = (1,2,2)$, $Q = (2,3,4)$.

3.2   Extension to n-space

If we are given n functions

$$t \mapsto x_j(t), \qquad j = 1, 2, \ldots, n; \qquad a \le t \le b,$$

defining a smooth curve C in n-space, and if $P_j(x_1, x_2, \ldots, x_n)$ are n continuous functions, we set

$$\int_C P_1 \, dx_1 + P_2 \, dx_2 + \cdots + P_n \, dx_n = \int_C \sum_{j=1}^{n} P_j \, dx_j$$

$$= \int_a^b \left\{ \sum_{j=1}^{n} P_j[x_1(t), x_2(t), \ldots, x_n(t)] x_j'(t) \right\} dt. \tag{6}$$

Everything said above applies to this line integral, with obvious modifications. We remark that the integrability conditions now read:

$$\frac{\partial P_i}{\partial x_j} = \frac{\partial P_j}{\partial x_i}, \qquad \text{for } i, j = 1, 2, \ldots, n; \qquad i \ne j.$$

There are $\frac{1}{2}n(n-1)$ of those.

PROBLEMS

1. Compute $\int_C x_1 \, dx_1 + x_2 \, dx_2 + \cdots + x_n \, dx_n$, where C is the curve $x_1 = a_1 t$, $x_2 = a_2 t, \ldots, x_n = a_n t$ for $t_0 \le t \le t_1$.

2. Compute $\int_C x_1^2 \, dx_1 + x_2^2 \, dx_2 + \cdots + x_n^2 \, dx_n$, where C is the curve $x_i = a_i t, i = 1, 2, \ldots, n$, for $t_0 \le t \le t_1$.

3. Compute $\displaystyle\int_C \frac{dx_1}{x_1} + \frac{dx_2}{x_2} + \cdots + \frac{dx_n}{x_n}$ from the point $P = (1, 1, \ldots, 1)$ to the point $Q = (2, 2, \ldots, 2)$.

4. Compute $\int_C e^{x_1} \, dx_1 + 2e^{2x_2} \, dx_2 + \cdots + ne^{nx_n} \, dx_n$ from the point $P = (0, 0, \ldots, 0)$ to the point $Q = (1, 1, \ldots, 1)$.

§4   Applications of line integrals to mechanics[‡]

4.1   Work

To apply line integrals such as

$$\int_C F \, dx + G \, dy + H \, dz \tag{1}$$

to mechanics, it is convenient to use vector language. We combine the three functions defining the curve C into one vector-valued function

$$t \mapsto \mathbf{x}(t) = x(t)\mathbf{i} + y(t)\mathbf{j} + z(t)\mathbf{k}, \qquad a \le t \le b, \tag{2}$$

[‡]Optional section.

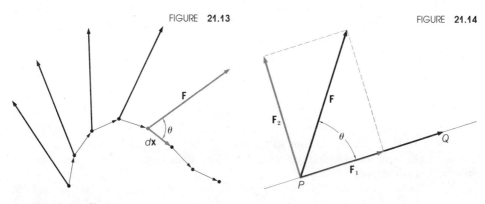

FIGURE 21.13   FIGURE 21.14

so that, formally,

$$dx = (x'(t)\mathbf{i} + y'(t)\mathbf{j} + z'(t)\mathbf{k})\, dt.$$

We combine the functions F, G, and H into one vector-valued function of the point $\mathbf{x} = x\mathbf{i} + y\mathbf{j} + z\mathbf{k}$:

$$\mathbf{F}(\mathbf{x}) = \mathbf{F}(x,y,z) = F(x,y,z)\mathbf{i} + G(x,y,z)\mathbf{j} + H(x,y,z)\mathbf{k}. \tag{3}$$

Recalling the definition of scalar products (see Chapter 18, §3.1), we write (1) in the form

$$\int_C (\mathbf{F},d\mathbf{x}). \tag{1'}$$

This notation suggests a mathematical myth depicted in Figure 21.13. The curve C consists of infinitely many, infinitely small directed segments; each determines an infinitely short vector $d\mathbf{x}$. At each segment take the corresponding value of $\mathbf{F}$ and form the scalar product

$$(\mathbf{F},d\mathbf{x}) = |\mathbf{F}|\,|d\mathbf{x}|\cos\theta, \tag{4}$$

where θ is the angle between $\mathbf{F}$ and $d\mathbf{x}$. The integral (1′) is the sum of all these infinitely small numbers.

Recall now (see Chapter 9, §5.2) that if a particle is moved the distance s along a line, subject to a constant force $\mathbf{F}$ in the direction of the line, then the work W done against the force is defined to be $-|\mathbf{F}|s$ if the particle moved in the direction of $\mathbf{F}$ and $|\mathbf{F}|s$ if it moved opposite to that direction.

If the particle moves along a directed segment $\overrightarrow{PQ}$, subject to a constant force $\mathbf{F}$, and the angle between $\mathbf{F}$ and $\overrightarrow{PQ}$ is θ (see Figure 21.14), the work W done against the force is then defined to be $-|\mathbf{F}|\,|PQ|\cos\theta$. If we denote by $\mathbf{s}$ the vector determined by $\overrightarrow{PQ}$, then

$$W = -(\mathbf{F},\mathbf{s}). \tag{5}$$

[We arrive at this definition by setting $\mathbf{F} = \mathbf{F}_1 + \mathbf{F}_2$ where $\mathbf{F}_1$ has the direction of the line $\overrightarrow{PQ}$ and $\mathbf{F}_2$ is perpendicular to this line. The sensible assumption that no work is done against $\mathbf{F}_2$ leads to the conclusion:

$$W = -|\mathbf{F}_1|\,|\mathbf{s}| \quad \text{if} \quad 0 \le \theta \le \frac{\pi}{2}, \qquad W = |\mathbf{F}_1|\,|\mathbf{s}| \quad \text{if} \quad \frac{\pi}{2} \le \theta \le \pi.$$

Since $|\mathbf{F}_1| = |\mathbf{F}| \cos \theta$ in the first case and $|\mathbf{F}_1| = -|\mathbf{F}| \cos \theta$ in the second, (5) follows.]

Suppose now that a particle moves in a **force field** described by the function (3). This means that when the particle is at the point with position vector $\mathbf{x}$, the force acting on it is $\mathbf{F}(\mathbf{x})$. If our particle is moved along the curve C described by (2), we define the work done against the field on the particle as

$$W = - \int_C (\mathbf{F}, d\mathbf{x}). \tag{6}$$

[This definition is justified by the mathematical myth told above: W is the sum of numbers (4), each of which represents the work done against the force along an infinitely small segment of C.]

We know (see §1.1) that the integral (6) does not depend on the parametric representation used to define C. This means that the work does not depend on the speed with which the particle moved along C.

EXAMPLES 1.   Near the surface of the earth, the gravity force acting at the point (x,y,z) is

$$\mathbf{F} = -mg\mathbf{k}$$

if we choose $z = 0$ as the surface of the earth, make $\mathbf{k}$ point up, and denote by g the acceleration of gravity and by m the mass of the particle. The work done by the gravitational field in moving the particle from the point (x_0,y_0,z_0) to the point (x_1,y_1,z_1) along the curve C is

$$- \int_C -mg \, dz = mg \int_C dz = mg(z_1 - z_0).$$

It depends only on the z-coordinates of the endpoints, that is, on the difference in heights.

2.   Suppose

$$\mathbf{F}(x,y,z) = \mathbf{i} + x\mathbf{j}.$$

Consider two curves with the same endpoints $(0,0,0)$ and $(0,\pi,0)$:

$$C_1: x = 0, \quad y = t, \quad z = 0, \quad 0 \le t \le \pi$$

and

$$C_2: x = \sin t, \quad y = t, \quad z = 0, \quad 0 \le t \le \pi.$$

The work done along these two curves is

$$\int_{C_1} (\mathbf{F}, d\mathbf{x}) = \int_{C_1} dx + x \, dy = \int_0^\pi 0 \, dt = 0$$

and

$$\int_{C_2} (\mathbf{F}, d\mathbf{x}) = \int_{C_2} dx + x \, dy = \int_0^\pi \cos t \, dt + \sin t \, dt = 2.$$

Thus the work depends on the path, not only on the endpoints.

PROBLEMS

In Problems 1 to 6 for each given field of force $\mathbf{F}$ and each given curve C, compute the work W needed to move against $\mathbf{F}$ over C.

1. $\mathbf{F} = 3\mathbf{i} + 4\mathbf{j} + 5\mathbf{k}$; C is the line segment from (x_0,y_0,z_0) to (x_1,y_1,z_1).
2. $\mathbf{F} = x\mathbf{i} + y\mathbf{j} + z\mathbf{k}$; C is the line segment from (x_0,y_0,z_0) to (x_1,y_1,z_1).
3. $\mathbf{F} = 3\mathbf{i} + 4\mathbf{j} + 5\mathbf{k}$; C is the broken line that goes from (x_0,y_0,z_0) to (x_1,y_1,z_1) and from there to (x_2,y_2,z_2).

4. **F** as in Problem 2, C as in Problem 3.

5. $\mathbf{F} = y\mathbf{i} + z\mathbf{j} + y\mathbf{k};$ C is the great circle cut out of the unit sphere by the plane $x = y$, traversed once in the direction that is clockwise when seen from the point $(-1/\sqrt{2}, -1/\sqrt{2}, 0)$.

6. $\mathbf{F} = x^2 y\mathbf{i} + x^2 z\mathbf{j} + yz^2\mathbf{k};$ C is the triangle from the origin to $(1,1,0)$, to $(0,0,1)$, and back to the origin.

4.2   Potential energy

A force field $\mathbf{F}(x,y,z)$ is called **conservative** if no work is done in moving a particle along a closed path:

$$\int_C (\mathbf{F}, d\mathbf{x}) = 0 \qquad \text{if } C \text{ is closed.}$$

For instance, the field in Example 1 in §4.1 is conservative, but the field in Example 2 is not.

We know (see Theorem 1 in §1.3 which is also valid in space) that a field is conservative if and only if the work done in moving a particle along a path depends only on the endpoints of the path, and if and only if there is a function with gradient **F**. Then there also is a function whose gradient is $-\mathbf{F}$. This function is

$$W(x,y,z) = -\int_{(x_0, y_0, z_0)}^{(x,y,z)} (\mathbf{F}, d\mathbf{x}), \tag{7}$$

the work done in moving a particle, along any path, from some fixed reference point $\mathbf{x}_0 = (x_0, y_0, z_0)$ to $\mathbf{x} = (x,y,z)$. That **F** is indeed the gradient of $-W$, that is, the components of the force are the partial derivatives

$$F = -\frac{\partial W}{\partial x}, \qquad G = -\frac{\partial W}{\partial y}, \qquad H = -\frac{\partial W}{\partial z}, \tag{8}$$

follows by the argument used in proving Theorem 1 in §1.3.

We call $W(x)$ the **potential** of the field, or the **potential energy** of a particle located at **x**. It is determined by the field only up to an arbitrary additive constant: it is up to us how to choose the point (x_0, y_0, z_0) where $W = 0$, and adding a constant to W does not disturb (8).

If the field has a potential, that is, is conservative, then the integrability conditions are satisfied:

$$F_y = G_x, \qquad F_z = H_x, \qquad G_z = H_y.$$

If these conditions hold, and we are in a simply connected region, there is a potential.

The level surfaces of W are called equipotentials. The force is perpendicular to the equipotentials. At a critical point of W, we have $\mathbf{F} = -\operatorname{grad} W = 0$, and the particle is in equilibrium.

EXAMPLES 1.   According to Newton's law of universal gravitation (Chapter 17, §5.3), a mass M located at the origin exerts on a particle with mass 1 located at (x,y,z) an attractive force **F** with the components

$$-\frac{\gamma M x}{(x^2 + y^2 + z^2)^{3/2}}, \qquad -\frac{\gamma M y}{(x^2 + y^2 + z^2)^{3/2}}, \qquad -\frac{\gamma M z}{(x^2 + y^2 + z^2)^{3/2}}.$$

Indeed, this $\mathbf{F}$ can be written as $-M\gamma r^{-3/2}(x\mathbf{i} + y\mathbf{j} + z\mathbf{k})$. Hence it has the direction pointing from (x,y,z) to the origin and the magnitude $M\gamma r^{-3/2}(x^2 + y^2 + z^2)^{1/2} = M\gamma/r$, as it should.

We check that the components of $\mathbf{F}$ are the negatives of the partial derivatives of

$$W(x,y,z) = \frac{\gamma M}{r} = \frac{\gamma M}{(x^2 + y^2 + z^2)^{1/2}}.$$

This function is therefore a potential of the gravitational field produced by the mass M.

2. If we restrict ourselves to a small region near the surface of the earth and choose a coordinate system as in Example 1, §4.1, the gravitational field has the potential

$$W(x,y,z) = mgz \qquad (z = \text{height above earth}),$$

in conformity with Chapter 9, §5.3. Indeed, $W_x = W_y = 0$ and $W_z = mg$.

PROBLEMS

1. Consider the gravitational force field due to two particles of masses m_1 and m_2, located at $\mathbf{x}_1$ and $\mathbf{x}_2$. Show that this field is conservative by finding a potential.
2. Consider the gravitational force field due to n particles of masses $m_1, m_2, \ldots, m_n$, located at $\mathbf{x}_1, \mathbf{x}_2, \ldots, \mathbf{x}_n$. Show that this field has a potential and is therefore conservative.
3. Is the field given by $\mathbf{F} = x\mathbf{i} + y\mathbf{j} + z\mathbf{k}$ conservative? If so, what is its potential?

4.3 Conservation of energy

The kinetic energy of a moving particle of mass m is defined as

$$K = \tfrac{1}{2}mv^2, \qquad v = \text{speed},$$

as in the case of rectilinear motion (see Chapter 9, §5.4). If t represents time, and the motion of our particle is described by $t \mapsto \mathbf{x}(t) = x(t)\mathbf{i} + y(t)\mathbf{j} + z(t)\mathbf{k}$, then the velocity vector is $\mathbf{x}'(t)$, and the speed is $|\mathbf{x}'(t)|$. Hence

$$K = \frac{1}{2} m|\mathbf{x}'(t)|^2 = \frac{m}{2}(x'(t)^2 + y'(t)^2 + z'(t)^2).$$

We can now state and prove the basic theorem on conservation of energy which contains the result of Chapter 9, §5.4, as a special case.

If a particle moves under the influence of a conservative force, its total energy $E = K + W$ (that is, the sum of kinetic and potential energies) remains constant.

Proof. We assume, of course, that the particle obeys Newton's law of motion

$$m\mathbf{x}''(t) = \mathbf{F}(\mathbf{x}(t)).$$

Since $\mathbf{F}$ derives from a potential W, we have $\mathbf{F} = -\operatorname{grad} W$. Hence we also have $m\mathbf{x}'' = -\operatorname{grad} W$ or, in components,

$$mx''(t) = -\left(\frac{\partial W}{\partial x}\right)_{\mathbf{x}=\mathbf{x}(t)}, \quad my''(t) = -\left(\frac{\partial W}{\partial y}\right)_{\mathbf{x}=\mathbf{x}(t)}, \quad mz''(t) = -\left(\frac{\partial W}{\partial z}\right)_{\mathbf{x}=\mathbf{x}(t)}. \quad (9)$$

Now, the energy of our particle at the time t is

$$E(t) = \frac{m}{2}(x'(t)^2 + y'(t)^2 + z'(t)^2) + W(x(t),y(t),z(t)).$$

Using the chain rule, we compute

$$E'(t) = mx'(t)x''(t) + my'(t)y''(t) + mz'(t)z''(t)$$

$$+ \left(\frac{\partial W}{\partial x}\right)_{\mathbf{x}(t)} x'(t) + \left(\frac{\partial W}{\partial y}\right)_{\mathbf{x}(t)} y'(t) + \left(\frac{\partial W}{\partial z}\right)_{\mathbf{x}(t)} z'(t)$$

$$= x'(t)\left\{mx''(t) + \left(\frac{\partial W}{\partial x}\right)_{\mathbf{x}(t)}\right\} + y'(t)\left\{my''(t) + \left(\frac{\partial W}{\partial y}\right)_{\mathbf{x}(t)}\right\}$$

$$+ z'(t)\left\{mz''(t) + \left(\frac{\partial W}{\partial z}\right)_{\mathbf{x}(t)}\right\} = 0,$$

by virtue of (9). Hence E is indeed constant.

This theorem is a special case of a general principle that dominates all of physics.

PROBLEMS

1. Verify that the motion of a projectile described in Chapter 17, §3.4, obeys the law of conservation of energy.

*2. Verify that a planet moving around the sun according to Kepler's laws (see Chapter 17, §5) obeys the law of conservation of energy.

THEORETICAL SUPPLEMENT

For 21 chapters we developed the main results and techniques of calculus relying, in some of the most crucial moments, on geometric intuition. In doing so, we followed the historical development of calculus. Now we want to show how calculus can be built up rigorously from properties of numbers. The resulting theory gives mathematicians and users of mathematics full assurance that calculus can never lead them astray.

The underlying theme of this supplement is a property of real numbers called the *least upper bound principle*. After explaining this principle in §1.6, we shall use it to derive the basic theorems of calculus.

§1 Real numbers

To develop calculus rigorously, we must lay the proper foundations, and this requires a renewed discussion of numbers (see Chapter 1).

1.1 Field postulates

In Chapter 1, §2.5 and §2.6, we said that real numbers (defined there as positive and negative nonterminating decimals and zero) obey the familiar rules of algebra. These rules all follow from relatively few basic postulates. We state first the postulates which describes addition and multiplication.

1. *For any two numbers a and b, there is a unique number called the **sum** of a and b and denoted by $a + b$.*
2. *$a + b = b + a$ for all numbers a and b.*
3. *$(a + b) + c = a + (b + c)$ for all numbers a, b, and c.*
4. *There is a number called **zero** and denoted by 0, such that $0 + a = a$ for all numbers a.*
5. *For every number a, there is a unique number called its **negative** and denoted by $(-a)$, such that $a + (-a) = 0$.*
6. *For any two numbers a and b, there is a unique number called the **product** of a and b and denoted by ab.*
7. *$ab = ba$ for all numbers a and b.*
8. *$(ab)c = a(bc)$ for all numbers a, b, and c.*
9. *$a(b + c) = ab + ac$ for all numbers a, b, and c.*
10. *There is a number called **one** and denoted by 1, such that $1 \cdot a = a$ for all numbers a.*
11. *$1 \neq 0$.*
12. *For every number $a \neq 0$, there is a unique number called its **reciprocal** and denoted by a^{-1}, such that $a(a^{-1}) = 1$.*

Postulates **1** to **12** are true if by "number" we mean "rational number" (positive or negative common fraction or zero). They are true if by "number" we mean real number. (Readers familiar with complex numbers will note that postulates **1** to **12** remain true if by "number" we mean complex number.)

Postulates **1** to **12** are called *field axioms,* and mathematicians abbreviate the statement that the rational numbers satisfy these axioms by saying that the rational numbers form a **field.**

Why does one go to the trouble of writing out such almost self-evident statements as **1** to **12**, and why does one dignify them by a special name? Because there are many different mathematical systems obeying the same postulates, and it is convenient to know which rules of algebra hold whenever statements **1** to **12** are true.

The field axioms do not mention integers (whole numbers) explicitly. We define a number to be an integer if it is 1 or obtained from 1 by addition and by taking negatives. Thus $1 + 1 = 2$, $2 + 1 = 3$, and so forth, are integers, as are -1, -2, -3, and so forth, and so is $0 = 1 + (-1)$. The integers do *not* form a field, since the reciprocal a^{-1} of an integer a is not an integer unless $a = 1$ or $a = -1$.

We agree to write $a + b + c$ instead of $(a + b) + c$ and abc instead of $(ab)c$ and similarly for sums and products of more than three terms. The associative laws **3** and **8** assure us that this will not lead to confusion. We note that in view of the commutative laws **2** and **7** we may write $b + a + c$ instead of $a + b + c$, bac instead of abc, and so forth. We also agree to write $b - a$ instead of $b + (-a)$ and b/a instead of $b(a^{-1})$. In view of the commutative laws **2** and **7**, we also have $-a + b = b + (-a) = b - a$, $a^{-1}b = ba^{-1} = b/a$. We note that $a^{-1} = 1/a$ for $a \neq 0$.

We already said that all familiar rules of algebra can be deduced from postulates **1** to **12**. In such a derivation one must justify each step by referring to the postulates or to some consequence of the postulates already established. We illustrate by giving a few examples.

EXAMPLES   1.   *Rule of signs.* $-0 = 0$, $-(-a) = a$ for all a.

Proof.   The first statement is true because -0 is, by **5**, the number x such that $0 + x = 0$, and we know that $0 + 0 = 0$ by **4**. The second statement means that $(-a) + a = 0$; this is so by **5** and **2**.

2.   *Subtraction.* $a + x = b$ if and only if $x = b - a$ $[=b + (-a)$, by definition$]$.

Proof.   If $a + x = b$, then $(a + x) + (-a) = b + (-a)$. But $b + (-a) = b - a$, by definition. Also, by **2**, **3**, **4**, and **5**, we have $(a + x) + (-a) = (x + a) + (-a) = x + (a + (-a)) = x + 0 = 0 + x = x$. Hence $x = b - a$.

Assume next that $x = b - a = b + (-a)$. Then, using **2**, **3**, **4**, and **5**, we have $a + x = a + (b + (-a)) = a + ((-a) + b) = (a + (-a)) + b = 0 + b = b$.

3.   *Division.*   Let $a \neq 0$. Then $ax = b$ if and only if $x = ba^{-1}$ $(=b/a$, by definition$)$.

Proof.   If $ax = b$, then $(ax)b^{-1} = ba^{-1}$. But by **7**, **8**, **10**, and **12**, we have $(ax)a^{-1} = (xa)a^{-1} = x(aa^{-1}) = x \cdot 1 = 1 \cdot x = x$. Hence $x = ba^{-1}$.

Assume next that $x = ba^{-1}$. Then $ax = a(ba^{-1}) = a(a^{-1}b) = (aa^{-1})b = 1 \cdot b = b$, where we used **7**, **8**, **10**, and **12**.

4.   *Multiplication by zero.*   For all a, $0a = a$.

Proof.   The distributive law **9** applied to the case $c = 0$ yields $a(b + 0) = ab + a0 = ab + 0a$, by **7**. But $a(b + 0) = a(0 + b) = ab$, by **2** and **4**. Hence $ab + 0a = ab$. Thus, by Example 2, $0a = ab - ab = ab + (-ab) = 0$, by **5**.

5.   *Cancellation law.*   If $ab = 0$, then either $a = 0$ or $b = 0$.

Proof.   Assume that $a \neq 0$, $b \neq 0$. We must show that $ab \neq 0$. Set $c = b^{-1}a^{-1}$. Then, using postulates **8**, **10**, and **12**, we have $(ab)c = a(bc) = a(b(b^{-1}a^{-1})) = a((bb^{-1})a^{-1}) = a(1 \cdot a^{-1}) = aa^{-1} = 1 \neq 0$, by **11**. This shows that $ab \neq 0$. For, if ab were 0, we should have $(ab)c = 0$, by Example 4.

6.   *Sign of a product.* $(-a)b = -(ab)$, $(-a)(-b) = ab$.

Proof.   The first statement means that $ab + ((-a)b) = 0$. It is so, because by **9** and **7** we have $ab + ((-a)b) = (a + (-a))b = 0b = 0$, the last step being a consequence of Example 4. To prove the second statement, we use the result just established and Example 1. We have $(-a)(-b) = -(a(-b)) = -((-b)a) = -(-ba) = ba = ab$.

PROBLEMS

Verify the following statements, valid for all a, b, c, and d, using the field postulates.

1. $(a + b) + (c + d) = (d + a) + (c + b)$.
2. $(a + b)(c + d) = (ac + bd) + (bc + ad)$.
3. $(ab + ac)d = a(bd + cd)$.
4. If $a \neq 0$ and $a = aa$, then $a = 1$.

5. $\dfrac{1}{1} = 1$.

6. $\dfrac{a}{1} = a$.

7. $\dfrac{1}{a^{-1}} = a$,   for $a \neq 0$.

8. $-\dfrac{1}{a} = \dfrac{1}{-a}$,   for $a \neq 0$.

9. $\dfrac{a}{-1} = -a$.

10. $\dfrac{-a}{-b} = \dfrac{a}{b}$,   for $b \neq 0$.

11. $\dfrac{1}{ab} = \dfrac{1}{a}\dfrac{1}{b}$,   for $a \neq 0$, $b \neq 0$.

12. $\dfrac{ab}{ac} = \dfrac{b}{c}$,   for $a \neq 0$, $c \neq 0$.

13. $\dfrac{a + b}{c} = \dfrac{a}{c} + \dfrac{b}{c}$, for $c \neq 0$.

14. $\dfrac{a}{b} + \dfrac{c}{d} = \dfrac{ad + bc}{bd}$, for $b \neq 0, d \neq 0$.

15. $(a + b)^2 = a^2 + 2ab + b^2$.
16. $(a + b)(a - b) = a^2 - b^2$.
17. $(-a)^3 = -a^3$.
18. $(-a)^4 = a^4$.

1.2 Order postulates

We recall next the postulates pertaining to the relation $<$ (less than). (These are rules **I** to **IV** of Chapter 1, §3.2.)

13. *If a and b are two numbers, then one of the three statements "$a = b$," "$a < b$," and "$b < a$" is true and the other two are false.*
14. *If a, b, c are three numbers, $a < b$ and $b < c$, then $a < c$.*
15. *If a, b, c, are three numbers, and $a < b$, then $a + c < b + c$.*
16. *If a, b, c are three numbers, $a < b$ and $0 < c$, then $ac < bc$.*

A system of numbers satisfying **1** to **16** is called an **ordered field.** Thus rational numbers form an ordered field and so do real numbers (but complex numbers do not).

Here are some consequences of postulates **1** to **16.** (We recall that $a > b$ means the same as $b < a$ and that $a \leq b$ means that either $a < b$ or $a = b$.)

$$\text{If } a > 0, b > 0, \text{ then } a + b > 0. \tag{1}$$

$$\text{If } a > 0, b > 0, \text{ then } ab > 0. \tag{2}$$

$$\text{If } a > 0, \text{ then } -a < 0. \tag{3}$$

$$\text{If } a \neq 0, \text{ then } a^2 > 0. \tag{4}$$

$$\text{If } a > 0, b < 0, \text{ then } ab < 0. \tag{5}$$

$$\text{If } a > 0, b > 0, \text{ then } a/b > 0. \tag{6}$$

$$\cdots -4 < -3 < -2 < -1 < 0 < 1 < 2 < 3 < 4 < \cdots. \tag{7}$$

These statements are quite obvious; nevertheless we shall derive them from postulates **1** to **16.** Assume that $a > 0$. Then $a + b > 0 + b = b$ (by **15**). If also $b > a$, then $a + b > 0$ (by **14**). This proves (1).

If $a > 0$ and $b > 0$, then $ab > 0b = 0$ (by **16**); this proves (2).

If $a > 0$, then $a \neq 0$ (by **13**); hence either $-a > 0$ or $-a < 0$ (again by **13**). If $-a > 0$, then $a + (-a) > 0$, by (1). This would mean $0 > 0$, which is impossible (by **13**). Hence (3) holds.

If $a \neq 0$, then either $a > 0$ or $a < 0$. If $a > 0$, then $a^2 = aa > 0$, by (2), and if $a < 0$, then $-a > 0$, by (3), and $a^2 = (-a)(-a) > 0$, by (2). This proves (4).

If $b < 0$, then $-b > 0$, by (3), so that, if also $a > 0$, then $a(-b) = -ab > 0$, by (2), and $ab = -(-ab) < 0$, by (3). This proves (5).

If $a > 0$ and $b > 0$, then $a/b \neq 0$. For if $a/b = 0$, then $0 = (a/b)b = a$. We cannot have $a/b < 0$, for if $a/b < 0$, then $a = (a/b)b < 0$, by (5). Therefore, $a/b > 0$ and (6) is proved.

Since $1 \neq 0$ (by **11**) and $1^2 = 1$, we have $1 > 0$, by (4). Hence $2 = 1 + 1 > 0 + 1 = 1$ (by **15**). Similarly, $3 = 2 + 1 > 2$, $4 = 3 + 1 > 3$, and so forth. Also $-1 < 0$, by (3), $-2 < -1$, and so on. This proves (7).

It follows from (7) that $1 \neq 2$, $2 \neq 3$, $1 \neq 3$, $1 \neq 4$, and so on. It is interesting to note that, if we use only the field axioms **1** to **12,** we cannot prove that $1 \neq 3$.

PROBLEMS

Verify the following statements, valid for all numbers a, b, c, and d, using the ordered field postulates.

1. If $a > 0$, $b > 0$, $c > 0$, then $a + b + c > 0$.
2. If $a > 0$, $b > 0$, $c > 0$, then $abc > 0$.
3. If $a > 0$, $b > 0$, $c < 0$, then $abc < 0$.
4. If $a \neq 0$, $b \neq 0$, then $a^2 + b^2 > 0$.
5. If $a < b$, then $a < \frac{1}{2}(a + b) < b$.
6. If $a > b$, $c > d$, then $(a + b)^2 + (c + d)^2 > 0$.

7. If $a > 0$, $b > 0$, $c > 0$, $d > 0$, then $\dfrac{a + b}{c + d} > 0$.

8. If $a > b > c > d > 0$, then $\dfrac{1}{a + b} < \dfrac{1}{c + d}$.

1.3   Archimedean postulate

Another order postulate, which we did not state explicitly in Chapter 1, sounds completely obvious, but it is very important.

17. *For every number a, there is an integer k such that $a < k$.*

We call this the **Archimedean postulate.** Archimedes stated an equivalent axiom in geometric terms; it was also tacitly used by Greek mathematicians who lived long before Archimedes. The Archimedean postulate has two useful consequences.

Corollary 1.   *If $0 \le a \le 1/n$ for all positive integers n, then $a = 0$.*

Proof.   By hypothesis, $a \ge 0$. Suppose $a > 0$. Then $1/a > 0$. By **17** there is an integer, call it s, such that $1/a < s$. Also $s > 0$, for otherwise we would have $1/a < 0$. Multiplying the inequality $s > 1/a$ by the positive number a/s, we obtain $a > 1/s$. Hence it is not true that $a \le 1/n$ for *all* positive integers n. The assumption that $a > 0$ is untenable.

Corollary 2.   *If $A > 1$ and $0 \le a \le A^{-n}$ for all positive integers n, then $a = 0$.*

Proof.   Since $A > 1$, we have $A = 1 + b$, $b > 0$. Hence

$$A^n = (1 + b)(1 + b) \ldots (1 + b) = 1 + nb + \text{other positive numbers} > nb.$$

Thus if $0 \le a \le A^{-n}$, then $0 \le a \le 1/nb$ and $0 \le ab \le 1/n$. If this holds for all positive integers n, then $ab = 0$, by Corollary 1. Hence $a = 0$, by the cancellation law (since $b \neq 0$).

EXAMPLES 1.   We use the Archimedean postulate to show that *every interval contains rational numbers and irrational numbers.*

Proof.   Let x be the midpoint of the interval in question; we assume that $x > 0$ and write $x = a.\alpha_1\alpha_2\alpha_3 \ldots$, an infinite decimal. Let l be the length of the interval, and let m be a positive integer such that $10^{-m} < l/2$. (Such an m exists. Otherwise we would have $0 \le l \le 10^{-n}$ for *all* positive integers n, and, by Corollary 2 to the Archimedean postulate, l would be 0. But the length of an interval must be positive.) Now set

$$y_1 = a + .\alpha_1\alpha_2 \ldots \alpha_m, \qquad y_2 = a + .\alpha_1\alpha_2 \ldots \alpha_m 10110111011110 \ldots.$$

Then $|y_1 - x| < 10^{-m}, |y_2 - x| < 10^{-m}$, so that both y_1 and y_2 are points in our interval. But y_1 is rational and y_2 is irrational. Cases where $x = 0$ or $x < 0$ can be treated similarly.

2. Let us verify a statement made in Chapter 1, §3.1: *the number $\frac{1}{3}$ is the only number which satisfies all inequalities*

$$0 < \tfrac{1}{3} \le 1, \qquad .3 < \tfrac{1}{3} \le .4, \qquad .33 < \tfrac{1}{3} \le .34, \qquad .333 < \tfrac{1}{3} \le .334 \qquad \text{and so forth.}$$

Proof. If y were another such number, then $\frac{1}{3}$ and y would lie in the same interval of length 1, of length $\frac{1}{10}$, of length $\frac{1}{100}$, and so forth. Hence we would have $|\frac{1}{3} - y| \le 10^{-m}$ for $m = 1, 2, 3, \ldots$. Then $|\frac{1}{3} - y| = 0$, by Corollary 2 to the Archimedean postulate, and $\frac{1}{3} = y$.

Like the previous postulates, **17** is valid for both rational and real numbers. We state (in §1.6) a postulate which distinguishes these two number systems. As a preparation we discuss bounds and then mathematical induction.

PROBLEMS

1. If $0 \le a \le 1/\sqrt{n}$ for all positive integers n, what can you conclude about a?
2. For which numbers x is it true that $0 \le x \le 10^{-j}$ for $j = 1, 2, 3, \ldots$?
3. For which numbers x is it true that $0 \le x \le 4(n + 1)^{-2}$ for all positive integers n?
4. For which numbers x is it true that $0 \le x^2 - 1 \le 2^{-j}$ for $j = 1, 2, 3, \ldots$?
5. Prove that $.5 = .4\overline{9}$, that is, show that $\frac{1}{2}$ is the only number x such that

$$\overset{n}{\overbrace{.4999\ldots 9}} \le x \le \overset{n}{\overbrace{.4999\ldots 9}} + 10^{-n-1}, \qquad n = 1, 2, 3, \ldots.$$

6. Prove that $.6\overline{9} = .7$.

1.4 Bounds

Let S be a **set** (collection) of real numbers. We say that "x is an **element** of S" or "x belongs to S" or "x is in S" to indicate that x is a member of the set. A set that contains no members is called **empty**.

A number c is called an **upper bound** for S if every x in S satisfies $x \le c$. One also calls such a c an upper bound of S, or one says that S has c for an upper bound. Whether c belongs to S or not is irrelevant.

A set S is called **bounded from above** if it has an upper bound; otherwise S is called **unbounded from above**.

A number c is called a **lower bound** for S if $c \le x$ for all x in S. If a lower bound exists, S is called **bounded from below**. A set is called **bounded** if it is bounded from above and from below. (The empty set is bounded, by convention.)

An upper bound for S, which is also an element of S, is called the greatest or largest element of S. A lower bound for S, which is also an element of S, is called the smallest or the least element of S.

A set of numbers may be considered a set of points on the number line. An upper (lower) bound for S is a point such that no element of S lies to its right (to its left).

EXAMPLES The interval $(-10, +\infty)$ is bounded from below but not from above. The interval $(-100, 100)$ is bounded. Any finite set, that is, a set with finitely many elements, is bounded.

If S consists of 1, 2, and -5, then 2 is an upper bound; so are 2.0001 and 10^{10}. If S is the set of all negative rationals, then 0 is an upper bound. So is every positive number. If S is the set of all integers, S has no upper bound.

Zero is the least element in the set of all nonnegative rational numbers. The interval $(-2,3)$ has no least element and no greatest element.

___

PROBLEMS

1. Is 1 an upper bound for the interval $[-3,1]$?
2. Is 12 an upper bound for the interval $(-1,0)$?
3. Is 3 a lower bound for the interval $(3,+\infty)$?
4. Is 2 a lower bound for the interval $(-\infty,2)$?
5. Is the set of all integers divisible by 3 bounded? Bounded from above? Bounded from below?
6. Is the set of all squares of all integers bounded? Bounded from above? Bounded from below?
7. Find the least elements and the largest element of the set $\{\frac{3}{1},\frac{3}{4},\frac{3}{5},\frac{4}{1},\frac{4}{3},\frac{4}{5},\frac{5}{1},\frac{5}{3},\frac{5}{4}\}$.
8. Does the interval $(0,2)$ have a least element? A largest element?
9. Does the infinite set $\{\frac{1}{2},\frac{2}{3},\frac{3}{4},\frac{4}{5},\ldots\}$ have a least element? A largest element? Is it bounded?
10. Does the set of rational numbers in the interval $[0,1)$ have a least element? A largest element? Is $\sqrt{2}$ an upper bound for this set?

1.5   Principle of least integer. Mathematical induction

I. *Let S be a nonempty set of integers. If S is bounded from below, S contains a least element.*

The statement is almost self-evident. It is assumed, perhaps in a different wording, in every axiomatic treatment of integers. We call **I** the **principle of least integer.** The principle has an immediate corollary.

I'. *Let S be a nonempty set of integers. If S is bounded from above, S contains a largest element.*

Proof.   Let T be the set of all negatives of the elements of S. Thus x is in T if and only if $-x$ is in S. The set T is a nonempty set of integers. Let c be an upper bound of S. Then $y \leq c$ for all y in S. Hence $-c \leq -y$ for all y in T; hence $-c$ is a lower bound for T. By **I**, T has a least element b. Then $-b$ is the greatest element of S.

The principle of least integer is the basis of proofs by **mathematical induction.** Let us explain it in an example.

We want to establish the formula for the sum of squares of the first m positive integers:

$$1^2 + 2^2 + 3^2 + \cdots + m^2 = \frac{m^3}{3} + \frac{m^2}{2} + \frac{m}{6}. \tag{8}$$

We claim that it is true for all positive integers m. If the claim is false, there is some positive integer for which the relation (8) is false: the set of positive integers m for which (8) is false is not empty. There is then the smallest integer, call it r, for which (8) is false (by the principle of least integer, since positive integers are bounded from below). Either $r = 1$ (case 1) or $r > 1$ (case 2). In case 2, we may set $r = k + 1$, where $k > 0$ is an integer.

We observe now that case 1 cannot occur. In other words: the relation to be established holds for $m = 1$. Indeed, for $m = 1$, relation (8) reads: $1 = \frac{1}{3} + \frac{1}{2} + \frac{1}{6}$.

Next, case 2 cannot occur either. For if it does, we have

$$1^2 + 2^2 + \cdots + k^2 + (k+1)^2 \neq \frac{(k+1)^3}{3} + \frac{(k+1)^2}{2} + \frac{k+1}{6} \qquad (9)$$

and

$$1^2 + 2^2 + \cdots + k^2 = \frac{k^3}{3} + \frac{k^2}{2} + \frac{k}{6}. \qquad (10)$$

Adding $(k+1)^2$ to both sides of (10), however, we obtain the correct equation,

$$1^2 + 2^2 + \cdots + k^2 + (k+1)^2 = \frac{k^3}{3} + \frac{k^2}{2} + \frac{k}{6} + (k+1)^2$$

and, after some manipulations,

$$1^2 + 2^2 + \cdots + k^2 + (k+1)^2 = \frac{(k+1)^3}{3} + \frac{(k+1)^2}{2} + \frac{k+1}{6}.$$

This correct result contradicts (9).

We conclude that (8) is true for all positive integers m.

The example just given explains a general method. If we want to prove that a certain statement about all positive integers is true, we may do so by carrying out two steps.

1. *First step.* Show that the statement is true for the integer 1.
2. *Second step.* Show that *if* the statement were true for any integer $k > 0$, *then* it would also be true for $k + 1$.

If the two steps are carried out, then there can be no positive integer for which the statement in question would be false. For, if there were such positive integers, one would be the smallest among them, call it r. But step 1 shows that r cannot be 1, and step 2 shows that r cannot be greater than 1.

EXAMPLE Prove, by induction, that if $0 < x < y$, and $m > 0$ is an integer, then $x^m < y^m$.

ANSWER *First step in induction proof.* For $m = 1$, the statement reads: if $0 < x < y$, then $x < y$. This is true.

Second step. Assume the statement to be true for some fixed integer $k > 0$. Hence, if $0 < x < y$, then $x^k < y^k$. But then $x^{k+1} = x^k x < y^k x$ and $y^k x < y^k y = y^{k+1}$, by **16**, and therefore $x^{k+1} < y^{k+1}$, by **14**. Thus the statement holds for $k + 1$.

This completes the proof.

Strictly speaking, the method of mathematical induction must be used whenever one proves a theorem about all positive integers. If the situation is very simple, however, it is customary (and legitimate) to replace the formal induction proof by a vague "and so on."

PROBLEMS

Prove, by mathematical induction, that the following statements are true for all positive integers n.

1. $1 + 2 + 3 + \cdots + n = \frac{1}{2}n(n+1)$.
2. $1 + q + q^2 + \cdots + q^{n-1} = (1 - q^n)/(1 - q)$, for any number $q \neq 1$.

3. $1 + 3 + 5 + \cdots + (2n - 1) = n^2$.

4. $1 \cdot 2 + 2 \cdot 3 + 3 \cdot 4 + \cdots + n(n + 1) = \frac{1}{3}n(n + 1)(n + 2)$.

5. $\dfrac{1}{1 \cdot 2} + \dfrac{1}{2 \cdot 3} + \dfrac{1}{3 \cdot 4} + \cdots + \dfrac{1}{n(n + 1)} = \dfrac{1}{n + 1}$.

6. $2 + 2^2 + 2^3 + \cdots + 2^n = 2(2^n - 1)$.

7. $1^3 + 3^3 + 5^3 + \cdots + (2n - 1)^3 = n^2(2n^2 - 1)$.

8. $3n^2 \geq 2n + 1$.

9. $4^n \geq n^2$.

10. If $a > -1$, then $(1 + a)^n \geq 1 + na$.

11. $\dfrac{1}{1^2} + \dfrac{1}{2^2} + \dfrac{1}{3^2} + \cdots + \dfrac{1}{n^2} < 3 - \dfrac{1}{n}$.

1.6 Least upper bound principle

Statements **I** and **I′** need not be true for sets of real numbers. Thus, if S is the set of all negative numbers, it has no largest element. But the set of all upper bounds for this S has a least element, namely, 0. This situation is typical. We state now the so-called **principle of least upper bound** or **completeness postulate** for real numbers. It is to be added to postulates **1** to **17**.

18. *A nonempty set of real numbers that is bounded from above has a least upper bound.*

This means that, if S is a nonempty set, and there are numbers that are greater than, or equal to, all members of S, then there is a smallest such number. This least upper bound may, or may not, belong to the set S. If all members of S are rational, the least upper bound may be irrational. (For example, the least upper bound of all rational numbers x satisfying $x^2 < 2$ is the irrational number $\sqrt{2}$.)

A corollary of **18** is

18′. *A nonempty set of real numbers that is bounded from below has a greatest lower bound.*

The reader is urged to derive **18′** from **18** following the pattern used above to derive **I′** from **I**.

Here is the geometric interpretation of **18**. Let S be a nonempty set of points on the number line. Let a be a point such that no point in S is to the right of a. Then there is a point b such that, first, no point of S is to the right of b and, second, if c is any point to the left of b, there is a point in S to the right of c.

That real numbers (defined as nonterminating decimals) satisfy postulates **1** to **17** is intuitively obvious, but the proof is not easy. The meaning of **18** may be difficult to grasp and is far from obvious. The proof, however, is not very difficult. Before presenting it, we emphasize that in developing calculus rigorously, postulate **18** must be used only a few times but always at crucial points in the argument. Just as mathematical induction is present, explicitly or implicitly, in all theorems about all integers, so the least upper bound principle is present in all significant theorems about real numbers.

Proof of postulate **18**. Let S be a nonempty set of real numbers, and let c be an upper bound for S. We shall compute the least upper bound of S.

By the Archimedean postulate **17**, there is an integer k such that $c \leq k$. This k is also an upper bound of S. The set of all integers that are upper bounds for S is itself bounded from below (any element of S is a lower bound). Hence the set of such integers has (by **I**) a smallest element. Call it $a + 1$. The integer a is not an upper bound for S, but $a + 1$ is. We consider the numbers $a + \frac{1}{10}$, $a + \frac{2}{10}$, ..., $a + \frac{9}{10}$ and find a digit α_1 (among 0, 1, ..., 9) such that $a + .\alpha_1$ is not an upper bound for S, but $a + .\alpha_1 + .1$ is. Continuing in this way, we obtain a real number

$$x = a + .\alpha_1\alpha_2\alpha_3 \ldots$$

such that: for all $k = 1, 2, 3, \ldots$,

$$a + .\alpha_1\alpha_2 \ldots \alpha_k \text{ is not an upper bound for } S \qquad (11)$$

and

$$a + .\alpha_1 \ldots \alpha_k + 10^{-k} \text{ is an upper bound for } S. \qquad (12)$$

(We could write $x = a.\alpha_1\alpha_2 \ldots$ if we knew that $a \geq 0$, which it need not be.)

Let y be any number belonging to S. Statement (12) implies that $y \leq a + .\alpha_1 \ldots \alpha_k + 10^{-k}$ (for all $k = 2, 3, \ldots$). Since $a + .\alpha_1 \ldots \alpha_k \leq x$, we have $y \leq x + 10^{-k}$ or $y - x \leq 10^{-k}$, $k = 2, 3, \ldots$. This means that $y - x \leq 0$, by Corollary 2 to **17** (see §1.3), so that, y being an arbitrary element of S, x is an upper bound for S.

Assume next that w is an upper bound for S. By (11) we have $w \geq a + .\alpha_1\alpha_2 \ldots \alpha_k$ for $k = 2, 3, \ldots$. For, otherwise, $a + .\alpha_2\alpha_2 \ldots \alpha_k$ would be an upper bound for S. Hence $w \geq x - 10^{-k}$ or $x - w \leq 10^{-k}$, for all positive integers k. This means that $x - w \leq 0$ or $x \leq w$. Hence x is the smallest upper bound for S.

[In the preceding proof we used the principle of least integer. It can be shown that this principle is a consequence of **18** and **17**. On the other hand, **17** is a consequence of **18** and **I**. Also, by rephrasing our definition of integers (see §1.1), we could derive **I**. We do not prove these statements here, since the proofs would be of no use to us later.]

It is customary to abbreviate postulates **1** to **18** by saying that the real numbers form a **complete, Archimedean, ordered field.**

PROBLEMS

1. Let S be the set of all rational numbers x with $x^2 + 4 < 6$. Find the least upper bound of S.
2. Let S be the set of all rational numbers z with $2z^3 - 1 < 15$. Find the least upper bound of S.
3. Let S be the set $\{\frac{1}{2}, \frac{1}{3}, \frac{1}{5}, \frac{1}{7}, \frac{1}{11}, \frac{1}{13}, \ldots\}$. Find the greatest lower bound of S.
4. Let S be the set of all real numbers whose decimal expansion starts with .12.... Find the least upper bound and greatest lower bound of S.
5. Let S be the set of all irrational numbers t with $1 < t^3 + 1 \leq 3$. Find the least upper bound and greatest lower bound of S.
6. Let S be the set of all irrational numbers v with $v^2 + v < 2$. Find the least upper bound and greatest lower bound of S.

1.7 Sums and products of real numbers

The whole stupendous structure of mathematical analysis rests on postulates **1** to **18**. So does most of the new mathematics that is being discovered or invented now. Some mathematicians prefer to treat postulates **1** to **18** as **axioms for the real number system,** that is, as assumed statements whose logical consequences we are investigating. Others like to think of real numbers as entities constructed out of rational numbers, say, by forming nonterminating decimals. One must then define addition and multiplication of real numbers so as to satisfy postulates 1 to 18. We indicate very briefly *how* this can be done.

We assume that the rational numbers are known and that the real numbers and the relation $<$ between real numbers are defined as in Chapter 1, §2.4 and §3.1. Postulates 13, 14, and 17 follow without serious difficulties. Also, and this is most important, we can prove 18 (see §1.6).

We now use the least upper bound principle 18 in defining addition and multiplication.

Let α and β be positive real numbers. Let A be the set of all rational numbers a with $0 < a < \alpha$, B the set of all rational numbers b with $0 < b < \beta$, S the set of all rational numbers $a + b$ with a in A and b in B, and P the set of all rational numbers ab with a in A and b in B. These sets are bounded. Now we prove that if α and β are both rational, then the least upper bound of S is $\alpha + \beta$ and that of P is $\alpha\beta$. If α and β are not both rational, then we *define* the sum $\alpha + \beta$ and the product $\alpha\beta$ as the least upper bounds of S and P, respectively.

Next, we prove that $1 \cdot \alpha = \alpha$ for all α, where 1 is the real number $.999 \ldots = .\overline{9}$, that for $0 < \alpha < \beta$ there is a unique real number x with $\alpha + x = \beta$, and that for $0 < \alpha$, $0 < \beta$ there is a unique real number y with $\alpha y = \beta$. (We obtain x as the least upper bound of the set of all positive rational c such that $a + c < \beta$ for all a in A, and we obtain y as the least upper bound of the set of all positive rational d with $ad < \beta$ for all a in A.) It is now not too difficult to define the sum and product of two real numbers which need not be positive, and it is possible to verify all remaining postulates.

Assuming postulates 1 to 18, it is not difficult to prove that the sum and product of two real positive numbers, say,

$$\alpha = a_0 a_1 a_2 a_3 \ldots \qquad \text{and} \qquad \beta = b_0 b_1 b_2 b_3 \ldots,$$

can be computed, with any desired degree of accuracy, by "rounding off."

Let $m \geq 1$ be an integer, and set

$$\alpha_m = a_0 a_1 a_2 \ldots a_m, \qquad \beta_m = b_0 b_1 b_2 b_3 \ldots b_m.$$

These two numbers are rational; we know how to compute $\alpha_m + \beta_m$ and $\alpha_m \beta_m$. Now

$$|(\alpha + \beta) - (\alpha_m + \beta_m)| \leq |\alpha - \alpha_m| + |\beta - \beta_m|$$

$$\leq |\underbrace{.00 \ldots 0}_{m} a_{m+1} a_{m+2} \ldots| + |\underbrace{.00 \ldots 0}_{m} b_{m+1} b_{m+2} \ldots| \leq 2 \cdot 10^{-m-1}.$$

This will be as small as we like if m is large enough. Similarly,

$$|\alpha\beta - \alpha_m \beta_m| = |\alpha\beta - \alpha_m \beta + \alpha_m \beta - \alpha_m \beta_m|$$

$$\leq |\beta| \, |\alpha - \alpha_m| + |\alpha_m| \, |\beta - \beta_m|$$

$$\leq (b + 1)10^{-m-1} + (a + 1)10^{-m-1}$$

will be arbitrarily small for m sufficiently large.

1.8   Eudoxos' theory of proportions

It is instructive to compare real numbers with Eudoxos' theory of proportions used in Greek mathematics. The difficulty, resolved by Eudoxos, is best grasped by asking oneself: what is the geometric meaning of the statement

$$\frac{|PQ|}{|RS|} = \frac{|P'Q'|}{|R'S'|}$$

if the segments PQ and RS need not be commensurable (that is, if the ratio $|PQ|/|RS|$ is not a rational number)?

The key observation is as follows. Let m and n be positive integers. Whether PQ and RS are commensurable or not, the statement $\dfrac{|PQ|}{|RS|} > \dfrac{m}{n}$ has a simple geometric meaning: a segment n times as long as PQ is longer than a segment m times as long as RS; in symbols: $n|PQ| > m|RS|$. Now Eudoxos calls two ratios of lengths equal if there are *no* integers m and n such that one ratio is greater than m/n and the other ratio is not.

The same definition applies to ratios of other geometric quantities: areas, volumes, and angles. Greek mathematicians compared only ratios of like quantities. They would not say, for instance, that the ratio of the length of a circle to that of its diameter equals the ratio of the area of the circle to that of a square erected on the radius.

The reader will note how close Eudoxos' definition is to our use of real numbers. We "know" a real number x if we know for which rational numbers r the inequality $x > r$ is true; this is the information conveyed by the decimal expansion of x. Eudoxos "knew" the ratio of two geometric quantities α and β if he knew for which integers m and n the quantity $m\alpha$ exceeded $n\beta$.

1.9   Different viewpoints

Readers interested in the foundations of mathematics should know that some mathematicians (called *intuitionists* or *constructivists*) reject the theory of real numbers summarized in postulates **1** to **18**. They contend that a statement about a number x is meaningful only if it contains an actual recipe for calculating this number. The proof for the existence of a least upper bound for any bounded set, given in §1.6, does not satisfy this condition. For it contains no prescription for deciding, at each stage, whether a rational number A is or is not an upper bound for the set S. But, says a "classical" mathematician, one thing is obvious: either A is a bound or it is not! This time-honored "principle of the excluded middle," however, is not accepted by the intuitionists.

To develop calculus in a way satisfactory to constructivists is possible but more laborious. The constructivist school of thought numbered among its adherents some of the greatest mathematicians, for instance, Kronecker, Poincaré, Brouwer, and Weyl.

A completely different approach has been developed recently by a group of mathematicians led by the late A. Robinson. They create a system of real numbers in which there are also "infinitely small" and "infinitely large numbers," and the Archimedean postulate **17** is not valid for all numbers. In Robinson's *nonstandard analysis* some of the mathematical myths which we related in the text (see Chapter 4, §1.5, Chapter 8, §1.7, Chapter 20, §1.1) become rigorous statements.

§2 Convergent sequences

As an illustration of the power of the least upper bound principle, and as a preparation of a discussion of continuous function, we establish some important theorems about infinite sequences (see Chapter 15, §1.1 and §1.2). We shall return to this subject in §6 of the supplement.

2.1 Convergence of monotone sequences

Throughout this supplement the letters i, j, and k stand for nonnegative integers. We recall that "**for nearly all** j" means "for all but finitely many j," that is, "for all j exceeding some integer N."

We recall (see Chapter 15, §1.2) that if $\{x_i\} = x_1, x_2, x_3, \ldots$ is a sequence of real numbers, and x a number, the statement "$\{x_i\}$ **converges** to x" or, in symbols, "$\lim_{i \to \infty} x_i = x$" means that, for every $\epsilon > 0$, $|x - x_i| \leq \epsilon$ for nearly all i.

We also recall that the sequence $\alpha_1, \alpha_2, \alpha_3, \ldots$ is called **increasing** if $\alpha_1 < \alpha_2 < \alpha_3 < \cdots$, **nondecreasing** if $\alpha_1 \leq \alpha_2 \leq \alpha_3 \leq \cdots$. Similarly, $\{\alpha_i\}$ is called **decreasing** if $\alpha_i > \alpha_{i+1}$ for all i, **nonincreasing** if $\alpha_i \geq \alpha_{i+1}$ for all i. A sequence that is either nondecreasing or nonincreasing is called **monotone.**

A sequence $\{\alpha_i\}$ is called **bounded from above** if there is a number M such that $\alpha_i \leq M$ for all i, **bounded from below** if there is a number m such that $m \leq \alpha_i$ for all i. A **bounded** sequence is one bounded from both above and below.

Theorem A. *A bounded nondecreasing sequence $\alpha_1, \alpha_2, \alpha_3, \ldots$ has a limit; the limit is the smallest number that is not less than any α_i.*

This basic result contains Theorem 3 in Chapter 15, §1.4, stated there without proof.

Proof. Let α be the least upper bound of the set of numbers $\{\alpha_1, \alpha_2, \alpha_3, \ldots\}$. We show that $\lim_{i \to \infty} \alpha_i = \alpha$.

LEOPOLD KRONECKER (1823–1891), a distinguished algebraist, summarized his attitude toward the foundations of mathematics in the epigram: "The good Lord made positive integers, everything else is the handiwork of men." He severely criticized Cantor's theory of infinite sets.

HENRI POINCARÉ (1857–1912). For several decades Poincaré's restless intellect dominated mathematics and mathematical physics. He created, or transformed, several branches of mathematics, and he arrived at the equations of special relativity independently of Einstein. Poincaré also wrote several popular books about science that earned him a seat in the literary Académie Francaise.

L. E. J. BROUWER (1881–1967) did some of the fundamental work in topology, a modern branch of geometry. Then the Dutch mathematician turned to the foundations of mathematics and became the leader of the "intuitionists."

HERMANN WEYL (1885–1955) was one of the few truly universal mathematicians of our time; his interests ranged from philosophy to physics and comprised all of mathematics. One of his many famous books (*Symmetry*, Princeton, 1952) can even be enjoyed by the general reader.

Weyl left his native Germany after Hitler came to power and accepted a chair at the Institute for Advanced Study in Princeton, N. J.

ABRAHAM ROBINSON (1918–1974) began his career as an engineer. His early work was in aerodynamics. Later he turned to algebra, mathematical logic, and nonstandard analysis. Robinson received his Ph.D. in London; he worked at the Royal Aircraft Establishment and taught in Canada, Israel, and the United States.

Let $\epsilon > 0$. Assume that $\alpha_i \leq \alpha - \epsilon$ for infinitely many i. Then, for any given j, there is an k with $j < k$ and $\alpha_k \leq \alpha - \epsilon$; we may conclude that $\alpha_j \leq \alpha_i \leq \alpha - \epsilon$. Hence our assumption implies that $\alpha_j \leq \alpha - \epsilon$ for *all* j, and $\alpha - \epsilon$ is an upper bound for $\{\alpha_1, \alpha_2, \ldots\}$. That is absurd since α is the least upper bound. We conclude that $\alpha_i > \alpha - \epsilon$ for nearly all i. But $\alpha_i \leq \alpha$ for *all* i. Hence $|\alpha_i - \alpha| \leq \epsilon$ for nearly all i. Since ϵ was an arbitrary positive number, $\lim_{i \to \infty} \alpha_i = \alpha$.

Theorem A′. *A bounded nonincreasing sequence α_1, α_2, α_3, ... has a limit. The limit is the largest number that is not greater than any α_i.*

This can either be proved similarly or deduced from Theorem A.

PROBLEMS

1. Prove Theorem A′ *directly*. Modify the proof given for nondecreasing sequences.
2. Deduce Theorem A′ from Theorem A.

2.2   The Bolzano-Weierstrass theorem

If $\{x_i\} = x_1, x_2, x_3, \ldots$ is an infinite sequence, and $j_1 < j_2 < j_3 < \cdots$ is an infinite sequence of positive integers, then $\{x_{j_n}\} = x_{j_1}, x_{j_2}, x_{j_3}, \ldots$ is called a **subsequence** of $\{x_j\}$.

A bounded sequence that is not monotone need not converge. For instance, the sequence

$$\tfrac{1}{2}, \ -\tfrac{1}{2}\tfrac{2}{3}, \ -\tfrac{2}{3}\tfrac{3}{4}, \ -\tfrac{3}{4}\tfrac{4}{5}, \ -\tfrac{4}{5}\tfrac{5}{6}, \ldots$$

has no limit. But this sequence contains convergent subsequences, for instance, the sequence $\tfrac{1}{2}, \tfrac{2}{3}, \tfrac{3}{4}, \tfrac{4}{5}, \ldots$, which converges to 1, and the sequence $-\tfrac{1}{2}, -\tfrac{2}{3}, -\tfrac{3}{4}, -\tfrac{4}{5}, \ldots$, which converges to -1. This is no accident, as the following theorem shows.

Theorem B (Bolzano-Weierstrass). *From every bounded sequence of real numbers, one can select a convergent subsequence.*

Proof. Let $x_1, x_2, x_3, \ldots$ be a sequence and M a number such that $|x_j| \leq M$ for all j. Let S be the set of all numbers s with the property: there are infinitely many j with $s \leq x_j$. The set S is not empty—it contains $(-M)$. The set S is bounded from above —the number $M + 1$ is an upper bound. Let u be the least upper bound of S. We claim that, for every $\epsilon > 0$,

$$\text{there are infinitely many } j \text{ with } u - \epsilon \leq x_j \leq u + \epsilon. \tag{1}$$

Indeed, assume that, for some $\epsilon > 0$, (1) is false. Then, for this ϵ, $u - \epsilon \leq x_j \leq u + \epsilon$ for only finitely many j. Also, $u + \epsilon$ is not in S (since u is an upper bound for S) so that $u + \epsilon \leq x_j$ for only finitely many j. Therefore no point y with $y \geq u - \epsilon$ belongs to S. Hence $u - \epsilon$ is an upper bound for S, and u is not the least upper bound. This is a contradiction.

We proved (1). Applying this to $\epsilon = 1$, $\epsilon = \tfrac{1}{2}$, $\epsilon = \tfrac{1}{3}, \ldots$, we can find positive integers $j_1 < j_2 < j_3 < \cdots$ such that $|x_{j_1} - u| \leq 1$, $|x_{j_2} - u| \leq \tfrac{1}{2}$, $|x_{j_3} - u| \leq \tfrac{1}{3}$, and so forth. The sequence $x_{j_1}, x_{j_2}, x_{j_3}, \ldots$ (a subsequence of $x_1, x_2, x_3, \ldots$) converges to u.

2.3 Sequences of points‡

In this supplement the word "point" will usually denote a point in the plane. We use a Cartesian coordinate system, and we identify a point with its position vector $\mathbf{x}$, and with its coordinate pair (x,y). We restrict ourselves to the plane *only* to simplify writing. Everything remains valid in spaces of n dimensions, for all n.

A sequence of points $\mathbf{x}_1 = (x_1, y_1)$, $\mathbf{x}_2 = (x_2, y_2)$, $\mathbf{x}_3 = (x_3, y_3)$, ... is called **bounded** if the sequence of numbers $|\mathbf{x}_1|, |\mathbf{x}_2|, |\mathbf{x}_3|, \ldots$ is bounded. Since $|\mathbf{x}_j|^2 = x_j^2 + y_j^2$, we have: $|x_j| \leq |\mathbf{x}_j|, |y_j| \leq |\mathbf{x}_j|, |\mathbf{x}_j| \leq |x_j| + |y_j|$. We conclude that the sequence of points $\{\mathbf{x}_j\}$ is bounded if and only if *both* sequences of numbers, $\{x_j\}$ and $\{y_j\}$, are.

A sequence of points $\{\mathbf{x}_j\}$ is said to **converge** to $\mathbf{a} = (a,b)$ (or to have the **limit a**) if the sequence of numbers $\{|\mathbf{x}_j - \mathbf{a}|\}$ converges to 0. We conclude, as before, that $\{\mathbf{x}_j\}$ converges to $\mathbf{a}$ if and only if the *two* sequences of numbers, $\{x_j\}$ and $\{y_j\}$, converge to a and b, respectively.

Theorem C (*The Bolzano-Weierstrass theorem in the plane*). *Every bounded sequence of points in the plane contains a convergent subsequence.*

Proof. Let $\{\mathbf{x}_j\} = \{(x_j, y_j)\}$ be a bounded sequence. Select a subsequence for which the sequence of first coordinates converges—this is possible by Theorem B. From this subsequence select a subsequence for which the second coordinates also converge.

§3 Continuous functions

In this section we establish the basic properties of continuous functions which, in the main text, we have accepted as geometrically obvious. We recall (see Chapter 3, §3.2) that a function $f(x)$, defined at $x = x_0$, is said to be **continuous** at a point x_0 if, for every $\epsilon > 0$, there is a $\delta > 0$ such that, for every x with $|x - x_0| < \delta$, for which $f(x)$ is defined, we have $|f(x) - f(x_0)| < \epsilon$. An equivalent statement is: for every $\epsilon > 0$, $|f(x) - f(x_0)| < \epsilon$ for all x near x_0 (at which f is defined). "**Near** x_0" means: in some interval with x_0 as midpoint.

Similar definitions are given for functions of several variables (see Chapter 19, §1.3).

3.1 Continuity of sums, differences, products, and quotients

We proceed to prove Theorem 1 from Chapter 3, §3.3. The theorem asserts that *if two functions, f and g, are continuous at a point x_0, so are $f + g$, $f - g$, fg, and, assuming that $g(x_0) \neq 0$, also f/g.*

The proof does *not* involve the least upper bound principle. The theorem and the proof extend at once to functions of two (or more) variables.

The proof is divided into several steps. Throughout the proof, ϵ denotes a given positive number.

(i) Let f be a function continuous at x_0, and let c be a constant. Then the function cf is continuous at x_0.

‡This subsection will be used only in §7.

Proof. If $c = 0$, the function $cf(x)$ is the constant 0 and hence continuous everywhere (see Example 2 in Chapter 3, §3.2). We assume now that $c \neq 0$. Since f is continuous at x_0, there is a δ such that $|f(x) - f(x_0)| < |c|^{-1}\epsilon$ for $|x - x_0| < \delta$. For this δ, $|cf(x) - cf(x_0)| = |c|\,|f(x) - f(x_0)| < \epsilon$ provided that $|x - x_0| < \delta$.

(ii). If f and g are continuous at x_0, so is $f + g$.

Proof. Since f is continuous at x_0, there is a δ_1 such that $|f(x) - f(x_0)| < \epsilon/2$ for $|x - x_0| < \delta_1$. Similarly, since g is continuous at x_0, there is a δ_2 such that $|g(x) - g(x_0)| < \epsilon/2$ for $|x - x_0| < \delta_2$. Let δ be the smaller of the two numbers δ_1 and δ_2. Then, for $|x - x_0| < \delta$, we have that

$$|f(x) + g(x) - f(x_0) - g(x_0)| \leq |f(x) - f(x_0)| + |g(x) - g(x_0)| < \frac{\epsilon}{2} + \frac{\epsilon}{2} = \epsilon.$$

(iii). If f and g are continuous at x_0, so is $f - g$.

Proof. The function $-g = (-1)g$ is continuous at x_0, by (i). The function $f - g = f + (-g)$ is continuous at x_0, by (ii).

(iv). Let f and g be continuous at x_0, and assume that $f(x_0) = g(x_0) = 0$. Then fg is continuous at x_0.

Proof. There is a $\delta_1 > 0$ such that $|f(x)| = |f(x) - f(x_0)| < \sqrt{\epsilon}$ for $|x - x_0| < \delta_1$, and a $\delta_2 > 0$ such that $|g(x)| < \sqrt{\epsilon}$ for $|x - x_0| < \delta_2$. If $|x - x_0| < \delta$, the smaller of the numbers δ_1 and δ_2, then $|f(x)g(x) - f(x_0)g(x_0)| = |f(x)g(x)| < \sqrt{\epsilon}\sqrt{\epsilon} = \epsilon$.

(v). If f and g are continuous at x_0, so is fg.

Proof. Set $f_1(x) = f(x) - f(x_0)$, $g_1(x) = g(x) - g(x_0)$, and $\phi = f_1 g_1$. We have that $f(x)g(x) = \phi(x) + f(x_0)g(x) + g(x_0)f(x) - f(x_0)g(x_0)$. Now f_1 and g_1 are continuous at x_0, by (iii). So are the functions $f(x_0)g(x)$ and $g(x_0)f(x)$, by (i). So is the function $\phi(x)$, by (iv). So is, finally, the function fg by (ii), since it is a sum of continuous functions.

(vi). Assume that f is continuous at x_0 and $f(x_0) = 1$. Then $1/f$ is continuous at x_0.

Proof. There is a $\delta_1 > 0$ such that $|f(x) - 1| < \frac{1}{2}$ for $|x - x_0| < \delta_1$ and a δ_2 such that $|f(x) - 1| < \epsilon/2$ for $|x - x_0| < \delta_2$. If $|x - x_0| < \delta$, the smaller of the numbers δ_1 and δ_2, then $|f(x)| > \frac{1}{2}$, and therefore

$$\left|\frac{1}{f(x)} - 1\right| = \frac{|1 - f(x)|}{|f(x)|} < \frac{\frac{1}{2}\epsilon}{\frac{1}{2}} = \epsilon.$$

(vii). If f is continuous at x_0 and $f(x_0) \neq 0$, then $1/f$ is continuous at x_0.

Proof. The function $[1/f(x_0)]f(x)$ is continuous at x_0 by (i), and its value at x_0 is 1. Hence its reciprocal, the function $f(x_0)/f(x)$, is continuous at x_0 by (vi), and so is the function $1/f(x) = [1/f(x_0)][f(x_0)/f(x)]$, again by (i).

(viii). Let f and g be continuous at x_0, and $g(x_0) \neq 0$. Then f/g is continuous at x_0.

Proof. The function $1/g$ is continuous at x_0, by (vii). So is $f/g = (1/g)f$, by (v).

The desired theorem follows from steps (ii), (iii), (v), and (viii).

3.2   The intermediate value theorem

The theorem (Theorem 4 in Chapter 3, §3.4) reads as follows.

Let $f(x)$ be defined and continuous for $a \leq x \leq b$. Set $\alpha = f(a)$, $\beta = f(b)$, and let $\alpha \neq \beta$. Let γ be a number such that either $\alpha < \gamma < \beta$ or $\beta < \gamma < \alpha$. Then there is a number c in the interval (a,b) with $f(c) = \gamma$.

The proof involves directly the least upper bound principle (postulate **18**).

We assume that $\alpha < \gamma < \beta$. Let S denote the set of all numbers x such that $a \leq x \leq b$ and $f(x) < \gamma$. The set S is not empty, since it contains the number a; it is bounded from above, because b is an upper bound for S. Hence there is a number, call it c, which is the least upper bound for S. This c is the smallest number with the property: if $f(x) < \gamma$, then $x \leq c$. Clearly $a \leq c$ (for a belongs to S) and $c \leq b$ (for b is an upper bound for S).

Three cases are conceivable: $f(c) < \gamma$, $f(c) > \gamma$, and $f(c) = \gamma$. We shall prove that the first two cases lead to contradictions. Therefore the third must take place, so that $f(c) = \gamma$ and the theorem is proved.

Assume that $f(c) < \gamma$. Then $f(c) < \beta$; hence $c \neq b$ and $c < b$. Since f is continuous at c, we have $f(x) < \gamma$ near c (by continuity). Therefore there are points to the right of c where $f < \gamma$, but all such points belong to S. Since c is an upper bound for S, such points cannot lie to the right of c. This is a contradiction.

Assume next that $f(c) > \gamma$. Thus $f(c) > \alpha$; hence $c \neq a$ and $c > a$. Since f is continuous at c, we have $f(x) > \gamma$ near c (by continuity). This means that there is a number c_1 such that $a < c_1 < c$ and $f(x) > \gamma$ for every x with $c_1 < x \leq c$. Since no point of S (the set where $f < \gamma$) lies to the right of c, no point of S lies to the right of c_1. Thus c_1 is an upper bound for S and is smaller than c. But c was the smallest upper bound for S. This is a contradiction.

REMARK   In Chapter 3, §3.4, we concluded that the intermediate value theorem held by appealing to geometric intuition. The graph of a continuous function consists of one piece; if it has a point below a line $y = c$ and a point above this line, there must be a point on the graph at which the graph crosses the line $y = c$. What can be simpler than this argument? Why complicate matters by appealing to the least upper bound principle and by giving a rather involved indirect proof?

The same question can be asked about many "geometrically obvious" theorems for which mathematicians nevertheless give analytic proofs.

The answer is: our geometric intuition is an invaluable guide, but, unfortunately, not an infallible one. The analytic proofs do not replace intuition; they reinforce it. Only our intuition and imagination, applied to particular cases, can suggest general theorems. Only a rigorous proof can assure us that intuition did not lead us astray.

3.3   Continuity of inverse functions

Let $y = f(x)$ be a continuous increasing function defined for $a \leq x \leq b$, and set $\alpha = f(a), \beta = f(b)$. Let $x = g(y), \alpha \leq y \leq \beta$ be the inverse function, so that $f(g(y)) = y$, $g(f(x)) = x$. Then g is continuous.

This is part (2) of Theorem 1 in Chapter 6, §3.4. We recall that the existence of the inverse function [part (1) of that theorem] follows at once from the intermediate value theorem. We also note that g is increasing. Indeed, if $y_1 < y_2$, and if we had $g(y_1) \geq g(y_2)$, then we would also have $f(g(y_1)) \geq f(g(y_2))$ or $y_1 \geq y_2$, which is absurd.

FIGURE S.1

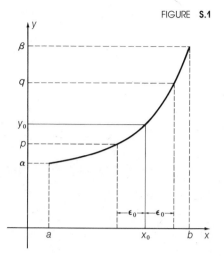

Proof.   Let y_0 be a number such that $\alpha < y_0 < \beta$. We shall prove that g is continuous at y_0. Since f takes on all values between α and β, there is a number x_0 such that $f(x_0) = y_0$. Let $\epsilon > 0$ be given. We choose an $\epsilon_0 > 0$ such that $\epsilon_0 < \epsilon$ and $a < x_0 - \epsilon_0 < x_0 + \epsilon_0 < b$. Set $f(x_0 - \epsilon_0) = p$, $f(x_0 + \epsilon_0) = q$. Since f is increasing, $\alpha < p < y_0 < q < \beta$; see Figure S.1. Since g is increasing we have that, for every y with $p < y < q$, $g(y)$ lies between $x_0 - \epsilon_0$ and $x_0 + \epsilon_0$. For such a y, therefore, $|g(y) - g(y_0)| < \epsilon_0 < \epsilon$. Now let $\delta > 0$ be the smaller of the two numbers $q - y_0$, $y_0 - p$. If $|y - y_0| < \delta$, we have $|g(y) - g(y_0)| < \epsilon$. The existence of such a δ for every $\epsilon > 0$ proves the continuity of g at y_0.

It remains to show that g is continuous at $y = \alpha$ and $y = \beta$. The proof is similar to the one just given and is left to the reader.

PROBLEMS

1. Under the hypotheses of the theorem considered here, prove that g is continuous at $y = \alpha$.
2. Under the same hypotheses, show that g is continuous at $y = \beta$.

3.4   Derivatives of inverse functions

We give now an analytic proof of Theorem 2, in Chapter 5, §1.2, about derivatives of inverse functions. (The theorem and proof remain valid at the endpoints of the interval considered, for one-sided derivatives.)

We consider two continuous increasing (or decreasing) functions $y = f(x)$ and $x = g(y)$ inverse to each other, so that $f(g(y)) = y$, $g(f(x)) = x$. Assume that $f'(x_0)$ exists and $f'(x_0) \neq 0$. Let $y_0 = f(x_0)$. We shall show that g has a derivative at y_0 and that $g'(y_0) = 1/f'(x_0)$.

Set

$$k(h) = g(y_0 + h) - g(y_0) = g(y_0 + h) - x_0.$$

Then

$$y_0 + h = f(x_0 + k(h)),$$

$k(h)$ is a continuous function, $k(h) \neq 0$ for $h \neq 0$, and $k(0) = 0$. We can verify that

$$\lim_{h \to 0} \frac{f(x_0 + k(h)) - f(x_0)}{k(h)} = f'(x_0), \tag{1}$$

so that

$$\frac{1}{f'(x_0)} = \lim_{h \to 0} \frac{k(h)}{f(x_0 + k(h)) - f(x_0)} = \lim_{h \to 0} \frac{g(y_0 + h) - g(y_0)}{(y_0 + h) - y_0}$$

$$= \lim_{h \to 0} \frac{g(y_0 + h) - g(y_0)}{h} = g'(y_0).$$

In Chapter 6, §1.2, we obtained the same result geometrically.

PROBLEMS

1. Verify statement (1).
2. Carry out the proof of the extension of the theorem mentioned in the parenthetical remark in the beginning of this subsection.

3.5 The maximum theorem

We first establish

Lemma 1. *Let $f(x)$ be defined and continuous in a finite closed interval $[a,b]$, that is, for $a \leq x \leq b$. Then f is bounded.*

Before proving this, we note that the conditions on the interval are essential. The function $f(x) = x$ is continuous in the infinite interval $-\infty < x < +\infty$ but not bounded in this interval. The function $f(x) = 1/x$ is continuous but not bounded in the open finite interval $0 < x < 1$.

Proof.   Assume that f is continuous but unbounded in $[a,b]$. Then, for every $j = 1$, $2, 3, \ldots$, there is an x_j with $a \leq x_j \leq b$, $|f(x_j)| > j$. By Theorem B in §2, there is a convergent subsequence $x_{j_1}, x_{j_2}, x_{j_3}, \ldots$. Set $\xi = \lim_{i \to \infty} x_{j_i}$. Then $a \leq \xi \leq b$. Since f is continuous, there is a $\delta > 0$ such that $|f(x) - f(\xi)| < 1$ if $|x - \xi| < \delta$. But $|x_{j_i} - \xi| < \delta$ for nearly all i. We conclude that $|f(x_{j_i}) - f(\xi)| < 1$ for nearly all i; hence $|f(x_{j_i})| \leq 1 + |f(\xi)|$ for nearly all i. Since $j < |f(x_j)|$ for all j, we conclude that $j_i \leq 1 + |f(\xi)|$ for nearly all i. This is absurd; hence the assumption that f is unbounded is untenable.

Now we can prove Theorem 1 in Chapter 5, §1.1. It will suffice to prove the following statement.

A continuous function $f(x)$ defined on a finite closed interval $[a,b]$ has a maximum.

Proof.   By Lemma 1, the set of values taken on by f in $[a,b]$ is bounded; let M be the least upper bound of this set. This means: $M \geq f(x)$ for $a \leq x \leq b$, and M is the smallest number with this property. We must show that there is a ξ, $a \leq \xi \leq b$, with $f(\xi) = M$. Assume there is no such ξ. Then $M - f(x) > 0$ for $a \leq x \leq b$; hence $g(x) = 1/[M - f(x)]$ is continuous for $a \leq x \leq b$, Lemma 1 applies, and there is a positive number A such that

$$g(x) = \frac{1}{M - f(x)} \leq A, \quad \text{for all } x, \quad a \leq x \leq b.$$

This implies that $M - f(x) \geq 1/A$ or $f(x) \leq M - (1/A)$ in $[a,b]$. But this contradicts the definition of M.

The existence of a minimum is proved similarly.

We recall (see Chapter 14, §1.1) that the maximum theorem implies Rolle's theorem, the mean value theorem, and hence also the theorems about functions with nonnegative and positive derivatives and the theorem that a function with a derivative 0 at every point is a constant (see Chapter 14, §1.4).

PROBLEMS

1. Prove *directly* that a function $f(x)$, continuous on a finite closed interval, has a minimum.
2. Deduce the statement in Problem 1 from the statement proved in §3.5 about the existence of a maximum.
3. Prove that if $f(x)$ is nondecreasing for $a < x < b$, and $f'(x)$ exists for all x, $a < x < b$, then $f'(x) \geq 0$. (This is the converse of Theorem 2 in Chapter 5, §2.2.)

3.6   Uniform continuity‡

A function $x \mapsto f(x)$ defined on an interval is called **uniformly continuous** in this interval if, given any positive number $\epsilon > 0$, one can find a positive number δ such that: if x_1 and x_2 are any two points in the interval and $|x_1 - x_2| < \delta$, then $|f(x_1) - f(x_2)| < \epsilon$.

A uniformly continuous function is always continuous. The converse need not be true. For instance, the function $f(x) = 1/x$ is continuous, but not uniformly continuous, in the interval $0 < x < 1$. Indeed, let n be an integer > 3, and set $x_1 = 1/n$, $x_2 = 2/n$. Then $0 < x_1 < 1$, $0 < x_2 < 1$, $|x_1 - x_2| = 1/n$, and $|f(x_1) - f(x_2)| = n/2$. Choosing n large enough, we can make $|x_1 - x_2|$ as small and $|f(x_1) - f(x_2)|$ as large as we like.

Theorem A. *Let $f(x)$ be defined and continuous in a closed finite interval $[a,b]$. Then $f(x)$ is uniformly continuous in this interval.*

Proof.   Assume f is not uniformly continuous. Then there is a number $\epsilon > 0$ such that for no $\delta > 0$ is it true that: if $|x_1 - x_2| < \delta$, then $|f(x_1) - f(x_2)| < \epsilon$. In other words: for this $\epsilon > 0$ and for every $\delta > 0$, we can find points x and $\hat{x}$ in our interval such that $|x - \hat{x}| < \delta$ and $|f(x) - f(\hat{x})| \geq \epsilon$. We do this for a sequence of positive numbers δ_1, δ_2, δ_3, ... converging to 0 and obtain two sequences of numbers x_j, $\hat{x}_j$ $[a,b]$ with

$$a \leq x_j \leq b, \qquad a \leq \hat{x}_j \leq b, \qquad \lim_{j \to \infty} |x_j - \hat{x}_j| = 0, \qquad |f(x_j) - f(\hat{x}_j)| \geq \epsilon,$$

for all j. Selecting, if need be, a subsequence, we may assume that the sequence x_j converges (by Theorem B in §2). Set $\xi = \lim_{j \to \infty} x_j$. Then $a \leq \xi \leq b$. Also $|\hat{x}_j - \xi| = |\hat{x}_j - x_j + x_j - \xi| \leq |\hat{x}_j - x_j| + |x_j - \xi|$ so that $\lim_{j \to \infty} \hat{x}_j = \xi$. But f is continuous at ξ, and if j is large enough, x_j and $\hat{x}_j$ are as close to ξ as we like. Hence if j is large enough, then $|f(x_j) - f(\xi)| < \epsilon/3$ and $|f(\hat{x}_j) - f(\xi)| < \epsilon/3$. Therefore

$$|f(x_j) - f(\hat{x}_j)| = |f(x_j) - f(\xi)| + |f(\xi) - f(\hat{x}_j)| \leq \frac{\epsilon}{3} + \frac{\epsilon}{3} < \epsilon.$$

This contradicts the way the numbers x_j and $\hat{x}_j$ have been selected. The assumption that $f(x)$ is not uniformly continuous is untenable.

‡This subsection will be used only in §7.

PROBLEMS

1. Prove *directly* that $f(x) = x^2$ is uniformly continuous for $-2 \leq x \leq 2$.
2. Prove that $f(x) = \sin x$ is uniformly continuous for $-\infty < x < +\infty$. [*Hint:* Write $\sin x_1 - \sin x_2$ as an integral.]
3. Prove that if a function $f(x)$ has in an interval the derivative $f'(x)$, and if $|f'(x)| \leq 1$ in the interval, then $f(x)$ is uniformly continuous.

3.7   Compact sets‡

We want to extend some of the theorems proved above to functions of two (or more) variables. It is advisable to consider functions defined not only on intervals but also on more general sets.

We recall that a set S of points (in the plane) is called **bounded** if it lies in some (two-dimensional) interval, that is, if the set of all numbers $|x|$, x in S, is bounded.

A set S of points (in the plane) is called **closed** if the following condition holds: if the sequence $\{x_j\}$ converges to x, and all x_j belong to S, then x also belongs to S.

A set S of points (in the plane) is called **compact** if, whenever $\{x_j\}$ is a sequence such that all x_j belong to S, one can find a subsequence $\{x_{j_k}\}$ which converges to a point x, also in S.

Theorem B.   *A set S of points (in the plane) is compact if and only if S is both bounded and closed.*

Proof.   If S is not bounded, then there is *no* number M such that $|x| \leq M$ for all x in S. Hence there are points x_1, x_2, x_3, ... in S with $|x_j| > j$. The sequence $\{x_j\}$ has no convergent subsequence. Thus S is not compact.

If S is not closed, there is a sequence $\{x_j\}$ of points in S that converges to a point x not in S. No subsequence of $\{x_j\}$ can converge to a point in S. Thus S is not compact.

If S is bounded, then every sequence of points in S is bounded and hence contains a convergent subsequence (by Theorem C in §2). If S is also closed, the limit of this subsequence is a point in S. Hence S is compact.

The definitions, and the theorem, extend to sets on the line, in ordinary space, or in n-space, for all n.

EXAMPLES 1.   Show that a closed two-dimensional interval, that is, the set of all $x = (x,y)$ with $a \leq x \leq b$, $c \leq y \leq d$, is compact.

ANSWER   The interval is clearly bounded. If x_j belongs to the interval for all j, and $\{x_j\}$ converges to x, then we have $a \leq x_j \leq b$, $c \leq y_i \leq d$ so that the limit $x = \lim_{j \to \infty} x_j$ and $y = \lim_{j \to \infty} y_j$ satisfy $a \leq x \leq b$, $c \leq y \leq d$. Hence $x = (x,y)$ belongs to the interval. The set is closed.

2.   Show that if $S_1, S_2, \ldots, S_n$ are finitely many compact sets, and S their union (the set consisting of all points belonging to any one of the sets $S_1, S_2, \ldots, S_n$), then S is compact.

ANSWER   Let $\{x_j\}$ be an infinite sequence of points of S. Then an infinite subsequence must belong to one of the sets $S_1, S_2, \ldots, S_n$, say, to S_1. Since S_1 is compact, a subsequence of this subsequence converges to a point of S_1, that is, to a point of S.

The definition of maximum value, minimum value, and uniform continuity extend at once to functions of several variables defined on some set.

‡This subsection will be used only in §7.

Theorem C. *If $f(x,y) = f(\mathbf{x})$ is a continuous function defined on a compact set S in the plane, then f is bounded on S, attains on S its maximum and minimum values, and is uniformly continuous on S.*

This is proved exactly as Lemma 1, the maximum theorem, and Theorem A in §3.5 and §3.6. A theorem analogous to Theorem C holds for functions of n variables, for any n.

PROBLEMS

1. Prove that the union of finitely many closed sets is closed.
2. Prove that the intersection of any number of closed sets is closed. (The intersection of sets $S_1, S_2, \ldots$ is the set of points that belong to each of the sets $S_1, S_2, \ldots$.)
3. Prove that the intersection of any number of compact sets is compact.
4. Prove directly that $f(x,y) = xy + x^2$ is uniformly continuous for $0 \leq x \leq 1$, $0 \leq y \leq 1$.

§4 Integrals

We proceed to establish those theorems about integrals (of functions of one variable) which have been accepted without proof in the main text.

4.1 Upper and lower integrals

We recall the definition of a step function and of the integral of a step function (Chapter 8, §§1.5 and 1.6). For step functions f and g, it is quite obvious that

$$\int_a^b f(x)\,dx = \alpha(b-a), \qquad \text{if } f(x) = \alpha, \quad \text{a constant} \tag{1}$$

$$\int_a^b f(x)\,dx = \int_a^c f(x)\,dx + \int_c^b f(x)\,dx, \qquad \text{if } a < c < b \tag{2}$$

$$\int_a^b f(x)\,dx \leq \int_a^b g(x)\,dx, \qquad \text{if } f \leq g, \quad a < b. \tag{3}$$

Here (a,b) is a finite interval, and $f \leq g$ means that $f(x) \leq g(x)$ for all x, $a < x < b$.

Now let $f(x)$, $a < x < b$, be any bounded function. Then there is a number M such that $-M \leq f \leq M$. The set of numbers $\int_a^b \phi(x)\,dx$, ϕ a step function such that $\phi \leq f$, is not empty [it contains the number $\int_a^b (-M)\,dx = -M(b-a)$] and is bounded from above [the number $\int_a^b M\,dx = M(b-a)$ is an upper bound]. We define the **lower integral** of f as

$$\int_a^b f(x)\,dx = \text{least upper bound of } \int_a^b \phi(x)\,dx, \phi \text{ a step function}, \qquad \phi \leq f. \tag{4}$$

The **upper integral** is defined similarly:

$$\overline{\int_a^b} f(x)\,dx = \text{greatest lower bound of } \int_a^b \psi(x)\,dx, \psi \text{ a step function}, \qquad f \leq \psi. \tag{5}$$

Lemma 1.   *For bounded functions f and g,*

$$\underline{\int_{a}^{b}} f(x)\, dx = \alpha, \qquad \text{if } f(x) = \alpha, \quad \text{a constant} \tag{6}$$

$$\underline{\int_{a}^{b}} f(x)\, dx = \underline{\int_{a}^{c}} f(x)\, dx + \underline{\int_{c}^{b}} f(x)\, dx, \qquad \text{if } a < c < b, \tag{7}$$

$$\underline{\int_{a}^{b}} f(x)\, dx \le \underline{\int_{a}^{b}} g(x)\, dx, \qquad \text{if } f \le g, \quad a < b. \tag{8}$$

Proof.   We verify easily that

$$\int_{a}^{b} f(x)\, dx = \underline{\int_{a}^{b}} f(x)\, dx = \overline{\int_{a}^{b}} f(x)\, dx \text{ for a step function } f. \tag{9}$$

Hence (6) follows from (1).

Next, given a bounded function f and an $\epsilon > 0$, there is a step function ϕ_0 such that

$$\phi_0 \le f \qquad \text{and} \qquad \underline{\int_{a}^{b}} f(x)\, dx - \epsilon \le \int_{a}^{b} \phi_0(x)\, dx \le \underline{\int_{a}^{b}} f(x)\, dx. \tag{10}$$

Indeed, if for some positive ϵ there would be no such ϕ, then $\int_a^b f\, dx - \epsilon$ would be an upper bound for all numbers $\int_a^b \phi(x)\, dx$, ϕ a step function with $\phi \le f$. This would contradict the definition (5) of the lower integral.

Now we establish (8). Let $\epsilon > 0$ be given. By the remark just made, there are step functions ϕ_1 and ϕ_2, defined in (a,c) and (c,b), respectively, such that $\phi_1(x) \le f(x)$ for $a < x < c$, $\phi_2(x) \le f(x)$ for $c < x < b$, and

$$\underline{\int_{a}^{c}} f(x)\, dx \le \int_{a}^{c} \phi_1(x)\, dx + \epsilon, \qquad \underline{\int_{c}^{b}} f(x)\, dx \le \int_{c}^{b} \phi_2(x)\, dx + \epsilon.$$

Define the step function $\phi(x)$ by $\phi(x) = \phi_1(x)$ for $a < x < c$, $\phi(x) = \phi_2(x)$ for $c < x < b$. Then ϕ is a step function defined in (a,b) and $\phi \le f$. Using the above inequalities, relation (2) for step functions, and the definition of the lower integral, we have

$$\underline{\int_{a}^{c}} f(x)\, dx + \underline{\int_{c}^{b}} f(x)\, dx \le \int_{a}^{c} \phi_1(x)\, dx + \int_{c}^{b} \phi_2(x)\, dx + 2\epsilon$$

$$= \int_{a}^{c} \phi(x)\, dx + \int_{c}^{b} \phi(x)\, dx + 2\epsilon$$

$$= \int_{a}^{b} \phi(x)\, dx + 2\epsilon \le \underline{\int_{a}^{b}} f(x)\, dx + 2\epsilon.$$

Since ϵ could be any positive number, this means that

$$\underline{\int_{a}^{c}} f(x)\, dx + \underline{\int_{c}^{b}} f(x)\, dx \le \underline{\int_{a}^{b}} f(x)\, dx. \tag{11}$$

On the other hand, there is a step function $\phi_3 \le f$ with

$$\underline{\int_{a}^{b}} f(x)\, dx \le \int_{a}^{b} \phi_3(x)\, dx + \epsilon.$$

Using (2), we find that this becomes

$$\int_{-a}^{b} f(x)\, dx \leq \int_{a}^{c} \phi_3(x)\, dx + \int_{c}^{b} \phi_3(x)\, dx + \epsilon \leq \int_{-a}^{c} f(x)\, dx + \int_{-c}^{b} f(x)\, dx + \epsilon,$$

and, since ϵ is arbitrary,

$$\int_{-a}^{b} f(x)\, dx \leq \int_{-a}^{c} f(x)\, dx + \int_{-c}^{b} f(x)\, dx. \tag{12}$$

Together with the reverse inequality (11) already established, this proves statement (7).

To prove (8), let ϵ be again a given positive number. There is a step function $\phi \leq f$ with

$$\int_{-a}^{b} f(x)\, dx \leq \int_{a}^{b} \phi(x)\, dx + \epsilon.$$

But since $f \leq g$, we have $\phi \leq g$, and hence

$$\int_{a}^{b} \phi(x)\, dx \leq \int_{-a}^{b} g(x)\, dx,$$

so that

$$\int_{-a}^{b} f(x)\, dx \leq \int_{-a}^{b} g(x)\, dx + \epsilon.$$

Since ϵ is arbitrary, we get the desired inequality (8).

One can, in the same way, establish statements analogous to (6), (7), and (8) for upper integrals.

PROBLEMS

1. Prove statement (9).
2. Prove (6), (7), and (8) for upper integrals.

4.2   Proof of the basic theorems

Theorem 1 in Chapter 8, §1.6, asserts that if f is bounded and piecewise continuous, then there is exactly one number A such that $\int_a^b \phi\, dx \leq A \leq \int_a^b \psi\, dx$ for all step functions ϕ and ψ with $\phi \leq f \leq \psi$. This statement is equivalent to the equality

$$\int_{-a}^{b} f(x)\, dx = \int_{a}^{\bar{b}} f(x)\, dx, \qquad \text{if } f \text{ is bounded and piecewise continuous.} \tag{13}$$

We shall prove this here.

We recall the proof, given in Chapter 8, §5, of the first part of the fundamental theorem of calculus. The proof used *only* the formula for the integral of a constant and the additivity and monotonicity properties of the integral. In view of Lemma 1 and the corresponding statement for upper integrals, we can repeat the proof *verbatim* and conclude that the functions $\underline{G}(x)$ and $\bar{G}(x)$ defined by

$$\underline{G}(a) = \bar{G}(a) = 0, \quad \underline{G}(x) = \int_{-a}^{x} f(t)\, dt, \quad \bar{G}(x) = \int_{a}^{\bar{x}} f(t)\, dt, \quad \text{for } a < t < b \tag{14}$$

are continuous for $a \leq x \leq b$, and at every point x $(a < x < b)$ at which f is continuous, we have $\underline{G}'(x) = \bar{G}'(x) = f(x)$.

Assume now that f is piecewise continuous, and set $H(x) = \overline{G}(x) - \underline{G}(x)$. Then $H(x)$ is continuous for $a \le x \le b$, $H(a) = 0$ and

$$H'(x) = \overline{G}'(x) - \underline{G}'(x) = f(x) - f(x) = 0$$

for all x, $a < x < b$, except perhaps the finitely many values x at which f fails to be continuous. Hence $H(x)$ is constant (by Theorem 1 in Chapter 6, §4.1; see also Chapter 14, §1.4). Thus $H(b) = H(a) = 0$ or $\overline{G}(b) = \underline{G}(b)$, which is the same as (13).

The common value of $\underline{\int_a^b} f \, dx$ and $\overline{\int_a^b} f \, dx$ is called the Riemann integral of f and is denoted by $\int_a^b f(x) \, dx$. It now follows from (13), (6), (7), and (8) that statements (1), (2), and (3) hold for all bounded, piecewise continuous functions f and g. This is Theorem 2 in Chapter 8, §1.6.

REMARK   The monotonicity property (3) of integrals implies that

$$\left| \int_a^b f(x) \, dx \right| \le \int_a^b |f(x)| \, dx \le M(b - a)$$

provided that

$$|f(x)| \le M, \qquad \text{for } a \le x \le b.$$

The proof is left to the reader.

PROBLEMS

1. Carry out the proof of the analog of the fundamental theorem of calculus for the functions defined by (14).
2. Prove the statement in the Remark.

4.3   Riemann integrable functions

A bounded function $f(x)$ such that $\underline{\int_a^b} f(x) \, dx = \overline{\int_a^b} f(x) \, dx$ is called **Riemann integrable.** A necessary and sufficient condition is that, for every $\epsilon > 0$, there should exist step functions ϕ and ψ with $\phi \le f \le \psi$ and $\int_a^b \psi \, dx - \int_a^b \phi \, dx < \epsilon$.

The argument in Chapter 8, §1.5, shows that *all bounded monotone functions are Riemann integrable.* Since a monotone function defined in a finite interval can have infinitely many discontinuity points, there are Riemann integrable functions which are not piecewise continuous.

An example of a function which is *not* Riemann integrable is $f(x) = 0$ for x rational, $f(x) = 1$ for x irrational. For this f, $\underline{\int_0^1} f(x) \, dx = 0$, $\overline{\int_0^1} f(x) \, dx = 1$.

PROBLEMS

1. Give an example of a monotone function $f(x)$ defined for $0 < x < 1$ which is not piecewise continuous. [*Hint:* Define f as constant in the intervals $(\frac{1}{2}, 0)$, $(\frac{1}{4}, \frac{1}{2})$, $(\frac{1}{8}, \frac{1}{4})$, $(\frac{1}{16}, \frac{1}{8})$, ... and so forth.]
2. Verify the statement made in §4.3 about the necessary and sufficient conditions for Riemann integrability.
3. Verify the statement made in §4.3 about the function $f(x) = 0$ for x rational, $f(x) = 1$ for x irrational.

4.4 The truncation error in numerical integration

In this subsection, we prove the estimates for the error in the trapezoidal rule and in Simpson's rule stated in §1 of Chapter 13.

Let $f(x)$ be a continuous function defined in the finite interval $[a,b]$, and let $\phi(x)$ be a *linear* function that coincides with $f(x)$ at the endpoints; that is, it satisfies the conditions $\phi(a) = f(a)$, $\phi(b) = f(b)$. We set

$$E^* = \int_a^b f(t)\,dt - \int_a^b \phi(t)\,dt$$

and want to estimate E^*, assuming that $f(x)$ has a continuous second derivative $f''(x)$ in $[a,b]$, and

$$|f''(x)| \leq M_2. \tag{15}$$

To achieve this, we set

$$g(x) = \int_a^x f(t)\,dt - \int_a^x \phi(t)\,dt + \left\{ 2\left(\frac{x-a}{b-a}\right)^3 - 3\left(\frac{x-a}{b-a}\right)^2 \right\} E^*.$$

We have $g(a) = 0$, $g(b) = 0$, so that by Rolle's theorem (see Chapter 14, §1.1) there is a point c with $a < c < b$ and $g'(c) = 0$. We compute that $g'(a) = g'(b) = 0$. Hence, again by Rolle's theorem, there are points c_1 and c_2 such that $a < c_1 < c$, $c < c_2 < b$, and $g''(c_1) = 0$, $g''(c_2) = 0$. We conclude, once more by Rolle, that there is a point y between c_1 and c_2 with $g'''(y) = 0$. Thus, since $\phi''(x) = 0$ for all x, we have

$$0 = g'''(y) = f''(y) + \frac{12}{(b-a)^3} E^*.$$

Noting (15), we conclude that

$$|E^*| \leq \frac{(b-a)^3}{12} M_2. \tag{16}$$

Suppose next that we divide the interval $[a,b]$ into n equal parts, each of length h, and use the trapezoidal rule to obtain an approximate value for the integral $\int_a^b f(t)\,dt$. This means that, in each subinterval, we replace f by a linear function that coincides with f at the endpoints, compute the integral of this linear function over the subinterval, and add the n numbers so obtained. According to the result established above, the error committed in computing the integral over a subinterval is, in absolute value, at most $h^3 M_2/12$. The total absolute error, E, is therefore not greater than

$$\frac{nh^3 M_2}{12} = (b-a)h^2 \frac{M_2}{12}.$$

This is the estimate stated in §1.4 of Chapter 13.

Let f be as before, except that now we assume that f has in $[a,b]$ a continuous fourth derivative $f^{(4)}$ and

$$|f^{(4)}(x)| \leq M_4. \tag{17}$$

Now we denote by $\phi(x)$ a *quadratic* polynomial which coincides with $f(x)$ at the endpoints and at the midpoint of the interval considered; that is, it satisfies the conditions

$$\phi(a) = f(a), \qquad \phi\left(\frac{a+b}{2}\right) = f\left(\frac{a+b}{2}\right), \qquad \phi(b) = f(b).$$

We set

$$E^* = \int_a^b f(t) \, dt - \int_a^b \phi(t) \, dt$$

and proceed to estimate E^*.
    We consider first the function

$$q(x) = (x - a)\left(x - \frac{a+b}{2}\right)(x - b)$$

and compute that

$$q(a) = q\left(\frac{a+b}{2}\right) = q(b) = 0, \qquad \int_a^b q(t) \, dt = 0, \qquad \int_a^{(a+b)/2} q(t) \, dt > 0.$$

It follows that, for every choice of the number s, the function $\phi_0(x) = \phi(x) + sq(x)$ satisfies the conditions

$$\phi_0(a) - f(a) = \phi_0\left(\frac{a+b}{2}\right) - f\left(\frac{a+b}{2}\right) = \phi_0(b) - f(b) = 0$$

and

$$E^* = \int_a^b f(t) \, dt - \int_a^b \phi_0(t) \, dt.$$

Also, s can be chosen so that

$$\int_a^{(a+b)/2} f(t) \, dt = \int_a^{(a+b)/2} \phi_0(t) \, dt.$$

Next, we form the fifth-degree polynomial

$$p(x) = \frac{4(x - a)^2\left(x - \frac{a+b}{2}\right)^2(7b - a - 6x)}{(b - a)^5}$$

and verify that it satisfies the conditions

$$p(a) = p'(a) = p\left(\frac{a+b}{2}\right) = p'\left(\frac{a+b}{2}\right) = p'(b) = 0, \qquad p(b) = 1.$$

Finally, we consider the function

$$g(x) = \int_a^x f(t) \, dt - \int_a^x \phi_0(t) \, dt - E^* p(x).$$

We verify that

$$g(a) = g\left(\frac{a+b}{2}\right) = g(b) = 0$$

and also

$$g'(a) = g'\left(\frac{a+b}{2}\right) = g'(b) = 0. \qquad (18)$$

By repeated applications of Rolle's theorem, we conclude that g' is 0 at a point between a and $(a + b)/2$ and at a point between $(a + b)/2$ and b, so that, in view of (18), the function g'' is 0 at at least four points in our interval, the function g''' is 0 at at least three points, the function $g^{(4)}$ is 0 at at least two points, and, finally, the function $g^{(5)}$ is 0 at at least one point. Call this point y. Since ϕ_0 is a cubic polynomial, $\phi_0^{(4)}(x) = 0$ for all x. Also $p^{(5)}(x) = -2880(b - a)^{-5}$. Thus

$$0 = g^{(5)}(y) = f^{(4)}(y) - \frac{2880E^*}{(b - a)^5},$$

and, by (17), we obtain the estimate

$$|E^*| \leq \frac{(b - a)^5 M_4}{2880}.$$

Now if we divide $[a,b]$ into $2m$ equal intervals (each of length h) and compute the integral $\int_a^b f(t)\, dt$ by Simpson's rule, we are using in each of the m pairs of successive intervals the construction described above. The absolute error over each pair of subintervals is, therefore, at most $(2h)^5 M_4/2880$, and the total absolute error does not exceed

$$\frac{m(2h)^5 M_4}{2880} = (b - a)\frac{h^4 M_4}{180}.$$

This is the estimate stated in §1.4 of Chapter 13.

§5   A calculus approach to trigonometric functions

Using calculus, one can arrive at the main properties of trigonometric functions in a way which is both faster and more elegant than the traditional geometric approach outlined in Chapter 10, §1. We must begin, however, with an inverse trigonometric function.

5.1   Integral formulas for the arc cosine

Let P be a point with coordinates (x,y) located on the unit circle $(x^2 + y^2 = 1)$ and such that $y > 0$, and let Q be the point $(0,1)$ and θ the length of the circular arc PQ. By the length formula (see Chapter 9, §3.2)

$$\theta = \int_x^1 \frac{dt}{\sqrt{1 - t^2}}.$$

On the other hand, by the definition of the cosine function (see Chapter 10, §1.3), $x = \cos \theta$. Since $0 < \theta < \pi$, we conclude that $\theta = \text{arc cos } x$. Thus

$$\text{arc cos } x = \int_x^1 \frac{dt}{\sqrt{1 - t^2}}. \tag{1}$$

(In Chapter 10, §3.1, we obtained this formula in a different way using the formula for differentiating $\cos \theta$.)

FIGURE S.2

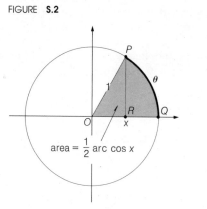

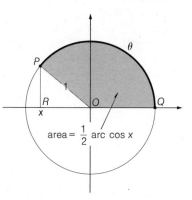

The area A of the circular sector POQ is $A = \frac{1}{2}(1)^2\theta = \frac{1}{2}\theta$ (remember that $|OP| = |OQ| = 1$). The area B under the circular arc PQ and above the x-axis is

$$B = \int_x^1 \sqrt{1 - t^2}\, dt.$$

Let R be the point $(x,0)$ (see Figure S.2). If $x > 0$, then $A = B +$ (the area of the triangle OPR); if $x = 0$, then $A = B$; if $x < 0$, then $A = B -$ (the area of OPR). In all cases we obtain

$$\tfrac{1}{2}\theta = \tfrac{1}{2}x\sqrt{1 - x^2} + \int_x^1 \sqrt{1 - t^2}\, dt$$

so that

$$\text{arc cos } x = x\sqrt{x^2 - 1} + 2\int_x^1 \sqrt{1 - t^2}\, dt. \tag{2}$$

REMARK   The integral in (1) is improper. That the right sides of (1) and (2) coincide can also be seen as follows: both functions have the derivative $-1/\sqrt{1 - x^2}$, and both are 0 for $x = 1$.

5.2   The functions $\cos\theta$ and $\sin\theta$

Let us now consider (1) or (2) as the *definition* of the function $\theta = \text{arc cos } x$. By (2) and the fundamental theorem of calculus, we conclude that arc cos x is continuous for $-1 \le x \le 1$, that arc cos $1 = 0$ and arc cos $(-1) = 2\int_{-1}^1 \sqrt{1 - t^2}\, dt = \pi$, and that

$$\frac{d \text{ arc cos } x}{dx} = -\frac{1}{\sqrt{1 - x^2}}, \qquad \text{for } -1 < x < 1. \tag{3}$$

Thus arc cos x is a decreasing function.

Now we *define* the function $x = \cos\theta$, $0 \le \theta \le \pi$, as the function inverse to arc cos x. By the theorems on inverse functions, $\cos\theta$ is continuous for $0 \le \theta \le \pi$, $\cos 0 = 1$, $\cos\pi = -1$, and, for $0 < \theta < \pi$,

$$\frac{d\cos\theta}{d\theta} = \frac{dx}{d\theta} = \frac{1}{d\theta/dx} = \frac{1}{d \text{ arc cos } x/dx} = -\sqrt{1 - x^2} = -\sqrt{1 - \cos^2\theta}. \tag{4}$$

We *define*: $\sin\theta = \sqrt{1 - \cos^2\theta}$, $0 \le \theta \le \pi$. Then $\sin\theta$ is continuous, $\sin 0 = 0$, $\sin\pi = 0$, and, by (4),

$$\frac{d\cos\theta}{d\theta} = -\sin\theta. \tag{5}$$

Also, by (5) and the chain rule,

$$\frac{d\sin\theta}{d\theta} = \frac{d\sqrt{1 - \cos^2\theta}}{d\theta} = \frac{1}{2\sqrt{1 - \cos^2\theta}} \frac{d(1 - \cos^2\theta)}{d\theta}$$

$$= \frac{1}{2\sqrt{1 - \cos^2\theta}} 2\cos\theta \sin\theta$$

or

$$\frac{d\sin\theta}{d\theta} = \cos\theta. \tag{6}$$

Now we *define* $\cos\theta$ and $\sin\theta$ in the interval $[-\pi, 0]$ by requiring that $\cos(-\theta) = \cos\theta$, $\sin(-\theta) = -\sin\theta$. Then we *extend the definition* to all θ by requiring that $\cos\theta$ and $\sin\theta$ be periodic with period 2π. The functions so defined are continuous for all θ. Using Theorem 1 in Chapter 6, §5.2, we can verify that (5) and (6), originally proved only for $0 < \theta < \pi$, hold for all θ.

In Chapter 10, §1.6, we gave a geometric proof of the addition theorem for sines and cosines, and in §2.2 of that chapter we used the addition theorem to obtain the differentiation rules (5) and (6). Having obtained these rules independently of the addition theorem, we can now give an analytic proof of that theorem. (Another version of this proof is given in Chapter 12, §2.1, Example 2.)

Let ψ be a fixed number. For every ϕ, set

$$f(\phi) = \cos(\phi + \psi) - \cos\phi \cos\psi + \sin\phi \sin\psi,$$

$$g(\phi) = \sin(\phi + \psi) - \sin\phi \cos\psi - \cos\phi \sin\psi.$$

We must show that $f(\phi) = g(\phi) = 0$ for all ϕ. It will suffice to show that $h(\phi) = f(\phi)^2 + g(\phi)^2$ is 0 for all ϕ.

Using (5) and (6), we compute that $f'(\phi) = -g(\phi)$, $g'(\phi) = f(\phi)$. Hence

$$h'(\phi) = 2f(\phi)f'(\phi) + 2g(\phi)g'(\phi) = -2f(\phi)g(\phi) + 2g(\phi)f(\phi) = 0,$$

for all ϕ. Thus $h(\phi) = h(0)$, but we see easily that $f(0) = g(0) = 0$, so that $h(0) = 0$.

REMARK Another approach to trigonometric functions can be based on the integral formula (17) in Chapter 10, §3.2, for the arc tangent.

PROBLEMS

1. In Chapter 10, §2, the differentiation rules for sine and cosine have been derived from the fact that $\lim_{h\to 0}(\sin h/h) = 1$ and $\lim_{h\to 0}(\cos h - 1)/h = 0$. Derive these limit relations from (5) and (6).

2. Derive (5) and (6) from formulas expressing $\sin\theta$ and $\cos\theta$ in terms of $\tan(\theta/2)$ and from the integral formula for the arc tangent [see Chapter 10, §1.4, Example 4, and §3.2, Equation (17)].

FIGURE S.3   FIGURE S.4

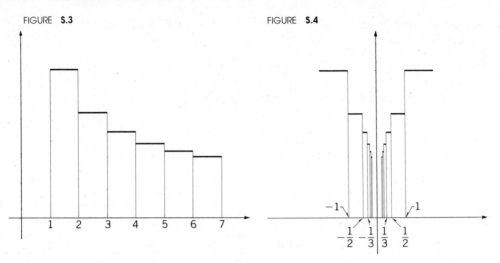

§6 Infinite sequences and series

We proceed to establish some theorems stated without proof in Chapter 15.

6.1 Limits of sequences and limits of functions

It is possible to reduce the concept of limits of sequences to that of limits of functions. For every real number x, let the symbol $[x]$ denote the greatest integer not exceeding x (thus, for instance, $[2] = 2$, $[2.001] = 2$, $[\pi] = 3$). Now let $a_1, a_2, a_3, \ldots$ be a sequence. We associate with it the function

$$x \mapsto a_{[x]}$$

defined for all $x \geq 1$. It is a step function that takes on the value a_1 for $1 \leq x < 2$, the value a_2 for $2 \leq x < 3$, the value a_3 for $3 \leq x < 4$, and so forth.

We can also associate with our sequence the function

$$x \mapsto a_{[|1/x|]}$$

defined for all values of $x \neq 0$. This function takes on the value a_1 for $x \leq -1$ and $x \geq 1$, the value a_2 for $-1 < x \leq -\frac{1}{2}$ and $\frac{1}{2} \leq x < 1$, the value a_3 for $-\frac{1}{2} < x \leq \frac{1}{3}$ and $\frac{1}{3} \leq x < \frac{1}{2}$, and so forth.

The graphs of the functions $x \mapsto a_{[x]}$ and $x \mapsto a_{[|1/x|]}$ for the sequence $a_j = 5/\sqrt{j}$ are shown in Figures S.3 and S.4.

Recalling the definitions of limits of functions (Chapter 3, §4.2 and §4.4, Chapter 5, §3.1 and §3.2), we see that the three statements

$$\lim_{i \to \infty} a_i = \alpha, \qquad \lim_{x \to +\infty} a_{[x]} = \alpha, \qquad \lim_{x \to 0} a_{[|1/x|]} = \alpha$$

are *equivalent*.

We can now transfer to limits of sequences theorems previously proved about limits of functions. This establishes Theorems 1 and 2 in Chapter 15, §1.3.

One can also prove these theorems directly. In case of Theorem 1, one may repeat almost verbatim the argument in §3.1 of this supplement. We now prove Theorem 2.

We are given that $f(x)$ is continuous at $x = \alpha$ and that $\lim_{n \to \infty} a_n = \alpha$. We set $\beta = f(\alpha)$, $b_n = f(a_n)$, and show that $\lim_{n \to \infty} b_n = \beta$.

Let $\epsilon > 0$ be given. Since f is continuous at α, there is a $\delta > 0$ such that $|f(x) - \beta| < \epsilon$ for $|x - \alpha| < \delta$. Since the sequence $\{a_n\}$ converges to α, there is an $N > 0$ such that $|a_n - \alpha| < \delta$ for $n > N$. Hence, for $n > N$, we have $|b_n - \beta| = |f(a_n) - \beta| < \epsilon$. This proves the assertion.

REMARK In Chapter 15, §2, we used convergent sequences to assign a value to certain infinite sums. In a similar way we can assign a value to certain **infinite products**. If $a_1, a_2, a_3, \ldots$ are numbers, and if the limit

$$\lim_{n \to \infty} (a_1 a_2 a_3 \ldots a_n) = p \neq 0$$

exists, we say that the infinite product $a_1 a_2 a_3 \ldots$ converges to p and write $a_1 a_2 a_3 \ldots = p$. An example is the **Wallis' infinite product** for $\pi/2$:

$$\frac{\pi}{2} = \frac{2}{1} \cdot \frac{2}{3} \cdot \frac{4}{3} \cdot \frac{4}{5} \cdot \frac{6}{5} \cdot \frac{6}{7} \cdot \frac{8}{7} \ldots .$$

For the proof, see Problems 3 and 4.

PROBLEMS

1. Prove Theorem 1 in Chapter 15, §1.3, by the method used in §3.1 of this supplement.
2. Let $f(x)$ be a function defined for $|x - a| < 1$. Assume that for *every* sequence $\{x_i\}$, such that $|x_i - a| < 1$ and $\lim_{i \to \infty} x_i = a$, we have $\lim_{i \to \infty} f(x_i) = \alpha$. Prove that $\lim_{x \to a} f(x) = \alpha$.
3. Set $I_n = \int_0^{\pi/2} \sin^n x \, dx$, $n = 1, 2, \ldots$, and show that

$$\lim_{k \to \infty} \left(\frac{I_{2k}}{I_{2k+1}} \right) = 1.$$

[*Hint:* First show that $I_{2k+1} < I_{2k} < I_{2k-1}$ and $1 < I_{2k}/I_{2k+1} < I_{2k-1}/I_{2k+1}$, then use the reduction formula (14) in Chapter 13, §2.4.]

4. Prove that

$$\frac{\pi}{2} = \lim_{k \to \infty} \frac{2}{1} \cdot \frac{2}{3} \cdot \frac{4}{3} \cdot \frac{4}{5} \cdot \frac{6}{5} \cdot \frac{6}{7} \cdots \frac{2k}{2k - 1} \cdot \frac{2k}{2k + 1}.$$

[*Hint:* Use relations (13) and (15) in Chapter 13, §2.4, and the result of Problem 1 above.]

6.2   Rearranging terms in a convergent series

We consider next a convergent infinite series

$$\sum_{j=0}^{\infty} a_j \tag{1}$$

and investigate the effect of rearranging the terms. We may assume that no a_j is 0, since such terms could be dropped. Since the series converges, we have (see Theorem 1 in Chapter 15, §2.3) that $\lim_{j \to \infty} a_j = 0$. It follows that, for every number $T > 0$, nearly all

---

JOHN WALLIS (1616–1703) was professor of mathematics at Oxford and one of the precursors of calculus.

a_j satisfy $|a_j| < T$. Therefore we can collect all positive terms among the a_j and arrange them in decreasing order. We call these terms α_0, α_1, α_2, ... Thus

$$\alpha_0 \geq \alpha_1 \geq \alpha_2 \geq \cdots, \qquad \alpha_j > 0, \qquad \lim_{j\to\infty} \alpha_j = 0.$$

Similarly, we denote the negative numbers among the a_j, arranged in descending order of absolute values, by $-\beta_0$, $-\beta_1$, $-\beta_2$, ... We have

$$\beta_0 \geq \beta_1 \geq \beta_2 \geq \cdots, \qquad \beta_j > 0, \qquad \lim_{j\to\infty} \beta_j = 0.$$

Set

$$P = \sum_{j=0}^{\infty} \alpha_j, \qquad N = \sum_{j=0}^{\infty} \beta_j.$$

(If there are no α's, set $P = 0$; if there are no β's, set $N = 0$.) Of course, we must admit $+\infty$ as a possible value for P and N. However, either $P = +\infty$ and $N = +\infty$, or $P < +\infty$ and $N < +\infty$. (In other words, the series of positive terms and the series of negative terms are either both divergent or both convergent.)

Indeed, assume that $P = +\infty$, $N < +\infty$. For every m, we have $\Sigma_{j=0}^{m} a_j = S_m - T_m$, where S_m is the sum of all positive numbers among $a_0, a_1, \ldots, a_m$ and $-T_m$ is the sum of all negative numbers among $a_0, a_1, \ldots, a_m$. Clearly $0 \leq T_m \leq N$, so that $\Sigma_{j=0}^{m} a_j \geq S_m - N$. But if m is large enough, the finite sequence $a_0, a_1, \ldots, a_m$ will contain as many terms $\alpha_0, \alpha_1, \alpha_2, \ldots$ as we like, and S_m will be as large as we like. Hence so will $\Sigma_{j=0}^{m} a_j$. This contradicts the assumption that (1) converges.

We showed that the case $P = +\infty$, $N < +\infty$ is impossible. The case $P < +\infty$ is also impossible for the same reason.

We show next that, if $P < +\infty$, $N < +\infty$, then $\Sigma_{j=0}^{\infty} a_j = P - N$. Indeed, if we use the same notations as before, we observe that for large m the sum S_m will be as close to P as we like, and the sum T_m will be as close to N as we like.

Finally, we note that $\Sigma_{j=0}^{\infty} |a_j| = P + N$. (In other words, the series $\Sigma_{j=0}^{\infty} a_j$ converges absolutely if and only if $P + N < +\infty$.) The proof is left to the reader.

The preceding remarks constitute a proof of the first part of Theorem 10 in Chapter 15, §2.9: *an absolutely convergent series remains convergent if its terms are rearranged; the rearrangement does not affect the sum of the series.*

We assume now that (1) is not absolutely convergent, so that

$$P = +\infty, \qquad N = +\infty.$$

We show first how to arrange the terms α_k and $-\beta_j$ into a divergent series.

Since $\Sigma_{k=0}^{\infty} \alpha_k = +\infty$, we can divide this series into blocks as follows:

$$\underbrace{\alpha_0 + \alpha_1 + \alpha_2 + \cdots + \alpha_{n_1}}_{>1} + \underbrace{\alpha_{n_1+1} + \alpha_{n_1+2} + \cdots + \alpha_{n_2}}_{>\beta_1 + 1}$$

$$+ \underbrace{\alpha_{n_2+1} + \cdots + \alpha_{n_3}}_{>\beta_2 + 1} + \underbrace{\alpha_{n_3+1} + \cdots + \alpha_{n_4}}_{>\beta_3 + 1} + \cdots.$$

The series

$$\underbrace{\alpha_0 + \alpha_1 + \cdots + \alpha_{n_1}}_{} + \underbrace{(-\beta_1) + \alpha_{n+1} + \cdots + \alpha_{n_2}}_{}$$

$$+ \underbrace{(-\beta_2) + \alpha_{n_2+1} + \cdots + \alpha_{n_3}}_{} + \underbrace{(-\beta_3) + \alpha_{n_3+1} + \cdots + \alpha_{n_4}}_{} + \cdots \qquad (2)$$

is a rearrangement of the original series (1) and diverges.

Now let s be any given number. The terms $\alpha_0, \alpha_1, \ldots, -\beta_0, -\beta_1, \ldots$ can be arranged into a series that converges to s:

$$\underbrace{\alpha_0 + \alpha_1 + \cdots + \alpha_{n_1}}_{\sigma_1} \underbrace{- \beta_0 - \beta_1 - \cdots - \beta_{m_1}}_{\tau_1} + \underbrace{\alpha_{n_1+1} + \cdots + \alpha_{n_2}}_{\sigma_2}$$

$$\underbrace{- \beta_{m_1+1} - \cdots - \beta_{m_2}}_{\tau_2} + \underbrace{\alpha_{n_2+1} + \cdots + \alpha_{n_3}}_{\sigma_3} \underbrace{- \beta_{m_2+1} - \cdots - \beta_{m_3}}_{\tau_3} + \alpha_{n_3+1} + \cdots \qquad (3)$$

Here the numbers $n_1, m_1, n_2, m_2, \ldots$ are chosen as the *smallest* numbers satisfying the conditions: $\sigma_1 > s$, $\sigma_1 + \tau_1 < s$, $\sigma_1 + \tau_1 + \sigma_2 > s$, $\sigma_1 + \tau_1 + \sigma_2 + \tau_2 < s$, $\sigma_1 + \tau_1 + \sigma_2 + \tau_2 + \sigma_3 > s, \ldots$ (One can find such numbers since the series $\alpha_1 + \alpha_2 + \alpha_3 + \cdots$ and $\beta_1 + \beta_2 + \beta_3 + \cdots$ diverge to $+\infty$, and so does every series obtained from either of those two by dropping finitely many terms.) The partial sums of the series (3) oscillate around s.

Also, since n_1 is the smallest number such that $\sigma_1 > s$, we have $|\sigma_1 - s| \le \alpha_{n_1}$. Similarly, since m_1 is the smallest number such that $\sigma_1 + \tau_1 < s$, we must have $|(\sigma_1 + \tau_1) - s| \le \beta_{m_1}$. For the same reason $|(\sigma_1 + \tau_1 + \sigma_2) - s| \le \alpha_{n_2}$, and $|(\sigma_1 + \tau_1 + \sigma_2 + \tau_2) - s| \le \beta_{m_2}$, and so forth. Since $\lim_{j\to\infty} \alpha_j = \lim_{j\to\infty} \beta_j = 0$, we conclude that

$$\sigma_1 + \tau_1 + \sigma_2 + \tau_2 + \sigma_3 + \cdots = s. \qquad (4)$$

Hence

$$\lim_{i\to\infty} \sigma_i = 0, \qquad \lim_{i\to\infty} \tau_i = 0.$$

Now, a partial sum of (3) can be written either as

$$\sigma_1 + \tau_1 + \cdots + \sigma_n + \tau_n + r, \qquad \text{with } |r| < \sigma_{n+1}$$

or as

$$\sigma_1 + \tau_1 + \cdots + \sigma_n + r, \qquad \text{with } |r| < \tau_n.$$

It follows that the series (3), a rearrangement of (1), converges to s.

We have now established the second part of Theorem 10 in Chapter 15, §2.9: *a series that converges, but not absolutely, can be rearranged so as to diverge or so as to converge to any given number.*

In view of this theorem, an absolutely convergent series is said to converge **unconditionally**; a series that converges, but not absolutely, is said to converge **conditionally.**

PROBLEMS

1. Show that $P < +\infty$ and $N < +\infty$ if $|a_0| + |a_1| + |a_2| + \cdots$ converges.
2. Show that $|a_0| + |a_1| + |a_2| + \cdots$ converges if $P < +\infty$ and $N < +\infty$.
3. Rearrange the terms of the series $1 - \frac{1}{2} + \frac{1}{3} - \frac{1}{4} + \cdots$ so as to form a divergent series.
4. Rearrange the terms of the series $1 - \frac{1}{2} + \frac{1}{3} - \frac{1}{4} + \cdots$ so as to form a series converging to 1.

6.3   Uniform convergence

Our next aim is to establish the theorems concerning the radius of convergence of power series, stated in Chapter 15, §3. As a preparation we introduce a concept of wider significance.

Let $f_1(x), f_2(x), \ldots$ be a sequence of functions defined in some interval I. We say that the sequence $f_1, f_2, \ldots$ **converges uniformly** in I, to a function $f(x)$, and we write

$$\lim_{j \to \infty} f_j(x) = f(x), \qquad \text{uniformly for } x \text{ in } I,$$

if the following holds. *For every $\epsilon > 0$, there is an N such that $|f_j(x) - f(x)| < \epsilon$ for $j > N$ and for all x in I.*

EXAMPLES   1. $\lim_{j \to \infty} x^j = 0$ uniformly for $|x| \leq \frac{1}{2}$.

Indeed, for $|x| \leq \frac{1}{2}$, we have $|x^j| \leq 2^{-j}$ so that $|x^j| < \epsilon$ wherever $2^{-j} < \epsilon$, that is, whenever $j > -\log 2/\log \epsilon$.

2. $\lim_{j \to \infty} x^j = 0$ for $0 < x < 1$ but not uniformly.

Indeed, if j is given, however large, we can find an x in the interval $(0,1)$ with x^j as close to 1 as desired. For example, we choose a small positive number η and set $x = (1 - \eta)^{1/j}$.

Theorem A.   *Let the functions $f_1(x), f_2(x), \ldots$ be continuous for $a \leq x \leq b$, and assume that $\lim_{j \to \infty} f_j(x) = f(x)$ uniformly in $a \leq x \leq b$. Then $f(x)$ is continuous and, for $a \leq x \leq b$,*

$$\lim_{j \to \infty} \int_a^x f_j(t)\, dt = \int_a^x f(t)\, dt.$$

Proof.   Let x_0 be a point in $[a,b]$; we show that $f(x)$ is continuous at x_0. Let $\epsilon > 0$ be given; we shall find a $\delta > 0$ such that $|f(x_0) - f(x)| < \epsilon$ for all x in $[a,b]$ with $|x - x_0| < \delta$. By uniform convergence, there is a j such that $|f(x) - f_j(x)| < \epsilon/3$ for all x in $[a,b]$. Since f_j is continuous at x_0, there is a $\delta > 0$ such that $|f_j(x_0) - f_j(x)| < \epsilon/3$ for x in $[a,b]$ and $|x - x_0| < \delta$. For such x we have

$$|f(x) - f(x_0)| = |[f(x) - f_j(x)] + [f_j(x) - f_j(x_0)] + [f_j(x_0) - f(x_0)]|$$

$$\leq |f(x) - f_j(x)| + |f_j(x) - f_j(x_0)| + |f_j(x_0) - f(x_0)| < \frac{\epsilon}{3} + \frac{\epsilon}{3} + \frac{\epsilon}{3} = \epsilon.$$

Let $\epsilon > 0$ be given, and now let j be so large that $|f_j - f| < \epsilon/(b - a)$ in $[a,b]$. Then, by the Remark in §4.2,

$$\left| \int_a^x f_j(t)\, dt - \int_a^x f(t)\, dt \right| = \left| \int_a^x [f_j(t) - f(t)]\, dt \right| \leq \frac{\epsilon}{b - a} |x - a| \leq \epsilon,$$

for $a \leq x \leq b$, which proves the second assertion.

From sequences one can pass to series. A series of functions $\sum_{n=0}^{\infty} \phi_j(x)$ **converges uniformly** in an interval I to a function $\Phi(x)$ if the sequence of partial sums converges uniformly to Φ:

$$\lim_{j \to \infty} \sum_{n=0}^{j} \phi_n(x) = \Phi(x), \qquad \text{uniformly for } x \text{ in } I;$$

that is, if for every ϵ, there is an N such that

$$\left| \sum_{n=0}^{j} \phi_n(x) - \Phi(x) \right| < \epsilon, \qquad \text{for } x \text{ in } I, \quad j \geq N.$$

It follows from Theorem A that *if all terms in the uniformly convergent series* $\Sigma \, \phi_j(x)$ *are continuous, then so is the sum, and that the series can be integrated term by term:*

$$\sum_{n=0}^{\infty} \int_{x_0}^{x} \phi_n(t) \, dt = \int_{x_0}^{x} \left(\sum_{n=0}^{\infty} \phi_n(t) \right) dt,$$

provided that x_0 and x lie in the interval of uniform convergence.

PROBLEMS

1. Let A be any positive number. Show that the sequence $\{x^j/j!\}$ converges to 0 uniformly for $|x| < A$.
2. Give an example of a uniformly convergent sequence of discontinuous functions, with a continuous limit.

6.4   Radius of convergence

We proceed to prove the assertions about power series made without proof in Chapter 15, §3 (Theorems 1 and 2). To simplify writing, we consider only power series about $x_0 = 0$; this is clearly sufficient.

The basic result in the theory of power series is

Lemma 1   (Abel's lemma).   *If the power series*

$$\sum_{n=0}^{\infty} a_n x^n \tag{5}$$

converges for some $x = \rho \neq 0$, *then the series converges absolutely for* $|x| < |\rho|$, *and there is a constant M such that*

$$|a_n| \leq \frac{M}{|\rho|^n}, \qquad \text{for } n = 0, 1, 2, \dots. \tag{6}$$

Proof.   If $\Sigma_{n=0}^{\infty} a_n \rho^n$ converges, then $\lim_{n \to \infty} a_n \rho^n = 0$ (by Theorem 1 in Chapter 15, §3.1). Therefore, by the definition of limit, $|a_n \rho^n| < 1$ for nearly all n, say, for $n > N$. Let M be the largest of the numbers $|a_0|, |a_1 \rho|, |a_2 \rho^2|, \dots, |a_N \rho^N|$, and 1. Then $|a_n \rho^n| \leq M$ or $|a_n| \, |\rho^n| \leq M$, for all n. Since $\rho \neq 0$, assertion (6) follows.

Now let x be a number such that $|x| < |\rho|$. Then $|x/\rho| < 1$ and, by (6), $|a_n x^n| \leq M|x/\rho|^n$. Thus the terms of (5) are, in absolute value, not greater than those of the convergent series

$$M + M \left| \frac{x}{\rho} \right| + M \left| \frac{x}{\rho} \right|^2 + \cdots.$$

Hence the series $\Sigma \, |a_n x^n|$ converges (by Theorem 5 in Chapter 15, §2.4) and (1) converges absolutely (compare Theorem 9 in Chapter 15, §2.9).

Lemma 2.   *If the power series (5) converges for some but not all $x \neq 0$, then there is a number $R > 0$ (the radius of convergence) such that the series (5) converges absolutely for $|x| < R$ and diverges for $|x| > R$.*

Proof.   Assume that (5) converges for $x = \rho_1 \neq 0$ and diverges for $x = \rho_2$. Then (5) converges for all x such that $|x| < |\rho_1|$ and diverges for all $|x| > |\rho_2|$. This follows from Abel's lemma. Let S be the set of all positive numbers ρ such that (5) converges for $|x| < \rho$. This set is not empty: it contains $|\rho_1|$. It is bounded: ρ_2 is an upper bound. Let R be the least upper bound of S. If $|x_0| < R$, then $\sum_{n=0}^{\infty} a_n x_0^n$ converges, for otherwise $|x_0|$ would be an upper bound for S. By the same token, for every $x_1, |x_0| < |x_1| < R$, the series $\sum_{n=0}^{\infty} a_n x_1^n$ converges, hence $\sum_{n=0}^{\infty} a_n x_0^n$ converges absolutely, by Abel's lemma. Assume next that $|x_2| > R$. Then $\sum_{n=0}^{\infty} a_n x_2^n$ diverges. For, if this series would converge, $|x_2|$ would belong to S, by Abel's lemma.

We see that R is the desired radius of convergence.

If the series converges for all values of x, one sets $R = +\infty$; if it diverges for all $x \neq 0$, one sets $R = 0$.

Lemma 3.   *The three power series*

$$\sum_{n=0}^{\infty} a_n x^n, \qquad \sum_{n=1}^{\infty} n a_n x^{n-1}, \qquad \sum_{n=0}^{\infty} \frac{a_n}{n+1} x^{n+1} \tag{7}$$

have equal radii of convergence.

Proof.   Denote the three radii of convergence by R, R_*, and R^*, respectively. (These symbols may be nonnegative numbers or $+\infty$.)

If $R > 0$, let ρ be a number such that $0 < \rho < R$. Then, by Abel's lemma, we have inequality (6) and, for $|x| < \rho$,

$$|n a_n x^{n-1}| \leq n \frac{M}{\rho} \left(\frac{|x|}{\rho} \right)^{n-1}, \qquad \left| \frac{a_n x^{n+1}}{n+1} \right| \leq \frac{M|x|}{(n+1)} \left(\frac{|x|}{\rho} \right)^n \leq M|x| \left(\frac{x}{\rho} \right)^n$$

so that the terms of the second and third series in (3) are in absolute value not greater than the corresponding terms of the convergent series

$$\frac{M}{\rho} + 2 \frac{M}{\rho} \frac{|x|}{\rho} + 3 \frac{M}{\rho} \frac{|x|^2}{\rho^2} + \cdots = \frac{M}{\rho} \left(1 - \frac{|x|}{\rho} \right)^{-2}$$

and

$$M|x| + M|x| \frac{|x|}{\rho} + M|x| \left(\frac{|x|}{\rho} \right)^2 + \cdots = M|x| \left(1 - \frac{|x|}{\rho} \right)^{-1},$$

respectively. We apply Theorem 5 in Chapter 15, §2.4, and conclude that the second and third series (7) converge for $|x| < \rho < R$, hence for *all* $|x| < R$. Therefore $R^* \geq R$ and $R_* \geq R$.

But the first series in (7) is obtained from the third in the same way as the second is obtained from the first, namely, by term-by-term differentiation. Therefore we also have $R \geq R^*$, hence $R \geq R^*$. By the same arguments $R = R_*$.

Lemma 4.   *Assume that $\sum_{n=0}^{\infty} a_n x^n$ converges for $|x| < A$ (for some $A > 0$), and set*

$$f(x) = \sum_{n=0}^{\infty} a_n x^n. \tag{8}$$

Then the series converges uniformly in every interval $|x| \le B < A$, and $f(x)$ is continuous for $|x| < A$.

Proof. By Abel's lemma, there is a number M such that $|a_n| \le MA^{-n}$ for all n. For $|x| \le B < A$, we have that

$$\left| \sum_{n=0}^{\infty} a_n x^n - \sum_{n=0}^{j} a_n x^n \right| = \left| \sum_{n=j+1}^{\infty} a_n x^n \right| \le \sum_{n=j+1}^{\infty} |a_n x^n|$$

$$\le \sum_{n=j+1}^{\infty} MA^{-n}B^n = M\left(\frac{B}{A}\right)^{j+1}\left[1 + \left(\frac{B}{A}\right) + \left(\frac{B}{A}\right)^2 + \cdots\right]$$

$$= M\left(\frac{B}{A}\right)^{j+1}\left[1 - \left(\frac{B}{A}\right)\right]^{-1},$$

which will be as small as we like if j is large enough (recall that $0 < B/A < 1$). This proves the first assertion. The second follows from the results of §6.3.

Lemma 5. *Under the hypothesis of Lemma 4,*

$$\int_0^x f(t)\, dt = \sum_{n=0}^{\infty} \frac{a_n x^{n+1}}{n+1}, \tag{9}$$

for $|x| < A$.

Proof. Apply Lemma 4 and the results of §6.3.

Lemma 6. *Under the hypothesis of Lemma 4,*

$$f'(x) = \sum_{n=1}^{\infty} n a_n x^{n-1}.$$

Proof. Set, for $|x| < A$, $g(x) = \sum_{n=1}^{\infty} n a_n x^{n-1}$, the series being convergent, by Lemma 3, and $g(x)$ being continuous, by Lemma 4. By Lemma 5, we have

$$\int_0^x g(t)\, dt = \sum_{n=1}^{\infty} a_n x^n = f(x) - a_0.$$

Hence $g(x) = f'(x)$ by the fundamental theorem of calculus.

Now we proved Theorems 1 and 2 of Chapter 15, §3, for $x_0 = 0$.

PROBLEMS

1. Let k be a fixed positive integer. Prove that the series

$$\sum_{n=0}^{\infty} a_n x^n \quad \text{and} \quad \sum_{n=k+1}^{\infty} n(n-1)(n-2)\ldots(n-k)x^{n-k-1}$$

have the same radius of convergence.

2. State and prove the analog of Abel's lemma for a power series $\sum_{n=0}^{\infty} a_n(x - x_0)^n$.

3. State and prove the analogs of Lemmas 2 to 6 for power series about x_0.

§7 Integration in two variables

In this section we establish several theorems about integrals stated in Chapters 19 and 20 without proof. Everything said here extends with obvious modifications to functions of more than two variables.

7.1 Integrals depending on a parameter

The following lemma will be needed later.

Lemma 1. *Let $f(x,y)$ be continuous for $a \leq x \leq b$, $c \leq y \leq d$, and set*

$$g(x) = \int_c^d f(x,y)\, dy.$$

Then $g(x)$ is continuous for $a \leq x \leq b$.

The proof depends on the concept of uniform continuity (see §3.6, especially Example 1 and Theorem C).

Let x_0 be a point such that $a \leq x_0 \leq b$. We show that for every $\epsilon > 0$ there is a $\delta > 0$ such that for all x, $a \leq x \leq b$, with $|x - x_0| < \delta$, we have $|g(x) - g(x_0)| < \epsilon$. Indeed, since $f(x,y)$ is uniformly continuous for $a \leq x \leq b$, $c \leq y \leq d$, there is a $\delta > 0$ such that $|f(x,y) - f(x_0,y)| < \epsilon/(c - d)$ for all $|x - x_0| < \delta$ (and, of course, for $a \leq x \leq b$, $c \leq y \leq d$). Hence, by the Remark in §4.2,

$$|g(x) - g(x_0)| = \left| \int_c^d f(x,y)\, dy - \int_c^d f(x_0,y)\, dy \right|$$

$$= \left| \int_c^d [f(x,y) - f(x_0,y)]\, dy \right| \leq \int_c^d |f(x,y) - f(x_0,y)|\, dy$$

$$\leq \left[\frac{\epsilon}{(d - c)} \right] (d - c) = \epsilon.$$

This proves the assertion.

7.2 Differentiation under the integral sign

Lemma 2. *Consider a continuous function $f(x,y)$ with a continuous partial derivative $f_x(x,y)$, for $a \leq x \leq b$, $c \leq y \leq d$. Set*

$$g(x) = \int_c^d f(x,y)\, dy. \tag{1}$$

Then

$$g'(x) = \int_c^d f_x(x,y)\, dy, \qquad a < x < b. \tag{2}$$

Proof. We observe that $f_x(x,y)$ is uniformly continuous for $a \leq x \leq b$, $c \leq y \leq d$. In particular, for every $\epsilon > 0$, there is a number $\delta > 0$ such that

$$|f_x(x + \xi, y) - f_x(x,y)| < \epsilon, \qquad \text{if } |\xi| < \delta. \tag{3}$$

Given a point (x,y) and a number $h \neq 0$ there is, by the mean value theorem, a number ξ depending on x, y, and h such that

$$f(x + h,y) = f(x,y) + hf_x(x + \xi,y), \qquad |\xi| < |h|. \tag{4}$$

Therefore

$$\left| \frac{1}{h} \{F(x + h) - F(x)\} - \int_d^c f_x(x,y)\, dy \right|$$

$$= \left| \frac{1}{h} \left\{ \int_c^d f(x + h,y)\, dy - \int_c^d f(x,y)\, dy \right\} - \int_c^d f_x(x,y)\, dy \right|$$

$$= \left| \int_c^d \left\{ \frac{f(x + h,y) - f(x,y)}{h} - f_x(x,y) \right\} dy \right|$$

$$= \left| \int_c^d \{f_x(x + \xi,y) - f_x(x,y)\}\, dy \right|$$

$$\leq \int_c^d |f_x(x + \xi,y) - f_x(x,y)|\, dy \leq (d - c)\epsilon,$$

if $|h| \leq \delta$, δ being determined from (3). This means that

$$\lim_{h \to 0} \frac{F(x + h) - F(x)}{h} = \int_c^d F_x(x,y)\, dy,$$

which proves (2).

Theorem 2 in Chapter 19, §3.2, follows from Lemma 2 and Lemma 1.

7.3   Upper and lower double integrals

We proceed to establish the existence of double integrals and the fact that double integrals can be computed like repeated integrals. In other words, we shall prove Theorems 1 and 2 of Chapter 20, §2.3. We shall work with a fixed (finite) interval

$$I: a < x < b, \qquad c < y < d. \tag{5}$$

Step functions of two variables, and their integrals, have been defined in Chapter 20, §2.2. For step functions, f and g, the following properties are obviously true:

$$\iint_I f(x,y)\, dx\, dy = K(b - a)(d - c), \qquad \text{if } f(x,y) = K, \quad \text{a constant}, \tag{6}$$

$$\iint_I f(x,y)\, dx\, dy = \iint_{\substack{a<x<t \\ c<y<d}} f\, dx\, dy + \iint_{\substack{t<x<b \\ c<y<d}} f\, dx\, dy, \quad \text{for } a < t < b, \tag{7}$$

$$\iint_I f(x,y)\, dx\, dy = \iint_{\substack{a<x<b \\ c<y<s}} f\, dx\, dy + \iint_{\substack{a<x<b \\ s<y<d}} f\, dx\, dy, \quad \text{for } c < s < d, \tag{8}$$

$$\iint_I f(x,y)\, dx\, dy \leq \iint_I g(x,y)\, dx\, dy, \qquad \text{if } f(x,y) \leq g(x,y) \text{ in } I. \tag{9}$$

As in §4.1, we define, for every bounded function $f(x,y)$ defined in I, the **lower** and **upper double integrals:**

$$\underline{\iint_I} f\, dx\, dy = \text{least upper bound of } \iint_I \phi\, dx\, dy,$$

$$\text{for all step functions } \phi(x,y) \text{ with } \phi \leq f \text{ in } I, \tag{10}$$

and

$$\overline{\iint_I} f\, dx\, dy = \text{greatest lower bound of } \iint_I \psi\, dx\, dy,$$

$$\text{for all step functions } \psi(x,y) \text{ with } f \leq \psi \text{ in } I. \tag{11}$$

As in §4.1, we obtain, from (6), (7), (8), (9), (10), and (11) the relations

$$\underline{\iint_I} f\, dx\, dy = K(b-a)(d-c), \qquad \text{if } f(x,y) = K, \quad \text{a constant}, \tag{12}$$

$$\underline{\iint_I} f\, dx\, dy = \underline{\iint_{\substack{a<x<t \\ c<y<d}}} f\, dx\, dy + \underline{\iint_{\substack{t<x<b \\ c<y<d}}} f\, dx\, dy, \quad \text{for } a<t<b, \tag{13}$$

$$\underline{\iint_I} f\, dx\, dy = \underline{\iint_{\substack{a<x<b \\ c<y<s}}} f\, dx\, dy + \underline{\iint_{\substack{a<x<b \\ s<y<d}}} f\, dx\, dy, \quad \text{for } c<s<d, \tag{14}$$

$$\underline{\iint_I} f\, dx\, dy \leq \underline{\iint_I} g\, dx\, dy, \qquad \text{if } f \leq g \text{ in } I. \tag{15}$$

These relations hold for all bounded functions f and g, and similar relations are valid for upper integrals.

Relations (12) and (15) imply that

$$\left| \underline{\iint_I} f(x,y)\, dx\, dy \right| \leq M(b-a)(d-c), \quad \text{if } |f(x,y)| \leq M \text{ in } I. \tag{16}$$

We can verify that the integral of a step function of two variables is also its lower and upper integrals.

Lemma 3. *Let $f(y)$ be a bounded function defined for $c < y < d$.* (Such a function may be considered as a function of two variables, x and y.) *Then*

$$\underline{\iint_I} f(y)\, dx\, dy = (b-a) \underline{\int_c^d} f(y)\, dy. \tag{17}$$

Similar lemmas are valid for upper integrals and for functions of x.

Proof of Lemma 3.   Let $\phi(x,y)$ be a step function defined in I and satisfying $\phi(x,y) \leq f(y)$. There is a step function $\Phi(y)$, $c < y < d$, such that $\phi(x,y) \leq \Phi(y) \leq f(y)$. For instance, we may set $\Phi(y) = $ largest value of $\phi(x,y)$ for $a < x < b$ (disregarding the values of ϕ at points of discontinuity). We conclude that the left side of (17) is the least upper bound of

$$\iint_I \Phi(y)\, dx\, dy, \tag{18}$$

for all step functions $\Phi(y)$ with $\Phi \le f$. But for a step function $\Phi(y)$, it is easy to see that the double integral (18) equals

$$(b - a) \int_c^d \Phi(y)\, dy, \tag{18'}$$

and the least upper bound of (18'), for all step functions Φ with $\Phi \le f$, is the right side of (17); see §4.1.

PROBLEMS

1. Derive relations (12) to (15) from the relations (6) to (11).
2. Prove statement (16).
3. State and prove the analog of Lemma 3 for lower integrals and for functions of x alone.
4. State and prove the analog of Lemma 3 for upper integrals.

7.4   Double integrals and iterated integrals

Let $f(x,y)$ be a bounded function defined in I, and set

$$F(t) = \underline{\iint}_{\substack{a < x < t \\ c < y < d}} f(x,y)\, dx\, dy, \qquad \text{for } a < t \le b, \tag{19}$$

$$F(a) = 0. \tag{20}$$

Lemma 4. *Let $f(x,y)$ be bounded and piecewise continuous in I. Then* (i) *$F(t)$ is continuous for $a \le t \le b$,* (ii) *$F'(t)$ exists at all but finitely many points t, $a < t < b$,*

$$F'(t) = \int_c^d f(t,y)\, dy, \tag{21}$$

and $F'(t)$ is bounded and piecewise continuous.

The proof of Lemma 4 will be found in §7.5 and §7.6. Assuming the lemma, we apply the fundamental theorem of calculus to $F(t)$ and conclude that (for a bounded and piecewise continuous f)

$$\underline{\iint}_I f(x,y)\, dx\, dy = F(b) = \int_a^b F'(t)\, dt + F(a) = \int_a^b \left\{ \int_c^d f(x,y)\, dy \right\} dx$$

or

$$\underline{\iint}_I f(x,y)\, dx\, dy = \int_a^b \left\{ \int_c^d f(x,y)\, dy \right\} dx. \tag{22}$$

There is a lemma similar to Lemma 4 for upper integrals; it leads to the identity

$$\overline{\iint}_I f(x,y)\, dx\, dy = \int_a^b \left\{ \int_c^d f(x,y)\, dy \right\} dx. \tag{23}$$

Thus, for a bounded piecewise continuous function $f(x,y)$, the upper integral and lower integral coincide; both are equal to the iterated integral $\int_a^b \{ \int_c^d f(x,y)\, dy \}\, dx$. This number is, therefore, the only number between $\iint_I \phi\, dx\, dy$ and $\iint_I \psi\, dx\, dy$, for any two step functions ϕ and ψ with $\phi \le f \le \psi$. This proves Theorem 1 in Chapter 20, §2.3.

We prove in the same way that, for a bounded piecewise continuous f,

$$\underline{\iint_I} f \, dx \, dy = \overline{\iint_I} f \, dx \, dy = \int_c^d \left\{ \int_a^b f(x,y) \, dx \right\} dy. \tag{24}$$

Relations (22), (23), and (24) imply Theorem 2 in Chapter 20, §2.3.

REMARK   A bounded function $f(x,y)$ for which the upper and lower integrals (over an interval) coincide is called **Riemann integrable** (over this interval). The common value of the upper and lower integral is called the Riemann integral of f. This shows that the Riemann integral can be approximated arbitrarily closely by Riemann sums (see Chapter 20, §2.4).

Properties (6), (7), (8), and (9) hold for all Riemann integrable functions. In Chapter 20, §1.4, we proved them using iterated integrals.

PROBLEMS

1. Prove relation (23).
2. Prove relations (24).

7.5   Proof of the key lemma

The lemma in question is Lemma 4. We prove it first assuming that $f(x,y)$ is continuous for $a \leq x \leq b$, $c \leq y \leq d$, and thus uniformly continuous (see §3.7).

Since f is bounded, there is a constant M such that

$$|f| \leq M, \qquad \text{in } I. \tag{25}$$

By (19) we have, for $a \leq t < \tau < d$,

$$|F(\tau) - F(t)| \leq \left| \underline{\iint}_{\substack{t < x < \tau \\ c < y < d}} f(x,y) \, dx \, dy \right|,$$

so that, by (16) applied to the interval $t \leq x \leq \tau$, $c \leq y \leq d$,

$$|F(\tau) - F(t)| \leq M(d - c)(\tau - t).$$

Hence $F(t)$ is continuous. [Indeed, for $\epsilon > 0$, set $\delta = \epsilon/M(d - c)$. Then $|F(t_1) - F(t_2)| < \epsilon$ for $|t_1 - t_2| < \delta$.] This proves (i).

We note next that the right side of (21) is bounded by the Remark in §4.3,

$$\left| \int_c^d f(x,y) \, dy \right| \leq M(d - c).$$

Note that the proof of the continuity of $F(t)$ and of the boundedness of $F'(t)$, given by (21), remain valid for *all* bounded f, continuous or not.

We consider next the difference quotient $[F(t + h) - F(t)]/h$ for some fixed t, $a < t < b$, and for $h \neq 0$, $|h|$ small. For the sake of definiteness assume that $h > 0$. Let $\epsilon > 0$ be given, and set $\epsilon_1 = \epsilon/(d - c)$. Since f is uniformly continuous in I, there is a $\delta > 0$ such that

$$|f(x_1,y) - f(x_2,y)| < \epsilon_1, \qquad \text{for } |x_1 - x_2| < \delta \tag{26}$$

and for all y, $c < y < d$. We assume that $0 < h < \delta$ and denote by R the interval

$$R: t < x < t + h, \qquad c < y < d. \tag{27}$$

Now

$$\frac{1}{h}[F(t+h) - F(h)] = \frac{1}{h}\underline{\iint_R} f(x,y)\, dx\, dy \qquad\qquad \text{[by (19) and (13)]}$$

$$= \frac{1}{h}\underline{\iint_R} \{f(t,y) + [f(x,y) - f(t,y)]\}\, dx\, dy$$

$$\leq \frac{1}{h}\underline{\iint_R} \{f(t,y) + \epsilon_1\}\, dx\, dy \qquad\qquad \text{[by (26) and (15)]}$$

$$= \int_c^d \{f(t,y) + \epsilon_1\}\, dy \qquad\qquad \text{[by Lemma 3]}$$

$$= \int_{\underline{c}}^d \{f(t,y) + \epsilon_1\}\, dy \qquad\qquad \text{[by §4.2]}$$

$$= \int_c^d f(t,y)\, dy + (d - c)\,\epsilon_1 = \int_c^d f(t,y)\, dy + \epsilon.$$

We prove similarly that

$$\frac{1}{h}[F(t+h) - f(t)] \geq \int_c^d f(t,y)\, dy - \epsilon,$$

and we can obtain a corresponding result for $h < 0$. Hence

$$\left| \frac{F(t+h) - F(t)}{h} - \int_c^d f(t,y)\, dy \right| \leq \epsilon$$

if $|h| < \delta$. In other words, (21) holds.

The continuity of $F'(t)$ now follows from Lemma 1 in §1. This completes the proof of Lemma 4 for continuous f.

7.6 Conclusion of the proof

We proceed to prove Lemma 4 for the case when $f(x,y)$ is bounded [satisfies (25)] and piecewise continuous. For the sake of definiteness we assume that

$$f(x,y) \text{ is continuous for all } (x,y) \text{ with } a \leq x \leq b,$$
$$c \leq y \leq d, \text{ except for } x = \xi \text{ and for } x = w(y), \text{ where}$$
$$\xi \text{ is a given point, } a < \xi < b, \text{ and } w(x) \text{ a continuous} \qquad (28)$$
$$\text{increasing function defined for } a \leq x \leq b \text{ and}$$
$$\text{satisfying } c < w(x) < d; \text{ see Figure S.5.}$$

Under this assumption we shall show that (21) holds for $a < t < \xi$ and for $\xi < t < b$ and that $F'(t)$ is continuous for $a < t < b$, $t \neq \xi$.

Let a t, $a < t < b$ and $t \neq \xi$, and an $\epsilon > 0$ be given. We must exhibit a $\delta > 0$ such that, for $0 < |h| < \delta$, we have

$$\left| \frac{F(t+h) - F(h)}{h} - \int_c^d f(t,y)\, dy \right| < \epsilon \qquad (29)$$

and

$$\left| \int_c^d f(t+h,y)\, dy - \int_c^d f(t,y)\, dy \right| < \epsilon. \qquad (30)$$

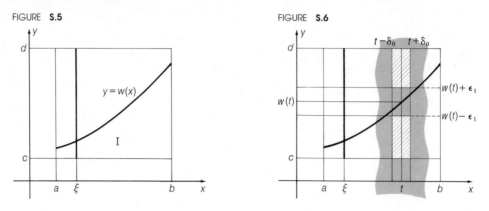

FIGURE **S.5**

FIGURE **S.6**

We set $\epsilon_1 = \epsilon/12M$ and choose a $\delta_0 > 0$ such that $\delta_0 < |t - \xi|$ and

$$|w(x) - w(t)| < \epsilon_1, \qquad \text{for } |x - t| < \delta_0 \tag{31}$$

(this is possible since w is continuous). Now $f(x,y)$ is continuous in the closed intervals

$$t - \delta_0 \leq x \leq t + \delta_0 \qquad \text{and} \qquad c \leq y \leq w(t) - \epsilon_1 \tag{32}$$

and

$$t - \delta_0 \leq x \leq t + \delta_0 \qquad \text{and} \qquad w(t) + \epsilon_1 \leq y \leq d. \tag{33}$$

These are the shaded intervals in Figure S.6. For $t - \delta_0 < s < t + \delta_0$, consider the three intervals

$$\alpha: t - \delta_0 < x < s, \qquad c < y < w(t) - \epsilon_1,$$

$$\beta: t - \delta_0 < x < s, \qquad w(t) - \epsilon_1 < y < w(t) + \epsilon_1,$$

$$\gamma: t - \delta_0 < x < s, \qquad w(t) + \epsilon_1 < y < d,$$

and let $A(s)$, $B(s)$, and $C(s)$ be the lower integrals of $f(x,y)$ over α, β, and γ, respectively. Then, by (19), (13), and (14),

$$F(s) = A(s) + B(s) + C(s) + \text{const.}, \tag{34}$$

where the constant is the lower integral of f over the interval $a < x < t - \delta_0$, $c < y < d$. Hence

$$\frac{F(t + h) - F(t)}{h} = \frac{A(t + h) - A(t)}{h} + \frac{B(t + h) - B(t)}{h} + \frac{C(t + h) - C(t)}{h}. \tag{35}$$

By (13), $B(t + h) - B(t)$ is the lower integral of f over the interval

$$\begin{cases} t < x < t + h \qquad (\text{or } t + h < x < t \text{ if } h < 0), \\ w(t) - \epsilon_1 < y < w(t) + \epsilon_1. \end{cases}$$

Noting (25) and (16), we conclude that

$$|B(t + h) - B(t)| \leq M|h|(2\epsilon_1) = |h|\frac{\epsilon}{6}. \tag{36}$$

Also,

$$\left| \int_{w(t)-\epsilon_1}^{w(t)+\epsilon_1} f(t,y)\, dy \right| \le M(2\epsilon_1) = \frac{\epsilon}{6}, \tag{37}$$

so that, by (36) and (37),

$$\left| \frac{B(t+h) - B(t)}{h} - \int_{w(t)-\epsilon_1}^{w(t)+\epsilon_1} f(t,y)\, dy \right| \le \frac{\epsilon}{3}. \tag{38}$$

On the other hand, applying the results of §7.5 to the intervals (32) and (33), we can find a δ, with $0 < \delta < \delta_0$, such that

$$\left| \frac{A(t+h) - A(t)}{h} - \int_c^{w(t)-\epsilon_1} f(t,y)\, dy \right| < \frac{\epsilon}{3}, \tag{39}$$

$$\left| \frac{C(t+h) - C(t)}{h} - \int_{w(t)+\epsilon_1}^{d} f(t,y)\, dy \right| < \frac{\epsilon}{3}, \tag{40}$$

for $0 < |h| < \delta$. For such h, we conclude from (35), (38), (39), and (40) that (29) holds. This establishes (21).

Applying Lemma 1 in §7.1 to the intervals (32) and (33), we can find a δ, with $0 < \delta < \delta_0$, such that

$$\left| \int_c^{w(t)-\epsilon_1} f(t+h,y)\, dy - \int_c^{w(t)-\epsilon_1} f(t,y)\, dy \right| < \frac{\epsilon}{3}, \tag{41}$$

$$\left| \int_{w(t)+\epsilon_1}^{d} f(t+h,y)\, dy - \int_{w(t)+\epsilon_1}^{d} f(t,y)\, dy \right| < \frac{\epsilon}{3} \tag{42}$$

for $|h| < \delta$. On the other hand, by (37), and a similar inequality with t replaced by $t + h$, we have

$$\left| \int_{w(t)-\epsilon_1}^{w(t)+\epsilon_1} f(t+h,y)\, dy - \int_{w(t)-\epsilon_1}^{w(t)+\epsilon_1} f(t,y)\, dy \right| \le \frac{\epsilon}{3}. \tag{43}$$

From (41), (42), and (43) inequality (30) follows. This establishes the continuity of $F'(t)$, for $t \ne \xi$, and completes the proof of Lemma 4 under the assumption (28).

A similar argument can be made for every piecewise continuous function (see the definition of piecewise continuity in Chapter 20, §2.1).

7.7   On the definition of area

The analytic definition of area A of a plane region D reads

$$A(D) = A = \iint_D dx\, dy = \iint_I \chi_D(x,y)\, dx\, dy; \tag{44}$$

see Chapter 20, §1.4. It is assumed that D is bounded and its characteristic function χ_D (which equals 1 on D and 0 outside D) is piecewise continuous. The letter I in (44) denotes some interval containing D; the choice of I does not affect the value of A. We saw in Chapter 20, §2.4, that (44) is essentially the area formula of Chapter 9, §1.1.

Assume that we introduce another Cartesian coordinate system (X, Y), and let $\widehat{\chi}_D$ denote the characteristic function of D in the new system. Then $\widehat{\chi}_D$ is again piecewise continuous (the definition of piecewise continuity in Chapter 20, §2.1, insures this). Is the area $\widehat{A}$ computed by the formula

$$\widehat{A}(D) = \iint_I \widehat{\chi}_D(X, Y) \, dX \, dY$$

equal to $A(D)$? In other words, is the area of D independent of the position of D?

Our geometric intuition tells us: yes, of course. Let us justify the intuition by a proof. The proof involves several steps.

(i) If D is a rectangle with sides a, b, then $A(D) = ab$, no matter how D is situated.

This is verified by an elementary calculation.

(ii) Call D an *elementary region* if it can be decomposed into finitely many nonoverlapping rectangles $R_1, R_2, \ldots, R_n$. Then $A(D) = A(R_1) + \cdots + A(R_n)$.

This follows by relations (15) and (16) in Chapter 20, §1.4. Hence $A(D)$ is independent of position.

(iii) If D' and D'' are two bounded sets with piecewise continuous characteristic functions, and if D' is contained in D'', then $A(D') \leq A(D'')$.

Proof.   Note that $\chi_{D'} \leq \chi_{D''}$, and apply relation (13) in Chapter 20, §1.4.

(iv) Let D be a bounded set with piecewise continuous characteristic function, and let ϵ be any given positive number. Then there are elementary regions D_1 and D_2 such that D_1 is contained in D and D is contained in D_2, and $A(D_2) - A(D_1) \leq \epsilon$.

Proof.   By the theory of double integrals expounded in §§7.3–7.6, there are step functions ϕ and ψ such that $\phi \leq \chi_D \leq \psi$ and

$$\iint_I \phi \, dx \, dy \leq \iint_I \chi_D \, dx \, dy = A(D) \leq \iint_I \psi \, dx \, dy,$$

$$\iint_I \psi \, dx \, dy - \iint_I \phi \, dx \, dy \leq \epsilon.$$

We may assume that ϕ and ψ have the same intervals of constancy: $R_1, R_2, \ldots, R_N$.

Define two new step functions $\widehat{\phi}$ and $\widehat{\psi}$ by setting: $\widehat{\phi} = 0$ in R_j if R_j contains some point not in D, $\widehat{\phi} = 1$ if R_j is contained in D; $\widehat{\psi} = 0$ in R_j if R_j contains no points of D, $\widehat{\psi} = 1$ if R_j contains some point of D. Then, as the reader should verify, $\phi \leq \widehat{\phi} \leq \chi_D \leq \widehat{\psi} \leq \psi$. Hence

$$\iint_I \widehat{\phi} \, dx \, dy \leq A(D) \leq \iint_I \widehat{\psi} \, dx \, dy, \qquad \iint_I \widehat{\psi} \, dx \, dy - \iint_I \widehat{\phi} \, dx \, dy \leq \epsilon.$$

Now $\widehat{\phi}$ and $\widehat{\psi}$ are characteristic functions of elementary regions D_1 and D_2 having the required properties.

(v) Let D be as in step (iv). Then $A(D) = \widehat{A}(D)$.

Proof.   Let D_1 and D_2 be as in step (iv). By step (iii), $\widehat{A}(D_1) \leq \widehat{A}(D) \leq \widehat{A}(D_2)$. By step (ii), $\widehat{A}(D_1) = A(D_1)$, $\widehat{A}(D_2) = A(D_2)$. Since both numbers, $A(D)$ and $\widehat{A}(D)$, lie between $A(D_1)$ and $A(D_2)$, we have that $|A(D) - \widehat{A}(D)| \leq \epsilon$. Since ϵ was arbitrary, $A(D) = \widehat{A}(D)$.

7.8   Polar coordinates

We conclude by proving the formula for integrating in polar coordinates, given in Chapter 20, §3.2.

Let R be a positive number; we consider in this subsection only such bounded piecewise continuous functions $f(x,y)$ which are 0 for $R > 0$ and for which $(r,\theta) \mapsto f(r\cos\theta, r\sin\theta)$ is a bounded piecewise continuous function of (r,θ) in $0 < r < R$, $0 < \theta < 2\pi$. For such f, set

$$T[f] = \int_{x^2+y^2<R^2}\!\!\int f(x,y)\,dx\,dy, \qquad T_*[f] = \int_0^{2\pi}\!\int_0^R f(r\cos\theta, r\sin\theta)r\,dr\,d\theta.$$

We must show that

$$T[f] = T_*[f]. \tag{45}$$

Since $T_*[f]$ is computed by carrying out two single integrations, we can show that, if $f_1 \le f_2$, then $T_*[f_1] \le T_*[f_2]$. Now if f is given, we can find step functions ϕ and ψ with $\phi \le f \le \psi$, $T[\phi] \le T[f] \le T[\psi]$, and $T[\psi] - T[\phi]$ as small as we like. By the remark just made, we have the inequality $T_*[\phi] \le T_*[f] \le T_*[\psi]$. We conclude that (45) holds in general if it holds for step functions.

Since $T_*[f]$ is an iterated integral, we can show that

$$T_*[cf] = cT_*[f], \qquad c \text{ a constant,}$$

and

$$T_*[f_1 + f_2] = T_*[f_1] + T_*[f_2].$$

Every step function is a sum of functions of the form $c\chi_I$, I an interval. Hence it suffices to prove (45) when f is the characteristic function of an interval, that is, when $f = 1$ in an interval and $f = 0$ outside the interval. This can be done by a direct calculation.

[It pays to note that every interval can be decomposed into intervals lying in a quadrant; an interval lying in a quadrant can be represented as an interval with one side on a coordinate axis from which another such interval has been removed; an interval with one side on an axis can be represented as an interval with two sides on axes from which another such interval has been removed; and that an interval with two sides on the axes can be decomposed into two right triangles with one vertex at the origin and the two legs parallel to the axes. Hence (45) needs to be proved only for characteristic functions of such triangles. This is easy—see Example 4, Chapter 16, §1.4.]

REMARK We can now prove rigorously the area formula (13) in Chapter 16, §1.4. Let D be the region bounded by the map $\theta = \alpha$ and $\theta = \beta > \alpha$ and the arc $r = f(\theta)$, $\alpha \le \theta \le \beta$. [Here r and θ are polar coordinates, $0 \le \alpha < \beta \le 2\pi$, f is continuous and positive.] Then the area A of D is

$$A = \iint_D dx\,dy = \int_\alpha^\beta \int_0^{f(\theta)} r\,dr\,d\theta$$

$$= \int_\alpha^\beta \frac{1}{2}r^2 \Big|_0^{f(\theta)} d\theta = \frac{1}{2}\int_\alpha^\beta f(\theta)^2\,d\theta,$$

as asserted. (Another proof will be found in Chapter 21, §2.2.)

TABLES

Table 1   Table of integrals*

___

Forms containing ($a + bu$)

1. $\int \dfrac{du}{(a + bu)} = \dfrac{1}{b} \log|a + bu|.$

2. $\int \dfrac{du}{(a + bu)^n} = -\dfrac{1}{(n - 1)b(a + bu)^{n-1}}.$

3. $\int \dfrac{u\,du}{(a + bu)} = \dfrac{(a + bu)}{b^2} - \dfrac{a}{b^2} \log|a + bu|.$

4. $\int \dfrac{u\,du}{(a + bu)^2} = \dfrac{a}{b^2(a + bu)} + \dfrac{1}{b^2} \log|a + bu|.$

5. $\int \dfrac{u\,du}{(a + bu)^n} = -\dfrac{1}{(n - 2)b^2(a + bu)^{n-2}} + \dfrac{a}{(n - 1)b^2(a + bu)^{n-1}}.$

6. $\int \dfrac{u^2\,du}{(a + bu)} = \dfrac{(a + bu)^2}{2b^3} - \dfrac{2a(a + bu)}{b^3} + \dfrac{a^2}{b^3} \log|a + bu|.$

7. $\int \dfrac{u^2\,du}{(a + bu)^2} = \dfrac{(a + bu)}{b^3} - \dfrac{a^2}{b^3(a + bu)} - \dfrac{2a}{b^3} \log|a + bu|.$

8. $\int \dfrac{du}{u(a + bu)} = -\dfrac{1}{a} \log \left| \dfrac{a + bu}{u} \right|.$

9. $\int \dfrac{du}{u(a + bu)^2} = -\dfrac{bu}{a^2(a + bu)} - \dfrac{1}{a^2} \log \left| \dfrac{a + bu}{u} \right|.$

10. $\int \dfrac{du}{u^2(a + bu)} = -\dfrac{1}{au} + \dfrac{b}{a^2} \log \left| \dfrac{a + bu}{u} \right|.$

11. $\int \dfrac{du}{u^2(a + bu)^2} = -\dfrac{b}{a^2(a + bu)} - \dfrac{1}{a^2u} + \dfrac{2b}{a^3} \log \left| \dfrac{a + bu}{u} \right|.$

Forms containing $\sqrt{a + bu}$

12. $\int \dfrac{du}{\sqrt{a + bu}} = \dfrac{2\sqrt{a + bu}}{b}.$

13a. $\int \dfrac{du}{u\sqrt{a + bu}} = \dfrac{1}{\sqrt{a}} \log \left| \dfrac{\sqrt{a + bu} - \sqrt{a}}{\sqrt{a + bu} + \sqrt{a}} \right|, \quad a > 0.$

13b. $\int \dfrac{du}{u\sqrt{a + bu}} = \dfrac{2}{\sqrt{-a}} \arctan \left(\dfrac{\sqrt{a + bu}}{\sqrt{-a}} \right), \quad a < 0.$

14. $\int \dfrac{du}{u^n\sqrt{a + bu}} = -\dfrac{\sqrt{a + bu}}{(n - 1)au^{n-1}} - \dfrac{(2n - 3)b}{2(n - 1)a} \int \dfrac{du}{u^{n-1}\sqrt{a + bu}}.$

*The constant of integration is omitted. The formulas are valid whenever meaningful.

Table 1   713

15. $\int \dfrac{u^n\,du}{\sqrt{a+bu}} = \dfrac{2u^n\sqrt{a+bu}}{(2n+1)b} - \dfrac{2na}{(2n+1)b}\int\dfrac{u^{n-1}\,du}{\sqrt{a+bu}}.$

16. $\int \sqrt{a+bu}\,du = \dfrac{2(a+bu)^{3/2}}{3b}.$

17a. $\int \dfrac{\sqrt{a+bu}}{u}\,du = 2\sqrt{a+bu} + \sqrt{a}\log\left|\dfrac{\sqrt{a+bu}-\sqrt{a}}{\sqrt{a+bu}+\sqrt{a}}\right|,\quad a>0.$

17b. $\int \dfrac{\sqrt{a+bu}}{u}\,du = 2\sqrt{a+bu} - 2\sqrt{-a}\arctan\left(\dfrac{\sqrt{a+bu}}{\sqrt{-a}}\right),\quad a<0.$

18. $\int \dfrac{\sqrt{a+bu}}{u^n}\,du = -\dfrac{(a+bu)^{3/2}}{(n-1)au^{n-1}} - \dfrac{(2n-5)b}{2(n-1)a}\int\dfrac{\sqrt{a+bu}}{u^{n-1}}\,du.$

19. $\int u^n\sqrt{a+bu}\,du = \dfrac{2u^n(a+bu)^{3/2}}{(2n+3)b} - \dfrac{2na}{(2n+3)b}\int u^{n-1}\sqrt{a+bu}\,du.$

Forms containing (a^2+u^2)

20. $\int \dfrac{du}{(a^2+u^2)} = \dfrac{1}{a}\arctan\dfrac{u}{a}.$

21. $\int \dfrac{du}{(a^2+u^2)^n} = \dfrac{u}{2(n-1)a^2(a^2+u^2)^{n-1}} + \dfrac{(2n-3)}{2(n-1)a^2}\int\dfrac{du}{(a^2+u^2)^{n-1}}.$

22. $\int \dfrac{u\,du}{(a^2+u^2)} = \dfrac{1}{2}\log|a^2+u^2|.$

23. $\int \dfrac{u\,du}{(a^2+u^2)^n} = -\dfrac{1}{2(n-1)(a^2+u^2)^{n-1}}.$

24. $\int \dfrac{u^2\,du}{(a^2+u^2)} = u - a\arctan\dfrac{u}{a}.$

25. $\int \dfrac{u^2\,du}{(a^2+u^2)^n} = -\dfrac{u}{2(n-1)(a^2+u^2)^{n-1}} + \dfrac{1}{2(n-1)}\int\dfrac{du}{(a^2+u^2)^{n-1}}.$

26. $\int \dfrac{du}{u(a^2+u^2)} = \dfrac{1}{2a^2}\log\left|\dfrac{u^2}{a^2+u^2}\right|.$

27. $\int \dfrac{du}{u^2(a^2+u^2)} = -\dfrac{1}{a^2u} - \dfrac{1}{a^3}\arctan\dfrac{u}{a}.$

Forms containing (a^2-u^2)

28. $\int \dfrac{du}{a^2-u^2} = \dfrac{1}{2a}\log\left|\dfrac{a+u}{a-u}\right|.$

29. $\int \dfrac{du}{(a^2-u^2)^n} = \dfrac{u}{2(n-1)a^2(a^2-u^2)^{n-1}} + \dfrac{2n-3}{2(n-1)a^2}\int\dfrac{du}{(a^2-u^2)^{n-1}}.$

30. $\int \dfrac{u\,du}{(a^2 - u^2)} = -\dfrac{1}{2}\log|a^2 - u^2|.$

31. $\int \dfrac{u\,du}{(a^2 - u^2)^n} = \dfrac{1}{2(n-1)(a^2 - u^2)^{n-1}}.$

32. $\int \dfrac{u^2\,du}{(a^2 - u^2)} = -u + \dfrac{a}{2}\log\left|\dfrac{a+u}{a-u}\right|.$

33. $\int \dfrac{u^2\,du}{(a^2 - u^2)^n} = \dfrac{u}{2(n-1)(a^2 - u^2)^{n-1}} - \dfrac{1}{2(n-1)}\int \dfrac{du}{(a^2 - u^2)^{n-1}}.$

34. $\int \dfrac{du}{u(a^2 - u^2)} = \dfrac{1}{2a^2}\log\left|\dfrac{u^2}{a^2 - u^2}\right|.$

35. $\int \dfrac{du}{u^2(a^2 - u^2)} = -\dfrac{1}{a^2 u} + \dfrac{1}{2a^3}\log\left|\dfrac{a+u}{a-u}\right|.$

Forms containing $\sqrt{a^2 + u^2}$

36. $\int \dfrac{du}{\sqrt{a^2 + u^2}} = \log|u + \sqrt{a^2 + u^2}|.$

37. $\int \dfrac{u^n\,du}{\sqrt{a^2 + u^2}} = \dfrac{u^{n-1}\sqrt{a^2 + u^2}}{n} - \dfrac{(n-1)a^2}{n}\int \dfrac{u^{n-2}\,du}{\sqrt{a^2 + u^2}}.$

38. $\int \dfrac{du}{u\sqrt{a^2 + u^2}} = -\dfrac{1}{a}\log\left|\dfrac{a + \sqrt{a^2 + u^2}}{u}\right|.$

39. $\int \dfrac{du}{u^n\sqrt{a^2 + u^2}} = -\dfrac{\sqrt{a^2 + u^2}}{(n-1)a^2 u^{n-1}} - \dfrac{(n-2)}{(n-1)a^2}\int \dfrac{du}{u^{n-2}\sqrt{a^2 + u^2}}.$

40. $\int \sqrt{a^2 + u^2}\,du = \dfrac{1}{2}u\sqrt{a^2 + u^2} + \dfrac{a^2}{2}\log|u + \sqrt{a^2 + u^2}|.$

41. $\int u^n\sqrt{a^2 + u^2}\,du = \dfrac{u^{n-1}(a^2 + u^2)^{3/2}}{(n+2)} - \dfrac{(n-1)a^2}{(n+2)}\int u^{n-2}\sqrt{a^2 + u^2}\,du.$

42. $\int \dfrac{\sqrt{a^2 + u^2}}{u}\,du = \sqrt{a^2 + u^2} - a\log\left|\dfrac{a + \sqrt{a^2 + u^2}}{u}\right|.$

43. $\int \dfrac{\sqrt{a^2 + u^2}}{u^n}\,du = -\dfrac{(a^2 + u^2)^{3/2}}{(n-1)a^2 u^{n-1}} - \dfrac{(n-4)}{(n-1)a^2}\int \dfrac{\sqrt{a^2 + u^2}}{u^{n-2}}\,du.$

44. $\int \dfrac{du}{(a^2 + u^2)^{3/2}} = \dfrac{u}{a^2\sqrt{a^2 + u^2}}.$

45. $\int \dfrac{u\,du}{(a^2 + u^2)^{3/2}} = -\dfrac{1}{\sqrt{a^2 + u^2}}.$

46. $\int \dfrac{u^2\,du}{(a^2 + u^2)^{3/2}} = -\dfrac{u}{\sqrt{a^2 + u^2}} + \log|u + \sqrt{a^2 + u^2}|.$

Table 1 715

47. $\int \dfrac{u^n \, du}{(a^2 + u^2)^{3/2}} = \dfrac{u^{n-1}}{(n-2)\sqrt{a^2 + u^2}} - \dfrac{(n-1)a^2}{(n-2)} \int \dfrac{u^{n-2} \, du}{(a^2 + u^2)^{3/2}}.$

48. $\int \dfrac{du}{u(a^2 + u^2)^{3/2}} = \dfrac{1}{a^2 \sqrt{a^2 + u^2}} - \dfrac{1}{a^3} \log \left| \dfrac{a + \sqrt{a^2 + u^2}}{u} \right|.$

49. $\int \dfrac{du}{u^n (a^2 + u^2)^{3/2}} = -\dfrac{1}{(n-1)a^2 u^{n-1} \sqrt{a^2 + u^2}} - \dfrac{n}{(n-1)a^2} \int \dfrac{du}{u^{n-2}(a^2 + u^2)^{3/2}}.$

50. $\int (a^2 + u^2)^{3/2} \, du = \dfrac{1}{4} u(a^2 + u^2)^{3/2} + \dfrac{3a^2}{8} u \sqrt{a^2 + u^2} + \dfrac{3a^4}{8} \log|u + \sqrt{a^2 + u^2}|.$

51. $\int u^n (a^2 + u^2)^{3/2} \, du = \dfrac{u^{n-1}(a^2 + u^2)^{5/2}}{(n+4)} - \dfrac{(n-1)a^2}{(n+4)} \int u^{n-2}(u^2 + a^2)^{3/2} \, du.$

52. $\int \dfrac{(a^2 + u^2)^{3/2}}{u} \, du = \dfrac{1}{3}(a^2 + u^2)^{3/2} + a^2 \sqrt{a^2 + u^2} - a^3 \log \left| \dfrac{a + \sqrt{a^2 + u^2}}{u} \right|.$

53. $\int \dfrac{(a^2 + u^2)^{3/2}}{u^n} \, du = -\dfrac{(a^2 + u^2)^{5/2}}{(n-1)a^2 u^{n-1}} - \dfrac{(n-6)}{(n-1)a^2} \int \dfrac{(a^2 + u^2)^{3/2}}{u^{n-2}} \, du.$

Forms containing $\sqrt{a^2 - u^2}$

54. $\int \dfrac{du}{\sqrt{a^2 - u^2}} = \arcsin \dfrac{u}{a}.$

55. $\int \dfrac{u^n \, du}{\sqrt{a^2 - u^2}} = -\dfrac{u^{n-1}\sqrt{a^2 - u^2}}{n} + \dfrac{(n-1)a^2}{n} \int \dfrac{u^{n-2} \, du}{\sqrt{a^2 - u^2}}.$

56. $\int \dfrac{du}{u\sqrt{a^2 - u^2}} = -\dfrac{1}{a} \log \left| \dfrac{a + \sqrt{a^2 - u^2}}{u} \right|.$

57. $\int \dfrac{du}{u^n \sqrt{a^2 - u^2}} = -\dfrac{\sqrt{a^2 - u^2}}{(n-1)a^2 u^{n-1}} + \dfrac{(n-2)}{(n-1)a^2} \int \dfrac{du}{u^{n-2}\sqrt{a^2 - u^2}}.$

58. $\int \sqrt{a^2 - u^2} \, du = \dfrac{1}{2} u \sqrt{a^2 - u^2} + \dfrac{a^2}{2} \arcsin \dfrac{u}{a}.$

59. $\int u^n \sqrt{a^2 - u^2} \, du = -\dfrac{u^{n-1}(a^2 - u^2)^{3/2}}{(n+2)} + \dfrac{(n-1)a^2}{(n+2)} \int u^{n-2}\sqrt{a^2 - u^2} \, du.$

60. $\int \dfrac{\sqrt{a^2 - u^2}}{u} \, du = \sqrt{a^2 - u^2} - a \log \left| \dfrac{a + \sqrt{a^2 - u^2}}{u} \right|.$

61. $\int \dfrac{\sqrt{a^2 - u^2}}{u^n} \, du = -\dfrac{(a^2 - u^2)^{3/2}}{(n-1)a^2 u^{n-1}} + \dfrac{(n-4)}{(n-1)a^2} \int \dfrac{\sqrt{a^2 - u^2}}{u^{n-2}} \, du.$

62. $\int \dfrac{du}{(a^2 - u^2)^{3/2}} = \dfrac{u}{a^2 \sqrt{a^2 - u^2}}.$

63. $\int \dfrac{u \, du}{(a^2 - u^2)^{3/2}} = \dfrac{1}{\sqrt{a^2 - u^2}}.$

64. $\int \dfrac{u^2\, du}{(a^2 - u^2)^{3/2}} = \dfrac{u}{\sqrt{a^2 - u^2}} - \arcsin \dfrac{u}{a}.$

65. $\int \dfrac{u^n\, du}{(a^2 - u^2)^{3/2}} = -\dfrac{u^{n-1}}{(n-2)\sqrt{a^2 - u^2}} + \dfrac{(n-1)a^2}{(n-2)} \int \dfrac{u^{n-2}\, du}{(a^2 - u^2)^{3/2}}.$

66. $\int \dfrac{du}{u(a^2 - u^2)^{3/2}} = \dfrac{1}{a^2 \sqrt{a^2 - u^2}} - \dfrac{1}{a^3} \log \left| \dfrac{a + \sqrt{a^2 - u^2}}{u} \right|.$

67. $\int \dfrac{du}{u^n(a^2 - u^2)^{3/2}} = -\dfrac{1}{(n-1)a^2 u^{n-1}\sqrt{a^2 - u^2}} + \dfrac{n}{(n-1)a^2} \int \dfrac{du}{u^{n-2}(a^2 - u^2)^{3/2}}.$

68. $\int (a^2 - u^2)^{3/2}\, du = \dfrac{1}{4} u(a^2 - u^2)^{3/2} + \dfrac{3a^2}{8} u \sqrt{a^2 - u^2} + \dfrac{3a^4}{8} \arcsin \dfrac{u}{a}.$

69. $\int u^n(a^2 - u^2)^{3/2}\, du = -\dfrac{u^{n-1}(a^2 - u^2)^{5/2}}{(n+4)} + \dfrac{(n-1)a^2}{(n+4)} \int u^{n-2}(a^2 - u^2)^{3/2}\, du.$

70. $\int \dfrac{(a^2 - u^2)^{3/2}}{u}\, du = \dfrac{1}{3}(a^2 - u^2)^{3/2} + a^2 \sqrt{a^2 - u^2} - a^3 \log \left| \dfrac{a + \sqrt{a^2 - u^2}}{u} \right|.$

71. $\int \dfrac{(a^2 - u^2)^{3/2}}{u^n}\, du = -\dfrac{(a^2 - u^2)^{5/2}}{(n-1)a^2 u^{n-1}} + \dfrac{(n-6)}{(n-1)a^2} \int \dfrac{(a^2 - u^2)^{3/2}}{u^{n-2}}\, du.$

Forms containing $\sqrt{u^2 - a^2}$

72. $\int \dfrac{du}{\sqrt{u^2 - a^2}} = \log|u + \sqrt{u^2 - a^2}|.$

73. $\int \dfrac{u^n\, du}{\sqrt{u^2 - a^2}} = \dfrac{1}{n} u^{n-1}\sqrt{u^2 - a^2} + \dfrac{(n-1)a^2}{n} \int \dfrac{u^{n-2}\, du}{\sqrt{u^2 - a^2}}.$

74. $\int \dfrac{du}{u\sqrt{u^2 - a^2}} = \dfrac{1}{a} \operatorname{arc\,sec} \left| \dfrac{u}{a} \right|.$

75. $\int \dfrac{du}{u^n \sqrt{u^2 - a^2}} = \dfrac{\sqrt{u^2 - a^2}}{(n-1)a^2 u^{n-1}} + \dfrac{(n-2)}{(n-1)a^2} \int \dfrac{du}{u^{n-2}\sqrt{u^2 - a^2}}.$

76. $\int \sqrt{u^2 - a^2}\, du = \dfrac{1}{2} u \sqrt{u^2 - a^2} - \dfrac{a^2}{2} \log|u + \sqrt{u^2 - a^2}|.$

77. $\int u^n \sqrt{u^2 - a^2}\, du = \dfrac{u^{n-1}(u^2 - a^2)^{3/2}}{(n+2)} + \dfrac{(n-1)a^2}{(n+2)} \int u^{n-2}\sqrt{u^2 - a^2}\, du.$

78. $\int \dfrac{\sqrt{u^2 - a^2}}{u}\, du = \sqrt{u^2 - a^2} - a \operatorname{arc\,sec} \left| \dfrac{u}{a} \right|.$

79. $\int \dfrac{\sqrt{u^2 - a^2}}{u^n}\, du = \dfrac{(u^2 - a^2)^{3/2}}{(n-1)a^2 u^{n-1}} + \dfrac{(n-4)}{(n-1)a^2} \int \dfrac{\sqrt{u^2 - a^2}}{u^{n-2}}\, du.$

80. $\int \dfrac{du}{(u^2 - a^2)^{3/2}} = -\dfrac{u}{a^2 \sqrt{u^2 - a^2}}.$

Table 1   717

81. $\displaystyle\int \frac{u\,du}{(u^2 - a^2)^{3/2}} = -\frac{1}{\sqrt{u^2 - a^2}}.$

82. $\displaystyle\int \frac{u^2\,du}{(u^2 - a^2)^{3/2}} = -\frac{u}{\sqrt{u^2 - a^2}} + \log|u + \sqrt{u^2 - a^2}|.$

83. $\displaystyle\int \frac{u^n\,du}{(u^2 - a^2)^{3/2}} = \frac{u^{n-1}}{(n - 2)\sqrt{u^2 - a^2}} + \frac{(n - 1)a^2}{(n - 2)}\int \frac{u^{n-2}\,du}{(u^2 - a^2)^{3/2}}.$

84. $\displaystyle\int \frac{du}{u(u^2 - a^2)^{3/2}} = -\frac{1}{a^2\sqrt{u^2 - a^2}} - \frac{1}{a^3}\,\text{arc sec}\,\left|\frac{u}{a}\right|.$

85. $\displaystyle\int \frac{du}{u^n(u^2 - a^2)^{3/2}} = \frac{1}{(n - 1)a^2 u^{n-1}\sqrt{u^2 - a^2}} + \frac{n}{(n - 1)a^2}\int \frac{du}{u^{n-2}(u^2 - a^2)^{3/2}}.$

86. $\displaystyle\int (u^2 - a^2)^{3/2}\,du = \frac{1}{4}u(u^2 - a^2)^{3/2} - \frac{3a^2}{8}u\sqrt{u^2 - a^2} + \frac{3a^4}{8}\log|u + \sqrt{u^2 - a^2}|.$

87. $\displaystyle\int u^n(u^2 - a^2)^{3/2}\,du = \frac{u^{n-1}(u^2 - a^2)^{5/2}}{(n + 4)} + \frac{(n - 1)a^2}{(n + 4)}\int u^{n-2}(u^2 - a^2)^{3/2}\,du.$

88. $\displaystyle\int \frac{(u^2 - a^2)^{3/2}}{u}\,du = \frac{1}{3}(u^2 - a^2)^{3/2} - a^2\sqrt{u^2 - a^2} + a^3\,\text{arc sec}\,\left|\frac{u}{a}\right|.$

89. $\displaystyle\int \frac{(u^2 - a^2)^{3/2}}{u^n}\,du = \frac{(u^2 - a^2)^{5/2}}{(n - 1)a^2 u^{n-1}} + \frac{(n - 6)}{(n - 1)a^2}\int \frac{(u^2 - a^2)^{3/2}}{u^{n-2}}\,du.$

Trigonometric forms

90. $\displaystyle\int \sin u\,du = -\cos u.$

91. $\displaystyle\int \sin^2 u\,du = \tfrac{1}{2}u - \tfrac{1}{2}\sin u \cos u.$

92. $\displaystyle\int \sin^n u\,du = -\frac{1}{n}\sin^{n-1} u \cos u + \frac{(n - 1)}{n}\int \sin^{n-2} u\,du.$

93. $\displaystyle\int u^n \sin u\,du = -u^n \cos u + n\int u^{n-1}\cos u\,du.$

94. $\displaystyle\int \sin mu \sin nu\,du = -\frac{\sin(m + n)u}{2(m + n)} + \frac{\sin(m - n)u}{2(m - n)}.$

95a. $\displaystyle\int \frac{du}{(a + b\sin u)} = \frac{2}{\sqrt{a^2 - b^2}}\,\text{arc tan}\left(\frac{a\tan\frac{u}{2} + b}{\sqrt{a^2 - b^2}}\right),\quad a^2 > b^2.$

95b. $\displaystyle\int \frac{du}{(a + b\sin u)} = \frac{1}{\sqrt{b^2 - a^2}}\log\left|\frac{a\tan\frac{u}{2} + b - \sqrt{b^2 - a^2}}{a\tan\frac{u}{2} + b + \sqrt{b^2 - a^2}}\right|,\quad b^2 > a^2.$

96. $\displaystyle\int \cos u\,du = \sin u.$

97. $\displaystyle\int \cos^2 u\,du = \tfrac{1}{2}u + \tfrac{1}{2}\sin u \cos u.$

98. $\int \cos^n u \, du = \dfrac{1}{n} \cos^{n-1} u \sin u + \dfrac{(n-1)}{n} \int \cos^{n-2} u \, du.$

99. $\int u^n \cos u \, du = u^n \sin u - n \int u^{n-1} \sin u \, du.$

100. $\int \cos mu \cos nu \, du = \dfrac{\sin(m+n)u}{2(m+n)} + \dfrac{\sin(m-n)u}{2(m-n)}.$

101a. $\int \dfrac{du}{(a + b \cos u)} = \dfrac{2}{\sqrt{a^2 - b^2}} \arctan\left(\dfrac{(a-b)\tan \frac{u}{2}}{\sqrt{a^2 - b^2}}\right), \quad a^2 > b^2.$

101b. $\int \dfrac{du}{(a + b \cos u)} = \dfrac{1}{\sqrt{b^2 - a^2}} \log \left| \dfrac{(b-a)\tan \frac{u}{2} + \sqrt{b^2 - a^2}}{(b-a)\tan \frac{u}{2} - \sqrt{b^2 - a^2}} \right|, \quad b^2 > a^2.$

102. $\int \sin u \cos u \, du = \frac{1}{2} \sin^2 u.$

103a. $\int \sin^m u \cos^n u \, du = \dfrac{\sin^{m+1} u \cos^{n-1} u}{(m+n)} + \dfrac{(n-1)}{(m+n)} \int \sin^m u \cos^{n-2} u \, du.$

103b. $\int \sin^m u \cos^n u \, du = -\dfrac{\sin^{m-1} u \cos^{n+1} u}{(m+n)} + \dfrac{(m-1)}{(m+n)} \int \sin^{m-2} u \cos^n u \, du.$

104. $\int \sin mu \cos nu \, du = -\dfrac{\cos(m+n)u}{2(m+n)} - \dfrac{\cos(m-n)u}{2(m-n)}.$

105. $\int \tan u \, du = \log|\sec u|.$

106. $\int \tan^2 u \, du = \tan u - u.$

107. $\int \tan^n u \, du = \dfrac{1}{(n-1)} \tan^{n-1} u - \int \tan^{n-2} u \, du.$

108. $\int \sec u \, du = \log|\sec u + \tan u|.$

109. $\int \sec^2 u \, du = \tan u.$

110. $\int \sec^n u \, du = \dfrac{1}{(n-1)} \sec^{n-2} u \tan u + \dfrac{(n-2)}{(n-1)} \int \sec^{n-2} u \, du.$

111. $\int \sec u \tan u \, du = \sec u.$

112a. $\int \sec^m u \tan^n u \, du = \dfrac{\sec^m u \tan^{n-1} u}{(m+n-1)} - \dfrac{(n-1)}{(m+n-1)} \int \sec^m u \tan^{n-2} u \, du.$

112b. $\int \sec^m u \tan^n u \, du = \dfrac{\sec^{m-2} u \tan^{n+1} u}{(m+n-1)} + \dfrac{(m-2)}{(m+n-1)} \int \sec^{m-2} u \tan^n u \, du.$

113. $\int \cot u \, du = \log|\sin u|.$

114. $\int \cot^2 u \, du = -\cot u - u.$

115. $\int \cot^n u \, du = -\dfrac{1}{(n-1)} \cot^{n-1} u - \int \cot^{n-2} u \, du.$

Table 1 719

116. $\int \csc u\, du = -\log|\csc u + \cot u|.$

117. $\int \csc^2 u\, du = -\cot u.$

118. $\int \csc^n u\, du = -\dfrac{1}{(n-1)} \csc^{n-2} u \cot u + \dfrac{(n-2)}{(n-1)} \int \csc^{n-2} u\, du.$

119. $\int \csc u \cot u\, du = -\csc u.$

120a. $\int \csc^m u \cot^n u\, du = -\dfrac{\csc^m u \cot^{n-1} u}{(m+n-1)} - \dfrac{(n-1)}{(m+n-1)} \int \csc^m u \cot^{n-2} u\, du.$

120b. $\int \csc^m u \cot^n u\, du = -\dfrac{\csc^{m-2} u \cot^{n+1} u}{(m+n-1)} + \dfrac{(m-2)}{(m+n-1)} \int \csc^{m-2} u \cot^n u\, du.$

Inverse trigonometric forms

121. $\int \arcsin u\, du = u \arcsin u + \sqrt{1-u^2}.$

122. $\int \arccos u\, du = u \arccos u - \sqrt{1-u^2}.$

123. $\int \arctan u = u \arctan u - \tfrac{1}{2}\log|1+u^2|.$

124. $\int \operatorname{arc\,sec} u\, du = u \operatorname{arc\,sec} u - \log|u + \sqrt{u^2-1}|.$

125. $\int \operatorname{arc\,cot} u\, du = u \operatorname{arc\,cot} u + \tfrac{1}{2}\log|1+u^2|.$

126. $\int \operatorname{arc\,csc} u\, du = u \operatorname{arc\,csc} u + \log|u + \sqrt{u^2-1}|.$

Exponential and logarithmic forms

127. $\int e^u\, du = e^u.$

128. $\int a^u\, du = \dfrac{a^u}{\log a}.$

129. $\int u^n e^u\, du = u^n e^u - n \int u^{n-1} e^u\, du.$

130. $\int e^u \sin au\, du = \dfrac{e^u}{(1+a^2)}(\sin au - a \cos au).$

131. $\int e^u \sin^n au\, du = \dfrac{e^u \sin^{n-1} au}{(1+n^2 a^2)}(\sin au - na \cos au) + \dfrac{n(n-1)a^2}{(1+n^2 a^2)} \int e^u \sin^{n-2} au\, du.$

132. $\int e^u \cos au\, du = \dfrac{e^u}{(1+a^2)}(a \sin au + \cos au).$

133. $\int e^u \cos^n au\, du = \dfrac{e^u \cos^{n-1} au}{(1+n^2 a^2)}(\cos au + na \sin au) + \dfrac{n(n-1)a^2}{(1+n^2 a^2)} \int e^u \cos^{n-2} au\, du.$

134. $\int \log u\, du = u \log u - u.$

135. $\int \dfrac{\log u}{u}\, du = \dfrac{1}{2}(\log u)^2.$

136. $\int u^n \log u \, du = -\dfrac{u^{n+1}}{(n+1)^2} + \dfrac{u^{n+1}}{(n+1)} \log u.$

137. $\int u^n (\log u)^m \, du = \dfrac{u^{n+1}(\log u)^m}{(n+1)} - \dfrac{m}{(n+1)} \int u^n (\log u)^{m-1} \, du.$

138. $\int \dfrac{du}{u \log u} = \log|\log u|.$

139. $\int \dfrac{du}{u(\log u)^n} = -\dfrac{1}{(n-1)(\log u)^{n-1}}.$

Hyperbolic forms

140. $\int \sinh u \, du = \cosh u.$

141. $\int \cosh u \, du = \sinh u.$

142. $\int \tanh u \, du = \log|\cosh u|.$

143. $\int \operatorname{sech} u \, du = \operatorname{arc\,tan}(\sinh u).$

144. $\int \operatorname{sech}^2 u \, du = \tanh u.$

145. $\int \operatorname{sech} u \tanh u \, du = -\operatorname{sech} u.$

146. $\int \coth u \, du = \log|\sinh u|.$

147. $\int \operatorname{csch} u \, du = \log \left| \tanh \dfrac{u}{2} \right|.$

148. $\int \operatorname{csch}^2 u \, du = -\coth u.$

149. $\int \coth u \operatorname{csch} u \, du = -\operatorname{csch} u.$

Table 2   721

Table 2   Powers, roots, and reciprocals of numbers 1–100*

N	N^2	$\sqrt{N}$	$\sqrt{10N}$	N^3	$\sqrt[3]{N}$	$\sqrt[3]{10N}$	$\sqrt[3]{100N}$	$1000/N$
1	1	1.00 000	3.16 228	1	1.00 000	2.15 443	4.64 159	1000.00
2	4	1.41 421	4.47 214	8	1.25 992	2.71 442	5.84 804	500.00 0
3	9	1.73 205	5.47 723	27	1.44 225	3.10 723	6.69 433	333.33 3
4	16	2.00 000	6.32 456	64	1.58 740	3.41 995	7.36 806	250.00 0
5	25	2.23 607	7.07 107	125	1.70 998	3.68 403	7.93 701	200.00 0
6	36	2.44 949	7.74 597	216	1.81 712	3.91 487	8.43 433	166.66 7
7	49	2.64 575	8.36 660	343	1.91 293	4.12 129	8.87 904	142.85 7
8	64	2.82 843	8.94 427	512	2.00 000	4.30 887	9.28 318	125.00 0
9	81	3.00 000	9.48 683	729	2.08 008	4.48 140	9.65 489	111.11 1
10	100	3.16 228	10.00 00	1 000	2.15 443	4.64 159	10.00 00	100.00 0
11	121	3.31 662	10.48 81	1 331	2.22 398	4.79 142	10.32 28	90.90 91
12	144	3.46 410	10.95 45	1 728	2.28 943	4.93 242	10.62 66	83.33 33
13	169	3.60 555	11.40 18	2 197	2.35 133	5.06 580	10.91 39	76.92 31
14	196	3.74 166	11.83 22	2 744	2.41 014	5.19 249	11.18 69	71.42 86
15	225	3.87 298	12.24 74	3 375	2.46 621	5.31 329	11.44 71	66.66 67
16	256	4.00 000	12.64 91	4 096	2.51 984	5.42 884	11.69 61	62.50 00
17	289	4.12 311	13.03 84	4 913	2.57 128	5.53 966	11.93 48	58.82 35
18	324	4.24 264	13.41 64	5 832	2.62 074	5.64 622	12.16 44	55.55 56
19	361	4.35 890	13.78 40	6 859	2.66 840	5.74 890	12.38 56	52.63 16
20	400	4.47 214	14.14 21	8 000	2.71 442	5.84 804	12.59 92	50.00 00
21	441	4.58 258	14.49 14	9 261	2.75 892	5.94 392	12.80 58	47.61 90
22	484	4.69 042	14.83 24	10 648	2.80 204	6.03 681	13.00 59	45.45 45
23	529	4.79 583	15.16 58	12 167	2.84 387	6.12 693	13.20 01	43.47 83
24	576	4.89 898	15.49 19	13 824	2.88 450	6.21 446	13.38 87	41.66 67
25	625	5.00 000	15.81 14	15 625	2.92 402	6.29 961	13.57 21	40.00 00
26	676	5.09 902	16.12 45	17 576	2.96 250	6.38 250	13.75 07	38.46 15
27	729	5.19 615	16.43 17	19 683	3.00 000	6.46 330	13.92 48	37.03 70
28	784	5.29 150	16.73 32	21 952	3.03 659	6.54 213	14.09 46	35.71 43
29	841	5.38 516	17.02 94	24 389	3.07 232	6.61 911	14.26 04	34.48 28
30	900	5.47 723	17.32 05	27 000	3.10 723	6.69 433	14.42 25	33.33 33
31	961	5.56 776	17.60 68	29 791	3.14 138	6.76 790	14.58 10	32.25 81
32	1 024	5.65 685	17.88 85	32 768	3.17 480	6.83 990	14.73 61	31.25 00
33	1 089	5.74 456	18.16 59	35 937	3.20 753	6.91 042	14.88 81	30.30 30
34	1 156	5.83 095	18.43 91	39 304	3.23 961	6.97 953	15.03 69	29.41 18
35	1 225	5.91 608	18.70 83	42 875	3.27 107	7.04 730	15.18 29	28.57 14
36	1 296	6.00 000	18.97 37	46 656	3.30 193	7.11 379	15.32 62	27.77 78
37	1 369	6.08 276	19.23 54	50 653	3.33 222	7.17 905	15.46 68	27.02 70
38	1 444	6.16 441	19.49 36	54 872	3.36 198	7.24 316	15.60 49	26.31 58
39	1 521	6.24 500	19.74 84	59 319	3.39 121	7.30 614	15.74 06	25.64 10
40	1 600	6.32 456	20.00 00	64 000	3.41 995	7.36 806	15.87 40	25.00 00
41	1 681	6.40 312	20.24 85	68 921	3.44 822	7.42 896	16.00 52	24.39 02
42	1 764	6.48 074	20.49 39	74 088	3.47 603	7.48 887	16.13 43	23.80 95
43	1 849	6.55 744	20.73 64	79 507	3.50 340	7.54 784	16.26 13	23.25 58
44	1 936	6.63 325	20.97 62	85 184	3.53 035	7.60 590	16.38 64	22.72 73
45	2 025	6.70 820	21.21 32	91 125	3.55 689	7.66 309	16.50 96	22.22 22
46	2 116	6.78 233	21.44 76	97 336	3.58 305	7.71 944	16.63 10	21.73 91
47	2 209	6.85 565	21.67 95	103 823	3.60 883	7.77 498	16.75 07	21.27 66
48	2 304	6.92 820	21.90 89	110 592	3.63 424	7.82 974	16.86 87	20.83 33
49	2 401	7.00 000	22.13 59	117 649	3.65 931	7.88 374	16.98 50	20.40 82
50	2 500	7.07 107	22.36 07	125 000	3.68 403	7.93 701	17.09 98	20.00 00

*From *Rinehart Mathematical Tables, Formulas and Curves*, compiled by Harold D. Larsen. Copyright © 1948, 1953 by Harold D. Larsen. Reprinted by permission of Holt, Rinehart and Winston.

Table 2 (continued)   Powers, roots, and reciprocals of numbers 1–100

N	N^2	$\sqrt{N}$	$\sqrt{10N}$	N^3	$\sqrt[3]{N}$	$\sqrt[3]{10N}$	$\sqrt[3]{100N}$	$1000/N$
51	2 601	7.14 143	22.58 32	132 651	3.70 843	7.98 957	17.21 30	19.60 78
52	2 704	7.21 110	22.80 35	140 608	3.73 251	8.04 145	17.32 48	19.23 08
53	2 809	7.28 011	23.02 17	148 877	3.75 629	8.09 267	17.43 51	18.86 79
54	2 916	7.34 847	23.23 79	157 464	3.77 976	8.14 325	17.54 41	18.51 85
55	3 025	7.41 620	23.45 21	166 375	3.80 295	8.19 321	17.65 17	18.18 18
56	3 136	7.48 331	23.66 43	175 616	3.82 586	8.24 257	17.75 81	17.85 71
57	3 249	7.54 983	23.87 47	185 193	3.84 850	8.29 134	17.86 32	17.54 39
58	3 364	7.61 577	24.08 32	195 112	3.87 088	8.33 955	17.96 70	17.24 14
59	3 481	7.68 115	24.28 99	205 379	3.89 300	8.38 721	18.06 97	16.94 92
60	3 600	7.74 597	24.49 49	216 000	3.91 487	8.43 433	18.17 12	16.66 67
61	3 721	7.81 025	24.69 82	226 981	3.93 650	8.48 093	18.27 16	16.39 34
62	3 844	7.87 401	24.89 98	238 328	3.95 789	8.52 702	18.37 09	16.12 90
63	3 969	7.93 725	25.09 98	250 047	3.97 906	8.57 262	18.46 91	15.87 30
64	4 096	8.00 000	25.29 82	262 144	4.00 000	8.61 774	18.56 64	15.62 50
65	4 225	8.06 226	25.49 51	274 625	4.02 073	8.66 239	18.66 26	15.38 46
66	4 356	8.12 404	25.69 05	287 496	4.04 124	8.70 659	18.75 78	15.15 15
67	4 489	8.18 535	25.88 44	300 763	4.06 155	8.75 034	18.85 20	14.92 54
68	4 624	8.24 621	26.07 68	314 432	4.08 166	8.79 366	18.94 54	14.70 59
69	4 761	8.30 662	26.26 79	328 509	4.10 157	8.83 656	19.03 78	14.49 28
70	4 900	8.36 660	26.45 75	343 000	4.12 129	8.87 904	19.12 93	14.28 57
71	5 041	8.42 615	26.64 58	357 911	4.14 082	8.92 112	19.22 00	14.08 45
72	5 184	8.48 528	26.83 28	373 248	4.16 017	8.96 281	19.30 98	13.88 89
73	5 329	8.54 400	27.01 85	389 017	4.17 934	9.00 411	19.39 88	13.69 86
74	5 476	8.60 233	27.20 29	405 224	4.19 834	9.04 504	19.48 70	13.51 35
75	5 625	8.66 025	27.38 61	421 875	4.21 716	9.08 560	19.57 43	13.33 33
76	5 776	8.71 780	27.56 81	438 976	4.23 582	9.12 581	19.66 10	13.15 79
77	5 929	8.77 496	27.74 89	456 533	4.25 432	9.16 566	19.74 68	12.98 70
78	6 084	8.83 176	27.92 85	474 552	4.27 266	9.20 516	19.83 19	12.82 05
79	6 241	8.88 819	28.10 69	493 039	4.29 084	9.24 434	19.91 63	12.65 82
80	6 400	8.94 427	28.28 43	512 000	4.30 887	9.28 318	20.00 00	12.50 00
81	6 561	9.00 000	28.46 05	531 441	4.32 675	9.32 170	20.08 30	12.34 57
82	6 724	9.05 539	28.63 56	551 368	4.34 448	9.35 990	20.16 53	12.19 51
83	6 889	9.11 043	28.80 97	571 787	4.36 207	9.39 780	20.24 69	12.04 82
84	7 056	9.16 515	28.98 28	592 704	4.37 952	9.43 539	20.32 79	11.90 48
85	7 225	9.21 954	29.15 48	614 125	4.39 683	9.47 268	20.40 83	11.76 47
86	7 396	9.27 362	29.32 58	636 056	4.41 400	9.50 969	20.48 80	11.62 79
87	7 569	9.32 738	29.49 58	658 503	4.43 105	9.54 640	20.56 71	11.49 43
88	7 744	9.38 083	29.66 48	681 472	4.44 796	9.58 284	20.64 56	11.36 36
89	7 921	9.43 398	29.83 29	704 969	4.46 475	9.61 900	20.72 35	11.23 60
90	8 100	9.48 683	30.00 00	729 000	4.48 140	9.65 489	20.80 08	11.11 11
91	8 281	9.53 939	30.16 62	753 571	4.49 794	9.69 052	20.87 76	10.98 90
92	8 464	9.59 166	30.33 15	778 688	4.51 436	9.72 589	20.95 38	10.86 96
93	8 649	9.64 365	30.49 59	804 357	4.53 065	9.76 100	21.02 94	10.75 27
94	8.836	9.69 536	30.65 94	830 684	4.54 684	9.79 586	21.10 45	10.63 83
95	9 025	9.74 679	30.82 21	857 375	4.56 290	9.83 048	21.17 91	10.52 63
96	9 216	9.79 796	30.98 39	884 736	4.57 886	9.86 485	21.25 32	10.41 67
97	9 409	9.84 886	31.14 48	912 673	4.59 470	9.89 898	21.32 67	10.30 93
98	9 604	9.89 949	31.30 50	941 192	4.61 044	9.93 288	21.39 97	10.20 41
99	9 801	9.94 987	31.46 43	970 299	4.62 607	9.96 655	21.47 23	10.10 10
100	10 000	10.00 000	31.62 28	1 000 000	4.64 159	10.00 000	21.54 43	10.00 00

Table 3   723

Table 3   Trigonometric functions of numbers 0–1.6

x	sin x	tan x	cot x	cos x	x	sin x	tan x	cot x	cos x
0.00	.00000	.00000	—	1.00000	**0.40**	.38942	.42279	2.3652	.92106
.01	.01000	.01000	99.997	0.99995	.41	.39861	.43463	2.3008	.91712
.02	.02000	.02000	49.993	.99980	.42	.40776	.44657	2.2393	.91309
.03	.03000	.03001	33.323	.99955	.43	.41687	.45862	2.1804	.90897
.04	.03999	.04002	24.987	.99920	.44	.42594	.47078	2.1241	.90475
.05	.04998	.05004	19.983	.99875	.45	.43497	.48306	2.0702	.90045
.06	.05996	.06007	16.647	.99820	.46	.44395	.49545	2.0184	.89605
.07	.06994	.07011	14.262	.99755	.47	.45289	.50797	1.9686	.89157
.08	.07991	.08017	12.473	.99680	.48	.46178	.52061	1.9208	.88699
.09	.08988	.09024	11.081	.99595	.49	.47063	.53339	1.8748	.88233
0.10	.09983	.10033	9.9666	.99500	**0.50**	.47943	.54630	1.8305	.87758
.11	.10978	.11045	9.0542	.99396	.51	.48818	.55936	1.7878	.87274
.12	.11971	.12058	8.2933	.99281	.52	.49688	.57256	1.7465	.86782
.13	.12963	.13074	7.6489	.99156	.53	.50553	.58592	1.7067	.86281
.14	.13954	.14092	7.0961	.99022	.54	.51414	.59943	1.6683	.85771
.15	.14944	.15114	6.6166	.98877	.55	.52269	.61311	1.6310	.85252
.16	.15932	.16138	6.1966	.98723	.56	.53119	.62695	1.5950	.84726
.17	.16918	.17166	5.8256	.98558	.57	.53963	.64097	1.5601	.84190
.18	.17903	.18197	5.4954	.98384	.58	.54802	.65517	1.5263	.83646
.19	.18886	.19232	5.1997	.98200	.59	.55636	.66956	1.4935	.83094
0.20	.19867	.20271	4.9332	.98007	**0.60**	.56464	.68414	1.4617	.82534
.21	.20846	.21314	4.6917	.97803	.61	.57287	.69892	1.4308	.81965
.22	.21823	.22362	4.4719	.97590	.62	.58104	.71391	1.4007	.81388
.23	.22798	.23414	4.2709	.97367	.63	.58914	.72911	1.3715	.80803
.24	.23770	.24472	4.0864	.97134	.64	.59720	.74454	1.3431	.80210
.25	.24740	.25534	3.9163	.96891	.65	.60519	.76020	1.3154	.79608
.26	.25708	.26602	3.7591	.96639	.66	.61312	.77610	1.2885	.78999
.27	.26673	.27676	3.6133	.96377	.67	.62099	.79225	1.2622	.78382
.28	.27636	.28755	3.4776	.96106	.68	.62879	.80866	1.2366	.77757
.29	.28595	.29841	3.3511	.95824	.69	.63654	.82534	1.2116	.77125
0.30	.29552	.30934	3.2327	.95534	**0.70**	.64422	.84229	1.1872	.76484
.31	.30506	.32033	3.1218	.95233	.71	.65183	.85953	1.1634	.75836
.32	.31457	.33139	3.0176	.94924	.72	.65938	.87707	1.1402	.75181
.33	.32404	.34252	2.9195	.94604	.73	.66687	.89492	1.1174	.74517
.34	.33349	.35374	2.8270	.94275	.74	.67429	.91309	1.0952	.73847
.35	.34290	.36503	2.7395	.93937	.75	.68164	.93160	1.0734	.73169
.36	.35227	.37640	2.6567	.93590	.76	.68892	.95045	1.0521	.72484
.37	.36162	.38786	2.5782	.93233	.77	.69614	.96967	1.0313	.71791
.38	.37092	.39941	2.5037	.92866	.78	.70328	.98926	1.0109	.71091
.39	.38019	.41105	2.4328	.92491	.79	.71035	1.0092	.99084	.70385
x	sin x	tan x	cot x	cos x	x	sin x	tan x	cot x	cos x

Table 3 (continued)   Trigonometric functions of numbers 0–1.6

x	$\sin x$	$\tan x$	$\cot x$	$\cos x$	x	$\sin x$	$\tan x$	$\cot x$	$\cos x$
0.80	.71736	1.0296	.97121	.69671	**1.20**	.93204	2.5722	.38878	.36236
.81	.72429	1.0505	.95197	.68950	1.21	.93562	2.6503	.37731	.35302
.82	.73115	1.0717	.93309	.68222	1.22	.93910	2.7328	.36593	.34365
.83	.73793	1.0934	.91455	.67488	1.23	.94249	2.8198	.35463	.33424
.84	.74464	1.1156	.89635	.66746	1.24	.94578	2.9119	.34341	.32480
.85	.75128	1.1383	.87848	.65998	1.25	.94898	3.0096	.33227	.31532
.86	.75784	1.1616	.86091	.65244	1.26	.95209	3.1133	.32121	.30582
.87	.76433	1.1853	.84365	.64483	1.27	.95510	3.2236	.31021	.29628
.88	.77074	1.2097	.82668	.63715	1.28	.95802	3.3413	.29928	.28672
.89	.77707	1.2346	.80998	.62941	1.29	.96084	3.4672	.28842	.27712
0.90	.78333	1.2602	.79355	.62161	**1.30**	.96356	3.6021	.27762	.26750
.91	.78950	1.2864	.77738	.61375	1.31	.96618	3.7471	.26687	.25785
.92	.79560	1.3133	.76146	.60582	1.32	.96872	3.9033	.25619	.24818
.93	.80162	1.3409	.74578	.59783	1.33	.97115	4.0723	.24556	.23848
.94	.80756	1.3692	.73034	.58979	1.34	.97348	4.2556	.23498	.22875
.95	.81342	1.3984	.71511	.58168	1.35	.97572	4.4552	.22446	.21901
.96	.81919	1.4284	.70010	.57352	1.36	.97786	4.6734	.21398	.20924
.97	.82489	1.4592	.68531	.56530	1.37	.97991	4.9131	.20354	.19945
.98	.83050	1.4910	.67071	.55702	1.38	.98185	5.1774	.19315	.18964
.99	.83603	1.5237	.65631	.54869	1.39	.98370	5.4707	.18279	.17981
1.00	.84147	1.5574	.64209	.54030	**1.40**	.98545	5.7979	.17248	.16997
1.01	.84683	1.5922	.62806	.53186	1.41	.98710	6.1654	.16220	.16010
1.02	.85211	1.6281	.61420	.52337	1.42	.98865	6.5811	.15195	.15023
1.03	.85730	1.6652	.60051	.51482	1.43	.99010	7.0555	.14173	.14033
1.04	.86240	1.7036	.58699	.50622	1.44	.99146	7.6018	.13155	.13042
1.05	.86742	1.7433	.57362	.49757	1.45	.99271	8.2381	.12139	.12050
1.06	.87236	1.7844	.56040	.48887	1.46	.99387	8.9886	.11125	.11057
1.07	.87720	1.8270	.54734	.48012	1.47	.99492	9.8874	.10114	.10063
1.08	.88196	1.8712	.53441	.47133	1.48	.99588	10.983	.09105	.09067
1.09	.88663	1.9171	.52162	.46249	1.49	.99674	12.350	.08097	.08071
1.10	.89121	1.9648	.50897	.45360	**1.50**	.99749	14.101	.07091	.07074
1.11	.89570	2.0143	.49644	.44466	1.51	.99815	16.428	.06087	.06076
1.12	.90010	2.0660	.48404	.43568	1.52	.99871	19.670	.05084	.05077
1.13	.90441	2.1198	.47175	.42666	1.53	.99917	24.498	.04082	.04079
1.14	.90863	2.1759	.45959	.41759	1.54	.99953	32.461	.03081	.03079
1.15	.91276	2.2345	.44753	.40849	1.55	.99978	48.078	.02080	.02079
1.16	.91680	2.2958	.43558	.39934	1.56	.99994	92.621	.01080	.01080
1.17	.92075	2.3600	.42373	.39015	1.57	1.00000	1255.8	.00080	.00080
1.18	.92461	2.4273	.41199	.38092	1.58	.99996	−108.65	−.00920	−.00920
1.19	.92837	2.4979	.40034	.37166	1.59	.99982	−52.067	−.01921	−.01920
					1.60	.99957	−34.233	−.02921	−.02920
x	$\sin x$	$\tan x$	$\cot x$	$\cos x$					

Table 4   725

Table 4   Trigonometric functions. Arguments in degrees

Degrees	Radians	Sine	Tangent	Cotangent	Cosine		
0	0	0	0	———	1.0000	1.5708	**90**
1	.0175	.0175	.0175	57.290	.9998	1.5533	89
2	.0349	.0349	.0349	28.636	.9994	1.5359	88
3	.0524	.0523	.0524	19.081	.9986	1.5184	87
4	.0698	.0698	.0699	14.301	.9976	1.5010	86
5	.0873	.0872	.0875	11.430	.9962	1.4835	**85**
6	.1047	.1045	.1051	9.5144	.9945	1.4661	84
7	.1222	.1219	.1228	8.1443	.9925	1.4486	83
8	.1396	.1392	.1405	7.1154	.9903	1.4312	82
9	.1571	.1564	.1584	6.3138	.9877	1.4137	81
10	.1745	.1736	.1763	5.6713	.9848	1.3963	**80**
11	.1920	.1908	.1944	5.1446	.9816	1.3788	79
12	.2094	.2079	.2126	4.7046	.9781	1.3614	78
13	.2269	.2250	.2309	4.3315	.9744	1.3439	77
14	.2443	.2419	.2493	4.0108	.9703	1.3265	76
15	.2618	.2588	.2679	3.7321	.9659	1.3090	**75**
16	.2793	.2756	.2867	3.4874	.9613	1.2915	74
17	.2967	.2924	.3057	3.2709	.9563	L.2741	73
18	.3142	.3090	.3249	3.0777	.9511	1.2566	72
19	.3316	.3256	.3443	2.9042	.9455	1.2392	71
20	.3491	.3420	.3640	2.7475	.9397	1.2217	**70**
21	.3665	.3584	.3839	2.6051	.9336	1.2043	69
22	.3840	.3746	.4040	2.4751	.9272	1.1868	68
23	.4014	.3907	.4245	2.3559	.9205	1.1694	67
24	.4189	.4067	.4452	2.2460	.9135	1.1519	66
25	.4363	.4226	.4663	2.1445	.9063	1.1345	**65**
26	.4538	.4384	.4877	2.0503	.8988	1.1170	64
27	.4712	.4540	.5095	1.9626	.8910	1.0996	63
28	.4887	.4695	.5317	1.8807	.8829	1.0821	62
29	.5061	.4848	.5543	1.8040	.8746	1.0647	61
30	.5236	.5000	.5774	1.7321	.8660	1.0472	**60**
31	.5411	.5150	.6009	1.6643	.8572	1.0297	59
32	.5585	.5299	.6249	1.6003	.8480	1.0123	58
33	.5760	.5446	.6494	1.5399	.8387	.9948	57
34	.5934	.5592	.6745	1.4826	.8290	.9774	56
35	.6109	.5736	.7002	1.4281	.8192	.9599	**55**
36	.6283	.5878	.7265	1.3764	.8090	.9425	54
37	.6458	.6018	.7536	1.3270	.7986	.9250	53
38	.6632	.6157	.7813	1.2799	.7880	.9076	52
39	.6807	.6293	.8098	1.2349	.7771	.8901	51
40	.6981	.6428	.8391	1.1918	.7660	.8727	**50**
41	.7156	.6561	.8693	1.1504	.7547	.8552	49
42	.7330	.6691	.9004	1.1106	.7431	.8378	48
43	.7505	.6820	.9325	1.0724	.7314	.8203	47
44	.7679	.6947	.9657	1.0355	.7193	.8029	46
45	.7854	.7071	1.0000	1.0000	.7071	.7854	**45**
		Cosine	Cotangent	Tangent	Sine	Radians	Degrees

Table 5 Natural logarithms of numbers 1.00–10.09

Logarithms of numbers outside this range can be computed by using the functional equation ($\log ab = \log a + \log b$) and the following values of the logarithmic function: $\log .1 = .6974$–3; $\log .01 = .3948$–5; $\log .001 = .0922$–7; $\log .0001 = .7897$–10; $\log .00001 = .4871$–12; $\log .000\,001 = .1845$–14.

N	.00	.01	.02	.03	.04	.05	.06	.07	.08	.09
1.0	0.0000	0.0100	0.0198	0.0296	0.0392	0.0488	0.0583	0.0677	0.0770	0.0862
1.1	0.0953	0.1044	0.1133	0.1222	0.1310	0.1398	0.1484	0.1570	0.1655	0.1740
1.2	0.1823	0.1906	0.1989	0.2070	0.2151	0.2231	0.2311	0.2390	0.2469	0.2546
1.3	0.2624	0.2700	0.2776	0.2852	0.2927	0.3001	0.3075	0.3148	0.3221	0.3293
1.4	0.3365	0.3436	0.3507	0.3577	0.3646	0.3716	0.3784	0.3853	0.3920	0.3988
1.5	0.4055	0.4121	0.4187	0.4253	0.4318	0.4383	0.4447	0.4511	0.4574	0.4637
1.6	0.4700	0.4762	0.4824	0.4886	0.4947	0.5008	0.5068	0.5128	0.5188	0.5247
1.7	0.5306	0.5365	0.5423	0.5481	0.5539	0.5596	0.5653	0.5710	0.5766	0.5822
1.8	0.5878	0.5933	0.5988	0.6043	0.6098	0.6152	0.6206	0.6259	0.6313	0.6366
1.9	0.6419	0.6471	0.6523	0.6575	0.6627	0.6678	0.6729	0.6780	0.6831	0.6881
2.0	0.6931	0.6981	0.7031	0.7080	0.7129	0.7178	0.7227	0.7275	0.7324	0.7372
2.1	0.7419	0.7467	0.7514	0.7561	0.7608	0.7655	0.7701	0.7747	0.7793	0.7839
2.2	0.7885	0.7930	0.7975	0.8020	0.8065	0.8109	0.8154	0.8198	0.8242	0.8286
2.3	0.8329	0.8372	0.8416	0.8459	0.8502	0.8544	0.8587	0.8629	0.8671	0.8713
2.4	0.8755	0.8796	0.8838	0.8879	0.8920	0.8961	0.9002	0.9042	0.9083	0.9123
2.5	0.9163	0.9203	0.9243	0.9282	0.9322	0.9361	0.9400	0.9439	0.9478	0.9517
2.6	0.9555	0.9594	0.9632	0.9670	0.9708	0.9746	0.9783	0.9821	0.9858	0.9895
2.7	0.9933	0.9969	1.0006	1.0043	1.0080	1.0116	1.0152	1.0188	1.0225	1.0260
2.8	1.0296	1.0332	1.0367	1.0403	1.0438	1.0473	1.0508	1.0543	1.0578	1.0613
2.9	1.0647	1.0682	1.0716	1.0750	1.0784	1.0818	1.0852	1.0886	1.0919	1.0953
3.0	1.0986	1.1019	1.1053	1.1086	1.1119	1.1151	1.1184	1.1217	1.1249	1.1282
3.1	1.1314	1.1346	1.1378	1.1410	1.1442	1.1474	1.1506	1.1537	1.1569	1.1600
3.2	1.1632	1.1663	1.1694	1.1725	1.1756	1.1787	1.1817	1.1848	1.1878	1.1909
3.3	1.1939	1.1969	1.2000	1.2030	1.2060	1.2090	1.2119	1.2149	1.2179	1.2208
3.4	1.2238	1.2267	1.2296	1.2326	1.2355	1.2384	1.2413	1.2442	1.2470	1.2499
3.5	1.2528	1.2556	1.2585	1.2613	1.2641	1.2669	1.2698	1.2726	1.2754	1.2782
3.6	1.2809	1.2837	1.2865	1.2892	1.2920	1.2947	1.2975	1.3002	1.3029	1.3056
3.7	1.3083	1.3110	1.3137	1.3164	1.3191	1.3218	1.3244	1.3271	1.3297	1.3324
3.8	1.3350	1.3376	1.3403	1.3429	1.3455	1.3481	1.3507	1.3533	1.3558	1.3584
3.9	1.3610	1.3635	1.3661	1.3686	1.3712	1.3737	1.3762	1.3788	1.3813	1.3838
4.0	1.3863	1.3888	1.3913	1.3938	1.3962	1.3987	1.4012	1.4036	1.4061	1.4085
4.1	1.4110	1.4134	1.4159	1.4183	1.4207	1.4231	1.4255	1.4279	1.4303	1.4327
4.2	1.4351	1.4375	1.4398	1.4422	1.4446	1.4469	1.4493	1.4516	1.4540	1.4563
4.3	1.4586	1.4609	1.4633	1.4656	1.4679	1.4702	1.4725	1.4748	1.4770	1.4793
4.4	1.4816	1.4839	1.4861	1.4884	1.4907	1.4929	1.4951	1.4974	1.4996	1.5019
4.5	1.5041	1.5063	1.5085	1.5107	1.5129	1.5151	1.5173	1.5195	1.5217	1.5239
4.6	1.5261	1.5282	1.5304	1.5326	1.5347	1.5369	1.5390	1.5412	1.5433	1.5454
4.7	1.5476	1.5497	1.5518	1.5539	1.5560	1.5581	1.5602	1.5623	1.5644	1.5665
4.8	1.5686	1.5707	1.5728	1.5748	1.5769	1.5790	1.5810	1.5831	1.5851	1.5872
4.9	1.5892	1.5913	1.5933	1.5953	1.5974	1.5994	1.6014	1.6034	1.6054	1.6074
5.0	1.6094	1.6114	1.6134	1.6154	1.6174	1.6194	1.6214	1.6233	1.6253	1.6273
5.1	1.6292	1.6312	1.6332	1.6351	1.6371	1.6390	1.6409	1.6429	1.6448	1.6467
5.2	1.6487	1.6506	1.6525	1.6544	1.6563	1.6582	1.6601	1.6620	1.6639	1.6658
5.3	1.6677	1.6696	1.6715	1.6734	1.6752	1.6771	1.6790	1.6808	1.6827	1.6845
5.4	1.6864	1.6882	1.6901	1.6919	1.6938	1.6956	1.6974	1.6993	1.7011	1.7029
5.5	1.7047	1.7066	1.7084	1.7102	1.7120	1.7138	1.7156	1.7174	1.7192	1.7210
N	.00	.01	.02	.03	.04	.05	.06	.07	.08	.09

Table 5 727

Table 5 (continued) Natural logarithms of numbers 1.00–10.09

N	.00	.01	.02	.03	.04	.05	.06	.07	.08	.09
5.5	1.7047	1.7066	1.7084	1.7102	1.7120	1.7138	1.7156	1.7174	1.7192	1.7210
5.6	1.7228	1.7246	1.7263	1.7281	1.7299	1.7317	1.7334	1.7352	1.7370	1.7387
5.7	1.7405	1.7422	1.7440	1.7457	1.7475	1.7492	1.7509	1.7527	1.7544	1.7561
5.8	1.7579	1.7596	1.7613	1.7630	1.7647	1.7664	1.7681	1.7699	1.7716	1.7733
5.9	1.7750	1.7766	1.7783	1.7800	1.7817	1.7834	1.7851	1.7867	1.7884	1.7901
6.0	1.7918	1.7934	1.7951	1.7967	1.7984	1.8001	1.8017	1.8034	1.8050	1.8066
6.1	1.8083	1.8099	1.8116	1.8132	1.8148	1.8165	1.8181	1.8197	1.8213	1.8229
6.2	1.8245	1.8262	1.8278	1.8294	1.8310	1.8326	1.8342	1.8358	1.8374	1.8390
6.3	1.8405	1.8421	1.8437	1.8453	1.8469	1.8485	1.8500	1.8516	1.8532	1.8547
6.4	1.8563	1.8579	1.8594	1.8610	1.8625	1.8641	1.8656	1.8672	1.8687	1.8703
6.5	1.8718	1.8733	1.8749	1.8764	1.8779	1.8795	1.8810	1.8825	1.8840	1.8856
6.6	1.8871	1.8886	1.8901	1.8916	1.8931	1.8946	1.8961	1.8976	1.8991	1.9006
6.7	1.9021	1.9036	1.9051	1.9066	1.9081	1.9095	1.9110	1.9125	1.9140	1.9155
6.8	1.9169	1.9184	1.9199	1.9213	1.9228	1.9242	1.9257	1.9272	1.9286	1.9301
6.9	1.9315	1.9330	1.9344	1.9359	1.9373	1.9387	1.9402	1.9416	1.9430	1.9445
7.0	1.9459	1.9473	1.9488	1.9502	1.9516	1.9530	1.9544	1.9559	1.9573	1.9587
7.1	1.9601	1.9615	1.9629	1.9643	1.9657	1.9671	1.9685	1.9699	1.9713	1.9727
7.2	1.9741	1.9755	1.9769	1.9782	1.9796	1.9810	1.9824	1.9838	1.9851	1.9865
7.3	1.9879	1.9892	1.9906	1.9920	1.9933	1.9947	1.9961	1.9974	1.9988	2.0001
7.4	2.0015	2.0028	2.0042	2.0055	2.0069	2.0082	2.0096	2.0109	2.0122	2.0136
7.5	2.0149	2.0162	2.0176	2.0189	2.0202	2.0215	2.0229	2.0242	2.0255	2.0268
7.6	2.0281	2.0295	2.0308	2.0321	2.0334	2.0347	2.0360	2.0373	2.0386	2.0399
7.7	2.0412	2.0425	2.0438	2.0451	2.0464	2.0477	2.0490	2.0503	2.0516	2.0528
7.8	2.0541	2.0554	2.0567	2.0580	2.0592	2.0605	2.0618	2.0631	2.0643	2.0656
7.9	2.0669	2.0681	2.0694	2.0707	2.0719	2.0732	2.0744	2.0757	2.0769	2.0782
8.0	2.0794	2.0807	2.0819	2.0832	2.0844	2.0857	2.0869	2.0882	2.0894	2.0906
8.1	2.0919	2.0931	2.0943	2.0956	2.0968	2.0980	2.0992	2.1005	2.1017	2.1029
8.2	2.1041	2.1054	2.1066	2.1078	2.1090	2.1102	2.1114	2.1126	2.1138	2.1150
8.3	2.1163	2.1175	2.1187	2.1199	2.1211	2.1223	2.1235	2.1247	2.1258	2.1270
8.4	2.1282	2.1294	2.1306	2.1318	2.1330	2.1342	2.1353	2.1365	2.1377	2.1389
8.5	2.1401	2.1412	2.1424	2.1436	2.1448	2.1459	2.1471	2.1483	2.1494	2.1506
8.6	2.1518	2.1529	2.1541	2.1552	2.1564	2.1576	2.1587	2.1599	2.1610	2.1622
8.7	2.1633	2.1645	2.1656	2.1668	2.1679	2.1691	2.1702	2.1713	2.1725	2.1736
8.8	2.1748	2.1759	2.1770	2.1782	2.1793	2.1804	2.1815	2.1827	2.1838	2.1849
8.9	2.1861	2.1872	2.1883	2.1894	2.1905	2.1917	2.1928	2.1939	2.1950	2.1961
9.0	2.1972	2.1983	2.1994	2.2006	2.2017	2.2028	2.2039	2.2050	2.2061	2.2072
9.1	2.2083	2.2094	2.2105	2.2116	2.2127	2.2138	2.2148	2.2159	2.2170	2.2181
9.2	2.2192	2.2203	2.2214	2.2225	2.2235	2.2246	2.2257	2.2268	2.2279	2.2289
9.3	2.2300	2.2311	2.2322	2.2332	2.2343	2.2354	2.2364	2.2375	2.2386	2.2396
9.4	2.2407	2.2418	2.2428	2.2439	2.2450	2.2460	2.2471	2.2481	2.2492	2.2502
9.5	2.2513	2.2523	2.2534	2.2544	2.2555	2.2565	2.2576	2.2586	2.2597	2.2607
9.6	2.2618	2.2628	2.2638	2.2649	2.2659	2.2670	2.2680	2.2690	2.2701	2.2711
9.7	2.2721	2.2732	2.2742	2.2752	2.2762	2.2773	2.2783	2.2793	2.2803	2.2814
9.8	2.2824	2.2834	2.2844	2.2854	2.2865	2.2875	2.2885	2.2895	2.2905	2.2915
9.9	2.2925	2.2935	2.2946	2.2956	2.2966	2.2976	2.2986	2.2996	2.3006	2.3016
10.0	2.3026	2.3036	2.3046	2.3056	2.3066	2.3076	2.3086	2.3096	2.3106	2.3115
N	.00	.01	.02	.03	.04	.05	.06	.07	.08	.09

Table 6   Mantissas of common logarithms of numbers 100–999

N	0	1	2	3	4	5	6	7	8	9
10	0000	0043	0086	0128	0170	0212	0253	0294	0334	0374
11	0414	0453	0492	0531	0569	0607	0645	0682	0719	0755
12	0792	0828	0864	0899	0934	0969	1004	1038	1072	1106
13	1139	1173	1206	1239	1271	1303	1335	1367	1399	1430
14	1461	1492	1523	1553	1584	1614	1644	1673	1703	1732
15	1761	1790	1818	1847	1875	1903	1931	1959	1987	2014
16	2041	2068	2095	2122	2148	2175	2201	2227	2253	2279
17	2304	2330	2355	2380	2405	2430	2455	2480	2504	2529
18	2553	2577	2601	2625	2648	2672	2695	2718	2742	2765
19	2788	2810	2833	2856	2878	2900	2923	2945	2967	2989
20	3010	3032	3054	3075	3096	3118	3139	3160	3181	3201
21	3222	3243	3263	3284	3304	3324	3345	3365	3385	3404
22	3424	3444	3464	3483	3502	3522	3541	3560	3579	3598
23	3617	3636	3655	3674	3692	3711	3729	3747	3766	3784
24	3802	3820	3838	3856	3874	3892	3909	3927	3945	3962
25	3979	3997	4014	4031	4048	4065	4082	4099	4116	4133
26	4150	4166	4183	4200	4216	4232	4249	4265	4281	4298
27	4314	4330	4346	4362	4378	4393	4409	4425	4440	4456
28	4472	4487	4502	4518	4533	4548	4564	4579	4594	4609
29	4624	4639	4654	4669	4683	4698	4713	4728	4742	4757
30	4771	4786	4800	4814	4829	4843	4857	4871	4886	4900
31	4914	4928	4942	4955	4969	4983	4997	5011	5024	5038
32	5051	5065	5079	5092	5105	5119	5132	5145	5159	5172
33	5185	5198	5211	5224	5237	5250	5263	5276	5289	5302
34	5315	5328	5340	5353	5366	5378	5391	5403	5416	5428
35	5441	5453	5465	5478	5490	5502	5514	5527	5539	5551
36	5563	5575	5587	5599	5611	5623	5635	5647	5658	5670
37	5682	5694	5705	5717	5729	5740	5752	5763	5775	5786
38	5798	5809	5821	5832	5843	5855	5866	5877	5888	5899
39	5911	5922	5933	5944	5955	5966	5977	5988	5999	6010
40	6021	6031	6042	6053	6064	6075	6085	6096	6107	6117
41	6128	6138	6149	6160	6170	6180	6191	6201	6212	6222
42	6232	6243	6253	6263	6274	6284	6294	6304	6314	6325
43	6335	6345	6355	6365	6375	6385	6395	6405	6415	6425
44	6435	6444	6454	6464	6474	6484	6493	6503	6513	6522
45	6532	6542	6551	6561	6571	6580	6590	6599	6609	6618
46	6628	6637	6646	6656	6665	6675	6684	6693	6702	6712
47	6721	6730	6739	6749	6758	6767	6776	6785	6794	6803
48	6812	6821	6830	6839	6848	6857	6866	6875	6884	6893
49	6902	6911	6920	6928	6937	6946	6955	6964	6972	6981
50	6990	6998	7007	7016	7024	7033	7042	7050	7059	7067
51	7076	7084	7093	7101	7110	7118	7126	7135	7143	7152
52	7160	7168	7177	7185	7193	7202	7210	7218	7226	7235
53	7243	7251	7259	7267	7275	7284	7292	7300	7308	7316
54	7324	7332	7340	7348	7356	7364	7372	7380	7388	7396

Table 6 729

Table 6 (continued)   Mantissas of common logarithms of numbers 100–999

N	0	1	2	3	4	5	6	7	8	9
55	7404	7412	7419	7427	7435	7443	7451	7459	7466	7474
56	7482	7490	7497	7505	7513	7520	7528	7536	7543	7551
57	7559	7566	7574	7582	7589	7597	7604	7612	7619	7627
58	7634	7642	7649	7657	7664	7672	7679	7686	7694	7701
59	7709	7716	7723	7731	7738	7745	7752	7760	7767	7774
60	7782	7789	7796	7803	7810	7818	7825	7832	7839	7846
61	7853	7860	7868	7875	7882	7889	7896	7903	7910	7917
62	7924	7931	7938	7945	7952	7959	7966	7973	7980	7987
63	7993	8000	8007	8014	8021	8028	8035	8041	8048	8055
64	8062	8069	8075	8082	8089	8096	8102	8109	8116	8122
65	8129	8136	8142	8149	8156	8162	8169	8176	8182	8189
66	8195	8202	8209	8215	8222	8228	8235	8241	8248	8254
67	8261	8267	8274	8280	8287	8293	8299	8306	8312	8319
68	8325	8331	8338	8344	8351	8357	8363	8370	8376	8382
69	8388	8395	8401	8407	8414	8420	8426	8432	8439	8445
70	8451	8457	8463	8470	8476	8482	8488	8494	8500	8506
71	8513	8519	8525	8531	8537	8543	8549	8555	8561	8567
72	8573	8579	8585	8591	8597	8603	8609	8615	8621	8627
73	8633	8639	8645	8651	8657	8663	8669	8675	8681	8686
74	8692	8698	8704	8710	8716	8722	8727	8733	8739	8745
75	8751	8756	8762	8768	8774	8779	8785	8791	8797	8802
76	8808	8814	8820	8825	8831	8837	8842	8848	8854	8859
77	8865	8871	8876	8882	8887	8893	8899	8904	8910	8915
78	8921	8927	8932	8938	8943	8949	8954	8960	8965	8971
79	8976	8982	8987	8993	8998	9004	9009	9015	9020	9025
80	9031	9036	9042	9047	9053	9058	9063	9069	9074	9079
81	9085	9090	9096	9101	9106	9112	9117	9122	9128	9133
82	9138	9143	9149	9154	9159	9165	9170	9175	9180	9186
83	9191	9196	9201	9206	9212	9217	9222	9227	9232	9238
84	9243	9248	9253	9258	9263	9269	9274	9279	9284	9289
85	9294	9299	9304	9309	9315	9320	9325	9330	9335	9340
86	9345	9350	9355	9360	9365	9370	9375	9380	9385	9390
87	9395	9400	9405	9410	9415	9420	9425	9430	9435	9440
88	9445	9450	9455	9460	9465	9469	9474	9479	9484	9489
89	9494	9499	9504	9509	9513	9518	9523	9528	9533	9538
90	9542	9547	9552	9557	9562	9566	9571	9576	9581	9586
91	9590	9595	9600	9605	9609	9614	9619	9624	9628	9633
92	9638	9643	9647	9652	9657	9661	9666	9671	9675	9680
93	9685	9689	9694	9699	9703	9708	9713	9717	9722	9727
94	9731	9736	9741	9745	9750	9754	9759	9763	9768	9773
95	9777	9782	9786	9791	9795	9800	9805	9809	9814	9818
96	9823	9827	9832	9836	9841	9845	9850	9854	9859	9863
97	9868	9872	9877	9881	9886	9890	9894	9899	9903	9908
98	9912	9917	9921	9926	9930	9934	9939	9943	9948	9952
99	9956	9961	9965	9969	9974	9978	9983	9987	9991	9996

Table 7 The exponential function and its reciprocal in the range 0–6

x	e^x	e^{-x}	x	e^x	e^{-x}
0.0	1.0000	1.0000	**3.0**	20.086	.04979
0.1	1.1052	.90484	3.1	22.198	.04505
0.2	1.2214	.81873	3.2	24.533	.04076
0.3	1.3499	.74082	3.3	27.113	.03688
0.4	1.4918	.67032	3.4	29.964	.03337
0.5	1.6487	.60653	3.5	33.115	.03020
0.6	1.8221	.54881	3.6	36.598	.02732
0.7	2.0138	.49659	3.7	40.447	.02472
0.8	2.2255	.44933	3.8	44.701	.02237
0.9	2.4596	.40657	3.9	49.402	.02024
1.0	2.7183	.36788	**4.0**	54.598	.01832
1.1	3.0042	.33287	4.1	60.340	.01657
1.2	3.3201	.30119	4.2	66.686	.01500
1.3	3.6693	.27253	4.3	73.700	.01357
1.4	4.0552	.24660	4.4	81.451	.01228
1.5	4.4817	.22313	4.5	90.017	.01111
1.6	4.9530	.20190	4.6	99.484	.01005
1.7	5.4739	.18268	4.7	109.95	.00910
1.8	6.0496	.16530	4.8	121.51	.00823
1.9	6.6859	.14957	4.9	134.29	.00745
2.0	7.3891	.13534	**5.0**	148.41	.00674
2.1	8.1662	.12246	5.1	164.02	.00610
2.2	9.0250	.11080	5.2	181.27	.00552
2.3	9.9742	.10026	5.3	200.34	.00499
2.4	11.023	.09072	5.4	221.41	.00452
2.5	12.182	.08208	5.5	244.69	.00409
2.6	13.464	.07427	5.6	270.43	.00370
2.7	14.880	.06721	5.7	298.87	.00335
2.8	16.445	.06081	5.8	330.30	.00303
2.9	18.174	.05502	5.9	365.04	.00274
3.0	20.086	.04979	**6.0**	403.43	.00248

Table 8   731

Table 8   The Greek alphabet

Letters		Names		Letters		Names
A	α	alpha		N	ν	nu
B	β	beta		Ξ	ξ	xi
Γ	γ	gamma		O	o	omicron
Δ	δ	delta		Π	π	pi
E	ϵ	epsilon		P	ρ	rho
Z	ζ	zeta		Σ	σ	sigma
H	η	eta		T	τ	tau
Θ	θ	theta		Υ	υ	upsilon
I	ι	iota		Φ	ϕ	phi
K	κ	kappa		X	χ	chi
Λ	λ	lambda		Ψ	ψ	psi
M	μ	mu		Ω	ω	omega

ANSWERS TO
ODD-NUMBERED PROBLEMS

Chapter 1

§2.3

1. $2.\overline{9}$.

3. $.24\overline{9}$.

5. $4.\overline{6}$.

7. $.\overline{36}$.

§2.7

1. Yes.
3. Yes.
5. Yes.
7. Yes.
9. Yes.
11,13,15. Expand left-hand side.
17. Expand right-hand side.
19,21. Simplify left-hand side.
23. 64.

25. 64.
27. 16.
29. 1.
31. 10.
33. No.
35. No.
37. Yes.
39. a^5b^5.
41. b^6c^{12}.
43. $-a^9b^3$.
45. $a^{27}b^9$.
47. 64.
49. $\frac{1}{16}$.
51. $\frac{1}{4}$.

§2.8

1. 4.
3. .7.
5. $\frac{1}{81}$.
7. 10.

9. 16.
11. a^2b^3.
13. $a^2bc^{1/2}$.
15. $-a^{-1}$.
17. $2a^{1/2}b^2$.
19. $a^{(4n+7)/2}b^{(-4n+1)/2}$.
21. $2x^2$.
23. $-2x^2y$.
25. $2x^{-2}y^2$.
27. x^3.
29. $x^{-2}y^{1/2}$.

§3.1

1. Yes.
3. Yes.
5. Yes.
7. Yes.
9. Yes.
11. $x > y$.
13. $x = y$.

733

§3.2

1. $a = b$.
3. $a = b = c$.
5. None.
7. No.
9. No.
11. Use Hint or expand the expression $(x - 1)^2 \geq 0$, divide by x, and rearrange terms.

§3.3

1. $x > 4$.
3. $x \leq -\frac{1}{4}$.
5. No solution.
7. $x > 7$.
9. $-2 \leq x \leq 1$.
11. $-2 < x < -1$.

§3.4

1. 3.
3. 27.
5. 256.
7. 1.
9. $|x - 2|$.
11. $|x - a|$.
13. $|x - 3| < 2$.
15. $|x + 1| > 3$.
17. The set of numbers whose distance from 4 is less than or equal to 2.
19. The set of numbers whose distance from -1 is greater than or equal to 4.
21. The set of numbers whose distance from -3 is less than or equal to 4 but greater than 1.
23. $x = 1$, $x = -3$.
25. $x = \frac{7}{2}$, $x = -\frac{1}{2}$.
27. $x = 0$, $x = -1$.
29. $x = \pm 3$, $x = \pm \sqrt{7}$.
31. $-1 < x < 3$.
33. $\frac{1}{3} \leq x \leq \frac{7}{3}$.
35. $x > \frac{4}{3}$ or $x < -\frac{8}{3}$.
37. $\frac{3}{2} < x < \frac{5}{2}$ or $\frac{7}{6} < x < \frac{3}{2}$.

§3.5

1. Yes.
3. Yes.
5. Yes.
7. -1.
9. $(1, 1.1]$.

Chapter 2

§1.1

1, 3, 5, 7.

9.

11.

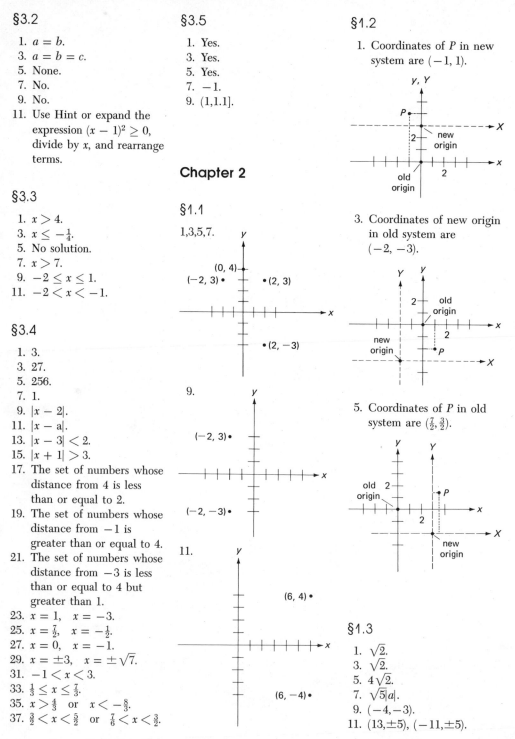

§1.2

1. Coordinates of P in new system are $(-1, 1)$.

3. Coordinates of new origin in old system are $(-2, -3)$.

5. Coordinates of P in old system are $(\frac{7}{2}, \frac{3}{2})$.

§1.3

1. $\sqrt{2}$.
3. $\sqrt{2}$.
5. $4\sqrt{2}$.
7. $\sqrt{5}|a|$.
9. $(-4, -3)$.
11. $(13, \pm 5)$, $(-11, \pm 5)$.

§2.1

1. 1.
3. 14.
5. 2.
7. $-\frac{2}{5}$.
9. $\frac{3}{2}$.
11. $-\frac{1}{5}$.
13. Yes.

§2.2

1. $-\frac{1}{2}$.
3. $\frac{2}{3}$.
5. $-\frac{4}{3}$.
7. Undefined.
9. l_1.
11. $-\frac{1}{3}$, $-\frac{1}{4}$.
13. -3, $-\frac{3}{2}$.
15. 0, 0.
17. -5, none.
19. $-C/A$, $-C/B$.
21.

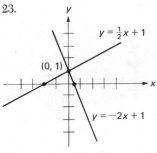

23.

25. $y = 4x + 5$.
27. $y = \frac{3}{4}x - 45$.
29. $y = -\frac{1}{2}x + 6$.
31. $y = -\frac{4}{9}x - \frac{2}{3}$.
33. $y = \frac{1}{12}x + 2$.

35. $y = -6x + 3$.
37. $y = 1$.
39. $y = 10x - 50$.
41. $y = 3x + 2$.
43. $\frac{3}{2}$, $\frac{1}{2}$.
45. $(1, 1)$.

§2.3

1. $y = \frac{2}{3}x$.
3. 3.
5. ± 2.
7. $y = -3x + 15$.
9. $y = \frac{3}{2}x$.
11. $y = \frac{6}{7}x - \frac{17}{7}$.

§3.1

1. $(x - 0)^2 + (y - 3)^2 = 2^2$.
3. $(x - 1)^2 + (y - 6)^2 = 5^2$.
5. $(x - 1)^2 + (y - 0)^2 = 4^2$.
7. Circle of radius 1 and center $(-1, -2)$.
9. No solution.
11. Circle of radius 1 and center $(-\frac{1}{2}, \frac{1}{2})$.
13. The equation of the circle is $(x - 0)^2 + (y - b)^2 = (\sqrt{1 + b^2})^2$. Note that when $y = 0$, $x = \pm 1$. The center is at $(0, b)$.
15. The center of the circle lies on the perpendicular bisector of the line segment joining the points (a, b) and (c, d). Suppose $b \neq d$. Then $(x - x_1)^2 + (y - y_1)^2 = (a - x_1)^2 + (b - y_1)^2$ where
$$y_1 = \frac{b + d}{2} - \left(\frac{c - a}{d - b}\right)$$
$$\times \left(x_1 - \frac{a + c}{2}\right) \text{ and } x_1 \text{ is}$$
arbitrary. If $b = d$, $a \neq c$, then

$$\left(x - \frac{a + c}{2}\right)^2 + (y - y_1)^2$$
$$= \left(\frac{c - a}{2}\right)^2 + (y_1 - b)^2,$$
where y_1 is arbitrary.

§3.2

1. -2.
3. $y = \frac{2}{7}x + \frac{43}{7}$.
5. $(16, 5)$.
7. $y = \dfrac{\sqrt{2}}{4}x - \sqrt{2}$,

$y = -\dfrac{\sqrt{2}}{4}x + \sqrt{2}$.

§4.2

1. Focus $(0, 2)$; directrix $y = -2$.

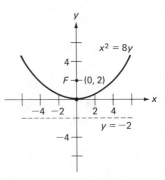

3. Focus $(1, 0)$; directrix $x = -1$.

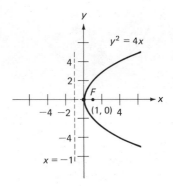

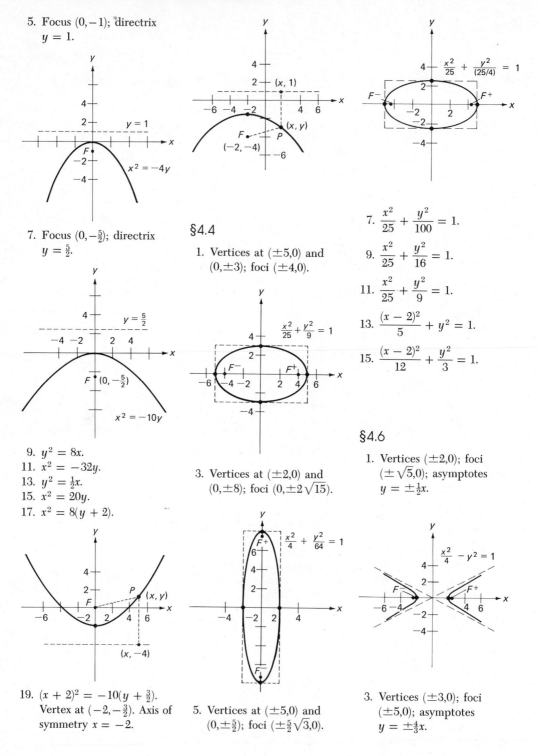

5. Focus $(0,-1)$; directrix $y = 1$.

7. Focus $(0,-\frac{5}{2})$; directrix $y = \frac{5}{2}$.

9. $y^2 = 8x$.
11. $x^2 = -32y$.
13. $y^2 = \frac{1}{2}x$.
15. $x^2 = 20y$.
17. $x^2 = 8(y + 2)$.

19. $(x + 2)^2 = -10(y + \frac{3}{2})$. Vertex at $(-2,-\frac{3}{2})$. Axis of symmetry $x = -2$.

§4.4

1. Vertices at $(\pm 5,0)$ and $(0,\pm 3)$; foci $(\pm 4,0)$.

3. Vertices at $(\pm 2,0)$ and $(0,\pm 8)$; foci $(0,\pm 2\sqrt{15})$.

5. Vertices at $(\pm 5,0)$ and $(0,\pm\frac{5}{2})$; foci $(\pm\frac{5}{2}\sqrt{3},0)$.

7. $\dfrac{x^2}{25} + \dfrac{y^2}{100} = 1.$

9. $\dfrac{x^2}{25} + \dfrac{y^2}{16} = 1.$

11. $\dfrac{x^2}{25} + \dfrac{y^2}{9} = 1.$

13. $\dfrac{(x - 2)^2}{5} + y^2 = 1.$

15. $\dfrac{(x - 2)^2}{12} + \dfrac{y^2}{3} = 1.$

§4.6

1. Vertices $(\pm 2,0)$; foci $(\pm\sqrt{5},0)$; asymptotes $y = \pm\frac{1}{2}x$.

3. Vertices $(\pm 3,0)$; foci $(\pm 5,0)$; asymptotes $y = \pm\frac{4}{3}x$.

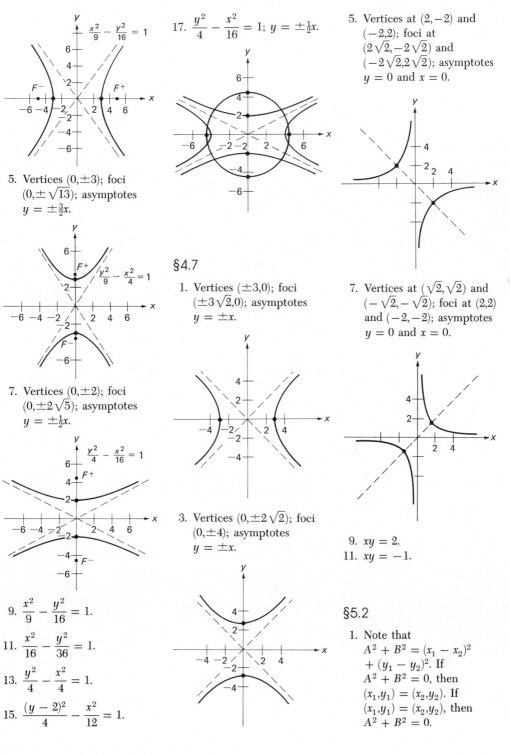

$\dfrac{x^2}{9} - \dfrac{y^2}{16} = 1$

5. Vertices $(0,\pm 3)$; foci $(0,\pm \sqrt{13})$; asymptotes $y = \pm \tfrac{3}{2}x$.

$\dfrac{y^2}{9} - \dfrac{x^2}{4} = 1$

7. Vertices $(0,\pm 2)$; foci $(0,\pm 2\sqrt{5})$; asymptotes $y = \pm \tfrac{1}{2}x$.

$\dfrac{y^2}{4} - \dfrac{x^2}{16} = 1$

9. $\dfrac{x^2}{9} - \dfrac{y^2}{16} = 1$.

11. $\dfrac{x^2}{16} - \dfrac{y^2}{36} = 1$.

13. $\dfrac{y^2}{4} - \dfrac{x^2}{4} = 1$.

15. $\dfrac{(y-2)^2}{4} - \dfrac{x^2}{12} = 1$.

17. $\dfrac{y^2}{4} - \dfrac{x^2}{16} = 1$; $y = \pm\tfrac{1}{2}x$.

§4.7

1. Vertices $(\pm 3,0)$; foci $(\pm 3\sqrt{2},0)$; asymptotes $y = \pm x$.

3. Vertices $(0,\pm 2\sqrt{2})$; foci $(0,\pm 4)$; asymptotes $y = \pm x$.

5. Vertices at $(2,-2)$ and $(-2,2)$; foci at $(2\sqrt{2},-2\sqrt{2})$ and $(-2\sqrt{2},2\sqrt{2})$; asymptotes $y = 0$ and $x = 0$.

7. Vertices at $(\sqrt{2},\sqrt{2})$ and $(-\sqrt{2},-\sqrt{2})$; foci at $(2,2)$ and $(-2,-2)$; asymptotes $y = 0$ and $x = 0$.

9. $xy = 2$.
11. $xy = -1$.

§5.2

1. Note that
$A^2 + B^2 = (x_1 - x_2)^2 + (y_1 - y_2)^2$. If
$A^2 + B^2 = 0$, then
$(x_1,y_1) = (x_2,y_2)$. If
$(x_1,y_1) = (x_2,y_2)$, then
$A^2 + B^2 = 0$.

Chapter 3

§1.2

1. 97; $\frac{7}{2}$; 31.

3. 0; $\frac{20}{17}$; $\frac{\sqrt{2}+1}{3}$.

5. 1; 8; -2.

7. $\frac{163}{2}$; 2; $\frac{99}{2}$.

9. 5; $\frac{201}{8}$; $\frac{36+\sqrt{2}}{4}$.

11. $-1 \le x \le 1$.

13. All $x \ge \frac{2}{3}$ except $x = 1$.

15. $x \ge -1$; all $x \ge -1$
    except $x = 3$.

§1.3

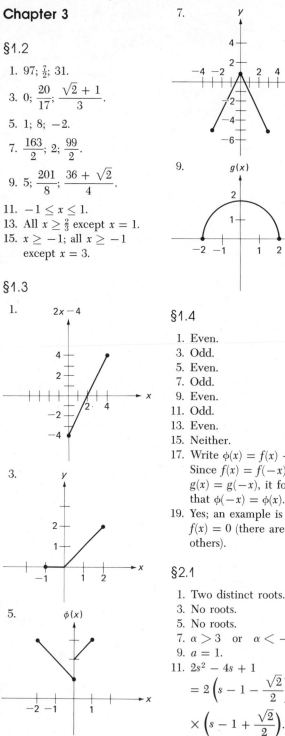

1. $2x - 4$

3.

5. $\phi(x)$

7.

9. $g(x)$

§1.4

1. Even.
3. Odd.
5. Even.
7. Odd.
9. Even.
11. Odd.
13. Even.
15. Neither.
17. Write $\phi(x) = f(x) + g(x)$.
    Since $f(x) = f(-x)$,
    $g(x) = g(-x)$, it follows
    that $\phi(-x) = \phi(x)$.
19. Yes; an example is
    $f(x) = 0$ (there are no
    others).

§2.1

1. Two distinct roots.
3. No roots.
5. No roots.
7. $\alpha > 3$ or $\alpha < -3$.
9. $a = 1$.
11. $2s^2 - 4s + 1$
$$= 2\left(s - 1 - \frac{\sqrt{2}}{2}\right)$$
$$\times \left(s - 1 + \frac{\sqrt{2}}{2}\right).$$

§2.2

1. Vertex $(0, -4)$; axis $x = 0$.

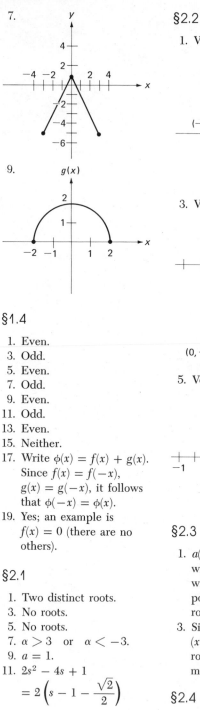

3. Vertex $(3, 0)$; axis $x = 3$.

5. Vertex $(\frac{3}{2}, \frac{1}{4})$; axis $x = \frac{3}{2}$.

§2.3

1. $a(x - 1)^2(x + 1)xg(x)$,
   where a is any number
   with $a \ne 0$ and $g(x)$ is any
   polynomial without real
   roots.

3. Since $x^4 \ne -2$ for any x,
   $(x^4 + 2)$ contains no real
   roots. $x = 2$ is a root of
   multiplicity 10.

§2.4

1. $x \mapsto \dfrac{x - 3}{x + 2}$.

3. $f(x)$ already in simplest form (there are no common roots).

5. $\phi(z) = \dfrac{1}{z-4}$.

§3.2

1. $\delta = \frac{1}{90}$.
3. $\delta = \frac{1}{10}$.
5. $\delta = \frac{5}{2}\epsilon$.
7. $\delta = \sqrt{1+\epsilon} - 1$.
9. Any δ such that $\delta < 1$ and $\delta < \frac{1}{6}\epsilon$.
11. Any δ such that $\delta < 1$ and $\delta < 2\epsilon$.

§3.3

1. $\phi(x)\psi(x)$ is continuous, $\psi(x) + \phi(x)\psi(x)$ is continuous, and $\phi(x) + \psi(x) + \phi(x)\psi(x)$ is continuous, all by Theorem 1.
3. $1 + x^2$ is continuous by Theorem 2. Since $1 + x^2 \neq 0$, $\dfrac{\phi(x)}{1+x^2}$ is continuous by Theorem 1.
5. $1 + x^2$ is continuous by Theorem 2. x^{100} is continuous by repeated application of Theorem 1. $\phi(x)\psi(x)$ is continuous, and $\phi(x)\psi(x)x^{100}$ is continuous, both by Theorem 1. Hence $\phi(x)\psi(x)x^{100}/(1+x^2)$ is continuous by Theorem 1 since $(1+x^2)$ is never zero.
7. Yes; no breaks in graph.

§3.4

1. f has at least one zero.

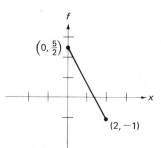

3. g has at least four zeros.

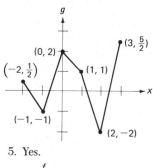

5. Yes.

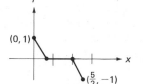

§4.2

1. -1.
3. 6.
5. 1.
7. -2.
9. 8.
11. Limit does not exist.

§4.3

1. -9.
3. 14.
5. $-\frac{7}{8}$.
7. Limit does not exist.
9. -1.
11. $\frac{5}{4}$.
13. Limit does not exist.

§4.4

1. (a) 0; (b) 0.

3. (a) $\frac{8}{5}$; (b) $\frac{8}{5}$.
5. 1.
7. No limit.

Chapter 4

§1.1

1. M approaches 4.

h	M
1	5
-1	3
.1	4.1
$-.1$	3.9
.01	4.01
$-.01$	3.99
.001	4.001
$-.001$	3.999

3. M approaches 1.

h	M
1	3
-1	-1
.1	1.2
$-.1$	.8
$-.01$	1.02
$-.01$	.98
.001	1.002
$-.001$	.998

5. $M = 10(x_0 - 1) + 5h$.
7. $M = 6x_0^2 + 2h(3x_0 + h)$.
9. $M = \dfrac{-1}{(4 + x_0)(4 + x_0 + h)}$.

§1.3

1. $f'(-1) = -6$.
3. $f'(4) = 5$.
5. $f'(2) = -1$.
7. $4x + y + 3 = 0$.
9. $x + 12y - 86 = 0$.
11. $3x - 4y - 1 = 0$.
13. $x - 4y + 19 = 0$.

§1.4

1. $f'(x) = -6x.$
3. $f'(x) = 4(x + 1).$
5. $f'(s) = 3s^2.$
7. $g'(x) = \dfrac{-2}{(2x - 3)^2}.$
9. $\dfrac{-2x}{(1 + x^2)^2}.$
11. $G'(x) = \dfrac{13}{(3x + 2)^2}.$

§1.5

1. 2.
3. $-1/x^2.$
5. $-3.$
7. 0.

§1.6

1. $x^2 + 1 = 5 + 4(x - 2) + (x - 2)^2;$ $r(x) = (x - 2).$
3. $3x^2 - x + 6 = 8 + 5(x - 1) + 3(x - 1)^2;$ $r(x) = 3(x - 1).$
5. $2x^2 - 4x + 1 = 1 - 4(x - 0) + 2x(x - 0);$ $r(x) = 2x.$
7. $\dfrac{1}{x + 1} = \frac{1}{3} - \frac{1}{9}(x - 2) + \dfrac{(x - 2)^2}{9(x + 1)};$ $r(x) = \dfrac{(x - 2)}{9(x + 1)}.$

§2.1

1. $f'.$
3. $f' - 2g'.$
5. $3\dfrac{df}{dx}.$
7. $-2\dfrac{df}{dx} + 4\dfrac{dg}{dx} - \dfrac{dh}{dx}.$
9. $4 + x.$
11. $-4x - 4.$
13. $1 + \dfrac{1}{x^2} - 8x.$

§2.2

1. $5x^4.$
3. $30x^9.$
5. $-x + 2.$
7. $-6t^2 + 6.$
9. $4t^3 + 3t^2 + 2t + 1.$
11. $12t^{11} - 18t^8 + 24t^5 - 18t^2.$
13. $-18.$
15. 48.
17. $8x + y - 13 = 0.$
19. $9x - y - 17 = 0;$ $x + 9y - 93 = 0.$
21. $(-4, -28).$

§2.3

1. $-15x^{-4}.$
3. $2x^{-3} + 6x^{-2}.$
5. $-8x^{-5} + 8x^{-3}.$
7. $-t^{-3} - 2t^{-2} + 4t.$
9. $-8t^{-9} + 12t^{-7} - 12t^{-5} + 8t^{-3}.$
11. $\dfrac{-1}{(x - 1)^2}.$
13. $\dfrac{-5}{(2x - 1)^2}.$
15. $\dfrac{2x(x + 1)}{(2x + 1)^2}.$
17. $\dfrac{-(1 + 8t)}{(1 + t + 4t^2)^2}.$
19. $\frac{2}{3}.$
21. $\frac{5}{27}.$
23. $x - y = 0.$

§2.4

1. $42x^5.$
3. $6x - 2.$
5. 48.
7. $18 + 96s.$
9. 52.
11. $\dfrac{64}{(1 + 4x)^3}.$

§3.1

1. 299 ft/sec.
3. 2 sec.
5. 5 ft/sec; $\frac{3}{2}$ sec.
7. 2 sec and 4 sec.
9. $x > \frac{1}{3}$ sec.

§3.2

1. 2 ft/sec;   2 ft/sec$^2.$
3. 0 ft/sec; $-\frac{1}{2}$ ft/sec$^2.$
5. 145 ft/sec;   84 ft/sec$^2.$
7. $t > 4$ sec; 0 sec $< t < 4$ sec.
9. $v(t) = (t - 3)(t - 5)$ and $a(t) = 2(t - 4).$ The particle starts at $s = 2$ ft and moves in the direction of increasing s to $s = 20$ ft where it stops at $t = 3$ sec. It then reverses direction and moves in the direction of decreasing s to $s = \frac{58}{3}$ ft where the acceleration becomes zero at $t = 4$ sec. The particle continues to move in the direction of decreasing s to $s = \frac{56}{3}$ ft where it stops again at $t = 5$ sec. The particle now reverses direction and moves in the direction of increasing s for all subsequent time.

§3.3

1. 64 ft/sec.
3. 3 sec;   96 ft/sec.
5. 1.6 sec, approx.

Chapter 5

§1.2

1. Maximum:
   $f(5) = 17;$ minimum:
   $f(2) = 8.$

3. Maximum:
$f(-2) = 32$;   minimum:
$f(2) = 0$.
5. Maximum:
$f(2) = 28$;   minimum:
$f(0) = 8$.
7. Maximum:
$\phi(\frac{1}{2}) = \frac{9}{7}$;   minimum:
$\phi(-4) = -\frac{9}{11}$.

§1.3

1. 1.
3. $\frac{1}{2}$.
5. 10, 10.
7. 250, 750.
9. 10 ft by 20 ft by 10 ft.
11. 256/9 square units.
13. Field should be a circle of area $250{,}000/\pi$ sq ft.
15. Faction favoring the vertical stripe wins. Vertical stripe is 1 ft wide. Flag is 4 ft by 3 ft.
17. Height $= 16$ in., width $= 8$ in.
19. For minimum area, cut wire into two pieces of length $4L/(\pi + 4)$ in. and $\pi L/(\pi + 4)$ in. Bend the first piece into a square, the second into a circle.
21. Length $= 12$ in., width $= 6$ in., height $= 8$ in.
23. $4\pi\sqrt{3}a^3/9$ cubic units.
25. Radius $=$ height $= 2$ in.
27. 125 units at $250 per unit.
29. $\frac{1}{3}$ of a mile from the weaker source. $1/(k+1)$ of a mile from the weaker source.
31. 1750° F.

§2.1

1. Increasing.
3. Nondecreasing.
5. None of these.

7. Decreasing on $(-\infty, -\frac{1}{2})$;   increasing on $(-\frac{1}{2}, +\infty)$.
9. Decreasing on $(-\infty, 0)$, $(0,1)$, and $(2, +\infty)$;   increasing on $(1,2)$.
11.

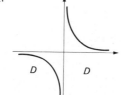

13.

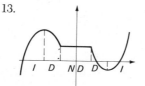

15.

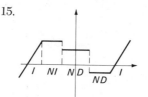

§2.3

1. $(1,3)$.
3. $(-\infty, +\infty)$.
5. $(0,2)$, $(5, +\infty)$.
7. $(-\infty, 1)$, $(2,3)$, $(4, +\infty)$.
9. $(-\infty, -3)$, $(-2,2)$, $(3, +\infty)$.
11. f increases in $(\frac{5}{2}, +\infty)$, decreases in $(-\infty, \frac{5}{2})$.
13. f increases for all x.
15. g increases in $(-\infty, 0)$, decreases in $(0, +\infty)$.
17. f increases in $(-\infty, -2)$ and $(3, +\infty)$, decreases in $(-2,3)$.
19. g increases in $(-\infty, \frac{1}{3})$ and $(1, +\infty)$, decreases in $(\frac{1}{3}, 1)$.

21. g increases in $(0, +\infty)$, decreases in $(-\infty, 0)$.
23. f increases in $(-\infty, -1)$ and $(-1, +\infty)$.

§2.4

1. f is convex for all x. No inflection points.
3. g is convex for $(-\infty, 0)$, concave for $(0, +\infty)$. Inflection point at $(0,0)$.
5. f is convex for $(-\infty, \frac{1}{2})$, concave for $(\frac{1}{2}, +\infty)$. Inflection point at $(\frac{1}{2}, -\frac{11}{2})$.
7. f is convex for $(0, +\infty)$, concave for $(-\infty, 0)$. No inflection points.
9. g is convex for $(-\infty, -\frac{2}{3}\sqrt{3})$ and $(\frac{2}{3}\sqrt{3}, +\infty)$, concave for $(-\frac{2}{3}\sqrt{3}, \frac{2}{3}\sqrt{3})$. Inflection points at $(\pm\frac{2}{3}\sqrt{3}, \frac{64}{9})$.
11. g is convex for $(-\infty, -1)$ and $(0, +\infty)$, concave for $(-1, 0)$. Inflection point at $(-1, 2)$.

§2.6

1. Minimum at $(-2, 0)$.
3. Maximum at $(\frac{1}{3}, \frac{4}{3})$.
5. No maxima or minima.
7. Maximum at $(2, 80)$;   minimum at $(-3, -45)$.
9. Minimum at $(2, 0)$.
11. Maximum at $(0, 81)$;   minima at $(\pm 3, 0)$.
13. Maxima at $(\pm 1, 12)$;   minima at $(0, 1)$ and $(\pm 2, -15)$.
15. Maximum at $(0, \frac{1}{9})$.
17. Maximum at $(0, 2)$.
19. Maximum at $(0, 1)$.
21. Maximum at $(0, 4)$;   minima at $(\pm 2, 0)$.

§3.1

1. $\lim\limits_{x \to -2} f(x) = +\infty$.

3. $\lim\limits_{x \to 0} f(x) = +\infty$.

5. $\lim\limits_{x \to 2^-} f(x) = -\infty$,

$\lim\limits_{x \to 2^+} f(x) = +\infty$.

7. $\lim\limits_{x \to 1} f(x) = +\infty$.

9. $\lim\limits_{x \to 4} f(x) = -\infty$.

§3.2

1. $+\infty, -\infty$.
3. $0, 0$.
5. $1,1$.
7. $1,1$.
9. $0, 0$.
11. $6,6$.
13. $2,2$.
15. $0,0$.
17. $+\infty, -\infty$.
19. $-\infty, +\infty$.

§3.3

1. Maximum at $(3,4)$.

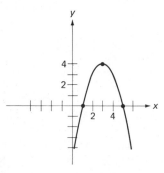

3. Minimum at $(2, -\frac{9}{4})$.

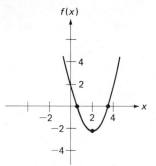

5. Maximum at $(1,2)$.
Minimum at $(-1,-2)$.
Inflection point at $(0,0)$.

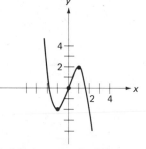

7. Maximum at $(-1,4)$.
Minimum at $(1,0)$.
Inflection point at $(0,2)$.

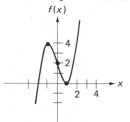

9. Maximum at $(-4,100)$.
Minimum at $(2,-8)$.
Inflection point at $(-1,46)$.

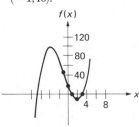

11. Minimum at $(3,-17)$.
Inflection points at $(0,10)$
and $(2,-6)$.

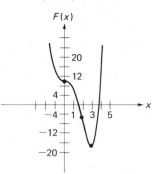

13. Horizontal tangent $y = 1$.
Vertical asymptote $x = 3$.

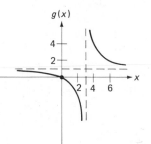

15. Maximum at $(0,2)$.
Inflection points at
$\left(\neq \dfrac{2}{\sqrt{3}}, \dfrac{3}{2} \right)$.
Horizontal asymptote
$y = 0$.

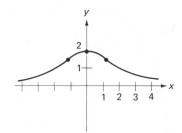

17. Maximum at $(3,\frac{9}{4})$.
Inflection point at $(5,2)$.
Vertical asymptote
$x = -1$.

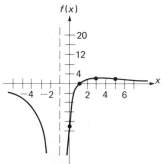

$f(x)$

19. Maximum at $(1,2)$.
Minimum at $(-1,0)$.
Inflection points at $(0,1)$,
$$\left(\sqrt{3}, \left(\frac{\sqrt{3}+1}{2}\right)^2\right), \text{ and}$$
$$\left(-\sqrt{3}, \left(\frac{\sqrt{3}-1}{2}\right)^2\right).$$

Horizontal asymptote
$s = 1$.

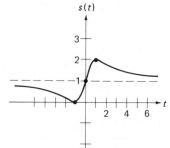

$s(t)$

21. Maximum at $(-1,-9)$.
Minimum at $(3,-1)$.
Vertical asymptote $u = 1$.

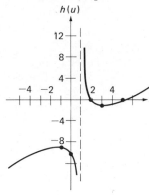

$h(u)$

Chapter 6

§1.1

1. Let $y = f(x)$. Then
$x = 2y - 10$.
3. Let $y = f(x)$. Then
$x = y^2$.
5. $x = \dfrac{y+1}{y-1}$.
7. Let $z = g(y)$. Then
$y = (z+1)^{-1/3}$.
9. Let $w = f(z)$. Then
$z = \left(1 - \dfrac{1}{w}\right)^{1/3}$.
11. $x \mapsto \sqrt[4]{x}$.
13. No inverse.
15. Let $y = x^2 + x - 1$. Then
$x = \frac{1}{2}(-1 - \sqrt{5+y})$.
17.

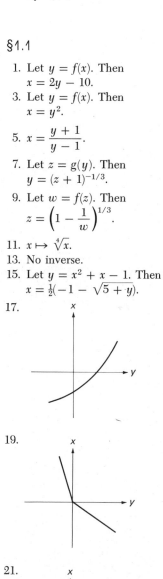

19.

21.

§1.2

1. $g'(3) = -4$.
3. $\psi'(1) = -\frac{3}{2}$.
5. $(dx/dy)_{y=3} = -1$,
$(dx/dy)_{y=0} = -\frac{1}{7}$.
7. (a) $k'(3) = \frac{1}{3}$,
$k'(7) = \frac{1}{5}$; (b) $l'(3) = -\frac{1}{3}$,
$l'(7) = -\frac{1}{5}$.
9. (a) $g'(0) = \frac{1}{18}$;
(b) $h'(0) = -\frac{1}{9}$;
(c) $k'(0) = \frac{1}{18}$.

§1.3

1. $dy/dx = \frac{3}{2}x^{-1/2}$.
3. $dy/dx = \frac{1}{3}x^{-2/3}$.
5. $dy/dx = \frac{3}{4}x^{-7/8}$.
7. $f'(x) = \frac{1}{4}x^{-1/2}\left(\dfrac{1+x}{x}\right)$.
9. $f'(x) = -x^{-4/3} - \frac{1}{3}x^{-7/6}$.
11. $g'(t) = \frac{1}{3}t^{-2/3} + 3 - \frac{1}{3}t^{-4/3}$.
13. $g'(t) = \frac{1}{10}t^{-9/10} + t^{-4/5}$
$+ \frac{1}{2}t^{-1/2}$.
15. $y = -\frac{1}{8}x + \frac{5}{2}$.
17. $y = 9x - 24$.

§2.1

1. $f \circ g = -12x + 8$,
$g \circ f = -12x + 7$.
3. $h \circ g = \dfrac{1}{x^2+1}$,
$h \circ g(-2) = \frac{1}{5}$,
$h(g(3)) = \frac{1}{10}$.
5. $f(g(2)) = 82$,
$g(f(-1)) = 27$,
$g(g(0)) = 102$.
7. $f \circ g = \sqrt{4x^2 + 12x + 13}$,
$g \circ f = 2\sqrt{x^2 + 4} + 3$,
$f(f(x)) = \sqrt{x^2 + 8}$.
9. $0 < x < 1$.
11. 2.
13. 3.
15. $\frac{1}{3}$.
17. $\dfrac{1}{\sqrt{3}}$.
19. 2.

§2.3

1. $y' = -18x(1 - 3x^2)^2$.
3. $y' = 4(2x + 3)$
   $\times (x^2 + 3x - 1)^3$.
5. $f' = 100(5x^4 - 12x^3 + 3x^2$
   $- 6x + 1)(x^5 - 3x^4$
   $+ x^3 - 3x^2 + x)^{99}$.
7. $g' = -24(x^3 - x - x^{-3}$
   $+ x^{-5})(x^4 - 2x^2 + 2x^{-2}$
   $- x^{-4})^{-7}$.
9. $\phi' = 3x(1 + 3x^2)^{-1/2}$.
11. $F' = \dfrac{(t^2 - 1)}{2t^2}\left(t + \dfrac{1}{t}\right)^{-1/2}$.
13. $F' = t(4 - t^2)^{-3/2}$.
15. $y' = -84x(1 + 4x)^2$
    $\times (1 - 3x)^3$.
17. $y' = \dfrac{x(9x^2 - 23)}{(x^2 - 4)^{1/2}}$.
19. $f' = \dfrac{8(2x - 1)}{(2x + 1)^3}$.
21. $f' = \dfrac{x(3x + 2)}{(1 + 2x)^{3/2}}$.
23. $\psi' = -(1 - 4z)^{-1/2}$
    $\times [1 + (1 - 4z)^{1/2}]^{-1/2}$.
25. $\psi' = \dfrac{-1}{\sqrt{1 + z}(1 + \sqrt{1 + z})^2}$.

§2.4

1. $f' = \frac{3}{4}x^{-1/4}$.
3. $f' = \frac{2}{5}(x^{-3/5} + 5 - x^{-7/5})$.
5. $g' = \frac{11}{2}t^{9/2}$.
7. $g' = -5t^{-7/2} - 8t^{-11/3}$
   $- 11t^{-15/4}$.
9. $y' = -3(x + 1)$
   $\times (x^2 + 2x + 3)^{-5/2}$.
11. $y' = x^2(24 - 5x^3)$
    $\times (8 - x^3)^{-1/3}$.
13. $F' = 2(2 + x^2)$
    $\times (4 - x^2)^{-5/2}$.
15. $y = \frac{5}{6}x + \frac{13}{6}$.
17. $y = \frac{3}{4}x + \frac{5}{4}$.
19. $2/\sqrt{5}, \ k/\sqrt{k^2 + 1}$.
21. 3.
23. $5\sqrt{5}$ ft.
25. 2.

27. 15 mi.
29. a^2 square units.
31. Height $= 6$ cm,
    radius $= 3\sqrt{2}$ cm.
33. Lay 250 ft on the ground
    and 1250 ft under water;
    lay all the line under
    water.

§3.1

1. 0.
3. -1.
5. $-\frac{3}{4}$.

§3.2

1. $\dfrac{dy}{dx} = \dfrac{x}{4y}$.
3. $\dfrac{dy}{dx} = -\dfrac{(1 + 4xy)}{2(y + x^2)}$.
5. $\dfrac{dy}{dx} = \dfrac{2xy(1 - y)}{(2x^2y - 2y - x^2)}$.
7. $\dfrac{dy}{dx} = \dfrac{y(y^3 - 2xy - 9x^2)}{x(2xy + 3x^2 - 4y^3)}$.
9. 4.
11. $-\frac{3}{4}$.

§4.2

1. $F(x) = -\dfrac{x^2}{2} + x + c$.
3. $F(x) = \frac{1}{5}x^5 + \frac{1}{3}x^3 + x + c$.
5. $F(x) = 2x^6 - 2x^4 + x^2$
   $- 3x + c$.
7. $G(t) = -\frac{1}{2}t^{-2} + c$.
9. $G(t) = -\frac{2}{3}t^{-3} + t^{-1}$
   $+ \frac{1}{2}t^2 + c$.
11. $G(t) = t - 4t^{-2}$
    $+ 8t^{-4} + c$.
13. $G(t) = \frac{4}{5}t^{5/2} + \frac{2}{7}t^{7/2} + c$.
15. $H(u) = \frac{1}{10}(u + 1)^{10} + c$.
17. $H(u) = \frac{1}{11}u^2 + 3u$
    $+ 4)^{11} + c$.

§4.3

1. $f(x) = \frac{1}{2}x + c$.
3. $g(t) = -\frac{1}{2}t^2 + c_1 t + c_2$.

5. $f(x) = \frac{1}{3}x^3 - \dfrac{1}{x} + c$.
7. $g(t) = \frac{2}{3}t^3 - \frac{3}{2}t^2 + t - \frac{1}{6}$.
9. $h(u) = u^3 + \frac{1}{2}u^2$
   $- \frac{7}{2}u + 2$.
11. $f(x) = \frac{1}{4}x^2 - 2$.

§5.1

1. $g'(1^-) = 3, \quad g'(1^+) = 2$.
3. $h'(-2^-) = 0; \ h'(-2^+)$
   does not exist; $h'(2^-)$
   does not exist;
   $h'(2^+) = 0$.
5. $f'(0^-) = 1, \quad f'(0^+) = 0$.
7. $h'(1^-) = 2, \quad h'(1^+) = 1$.

§5.2

1. $f'(x) = \frac{2}{3}x^{-1/3}$ for
   $x \neq 0; \ f'(0^-) = -\infty,$
   $f'(0^+) = +\infty$.
3. $f'(s) = \frac{3}{7}(s - 1)^{-4/7}$ for
   $s \neq 1; \ f'(1) = +\infty$.
5. $g'(u) = \frac{16}{3}(8u - 1)^{-1/3}$ for
   $u < \frac{1}{8}, g'(u) = 16(8u - 1)$
   for $u > \frac{1}{8}; \ g'(\frac{1}{8}^-) = -\infty,$
   $g'(\frac{1}{8}^+) = 0$.
7. $A'(t) = \frac{1}{6}t^{-2/3}(t^{1/3} + 1)^{-1/2}$
   for $t \neq 0$ and
   $t > -1; \ A'(0) = +\infty,$
   $A'(-1^+) = +\infty$.

Chapter 7

§1.1

1. 3 gm cm/sec^2.
3. 12 gm cm/sec^2; 14 gm
   cm/sec^2; 30 gm cm/sec^2.
5. $-2{,}916{,}000$ gm cm/sec^2;
   $-900{,}000$ gm cm/sec^2.

§1.3

1. 64 ft, 4 sec.
3. 2.9 sec, -101 ft/sec.
5. 56.6 ft/sec.

§1.4

1. Relativistic force: .012 gm cm/sec²;   Newtonian force: .001 gm cm/sec².

3. (i) When $v = .5c$, $t = 100$ sec; relativistic force $= (2\sqrt{3}/3)c \times 10^{-5}$ gm cm/sec²;   Newtonian force $= \frac{3}{4}c \times 10^{-5}$ gm cm/sec².

   (ii) When $v = .9c$, $t = 900$ sec; relativistic force $= 36.3c \times 10^{-7}$ gm cm/sec²;   Newtonian force $= 3c \times 10^{-7}$ gm cm/sec².

§2.1

1. $A(r) = \pi r^2, \dfrac{dA}{dr} = 2\pi r.$

3. $V(r) = \frac{4}{3}\pi r^3, \dfrac{dV}{dr} = 4\pi r^2.$

5. $V(h) = \pi(\frac{1}{2})^2 h$
   $= \dfrac{\pi}{4}h, \quad \dfrac{dV}{dh} = \dfrac{\pi}{4},$
   $\left.\dfrac{dV}{dh}\right|_{h=1} = \dfrac{\pi}{4}.$

§2.2

1. 9.3.
3. 3.952.
5. 7.007.
7. .1999.
9. $\sqrt{1+h} \approx 1 + \frac{1}{2}h.$

§2.3

1. $\dfrac{1}{25\pi}$ ft/min.

3. π sq in./sec.
5. 3 ft/min.
7. 2 lb/sq in./min.
9. 39 ft/min.

11. $\frac{64}{5}$ knots.
13. 4 ft/sec.
15. 46.5 ft/sec.
17. -30 ft/sec.

§3.1

1. $C'(x) = 10 + 2x$;
   $C'(50) = 110$;
   $C(51) - C(50) = 111$;
   error in using $C'(x)$ at level $x = 50$ is approximately 1 percent.

3. $C'(x) = 10 - 60x + 3x^2$;
   $\dfrac{C(x)}{x} = \dfrac{4000}{x} + 10 - 30x + x^2;\quad x = 20.$

§3.2

1. $x = 40$; $\pi(40) = \$400$.

§3.3

1. $x = 50$.
3. $x = 40$; $\pi(40) = \$100{,}000$.

Chapter 8

§1.1

1. Area $= \frac{45}{2}$.

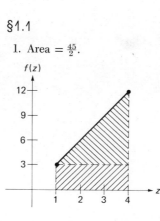
$f(z)$

3. Area $= 1$.

$f(y)$

5. Area $= \frac{7}{2}$.

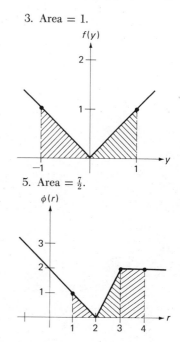
$\phi(r)$

§1.2

1. Area $= -24$.

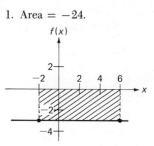
$f(x)$

3. Area $= -10$.

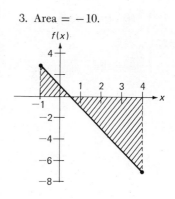
$f(x)$

5. Area $= -3$.

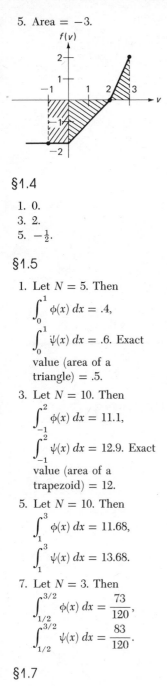

§1.4

1. 0.
3. 2.
5. $-\frac{1}{2}$.

§1.5

1. Let $N = 5$. Then
$$\int_0^1 \phi(x)\,dx = .4,$$
$$\int_0^1 \psi(x)\,dx = .6. \text{ Exact}$$
value (area of a triangle) $= .5$.

3. Let $N = 10$. Then
$$\int_{-1}^2 \phi(x)\,dx = 11.1,$$
$$\int_{-1}^2 \psi(x)\,dx = 12.9. \text{ Exact}$$
value (area of a trapezoid) $= 12$.

5. Let $N = 10$. Then
$$\int_1^3 \phi(x)\,dx = 11.68,$$
$$\int_1^3 \psi(x)\,dx = 13.68.$$

7. Let $N = 3$. Then
$$\int_{1/2}^{3/2} \phi(x)\,dx = \frac{73}{120},$$
$$\int_{1/2}^{3/2} \psi(x)\,dx = \frac{83}{120}.$$

§1.7

1. Area using left endpoints is 1.4; area using right endpoints is 1.6.
3. .33.

§2.1

1. Area of "shorter" rectangle is ht; area of "taller" rectangle is $h(t+h)$. $A'(t) = t$. Desired area is $t^2/2$.
3. Area of "shorter" rectangle is $h\sqrt{t}$; area of "taller" rectangle is $h\sqrt{t+h}$. $A'(t) = \sqrt{t}$. Desired area is $\frac{2}{3}t^{3/2}$.

§2.3

1. 15.
3. $\frac{9}{2}$.
5. 12.
7. $\frac{1}{12}$.
9. $-\frac{40}{3}$.
11. $\frac{1}{2}$.
13. $\frac{5}{2}$.
15. $\dfrac{n}{n+1}(2^{n+1}-1)$.
17. $\frac{15}{4}$.
19. $-\frac{13}{8}$.
21. $\frac{26}{15}$.

§2.4

1. $\frac{1}{2}x^2 + c$.
3. $x + \frac{1}{2}x^2 + c$.
5. $\frac{1}{2}z^2 + \frac{1}{6}z^6 + c$.
7. $\frac{1}{2}t^2 + \frac{1}{3}t^3 + \frac{1}{4}t^4 + c$.
9. $x^2 + 4x + c$.
11. $7x^6$.
13. $3(t+2)(t^2 + 4t + 1)^{1/2}$.
15. 16.
17. $\frac{6}{5}u^{5/3} - \frac{1}{6}u^6 - \frac{31}{30}$.
19. $\frac{2}{15}(t^{3/2}-1)^5$.

§2.5

1. $v(64) \cong 1375\text{ ft/sec}$; $s(64) \cong 22{,}160\text{ ft}$.
3. Each small box has area 1. Approximate velocities are found by counting boxes. We find

t	$v(t)$
1	30
2	80
3	130
4	200
5	280
6	380
7	500
8	650
9	830
10	1050

The area under the graph of the velocity from 0 to 10 is the distance traveled after 10 sec. It is approximately 3600 ft.

§3.2

1. 4.
3. 2.
5. $-\frac{1}{3}$.
7. 2.
9. $\frac{47}{60}$.
11. 6.
13. -56.
15. $\frac{89}{3}$.
17. $\frac{16}{3}$.
19. $\frac{216}{35}$.

§3.3

1. $-dx$.
3. $\dfrac{1}{2\sqrt{x}}\,dx$.
5. $-2x\,dx$.
7. $\frac{3}{2}x(x-2)$
$\times (x^3 - 2x^2 + 1)^{-1/2}dx$.
9. $y = \frac{2}{3}x^{3/2} - 18$.
11. $f(2) = 32$.
13. $g(-1) = 0$.

15. $f(u) = (u^2 + 1)^{-3/2}$.

17. $f(x) = \frac{2}{5}x^{5/2}$
   $- \frac{2}{7}(x + 1)^7 + c$.

19. $f(2) = 1$.

21. $f(x) = \frac{1}{2}x + 1$;
   $g(x) = \frac{1}{2}x^2 - \frac{1}{2}x + 1$.

23. Use the quotient formula for derivatives and the definition for differential.

25. $\frac{dy}{dx} = \frac{3}{2}(u^2 + 1)$
   $\times \left(\frac{1}{\sqrt{x + 1}} + 1 \right)$.

27. $\frac{dy}{dx} = ux^{-3/2}(x - 1)$
   $\times \left(\frac{1}{\sqrt{1 + u^2}} + \frac{u^2}{\sqrt{1 + u^4}} \right)$.

29. $\frac{dy}{dx}$
   $= \frac{3x(w + 3)(u - 1)(x^2 + 1)}{u^{1/2}(w + 1)^{3/2}}$.

§3.4

1. $-\frac{1}{10}(1 - 2x)^5 + c$.

3. $-\frac{2}{9}(2 - 3x)^{3/2} + c$.

5. 30.

7. $-\frac{1}{(2 + x)} + c$.

9. $-\frac{1}{3}(2 - x^2)^{3/2} + c$.

11. $\frac{3}{20}$.

13. $\frac{45}{8}$.

15. $(a^2 + x^2)^{1/2} + c$.

17. $\frac{1}{2a(1 - n)}$
   $\times (1 + ax^2)^{1-n} + c$.

19. $-\frac{3}{16}(2 - x^4)^{4/3} + c$.

21. $\frac{2}{3}(1 + 2x + 3x^2)^{3/2} + c$.

23. $\frac{1}{48}$.

25. $\frac{2}{3}\left(x^2 - \frac{1}{x}\right)^{3/2} + c$.

27. $\frac{2}{r + 1}(1 + \sqrt{x})^{r+1} + c$.

§3.6

1. 0.

3. 0.

5. 2.

7. $\frac{2}{5}a^5$.

§4.1

1. $\frac{3}{2}$.

3. $+\infty$.

5. $2\sqrt{2}$.

7. 2.

9. -3.

11. $\frac{1}{(1 - r)}a^{1-r}$.

§4.2

1. $\frac{1}{2}$.

3. 3.

5. $\frac{1}{2}$.

7. $\frac{3}{2}$.

9. $+\infty$.

§4.3

1. $+\infty$.

3. $+\infty$.

5. $-\infty$.

7. $+\infty$.

9. $+\infty$.

Chapter 9

§1.1

1. $A = \frac{32}{3}$.

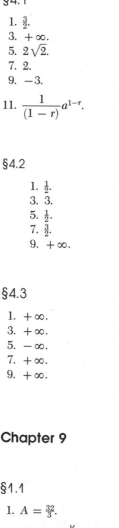

3. $A = 16$.

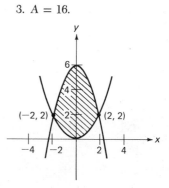

5. $A = \frac{1}{12}$.

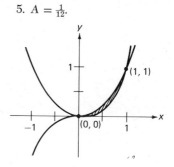

7. $A = \frac{81}{2}$.

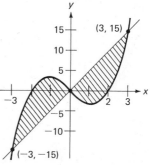

9. $A = \frac{9}{2}$.

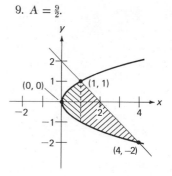

11. $A = \frac{5}{2}$.

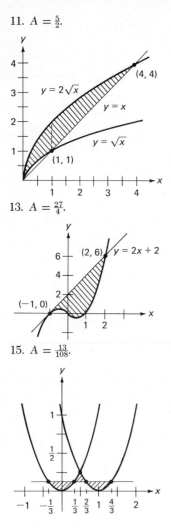

13. $A = \frac{27}{4}$.

15. $A = \frac{13}{108}$.

§2.2

1. $\pi R^2 H$.

3. $\pi h^2 \left(a - \frac{h}{3} \right)$.

5. $\frac{1}{3}\pi H(R^2 + r^2 + Rr)$.
7. 2.
9. $16\pi/5$.
11. $\frac{1}{3}$.

§2.3

1. $8\pi/15$.
3. $127\pi/7$.

5. 256π.
7. $\pi/2$.
9. $\pi(2 - \sqrt{2})/3$.
11. $8\pi/15$.
13. $176\pi/15$.
15. 6π.

§2.4

1. 8π; $32\pi/5$.
3. $128\pi/3$.
5. $\frac{4}{3}\pi a^3$.
7. $3\pi/10$.
9. 4π.
11. $5\pi/6$.

§3.1

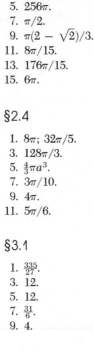

1. $\frac{335}{27}$.
3. 12.
5. 12.
7. $\frac{31}{6}$.
9. 4.

§3.2

1. $A(a,b) = \frac{1}{2}rL(a,b)$; $L(-r,r) = \pi r$.

3. $A(a,b) = \int_a^b \sqrt{r^2 - x^2}\, dx$
$- \frac{1}{2}b\sqrt{r^2 - b^2} + \frac{1}{2}a\sqrt{r^2 - a^2}$.

5. Use Problem 4, set $a = 0$, substitute for $A(0,r)$, and integrate.

§4.1

1. $\alpha = 0$.
3. Yes.
5. Total mass $= 1$; $\alpha = \frac{3}{4}$.

§4.2

1. $\alpha = \frac{7}{6}$.
3. $f(x) < 0$ for $-1 \le x < -\frac{1}{2}$; not a distribution function.
5. $\alpha = -\frac{1}{2}$.
7. $\alpha = 0$.

§4.3

1. 236.
3. 450,000.

5. $\sigma^2 = \frac{1}{18}$; $\sigma = \frac{\sqrt{2}}{6}$.

7. $\sigma^2 = \frac{3}{20}$; $\sigma = \frac{1}{10}\sqrt{15}$.

9. $\sigma = \frac{a}{\sqrt{3}}$; σ increases as a increases; yes.

§5.1

1. -6 gm cm/sec^2.
3. 19.6 cm.

§5.3

1. .09 gm cm^2/sec^2.
3. 67.5 gm cm^2/sec^2.
5. 2.25 gm cm^2/sec^2.

§5.5

1. $200\sqrt{15}$ mph ≈ 770 mph.
3. $\frac{529}{8} m$ gm cm^2/sec^2.
5. (a) $v(t) = 3t^2$.
   (b) $K(t) = \frac{9}{2}t^4$.
   (c) $s(t) = t^3 + s_0$, where s_0 is the position of the particle at time $t = 0$.
   (d) $F(s) = 6(s - s_0)^{1/3}$.
   (e) $W(s) = -\frac{9}{2}(s - s_0)^{4/3}$.
   (f) $E = W + K = 0$.

§5.7

1. About 2.37×10^5 cm/sec.
3. About 6.16×10^7 cm/sec.

5. $\frac{dv_0}{dR} = -\sqrt{\frac{\gamma M}{2R^3}}$.

Chapter 10

§1.1

1. $\pi/18$.
3. $2\pi/15$.
5. $\pi/4$.

7. $\pi/2$.
9. $5\pi/6$.
11. $22.5°$.
13. $45°$.
15. $120°$.
17. $210°$.
19. $315°$.
21. $\pi/8640$.

§1.2

1. $\theta = 360° + 120°$.

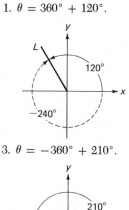

3. $\theta = -360° + 210°$.

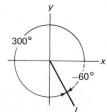

5. $\theta = 3 \cdot 360° + 300°$.

7. $\theta = 2\pi + \dfrac{\pi}{3}$.

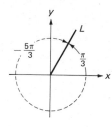

9. $\theta = -2\pi + \dfrac{3\pi}{4}$.

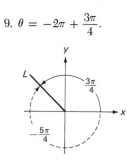

§1.3

1. Proceed as in Example 1.
3. Proceed as in Example 2.
5. $\sqrt{3}/2$.
7. $\sqrt{2}/2$.
9. $\sqrt{2}/2$.
11. $-1/2$.
13. $-\sqrt{3}/2$.
15. $+1/2$.
17. $-.9511$.
19. $.3907$.
21. $.6561$.
23. $-.3584$.

§1.4

1.

θ	$\tan\theta$	$\cot\theta$
$0 = 0°$	0	$\pm\infty$
$\dfrac{\pi}{6} = 30°$	$\dfrac{\sqrt{3}}{3}$	$\sqrt{3}$
$\dfrac{\pi}{4} = 45°$	1	1
$\dfrac{\pi}{3} = 60°$	$\sqrt{3}$	$\dfrac{\sqrt{3}}{3}$
$\dfrac{\pi}{2} = 90°$	$\pm\infty$	0
$\pi = 180°$	0	$\pm\infty$
$\dfrac{3\pi}{2} = 270°$	$\pm\infty$	0
$2\pi = 360°$	0	$\pm\infty$

3. Express the cotangent in terms of sines and cosines.
5. Express the cosecant in terms of the sine.
7,9. Express the given identities in terms of sines and/or cosines.
11. -1.
13. 2.
15. $-\sqrt{3}$.
17. $-\sqrt{3}$.
19. $2\sqrt{3}/3$.

§2.2

1. $2\cos(2x + 1)$.
3. $-2(x + 2)$
   $\times \sin(x^2 + 4x + 1)$.
5. $\dfrac{x}{\sqrt{1 + x^2}}\cos\sqrt{1 + x^2}$.
7. $\dfrac{\sin\theta}{2\sqrt{1 - \cos\theta}}$.
9. $2x\sin(4x^2 - 1)$
   $+ 8x^3\cos(4x^2 - 1)$.
11. $\dfrac{2\sin\theta}{(1 + \cos\theta)^2}$.
13. $-4\cos 4t\sin(2\sin 4t)$.
15. $-\dfrac{6}{(1 + t)^2}\sin^2\left(\dfrac{1 - t}{1 + t}\right)$
    $\times\cos\left(\dfrac{1 - t}{1 + t}\right)$.
17. Maxima at $x = 0$, $x = \dfrac{\pi}{2}$;

    minimum at $x = \dfrac{\pi}{4}$;

    inflection points at $x = \dfrac{\pi}{8}$,

    $x = \dfrac{3\pi}{8}$.

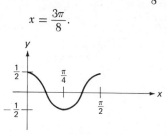

19. Maximum at $x = \dfrac{7\pi}{2}$;

minimum at $x = \dfrac{3\pi}{2}$;

inflection points at $x = \dfrac{\pi}{2}$,

$x = \dfrac{5\pi}{2}$.

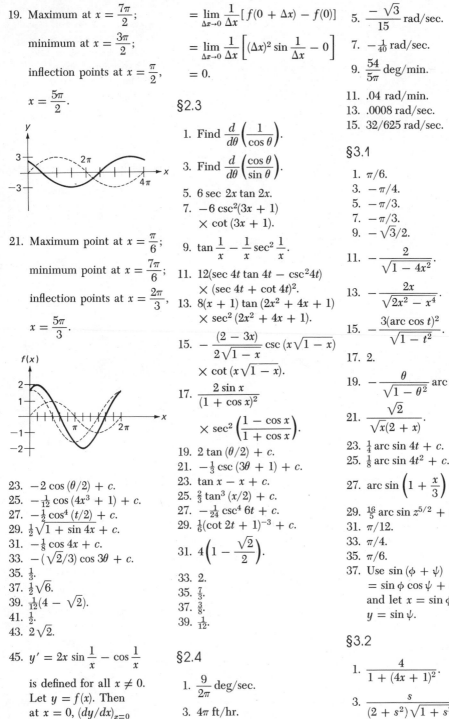

21. Maximum point at $x = \dfrac{\pi}{6}$;

minimum point at $x = \dfrac{7\pi}{6}$;

inflection points at $x = \dfrac{2\pi}{3}$,

$x = \dfrac{5\pi}{3}$.

23. $-2 \cos(\theta/2) + c.$
25. $-\frac{1}{12} \cos(4x^3 + 1) + c.$
27. $-\frac{1}{2} \cos^4(t/2) + c.$
29. $\frac{1}{2}\sqrt{1 + \sin 4x} + c.$
31. $-\frac{1}{8} \cos 4x + c.$
33. $-(\sqrt{2}/3) \cos 3\theta + c.$
35. $\frac{1}{3}.$
37. $\frac{1}{2}\sqrt{6}.$
39. $\frac{1}{12}(4 - \sqrt{2}).$
41. $\frac{1}{2}.$
43. $2\sqrt{2}.$

45. $y' = 2x \sin \dfrac{1}{x} - \cos \dfrac{1}{x}$

is defined for all $x \neq 0$.
Let $y = f(x)$. Then
at $x = 0$, $(dy/dx)_{x=0}$

$= \lim_{\Delta x \to 0} \dfrac{1}{\Delta x}[f(0 + \Delta x) - f(0)]$

$= \lim_{\Delta x \to 0} \dfrac{1}{\Delta x}\left[(\Delta x)^2 \sin \dfrac{1}{\Delta x} - 0\right]$

$= 0.$

§2.3

1. Find $\dfrac{d}{d\theta}\left(\dfrac{1}{\cos \theta}\right).$

3. Find $\dfrac{d}{d\theta}\left(\dfrac{\cos \theta}{\sin \theta}\right).$

5. $6 \sec 2x \tan 2x.$
7. $-6 \csc^2(3x + 1)$
   $\times \cot(3x + 1).$

9. $\tan \dfrac{1}{x} - \dfrac{1}{x}\sec^2 \dfrac{1}{x}.$

11. $12(\sec 4t \tan 4t - \csc^2 4t)$
    $\times (\sec 4t + \cot 4t)^2.$
13. $8(x + 1) \tan(2x^2 + 4x + 1)$
    $\times \sec^2(2x^2 + 4x + 1).$

15. $-\dfrac{(2 - 3x)}{2\sqrt{1 - x}} \csc(x\sqrt{1 - x})$
    $\times \cot(x\sqrt{1 - x}).$

17. $\dfrac{2 \sin x}{(1 + \cos x)^2}$
    $\times \sec^2\left(\dfrac{1 - \cos x}{1 + \cos x}\right).$

19. $2 \tan(\theta/2) + c.$
21. $-\frac{1}{3} \csc(3\theta + 1) + c.$
23. $\tan x - x + c.$
25. $\frac{2}{3} \tan^3(x/2) + c.$
27. $-\frac{1}{24} \csc^4 6t + c.$
29. $\frac{1}{6}(\cot 2t + 1)^{-3} + c.$

31. $4\left(1 - \dfrac{\sqrt{2}}{2}\right).$

33. $2.$
35. $\frac{7}{3}.$
37. $\frac{3}{8}.$
39. $\frac{1}{12}.$

§2.4

1. $\dfrac{9}{2\pi}$ deg/sec.

3. 4π ft/hr.

5. $\dfrac{-\sqrt{3}}{15}$ rad/sec.

7. $-\frac{1}{40}$ rad/sec.

9. $\dfrac{54}{5\pi}$ deg/min.

11. .04 rad/min.
13. .0008 rad/sec.
15. 32/625 rad/sec.

§3.1

1. $\pi/6.$
3. $-\pi/4.$
5. $-\pi/3.$
7. $-\pi/3.$
9. $-\sqrt{3}/2.$

11. $-\dfrac{2}{\sqrt{1 - 4x^2}}.$

13. $-\dfrac{2x}{\sqrt{2x^2 - x^4}}.$

15. $-\dfrac{3(\arccos t)^2}{\sqrt{1 - t^2}}.$

17. $2.$

19. $-\dfrac{\theta}{\sqrt{1 - \theta^2}}\arccos\theta - 1.$

21. $\dfrac{\sqrt{2}}{\sqrt{x}(2 + x)}.$

23. $\frac{1}{4} \arcsin 4t + c.$
25. $\frac{1}{8} \arcsin 4t^2 + c.$

27. $\arcsin\left(1 + \dfrac{x}{3}\right) + c.$

29. $\frac{16}{5} \arcsin z^{5/2} + c.$
31. $\pi/12.$
33. $\pi/4.$
35. $\pi/6.$
37. Use $\sin(\phi + \psi)$
    $= \sin \phi \cos \psi + \cos \phi \sin \psi,$
    and let $x = \sin \phi,$
    $y = \sin \psi.$

§3.2

1. $\dfrac{4}{1 + (4x + 1)^2}.$

3. $\dfrac{s}{(2 + s^2)\sqrt{1 + s^2}}.$

5. $\arctan \dfrac{1}{\sqrt{x}} - \dfrac{\sqrt{x}}{2(1 + x)}$.

7. $\dfrac{1}{\sqrt{1 - 4z^2}}$

$\times \dfrac{1}{[1 + (\arcsin 2z)^2]}$.

9. $\frac{1}{3}\arctan 3t + c$.

11. $\frac{1}{2}\arctan (2t - 1) + c$.

13. $\frac{1}{4}\arctan (x - \frac{3}{2}) + c$.

15. $\frac{2}{3}\arctan z^{3/2} + c$.

17. $\pi/16$.

19. 0.

21. Use $\tan (\phi + \psi)$
$= \dfrac{\tan \phi + \tan \psi}{1 - \tan \phi \tan \psi}$,
and let $x = \tan \phi$,
$y = \tan \psi$.

23. Differentiate both sides.

25. Use $\sin 2\theta = \dfrac{2 \tan \theta}{1 + \tan^2 \theta}$,
and let $\theta = \arctan u$.

27. Use the result in Problem
25 and the half-angle
formula for the sine. Treat
$u \geq 0$ and $u < 0$
separately.

29. Take tangent of both
sides, and use the addition
theorem for the tangent.

§3.3

1. Let $\theta = \operatorname{arc\,cot} x$, and use
the theorem on
differentiating inverse
functions.

3,5,7. Differentiate both sides.

9. $\dfrac{1}{2|x| \sqrt{x - 1}}$.

11. $\dfrac{-2x}{1 + x^4}$.

13. $\dfrac{t}{|t|(1 + t^2)}$.

15. $\dfrac{-3(\operatorname{arc\,cot} \sqrt{z})^{1/2}}{4\sqrt{z}(1 + z)}$.

17. $\dfrac{1}{1 + x^2}$.

19. $\dfrac{\sec^2 x}{1 + 4\tan^2 x}$.

21. $\frac{1}{5}\operatorname{arc\,sec} |\frac{3}{5}x| + c$.

23. $\pi/3$.

25. $\pi/6$.

Chapter 11

§1.1

1. $y' = \dfrac{2}{3 + 2x}$.

3. $f'(x) = \dfrac{4x}{1 + 2x^2}$.

5. $f'(x) = \dfrac{1}{x \log x}$.

7. $y' = \dfrac{4}{x}(\log 2x)^3$.

9. $g'(t) = 3 \cot 3t$.

11. $g'(t) = \dfrac{1 - 2\sin 2t}{t + \cos 2t}$.

13. $\frac{1}{2}\log |2x + 5| + c$.

15. $\frac{1}{2}\log |x^2 + 4x + 1| + c$.

17. $-\frac{1}{3}\log |1 - 6x + 3x^2 - x^3| + c$.

19. $\frac{1}{2}(\log x)^2 + c$.

21. $-\frac{1}{2}\log |1 + 2\cos x| + c$.

23. $\log |\sec x + \tan x| + c$.

§1.2

1. 2.3026.

3. .1823.

5. -2.4848.

7. .4621.

9. $y' = -\dfrac{2}{(2x + 1)}$.

11. $y' = \dfrac{4x}{x^4 - 1}$.

13. $f'(x) = \dfrac{1}{1 - x^2}$.

15. $f'(x) = \dfrac{16x(x^2 - 11)}{(x^2 + 4)(4x^2 + 1)}$.

17. $g'(t) = \cot t - 2\tan 2t$.

19. $g'(t) = \dfrac{\sec^2 t}{\tan t(1 + \tan t)}$.

§1.3

1. $\displaystyle\int_1^{2.6} \dfrac{dx}{x} \approx .96$,

$\displaystyle\int_1^{2.8} \dfrac{dx}{x} \approx 1.03$. Hence
$2.6 < e < 2.8$.

§2.1

1. Consider $\log (a^{x+y})$, and
rearrange terms.

3. Consider $\log (a^x)^y$, and
rearrange terms.

§2.2

1.·4.

3. $\frac{1}{2}$.

5. -2.

7. 81.

9. 2.

11. $f'(1) = .4343$.

13. $f'(2) = 3.607$.

15. Use $\log_b a = \dfrac{\log a}{\log b}$, and
let $a = 1$.

17. Use $\log_b \bar{a} = \dfrac{\log \bar{a}}{\log b}$, and
let $\bar{a} = a^x$.

§2.3

1. 31.4.

3. .707.

5. .00458.

7. 28.2.

9. 1.48.

11. .68613.

§3.1

1. $y' = 3e^{3x}$.

3. $y' = (x - 2)e^{(x^2 - 4x + 3)}$.

5. $f'(x) = 2x(1 + x)e^{2x}$.

7. $g'(t) = 2(\cos 2t)e^{\sin 2t}$.

9. $g'(t) = \dfrac{4}{1 + 16t^2}e^{\operatorname{arc\,tan} 4t}$.

11. $g'(t) = \dfrac{e^t}{1 + e^{2t}}$.

13. $y' = \dfrac{e^x}{2(e^x + 1)}$.

15. $f'(x) =$
$-2 \sin (2(2x + 1))e^{\cos^2(2x+1)}$

17. $f'(x) = \dfrac{e^{2x}(e^{2x} - 2)}{(e^{2x} - 1)^{3/2}}$.

19. $(e - 1)/2e$.

21. $\log (e^x + 1) + c$.

23. 0.

25. $\frac{1}{21}(1 + 3e^x)^7 + c$.

27. $\log \dfrac{(e^2 + 1)}{2e}$.

29. $\frac{1}{4}e^{4\tan t} + c$.

31. $\sin e^t + c$.

33. $2e^{t/2} + c$.

35. No maximum, minimum, or inflection points.

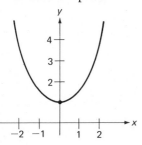

37. Minimum at $x = 0$, $y = 1$; no inflection points.

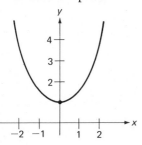

39. Maximum at $x = 1$,
$y = \dfrac{1}{e} \approx .368$;   inflection
point at $x = 2$,
$y = \dfrac{2}{e^2} \approx .271$.

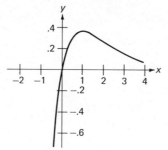

41. Maximum at $x = 0$,
$y = 1$;   inflection points at
$x = \pm.707$, $y \approx .607$.

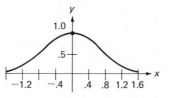

43. Substitute y, y', and y'' into the differential equation.

45. Substitute y, y', and y'' into the differential equation.

§3.2

1. $y' = (4 \log 2)4^{2x}$.

3. $y' = \dfrac{-(\log 3)3^{1/x}}{x^2}$.

5. $f'(x) = \dfrac{x(\log 10)10^{\sqrt{x^2-4}}}{\sqrt{x^2 - 4}}$.

7. $f'(x) = \frac{1}{2}(\log 2)$
$\times \left(\cos \dfrac{x}{2}\right) 2^{\sin x/2}$.

9. $\dfrac{1}{\log 16} 16^x + c$.

11. $\dfrac{1}{3 \log 10} 10^{(x^3+3x+1)} + c$.

13. $\dfrac{2^x e^x}{(1 + \log 2)} + c$.

15. $\dfrac{1}{\log 4} 4^{\sin x} + c$.

§3.3

1. $y' = -2(1 + \log x)x^{-2x}$.

3. $y' = \dfrac{x^{\sqrt{x}}}{\sqrt{x}}(1 + \log \sqrt{x})$.

5. $y' = \dfrac{-1}{(1 + e^x)^x}$
$\times \left(\log (1 + e^x) + \dfrac{xe^x}{(1 + e^-)}\right)$.

7. $y' = \dfrac{(2x + 1)^8}{(x + 4)^2(x + 1)^4}$
$\times \left(\dfrac{16}{2x + 1} - \dfrac{2}{x + 4} - \dfrac{4}{x + 1}\right)$.

9. $y' = \frac{1}{2} \sqrt{(x - 1)(x - 2)(x - 3)}$
$\times \left(\dfrac{1}{x - 1} + \dfrac{1}{x - 2} + \dfrac{1}{x - 3}\right)$.

§3.4

1. $+\infty$.

3. 0.

5. 0.

7. 1.

9. 0.

11. 1.

§3.5

1. Use definitions for $\sinh x$, $\cosh x$, and $\tanh x$, and set $x = 0$.

3. Use definition for $\cosh x$.

5. Divide $\sinh^2 x - \cosh^2 x = 1$ by $\cosh^2 x$.

7. Use definitions for $\sinh (x + y)$, $\sinh x$, $\cosh x$, $\sinh y$, and $\cosh y$, and show both sides of identity are equal.

9. Let $\tanh (x + y)$
$= \dfrac{\sinh (x + y)}{\cosh (x + y)}$, and
use addition theorems
for $\sinh (x + y)$ and
$\cosh (x + y)$.

11. 1.

13. 1.

15. 1.

17. 0.
19. $+\infty$.
21.

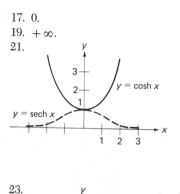

23.

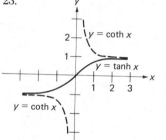

25. Let $\cosh x = \frac{1}{2}(e^x + e^{-x})$ and differentiate.

27. Let $\coth x = \dfrac{1}{\tanh x}$ and differentiate.

29. Let $\operatorname{csch} x = \dfrac{1}{\sinh x}$ and differentiate.

31. $y' = 4 \sinh 4x$.

33. $y' = \frac{1}{2} \operatorname{sech}^2 \frac{x}{2}$.

35. $f'(x) = -8 \operatorname{sech}^4 2x \times \tanh 2x$.

37. $f'(x) = -\dfrac{1}{2\sqrt{x}} \times \operatorname{csch} \sqrt{x} \coth \sqrt{x}$.

39. $g'(t) = -4 \tanh 4t$.
41. $\frac{1}{3}\cosh(3x + 2) + c$.
43. $\frac{1}{6}\coth(1 - 6x) + c$.
45. $-\tanh \dfrac{1}{x} + c$.
47. $\frac{1}{6}\cosh^3 2z + c$.
49. $\log|\cosh z| + c$.

Chapter 12

§1.1

1. $g(t) = -2e^t$.
3. $g(t) = 6e^{-\frac{1}{2}t}$;
   $g(2) = 6e^{-1}$.
5. $f(x) = e^{\frac{1}{3}(9-2x)}$; $f(0) = e^3$.
7. $f(x) = e^{x^2}$.
9. $f(x) = e^{\sqrt{x}-2}$.

§1.3

1. $e^{1/2}$.
3. e^2.
5. e.
7. e^{-1}.

§1.4

1. 62,500 bacteria, approx.

§1.5

1. 3.32 years, approx.

§2.1

1. $f(x) = -\cos x + \sin x$.
3. $y = 8 \cos x + 4 \sin x$.
5. $f(x) = -2 \cos x$.
7. Let $\phi(x) = \cos(x + y) - \cos x \cos y + \sin x \sin y$, and proceed as in Example 2.

§2.2

1. $f(x) = \cos 4x + \frac{1}{4}\sin 4x$.
3. $y = \frac{1}{3}\cos 3x - \sin 3x$.
5. $f(x) = \frac{1}{10}\sqrt{2}\cos \frac{5}{4}x + \frac{9}{10}\sqrt{2}\sin \frac{5}{4}x$.

§2.3

1. $k = \frac{5}{9}$ gm/sec^2.
3. $v(t) = \sqrt{3}\sin \sqrt{3}t - \sqrt{3}\cos \sqrt{3}t$.
5. $s(t) = \cos t/2 + 3 \sin t/2$;
   $A = \sqrt{10}$.

§3.1

1. $f'(x) = -2xe^{-x^2}$,
   $f''(x) = -2(1 - 2x^2)e^{-x^2}$.
   Use tests for establishing maxima and inflection points.

§4.1

1. $f(x) = 5e^x - 4$.
3. $f(x) = (2 + x)e^{\frac{1}{2}x}$.
5. $f(x) = -\dfrac{1}{3e^{3x}(1 + e^{3x})}$.

§4.4

1. $f(t) = 2 \cosh 3t + \frac{2}{3}\sinh 3t$.
3. $f(t) = \cosh t + \sinh t$.
5. Let $f(t) = F(\omega t)$, and use the fact that the general solution of $F''(t) - F(t)$ is $F(t) = a \cosh t + b \sinh t$.

Chapter 13

§1.4

1. $\frac{19}{4} = 4.750$, $\frac{14}{3} \approx 4.667$;
   $\frac{1}{12} \approx .083$,   0.
3. $\dfrac{893}{560} \approx 1.59464$,
   $\dfrac{4063}{2520} \approx 1.61230$;
   $\dfrac{1}{8} = .12500$,
   $\dfrac{1}{40} = .02500$.
5. .56490,   .55494;
   .01662,   .00008.
7. 7.7697,   7.7335;
   .0625,   .0006.

§2.1

1. $\frac{1}{224}(1 + 2x)^7(14x - 1) + C$.
3. $\frac{1}{14}$.

5. $2x \sin \dfrac{x}{2} + 4 \cos \dfrac{x}{2} + C.$

7. $\frac{1}{9}.$

9. $\frac{2}{15}(1 + x)^{3/2}(3x - 2) + C.$

11. $0.$

13. $\pi - 2.$

15. $\frac{1}{2}(\sin x^2 - x^2 \cos x^2) + C.$

17. $-\dfrac{(5 - 2x)^{3/2}}{21}$
$\times (3x^2 + 6x + 10) + C.$

§2.2

1. $\frac{1}{2}(2x + 1) \log (2x + 1)$
$- x + C.$

3. $\frac{1}{2}x^2 \arctan x^2$
$- \frac{1}{4}\log (1 + x^4) + C.$

5. $\frac{1}{3}(x^3 - 8) \arcsin (x^3 - 8)$
$+ \frac{1}{3}\sqrt{1 - (x^3 - 8)^2} + C.$

7. $\frac{1}{26}e^x(\sin 5x - 5 \cos 5x)$
$+ C.$

9. $\frac{1}{10}e^{2x}(\cos 4x + 2 \sin 4x)$
$+ C.$

11. $\frac{1}{4}(2x^2 - 1) \arcsin x$
$+ \dfrac{x}{4}\sqrt{1 - x^2} + C.$

13. $\dfrac{\pi - 1}{2}.$

15. $\dfrac{xe^x}{1 + a^2}(\cos ax$
$+ a \sin ax)$
$- \dfrac{e^x}{(1 + a^2)^2}$
$[2a \sin ax$
$+ (1 - a^2)$
$\cos ax] + C.$

§2.3

1. $-e^{-x}(x^2 + 2x + 2) + C.$

3. $(8/e)(2e - 5).$

5. $(1/8e^2)(3e^2 - 19).$

7. $2x \cos x + (x^2 - 2) \sin x$
$+ C.$

9. $8x \sin (x/2)$
$- 2(x^2 - 8) \cos (x/2) + C.$

11. $3(x^2 - 2) \sin x$
$- x(x^2 - 6) \cos x + C.$

13. $\frac{1}{5}.$

15. $x(\log x)^3 - 3x(\log x)^2$
$+ 6x \log x - 6x + C.$

17. $-\dfrac{1}{2x^2}(\log x)^2 - \dfrac{1}{2x^2}(\log x)$
$-\dfrac{1}{4x^2} + C.$

§2.4

1. $\dfrac{x}{2(1 + 9x^2)}$
$+ \frac{1}{6}\arctan 3x + C.$

3. $\dfrac{x}{4(1 + x^2)^2} + \dfrac{3x}{8(1 + x^2)}$
$+ \frac{3}{8}\arctan x + C.$

5. $-\frac{1}{7}\sin^2 x \cos^5 x$
$- \frac{2}{35}\cos^5 x + C.$

7. $\frac{1}{10}\sin 2x \cos^4 2x$
$+ \frac{2}{15}\sin 2x \cos^2 2x$
$+ \frac{4}{15}\sin 2x + C.$

9. $-\frac{1}{32}\sin^2 4x \cos^6 4x$
$-\frac{1}{96}\cos^6 4x + C.$

11. $5\pi/12.$

§3.1

1. $\frac{2}{5}.$

3. $-\frac{1}{7}\log |8 - 7x| + C.$

5. $\frac{2}{3}\log 2.$

7. $\arctan (x + 2) + C.$

9. $\arctan (2x + 1) + C.$

11. $\frac{1}{5}\arctan \frac{2}{5}(x - \frac{3}{2}) + C.$

13. $\dfrac{(x + 3)}{2(x^2 + 6x + 10)}$
$+ \frac{1}{2}\arctan (x + 3) + C.$

15. $\dfrac{(2x + 1)}{2(2x^2 + 2x + 1)}$
$+ \arctan (2x + 1) + C.$

17. $\dfrac{(x + 4)}{4(x^2 + 16x + 17)^2}$
$+ \dfrac{3(x + 4)}{8(x^2 + 16x + 17)}$
$+ \frac{3}{8}\arctan (x + 4) + C.$

19. $\frac{3}{8}\log 2 + \pi/4.$

21. $\frac{3}{2}\log |x^3 + 4x + 5|$
$-7 \tan (x + 2) + C.$

23. $\pi/4 - \frac{1}{2}\log 2.$

25. $-\dfrac{(5x + 3)}{2(2x^2 + 2x + 1)}$
$- \frac{5}{2}\arctan (2x + 1) + C.$

27. $-\dfrac{(3x + 13)}{4(x^2 + 8x + 17)^2}$
$-\dfrac{9(x + 4)}{8(x^2 + 8x + 17)}$
$- \frac{9}{8}\arctan (x + 4) + C.$

§3.3

1. $\frac{2}{3}\log |x + 2|$
$+ \frac{7}{3}\log |x - 4| + C.$

3. $\log \dfrac{x^2(x - 2)^4}{(x - 1)^6} + C.$

5. $\frac{1}{2}\log \left| \dfrac{(x - 1)(x - 2)^2}{(x + 1)} \right| + C.$

7. $-\dfrac{4}{(x + 2)^2} + \dfrac{12}{(x + 2)}$
$+ 3 \log |x + 2| + C.$

9. $\log \left| \dfrac{x^2 + x + 1}{x} \right| + C.$

11. $\dfrac{x^2}{2} + \frac{1}{2}\log |x^2(x^2 + 1)|$
$+ \arctan x + C.$

13. $2 \log x + \dfrac{5}{x} - \dfrac{1}{x + 1} + C.$

15. $-\dfrac{(2 + x)}{2(x^2 + 4)}$
$+ \frac{1}{4}\arctan \dfrac{x}{2} + C.$

17. $-\dfrac{1}{(x - 1)}$
$+ \log \left| \dfrac{x^2 + x + 2}{(x - 1)^2} \right| + C.$

§4.2

1. $\frac{1}{2}\log |e^{2x} - 1| + C.$

3. $-\dfrac{1}{e^x + 1} + C.$

5. $\frac{1}{5}x - \frac{1}{10}\log(e^{2x} + 4e^x + 5)$
   $- \frac{2}{5}\arctan(e^x + 2) + C.$

7. $-\dfrac{2}{(e^x + 1)} + C.$

9. $-x - \dfrac{2}{(e^x + 1)}$
   $+ 2\log(e^x + 1) + C.$

$\left|\tan\dfrac{(x + \theta)}{2}\right| + C$ where

$\theta = \arcsin\dfrac{a}{\sqrt{a^2 + b^2}}.$

13. $\dfrac{1}{ab}\arctan\left(\dfrac{b}{a}\tan x\right) + C.$

§4.7

1. $\frac{1}{4}\sin 2x + \frac{1}{12}\sin 6x + C.$
3. $\frac{1}{8}\sin 4x - \frac{1}{16}\sin 8x + C.$
5. $-\frac{1}{2}\cos x - \frac{1}{18}\cos 9x + C.$
7. $\frac{3}{4}\cos x - \frac{1}{12}\cos 3x$
   $- \frac{1}{20}\cos 5x$
   $+ \frac{3}{28}\cos 7x + C.$
9. $\pi/8.$

§4.3

1. $\frac{3}{16}(4x + 1)^{4/3} + C.$
3. $\frac{2}{5}\sqrt{x - 3}(24 + 4x + x^2)$
   $+ C.$
5. $-\frac{1}{10}(1 - 2x)^{5/4}$
   $+ \frac{1}{9}(1 - 2x)^{9/4}$
   $- \frac{1}{26}(1 - 2x)^{13/4} + C.$
7. $-x + 8\sqrt{x}$
   $- 16\log|\sqrt{x} + 2| + C.$
9. $-\sqrt{x(1 - x)}$
   $+ \arctan\sqrt{\dfrac{x}{1 - x}} + C.$
11. $4\sqrt{\dfrac{1 + x}{1 - x}}$
   $+ \sqrt{(1 + x)(1 - x)}$
   $- 6\arctan\sqrt{\dfrac{1 + x}{1 - x}} + C.$

§4.4

1. $\tan x/2 + C.$
3. $\frac{1}{2}\arctan(\frac{1}{2}\tan x/2) + C.$
5. $x + \dfrac{2}{1 + \tan x/2} + C$
   or
   $x + \tan\left(\dfrac{\pi}{4} - \dfrac{x}{2}\right) + C'.$
7. $\frac{1}{2}\tan x/2 - \frac{1}{6}\tan^3 x/2 + C.$
9. $x/5 - \frac{3}{20}\log$
   $\left|\dfrac{\tan x/2 + 2}{\tan x/2 - 2}\right| + C.$
11. $\dfrac{1}{\sqrt{a^2 + b^2}}\log|\csc(x + \theta)$
   $- \cot(x + \theta)| + C$
   or
   $\dfrac{1}{\sqrt{a^2 + b^2}}\log$

§4.5

1. $\arcsin x + C.$
3. $\frac{1}{2}\arcsin 2x + C.$
5. $-\dfrac{\sqrt{1 - x^2}}{x}$
   $- \arcsin x + C.$
7. $-\frac{1}{2}x\sqrt{36 - x^2}$
   $+ 18\arcsin x/6 + C.$
9. $\log|x + \sqrt{x^2 - 16}| + C.$
11. $\frac{1}{3}\log|3x + \sqrt{9x^2 - 1}| + C.$
13. $\frac{1}{5}\operatorname{arc\,sec}|x/5| + C.$
15. $\sqrt{x^2 - 9}$
   $- 3\operatorname{arc\,sec}|x/3| + C.$
17. $\log|x + \sqrt{x^2 + 16}| + C.$
19. $-\dfrac{\sqrt{x^2 + 1}}{x} + C.$
21. $\frac{1}{2}x\sqrt{x^2 + 9}$
   $+ \frac{9}{2}\log|x + \sqrt{x^2 + 9}|$
   $+ C.$
23. $-\dfrac{x}{\sqrt{x^2 + 1}}$
   $+ \log|x + \sqrt{x^2 + 1}| + C.$

§4.6

1. $\arcsin\left(\dfrac{x - 4}{4}\right) + C.$
3. $\dfrac{(x - 2)}{\sqrt{x^2 - 4x + 5}} + C.$
5. $\sqrt{x^2 - 6x + 10}$
   $+ 3\log|x - 3$
   $+ \sqrt{x^2 - 6x + 10}| + C.$
7. $-\arcsin\left(\dfrac{3x + 1}{4x}\right) + C.$

§5

1. $\frac{1}{9}(2 + 3x) - \frac{2}{9}\log|2 + 3x|$
   $+ C.$
3. $\dfrac{x}{4(2 - x)}$
   $- \frac{1}{4}\log\left|\dfrac{2 - x}{x}\right| + C.$
5. $\dfrac{(x - 9)^{3/2}}{9x} - \dfrac{\sqrt{x - 9}}{9}$
   $+ \frac{1}{3}\arctan\left(\dfrac{\sqrt{x - 9}}{2}\right)$
   $+ C.$
7. $\frac{1}{2}t^2 - \frac{1}{2}\arctan t^2 + C.$
9. $-\dfrac{2}{\sqrt{t}} + \log\left|\dfrac{1 + \sqrt{t}}{1 - \sqrt{t}}\right|$
   $+ C.$
11. $-\dfrac{1}{9x\sqrt{9 + x^2}}$
   $- \dfrac{2x}{81\sqrt{9 + x^2}} + C.$
13. $-\dfrac{\sqrt{9 - 16x^2}}{9x} + C.$
15. $\frac{1}{6}x(1 - x^2)^{5/2}$
   $+ \frac{1}{24}x(1 - x^2)^{3/2}$
   $+ \frac{1}{16}x\sqrt{1 - x^2}$
   $+ \frac{1}{16}\arcsin x + C.$
17. $\dfrac{s^3}{2\sqrt{s^2 - 9}} - \dfrac{27s}{2\sqrt{s^2 - 9}}$
   $+ \dfrac{27}{2}\log|s + \sqrt{s^2 - 9}|$
   $+ C.$
19. $\frac{1}{4}\arctan\left(\dfrac{5\tan s - 3}{4}\right)$
   $+ C.$

21. $\frac{1}{12}\cos^3 3x \sin 3x$
$+ \frac{1}{8}\sin 3x \cos 3x + \frac{3}{8}x + C.$

23. $\frac{1}{6}\sin^4 x \cos^2 x$
$+ \frac{1}{12}\sin^4 x + C.$

25. $\frac{1}{20}\sec^4 4x \tan 4x$
$+ \frac{1}{15}\sec^2 4x \tan 4x$
$+ \frac{2}{15}\tan 4x + C.$

27. $\frac{1}{5}\sec^3 t \tan^2 t$
$- \frac{2}{15}\sec t \tan^2 t$
$+ \frac{2}{15}\sec t + C.$

29. arc sin $t^2/2$
$+ \frac{1}{2}\sqrt{4 - t^4} + C.$

31. $\frac{1}{2}x^3e^{2x} - \frac{3}{4}x^2e^{2x} + \frac{3}{4}xe^{2x}$
$- \frac{3}{8}e^{2x} + C.$

33. $\frac{1}{148}e^{4x}(\cos^3 8x$
$+ 6\cos^2 8x \sin 8x$
$+ \frac{48}{5}\sin 8x$
$+ \frac{24}{5}\cos 8x) + C.$

35. $\frac{1}{3}x^3((\log x/2)^3 - (\log x/2)^2$
$+ \frac{2}{3}\log x/2 - \frac{2}{9}) + C.$

37. $-\frac{1}{4}$ arc tan (sinh (1 − 4s))
$+ C.$

Chapter 14

§1.1

1. $1 < \xi < 4$, that is, all values between 1 and 4.
3. $e^2(e - 1)$.
5. $1 + \log(e - 1)$.
7. $0, \pm\sqrt{2}/2$.
9. $\sqrt{3}$.
11. $f'(x)$ does not have a derivative at $x = 0$.
13. $f(x)$ is not continuous in the interval (0,2).

§1.2

1. $1 < \xi < 3$, that is, all values between 1 and 3.
3. $\sqrt[3]{3}$.
5. $\pi/2$.

§1.3

1. From Example 1 in the text we have
$\sqrt{1 + x} = 1$
$+ x/2 + R(x)$, where
$|R(x)| < \frac{1}{8}x^2$ for $x \geq 0$
and $|R(x)| < x^2$ for
$-\frac{3}{4} \leq x \leq 0$. Write
$\sqrt{5} = 2\sqrt{1 + \frac{1}{4}}$. Let
$x = \frac{1}{4} > 0$. Then
$\sqrt{5} \approx 2.25$, and the maximum error committed is $2(\frac{1}{8})(\frac{1}{4})^2 = \frac{1}{64} < .02$.

3. $(.044)^{1/2} = .2\sqrt{1 + .1}$
$\approx .21000; \quad (.2)(\frac{1}{8})(.1)^2$
$= 1/4000 < .00025$.

5. $\sqrt{.0062} = .08\sqrt{1 - \frac{1}{32}}$
$\approx .07878; \quad .08(-\frac{1}{32})^2$
$= 1/12,800 < .00008$.

7. Use Equation (11) with
$f(x) = e^{-x}, x_0 = 0$, and
$x = .04$. Since $f''(x)$ is
largest at $x = 0$, let $\xi = 0$.
Then $e^{-.04} \approx .9600$, and
the maximum error committed is
$\frac{1}{2}f''(0)(.04 - 0)^2 = .0008$.

9. $(.03)^{1/3} \approx .31111;$
$\frac{1}{2}f''(.027)(.03 - .027)^2$
$= \frac{1}{2430} < .00042.$

11. Let $f(x) = \log x, x_0 = 1.00$,
$x = 1.04$, and $\xi = 1.00$.
Then $\log 1.04 \approx .0400$,
and $\frac{1}{2}f''(1)(1.04$
$- 1.00)^2 = .0008$.

§2.1

1. From Example 1 in the text we have
$\sqrt{1 + x} = 1 + \frac{1}{2}x - \frac{1}{8}x^2$
$+ R(x)$, where
$R(x) \leq \frac{1}{16}x^3$
for $x \geq 0$ and
$|R(x)| \leq |2x^3|$ for
$-\frac{3}{4} \leq x \leq 0$. Write

$\sqrt{110} = 10\sqrt{1 + \frac{1}{10}}$. Let
$x = \frac{1}{10} > 0$. Then
$\sqrt{110} \approx 10.4875$, and
the maximum error committed is $10(\frac{1}{16})(\frac{1}{10})^3$
$= .000625 < .0007$.

3. $(.164)^{1/2} \approx .4049688;$
$.4(\frac{1}{16})(.025)^3 < .0000004.$

5. $\sqrt{.00038} \approx .019494;$
$|(.02)(2)(-.05)^3|$
$= .000005.$

7. Use Equation (6) with
$f(x) = e^{-x}, x_0 = 0$, and
$x = .05$. Since $|f'''(x)|$ is
largest at $x = 0$, let
$\xi = 0$. Then $e^{-.05}$
$\approx .95125$, and the maximum error committed is
$\frac{1}{6}|f'''(0)|(.05 - 0)^3$
$= \frac{125}{6}10^{-6} < .00003.$

9. Let $f(x) = \log x, x_0 = 1$,
$x = 1.2$, and $\xi = 1$. Then
$\log(1.2) \approx .180$, and
$\frac{1}{6}|f'''(1)|(1.2 - 1)^3$
$= \frac{8}{3}10^{-3} < .003.$

11. Let $f(x) = x^{1/3}, x_0 = 27$,
$x = 30$, and $\xi = 27$. Then
$\sqrt[3]{30} \approx 3.1070$, and
$\frac{1}{6}|f'''(27)|(30 - 27)^3$
$= 10/(2 \cdot 3^9) < .0003.$

§2.2

1. $P(x) = 1 - x + x^2 - x^3$.
3. $P(x) = 1 - \frac{1}{2}x^2 + \frac{1}{24}x^4$.
5. $P(x) = 1 - x + \frac{1}{2}x^2 - \frac{1}{6}x^3$
$+ \frac{1}{24}x^4$.
7. $P(x) = 1 + 4(x - 1)$
$+ 6(x - 1)^2 + 4(x - 1)^3$
$+ (x - 1)^4$.

9. $P(x) = \frac{1}{2} + \frac{\sqrt{3}}{2}\left(x - \frac{\pi}{6}\right)$
$- \frac{1}{4}\left(x - \frac{\pi}{6}\right)^2$
$- \frac{\sqrt{3}}{12}\left(x - \frac{\pi}{6}\right)^3.$

11. $P(x) = (x - 1) - \frac{1}{2}(x - 1)^2$
$+ \frac{1}{3}(x - 1)^3 - \frac{1}{4}(x - 1)^4.$

13. $P(x) = 1 + x + \frac{1}{2}x^2 + \frac{1}{6}x^3$
    $+ \frac{1}{24}x^4$, $|R(x)| \le e/120$.
15. $P(x) = x - \frac{1}{2}x^2 + \frac{1}{3}x^3$
    $- \frac{1}{4}x^4$, $|R(x)| \le \frac{1}{160}$.
17. $P(x) = 1 - \frac{3}{2}x + \frac{15}{8}x^2$
    $- \frac{35}{48}x^3$, $|R(x)| \le 35/1152$.

§2.3

1. $\sqrt{1.2} \approx 1.0950$;
   $R = .0005$.
3. $\sqrt{1.4} \approx 1.184$; $R = .001$.
5. $\sqrt{2} \approx 1.40$;
   $R = \frac{7}{256} < .03$.
7. $\sqrt[3]{.9} \approx .9656$;
   $R = |\frac{5}{81}(.729)^{-8/3}(-.1)^3|$
   $< .0002$.
9. $(1.1)^{.2} \approx 1.019248$;
   $R = (1.68)10^{-6} < .000002$.
11. In Equation (16) replace
    α with $n - 1$ and j with
    $j - 1$. Then write out
    $\frac{n}{j}\binom{n-1}{j-1}$, and show that
    it is the same as $\binom{n}{j}$.

13. Multiply $\binom{n}{j-1}$ by $\frac{j}{j}$.
    Then $\binom{n}{j-1} + \binom{n}{j}$
    $= \binom{n}{j-1}\frac{j}{j} + \binom{n}{j}$
    $\frac{n(n-1)(n-2)\cdots}{1 \cdot 2 \cdot 3 \cdots}$
    $\times \frac{\cdots(n+1-j+1)(n+1)}{\cdots(j-1)j}$
    $= \binom{n+1}{j}$.
15. Use the binomial theorem
    with $x = -1$. Then $0 = 1$
    $+ \binom{n}{1}(-1) + \binom{n}{2}(-1)^2$
    $+ \binom{n}{3}(-1)^3 + \cdots$
    $+ \binom{n}{n}(-1)^n$. But $\binom{n}{0} = 1$.

§2.4

1. 1.221.
3. 1.492.
5. .1987.
7. .9950.

§2.5

1. .09533, $|R| \le .00003$.
3. .42, $|R| \le .02$.
5. 1.08, $|R| \le .02$.
7. .4053, $|R| \le .0002$.
9. $f(x) = 1 - x + x^2 - x^3$
   $+ \cdots + (-1)^n x^n + R(x)$;
   $|R(x)| \le \left|\frac{x^{n+1}}{(1+\xi)^{n+1}}\right|$,
   ξ between 0 and x. Tay-
   lor's remainder depends on
   ξ; the remainder given in
   Equation (25) does not.

§2.6

1. .099667, $|R| \le .000002$.
3. .465, $|R| \le .002$.

§2.7

1. 24.
3. $\frac{4}{5}$.
5. $-\frac{115}{144}$.
7. $\frac{31}{2}$.
9. $\frac{561}{384}$.
11. $\displaystyle\sum_{k=1}^{6} 2k$.
13. $\displaystyle\sum_{k=0}^{5} \frac{1}{2+4k}$.
15. $\displaystyle\sum_{k=1}^{5} \frac{(2k-1)}{3k}$.
17. $\displaystyle\sum_{k=1}^{5} \frac{(2k-1)^2(2k)^2}{(2k+1)(2k+2)}$.
19. Expand the summation
    and collect terms.

21. Expand the left-hand side
    and regroup terms.
23. Write out the sum. Apply
    the result of Problem 13,
    §2.3, to each term except
    the first one. Then
    $$\binom{n+1}{0} = \binom{n}{0} = 1,$$
    and all terms except the
    last cancel in pairs.
25. Write out the sum.
    Rewrite the result in
    Problem 13, §2.3, as
    $$\binom{n}{j-1} = \binom{n+1}{j} - \binom{n}{j}.$$
    Replace j by $(k+1)$.
    Apply this to each term in
    the sum. Everything drops
    out except $\binom{k+n+1}{k+1}$
    and $\binom{k}{k+1}$. But $\binom{k}{k+1}$
    $= 0$.

§3.1

1. $\frac{7}{5}$.
3. 3.
5. 0.
7. $\frac{1}{6}$.
9. 1.
11. 4.

§3.2

1. $\frac{1}{2}$.
3. $-\infty$.
5. $+\infty$.
7. 0.
9. 1.

§3.3

1. 1.
3. $+\infty$.
5. 0.
7. 1.
9. 1.

11. e.

13. 1.

15. 1.

Chapter 15

§1.1

1. 0,2,6,12,20.

3. $1, -\frac{1}{2} \cdot \frac{1}{3}, -\frac{1}{4} \cdot \frac{1}{5}$.

5. 2,1,4,11,22.

7. $0,1,-8,27,-64$.

9. 3,4,5.

11. $b_n = (n + 2)^2$,
$n = -1,0, \ldots$.

13. $a_k = 4k, \quad k = 0,1, \ldots$.

§1.2

1. $a_n = \dfrac{1}{n^3}, \quad n = 1,2, \ldots$;
0.

3. Diverges to $+\infty$.

5. Does not converge.

7. $\frac{3}{2}$.

9. $a_n = \dfrac{2n + 1}{n + 1}$,
$n = 1,2, \ldots$; \qquad 2.

11. 0.

13. $a_n = \dfrac{n^2 + 1}{n^2 + 2}$,
$n = 1,2, \ldots$; \qquad 1.

15. $a_n = 1 + \dfrac{1}{10^n}$,
$n = 1,2, \ldots$; \qquad 1.

§1.3

1. 1.

3. $-\infty$.

5. 1.

7. 4.

9. Does not converge.

11. -4.

13. -2.

15. $+\infty$.

17. -1.

19. 0.

21. Use Theorem 2 with
$f(x) = q^x, a_n = 1/n$.

§1.4

1. Nondecreasing; $\frac{2}{3}$.

3. Neither; 0.

5. Neither;
unbounded.

7. Nondecreasing; 1.

9. Nonincreasing; 0.

11. Nonincreasing; 0.

13. Nonincreasing; 0.

§2.2

1. $S_0 = 0, \quad S_1 = \frac{1}{2}, \quad S_2 = \frac{7}{6}$,
$S_3 = \frac{23}{12}$.

3. $S_0 = 1, \quad S_1 = \frac{3}{2}, \quad S_2 = \frac{17}{10}$,
$S_3 = \frac{9}{5}$.

5. $S_1 = \frac{1}{2}, \quad S_2 = \frac{7}{8}, \quad S_3 = \frac{57}{48}$,
$S_4 = \frac{561}{384}$.

7. Let $S_n = \displaystyle\sum_{j=0}^{n} a_j$. Then
$S_n = S_{n-1} + a_n$,
$a_n = S_n - S_{n-1}$. Since
$$S_n = \frac{n}{n + 2},$$
$$S_{n-1} = \frac{n - 1}{n + 1}$$
and $a_n = \dfrac{2}{(n + 1)(n + 2)}$.

Hence the infinite series is
$$\sum_{n=0}^{\infty} \frac{2}{(n + 1)(n + 2)}.$$

§2.3

1. May converge.

3. May converge.

5. Diverges.

7. Diverges.

9. May converge.

§2.4

1. Diverges.

3. Diverges.

5. Diverges.

7. Diverges.

9. Converges.

11. Diverges.

13. Converges.

15. Diverges.

17. Converges.

§2.5

1. $\frac{4}{9}$.

3. $\frac{2}{11}$.

5. $\frac{128}{99}$.

7. $\frac{151}{75}$.

9. $\frac{179}{825}$.

11. $\frac{121}{555}$.

§2.6

1. Converges.

3. Converges.

5. Converges.

7. Converges.

9. Converges.

11. Diverges.

13. Diverges.

15. Converges.

§2.7

1. Converges.

3. Ratio test fails.

5. Ratio test fails.

7. Converges.

9. Converges.

11. Converges.

13. Ratio test fails.

15. Diverges.

§2.8

1. Converges.

3. Diverges.

5. Converges.

7. Diverges.
9. Converges.
11. Converges.

§2.9

1. Absolutely convergent.
3. Conditionally convergent.
5. Conditionally convergent.
7. Conditionally convergent.
9. Absolutely convergent.
11. Absolutely convergent.
13. Divergent.

§3.1

1. $-1 < x < 1$.
3. $-\infty < x < \infty$.
5. $-1 \le x \le 1$.
7. $-1 \le x \le 1$.
9. $x = 0$.
11. $3 < x < 5$.
13. $-3 < x \le -1$.
15. $-2 < x < 4$.
17. $-\infty < x < \infty$.
19. $-\frac{5}{2} < x < -\frac{3}{2}$.
21. Apply ratio test to

$$\sum_{n=0}^{\infty} a_n (x - x_0)^n.$$

23. $R^{1/k}$.

§3.2

1. $\dfrac{1}{1-x} = \sum_{n=0}^{\infty} x^n,$

$-1 < x < 1$.

3. $\dfrac{1}{(1-x)^2} = \sum_{n=1}^{\infty} n x^{n-1},$

$-1 < x < +1$.

5. $\sin \dfrac{x}{2}$

$$= \sum_{n=1}^{\infty} \frac{(-1)^{n+1}}{2^{2n-1}(2n-1)!} x^{2n-1},$$

$-\infty < x < +\infty$.

7. $\log\left(1 + \dfrac{x}{2}\right)$

$$= \sum_{n=1}^{\infty} \frac{(-1)^{n+1}}{n\,2^n} x^n,$$

$-2 < x < 2$.

9. $\sin^2 x$

$$= \sum_{n=1}^{\infty} (-1)^{n+1} \frac{2^{2n-1}}{(2n)!} x^{2n},$$

$-\infty < x < +\infty$.

11. $e^{-x/2} = \displaystyle\sum_{n=0}^{\infty} \frac{(-1)^n}{2^n(n!)} x^n,$

$-\infty < x < +\infty$.

13. $4^x = \displaystyle\sum_{n=0}^{\infty} \frac{(\log 4)^n}{n!} x^n,$

$-\infty < x < +\infty$.

15. $\arctan 2x$

$$= \sum_{n=0}^{\infty} (-1)^n \frac{2^{2n+1}}{(2n+1)} x^{2n+1},$$

$-\frac{1}{2} < x < \frac{1}{2}$.

17. $\displaystyle\int_0^x \frac{dt}{1+t^3}$

$$= \sum_{n=0}^{\infty} \frac{(-1)^n}{(3n+1)} x^{3n+1},$$

$-1 < x < 1$.

19. $\displaystyle\int_0^x \frac{\log(1+t)}{t}\,dt$

$$= \sum_{n=1}^{\infty} \frac{(-1)^{n+1}}{n^2} x^n,$$

$-1 < x < 1$.

§3.3

1. $(x+1)^{-1/3} = \displaystyle\sum_{n=0}^{\infty} \binom{-\frac{1}{3}}{n} x^n,$

$|x| < 1$.

3. $(x^2+1)^{2/5}$

$$= \sum_{n=0}^{\infty} \binom{\frac{2}{5}}{n} x^{2n},$$

$|x| < 1$.

5. $(x+4)^{-3/2}$

$$= \sum_{n=0}^{\infty} \binom{-\frac{3}{2}}{n} \frac{x^n}{8(4^n)},$$

$|x| < 4$.

7. $\displaystyle\int_0^x (1+u^2)^{3/2}\,du$

$$= \sum_{n=0}^{\infty} \binom{\frac{3}{2}}{n} \frac{x^{2n+1}}{(2n+1)},$$

$|x| < 1$.

9. $\dfrac{2}{x} \arcsin \dfrac{x}{2}$

$$= \sum_{n=0}^{\infty} (-1)^n \binom{-\frac{1}{2}}{n} \frac{x^{2n}}{(2n+1)2^{2n}},$$

$|x| < 2$.

§3.4

1. $\log x = (x-1) - \frac{1}{2}(x-1)^2$
$+ \frac{1}{3}(x-1)^3 + \cdots$.
3. $\sqrt{x} = 3 + \frac{1}{6}(x-9)$
$- \frac{1}{216}(x-9)^2 + \cdots$.
5. $x^{3/2} = 8 + 3(x-4)$
$+ \frac{3}{16}(x-4)^2 + \cdots$.
7. $e^{-2x} = e^{-2} - 2e^{-2}(x-1)$
$+ 2e^{-2}(x-1)^2 + \cdots$.
9. $\sin x = \frac{1}{2} + \frac{\sqrt{3}}{2}\left(x - \frac{\pi}{6}\right)$
$- \frac{1}{4}\left(x - \frac{\pi}{6}\right)^2 + \cdots$.
11. $\tan x = x + \frac{1}{3}x^3 + \frac{2}{15}x^5$
$+ \cdots$.
13. $\log(1 + e^x) = \log 2 + \frac{1}{2}x$
$+ \frac{1}{8}x^2 + \cdots$.
15. $\dfrac{\sin x}{1+x}$
$= x - x^2 + \frac{5}{6}x^3 + \cdots$.

17. $\dfrac{\arctan x}{1 + x}$

$= x - x^2 + \tfrac{2}{3}x^3 + \cdots$.

19. $e^{-x}\tan x$

$= x - x^2 + \tfrac{5}{6}x^3 + \cdots$.

21. Use an induction argument.

23. If $f(x) = A_0 + A_1 x + A_2 x^2 + \cdots$, then
$f(x) - f(-x) = 2A_1 x + 2A_3 x^3 + \cdots = 0$ if and only if
$A_1 = A_3 = \cdots = 0$,
and $f(x) + f(-x) = 2A_0 + 2A_2 x^2 + \cdots = 0$ if and only if
$A_0 = A_2 = \cdots = 0$.

Chapter 16

§1.1

1. $(0,2)$.
3. $(0,-3)$.
5. $(-\tfrac{3}{2},\sqrt{3}/2)$.
7. $(\tfrac{1}{2},\pi/2)$.
9. $(2,5\pi/6)$.
11. $(2,5\pi/3)$.
13. $\sqrt{3}$.
15. $\sqrt{13}$.
17. $C \approx 54°20'$.
19. $a = \sqrt{5 - 2\sqrt{2}} \approx 1.47$.

§1.2

1. A circle of radius 2 with center at the origin. A circle of radius $\tfrac{1}{2}$ with center at the origin.
3. A ray emanating from the origin making an angle of $120°$ with the positive x-axis. A ray emanating from the origin making an angle of $240°$ with the positive x-axis.
5. $r^2 - 4r\sin\theta + 3 = 0$.
7. $r(\sin\theta - 4\cos\theta) = 2$.

9. $x - 2y + 4 = 0$.
11. $(x - 1)^2 + y^2 = 4$;   circle of radius 2 with center at $(1,0)$.
13. $(x - \tfrac{1}{2})^2 + (y - \tfrac{1}{2})^2 = \tfrac{1}{2}$;   circle of radius $1/\sqrt{2}$ with center at $(\tfrac{1}{2},\tfrac{1}{2})$.
15. $x = \tfrac{1}{2}(y^2 - 1)$;   parabola.

§1.3

1. $r = 2\sin\theta$.

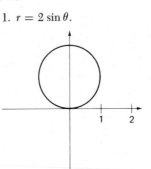

3. $r = \sin 3\theta$.

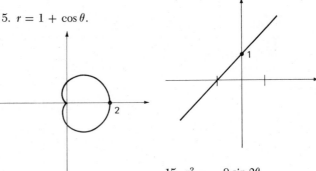

5. $r = 1 + \cos\theta$.

7. $r = 2 - \sin\theta$.

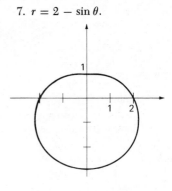

9. $r = 2(1 + 2\cos\theta)$.

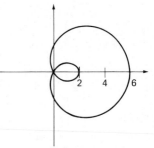

11. $r = \pi/\theta$.

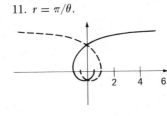

13. $r = \dfrac{1}{\sin\theta - \cos\theta}$.

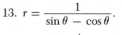

15. $r^2 = -9\sin 2\theta$.

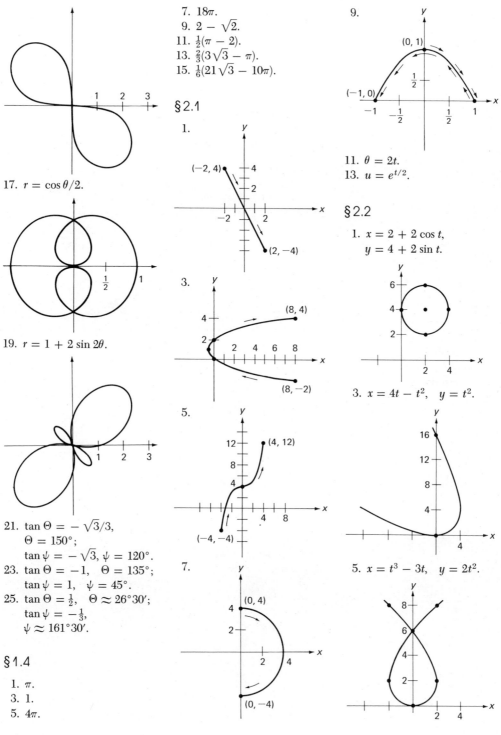

17. $r = \cos\theta/2$.

19. $r = 1 + 2\sin 2\theta$.

21. $\tan\Theta = -\sqrt{3}/3$,
    $\Theta = 150°$;
    $\tan\psi = -\sqrt{3}, \psi = 120°$.
23. $\tan\Theta = -1, \quad \Theta = 135°$;
    $\tan\psi = 1, \quad \psi = 45°$.
25. $\tan\Theta = \frac{1}{2}, \quad \Theta \approx 26°30'$;
    $\tan\psi = -\frac{1}{3}$,
    $\psi \approx 161°30'$.

§1.4

1. π.
3. 1.
5. 4π.

7. 18π.
9. $2 - \sqrt{2}$.
11. $\frac{1}{2}(\pi - 2)$.
13. $\frac{2}{3}(3\sqrt{3} - \pi)$.
15. $\frac{1}{6}(21\sqrt{3} - 10\pi)$.

§2.1

1.

(−2, 4)

(2, −4)

3.

(8, 4)

(8, −2)

5.

(4, 12)

(−4, −4)

7.

(0, 4)

(0, −4)

9.

(0, 1)

(−1, 0)

11. $\theta = 2t$.
13. $u = e^{t/2}$.

§2.2

1. $x = 2 + 2\cos t$,
   $y = 4 + 2\sin t$.

3. $x = 4t - t^2, \quad y = t^2$.

5. $x = t^3 - 3t, \quad y = 2t^2$.

7. Use the chain rule, the theorem for the derivative of an inverse function, and the rule for the differentiation of a quotient.

9. $\dfrac{dy}{dx} = \dfrac{3t^2}{2t - 1}$,

$\dfrac{d^2y}{dx^2} = \dfrac{6t(t - 1)}{(2t - 1)^3}$.

§2.3

1. $\frac{3}{2}$.
3. $122/27$.
5. $\frac{3}{2}$.
7. 8π (circle traversed twice).
9. 6.
11. $3\pi/2$.
13. Integrate t from 0 to $\pi/2$, multiply by 4 (symmetry), simplify, and use the substitution $t = \pi/2 - \psi$.

§2.4

1. See figure. Let $R = (x,y)$ denote the tracing point.

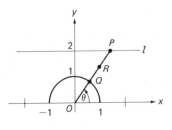

$x = \frac{1}{2}\cos\theta + \cot\theta$,
$y = \frac{1}{2}\sin\theta + 1$ for $0 < \theta < \pi$.

3. See figure. The tracing point R (x,y) is initially at $P = (0, a - b)$.

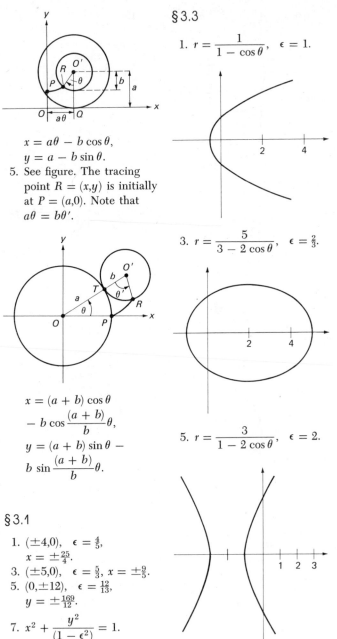

$x = a\theta - b\cos\theta$,
$y = a - b\sin\theta$.

5. See figure. The tracing point $R = (x,y)$ is initially at $P = (a,0)$. Note that $a\theta = b\theta'$.

$x = (a + b)\cos\theta - b\cos\dfrac{(a + b)}{b}\theta$,

$y = (a + b)\sin\theta - b\sin\dfrac{(a + b)}{b}\theta$.

§3.1

1. $(\pm 4, 0)$, $\epsilon = \frac{4}{5}$, $x = \pm\frac{25}{4}$.
3. $(\pm 5, 0)$, $\epsilon = \frac{5}{3}$, $x = \pm\frac{9}{5}$.
5. $(0, \pm 12)$, $\epsilon = \frac{12}{13}$, $y = \pm\frac{169}{12}$.
7. $x^2 + \dfrac{y^2}{(1 - \epsilon^2)} = 1$.

§3.2

1. $\dfrac{(x - 2)^2}{4} + \dfrac{(y - 4)^2}{3} = 1$.

§3.3

1. $r = \dfrac{1}{1 - \cos\theta}$, $\epsilon = 1$.

3. $r = \dfrac{5}{3 - 2\cos\theta}$, $\epsilon = \frac{2}{3}$.

5. $r = \dfrac{3}{1 - 2\cos\theta}$, $\epsilon = 2$.

7. Find the vertices, and determine the midpoint. Write the given equation in rectangular coordinates.

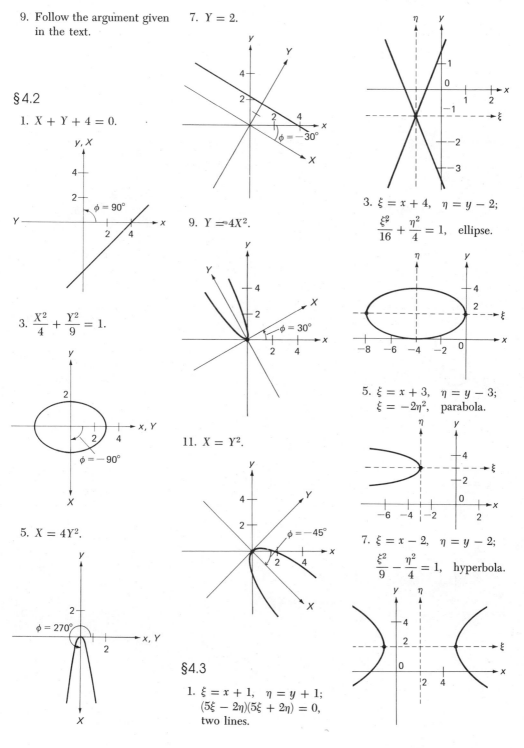

9. Follow the argument given in the text.

§4.2

1. $X + Y + 4 = 0$.

3. $\dfrac{X^2}{4} + \dfrac{Y^2}{9} = 1$.

5. $X = 4Y^2$.

7. $Y = 2$.

9. $Y = 4X^2$.

11. $X = Y^2$.

§4.3

1. $\xi = x + 1$, $\eta = y + 1$; $(5\xi - 2\eta)(5\xi + 2\eta) = 0$, two lines.

3. $\xi = x + 4$, $\eta = y - 2$; $\dfrac{\xi^2}{16} + \dfrac{\eta^2}{4} = 1$, ellipse.

5. $\xi = x + 3$, $\eta = y - 3$; $\xi = -2\eta^2$, parabola.

7. $\xi = x - 2$, $\eta = y - 2$; $\dfrac{\xi^2}{9} - \dfrac{\eta^2}{4} = 1$, hyperbola.

9. $\xi = x + 2$, $\eta = y - 4$;

$$\xi^2 + \frac{\eta^2}{4} = -1,$$

empty set.

§4.4

1. $\phi = 30°$, $\dfrac{(X - 4)^2}{16}$

$+ \dfrac{Y^2}{4} = 1$;

$\xi = x - 4$, $\eta = Y$,

$\dfrac{\xi^2}{16} + \dfrac{\eta^2}{4} = 1$; ellipse.

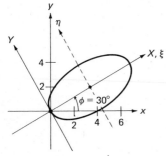

3. $\phi = -30°$,

$$\dfrac{Y^2}{4} - (X + 2)^2 = 1;$$

$\xi = X + 2$, $\eta = Y$,

$\dfrac{\eta^2}{4} - \xi^2 = 1$; hyperbola.

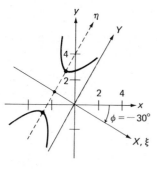

5. $\phi = 45°$,

$Y - 2 = (X - 2)^2$;

$\xi = X - 2$, $\eta = Y - 2$,

$\xi^2 = \eta$; parabola.

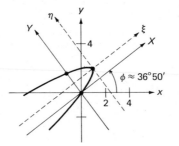

7. $\phi \approx 36°50'$,

$X - 2 = -2(Y - 1)^2$;

$\xi = X - 2$, $\eta = Y - 1$,

$\xi = -2\eta^2$; parabola

§4.5

1. $D = 0$; one line.
3. $D = 0$; two parallel lines.
5. $D = 0$, empty set.
7. $D = -4$; hyperbola.

Chapter 17

§1.1

1. Yes.
3. Yes.
5. $T = (3,4)$.
7. $C = (-3,1)$.
9. $R = (-1,2)$.

§1.2

1. $\overrightarrow{PQ} = \langle 1,3 \rangle$, $|PQ| = \sqrt{10}$.
3. $\overrightarrow{PQ} = \langle 4,4 \rangle$, $|PQ| = 4\sqrt{2}$.
5. $\overrightarrow{PQ} = \langle 5,2 \rangle$, $|PQ| = \sqrt{29}$.
7. $\overrightarrow{PQ} = \langle 6,-4 \rangle$, $|PQ| = 2\sqrt{13}$.
9. $\overrightarrow{RS} = \langle 5,2 \rangle$, $S = (6,-1)$.

§1.3

1. $\mathbf{a} + \mathbf{b} = \langle 4,-4 \rangle$, $4\mathbf{a} = \langle 4,-8 \rangle$, $(-5)\mathbf{b} = \langle -15,10 \rangle$.
3. $\mathbf{a} - \mathbf{b} = \langle -2,-2 \rangle$, $\frac{1}{4}(\mathbf{a} - \mathbf{b}) = \langle -\frac{1}{2},-\frac{1}{2} \rangle$, $-2(\mathbf{a} - \mathbf{b}) = \langle 4,4 \rangle$.
5. $2\mathbf{a} - \mathbf{b} = \langle -1,0 \rangle$.

§1.4

1. $\langle 2,38 \rangle$.
3. $\langle 1,8 \rangle$.
5. $\langle -6,-5 \rangle$.
7. $\mathbf{a} = -\mathbf{b} - 2\mathbf{c}$.
9. $\mathbf{a} = -\frac{1}{2}\mathbf{c}$, $\mathbf{b} = \frac{1}{2}\mathbf{c}$.
11. Proof of **2**.
   $\mathbf{a} + \mathbf{b} = \langle a_1 + b_1,$
   $a_2 + b_2 \rangle = \langle b_1 + a_1,$
   $b_2 + a_2 \rangle = \mathbf{b} + \mathbf{a}$.
   Proof of **3**.
   $(\mathbf{a} + \mathbf{b}) + \mathbf{c} = \langle a_1 + b_1,$
   $a_2 + b_2 \rangle + \langle c_1,c_2 \rangle$
   $= \langle a_1 + b_1 + c_1,$
   $a_2 + b_2 + c_2 \rangle$
   $= \langle a_1 + (b_1 + c_1),$
   $a_2 + (b_2 + c_2) \rangle$
   $= \mathbf{a} + (\mathbf{b} + \mathbf{c})$.
13. Let $\mathbf{x}$ denote the negative vector of $\mathbf{a}$. Then
   $\mathbf{a} + \mathbf{x} = \mathbf{0}$,
   $\langle a_1 + x_1, a_2 + x_2 \rangle = \langle 0,0 \rangle$,
   and $x_1 = -a_1$, $x_2 = -a_2$.
   Hence $\mathbf{x} = (-1)\mathbf{a} = -\mathbf{a}$.

§1.5

1. $\mathbf{a} = -\mathbf{i} + 5\mathbf{j}$.
3. $\mathbf{a} + 4\mathbf{b} = 18\mathbf{i} + 7\mathbf{j}$, $\mathbf{b} - 3\mathbf{a} = -2\mathbf{i} - 8\mathbf{j}$.

5. $\langle 0,5 \rangle$.

7. $\langle 4,-26 \rangle$.

9. $\mathbf{a} = \langle \frac{1}{2},\frac{1}{2} \rangle$, $\mathbf{b} = \langle \frac{1}{2},-\frac{1}{2} \rangle$.

11. $a_1\mathbf{i} + a_2\mathbf{j}$ is the vector $\mathbf{a}$
with components $\langle a_1,a_2 \rangle$.
By definition of length,
$|\mathbf{a}| = |a_1\mathbf{i} + a_2\mathbf{j}|$
$= \sqrt{a_1^2 + a_2^2}$.

§1.6

1. $|\mathbf{P}| = 1$, $\theta = 30°$.

3. $|\mathbf{R}| = 2$, $\theta = 120°$.

5. Yes.

7. No.

9. $\mathbf{R} = (-\frac{3}{2},\frac{7}{2})$.

11. $\mathbf{R} = (0,7)$.

13. $\mathbf{R} = \langle -2,-8 \rangle$.

§1.7

1. $x = 3t + 1$, $y = t + 2$.

3. $x = 3t - 2$, $y = 6t - 2$.

5. $x = -3t + 5$, $y = -3t$.

7. $x = t$, $y = 0$.

9. $x = -2t$, $y = t - 1$.

11. $x = 3t$, $y = -2t + 2$.

§2.1

1. $\mathbf{f}(t) = \{1 + \frac{1}{9}t^2\}\mathbf{i}$
$+ \{1 + \frac{2}{9}t^2 + \frac{1}{81}t^4\}\mathbf{j}$.

3. $\mathbf{f}(t) = \frac{\sqrt{2}}{2}(t - 2)^{3/2}\mathbf{i}$
$+ \frac{\sqrt{2}}{4}(t - 2)^{9/2}\mathbf{j}$.

5. $\mathbf{f}(t) = \{(t + 3)^2 + 1\}\mathbf{i}$
$+ \log\{(t + 3)^2 + 1\}\mathbf{j}$.

7. $\mathbf{f}(t) = \{(t/2)^{2/3} - 1\}\mathbf{i}$
$+ \frac{2}{3}\{(t/2)^{2/3} - 1\}^{3/2}\mathbf{j}$.

§2.2

1. $-\sin 2t\mathbf{i} + \cos t\mathbf{j}$.

3. $\frac{3}{2}t^{1/2}\mathbf{i} - (t \sec t \tan t$
$+ \sec t)\mathbf{j}$.

5. $\frac{15}{4}t^{1/2}\mathbf{i}$.

7. $-\frac{1}{4}t^{-3/2}\mathbf{i} + e^t\mathbf{j}$.

9. $\mathbf{f}'(0) = \mathbf{i} + 3\mathbf{j}$.

11. $\frac{1}{24}\mathbf{i} - 2\log\frac{5}{4}\mathbf{j}$.

13. $\log 2\mathbf{i} + \log\frac{3}{2}\mathbf{j}$.

15. $\frac{1}{3}\{(\pi + 1)^{3/2} - 1\}\mathbf{i}$.

17. $\mathbf{i} + \{2\log 2 - 1\}\mathbf{j}$.

19. $\mathbf{f}(t) = (\frac{1}{12}t^4 + \frac{11}{12}t + 1)\mathbf{i}$
$+ (-2t + 1)\mathbf{j}$.

21. Use the fundamental
theorem of calculus for
scalar-valued functions.

§2.3

1. $\mathbf{T}(t) = -\dfrac{\sin t}{\sqrt{1 + \sin^2 t}}\mathbf{i}$
$+ \dfrac{1}{\sqrt{1 + \sin^2 t}}\mathbf{j}$;
$\mathbf{N}(t) = -\dfrac{1}{\sqrt{1 + \sin^2 t}}\mathbf{i}$
$- \dfrac{\sin t}{\sqrt{1 + \sin^2 t}}\mathbf{j}$.

3. $\mathbf{T}(t) = \dfrac{3t^2}{|t|\sqrt{9t^2 + 4}}\mathbf{i}$
$- \dfrac{2t}{|t|\sqrt{9t^2 + 4}}\mathbf{j}$;
$\mathbf{N}(t) = \dfrac{2t}{|t|\sqrt{9t^2 + 4}}\mathbf{i}$
$+ \dfrac{3t^2}{|t|\sqrt{9t^2 + 4}}\mathbf{j}$.

5. $\mathbf{T}(t) = \dfrac{\sin^2 t \cos t}{|\sin t \cos t|}\mathbf{i}$
$- \dfrac{\cos^2 t \sin t}{|\sin t \cos t|}\mathbf{j}$;
$\mathbf{N}(t) = \dfrac{\cos^2 t \sin t}{|\sin t \cos t|}\mathbf{i}$
$+ \dfrac{\sin^2 t \cos t}{|\sin t \cos t|}\mathbf{j}$.

7. $\mathbf{T}(t) = \dfrac{1}{\sqrt{1 + 9t^4}}\mathbf{i}$
$+ \dfrac{3t^2}{\sqrt{1 + 9t^4}}\mathbf{j}$;
$\mathbf{N}(t) = -\dfrac{3t^2}{\sqrt{1 + 9t^4}}\mathbf{i}$
$+ \dfrac{1}{\sqrt{1 + 9t^4}}\mathbf{j}$.

§2.4

1. $\mathbf{f}(s) = \dfrac{2}{\sqrt{5}}s\mathbf{i}$
$+ \left(-\dfrac{s}{\sqrt{5}} + 1\right)\mathbf{j}$.

3. $\mathbf{f}(s) = \{(\frac{3}{2}s + 1)^{2/3} - 1\}\mathbf{i}$
$+ \frac{2}{3}\{(\frac{3}{2}s + 1)^{2/3} - 1\}\mathbf{j}$.

5. $\mathbf{f}(s) = \{(\frac{3}{4}s + 1)^{2/3} - 1\}\mathbf{i}$
$+ \displaystyle\int_0^{\{(\frac{3}{4}s+1)^{2/3}-1\}} \sqrt{3 + 4u}\, du\,\mathbf{j}$
or $\mathbf{f}(s) = \{(\frac{3}{4}s + 1)^{2/3}$
$- 1\}\mathbf{i} + [\frac{1}{6}\{4(\frac{3}{4}s + 1)^{2/3}$
$- 1\}^{3/2} - \frac{1}{6}(3)^{3/2}]\mathbf{j}$.

§3.1

1. $\mathbf{v}(t) = \mathbf{i} + t\mathbf{j}$; $|\mathbf{v}(t)|$
$= \sqrt{1 + t^2}$; $\mathbf{a}(t) = \mathbf{j}$.

3. $\mathbf{v}(t) = -2\sin t\mathbf{i} - 3\sin t\mathbf{j}$;
$|\mathbf{v}(t)| = \sqrt{13}|\sin t|$;
$\mathbf{a}(t) = -2\cos t\mathbf{i} - 3\cos t\mathbf{j}$.

5. $\mathbf{v}(t) = t(t^2 + 1)^{-1/2}\mathbf{i}$
$- t(t^2 + 1)^{-3/2}\mathbf{j}$; $|\mathbf{v}(t)|$
$= |t|(t^4 + 2t^2 + 2)^{1/2}$
$\times (t^2 + 1)^{-3/2}$;
$\mathbf{a}(t) = (t^2 + 1)^{-3/2}\mathbf{i}$
$+ (2t^2 - 1)(t^2 + 1)^{-5/2}\mathbf{j}$.

7. $v_x = 16$ ft/sec;
$\mathbf{a} = -64\mathbf{i}$ ft/sec^2.

9. $\mathbf{v} = \dfrac{5}{\sqrt{2}}\mathbf{i} + \dfrac{5}{\sqrt{2}}\mathbf{j}$ ft/sec;
$\mathbf{a} = -\frac{25}{8}\mathbf{i} + \frac{25}{8}\mathbf{j}$ ft/sec^2.

§3.4

1. Assume projectile is shot
from point $(0,0)$ at time
$t = 0$. Then $\gamma = \delta = 0$ and
$x(t) = \alpha t$, $y(t) = -\frac{1}{2}gt^2$
$+ \beta t$. The projectile
strikes the ground when
$y(t) = 0$, that is, when
$t = 0$ or $t = 2\beta/g$. Hence
$\sigma_0 = \{(\dot{x}(0))^2 +$
$(\dot{y}(0))^2\}^{1/2}$
$= \{(\dot{x}(2\beta/g))^2 +$
$(\dot{y}(2\beta/g))^2\}^{1/2} = \sqrt{\alpha^2 + \beta^2}$
$= \sigma_1$.

3. Assume projectile is shot from point $(0,0)$ at time $t = 0$. Then $x(t) = \alpha t$, $y(t) = -\frac{1}{2}gt^2 + \beta t$.
(a) The projectile reaches its maximum distance when $t = 2\beta/g$. (See Problem 1.) Hence $r = x(2\beta/g) = \alpha \cdot 2\beta/g = 2\alpha\beta/g$. The projectile attains its maximum height when $t = \beta/g$. (See Problem 2.) Hence $h = y(\beta/g) = -\frac{1}{2}g(\beta^2/g^2) + \beta \cdot \beta/g = \beta^2/2g$. (b) Draw a figure, and note that $\tan \phi = 2h/r$. Substituting for h and r, $\tan \phi = \beta/2\alpha$. But $\tan \theta = \dot{y}(0)/\dot{x}(0) = \beta/\alpha$.

5. 32 ft.

7. $\mathbf{v} = 400\mathbf{i} + 80\mathbf{j}$ ft/sec.

§4.1

1. $\kappa(x) = \dfrac{2}{[1 + 4(x - 1)^2]^{3/2}}$;
$\kappa(x) > 0$ for all x.

3. $\kappa(x) = \dfrac{2x^3}{(1 + x^4)^{3/2}}$; $\kappa(x)$
> 0 for $0 < x < +\infty$;
$x = 0$ is an inflection point.

5. $\kappa(t) = \dfrac{-4t}{(t^4 + 4)^{3/2}}$;
$\kappa(t) > $ for $t < 0$,
$\kappa(t) < 0$ for $t > 0$; $t = 0$ is an inflection point.

7. $\kappa(t) = -\dfrac{1}{\sqrt{2}}e^{-t}$;
$\kappa(t) < 0$ for all t.

9. $\kappa(x) = \cos x$. Curvature has a local maximum at $(0,0)$. There are no minima in the interval $-\pi/2 < x < \pi/2$.

11. $\kappa = 0$ at $(1,1)$.

§4.2

1. $a_T = \dfrac{4t}{\sqrt{1 + 4t^2}}$,
$a_N = \dfrac{2}{\sqrt{1 + 4t^2}}$.

3. $a_T = \dfrac{-1}{t^2\sqrt{1 + t^2}}$,
$a_N = -\dfrac{1}{t\sqrt{1 + t^2}}$.

5. $a_T = \dfrac{(e^{2t} - e^{-2t})}{(e^{2t} + e^{-2t})^{1/2}}$,
$a_N = \dfrac{2}{(e^{2t} + e^{-2t})^{1/2}}$.

§4.4

1. For the inclined plane $x(s) = s \cos \theta$, $y(s) = s \sin \theta$. Hence the curvature $\kappa = 0$ and $dx/ds = \cos \theta$. The second equation (11) in §4.3 becomes $\gamma = mg \cos \theta$, that is, the magnitude of the force depends only on θ.

§4.5

1. Rate of change of period bigger on moon.

§5.4

1. (a) Use Equation (1) with $r = \beta e^\theta$. Then $\beta^2 e^{2\theta}\, d\theta/dt = 2\alpha$. The solution of this differential equation is $\frac{1}{2}(\beta e^\theta)^2 = 2\alpha t + c$ or $\frac{1}{2}r^2 = 2\alpha t + c$. Hence $r(t) = (4\alpha t + r_0^2)^{1/2}$ where $r_0 = r(0)$. (b) Since $r = \beta e^\theta$, $\dot{r} = \beta e^\theta \dot{\theta}$. Hence $\dot{r} = (\beta e^\theta)\dot{\theta} = (r)(2\alpha/r^2) = 2\alpha/r$, where we use Equation (1). Also

$\ddot{r} = -(2\alpha/r^2)\dot{r}$
$= -(2\alpha/r^2)(2\alpha/r)$
$= -4\alpha^2/r^3$. Hence, from §4.1, $|\mathbf{a}| = |\ddot{r} - r\dot{\theta}^2|$
$= |\ddot{r} - (1/r^3)(r^4\dot{\theta}^2)|$
$= |\ddot{r} - (4\alpha^2/r^3) = 8\alpha^2/r^3$.

Chapter 18

§1.1

1,3,5.

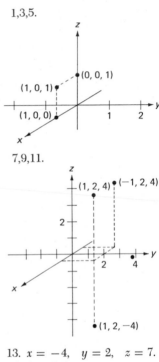

7,9,11.

13. $x = -4$, $y = 2$, $z = 7$.

15. $(1, -6, 1)$.

17. $(1, -2, -10)$.

§1.2

1. Sphere with center at $(2, -1, 4)$ and radius of 4.

3. Center at $(2, -3, \frac{1}{2})$; radius of 1.

5. Radius is 3; center at $(0, 0, 1)$.

7. $(x - 1)^2 + (y + 2)^2 + (z - 2)^2 = 6$.

9. $z = \pm 6$, that is, two planes.

§1.3

1. $(\sqrt{2}, \sqrt{2}, 0)$.
3. $(2,0,2)$.
5. $(\sqrt{3}/2, \frac{1}{2}, -4)$.
7. $(0, -2, 3)$.
9. $(2\sqrt{2}, -\pi/4, -3)$.
11. $(4, -\pi/3, 2)$.
13. $(\sqrt{5}, \text{arc tan } 2, 3)$.
15. $(4, -15°, 1)$.

§1.4

1. $z^2 = 4r^2$,
   $z^2 = 4(x^2 + y^2)$.
3. $z = 2r^2$, $z = 2(x^2 + y^2)$.
5. $4r^2 + z^2 = 100$,
   $4(x^2 + y^2) + z^2 = 100$.
7. $z = e^{-|r|}$, $z = e^{-\sqrt{x^2+y^2}}$.

§1.5

1. $(\sqrt{2}, 0, \sqrt{2})$.
3. $(0, 0, \sqrt{2})$.
5. $(0, -\sqrt{2}, 0)$.
7. $(\sqrt{3}/2, -\frac{3}{2}, 1)$.
9. $(1, 0, \pi/2)$.
11. $(\sqrt{2}, \pi/4, \pi/2)$.
13. $(4, \pi/6, \pi/3)$.
15. $(4, -\pi/6, \pi/6)$.

§2.1

1. $R = (7,4,2)$.
3. $S = (4,6,6)$.
5. $\overrightarrow{PQ} = \langle 2, -4, 6 \rangle$;
   $|PQ| = 2\sqrt{14}$.
7. $\langle 1, -\frac{1}{2}, \frac{3}{2} \rangle$; $\langle 15, -21, -18 \rangle$.
9. $\langle 18, -5, 33 \rangle$;
   $18\mathbf{i} - 5\mathbf{j} + 33\mathbf{k}$.
11. $\mathbf{a} = \mathbf{i} + \mathbf{j} = \langle 1,1,0 \rangle$,
    $\mathbf{b} = \mathbf{i} + \mathbf{j} + \mathbf{k} = \langle 1,1,1 \rangle$,
    $\mathbf{c} = \mathbf{i} + \mathbf{k} = \langle 1,0,1 \rangle$.

§2.2

1. $x = 3 + 2t$, $y = 4$,
   $z = 5 - 2t$.
3. $x = 1 - 3t$, $y = 4 + 4t$,
   $z = 2 + 2t$.
5. $x = -3 + 15t$,
   $y = 2 - 6t$, $z = 6 - 4t$.
7. $(0,4,8)$, $(8,4,0)$; line does not meet the xz plane.
9. $(2,0,4)$.
11. Yes.
13. $(\frac{3}{2}, 1, -\frac{7}{2})$.
15. $P = (a, 0, 6)$,
    $Q = (2 - a, 4, 0)$, where a is any number.

§2.3

1. $\mathbf{T}(1) = \frac{1}{2}(\sqrt{2}\mathbf{i} + \mathbf{j} + \mathbf{k})$.
3. $\mathbf{T}\left(\dfrac{\pi}{2}\right) = \dfrac{1}{\sqrt{\pi^2 + 8}}$
   $\times (-\pi\mathbf{i} + 2\mathbf{j} + 2\mathbf{k})$.
5. e.
7. $\frac{52}{3}$.
9. $\mathbf{v}(t) = 2\mathbf{i} + 2t\mathbf{j} + 2\mathbf{k}$;
   $|\mathbf{v}(t)| = 2\sqrt{2 + t^2}$;
   $\mathbf{a}(t) = 2\mathbf{j}$.
11. $\mathbf{v}(t) = e^t(\cos t - \sin t)\mathbf{i}$
    $+ e^t(\sin t + \cos t)\mathbf{j} + e^t\mathbf{k}$;
    $|\mathbf{v}(t)| = \sqrt{3}e^t$; $\mathbf{a}(t)$
    $= -2e^t \sin t\mathbf{i} + 2e^t \cos t\mathbf{j}$
    $+ e^t\mathbf{k}$.
13. Parabola.
15. Right-handed circular helix.

§3.1

1. 8.
3. 20.
5. 120°.
7. $\text{arc cos } \frac{13}{21} \approx 52°$.
9. $\alpha = 1$, $\alpha = -\frac{1}{6}$.

§3.2

1. Expand $(\mathbf{a} + \mathbf{b}, \mathbf{a} + \mathbf{b})$ using definition of inner product and group terms.

3. $\mathbf{a} = 2\mathbf{i} + 2\mathbf{j} + 2\mathbf{k}$.
5. No.
7. $\text{arc cos } \frac{5}{7} \approx 44°$.
9. $45°$.
11. $\text{arc cos } \sqrt{\frac{2}{3}} \approx 35°$.

§3.3

1. $\cos \alpha_1 = 2/\sqrt{29}$,
   $\cos \alpha_2 = -3/\sqrt{29}$,
   $\cos \alpha_3 = 4/\sqrt{29}$.
3. $\cos \alpha_1 = -\frac{3}{5}$, $\cos \alpha_2 = \frac{4}{5}$,
   $\cos \alpha_3 = 0$.
5. $\cos \alpha_1 = 1/\sqrt{21}$,
   $\cos \alpha_2 = -2/\sqrt{21}$,
   $\cos \alpha_3 = 4/\sqrt{21}$.
7. $\alpha_1 = 90°$, $\alpha_2 = 20°$,
   $\alpha_3 = 70°$.
9. $\alpha_1 = \text{arc cos } \sqrt{3}/3 \approx 55°$.
11. $\theta = \text{arc cos } \frac{12}{25} \approx 61°$.
13. Verify that $\cos \theta$
    $= \cos \alpha_1 \cos \beta_1$
    $+ \cos \alpha_2 \cos \beta_2$
    $+ \cos \alpha_3 \cos \beta_3 = 0$,
    where α_1, α_2, α_3 are direction angles for $\mathbf{a}$ and β_1, β_2, β_3 are direction angles for $\mathbf{b}$.
15. $\cos \alpha_1 = \sin \phi \cos \theta$,
    $\cos \alpha_2 = \sin \phi \sin \theta$,
    $\cos \alpha_3 = \cos \phi$.

§4.1

1. $\alpha_1 \approx 48°$, $\alpha_2 \approx 71°$,
   $\alpha_3 \approx 48°$; $d = \frac{1}{2}$.
3. $\alpha_1 \approx 96°$, $\alpha_2 \approx 116°$,
   $\alpha_3 \approx 27°$; $d = 1$.
5. $\alpha_1 \approx 65°$, $\alpha_2 \approx 124°$,
   $\alpha_3 \approx 135°$; $d = 0$.
7. $5x - 5y - 6z + 5 = 0$.
9. $3x + 2y - z = 0$.
11. $\sqrt{42}$.
13. $x + 4y - 3z = 0$.

§4.2

1. $2x - 3y - 2 = 0$.
3. $x - 20y + 7z + 12 = 0$.

5. $a = 2$.

7. Parallel.

9. Not parallel. Let $x = t$. Then $y = -t$, $z = \frac{3}{2}$.

11. $y - 2x + 2 = 0$, $6x + z - 12 = 0$.

13. $5x + 6y - 35 = 0$, $y + 5z - 30 = 0$.

15. $3x - 4y - 10 = 0$, $7y + 3z - 5 = 0$.

17. $x = t$, $y = t$, $z = t$.

19. $x = 2t$, $y = t$, $z = \frac{9}{4}t$.

21. The points are $(\frac{1}{2}t - \frac{5}{2}, \frac{1}{2}t - \frac{3}{2}, t)$ for all t.

§4.3

1. $\phi \approx 71°$.

3. $\phi \approx 48°$.

5. Skew.

7. Parallel.

9. l_1 and l_2 intersect at $(3, -1, 4)$.

11. Skew.

13. l and σ intersect at $(2, 3, 4)$.

15. l is parallel to the plane σ.

17. l intersects σ at the point $(28, 14, 7)$.

19. l intersects σ at the point $(-\frac{2}{3}, \frac{1}{3}, \frac{7}{6})$.

§5.1

1. Parabolic cylinder.

3. Elliptic cylinder.

5. Hyperbolic cylinder.

7. Circular cone.

9. Parabolic cylinder.

11. Hyperbolic cylinder.

§5.2

1. Set $z = \gamma$ in Equation (3). This gives the elliptic cylinder $x^2/a^2 + y^2/b^2 - \gamma^2 = 0$. The intersection of the plane $z = \gamma \neq 0$ and the elliptic cylinder gives an ellipse. When $\gamma = 0$, the plane $z = 0$ and the line $x^2/a^2 + y^2/b^2 = 0$ intersect at the point $(0,0,0)$.

3. Elliptic cone.

§5.3

1. Prolate spheroid.

3. General ellipsoid.

5. General hyperboloid of two sheets.

7. General hyperboloid of one sheet.

9. Prolate spheroid.

11. Hyperboloid of revolution of two sheets.

§5.4

1. Elliptic paraboloid.

3. Hyperbolic paraboloid.

5. Elliptic paraboloid.

§5.5

1. Use Equations (10) and (11). Then $x = \frac{1}{2}(\sqrt{3} + t)$, $y = \frac{1}{2}(1 - \sqrt{3}t)$, $z = \frac{1}{10}t$, and $x = \frac{1}{2}(\sqrt{3} - t)$, $y = \frac{1}{2}(1 + \sqrt{3}t)$, $z = \frac{1}{10}t$.

3. Use Equations (12) and (13). Then $x = \frac{1}{2}t$, $y = -(1/\sqrt{5})t$, $z = 0$, and $x = \frac{1}{2}t$, $y = (1/\sqrt{5})t$, $z = 0$.

5. If the lines are parallel for $\alpha \neq \beta$, we must have the parametric representation of each coordinate on the two lines in the same ratio. Thus we have
$$\frac{a \cos \alpha + at \sin \alpha}{a \cos \beta + at \sin \beta}$$
$$= \frac{b \sin \alpha - bt \cos \alpha}{b \sin \beta - bt \cos \beta} = \frac{t}{t}$$
or $(\cos \alpha - \cos \beta)$
$= (\sin \beta - \sin \alpha)t$ and
$(\sin \alpha - \sin \beta)$
$= (\cos \alpha - \cos \beta)t$ or
$(\cos \alpha - \cos \beta)$
$= -(\cos \alpha - \cos \beta)t^2$.
This is valid for all t only if $\cos \alpha = \cos \beta$. Also, from one of the earlier equations, $\sin \alpha = \sin \beta$. Hence $\alpha = \beta$ if the lines are not skew.

7. Proceed as in Problem 5. If the lines are not skew, we must have the equations with $\alpha = A$ and $\alpha = B$ proportional. Thus we have $\dfrac{A + t}{B + t} = \dfrac{A - t}{B - t}$
$= \dfrac{A}{B}$ for $B \neq 0$, or
$B(A + t) = A(B + t)$ or $Bt = At$ for all t and hence $A = B$. If $B = 0$, reverse the ratio.

§6.2

1. $S = (0, 0, 1)$.

3. $\mathbf{a} \times \mathbf{b} = \mathbf{i}$.

5. $\mathbf{a} \times \mathbf{b} = 12\mathbf{k}$, $\mathbf{b} \times \mathbf{a} = -12\mathbf{k}$.

7. 4.

§6.3

1. Use Theorem 1.

3. Use Theorem 1.

5. $-10\mathbf{i} + 7\mathbf{j} + \mathbf{k}$.

7. $-2\mathbf{i} + \mathbf{j} - \mathbf{k}$.

9. $-26\mathbf{i} - \mathbf{j} + 7\mathbf{k}$.

11. $4\sqrt{14}$.

13. $(\sqrt{2}/6)(-\mathbf{i} + 4\mathbf{j} - \mathbf{k})$.

15. 7.

§6.4

1. 4.

3. 32.

5. Expand each identity using the definitions of cross product and dot product.

7. Expand both sides using the definitions of cross product and dot (inner) product.

9. Let $\mathbf{a} = a_1\mathbf{i} + a_2\mathbf{j} + a_3\mathbf{k}$, and expand both sides.

11. Let $\mathbf{a} = a_1\mathbf{i} + a_2\mathbf{j} + a_3\mathbf{k}$, $\mathbf{b} = b_1\mathbf{i} + b_2\mathbf{j} + b_3\mathbf{k}$, and expand both sides.

Chapter 19

§1.1

1. $g(x,y)$ defined for all (x,y) such that $y \leq 2x + 3$.
3. $g(x,y)$ defined for all (x,y) such that $x^2 - y^2 > 9$.
5. $f(2,1) = 25$; $f(-3,0) = -27$.
7. $F(2x,y^2) = 2x + y^4$; $F(x^2, \sqrt{x + y}) = x^2 + x + y$.
9. $F(e^x, e^{-x}) = \dfrac{e^{3x}}{e^{6x} + 1}$; $F(e^x, 2e^x) = \frac{8}{9}e^{3x}$.
11. $f(x,f(x,y)) = \dfrac{x + y}{x + y + 1}$; $f(f(x,y), f(x,y)) = \frac{1}{2}$.
13. Note that $\tan u = x/y$. Hence $\sin u = x/\sqrt{x^2 + y^2}$, $\cos u = y/\sqrt{x^2 + y^2}$. Now substitute for $\sqrt{v}$, $\sin u$, $\cos u$, and verify.

§1.2

1. Open.
3. Open.
5. Not open.
7. The set of all (x,y) such that $y = -x$.
9. The set is its own boundary.
11. Yes. The set of all (x,y) such that $x^2 + y^2 = 1$.
13. Continuous.

15. Discontinuous when $xy = -1$.
17. Discontinuous when $x = 0$, $y = 0$.
19. Discontinuous when $y = \dfrac{2n + 1}{2}\pi$, $n = 0,\pm 1,\pm 2, \ldots$.
21. Discontinuous at $(0,0)$.
23. Continuous.
25. Continuous.

§1.3

1. $f(1,-2,3) = -5$; $f(-1,0,4) = -4$.
3. $f(a, -2a, a^2) = 2a^2(4a - 1)$; $f(a + b, b^2, a - b) = 3b^3 - 4b^2 - ab^2 + 4a^2$.
5. $u(\pi, \pi/2, 0, \pi/4) - (\sqrt{2}/2)e^{\pi^2/2}$.
7. $F\left(\dfrac{3}{x^2 + 1}, \dfrac{2}{x^2 + 1}, \dfrac{1}{x^2 + 1}\right) = \dfrac{6}{5(x^2 + 1)^2}$.
9. Defined and continuous for all (x,y,z).
11. Defined and continuous everywhere except at $x = 0$, $y = 0$, $z = 0$.
13. Defined and continuous everywhere except for $x^2 + y^2 + z^2 = t^2$.

§2.1

1. $\partial f/\partial x = 1$, $\partial f/\partial y = 2$.
3. $\partial f/\partial x = 2y^2(xy - 1)$, $\partial f/\partial y = xy(3xy - 4)$.
5. $\partial f/\partial x = 2x\cos(x^2 + y)$, $\partial f/\partial y = \cos(x^2 + y)$.
7. $\partial f/\partial x = -1/x^2y^2$, $\partial f/\partial y = -(1/y)((1/x^2y) + 1)$.
9. $\partial f/\partial x = (2 - 1/x^2)e^{x^2+y^2}$, $\partial f/\partial y = (2y/x)e^{x^2+y^2}$.

11. $\partial f/\partial x = 1/x$, $\partial f/\partial y = 2\csc 2y$.
13. $f_x(-1,2) = -64$, $f_y(-1,2) = 80$.
15. $f_x(1,0) = 1/\sqrt{2}$, $f_y(1,0) = 0$.
17. $g(x) = \log(x\tan 3)$, $g'(2) = \frac{1}{2}$; $(\partial f/\partial x)_{(2,3)} = \frac{1}{2}$.
19. $z_x = x/\sqrt{x^2 + y^2}$, $z_y = y/\sqrt{x^2 + y^2}$. Substitute and verify.

§2.2

1. $2x + 2y - z = 2$; $x = 1 + 2t$, $y = 1 + 2t$, $z = 2 - t$.
3. $6x + 8y - z = 33$; $x = 3 + 6t$, $y = 2 + 8t$, $z = 1 - t$.
5. $2x + 2y + 3z = 7$; $x = 1 - \frac{2}{3}t$, $y = 1 - \frac{2}{3}t$, $z = 1 - t$.
7. $2x + 2y + z = 3$; $x = \frac{2}{3} - 2t$, $y = \frac{2}{3} - 2t$, $z = \frac{1}{3} - t$.
9. The normal vectors have components $\langle -1,1,-1 \rangle$ and $\langle \frac{5}{6}, -\frac{1}{6}, -1 \rangle$, respectively, and their inner product is zero.
11. The upper half of the ellipsoid is the graph of the function $z = c\sqrt{1 - (x/a)^2 - (y/b)^2}$. The equation of the tangent plane at (x_0, y_0, z_0) is $z = z_0 - (c^2x_0/a^2z_0)(x - x_0) - (c^2y_0/b^2z_0)(y - y_0)$. Collecting terms, we obtain $xx_0/a^2 + yy_0/b^2 + zz_0/c^2 = z_0^2 + x_0^2/a^2 + y_0^2/b^2 = 1$.

§2.3

1. $\partial z/\partial u = 14u - 13v$; $\partial z/\partial v = 14v - 13u$.

3. $\partial z/\partial u = 2u$; $\partial z/\partial v = 0$.

5. $\partial z/\partial u = -(1/u^2)\cos v$;
$\partial z/\partial v = -(1/u)\sin v$.

7. 1,1.

9. 2,-2.

11. -17,3.

13. Let $F(x,y) = f(t)$ where
$t = x - 2y$. Then
$F_x(x,y) = (\partial f/\partial t)(\partial t/\partial x)$
$= f'(t)$ and
$F_y(x,y) = (\partial f/\partial t)(\partial t/\partial y)$
$= -2f'(t)$ by the chain
rule.
Substitute and verify.

15. Let $\phi(u,v) = f(t)$ where
$t = u + v$. Then
$\phi_u(u,v) = (\partial f/\partial t)(\partial t/\partial u)$
$= f'(t) > 0$ and
$\phi_v(u,v) = (\partial f/\partial t)(\partial t/\partial v)$
$= f'(t) > 0$ by the chain
rule. Hence
$\phi_u + \phi_v = 2f'(t) > 0$.

§2.4

1. $dz/ds = 6\cos 4x - 8\sin 4x$.

3. $dz/ds = 2(e^{2s} - e^{-2s})/(e^{2s}$
$+ e^{-2s}) = 2\tanh 2s$.

5. $dz/ds = 2\cot 2s$.

7. (a) grad $F = 2ye^{2x}\mathbf{i} + e^{2x}\mathbf{j}$,
(b) no critical points,
(c) $(1/\sqrt{65})(8\mathbf{i} + \mathbf{j})$,
(d) $dF/ds = -\frac{7}{2}\sqrt{2}$.

9. (a) grad $F = 2(x - y + 3)\mathbf{i}$
$- 2(x - 4y)\mathbf{j}$,
(b) $(-4,-1)$,
(c) $(1/\sqrt{5})(2\mathbf{i} + \mathbf{j})$,
(d) $dF/ds = -2(2 + \sqrt{3})$.

§2.5

1. $3x^2 - 6y^2 = 168$,
$\mathbf{v} = \mathbf{i} + 2\mathbf{j}$.

3. $x^2 + y^2 - 6x + 4y = -5$,
$\mathbf{v} = \mathbf{i} - \mathbf{j}$.

5. $e^x\cos y = 1$, $\mathbf{v} = \mathbf{j}$.

§2.6

1. $f_x = yz$, $f_y = xz$,
$f_z = xy$.

3. $f_x = 2x\tan yz$,
$f_y = x^2z\sec^2 yz$,
$f_z = x^2y\sec^2 yz$.

5. $f_x = -y\sin xy\cos yz$,
$f_y = -x\sin xy\cos yz$
$- z\sin yz\cos xy$,
$f_z = -y\sin yz\cos xy$.

7. $f_x = \dfrac{2x}{y^2 + z^2}$,
$f_y = -2y\dfrac{(x^2 + z^2)}{(y^2 + z^2)^2}$,
$f_z = 2z\dfrac{(y^2 - x^2)}{(y^2 + z^2)^2}$.

9. $f_x(1,2,-1) = 1/e$,
$f_y(1,2,-1) = 0$,
$f_z(1,2,-1) = 3/e$.

11. $G_x = yzt^2$, $G_y = xzt^2$,
$G_z = xyt^2$, $G_t = 2xyzt$.

13. $G_x =$
$-x(x^2 + y^2 + z^2$
$- t^2)^{-3/2}$,
$G_y =$
$-y(x^2 + y^2 + z^2$
$- t^2)^{-3/2}$,
$G_z =$
$-z(x^2 + y^2 + z^2$
$- t^2)^{-3/2}$,
$G_t =$
$t(x^2 + y^2 + z^2 - t^2)^{-3/2}$.

15. $\partial w/\partial r = 2r$,
$\partial w/\partial s = 2te^{2st}$,
$\partial w/\partial t = 2se^{2st}$.

17. $\dfrac{\partial w}{\partial r}$
$= \dfrac{8s^2(r + t)}{(r^2 + 4s^2 + t^2 + 2rt)^2}$,
$\dfrac{\partial w}{\partial s}$
$= \dfrac{-8s(r + t)^2}{(r^2 + 4s^2 + t^2 + 2rt)^2}$,
$\dfrac{\partial w}{\partial t}$
$= \dfrac{8s^2(r + t)}{(r^2 + 4s^2 + t^2 + 2rt)^2}$.

19. $\partial w/\partial r = 2se^{2rs}$
$(\cos t + \sin t + \cos t\sin t)$,
$\partial w/\partial s = 2re^{2rs}$
$(\cos t + \sin t + \cos t\sin t)$,
$\partial w/\partial t = e^{2rs}$
$(\cos t - \sin t + \cos 2t)$.

§2.7

1. $df = (3x^2 + 4xy + 2y^2)\,dx$
$+ (3y^2 + 4xy + 2x^2)\,dy$.

3. $df = \dfrac{dx}{x + 1} - \dfrac{dy}{y + 1}$.

5. $df = \dfrac{(dx + dy)}{2\sqrt{1 - (x + y)^2}}$.

7. 4,2.

9. $dF = vw(2u + 4v + w)\,du$
$+ uw(u + 8v + w)\,dv$
$+ uv(u + 4v + 2w)\,dw$.

11. $dF = \dfrac{1}{u^2v + v^2w + uw^2}$
$[(2uv + w^2)\,du$
$+ (u^2 + 2vw)\,dv$
$+ (v^2 + 2uw)\,dw]$.

13. $dF = \dfrac{1}{1 + u^2v^2w^2}$
$(vw\,du + uw\,dv + uv\,dw)$.

15. 0, 0, 0.

17. Yes; yes.

§2.8

1. (a) grad $F =$
$2x\mathbf{i} - 4y\mathbf{j} + 2z\mathbf{k}$,
(b) critical point at $(0,0,0)$,
(c) $(1/\sqrt{3})(\mathbf{i} - \mathbf{j} + \mathbf{k})$,
(d) $4\sqrt{3}$, (e) $\sqrt{2}$.

3. (a) grad $F = (y - z + 1)\mathbf{i}$
$+ (x + z)\mathbf{j} + (y - x)\mathbf{k}$,
(b) critical point at
$(-\frac{1}{2}, -\frac{1}{2}, \frac{1}{2})$,
(c) $\frac{1}{3}(2\mathbf{i} + 2\mathbf{j} + \mathbf{k})$,
(d) 3, (e) $\sqrt{3}$.

5. (a) grad $F = (3x^2 - 12)\mathbf{i}$
$+ (z - 2y)\mathbf{j} + y\mathbf{k}$,

(b) critical points at $(2,0,0)$ and $(-2,0,0)$,

(c) $(1\sqrt{5})(-2\mathbf{j} + \mathbf{k})$,

(d) $\sqrt{5}$,   (e) $-\frac{2}{3}$.

7. $-4/25$.

§2.9

1. Homogeneous, degree 3.
3. Not homogeneous.
5. Not homogeneous.
7. Homogeneous, degree $\frac{4}{3}$.
9. We are given that $f(x_1, x_2, \ldots, x_n)$ is homogeneous of degree α. Hence $f(tx_1, tx_2, \ldots, tx_n) = t^\alpha f(x_1, x_2, \ldots, x_n)$. Differentiate both sides with respect to t using the chain rule, and set $t = 1$. The desired equation follows.

§3.1

1. $f_{xy} = f_{yx} = 3$.
3. $f_{xy} = f_{yx} = -2xy/(x^2 + y^2)^2$.
5. $f_{xy} = f_{yx} = 2(x - y)/(x + y)^3$.
7. $f_{xy} = f_{yx} = 8\sec^2 2x \sec 4y \tan 4y$.
9. $f_{xy} = f_{yx} = 4xy \sec (x^2 + y^2)[\tan^2 (x^2 + y^2) + \sec^2 (x^2 + y^2)]$.

11. $\dfrac{\partial}{\partial x} f = \dfrac{\partial}{\partial x} \sum a_j(x)b_j(y)$

$= \sum \dfrac{\partial}{\partial x}(a_j(x)b_j(y))$

$= \sum a_j'(x)b_j(y)$. Similarly,

$\dfrac{\partial}{\partial y} f_x = \sum a_j'(x)b_j'(y)$.

Repeat argument, and show

that $\dfrac{\partial}{\partial y} f = \sum a_j(x)b_j'(x)$

and $\dfrac{\partial}{\partial x} f_y = \sum a_j'(x)b_j'(y)$.

§3.2

1. $3x^2 + 1$.
3. $-\dfrac{1}{(x + 2)(x + 1)}$.
5. $-(\pi^2/8) \sin x$.
7. $\dfrac{1}{x}(\cos ex - \cos x)$.
9. $\dfrac{1}{x} \sin x^2$.
11. $\frac{1}{2}x^2(5x^2 - 3)$.

§3.3

1. $f(x,y) = x^2 - 4xy + 3y^2 + y$.
3. $f(x,y) = e^x \cos y + e^y \sin x - 1$.
5. $\phi(x,y) = x^3 y + 3x^2 y^2 - xy^3 + x + C$.
7. $\phi(x,y) = x \sin (x + y) + \cos (x + y) + x + C$.

§3.5

1. $f_{xxx} = 24,\quad f_{xxy} = 8$, $f_{xyy} = 4,\quad f_{yyy} = 48$.
3. $f_{xxx} = -y^3 \cos xy$, $f_{xxy} = -y(2 \sin xy + xy \cos xy)$, $f_{xyy} = -x(2 \sin xy + xy \cos xy)$, $f_{yyy} = -x^3 \cos xy$.
5. $f_{xxxx} = 16y^4 e^{2xy}$, $f_{xxxy} = 8y^2 e^{2xy}(3 + 2xy)$, $f_{xxyy} = 8e^{2xy}(1 + 4xy + 2x^2 y^2)$, $f_{xyyy} = 8x^2 e^{2xy}(3 + 2xy)$, $f_{yyyy} = 16y^4 e^{2xy}$.

§3.6

1. $f_{xx} = 0,\quad f_{xy} = 2yz^3$, $f_{xz} = 3y^2 z^2,\quad f_{yy} = 2xz^3$, $f_{yz} = 6xyz^2,\quad f_{zz} = 6xy^2 z$.
3. $f_{xxx} = 0,\quad f_{xxy} = 4$, $f_{xxz} = 4,\quad f_{xyy} = 4$, $f_{xyz} = 0,\quad f_{xzz} = 4$, $f_{yyy} = 0,\quad f_{yyz} = 4$, $f_{yzz} = 4,\quad f_{zzz} = 0$.

5. $f_{xxx} = 0,\quad f_{xxy} = 0$, $f_{xxz} = 0$, $f_{xyy} = -\cos y \sin z$, $f_{xyz} = -\sin y \cos z$, $f_{xzz} = -\cos y \sin z$, $f_{yyy} = x \sin y \sin z$, $f_{yyz} = -x \cos y \cos z$, $f_{yzz} = x \sin y \sin z$, $f_{zzz} = -x \cos y \cos z$.
7. 2^{-n}.
9. $f = \frac{1}{3}(x^3 + y^3 + z^3) + xyz + \frac{2}{3}$.
11. $f = xe^{yz} + ye^{xy}$.
13. $f = \log xyz + xyz$.
15. $f = x \sin yz + 1$.

§4.1

1. Strict minimum at $(\frac{1}{2}, -\frac{1}{2})$.
3. Saddle point at $(0,0)$.
5. Strict minimum at $(-2, -1)$.
7. Saddle point at $(2, -2)$.
9. Strict minimum at $(1,1)$, strict maximum at $(-1, -1)$, saddle points at $(1, -1)$ and $(-1, 1)$.

§4.2

1. Maximum $f = 8$ at $(0,4)$; minimum $f = -13$ at $(2,1)$.
3. Maximum $f = 28$ at $(-2,3)$; minimum $f = -3$ at $(-1, -2)$.
5. Maximum $f = 1$ at $(1,0)$ and $(0,1)$; minimum $f = -\frac{1}{2}$ at $(\frac{1}{2}, \frac{1}{2})$.
7. 100, 100, 100.
9. Cube $(x = y = z)$.
11. Square base $(x = y = 6\,\text{ft})$, height $(z = 8\,\text{ft})$.
13. $(0,0,-1)$.
15. $\frac{8}{9}\sqrt{3}abc$.

Chapter 20

§1.1

1. 14.
3. $\frac{140}{9}$.
5. $\frac{1}{2}(e-1)^2$.
7. 972.
9. $\frac{1}{2}(1-1/e)$.
11. $\frac{1}{2}e-1$.

§1.2

1. -4.
3. 0.
5. $-\frac{1}{2}$.

§1.3

1. $\frac{1}{15}$.
3. 0.
5. $\pi^2/8$.
7. $\frac{2}{3}\pi abc$.
9. $\frac{1}{3}(2\cos 1 - \cos 2 - \cos 4)$.

§1.4

1. $\displaystyle\iint_D \alpha f\, dx\, dy$

$\displaystyle = \iint_{\substack{a<x<b \\ c<y<d}} \alpha f(x,y)\, dx\, dy$

$\displaystyle = \int_a^b \left\{ \int_c^d \alpha f(x,y)\, dx \right\} dy$

$\displaystyle = \int_a^b \alpha \left\{ \int_c^d f(x,y)\, dx \right\} dy$

$\displaystyle = \alpha \int_a^b \left\{ \int_c^d f(x,y)\, dx \right\} dy$

$\displaystyle = \alpha \iint_{\substack{a<x<b \\ c<y<d}} f(x,y)\, dx\, dy$

$\displaystyle = \alpha \iint_D f\, dx\, dy$.

§2.1

1. Yes. For $xy \neq 0$,
   $x^2 + y^2 + 6$ and xy are

clearly continuous
functions, and so is their
ratio. The only points of
discontinuity are those for
which $xy = 0$ where the
function jumps to 77. But
this set is merely the
union of the lines $x = 0$
and $y = 0$.
3. Yes, since $f + g$, $f - g$,
   and fg can be
   discontinuous only at
   points where $f(x,y)$ or
   $g(x,y)$ is discontinuous.
5. Yes, since $\sin f$ can be
   discontinuous only at
   points where $f(x,y)$ is
   discontinuous.

§2.2

1. The interval I may be
   broken into a set of
   intervals $\{I_i\}$ and a set
   $\{J_j\}$. On I_i, ϕ has value f_i
   and on J_j, ψ has value g_j.
   Define $M_{ij} = I_i \cap J_j$.
   $\{M_{ij}\}$ is an interval
   decomposition of the
   domain and on M_{ij},
   $\phi + \psi = f_i + g_j$, and
   $\phi\psi = f_i g_j$.

§2.4

1. Riemann sum = 64;
   exact value = 64.
3. Riemann sum = 155.625;
   exact value = 157.5.
5. Riemann sum = 124;
   exact value = 128.

§3.1

1. $\frac{13}{3}$.
3. $\sqrt{2} + \log(1 + \sqrt{2})$.
5. 9.
7. 32.
9. 4π.

§3.2

1. $\pi/6$.
3. 16.
5. $(\pi/4)(R_2^4 - R_1^4)(A + C)$.
7. $2\pi^4$.
9. $3\pi(2\sqrt{2} - 1)$.

§3.3

1. $A = 2\pi \displaystyle\int_{x_0}^{x_1} \phi(x)$
   $\sqrt{1 + [\phi'(x)]^2}\, dx$.
3. $A = 2\pi \displaystyle\int_{y_0}^{y_1} \phi(y)$
   $\sqrt{1 + [\phi'(y)]^2}\, dy$.
5. $13\pi/3$.
7. $(\pi/9)(17\sqrt{17} - 1)$.
9. $(\pi/2)(15 + \log 2)$.
11. $\pi\left[2\sqrt{5} - \sqrt{2} \right.$
    $\left. + \log\left(\dfrac{2 + \sqrt{5}}{1 + \sqrt{2}}\right)\right]$.
13. $\dfrac{\pi}{6}[3\sqrt{2} + \log(3 + \sqrt{10})]$.

§3.4

1. Integral diverges for
   $\alpha \geq 2$. For $\alpha < 2$, integral
   is equal to $2\pi/(2 - \alpha)$.
3. Diverges.
5. Diverges.

§3.5

1. $V = \pi$; A
   $\displaystyle = 2\pi \int_1^\infty \frac{1}{x}\sqrt{1 + \frac{1}{x^4}}\, dx$
   $\displaystyle > 2\pi \int_1^\infty \frac{1}{x}\, dx = +\infty$.
3. Diverges.
5. π.

§3.6

1. Let $x^2 = u$,
   $du/\sqrt{u} = 2\, dx$. Then the
   integral becomes

$$2 \int_0^\infty e^{-x^2} \, dx$$

$$= \int_{-\infty}^\infty e^{-x^2} \, dx = \sqrt{\pi},$$

where we use Equation (19).

§4.1

1. 1.
3. $e - \frac{5}{2}$.
5. 0.
7. 1157/105.
9. 64.
11. $\frac{7}{2}$.
13. $\frac{1}{8}(\pi - 2)$.

§4.2

1. 4.
3. $\frac{4}{3}\pi a^3$.
5. $(\pi/2)a^4$.
7. $\frac{4}{3}\pi abc$.

§4.3

1. $(244/3)\pi$.
3. $(64/9)(3\pi - 4)$.
5. $\frac{3}{4}\pi$.
7. 0.

§4.5

1. $128\pi/15$.
3. $9\sqrt{3}\pi$.
5. $\alpha > 3$.
7. $9\sqrt{3}\pi$.
9. π.

§5.1

1. $(\pi/3)R^6$.
3. $(\pi h/6)[(1 + h^2)^{3/2} - h^3]$.
5. $\frac{2}{3}$.
7. $\frac{1}{12}a^3$.

§5.2

1. Place axis of cylinder along positive z-axis and base in xy plane. Center of mass at $(0,0,2h/3)$.
3. Place the corner of the square at the origin in the xy plane. Take two sides along positive coordinate axes. Center of mass at $(5a/8, 5a/8, 0)$.
5. $(0, -1/3\pi, 4\pi)$.
7. $(17/15, 79/30)$.
9. $(2/\log 3, 1/(3 \log 3))$.
11. $(0,0)$.
13. $(4, 17/5)$.
15. Place axis of hemisphere along positive z-axis and base in xy plane. Centroid at $(0,0,3R/8)$.
17. Place axis of cone along positive z-axis and vertex at origin. Centroid at $(0,0,\frac{8}{3}(2 - \sqrt{3}))$.
19. $((3\sqrt{2}/8)R, 0, 0)$.
21. $\left(\dfrac{128 - 3\pi}{8(32 - \pi)}, \dfrac{1}{2}, \dfrac{1}{2}\right)$.
23. $(2/\pi, 2/\pi)$.
25. $\left(0, \dfrac{e^4 + 4e^2 - 1}{4e(e^2 - 1)}\right)$.
27. $(\frac{5}{6}, \frac{1}{2}, \frac{1}{6})$.

§5.3

1. $V = \pi r^2 H$.
3. $V = \pi c(b^2 - a^2)$.
5. Let the vertices of an equilateral triangle of side a be located at $(0,0)$, $(\sqrt{3}/2)a, a/2)$, and $((\sqrt{3}/2)a, -a/2)$. Then the centroid has coordinates $(a/\sqrt{3}, 0)$.

7. $S = \pi r \sqrt{r^2 + h^2}$, where r is the radius and h the height.
9. If the line segment has length r and touches l, the theorem says that the area of a circle of radius r is πr^2.

Chapter 21

§1.1

1. 10.
3. 21.
5. $\dfrac{\sqrt{2}}{4}(4 + \pi) - \frac{1}{4}(6 + \pi)$.
7. $\frac{1}{6}(4 + 3\pi)$.

§1.2

1. $\frac{4}{3}$.
3. $\frac{4}{9}(3 + \sqrt{2})$.
5. 3.

§1.3

1. $A(x_1 - x_0) + B(y_1 - y_0)$.
3. 25.
5. -2.
7. $16e - 1$.

§2.1

1. 0.
3. $\frac{1}{2}$.
5. $2(1 - 2 \cos 1 + 2 \sin 1 + \cos 2 - \sin 2)$.

§2.2

1. πab.
3. 29π.
5. $62/5$.

§2.3

1. $\iint_D \frac{\partial Q}{\partial x}\, dx\, dy$

$= \int_{f(a)}^{c} \int_{a}^{g(y)} \frac{\partial Q}{\partial x}\, dx\, dy$

$= \int_{f(a)}^{c} [Q(g(y),y)$

$\quad - Q(a,y)]\, dy.$ Since dy

$= 0$ on C_2, $\int_c Q\, dy$

$= \int_{f(a)}^{c} \phi(g(y),y)\, dy$

$\quad + \int_{c}^{f(a)} Q(a,y)\, dy.$

§3.1

1. 0.
3. $\frac{1}{6}(\sqrt{2} + 4)$.

5. 10.
7. $-\frac{1}{2}$.
9. $A(x_1 - x_0) + B(y_1 - y_0)$
   $\quad + C(z_1 - z_0)$.
11. 16.
13. 1.

§3.2

1. $\frac{1}{2}(a_1^2 + a_2^2 + \cdots + a_n^2)$
   $\quad \times (t_1^2 - t_0^2)$.
3. $n \log 2$.

§4.1

1. $3(x_1 - x_0) + 4(y_1 - y_0)$
   $\quad + 5(z_1 - z_0)$.
3. $3(x_2 - x_0) + 4(y_2 - y_0)$
   $\quad + 5(z_2 - z_0)$.
5. 0.

§4.2

1. $W = \gamma M \left(\dfrac{m_1}{r_1} + \dfrac{m_2}{r_2} \right)$.

3. Yes.
   $W = -\frac{1}{2}(x^2 + y^2 + z^2)$.

§4.3

1. $x(t) = \alpha t + \gamma$,
   $x'(t) = \alpha$ and
   $y(t) = -\frac{1}{2}gt^2 + \beta t + \delta$,
   $y'(t) = -gt + \beta$. Hence
   $K = \frac{1}{2}mv^2 = \frac{1}{2}m(x'(t)^2$
   $\quad + y'(t)^2) = \frac{1}{2}m(\alpha^2 +$
   $\beta^2 - 2g\beta t + g^2t^2)$. Also
   $W = mgy = \frac{1}{2}m(2g\delta$
   $\quad + 2g\beta t - g^2t^2)$. Therefore
   $E = K + W = \frac{1}{2}m(\alpha^2$
   $\quad + \beta^2 + 2g\delta)$ and $E' = 0$.

INDEX